THE STANDARD ABBREVIATIONS

In this book, several units find their names less drastically abbreviated than in the technical literature; this slight redundancy appears to facilitate the reading considerably. The standard abbreviations are listed below; note that they are not followed by a period.

ampere	A	horsepower	hp
Ångstrom	Å	hour	h
astronomical unit	AU	inch	in
atmosphere	atm	joule	J
atomic mass unit	u	kelvin	K
becquerel	Bq	Kelvin degree	K
bel	B	light-year	ly
British thermal unit	Btu	liter	l
calorie	cal	meter	m
Celsius degree	C°	mile	mi
coulomb	C	minute	min
curie	Ci	mole	mol
degree Celsius	°C	newton	N
degree Fahrenheit	°F	ohm	Ω
degree Kelvin (=kelvin)	K	pascal	Pa
dyne	dyn	pound	lb
electron-volt	eV	quart	qt
erg	erg	radian	rad
Fahrenheit degree	F°	second	s
gallon	gal	tesla	T
gauss	G	volt	V
gram	g	watt	W
henry	H	watt-hour	Wh
hertz	Hz	weber	Wb

ELEMENTS OF
PHYSICS

ELEMENTS OF
PHYSICS

MARCEL WELLNER

Syracuse University
Syracuse, New York

SPRINGER SCIENCE+BUSINESS MEDIA, LLC

Library of Congress Cataloging-in-Publication Data

Wellner, Marcel.
    Elements of physics / Marcel Wellner.
        p.   cm.
    Includes bibliographical references and index.
    ISBN 978-1-4613-6723-9   ISBN 978-1-4615-3860-8 (eBook)
    DOI 10.1007/978-1-4615-3860-8
    1. Physics.   I. Title.
    QC21.2.W437 1991
    530--dc20   91-20745
                                                              CIP

Cover illustration: A spacecraft orbiting a planet is a
prime display of classical mechanics; the radio pulses it
sends home epitomize electromagnetism. Mechanics and
electromagnetism are two main topics of this book.

ISBN 978-1-4613-6723-9

© 1991 Springer Science+Business Media New York
Originally published by Plenum Press, New York in 1991
Softcover reprint of the hardcover 1st edition 1991

To the memory of my parents,

Jules and Lucie Wellner-Rapoport,

victims of the Holocaust.

PREFACE

To the Instructor

We are seeing an increased need for a one-year survey of physics, at the calculus level, and with the inclusion of some modern physics. A growing number of students—in engineering as well as in the sciences—must take early technical courses that demand a reasonable familiarity with physics *as a whole*.

The present book is a response to that need. The author is well aware that introductory physics cannot be compressed or pruned ad infinitum; nevertheless, the one-year goal may yet be reachable.

A slim volume does not seem to be the answer. Rather than compressing or pruning, I have tried to work towards a smoother exposition. To that end a variety of devices—not necessarily bulk-saving—have been enlisted: a liberal use of line drawings; a modest number of chapters, but each fairly broad, in the hope of improving the continuity of topics; the avoidance of technical busy-work; the serious revival, in modern form, of some old approaches like field lines and—yes—magnetic poles; the free use of chapter end notes to reduce digressions in the main text; and formula summaries at the end of each chapter in order to help avoid unproductive leafing through the book, as well as to facilitate review.

A calculus course, concurrent with the present material, is assumed; but how much previous exposure to calculus does our treatment require? In principle, none. The early concepts are explained as needed, although not in a mathematically rigorous manner.

While the *language* of calculus is indispensable here, its manipulative power will, with some regret, be left pretty much unexploited; calculus-centered exercises, seductive though they are, would not help us accomplish our mission.

Suggested scheduling. How much material should be covered in one term? Some possible apportionments of the 28 chapters (24 without the modern physics) are indicated in the table below.

	With modern physics	Without modern physics
2 terms	$14 + 14$	$12 + 12$
3 terms	$9 + 10 + 9$	$9 + 8 + 7$

Enough problems are provided for three full semesters, if desirable.

To the Student

The message of this book is mostly self-explanatory; but one of its aspects is not obvious, and you need to be alerted to it in advance. It consists of an attempt to promote five habits of thought that physicists cultivate, and that you will find rewarding to acquire.

The principal one is the use of *idealized models.* Instead of attacking a difficult realistic problem, it is far better to start with a slightly unrealistic one that is, however, easy to solve. Very often, the harder problem can thereafter be mastered. In these pages you will encounter many of the idealized concepts that have led to the amazing success of physics in unraveling the real world: point particles, ideal springs, frictionless surfaces, perfectly circular orbits, ideal gases, infinitely thin wires, plane waves, and many others.

Next in order of importance is the repeated harnessing of just *a few basic principles*—especially the conservation of energy—to analyze new situations of any kind.

Among other useful strategies illustrated here, a popular one is the *search for possible symmetries* in a problem. A symmetry is an equivalence between different directions (for example, right and left), or between different locations or different objects. When valid, a symmetry can sometimes simplify a problem to the point where its solution becomes obvious without calculation.

Finally, a pair of mutually related tools should be mentioned. One is *the art of making rough estimates*, a valuable skill that deliberately renounces all unnecessary precision. (That skill, to be sure, is only acquired gradually with maturity and experience.) The other is *dimensional analysis*, which amazingly enough can produce detailed formulas merely from a consistency requirement involving physical units (kilogram, meter, second, etc.)

Admittedly, the foregoing descriptions are far too sketchy to be completely meaningful to you at the present stage; but their purpose will be accomplished if your recognition is evoked at the appropriate place.

A few words are in order, as well, about some of the reference materials in this book. If you are aware of their existence they may save you a trip to the library, perhaps in connection with some of your other courses. They consist of appendices (on conversion factors, mathematical formulas, etc.); also, interspersed throughout the book you will find tables and "spectra," or charts, of typical and extreme magnitudes for physical quantities observed in the real world. These tables and spectra may be located through Appendix VII or through the book's alphabetical index.

To end this preface, a remark on "he or she."

May the women readers of this book forgive my stylistic use of the generic "he"; no barrier is intended. Physical science badly needs the talent of women and men alike; its beauty and adventure invite both.

Acknowledgments

Without the unflagging support and participation of my wife Magdeleine, this book would surely not exist. I am grateful to Syracuse University for providing the environment needed by a project of this magnitude. And over the years, superseding any pedagogical doctrine, my students have been my teachers.

Certain milestones of physics text writing are bound to have shaped much of what I have done here. I refer especially to the original *Principles of Physics* by Sears, and its successor by Sears, Zemansky, and Young; also, in its early editions, *Fundamentals of Physics* by Halliday and Resnick; and *Physics* by Tipler. Any of these authors may profitably be consulted for much that is missing in the present pages.

I am greatly indebted to the constructive reading of the entire manuscript by A. L. Ford (Texas A & M University), Michael Lieber (University of Arkansas), Edward Pollack (University of Connecticut), John E. Potts (University of Michigan), C. W. Price (Millersville University), and Daniel Knight, the latter with particular attention to the examples and problems.

Selected portions of the book have benefited from the suggestions of Herbert Berry, Harvey Kaplan, Carl Rosenzweig, Eric Schiff, John Trischka (all of Syracuse University), John Peterson (Fresno City College), and Daniel Wellner (Cornell University Medical School).

The later stages of this enterprise have owed much to the patience and professionalism of Amelia McNamara and Kenneth Schubach at Plenum Publishing Corporation.

No amount of consultation will remove all flaws and errors; for these I must take full responsibility.

M. W.

CONTENTS

CHAPTER 1. **MEASUREMENT** 1

A. Theories versus Data 1
B. Precision in Numbers 2
C. Distance 4
D. Time 8
Condensed Checklist 10
True or False 11
Problems 11

CHAPTER 2. **MOTION** 13

A. Uniform Motion 13
B. Accelerated Motion (One-Dimensional) . . . 15
C. Vertical Free Fall 18
D. Vectors 23
E. Motion in a Plane 30
F. Projectiles 34
Condensed Checklist 38
True or False 39
Problems 39

CHAPTER 3. **FORCE** 45

A. The Law of Inertia
   (Newton's First Law) 45
B. Force and Acceleration
   (Newton's Second Law) 46
C. Combining Forces 50
D. Action and Reaction
   (Newton's Third Law) 53
E. Pulleys and Surfaces 56

F. Effective Gravity 62
Note 64
Condensed Checklist 64
True or False 65
Problems 65

CHAPTER 4. **ENERGY** 75

A. Conservation of Energy in Free Fall 75
B. Work 78
C. Work and the Energy of a Particle 83
D. Work and the Energy of a System 90
E. Conservation of Total Energy 95
Note 96
Condensed Checklist 97
True or False 97
Problems 98

CHAPTER 5. **MOMENTUM** 103

A. Conservation of Momentum 103
B. Two-Body Collisions 106
C. Force from a Jet 110
D. The Center of Mass 111
Note 117
Condensed Checklist 118
True or False 118
Problems 118

CHAPTER 6. **CIRCULAR MOTION** 125

A. Angular Variables 125
B. Uniform Circular Motion 127

C. Nonuniform Circular Motion 134
Notes 136
Condensed Checklist 137
True or False 138
Problems 138

CHAPTER 7. ROTATION AND TORQUE 143

A. Rotation About a Fixed Axis 143
B. Torques in Terms of Forces. 150
C. Rotational Equilibrium 154
D. Rotation and the Center of Mass 158
E. Angular Momentum 160
Notes 164
Condensed Checklist 166
True or False 167
Problems 167

CHAPTER 8. GRAVITY 175

A. Astronomical Comments 175
B. Universal Gravitation 177
C. Orbits 182
D. Gravitational Potential Energy 185
Notes 188
Condensed Checklist 189
True or False 189
Problems 189

CHAPTER 9. FLUIDS 193

A. Pressure and Density. 193
B. Archimedes' and Pascal's Principles. 198
Note 202
Condensed Checklist 204
True or False 205
Problems 205

CHAPTER 10. HEAT 209

A. Temperature 209
B. Heat as Transferred Energy 215
C. Specific Heat 217

D. Phase Transitions 219
Notes 221
Condensed Checklist 222
True or False 223
Problems 223

**CHAPTER 11. MOLECULES AND
          GASES** 227

A. The Existence of Molecules. 227
B. Atomic Masses and the Kilomole 229
C. The Ideal-Gas Law 233
D. Kinetic Theory. 236
E. Internal Energy and Heat Capacity 241
Notes 243
Condensed Checklist 244
True or False 244
Problems 245

CHAPTER 12. WORK FROM HEAT 249

A. Heat and the Conservation of Energy
        (First Law of Thermodynamics) . . . 249
B. Heat Engines and the Second Law of
        Thermodynamics 256
C. Entropy and the Waste of Energy 262
D. The Logic of this Chapter 266
Condensed Checklist 266
True or False 267
Problems 268

CHAPTER 13. VIBRATIONS 273

A. Hooke's Law and the Ideal Spring 273
B. Harmonic Motion 278
C. Superposition and Sound 284
D. Resonance 288
Condensed Checklist 292
True or False 292
Problems 293

CHAPTER 14. WAVES 297

A. Traveling Waves in Strings 297
B. Classifying Waves 303

C. Sound in Gases............................ 304
D. Standing Waves 308
Notes 312
Condensed Checklist 315
True or False.............................. 315
Problems 316

CHAPTER 15. ELECTRIC CHARGE AND
                       ITS FIELD 319

A. The Electric Force...................... 319
B. The Electric Field 326
C. The Electric Flux and Gauss' Law 330
Note....................................... 335
Condensed Checklist 337
True or False.............................. 337
Problems 338

CHAPTER 16. THE ELECTRIC
                       POTENTIAL 343

A. Introducing the Potential............... 343
B. Single Conductors in Equilibrium 349
C. Capacitors 353
D. Energy of the Electric Field............ 358
E. Particles Controlled by Electric Fields..... 359
Note....................................... 363
Condensed Checklist 363
True or False.............................. 364
Problems 365

CHAPTER 17. DIRECT CURRENTS...... 371

A. Ohm's Law 371
B. Electric Power......................... 373
C. Circuits 375
D. Conduction 381
Notes 384
Condensed Checklist 385
True or False.............................. 385
Problems 386

CHAPTER 18. MAGNETIC FORCES 391

A. Magnetic Poles and Fields 391
B. Magnetic Force on a Moving Charge..... 397
C. Magnetic Force on a Current 407
Notes 412
Condensed Checklist 413
True or False.............................. 414
Problems 415

CHAPTER 19. MAGNETIC FIELDS FROM
                       CURRENTS 421

A. The Law of Biot and Savart............. 421
B. Proof by Action and Reaction 428
C. Ampère's Law 429
D. Magnetic Poles: An Illusion 436
Notes 439
Condensed Checklist 442
True or False.............................. 442
Problems 443

CHAPTER 20. MAGNETIC INDUCTION 449

A. Electric Field Seen by a Moving Object 449
B. Moving Circuits........................ 451
C. Faraday's Law of Induction 454
D. Application to Coils................... 456
E. The Induced Electric Field 459
Note....................................... 462
Condensed Checklist 463
True or False.............................. 464
Problems 464

CHAPTER 21. INDUCTANCE AND
                       ALTERNATING CURRENTS 469

A. Self-Induction 470
B. Natural Oscillations in a Circuit......... 475
C. Alternating Currents 478
Notes 483

Condensed Checklist 484
True or False 485
Problems 485

CHAPTER 22. ELECTROMAGNETIC
                        WAVES 489

A. Theory of Electromagnetic Waves 489
B. Energy and Momentum in a Wave 495
C. The Electromagnetic Spectrum. 500
Note 505
Condensed Checklist 505
True or False 506
Problems 506

CHAPTER 23. GEOMETRICAL OPTICS   511

A. Wave Fronts and Rays 511
B. Refraction of a Plane Wave. 512
C. Reflection of a Plane Wave 517
D. Image Formation by Lenses 519
E. Image Formation by Mirrors 528
Notes 533
Condensed Checklist 535
True or False 536
Problems 537

CHAPTER 24. WAVE OPTICS 543

A. Interference by Reflection 544
B. Diffraction from Slits 552
C. Noncoherent Light. 560
D. Limits of Resolution 562
E. Polarization 565
Notes 569
Condensed Checklist 569
True or False 570
Problems 571

CHAPTER 25. RELATIVITY 577

A. The Michelson–Morley Experiment. 577
B. The Postulates of Relativity 581

C. The Equivalence of Mass and Energy 582
D. Time Dilation 585
E. The Logic of this Chapter 589
Notes 589
Condensed Checklist 590
True or False 591
Problems 591

CHAPTER 26. WAVES VERSUS
                        PARTICLES. 595

A. Photoemission 596
B. X Rays 601
C. The Compton Effect 603
D. Electron Waves 606
E. The Role of Probability 608
Note 611
Condensed Checklist 612
True or False 612
Problems 613

CHAPTER 27. ATOMIC STRUCTURE. . . 617

A. Energy Levels. 617
B. Bohr's Model of the Hydrogen Atom 624
C. Electron States and Their Four
                 Quantum Numbers 628
D. The Exclusion Principle and the
                 Periodic Table 633
Notes 640
Condensed Checklist 642
True or False 643
Problems 643

CHAPTER 28. THE NUCLEUS. 647

A. The Stable Nuclei 648
B. Radioactivity 652
C. Nuclear Reactions 658
Notes 663
Condensed Checklist 664

True or False 665
Problems 665

APPENDICES 669

  I. **Energy and Power Spectra** 669

  II. **The Vector Cross Product** 671

  III. **Mathematical Formulas** 673

  IV. **Conversion Factors** 675

  V. **Numerical Constants** 677

  VI. **The Greek Alphabet** 679

  VII. **List of Tables and Spectra** 681

INDEX 683

CHAPTER 1

MEASUREMENT

Can Nature be understood on the basis of logic? Physics teaches us that the answer is, to a great extent, "yes," and that the relevant logic includes a great deal of mathematics. The quantitative laws of physics are expressed by *mathematical equations*, whose symbols are directly or indirectly related to *the numerical results of measurements*.

We are all familiar with the act of measurement. It is a necessity in most businesses, industries, and crafts. The purpose of measurement in science is, however, rather different from what it is in other activities.

We begin this chapter by sketching the role of measurement in establishing the physical laws. We then discuss the concept of *precision*, as it applies to numerical data; that concept will lead to a set of practical rules for handling numbers when solving problems.

In order to conduct measurements and communicate them to others, we must agree on the *units* in which they shall be expressed. Physics uses a great variety of units; two basic ones, the *meter* and the *second*, are selected for brief descriptions, which conclude the chapter. The other units will be introduced later, as they are needed.

A. Theories versus Data

We first consider how a theory can emerge from a set of data, or be destroyed by it.

i. Qualitative and Quantitative Facts

Suppose we originally are in possession of a body of facts concerning the physical world. Some of these facts may not be expressible in numbers, but rather in terms of the existence of some things or phenomena; these are called **qualitative facts**. Examples are the existence of solids, liquids, and gases, or the existence of radio waves, etc. No measurement need be involved in the recognition of qualitative facts.

Some other known facts do depend on measurement, and are called **quantitative facts**, **numerical data**, or simply **data**; examples are the density of certain substances, the frequency of certain waves, etc.

ii. Theoretical Explanation

Prompted by the known facts or just by his imagination, someone may set up a logical structure called a **theory**, or sometimes more modestly a **model**. This involves a set of assumptions, which must be simpler than the facts he is trying to explain. From these assumptions he logically deduces a set of qualitative, and also quantitative results; these should be in agreement with the known facts. If he gets gross disagreement, the theory must be discarded; if he gets fair but not perfect agreement, the theory can serve until a better one is found, or else it can be modified.

Theories can sometimes be improved if one tampers with the assumptions, but making the assumptions more complicated to fit a few more facts is not usually considered an improvement. (Here we see that human judgement and even taste may be needed in evaluating theories.)

Occasionally the numbers calculated from the theory agree spectacularly with the measured data, of which there may be plenty; such a theory then becomes well established. Sometimes a theory even explains many more facts than its designer intended, and thus the theory may be said to have a life of its own, or to be more clever than its designer. Why such lucky windfalls ever happen is not entirely clear, but they have happened repeatedly in the past.

iii. Predictive Power of a Theory

A theory that explains the known facts still must pass another test, that of **prediction**. From the theory one must be able to calculate the results of experiments not yet performed. If those experiments are subsequently made and agreement is obtained, the theory emerges much stronger. Correct prediction will often convince where mere explanation will not.

iv. Evolution of a Theory

Two rival theories, both wrong, may have to be taken seriously because nothing better is available. One theory explains, say, experiment A, while giving the wrong result for experiment B. The second theory explains B but contradicts A. One must live with this situation until it is resolved by the birth of a "supertheory" that explains both A and B. Such a supertheory is exemplified by quantum mechanics, which reconciles two well-established but seemingly contradictory properties of light—its wavelike and particlelike aspects.

Even the best-established theory may eventually fail to agree with certain new experiments. No matter how many experiments are made, it is forever impossible to prove that a theory is absolutely right. This is because it is impossible to perform every conceivable experiment, and no measurement can be carried out to perfect accuracy. On the other hand, even a single measurement can overthrow the most impressive theory if it disagrees with it.

Fortunately, experience has shown that when a well-established and widely tested theory must be discarded, it is often in favor of a theory that retains the important features of the old one, so that the insight gained from the old theory is not destroyed by the new one. It must be added that, so far, the ultimate physical theory has not been written, and perhaps it never will be.

v. The Theories Studied in This Textbook

This book, like all physics texts, can show little of the historic interplay between data and theories. Discarded theories are legion, and the reasons for discarding them are often quite subtle. In the space and time available we must confine ourselves to the best-established aspects of physics. Although several of the theories we shall study are now known to present an oversimplified view, they have survived to this day in their essentials. We shall attempt to indicate their limitations at the appropriate places.

vi. The Details of Science

Some aspects of science are so general and profound that they verge on philosophy. They must nevertheless be given some consideration by every practicing scientist before the plunge is taken into technical details. The reason is simple—before doing something, one must know why one wants to do it.

In particular, the practical minutiae of measurement are directly motivated by generalities such as we have been discussing. This book is devoted to the specifics, however, and we now turn to them.

B. Precision in Numbers

These are two ways in which science generates numbers: as the result of measurement, or as the result of calculation. In either case, how much precision is reasonable in recording a number? While using this book, and in particular while solving its problems, the student will face many such (minor) decisions; the following are rough guidelines.

i. Significant Figures

The amount of physical information contained in a number depends on how many significant figures are in that number. For example, the two numbers 234.628 and 0.003 467 30, although very different, contain a comparable amount of information. They both have six significant figures, counted as follows. First, discard all the leading zeros. (There are none in 234.628, and three in 0.003 467 30; do not worry about the decimal point.) Then count the remaining digits. (Thus, a trailing zero, such as occurs in 0.003 467 30, is a significant figure.) One would say of both examples above that they are precise to six significant figures (or to six figures, or to six places, or to about one part in 1 000 000). The last significant figure (i.e., the last written digit) in a physical measurement is meant as a best guess but must be considered uncertain by at least half a unit or so either way.

Illustrations: 27.3 means "probably between 27.25 and 27.35"; 3.60 means "probably between 3.595 and 3.605." The respective **uncertainties**, or **limits of precision**,* are ± 0.1 and ± 0.01.

(Caution: Counting significant figures is only a rough index of precision. For example, 0.999 and 1.000 are nearly the same quantity, and both are known to within 0.001. Yet, 0.999 has three figures while 1.000 has four.)

ii. Power-of-Ten Notation

Suppose a certain number has been rounded off to the nearest 1000, so as to read 48 000. How do we convey the fact that there are only two significant figures, namely, the digits 4 and 8? Our prescription gives a count of 5 significant figures, which is wrong. One way out is to replace 48 000 by the notation 48×10^3, or 4.8×10^4, with the convention that a factor of the form 10^n carries no significant figures. Thus, as another example, the number 250 might be represented, to the nearest 10, by 25×10^1 (two significant figures). It must be said that in practice this notation is sometimes given up when the precision is made clear from the context.

Aside from making the precision clear, the main use and advantage of the power-of-ten notation is in representing compactly very large or very small numbers, such as

$$300\ 000\ 000 = 3 \times 10^8 \qquad \text{(to the nearest } 10^8\text{)}$$

or

$$0.000\ 009\ 8 = 9.8 \times 10^{-6}$$

A standard place for the decimal point is recommended, namely, after the first significant

* There appears to be no *precise* definition of "precision."

figure: thus, 2.34×10^5 rather than 23.4×10^4 or 0.234×10^6, although all three are equal and carry the same information.

In order to emphasize the approximate nature of an equality we may use the sign $\approx$, as in

$$\pi \approx 3.14$$

iii. Precision in Physics Problems

In this book, unless otherwise indicated, it will rarely be necessary to go beyond three figures; two will often be sufficient. As a rule of thumb, *the precision of the answer should match the precision of the data included in the question.* Beware of electronic calculators, which can mindlessly supply a great excess of significant figures. Excessive precision is misleading because it appears to be physically significant but actually is not; it should be counted as a full-blown mistake.

Symbolic Formulas: A Necessity

A symbolic formula is greatly superior to a numerical calculation, quite aside from questions of precision. A formula is in reality the solution, once and for all, to an infinity of similar problems. It confers a kind of understanding that cannot be achieved with numbers; it also makes faulty or clumsy reasoning much easier to trace and to correct than would be possible in a numerical solution.

When working with symbolic formulas or equations it is best to treat them at first as exact. In this way the mathematics can be kept simple and powerful. Any approximate numbers may then be substituted into the end result after it has already been simplified as much as possible. A trivial fable illustrates one aspect of this: a student inserted the approximate numbers

$$a = 1.00, \qquad b = 1.01$$

into the formula

$$x = \frac{a^2 - 2ab + b^2}{(a-b)^2} \qquad (1.B.1)$$

His result for x had very low precision. (Why? Try

it out.) But a little "exact mathematics" would have shown him that Eq. (1.B.1) implies

$$x = 1$$

to perfect precision, no matter what a and b are, as long as $a \neq b$. Moral: *simplify the formula before inserting the numbers.*

C. Distance

Quantitative physics was founded on the measurement of distances, time intervals, and weights. These three kinds of measurement have been for a long time the easiest and the most reliable. To this day, distance and time have retained their fundamental status and are at the basis of our understanding of natural phenomena. The average person has a good grasp of these two concepts and of the daily-life measurements corresponding to them. In recent history, however, the range of these measurements has been extended prodigiously outside the daily-life area, both towards the very large and towards the very small. Our discussion in this section is intended as a reminder of that fact. We also take this opportunity to bring up the essential question of units.

What about the third traditional quantity, weight? It has been found that a related, but different concept, that of mass, is much more fundamental. Mass and its relation to weight will be studied at a later stage (Chap. 3, Force).

i. Working Definition of Distance

Physical quantities are often defined in terms of a *prescription* for measuring them. One prescription for measuring the distance between two motionless points is the familiar one of joining them by the edge of a ruler and noting the difference between those readings on the ruler that coincide with the points. The "points" in question are never localized with perfect precision: they are small objects or marks. The existence of rulers should not be taken for granted: it depends on the existence of rigid objects, i.e., objects that can be casually carried from place to place without undergoing any detectable change. We owe this fortunate property to the fact that atoms exist, and that, in solids, they are attached to one another by substantial forces.

Measurement by means of a ruler becomes impossible in the most interesting cases: when the objects involved are invisible, or moving, or very far away, or very close together. In such cases, e.g., those involving astronomical bodies or atomic particles, indirect methods are used. Their ingenuity and success are among the greatest triumphs of physics. Illustrations will be discussed as the occasion arises.

In principle, a unit of distance, or length, can be constructed by referring all distance measurements to a single ruler of good quality, bearing two special but arbitrary marks and called the **standard** (or prototype) **ruler**. The distance between these two marks is declared to be one unit, and all other rulers must be directly or indirectly compared to the standard in order to give compatible measurements.

ii. The Metric System

The standard of length internationally used in science is the **meter** (official abbreviation: m). Until recently the prototype meter was a carefully built engraved metal bar kept at the International Bureau of Weights and Measures in Sèvres, France, and made to be used at a specific temperature. That prototype, as well as its duplicates scattered around the world, are still valuable, but have been superseded in precision and reliability by standard light beams, which carry an extremely fine wave pattern of known spacing. In order to make use of those patterns in a length measurement one needs an instrument called an **interferometer**, to be discussed in Chap. 24 (Wave Optics). Interferometric methods provide a standard meter that is precise to better than one part in 10^8. Physical applications requiring such an enormous precision are rare and specialized; a good commercial length scale is sufficient in the vast majority of laboratory measurements. *For our purposes, the working unit of 1 meter is obtainable from any such scale.*

International Definition

The meter was originally chosen (during the French Revolution) in such a way that 4×10^7 meters, as exactly as possible, should equal the Earth's circumference along a meridian at sea level. According to present-day surveys, and in terms of today's meter, that value for the Earth's circumference is still good to one part in 10^4.

As we have just mentioned, the contemporary meter is defined in terms of the properties of light (or of electromagnetic waves). In order to be meaningful to the reader of this book, the official definition will be quoted in Chap. 25 (Relativity), where it properly belongs.

[Those of us who are accustomed to British units might note the relation

$$1 \text{ yard} = 0.9 \text{ meter} \qquad \text{(approximately)}$$

The inch is exactly defined as 0.0254 m.]

Larger and smaller units are conveniently derived from the meter through multiplication or division by powers of 10. These auxiliary units are labeled by prefixes, as shown in Table 1.1.

The prefixes of Table 1.1 are used in front of a variety of units. Here are some illustrations:

$$1 \text{ kilometer} = 1 \text{ km} = 10^3 \text{ meters} = 10^3 \text{ m}$$

$$1 \text{ microsecond} = 1 \ \mu s = 10^{-6} \text{ second} = 10^{-6} \text{ s}$$

Conversion between Units

Tables of conversion between metric, British, and other units may be found in Appendix IV.

iii. Dealing with Units

The name of a unit, such as centimeter, degree of angle, minute of time, volt, or one of countless others, must already be implied or furnished with the scale to indicate what is being measured. The unit also specifies a standard of comparison for what is being measured.

A reading on a scale is thus given in terms of the appropriate number, followed by the name of the unit, for example 21.1 centimeters, -38.9 degrees Celsius, etc. (The abbreviations, like cm for centimeter, get no plural form and no abbreviative period.)

There exists a set of recommended units (the so-called SI, or **système international**, of which the metric system is a part), and we shall ordinarily make use of them, unless alternative units are much more convenient or much better known. Sometimes it is advisable to use a less abbreviated version than the standard one. For example, it is better to use the symbol "sec" rather than "s" for the second of time in a context where s denotes a length, which it often does. Similarly, "meter" is better than "m" if the symbol m is used for a mass.

Should Units Appear in Formulas?

The answer is, "preferably no." Expressions such as

$$\text{time} = t \text{ sec}$$

lead to confusion. The recommended practice is

$$\text{time} = t$$

where *the symbol t includes the units*, as in $t = 6$ sec. This has the virtue that one may switch units without switching symbols: the formula may be used in any system of units. For example,

$$t = 6 \text{ sec}, \qquad t = 0.1 \text{ min}$$

are mutually consistent statements.

Table 1.1. Power-of-Ten Multipliers

Prefix	Symbol = (multiplying factor)	Prefix	Symbol = (multiplying factor)
deca[a]	da = 10	deci[b]	d = 10^{-1}
hecto[a]	h = 10^2	centi[c]	c = 10^{-2}
kilo	k = 10^3	milli	m = 10^{-3}
mega	M = 10^6	micro[d]	$\mu = 10^{-6}$
giga	G = 10^9	nano	n = 10^{-9}
tera	T = 10^{12}	pico	p = 10^{-12}
peta[a]	P = 10^{15}	femto	f = 10^{-15}
exa[a]	E = 10^{18}	atto	a = 10^{-18}

[a] Seldom used in physics.
[b] Seldom used in physics, except in "decibel."
[c] Seldom used in physics, except in "centimeter" and "centipoise."
[d] μ is the lower-case Greek letter "mu."

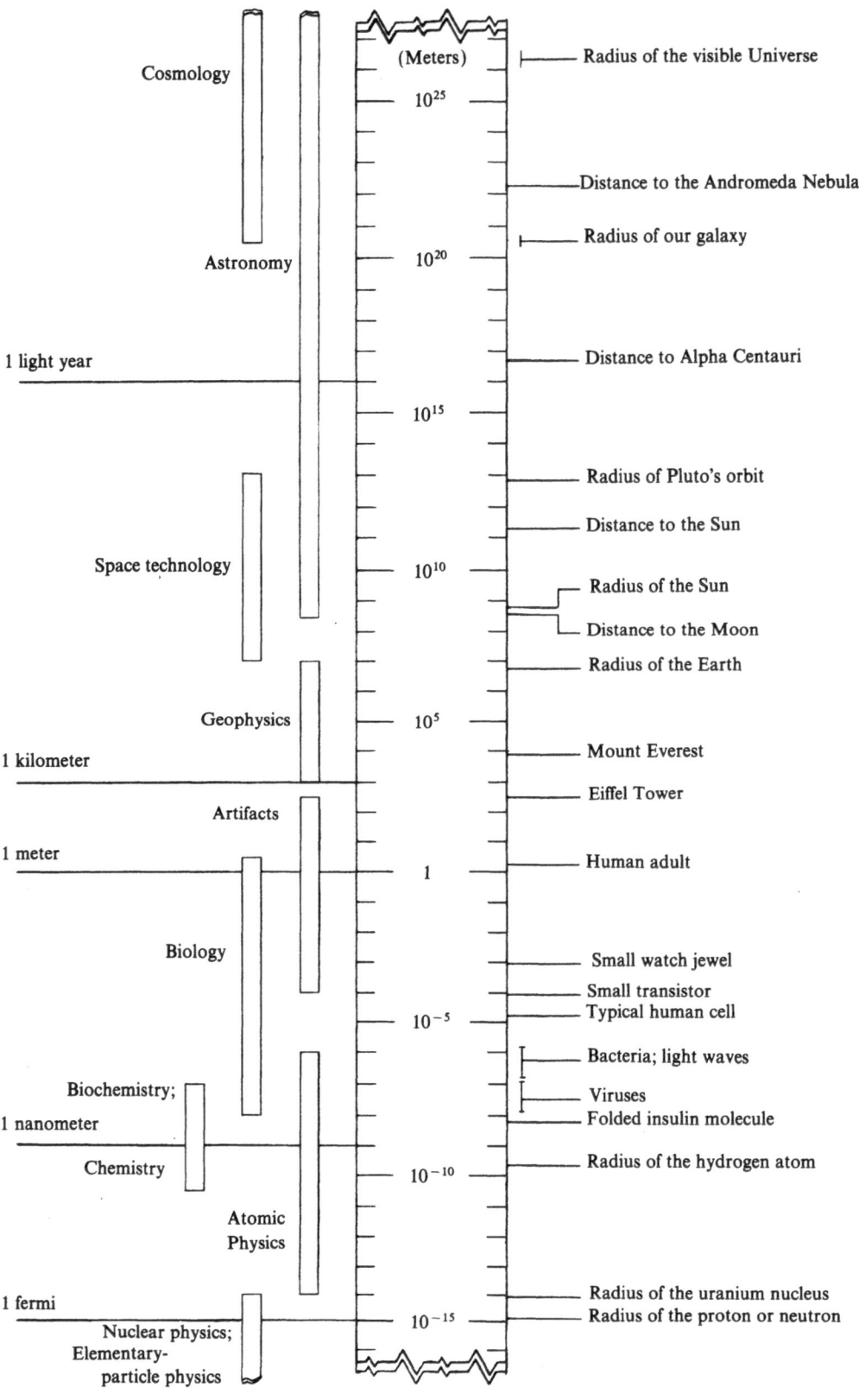

iv. A Spectrum* of Distances

Figure 1.1 attempts to convey the range of distances that have been measured. A few minutes' contemplation of that diagram can be useful in obtaining a feel for the domain of exploration open to physical science.

v. Logarithmic Scales

The reader will observe that the scale of Fig. 1.1 is nonuniform in a special way, so that the very small and the very large can be represented in the same picture. Such so-called **logarithmic scales** will be used occasionally throughout this book. On a logarithmic scale, *successive powers of ten are plotted at equal distance intervals*. More specifically, at what point should a given number N be plotted? Our special reference point on the scale will be where we plot the number 1 (we shall not be able to plot zero). Let D be the actual distance on the paper, measured along the scale, between the number 1 and the number N. Then the rule is: make D proportional to the logarithm of N. Thus, for some proportionality constant s,

$$D = s \log N \qquad (1.C.1)$$

(This formula uses base-ten logarithms.) Taking $N = 10$ as a special case, we see that the fixed distance s is that between the number 1 and the number 10. More generally, we see that if $N = 10$, 100, 1000, etc., then $D = s$, $2s$, $3s$, etc.

---

* A spectrum ("something to look at") is any displayed sequence. This word found its first physical use in optics.

Example 1.1. Plotting Some Points. Consider a horizontal logarithmic scale showing the points 1, 10, 100, etc., plotted from left to right at intervals $s = 1.0$ cm. Where do we plot the numbers 0.1, 50, 0, −50?

We have for $\log N$:

$$\log 0.1 = -1$$
$$\log 50 \approx 1.7$$
$$\log 0 = -\infty \qquad \text{(minus infinity)}$$
$$\log(-50) = \text{(not a real number)}$$

Hence the plotted points are located as follows. For $N = 0.1$,

$$D = (1.0 \text{ cm})(-1) = \underline{-1.0 \text{ cm}}$$

(that is, 1.0 cm to the left of the number 1); for $N = 50$,

$$D = (1.0 \text{ cm})(1.7) = \underline{1.7 \text{ cm}}$$

(that is, 1.7 cm to the right of the number 1).
The numbers 0 and −50 <u>cannot be plotted</u>.

It must be kept in mind that a logarithmic scale has no origin and no negative values.

vi. Position Along a Path

Studying the motion of an object means, first of all, recording where it is located at various times. To make this task as simple as possible we shall, whenever we can, represent an object by a single

---

Figure 1.1. Some measured distances (logarithmic scale). This shows the approximate numerical value, in meters, of various observed distances. (Some other common units are shown on the left.) A small vertical bar denotes an uncertainty or an allowed range in the possible sizes. "Distances to" are meant as measured from Earth. The Andromeda nebula is our neighboring galaxy. Our own galaxy is the Milky Way. Alpha Centauri is the nearest known star (apart from the Sun). Pluto is probably the outermost planet, so that its orbit about circumscribes the solar system. "Light waves" refers to the approximate range of wavelengths for visible light; details will be found in Chap. 22, whose spectra can profitably be compared with this one. The proton is the nucleus of the hydrogen atom. Much smaller structures than any indicated here must surely exist. For example, to date, the electron's measured size is consistent with zero. The fields of human activity shown on the left are not really as sharply limited as would appear from the chart. For example, cosmology legitimately deals with nuclei. This chart is a good place to introduce one of the physicist's favorite expressions, "order of magnitude." Two numbers are said to be of the same order of magnitude if they are appreciably less than a factor of 10 apart. Thus, the Moon's orbit and the Sun's radius have the same order of magnitude (one is about half the other); the Earth's radius is two orders of magnitude smaller than either of those. The order of magnitude is itself not a sharply defined concept.

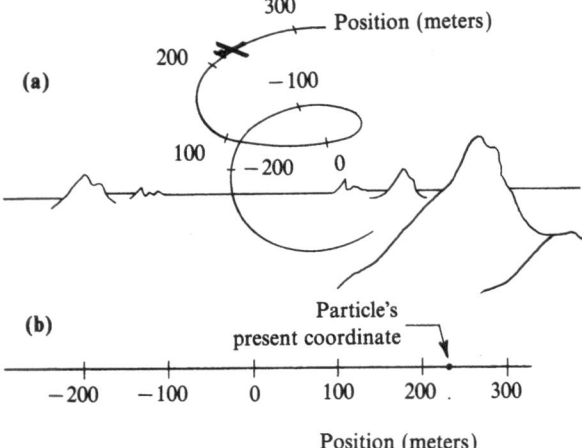

Figure 1.2. Coordinate of a particle along its path. We can think of an abstract coordinate curve set along the path of a particle. In (a) the particle (here a glider) is shown semirealistically with its coordinate curve. If we are interested in distances along the path rather than in the actual shape of the path, then (b) is a more convenient representation.

point. This may seem outrageously inappropriate if the object is complex. Still, even a Swiss watch may be represented by no more than a point if one wishes to study certain aspects of its motion, let us say as it is being dropped.

We shall not worry, for the time being, about which point of an actual object is the most representative one.* If the object is symmetric to such an extent that it has an obvious center (as is the case for a sphere, cube, or parallelopiped made of a uniform material), then we naturally pick that center as the representative point. In many cases a point is too simple a concept to represent an object, even in physics, and a more complicated description is needed. In the first few chapters of this book we shall study only those situations where an object can be described by a single point. This leads to the following definition:

> A **particle** is an object whose position is adequately described as that of a point.
>
> (1.C.2)

The smaller the object, the more realistically it can be called a particle.

Suppose a particle moves along a known path or trajectory, Fig. 1.2. Then its position along the path can be described by (1) setting up in advance a dis-

tance scale along the path, and (2) noting the scale marking that coincides with the particle at any time. That scale marking is called the **coordinate** of the particle, and the scale itself is a coordinate curve. A straight coordinate line is called a **coordinate axis**.

In theoretical work we merely imagine such a procedure, noting that we could in principle execute it if we had to. The favorite coordinate axes of physics are abstract, and we fortunately do not have to fill all space with real-life meter sticks.

D. Time

We saw that, as far as the physicist is concerned, **distance** means "what is measured with a ruler." Similarly, **time** means "what is measured with a clock." A perfect clock is a device that (1) repeats an identical pattern of behavior, or cycle (such as the back-and-forth swinging of a pendulum) indefinitely many times in succession; (2) can be made in many identical copies that can be distributed to different locations and work independently of each other; (3) always continues to agree with all its copies; that is, they continue to perform their cycles simultaneously.

No actual device is a perfect clock, and a real clock's limitations show up when it and its "clones" begin to get out of step.

i. Working Definition of Time; The Second

For many centuries the prototype clock was the spinning Earth itself; the basic unit of time was the ordinary or solar day. (A given meridian of the Earth faces the Sun at intervals of one solar day.) The spinning Earth does not quite satisfy criteria (2) and (3) of a perfect clock, but it almost does, since the sky if full of other such "devices," not identical, but similar—the orbiting and spinning astronomical bodies. When suitably subdivided, their cyclic behavior does, in most cases, maintain a good agreement with that of the Earth.

When it became recognized that the solar day is not absolutely constant during the year, the unit of time was redefined as the mean, over as many years as possible, of the solar day. A more recent definition involved the Earth's orbit as a whole ("ephemeris time").

---

* In Chap. 5 such a point (the center of mass) will be specified.

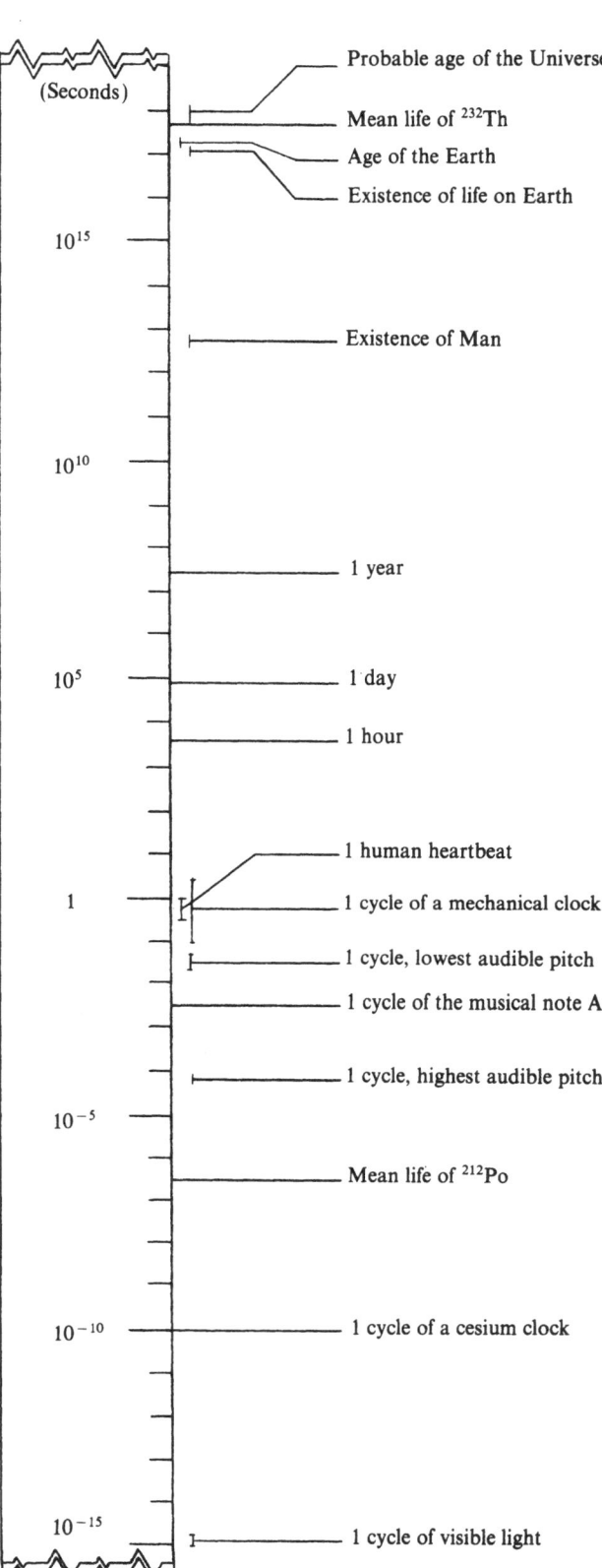

Figure 1.3. Some measured time intervals (logarithmic scale). The various durations shown here are representative of the orders of magnitude accessible to direct or indirect measurement. In contrast to distances, time intervals cannot well be grouped in terms of scientific disciplines. For example, one of the stablest radioactive elements, thorium 232, has a half-life (see Chap. 28) comparable to the age of the Universe, while a related element, polonium 212, is so unstable that its nuclei exist on the average for 4×10^{-7} sec. For further points of comparison, see the circular motion spectrum (Chap. 6), the acoustic spectrum (Chap. 13), and the electromagnetic radiation spectrum (Chap. 22), keeping in mind that frequencies rather than time intervals are displayed in the latter two spectra. There is no hint, in Nature, of anything like a shortest characteristic time. Radioactive particles with arbitrarily short mean lives, as well as radiation with arbitrarily short cycles of vibration appear to be possible.

Instrumentation has now progressed to the point where an atomic standard is much more precise (by at least three significant figures) than an average based on the spinning Earth. In Chap. 27 we shall see that atoms can emit accurately periodic electromagnetic waves and hence that certain well-chosen atoms can constitute a clock.

For most scientific purposes, however, the **mean solar day** provides an accurate enough standard. Subdivisions are defined as follows (with the abbreviations hour = h, minute = min, second = sec):

$$1 \text{ mean solar day} \approx 24 \text{ h} \qquad (\text{to 1 part in } 10^9)$$
$$1 \text{ h} = 60 \text{ min} \qquad (\text{exactly}) \quad (1.\text{D}.1)$$
$$1 \text{ min} = 60 \text{ sec} \qquad (\text{exactly})$$

The time elapsed between two clock readings is called a **time interval**, or **duration**. If the first reading is t_1 and the later reading is t_2, then the interval is the difference $t_2 - t_1$.

Current International Definition

The official definition of the second, in terms of atomic radiation, will be quoted in Chap. 27, after the necessary rationale has been explained.

ii. A Spectrum of Time Intervals

Figure 1.3 shows on a logarithmic scale the range of time intervals that are of interest to the physicist and which have been measured by a variety of methods.

Condensed Checklist

A "condensed checklist" will be provided at the end of each chapter. The main purpose of these lists is to serve as reminders for doing the problems: leafing through a chapter to locate a formula is no help to learning. The checklist also serves to review the subject matter: is each formula thoroughly understood? If so, its physical context has probably been grasped. Physics formulas must be memorized effortlessly, through their comprehension, or not at all.

Statement	Reference
Significant figures: all but the leading zeros	Sec. B i
Precision: measured by the number of significant figures	
Power-of-ten notation: $\Box \, . \, \Box \, \Box \, ... \times 10^{\Box}$	Sec. B ii
Physics problems: find the general (symbolic) solution before inserting the data; the precision of the answer depends on the precision of the data.	Sec. B iv
Distance (definition): the difference between two readings on an appropriate ruler;	Sec. C i
alternative definition, in terms of the delay in a light signal:	Chap. 25
The meter:	
Defined from the second and the speed of light	Chap. 25
Earth's circumference $\approx 4.000 \times 10^7$ meters	
1 inch $= 0.0254$ meter (exactly)	Sec. C ii
Decimal prefixes	Table 1.1
Conversion factors between units (table)	Appendix IV
Spectrum of distances	Fig. 1.1
Logarithmic plot of N: $D = s \log N$	(1.C.1)
Particle (definition): An object whose position is adequately described as that of a point	(1.C.2)
Position coordinate (definition): number that labels a marking in space.	Sec. C vi
Coordinate curve (definition): a curve in space, along which coordinates are marked	
Coordinate axis (definition): a straight coordinate line	
Time (definition): what is measured with a clock.	Sec. D, introduction

The second:

Atomic definition   Chap. 27

Approximate definition:

1 mean solar day $\approx$ 24.000 000 0 h

$$1 \text{ h} = 60 \text{ min} \qquad (1.D.1)$$

$$1 \text{ min} = 60 \text{ sec}$$

Spectrum of time intervals   Fig. 1.3

True or False[†]

1. Extensive measurements are needed before a good theory can be formulated.

2. In science, one should avoid excessive precision when recording a measurement.

3. All the following have the same number of significant figures: 208, 0.0208, 2.08×10^5, 28.0, 0.0280.

4. The number 32 is given to less precision than 0.032.

5. When multiplied by 10^6, an imprecise number becomes even less precise.

6. Zero is the smallest number that can be plotted on a logarithmic scale.

7. On a logarithmic scale, three points representing 1/3, 3, 27 define two equal intervals.

Problems[‡]

Section B: Precision in Numbers

*1.1. How many significant figures are there in the following numbers? (a) 3.1416; (b) 1492; (c) 0.001; (d) 1; (e) 0; (f) 2000; (g) 137.03602; (h) 0.000; (i) 2000.0; (j) 0.010×10^{34}.

1.2. How many significant figures are there in the following numbers? (a) 1800; (b) 1×10^3; (c) 20×10^{-6}; (d) 0.00×10^{48}; (e) 256.0; (f) 0×10^{-2}; (g) 0.00500; (h) 20.0; (i) 909.090; (j) 1.0001.

[In Problems 1.3–1.5, both excessive and insufficient precision should be counted as wrong; all quantities shown (other than powers of 10) are approximate. Hint for Problems 1.3 and 1.4: How large is the uncertainty in *each* term? Hint for Problem 1.5: Take the overall precision to be that of the *less* precise factor, as indicated by the number of significant figures.]

*1.3. Evaluate: (a) $35.6 + 0.439$; (b) $0 + 0.123$; (c) $0.0 - 0.518$; (d) $3 \times 10^8 + 6$; (e) $3 \times 10^2 - 142$; (f) $3 \times 10^2 - 300$; (g) $520 \times 10^{-7} - 0.52 \times 10^{-4}$; (h) $2 + 2.0$; (i) $2 - 2.0$; (j) $0 + 0.0$.

1.4. Evaluate: (a) $0 - 0.0$; (b) $0.50 + 0.5$; (c) $0.50 - 0.5$; (d) $0.048 \times 10^6 - 480 \times 10^2$; (e) $6 \times 10^4 + 1 \times 10^3$; (f) $60 \times 10^4 + 1 \times 10^3$; (g) $1 + 0.1 + 0.01$; (h) $3.1 - 3.2 + 0.10$; (i) $100 + 1$; (j) $1 \times 10^2 + 10$.

1.5. Evaluate: (a) $3.14 \times 1.3 \times 10^7$; (b) 999×1.001; (c) 251×302; (d) $(1.661 \times 10^{-27}) \times (6.022 \times 10^{26})$; (e) $\dfrac{1.602 \times 10^{-19}}{9.110 \times 10^{-31}}$.

*1.6. An egg fell from a height of 52 meters without breaking. Then it fell one additional millimeter and was completely smashed. By how many millimeters did the egg fall during these events?

*1.7. Euclidean geometry teaches that the circumference of a circle is π times its diameter, with $\pi \approx 3.14159265$. If we tried to verify this result by measurement, using objects found in the average home or stationery store, to how many figures might we expect to check π? Give a brief justification of your answer.

Sections C: Distance, and D: Time

1.8. Draw a logarithmic scale where the numbers 1 and 10 are plotted 5 cm apart. On your scale, show the numbers 0.2, 0.5, 2.0, 3.0, 4.0, 6.0.

*1.9. On the scale of Problem 1.8, which numbers are located 1.0, 2.0, 3.0 cm to the right of the number 1? Which numbers are located 1.0, 2.0, 3.0 cm to the left of the number 1?

1.10. If some positive numbers P, Q, R, S satisfy the proportion $P/Q = R/S$, it means that the distance between P and Q is the same as that between R and S on a logarithmic scale. Prove this mathematically.

*1.11. Using Fig. 1.1 and the result of Problem 1.10, (a) complete the approximate statement "Man is to Mount Everest as _____ is to Man"; (b) find at least three more proportions of that type in Fig. 1.1 and (c) at least two in Fig. 1.3.

Answers to True or False

1. False (a guess may work; measurements are needed for confirmation).

2. True.

3. True.

4. False.

[†] The answers are found after the Problems section.
[‡] An asterisk indicates that an answer or hint is given after the Problems section.

5. False.

6. False (cannot be plotted).

7. True.

Answers or Hints to Problems

1.1. (a) 5; (b) 4; (c) 1; (d) 1; (e) 0; (f) 1; (g) 8; (h) 0; (i) 5; (j) 2.

1.3. (a) 36.0; (b) 0; (c) -0.5; (d) 3×10^8; (e) 2×10^2; (f) 0×10^2; (g) 0×10^{-7}; (h) 4; (i) 0; (j) 0.

1.6. 52×10^3 (52001 is wrong).

1.7. With string or ruler, you should not expect much better than between 1% and 10% accuracy.

1.9. 1.58, 2.51, 3.98; 0.631, 0.398, 0.251.

1.11. (a) A small transistor.

CHAPTER 2

MOTION

Space and time are the arena in which all physical events take place. An object's position in space will generally change with time; this obvious phenomenon, **motion,** is used as our first clue for deducing the laws of physics. The study of motion is traditionally called **kinematics.**

This chapter deals with two main aspects of motion, namely, **velocity** and **acceleration,** and demonstrates how these quantities are used to predict the future trajectory of a particle from its present condition. Particular attention is paid here to the motion of a **freely falling object.** This chapter does not do justice to two important types of motion—circular and oscillatory—which will be treated separately later on.

The following material offers a self-contained review of the needed **differential calculus** and **vector analysis.** We shall at first only learn their basic language, whose effect is to make physics immeasurably easier to explain and to understand. The powerful techniques of these disciplines will remain mostly unexploited in this book; they should be learned in a calculus course.

A. Uniform Motion

A particle P is said to undergo *uniform motion* whenever the following two conditions occur together:

> (1) The trajectory of P is a straight line.
> (2) The distance traveled by P is proportional to the time elapsed.

$$(2.A.1)$$

By first examining this simplest of all motions we prepare ourselves to deal with far more complicated ones.

i. The Equation for Uniform Motion

Suppose, as in Fig. 2.1, that P moves along the x axis (say in the positive direction for definiteness). At times zero and t, P has coordinates x_0 and x, respectively. Then the distance covered in time t is $x - x_0$, and we have the proportionality

$$\boxed{x - x_0 = v_x t} \qquad (v_x = \text{const}) \quad (2.A.2)$$

The proportionality constant v_x is called the **velocity** of P in the x direction; the coordinate x_0 is called the **initial value** of x.

Questions of Sign

Let us call the motion **forward** if v_x is positive, and **backward** if v_x is negative. [In the latter case, $x - x_0$ in Eq. (2.A.2) becomes more and more negative.] The **distance** covered, l, is the absolute value of $x - x_0$:

$$l = |x - x_0| \qquad (2.A.3)$$

correctly defined for both directions of motion; "distance" always denotes a positive number.

Similarly, one defines a particle's **speed** v (without a subscript) as the *absolute value of the velocity*:

$$v = |v_x| \qquad (2.A.4)$$

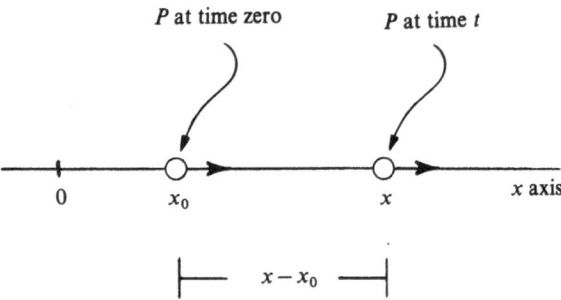

Figure 2.1. A particle (small circle), shown at two different times.

Equations (2.A.3) and (2.A.4) lead us to the simplest and most familiar statement of uniform motion. For positive t, Eq. (2.A.2) gives

$$l = vt \quad \text{or} \quad v = \frac{l}{t} \qquad (2.A.5)$$

Speed is distance traveled per unit time. Thus, the SI unit of speed (or velocity) is the meter/sec.

Note that Eqs. (2.A.5) carry less information than (2.A.2) since they do not indicate whether the motion is forward or backward.

> **Example 2.1. An Initial-Value Problem with Negative Velocity.** A runner, known for his steadiness of pace (7 km/h), passes the 16-km marker at 2 p.m.; he is heading toward the 15-km marker. Where is he expected at 4 p.m.?
>
> Equation (2.A.2) gives
>
> $$x = x_0 + v_x t \qquad (2.A.6)$$
>
> where x is the expected coordinate. Here the velocity is $v_x = -7$ km/h, and the initial value of x is $x_0 = 16$ km. Equation (2.A.6) becomes
>
> $$x = 16 \text{ km} + \left(-7 \frac{\text{km}}{\text{h}} \right) (4-2) \text{ h}$$
>
> $$= \underline{\underline{2 \text{ km}}}$$

The runner is expected at the 2-km marker.

ii. A Spectrum of Speeds

Uniform motion is rarely observed in the real world. Nevertheless, in a few cases of great interest one does observe close to perfectly uniform motion.

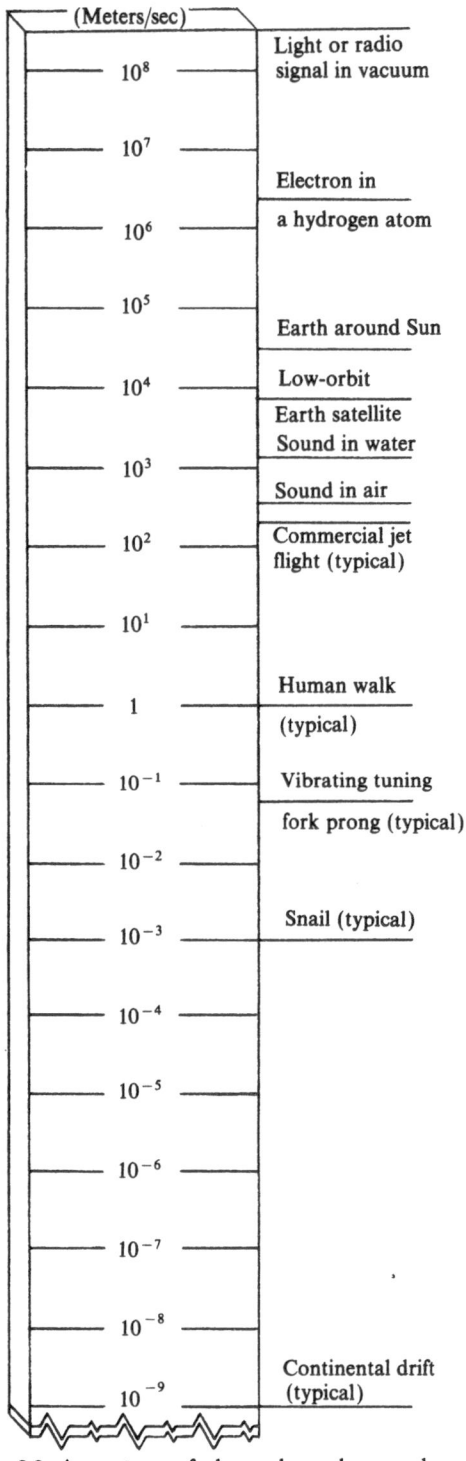

Figure 2.2. A spectrum of observed speeds, on a logarithmic scale. Since some of these examples involve curved trajectories, we are not always dealing with uniform motion. Nevertheless, in those cases the speed v is still defined as the length of (curved) path covered in unit time. The speed of light (3.0×10^8 meters/sec) constitutes an absolute upper limit. The speed of the Earth is with the Sun considered at rest; the listed sound speeds are as measured at room temperature.

This happens in particular for signals carried by waves traveling in a uniform medium, i.e., in an environment that has identical properties at all points. (A perfect vacuum is the prime example of such an environment; a reasonably homogeneous substance can often also be considered uniform.)

Such wave signals are not tangible objects in the usual sense, but their passage can be detected and at any single time they occupy a limited region of space, so that their motion can be treated as that of a particle.

No signal or object has ever been observed to travel faster than a light or radio signal in vacuum. This fact has great fundamental significance and will be discussed later in connection with the theory of special relativity.

A few representative speeds are shown in Fig. 2.2.

iii. Velocity and Graphs

Figure 2.3 shows x plotted against t in the case of a particle in uniform motion. In that figure, *the slope of the graph represents the velocity of the particle*:

$$\boxed{\text{Slope} = v_x} \qquad (2.A.7)$$

We let Δx = forward distance traveled = **displacement**, Δt = time elapsed, so that

$$\boxed{v_x = \frac{\Delta x}{\Delta t}} \quad \text{(a constant number)} \quad (2.A.8)$$

This is by definition the slope of a linear graph. In the backward case, both sides of (2.A.8) change sign, as they should.

B. Accelerated Motion (One-Dimensional)

Motion along a straight line (rectilinear motion) is further examined in this section. However, we now consider a velocity that changes in the course of time; the motion is called *accelerated*.

Everyone, especially a car driver, has a reasonably good intuitive feel for variable velocities. Nevertheless, considerable care is required to formulate a quantitative definition. This will be done graphically at first.

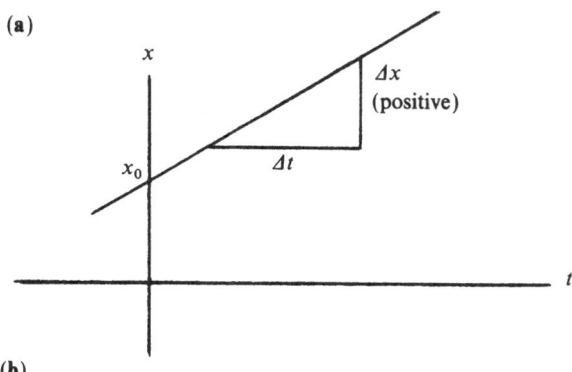
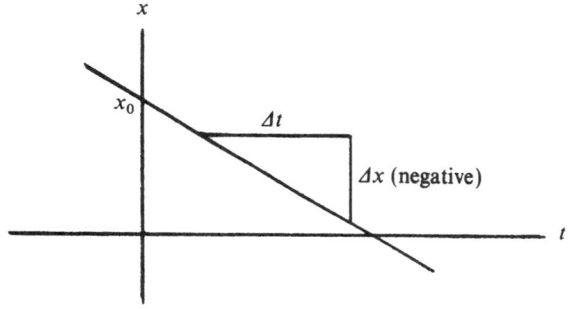
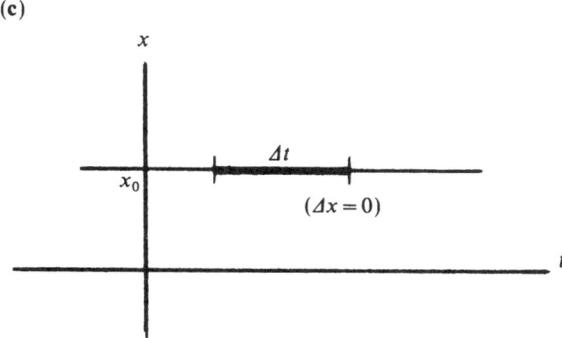

Figure 2.3. Position against time for a uniformly moving particle. The slope is the velocity. (a) Forward motion; (b) Backward motion; (c) No motion.

To avoid confusion, we must keep in mind that *a particle may have a straight path in space, while the graph of its position against time is curved.*

i. Slope of a Curve*

We have seen that, in the case of uniform motion, the velocity happens to be the slope of an appropriate straight-line graph. Can a slope be meaningfully defined if the graph is curved, i.e., if it

* This subsection may be omitted by anyone who has taken a course in calculus.

represents accelerated motion? To obtain such a definition, we start from the assumption that:

> An infinitesimal portion of a curve
> has its own slope (the local slope).   (2.B.1)

This local slope can be determined as that of a straight line if we accept the following statement:

> A small enough piece of curve cannot be
> distinguished from a segment of straight line.
>
> (2.B.2)

This is illustrated in Fig. 2.4.

Let us work within such a small region of the curve. Then we have

$$\text{(Slope of } x \text{ against } t) = \frac{\Delta x}{\Delta t} \quad \text{(small } \Delta t)$$

$$(2.B.3)$$

The smaller we make Δt, the smaller Δx will become, and the more accurately we shall match the slope of an infinitesimal segment of the curve. Hence we let Δt approach (but not reach) zero:

$$\Delta t \to 0$$

Correspondingly, Δx will also approach zero in just

such a way that the ratio $\Delta x / \Delta t$ gives us the local slope. In mathematical notation, we have

$$\text{(Slope of } x \text{ against } t) = \lim_{\Delta t \to 0} \frac{\Delta x}{\Delta t} \quad (2.B.4)$$

(Read: "The limit, as Δt approaches zero, of $\Delta x / \Delta t$".) This is abbreviated as follows:

$$\text{(Slope of } x \text{ against } t) = \frac{dx}{dt} \quad (2.B.5)$$

where dx/dt is called the **derivative** of x with respect to t.

It is essential to keep in mind that the derivative is an exact, rather than an approximate concept.

The slope of a mathematically given curve can in principle be calculated at any point. Figure 2.4 shows that *the local slope dx/dt of a curve at some point A is the slope of the straight line that is tangent to the curve at A.*

Example 2.2. Slope of a Parabola. Suppose x depends on t according to

$$x = Kt^2 \quad (2.B.6)$$

for some constant K. The graph of this function, Fig. 2.5, is called a **parabola**. Find its slope at an arbitrary t.

We must choose some value of t, and then increase it by a small amount Δt:

$$t \to t + \Delta t$$

The modified value of x is

$$x + \Delta x = K(t + \Delta t)^2$$
$$= K[t^2 + 2t\,\Delta t + (\Delta t)^2]$$

To find Δx, subtract, from the above, the corresponding sides of Eq. (2.B.6), with the result

$$\Delta x = K[2t\,\Delta t + (\Delta t)^2]$$

We need the ratio $\Delta x / \Delta t$:

$$\frac{\Delta x}{\Delta t} = K(2t + \Delta t)$$

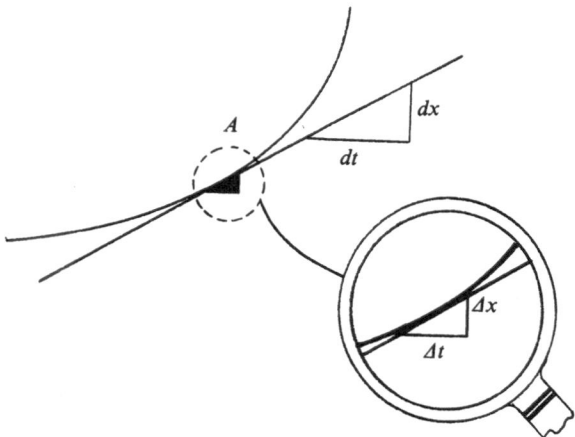

Figure 2.4. A small segment of curve (at A), when magnified, can hardly be told apart from a segment of straight line; see statement (2.B.2). The slope at A is $\Delta x/\Delta t$ in the limit when Δt, Δx (shown in the magnified view) are very small. This slope is denoted by dx/dt. The so-called differentials dx and dt can be visualized as the rectangular sides of a triangle along the line tangent to the curve at A.

Finally, let $\Delta t \to 0$:

$$\frac{dx}{dt} = K(2t + 0) = \underline{\underline{2Kt}}$$

the desired result. Note that this slope vanishes at $t = 0$, and keeps increasing for increasing t; it is positive or negative according to the sign of t. This can also be seen in Fig. 2.5.

Calculus provides a repertoire of ready-made formulas such as the one we have obtained,

$$\frac{d(Kt^2)}{dt} = 2Kt \qquad (2.B.7)$$

(Some further formulas are listed in Appendix III.) By using that repertoire one avoids having to do the type of step-by-step calculations outlined above.

Similarly to (2.B.7), one has

$$\frac{d(Kt)}{dt} = K, \qquad \frac{dK}{dt} = 0 \qquad (K = \text{const}) \quad (2.B.8)$$

It is useful to know, also, that the derivative of a sum may be calculated term by term: for any functions f, g, h,... we have

$$\frac{d(f + g + h + \cdots)}{dt} = \frac{df}{dt} + \frac{dg}{dt} + \frac{dh}{dt} + \cdots \quad (2.B.9)$$

ii. Instantaneous Velocity

Consider a particle P that occupies position x at time t. It is natural to define the (instantaneous) velocity v_x in a manner consistent with the uniform-motion formula (2.A.8):

$$v_x = \lim_{\Delta t \to 0} \frac{\Delta x}{\Delta t}$$

or

$$\boxed{v_x = \frac{dx}{dt}} \qquad (2.B.10)$$

In words: *the velocity is the time rate of change of position.* From the preceding subsection, it is clear that v_x *is also the slope of the graph of x against t.*

The speed is still defined as the absolute value of the velocity,

$$v = |v_x| \qquad (2.B.11)$$

Example 2.3. The Quadratic Time Dependence Again. A certain marble, clocked while rolling down an inclined plane, is found to have a downhill coordinate x, measured in meters along the incline, of

$$x = 3t^2 - 2t + 1 \qquad (2.B.12)$$

where t is measured in seconds. Find the marble's velocity at time $t = 2$ sec.

Differentiating (2.B.12) with respect to t, and using results (2.B.7), (2.B.8), (2.B.9), we have, in units of meter/sec,

$$v_x = \frac{dx}{dt} = (2)(3)t + (-2) + 0$$

or, at $t = 2$ sec,

$$v_x = 10 \, \underline{\underline{\frac{\text{meters}}{\text{sec}}}}$$

iii. Acceleration

In dealing with a variable velocity it becomes meaningful to ask at what rate the velocity itself changes.

The previous subsections already contain all the necessary tools to calculate a rate of change. These can be applied to changing velocities.

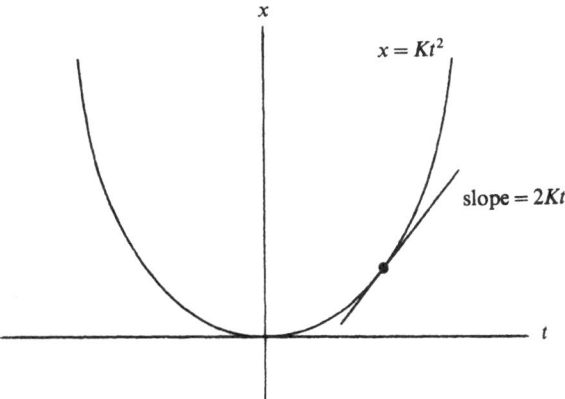

Figure 2.5. A parabola and its slope.

Table 2.1. Definitions of velocity and acceleration.

One-dimensional velocity	One-dimensional acceleration	
The velocity is the time rate of change of the position.	The acceleration is the time rate of change of the velocity.	(2.B.13)
Velocity is the slope of position-against-time.	Acceleration is the slope of velocity-against-time.	(2.B.14)
$v_x = \lim\limits_{\Delta t \to 0} \dfrac{\Delta x}{\Delta t} = \dfrac{dx}{dt}$	$a_x = \lim\limits_{\Delta t \to 0} \dfrac{\Delta v_x}{\Delta t} = \dfrac{dv_x}{dt}$	(2.B.15)

In analogy to v_x, one defines an **acceleration** a_x in the x direction. The left column of Table 2.1 recalls the various ways of defining v_x, and the right column lists the corresponding statements for a_x. This last relation allows us to define the acceleration in terms of the position. We have, according to (2.B.15),

$$a_x = \frac{d\left(\dfrac{dx}{dt}\right)}{dt} \qquad (2.B.16)$$

This is denoted for short by

$$a_x = \frac{d^2 x}{dt^2} \qquad (2.B.17)$$

The acceleration is the second time derivative of the position coordinate.

Figure 2.6 illustrates how x, v_x, a_x are plotted from one another.

It is probably becoming clear to the reader that he can, if he wishes, define rates of change in infinite succession. In fact, some branches of engineering make use of the time rate of change of acceleration (called the **jerk**). In this book there is no need to go beyond the acceleration.

Units

Any one of the statements (2.B.13)–(2.B.15) in the right-hand column shows that the units of acceleration are those of velocity/time. Hence the SI unit of acceleration is

$$\frac{1\left(\dfrac{\text{meter}}{\text{sec}}\right)}{1 \text{ sec}} = 1 \frac{\text{meter}}{\text{sec}^2} \qquad (2.B.18)$$

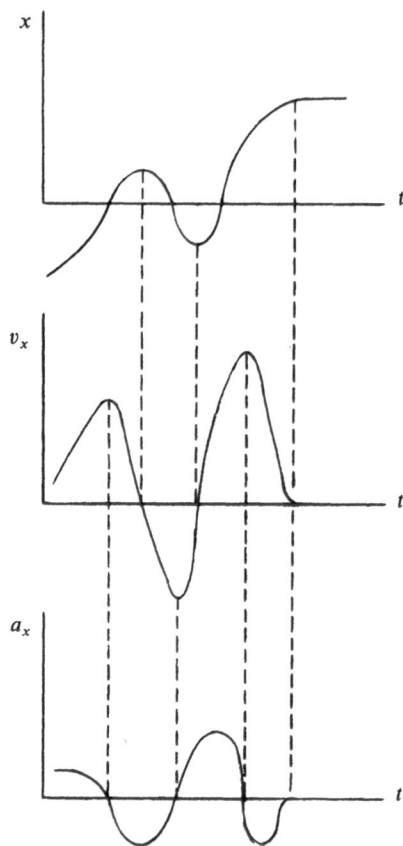

Figure 2.6. Plotting v_x from x, and a_x from v_x. Note how a horizontal slope in x corresponds to a zero value in v_x, and similarly from v_x to a_x.

[We see that a unit is treated algebraically like any number; Eq. (2.B.18) is analogous to the familiar relation $(a/b)/b = a/b^2$. If, for example, a velocity, given in km/h, is known to change by a certain amount every second, we must face the combination $1(\text{km/h})/\text{sec} = 1 \text{ km/(h sec)} = (1/3600)(\text{km/sec}^2)$, where we have used $1 \text{ h} = 3600$ sec.]

Case of Zero Acceleration

If $a_x = 0$ at all times, then we have $v_x = \text{const}$, and we are back to the case of uniform motion.

C. Vertical Free Fall

The simplest possible time dependence for a velocity v_x occurs when no dependence exists at all, $v_x = \text{const}$ (uniform motion). The next simplest case, **uniformly accelerated motion** (acceleration = const), happens to be exhibited in nature by freely falling objects.

i. The Acceleration of Gravity

On earth, any object released from rest in vacuum (or *when air resistance can be neglected*) falls vertically down in such a way that *its velocity is in proportion to the time elapsed*:

$$v_{\text{down}} = gt \qquad (2.\text{C}.1)$$

The proportionality constant, g, can be seen from this relation to have the SI units of velocity/time, or meter/sec², it is called the **acceleration of gravity** and is measured to be

$$g = 9.8 \, \frac{\text{meters}}{\text{sec}^2} \qquad (2.\text{C}.2)$$

at sea level; it varies slightly with the geographic location and with the altitude, but, at a given location, it is rigorously—in our model without air drag—the same for all objects:

> The acceleration of free fall is independent of the object's size, shape, chemical composition, and, especially, weight. $\qquad (2.\text{C}.3)$

This extraordinary observation is due to Galileo Galilei (1564–1642) who, according to legend, demonstrated its validity by dropping two unequal weights from the leaning tower of Pisa, his native city. After three centuries, statement (2.C.3), now called the **principle of equivalence**, has become a cornerstone of Einstein's general theory of relativity.

For the purpose of this book, the value

$$g \approx 10 \, \frac{\text{meters}}{\text{sec}^2} \qquad (2.\text{C}.4)$$

although slightly wrong to two significant figures, is convenient and usually adequate. Figure 2.7 shows the value of g at the surface of some astronomical bodies.

How is g measured? One can, for example, apply high-speed photography to an object falling along a vertical distance scale. However, the simplest and

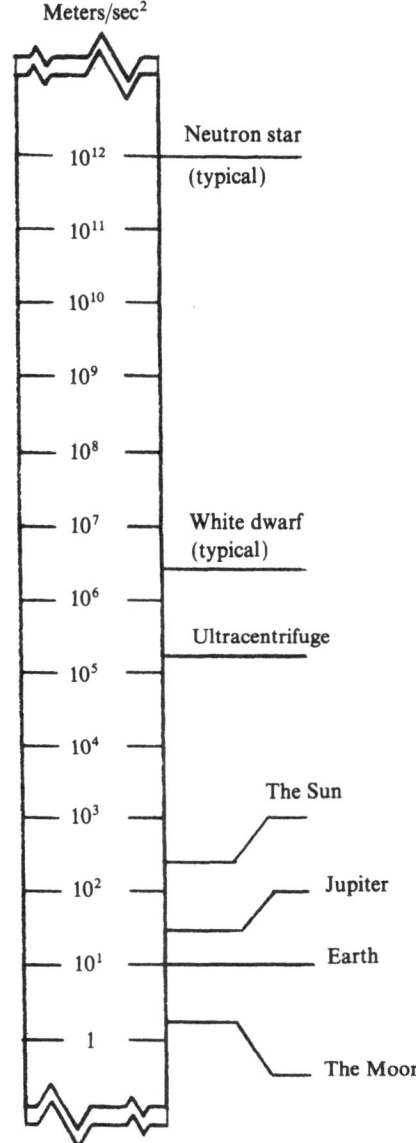

Figure 2.7. A spectrum of g's. The acceleration of gravity at the surface of some astronomical bodies is shown on a logarithmic scale. A white dwarf is a star highly compacted (to planetary size) by its own gravity. Neutron stars, which almost certainly exist, are even smaller, their interiors consisting of solidly packed neutrons. There is no obvious theoretical limit to the size of g. The simulated gravity in a good laboratory ultracentrifuge is shown for comparison, see also Sec. F of the next chapter.

most accurate methods *infer g* from a partially controlled fall: they involve rolling an object down an inclined plane; using strings and pulleys; or best of all, observing the oscillations of a pendulum. A discussion of these methods must be postponed until the concept of force has been introduced.

The value of g at a given location can be

measured to better than one part in 10^6. The difference in g that *would* exist between two bodies of different composition if the equivalence principle were wrong can be measured more accurately still—in some cases to better than $10^{-11}g$. Even such an amazing sensitivity fails to detect any deviation from equivalence.

Nonzero Initial Velocity

Figure 2.8 shows v_{down} plotted against t, Eq. (2.C.1). If the launching occurs at a time other than zero, or if the object is thrown vertically at time zero, the graph of Fig. 2.9 is applicable; it fits the case of a *downward throw*. To modify the graph for an *upward throw*, it should be parallel-shifted to below the origin (reader, please verify; note then the initially decreasing *speed*, a **deceleration**).

The general equation is

$$v_{down} = v_{0,down} + gt \qquad (2.C.5)$$

which is basic for uniformly accelerated motion.

To verify that g is, technically speaking, the downward acceleration, we take the time derivative on both sides of (2.C.5):

$$\frac{dv_{down}}{dt} = 0 + g$$

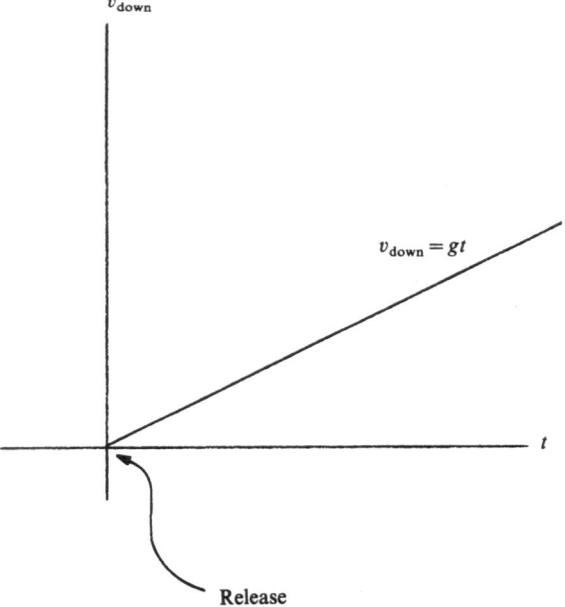

Figure 2.8. Downward velocity as a function of time. The falling object is at rest when $t = 0$.

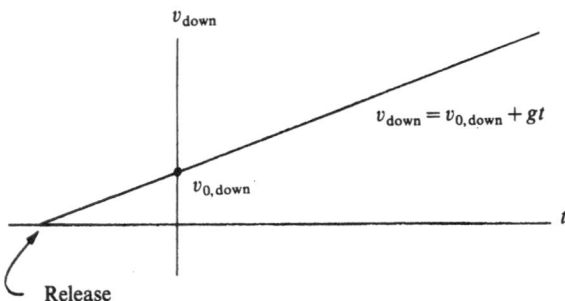

Figure 2.9. This is how an earlier release modifies Fig. 2.8. The positive t portion of the graph could result from a downward throw, with speed v_{0x}, at time zero.

or

$$\boxed{a_{down} = g} \qquad (2.C.6)$$

ii. Uniformly Accelerated Motion

Given the initial velocity of a falling particle, we can predict its velocity at any later time, through Eq. (2.C.5). Given the particle's initial altitude, we want to predict its later altitudes as well. *The particle's later motion would then be completely determined from the initial data.* Here we solve this problem for any constant a_x.

Let a particle P move along the x axis with a velocity [see Eq. (2.C.5)]

$$\boxed{v_x = v_{0x} + a_x t} \qquad (a_x = \text{const}) \quad (2.C.7)$$

Where is P located at any given t, if its initial coordinate is x_0? This question will be answered in two steps.*

1. *Guessing the x Coordinate.* Consider a time interval t, starting at time zero. The velocity of P is v_{0x} at the beginning of that interval, and v_x at the end. The average of these two values is

$$v_{av} = \tfrac{1}{2}(v_{0x} + v_x) = \tfrac{1}{2}[v_{0x} + (v_{0x} + a_x t)]$$
$$= v_{0x} + \tfrac{1}{2}a_x t$$

Perhaps the displacement of P during time t can be computed naively as in the case of uniform motion,

---

* The student who feels at ease with integration will prefer it as a far more direct method for getting from (2.C.7) to (2.C.9).

but using the average velocity. If this is legitimate, we have

$$x - x_0 \overset{?}{=} v_{av} t$$

which becomes

$$x - x_0 \overset{?}{=} v_{0x} t + \tfrac{1}{2} a_x t^2 \qquad (2.C.8)$$

2. Verifying the Guess. A guess is a risky substitute for a derivation. To check Eq. (2.C.8), we take the time derivative on both sides. The result is [recall Eqs. (2.B.7), (2.B.8), (2.B.9)]

$$\frac{dx}{dt} - 0 \overset{?}{=} v_{0x} + a_x t$$

or

$$v_x \overset{?}{=} v_{0x} + a_x t$$

This is indeed the correct velocity at all times t. We safely conclude from (2.C.8) that x is given by

$$x = x_0 + v_{0x} t + \tfrac{1}{2} a_x t^2 \qquad (2.C.9)$$

How Does the Velocity Depend on Position?

We already know how v_x depends on t. To find how it depends on x, we eliminate t between Eqs. (2.C.7) and (2.C.9). [An easy method: square both sides of (2.C.7), divide by $2a_x$, and subtract the corresponding sides of (2.C.9).] The result may be written

$$v_x^2 - v_{0x}^2 = 2a_x(x - x_0) \qquad (2.C.10)$$

The square of the velocity increases by an amount proportional to the displacement.

[An important caution: the reader must keep in mind that the formulas of this section are not general, but valid only when $a_x = \text{const.}$]

iii. Free Fall Calculations

Let us apply the preceding results to a freely falling particle P. Let z denote the **altitude** of P above some fixed reference level such as the ground

(z is the physicist's favorite symbol for a vertical coordinate). The downward direction is then negative, and

$$a_z = -g \qquad (2.C.11)$$

In this notation, Eqs. (2.C.7), (2.C.9), (2.C.10) become

$$v_z = v_{0z} - gt \qquad (2.C.12)$$

$$z = z_0 + v_{0z} t - \tfrac{1}{2} g t^2 \qquad (2.C.13)$$

$$v_z^2 - v_{0z}^2 = -2g(z - z_0) \qquad (2.C.14)$$

The velocity and altitude, Eqs. (2.C.12) and (2.C.13), are plotted in Fig. 2.10.

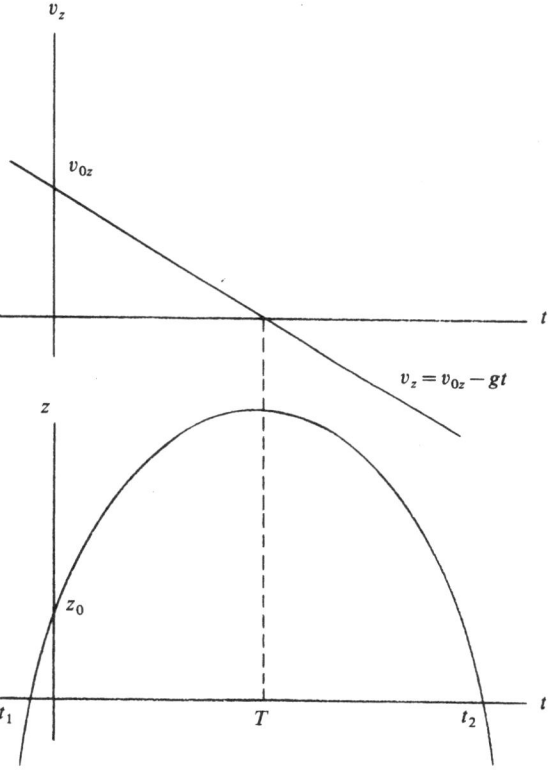

Figure 2.10. Velocity (top) and altitude (bottom) under vertical free fall. This figure illustrates an *upward* velocity at time zero. The parabola's shape is symmetric about the time T at which the highest altitude is reached. (To verify this easily, choose the origin of time such that $T=0$, which means $v_{0z} = 0$.) Hence, in this figure, we have $T - t_1 = t_2 - T$.

Example 2.4. Time of Flight and Maximum Altitude. If a stone is thrown upwards, from ground level, with an initial speed of 12 m/s, (a) how much time does the stone need to reach its highest altitude? (b) How high does it rise? (c) After how much time does it fall back to the ground?

(a) Time of rise: At the highest point, the velocity vanishes, $v_z = 0$. Equation (2.C.12) gives

$$0 = v_{0z} - gt$$

so that

$$t = \frac{v_{0z}}{g} = \frac{12}{10} \sec = 1.2 \sec \qquad (2.\text{C}.15)$$

(b) Maximum altitude: We could insert the time of rise, (a) above, into Eq. (2.C.13) to find z. More directly, from (2.C.14) with $z_0 = 0$, $z = z_{max} =$ maximum altitude, and $v_z = 0$:

$$0^2 - v_{0z}^2 = -2g(z - 0)$$

so that

$$z_{max} = \frac{v_{0z}^2}{2g} \qquad (2.\text{C}.16)$$

$$= \frac{(12)^2}{(2)(10)} \text{ meters} = 7.2 \text{ meters}$$

(c) Total time of flight: The symmetry shown in Fig. 2.10 implies that the time of fall equals the time of rise:

$$\text{Total time} = (2)(\text{time of rise}) = 2.4 \sec$$

Alternatively (without using symmetry), apply Eq. (2.C.13) with $z_0 = 0$ (ground level launching) and $z = 0$ (ground level arrival). We have

$$0 = 0 + v_{0z}t - \tfrac{1}{2}gt^2$$

with solutions

$$t = 0$$

(the stone has not yet left the ground) and

$$t = \frac{2v_{0z}}{g} \qquad (2.\text{C}.17)$$

$$= \frac{(2)(12)}{10} \sec = 2.4 \sec$$

as before.

Example 2.5. Time for a Less-Than-Maximum Rise. How much time does the stone of the preceding example need in order to reach an altitude of 2.2 meters?

Equation (2.C.13) is quadratic in t. It has two solutions,

$$t = \frac{v_{0z} \pm [v_{0z}^2 - 2g(z - z_0)]^{1/2}}{g} \qquad (2.\text{C}.18)$$

$$= \frac{12 \pm [(12)^2 - (2)(10)(2.2)]^{1/2}}{10} \sec$$

$$= \begin{cases} 2.2 \sec \\ \text{or} \\ 0.2 \sec \end{cases}$$

How should the double solution be interpreted? The answer is that the lesser time is needed to reach the required altitude on the way up; the same altitude is revisited later, on the way down. The reader is invited to sketch a graph of z against t to visualize the situation better.

iv. Dimensional Analysis

In our calculations, the units of meter and second are getting combined in a variety of ways. A surprising amount of information can be extracted simply from requiring that all units be combined consistently.

If two physical quantities satisfy an equality such as

$$A = B$$

then they must be expressible in the same units. Equivalently stated, A and B, have the same dimensions.

This obvious and trivial fact is of great help in

verifying, remembering, and even deriving physical laws. Two examples will make this clear.

Example 2.6. The Displacement Formula. In an emergency (an urgent spaceship maneuver or a closed-book examination), someone becomes unsure of the distance x covered in time t when starting from rest, under a constant acceleration a_x. Is it $x = \frac{1}{2}a_x^2 t$ or $x = \frac{1}{2}a_x t^2$?

Let us use the metric system. Then x is expressed in meters, for example,

$$x = 5 \text{ meters}$$

Dimensional analysis is not concerned with numerical values. Hence we write

$$x \sim \text{meter}$$

(read: "x has the units of meter"). Similarly,

$$a_x \sim \frac{\text{meter}}{\text{sec}^2}$$

$$t \sim \text{sec}$$

Pure numbers have no units, and hence they are all equivalent to each other in this method, for example,

$$\tfrac{1}{2} \sim 1, \qquad \pi \sim 1, \qquad \text{etc.}$$

To test the formula $x \stackrel{?}{=} \frac{1}{2}a_x^2 t$, insert the units:

$$\text{meter} \stackrel{?}{\sim} (1)\left(\frac{\text{meter}}{\text{sec}^2}\right)^2 (\text{sec})$$

or

$$\text{meter} \stackrel{?}{\sim} \frac{\text{meter}^2}{(\text{sec}^3)}$$

which is wrong. On the other hand, $x \stackrel{?}{=} \frac{1}{2}a_x t^2$ gives

$$\text{meter} \stackrel{?}{\sim} (1)\left(\frac{\text{meter}}{\text{sec}^2}\right) (\text{sec})^2$$

or

$$\underline{\text{meter} \sim \text{meter}}$$

which verifies the formula. This method is, of course, incapable of checking purely numerical factors such as 1/2.

Example 2.7. The Maximum-Altitude Formula. How high does a stone rise if thrown upwards at speed v_{0z}? "Derive" the formula by dimensional analysis.

———

Assume that the result, z, depends only on v_{0z} and on g. (Admittedly, this is quite an educated assumption, which in fact largely solves the problem.) Set

$$z \sim \text{meter}$$

$$v_{0z} \sim \frac{\text{meter}}{\text{sec}}$$

$$g \sim \frac{\text{meter}}{\text{sec}^2}$$

We must combine g and v_{0z} so as to get z. Hence we try

$$z \sim v^\alpha g^\beta \qquad\qquad (2.C.19)$$

for some unknown exponents α, β. This gives

$$\text{meter} \sim \left(\frac{\text{meter}}{\text{sec}}\right)^\alpha \left(\frac{\text{meter}}{\text{sec}^2}\right)^\beta$$

or

$$\text{meter} \sim \frac{\text{meter}^{\alpha+\beta}}{\text{sec}^{\alpha+2\beta}}$$

Comparing both sides of this equation, we conclude that

$$\alpha + \beta = 1$$

$$\alpha + 2\beta = 0$$

The solution is $\alpha = 2$, $\beta = 1$. Therefore (2.C.19) reads

$$z \sim \frac{v_{0z}^2}{g}$$

This is as close as dimensional analysis will come to giving use Eq. (2.C.16),

$$z = \frac{v_{0z}^2}{2g}$$

D. Vectors

In this second half of the chapter (Secs. D, E, F) we broaden our expertise from one-dimensional to (mostly) two-dimensional. Real-life objects are seldom considerate enough to move along straight lines, and therefore we must analyze *curved trajectories*. For that purpose there exists a convenient formalism, called **vector calculus** or **vector analysis**, whose rudiments are given in this section. Vector calculus, more or less in the form presented here, was developed by Josiah Willard Gibbs (1839–1903).

i. Displacements

Consider any two fixed points A, B in space; see Fig. 2.11. If a particle starts at A and follows any path to B, the net result of the motion (namely, location A followed by location B) depends on A and B, but not, of course, on the details of the intermediate path. The net result is called a **displacement**, or **displacement vector**; it is represented pictorially by a straight arrow whose tail is at A and whose head is at B. More abstractly, the displacement vector is symbolized by the notation $\overrightarrow{AB}$. The length of $\overrightarrow{AB}$ (that is, the length of the arrow) is called the **magnitude** or **absolute value** of $\overrightarrow{AB}$ and is denoted by $|\overrightarrow{AB}|$. For example, if A and B are located 1 km apart (straight-line distance), then $|\overrightarrow{AB}| = 1$ km. The vector $\overrightarrow{AB}$ carries more information (direction as well as magnitude) than an ordinary number such as $|\overrightarrow{AB}|$. Displacements will serve as models for all other vectors to be met later in this book, insofar as they possess the following special **addition property**.

Two successive displacements $\overrightarrow{AB}$ and $\overrightarrow{BC}$ can be considered as a single overall displacement $\overrightarrow{AC}$. This displacement from A to C is called the **vector sum**, or simply the **sum**, of the two individual displacements:

$$\boxed{\overrightarrow{AC} = \overrightarrow{AB} + \overrightarrow{BC}} \qquad (2.\text{D}.1)$$

The relation is illustrated in Fig. 2.12 and gives rise to the "head-to-tail method" of adding two vectors. Warning: the figure shows that, in general, no such equality is true for the magnitudes,

$$|\overrightarrow{AC}| \neq |\overrightarrow{AB}| + |\overrightarrow{BC}|$$

The equality holds only if $\overrightarrow{AB}$ and $\overrightarrow{BC}$ are parallel.

To obtain the correct magnitude, apply the "cosine rule" to the angle α shown in Fig. 2.12b:

$$|\overrightarrow{AC}|^2 = |\overrightarrow{AB}|^2 + |\overrightarrow{BC}|^2 - 2\,|\overrightarrow{AB}|\,|\overrightarrow{BC}|\cos\alpha$$
$$(2.\text{D}.2)$$

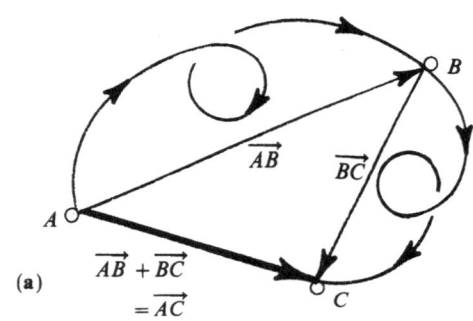

(a)

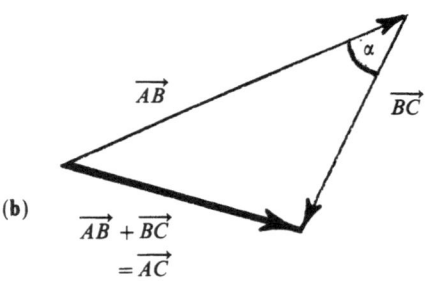

(b)

Figure 2.12. Two successive displacements (vector addition). (a) The actual paths with their corresponding vectors; (b) The vectors without the irrelevancies. The net displacement is labeled $\overrightarrow{AB} + \overrightarrow{BC}$. Part (b) of this figure should serve as a set of instructions for adding two vectors $\overrightarrow{AB}$ and $\overrightarrow{BC}$: put the tail of $\overrightarrow{BC}$ at the head of $\overrightarrow{AB}$; the sum $\overrightarrow{AB} + \overrightarrow{BC}$ then extends from the tail of $\overrightarrow{AB}$ to the head of $\overrightarrow{BC}$. It is clear that the *magnitudes* (lengths) of $\overrightarrow{AB}$ and $\overrightarrow{BC}$ do not add up to give that of $\overrightarrow{AB} + \overrightarrow{BC}$.

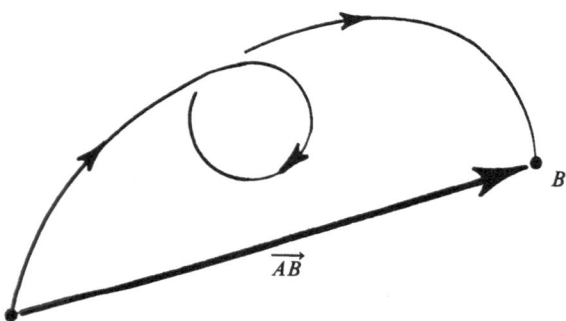

Figure 2.11. The displacement vector. The vector $\overrightarrow{AB}$ (heavy arrow from A to B) summarizes the net displacement without reference to the details of the curved path from A to B.

ii. Operations with Vectors

For our purposes, *a vector is any mathematical entity that has magnitude and direction, and that follows the same calculational rules as displacements.* Pictorially, and with a single exception (the "zero vector"), a vector can always be represented by a straight arrow whose length represents its magnitude. Of course, vector diagrams may be freely scaled up or down to permit a convenient representation on paper.

The basic rules of vector algebra are as follows:

1. Vector Equality. Two vectors are said to be *equal* if and only if they have the same magnitude and the same direction. In Fig. 2.13, we have $\overrightarrow{AB} = \overrightarrow{CD}$. Thus, by definition, *a vector remains unchanged if shifted parallel to itself.*

Notation: Since $\overrightarrow{AB}$ and $\overrightarrow{CD}$ are the same vector, they can be denoted by the same symbol—one that makes no reference to the end points A, B, C, D. For example,

$$\overrightarrow{AB} = \overrightarrow{CD} = \mathbf{p}$$

(Boldface type will be reserved for vectors. In handwritten work, use the notation $\vec{p}$.) Figure 2.14 illustrates some cases of vector equality and inequality.

2. Vector Addition. Vectors must be *added* like successive displacements. Figure 2.15 illustrates the vector sum

$$\mathbf{p} + \mathbf{q} = \mathbf{r}$$

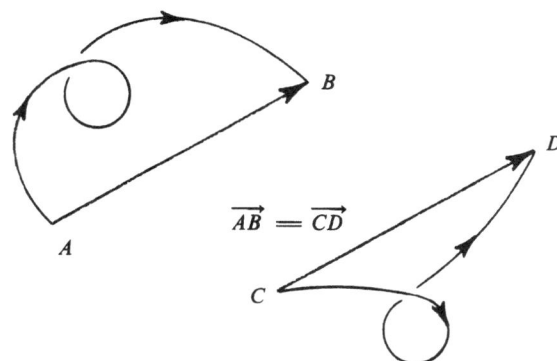

Figure 2.13. Two equal displacement vectors, $\overrightarrow{AB} = \overrightarrow{CD}$. The points A, B, C, D are at four distant locations.

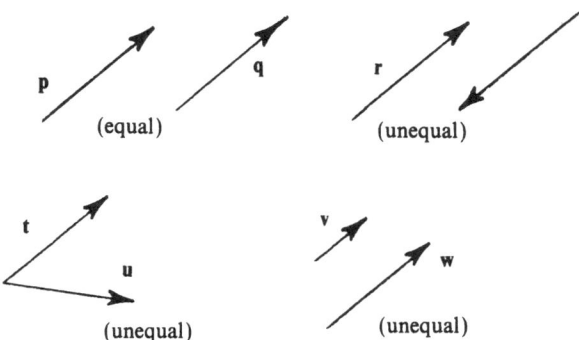

Figure 2.14. Equality and inequality among vectors. In these examples, $\mathbf{p} = \mathbf{q}$, $\mathbf{r} \neq \mathbf{s}$, $\mathbf{t} \neq \mathbf{u}$, $\mathbf{v} \neq \mathbf{w}$; however, $\mathbf{s} = -\mathbf{r}$.

and shows the commutative property

$$\mathbf{p} + \mathbf{q} = \mathbf{q} + \mathbf{p}$$

It also shows a "parallelogram method" of adding two vectors, which is equivalent to the "head-to-tail method."

3. Negative of a Vector. The vector $-\mathbf{p}$ has the same magnitude as $\mathbf{p}$ but points in the opposite direction. (A magnitude is never negative.) In Fig. 2.14, we have $\mathbf{s} = -\mathbf{r}$. In the case of displacements, we always have $\overrightarrow{BA} = -\overrightarrow{AB}$.

4. Zero Vector. We see that the sum

$$\mathbf{p} + (-\mathbf{p})$$

is a vector joining a point to itself (use the head-to-tail method). Hence this resultant vector has zero length and no well-defined direction. In fact, it cannot be drawn as an arrow, and is called the **zero vector, 0**:

$$\mathbf{p} + (-\mathbf{p}) = \mathbf{0}$$

The symbol **0** is often written as an ordinary zero without great risk of confusion.

5. Vector Subtraction. The difference $\mathbf{p} - \mathbf{q}$ is defined as

$$\mathbf{p} - \mathbf{q} = \mathbf{p} + (-\mathbf{q})$$

Hence it is reduced to the sum of two vectors. Figure 2.16 shows the construction for finding $\mathbf{p} - \mathbf{q}$ when $\mathbf{p}$ and $\mathbf{q}$ are given.

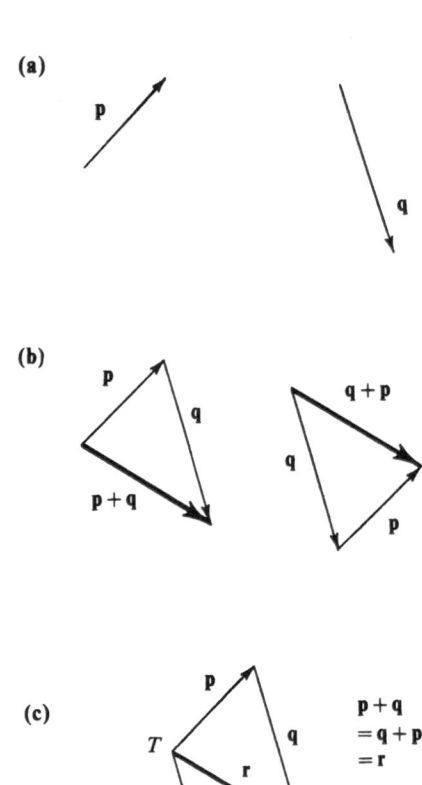

(a)

(b)

(c)

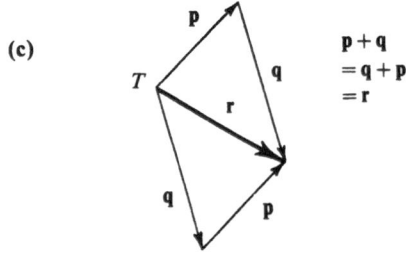

(d)

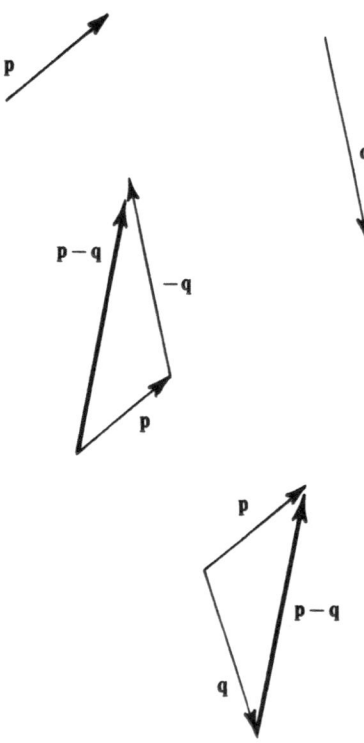

Figure 2.16. Vector subtraction. If **p** and **q** are given vectors, this figure shows two equivalent methods of constructing the vector difference **p** − **q**. Note that **p** − **q** would correspond to the unused diagonal in Fig. 2.15c.

As a special case, we have

$$\mathbf{p} - \mathbf{p} = \mathbf{p} + (-\mathbf{p}) = 0$$

surely a reasonable result.

6. Multiplying a Vector by a Number.* If n is a positive number and **p** is a vector, then $n\mathbf{p}$ is a new vector whose magnitude is $n\,|\mathbf{p}|$ and which has the same direction as **p**. To multiply **p** by a negative number, consider $(-n)\mathbf{p}$, where n is still positive. This product is naturally defined as

$$(-n)\mathbf{p} = -(n\mathbf{p})$$

i.e., the negative of $n\mathbf{p}$. Figure 2.17 shows some examples of a number multiplying a vector. Note that $(0)(\mathbf{p}) = \mathbf{0}$.

The same distributive properties are valid as in

Figure 2.15. Vector addition again. In (a), two vectors **p** and **q** are given. In (b) they are added according to the prescription of Fig. 2.12, but in two different orders **p** + **q** and **q** + **p**. Putting these two triangles against each other, as in (c), demonstrates that the result is the same irrespective of the order of terms; this part of the figure also illustrates a "tail-to-tail," or "parallelogram" method for adding two vectors **p** and **q** together at T; complete the parallelogram; the diagonal extending out of T then provides the vector sum **p** + **q**. Part (d) of this figure shows particularly simple cases, in which the magnitudes add or subtract:

$$|\mathbf{r} + \mathbf{s}| = |\mathbf{r}| + |\mathbf{s}|, \qquad |\mathbf{t} + \mathbf{u}| = |\mathbf{t}| - |\mathbf{u}|, \qquad |\mathbf{v} - \mathbf{v}| = 0$$

* Ordinary numbers are sometimes referred to as **scalars** when they are contrasted to vectors. Since the term "scalar" has acquired a more technical connotation it is better avoided in this context.

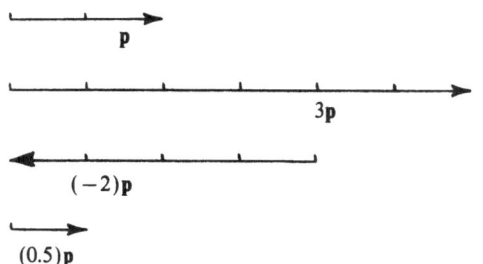

Figure 2.17. Multiplying an ordinary number by a vector. For a given **p**, this example shows **p** [or $(1)(\mathbf{p})$], $3\mathbf{p}$, $(-2)\mathbf{p}$, and $(0.5)\mathbf{p}$.

ordinary arithmetic or algebra: if m, n are any two numbers, then

$$m\mathbf{p} + n\mathbf{p} = (m+n)\mathbf{p}$$

If **p** and **q** are any two vectors, then

$$n\mathbf{p} + n\mathbf{q} = n(\mathbf{p}+\mathbf{q})$$

These rules are welcome, allowing us, for example, to write

$$\mathbf{p} + \mathbf{p} = 2\mathbf{p}$$

7. Multiplying a Vector by a Vector. We shall refrain until Chap. 4 (Energy) from such a novel operation.

8. Meaningless Expressions. Equalities such as $n = \mathbf{p}$, where n is a number and **p** is a vector, are totally meaningless. Similarly, it makes no sense to mix numbers and vectors in a sum, as in $n + \mathbf{p}$.

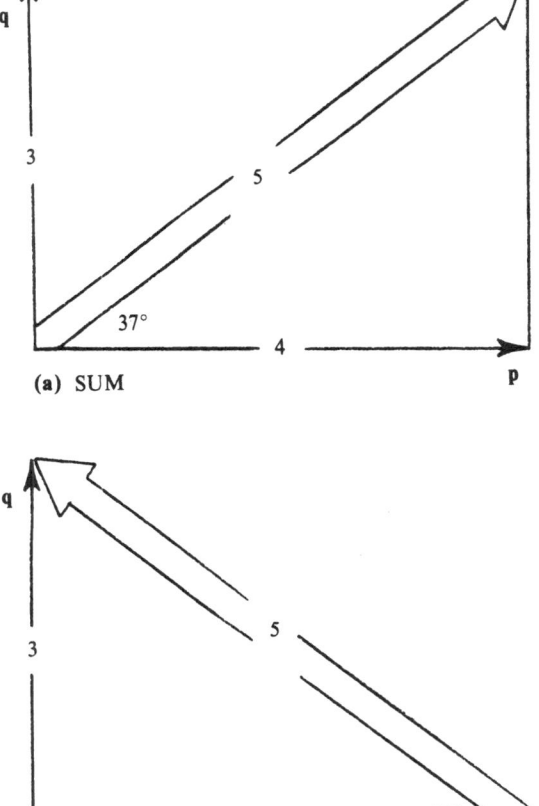

(a) SUM

(b) DIFFERENCE

Figure 2.18. (a) How to find the sum of an eastward (**p**) and a northward (**q**) vector with respective magnitudes 4 and 3. The resultant magnitude is $[(3)^2 + (4)^2]^{1/2} = 5$ from Pythagoras's theorem; the resultant direction can be found by trigonometry from either one of the triangles shown. For example, if θ is the resultant angle with respect to east, then $\tan\theta = 3/4$, or $\theta \approx 37°$. Equivalently, from the same triangle, $\sin\theta = 5/3$ or $\theta \approx 37°$. (b) How to find the difference $\mathbf{q} - \mathbf{p}$ between those same vectors. Calculations are particularly simple here owing to the orthogonality of **p** and **q**. In particular, note that $|\mathbf{q}-\mathbf{p}| = |\mathbf{p}+\mathbf{q}|$, a fact not usually true.

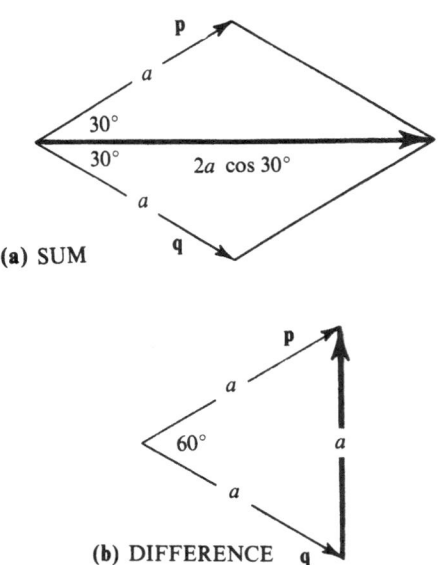

(a) SUM

(b) DIFFERENCE

Figure 2.19. (a) We are given **p** and **q** ($|\mathbf{p}| = |\mathbf{q}| = a$), pointing 30° north and south of east, respectively. The resultant direction is found to be due east *without calculations*, because the figure is symmetric about the east–west line. Symmetry is an important shortcut in working out problems and should be used whenever possible. The resultant magnitude is $|\mathbf{p}+\mathbf{q}| = 2a\cos 30° \approx 1.8a$. (b) The difference $\mathbf{p} - \mathbf{q}$ points due north, being the other diagonal of the rhombus shown in (a). Its magnitude is $|\mathbf{p}-\mathbf{q}| = a$, since the 60° angle, together with $|\mathbf{p}| = |\mathbf{q}|$, implies an equilateral triangle.

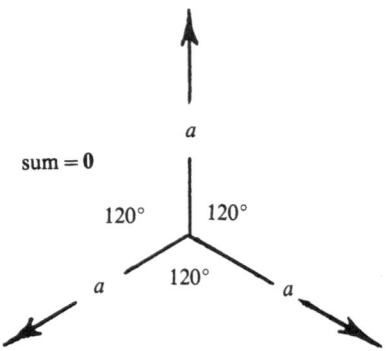

Figure 2.20. Getting the complete result from symmetry. If this figure is rotated by 120°, it does not change. The same property must be true for the sum of the three vectors shown. The only vector that does not change when rotated is the zero vector.

iii. Some Illustrations

The most direct method of obtaining numerical results when adding or subtracting vectors consists of making a scale drawing and measuring the required magnitudes and angles. This, however, is often slow and inaccurate when compared to methods using trigonometry or symmetry considerations. Some examples of vector calculations are shown in Figs. 2.18–2.21.

One of the best methods, using rectangular components, deserves special study and is discussed in the next subsection.

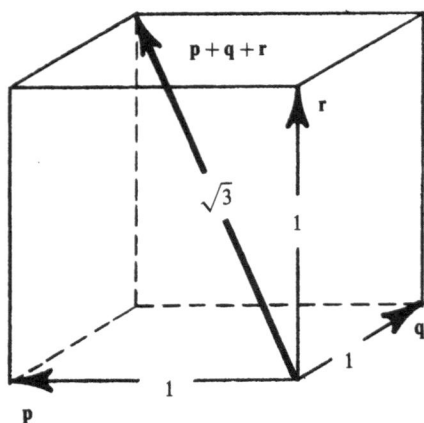

Figure 2.21. A three-dimensional illustration. What is the sum $\mathbf{p} + \mathbf{q} + \mathbf{r}$, if $\mathbf{p}$, $\mathbf{q}$, $\mathbf{r}$ are unit vectors along the sides of a cube? We may, for example, construct $\mathbf{p} + \mathbf{q}$, and then add $\mathbf{r}$ to the result. The total sum points along one of the cube's main diagonals, and its magnitude is $\sqrt{3}$.

iv. Rectangular Components

A convenient way of specifying a vector $\mathbf{p}$ is shown in the two-dimensional example of Fig. 2.22. The procedure is to set up a pair of rectangular axes x, y and to record the perpendicular projections p_x, p_y of $\mathbf{p}$ along the two axes. The sign of a projection must be carefully watched: it is positive or negative according to whether the projection points along the positive or negative direction of the relevant axis. The quantities p_x, p_y are ordinary numbers and are called the (rectangular) components of $\mathbf{p}$. If we know the components p_x and p_y we can easily find the magnitude and the direction of $\mathbf{p}$, as follows.

1. The Magnitude. Apply Pythagoras' theorem to any triangle in Fig. 2.22. The result is

$$\boxed{|\mathbf{p}| = (p_x^2 + p_y^2)^{1/2}} \qquad (2.D.3)$$

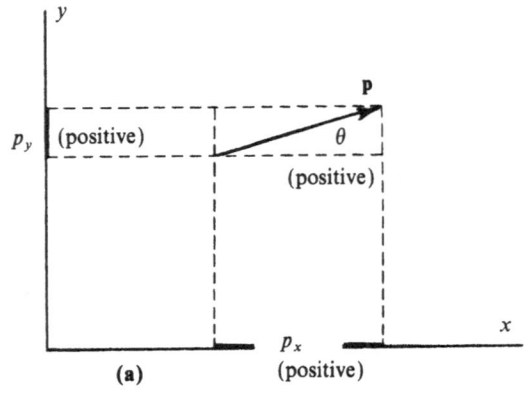

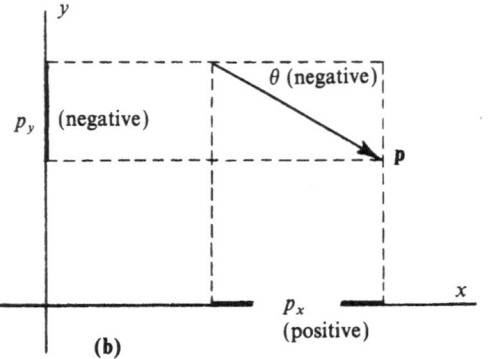

Figure 2.22. Rectangular components of a vector. In this figure the projections p_x and p_y are the components of $\mathbf{p}$; In (a), both p_x and p_y are positive numbers; in (b), p_x is positive, p_y is negative. This is indicated by the mentions (positive), (negative) rather than by an arrow, since we must keep in mind that p_x and p_y are not vectors. Knowing p_x and p_y enables one to reconstruct $\mathbf{p}$, as explained by Eqs. (2.D.3), (2.D.4).

2. The Direction. Let θ be the angle between **p** and the x axis. Then Fig. 2.22 gives

$$\tan \theta = \frac{p_y}{p_x} \qquad (2.D.4)$$

from which θ can be determined. [Conversely, if $|\mathbf{p}|$ and θ are given, can we calculate p_x and p_y? Figure 2.22 shows that

$$p_x = |\mathbf{p}| \cos \theta, \qquad p_y = |\mathbf{p}| \sin \theta \quad (2.D.5)$$

As a consistency check insert (2.D.5) into (2.D.3) and (2.D.4). It is important to keep in mind that a vector in a plane always contains two items of information, which can be presented in a variety of ways, such as prescribing $|\mathbf{p}|$ and θ, or p_x and p_y. Similarly, in three-dimensional space, a vector stands for three items of information.]

Components are most useful for adding and subtracting vectors. Figure 2.23 shows that, if

$$\mathbf{p} + \mathbf{q} = \mathbf{r}$$

(a vector equality), then also

$$p_x + q_x = r_x$$

(an equality between numbers). Similarly,

$$p_y + q_y = r_y$$

Thus, a vector sum is reduced to ordinary algebraic sums. Vector differences behave in a corresponding way.

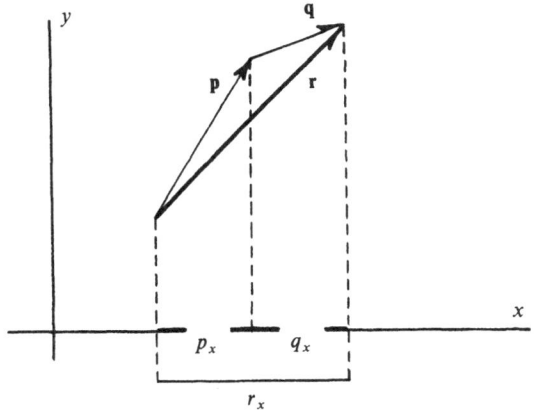

Figure 2.23. Adding components. If $\mathbf{r} = \mathbf{p} + \mathbf{q}$, then also $r_x = p_x + q_x$; verify, on this figure, the corresponding statement for r_y.

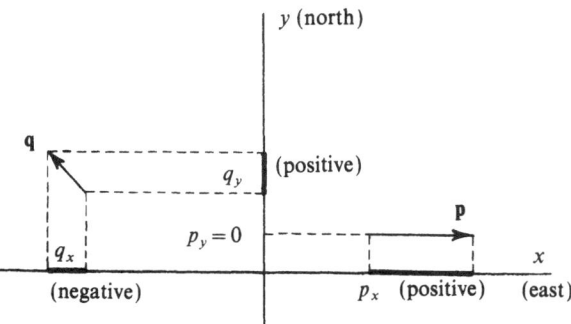

Figure 2.24. Adding two vectors **p**, **q** by the method of components, see Example 2.8. This figure shows the components to be added. The reader may wish to supply a sketch of the resultant.

Example 2.8. Magnitude and Direction of a Vector Sum. Given, two vectors **p**, **q**; assume that **p** has magnitude 2.0 and points due east; **q** has magnitude 1.0 and points north-west. Find the magnitude and the direction of the resultant,

$$\mathbf{r} = \mathbf{p} + \mathbf{q}$$

by using components along an eastward x axis and a northward y axis.

From Fig. 2.24 we find

$$
\begin{aligned}
p_x &&= 2.0 \\
q_x = -(1.0) \cos 45^\circ &&= -0.71 \\
\hline
r_x = p_x + q_x &&= 1.3
\end{aligned}
$$

$$
\begin{aligned}
p_y &&= 0.0 \\
q_y = (1.0) \cos 45^\circ &&= +0.71 \\
\hline
r_y = p_y + q_y &&= 0.71
\end{aligned}
$$

Magnitude: From Eq. (2.D.3), we have

$$|\mathbf{r}| = (r_x^2 + r_y^2)^{1/2} = [(1.3)^2 + (0.71)^2]^{1/2}$$
$$= \underline{1.5}$$

Direction: The angle θ between **r** and the x axis satisfies

$$\tan \theta = \frac{r_y}{r_x} = \frac{0.71}{1.3} = 0.55$$

so that

$$\underline{\theta = 29^\circ}$$

Thus, **r** points 29° north of east, or, in more conventional language, $90° - 29° = 61°$ east of north.

[Note the simplification due to $p_y = 0$. This comes from having one of the axes (x) parallel to one of the vectors (**p**). It is often advantageous to choose one's axes so that as many of the given vectors as possible are parallel to one of the axes. An alternative strategy exists if one can guess in advance, from the symmetry of the situation, how the resultant will point, as in Fig. 2.19. In such a case it may be more efficient to have one of the axes in that direction.]

Although the preceding example deals with only two vectors, it clearly shows that components will provide a safe and systematic way of adding many vectors, since a vector sum

$$\mathbf{p} + \mathbf{q} + \mathbf{r} + \cdots$$

will reduce in a plane to the calculation of two ordinary sums

$$p_x + q_x + r_x + \cdots, \qquad p_y + q_y + r_y + \cdots$$

A three-dimensional illustration of rectangular components is shown in Fig. 2.25. The trigonometry needed to relate the direction to the components is somewhat more involved than in the two-dimensional case, and we need not bother with it here. The magnitude of **p** in Fig. 2.25 is given by

$$|\mathbf{p}| = (p_x^2 + p_y^2 + p_z^2)^{1/2} \qquad (2.\text{D}.6)$$

a straightforward extension of (2.D.3).

E. Motion in a Plane

Our task in this section is to develop a simple way of working with velocity and acceleration when motion is not confined to a straight line. Vectors provide a natural language for doing this. For simplicity, we ordinarily assume all trajectories to occur in a single plane (two-dimensional motion), but the same concepts can be applied to three-dimensional motion as well.

i. The Position Vector

In a plane, the position of a particle P is specified by its rectangular coordinates x, y. Here we consider an alternative description: it consists of drawing the **position vector r**, defined as the vector whose tail is at the origin of coordinates, and whose head is at the location of P; see Fig. 2.26. Note that:

The vector **r** and the two numbers (x, y) contain exactly the same information.

$$(2.\text{E}.1)$$

In fact, as the figure shows, x, y are the **rectangular components** of **r**:

$$\boxed{r_x = x, \qquad r_y = y} \qquad (2.\text{E}.2)$$

ii. The Velocity Vector

The one-dimensional velocity was defined as

$$v_x = \frac{dx}{dt} \qquad (2.\text{E}.3)$$

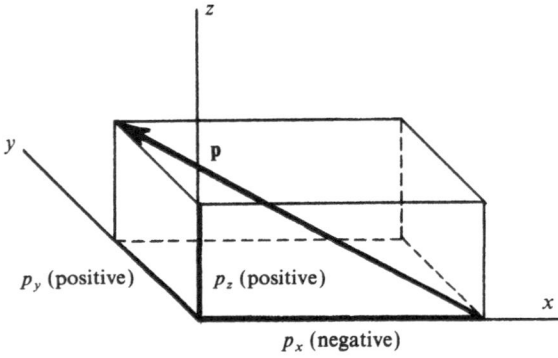

Figure 2.25. Rectangular components in three dimensions. In this example p_y and p_z are positive, while p_x is negative.

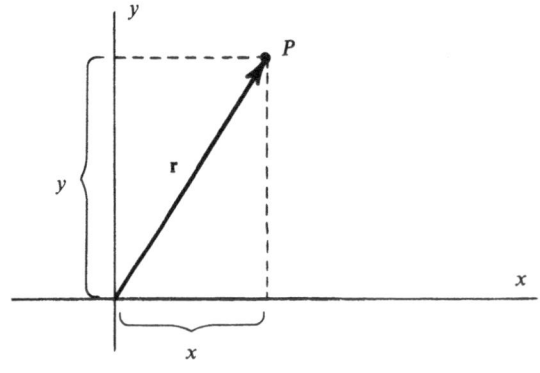

Figure 2.26. The position vector **r** for a particle P.

This serves as a model for defining the **velocity vector**

$$\mathbf{v} = \frac{d\mathbf{r}}{dt} \qquad (2.E.4)$$

as *the time rate of change of the position vector.* We must now examine in detail what is meant by that concept.

Just as dx/dt stands for

$$\frac{dx}{dt} = \lim_{\Delta t \to 0} \frac{\Delta x}{\Delta t} \qquad (2.E.5)$$

where Δx is the increase in x during Δt, we have for $d\mathbf{r}/dt$

$$\frac{d\mathbf{r}}{dt} = \lim_{\Delta t \to 0} \frac{\Delta \mathbf{r}}{\Delta t} \qquad (2.E.6)$$

Here $\Delta \mathbf{r}$ is the **displacement vector** corresponding to a change in the particle's position

$$\mathbf{r} \to \mathbf{r} + \Delta \mathbf{r} \qquad (2.E.7)$$

during Δt; this construction is illustrated in Fig. 2.27.

Three features of $\mathbf{v}$ contribute to its usefulness, and must be kept in mind.

1. Direction of $\mathbf{v}$. Figure 2.28 shows that $\mathbf{v}$ *points forward along the particle's trajectory.* More precisely, the direction of $\mathbf{v}$ is that of the trajectory's tangent, drawn at the point occupied by the particle.

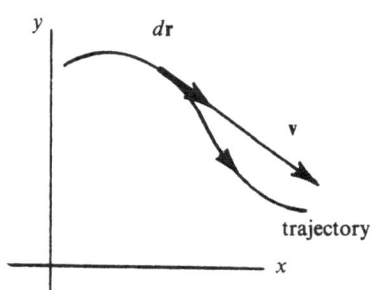

Figure 2.28. Since $\mathbf{v} = d\mathbf{r}/dt$, $\mathbf{v}$ has the direction of $d\mathbf{r}$. Therefore it points tangentially forward along the trajectory.

2. Magnitude of $\mathbf{v}$. This quantity, denoted by v and better known as the *speed*, equals *the length of path traveled per unit time.*

$$v = |\mathbf{v}| \qquad (2.E.8)$$

$$= \frac{|d\mathbf{r}|}{dt} \qquad (2.E.9)$$

where $|d\mathbf{r}|$ is the infinitesimal distance traveled during dt (we assume a positive time increase dt). Equation (2.E.8) follows directly from taking magnitudes on both sides of definition (2.E.4).

3. Components of $\mathbf{v}$. Since the x component of $\mathbf{r}$ is just x, it follows that the x component of $d\mathbf{r}/dt$ is dx/dt. (The rule is: *Differentiating a vector means differentiating its components.* This goes back to the fact that a derivative amounts to a difference, and that taking the difference between two vectors means taking the difference between their components.)

Taking the x component on both sides of (2.E.4), we find

$$v_x = \frac{dx}{dt} \qquad (2.E.10a)$$

Similarly,

$$v_y = \frac{dy}{dt} \qquad (2.E.10b)$$

(These are already familiar equations, each fitting the special case of rectilinear motion along the appropriate axis.)

In connection with rectangular components, we recall that the magnitude v is expressible as

$$v = (v_x^2 + v_y^2)^{1/2} \qquad (2.E.11)$$

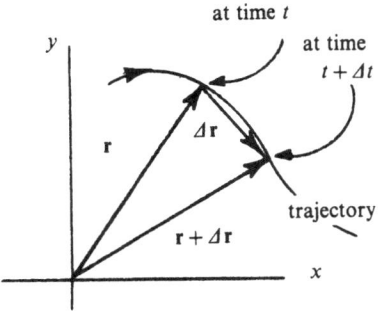

Figure 2.27. The displacement vector $\Delta \mathbf{r}$.

iii. Combined and Relative Velocities

It is often natural to view a motion as being made of two combined motions.

> **Example 2.9. Parallel Motions.** A train (T) moves north at a velocity $v_{TG} = 100$ km/h with respect to the ground (G); inside the train, a passenger (P), facing north, walks at a velocity $v_{PT} = 2$ km/h with respect to the train. What is his "real" velocity v_{PG} (i.e., with respect to the ground outside)?
>
> _____
>
> The answer is 102 km/h;
>
> more generally,
>
> $$v_{PG} = v_{TG} + v_{PT} \qquad (2.E.12)$$
>
> One calls v_{PT} the **relative velocity** of P with respect to T.

We next analyze a situation where not all velocities are in the same direction. Suppose a flat board (B), lying on a table top, is being shifted parallel to itself at a velocity $\mathbf{v}_{BT}$ relative to the table (T). An ant (A) is crawling on the board at a velocity $\mathbf{v}_{AB}$ relative to the board. What velocity $\mathbf{v}_{AT}$ does the ant have relative to the table?

This problem is most easily solved by first considering the relevant position coordinates; see Fig. 2.29. Let a coordinate system, with origin T, be drawn on the table; another one, with origin B, is drawn on the board. All positions and velocities

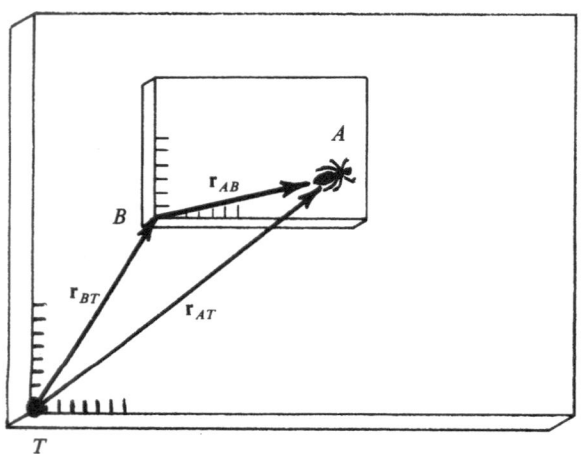

Figure 2.29. The "true" position of A ($\mathbf{r}_{AT}$), and its relative position $\mathbf{r}_{AB}$ relative to B.

relative to the table will, for convenience, be called "true." The vector $\mathbf{r}_{BT} = \overrightarrow{TB}$ is then the true position of point B; it therefore indicates the position of the whole board. Similarly, the vector $\mathbf{r}_{AT} = \overrightarrow{TA}$ is the true position of the ant. Finally, the vector $\mathbf{r}_{AB} = \overrightarrow{BA}$ is the *position of the ant relative to the board*. We see from the figure that

$$\mathbf{r}_{AT} = \mathbf{r}_{BT} + \mathbf{r}_{AB} \qquad (2.E.13)$$

Taking the derivative on both sides of this vector sum yields immediately

$$\boxed{\mathbf{v}_{AT} = \mathbf{v}_{BT} + \mathbf{v}_{AB}} \qquad (2.E.14)$$

where $\mathbf{v}_{AT}$ is the true velocity of A, $\mathbf{v}_{BT}$ is the true velocity of B, and $\mathbf{v}_{AB}$ is the *velocity of A relative to B*. Equation (2.E.14) is most easily remembered through its special case (2.E.12).

> **Example 2.10. Swimming in a Water Current.** In Fig. 2.30 a swimmer (A), whose head is pointing north, is achieving a water speed of 1.00 km/h. He is, however, in an eastward current of water (B), of speed 0.75 km/h. What is, relative to the shore (C), his speed and direction of motion?
>
> _____
>
> The required velocity is
>
> $$\mathbf{v}_{AC} = \mathbf{v}_{BC} + \mathbf{v}_{AB}$$
>
> where $\mathbf{v}_{BC}$ is 0.75 km/h east, and $\mathbf{v}_{AB}$ is 1.00 km/h north. From the figure we find
>
> $$v_{AC} = (v_{BC}^2 + v_{AB}^2)^{1/2}$$
> $$= [(0.75)^2 + (1.00)^2]^{1/2} \frac{\text{km}}{\text{h}} = 1.25 \frac{\text{km}}{\text{h}}$$
>
> The direction θ (east of north) is given by
>
> $$\tan \theta = \frac{v_{BC}}{v_{AB}} = 0.75$$
>
> so that
>
> $$\theta = 37°$$

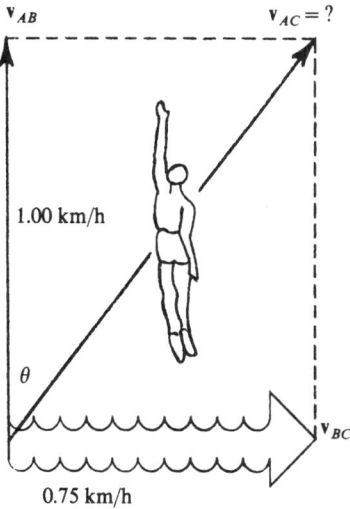

Figure 2.30. The swimmer in a cross current. His relative velocity with respect to the water is $\mathbf{v}_{AB}$, the velocity of the current is $\mathbf{v}_{BC}$, and the swimmer's resultant ("true") velocity is $\mathbf{v}_{AC}$.

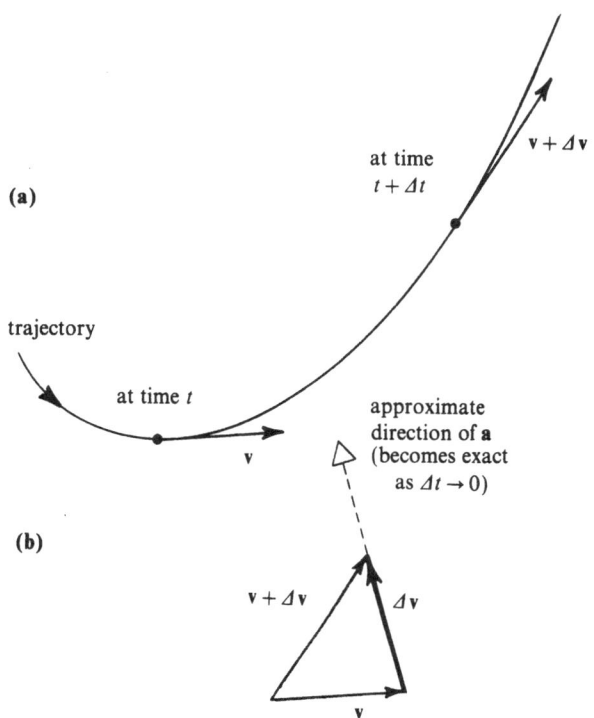

Figure 2.31. (a) The velocity vector at two different times. (b) The change $\varDelta\mathbf{v}$ in the velocity vector. In the limit of $\varDelta t \rightarrow 0$, $\varDelta\mathbf{v}/\varDelta t$ becomes the acceleration vector $\mathbf{a}$.

iv. The Acceleration Vector

For motion along the x axis, acceleration was defined as $a_x = dv_x/dt$. Similarly, for motion along a curve, the **acceleration vector** is defined as

$$\mathbf{a} = \frac{d\mathbf{v}}{dt} \qquad (2.E.15)$$

the time rate of change of the velocity vector. We have already learned, in connection with $\mathbf{v} = d\mathbf{r}/dt$, how to take the derivative of a vector. The same construction is applicable to $\mathbf{a}$, and is illustrated in Fig. 2.31. An interesting property of $\mathbf{a}$ can be seen: unlike $\mathbf{v}$, it does not necessarily point in the direction of motion. We shall encounter this feature in our study of projectiles (next section) and of circular motion (Chap. 6).

Equivalently to (2.E.15), we can write

$$\mathbf{a} = \frac{d^2\mathbf{r}}{dt^2} \qquad (2.E.16)$$

Summary

The vector definitions of position, velocity, and acceleration are shown in Table 2.2 together with their (already familiar) x component. This table carries an important implication: *the motion of a particle* (left column) *can be mathematically separated into an x motion* (right column) *and a y motion.* Each of these can be discussed as if it took place, by itself, along its own axis. Therefore our previous study of rectilinear motion can be taken over and applied without change to the most complicated orbits, provided we consider one component at a time. This principle will be of great help to us in the next section, where we study projectiles.

Table 2.2. Vector Language versus One-Component Language.

Vector definition	x component (y component: similar)	
$\mathbf{r}$	x	(2.E.17)
$\mathbf{v} = \dfrac{d\mathbf{r}}{dt}$	$v_x = \dfrac{dx}{dt}$	(2.E.18)
$\mathbf{a} = \dfrac{d\mathbf{v}}{dt} = \dfrac{d\mathbf{r}}{dt^2}$	$a_x = \dfrac{dv_x}{dt} = \dfrac{d^2x}{dt^2}$	(2.E.19)

A Special Case: Uniform Motion (Revisited)

Here the vector formulas are easily written in analogy with their one-dimensional counterparts (Sec. A). The speed and direction of motion are determined by

$$\mathbf{v} = \frac{d\mathbf{r}}{dt} = \frac{\Delta\mathbf{r}}{\Delta t} = \text{constant vector} \qquad (2.\text{E}.20)$$

(finite increments $\Delta\mathbf{r}$, Δt are permitted here); we have

$$\mathbf{a} = \mathbf{0} \qquad (2.\text{E}.21)$$

The position changes with time according to

$$\mathbf{r} = \mathbf{r}_0 + \mathbf{v}t \qquad (2.\text{E}.22)$$

F. Projectiles

We are now fully equipped to extend the laws of free fall to the nonvertical case.

i. Nonvertical Free Fall

Suppose a particle, here called a **projectile**, is thrown in an arbitrary direction under idealized conditions (air effects are negligible and the trajectory is much smaller than planetary size). The projectile's motion is still surprisingly simple to describe. We first do so without equations.

Observation shows that

> The whole trajectory lies in a vertical plane.   (2.F.1)

Figure 2.32 illustrates how to select two convenient coordinate axes, x (horizontal) and z (vertically up) in order to describe the motion. In terms of these coordinates, it is further observed that

> The horizontal coordinate exhibits uniform motion.

(2.F.2)

Finally,

> The vertical coordinate exhibits the same constant acceleration, $-g$, that characterizes vertical free fall.

(2.F.3)

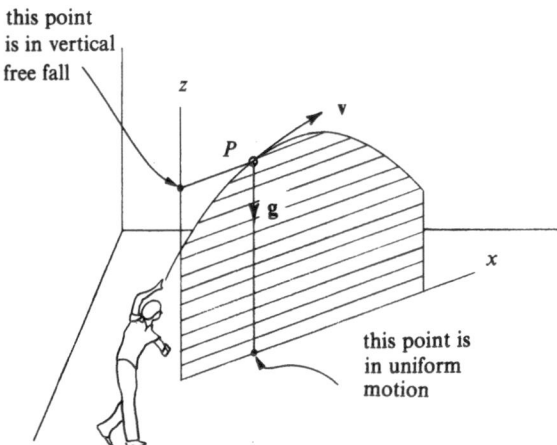

Figure 2.32. The trajectory of a projectile P is always contained in a vertical plane (shaded). Hence only two axes, x and z, are needed to describe its shape. The acceleration vector $\mathbf{g}$ is always vertically down, with constant magnitude. The velocity vector $\mathbf{v}$ is shown along the trajectory. The x and z coordinates of the projectile are in uniform motion and vertical free fall, respectively.

These observations, combined with the equivalence principle discussed in Sec. C, are condensed as follows:

> The acceleration vector $\mathbf{g}$ points downward, and has constant magnitude; $\mathbf{g}$ does not depend on the nature of the projectile.

(2.F.4)

ii. Equations for Projectile Motion

Statement (2.F.4), summarizing the behavior of a projectile, must now be converted into equations involving the projectile's position $\mathbf{r}$ and velocity $\mathbf{v}$. These equations can then be used to make quantitative predictions.

Position

To find $\mathbf{r}$ as a function of time, we use the information that *the acceleration is constant*. In the one-dimensional case, Eq. (2.C.9), this led to

$$x = x_0 + v_{0x}t + \tfrac{1}{2}a_x t^2 \qquad (a_x = \text{const}) \qquad (2.\text{F}.5)$$

The corresponding result in a plane can be

expressed in terms of vectors. We expect to find the modifications

$x \to \mathbf{r}$ (position at time t)

$x_0 \to \mathbf{r}_0$ (initial position)

$v_{0x} \to \mathbf{v}_0$ (initial velocity)   (2.F.6)

$a_x \to \mathbf{a} = \mathbf{g}$ (constant acceleration)

(Caution: Note the sign convention $\mathbf{a} = +\mathbf{g}$: *the downward vector* $+\mathbf{g}$ *is the acceleration of gravity*; we still have $a_z = -g$ if the z axis points upward.) Equation (2.F.5) is therefore replaced by

$$\boxed{\mathbf{r} = \mathbf{r}_0 + \mathbf{v}_0 t + \tfrac{1}{2}\mathbf{g}t^2} \qquad (2.F.7)$$

This formula is only a guess so far, but it is quickly checked. Let us look at the x and z components of (2.F.7), noting that $\mathbf{g}$ has the components $g_x = 0$, $g_z = -g$.

For the x component we have

$$\boxed{x = x_0 + v_{0x}t} \qquad (2.F.8)$$

as required by statement (2.F.2).

For the z component we have

$$\boxed{z = z_0 + v_{0z}t - \tfrac{1}{2}gt^2} \qquad (2.F.9)$$

as required by (2.F.3).

Velocity

A similar, but simpler argument leads to the velocity equation,

$$\boxed{\mathbf{v} = \mathbf{v}_0 + \mathbf{g}t} \qquad (2.F.10)$$

with components

$$v_x = v_{0x} \qquad (2.F.11)$$

$$v_z = v_{0z} - gt \qquad (2.F.12)$$

iii. Shape of the Trajectory: A Parabola

Eliminating t between Eqs. (2.F.8) and (2.F.9) gives a relation between x and z: we can plot the

trajectory in space. For simplicity, let us launch the projectile at the origin, $x_0 = z_0 = 0$. (This may affect the trajectory's location, but should not affect its shape.)

Then (2.F.8) and (2.F.9) read

$$x = v_{0x}t \qquad (2.F.13)$$

$$z = v_{0z}t - \tfrac{1}{2}gt^2 \qquad (2.F.14)$$

Using (2.F.13) to express t in terms of x, we get from (2.F.14)

$$z = \left(\frac{v_{0z}}{v_{0x}}\right)x - \left(\frac{g}{2v_{0x}^2}\right)x^2 \qquad (2.F.15)$$

The quantities in parentheses are constants, and the graph of z against x is then a **parabola**, as was the earlier graph of z against t. (On a planet with negligible gravity, $g = 0$, *the parabola would turn into a straight line*.)

iv. Two Trajectories Compared (with and without Gravity)

Let us assume some fixed initial conditions x_0, z_0, v_{0x}, v_{0z}. The free-fall equations (2.F.8), (2.F.9) can then be written in terms of what would happen without gravity. For $g = 0$, uniform motion would prevail:

$$x_{NG} = x_0 + v_{0x}t \qquad (2.F.16)$$

$$z_{NG} = z_0 + v_{0z}t \qquad (2.F.17)$$

where the subscript NG stands for "no gravity." Then, at any given time, (2.F.8) reads

$$x = x_{NG} \qquad (2.F.18)$$

Equation (2.F.9) becomes

$$z = z_{NG} - \tfrac{1}{2}gt^2 \qquad (2.F.19)$$

The contents of these two equations may be expressed as follows:

> For given initial conditions, and at any given time, a projectile is vertically below the position it would occupy if there were no gravity; it is lower than this by a distance $\tfrac{1}{2}gt^2$.   (2.F.20)

This behavior is illustrated in Fig. 2.33.

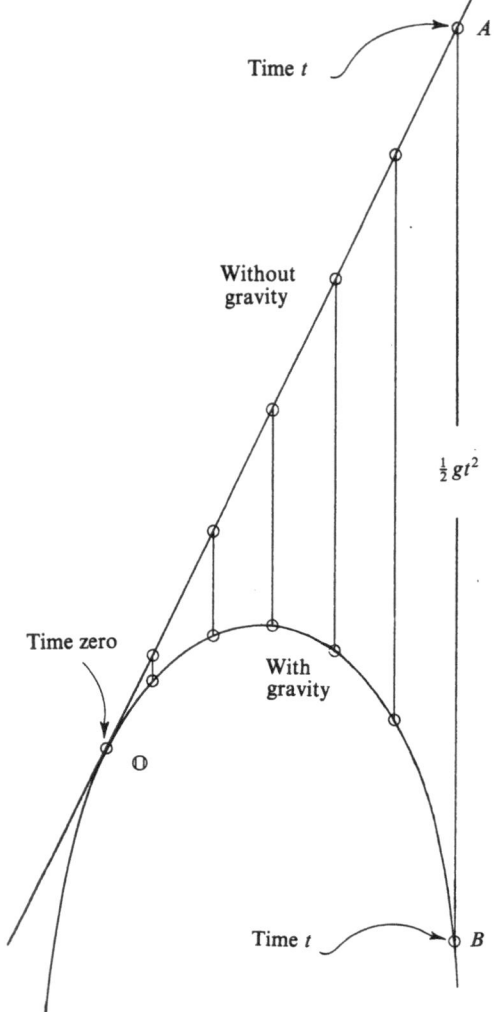

Time *t*

Without gravity

$\frac{1}{2}gt^2$

Time zero

With gravity

O

Time *t*

Figure 2.33. A projectile is launched at point $\mathbb{O}$ with given initial conditions. Its trajectory under gravity ($\mathbb{O}B$) is compared with the trajectory ($\mathbb{O}A$) that it would follow in the absence of gravity. The comparison is summarized by statement (2.F.20).

The Monkey and the Slingshot

This is a classic physics fable, based on statement (2.F.20). A hunter aims his slingshot at a monkey hanging from a branch. The hunter forgets that gravity will curve the path of his stone, and he aims along a straight line. The monkey, upon hearing the snap of the slingshot, releases the branch and falls straight down. Nevertheless, he gets hit by the stone. The explanation is as follows; see Fig. 2.33. Let *t* be the time needed for the stone to arrive at the *x* coordinate of the monkey. At this time, in the absence of gravity, the stone would be at *A*. Actually, the stone is at *B*, a distance $gt^2/2$ lower.

But this is also the monkey's position since he fell a distance of $gt^2/2$ from rest, starting from *A* at $t = 0$. (We neglect the delay of the sound signal that alerted him.)

v. Examples

Example 2.11. An Initial-Value Problem. A paperweight is thrown out of a window at a speed of 6.0 meters/sec, in a direction 30° above horizontal. After 1.5 sec, what is the paperweight's (a) speed, (b) direction of motion, and (c) position, assuming it is still in flight?

First, resolve the initial conditions into vertical and horizontal components. Let, at any time, θ be the direction of motion above the horizontal. Then we have

$$v_x = v \cos \theta, \qquad v_z = v \sin \theta$$

In particular, at $t = 0$, we have

$$v_{0x} = v_0 \cos \theta_0, \qquad v_{0z} = v_0 \sin \theta_0$$

where

$$v_0 = 6.0 \text{ meters/sec}, \qquad \theta_0 = 30°$$

so that

$$v_{0x} = \frac{5.2 \text{ meters}}{\text{sec}}, \qquad v_{0z} = \frac{3.0 \text{ meters}}{\text{sec}}$$

(a) *Speed.* The *x* velocity is constant:

$$v_x = v_{0x} = 5.2 \frac{\text{meters}}{\text{sec}}$$

The *z* velocity is [see Eq. (2.F.12)]

$$v_z = v_{0z} - gt$$

$$= [3.0 - (10)(1.5)] \frac{\text{meters}}{\text{sec}}$$

$$= -12 \frac{\text{meters}}{\text{sec}}$$

meaning 12 m/s downward. The speed is

$$v = (v_x^2 + v_z^2)^{1/2} = [(5.2)^2 + (12)^2]^{1/2} \frac{\text{meters}}{\text{sec}}$$

$$= 13 \frac{\text{meters}}{\text{sec}}$$

(b) *Direction of Motion.* This is the direction of **v**. As in Eq. (2.D.4),

$$\tan \theta = \frac{v_z}{v_x} = \frac{-12}{5.2} = -2.3$$

so that

$$\underline{\theta = -66.5°}$$

that is, 66.5° below the horizontal.

(c) *Position.* Let the launching point be the origin. Then, from Eq. (2.F.8), the horizontal distance traveled is

$$x = 0 + v_{0x}t$$

$$= (5.2)(1.5) \text{ meters}$$

$$= 7.8 \text{ meters}$$

The vertical rise is, according to (2.F.9),

$$z = 0 + v_{0z}t - \tfrac{1}{2}gt^2$$

$$= [(3.0)(1.5) - (\tfrac{1}{2})(10)(1.5)^2] \text{ meters}$$

$$= \underline{-6.8 \text{ meters}}$$

meaning that the paperweight is 6.8 meters below window level. The trajectory is plotted in Fig. 2.34.

Example 2.12. The Range of a Water Spout. A garden hose, with its nozzle at ground level, projects a stream of water at 37° to the horizontal. The water leaves the hose at a speed of 8.0 meters/sec. How far away does it reach the ground, which is assumed horizontal?

Consider a small portion of water as a particle, and assume it undergoes free fall. It leaves the ground with initial x and z velocities

$$v_{0x} = v_0 \cos \theta_0, \qquad v_{0z} = v_0 \sin \theta_0$$

where

$$v_0 = 8.0 \text{ meters/sec}, \qquad \theta_0 = 37°$$

First we find the total time of flight. This is of course synonymous with the total time taken by the vertical component of the motion. We have previously solved this problem in terms of v_{0z} [see Eq. (2.C.17)]:

$$t_{\text{flight}} = \frac{2v_{0z}}{g}$$

Next we use the constant horizontal velocity to find the range:

$$x = v_{0x} t_{\text{flight}} = \frac{2v_{0x} v_{0z}}{g}$$

or

$$x = \frac{2v_0^2 \cos \theta_0 \sin \theta_0}{g} \qquad (2.F.21)$$

In this case,

$$x = \frac{(2)(8.0)^2 (0.80)(0.60)}{10} \text{ meters}$$

$$= \underline{6.2 \text{ meters}}$$

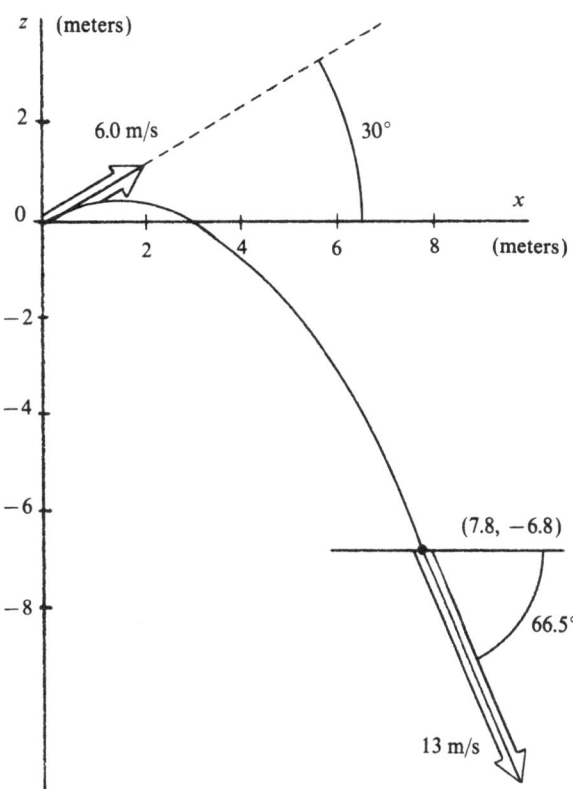

Figure 2.34. The flying paperweight. Example 2.11 should be carefully followed on this graph.

vi. Galilean Relativity

Setting calculations aside, we conclude the chapter on a more philosophical note: free fall tells us less about our environment than we might hope.

Suppose a prisoner, kept in a uniformly traveling, windowless railroad car, wants to use the laws of free fall to deduce his speed. He marks a spot A on the ceiling, and another one, B, on the floor vertically below A. He then releases a penny at A and measures how far from B it hits the floor. Will that measurement reveal the train's speed?

The answer is no, because the penny will strike the exact spot B, *just as though the train were at rest*; Fig. 2.35 illustrates what happens. More generally,

> The observed laws of free fall are the same whether the laboratory is uniformly moving or at rest.

(2.F.22)

This is true even if the laboratory has a vertical component of velocity; for example, the free-fall measurements might be conducted inside a *uniformly* rising elevator. (The laboratory cannot rotate, however, without those measurements being affected. If it does rotate, most of its points can be shown to move nonuniformly.)

Statement (2.F.22) is called the principle of **Galilean relativity**; equivalently, it says that *a uniform motion of the environment cannot be detected by projectile experiments.* We need not worry about the formal proof of (2.F.22).

Extensions of Galilean Relativity

The principle remains valid even for objects interacting through mechanical devices, collisions, etc., rather than just under the influence of gravity.

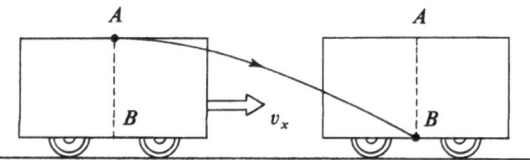

Figure 2.35. The solid curve is the "true" trajectory of a penny falling from A to B in a uniformly moving railroad car. The penny's behavior is explained by the fact that its initial x velocity is v_x, the same as the car's.

The conclusion is that *a mechanical experiment conducted entirely within a uniformly moving laboratory yields the same results as if the laboratory were at rest.* The aerodynamic wind tunnel is an outstanding application of this principle. When a stationary airplane is tested in an airstream of velocity $-\mathbf{v}$, all measurements are also applicable to an airplane of velocity $+\mathbf{v}$ in still air.

The ultimate extension of Galilean relativity is found in **Einsteinian relativity**: according to it, there exists no phenomenon of any kind which, when observed entirely within a uniformly moving environment, is able to reveal such a motion.

Condensed Checklist

(The vector formulas should serve as reminders for their one-dimensional versions as well.)

General motion:

$$\mathbf{v} = \frac{d\mathbf{r}}{dt} \qquad (2.E.4)$$

$$\mathbf{a} = \frac{d\mathbf{v}}{dt} = \frac{d^2\mathbf{r}}{dt^2} \qquad (2.E.15), (2.E.16)$$

Uniform motion:

$$\mathbf{a} = 0 \qquad (2.E.21)$$

$$\mathbf{v} = \frac{\Delta\mathbf{r}}{\Delta t} = \text{constant vector} \qquad (2.E.20)$$

$$\mathbf{r} = \mathbf{r}_0 + \mathbf{v}t \qquad (2.E.22)$$

Uniformly accelerated motion:

$$\mathbf{a} = \text{constant vector} \qquad (2.F.6)^*$$

$$\mathbf{v} = \mathbf{v}_0 + \mathbf{a}t \qquad (2.F.10)^*$$

$$\mathbf{r} = \mathbf{r}_0 + \mathbf{v}_0 t + \tfrac{1}{2}\mathbf{a}t^2 \qquad (2.F.7)^*$$

$$v_x^2 - v_{0x}^2 = 2a_x(x - x_0) \qquad (2.C.10)^*$$

(true for each component)

Free fall:

$$\mathbf{a} = \mathbf{g}, \qquad g = 9.8 \frac{\text{meters}}{\text{sec}^2}$$

(2.F.6), (2.C.2)

$$a_z = -g \qquad (2.C.11)$$

$$z_{\max} = v_{0z}^2/2g \qquad (2.C.16)$$

* Take $\mathbf{g} = \mathbf{a}$ for generality.

Combining velocities:

$$\mathbf{v}_{AC} = \mathbf{v}_{AB} + \mathbf{v}_{BC} \qquad (2.E.14)$$

$$(\mathbf{v}_{AB}: A \text{ relative to } B, \text{ etc.})$$

Vector calculations:

(two dimensional; θ = angle with x axis)

$$|\mathbf{p}| = (p_x^2 + p_y^2)^{1/2} \qquad (2.D.3)$$

$$\tan \theta = p_y/p_x \qquad (2.D.4)$$

$$p_x = |\mathbf{p}| \cos \theta$$
$$\qquad\qquad\qquad (2.D.5)$$
$$p_y = |\mathbf{p}| \sin \theta$$

True or False

1. A stone that has been thrown vertically upward reverses its acceleration as it reaches the top of its trajectory.

2. A freely falling particle has zero jerk.

3. If a flea doubles the speed at which it takes off vertically, then it approximately quadruples the height it reaches.

4. The distance traveled during one second by a vertically falling stone increases by the same amount each second.

5. If two objects are dropped from rest one after the other from the same point, the distance between them increases uniformly in time.

6. If their directions are properly chosen, two vectors of unequal magnitudes can add up to zero.

7. Position is a vector, but time is not.

8. The time rate of change of a vector is always a vector.

9. Any calculation using vectors can also be done without them, although perhaps more laboriously.

10. Because of the current, a given swimmer needs more time to cross a river than an equal width of still water.

11. A projectile's acceleration and velocity vectors can be mutually perpendicular.

12. If one horizontal distance AB is longer than another horizontal distance CD, then a projectile's trajectory must reach higher when it is thrown from A to B than when it is thrown from C to D.

13. If a film portraying a projectile in motion were run backwards, this fact could be detected from the apparent violation of the laws of free fall.

14. A vertical free-fall trajectory generally appears as a parabola to a uniformly moving observer.

15. If statement 14 is true, it constitutes an exception to Galilean relativity.

Problems

Section A: Uniform Motion

2.1. A car, traveling at constant velocity, passes the 240-km marker at 9:00 a.m. and the 400-km marker at 11:00 a.m. (a) When was it at the 0-km marker? (b) Make a graph of the car's position against time.

2.2. A thunderclap is heard 5 sec after the flash of lightning is seen. How far away did the lightning occur? (The visual delay is negligible, since the speed of the light signal is about 10^6 times greater than that of the sound signal.)

*2.3. Express some approximate speeds of Fig. 2.2 in the following units: mile/sec (light signal); mile/h (Earth in its orbit, sound in air); inch/min (snail); inch/year (continental drift).

2.4. A paper bag is burst, and the echo, coming from a distant wall, is heard after 0.8 sec. How far is the wall? (Speed of sound = 340 meters/sec.)

2.5. How many times around the Earth could a radio signal go in one second? (Speed of radio waves = 3.0×10^8 meters/sec.)

*2.6. Two locomotives, each moving at 10 km/h, are on a head-on collision course. When they are still 1.0 meter apart, a fly takes off from the front surface of locomotive No. 1, and speeds straight towards locomotive No. 2. Upon meeting No. 2, the fly heads back to No. 1, then back to No. 2, etc. and in the end does not escape the crash. Assuming a constant speed of 20 km/h for the fly, what total back-and-forth distance does it cover?

2.7. A spacecraft splashes down in the sea and is detected by an underwater microphone 11.1 sec earlier than by an air microphone, which is just above the same location. How far away is the splashdown? (Speed of sound = 1450 meters/sec in water, 340 meters/sec in air.)

2.8. Two runners, A and B, leave the starting line at time zero. Runner A maintains a speed of 6 km/h for 10 min, then 3 km/h for 10 min, then 9 km/h for 5 min. He gets to the finish at the same time as B, who has kept up a constant speed v_B. (a) Plot the positions of A and B against time (with a common set of axes). (b) Find v_B. (c) During the race, who overtakes whom, where, and when?

*2.9. Two cars, initially 200 km apart, drive towards each other at speeds of 40 km/h and 60 km/h respectively. (a) Where and when do they pass each other? (b) Show the solution as the intersection of two linear graphs.

2.10. Two trains are moving uniformly. Train No. 1, whose speed is 100 km/h, leaves the station at 9:00 a.m.

Train No. 2, whose speed is 120 km/h, leaves the same station, on the same track, at 10:00 a.m. Where and when will the two trains collide?

Sections B: Accelerated Motion (One-Dimensional), and C: Vertical Free Fall

*2.11. A certain croquet ball takes off at a speed of 10 meter/sec after a contact of only 5.0×10^{-2} sec with the mallet. Estimate the ball's acceleration while it is being hit.

*2.12. Express $g = 9.8$ meter/sec^2 in the following units: ft/sec^2; miles/(min sec); km/h^2; μm/(μs)^2.

2.13. A train car is coasting along a straight horizontal track with a speed of 15 km/h. A locomotive catches up with it and pushes it with a constant acceleration of 0.01 g, until a speed of 51 km/h is reached. How many minutes did the push last?

*2.14. A model railroad car on a straight track passes a switch at a speed of 20 cm/sec. While keeping a constant forward acceleration of 2 cm/sec^2, it proceeds in 5 sec to another switch. What is the distance between both switches?

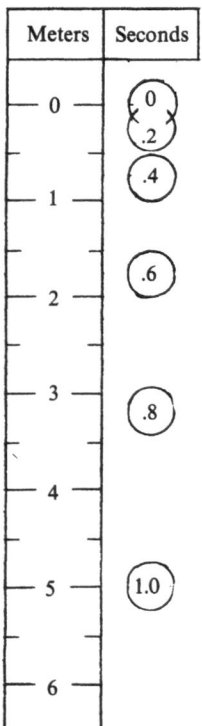

Meters	Seconds
0	0
	.2
	.4
1	
	.6
2	
	.8
3	
4	
5	1.0
6	

2.15. In the adjoining figure, a digital clock, falling beside a vertical meter stick, is represented at intervals of

0.2 sec. From these data, plot the six points of altitude against time; also show the theoretical curve.

*2.16. How high does a stone rise if thrown upwards at a speed of 4 meters/sec?

2.17. An apple, falling vertically from a tree, strikes the ground at a speed of 8 meters/sec. How high was it hanging?

2.18. To avoid hitting a stationary car, a truck going at 36 km/h must brake to a stop within 50 meters. How many g of uniform deceleration does this require?

*2.19. The tail of a rocket rising with constant acceleration passes the top of the launch tower with a speed of 25 meters/sec. After 2 more seconds that tail is 90 meters higher. How many g of acceleration does the rocket have?

2.20. Some trained dolphins can jump more than 5 meters above the water. How much speed, in km/h, must they pick up in the water to achieve this?

2.21. It took a vertically rising, uniformly accelerated rocket 20 sec to reach a velocity of 20 meters/sec. How many seconds did it require to reach 10 meters/sec?

2.22. On planet X, a stone falling from rest acquires a speed of 10 meters/sec after going down 10 meters. What speed does it have after going down 20 meters?

2.23. On planet Y, a stone has a speed of 20 meters/sec after a vertical 20-meter fall from rest. What is the acceleration of gravity g_Y on planet Y, expressed in Earth g?

*2.24. On planet Z, a stone has fallen 5 meters from rest when its speed is 5 meters/sec. How far down from its starting point is it when its speed is 15 meters/sec?

2.25. Roughly by what distance l should a particle fall from rest on a neutron star (see Fig. 2.7) to acquire a speed of 10 meters/sec? To what is l comparable in Fig. 1.1 of Chap. 1?

*2.26. A uniformly accelerated rocket with rectilinear trajectory has a speed of 100 meters/sec when observed 10 sec after launch. How fast was it going just 1 sec after launch?

*2.27. Starting from rest, a ball rolls down a slope at constant acceleration. If 2 sec are needed to travel over the first 2 meters, how many seconds are needed for the first 3 meters?

2.28. Starting from rest, a ball rolls down a slope at constant acceleration. If the first 3 meters are covered in 3 sec, what distance was covered after only 2 sec?

2.29. A departing jet airplane accelerates uniformly on the runway so as to reach a speed of 200 km/h in 10 sec. What is its acceleration in units of 1 g?

*2.30. A departing jet airplane has only 500 meters of runway, and must reach a speed of 180 km/h in order to take off. What minimum acceleration, in meters/sec², should it be capable of?

2.31. A handgun barrel, 10 cm long, fires a bullet at a speed of 400 meters/sec. What was the bullet's acceleration, assuming it to be constant?

*2.32. Find the velocity v_x of a particle as a function of the time t, if its position is given by (a) $x = kt^3$; (b) $x = 1/(pt + q)$, where k, p, q are constants.

2.33. Find the acceleration a_x in Problem 2.32 for both cases (a) and (b).

2.34. An object falls from rest at time zero. Between times t and $t + 4$ sec it goes down 100 meters. Find t.

*2.35. A steel ball is dropped vertically from a height of 5 cm on a smooth marble surface. The impact takes negligible time and the ball keeps bouncing back to the same height. About how many bounces occur every 10 sec?

2.36. Two stones are dropped from a high tower at an interval of 1 sec. How far apart are the stones 2 sec after the first one was dropped? What is the velocity of stone No. 2 relative to stone No. 1 (i.e., the difference between their velocities) at that time? What are the corresponding answers 3 sec after the first stone was dropped? If the tower is 100 meters high, at what time interval do the two stones strike the ground?

*2.37. A balloon descends vertically with constant speed v_B. At altitude h it releases some ballast. How much time does the ballast take to reach the ground? (Express your solution in terms of v_B, h, and g.)

2.38. Show in detail the algebra for deriving the velocity-versus-position result (2.C.10).

2.39. Plot v_z against t, and z against t, in such a way as to illustrate Examples 2.4 and 2.5.

*2.40. One light-year is the distance traveled in one year by a light signal (whose speed is 3×10^8 meters/sec). (a) How far, in meters, is a light-year? (b) How fast, in meters/sec, is a light-year per year? (c) How many g is 1 light-year/year²?

*2.41. A toy balloon takes off and keeps a constant vertical speed of 2 meters/sec. A time T later, a stone is thrown vertically up from the same spot at a speed of 12 meters/sec. It then barely succeeds in touching the bottom of the balloon. (a) Find T. (b) Draw, on a common grid, the graphs for the two objects' altitudes against time.

2.42. The **average velocity** between times 0 and t could be defined as $v_{av} = \frac{1}{2}(v_0 + v)$ or as $v_{av} = (x - x_0)/t$. Show that both definitions give the same result in the uniformly

accelerated case. (Caution: in more general cases they are not equivalent.)

2.43. A stone is thrown vertically upward at a speed of 2 meters/sec from a point 0.60 meter above the ground. (a) At what speed does it eventually hit the ground? (b) What would be the answer if the stone had been thrown downward, also at 2 meters/sec?

*2.44. A fountain is designed to produce a 5-meter vertical jet. If this jet starts off with a square cross section 2 cm on the side, what volume of water, in liters, must be pumped every minute? (1 liter = 10^{-3} meter³.) [First convince yourself that (water output per unit time) = (speed) × (cross section).]

2.45. A late traveler, 10 meters behind a train when it departs, races at a speed of 5 meters/sec to catch up. The train accelerates at 0.45 meter/sec². (a) Over what distance must the traveler run to catch the tail end of the train? (b) How much time will this take? (c) What is the meaning of the two mathematical solutions?

*2.46. In Problem 2.45, how low a speed could the traveler have afforded, while still catching the end of the train?

Sections D: Vectors, and E: Motion in a Plane

Each vector problem requires a sketch.

*2.47. Vector **p** has components $p_x = 4$, $p_y = -3$; vector **q** has components $q_x = 6$, $q_y = 3$. Find the magnitude and direction of $\mathbf{p} + \mathbf{q}$.

2.48. If, in Fig. 2.22a, we have $|\mathbf{p}| = 5$ and $\theta = 30°$, find p_x and p_y.

2.49. If, in Fig. 2.22b, we have $|\mathbf{p}| = 3$ and $\theta = -37°$, find p_x and p_y.

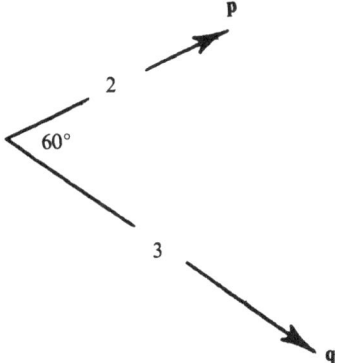

*2.50. Two vectors **p** and **q**, shown in the adjoining figure, have magnitudes 2 and 3, respectively, and the angle between them is 60°. (a) Use the cosine rule to determine $|\mathbf{p} + \mathbf{q}|$. (b) Use the cosine rule to determine $|\mathbf{p} - \mathbf{q}|$.

2.51. A certain balloon would rise at 3 meters/sec in still air. Actually, there is a horizontal breeze of 4 meters/sec. At what angle to the vertical does the balloon rise, and what is its speed?

*2.52. A passenger looks out of the window of an airplane traveling due north at 400 km/h. He sees another airplane whose actual velocity is 300 km/h due east. What is, in magnitude and direction, the apparent velocity of this second airplane, as seen by our observer?

2.53. At what angle to the vertical should one tilt one's umbrella when walking at 1 meter/sec in a drizzle whose drops fall vertically at a constant 4 meters/sec? (The tiny drops do not accelerate, because of air resistance.)

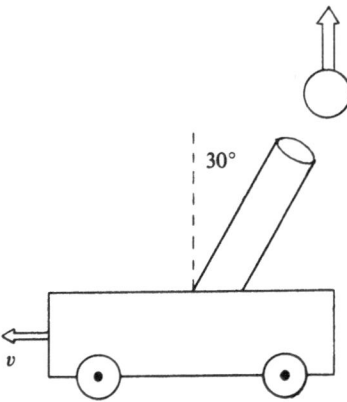

*2.54. A spring gun, mounted on a cart with speed v, is aimed in the backward direction at an angle of 30° to the vertical. (See the figure.) Its muzzle velocity is 2.0 meters/sec. Find how v must be adjusted so that the projectile shoots vertically up.

2.55. In order for an airplane to travel due north at a speed of 500 km/h in an east-to-west crosswind of 50 km/h, what should be its air velocity (i.e., the velocity relative to the air) in direction and magnitude?

2.56. An airplane whose air velocity is to the northwest at 71 km/h is flying in a 50-km/h wind, which comes from the west. How far, and in what direction, will the airplane travel in 5 hours?

*2.57. A swimmer, who takes pride in his constant 0.5-meter/sec stroke, crosses a 100-meter-wide river at right angles to the banks, but feels as tired as if he had swum 140 meters. What speed of water current would explain this fact?

2.58. A boat's speed gauge, dipped in the water, reads as if the boat had a speed of 8 knots due south. The boat's wind gauge seems to indicate a wind out of the west at 10 knots. Actually, the boat is in an ocean current that travels 60° east of north at 16 knots. What is the wind's speed (in knots) and what is its direction with respect to land? (1 knot = 1 nautical mile/h ≈ 0.514 meter/sec.)

Section F: Projectiles

2.59. Plot on the same (x, y) coordinate system the position of a particle falling from rest, and that of a particle thrown horizontally at a speed of 1 meter/sec. Both particles are released at time zero from the origin. Show positions at times 0 sec, 1 sec, 2 sec, and 3 sec. Compress the x or y scale so that the graph can be drawn and interpreted comfortably. How are the positions of the two particles related to each other at the same time?

*2.60. A marble rolls off a horizontal table top at a speed of 0.2 meter/sec. After 0.1 sec, at what angle is the tangent to the marble's trajectory dipping below the horizontal?

2.61. A spring gun with muzzle speed 25 meters/sec is aimed at 37° above horizontal. What is the projectile's velocity (speed and direction) 3 sec after it has been shot, assuming it is still in the air?

*2.62. A spring gun with muzzle speed 25 meters/sec is vertically mounted on a carriage that moves uniformly along a straight horizontal track at 8 meters/sec. (a) Where is the projectile (in horizontal and vertical position) in relation to the gun, 1 sec after being shot? (b) What angle does the trajectory (relative to the ground) make with the horizontal at that time? (c) How far ahead or behind the gun does the projectile eventually fall back?

*2.63. In an elevator that descends uniformly at 1.0 meter/sec, a passenger drops an object from a height of 2.0 meters above the elevator's floor. How much time elapses before the floor is reached?

2.64. A certain kangaroo progresses by regular jumps, 1 meter high, with 4 meters between bounces. He spends almost no time touching the ground. How fast, in km/h, does he travel in level country?

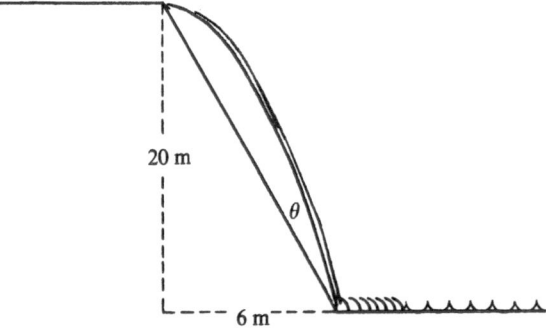

*2.65. A dam is 20 meters high and has a base that extends 6 meters beyond the top (see figure). The overflowing water falls at the exact foot of the dam. What is its horizontal speed as it leaves the top of the dam?

2.66. In the figure of Problem 2.65, find the angle θ between the dam (a flat sloping wall) and the bottom of the waterfall.

2.67. A marble rolls off a horizontal table top, which is 1.250 meters above a horizontal floor, at a speed of 3.000 meters/sec. After 0.500 sec, is the marble still falling?

*2.68. The jet of a fountain leaves the pool surface at an angle of 15° off the vertical, and forms an arch 5 meters wide at the base. (a) What is the water's initial speed? (b) How high is the fountain's arch? (c) How long does a small portion of water stay in the air?

2.69. A bullet is shot horizontally at a speed of 300 meter/sec. How far does it travel before being 1 cm below the line of aim?

*2.70. An elevator is rising at constant velocity. In it, a woman drops her purse from a height of 1.0 meter above the elevator's floor; when it strikes the floor it is as high above street level as when it was released. (a) Sketch qualitatively the combined graphs of altitude against time for the purse and for the elevator floor. (b) How fast is the elevator rising?

*2.71. The action of a science-fiction film, which involves much jumping and throwing objects, takes place on the Moon, where the acceleration of gravity is about 1/6 that of Earth. The filming takes place on Earth, however. By what factor should the replay be slowed down or speeded up to simulate the lunar environment?

2.72. The fountain of problem 2.44 is redesigned to produce an arch, 5 meters high and 5 meters wide at the base. The starting cross section is still 2 cm × 2 cm. How many liters must be pumped per minute?

*2.73. At what angle θ to the horizontal must a cannon be aimed to have its maximum range on level ground?

2.74. A gun with a muzzle speed of 600 meters/sec must bombard a spot that is 10 km away and at the same altitude. There are two angles θ_1, θ_2 above horizontal at which the gun could be aimed. What are they?

2.75. A gun with muzzle speed v_0 bombards a spot that is 9 km away and at the same altitude. There is only one angle of aim at which it can do this. Find v_0. (Cf. Problem 2.73.)

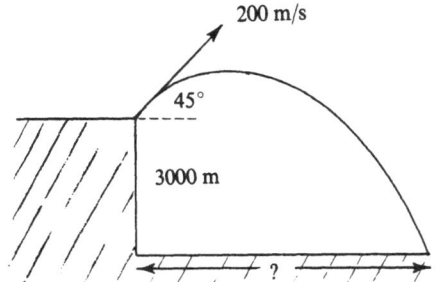

200 m/s

45°

3000 m

?

2.76. A cannon, aimed 45° above horizontal, and located on a 3000-meter-high plateau, has a muzzle speed of 200 meters/sec. How far, horizontally, would the plain be bombarded if air effects were negligible? (See the figure.)

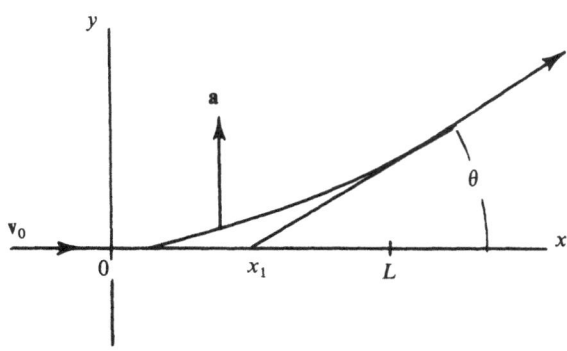

*2.77. In an electronic image tube, an electron is projected along the x axis with an initial speed v_0; see the figure. As long as its x coordinate is between 0 and L, the electron has a constant acceleration a in the y direction. Elsewhere it has a uniform velocity. In terms of v_0, a, and L, what is the final angle θ of the trajectory with the x axis? What is the coordinate x_1, on the x axis, from where the electron appears to come if one makes a straight-line extrapolation of its emerging trajectory?

Answers to True or False

1. False (v_z is reversed, but a_z never changes).
2. True [see between Fig. 2.6 and Eq. (2.B.18)].
3. True [see Eq. (2.C.16)].
4. True [the average velocity, v_{av}, increases linearly with time; see between Eqs. (2.C.7) and (2.C.8))].
5. True [their relative velocity is a constant, equal to $g \times$ (difference in starting time)].
6. False.
7. True.
8. True.
9. True.
10. False (apply Galilean relativity: let the water be motionless while the shores are moving).
11. True (at the top of the trajectory).
12. False (a fast enough trajectory is as flat as desired).
13. False [free-fall behavior, Eq. (2.F.7), is unchanged under the replacement $t \to -t$, provided $\mathbf{v}_0$ is replaced by $-\mathbf{v}_0$: all velocities are reversed in the film, but not $\mathbf{g}$].
14. True (v_{0x} is no longer zero).
15. False (the moving observer does not consider the fall to be "vertical").

Answers or Hints to Problems

2.3. 1 in./year for the continental drift.

2.6. Requires no knowledge of infinite series.

2.9. The slower and faster cars travel 80 and 120 km, respectively.

2.11. 200 meters/sec^2, assuming a constant acceleration.

2.12. $9.8 \times 10^{-6} \ \mu\mathrm{m}/(\mu\mathrm{s})^2$.

2.14. 125 cm.

2.16. 0.8 meter.

2.19. $2g$.

2.24. 45 meters.

2.26. First find the acceleration.

2.27. Write $x = \frac{1}{2} a_x t^2$ for $x = 2$ meters, $x = 3$ meters; eliminate a_x.

2.30. 2.5.

2.32. (a) $v_x = 3kt^2$; (b) $v_x = -p/(pt + q)^2$.

2.35. 50.

2.37. $[(v_B^2 + 2gh)^{1/2} - v_B]/g$.

2.40. (c) About 1.

2.41. Hint: When in contact, balloon and stone have the same velocity.

2.44. 240 liters.

2.46. Hint: The traveler reaches the train when both have the same velocity.

2.47. 10; parallel to the x axis, in the positive direction.

2.50. (a) $\sqrt{19}$; (b) $\sqrt{7}$.

2.52. 500 km/h, 37° east of south.

2.54. The projectile's "true" x velocity must vanish. Answer: $v = 1.0$ meter/sec.

2.57. Hint: Swimmer's speed relative to the water = $(140/100)(0.5)$ meter/sec. Answer: Current speed ≈ 0.35 meter/sec.

2.60. About 80°.

2.62. (c) It falls back right on top of the gun.

2.63. Use Galilean relativity.

2.65. First find the time of fall. Answer: Horizontal speed $= 3$ meters/sec.

2.68. See Eq. (2.F.21); the formula $2\cos\theta_0 \sin\theta_0 = \sin 2\theta_0$ is convenient. (a) 10 meters/sec.

2.70. (b) First use Galilean relativity to obtain the time of fall.

2.71. Slowed down by a factor of about $\sqrt{6} \approx 2.5$.

2.73. 45°. (You may use $2\cos\theta_0 \sin\theta_0 = \sin 2\theta_0$ instead of calculus.)

2.77. $\tan\theta = aL/v_0^2$; $x_1 = L/2$.

CHAPTER 3

FORCE

What causes anything to move as it does? This is one of the most fundamental questions of physics, and the science of **dynamics** has been developed to deal with it; this chapter is an introduction to that subject. In this area our thinking has been shaped almost entirely by one man, Sir Isaac Newton (1642–1727).

Broadly speaking, two unrelated factors must be distinguished, which together dictate the actual motion of an object. They are (1) the manner in which it has been launched (its **initial conditions**), and (2) the *forces* that act on it at all times during the motion. The emphasis in this chapter is on effects of type (2); we recall that effects of type (1) were explored in the preceding chapter in connection with projectiles.

Many of us still harbor the intuitive—but wrong—notion that force determines velocity. According to Newtonian physics, *acceleration, not velocity, is proportional to force*.

In what follows, force will be defined quantitatively; we shall learn how forces are measured, how several forces combine, and how they affect the behavior of a particle to which they are applied. (Entire objects will often be treated as if they were simple particles; a more realistic point of view will be taken in later chapters.)

Forces may be acting even when no acceleration—or even no motion—occurs at all. This can happen when the existing forces cancel one another; several such **equilibrium** situations will be examined in this chapter.

The concept of force will recur throughout our study of physics; it is needed in order to understand most natural phenomena, from cosmology down to the subnuclear world.

Mechanics (= kinematics + dynamics) combines motion and force into a comprehensive study. It forms a substantial part of this book's subject matter.

A. The Law of Inertia (Newton's First Law)

As we have seen, the motion of a particle must be referred to a coordinate system, also called a **frame of reference**, or simply a **frame**. Clearly, we must worry about whether the frame we are using has a motion of its own. An erratically moving frame would lead to a confusing description of a particle's motion, and would hide the laws of physics rather than reveal them. Therefore certain frames are to be preferred. As an illustration, the altitude of a freely falling particle should not be measured along the mast of a tossing sailboat (a poor frame of reference), but if measured along the wall of a physics laboratory, the altitude would follow the rules of the preceding chapter.

Newton's first law, which we now formulate, should be considered as *a recipe for selecting an appropriate frame of reference*, rather than as an ordinary physical law.

i. Inertial Frames

Of Newton's three laws of motion, the first one, also called the **law of inertia**, asserts that

> An isolated particle maintains a uniform motion.

(3.A.1)

There is more to this statement than meets the eye: it is a nontrivial task to make sure that a particle is truly isolated. Projectiles in terrestrial laboratories cannot be isolated (gravitationally) from the Earth. Our best-isolated "particles" are the stars (including the Sun, on whose motion the planets have a negligible effect); huge interstellar distances are a sufficient reason for that conclusion. We note, however, that a residual gravitational interaction still affects well-separated stars—no isolation is perfect.

The important aspect of (3.A.1) is that it says something about our frame of reference. If we have verified (3.A.1) for a certain particle (say a distant star), then a colleague of ours, whose coordinate axes are rotating or accelerating, will find the statement untrue: he will claim that the star is moving in a circle around him, or is otherwise accelerating. Thus again, there are "good" and "poor" frames, and Newton's first law gives us a procedure for selecting the good ones, called **inertial**:

> An inertial frame is one with respect to which Newton's first law is valid.

$$(3.A.2)$$

An Earthbound system does rotate and accelerate relatively to the stars, but gently enough so that *ordinary terrestrial laboratories are inertial for most practical purposes.* Unless specified otherwise, all measurements in physics are assumed to be performed with respect to an inertial system.

ii. The Horizontal Frictionless Plane

A uniformly moving particle (for example, a particle at rest) behaves *as if* isolated: the effects of gravity, friction, pressure, etc., must somehow be canceling each other out. An ordinary book lying on an ordinary table is an example of this: the effects of the table and of gravity cancel out.

In order to cancel out gravity, while still permitting free horizontal motion, one may place a particle on a **frictionless plane** or track. In practice, such a device can only be approximated, for example, through lubrication, or through the use of an air table; see Fig. 3.1. Rollers or wheels are another possibility, but their rotation must be ignored in a particle picture. Suspension from a very long wire is an interesting substitute for a supporting plane.

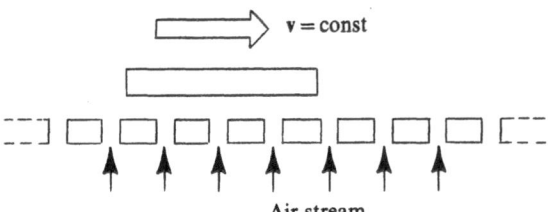

Figure 3.1. An air table. The glider is levitated by air injected from below through holes in the table surface.

On a successful horizontal frictionless plane, a sliding object maintains its initial velocity after being launched; Newton's first law is used to test the plane.

B. Force and Acceleration (Newton's Second Law)

Newton succeeded in defining a quantity —**force**—which represents the external influences on the motion of a particle.

Two approaches to force are possible, and both are used in this book. They consist of (1) diagnosing a force by its effect, namely, by the *acceleration* of the particle that experiences the force; or (2) examining the particle's surroundings in order to identify existing forces, such as contact with another object, gravitational attraction to the Earth, electric or magnetic effects from laboratory equipment, etc.

Approach (1) is used to *define* force; it is very simple and completely general, because it requires no itemization of the environment. But we cannot explore the world entirely with definitions, and therefore approach (2) is the one that yields the real physics. Throughout this book, a variety of forces will be studied in detail; all of them should, however, be describable in terms of the acceleration of a test particle.

This section is organized as follows. First, we select a standard particle, whose **mass** is the one relevant property. Next, from the particle's acceleration, we infer the total force acting on it; this relation is Newton's second law. One of its applications is the concept of **weight**, which must be carefully distinguished from mass.

The acceleration experiments described here are rather elaborate and difficult to perform accurately. Their purpose, as far as we are concerned, is only to illustrate principles. For the practical, accurate measurement of forces and masses, static methods are used; they are described in the next section.

i. A Standard of Mass

A special object, the **standard of mass**, has been selected by international consensus as the model to which all particles are compared in terms of their response to a force. That standard is *defined* to possess a mass of exactly 1 kilogram ($=1$ kg $=$ 1000 grams). It consists of a platinum cylinder, housed at the International Bureau of Weights and Measures, Sèvres, France. Many copies of it have been made for distribution to other countries. Although they are very expensive and are touched as little as possible, it is philosophically important to know that several (say n) standard kilograms can *in principle* be welded together to form a mass of n kg; they can also be cut into equal pieces to form known fractions of a kilogram. In short, these masses are, by definition, additive; thus, we can construct an object of any known mass m.

ii. The Definition of Force

Why is it reasonable to use the acceleration as a measure of force? The argument goes as follows. When left entirely to itself, a particle exhibits zero acceleration (law of inertia). Hence a nonzero acceleration implies an external influence on the motion; this influence is what we call a force.

Quantitatively, the **total force vector** $\mathbf{F}_{tot}$ acting on a particle with mass m and acceleration $\mathbf{a}$ is defined as

$$\boxed{\mathbf{F}_{tot} = m\mathbf{a}} \qquad (3.B.1)$$

(Newton's second law, or law of acceleration). Equation (3.B.1) calls for several comments.

The First Law as a Special Case

We see that $\mathbf{F}_{tot} = \mathbf{0}$ is equivalent to $\mathbf{a} = \mathbf{0}$, as it must be.

Mass Is Inertia

We shall see later that a given, controlled force (say from a spring or rubber band) can be tried on objects of different masses. Then (3.B.1) shows that a greater mass m goes with a smaller acceleration $|\mathbf{a}|$. Thus, mass constitutes **inertia**; that is, it tends to oppose attempts at acceleration.

Units

Replacing the symbols m, $\mathbf{a}$ in (3.B.1) by their SI units, we see that $\mathbf{F}_{tot}$ is expressed in kg meter/sec². For short, one defines

$$1 \text{ newton} = 1 \text{ N} = \frac{1 \text{ kg meter}}{\text{sec}^2} \qquad (3.B.2)$$

[Alternative units: One has 1 newton $=$ 10^5 dynes; the latter unit belongs to the c.g.s., or centimeter-gram-second system. The British gravitational system makes use of the **pound force** (lb or lbf), given by 1 lb ≈ 4.45 newtons.]

iii. The Spring Scale

A spring scale (or spring balance) accomplishes two things: it can be made to exert a *reproducible force* $\mathbf{F}$, and it also measures $\mathbf{F}$. Figure 3.2 shows the calibration of such a scale; the procedure utilizes a horizontal frictionless plane, along which a controlled acceleration $\mathbf{a}$ is imparted to a chosen mass m.

We clearly do not need or want a theory of springs at this stage: the calibration is done empiri-

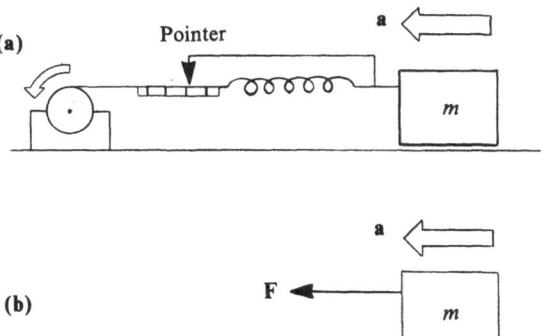

Figure 3.2. (a) Calibration of a spring scale on a frictionless table. The acceleration, $\mathbf{a}$, of m is controlled by the motor at the left. The pointer reading, F, depends on the stretch of the spring in some initially unspecified way; the scale is then marked so that $F = ma$. (Both m and a are known.) The procedure is cumbersome and inaccurate, but constitutes an essential **thought experiment**. (b) The direction of $\mathbf{F}$ is that of the spring. It is understood that $\mathbf{F}$ is the *total* force on the test object.

cally, using only the observed acceleration. A good spring scale, if much lighter than the mass being accelerated, displays in its readings a wonderful consistency with Newton's second law. For example, it will read an unchanged **F** when **a** is doubled while *m* is halved. This property is, in fact, the test of a "good" spring scale; it can then be called a "forcemeter." Having been calibrated, it can be used statically, i.e., without any motion at all; how and why this is done will be explained in the next section.

iv. Comparing Masses Dynamically*

It would be unfortunate if we had to dissect platinum standards in order to put together a desired mass. Actually, a forcemeter such as a spring scale allows us to assign a mass to arbitrary objects. Newton's law is "turned around" to yield, in terms of magnitudes, the amount

$$m = \frac{F}{a} \qquad (3.B.3)$$

This defines the mass *m* of any object being accelerated; *F* is read on a forcemeter previously calibrated by means of a standard mass; see Fig. 3.3.

By this method, or by more convenient and accurate ones such as weighing (discussed later), adequate mass standards can be made of inexpensive alloys such as brass. We note that 1 liter ($= 10^{-3}$ meter^3) of water has almost exactly a mass of 1 kg. This is not a coincidence: historically the kilogram was first defined in terms of that amount of water. It is observed, through the measurement of accelerations and the use of (3.B.3), that several objects can be tied together to yield an overall mass that is just the sum of the objects' individual masses. (For the platinum standards this was just a chosen prescription, of course.) Thus, mass is **additive**. (It is also considered, in a intuitive sense, to represent an **amount of matter**; when, in a metric-system country, we buy 10 kg of potatoes, we are not interested in their inertial properties.)

Conservation of Mass

Imagine an object to which nothing is being added from outside, and from which nothing is

* "Dynamically" = by subjecting to accelerations.

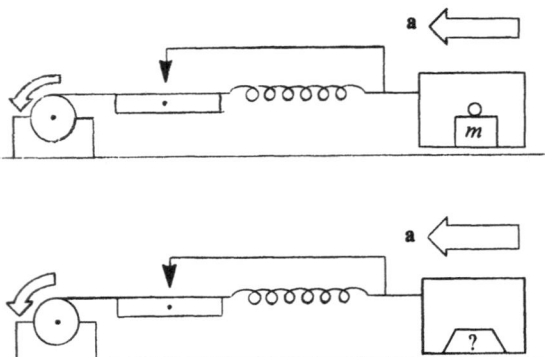

Figure 3.3. The known and unknown masses are declared to be equal: they experience the same acceleration when acted on by the same force.

allowed to escape (a "leakproof system"). It is known, ever since the meticulous weighing experiments of Antoine Lavoisier (1743–1794), that such an object keeps a constant mass, even if drastic changes (such as chemical reactions) are allowed to take place internally. This is the principle of **conservation of mass**.

v. A Spectrum of Masses

Figure 3.4 shows, on a logarithmic scale, various observed and calculated masses.

vi. Weight

In the laboratory, perhaps the most important of all mechanical force vectors is that which accelerates a falling object towards the center of the Earth. It is called the **force of gravity**, or equivalently, the object's *weight vector*, and will be denoted by **w**. Its value, for a mass *m*, is found at a glance if we set, in (3.B.1), $\mathbf{F}_{tot} = \mathbf{w}$ and $\mathbf{a} = \mathbf{g}$. The result is

$$\mathbf{w} = m\mathbf{g} \qquad (3.B.4)$$

This is a downward vector whose magnitude,

$$w = mg \qquad (3.B.5)$$

is called the object's **weight**. For example, a 1-kg mass has, on Earth, a weight

$$w_{1\,kg} = (1 \text{ kg})\left(9.8 \frac{\text{meters}}{\text{sec}^2}\right) = 9.8 \text{ newtons}$$

$$(3.B.6)$$

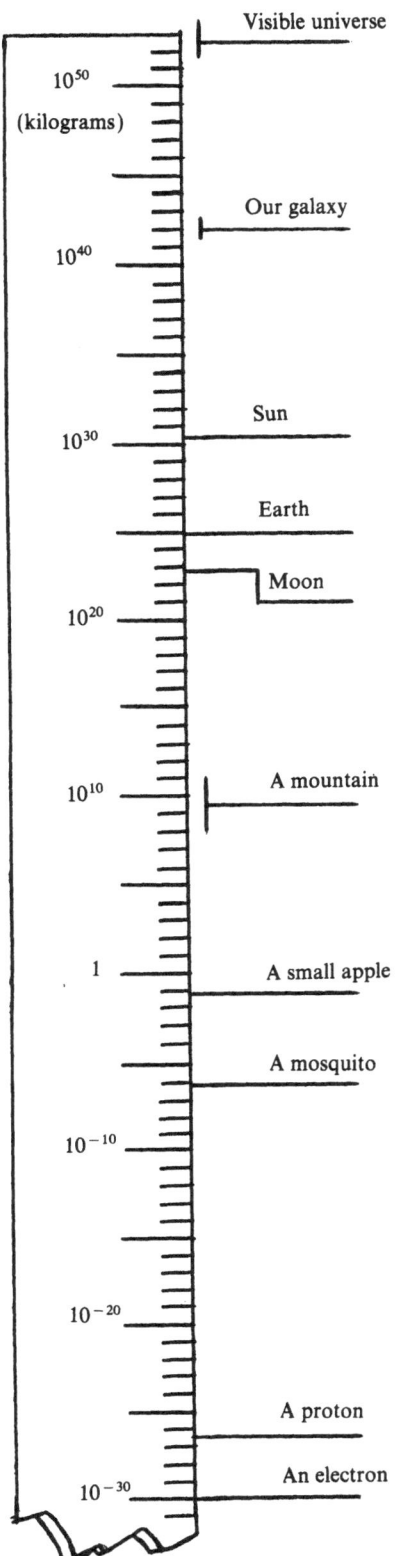

Figure 3.4. Typical masses, from the whole Universe down. There is no lower limit: as far as its known, several particles (neutrinos, photons) have zero rest mass.

(Thus, one liter of water weighs about 9.8 newtons.) On the Moon, its weight would be about 1/6 of that amount, since $g_{\text{Moon}} \approx \frac{1}{6} g_{\text{Earth}}$. Sufficiently far from astronomical bodies all objects have zero weight.

To summarize:

> The total force of gravity on an object is called its weight, and is proportional to its mass; weight varies with the local value of g, whereas mass is independent of location. Also, since **g** is independent of the falling objects's velocity, so is **w**: weight does not depend on motion.

(3.B.7)

vii. A Spectrum of Forces

Some forces arise from **contact** between two objects, or from contact between two parts of the same object. In other cases, the effect is exerted at a distance. The **gravitational** force is the prime example of a force exerted without a material transmitter. Two other such forces are easily demonstrated, namely, the **electric** and **magnetic** ones. (We shall see later that both these forces are different aspects of the same physical interaction.) Our study of physics will show that even the so-called contact forces are in reality forces at a distance. They turn out to be entirely due to the electric and magnetic forces between atoms, or rather between the electrons and nuclei of which the atoms are made.

Yet another kind of force, the *nuclear* one, is much better hidden, although it typically dwarfs the other kinds in magnitude. However, it acts only over very short distances, and thus is sheltered from observation by the long-range electric forces around it. (There is a force of even shorter range than the nuclear one, the so-called "**weak force**." Its existence has been inferred from a radioactive process called beta decay.)

Contemporary physics is progressively unifying this wealth of phenomena. Different-looking forces can have their properties deduced from each other. For example, we shall ourselves deduce some magnetic properties from electric ones. As to the nuclear force, it stems from one that is even stronger and better hidden, and which acts primarily between **quarks**, subparticles of which the proton and neutron are, at present, thought to be made.

Figure 3.5 will give an idea of the magnitudes characterizing some forces. It should be clear that most of them have not been measured by anything so direct as a system of springs or weights.

C. Combining Forces

Spring scales allow us to study the effect of several forces applied simultaneously to a single particle, as in Fig. 3.6a. One observes in practice that such a combination of forces $\mathbf{F}_1$, $\mathbf{F}_2$,... can be equivalently replaced by a single total force $\mathbf{F}_{tot}$ (the **resultant force**, or **net force**); if all forces have been included, $\mathbf{F}_{tot}$ is then the force that must be used in Newton's second law (3.B.1).

If the resultant force is zero, $\mathbf{F}_{tot} = \mathbf{0}$, the situation is called an **equilibrium of forces**. It implies a zero acceleration and therefore a constant velocity. Of particular interest is the case of **static equilibrium**, where no motion exists at any time. For example, every brick in a building, although subjected to a set of large forces, is in static equilibrium; the resultant force on each brick must therefore be zero.

i. Vector Addition of Forces

It is experimentally observed that a resultant force is the **vector sum** of the individual forces:

$$\mathbf{F}_{tot} = \mathbf{F}_1 + \mathbf{F}_2 + \cdots \qquad (3.C.1)$$

as Fig. 3.6 illustrates in detail.

An important special case is the equilibrium between two equal but opposite forces acting on

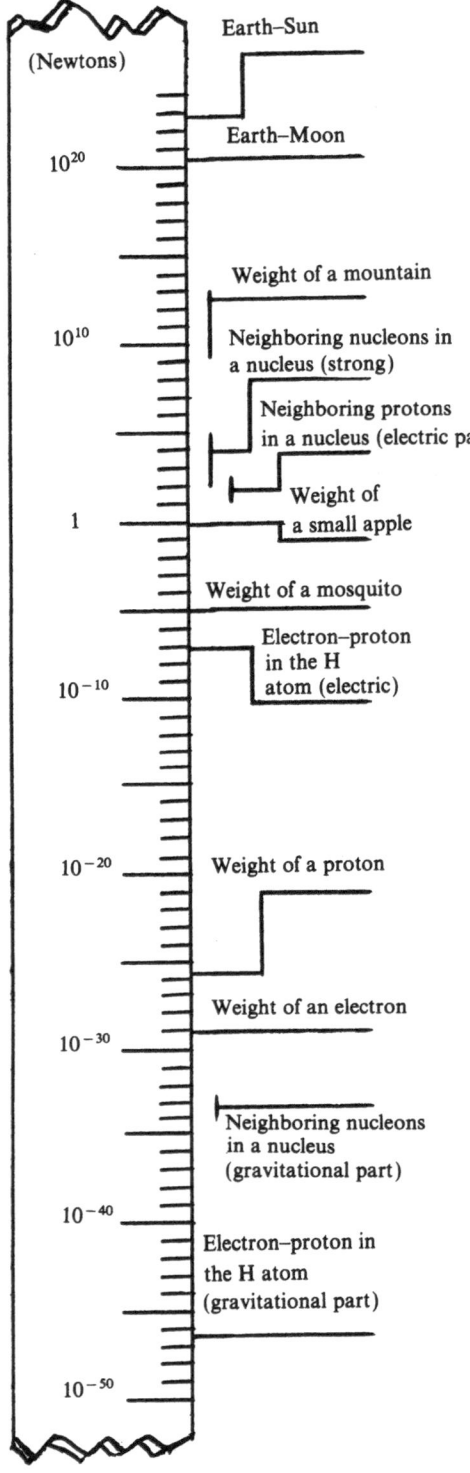

Figure 3.5. Typical forces. Weights are understood to be as measured on Earth. Pairs of objects are listed to indicate the force they exert on each other. The force between two neighboring nucleons (i.e., protons or neutrons) is almost wholly of a kind called "strong" or nuclear. The remaining fraction of it is mostly electric, and a far smaller fraction yet is gravitational.

These orders of magnitude show a staggering discrepancy. Note also that the weight of an apple (indeed, a whole basket of them) could be supported by the force between two single nucleons. We could envision a sturdy nucleon chain, much finer than a single atom. Unfortunately no one knows how to make such chains, or even whether they are possible. The forces that hold atoms together (see the electron–proton force) are almost entirely electric, but here again a minute fraction of the force must be gravitational. Magnetic forces in atoms and nuclei are not shown on this chart, but they are typically not much smaller than the electric ones.

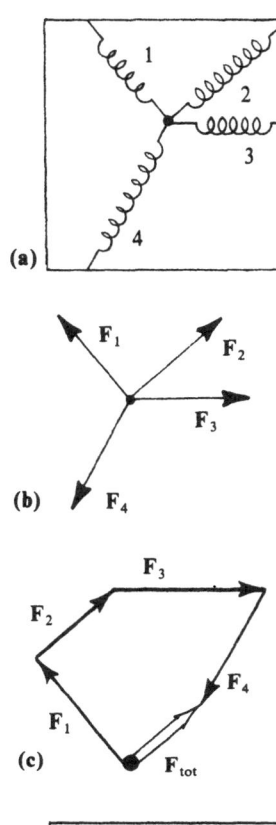

(a)

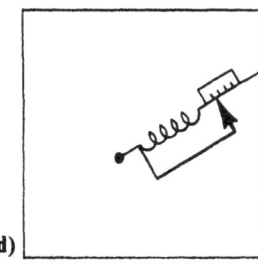

(b)

(c)

(d)

Figure 3.6. (a) Several forces acting on a particle. (b) Their vector representation. (c) Their resultant, $\mathbf{F}_{tot}$. (d) A spring scale replaces springs 1, 2, 3, 4, and measures $\mathbf{F}_{tot}$: the acceleration of the particle is as in (a). (Note: this is a "thought experiment," difficult to perform in practice.)

the same particle; see Fig. 3.7. Here we have $\mathbf{F}_1 + \mathbf{F}_2 = 0$, and we find experimentally (from $\mathbf{a} = 0$) that $\mathbf{F}_{tot} = 0$ as required by (3.C.1). The result could have been predicted from **symmetry**, which is a useful shortcut in many problems. (In this case, a nonzero $\mathbf{F}_{tot}$ would have to point in some unique direction, defined by $\mathbf{F}_1$ and $\mathbf{F}_2$; but no such favored direction exists.)

More generally, for nonparallel forces, assertion (3.C.1) is accurately verified by static experiments such as the one illustrated in Fig. 3.8. We have here an eminently practical method, in contrast to that of Fig. 3.6.

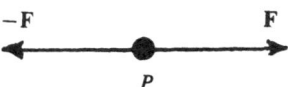

Figure 3.7. Two equal and opposite forces on the same particle P. The resultant vanishes.

ii. Examples

Example 3.1. Static Weighing. Figure 3.9a shows an object of mass $m = 5\,\mathrm{kg}$ suspended at rest from a spring balance. How much force, F, does the balance indicate?

Figure 3.9b shows the forces acting on m. Equilibrium means

$$\mathbf{F}_{tot} = \mathbf{0}$$

or, in the upward direction,

$$F - w = 0$$

Thus, the balance indicates the object's *weight*,

$$F = w = mg$$

$$= (5)(10) \text{ newtons} = \underline{50 \text{ newtons}}$$

[The elementary nature of this procedure obscures its basic importance. Static weighing owes its accuracy to the fact that it is a measurement by

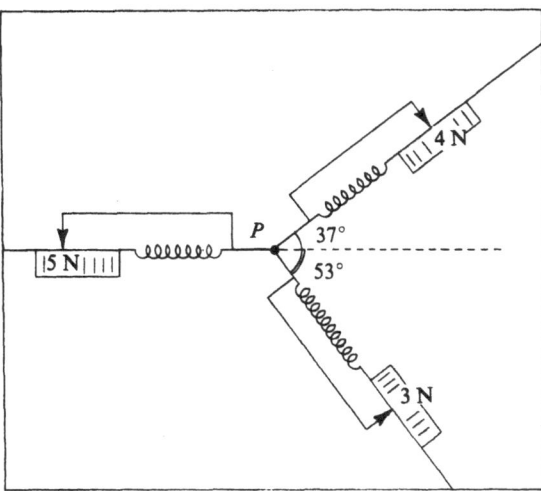

Figure 3.8. Top view of a horizontal frictionless plane on which P is held in static equilibrium. This experiment should verify the vector addition of forces. Does it?

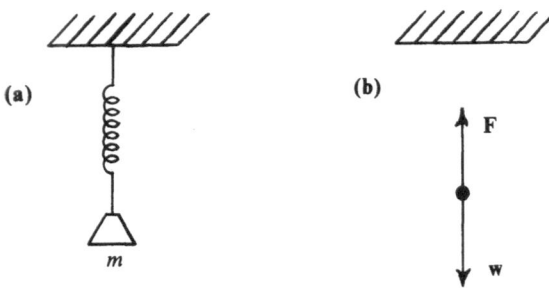

Figure 3.9. Static weighing. This is a more practical way of measuring mass (from $m = w/g$) than by actual acceleration.

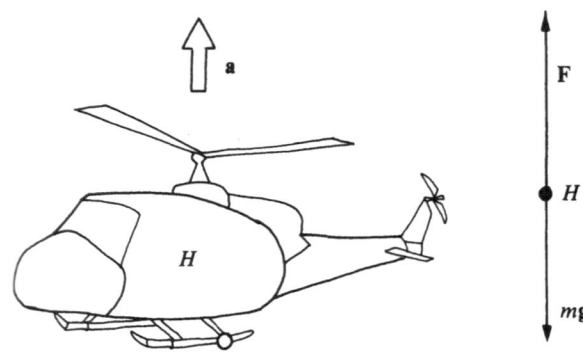

Figure 3.10. The vertically accelerating helicopter, as a helicopter and as a particle.

cancellation, also called a **null experiment**: a certain effect, here the acceleration, is observed to vanish. Many kinds of force, besides weight, are determined in this way; one spring can be calibrated against another, or a rocket engine can have its **thrust** (= force exerted on the engine by its exhaust flame) measured statically on a test bench. Null experiments of great sophistication are used in many areas of physics.]

Example 3.2. Unbalanced Vertical Forces. A helicopter of mass 1500 kg is rising with an acceleration of 2 meters/sec². What is its lift (= upward force due to the air)? By what percentage does the lift exceed the weight?

Let **F** = lift vector, $m = 1500$ kg, $a = 2$ meters/sec². Then Fig. 3.10 shows the forces on the helicopter; $m\mathbf{g}$ is its weight vector ($g = 10$ meters/sec²). We have

$$F_{tot} = ma$$

or, from the figure,

$$F - mg = ma$$

$$F = m(a + g)$$

$$= (1500)(2 + 10) \text{ newtons}$$

$$= \underline{18\,000 \text{ newtons}}$$

The excess lift, relative to the weight, is

$$\frac{F - mg}{mg} = \frac{ma}{mg} = \frac{2}{10}$$

(the value of m does not enter). Thus, the lift exceeds the weight by $\underline{20\%}$.

Example 3.3. A Rocket Problem (Unbalanced Nonparallel Forces). During propulsion, must a rocket's nose point in the direction of motion? Must it at least point in the direction of acceleration? The answer to both questions is no, as illustrated in Fig. 3.11a. A certain rocket of mass $m = 2000$ kg is just above the atmosphere, and must be kept in a horizontal trajectory with an acceleration $a = 30$ meters/sec²; we can take $g \approx 10$ meters/sec². Find (a) the angle θ which the rocket's body must make with the horizontal (i.e., the "attitude angle"), and (b) the magnitude F of the engine's thrust.

Figure 3.11b shows the two forces on the rocket: **F** and the weight $\mathbf{w} = m\mathbf{g}$. Their sum $\mathbf{F}_{tot} = \mathbf{F} + m\mathbf{g}$ must be horizontal. In terms of vertical and horizontal components, Newton's second law, $\mathbf{F}_{tot} = m\mathbf{a}$ gives

$$F \sin \theta - mg = 0 \qquad (3.C.2)$$

$$F \cos \theta = ma \qquad (3.C.3)$$

(a) *Attitude angle.* Eliminating F, we obtain

$$\tan \theta = \frac{g}{a} = \frac{10}{30}$$

or

$$\theta = \underline{18°}$$

(The rocket's mass is not involved.)

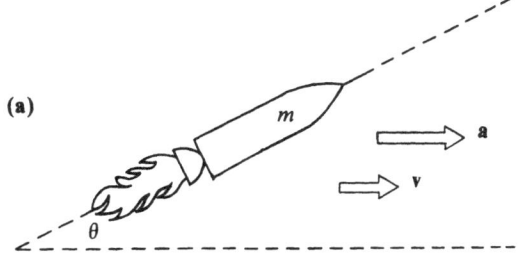

(a)

(b)

Figure 3.11. A horizontally moving and accelerating rocket.

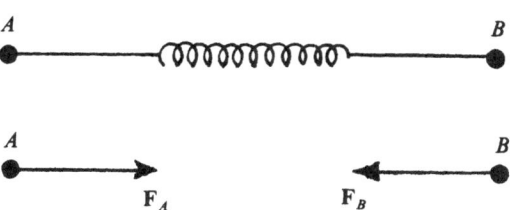

Figure 3.12. Action and reaction: here $\mathbf{F}_A = -\mathbf{F}_B$ by symmetry.

(b) Thrust. Knowing θ, we can solve either (3.C.2) or (3.C.3) for F. Alternatively, we can eliminate θ through $\cos^2\theta + \sin^2\theta = 1$, which amounts to using the Pythagorean theorem in Fig. 3.11b. The result is

$$F = m(a^2 + g^2)^{1/2}$$
$$= (2000)[(30)^2 + (10)^2]^{1/2} \text{ newtons}$$
$$= \underline{63\,000 \text{ newtons}}$$

D. Action and Reaction (Newton's Third Law)

So far, we have considered the effect of one or several forces acting on a single particle, while ignoring the objects that are exerting those forces. In the case of a particle pair A, B, the force on A might be due to B. We now ask how B itself is affected by the force it exerts.

i. Stating Newton's Third Law

Figure 3.12 shows two massive objects A, B connected by a light stretched spring. (A and B are therefore accelerating towards each other.) We consider A, B as two particles that exert forces on each other through the intermediary of the spring, itself not counted as a particle. This is a severe but useful idealization, which provides a model for an action at a distance between A and B.

If the spring is built symmetrically, i.e., if its two ends are indistinguishable, then they must exert forces which are equal in magnitude and opposite in direction:

$$\boxed{\mathbf{F}_A = -\mathbf{F}_B} \qquad (3.D.1)$$

$\mathbf{F}_B$, the **action**, is said to be exerted by A on B; $\mathbf{F}_A$ is its **reaction**, exerted by B on A. Of course, which of these two forces is considered the action, and which the reaction, is entirely a matter of choice.

The discussion becomes more interesting if we ask: "Could the spring be made with one end 'stronger' than the other, in such a way that $\mathbf{F}_A \neq -\mathbf{F}_B$?" Common experience shows this to be impossible, see the experiment of Fig. 3.13. If the box were acted on by unbalanced forces, it could forever accelerate to one side. This is never observed to happen.

Equation (3.D.1), which states that *action and reaction are equal and opposite*, is postulated to

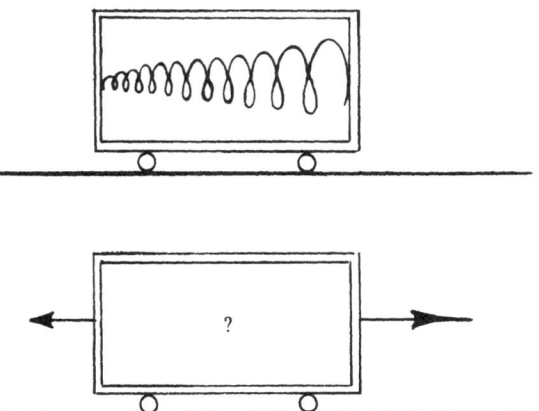

Figure 3.13. A hypothetical spring, of negligible mass, which maintains unbalanced forces on the box. If this were possible, all our transportation problems would be solved without the help of engines. We could also lift ourselves by our bootstraps, and we could push our cars forward while sitting inside.

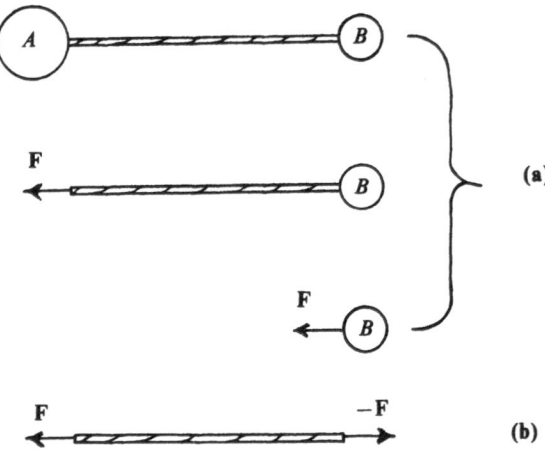

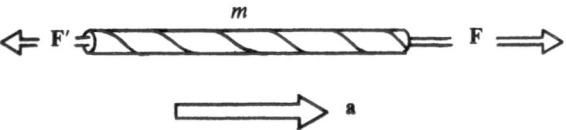

Figure 3.16. A piece of real string as an imperfect transmitter of force. Under any nonzero acceleration, we have $F' \neq F$ unless $m = 0$.

Figure 3.14. (a) The string "transmits" a force **F**: A exerts **F** on the string, and the latter exerts the same **F** on B. (b) Newton's third law implies that B exerts $-$**F** on its end of the string.

hold for any kind of force, and is known as **Newton's third law of motion**. It necessarily involves two particles, and must be carefully distinguished from the cancellation property of Fig. 3.7, which deals with a single particle.

As an illustration, the 5 kg object of Example 3.1 exerts a *downward* force of 50 newtons on the spring from which it is suspended (the reaction to **F**); it also exerts an *upward* force of 50 newtons on the Earth as a whole (reaction to **w**).

ii. Transmission of a Force (Strings and Rods)

When attached to two particles A and B, any light, flexible object (string, belt, or chain) will transmit a pull; any rigid one, such as a connecting rod, will transmit a pull or a push.

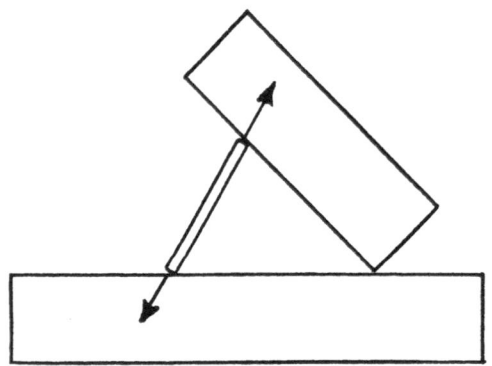

Figure 3.15. Alignment of the forces exerted by a light incompressible rod.

Figure 3.14 shows, for a string, what is meant by "transmitting a force." (The magnitude of that force is called the string's **tension**.)

In this chapter, strings and rods are assumed to be inextensible; rods are assumed incompressible. The forces they exert are always considered to be along the direction joining their two end points; see Fig. 3.15. (If attached at more than two points, a rod becomes a **lever**, capable of exerting transverse forces. This situation will not be considered until Chapter 7.)

Let us analyze the performance of a piece of real string as a force transmitter. Figure 3.16 shows a string segment of mass m. Assume for simplicity that the forces and the acceleration are along the length of the string, and that we are in a gravity-free environment. Let the string segment have a tension F at one end, and F' at the other. Then Newton's second law gives

$$F - F' = ma$$

where a is the string's acceleration. For a given a, and in the limit where $m \to 0$, we get

$$F = F'$$

so that the tension is then faithfully transmitted; the string may be called **ideal**. This argument, which shows that *the ideal string must have zero mass*, applies also to force transmitters other than strings; see, for example, the negligible-mass springs of Figs. 3.12 and 3.13.

Example 3.4. A Tightrope. In Fig. 3.17 a tightrope artist, of weight w, is standing at midpoint on a rope of length $2l$ and negligible weight. That midpoint is dipping by a height h below the ends of the rope, which are at equal levels. How much force, F, does each end of the rope exert on its support?

First we note that F is just the rope's tension. In Fig. 3.17b, the performer is shown as a particle. We can exploit the symmetry of the situation by choosing our axes horizontally (x) and vertically (y). Vertical equilibrium reads $(F_{\text{tot}})_y = 0$, or

$$2F_y - w = 0 \qquad (3.D.2)$$

where F_y is indicated in the figure. From the similarity of the triangles in parts (a) and (b) of the figure, we have

$$\frac{F}{l} = \frac{F_y}{h} \qquad (3.D.3)$$

Eliminating F_y between (3.D.2) and (3.D.3) results in

$$F = \left(\frac{l}{2h}\right) w \qquad (3.D.4)$$

If the rope is very tight $(h \ll l)$, a moderate weight w can generate an enormous tension F.

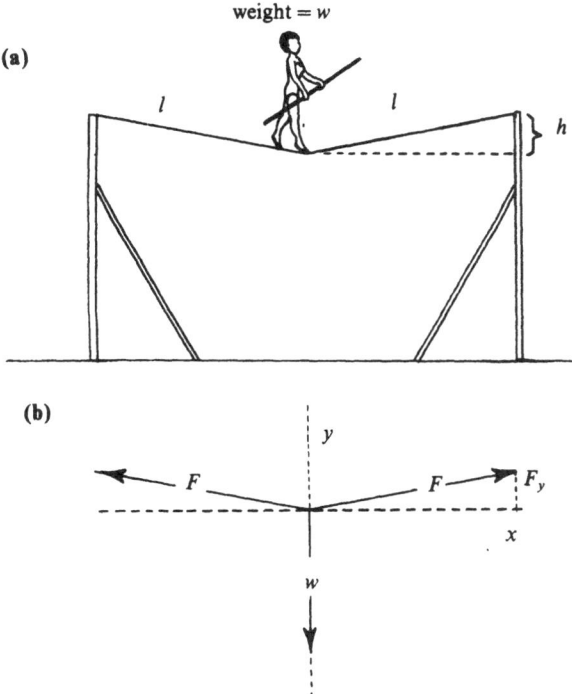

(a)

weight $= w$

(b)

Figure 3.17. (a) The tightrope artist of Example 3.4. (b) The performer as a particle. Note how our choice of axes respects the symmetry of the situation. We never need to state the horizontal equilibrium condition to find the tension F.

(a)

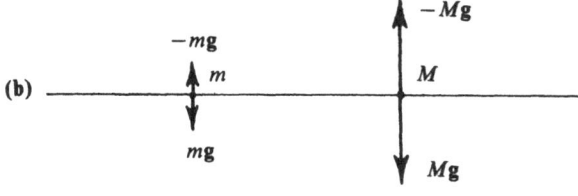

(b)

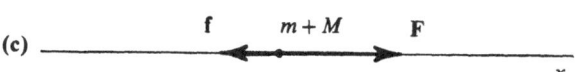

(c)

(d)

Figure 3.18. (a) Accelerating train. (b) Locomotive and car as separate particles (horizontal forces not shown). The weight vectors $m\mathbf{g}$, $M\mathbf{g}$, are canceled by upward forces $-m\mathbf{g}$, $-M\mathbf{g}$ due to the track. (Note: the forces on the track are never shown here; all forces are on the vehicles). (c) The whole train as one particle (vertical forces not shown). (d) Locomotive and car as separate particles (vertical forces not shown).

iii. An Example for Review

The following example is mathematically trivial, but it illustrates several aspects of problem solving, and should be studied carefully.

Example 3.5. Two Connected Masses. On a straight horizontal track, a locomotive of mass $M = 70\,000$ kg pulls a freight car of mass $m = 20\,000$ kg with an acceleration $a = 0.3$ meter/sec^2; see Fig. 3.18a. The track exerts a backward frictional force $f = 16\,000$ newtons on the freight car. Find (a) the wheel traction F (=forward frictional force of the track on M), and (b) the towing force F' of M on m. [Note: all horizontal propelling and retarding forces on a rolling vehicle are exerted *by the road*, at its point of contact with the wheels. (We neglect air effects). The detailed laws of friction will be discussed later on; they are not needed in the present example.]

Figure 3.18b shows only the vertical forces on each vehicle; Fig. 3.18c shows only the horizontal ones. There are no other forces. Since the vertical component of acceleration is zero for each vehicle, all vertical forces cancel out, and from now on they will be ignored.

Which part of the system is selected as a "particle" is a matter of problem-solving strategy. Any choice is correct, provided all forces are taken into account. In Fig. 3.18c, the whole train is a particle of mass $m + M$; in Fig. 3.18d we have two particles, of masses m and M. Note the action–reaction pair $\mathbf{F}'$ and $-\mathbf{F}'$ at the ends of the connecting bar.

The solution consists of writing Newton's second law (its x component) for each particle; they all have the same acceleration $a_x = a$. For $m + M$:

$$F - f = (m + M)\,a \qquad (3.D.5)$$

For m:

$$F' - f = ma \qquad (3.D.6)$$

For M:

$$F - F' = Ma \qquad (3.D.7)$$

As a consistency check, note that (3.D.5) follows from (3.D.6) and (3.D.7); one of our equations, say (3.D.7), is superfluous. Solving (3.D.5) for F, we have

$$F = (m + M)\,a + f$$
$$= [(20\,000 + 70\,000)(0.3) + 16\,000]\ \text{newtons}$$
$$= \underline{43\,000\ \text{newtons}}$$

From (3.D.6),

$$F' = ma + f$$
$$= [(20\,000)(0.3) + 16\,000]\ \text{newtons}$$
$$= \underline{22\,000\ \text{newtons}}$$

E. Pulleys and Surfaces

The present section illustrates some simple means of exerting forces: **pulleys, frictionless surfaces**, and **rough surfaces**.

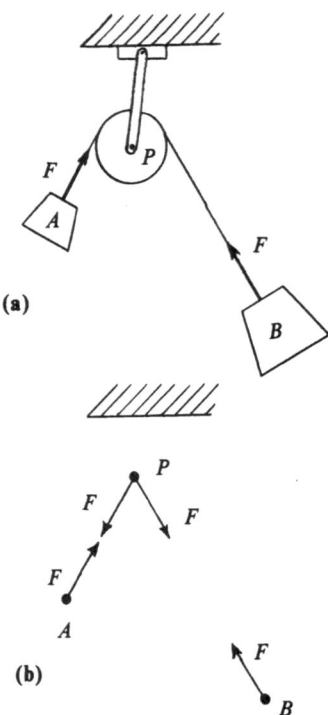

(a)

(b)

Figure 3.19. (a) Properties of an ideal pulley P. A force is transmitted between objects A and B with unchanged magnitude F but modified direction. (b) The law of action and reaction, valid separately between A and P, and between B and P. Objects A and B cannot be considered to exert a force directly on each other. (Only the forces exerted by the string are shown here.)

i. The Pulley

This is a convenient device for transmitting a force "around a corner." The ideal pulley (in practice, a light pulley with a low-friction bearing), is such that a string looped around it has the same tension on both sides, no matter what the pulley's state of motion might be; see Fig. 3.19. The size of the pulley is not relevant. The law of action and reaction cannot be applied to the forces transmitted around a pulley, *unless the pulley itself is included as one of the force-exerting bodies*. This is also demonstrated in Fig. 3.19.

Example 3.6. The Block and Tackle. The arrangement is shown in Fig. 3.20. All straight segments of rope are vertical. The large pulley-plus-weight system has a total weight W. What weight, w, on the free end of the rope, will produce equilibrium (static equilibrium or a uniform vertical motion)?

Vertical equilibrium for the small weight requires that the rope tension be w; all forces on the large system of weight W are shown. Vertical equilibrium for that system reads

$$4w - W = 0$$

so that

$$w = \frac{W}{4}$$

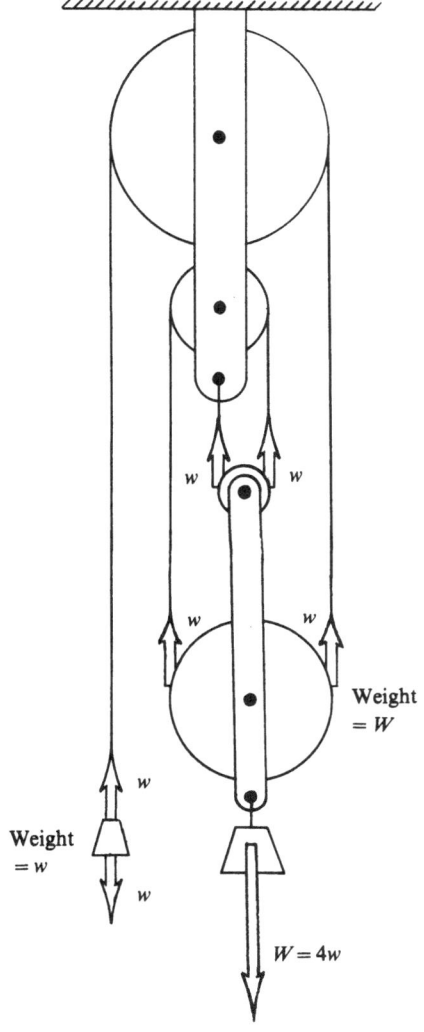

Figure 3.20. A block-and-tackle arrangement. The assembly of pulleys + weight on the lower right is hanging from four segments of rope. The rope is uninterrupted, from its point of attachment on the upper assembly to the small weight w. Therefore the tension is the same in every segment.

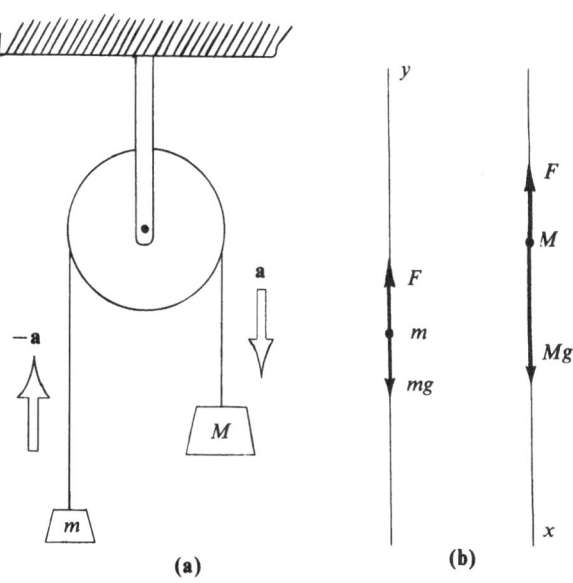

Figure 3.21. Atwood's machine.

Mechanical Advantage

A machine that can balance a movable weight W in equilibrium against another movable weight w is said to have a **mechanical advantage** W/w (equal to 4 in the preceding example). With an increased number of pulleys the mechanical advantage can of course be made much larger. Unlike energy (next chapter), force will always be cheap.

Our next example does not involve equilibrium.

Example 3.7. Atwood's Machine. Two hanging masses, m and M, are connected by an inextensible string which loops over a frictionless pulley, as shown in Fig. 3.21. The string and pulley have negligible mass. Given m and M, what is their acceleration? Like the inclined plane, this device can be used to observe gravity in "slow motion."

For definiteness, let $M > m$. Suppose M has a downward acceleration a; then m has an upward acceleration of the same magnitude. Figure 3.21a shows the physical setup; Fig. 3.21b shows all the forces acting on each particle. The tension in the string is F, and is equal on both sides; note that the two vectors **F** in the figure do *not* constitute an action–reaction pair.

It is convenient (but not necessary) to take the x and y axes in opposite directions, as shown in the figure. Then

$$a_x = a_y = a$$

Applying Newton's second law successively to m and to M, we have

$$F - mg = ma$$
$$Mg - F = Ma$$

Eliminating F, for example by adding the corresponding sides of these equations, results in

$$(M - m)g = (M + m)a$$

or

$$a = \frac{M - m}{M + m}g \qquad (3.E.1)$$

Here gravity is "diluted" by the dimensionless factor $(M - m)/(M + m)$. (Special cases: If $M = m$, then $a = 0$ as we should have by symmetry; note that a changes sign under the exchange $m \leftrightarrow M$. If $m = 0$, then $a = g$, as we should have, since this is just free fall.)

ii. The Frictionless Surface

If an object is in contact with a flat rigid surface (see Fig. 3.22), two kinds of force can be exerted by that surface on the object: a force $\mathbf{N}$, perpendicular (or **normal**) to the surface, and another force $\mathbf{F}_{\parallel}$, parallel to it. Any force vector $\mathbf{F}$ exerted by the surface can be written as the sum of two such perpendicular vectors, and therefore there is no need to list any other kind of force between surface and object. If one has $\mathbf{F}_{\parallel} = 0$ for any motion of the object, then the surface is called **frictionless**.

> A frictionless surface is one that can exert only a normal force. (3.E.2)

The magnitude N of that force is *self-adjusting* in such a way that any motion of the object into the surface is prevented. Two examples follow.

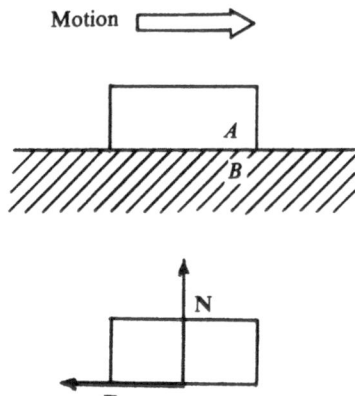

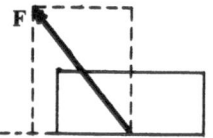

Figure 3.22. Forces exerted on a block A (moving or stationary) by a flat surface B: there is in general a **normal** force $\mathbf{N}$ and a **frictional** force $\mathbf{F}_{\parallel}$ (for the subscript $_{\parallel}$, read "parallel"). Any surface force $\mathbf{F}$ can be said to result from these two types of force. If $\mathbf{F}_{\parallel}$ always vanishes, B (or, more properly, the contact between A and B) is said to be frictionless.

Example 3.8. A Frictionless Inclined Plane. A block of given mass m slides down a frictionless plane which makes an angle θ with the horizontal; see Fig. 3.23. What is the magnitude of the block's acceleration, a?

We have, for the force on the block,

$$\mathbf{F}_{tot} = m\mathbf{a} \qquad (3.E.3)$$

We know that $\mathbf{a}$ must be along the inclined plane; this we choose to be our x direction, as shown. Then the x component of (3.E.3) is

$$(F_{tot})_x = ma_x$$

or, from the force diagram

$$-mg \sin \theta = ma_x$$

so that

$$a_x = -g \sin \theta$$

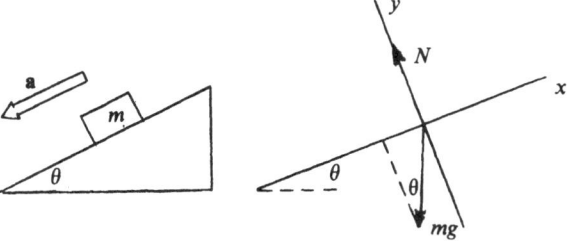

Figure 3.23. The block of Example 3.8, sliding on a frictionless plane. Note the choice of x and y axes in the force diagram for the block.

Thus, we obtain

$$a = g \sin \theta \qquad (3.E.4)$$

(Gravity becomes "diluted" by the dimensionless factor $\sin \theta$; the mass m is not relevant.) Note that $a(\theta = 0) = 0$ and $a(\theta = 90°) = g$; why were these values to be expected?

Example 3.9. Two Frictionless Surfaces. A child's wagon, seen from the side in Fig. 3.24a, consists of a light board and two pairs of heavy wheels. Each pair has weight w. The wagon must be stored at an angle θ to the horizontal because its box is too small. In terms of θ and w, what force N does the left pair of wheels exert on the wall of the box? (Ignore friction.)

Let us choose that left pair of wheels as the particle to be considered. The force diagram, with the x and y axes chosen horizontal and vertical, is shown in Fig. 3.24b. Since the wall of the box amounts to a frictionless surface, the force **N** is normal to it. Let F be the unknown magnitude of the force transmitted along the board. One equilibrium condition, $(F_{\text{tot}})_x = 0$, gives

$$N - F \cos \theta = 0 \qquad (3.E.5)$$

The other condition, $(F_{\text{tot}})_y = 0$, gives

$$F \sin \theta - w = 0 \qquad (3.E.6)$$

Eliminating F between these two equations, we obtain

$$N = w \cot \theta \qquad . \qquad (3.E.7)$$

We note that, when $\theta \to 0$, $\cot \theta$ becomes infinite. If the box is *nearly* large enough, it or the wagon will surely bend or get broken. If the wagon *as a whole* is considered as the particle, we see that the right pair of wheels exerts on the right wall a force of the same magnitude N as in (3.E.7). This follows from $(F_{\text{tot}})_x = 0$, applied to the whole wagon.

iii. Dry Friction (Kinetic)

Material surfaces are never perfectly smooth. When a surface A slides on a surface B (see Fig. 3.25), many microscopic contact forces due to the peaks and valleys combine into an observed resultant $\mathbf{F}_K$, parallel to the overall surface, and called the force of **kinetic** (sliding) friction. In the case of Fig. 3.25, $\mathbf{F}_K$ would be the force exerted by B on A.

Increasing the smoothness sometimes increases $\mathbf{F}_K$ rather than decreasing it, a fact attributed to the closer contact which is then achieved between the surfaces. The presence of tiny "cold-welded" spots, which are made and broken during the motion, is believed to contribute to $\mathbf{F}_K$.

The properties of $\mathbf{F}_K$ are difficult to predict from basic theory, but, if the force and velocity conditions are not too extreme, $\mathbf{F}_K$ turns out to obey the following simple but approximate empirical rules,

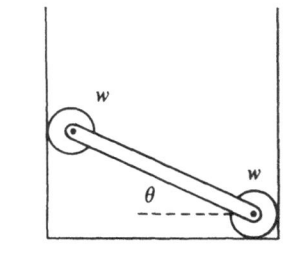

(a)

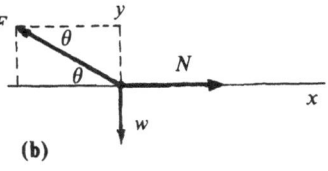

(b)

Figure 3.24. (a) A child's wagon in a box that is too small. The box is rectangular and rests on horizontal ground. (b) The left pair of wheels, as a particle.

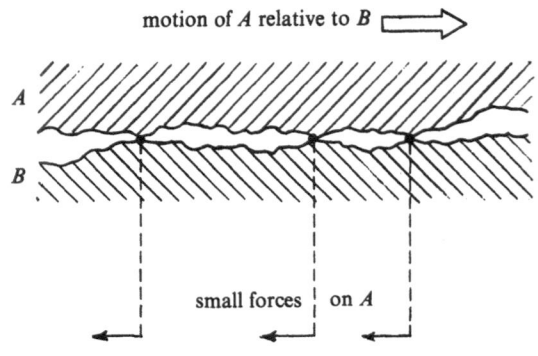

Figure 3.25. The microscopic origin of frictional forces. At some points of contact, especially when at rest, temporary welding may occur.

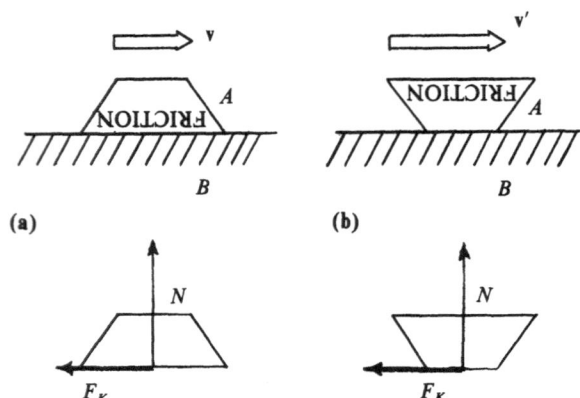

Figure 3.26. Area and speed independence of kinetic friction. In this figure, the top and bottom of a piece of material are similar except for their area; F_K is the same in (a) and (b).

established by Charles Augustin de Coulomb* (1736–1806).

For a surface A, sliding on a surface B at a nonzero velocity $\mathbf{v}$:

1. The force $\mathbf{F}_K$ exerted by B on A points oppositely to $\mathbf{v}$.
2. The magnitude F_K is independent of v.
3. If each surface involved is made of a given material, prepared to a given degree of smoothness, and if the normal force N has a fixed value, then F_K is independent of the area of contact; see Fig. 3.26.
4. The magnitudes F_K and N are proportional to each other:

$$F_K = \mu_K N \qquad (3.E.8)$$

The constant μ_K is called the **coefficient of kinetic friction** for the pair of surfaces involved. We see that its units are newton/newton $= 1$, so that μ_K is a "pure," or dimensionless number; it depends only on the materials of which the surfaces are made, on their smoothness, and on their cleanliness. It can range from as little as 0.05 if slippery materials such as Teflon are used, to as much as 1 or more, as for sandpaper on sandpaper.

(Caution: Most lubricants are liquids, and hence *the frictional properties of lubricated surfaces do not obey the above rules of solid friction.*)

---

* Much better known for his research in electricity and magnetism; see Chaps. 15 and 18.

Example 3.10. The Inclined Plane with Kinetic Friction. What does the acceleration of Example 3.8 become if the block and plane have a given coefficient of kinetic friction μ_K?

In this case the forces on the block are given by Fig. 3.27, with the same choice of axes as before. Let F_K be the force of friction, so that $F_K = \mu_K N$. We have again

$$\mathbf{F}_{tot} = m\mathbf{a} \qquad (3.E.9)$$

with $\mathbf{a}$ along the inclined plane. In the x direction, (3.E.9) gives

$$(F_{tot})_x = ma_x$$

or, from the figure,

$$-mg \sin \theta + \mu_K N = ma_x \qquad (3.E.10)$$

In the y direction,

$$(F_{tot})_y = ma_y$$

or

$$N - mg \cos \theta = 0 \qquad (3.E.11)$$

From Eqs. (3.E.10) and (3.E.11), we can eliminate N and find a_x:

$$a_x = -g \sin \theta + \mu_K g \cos \theta$$

or

$$a_x = g(-\sin\theta + \mu_K\cos\theta) \quad (3.E.12)$$

Again the mass does not enter the result. If μ_K is large enough, the right-hand side of (3.E.12) becomes positive. Then the acceleration a_x is *upwards* along the plane. This means that a block sliding down will decelerate to a stop. At equilibrium ($a_x = 0$), we have $\mu_K = \tan\theta$, so that *the coefficient of kinetic friction equals the slope at which the block will slide down with constant speed.*

iv. Dry Friction (Static)

Consider a fixed horizontal surface B and another surface A, able to slide on B, but temporarily at rest. Let us exert on A a small horizontal force $\mathbf{F}$; see Fig. 3.28. Surface A will not start sliding until F exceeds a critical value F_{cr}. This is the minimum force required to break the "sticking" of A and B. Since equilibrium prevails for all forces $F < F_{cr}$, we conclude that, in this range, B exerts on A a force $\mathbf{F}_x$ (the force of *static friction*), parallel to the surface, and such that

$$\mathbf{F}_S = -\mathbf{F}$$

Hence $\mathbf{F}_S$ is a *self-adjusting* force which prevents motion along the surface and which is subject to the limitation

$$F_S < F_{cr}$$

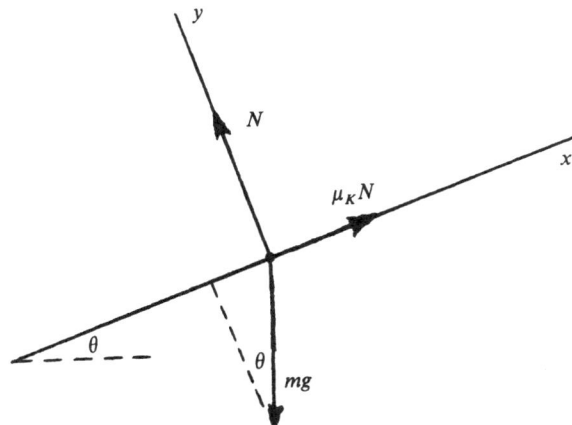

Figure 3.27. Force diagram for the same block as in Fig. 3.23. In this case, however, the contact has a coefficient of kinetic friction μ_K.

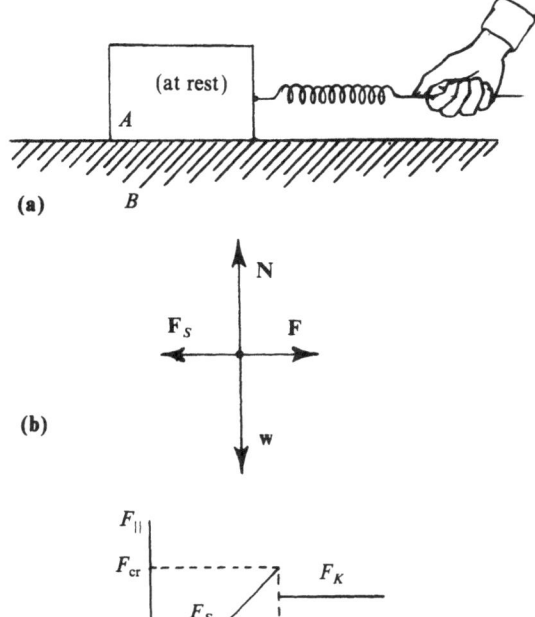

Figure 3.28. (a) A small horizontal pull $\mathbf{F}$ is applied to A. (b) All the forces on A are shown: $\mathbf{F}$ is exactly canceled by the force of static friction $\mathbf{F}_S$. (c) The frictional force $F_{\parallel}$, exerted by B on A, is plotted as F is gradually increased. When F exceeds F_{cr}, the block A is set in motion and F_S is suddenly reduced, becoming F_K.

As in the case of kinetic friction, the critical force F_{cr} is independent of the area of contact and proportional to the normal foce N:

$$\boxed{F_{cr} = \mu_S N} \quad (3.E.13)$$

where μ_S is the **coefficient of static friction** for the substances of A and B.

The static coefficient μ_S is always larger, typically by 10%–30%, than the kinetic one μ_K for the same pair of surfaces. Hence the force of friction abruptly decreases as an object is set in sliding motion.

A violin string, transversely driven by the bow, provides an example of alternation between sticking and slipping, i.e., between the operation of μ_S and μ_K, hundreds of times per second. These coefficients are empirically adjusted by the application of rosin to the horsehair of the bow.

Does static friction always prevent motion? Not in the case of **rolling without slipping**, shown in

Fig. 3.29. The force of static friction, F_S, does prevent relative motion between a wheel and its supporting surface *at the point of contact*, but does not prevent the wheel from turning. (Here we shall not examine in detail the more complicated phenomenon of "rolling friction," not to be confused with F_S above, and which arises from a nonzero area of contact associated with a slight deformation of the surfaces.)

F. Effective Gravity

Feelings of weightlessness or heaviness are a familiar feature of roller coaster rides. The physical phenomenon is a real one, and in the following we ask how the acceleration of free fall and the weight of an object are modified when measured by instruments that have an acceleration of their own.

i. An Accelerating Environment

Suppose the acceleration of a freely falling ball P is measured by an observer A in an airplane that is itself accelerating. Then the resulting measurement will not be the true acceleration of P; see Fig. 3.30. Rather, what will be observed is the acceleration of P relative to A.

Let $\mathbf{g}$ be the true free-ball acceleration of P

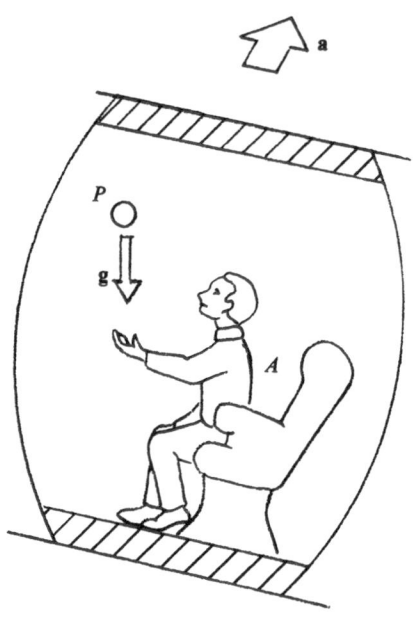

Figure 3.30. A freely falling particle P is observed from an accelerating environment. Observer A's laboratory has an acceleration $\mathbf{a}$.

(measured relative to ground); let $\mathbf{a}$ be the acceleration of observer A. If A measures P's acceleration, he will find an **effective gravity** $\mathbf{g}_{\text{eff}}$ (the apparent gravity felt in the airplane).

How does $\mathbf{g}_{\text{eff}}$ depend on $\mathbf{g}$ and on $\mathbf{a}$? All that needs to be realized is that relative accelerations combine like relative velocities. In Sec. E iii of Chap. 2 we learned that

$$\mathbf{v}_{PA} = \mathbf{v}_{PG} - \mathbf{v}_{AG}$$

where $\mathbf{v}_{PA}$ is the velocity of P relative to A, while $\mathbf{v}_{PG}$ and $\mathbf{v}_{AG}$ are P's and A's "true" velocities. The time derivative of this relation yields the corresponding result for the accelerations:

$$\boxed{\mathbf{g}_{\text{eff}} = \mathbf{g} - \mathbf{a}} \qquad (3.\text{F}.1)$$

Thus, speaking in terms of acceleration vectors,

> The effective gravity equals the true gravity less the observer's acceleration. $\qquad (3.\text{F}.2)$

The mass and other properties of P do not enter into these considerations. Therefore the principle of equivalence still holds in an accelerating

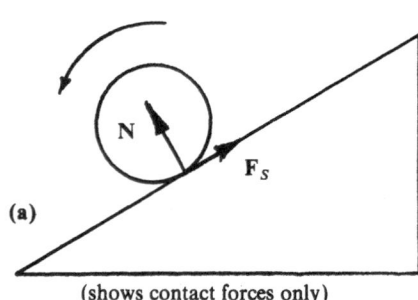

(a)

(shows contact forces only)

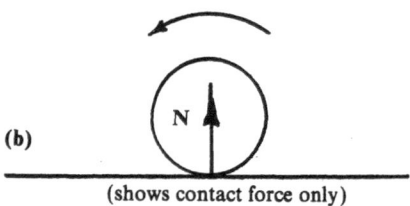

(b)

(shows contact force only)

Figure 3.29. Rolling without slipping. There can be static friction, as in (a), or no friction at all, as in (b). The weight vectors of the rolling objects are not shown.

laboratory. [For a statement of that principle, refer to (2.C.3) in the preceding chapter.]

Static weights will appear modified as well. If, as in Fig. 3.31, P is not freely falling, but is attached to A's airplane-laboratory (perhaps through the intermediary of a spring scale), that laboratory will be exerting on P a force, $-\mathbf{w}_{\text{eff}}$, to compensate for P's **effective weight**, $+\mathbf{w}_{\text{eff}}$. In addition, ordinary gravity still exerts a force $\mathbf{w}$, the **true weight** of P. Since P is assumed to have the same acceleration $\mathbf{a}$ as observer A, Newton's second law

$$\mathbf{F}_{\text{tot}} = m\mathbf{a}$$

applied to P, reads

$$-\mathbf{w}_{\text{eff}} + \mathbf{w} = m\mathbf{a}$$

or

$$\boxed{\mathbf{w}_{\text{eff}} = \mathbf{w} - m\mathbf{a}} \qquad (3.\text{F}.3)$$

[The reader should verify that the same result is formally obtained if one multiplies Eq. (3.F.1) by m.] In terms of vectors:

> The effective weight is the true weight diminished by the force needed to follow the observer. $\qquad (3.\text{F}.4)$

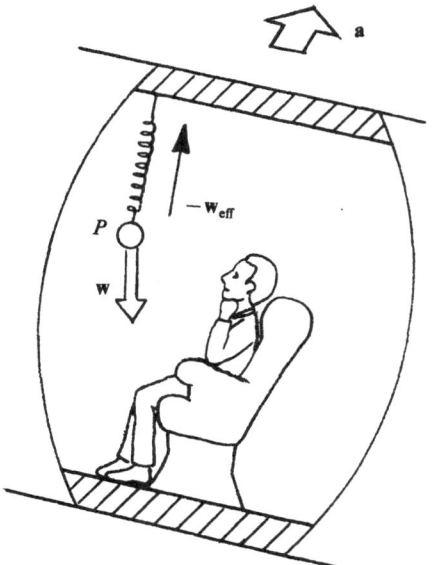

Figure 3.31. By means of a spring scale, a particle is weighed in an accelerating environment.

We conclude that gravity in both its aspects (acceleration and weight) is effectively modified by an acceleration of the environment.

ii. Examples

Example 3.11. The Freely Falling Elevator. (This is a rather trivial case, but of considerable interest.) What is the effective gravity $\mathbf{g}_{\text{eff}}$ in a freely falling elevator?

Here we have $\mathbf{a} = \mathbf{g}$, and therefore Eq. (3.F.1) gives

$$\mathbf{g}_{\text{eff}} = \mathbf{g} - \mathbf{g} = \underline{\underline{0}}$$

The freely falling elevator behaves like a gravity-free environment; so does a freely falling (or orbiting) vehicle anywhere in space. (In Chap. 6 we shall see how, by spinning the vehicle, we can reintroduce an apparent gravity.)

Example 3.12. Being Launched from an Aircraft Carrier. The specifications of a certain navy fighter plane call for it to be catapulted horizontally (with engines turned on) so as to reach a speed of 58 meter/sec in 2.5 sec. To how many g of effective gravity is the pilot subjected?

The horizontal acceleration (which we assume to be constant during launch) is

$$a = \frac{\varDelta v}{\varDelta t} = \frac{v_{\text{final}} - 0}{\varDelta t} = \frac{58 - 0}{2.5} \frac{\text{meter}}{\text{sec}^2}$$

$$= 23 \frac{\text{meter}}{\text{sec}^2}$$

The vector diagram for $\mathbf{g}_{\text{eff}} = \mathbf{g} - \mathbf{a}$ is shown in Fig. 3.32. We see that, in magnitude,

$$g_{\text{eff}} = (a^2 + g^2)^{1/2} = [(23)^2 + (10)^2]^{1/2} \frac{\text{meters}}{\text{sec}^2}$$

$$= 25 \frac{\text{meters}}{\text{sec}^2} = \underline{\underline{2.5g}}$$

(The figure also shows the "effective downward" direction, as experienced by the pilot.)

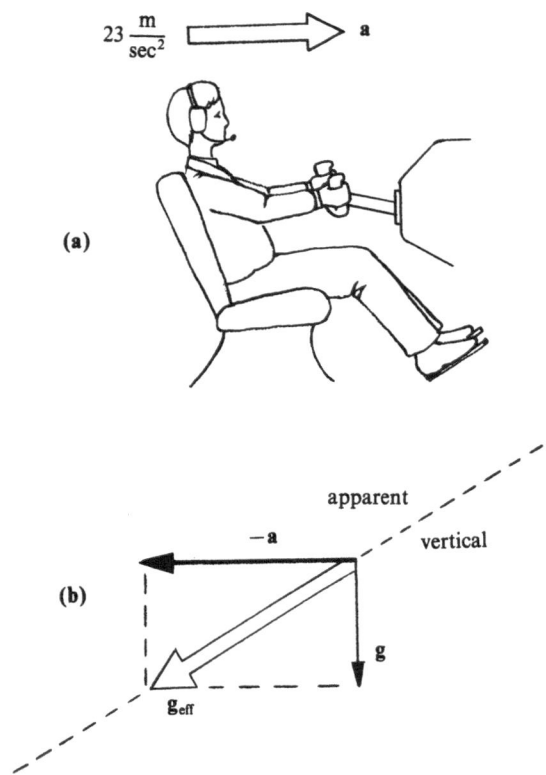

Figure 3.32. Effective gravity, g_{eff}, in an airplane during a horizontal launch.

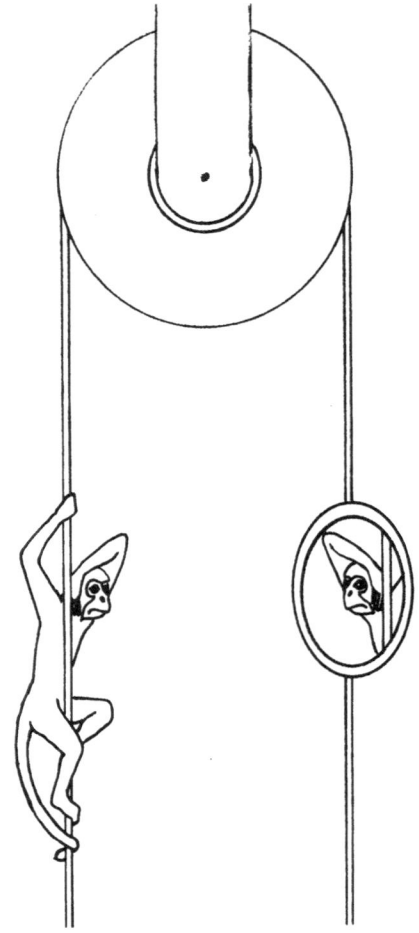

Figure 3.33. The monkey and the mirror.

Note

Initial-Value Problems and Determinism

If the force $\mathbf{F}_{tot}$ on a certain particle is given as a function of time, then *the particle's initial position and velocity completely determine its subsequent motion.* This is sometimes illustrated by the fable of the monkey and the mirror, Fig. 3.33. Here we have an Atwood machine featuring two equal masses $m = M$, one a monkey, and the other a mirror hanging next to the monkey. The system is initially at rest. The monkey becomes tired of seeing his image and climbs or descends the rope, but the mirror remains forever next to him. The explanation is as follows. If both coordinates are measured upwards, the mirror and the monkey start their motion with the same initial conditions; the total force acting on each is the same (same weight and same tension). Hence their motion must be identical since it is uniquely determined.

Initial-value problems have a deeper significance than might appear from this illustration. On a cosmic scale, Newton viewed the gravitational force on an astronomical body as fully determined by its position relative to all the other ones. In such a model, where each star is just a particle, the future behavior of the Universe should be completely determined by the position and velocities of all its parts at any one time. This conclusion has led to the nineteenth-century philosophy of **determinism**, which is at the basis of classical physics. Since the advent of quantum mechanics (see Chap. 26), it has been recognized that this form of determinism is of limited validity. Although well verified in celestial mechanics, it breaks down in many other areas such as the atomic and subatomic worlds.

Condensed Checklist

Newton's laws:

I. $\mathbf{a} = \mathbf{0}$ or $\mathbf{v} = \text{const}$
    (isolated particle, inertial frame). (3.A.1)

II. $\boxed{\mathbf{F}_{tot} = m\mathbf{a}}$ (3.B.1)

III. $\boxed{\mathbf{F}_{A \text{ on } B} = -\mathbf{F}_{B \text{ on } A}}$ (3.D.1)

Weight:

$$\boxed{w = mg} \qquad (3.B.4)$$

Resultant:

$$F_{tot} = F_1 + F_2 + \cdots \qquad (3.C.1)$$

Equilibrium:

$$F_{tot} = 0 \qquad \text{Sec. C, introduction}$$

Friction:

$$F_K = \mu_K N, \qquad F_{cr} = \mu_S N$$
$$(\mu_K < \mu_S) \qquad (3.E.8),\ (3.E.13)$$

Effective gravity:

$$g_{eff} = g - a \qquad (3.F.1)$$

Effective weight:

$$w_{eff} = w - ma \qquad (3.F.3)$$

True or False

1. A particle can be in equilibrium and yet moving.

2. As one racing car overtakes another (identical) racing car, the faster one must be acted on by a larger total force.

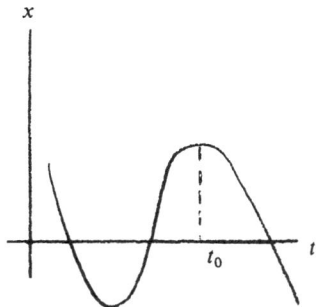

3. The adjoining graph of position against time implies a zero total force at time t_0.

4. At a pulling contest between two tractors, the steadily moving winner exerts, on the connecting chain, a larger force than the loser.

5. While just standing on his feet, a person exerts a zero net force on the whole Earth.

6. If a mass of 1 kg is subjected to a total force of magnitude 1 newton for 1 sec, its final velocity can be zero.

7. A total force of 1 newton, acting downwards on a 1-kg object, will always give it an acceleration equal to g.

8. A total force of 1 newton, acting in the $+x$ direction for 1 sec, will always displace a 1-kg object by 1 meter.

9. A total force of 1 newton, acting in the $+x$ direction for 1 sec, will always displace a 1-kg object by 1/2 meter if that object starts from rest.

10. A total force equal to the weight (on Earth) of a 10-kg mass is sufficient to accelerate a 1-kg mass from rest to a speed of about 100 meters/sec in one second.

11. Three forces, of magnitudes 5, 6, and 12 newtons, and acting at the same point, have a zero resultant if their mutual angles are suitably chosen.

12. The effective weight of an astronaut on take-off is proportional to the acceleration of his rocket.

Problems

Sections A: The Law of Inertia;
B: Force and Acceleration

3.1. The aggregate tension in (=force exerted by) the strings of a grand piano may amount to as much as 3×10^5 N. (a) What volume of water, in cubic meters, has the equivalent weight? (b) Express that force in pounds.

***3.2.** (a) What is the mass, in grams, of an apple whose weight on Earth is 1.00 newton? (b) What is the apple's weight in pounds? (c) What would be its mass, in grams, on the Moon? (d) How much, in newtons and in pounds, would it weigh on the Moon? ($g_{Moon} = 1.6$ meters/sec².)

3.3. A certain spring balance measures an applied *force* in units of 1 ounce (16 oz = 1 lb). Another spring balance measures a *suspended mass* in units of 1 gram. If used on the Moon, where $g_{Moon} = 1.6$ m/s², these scales might have to be relabeled. If we keep the units ounces and grams, (a) What should be the new value replacing the present 72-oz mark? (b) What should replace the present 2000-gram mark?

3.4. What is the weight on Earth, (a) in newtons, (b) in pounds, of 10.0 kg of potatoes?

***3.5.** What is the mass, in kilograms, of 20.0×10^3 lb of gravel, as weighed on Earth?

3.6. On the Moon, an astronaut picks up a 1.5-lb stone. (a) What is the mass, in kilograms, of his sample? (b) How much will it weigh, in pounds, when brought to Earth? ($g_{Moon} = 1.6$ meters/sec².)

3.7. A 15-kg object is observed to have an acceleration of 5 meters/sec². What is the total force acting on it?

3.8. A total, steady horizontal force of 3000 newtons is applied to a 1500-kg car, initially at rest. What is the car's speed after 5 sec?

3.9. On the runway, a certain 120 000-kg jetliner achieves a speed of 32 meters/sec in 8 sec, starting from rest. (a) What is its acceleration (assumed constant) in meters/sec², and in units of g? (b) What is the total force F of propulsion in newtons? (c) What fraction of the plane's weight does F represent? (d) How does F compare with the weight of a 75-kg passenger?

3.10. On horizontal ground, a truck is capable of reaching a 1-meter/sec² acceleration when empty (mass = 2000 kg). What is its maximum acceleration when it carries a 1 000-kg load?

*3.11. What is the shortest time in which a boy can raise a 1.5-kg fish a vertical distance of 2.5 meters from the water if his line has a breaking strength of 35 newtons? (Assume the fish leaves the water at zero speed.)

Sections C: Combining Forces;
D: Action and Reaction.
One-Dimensional

*3.12. In a tug-of-war game on level ground, ten winning pupils are pulling five teachers at a constant speed of 0.2 meter/sec. (a) Which team is exerting the greater force on the rope? (b) If, on the average, each pupil exerts a force of 150 newtons on that rope, what is its tension? (c) How much force does the average teacher exert on the rope? (d) How much horizontal force does the average pupil exert on the ground? Draw the horizontal-force diagrams for, (e) a single pupil as a particle (2 forces); (f) the losing team as a particle (2 forces); use the same scale in (e) and (f).

*3.13. An elevator moves upwards with a constant speed of 6 meters/sec. What force does a passenger, whose mass is 75 kg, exert on the floor of the elevator?

3.14. At a certain stage in his jump, a 150-lb parachutist experiences a deceleration of $2g$. Assuming all air resistance to be acting on the parachute rather than on the man himself, what force is exerted on him by his parachute lines?

*3.15. A 60-kg jumper leaves the ground vertically with a speed of 4 meters/sec. If it took him 0.8 sec to push himself off, what total force F (assumed constant) did his feet exert on the ground?

*3.16. An elevator, whose mass is 1600 kg when empty, is supported by a cable that is unsafe beyond a tension of 25 000 newtons. (a) What is the maximum safe upward acceleration without passengers? (b) What is the maximum safe mass of the passenger load at an upward acceleration of 1.5 meters/sec²?

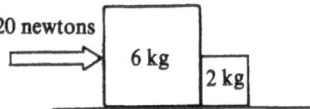

3.17. Two blocks, of masses 6 kg and 2 kg, are sliding on a horizontal frictionless plane; see figure. A horizontal force of 20 newtons is being applied to the larger block as shown. What is the force exerted by the larger block on the smaller one?

*3.18. A steel ball of weight w is suspended vertically below a block of wood of weight W by a light rod. After the system is dropped in vacuum, what is the tension (or compressive force) in the rod?

3.19. In the train arrangement of Fig. 3.18, let the three horizontal forces be in the ratio $F:F':f = 3:2:1$. Find the mass ratio M/m.

Sections C: Combining Forces;
D: Action and Reaction.
Two-Dimensional

*3.20. Three horses A, B, C, are tied to a common stake, each by a horizontal rope. Horse A pulls north, B pulls to 60° east of south, C pulls to 60° west of south. If each horse exerts a 2500-newton pull, find the magnitude and direction of the three horses' total pull on the stake.

3.21. A kite of known weight w is flying motionlessly at the end of a light nylon string, whose lower end is held fixed, and which makes an angle θ with the vertical. If the tension f in the string is given, find (a) the horizontal component, and (b) the vertical component of the wind's force on the kite.

3.22. A horizontal silk thread of length 16 cm is fastened at both ends. When a spider of weight 1.2×10^{-4} newton settles on the thread's midpoint, the latter sinks by 1.0 cm. What is then the tension in the thread?

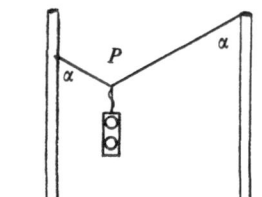

*3.23. A traffic signal of weight 8 lb hangs from point P as shown in the figure. The two slanted cables make equal angles $\alpha = 60°$ with the vertical. Find the tension in each of the three segments of cable. (Neglect the cables' weight.)

3.24. A northward-traveling train, applying the brakes on level ground, comes to rest in 100 sec from an original velocity of 36 km/h. Inside, a 2-kg bottle of beer is resting on a table. Assuming the deceleration to be constant, (a) what is the northward component of the force **F** exerted by the table on the bottle? (b) What is the direction of the total force **F**?

3.25. A pith ball of weight 5×10^{-6} newton is tied to a silk thread of negligible weight. What is the magnitude of a horizontal electric force that pushes the ball so that it hangs at 60° to the vertical?

*3.26. A container, weighing 2×10^4 lb, is hanging from a crane at the end of a 50-ft cable. When pulled horizontally by hand with a force of 100 lb, how far from its free-hanging position can the container be held?

3.27. Tugboats A and B must tow a ship at constant speed toward the west. Tug A pulls northwest with a force of 1410 lb. Tug B pulls 37° south of west. What is the magnitude of the force exerted by B?

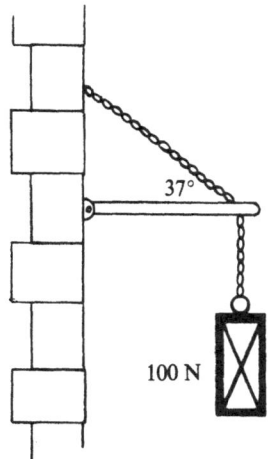

*3.28. A lantern of weight 100 newtons is suspended from the chain and hinged bar arrangement shown in the figure. Both chain and bar are much lighter than the lantern; the bar is horizontal. What is the tension in the slanted portion of the chain, if it makes an angle of 37° with the horizontal?

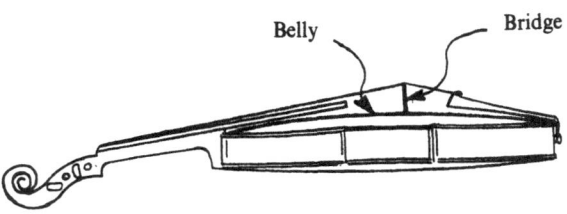

*3.29. From the scale figure, estimate the force exerted by the bridge on the belly of the violin, if the total tension in the strings is 68 lb.

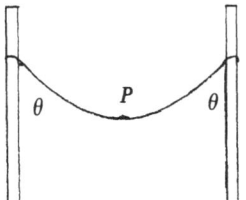

*3.30. A perfectly flexible rope of uniform thickness and material has total weight w. It hangs between two points at the same level, and makes an angle θ with the vertical at those points; see figure. In terms of w and θ, what is the tension at the lowest point P?

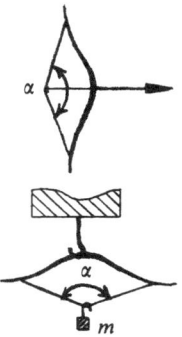

3.31. A bowstring is pulled in the middle until its two halves make an angle α with each other; see figure. That angle is bisected by the horizontal arrow, which has a mass of 0.20 kg. When released, the string gives the arrow an initial acceleration of 1.0×10^3 meters/sec². Later, the bow is suspended as shown. What mass m must be hung from the middle of the string to give it the same angle as before? (Neglect the weight of the bow and string.)

3.32. A train is moving along a straight level track. A passenger, who has suspended his camera from a hook by a long strap, estimates that it is hanging at a steady 10° to the vertical. What can he conclude about the magnitude of the train's acceleration? (See also Problem 3.72 further on.)

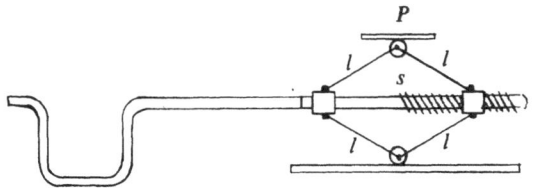

*3.33. A car jack, shown in the figure, consists of four hinged sections of equal length l; the horizontal diagonal is an adjustable length s of threaded bar. Given the lengths l and s, as well as the weight w supported at point P, what is the tension in the threaded bar?

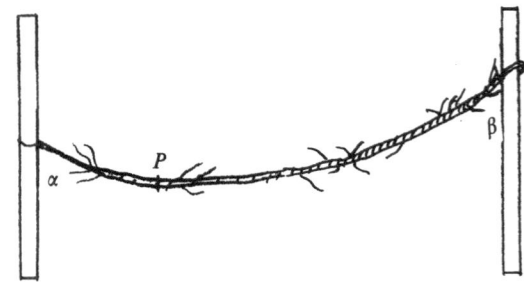

3.34. A perfectly flexible but irregularly made rope has total weight w. At its points of attachment, it makes angles α and β with the vertical, see figure. In terms of α, β, and w, what is the tension at the lowest point P? (Cf. Problem 3.30.)

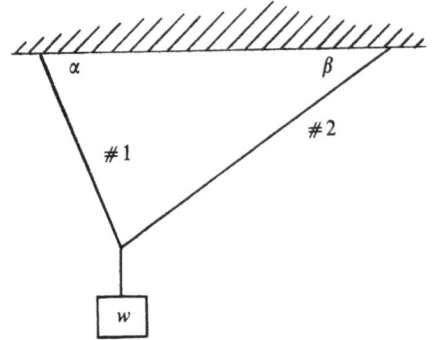

*3.35. A weight w is hanging from a horizontal ceiling through a weightless rope arrangement, as shown in the figure. Given w and the angles α and β, find the tensions F_1, F_2 in the corresponding rope segments. (Cf. Problem 3.23.)

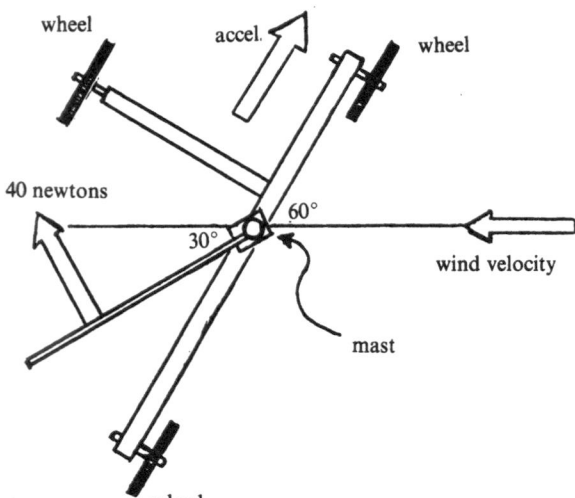

3.36. The figure shows a top view of a sailcart "tacking into the wind." The flat sail is held at 30° to the wind direction. Assume that a force of 40 newtons, due to the

wind, is acting normally to the sail. If the whole vehicle has a mass of 100 kg, and is frictionlessly constrained to move at 60° to the wind, what is its acceleration?

Section E: Pulleys and Surfaces. No Friction

3.37. Consider Fig. 3.24. Find, in magnitude, the force exerted by the right-hand pair of wheels (a) on the floor of the box; (b) on the whole box. (c) What angle does the force in (b) make with the horizontal? (Express your answers in terms of w and θ.)

*3.38. How many g of acceleration does a car have when coasting down an incline at 30° to the horizontal? (Ignore the wheels and use a frictionless-slide model.)

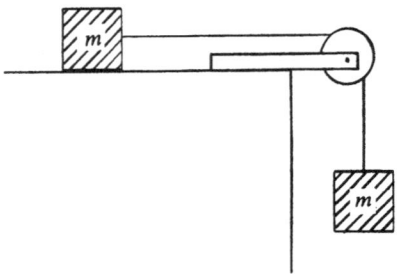

3.39. How large is the acceleration of the equal masses shown in the figure? (The upper mass slides on a horizontal frictionless plane; the string and pulley are ideal.)

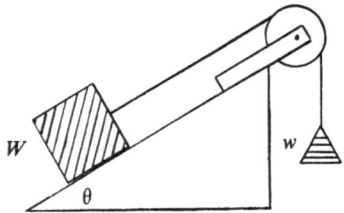

*3.40. A block of weight W slides at constant velocity on a frictionless plane, inclined at an angle θ to the horizontal; see the adjoining figure. If W and θ are given, (a) what is w, the weight hanging from the frictionless pulley in the figure? (b) What is, in magnitude, the normal force N of the plane on the block?

3.41. Assume the pulley of Fig. 3.19 to be in equilibrium. Show that the pulley's support, in this case a hinged bar, always bisects the angle between the two string segments. (Note: This is true even for an accelerating pulley if it is massless. The argument would be similar to the one illustrated in Fig. 3.16.)

3.42. Through an arrangement of frictionless pulley and horizontal rope (see figure), a man tries to propel a cart forward at constant velocity against a horizontal ground friction of 100 newtons. What force must he exert on the rope?

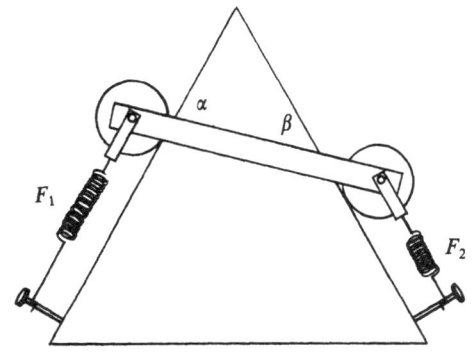

*3.43. In the equilibrium situation shown in the figure, the springs exert forces of magnitudes F_1 and F_2. The two massless, frictionless rollers are connected by a straight bar making angles α and β with the sides of the prism. The bar is not in contact with the prism. Given F_1, α, and β, find F_2.

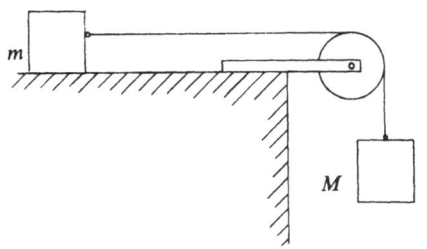

3.44. In the frictionless configuration of the figure, the vertically moving mass M is known. What should be the

horizontally moving mass m in order that M should fall with an acceleration $g/3$? (Assume an ideal string and pulley; cf. Problem 3.39.)

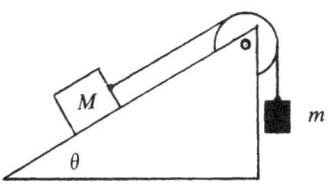

3.45. Two masses, m and M, are arranged as shown in the figure; m is hanging vertically. The plane, inclined at an angle θ to the horizontal, is frictionless. In terms of θ, g, M, and m, what is the acceleration of m? Under what condition is it downward? (Assume an ideal string and pulley.)

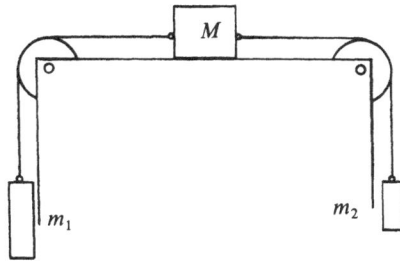

*3.46. Assume the arrangement shown in the figure to be totally frictionless. The motion of M is horizontal; m_1 and m_2 are hanging vertically. If g and the three masses M, m_1, m_2 are given, find the leftward acceleration of M. (Assume ideal pulleys and strings.)

*3.47. In Atwood's machine, Fig. 3.21, the string, the pulley, and its support are weightless. In terms of m, M, and g, what is the tension in the pulley's support, i.e., what weight does it experience? Does your result agree with your expectation when $m = M$?

3.48. In an accelerating dining car, the surface of the coffee in a passenger's cup makes a steady 5° angle with the horizontal. Assuming the train is on a level track, what is its acceleration? (Treat an arbitrary drop of coffee at the surface as being on a frictionless inclined plane; see also Problem 3.73 further on.)

*3.49. A truck of mass 4000 kg is coasting frictionlessly down an incline at 10° to the horizontal. The truck is being partially restrained by a tow cable parallel to the inclined surface. If the truck's forward acceleration is $0.05g$, what is the tension in the cable?

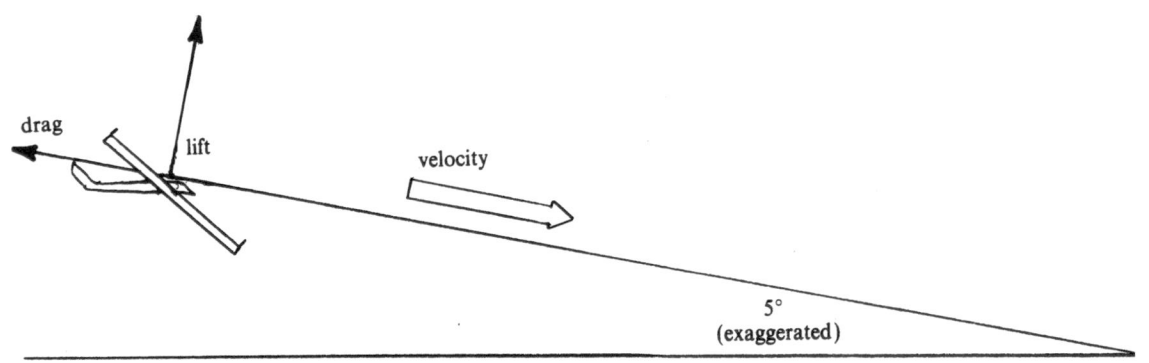

5°
(exaggerated)

3.50. A glider, weighing 1250 lb, comes in for a landing in still air, with a constant velocity of 35 mi/h at 5° below the horizontal (see the figure). Calculate the magnitude of the glider's lift (force in a vertical plane and perpendicular to the motion) and drag (backward force parallel to the motion), both in units of pounds.

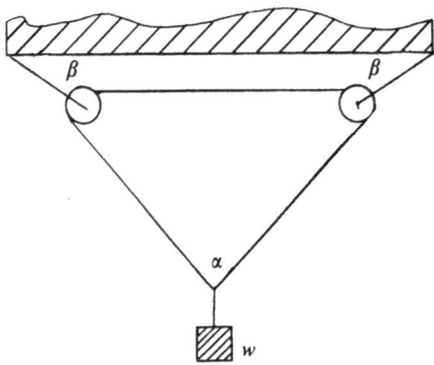

*3.51. A weight w hangs in equilibrium from a symmetric weightless, frictionless system of strings and pulleys as shown in the figure. If the angle α is given, find the angle β. (Cf. Problem 3.41.)

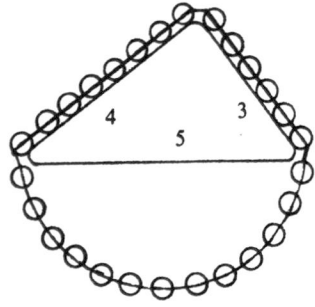

3.52. A fallacious perpetual-motion machine, analyzed by Simon Stevin (1548–1620), and shown here in the case of a 3:4:5 geometry with the long side horizontal, consists of equally spaced rollers on a closed chain. Since

there is always more weight on the left slope than on the right one, while the hanging portion is always approximately symmetric, how can the chain ever come to equilibrium? Demonstrate (as did Stevin) that equilibrium nevertheless will occur.

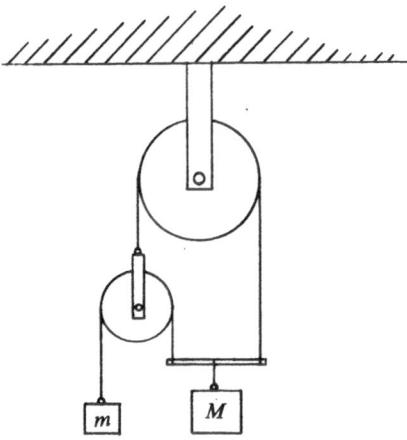

3.53. In the arrangement shown, what is, in terms of m, M, and g, the acceleration of the mass M? (Assume ideal strings and pulleys.)

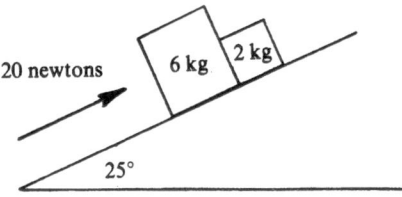

*3.54. Two blocks, of masses 6 kg and 2 kg, are sliding down a frictionless inclined plane; see figure. A force of 20 newtons, parallel to the incline, is applied to the larger block as shown. Calculate the magnitude of the force exerted by the larger block on the smaller one. (Cf. Problem 3.17.)

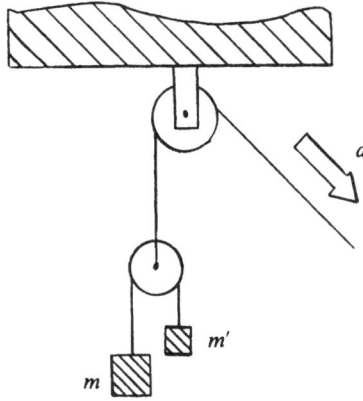

*3.55. In the arrangement shown, m is the larger of the two suspended masses m, m', and is initially at rest. What acceleration, a, must be given to the free end of the string in order to prevent (temporarily) m from falling? (Assume ideal strings and pulleys; express your answer in terms of m, m', and g.)

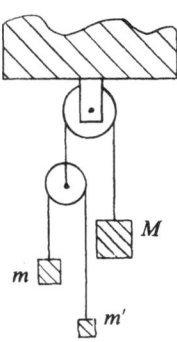

*3.56. In the compound Atwood machine shown, determine the downward acceleration, a, of M. [Assume ideal strings and pulleys; express your answer in terms of g, M, m, m'. You may find it helpful to introduce the so-called **reduced mass** $\mu = mm'/(m+m')$ of the mm' pair; cf. Problems 3.47 and 3.55.]

Section E: Pulleys and Surfaces. With Friction

*3.57. A block rests on a flat horizontal board, the coefficients of static and kinetic friction being 1.00 and 0.50. The board is then slowly tilted. (a) At what angle of inclination does the block start sliding? (b) What is then the magnitude of its acceleration?

*3.58. In a car, efficient braking requires that the brakes be applied to all four wheels. How is this consistent with the statement that the force of dry friction is independent of the surface area over which friction occurs?

3.59. An initially stationary block of mass 20.0 kg is being pulled along a level table top by a force that gradually increases from 10.0 newtons to 20.0 newtons. The coefficient of static and kinetic friction are 0.100 and 0.075. Show that the block will eventually start moving, and calculate its acceleration when it does.

3.60. A porcelain plate is resting on an unwrinkled linen tablecloth, the coefficient of static friction being 0.20. What is the minimum acceleration with which the tablecloth should be pulled (horizontally) in order to start sliding from under the plate?

*3.61. If, in Problem 3.60, the coefficient of kinetic friction is 0.15, what is the time available, after sliding starts, for pulling the tablecloth from under the plate if the latter's final velocity is not to exceed 12 cm/sec?

3.62. A porcelain plate is sliding at a speed of 12 cm/sec on a horizontal polished oak table top, the coefficient of kinetic friction being 0.05. How far will the plate slide?

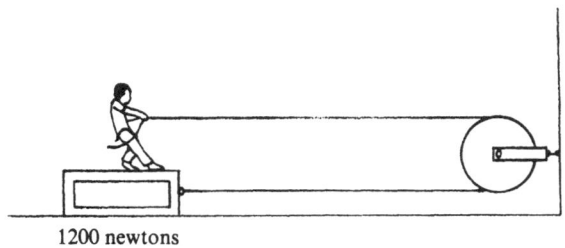

1200 newtons

3.63. A crate that weighs 1200 newtons rests on a horizontal floor, the (unknown) coefficient of static friction being μ_S. A boy climbs on the crate and succeeds in moving it by an arrangement of frictionless pulley and horizontal rope, as shown in the figure. He needs to exert a force F. Without the rope and pulley, he would have to push with a force $1.5F$. What is his weight?

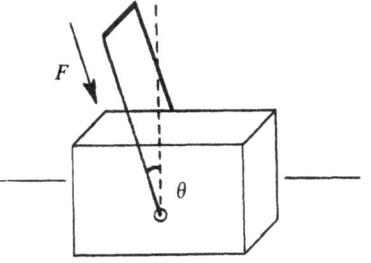

3.64. A block, resting on a horizontal surface with coefficient of static friction μ_S, is being pushed by a force F at a fixed angle θ to the vertical, through a light hinged handle; see figure. As F is increased, the block gets jammed harder and harder, and never moves. What is the largest angle θ which produces such a behavior?

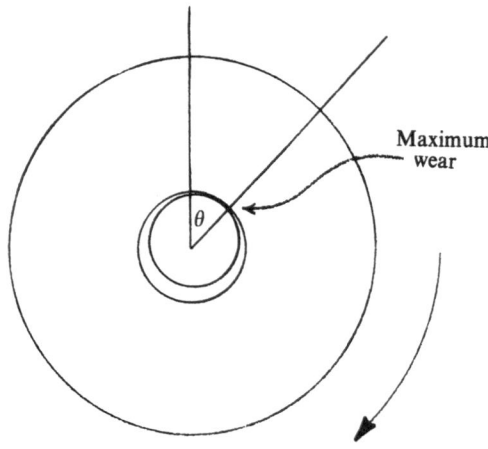

Maximum
wear

θ

*3.65. A wheel of given weight w spins, approximately steadily, on an unlubricated fixed horizontal axle; see figure. The fit is not perfectly tight. The coefficient μ_K of kinetic friction between wheel and axle is given. (a) In terms of w and μ_K, at what angle θ to the vertical will the axle get worn? (b) What are, in magnitude, the frictional and normal forces exerted by the axle on the wheel?

3.66. A certain car, propelled by its rear wheels, is capable of a 0.05g acceleration on a certain dry level road. (Attempting a higher acceleration will cause skidding.) Assuming that 1/3 the weight of the car rests on its rear wheels, what is the coefficient of static friction between tire and road?

3.67. If the car of Problem 3.66 weighs 1500 lb, by what percentage is the attainable acceleration increased if a 200-lb sandbag is placed over the rear axle? (Assume that the sandbag cannot slip, and that the weight on the front wheels is unchanged.)

*3.68. A certain car, with rear-wheel traction, has its weight equally distributed between front and rear axles. If the car has a mass M, by what percentage does the maximum acceleration, on a dry level road, increase if a load of mass m is securely placed midway between the four wheels?

Section F: Effective Gravity

3.69. Mr. X, who is on a diet, always weighs himself on a spring scale in his apartment house's elevator, when the latter starts down at an acceleration of 0.2g. By what percentage would you expect him to exceed his recommended weight? Explain.

3.70. Mrs. X is also on a diet, and always weighs herself in the elevator next to her husband, and in the same circumstances. But her scale works with a counterweight rather than with a spring. By what percentage would you expect her to exceed her recommended weight? Explain.

*3.71. At a certain instant of time, an elevator is moving downwards at a speed of 2.0 meters/sec with a downward acceleration of 0.10g. (a) What force does a passenger, whose mass is 60 kg, exert on the floor of the elevator? (b) By what percentage does that force differ from the passenger's weight? (c) What should be the downward acceleration, in units of g, in order that the passenger should exert his usual weight, but on the *ceiling* of the elevator?

3.72. Do Problem 3.32 in terms of effective gravity.

3.73. Do Problem 3.48 in terms of effective gravity.

Answers to True or False

1. True ($\mathbf{v} = $ const is enough).
2. False (accelerations are relevant, not velocities).
3. False ($v_x = 0$, but $a_x < 0$).
4. False (but the winner must have exerted a somewhat larger force at one time to accelerate the chain).
5. True (net force on person $= 0$, hence, by action–reaction,...).
6. True (with a suitable initial velocity).
7. False (should be 1 meter/sec^2).
8. False (depends on initial conditions).
9. True.
10. True (force $= 100$ newtons).
11. False (vectors of lengths 5, 6 cannot be combined to length 12).
12. False (proportional to acceleration $+g$).

Answers or Hints to Problems

3.2. (a) 102 g; (b) 0.225 lb; (c) 102 g; (d) 0.16 newton, 0.037 lb.

3.5. 9.07×10^3 kg.

3.11. 0.46 s.

3.12. (a) Both teams exert the same force.

3.13. 750 newtons, down.

3.15. First calculate the acceleration a; then use $F - mg = ma$.

3.16. (a) 5.8 meters/sec^2; (b) 610 kg.

3.18. Zero.

3.20. See Fig. 2.20 of Chap. 2.

3.23. Apply the equilibrium conditions to the knot at point P.

3.26. 0.25 ft. (Note: For small angle, $\sin \theta \approx \tan \theta$.)

3.28. Apply the equilibrium condition to the joint between bar and chain.

3.29. 28 lb (a 10% error is acceptable).

3.30. Require equilibrium for one half of the rope.

3.33. $ws/(4l^2 - s^2)^{1/2}$.

3.35. $F_1 = (w \cos \beta)/\sin(\alpha + \beta)$.

3.38. $(1/2) g$.

3.40. (a) $W \sin \theta$; (b) $W \cos \theta$.

3.43. Apply the equilibrium condition to each roller in terms of the tension F in the connecting bar; then eliminate F.

3.46. $[(m_1 - m_2)/(m_1 + m_2 + M)] g$.

3.47. $4\mu g$, where $\mu = (mM)/(m + M)$ is called the *reduced mass* of the mM pair.

3.49. 4.8×10^3 newtons.

3.51. $\beta = 45° - \alpha/4$.

3.54. The angle of the incline is irrelevant.

3.55. If the position of m is fixed, m' accelerates twice as much as the lower pulley. (Why?)

3.56. Use the acceleration of m or m' with respect to the lower pulley. Relative accelerations combine in the same way as relative velocities. (Why?)

3.57. (b) $g/2^{3/2} \approx 3.5$ meters/sec^2.

3.58. Hint: Four brakes *do* give about four times the maximum braking force of one brake.

3.61. 0.08 sec.

3.65. The wheel, considered as a particle, is in equilibrium.

3.68. Zero.

3.71. (a) 540 newtons, down; (b) 10% less; (c) $2g$.

CHAPTER 4

ENERGY

Energy, like force and motion, is an unspecialized concept, applicable to almost all fields of physics, and providing them with a common unit of measurement. Energy plays a starring role in the most abstract theories as well as in the most practical concerns. Everyone, of course, is aware of its economic, social, and military importance.

The present chapter attempts to illuminate only a small corner of that landscape. It examines just two kinds of energy, associated with a particle's motion (**kinetic energy**) and with a particle's altitude (**gravitational potential energy**). In the remainder of the book our view will become broader: we shall eventually consider **elastic energy** (e.g., the energy of springs), **thermal energy** (related to heat), **electromagnetic energy**, etc.

What makes energy worth talking about is the fact that it is conserved. Energy cannot increase in one place without correspondingly decreasing somewhere else. (One way of transferring energy is through **mechanical work**, a concept studied in what follows.)

Energy techniques are of great help in solving mechanics problems, a fact that the present chapter is meant to promote. Later we shall demonstrate the powerful problem-solving role of energy in many other areas of physics.

Ordinary numbers ("scalars") are easier to deal with than vectors, and energy is an ordinary number; *it has no direction.*

A. Conservation of Energy in Free Fall

i. Does Anything Remain Constant?

We have studied projectile motion rather thoroughly, and therefore it might seem unlikely that a fresh look at this phenomenon could yield a new insight. Such is nevertheless the case.

Let us concentrate first on the vertical (z) component of the motion. In that direction the variables of interest are the altitude z, the velocity component v_z, and the acceleration $a_z = -g$. We know that a_z is a constant, but both z and v_z change with time. Is it possible to find a new combination of these variables that remains constant throughout the falling motion? The answer is yes:

> The quantity $gz + \frac{1}{2}v_z^2$ does not change with time during free fall. (4.A.1)

To visualize this, imagine an object being thrown upward. As the altitude z increases, the speed v_z must decrease in just such a way that the decrease in $\frac{1}{2}v_z^2$ exactly compensates the increase in gz.

Statement (4.A.1) is not really new; we recall [Eq. (2.C.14) of Chap. 2] that

$$v_z^2 - v_{0z}^2 = -2g(z - z_0)$$

or, with time t on the left and time zero on the right,

$$gz + \frac{1}{2}v_z^2 = gz_0 + \frac{1}{2}v_{0z}^2 \qquad (4.A.2)$$

Here we have nothing but statement (4.A.1), referred to initial conditions.

This simple result will eventually lead us to one of the greatest generalizations of all physics—that of energy conservation.

As a small step in that direction, we first replace v_z^2 by the total squared *speed* $v^2 = v_x^2 + v_z^2$ ($x =$ horizontal coordinate of the projectile). This replacement is allowed in (4.A.1) by virtue of the following relations.

Vertical motion, statement (4.A.1):

$$gz + \tfrac{1}{2}v_z^2 = \text{const}$$

Horizontal motion:

$$\tfrac{1}{2}v_x^2 = \text{const}$$

Add both equations:

$$gz + \tfrac{1}{2}v^2 = \text{const} \qquad (4.A.3)$$

the desired result.

ii. Nomenclature and Comments

The combination of variables $gz + \tfrac{1}{2}v^2$ is important enough to warrant some new terminology. The property of not changing with time is called a **conservation law**. (Recall the conservation of mass as another example of such a law.) We note that if a conserved quantity is multiplied by any constant (i.e., by another conserved quantity) the result is still conserved. In particular, let us consider multiplication by the mass m of the projectile. The result,

$$\mathscr{E} = mgz + \tfrac{1}{2}mv^2 \qquad (4.A.4)$$

is called the projectile's **mechanical energy**.

Why is energy defined with a mass factor? The answer is that one wishes to generalize the idea of a conserved energy to larger and more complicated systems than just a falling particle. Then the energy, it turns out, needs to be **additive**; for example, two equal energy-bearing particles, taken together, will have twice the energy of each taken by itself. To be specific, consider two identical projectiles of mass m falling side by side without perturbing each other. Together, they can be considered as a single mass $2m$. As a consequence of the mass factor in (4.A.4), the energy of the combined mass will be twice that of a single mass.

Quite aside from additivity, we shall see that the mass factor is needed if conservation of energy is to be extended to general systems. Examples such as Atwood's machine further on, and the elastic collision, next chapter, exhibit this requirement.

Let us now go back to formula (4.A.4). Each of the two terms receives a name of its own. The quantity $\mathscr{U} = mgz$ is called the projectile's **gravitational potential energy** (an energy of position), while

$\mathscr{K} = \tfrac{1}{2}mv^2$ is called its **kinetic energy** (energy of motion). To summarize the definitions,

> Mechanical energy
> $\quad$ = potential energy
> $\qquad$ + kinetic energy
>
> or
>
> $$\mathscr{E} = \mathscr{U} + \mathscr{K} \qquad (4.A.5)$$
> $$\mathscr{U} = mgz \qquad (4.A.6)$$
> $$\mathscr{K} = \tfrac{1}{2}mv^2 \qquad (4.A.7)$$

The conservation of energy now reads simply $\mathscr{E} = \text{const}$, or, over any finite displacement,

$$\Delta\mathscr{E} = \Delta\mathscr{U} + \Delta\mathscr{K} = 0 \qquad (4.A.8)$$

(It is worth recalling that "Δ" means "final minus initial," as in $\Delta\mathscr{K} = \mathscr{K}_f - \mathscr{K}_i$, $\Delta t = t_f - t_i$, etc., whenever a time can be associated with the change.)

In connection with Eq. (4.A.8), we note once more that $\mathscr{U}$ and $\mathscr{K}$ are not individually conserved, although their sum is. We also note that the altitude z of a particle can only be specified relative to a chosen reference level (origin of the z axis). This is therefore also true for the potential energy.

> The potential energy mgz depends on the reference level with respect to which it is defined.

$$(4.A.9)$$

In any problem involving energy, we may choose the reference level to fit our convenience.

On the other hand, that choice does not enter into the value $-\Delta\mathscr{U}$ of a **potential drop** resulting from an **altitude drop** $h = -\Delta z$; we have

$$-\Delta\mathscr{U}\,(= \Delta\mathscr{K}) = mgh \qquad (4.A.10)$$

independently of the reference level.

A General Definition of Energy?

So far, we have seen two kinds of energy. In this book we shall encounter many more. Therefore, we are faced with a mild paradox: we cannot open our study of energy with its general definition. At the chapter's very end we shall be in a better position to do so.

iii. Units of Energy

There are several equivalent ways of determining the SI unit of energy.

(1) According to Eq. (4.A.6), we have

$$\mathcal{U} = (mg)(z) = (force)(distance)$$

Hence, in terms of units,

$$\mathcal{U} \sim \text{newton meter} \qquad (4.A.11)$$

or

$$\mathcal{U} \sim \frac{\text{kg meter}^2}{\text{sec}^2} \qquad (4.A.12)$$

(2) Let us look at the kinetic energy. Equation (4.A.7) gives

$$\mathcal{K} \sim \frac{\text{kg meter}^2}{\text{sec}^2} \qquad (4.A.13)$$

the same unit as for $\mathcal{U}$, Eq. (4.A.12), and properly so. The SI unit of energy has its own special name, the **joule**.*

$$1 \text{ joule} = 1 \text{ J} = 1 \text{ newton meter}$$

$$= 1 \frac{\text{kg meter}^2}{\text{sec}^2} \qquad (4.A.14)$$

(In the **c.g.s.** system one uses the **erg** as the unit of energy:

$$1 \text{ dyne centimeter} = 1 \frac{\text{gram cm}^2}{\text{sec}^2} = 1 \text{ erg}$$

so that

$$1 \text{ erg} = 10^{-7} \text{ joule}$$

In the so-called **British gravitational system**, the energy unit is the **foot-pound**:

$$1 \text{ ft lb} \approx 1.356 \text{ joules}$$

We shall encounter several other energy units later in this book.)

* After James Prescott Joule; see Chap. 10.

iv. Examples

Example 4.1. Changes in Energies. A 2-kg block falls from rest, off the edge of a 1-meter-high table. Using the floor as reference level, (a) What are the block's potential, kinetic, and total energies just as it begins to fall? (b) What are these quantities just before it hits the floor?

(a) *Initial energies.* Let z = vertical distance above the floor; let $m = 2$ kg. We shall indicate the initial and final situations by the subscripts i and f.

Potential energy (depends on reference level):

$$\mathcal{U}_i = mgz_i = (2)(10)(1) \text{ joules} = \underline{\underline{20 \text{ joules}}}$$

Kinetic energy (independent of reference level):

$$\mathcal{K}_i = \tfrac{1}{2}mv_i^2 = \underline{\underline{0}}$$

since $v_i = 0$.

Total energy (depends on reference level):

$$\mathcal{E}_i = \mathcal{U}_i + \mathcal{K}_i = (20 + 0) \text{ joules} = \underline{\underline{20 \text{ joules}}}$$

(b) *Final energies.* The total energy requires no calculation since it is conserved. We have

$$\mathcal{E}_f = \mathcal{E}_i = \underline{\underline{20 \text{ joules}}}$$

Turning to the potential energy, we find

$$\mathcal{U}_f = mgz_f = \underline{\underline{0}}$$

since $z_f = 0$. Next, we note that the final kinetic energy can be found without our having to calculate the final velocity, as follows. Conservation of energy means that

$$\mathcal{E}_f = \mathcal{E}_i$$

or

$$0 + \mathcal{K}_f = \mathcal{U}_i + 0$$

so that

$$\mathcal{K}_f = \mathcal{U}_i = \underline{\underline{20 \text{ joules}}} \qquad (4.A.15)$$

Example 4.2. An Artillery Problem. A gunshell of unspecified mass is fired at an unspecified angle to the vertical. If its muzzle speed is $v_i = 500$ meters/sec, at what (horizontal) speed v_f is it still traveling when it reaches its maximum altitude, which is $h = 8000$ meters?

This type of problem can be successfully handled with the methods of Chap. 2. However, the energy concept provides a fresh and far more efficient approach. We have

$$\mathcal{U}_f + \mathcal{K}_f = \mathcal{U}_i + \mathcal{K}_i$$

With the ground as reference level ($z_i = 0$, $z_f = h$), this reads

$$mgh + \tfrac{1}{2}mv_f^2 = 0 + \tfrac{1}{2}mv_i^2$$

Factoring out m, the shell's mass, and solving for v_f, we have

$$v_f = (v_i^2 - 2gh)^{1/2}$$

$$= [(500)^2 - (2)(10)(8000)]^{1/2} \frac{\text{meters}}{\text{sec}}$$

$$= \underline{\underline{300 \frac{\text{meters}}{\text{sec}}}}$$

(The information that h is maximal has not been used.)

v. Energy Conservation from Calculus

In order to obtain statement (4.A.1),

$$gz + \tfrac{1}{2}v_z^2 = \text{const}$$

we invoked the somewhat cumbersome algebra of Chap. 2. There is no need to do so; calculus provides a short and direct proof, as follows. Consider the time derivative

$$\frac{d}{dt}(gz + \tfrac{1}{2}v_z^2)$$

$$= g\frac{dz}{dt} + v_z\frac{dv_z}{dt}$$

$$= gv_z + v_z a_z = gv_z + (v_z)(-g) = 0$$

$$\text{(4.A.16)}$$

which is all that is required. Of course, it does help to know what quantity to differentiate in the first place.

B. Work

Physical events are often associated with the conversion of energy from one form into another. For example, in the case of a falling particle, some potential energy $\mathcal{U}$ is converted to kinetic energy $\mathcal{K}$.

The amount of energy converted during any part of the motion depends on the relevant force (the weight, in this case). It also depends on the particle's displacement (its altitude drop). The purpose of this section is to combine **force** and **displacement** into a new quantity, (mechanical) **work**, which tells us how much energy has been converted.

Loosely speaking, *work equals the product of force times displacement*; any force—not just gravity—may be involved in this definition.

i. Work Done by Gravity

Suppose a projectile falls between altitudes z_i and z_f. (For definiteness we may visualize z_i as higher than z_f.) The increase in kinetic energy is

$$\Delta\mathcal{K} = -\Delta\mathcal{U} = \mathcal{U}_i - \mathcal{U}_f$$

$$= mgz_i - mgz_f \qquad \text{(4.B.1)}$$

or

$$\Delta\mathcal{K} = mg(z_i - z_f) \qquad \text{(4.B.2)}$$

The right-hand side of this equation is called the **work**,

$$\boxed{\mathcal{W} = wh} \qquad \text{(4.B.3)}$$

done over an altitude drop

$$h = z_i - z_f \qquad \text{(4.B.4)}$$

by the force of gravity

$$w = mg \qquad \text{(4.B.5)}$$

We conclude from (4.B.2) that

$$\Delta\mathscr{K} = \mathscr{W} \qquad (4.B.6)$$

The kinetic energy of a projectile increases by an amount equal to the work done on it by gravity. Note that, on a rising projectile, gravity does negative work.

Our next task is to fashion (4.B.3) into a definition applicable to forces other than gravitational.

ii. Work Done by Any Constant Force

The work $\mathscr{W} = wh$ involves a force vector **w** and a displacement vector l; see Fig. 4.1; h is related to l by

$$h = l \cos \theta$$

Thus we have

$$\mathscr{W} = wl \cos \theta \qquad (4.B.7)$$

The effect of a force on a particle does not depend on whether the force is gravitational or not. The only relevant aspects of a force are its magnitude, its direction, and the particle to which it is applied. To emphasize that the work $\mathscr{W}$, Eq. (4.B.7), could have an origin other than gravitational, we set

$$\mathbf{w} = \mathbf{F}$$

Thus, Eq. (4.B.7) leads us to the following defini-

tions: the work $\mathscr{W}$ done by any constant force **F** during a net displacement l is given by

$$\boxed{\mathscr{W} = Fl \cos \theta} \qquad (4.B.8)$$

where θ is the angle between **F** and l when these vectors are displayed tail to tail. Note that the only information relevant to (4.B.8) consists of the vectors **F** and l, see Fig. 4.1b.

iii. The Dot Product of Two Vectors

The combination of variables on the right side of (4.B.8) finds many uses in physics, and accordingly a convenient notation has been devised for it:

$$Fl \cos \theta = \mathbf{F} \cdot l \qquad (4.B.9)$$

This expression is called the **dot product** (scalar product, inner product) of the two vectors **F** and l. The general definition of a dot product is as follows.

> The dot product of two arbitrary vectors **F** and l is the ordinary number
>
> $$\mathbf{F} \cdot l = Fl \cos \theta \qquad (4.B.10)$$
>
> where F, l are the vectors' magnitudes and θ is the angle between them when they are displayed tail to tail.

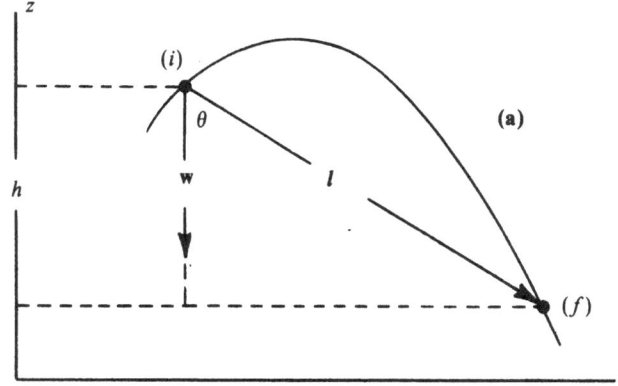

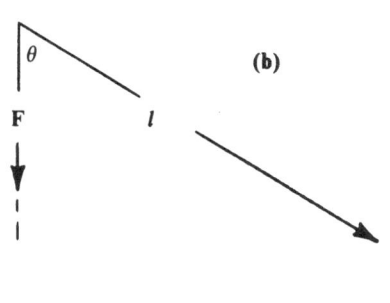

Figure 4.1. (a) Trajectory of a projectile falling through a height h. The kinetic energy gained between (i) and (f) depends only on the total force (the weight **w**) and the net displacement l. (b) The same process as (a), with the irrelevancies omitted. The force **w** is now called **F** to emphasize that the argument does not depend on the force's origin being gravity.

If this notation is used, definition (4.B.8) for the work of a constant force **F** over a net displacement *l* becomes

$$\boxed{\mathscr{W} = \mathbf{F} \cdot \boldsymbol{l}} \qquad (4.B.11)$$

For a projectile of weight vector **w**, as it completes a net displacement *l*, the changes in energy can now be expressed in the same elegant manner. Equation (4.B.6) reads

$$\Delta \mathscr{K} = \mathbf{w} \cdot \boldsymbol{l} \qquad (4.B.12)$$

or, from conservation,

$$-\Delta \mathscr{U} = \mathbf{w} \cdot \boldsymbol{l} \qquad (4.B.13)$$

Geometric Interpretations

The quantity $\mathbf{F} \cdot \boldsymbol{l}$ can be visualized in several ways, see Fig. 4.2. In Fig. 4.2a, we denote by $F_{\parallel}$ ("*F* parallel") the component of **F** in the direction of *l*. We see that $F_{\parallel} = F \cos \theta$, so that we may write

$$\mathbf{F} \cdot \boldsymbol{l} = F_{\parallel} l \qquad (4.B.14)$$

Similarly, in Fig. 4.2b we let $l_{\parallel}$ be the component of *l* along the direction of **F**. Then $l_{\parallel} = l \cos \theta$, and

$$\mathbf{F} \cdot \boldsymbol{l} = F l_{\parallel} \qquad (4.B.15)$$

To summarize, we have

$$\mathbf{F} \cdot \boldsymbol{l} = F l \cos \theta = F_{\parallel} l = F l_{\parallel} \qquad (4.B.16)$$

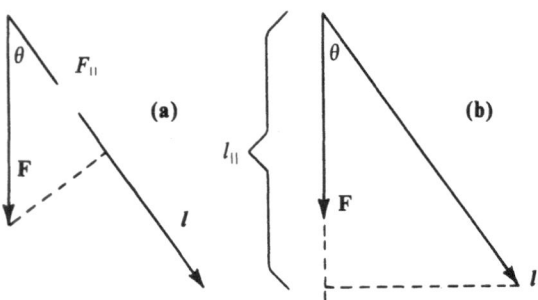

Figure 4.2. (a) Construction of the number $F_{\parallel}$ for use in the expression (4.B.14). In this case $F_{\parallel}$ is positive, but an obtuse θ would make $F_{\parallel}$ negative. (b) Construction of the number $l_{\parallel}$ for use in (4.B.15). This is actually the same prescription as in (a), the roles of **F** and *l* being interchanged.

Those are purely mathematical formulas, in which **F** and *l* are any vectors, not necessarily a force and a displacement.

[In physical terms, Eq. (4.B.14) appears reasonable. It says that *only the force component along the direction of the displacement "is of any use" in doing work.*]

Important Special Cases

If **F** and *l* point in the same direction, then

$$\mathbf{F} \cdot \boldsymbol{l} = F l \cos 0° = F l \qquad (4.B.17)$$

In particular, the dot product of a vector (for example **v**) with itself gives

$$\mathbf{v} \cdot \mathbf{v} = v^2 \qquad (4.B.18)$$

(just the square of its magnitude). If **F** and *l* point in opposite directions, then

$$\mathbf{F} \cdot \boldsymbol{l} = F l \cos 180° = -F l \qquad (4.B.19)$$

If **F** and *l* are mutually orthogonal, then

$$\mathbf{F} \cdot \boldsymbol{l} = F l \cos 90° = 0 \qquad (4.B.20)$$

(The relation $\mathbf{F} \cdot \boldsymbol{l} = 0$ is, in fact, a physicist's favorite way of expressing the orthogonality of two vectors.)

It is useful to keep in mind that a dot product is positive or negative according to whether the angle between the vectors is acute or obtuse; see Fig. 4.3.

Resemblance to an Ordinary Product

In many ways, we can treat the dot product $\mathbf{F} \cdot \boldsymbol{l}$ like the ordinary product of two numbers.

1. The two factors, **F** and *l*, play equal mathematical roles in the calculation and the dot product is commutative, $\mathbf{F} \cdot \boldsymbol{l} = \boldsymbol{l} \cdot \mathbf{F}$.

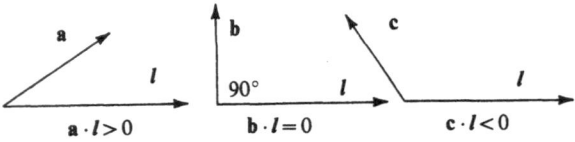

Figure 4.3. The sign of a dot product depends on how the enclosed angle compares with 90°.

2. If n is an ordinary number, then

$$(\mathbf{F}) \cdot (n\mathbf{l}) = n(\mathbf{F} \cdot \mathbf{l}) \qquad (4.B.21)$$

which may be seen from the fact that, for positive n, $n\mathbf{l}$ has the same direction as $\mathbf{l}$ but is n times as large. (How should we restate this explanation for negative n?)

3. Any vector sum $\mathbf{F}_1 + \mathbf{F}_2$ obeys the distributive rule

$$(\mathbf{F}_1 + \mathbf{F}_2) \cdot \mathbf{l} = \mathbf{F}_1 \cdot \mathbf{l} + \mathbf{F}_2 \cdot \mathbf{l} \quad (4.B.22)$$

[The mathematically inclined reader may see this from (4.B.16):

$$(\mathbf{F}_1 + \mathbf{F}_2) \cdot \mathbf{l} = (F_1 + F_2)_{\parallel} \, l$$

Now recall (Sec. D iv of Chap. 2) that adding two vectors means adding their components. Therefore

$$(F_1 + F_2)_{\parallel} = (F_1)_{\parallel} + (F_2)_{\parallel}$$

and we get (4.B.22).]

iv. Examples of Work Done by a Constant Force

Example 4.3. A Horse and Cart. A horse, pulling a cart over 1 km of straight road, exerts a constant forward force of 500 newtons. (a) How much work is done by the horse on the cart? (b) How much work is done by the cart on the horse?

———

(a) *Horse on cart*: Let $\mathbf{l}$ = displacement, $\mathbf{F}$ = force, both parallel in this case. Then the work done is

$$\mathscr{W} = \mathbf{F} \cdot \mathbf{l} = Fl$$
$$= (500 \text{ newtons})(1000 \text{ meters})$$
$$= 5 \times 10^5 \text{ joules}$$

(b) *Cart on horse*: The work is

$$\mathscr{W}' = (-\mathbf{F}) \cdot \mathbf{l} = -Fl = -5 \times 10^5 \text{ joules}$$

(Caution: We know nothing about the system's acceleration. It may be positive, negative, or zero according to frictional forces.)

Example 4.4. The Horse and Cart, Second Version. Let us modify the preceding example so that the horse exerts his 500 newtons at an upward angle $\theta = 37°$ to the road. The cart still covers a straight distance of 1 km. How much work is done by the horse on the cart?

———

Now we have

$$\mathscr{W} = \mathbf{F} \cdot \mathbf{l} = Fl \cos \theta$$
$$= (500 \text{ newtons})(1000 \text{ meters})(\cos 37°)$$
$$= 4 \times 10^5 \text{ joules}$$

v. Work Done by a Variable Force

If the force $\mathbf{F}$ varies with time and position, what happens to our recipe,

$$\mathscr{W} = \mathbf{F} \cdot \mathbf{l} \qquad (4.B.23)$$

for calculating the work? That relation can easily be adapted to the case of a variable force $\mathbf{F}$ if the displacement $\mathbf{l}$ is sufficiently small. Indeed, the following continuity (or smoothness) assumption is made about $\mathbf{F}$:

> F can be considered constant in magnitude and direction, provided it is observed during a very small displacement of the particle on which it acts.

$$(4.B.24)$$

To express the infinitesimal nature of the displacement, we replace the symbol $\mathbf{l}$ by $d\mathbf{r}$; the work done, which also becomes very small, will be denoted by $d\mathscr{W}$. Now Eq. (4.B.23) reads

$$\boxed{d\mathscr{W} = \mathbf{F} \cdot d\mathbf{r}} \qquad (4.B.25)$$

How does one calculate the work done by a variable force over a path that is not infinitesimal? Figure 4.4 indicates that, with enough patience, one might add up a very large number of small contributions,

$$\mathscr{W} = \mathbf{F}_1 \cdot d\mathbf{r}_1 + \mathbf{F}_2 \cdot d\mathbf{r}_2 + \cdots$$
$$= \sum_i \mathbf{F}_i \cdot d\mathbf{r}_i \qquad (4.B.26)$$

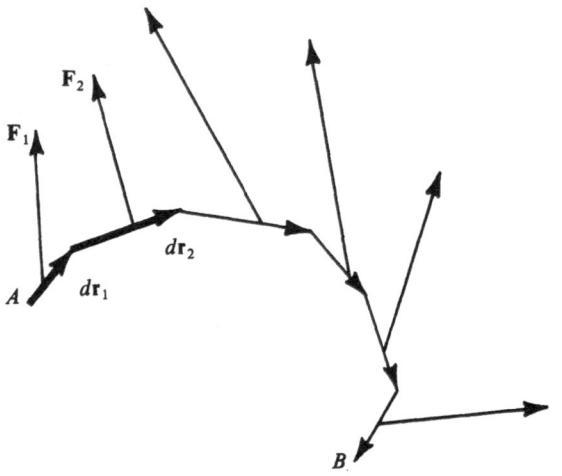

Figure 4.4. Calculating the work of a variable force over a path extending from A to B.

The symbol $\sum_i$ reads "sum, over all appropriate values of i, of ...." When taking the limit of infinitely many small terms, one replaces the symbol $\sum_i$ by $\int_A^B$ (which exhibits the beginning and end of the path). Thus, one writes

$$\mathscr{W} = \int_A^B \mathbf{F} \cdot d\mathbf{r} \qquad (4.B.27)$$

(Read: "$\mathscr{W}$ equals the **integral**, from A to B, of $\mathbf{F} \cdot d\mathbf{r}$.") As each $d\mathbf{r}$ tends to zero, and as the number of terms becomes infinitely large, the result is no longer an approximation, but *becomes exact*. In practice, no one can add an infinite number of terms, of course. Fortunately, integral calculus provides the exact result in many cases. (For further details and some mathematical rigor, see any good calculus text.) *Formula (4.B.27) is the general definition of work.*

Example 4.5. A Tangential Force. By exerting a forward force of constant magnitude 40 000 newtons, a tractor pulls a plow over a contoured path of total length 5 km. How much work is done by the tractor on the plow? (A force that remains parallel to the direction of travel is called **tangential**.)

We first note that $\mathbf{F}$ changes its direction. Therefore Eq. (4.B.23) is not applicable. However, $\mathbf{F}$ and $d\mathbf{r}$ are always parallel, and Eq. (4.B.27) becomes, with $|d\mathbf{r}| = ds$ (element of path length),

$$\mathscr{W} = \int_A^B F \, ds = F \int_A^B ds \qquad (4.B.28)$$

(Since F is constant, it can be factored out of the integral.)

Now $\int_A^B ds$ is just the total length of path, s_{tot}. Equation (4.B.28) gives

$$\mathscr{W} = F s_{\text{tot}} \qquad (4.B.29)$$

$$\mathscr{W} = (40\,000)(5000) \text{ joules} = \underline{2 \times 10^8 \text{ joules}}$$

[Compare (4.B.29) with (4.B.8) in the case where $\mathbf{F}_{\parallel} l$.]

vi. Power

Infinitesimal changes can be used to calculate rates. Let dt be the time needed for the displacement $d\mathbf{r}$ of a certain particle. Then its velocity is $\mathbf{v} = d\mathbf{r}/dt$. If both sides of (4.B.25) are divided by dt we obtain

$$\frac{d\mathscr{W}}{dt} = \mathbf{F} \cdot \mathbf{v} \qquad (4.B.30)$$

This quantity is called the **mechanical power** supplied to the particle by the force $\mathbf{F}$. Thus, *power is the rate of doing work; it is proportional to force and to velocity*.* Again, only the tangential component of a force is "useful" in this respect.

In terms of units, we see that

$$\text{power} = \frac{\text{work}}{\text{time}}$$

or

$$\text{power} \sim \frac{\text{joule}}{\text{sec}}$$

This, the SI unit of power, is called the **watt** (W) after James Watt, of steam-engine fame. Thus,

$$1 \text{ watt} = 1 \frac{\text{joule}}{\text{sec}} = 1 \frac{\text{kg meter}^2}{\text{sec}^3} \qquad (4.B.31)$$

---

* More generally (even outside mechanics) *power is a time rate of energy transfer.*

An alternative unit, still often used in engineering, is the **electric horsepower** (hp), more often just called the **horsepower**, and defined by

$$1 \text{ hp} = 746 \text{ watts} \tag{4.B.32}$$

exactly. (Convention has played a larger role in determining this figure than either electricity or the horse.)

Energy units may be expressed in terms of the watt:

$$1 \text{ watt second} = 1 \text{ joule} \tag{4.B.33}$$

$$1 \text{ kilowatt hour} = 1 \text{ kW h} = 3.6 \times 10^6 \text{ joules}$$

etc.; the kW h is of frequent commercial use. Next we consider two examples.

Example 4.6. The Horse and Cart, Third Version. Suppose the horse of Example 4.3 covers his 1 km in a time $t = 1$ h. What power $d\mathcal{W}/dt$ does he supply to the cart? Assume a constant velocity.

One method: Since work is done at a constant rate, we have

$$\frac{d\mathcal{W}}{dt} = \frac{\mathcal{W}}{t} = \frac{5 \times 10^5 \text{ joules}}{3600 \text{ sec}}$$

$$= \underline{140 \text{ watts}} = (140)\left(\frac{1}{746} \text{ hp}\right) = \underline{0.2 \text{ hp}}$$

Another method: Let $\mathbf{v}$ be the cart's velocity. Then, according to (4.B.30),

$$\frac{d\mathcal{W}}{dt} = \mathbf{F} \cdot \mathbf{v} = Fv$$

$$= (500 \text{ newtons})(1 \text{ km/h}) = \underline{140 \text{ watts}}$$

as before.

Example 4.7. The Niagara Falls. According to a tourist guide, the Niagara Falls are nearly 200 ft high and spill 30 million gallons each minute. (This is the portion not diverted for hydroelectric use.) How much power, in units of megawatt ($= 10^6$ watts) does this represent (i.e., how many megajoules of potential energy are converted to kinetic energy in one second)?

"Power," in this case, represents the total gravitational work (see Sec. B i) done per unit time on a large amount of water. The latter we shall consider as a collection of neighboring but individually falling particles, for which the initial and final positions are the top and bottom of the waterfall, respectively.

Consider a time interval $t = 1 \text{ min} = 60 \text{ sec}$. The potential energy converted is for a mass m of water,

$$\Delta\mathcal{U} = mgh$$

where

$$h = 200 \text{ ft} = (200)(0.3 \text{ meter})$$

and

$$g = 10 \text{ meters/sec}^2$$

For the mass we have

$$m = (\text{No. of liters})(1 \text{ kg})$$

$$= \frac{3 \times 10^7 \text{ gallons}}{1 \text{ liter}}(1 \text{ kg}) = (3 \times 10^7)(4)(1 \text{ kg})$$

Thus, the power is

$$\frac{d\mathcal{W}}{dt} = \frac{\Delta\mathcal{U}}{t} = \frac{mgh}{t}$$

$$= \frac{(3 \times 10^7)(4)(10)(200)(0.3)}{60} \text{ watts}$$

$$= 1 \times 10^9 \text{ watts} = \underline{1000 \text{ MW}}$$

(enough to supply 600 000 average homes).

C. Work and the Energy of a Particle

We now examine how the energy of a particle is affected if work is done on that particle by an arbitrary force.

As an illustration, let someone lift a suitcase; his hand does positive work on the suitcase. Work is an energy transfer, and we ask: where does the energy go? In this example, part becomes kinetic energy (the suitcase was initially at rest), and part becomes gravitational potential energy.

This type of energy budgeting will now be our topic; our plan of study is as follows. First, we consider the effect of the total force, $\mathbf{F}_{tot}$ (here, hand plus gravity). How does $\mathbf{F}_{tot}$ affect the kinetic energy? Next, we distinguish between the different contributions to $\mathbf{F}_{tot}$; we also ask how the combined potential and kinetic energy $\mathcal{U} + \mathcal{K}$ is affected.

Our results will be used, in the next section, for dealing with extended systems like rigid bodies and machines.

i. Kinetic Energy and Work

Suppose a particle is acted on by a constant total force $\mathbf{F}_{tot}$, while completing a net displacement l. To see what happens to the kinetic energy $\mathcal{K}$, we only need to look at the work

$$\mathcal{W}_{tot} = \mathbf{F}_{tot} \cdot l \qquad (4.\text{C}.1)$$

(the **total work** done on the particle). Our answer is that $\mathcal{K}$ increases by precisely that amount:

$$\boxed{\Delta \mathcal{K} = \mathcal{W}_{tot}} \qquad (4.\text{C}.2)$$

> The increase in the kinetic energy of a particle equals the work done by the total force acting on that particle ("total work").

$$(4.\text{C}.3)$$

This statement is usually referred to as the **work-energy theorem**; we must keep in mind that, in (4.C.2), $\mathcal{W}_{tot}$ is the work done, over l, by the combination of *all forces* (gravitational, frictional, etc.), that act on the particle.

[We no longer need to prove Eq. (4.C.2), because we have already done so in the case of a projectile. Is that case general enough? In other words, does Eq. (4.B.6), which states that $\Delta \mathcal{K} = \mathcal{W}$ for a purely gravitational force, imply (4.C.2), which is claimed for any kind of constant force? The answer is yes. As was already pointed out in *Sec. B ii*, what causes the force is not relevant; the mathematics uses only two of its properties, namely, that the force is the total one, and that it is constant. We also observe that (4.C.2) makes no reference to any special direction in space, such as the vertical.]

The Work-Energy Theorem for a Variable Force

What makes Eq. (4.C.2) particularly useful is that it remains valid even when the working force varies with time and position. [To see this, we consider several constant forces acting successively, as was illustrated in Fig. 4.4; each contributes a work $\mathcal{W}_1$, $\mathcal{W}_2$, etc. The overall change in $\mathcal{K}$ is then

$$\Delta \mathcal{K} = \mathcal{W}_1 + \mathcal{W}_2 + \cdots = \mathcal{W}_{tot} \qquad (4.\text{C}.4)$$

as in (4.C.2). Any variable force can be represented by sufficiently many of such steps.]

Example 4.8. Predicting a Final Speed. A particle of mass 4 kg has an initial speed of 3 meters/sec. A variable total force then does a work of 32 joules on that particle. What is its final speed?

Equation (4.C.2) gives

$$\mathcal{K}_f - \mathcal{K}_i = \mathcal{W}_{tot}$$

or

$$\tfrac{1}{2}mv_f^2 - \tfrac{1}{2}mv_i^2 = \mathcal{W}_{tot}$$

so that

$$v_f = \left(v_i^2 + \frac{2\mathcal{W}_{tot}}{m} \right)^{1/2}$$
$$= \left[(3)^2 + \frac{(2)(32)}{4} \right]^{1/2} \frac{\text{meters}}{\text{sec}} = 5 \frac{\text{meters}}{\text{sec}}$$

(Observe that spatial directions play no role in this type of problem.)

ii. Conservative and Nonconservative Forces (A Digression)

Before we can fully exploit the work-energy theorem, we must draw a distinction between two kinds of forces—those that conserve mechanical energy and those that do not.

As an illustration that does not involve free fall, consider a grandfather clock; it stores gravitational potential energy by means of its suspended mass. As the mass descends, work is done (mostly against

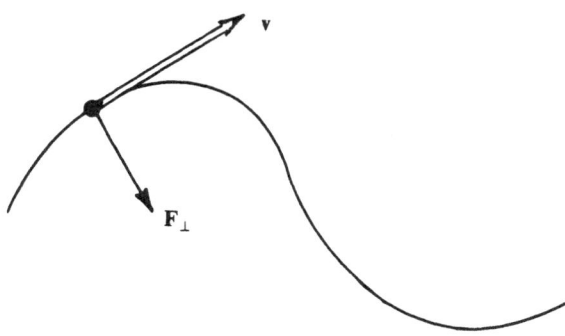

Figure 4.5. A transverse force, $\mathbf{F}_\perp$, is one that remains at right angles to the motion.

friction), and the stored energy is gradually given up. Because gravity can store energy and give it back, it is said to be conservative.

The precise definition of a **conservative force** $\mathbf{F}_{\text{cons}}$ is as follows. Suppose $\mathbf{F}_{\text{cons}}$ acts on a particle that travels from A to B. Then $\mathbf{F}_{\text{cons}}$ is conservative if

> There exists a potential energy function $\mathscr{U}$ (needed to indicate the amount of storage) such that the work $\mathscr{W}_{AB}$ done by $\mathbf{F}_{\text{cons}}$ is
> $$\mathscr{W}_{AB} = \mathscr{U}_A - \mathscr{U}_B \ (= -\Delta\mathscr{U})$$

(4.C.5)

Since $\mathscr{U}_A$ and $\mathscr{U}_B$ depend only on the end points, *the amount of work $\mathscr{W}_{AB}$ does not depend on the precise shape of the path or on how fast it is completed.* Conservative forces are not rare in physics; some will now be enumerated.

Transverse Forces

These are the simplest of all conservative forces. A transverse force, $\mathbf{F}_\perp$, is defined as one that acts on a particle at right angles to its velocity vector $\mathbf{v}$ (see Fig. 4.5):

$$\mathbf{F}_\perp \cdot \mathbf{v} = 0 \qquad (4.C.6)$$

It follows immediately [from Eq. (4.B.31)] that $\mathbf{F}_\perp$ transfers zero power, and therefore

> A transverse force does zero work. (4.C.7)

Statement (4.C.5) gives, for any two points A, B,

$$0 = -\Delta\mathscr{U} \qquad (4.C.8)$$

Thus, $\mathscr{U}$ does not change; it is constant in space and time, and may be taken to be zero.

> A transverse force contributes zero change to the potential energy. (4.C.9)

Figure 4.6 illustrates three transverse forces.

Gravity *(This chapter's prime illustration for a conservative force)*

Let us raise or lower a particle from A to B; the work done by *gravity alone* is $\mathscr{W}_{AB} = -\Delta\mathscr{U}$, independent of the detailed path between A and B. Thus, $\mathscr{W}_{AB} = -\Delta\mathscr{U}$ is valid not only under free fall,

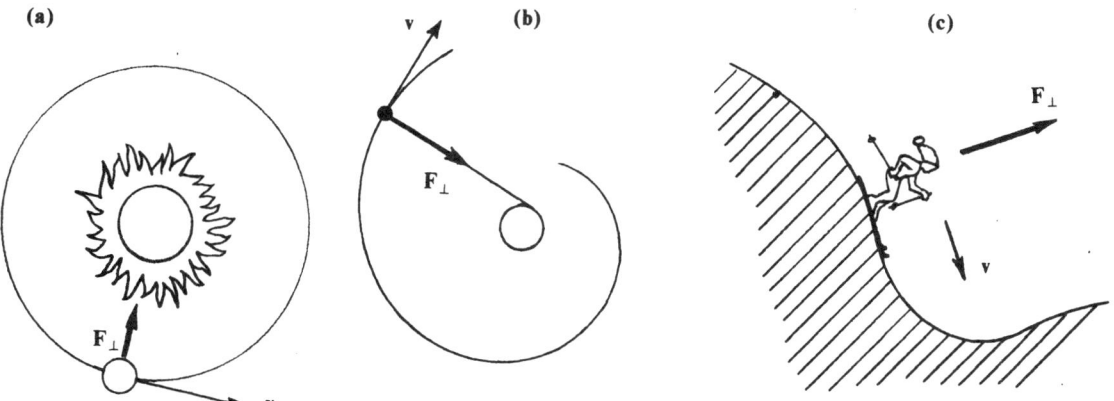

Figure 4.6. Some cases of transverse forces. (a) Attraction, towards the Sun, of a planet in circular orbit. (b) Force exerted by a string whose other end unwinds from a post. (c) Normal force exerted on a skier by a frictionless slope.

but in all circumstances; by statement (4.C.5), *gravity is a conservative force.*

[The path independence is simple to verify. Consider, as in Fig. 4.7, two alternative paths AXB (direct) and AYB (broken). For AXB we have

$$\mathscr{W}_{AXB} = \mathbf{w} \cdot \overrightarrow{AB}$$

($\mathbf{w}$ = weight of the particle). For AYB,

$$\mathscr{W}_{AYB} = \mathbf{w} \cdot \overrightarrow{AY} + \mathbf{w} \cdot \overrightarrow{YB}$$
$$= \mathbf{w} \cdot (\overrightarrow{AY} + \overrightarrow{YB}) = \mathbf{w} \cdot \overrightarrow{AB}$$

so that indeed $\mathscr{W}_{AXB} = \mathscr{W}_{AYB}$. Note the importance of having the same $\mathbf{w}$ over AY as over AB. Extending the argument to curved paths, as usual, is done by a limit of multiple broken paths.]

Other Conservative Forces

In Chap. 13 (Vibrations) we shall see that the force exerted by a stretched or compressed spring ("**elastic force**") is conservative.

Clocks, watches, and toys provide familiar illustrations of this type of energy storage. In a solid, sound waves carry half their energy as elastic potential energy; the other half is kinetic.

In the later chapters of this book we shall describe the conservative forces exerted by static **charges** or **magnets**. Examples of conservative forces are also found in the nuclear and subnuclear world. We keep in mind that in all these cases, "conservative" means that a potential energy can in principle be calculated.

Nonconservative Forces

Most mechanical forces are not conservative. If someone takes a book from the shelf and sets it on the table, his hand exerts a force that follows no reproducible mathematical pattern related to where the hand is located. Therefore there exists no potential energy function that could be associated with every position of the hand. The force exerted by the hand is **nonconservative**. The propulsive force of a vehicle is another such illustration. (To be sure, there does exist a storage of energy in a biological organism, or in a vehicle's fuel; but these energies cannot be described as a function $\mathscr{U}$ of position.)

Kinetic Friction

Kinetic friction is the classic example of a nonconservative force. Such a force $\mathbf{F}_K$, when exerted by a stationary surface on a moving particle P, always points *oppositely* to the motion of P. Suppose P travels from A to B, and then back to A, on a rough horizontal plane. Let l be the distance between A and B, and let $\mathbf{F}_K$ have constant magnitude $F_K \neq 0$. If we try to apply (4.C.5) to the roundtrip (A to A), we find

$$\mathscr{W}_{AA} = \mathscr{U}_A - \mathscr{U}_A = 0 \qquad (4.C.10)$$

Since F_K does negative work in both directions, the left side is

$$\mathscr{W}_{AA} = -F_K l - F_K l \neq 0$$

contradicting (4.C.10). The conclusion: One cannot associate a potential energy $\mathscr{U}$ with kinetic friction. (A force such as $\mathbf{F}_K$, which can only do negative total work on the participating objects, is called **dissipative**.)

iii. The Energy Budget of a Particle

As Fig. 4.8 illustrates, a particle P may be simultaneously subjected to several forces, some of which ($\mathbf{F}_{grav}$, $\mathbf{F}_{elastic}$) are known to be conservative, while others (together amounting to $\mathbf{F}_{sup}$, a "supplementary" force) are nonconservative, or perhaps of undetermined nature, possibly even conservative like the others. We shall now see that

The work done on P by $\mathbf{F}_{sup}$ is completely stored as kinetic and potential energy.

$$(4.C.11)$$

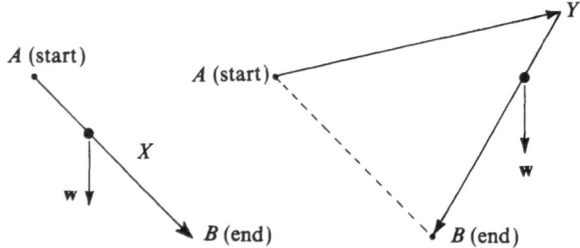

Figure 4.7. Two alternative routes AB for a particle of weight $\mathbf{w}$. (The particle might, for example, be carried by hand over the chosen path.) The work done by gravity is the same in both cases.

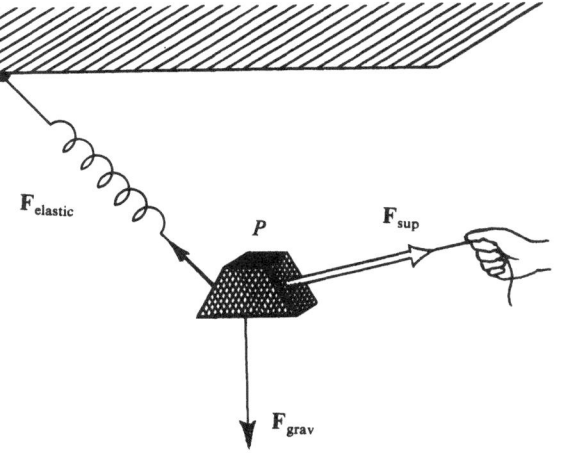

Figure 4.8. Conservative and nonconservative forces on a single particle.

The argument is as follows. Let $\mathbf{F}_{cons}$, $\mathbf{F}'_{cons}$, etc., be the set of conservative forces acting on P. Let

$$\mathbf{F}_{tot} = \mathbf{F}_{sup} + \mathbf{F}_{cons} + \mathbf{F}'_{cons} + \cdots \quad (4.C.12)$$

be the total force on P. Then, as P completes a small displacement $d\mathbf{r}$, Eq. (4.C.12) gives for the work done by $\mathbf{F}_{sup}$

$$\mathbf{F}_{sup} \cdot d\mathbf{r} = (-\mathbf{F}_{cons} \cdot d\mathbf{r} - \mathbf{F}'_{cons} \cdot d\mathbf{r} - \cdots) + \mathbf{F}_{tot} \cdot d\mathbf{r} \quad (4.C.13)$$

The left-hand side is the "supplementary work" $d\mathcal{W}_{sup}$; in the right-hand side, we recall from Eq. (4.C.4) that

$$\mathbf{F}_{cons} \cdot d\mathbf{r} = -d\mathcal{U} \quad (4.C.14)$$

and the work-energy theorem tells us that

$$\mathbf{F}_{tot} \cdot d\mathbf{r} = d\mathcal{K} \quad (4.C.15)$$

In conclusion, (4.C.13) becomes

$$d\mathcal{W}_{sup} = d(\mathcal{U} + \mathcal{U}' + \cdots + \mathcal{K}) \quad (4.C.16)$$

or, over a finite path length,

$$\mathcal{W}_{sup} = \Delta(\mathcal{U} + \mathcal{U}' + \cdots + \mathcal{K}) \quad (4.C.17)$$

as announced in (4.C.11). Here $\mathcal{U}$, $\mathcal{U}'$,... are the potential energies associated with the various con-

servative forces. Equation (4.C.17) is often referred to as the **energy principle**. It is, of course, not a conservation law unless $\mathcal{W}_{sup} = 0$.

iv. Examples with Conservation ($\mathcal{W}_{sup} = 0$)

The following are applications of Eq. (4.C.17) in the form

$$0 = \Delta(\mathcal{U} + \mathcal{K}) \quad (4.C.18)$$

where $\mathcal{U}$ is the gravitational potential energy. This equation is not affected by the presence of transverse forces, which do not contribute to any change in the potential energy.

Example 4.9. Centripetal Force. At an ice show, a skater S launches herself with speed v while tethered to a fixed point by a rope under tension; see Fig. 4.9. She then coasts in a circle and with negligible friction. Prove that her speed does not change.

———

Three forces are acting on S: a force $\mathbf{N}$, normal to the ice; the weight $\mathbf{w}$ of S, and the force $\mathbf{F}$ exerted by the rope. All three are transverse. Equation (4.C.18) becomes

$$\Delta \mathcal{K} = 0$$

so that

$$\underline{\underline{v = \text{const}}}$$

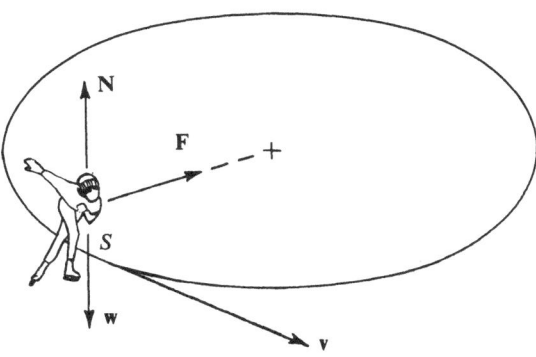

Figure 4.9. The skater as a particle: to the extent that friction is negligible, her speed remains constant.

[Here we have zero vertical acceleration, implying $\mathbf{N} + \mathbf{w} = \mathbf{0}$ (weight is canceled), and thus the total force on S is just $\mathbf{F}$; the latter is called a **centripetal** ($=$ center-seeking) force. *Circular motion under a purely centripetal force always proceeds at constant speed*; for details, see Chap. 6 further on.]

Example 4.10. A Roller Coaster. Figure 4.10 shows a frictionless roller coaster, in which a passenger car has a speed of 15 meters/sec at point A. What speed will it have at point B, if A and B are at heights of 32 meters and 12 meters, respectively?

Let us assume the whole car can be treated as a particle. It is subjected to only two forces: its weight, and the normal force $\mathbf{N}$ of the track. Since $\mathbf{N}$ is transverse, we can use Eq. (4.C.18), just as we would in the case of free fall. We have

$$(\mathcal{U} + \mathcal{K})_B = (\mathcal{U} + \mathcal{K})_A$$

or

$$mgh_B + \tfrac{1}{2}mv_B^2 = mgh_A + \tfrac{1}{2}mv_A^2$$

We note that the mass m factors out. Solving for v_B, we find

$$
\begin{aligned}
v_B &= [2g(h_A - h_B) + v_A^2]^{1/2} \\
&= [(2)(10)(32 - 12) + (15)^2]^{1/2} \text{ meters/sec} \\
&= \underline{\underline{25 \text{ meters/sec}}}
\end{aligned}
$$

A little reflection on the complexity of the motion, as well as on how incompletely it is specified, shows what a simple and powerful method we have here. In fact, it is hard to imagine how this type of problem could be solved without the energy principle.

Example 4.11. The Roller Coaster, Continued. In Fig. 4.10, will the car get over the hump at point C, which is at a height $h_C = 41$ meters? (The data are as in the preceding example.)

If the car could not go over the hump, it would reach a maximum height $h_{\max} < h_C$ before then. At $h_{\max}$, its velocity would reverse and therefore momentarily vanish. Thus, at maximum height, the car would have zero kinetic energy. Maximum height also means a maxi-

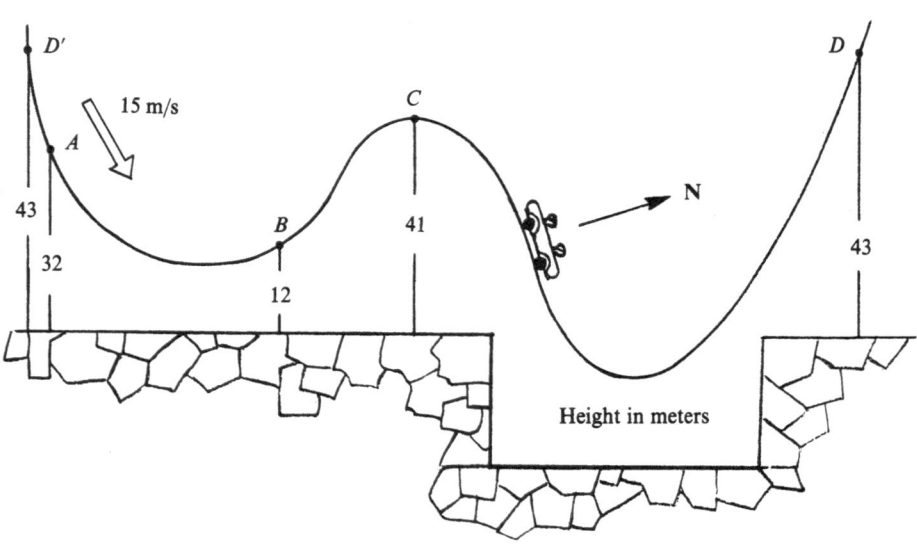

Figure 4.10. A frictionless roller coaster. The car, considered to be a particle, has a given speed at point A. In Example 4.10 we calculate its speed at B; whether it will get over the hump at C; and how high it will reach at D and D'. Only the nongravitational force $\mathbf{N}$ is shown; the weight vector of the car is omitted.

mum potential energy $\mathcal{U}_{max}$. Conservation gives, for $\mathcal{U} + \mathcal{K}$ at maximum height,

$$\mathcal{U}_{max} + 0 = (\mathcal{U} + \mathcal{K})_A$$

or

$$mgh_{max} = mgh_A + \tfrac{1}{2}mv_A^2$$

so that

$$h_{max} = h_A + \frac{v_A^2}{2g}$$

$$= \left[32 + \frac{(15)^2}{(2)(10)} \right] \text{meters} = \underline{\underline{43 \text{ meters}}}$$

This is higher than point C, so that the car will indeed get over the hump. In the figure, a further point D is shown where the car will in fact reach its allotted 43 meters, will stop, and will start reversing its motion. It will then go over the hump once more and continue until it reaches point D', again at 43 meters. It is clear that a perpetual back-and-forth motion will result. In practice any amount of friction, neglected here, will sooner or later trap the car at one of the two low spots.

v. Examples with Possible Nonconservation

The following cases involve, as before, gravitational and transverse forces, with, in addition, a (possibly) nonconservative force. The energy principle (4.C.17) is now

$$\mathcal{W}_{sup} = \Delta(\mathcal{U} + \mathcal{K}) \qquad (4.C.19)$$

It is sometimes advantageous to write it in its more basic form

$$\mathcal{W}_{tot} = \Delta\mathcal{K} \qquad (4.C.20)$$

(work-energy theorem).

Example 4.12. Lifting an Object. A student picks up a 4 kg book from a chair 50 cm high, walks around the room, and then *gently* sets the book on a table 100 cm high. How much work, $\mathcal{W}_{sup}$, does he do on the book?

In Eq. (4.C.19),

$$\mathcal{W}_{sup} = \Delta\mathcal{U} + \Delta\mathcal{K}$$

The initial and final kinetic energies vanish (the book is set down gently), and therefore $\Delta\mathcal{K} = 0$ just before it touches the table. Thus

$$\mathcal{W}_{sup} = \Delta\mathcal{U} = mgh \qquad (4.C.21)$$

where

$$m = 4 \text{ kg}, \qquad g = 10 \frac{\text{meters}}{\text{sec}^2}$$
$$h = (1.0 - 0.5) \text{ meter}$$

The result is

$$\mathcal{W}_{sup} = (4)(10)(1.0 - 0.5) \text{ joules} = \underline{\underline{20 \text{ joules}}}$$

What is the relevance, in this result, of having walked around the room? If we believe our calculation, the remarkable fact is that *there is no relevance*; the book's path need not be known in detail. The student's excess metabolic (chemical) energy, which has been spent in walking, does not enter into the *net* work on the book. Rather, it is mostly converted to heat—a statement which should become more meaningful in later chapters.

Example 4.13. A Playground Slide with Friction. A child X, whose mass is $m = 20$ kg, launches himself down with speed $v_i = 2$ meters/sec at the top of an inclined plane. After traveling down a length $l = 3$ meters of the incline, and lowering his altitude by $h = 1.5$ meters, he has an increased speed $v_f = 4$ meters/sec. Find the force F_K of kinetic friction (assumed constant) which acts on X.

From Eq. (4.C.17), the (negative) work done by $\mathbf{F}_K$ is

$$\mathcal{W}_{sup} = \Delta(\mathcal{U} + \mathcal{K})$$

or

$$-F_K l = \mathcal{U}_f - \mathcal{U}_i + \mathcal{K}_f - \mathcal{K}_i$$

Taking the bottom of the slide as reference level, this reads

$$-F_K l = 0 - mgh + \tfrac{1}{2}mv_f^2 - \tfrac{1}{2}mv_i^2$$

so that

$$F_K = \frac{m}{l}\left[gh + \tfrac{1}{2}(v_i^2 - v_f^2)\right]$$

$$= \frac{20}{3}\left[(10)(1.5) + \tfrac{1}{2}(2^2 - 4^2)\right] \text{ newtons}$$

$$= \underline{60 \text{ newtons}}$$

Example 4.14. Shooting a Pumpkin Seed. A wet seed, pinched between two fingers, shoots up vertically through a height $h = 4$ meters, see Fig. 4.11. If the seed is 2 mm thick and has a mass $m = 0.1$ gram, what force (assumed constant and nearly horizontal) does each finger exert on the seed? (Neglect friction.)

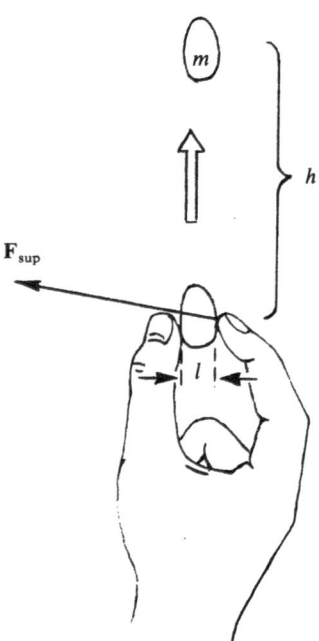

Figure 4.11. A pumpkin seed experiment; the "supplementary" force $\mathbf{F}_{\text{sup}}$ is exerted by the index finger; a similar force (not shown) is due to the thumb. The force is large and nearly—but not quite—horizontal. The seed's thickness is exaggerated for clarity.

Let the system under consideration be the seed, while gravity is the known conservative force. For definiteness, assume a stationary thumb and a moving index finger. The latter exerts a force F_{sup} and completes a distance $l = 2$ mm until it meets the thumb. Equation (4.C.19) reads

$$\mathscr{W}_{\text{sup}} = \Delta \mathscr{U} + \Delta \mathscr{K}$$

The left-hand side is more simply calculated from the motion of the finger than from that of the seed. The above reads

$$F_{\text{sup}} l = mgh + 0$$

since $\mathscr{K}_i = \mathscr{K}_f = 0$. We find

$$F_{\text{sup}} = \frac{mhg}{l} = \frac{(0.1 \times 10^{-3})(10)(4)}{2 \times 10^{-3}} \text{ newtons}$$

$$= \underline{2 \text{ newtons}}$$

The thumb must also exert a force of 2 newtons in the opposite direction, since the seed's horizontal acceleration is zero. (Would it make any difference if both fingers were assumed to move symmetrically, meeting halfway?)

An interesting aspect of the force $\mathbf{F}_{\text{sup}}$ is that its horizontal component—which accounts for most of the force—determines the work in our calculation, while we know that the small vertical component of the forces exerted by both fingers is what accelerates the seed. Both perceptions are correct.

D. Work and the Energy of a System

In this section we simply mean by "system" a set of particles, which, in general, are subject to mutual forces. Just as work done on a particle transfers energy to it, work done on any part of a system may be expected to transfer energy to that system. Here we propose to look into that type of energy transfer. When doing the bookkeeping on the energy of a system, one new feature arises: one must distinguish between work done by one part of the system on another part, as against work done on the system from the outside.

i. Mechanically Conservative Systems

Imagine a set of many particles interacting with each other. Ordinary objects fit that description: they are·composed of molecules, and maintain their shape owing to the mutual forces between the molecules. We assume for simplicity that each particle's potential energy is just gravitational. Thus, particles Nos. 1, 2, etc., have respective mechanical energies $\mathcal{K}_1 + \mathcal{U}_1$, $\mathcal{K}_2 + \mathcal{U}_2$, etc.

Let us do some nongravitational work $\mathcal{W}_{\text{sup}}$ on any part of such a system. In some (but not all) cases, its *total mechanical energy* will then increase by exactly the amount of work done:

$$\mathcal{W}_{\text{sup}} = \Delta(\mathcal{K}_1 + \mathcal{U}_1 + \mathcal{K}_2 + \mathcal{U}_2 + \cdots) \quad (4.D.1)$$

The system, therefore, stores an amount of energy $\mathcal{W}_{\text{sup}}$; we have here an **energy principle** analogous to (4.C.17). Equation (4.D.1) describes a **mechanically conservative system**.

Many systems fail to be mechanically conservative, for example, if kinetic friction occurs between some of their parts, or if they contain a motor with fuel or a battery. Living organisms are not mechanically conservative.

On the other hand, a system will satisfy Eq. (4.D.1) with purely gravitational terms $\mathcal{U}_1$, $\mathcal{U}_2$,..., if it just contains the following:

1. Rigid bodies;
2. Inextensible strings;
3. Frictionless pivots and parts which roll without slipping;
4. Frictionless sliding surfaces.

This permits frictionless wheels, levers, transmission belts and chains, pulleys, gears, cams, etc.—a mechanical engineer's dream.

The energy principle, (4.D.1), for a system involving (1)–(4) above, needs to be proved. The basic idea can be obtained from Fig. 4.12, illustrating the simplest case of a rigid object. Here two particles, 1 and 2, are connected by a light bar, through which they exert on one another a pair of action–reaction forces $\mathbf{F}$, $-\mathbf{F}$. An external, possibly nonconservative force $\mathbf{F}_{\text{sup}}$ is shown acting on particle No. 1. Each particle is also subject to a gravitational force, not shown in the figure. During a certain time interval, the work done by each of these forces is $\mathcal{W}_{\text{sup}}$ (done by $\mathbf{F}_{\text{sup}}$), $\mathcal{W}_1$ (done by $\mathbf{F}$), and $\mathcal{W}_2$ (done by $-\mathbf{F}$).

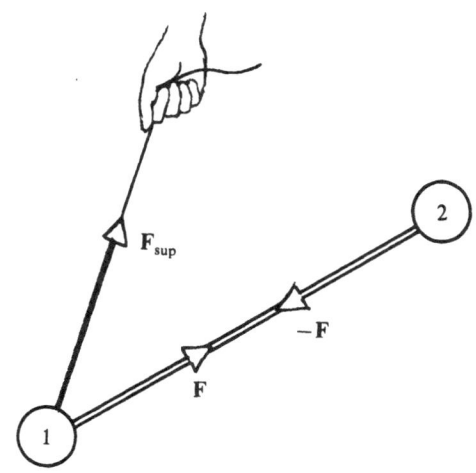

Figure 4.12. A system containing a rigid connecting bar.

The energy principle (4.C.17) for each of the two particles reads

$$\mathcal{W}_{\text{sup}} + \mathcal{W}_1 = \Delta(\mathcal{U}_1 + \mathcal{K}_1) \quad (4.D.2)$$

$$\mathcal{W}_2 = \Delta(\mathcal{U}_2 + \mathcal{K}_2) \quad (4.D.3)$$

Adding,

$$\mathcal{W}_{\text{sup}} + \mathcal{W}_1 + \mathcal{W}_2 = \Delta(\mathcal{U}_1 + \mathcal{K}_1 + \mathcal{U}_2 + \mathcal{K}_2) \quad (4.D.4)$$

The crux of the matter is that

$$\mathcal{W}_1 + \mathcal{W}_2 = 0 \quad (4.D.5)$$

i.e., *the rigid connecting bar supplies zero net work to the system*, and therefore Eq. (4.D.4) conforms to the energy principle (4.D.1). [A Note at the end of this chapter shows how to arrive at (4.D.5), which is not as obvious as it may seem, and which depends on the rigidity of the connection.]

ii. Examples with Conservation ($\mathcal{W}_{\text{sup}} = 0$)

Here we apply Eq. (4.D.1) in the form

$$\Delta(\mathcal{K}_1 + \mathcal{U}_1 + \mathcal{K}_2 + \mathcal{U}_2 + \cdots) = 0 \quad (4.D.6)$$

Example 4.15. Atwood's Machine. We are given two masses, m and M ($M > m$), suspended as shown in Fig. 4.13. The system is released from rest with m on the floor and M higher by an amount h. When M is about to strike the floor, it is found to have a speed v. Find h if v is known.

The conservation condition (4.D.6) between initial and final states can be written

$$(\mathscr{U} + \mathscr{K})_f = (\mathscr{U} + \mathscr{K})_i \qquad (4.D.7)$$

where $\mathscr{U}$ is the combined potential energy of the system, and $\mathscr{K}$ is its combined kinetic energy. We have

$$\mathscr{U}_i = \quad 0 \quad + \quad Mgh, \quad \mathscr{U}_f = \quad mgh \quad + \quad 0$$

$$\text{(for } m) \quad \text{(for } M) \qquad \text{(for } m) \quad \text{(for } M)$$

$$\mathscr{K}_i = \quad 0 \quad + \quad 0, \qquad \mathscr{K}_f = \tfrac{1}{2}mv^2 + \tfrac{1}{2}Mv^2$$

Equation (4.D.7) reads

$$mgh + \tfrac{1}{2}mv^2 + \tfrac{1}{2}Mv^2 = Mgh$$

Solving for h, we obtain

$$h = \frac{(m + M)\, v^2}{2(M - m)\, g}$$

The reader should verify that the acceleration obtained earlier in Eq. (3.E.1) of Chap. 3 yields the same result. We note that the string's tension does zero net work on the system, although it does nonzero work on each block.

Example 4.16. A Sliding Dumbbell. (This problem would be insoluble without the use of energy conservation.) A dumbell, Fig. 4.14, consists of two small but heavy spheres of equal mass m, connected by a straight light rigid bar of length l. When set at an angle θ against the (rounded) wall and floor arrangement shown in the figure, the dumbbell slips. At what speed v is it sent coasting along the floor? (Assume that everything happens in the plane of the figure, and that both spheres slide frictionlessly throughout the motion, starting from rest.)

Again we set, for the total energy of the dumbbell,

$$(\mathscr{U} + \mathscr{K})_f = (\mathscr{U} + \mathscr{K})_i \qquad (4.D.8)$$

If potential energies are defined with respect to the floor, we have

$$(\mathscr{U} + \mathscr{K})_f = 0 + \mathscr{K}_f = 2(\tfrac{1}{2}mv^2) = mv^2$$

(each mass contributes equally);

$$(\mathscr{U} + \mathscr{K})_i = \mathscr{U}_i + 0 = mgl \cos \theta$$

(only the upper mass contributes). Thus, Eq. (4.D.8) becomes

$$mv^2 = mgl \cos \theta$$

or

$$v = (gl \cos \theta)^{1/2}$$

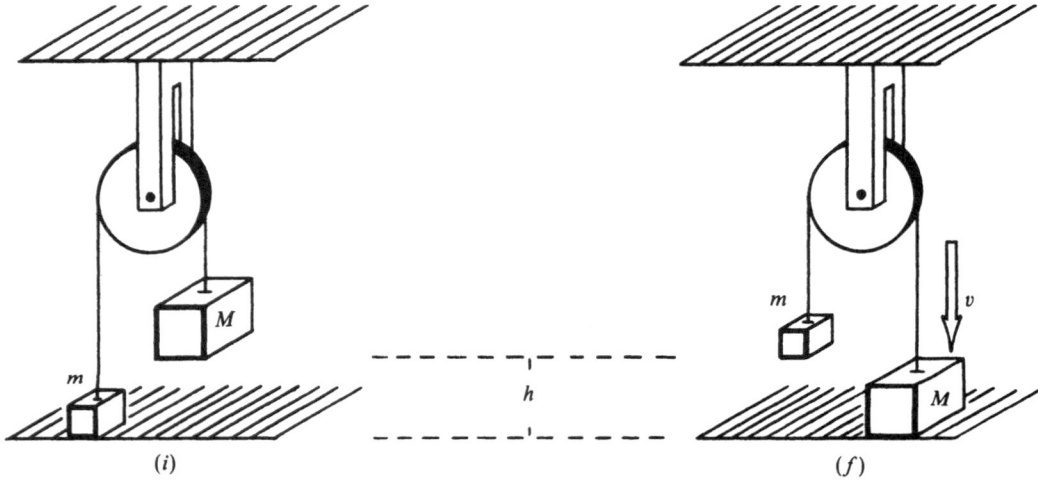

(i) $\qquad\qquad\qquad\qquad\qquad\qquad\qquad$ (f)

Figure 4.13. Atwood's machine as an energy-conserving system: (i) and (f) are the initial and final configurations considered in Example 4.15.

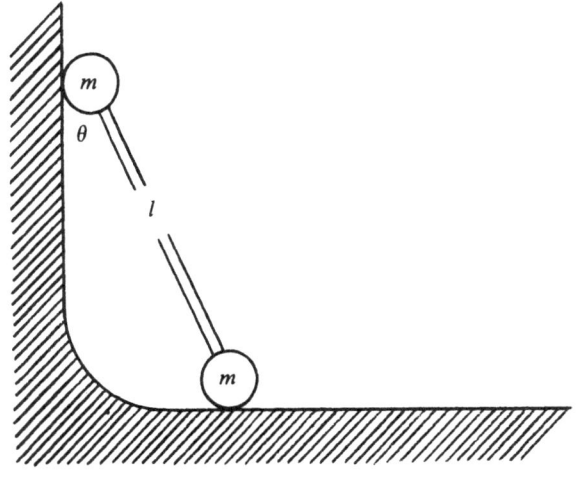

Figure 4.14. A dumbbell initially leans against a wall and slides horizontally after it slips down. Example 4.16 calculates its ultimate speed.

iii. Examples with Possible Nonconservation

Example 4.17. The Lever. This familiar "machine," shown in Fig. 4.15, consists of a light rigid bar, frictionlessly pivoted at a fixed point P called the **fulcrum**. Some external objects exert the forces $\mathbf{F}_1$ and $\mathbf{F}_2$ perpendicularly to the lever, in such a way that it remains in static equilibrium. If the corresponding "arms" have lengths R_1 and R_2, find the ratio F_2/F_1, or **mechanical advantage** of end No. 1 with respect to end No. 2.

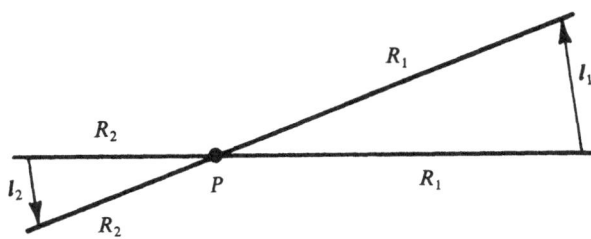

Figure 4.16. The lever of Fig. 4.15, slightly rotated. The point of application of $\mathbf{F}_1$ is displaced through a small l_1, parallel and opposite to $\mathbf{F}_1$, so that $\mathbf{F}_1 \cdot l_1 = -F_1 l_1$. Similarly, $\mathbf{F}_2 \cdot l_2 = +F_2 l_2$.

Let the whole system consist of just the lever itself, without the weights. In order to apply the energy principle, we use a trick: we can approximate static equilibrium as closely as we want by letting the lever move very slowly through a small angle, Fig. 4.16. Kinetic energy can then be neglected, and, for a light enough lever, so can potential energy. The energy principle, Eq. (4.D.1), becomes

$$\mathscr{W}_{\text{sup}} = 0$$

(conservation applies after all), or

$$-F_1 l_1 + F_2 l_2 = 0$$

Hence

$$\frac{F_2}{F_1} = \frac{l_1}{l_2} = \frac{R_1}{R_2} \qquad (4.\text{D}.9)$$

the last step from the similarity of the two triangles in Fig. 4.16.

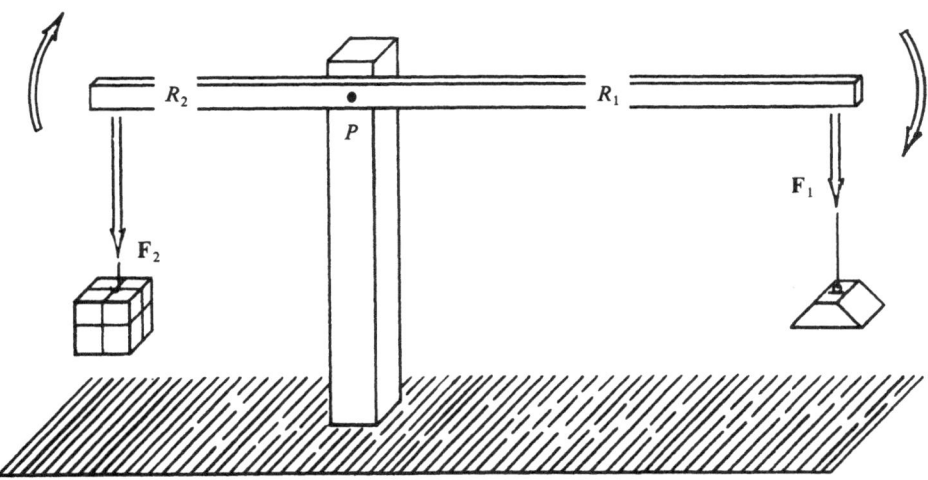

Figure 4.15. A lever. Two objects, considered external to the system, apply the forces F_1 and F_2 on the ends of the lever and perpendicularly to it.

Example 4.18. Launching by Counterweight. In testing the method of Fig. 4.17 for launching a glider, someone left the brakes on by mistake, and a final speed of only $v_f = 20$ meters/sec was achieved. The cliff's height is $h = 60$ meters; the masses are $M = 200$ kg for the glider and $m = 160$ kg for the counterweight. How much frictional force, F_{sup} (assumed constant) did the glider experience?

Once more, let $\mathcal{U}$ and $\mathcal{K}$ be combined for both particles. The energy principle is then

$$\mathcal{W}_{sup} = (\mathcal{U}_f + \mathcal{K}_f) - (\mathcal{U}_i + \mathcal{K}_i)$$

If potential energies are measured from the top of the cliff, this reads

$$-F_{sup}h = [(0 - mgh) + (\tfrac{1}{2}Mv^2 + \tfrac{1}{2}mv_f^2)]$$

$$- [(0 + 0) + (0 + 0)]$$

(The friction does negative work on the glider; note that m and M cover the same distance h, and have the same speed.) Solving for F_{sup},

$$F_{sup} = \frac{mgh - \tfrac{1}{2}(M + m)\,v_f^2}{h}$$

$$= \frac{(160)(10)(160) - \tfrac{1}{2}(200 + 160)(20)^2}{60} \text{ newtons}$$

$$= \underline{\underline{400 \text{ newtons}}}$$

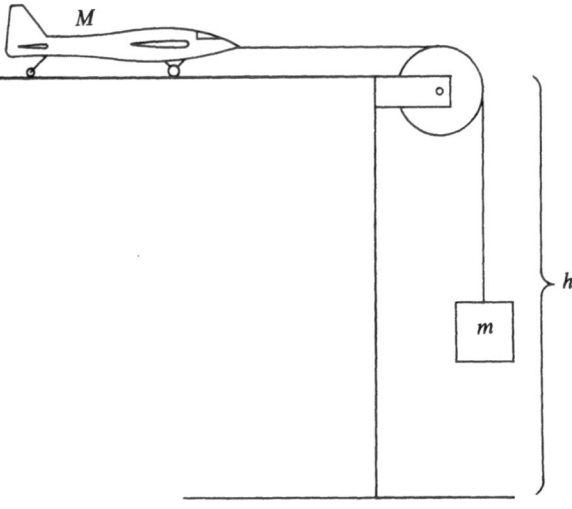

Figure 4.17. The glider is pulled by the cable and released just before it reaches the pulley (and as m hits bottom). The system starts from rest.

iv. Stability

The kinetic energy of a particle, $\tfrac{1}{2}mv^2$, cannot be negative. This has an interesting consequence for the behavior of mechanical systems. Consider a particle, initially held at rest on an incline, Fig. 4.18a. When released, the particle acquires some (necessarily positive) kinetic energy, and therefore its potential energy can only decrease. This simple observation remains true when applied to the sum of all kinetic or potential energies of any isolated system. Thus,

> An isolated system, released from rest, begins to change—if at all—so as to decrease its potential energy.

(4.D.10)

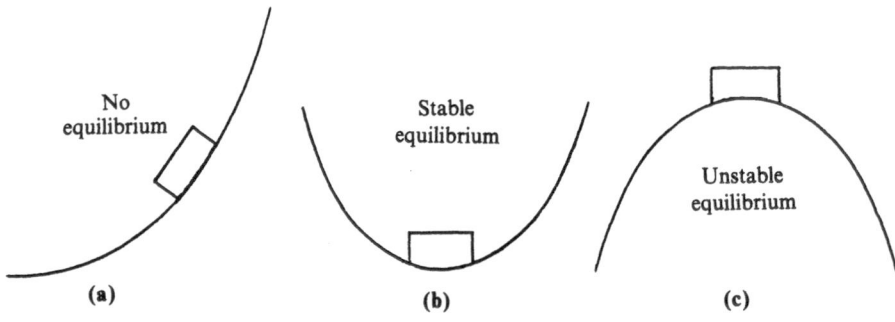

Figure 4.18. A particle is in stable or unstable equilibrium according to whether its potential energy is minimal or maximal.

What happens if the system already has a minimum value for its potential energy, as in Fig. 4.18b? Then its configuration cannot change at all; it is in so-called **stable equilibrium**. Moreover, if released from rest ever so little away from that minimum potential energy, the system will move toward it.

> A minimum in the total potential energy of an isolated system corresponds to a stable equilibrium.

$$(4.D.11)$$

On the other hand, suppose the potential energy is a maximum, as in Fig. 4.18c. Then, at least in this particular example, we see that the total force is zero, and no motion will develop. However, if the particle is released from rest just next to the maximum potential energy, then the latter, as we have seen, must be further decreased. The particle will therefore move *away* from that maximum. This is called a situation of **unstable equilibrium**.

> A maximum in the total potential energy of an isolated system corresponds to an unstable equilibrium.

$$(4.D.12)$$

(Borderline cases, as of a ball resting on a horizontal plane, are referred to as **neutral equilibrium**.)

E. Conservation of Total Energy

Let us examine yet again the mechanical-energy balance of a system:

$$\mathscr{W}_{\text{sup}} = \Delta(\mathscr{U} + \mathscr{K})\qquad(4.E.1)$$

Where does the supplementary work, $\mathscr{W}_{\text{sup}}$, come from? It is tempting to postulate that there is, somewhere, a supplementary energy, $\mathscr{E}_{\text{sup}}$, that is *depleted* by the exact amount $\mathscr{W}_{\text{sup}}$:

$$-\Delta\mathscr{E}_{\text{sup}} = \mathscr{W}_{\text{sup}}\qquad(4.E.2)$$

Combining (4.E.1) and (4.E.2) yields

$$\Delta(\mathscr{U} + \mathscr{K} + \mathscr{E}_{\text{sup}}) = 0\qquad(4.E.3)$$

that is to say, *the total energy* (including $\mathscr{E}_{\text{sup}}$) *is conserved*. As science has progressed, more and more kinds of energy $\mathscr{E}_{\text{sup}}$ have been discovered,

leading to a detailed verification of (4.E.3), which is one of the most basic laws of Nature.

In such a verification it is, of course, essential that all relevant energy reservoirs be taken into account. One way of making sure that no outside sources (or sinks) of energy have been forgotten in the bookkeeping is to prepare a system that is completely isolated from its surroundings. This is very hard to achieve, and much detailed physics is needed to determine what constitutes sufficient isolation. Nevertheless, the concept of an "isolated system" is a useful one and leads to a compact statement:

> The total energy of an isolated system is conserved.

$$(4.E.4)$$

What is the nature of the mysterious new term $\mathscr{E}_{\text{sup}}$? Sometimes it is quite familiar: it may just be the potential or kinetic energy of a neighboring system. In the case of friction, note that $\mathscr{W}_{\text{sup}}$, in Eq. (4.E.1), is negative; here $\mathscr{U} + \mathscr{K}$ *decreases* and $\mathscr{E}_{\text{sup}}$ *increases*. We shall learn in Chap. 10 that the mechanical energy drained away from a system by friction amounts to a so-called **heat transfer**; it helps raise the temperature of a substance. The new energy term $\mathscr{E}_{\text{sup}}$ is, in this case, called the **internal energy** of that substance, and really amounts to the mechanical energy of individual molecules as they vibrate or travel in a disorderly, random fashion.

Other contributions to $\mathscr{E}_{\text{sup}}$ will be studied in this book, notably the energy of **electric and magnetic fields**; also the huge hidden energy that a particle possesses, even when at rest, just by virtue of having a mass. The existence of this so-called **rest energy** was unsuspected until the advent of special relativity.

We can now conclude the chapter with the following definition.

General Definition of Energy

Because work has a simple mechanical meaning, while energy has multiple forms, the latter is customarily defined in terms of the former.

> Energy is a conserved quantity, portions of which can be transferred as work.

$$(4.E.5)$$

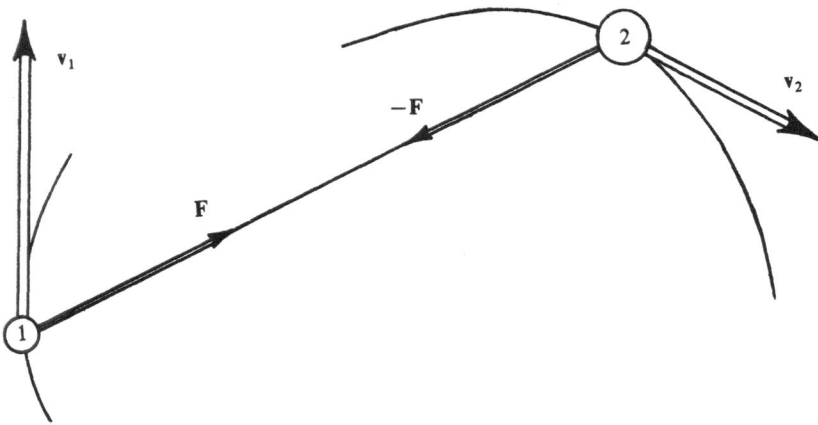

Figure 4.19. Two particles 1, 2, each shown with a portion of its trajectory, are connected by a light rigid bar. Together, the action-and-reaction forces **F** and −**F** do zero net work in such a case.

While somewhat abstract, this definition is perfectly sharp. It can be used in the laboratory, and nothing but energy fits it.

Energy and Power Spectra

An overview of energy and power must range over many fields of physics. Accordingly, the reader will find the relevant spectra in Appendix I rather than in this or any other specific chapter.

Note

Rigid Connections Contribute Zero Work

We want to prove that the rigid bar in Fig. 4.12 does zero work on its pair of particles; see Eq. (4.D.5). Equivalently, *it supplies zero power to them.*

Figure 4.19 shows the particles' motion; let their velocities be v_1 and v_2. The power supplied by **F** to particle 1 is [see Eq. (4.B.30)]

$$(\text{Power})_1 = \mathbf{F} \cdot \mathbf{v}_1$$

Similarly, −**F**, acting on particle 2, gives

$$(\text{Power})_2 = -\mathbf{F} \cdot \mathbf{v}_2$$

In total, we have

$$(\text{Power})_1 + (\text{Power})_2 = \mathbf{F} \cdot (\mathbf{v}_1 - \mathbf{v}_2) \quad (4.\text{N}.1)$$

Now the vectors **F** and $v_1 - v_2$ are at right angles to each

other. To see this, we reason as follows. The difference $v_1 - v_2$ is the velocity of 1 relative to 2. This means that $v_1 - v_2$ would be the velocity of 1, observed by someone moving along with 2. For that observer, particle 2 is at rest. The situation is depicted in Fig. 4.20: if 2 seems at rest, then a constant distance $|\overrightarrow{1,2}|$ means that 1 appears to describe a circle around 2, and therefore its apparent velocity $v_1 - v_2$ is exactly orthogonal to the direction of 1, 2, i.e., to the direction of **F**. Thus, we have

$$\mathbf{F} \cdot (\mathbf{v}_1 - \mathbf{v}_2) = 0$$

Going back to Eq. (4.N.1), we have

$$(\text{Power})_1 + (\text{Power})_2 = 0$$

which was to be shown.

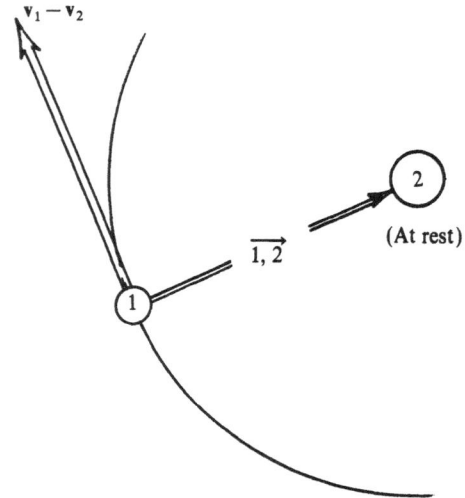

Figure 4.20. This is how the behavior in Fig. 4.19 appears to someone moving with particle 2.

Condensed Checklist

Dot product (any two vectors p, q):

$$\mathbf{p} \cdot \mathbf{q} = pq \cos \theta \qquad (\theta = \text{angle between} \quad (4.B.9)$$
$$\mathbf{p}, \mathbf{q}, \text{ tail to tail})$$

$$\mathbf{p} \cdot \mathbf{q} = p_{\parallel} q = p q_{\parallel} \qquad (4.B.14), (4.B.15)$$

$$\mathbf{p} \cdot \mathbf{p} = p^2 \qquad (4.B.18)$$

$$\left. \begin{array}{l} \mathbf{p} \cdot \mathbf{q} = 0 \\ \mathbf{p} \neq \mathbf{0}, \mathbf{q} \neq \mathbf{0} \end{array} \right\} \quad \text{means } \mathbf{p}, \mathbf{q} \text{ mutually orthogonal.}$$
$$(4.B.20)$$

Kinetic energy:

$$\boxed{\mathcal{K} = \tfrac{1}{2}mv^2} \qquad (4.A.7)$$

Potential energy:

$$\boxed{\mathcal{U} = mgz} \qquad \text{(uniform gravity)} \qquad (4.A.6)$$

Work:

$$\mathcal{W} = \mathbf{F} \cdot l \qquad \text{(constant } \mathbf{F}) \qquad (4.B.11)$$

$$d\mathcal{W} = \mathbf{F} \cdot d\mathbf{r} \qquad \text{(any } \mathbf{F}) \qquad (4.B.25)$$

$$\boxed{\mathcal{W} = \int_A^B \mathbf{F} \cdot d\mathbf{r}} \qquad (4.B.27)$$

Work done by a conservative force:

$$\mathcal{W}_{AB} = \mathcal{U}_A - \mathcal{U}_B = -\Delta \mathcal{U} \qquad (4.C.5)$$

Work-energy theorem:

$$\Delta \mathcal{K} = \mathcal{W}_{\text{tot}} \qquad (4.C.2)$$

$$d\mathcal{K} = \mathbf{F}_{\text{tot}} \cdot d\mathbf{r} \qquad (4.C.4)$$

Power:

$$\boxed{\frac{d\mathcal{W}}{dt} = \mathbf{F} \cdot \mathbf{v}} \qquad (4.B.30)$$

Conservation of mechanical energy:

$$(\mathcal{U} + \mathcal{K})_f = (\mathcal{U} + \mathcal{K})_i \qquad (4.C.18)$$

(under purely conservative forces).

Supplementary work on a system:

$$\boxed{\mathcal{W}_{\text{sup}} = \Delta(\mathcal{U} + \mathcal{K})} \qquad (4.C.17)$$

Units:

$$1 \text{ joule} = 1 \frac{\text{kilogram meter}^2}{\text{second}^2} \qquad (4.A.14)$$

$$1 \text{ watt} = 1 \frac{\text{joule}}{\text{second}} \qquad (4.B.31)$$

$$1 \text{ kilowatt hour} = 3.6 \times 10^6 \text{ joules} \qquad (4.B.33)$$

$$1 \text{ horsepower} = 746 \text{ watts} \qquad (4.B.32)$$

Energy and power spectra: Appendix I

True or False

1. The sum of a particle's kinetic and potential energies is sometimes not conserved.

2. The sum of a particle's kinetic and potential energies is conserved only in cases of free fall.

3. In an Atwood machine released from rest, the combined potential energy of both masses always decreases.

4. The force of gravity does negative work on a rising particle.

5. While an object slides on a rough, stationary surface, negative work is done on the object, and hence positive work is done on the surface.

6. The law of action and reaction implies that, if object A does a work $\mathcal{W}$ on object B, then B does a work $-\mathcal{W}$ on A.

7. An object's kinetic energy depends on the motion of the coordinate system in which it is measured.

8. If a particle's kinetic energy is increasing, it means that its potential energy must be decreasing.

9. In the case of a projectile, conservation of mechanical energy can be inferred from Newton's laws of motion.

10. Two equal-sized energies may point in opposite directions; they then cancel one another.

11. If $\mathbf{p} \cdot \mathbf{q} = 0$ we must have $\mathbf{p} = \mathbf{0}$ or $\mathbf{q} = \mathbf{0}$, or both.

12. A rocket engine that delivers a constant thrust delivers more power as the rocket picks up speed.

13. Kinetic friction is sometimes a transverse force.

14. If all forces acing on a particle are conservative, its total mechanical energy is conserved.

15. As an object travels between two given points A and B, the work done on it by a conservative force depends on the shape of the object's path.

Problems

Section A: Conservation of Energy in Free Fall; Includes Definitions of Kinetic and Potential Energies

*4.1. Find the kinetic energy, in joules, of (a) a paper clip of mass 1.5 g, falling at 5 meters/sec; (b) a person weighing 150 lb, running at 5 mi/h; (c) an airplane of mass 10 metric tons, flying at 200 km/h. (Note: 1 metric ton = 1000 kg.)

4.2. Find the speed of (a) a 50 000 metric ton battleship, cruising with a kinetic energy of 6×10^9 joules; (b) the Moon (mass $\approx 7 \times 10^{22}$ kg), whose kinetic energy in Earth orbit is about 4×10^{28} joules; (c) an electron (mass $\approx 9 \times 10^{-31}$ kg), orbiting with 1.6×10^{-18} joules of kinetic energy in a hydrogen atom. (Note: 1 metric ton = 1000 kg.)

4.3. Particles A and B have the same kinetic energy, but A's mass is 100 times that of B. If A moves at 100 meters/sec, what is the speed of B?

*4.4. Estimate the mass of an air molecule whose typical speed, at room temperature, is 500 meters/sec, while its kinetic energy is 6×10^{-21} joules.

4.5. Assuming that Edmund Hillary weighed between 100 and 200 lb, estimate, in joules, his potential energy relative to sea level when he conquered Mount Everest (height $\approx$ 30 000 ft).

4.6. A crate of mass 400 kg, initially at rest, is dropped from a height of 30 meters. What is its kinetic energy just before it hits the ground?

4.7. How should the answers to Example 4.1 be modified if the edge of the table is used as the reference level?

*4.8. A pingpong ball, initially at rest, is dropped from a height of 60 cm on a horizontal table, and bounces vertically up. If 20% of its kinetic energy was lost on impact, how high does it rebound? (Air friction is significant, but ignore it.)

4.9. Sputnik I, the first man-made satellite, had a mass of about 80 kg and circled the Earth about every 90 min in an orbit of less than 100 km altitude. Estimate its kinetic energy (a) in joules; (b) in kW h.

4.10. A pingpong ball of mass 2.0 g rolls off the edge of a table that is 1.00 meter high. When the ball strikes the floor, its speed is 5.00 meters/sec. At what speed did it leave the table? (Ignore air friction.)

4.11. A pingpong ball of mass m, traveling in the x direction at speed v_i, encounters a paddle that travels in the $-x$ direction at speed u. The ball rebounds in the $-x$ direction at speed v_f in such a way that its speed relative to the paddle is unchanged; the paddle does not appreciably change its own motion. Show that the impact has increased the ball's kinetic energy by an amount $\Delta \mathscr{K} = 2mv_{av}u$, where $v_{av} = \frac{1}{2}(v_i + v_f)$.

Section B: Work

*4.12. (a) How much work, in joules, do you do on a 5-g coin if you carry it up the stairs of the Empire State Building (height $\approx$ 400 meters)? (b) How much work would you do on it if you took the elevator?

4.13. A constant force vector of magnitude 5 newtons, acting on a particle whose net displacement is 3 meters long, does a work of 12 joules. Find the angle between the force and the displacement.

*4.14. A draftsman, tracing a square of side 3 cm on paper, holds his pen differently according to which side he draws. The frictional forces encountered by the pen are as follows: up, 250 dynes; to the right, 300 dynes; down, 450 dynes; to the left, 200 dynes. How much work, in ergs, does the pen do against friction?

*4.15. A mass of 20 kg, starting from rest, is pushed a distance of 5 meters up a frictionless incline at 30° to the horizontal. What is the minimum work required? (Find the force along the incline, and let work = force × distance; see Problem 4.23 for a more sophisticated approach.)

4.16. A tugboat, pulling to 37° west of north, is supplying 400 hp of towing power to a ship whose motion is due north at 0.5 km/h. What is, in newtons, the tension in the towing cable?

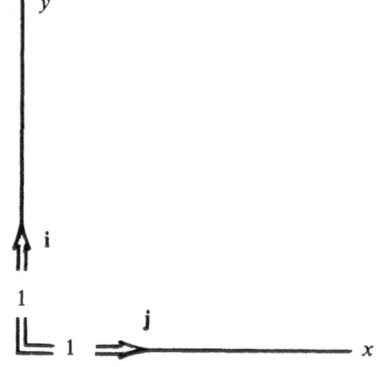

4.17. (a) Let $\mathbf{i}$, $\mathbf{j}$ be vectors of unit magnitude ("unit vectors") in the x and y directions, respectively; see the figure. Show that, for any vector $\mathbf{F}$ in the xy plane, one has $\mathbf{F} = F_x \mathbf{i} + F_y \mathbf{j}$, where F_x, F_y are the rectangular components of $\mathbf{F}$. (b) Hence show that for any two vectors $\mathbf{F}$, $\mathbf{l}$ in the xy plane, one has $\mathbf{F} \cdot \mathbf{l} = F_x l_x + F_y l_y$. (Note: In three dimensions, these results become $\mathbf{F} = F_x\mathbf{i} + F_y\mathbf{j} + F_z\mathbf{k}$, and $\mathbf{F} \cdot \mathbf{l} = F_x l_x + F_y l_y + F_z l_z$.)

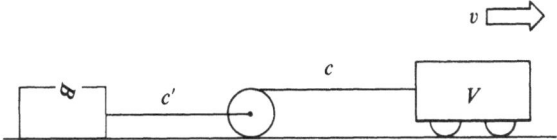

*4.18. A vehicle V, moving horizontally at constant speed of 2 meters/sec, pulls a circular drum by unwinding the horizontal cable c, as shown. The drum rolls without slipping and pulls a block B, against a kinetic frictional force of 8 newtons, by means of another horizontal cable c'. Assuming the cables to be massless and inextensible, (a) find the tension in c'; (b) find the speed of B; (c) find the power supplied to B through cable c'; (d) find the power supplied by V through cable c; (e) find the tension in cable c.

Section C: Work and the Energy of a Particle

*4.19. A 500-kg roller coaster car, initially moving at 6 meters/sec on a curved track, increases its altitude by an amount h, at which point the car reverses its motion. Assuming zero friction, find h.

4.20. A small girl, who weighs 50 lb, uses a light swing whose rope is 10 ft long. If her maximum speed is 6 ft/s, what is the height difference between her highest and lowest positions?

*4.21. A block of mass 4 kg is sent sliding at an initial speed of 2 meters/sec along a horizontal surface, against a constant frictional force F_K. If the block stops after 8 meters, find F_K.

*4.22. A 500-kg roller coaster car, moving initially at a speed of 12.0 meters/sec, travels a 50.0-meter length of curving track, decreasing its altitude by 10.0 meters in the process, and increasing its speed to 18.0 meters/sec. Calculate the frictional force (assumed constant) of the track. (Treat the car as a particle; use $g = 9.8$ SI units.)

4.23. Do Problem 4.15 by the potential energy method.

4.24. A particle of mass m is initially suspended at rest from a light inextensible string of length l. The particle is slowly pulled sideways by a horizontal force $\mathbf{F}$, until the string (still under tension) is at an angle θ to the vertical. How much work is done on the particle by (a) the force $\mathbf{F}$? (b) gravity? (c) the string? (Express your answers in terms of m, g, l, θ.)

4.25. A longitudinal force F is needed to raise a block of mass m slowly up a frictionless inclined plane of length l

and vertical height h. In terms of l and h, find the mechanical advantage w/F, where w is the weight of the block. (Consider the energy budget of the block in order to solve this problem.)

Sections D: Work and the Energy of a System; E: Conservation of Total Energy

4.26. In the equilibrium situation of Fig. 3.20 of Chap. 3, by what height does the large weight go down when the small one goes up by a height h? (Use the energy method.)

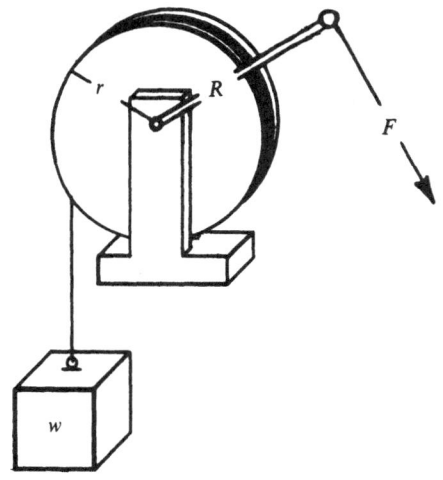

4.27. A winch (see the figure) consists of a drum of radius r, on which a light rope is wound. What perpendicular force F must be exerted on the end of the crank, whose length is R, in order to lift a weight w at a steady rate? (Neglect friction and the weight of the crank.)

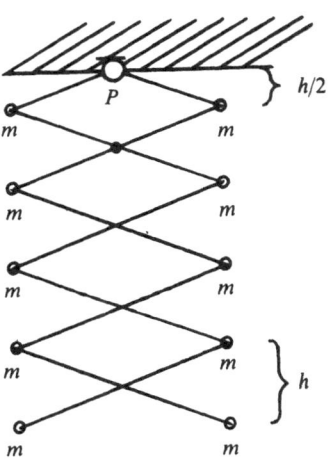

4.28. The light connecting bars in the figure are frictionlessly pivoted at each mass m, also wherever they cross, and at the ceiling suspension. All bar segments have equal length. If, initially, the successive masses are a height h apart, what minimum work must be done on point P to collapse the whole mechanism upwards against the ceiling? (Express your answer in terms of h, m, g.)

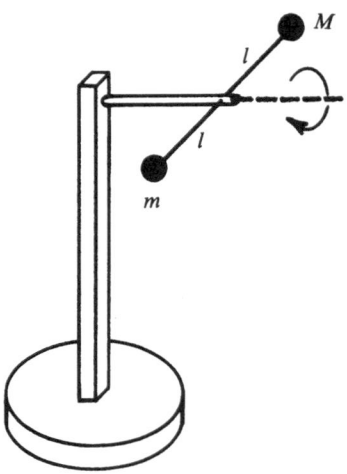

4.29. Two particles of masses m and M $(M > m)$ are connected by a light rigid bar of length $2l$ frictionlessly supported at its middle; see the figure. If the system is released from rest when horizontal, what is the speed of the masses (a) when m is vertically above M? (b) when the connecting bar is at 30° to the horizontal? (Express your results in terms of M, m, g, l; neglect friction.)

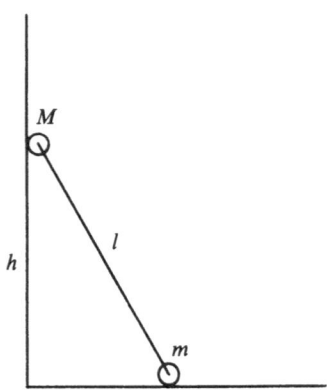

*4.30. Two particles of masses M and m are connected by a light rigid rod of length l; see the figure. Each particle slides, for the entire motion considered, on its frictionless track, vertical or horizontal as shown. The system is released from rest when M is at a height h above m. How fast is M moving as it reaches the same level as m? (Express your answer in terms of M, m, g, h, l.)

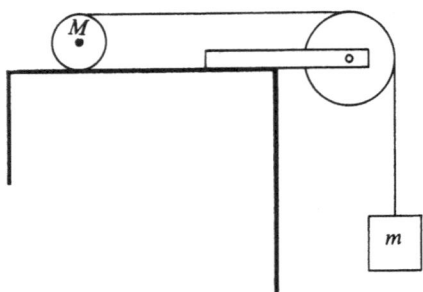

*4.31. In the figure, the wheel, which rolls horizontally without slipping, has negligible mass except for a particle of mass M at its center. A light string unwinds, passes over a light pulley, and terminates on a suspended particle of mass m. If the system is released at rest, by what height has m descended when its speed is v? (Express your answer in terms of M, m, g, v; neglect friction.)

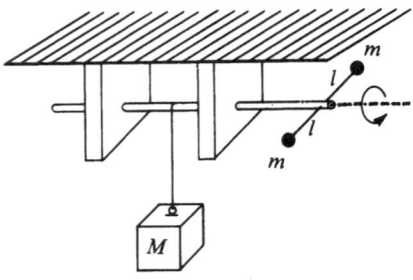

*4.32. In the frictionless device shown, a suspended mass M descends at negligible speed, unwinding an inextensible string from a thin shaft and causing two particles of equal masses m to whirl in a vertical plane. The light connecting bar has length $2l$. If the system is released from rest, find the speed of the whirling masses after M has gone down by a height h. (Express your answer in terms of M, m, g, l, h.)

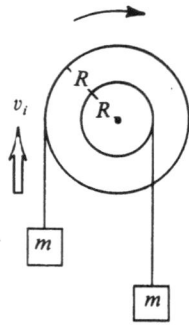

*4.33. In the system shown, the equal masses m are vertically suspended from two massless coupled pulleys. If the system is launched in such a way that the left mass rises at a speed v_i, how high will it get above its initial level? (The left string winds up, the right one unwinds; the pulley's radii are R and $2R$; neglect friction; express your answers in terms of m, g, v_i, R.)

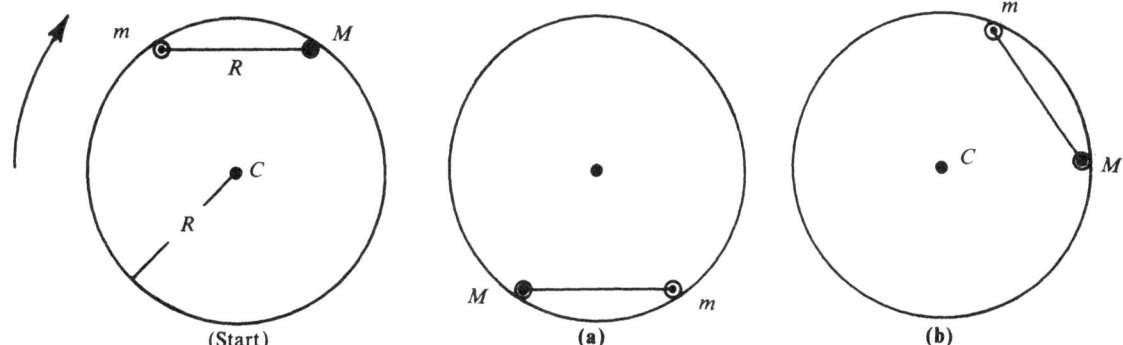

(Start)   (a)   (b)

4.34. Two particles, of masses m and M ($M>m$), are sliding as shown on a frictionless circular track of radius R (the center is at C). The light rigid connection between m and M also has length R. The track occupies a vertical plane. The system is released at rest when M is to the right of m. What is the particles' speed (a) when M is to the left of m? (b) when M is to the right of C? (Express your answer in terms of M, m, g, R.)

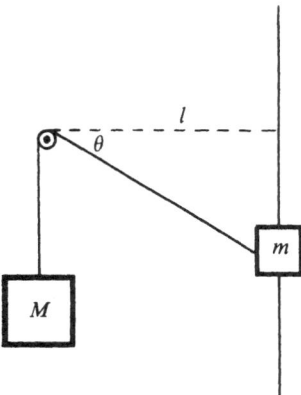

*4.35. In the system shown, the mass m is allowed to slide on a vertical track; the mass M hangs vertically below a small, light pulley at a distance l from the track; the string is light and inextensible. Let m be released from rest when its end of the string is at an angle θ to the horizontal, with m lower than the pulley. (a) Find m's speed when it reaches the level of the pulley. (Express your answer in terms of M, m, g, θ; neglect friction.) (b) Show that the problem has no solution unless $m/M \leqslant \tan(\theta/2)$.

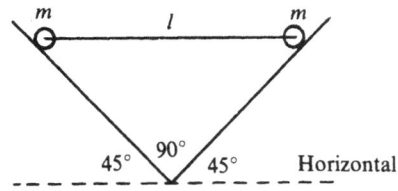

*4.36. Two particles of equal masses m are connected by a light rigid rod of length l, as shown. The system is initially at rest, in unstable equilibrium, on a V-shaped frictionless track in a vertical plane. A slight downward push is given to the left mass. What is its speed as it reaches the corner of the track?

Answers to True or False

1. True (e.g., under friction).
2. False.
3. True (kinetic energy increases).
4. True.
5. False (negative work on the object but zero work on the surface).
6. False (e.g., see item 5).
7. True (because v does).
8. False (external work is possible).
9. True.
10. False (energy has no direction).
11. False (they could be at 90° to one another).
12. True (power $\propto v$).
13. False.
14. True.
15. False.

Answers or Hints to Problems

4.1. (a) 1.9×10^{-2}; (b) 170; (c) 1.5×10^7.

4.4. 5×10^{-26} kg.

4.8. Note: It rebounds to a maximum potential energy (relative to the table) that is 20% less than when originally released. Why?

4.12. (b) Same answer as (a).

4.14. 3600.

4.15. 500 joules.

4.18. The drum's top moves twice as fast as its center. (c) 8 watts; (d) same answer as (c).

4.19. Note: $\mathcal{K} = 0$ where the motion reverses. Answer: 1.8 meters. (The 500 kg is irrelevant.)

4.21. 1 newton.

4.22. Use Eq. (4.C.19). Answer: 80 newtons.

4.30. The final speed of m is zero.

4.31. Note: M moves half as fast as m. Why?

4.32. The total potential energy of the two masses m is constant; why?

4.33. $5v_i^2/4g$.

4.35. M descends by a height $(l/\cos\theta) - l$; why? The final speed of M is zero; why?

(a) Answer: $(2gl)^{1/2}\left[\dfrac{M}{m}\left(\dfrac{1}{\cos\theta} - 1\right) - \tan\theta\right]^{1/2}$

(b) Consider the case where m barely makes it, meaning answer to (a) is 0.

4.36. $(2 - 2^{1/2})^{1/2}\,(gl)^{1/2}$.

CHAPTER 5

MOMENTUM

The search for conserved quantities is a favorite activity of physicists, because its rewards, in terms of understanding and predictive power, are so great. A modest glimpse of this was seen in the preceding chapter. There, conservation of energy enabled us to solve a whole new class of problems, sometimes featuring complicated systems.

In the present chapter a similar extension of our abilities will take place through another conservation law, that of the **momentum vector**.

For a single particle, momentum is defined as $m\mathbf{v}$, where m and $\mathbf{v}$ are the particle's mass and velocity. For a whole system of particles, the momentum is the vector sum of all the individual $m\mathbf{v}$. In the absence of any force, $\mathbf{v}$, as we know, is constant for a particle, and hence so is $m\mathbf{v}$. This is a trivial conservation law, which teaches us nothing new. What interests us here is a generalization to whole systems of *mutually interacting* particles.

In what follows we first establish, from Newton's laws of motion, that the *total momentum is conserved* in an isolated system; we later examine the **momentum transfer** into a system that is not isolated. These ideas will be applied to several topics:

1. Collisions;
2. The force exerted by a fluid stream, for example in a rocket engine;
3. The center of mass, in terms of which an object is legitimately represented as a particle;
4. In later chapters, the explanation of gas pressure as molecular impacts, and the phenomenon of radiation pressure.

Conservation of momentum not only follows from Newton's laws of motion; it is more fundamental than these laws. Today, even after the drastic modifications of Newton's laws brought about by modern physics, conservation of momentum is still intact.

A. Conservation of Momentum

i. Definition

The **momentum vector** of a particle (of mass m and velocity $\mathbf{v}$) is denoted by $\mathbf{p}$. It is defined as

$$\mathbf{p} = m\mathbf{v} \tag{5.A.1}$$

For a system of many particles (labeled 1, 2, 3,...), the **total momentum vector** is given by

$$\mathbf{p}_{\text{syst}} = m_1\mathbf{v}_1 + m_2\mathbf{v}_2 + \cdots = \sum_i m_i\mathbf{v}_i$$

$$\tag{5.A.2}$$

The relevant SI unit is the kg meter/sec.

As a one-particle illustration: a 0.50-kg pigeon flying at a speed of 4.0 meter/sec has a momentum of magnitude $p = (0.50)(4.0)$kg meter/sec $= 2.0$ kg meter/sec.

The vector $\mathbf{p}_{\text{syst}}$ fits our intuition for an "amount of motion." For example, if all speeds are doubled, $\mathbf{p}_{\text{syst}}$ is doubled; also, two identical particles traveling side by side have, together, twice the momentum of a single one.

Figure 5.1 illustrates the calculation of a system's momentum in a simple case.

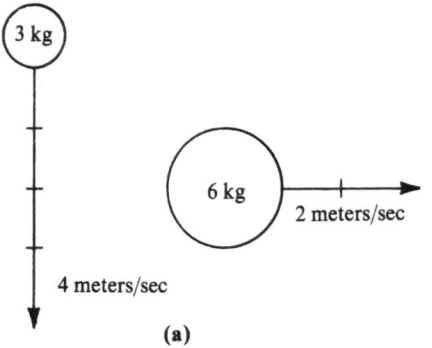

(a)

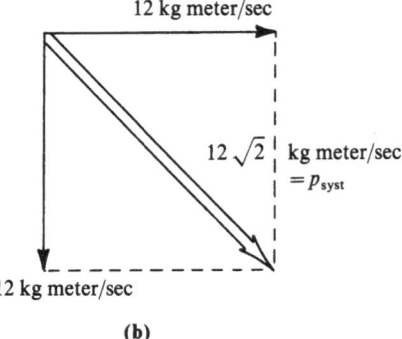

(b)

Figure 5.1. (a) Two particles shown with their masses and velocities. (b) The same particles' momentum vectors, and the system's combined momentum $\mathbf{p}_{\text{syst}}$.

Newton's Second Law, Revisited

This law acquires a particularly simple form if written in terms of **p**. We have

$$\mathbf{F}_{\text{tot}} = m\mathbf{a} = m \frac{d\mathbf{v}}{dt} = \frac{d}{dt}(m\mathbf{v})$$

(since m is constant), or

$$\boxed{\mathbf{F}_{\text{tot}} = \frac{d\mathbf{p}}{dt}} \qquad (5.A.3)$$

The total force on a particle equals the rate of change of its momentum.

We are now ready to verify that $\mathbf{p}_{\text{syst}}$ obeys a conservation law.

ii. Two Interacting Particles

Consider two particles, 1 and 2, of masses m_1 and m_2, moving with velocities $\mathbf{v}_1$ and $\mathbf{v}_2$. They exert a pair of action–reaction forces $\mathbf{F}_1$, $\mathbf{F}_2$ on each other, as shown in Fig. 5.2. (Therefore $\mathbf{v}_1$ and $\mathbf{v}_2$ are not necessarily constant.) We assume that *no external forces are acting on this system*, i.e., $\mathbf{F}_1$ and $\mathbf{F}_2$ are the only forces present.

It will now become apparent that the combined momentum of the system remains constant:

$$\frac{d\mathbf{p}_{\text{syst}}}{dt} = 0 \qquad (5.A.4)$$

where

$$\mathbf{p}_{\text{syst}} = \mathbf{p}_1 + \mathbf{p}_2 \qquad (5.A.5)$$

As a proof, take the time rate of change of (5.A.5):

$$\frac{d\mathbf{p}_{\text{syst}}}{dt} = \frac{d\mathbf{p}_1}{dt} + \frac{d\mathbf{p}_2}{dt}$$

$$= \mathbf{F}_1 + \mathbf{F}_2 = 0 \qquad (5.A.6)$$

by Newton's third law. Thus, although $\mathbf{p}_1$ and $\mathbf{p}_2$ may individually change, $\mathbf{p}_{\text{syst}}$ does not change.

This argument is so simple that it hides the power of our conclusion. We note that nothing is said about the exact nature of the mutual forces $\mathbf{F}_1$, $\mathbf{F}_2$ or about their behavior in time, which can be as complicated as we please. In particular, the total momentum is conserved even when those forces involve friction, as in Fig. 5.2b. (Mechanical energy is not conserved in such a case, as we recall from the preceding chapter.) To summarize: *the total momentum of a two-particle system is conserved when no external forces are present.*

iii. Many-Particle Systems

The two-particle result can easily be extended to systems of many particles, the argument being

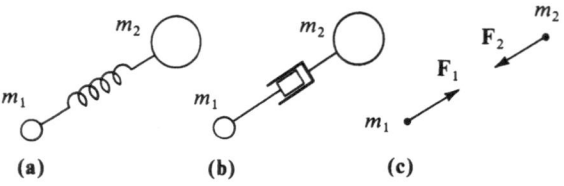

(a)   (b)   (c)

Figure 5.2. An isolated two-particle system. An arbitrary force is transmitted between them, for example by a spring (a), or a friction device (b). The action–reaction forces are shown in (c). The combined momentum is necessarily a constant vector.

essentially the same. Let us assume there are no external forces. This means that all the forces are exerted by the system's particles 1, 2, 3,... on one another. For example, the forces on particle 1 are just $\mathbf{F}_{2 \to 1}$ (exerted by particle 2), $\mathbf{F}_{3 \to 1}$ (exerted by particle 3), etc., as shown in Fig. 5.3.

Consider the momentum $\mathbf{p}_{\text{syst}}$ of the whole system,

$$\mathbf{p}_{\text{syst}} = \mathbf{p}_1 + \mathbf{p}_2 + \mathbf{p}_3 + \cdots$$

The rate of change of this vector is given by

$$\frac{d\mathbf{p}_{\text{syst}}}{dt} = \frac{d\mathbf{p}_1}{dt} + \frac{d\mathbf{p}_2}{dt} + \frac{d\mathbf{p}_3}{dt} + \cdots$$

$$= \mathbf{F}_1 + \mathbf{F}_2 + \mathbf{F}_3 + \cdots \tag{5.A.7}$$

According to (5.A.3), $\mathbf{F}_1$ is the total force on particle 1, etc. That is to say,

$$\mathbf{F}_1 = \mathbf{F}_{2 \to 1} + \mathbf{F}_{3 \to 1} + \cdots$$

etc. Thus the right-hand side of (5.A.7) is the sum of *all* the forces acting on *all* the particles.

A look at Fig. 5.3 shows that these forces can be grouped into action–reaction pairs. Equation (5.A.7) becomes

$$\frac{d\mathbf{p}_{\text{syst}}}{dt} = (\mathbf{F}_{2 \to 1} + \mathbf{F}_{1 \to 2})$$

$$+ (\mathbf{F}_{3 \to 1} + \mathbf{F}_{1 \to 3}) + \cdots \tag{5.A.8}$$

$$= 0 + 0 + \cdots \tag{5.A.9}$$

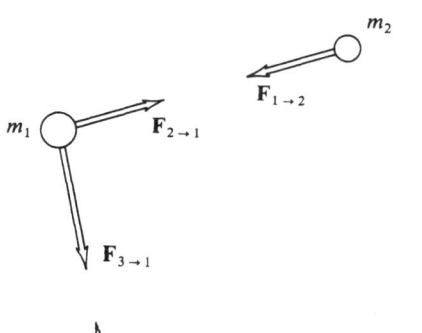

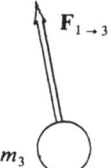

Figure 5.3. A system of many particles, of which only three are shown. All internal forces can be grouped into action–reaction pairs.

since each pair vanishes by virtue of Newton's third law. We conclude that $\mathbf{p}_{\text{syst}}$ does not change.

> The total momentum vector of an isolated system is conserved. $\tag{5.A.10}$

Here "isolated" means that there are no external forces.

iv. Nonconservation of Momentum

What happens if the system is not isolated? After the action–reaction pairing in Eq. (5.A.8) has been done, the external forces are left over. Instead of zero, Eq. (5.A.9) yields the sum, $\mathbf{F}_{\text{ext}}$, of these forces:

$$\boxed{\frac{d\mathbf{p}_{\text{syst}}}{dt} = \mathbf{F}_{\text{ext}}} \tag{5.A.11}$$

The rate of change of a system's total momentum equals the net *external* force acting on the system. $\tag{5.A.12}$

This allows practical calculations, since, although the internal forces of a system are often inaccessible to measurement, the external forces may be more easily found. (The objects of daily life are "systems" in the sense of this chapter.)

Since all internal forces cancel one another, "external force" is synonymous with "total force":

$$\mathbf{F}_{\text{ext}} = \mathbf{F}_{\text{tot}}$$

Equation (5.A.11) reads

$$\frac{d\mathbf{p}_{\text{syst}}}{dt} = \mathbf{F}_{\text{tot}} \tag{5.A.13}$$

which looks just like Newton's second law, (5.A.3), for a single particle. Thus, a *whole system can legitimately be represented as a particle*, at least as far as its momentum is concerned. By introducing the system's center of mass, in Sec. D of this chapter, we shall put the finishing touch on that argument.

B. Two-Body Collisions

A collision between two objects typically involves violent but short-lived forces about which little information is available; analysis might seem hopeless. Fortunately, conservation of momentum gives us simple and powerful predictions when applied to the system *composed of both interacting bodies.*

The outcome of a collision depends on what happens to the initially available kinetic energy. Some of it may be used up to heat the colliding objects, to break or permanently deform them, or to make them spin. Therefore, in order to solve a collision problem, some information is needed about the fate of the energy during that process.

Two extreme cases, in which simple conclusions are possible, will be discussed here. They are the **completely inelastic** case, in which two objects stick together after impact, and the **completely elastic** case, in which the total kinetic energy is the same before and after impact. Most real-life collisions are of an intermediate (that is, partially inelastic) nature.

Why does a collision conserve total momentum *in practice*? What happens is that the mutual force lasts a very short time but is very large. Thus, *during collision, the external forces* (gravity, air friction, etc.) *are negligible* compared to the mutual force.

To summarize: in what follows we avoid a detailed description of the collision forces, but we take advantage of the fact that, *just before* and *just after* collision, the total momentum is the same.

i. Completely Inelastic Collisions

This case is best treated by an example, which suggests a simple way to measure the speed of a bullet.

Example 5.1. Firing a Bullet into a Block. A bullet of mass $m = 3$ grams is fired horizontally into an initially stationary block of known mass $M = 3$ kg, which rests on a horizontal frictionless plane. After the impact, the block, with the imbedded bullet, is seen to move at a speed $v_f = 0.5$ meter/sec. (a) What was the speed v_i at which the bullet was fired? (b) How much mechanical energy is lost?

In this case, all momentum vectors are and remain horizontal. The conservation of total momentum states that

$$mv_i \;\; + \;\;\; 0 \;\; = \;\; (M+m)\,v_f$$

$$\qquad\uparrow \qquad\qquad \uparrow \qquad\qquad\qquad \uparrow$$

Initial momentum of bullet · Initial momentum of block · Final momentum of combined masses

$$(5.B.1)$$

a. Final speed: From Eq. (5.B.1) we find

$$v_i = \frac{M+m}{m}\, v_f \qquad (5.B.2)$$

$$= \left(\frac{3+0.003}{0.003}\right)(0.5)\,\frac{\text{meters}}{\text{sec}}$$

$$= \underline{500\,\frac{\text{meters}}{\text{sec}}}$$

Note: If M is much larger than m, as it is here, we have approximately $M + m \approx M$, and (5.B.2) becomes

$$v_i \approx \frac{M}{m}\, v_f \qquad (5.B.3)$$

Measuring the slow final speed v_f provides a practical way of measuring the bullet's speed v_i without elaborate equipment; in this connection see Problem 5.19.

b. Loss of mechanical energy: There is no potential energy to worry about. We have, for the initial and final kinetic energies,

$$\mathscr{K}_i = \tfrac{1}{2}mv_i^2 = \tfrac{1}{2}(0.003)(500)^2 \text{ joules}$$

$$= 380 \text{ joules}$$

$$\mathscr{K}_f = \tfrac{1}{2}(m+M)\,v_f^2 \approx \tfrac{1}{2}Mv_f^2$$

$$= \tfrac{1}{2}(3)(0.5)^2 \text{ joule} = 0.38 \text{ joule}$$

Thus, almost all the mechanical energy is lost, amounting to

$$\mathscr{K}_i - \mathscr{K}_f \approx \underline{380 \text{ joules}}$$

(The bullet, and to some extent the block, are observed to become hotter: the lost mechanical energy has been converted to "internal energy," as we shall see later.)

ii. Explosions

An explosion can be thought of as a completely inelastic collision in reverse. For example, consider a rifle, free to move horizontally. When the rifle fires a bullet, some nonmechanical (chemical) energy stored in the explosive is converted into kinetic energy of the final objects (the recoiling rifle and the bullet). Their combined horizontal momentum, however, is conserved; see Problem 5.10.

iii. Completely Elastic Collisions (General Description)

Two objects may collide in such a way that their combined kinetic energy is found unchanged after the event. Such collisions are called **completely elastic**, or just **elastic**. (We assume there is no appreciable spinning motion imparted by the collision; spinning would take up some of the kinetic energy.)

What happens *during* such an impact may be summarized as follows. The objects are momentarily compressed by large contact forces, and store elastic potential energy, as a spring would. The combined kinetic energy is correspondingly reduced. As the objects separate, they recover their original shape, and the potential energy is converted back into kinetic energy.

This scenario is not universal, however. In some important cases, no contact or deformation occurs at all; the two objects (for example, electrically charged particles) interact at a distance.

To a physicist, the real importance of elastic collisions arises from the behavior of atoms and subatomic particles, as well as of simple molecules. A wide range of collisions leave such particles unbroken and undeformed; also, friction does not exist for them individually. Therefore their collisions are, in many cases, elastic. Some important properties of gases can be understood on this basis; see Chap. 11.

In general, a collision between two particles may be represented as in Fig. 5.4: among the final and initial velocities, no two are aligned in the same direction.

We next demonstrate the basic feature of elastic collisions:

> The relative speed of two particles is the same before and after they collide elastically. (5.B.4)

We can see this most directly by considering first, as a special case, a system of two particles whose combined momentum vanishes, as illustrated in Fig. 5.5.

A System with Zero Momentum

Let the two particles have masses m, M, and corresponding velocities $\mathbf{v}, \mathbf{V}$. Zero combined momentum means, before impact, $m\mathbf{v}_i = -M\mathbf{V}_i$, or, in magnitude,

$$mv_i = MV_i \qquad (5.B.5)$$

Similarly, after impact, we have

$$mv_f = MV_f \qquad (5.B.6)$$

The combined kinetic energy, before and after impact, is

$$\mathscr{K}_i = \tfrac{1}{2}mv_i^2 + \tfrac{1}{2}MV_i^2 \qquad (5.B.7)$$

$$\mathscr{K}_f = \tfrac{1}{2}mv_f^2 + \tfrac{1}{2}MV_f^2 \qquad (5.B.8)$$

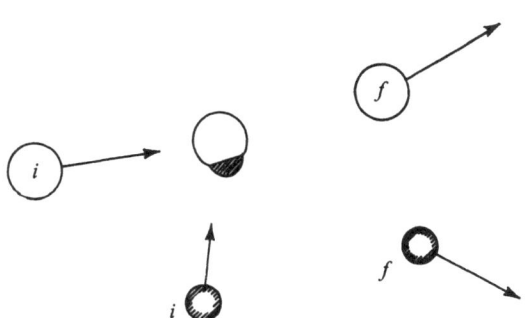

Figure 5.4. A typical two-particle collision. The arrows indicate their velocities. The labels *i* or *f* refer to the situations before or after impact, respectively. The particles are also shown, momentarily deformed, during the collision (which may be elastic).

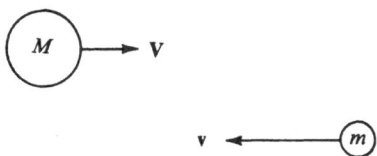

Figure 5.5. Two particles whose combined momentum vanishes. Since $M\mathbf{V} = -m\mathbf{v}$, the two velocity vectors must be pointing in opposite directions. (The typography should not mislead us: in this picture $v > V$.)

One way to have $\mathcal{K}_i = \mathcal{K}_f$ is to take

$$v_i = v_f, \qquad V_i = V_f \qquad (5.B.9)$$

Is this the only way? Suppose, for example, that, for a given v_i and V_i, we made v_f larger. Then, by Eq. (5.B.6), we have to make V_f larger as well. As a result, $\mathcal{K}_f$ would be larger than $\mathcal{K}_i$. It is clear, therefore, that Eq. (5.B.9) is the only possibility: *each speed is left unchanged by the impact*.

Some things do change, however: (1) the speed of approach turns into an (equal) speed of recession; (2) the line of travel is in general not the same before and after impact; see Fig. 5.6. Before the collision, the *relative* speed of approach is $v + V$; after the collision, the relative speed of recession is also $v + V$. (Remember that both v and V are speeds and therefore positive.)

Systems with Nonzero Momentum

In general, a two-particle system will have a combined momentum

$$\mathbf{p}_{syst} = m\mathbf{v}_i + M\mathbf{V}_i = m\mathbf{v}_f + M\mathbf{V}_f \quad (5.B.10)$$

which is different from zero. Nevertheless, we can convince ourselves that, even in such a case, the relative speed is still unchanged by the elastic collision:

$$\boxed{|\mathbf{v}_f - \mathbf{V}_f| = |\mathbf{v}_i - \mathbf{V}_i|} \qquad (5.B.11)$$

No new calculation is needed. Rather than discussing an entirely new system, let us look, from the point of view of a moving observer $\mathbb{O}$, at what we, who are stationary, consider to be a zero-momen-

tum case. Suppose observer $\mathbb{O}$ has a velocity $\mathbf{u}$. Then any velocity seen by us as $\mathbf{v}$ or $\mathbf{V}$ will be seen by $\mathbb{O}$ as $\mathbf{v} - \mathbf{u}$ or $\mathbf{V} - \mathbf{u}$. Therefore $\mathbb{O}$ will see *the same relative velocity* as we do:

$$(\mathbf{v} - \mathbf{u}) - (\mathbf{V} - \mathbf{u}) = \mathbf{v} - \mathbf{V}$$

and hence also the same relative speed $|\mathbf{v} - \mathbf{V}|$ as we do. We conclude that, for observer $\mathbb{O}$ as well as for us, the relative speed between the two particles is left unchanged by the impact. Since $\mathbb{O}$ can choose his own velocity $\mathbf{u}$ as he wishes, there is no restriction on the combined momentum he observes. [That momentum is $m(\mathbf{v} - \mathbf{u}) + M(\mathbf{V} - \mathbf{u}) = -(m + M)\mathbf{u}$.] Hence the zero-momentum restriction on our argument can be dropped.

iv. Elastic Head-on Collisions

A two-particle collision in which all initial and final motions occur along the same line will be referred to as a **head-on collision**. (Note: The definition includes "rear-end collisions.") Consider a pair of masses m, M moving along a line and having the respective velocities v, V, as shown in Fig. 5.7. It is easiest to think of both v and V as being positive ("to the right"), both before and after collision. A negative result must then be interpreted as a leftward motion.

Let us solve the following problem. If the initial velocities v_i, V_i are given, what are the final velocities v_f, V_f after a completely elastic collision?

Conservation of momentum (along the line of motion) gives

$$mv_f + MV_f = mv_i + MV_i \qquad (5.B.12)$$

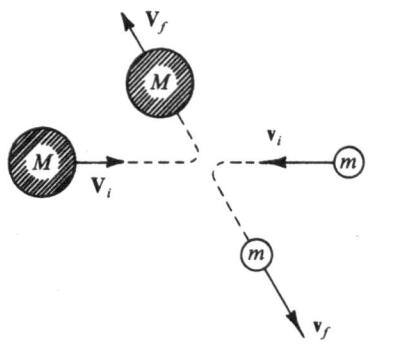

Figure 5.6. An elastic collision with zero total momentum.

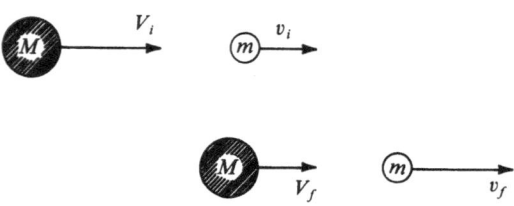

Figure 5.7. Two particles undergoing an elastic head-on collision (initial and final configurations). Note that, if the directions of motion are as shown here, we must have $V_i > v_i$ (otherwise no collision will occur), and $v_f > V_f$ (the particles are moving away from each other after collision).

Conservation of energy gives

$$\tfrac{1}{2}mv_f^2 + \tfrac{1}{2}MV_f^2 = \tfrac{1}{2}mv_i^2 + \tfrac{1}{2}MV_i^2 \quad (5.B.13)$$

In principle, the two equations (2.B.12), (2.B.13) can be solved for the two unknowns v_f, V_f. Equivalently, and much more simply, we can make use of the fact that the relative speed $|V - v|$ is unchanged by the collision. This means, according to Fig. 5.7, that

$$v_f - V_f = V_i - v_i \quad (5.B.14)$$

The set of equations (5.B.12), (5.B.14) is an easy one to solve for v_f and V_f, with the result

$$v_f = \frac{(m - M)\, v_i + 2MV_i}{m + M}$$
$$V_f = \frac{(M - m)\, V_i + 2mv_i}{m + M} \quad (5.B.15)$$

As well as yielding velocities, these formulas lend themselves to the calculation of masses, if enough velocities are known.

Example 5.2. A Mass Determination. A hydrogen atom, of mass $m = 1.7 \times 10^{-27}$ kg, collides head-on and elastically with another atom of unknown mass M, which is initially at rest. The hydrogen atom bounces back at 1/2 its original speed. Determine M.

The first equation (5.B.15), with $v_f = -\tfrac{1}{2}v_i$ and $V_i = 0$, becomes

$$-\frac{1}{2}v_i = \frac{(m - M)\, v_i}{m + M}$$

from which v_i can be factored out. Solving for M, we obtain

$$M = 3m = (3)(1.7 \times 10^{-27} \text{ kg}) = \underline{\underline{5.1 \times 10^{-27} \text{ kg}}}$$

Special Cases

Formulas (5.B.15) are particularly useful in special cases of collisions such as the following.

Equal Masses. If $m = M$, Eqs. (5.B.15) become

$$v_f = V_i, \qquad V_f = v_i$$

> In an elastic head-on collision, two equal masses simply exchange their velocities.

$$(5.B.16)$$

In a classic piece of billiard ball behavior, shown in Fig. 5.8, ball A, moving at speed v, hits a stationary ball B head-on. After collision, B moves at speed v in the same direction, while A is stationary.

Very Unequal Masses. Suppose M is much larger than m. Then Eqs. (5.B.15) can be approximated by

$$v_f \approx \frac{-Mv_i + 2MV_i}{M} = -v_i + 2V_i \quad (5.B.17)$$

$$v_f \approx \frac{MV_i}{M} = V_i \quad (5.B.18)$$

This last equation shows that the very massive particle hardly changes its velocity, a fact to be expected. *If the light particle is originally at rest* $(v_i = 0)$, then Eq. (5.B.17) has the interesting implication *that the massive particle imparts about twice its own velocity to the light one in an elastic head-on collision. If the very massive particle is originally at rest* $(V_i = 0$, v_i negative), then Eqs. (5.B.17) and (5.B.18) give

$$v_f \approx -v_i \quad (5.B.19)$$

$$V_f \approx V_i \quad (5.B.20)$$

(*The massive particle remains at rest; the light particle bounces back with its original speed. The*

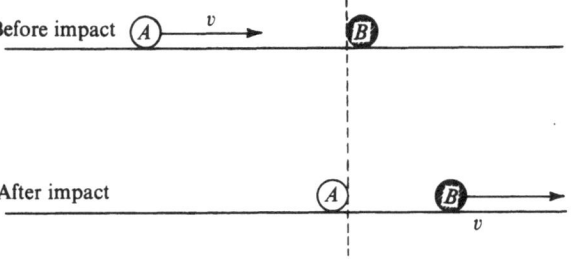

Figure 5.8. An initially moving billiard ball (A) impinging on an initially stationary one (B).

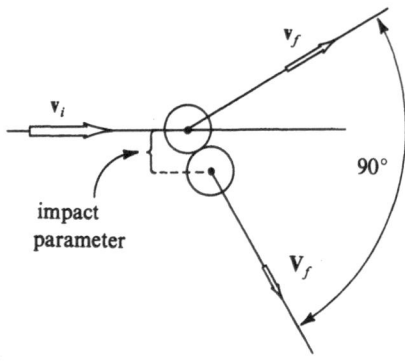

Figure 5.9. An off-center elastic collision between two equal particles. In nuclear laboratories, the 90° angle is a well-measured feature of, for example, elastic proton–proton collisions.

momentum transfer is $2mv_i$, but essentially all the kinetic energy remains in the light particle.)

v. A Right-Angle Emergence

A curious phenomenon, well known to billiard players and to nuclear physicists, is shown in Fig. 5.9. If an incoming particle strikes, elastically but off-center, *a stationary target of the same mass*, the two outgoing trajectories form a right angle. (Nevertheless, the emerging directions themselves are difficult to predict: they depend on the extent to which the incoming trajectory misses the exact center of the target; see "impact parameter" in the figure.)

To explain the right angle, it is convenient to look first at an elastic equal-mass collision with zero total momentum, Fig. 5.10. The initial velocities are then v_i and $-v_i$; the final velocities are v_f and $-v_f$; these four vectors all have the same magnitude; see Eq. (5.B.9).

We now need an observer who will see a stationary target. That observer must himself move with a

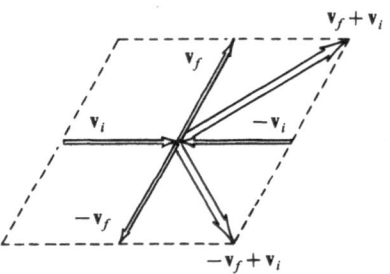

Figure 5.10. Geometric proof of perpendicularity between the final velocity vectors (wide arrows).

velocity $-v_i$ since he accompanies the target. He consequently sees final velocity vectors $v_f + v_i$ and $-v_f + v_i$. As the figure shows, these two vectors are along the diagonals of a rhombus, and hence are orthogonal, as claimed.

C. Force from a Jet

A steady jet of water, striking a turbine blade, pushes the latter with a steady force; a flaming gas jet, issuing from a rocket engine, can exert a steady force on that engine. These "gradual collisions" can be analyzed in terms of momentum transfers.

Bombardment by a stream of particles provides another application of these ideas. The pressure exerted by a gas on a piston is due to the repeated impacts of the gas molecules. This huge number of small "bumps" can fortunately be approximated by a steady overall force; the idea will be exploited in a later chapter (Chap. 11, Molecules and Gases).

The force exerted by a continuous stream of fluid is examined in the following example.

Example 5.3. The Thrust of a Rocket Engine. A rocket engine, kept at rest on a test bench, expels hot exhaust gas at a rate (mass per unit time) $\Delta m/\Delta t = 0.5$ kg/sec; the speed of the gas is $V = 200$ meter/sec. Find the **thrust** F (=force of gas on engine).

Consider a mass Δm of gas, expelled (say, in the x direction) during a time Δt. This is the system whose total momentum will be examined. It undergoes a force F_{ext}, which is the reaction to F. Hence, for the magnitudes, we have $F = F_{ext}$. The x component of Eq. (5.A.11) gives

$$F = \frac{\Delta p}{\Delta t} \qquad (5.C.1)$$

(Finite changes Δp, Δt can be used in a steady-state situation.) We now compute the right side of (5.C.1). The mass Δm of gas, starting at rest and ending at speed V, suffers a momentum change

$$\Delta p = p_f - p_i = (\Delta m)\,V - 0 \qquad (5.C.2)$$

so that

$$F = V \frac{\Delta m}{\Delta t} \qquad (5.C.3)$$

(For a variable rate of fuel burning, the expression

$$F = V \, dm/dt \qquad (5.C.4)$$

would have to be used.) Equation (5.C.3) gives, in this case,

$$F = (200)(0.5) \text{ newtons} = \underline{\underline{100 \text{ newtons}}}$$

$$(5.C.5)$$

Specific Impulse of a Flying Rocket

It would be unfortunate if the thrust formulas (5.C.3), (5.C.4) held true only in the test-bench case. Fortunately, they apply also to a moving rocket, provided V denotes the exhaust speed *relative to the engine*; V, which is usually not affected by the rocket's motion, is called the engine's **specific impulse** because it also happens to equal the magnitude of impulse (= net change in momentum) given to a unit exhaust mass:

$$\boxed{\text{Specific impulse} = \frac{\Delta p}{\Delta m} = V} \qquad (5.C.6)$$

according to Eq. (5.C.2). Each kind of rocket fuel, when optimally utilized, corresponds to a well-defined specific impulse. (In the formula $\Delta p/\Delta m$, the Δm must take into account all the combustion ingredients.) The fuels of modern astronautics can reach specific impulses of 3000 meters/sec or so, for example with liquid oxygen and liquid hydrogen as the reagents.*

A rocket engine works best *without* a surrounding atmosphere, which only impedes the expulsion of the jet gases.

Impinging Jets

The rocket situation of Example 5.3 can be visualized in reverse. Let a continuous stream of

---

*For Earth-based applications, (5.C.6) is often formulated in terms of exhaust weight rather than mass; the specific impulse is then a time rather than a speed.

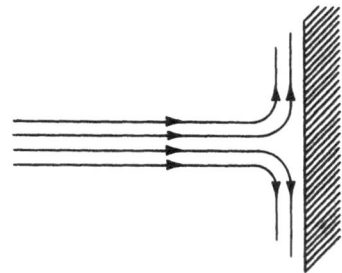

Figure 5.11. A fluid jet impinging on a stationary target (completely inelastic case).

mass Δm impinge at speed V on a fixed wall, and perpendicularly to it, during every time interval Δt. Assume this mass does not bounce, but adheres to or flows along its target (completely inelastic impact); see Fig. 5.11. What is the resulting force F on the wall? By the same argument as before, we have

$$\boxed{F = \frac{\Delta p}{\Delta t} = V \frac{\Delta m}{\Delta t}} \qquad (5.C.7)$$

(stream against wall, completely inelastic case).

D. The Center of Mass

This section fulfills a long-standing promise: to define the point whose motion best represents that of a whole system. For many purposes, the system can then be considered to have its total mass concentrated at that point, called the **center of mass**. This concept has many uses, some of which will be explored here and some in Chap. 7 (Rotation and Torque).

i. Definition

For any set of particles, one can define an average position; the center of mass, to be denoted by $\mathbb{C}$, is a natural modification of that idea.

First consider three identical particles, with x coordinates x_1, x_2, x_3. The x coordinate of their center of mass is defined as an ordinary average,

$$x_{\mathbb{C}} = \frac{x_1 + x_2 + x_3}{3} \qquad (5.D.1)$$

Next, suppose particles 2 and 3 occupy the same position. Then x_C becomes

$$x_C = \frac{x_1 + x_2 + x_2}{3} = \frac{(1)(x_1) + (2)(x_2)}{1 + 2} \quad (5.D.2)$$

as illustrated in Fig. 5.12. We note that the two combined particles can be considered as *a single particle of double mass*.

The general definition of x_C for a set of n particles with masses $m_1, m_2,..., m_n$ is a "weighted average" directly inspired from Eq. (5.D.2):

$$x_C = \frac{m_1 x_1 + m_2 x_2 + \cdots + m_n x_n}{m_1 + m_2 + \cdots + m_n} \quad (5.D.3)$$

or

$$x_C = \frac{\sum_i m_i x_i}{m_{syst}} \quad (5.D.4)$$

m_{syst} being the system's total mass. (Note how the units of mass cancel in this definition.) In a similar way, the y and z coordinates of the center of mass are defined as

$$y_C = \frac{\sum_i m_i y_i}{m_{syst}} \quad (5.D.5)$$

$$z_C = \frac{\sum_i m_i z_i}{m_{syst}} \quad (5.D.6)$$

An elegant and convenient way of summarizing this information consists of assigning to each par-

ticle a position vector, $\mathbf{r}_1$ for particle 1, $\mathbf{r}_2$ for particle 2, etc., as well as a position vector $\mathbf{r}_C$ for $\mathbb{C}$, as in Fig. 5.13. We then have

$$\mathbf{r}_C = \frac{m_1 \mathbf{r}_1 + m_2 \mathbf{r}_2 + \cdots}{m_1 + m_2 + \cdots} \quad (5.D.7)$$

or

$$\mathbf{r}_C = \frac{\sum_i m_i \mathbf{r}_i}{m_{syst}} \quad (5.D.8)$$

The rectangular components of (5.D.8) are just Eqs. (5.D.4), (5.D.5), (5.D.6).

Continuous Objects

The ordinary objects of daily life consist, in practice, of infinitely many infinitesimal particles. Here Eq. (5.D.8) must be interpreted as an integral:

$$\mathbf{r}_C = \frac{\int \mathbf{r} \, dm}{m_{syst}} \quad (5.D.9)$$

where $\mathbf{r}$ is the position vector of a typical infinitesimal mass dm, and the integral generally ranges over three dimensions. To evaluate the three integrals in (5.D.9) is a good exercise in applied mathematics, but it will receive no emphasis in this book. In what follows we illustrate some cases that do not require integration.

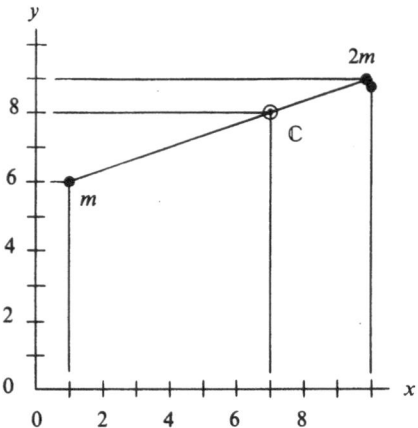

Figure 5.12. The construction (5.D.2) for two particles, of masses m and $2m$. In this two-dimensional illustration, that construction is applied to both the x and the y coordinate. The center of mass $\mathbb{C}$ is two-thirds of the way between the simple and double particles.

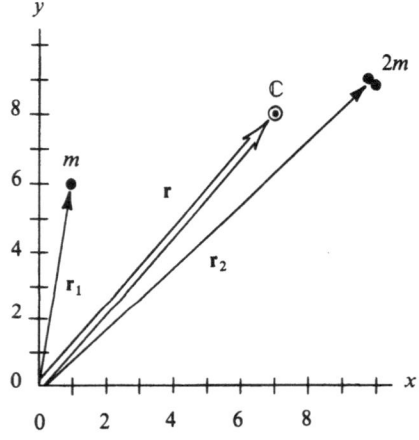

Figure 5.13. Position vectors for the example of Fig. 5.12.

ii. Finding the Center of Mass by Calculation

Given the positions and masses of a system of particles, where is its center of mass $\mathbb{C}$? In calculating the answer, three principles are particularly useful. We state them here without proof.

1. In relation to the system itself, the position of $\mathbb{C}$ is unique, and does not depend on the choice of coordinate axes. (For example, $\mathbb{C}$ for a set of two equal particles is always halfway between them.) Therefore *we are free to choose any convenient axes* in applying Eqs. (5.D.4), (5.D.5), (5.D.6).
2. For a symmetric system, $\mathbb{C}$ is at the center (or axis, or plane) of symmetry. This rule is especially useful when applied to *continuous objects* such as uniform straight bars, spheres, etc.
3. Suppose we must find $\mathbb{C}$ for a system consisting of several subsystems (not necessarily single particles). Then each subsystem may be thought of as concentrated at *its* own center of mass.

Some examples follow.

Example 5.4. A Four-Particle System. Four particles, with masses 1 kg, 1 kg, 2 kg, 6 kg, occupy the corners of a square of side 1 meter, as shown in Fig. 5.14a. Locate $\mathbb{C}$.

The system is symmetric about the dashed diagonal; therefore $\mathbb{C}$ lies on that diagonal. The subsystem with masses 1 kg, 1 kg can be concentrated at the center $\mathbb{O}$ of the square. The resulting system is shown in part (b) of the figure. Choosing an x axis along the dashed diagonal, with origin at $\mathbb{O}$, we have

$$x_{\mathbb{C}} = \frac{(-1/\sqrt{2})(2) + (0)(1+1) + (1/\sqrt{2})(6)}{2 + (1+1) + 6} \text{ meter}$$

$$= \underline{\underline{0.28 \text{ meter}}}$$

Thus, $\mathbb{C}$ lies 0.28 meter off center, towards the 6-kg corner.

Example 5.5. A Two-Bar System. Two equal uniform bars form a V-shape; see Fig. 5.15. Where is the center of mass $\mathbb{C}$ of that structure?

The figure shows the bars' individual centers of mass, $\mathbb{C}_A$ and $\mathbb{C}_B$, at the center of each bar. If the bars are assumed concentrated at these respective points, we obtain as the next step the point $\mathbb{C}$, <u>halfway between $\mathbb{C}_A$ and $\mathbb{C}_B$</u>.

Further Illustrations

Figure 5.16 shows the center of mass for some simple shapes. Readers with some proficiency in integral calculus may wish to check those results by using prescription (5.D.9).

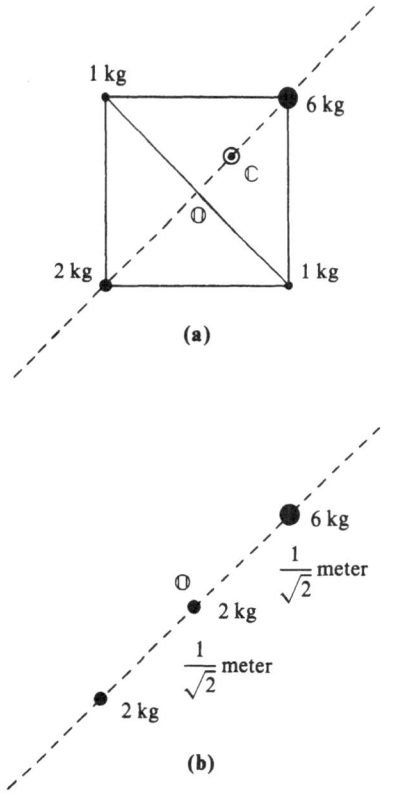

(a)

(b)

Figure 5.14. Finding $\mathbb{C}$ for a system of four particles.

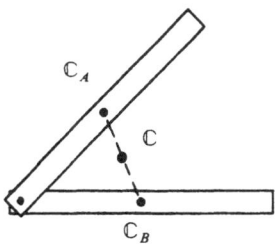

Figure 5.15. To locate the center of mass $\mathbb{C}$ of a two-piece object, each piece may be considered concentrated at its own center of mass $\mathbb{C}_A$, $\mathbb{C}_B$.

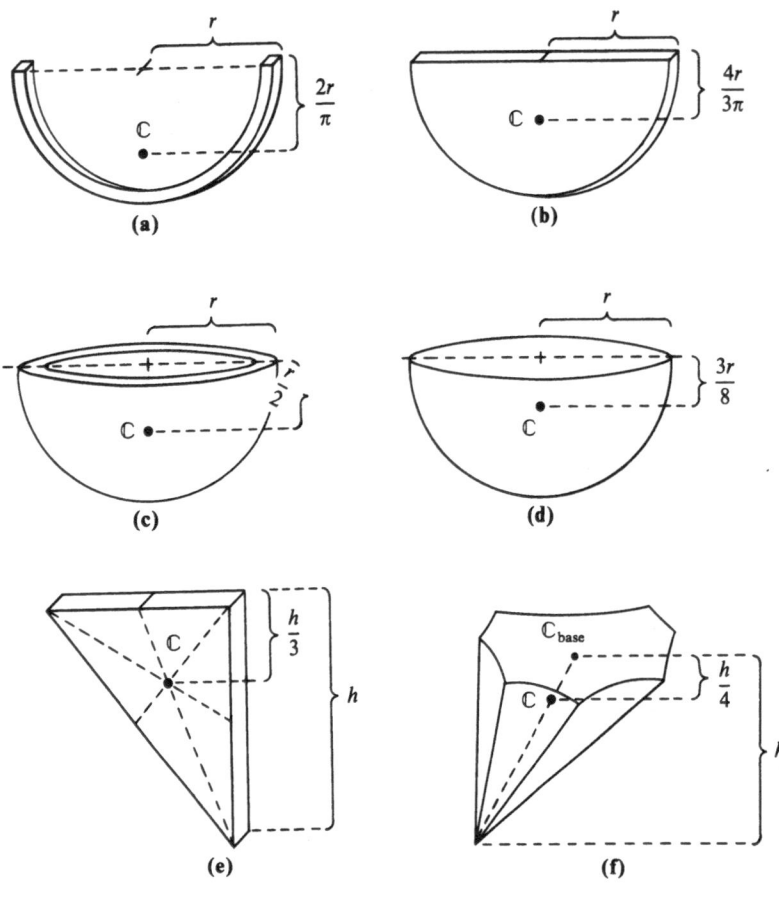

Figure 5.16. Position of the center of mass C for some partially asymmetric objects. Uniformity of the mass distribution is assumed. (a) Half of a thin circular hoop; (b) half of a thin circular disk; (c) thin hemispherical shell; (d) solid hemisphere; (e) flat triangular plate with arbitrary angles: C is at the intersection of the three medians; h is an altitude. [Note. A median joins a vertex with the midpoint of the opposite side. Length of a median: if a, b, c, are the sides and l is the median on a, then $l^2 = \frac{1}{2}(b^2 + c^2) - \frac{1}{4}a^2$]; (f) solid cone with a plane base of arbitrary shape; h is the altitude; C_{base} is the center of mass of the base, considered as a uniform flat plate.

iii. Motion of the Center of Mass

Science cannot be conducted through definitions alone, no matter how elegant. What is the center of mass good for? Here we demonstrate that C *moves according to Newton's law of acceleration for a single particle.*

Let $\mathbf{F}_{tot}$ be the total **external** force exerted on a system (forces exerted by one part of the system on another are correctly ignored in calculating $\mathbf{F}_{tot}$). Also, let m_{syst} be the mass of the whole system. Then Newton's second law,

$$\boxed{\mathbf{F}_{tot} = m_{syst}\mathbf{a}_C} \qquad (5.D.10)$$

is valid for the acceleration $\mathbf{a}_C$ of the center of mass. Just as in the single-particle case, where $\mathbf{F} = m\mathbf{a}$ is the time derivative of $\mathbf{p} = m\mathbf{v}$, Eq. (5.D.10) is the time derivative of

$$\boxed{\mathbf{p}_{syst} = \mathbf{m}_{syst}\mathbf{v}_C} \qquad (5.D.11)$$

Here $\mathbf{p}_{syst}$ is the total momentum of the system, and $\mathbf{v}_C$ is the velocity of the center of mass; we have $\mathbf{a}_C = d\mathbf{v}_C/dt$. These statements are proved in Note (ii) at the end of this chapter.

Application to Free Fall

The total external force on a freely falling system of particles 1, 2,... is

$$\mathbf{F}_{tot} = m_1\mathbf{g} + m_2\mathbf{g} + \cdots = m_{syst}\mathbf{g} \qquad (5.D.12)$$

where m_1, m_2,... are the individual masses. Therefore, from (5.D.11), the motion of the center of mass obeys

$$m_{syst}\mathbf{g} = m_{syst}\mathbf{a}_C$$

or

$$\boxed{\mathbf{a}_C = \mathbf{g}} \qquad (5.D.13)$$

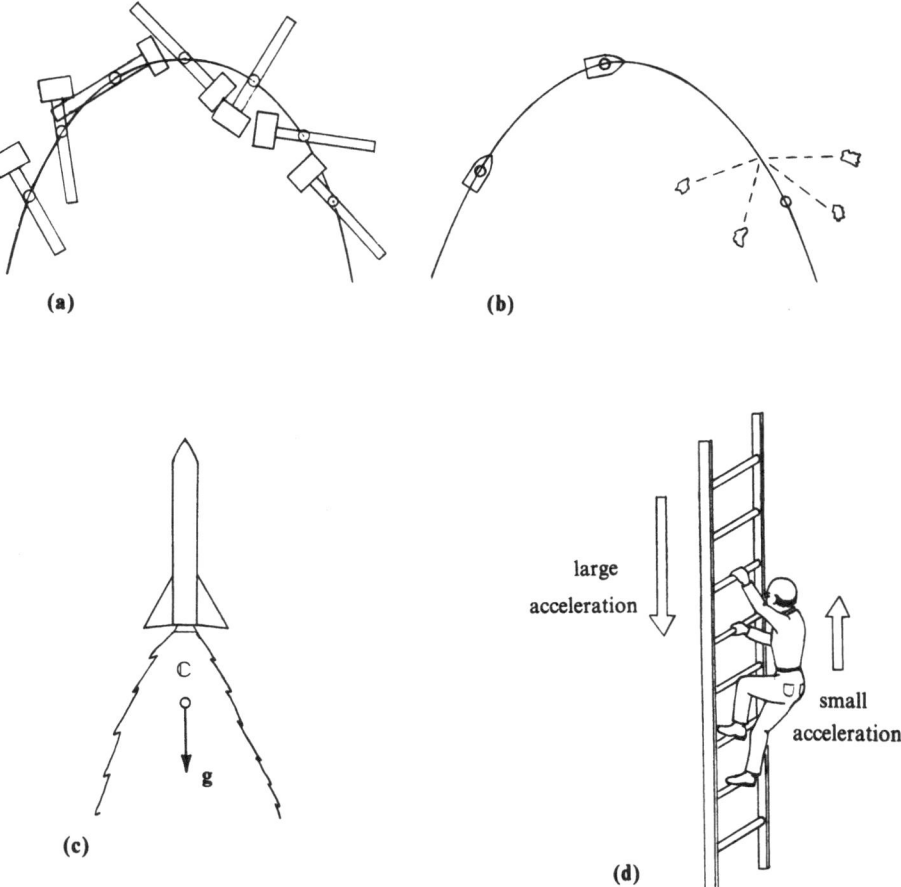

(a)

(b)

(c)

large
acceleration

small
acceleration

(d)

Figure 5.17. (a) The center of mass (small circle) follows a free-fall orbit. (b) The center of mass of an exploding shell continues on its free-fall orbit as if nothing were happening. As soon as a single fragment touches ground, the orbit of the center of mass is disturbed. (c) A rocket rising in vacuum. The center of mass of the combined rocket and trail of exhaust gas undergoes free fall: the combined system does not escape the laws of gravity. (d) An unsupported ladder can *in principle* be climbed successfully; although climber and ladder have a combined center of mass that is subject to ordinary free fall. The climber "running out of ladder" is analogous to a rocket running out of fuel.

The center of mass falls like a particle. In this sense complicated systems, with spin, moving parts, and internal interactions, are seen to satisfy the equivalence principle exactly. Figure 5.17 shows illustrations of that fact.

Example 5.6. An Exploding Shell. A gun fires a shell of mass $m = 75$ kg with muzzle speed $v_0 = 680$ meters/sec in a direction $\theta = 60°$ above horizontal, as illustrated in Fig. 5.18. The shell explodes in flight, and its two fragments, of masses $m_1 = 50$ kg and $m_2 = 25$ kg, land *simultaneously* at distances x_1 and x_2 horizontally and in a straight line from the gun. If $x_1 = 30$ km, find x_2. (Neglect air resistance and the mass of the explosive charge.)

When the fragments reach the ground, let their center of mass have coordinate X, equal to that of an intact projectile. We have

$$X = \frac{m_1 x_1 + m_2 x_2}{m} \tag{5.D.14}$$

giving

$$x_2 = \frac{mX - m_1 x_1}{m_2} \tag{5.D.15}$$

Here X is found by the methods of Chap. 2 to be

$$X = \frac{2v_{0x}v_{0z}}{g} = \frac{2v_0^2 \sin\theta \cos\theta}{g} \tag{5.D.16}$$

$$= \frac{(2)(680)^2 (\sin 60°)(\cos 60°)}{10} \text{ meters} = 40 \text{ km}$$

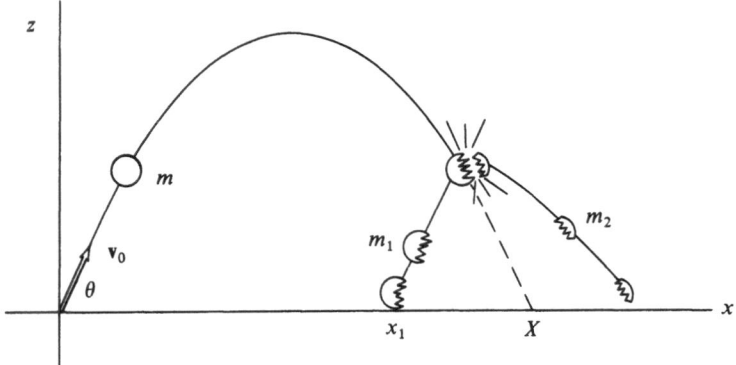

Figure 5.18. In this time sequence, a gunshell breaks up into two unequal pieces.

so that, from (5.D.15),

$$x_2 = \frac{(75)(40) - (50)(30)}{25} \text{ km} = \underline{\underline{60 \text{ km}}}$$

iv. Potential Energy and the Center of Mass

As far as its gravitational potential energy is concerned, an extended object may conveniently be treated as if it were wholly concentrated at its center of mass. That is to say, the gravitational potential energy $\mathcal{U}_{\text{syst}}$ of an extended system is given by

$$\boxed{\mathcal{U}_{\text{syst}} = m_{\text{syst}} g z_C} \qquad (5.D.17)$$

where m_{syst} is the total mass, g is the acceleration of gravity, and z_C is the vertical coordinate of the center of mass with respect to any fixed reference level.

The proof just consists of adding the individual potential energies $\mathcal{U}_1, \mathcal{U}_2,...$ of the system's particles, which have masses $m_1, m_2,...$ and vertical coordinates $z_1, z_2,...$. We have

$$\mathcal{U}_{\text{syst}} = m_1 g z_1 + m_2 g z_2 + \cdots$$
$$= g(m_1 z_1 + m_2 z_2 + \cdots) = (m_1 + m_2 + \cdots) g z_C$$

where Eq. (5.D.6) is used in the last step. Thus one obtains Eq. (5.D.17).

Application to Stability

If an object is in static equilibrium in such a way that its potential energy is a minimum, then the equilibrium is **stable**, as explained in Sec. D iv of Chap. 4. One illustration is shown in Fig. 5.19, where a statically suspended object has its center of mass C vertically below the point of suspension. Any amount of rotation would raise C and hence increase the potential energy. This provides a simple experimental way of determining a rigid body's center of mass. The body is suspended from two of its points P, P' in succession. Each time, a vertical line of body points is determined; the center of mass lies at the intersection of these two lines (or their extension) as illustrated in the figure. Another instance is seen in Fig. 5.20, which illustrates the rule that a rigid body at rest, whose center of mass is above an interior point of the basis which supports it, is in stable equilibrium. (The basis in question should be interpreted as the smallest convex figure that can be drawn around a horizontal-plane projection of all the points of support.)

Seemingly difficult problems may easily be solved if the potential energy of the center of mass is invoked.

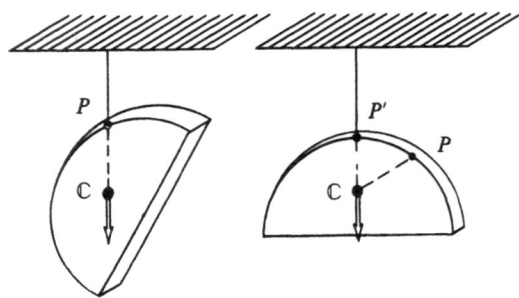

Figure 5.19. Stable equilibrium, and an experimental method for locating the center of mass C: the object is suspended in succession from two of its points, P, P'; each time a vertical line is drawn; C is at their intersection.

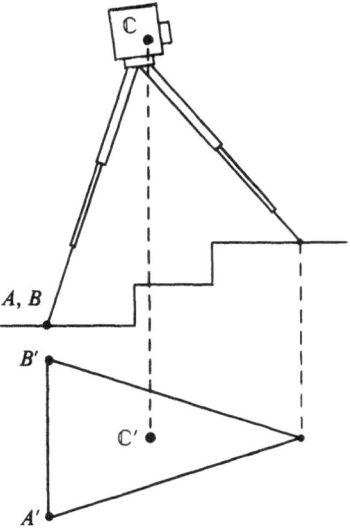

Figure 5.20. Another example of stable equilibrium. Point $\mathbb{C}'$, the vertical projection of $\mathbb{C}$, must be within the triangle whose corners are the projections of the tripod's feet on a horizontal plane. The explanation of this rule is that $\mathbb{C}$ would rise if the structure were pivoted, say about AB.

Example 5.7. Winding a Heavy Chain. A uniform chain of length l and mass m is suspended at point A from a horizontal cylinder whose circumference is also l; see Fig. 5.21. What is the minimum work, $\mathscr{W}$, needed to wind the chain completely around the cylinder? (Point A is initially at the same level as the cylinder's axis.)

We have

$$\mathscr{W} = \mathscr{U}_f - \mathscr{U}_i \qquad (5.D.18)$$

where $\mathscr{U}_i$, $\mathscr{U}_f$ are the initial and final potential energies of the chain. Choosing the reference

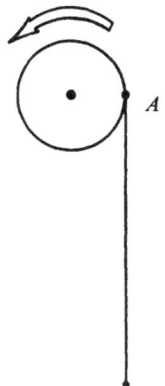

Figure 5.21. Winding a chain around a cylinder.

level to be the cylinder's axis, we have, for the altitude of the center of mass,

$$(z_\mathbb{C})_i = -\tfrac{1}{2}l$$

(here $\mathbb{C}$ is at the chain's midpoint), and

$$(z_\mathbb{C})_f = 0$$

since the wound chain is a circle with center at the zero level. Equation (5.D.18) becomes

$$\mathscr{W} = (mg)(0) - (mg)(-\tfrac{1}{2}l)$$

according to Eq. (5.D.17). Thus, we obtain

$$\mathscr{W} = \frac{mgl}{2} \qquad (5.D.19)$$

Note

Newton's Second Law for the Center of Mass

The proof of Eq. (5.D.10), $\mathbf{F}_{\text{tot}} = m_{\text{syst}} \mathbf{a}_\mathbb{C}$, proceeds in two steps.

1. We express the system's total momentum in "center-of-mass language." In our usual notation, we have

$$\mathbf{p}_{\text{syst}} = m_1 \mathbf{v}_1 + m_2 \mathbf{v}_2 + \cdots$$

$$= m_1 \frac{d\mathbf{r}_1}{dt} + m_2 \frac{d\mathbf{r}_2}{dt} + \cdots$$

$$= \frac{d}{dt}(m_1 \mathbf{r}_1 + m_2 \mathbf{r}_2 + \cdots)$$

Exploiting definition (5.D.7) for $\mathbf{r}_\mathbb{C}$, we obtain

$$\mathbf{p}_{\text{syst}} = \frac{d}{dt}(m_{\text{syst}} \mathbf{r}_\mathbb{C})$$

or, as stated in Eq. (5.D.11),

$$\boxed{\mathbf{p}_{\text{syst}} = m_{\text{syst}} \mathbf{v}_\mathbb{C}} \qquad (5.N.1)$$

where $\mathbf{v}_\mathbb{C} = d\mathbf{r}_\mathbb{C}/dt$.

2. We already possess an equation of motion for the complete system, namely, (5.A.13). Let us use it:

$$\mathbf{F}_{\text{tot}} = \frac{d\mathbf{p}_{\text{syst}}}{dt} \qquad (5.N.2)$$

With Eq. (5.N.1), this becomes

$$F_{tot} = m_{syst} \frac{d\mathbf{v}_C}{dt} = m_{syst} \mathbf{a}_C$$

as announced in Eq. (5.D.10).

Condensed Checklist

Momentum:

$$\boxed{\mathbf{p} = m\mathbf{v}} \qquad \mathbf{p}_{syst} = \sum_i m_i \mathbf{v}_i \qquad (5.A.1), (5.A.2)$$

Force:

$$\boxed{F_{tot} = \frac{d\mathbf{p}}{dt}} \qquad F_{ext} = \frac{d\mathbf{p}_{syst}}{dt} \qquad (5.A.3), (5.A.11)$$

Conservation:

$$\boxed{\Delta \mathbf{p}_{syst} = 0} \qquad if \quad F_{ext} = 0 \qquad (5.A.9)$$

Two-body collision:

$$m\mathbf{v}_f + M\mathbf{V}_f = m\mathbf{v}_i + M\mathbf{V}_i \qquad (5.B.10)$$

Elasticity:

$$|\mathbf{v}_f - \mathbf{V}_f| = |\mathbf{v}_i - \mathbf{V}_i| \qquad (5.B.11)$$

Elastic head-on case: $\qquad$ (5.B.15)

Force from a jet:

$$F = V \frac{dm}{dt} \qquad (completely\ inelastic) \qquad (5.C.4)$$

Center of mass:

$$\boxed{\mathbf{r}_C = \frac{\sum_i m_i \mathbf{r}_i}{m_{syst}}} \qquad (5.D.8)$$

$$\mathbf{p}_{syst} = m_{syst} \mathbf{v}_C \qquad (5.D.11)$$

$$\boxed{F_{tot} = m_{syst} \mathbf{a}_C} \qquad (5.D.10)$$

$$\boxed{\mathbf{a}_C = \mathbf{g}} \qquad (free\ fall) \qquad (5.D.13)$$

$$\mathcal{U}_{syst} = m_{syst} \, g z_C \qquad (5.D.17)$$

True or False

1. The momentum of a given particle appears as a different vector to two observers who are moving relatively to one another.

2. For the total momentum of a system to be conserved, it is necessary that all the forces on that system be conservative.

3. In a completely *elastic* collision between two particles of known mass, the final velocity vectors can be calculated from the initial velocity vectors.

4. In a completely *inelastic* collision between two particles of known mass, the final velocity vector can be calculated from the initial velocity vectors.

5. After a completely elastic two-body collision, the speed of each particle must be unchanged.

6. During a two-body collision, the momentum of each particle must change.

7. During a two-body collision, the energy of each particle must change.

8. If we know the total mass of a many-particle system, as well as its total kinetic energy, we can in principle calculate the magnitude of its total momentum.

9. A system whose center of mass is stationary must have zero total momentum.

10. In an Atwood machine (Chap. 3), the center of mass of the two suspended objects remains at rest.

Problems

Section A: Conservation of Momentum; B: Two-Body Collisions

*5.1. What is, in kg meter/sec, the momentum of a 2000-kg truck moving at 60 km/h?

5.2. A 0.5-gram beetle has a momentum of 1×10^{-5} kg meter/sec. How fast is it crawling?

5.3. A woman of mass 60 kg is running north at 2 meters/sec. Her 10-kg dog is running west at 9 meters/sec. Find the magnitude and the direction of their combined momentum.

5.4. A prospector, of mass 70 kg, carrying a 30-kg bag of gold, is sliding horizontally on smooth ice at a speed of 1 meter/sec. Observing open water ahead, and unable to brake, he throws his bag forward and barely succeeds in stopping himself. How fast was the bag moving after being thrown?

*5.5. A 200-lb inventor of physics problems, standing on smooth ice, is hit in the face by a 2-lb apple pie, traveling horizontally at 30 ft/sec. Estimate, in ft/sec, the speed imparted to the inventor.

*5.6. Two objects, of masses m and M, traveling on a horizontal frictionless plane, collide head-on with equal speeds v, then remain attached to each other. What is the speed of the combination? (Assume $M > m$; express your answer in terms of m, M, v.)

5.7. After throwing a 2-gram pingpong ball vertically upward, a player, using a firmly-held 200-gram paddle, strikes the ball forward when it is at the top of its trajectory. If the paddle moves at a steady 3 meters/sec, what is the speed of the ball after impact? (Assume an elastic collision.)

*5.8. (a) If $\mathcal{K}$ is a particle's kinetic energy, m its mass, and p the magnitude of its momentum, show that $p = (2m\mathcal{K})^{1/2}$. (b) If a certain electron, of mass 9.1×10^{-31} kg, has a kinetic energy of 1.6×10^{-16} joule, what is the magnitude of its momentum?

*5.9. How much kinetic energy is lost during the impact of Problem 5.6? (Express your answer as simply as you can in terms of m, M, v.)

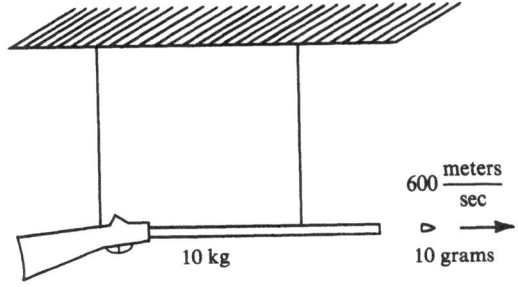

5.10. A 10-kg rifle, initially suspended at rest from vertical cords, as shown in the figure, fires a 10-gram bullet at a speed of 600 meters/sec. Find the recoil speed of the rifle just after the shot. (Neglect the external force needed to pull the trigger.)

*5.11. A railroad car A, of mass 50 metric tons, collides elastically with another car B, initially at rest. If the final speed of B is 1/2 the initial speed of A, what is the mass of B?

*5.12. Two pingpong balls, A and B, of unequal masses, are made to collide elastically head-on. Their initial speeds are both equal to 10 meters/sec. (a) If ball A bounces back at 20 meters/sec, what is the final velocity of ball B? (b) If B has a mass of 6 grams, what is the mass of A?

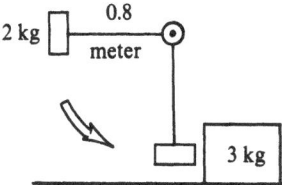

5.13. A hammer, of mass 2 kg and with a light handle of length 0.8 meter, is frictionlessly pivoted from a fixed point as shown in the figure. The hammer falls from a horizontal position where it was at rest, to a vertical one where it strikes a block of mass 3 kg resting on a horizontal frictionless surface. Assuming the impact to be elastic, (a) what speed does the block reach? (b) How high above its lowest position does the hammer bounce back?

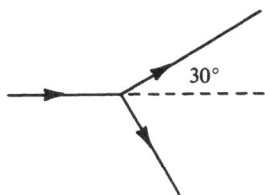

*5.14. A proton (mass $= 1.7 \times 10^{-27}$ kg), coming in at a speed of 3×10^7 meters/sec, strikes another, originally stationary proton. One of the two particles is observed to come out at an angle of 30° to the incoming direction, as shown. What is the velocity of each outcoming proton?

5.15. (a) Prove that the momentum of a particle increases by $\Delta\mathbf{p} = \mathbf{F}_{tot}\,\Delta t$ if a constant force $\mathbf{F}_{tot}$ acts on it during Δt. ($\mathbf{F}_{tot}\,\Delta t$ is called an **impulse**.) (b) A grand piano of mass 300 kg, on frictionless casters, and starting from rest, is being pushed on a smooth horizontal concert stage, first to the north for 4 sec, then to the west for 3 sec. If the force is always 200 newtons, what is, in magnitude and direction, the piano's velocity after these two pushes? (Use two impulses.)

*5.16. Two astronauts, A and B, initially floating at rest in a gravity-free region, are connected by a nylon rope. Astronaut A pulls until his speed relative to B is 3.0 meters/sec. If A's mass is 60 kg, while B's mass is 90 kg, what are their respective speeds?

*5.17. Derive Eq. (5.B.14), for the relative speeds in an elastic collision, from (5.B.12) and (5.B.13).

5.18. In the completely inelastic collision problem whose solution is Eq. (5.B.2), (a) calculate the kinetic energy loss (express your result in terms of m, M, and v_i); (b) show that, as $M \gg m$, nearly the whole kinetic energy is lost, i.e., $\mathcal{K}_f/\mathcal{K}_i \to 0$.

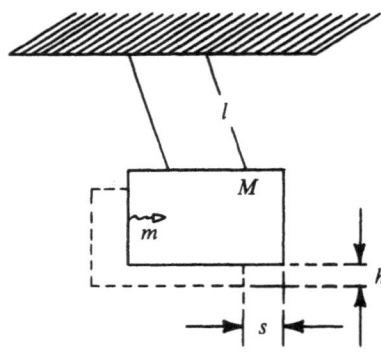

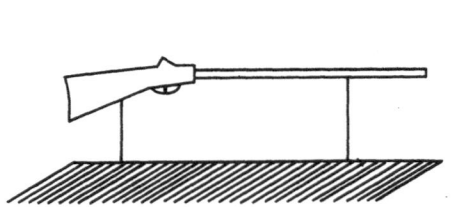

*5.19. (**A ballistic pendulum**). In the arrangement shown, a wooden block of mass M is initially suspended at rest from vertical cords of length l. A rifle fires a bullet of mass m horizontally at speed v into the block. (a) In terms of M, m, and v, find the speed of the block and bullet combination just after impact. (b) Find a relation between M, m, v, l, g, and h, where h is the maximum height to which the block will rise (use the energy method). (c) Find a geometric relation between l, h, and s, where s is the maximum horizontal deflection of the block. (d) Show that $h \approx s^2/2l$ if $h \ll l$. (e) Find a simple approximation for v in terms of m, M, l, s, and g when $h \ll l$, $m \ll M$. (f) Show that your answer is dimensionally correct.

*5.20. A stationary spaceship explodes into three fragments. Show that the three trajectories must be in the same plane. (No gravity is present.)

Section C: Force from a Jet

5.21. A fire hose sends 300 liters of water per minute horizontally at a speed of 15 meters/sec against a vertical wall. Assuming that all the water drips down along the wall, what force does the jet exert on the wall?

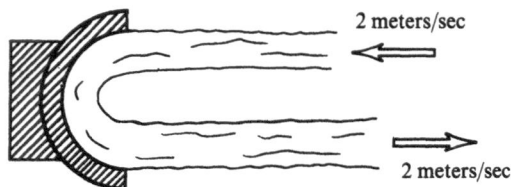

2 meters/sec

2 meters/sec

5.22. Some turbines have so-called **dished blades**, which send a jet of water back, as shown in the figure. If 30 kg of water comes in per second on such a blade (assumed stationary), and if the speed of both the incoming and outgoing water is about 2 meters/sec, find the force exerted by the water on the blade.

*5.23. What is the acceleration of a 250-kg rocket, in a gravity-free environment, if its specific impulse is 1200 meters/sec and fuel is expended at a rate of 2.0 kg every minute?

5.24. A ten-ton rocket (mass $= 1.0 \times 10^4$ kg) with a specific impulse of 2500 meters/sec, is designed to accelerate upward at 0.06 meter/sec^2 when launched. What mass of fuel must be consumed during the first second? Is it legitimate to assume a constant rocket mass during that time? Would your answer be significantly different if zero acceleration were assumed?

5.25. A certain 30-kg rocket reaches a vertical upward velocity of 20 meters/sec one second after launch from the ground. If fuel is burned at the rate of 0.50 kg/sec, find the specific impulse (assumed constant). Assume a constant mass for the rocket during the first second; why is this justified?

5.26. Owing to the burning of its fuel, a certain rocket, launched from rest in the absence of gravity, has a total mass that depends on time according to $m = (1400 \text{ kg}) \times e^{-t/T}$, where $T = 10$ min. Assuming a specific impulse of 1200 meters/sec, make a graph of the rocket's acceleration against time.

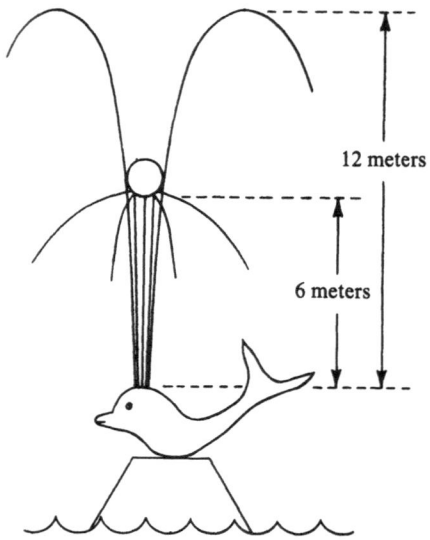

12 meters

6 meters

*5.27. When unimpeded, a certain fountain has a 12-meter vertical jet, and an output of 0.50 kg of water per second. A ball is then inserted into the jet, and kept in balance at a height of 6 meters, as shown. Assuming that half the water misses the ball, and that the other half makes a completely inelastic impact with it, find the ball's mass.

Section D: The Center of Mass

5.28. Three particles 1, 2, 3, are located on the x axis, at coordinates $x_1 = -4$ meters, $x_2 = +2$ meters, $x_3 = +10$ meters. Their respective masses are $m_1 = 1$ kg, $m_2 = 2$ kg, $m_3 = 3$ kg. Find the x coordinate of the center of mass for this system.

*5.29. The Earth's mass is 6.0×10^{24} kg; the Moon's is 7.3×10^{22} kg; the distance between their centers is about 60 terrestrial radii. How far from the Earth's center is the center of mass of the Earth–Moon system? (Express your answer in terrestrial radii.)

*5.30. Three particles are located in the xy plane. Their coordinates, in meters, are $(-2, +1)$, $(0, -2)$, $(+1, +1)$. Their respective masses are 2 g, 3 g, 4 g. Find the coordinates of the system's center of mass.

5.31. Do Example 5.4 with the origin of the x axis at the 2-kg corner.

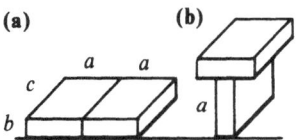

(a)   **(b)**

5.32. Two identical bricks, each of mass m and dimensions a, b, c, lie initially on the ground as shown in (a). What minimum work is needed to stack them as in (b)? (Express your answer in terms of m, g, a, b.)

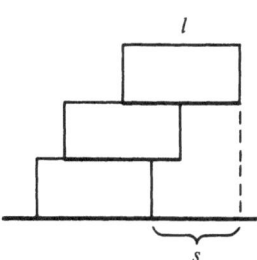

*5.33. Three identical rectangular blocks are piled up on a horizontal floor, as shown in the figure. (a) If each block has length l, what is the largest possible stable overhang s that the top block can have over the bottom one? (b) Find the corresponding result for four blocks. (c) How many blocks are needed to make s longer than one block? (Note: With enough blocks, the amount of overhang, surprisingly, can be made arbitrarily large.)

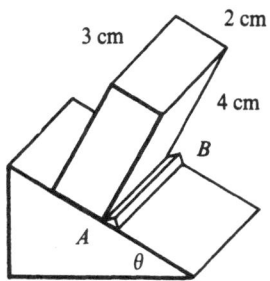

*5.34. A rectangular block of uniform composition and of dimensions 2.0 cm × 3.0 cm × 4.0 cm is resting on an inclined plane as shown. It is prevented from slipping by a small horizontal ridge AB. How large an angle θ can the inclined plane make with the horizontal before the block tips over?

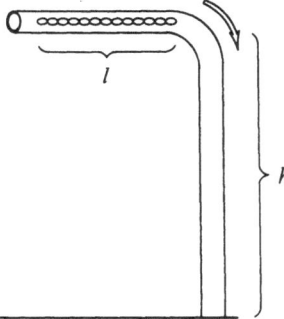

*5.35. A uniform chain of length l and mass m, initially at rest in the horizontal part of a frictionless tube as shown, receives a small nudge and falls into the vertical part, which has height h, with $h > l$. What is the speed of the chain when its leading end strikes the ground?

5.36. Prove principle No. 3 of Sec. D ii, concerning the center of mass of a set of objects.

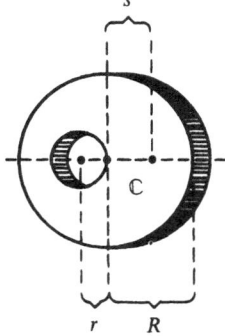

*5.37. Let $\mathbb{C}$ be the center of mass of a uniform circular disk of radius R with an eccentric circular hole of radius r, see the figure. If the edge of the hole goes through the center of the disk, find the distance s between $\mathbb{C}$ and the center of the disk.

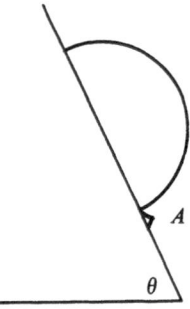

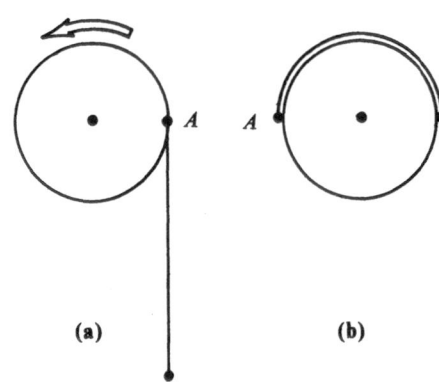

*5.38. A solid homogeneous hemisphere is lying on an inclined plane as shown, and is prevented from sliding by a small horizontal ridge at A. The plane makes an angle θ with the horizontal. How large can θ be made before the hemisphere tips over?

(a)   **(b)**

*5.41. A chain of length l and mass m is suspended, at point A as shown in (a), from a horizontal cylinder whose circumference is $2l$. Point A is at the same level as the cylinder's axis. In terms of m, l, g, what is the minimum work needed to wind the chain over the top half of the cylinder, as in (b)?

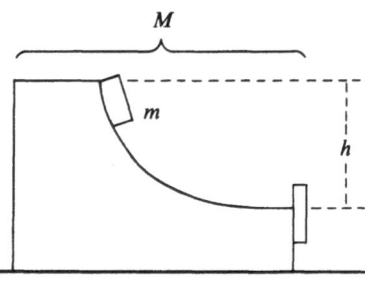

*5.39. A frictionless block of mass m is released from rest at the top of a slide of total height h, as shown in the figure. At the bottom, the horizontally moving block is stopped by a shock absorber, which does not allow it to bounce. The slide and block combination has a total mass M and rests on a horizontal plane with coefficient of kinetic friction μ_K. How far does the combined object move? (Express your answer in terms of m, M, g, h, μ_K. Static friction prevents motion of the slide before impact; assume the force of impact to be much larger than any force of friction.)

Answers to True or False

1. True.
2. False (true for energy).
3. False (unknown change in direction).
4. True.
5. False (true if the center of mass is at rest).
6. True (there is a nonzero force on each).
7. False (think of an elastic collision with the center of mass at rest).
8. False (internal relative motion may affect kinetic energy only).
9. True.
10. False (it goes down if released from rest).

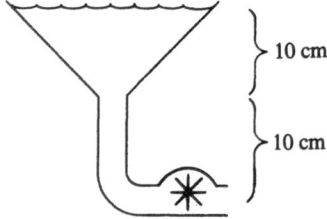

*5.40. A vertical conical funnel, shown in the figure, has a capacity of 1 liter and an altitude of 10 cm. Its pipe is 10 cm tall and of negligible capacity. What maximum work can we expect a small turbine to perform if it is connected to the base of the funnel, initially full of water, as it empties itself?

Answers or Hints to Problems

5.1. 3.3×10^4 kg meter/sec.

5.5. There is no need to calculate the masses.

5.6. $\dfrac{M-m}{M+m} v$.

5.8. (b) About 5×10^{35} kg meter/sec for the Sun.

5.9. $\dfrac{2mM}{M+m} v^2$.

5.11. Use the second Eq. (5.B.15).

5.12. (a) Zero; use Eq. (5.B.14); (b) 2 g.

5.14. The mass value is irrelevant, as well as which particle comes out at 30°.

5.16. 1.8 meters/sec for A.

5.17. In each equation, first put the terms in m on one side, those in M on the other.

5.19. (c) Use a right triangle of hypotenuse l and rectangular sides $l-h$ and s; (e) $v \approx (Ms/m)(g/l)^{1/2}$.

5.20. Assume the opposite and show that the momentum vector would not be conserved.

5.23. 0.16 meter/sec^2.

5.27. 0.28 kg.

5.29. At about 0.7 times the Earth's radius (i.e., inside the Earth).

5.30. $(0, 0)$.

5.33. Hint: Start with the top block, and work down. (c) 5.

5.34. $\arctan(1/2) \approx 27°$; the 3-cm width is irrelevant.

5.35. Use the center of mass and the energy principle.

5.37. Insert a small disk to complete the large one; use your knowledge of the resulting center of mass.

5.38. See item (d) of Fig. 5.16.

5.39. $(m/M)^2 (h/\mu_K)$.

5.40. Use item (f) of Fig. 5.16.

5.41. Use item (a) of Fig. 5.16.

CHAPTER 6

CIRCULAR MOTION

The circle is surely among the simplest of curves. Nevertheless, a particle in circular orbit necessarily experiences forces and accelerations that are far from trivial.

Circular motion plays an essential role in many areas of physics. It is, of course, directly relevant to the rotation of rigid objects, a study taken up in the next chapter. Later, we shall apply our knowledge of circular motion to the behavior of planets and satellites. At the opposite extreme in size, that knowledge will help us understand electron orbits in a simple model of the atom. On a laboratory scale, a beautiful illustration of circular motion can be offered by charged particles in the presence of a magnetic field. We note that few of the orbits found in Nature are exactly circular; but many of them can be idealized as such, with little loss in realism and great benefits in understanding.

Curiously enough, the study of circular motion has ramifications where no circular motion occurs at all, namely, in the theory of vibrating objects, as well as of alternating electric currents and of waves. All these applications are touched upon in the course of this book.

We shall approach circular motion as we approached free fall in Chap. 2. First comes a kinematic description, that is, one that makes no reference to the forces at play.

In this way we are able to relate the acceleration vector, **a**, to the speed and radius of the orbital motion. If the speed is constant, we shall obtain a famous result: **a** points toward the center of the orbit.

Once the acceleration is known, it is a simple matter to find the force responsible for it, through Newton's law $\mathbf{F}_{tot} = m\mathbf{a}$.

A. Angular Variables

The purpose of this section is to introduce convenient analogues to the concepts of position (x), velocity (v_x), and acceleration (a_x), which have served us so well in linear motion. In the circular case, the corresponding quantities are the angular position (θ, lower case Greek theta), the angular velocity (ω, lower case Greek omega), and the angular acceleration (α, lower case Greek alpha).

i. The Angular Position θ

Consider a particle P, constrained to move along the circumference of a given circle with radius r; see Fig. 6.1. That orbit can be used as a **coordinate curve** in order to specify the position of P. One direction of motion is arbitrarily chosen as positive, usually the counterclockwise direction as drawn on paper. (This convention is analogous, and just as arbitrary, as choosing left-to-right as the positive direction for a horizontal coordinate axis.) Some fixed point P_0 (usually the rightmost point) is chosen as the origin. The position of P can then be specified by the length of arc s between P_0 and P, together with an appropriate sign. Two positions are illustrated in Fig. 6.1: P, with a positive s; P', with a negative s'. If a particle goes around a complete turn or **revolution** (that is, an arc $\pm 2\pi r$), its original position is restored. Thus, a given position can be specified by infinitely many alternative coordinates. For example, the coordinates

$$s, \quad s + 2\pi r, \quad s - 2\pi r, \quad s + 4\pi r \quad (6.A.1)$$

etc., all refer to the same P. In practice one uses

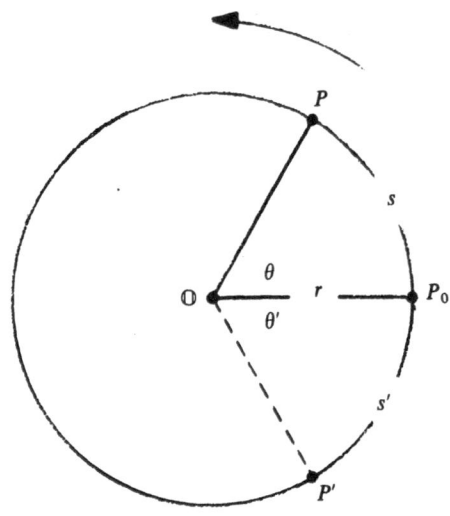

Figure 6.1. The arc s and angle θ equivalently label the position of P on a circle of given radius r. (Negative s' and θ' are featured by P'.)

whichever of these values appears to be most convenient.

Here we introduce a new but equivalent way of specifying the position of P, namely, through the angle θ shown in the figure:

$$\theta = \text{angle } P_0 \, \mathbb{O} \, P = \textbf{angular position of } P$$

This angle, with its vertex at the center $\mathbb{O}$ of the circle, is subtended by the arc s. Its sign is taken to be the same as that of s; in the figure, θ is positive, while θ' is negative.

ii. The Radian

How are θ and s related? In elementary geometry, it is shown that θ is proportional to the ratio between the arc s and the radius r of the circle:

$$\theta = (\text{const}) \frac{s}{r} \qquad (6.A.2)$$

The proportionality constant depends on the unit in which one chooses to express θ. The **radian** ($= 1$ rad) is defined as that unit of angle that makes the proportionality constant in Eq. (6.A.2) equal to 1:

$$\frac{\theta}{1 \text{ rad}} = \frac{s}{r} \qquad (6.A.3)$$

(the left side is just θ expressed in radians).

Let us evaluate the radian in terms of degrees. Consider a special case of Eq. (6.A.3) where $s = 2\pi r$ (a complete circumference). Correspondingly we must have $\theta = 360°$. Then (6.A.3) reads

$$\frac{360°}{1 \text{ rad}} = 2\pi \qquad (6.A.4)$$

so that

$$1 \text{ rad} = \frac{360°}{2\pi} \approx 57.3° \qquad (6.A.5)$$

A further simplification can now be achieved. We shall treat the radian as a *dimensionless quantity* equal to 1:

$$1 \text{ rad} = 1 \qquad (6.A.6)$$

Thus

If an angle is given without a unit, is it understood to be expressed in radians.

(6.A.7)

Equation (6.A.3) gives rise to

$$\boxed{\theta = \frac{s}{r}, \qquad s = r\theta} \qquad (6.A.8)$$

[The reader should learn to convert degrees and radians into each other effortlessly. In particular, Eq. (6.A.4) leads to the often-used equivalences $360° = 2\pi$, $180° = \pi$, $90° = \pi/2$, etc.; we also see that $\cos \pi = -1$, $\sin(\pi/2) = 1$, etc.]

Equations (6.A.8) imply that, *in the case of a unit circle* ($r = 1$), *the length of arc*, s, *equals the angle*, θ, when the latter is expressed in radians.

For arbitrary r, but with $\theta = 1$ rad $= 1$, Eqs. (6.A.8) state that:

One radian is the angle subtended at the center of the circle by an arc whose length equals the radius,

(6.A.9)

see Fig. 6.2.

Small-angle approximations are most simply done in terms of radians:

$$\sin \theta \approx \tan \theta \approx \theta \qquad (6.A.10)$$

$$\cos \theta \approx 1 - \tfrac{1}{2}\theta^2 \qquad (6.A.11)$$

(θ sufficiently small).

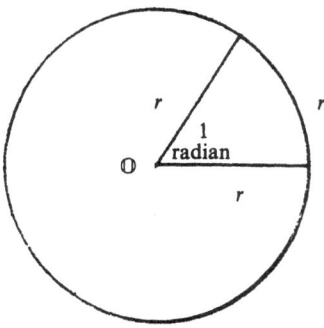

Figure 6.2. The angle subtended at O equals 1 radian in this figure.

iii. The Angular Velocity ω

Now that the position of P is well described, we need to specify how that position changes with time. We first recall the definition of **speed**:

> The speed, v, of a particle equals the absolute length of arc covered per unit time.

$$(6.A.12)$$

Symbolically,

$$v = \left| \frac{ds}{dt} \right| \qquad (6.A.13)$$

Rather than taking the absolute value, one may wish to keep the sign, in order to indicate whether the motion is counterclockwise or clockwise. To denote this we set

$$v_\theta = \frac{ds}{dt} = \pm v \qquad (6.A.14)$$

(+ for counterclockwise, − for clockwise motion), so that $v = |v_\theta|$. The quantity v_θ is the **tangential velocity**.

Just as v_θ is the rate of change of s, the **angular velocity** ω is defined as the rate of change of θ, where θ is expressed in radians:

$$\omega = \frac{d\theta}{dt} \qquad (6.A.15)$$

The angular velocity of P is the number of radians swept out per unit time by a radius attached to P. $\qquad (6.A.16)$

The quantity ω is the ratio of a pure number (obtained by counting radians) to an interval of time, say the second. Hence ω may be expressed in units of $1/\sec = \sec^{-1}$, the **inverse second**. That unit is also called the radian/second to emphasize that one is dealing with an angular velocity.

There exists a simple relation between ω and v_θ. Starting from Eq. (6.A.8),

$$s = r\theta$$

then noting that r is constant, and taking the rate of change on both sides, we have

$$\frac{ds}{dt} = r \frac{d\theta}{dt}$$

or

$$v_\theta = r\omega \qquad (6.A.17)$$

It should be kept in mind that our results and definitions so far refer to **instantaneous** values of v, v_θ, and ω. That is to say, these quantities do not have to be constant.

B. Uniform Circular Motion

In this simplest and most important case, the tangential velocity v_θ is constant by definition. We have

$$\frac{dv_\theta}{dt} = 0, \qquad \frac{dv}{dt} = 0, \qquad \frac{d\omega}{dt} = 0 \qquad (6.B.1)$$

On the other hand, the velocity vector **v** is not constant, because its direction changes with time. Therefore the acceleration vector,

$$\mathbf{a} = \frac{d\mathbf{v}}{dt} \qquad (6.B.2)$$

is nonzero. Our main task in this section is to determine **a** for a particle in uniform circular motion. It will be necessary to start with a few additional definitions.

i. The Frequency f and the Period T

The **frequency**, f, is defined as *the number of revolutions completed in unit time*. For any time interval Δt,

$$f = \frac{\text{number of revolutions during } \Delta t}{\Delta t} \qquad (6.B.3)$$

The **revolution per second**,* or rev/sec, is the appropriate unit for f; the **revolution per minute**, or rpm, is often used in engineering:

$$1 \text{ rpm} = \frac{1}{60}\frac{\text{rev}}{\text{sec}} \qquad (6.B.4)$$

Relation Between f and ω

During each revolution, 2π radians are covered, so that

$$\boxed{\omega = 2\pi f} \qquad (6.B.5)$$

Which is the more useful quantity, ω or f? Theoretical calculations in terms of ω are less cluttered by factors of 2π; however, practical results are easier to visualize in terms of f. (Note: In the technical literature, the lower-case Greek nu, v, is often preferred to f as a symbol for frequency.)

Caution: in the context of formula (6.B.5), the revolution is not an acceptable unit of angle; it is only a repetitive event. Attempts at setting $1 \text{ rev} = 360°$ are frustrated by (6.B.5), as in the following illustration. Let a particle revolve with a frequency $f = 1$ rev/sec, and thus with an angular frequency $\omega = 2\pi$ rad/sec. Inserting these quantities in Eq. (6.B.5) gives

$$2\pi\frac{\text{rad}}{\text{sec}} \overset{?}{=} 2\pi\frac{\text{rev}}{\text{sec}}$$

or $1 \text{ rad} \overset{?}{=} 1$ rev, an absurdity. Actually, "rev," in the above, is mathematically equivalent to the number 1. Very often, "rev/sec" is simply denoted by "sec^{-1}."

The **period** is defined as follows:

> The **period**, T, is the time required for one revolution. (6.B.6)

This quantity gives information equivalent to ω or to f.

* Also sometimes called the **hertz**, after Heinrich Hertz, the discoverer of radio waves; however, the hertz is better reserved for use in connection with vibrations and waves.

Relation Between f and T

Let us choose $\Delta t = T$ in Eq. (6.B.3). During that interval, the number of revolutions is 1. Therefore (6.B.3) becomes

$$\boxed{f = \frac{1}{T}} \qquad (6.B.7)$$

The frequency and the period are each other's inverses; T is expressed in seconds. Equivalently to (6.B.7), we have

$$\omega = \frac{2\pi}{T} \qquad (6.B.8)$$

See Eq. (6.B.5).

Example 6.1. A Spinning Bicycle Wheel. A bicycle wheel, of radius $r = 0.40$ meter, spins at a frequency $f = 120$ rpm while its axle is held fixed. (a) What is the wheel's period of revolution, T? (b) What is its angular velocity ω? (c) What is the speed v of a point P on the wheel's rim?

a. *Period.* We have

$$T = \frac{1}{f}$$

where

$$f = \frac{120}{60}\frac{\text{rev}}{\text{sec}} = 2\frac{\text{rev}}{\text{sec}}$$

Thus,

$$T = \frac{1}{2}\frac{\text{sec}}{\text{rev}} = 0.5\frac{\text{sec}}{\text{rev}} \ (=0.5 \text{ sec})$$

b. *Angular frequency.* Here we use

$$\omega = 2\pi f = (2\pi)(2)\frac{\text{rad}}{\text{sec}}$$

(note the change of label, rev → rad), or

$$\omega = 12\frac{\text{rad}}{\text{sec}} = 12 \text{ sec}^{-1}$$

c. *Speed.* We have

$$v = \omega r$$

$$= (12 \text{ sec}^{-1})(0.40 \text{ meter}) = 4.8 \frac{\text{meters}}{\text{sec}}$$

ii. The Velocity Vector v

The next step is to ask how the velocity vector **v** and the position vector **r** are related in the case of uniform circular motion. We recall that **r** is defined as having its tail at the origin of coordinates (chosen here as the center of the circle), and its head at the orbiting particle; **r** is sometimes called the "radius vector."

From Chap. 2 we recall the basic features of any velocity vector **v**: (1) the magnitude of **v** is always the speed $v = ds/dt$; (2) **v** always points forwards in a direction tangent to the trajectory.

Figure 6.3 illustrates a positive rotation and shows that, for a given **r**, the resulting **v** may be specified as follows:

1. The magnitude of **v** is $v = \omega r$; see Eq. (6.A.17);
2. The direction of **v** is always at 90° to the left of **r** in the plane of the motion. (In the clockwise case, "left" should be changed to "right.")

We conclude that $\mathbf{v} = d\mathbf{r}/dt$ is a vector of constant length, which rotates uniformly at the same angular velocity as **r**. Since a vector does not change when shifted parallel to itself, **v** may be drawn with its tail at the center of the circle. The diagram of Fig. 6.3 then assumes the appearance of

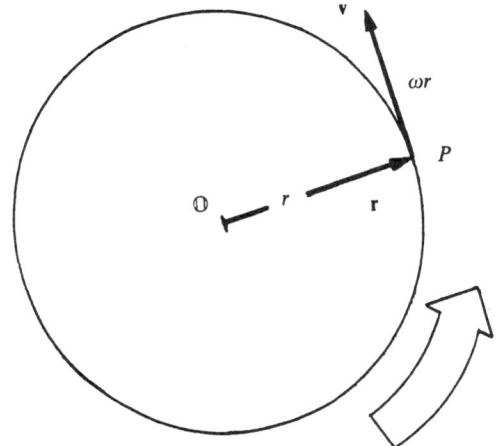

Figure 6.3. In uniform circular motion, **r** and **v** are both of constant magnitude; both rotate at the same angular velocity.

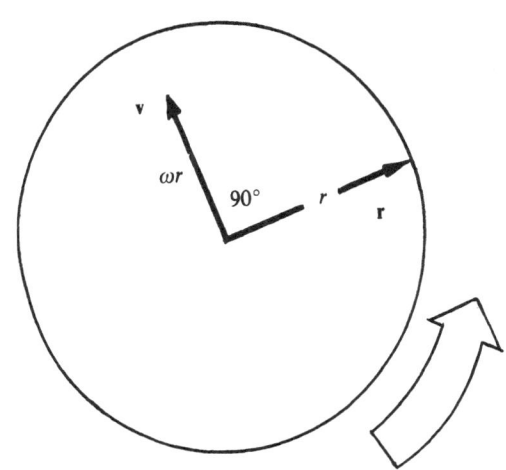

Figure 6.4. Another representation of Fig. 6.3. Here we see that **v** leads **r** by 90°.

Fig. 6.4. The description (1), (2) above applies of course equally well to this figure. To indicate the direction of **v** in a manner that does not depend on whether the rotation is in the positive or negative direction, one says that

$$\boxed{\mathbf{v} \text{ leads } \mathbf{r} \text{ by } 90°.} \qquad (6.B.9)$$

Phase Vectors

Since we often have to deal with rotating vectors such as **r** and **v**, it is economical to make the following definition:

> A **phase vector*** is a vector that has constant magnitude, and that rotates in a plane at a constant angular velocity. (6.B.10)

Thus, in the case of uniform circular motion, both **r** and **v** are phase vectors. They move in the same plane and with the same angular velocity. Unless otherwise stated, we shall assume phase vectors to rotate in the positive, or counterclockwise, direction.

iii. The Time Derivative of a Phase Vector

We have just painlessly solved an interesting vector problem.

---

* Also called **phasor**.

Problem. Let **r** be a phase vector with angular velocity ω; find its derivative $d\mathbf{r}/dt$.

Solution. As illustrated in Fig. 6.4,

$$\frac{d\mathbf{r}}{dt} \text{ is a phase vector of magnitude } \omega r;$$

$$\frac{d\mathbf{r}}{dt} \text{ leads } \mathbf{r} \text{ by } 90^\circ$$

In order to indicate that the problem is a general one, in which **r** need not be a particle's position, we restate it in terms of a different symbol.

Problem. Let *l* be *any phase vector whatsoever* whose angular velocity is ω; find its derivative, dl/dt.

Solution. (We must keep in mind that *l* has constant magnitude and that ω is constant.)

$$\frac{dl}{dt} \text{ is a phase vector of magnitude } \omega l;$$

$$\frac{dl}{dt} \text{ leads } l \text{ by } 90^\circ.$$

$$(6.B.11)$$

This describes the vector dl/dt completely.

The prescription is simple enough to allow us to take further and further derivatives without difficulty. Figure 6.5 shows *l*, dl/dt, d^2l/dt^2, d^3l/dt^3 on the same diagram. The whole picture must be visualized as rotating with a single angular velocity ω. A thorough familiarity with (6.B.11) is an asset in many physical and mathematical applications.

iv. The Centripetal Acceleration and Force

We are now ready to calculate the acceleration vector $\mathbf{a} = d\mathbf{v}/dt$ for uniform circular motion. Setting $l = \mathbf{v}$ in (6.B.11), we see that

$$\mathbf{a} \text{ is a phase vector of magnitude } a = \omega v. \quad (6.B.12)$$

$$\mathbf{a} \text{ leads } \mathbf{v} \text{ by } 90^\circ. \quad (6.B.13)$$

Since $v = \omega r$, (6.B.12) gives

$$a = \omega^2 r = \frac{v^2}{r} \quad (6.B.14)$$

Statement (6.B.13) is illustrated in Fig. 6.6 and shows that the acceleration points oppositely to the radius vector. The same situation is redrawn in Fig. 6.7 with the vector **a** originating on the particle P, in order to demonstrate that *the acceleration of P is toward the center of the circle*; it is called a *centripetal acceleration*. An acceleration orthogonal to the motion itself is nothing new to us; the situation at the top of a (parabolic) free-fall trajectory was an earlier example of this. The reader should become thoroughly comfortable with the idea that

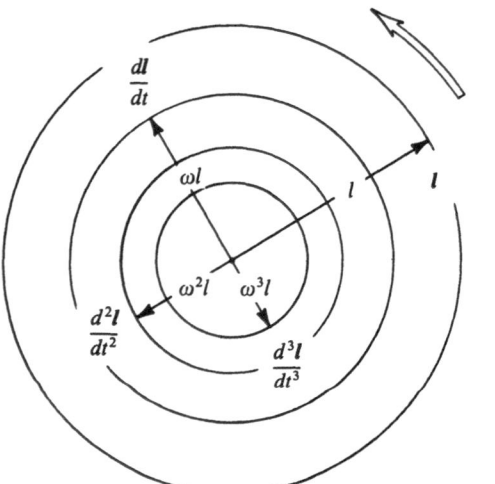

Figure 6.5. Successive time derivatives of a phase vector *l*. In our earlier diagrams, *l* was just **r**.

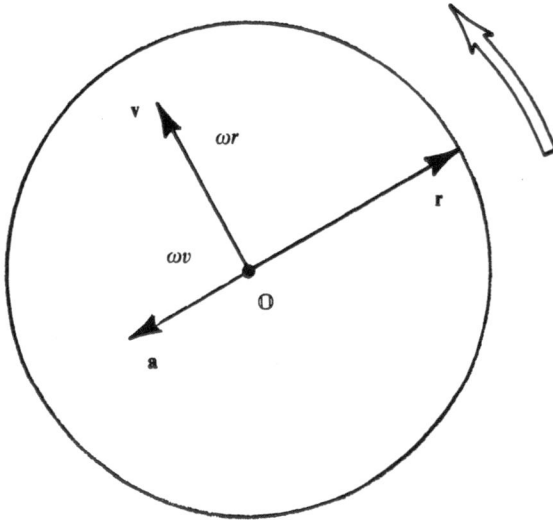

Figure 6.6. Two successive derivatives of the phase vector **r**.

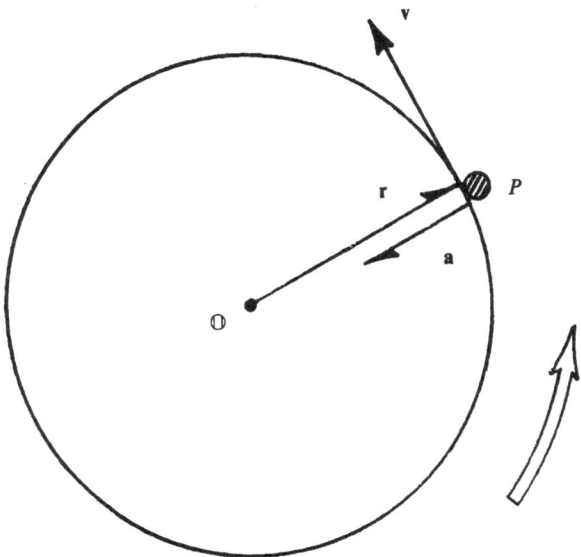

Figure 6.7. Another representation of Fig. 6.6. Here we see that **a** is a centripetal vector.

a constant speed is compatible with a nonzero acceleration. Formulas (6.B.14) are important and constitute the main result of this chapter. (Some judgment must be exercised as to whether $\omega^2 r$ or v^2/r is the most advantageous form in a given problem.)

If particle P has a mass m, Newton's second law gives the magnitude of the total force $\mathbf{F}_{tot}$ acting on it:

$$F_{tot} = m\omega^2 r = \frac{mv^2}{r} \qquad (6.B.15)$$

The direction of $\mathbf{F}_{tot}$ is toward the center of the orbit; $\mathbf{F}_{tot}$ is called a **centripetal force**. Whatever its origin (a frictionless track, a string, the gravitational attraction on a satellite, etc.), $\mathbf{F}_{tot}$ may be visualized as "responsible" for the uniform circular motion of P, or as "needed" to maintain that motion.

A Centrifugal Force?

Popular intuition associates circular motion with a force *away* from the center, or *centrifugal* force. As concerns $\mathbf{F}_{tot}$, this is the exact opposite of the truth. As concerns the reaction to $\mathbf{F}_{tot}$, that notion is sometimes correct; for example, while rounding a curve, a car subjects a passenger to a net centripetal force, but the passenger exerts a centrifugal force on

the car. The net force on the car (including friction from the road) must be centripetal.

v. Examples

Example 6.2. A Dynamo. The armature (rotating coil) of a certain dynamo has a frequency $f = 10$ rev/sec; its outer radius is $r = 16$ cm. Find the resulting force on a section of wire, of mass $m = 0.5$ gram, located at the coil's periphery.

According to (6.B.15), we have

$$F_{tot} = m\omega^2 r = m(2\pi f)^2\, r$$
$$= (0.5 \times 10^{-3})[(2\pi)(10)]^2(16 \times 10^{-2})\,\text{newton}$$
$$= \underline{0.3\ \text{newton}}$$

Example 6.3. A Low-Altitude Satellite. A satellite is launched so as to circle the Earth just above the atmosphere (at a few hundred kilometers' altitude). The satellite, which, after launch, is coasting without any further power consumption, undergoes uniform circular motion about the Earth's center. Assuming the sea-level value $g = 10$ meters/sec^2 for the satellite's centripetal acceleration, how long is its period of revolution, T?

Equation (6.B.14) gives

$$g = \omega^2 r \qquad (6.B.16)$$

where $r = 6.4 \times 10^6$ meters is the Earth's radius, to which a few hundred kilometers contribute a negligible amount. With Eq. (6.B.8) this becomes

$$g = \left(\frac{2\pi}{T}\right)^2 r \qquad (6.B.17)$$

Solving for T, we obtain

$$T = 2\pi \left(\frac{r}{g}\right)^{1/2} = 2\pi \left(\frac{6.4 \times 10^6}{10}\right)^{1/2}\ \text{sec} \qquad (6.B.18)$$
$$= 5.0 \times 10^3\ \text{sec} \approx \underline{84\ \text{min}} \qquad (6.B.19)$$

This is close to the figure actually realized by all low-altitude Earth satellites.

Example 6.4. Centripetal Acceleration at the Equator. Calculate, with respect to the Earth's center, the centripetal acceleration a of a building located at the Earth's equator.

The building's period of revolution is one day, $T = (24)(3600)$ sec; the radius of the building's orbit is the radius of the Earth, $r = 6.4 \times 10^6$ meters; the angular velocity is $\omega = 2\pi/T$. Hence Eq. (6.B.14) gives

$$a = \omega^2 r = \left(\frac{2\pi}{T}\right)^2 r$$

$$= \left[\frac{2\pi}{(24)(3600)}\right]^2 (6.4 \times 10^6) \frac{\text{meter}}{\text{sec}^2}$$

$$= 0.034 \frac{\text{meter}}{\text{sec}^2}$$

This is very small compared to $g = 10$ meters/sec². Therefore one would not expect the measurement of g to be much affected by the geographical latitude; see later, Example 6.7, for some further details.

Example 6.5. The Conical Pendulum. Figure 6.8 shows a particle of mass m suspended on a light string of length l, making a constant angle θ with the vertical. The particle undergoes uniform circular motion at a height h below the string's point of suspension. Gravity and the string provide the only forces needed to sustain the motion. Find the period of revolution T in terms of l, h, m, and g.

Newton's second law is

$$\mathbf{F}_{tot} = m\mathbf{a}$$

or

$$\mathbf{F} + m\mathbf{g} = m\mathbf{a} \qquad (6.B.20)$$

where $\mathbf{F}$ is the force provided by the string, $m\mathbf{g}$ is the weight, and $\mathbf{a}$ is the centripetal acceleration. The vertical component of (6.B.20) is

$$F\cos\theta - mg = 0 \qquad (6.B.21)$$

The centripetal component is

$$F\sin\theta = ma$$

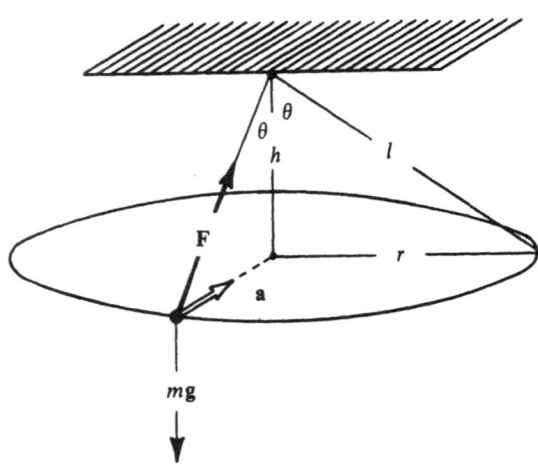

Figure 6.8. The conical pendulum.

or, with the help of Eq. (6.B.14),

$$F\sin\theta = m\omega^2 r \qquad (6.B.22)$$

where r is the radius of the orbit. Eliminating the tension F between (6.B.21) and (6.B.22) gives

$$g\tan\theta = \omega^2 r$$

With $\tan\theta = r/h$, this becomes

$$\frac{g}{h} = \omega^2, \qquad \omega = \left(\frac{g}{h}\right)^{1/2}$$

$$T = \frac{2\pi}{\omega} = 2\pi\left(\frac{h}{g}\right)^{1/2} \qquad (6.B.23)$$

quite a remarkable result since it depends neither on the string's length nor on the particle's mass. For example, the various conical pendulums shown in Fig. 6.9 have the same period and hence revolve together as if they constituted a rigid unit.

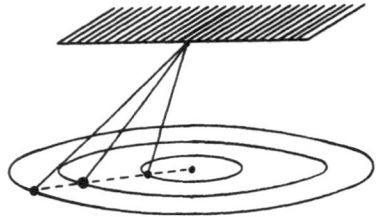

Figure 6.9. Several conical pendulums, suspended from the same point, and rotating at a common height, will move as if attached to a single invisible disk.

vi. A Spectrum of Circular Frequencies

Some illustrative periods and frequencies for circular motion are displayed in Fig. 6.10. On the whole, larger periods are associated with larger sizes.

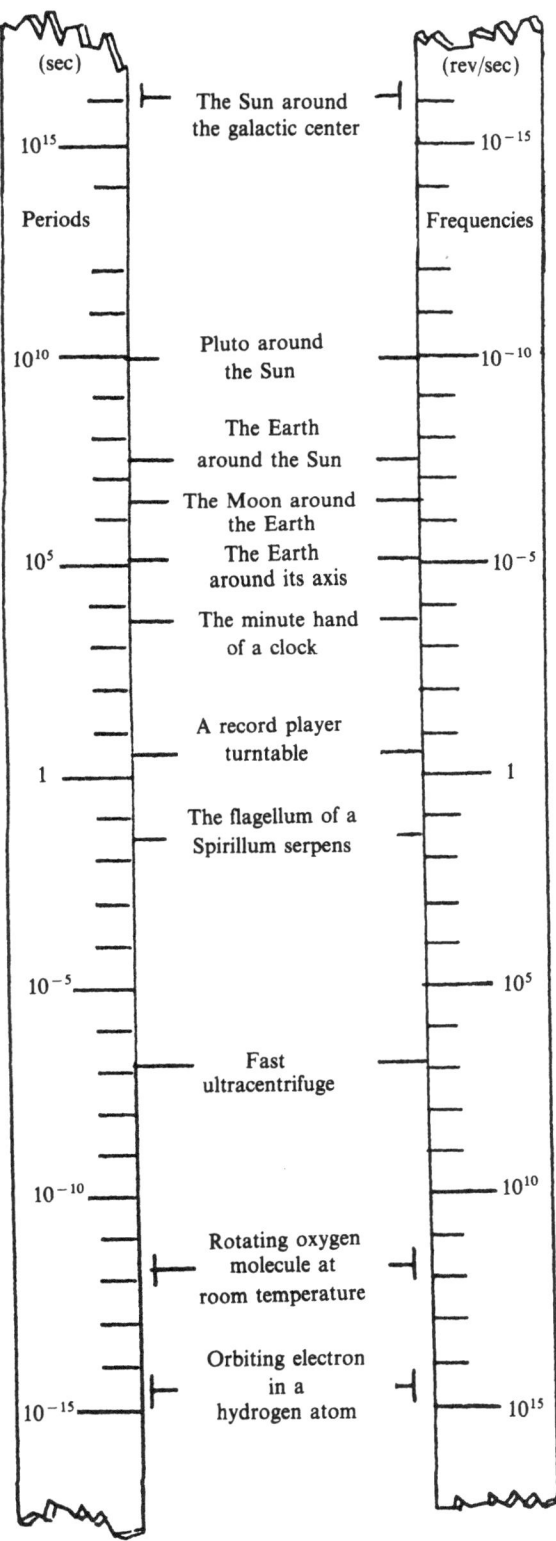

vii. Effective Gravity and Circular Motion

In Chap. 3 we saw how an accelerated environment creates the effect of a modified gravity. The rule is

$$\mathbf{g}_{\text{eff}} = \mathbf{g} - \mathbf{a} \qquad (6.B.24)$$

where $\mathbf{g}_{\text{eff}}$ is the effective acceleration of gravity, as observed in a laboratory whose own acceleration is $\mathbf{a}$; the true gravity is $\mathbf{g}$.

In one illustration, the freely falling elevator, the occupants eventually suffer a crash as the price of their temporary effective weightlessness. A more permanent method for modifying the effects of gravity is to subject one's environment to uniform circular motion. Gravity can thereby most easily be increased, as in the first example below, or it can be decreased, as in the second example.

Example 6.6. "Artificial Gravity" in a Space Station. Let us design a space station as a uniformly spinning cylinder, shown in Fig. 6.11. Supposing its radius to be $r = 15$ meters, what should be its period of rotation T to provide an Earthlike effective gravity at its cylindrical wall? (Neglect any true gravity that may be present—a legitimate neglect under free-fall conditions.)

Figure 6.10. A logarithmic spectrum of circular motion periods (left scale) or frequencies (right scale). Many of the motions listed are only approximately circular; several figures are typical rather than exact; the Sun's period around the Galaxy is rather imprecisely known. The laboratory ultracentrifuge features a typical high rotational frequency for a man-made object, slower ones being propellers, turbines, flywheels, etc. *Spirillum serpens* is a water-dwelling microorganism that rotates its tail, or flagellum, as a propeller at about 2400 rpm, while its body rotates inversely at about 800 rpm. The oxygen molecule, O_2, is a dumbbell-shaped assembly whose rotational energy is, on the average, a measure of the temperature of its surroundings.

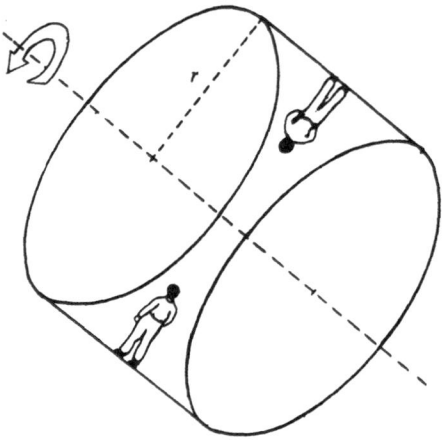

Figure 6.11. A spinning space station will provide artificial gravity.

Equation (6.B.24) gives

$$\mathbf{g}_{eff} = 0 - \mathbf{a}$$

Thus, the effective **g** is centrifugal, and equals in magnitude

$$g = a = \omega^2 r = \left(\frac{2\pi}{T}\right)^2 r \qquad (6.B.25)$$

Solving for T,

$$T = 2\pi \left(\frac{r}{g}\right)^{1/2} \qquad (6.B.26)$$

$$= 2\pi \left(\frac{15}{10}\right)^{1/2} \text{ sec} = \underline{7.7 \text{ sec}}$$

[The common laboratory centrifuge is another version of this example. It is often used to obtain faster settling (sedimentation) of a precipitate in a liquid medium. To all appearances, a centrifugal force acts on the suspended particles, but in reality there just is an *insufficient centripetal force* to keep them in a circular orbit.]

Example 6.7. Effective Gravity at the Equator. How much smaller is g_{eq}, the acceleration of gravity measured in a laboratory located on the Earth's equator, than the corresponding quantity at the poles, g_p? (Assume the Earth to be perfectly spherical.)

Since the poles are not subjected to the other latitudes' centripetal acceleration caused by the Earth's spin, we let $g_p = g$ and $g_{eq} = g_{eff}$. Then Eq. (6.B.24) gives, in the centripetal direction,

$$g_{eq} = g_p - a$$

where a was already calculated in Example (6.4): $a = 0.034$ meter/sec^2. We see that

$$g_p - g_{eq} = \underline{0.034 \text{ meter/sec}^2}$$

(spherical-Earth approximation)

The actually measured difference is 0.052 meter/sec^2. The additional 0.018 meter/sec^2 needed in our calculated value can be accounted for by the slight oblateness (flattening) of the Earth, itself an effect due to the Earth's spin. Oblateness diminishes the force of gravitational attraction because points on the equator are somewhat further from the Earth's center than the poles are. The motion of the Earth around the Sun may be totally ignored in the present example.

Surface of a Rotating Liquid

Consider a pail of water that has been put on a turntable. As the water rotates, its surface becomes concave because the effective gravity points differently at different locations in the pail. That intriguing effect is the topic of Note (ii) at the end of this chapter.

C. Nonuniform Circular Motion

For completeness, a few words should be said about circular motion with variable speed. In such a case the acceleration vector **a** is no longer purely centripetal. Figure 6.12 illustrates the fact that **a** has a tangential component a_θ and a radial component a_r.

Applications dealing with a_θ are left to the next chapter (Rotation and Torque), but the radial component a_r is of more immediate interest: it can be taken over unchanged from uniform circular motion, Eq. (6.B.14):

$$\boxed{|a_r| = \omega^2 r = \frac{v^2}{r}} \qquad (6.C.1)$$

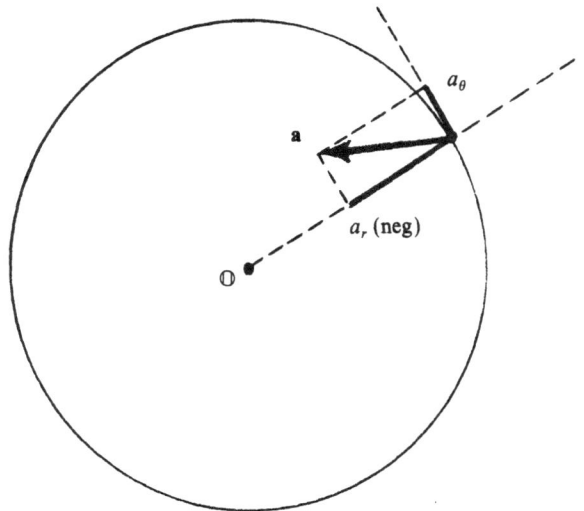

Figure 6.12. If the circular motion is nonuniform, the acceleration **a** has a nonzero tangential component a_θ as well as the usual centripetal component $|a_r|$. The notation a_r indicates the radial component, negative here because the radial direction is defined as positive outward.

where ω and v now depend on time in an arbitrary fashion. [If the sign is to be included, we must write

$$a_r = -\omega^2 r = -\frac{v^2}{r}$$

to indicate that the direction is opposite to **r**.]

The proof that (6.C.1) is still valid in the non-uniform case is given in a Note at the end of this chapter.

Example 6.8. A Vertical Circular Track. A particle P, of mass m, is released from rest on a frictionless circular track lying in a vertical plane; see Fig. 6.13. If the starting point A is level with the center of the circle, whose radius r is given, what (normal) force N does the track exert on P when it is at the lowest point B? (This example provides a good opportunity to combine *conservation of energy* with circular motion.)

At B, Newton's second law gives

$$\mathbf{F}_{tot} = m\mathbf{a} \qquad (6.C.2)$$

where **a** is the acceleration of P, and where $\mathbf{F}_{tot} = \mathbf{N} + m\mathbf{g}$. The centripetal component of (6.C.2) is

$$N - mg = m\frac{v^2}{r}$$

so that the required force is

$$N = m\left(\frac{v^2}{r} + g\right) \qquad (6.C.3)$$

To find v^2, we apply the energy principle:

Energy at B = energy at A

or, if potential energies are defined with respect to B,

$$0 + \tfrac{1}{2}mv^2 = mgr + 0$$

(The potential energy vanishes at B; the kinetic energy vanishes at A.) Solving this for v^2 and inserting in (6.C.3), we obtain

$$\underline{N = 3mg} \qquad (6.C.4)$$

At the time in question, the track must support three times the ordinary weight of P.

We see, incidentally, that there is zero tangential force (and hence also $a_\theta = 0$) at B in spite of the fact that we have a case of nonuniform circular motion. The tangential force and acceleration will not vanish at other points of the trajectory.

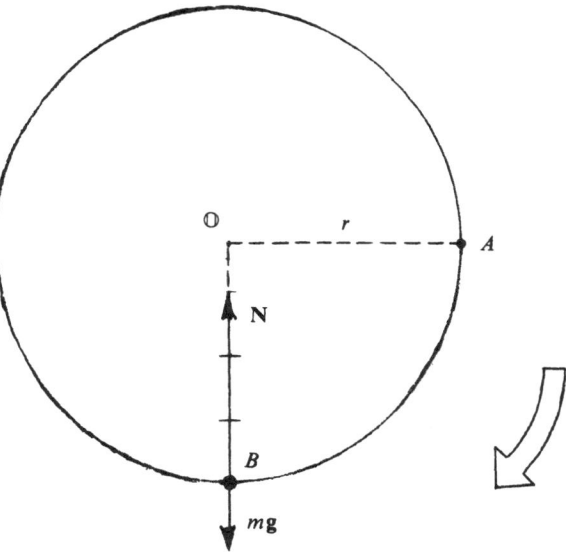

Figure 6.13. A particle of mass m, sliding frictionlessly along a vertical circle. If the starting point (from rest) is A, then the force between particle and track at B is 3 times the particle's usual weight. (Note: the particle could just be suspended from O by a light string or rod, in which case the arrangement is a **pendulum**.)

Notes

i. Acceleration in Nonuniform Circular Motion

In Sec. C it was stated that the **radial component** of acceleration is

$$a_r = -\frac{v^2}{r} \tag{6.N.1}$$

whether or not v is constant. An expression can also be found for the **tangential component** of acceleration. Not surprisingly, it is

$$a_\theta = \frac{d^2 s}{dt^2} \tag{6.N.2}$$

where the arc s is the position coordinate measured along the circumference.

While both relations, (6.N.1) and (6.N.2), are eminently plausible, they still must be proved. We shall do so by a "trick method," whose main ingredient is the differentiation rule for the dot product.

Let $\mathbf{f}$, $\mathbf{g}$ be two arbitrary time-varying vectors. Then we have

$$\frac{d}{dt}(\mathbf{f} \cdot \mathbf{g}) = \frac{d\mathbf{f}}{dt} \cdot \mathbf{g} + \mathbf{f} \cdot \frac{d\mathbf{g}}{dt} \tag{6.N.3}$$

just as if $\mathbf{f} \cdot \mathbf{g}$ were an ordinary product. [To obtain this rule, start as usual with a finite increment, $\Delta(\mathbf{f} \cdot \mathbf{g}) = (\mathbf{f} + \Delta\mathbf{f}) \cdot (\mathbf{g} + \Delta\mathbf{g}) - \mathbf{f} \cdot \mathbf{g}$, divide by Δt, and take the limit for $\Delta t \to 0$.]

Radial Acceleration

We know that $\mathbf{v}$ is always tangential to the circle, so that

$$\mathbf{v} \cdot \mathbf{r} = 0^* \tag{6.N.4}$$

Differentiating with respect to time,

$$\frac{d\mathbf{v}}{dt} \cdot \mathbf{r} + \mathbf{v} \cdot \frac{d\mathbf{r}}{dt} = 0$$

or

$$\mathbf{a} \cdot \mathbf{r} + \mathbf{v} \cdot \mathbf{v} = 0$$

---

*As a check, take the time derivative of $\mathbf{r} \cdot \mathbf{r}$ in the case of circular motion (not necessarily uniform); use (6.N.3).

or, using a component interpretation, see Eqs. (4.B.14) and (4.B.15) of Chap. 4,

$$(a_r)(r) + v^2 = 0$$

This gives the desired Eq. (6.N.1).

Tangential Acceleration

Here we start from the dot-product square

$$\mathbf{v} \cdot \mathbf{v} = v^2$$

Applying d/dt to both sides, we obtain

$$2\mathbf{v} \cdot \frac{d\mathbf{v}}{dt} = 2v \frac{dv}{dt}$$

or

$$\mathbf{v} \cdot \mathbf{a} = v \frac{dv}{dt} \tag{6.N.5}$$

A component interpretation of the left-hand side is

$$\mathbf{v} \cdot \mathbf{a} = v a_v = v a_\theta$$

where a_v is the component of $\mathbf{a}$ in the direction of $\mathbf{v}$, so that a_v and a_θ have the same meaning. The right-hand side of (6.N.5) is

$$v \frac{dv}{dt} = v \frac{d^2 s}{dt^2}$$

Thus, (6.N.5) becomes

$$v a_\theta = v \frac{d^2 s}{dt^2}$$

leading to Eq. (6.N.2).

ii. The Rotating-Pail Experiment ("Newton's Pail")

Imagine a philosophically inclined prisoner in a windowless cell that gently rotates about a vertical axis. If asked whether his cell is rotating or keeping a fixed orientation, he might answer: "This is a question that has no meaning except with reference to the outside world, from which I am isolated. Therefore, there is no way for me to decide." However, a pail of water, standing on the floor, would dramatically refute him, as was first pointed out by Newton. (Many other centrifuge-type effects would be equally decisive, of course.)

If a pail of water is steadily spun about its vertical axis, until the whole fluid mass rotates as a unit, the surface of the water will assume a smooth concave shape. That shape turns out to be a **paraboloid of revolution**: all cross sections that contain the axis are identical parabolas; see Fig. 6.14. To prove this, we first calculate the angle θ between the water's surface and the horizontal, at a distance r from the axis. We treat a surface drop of mass m

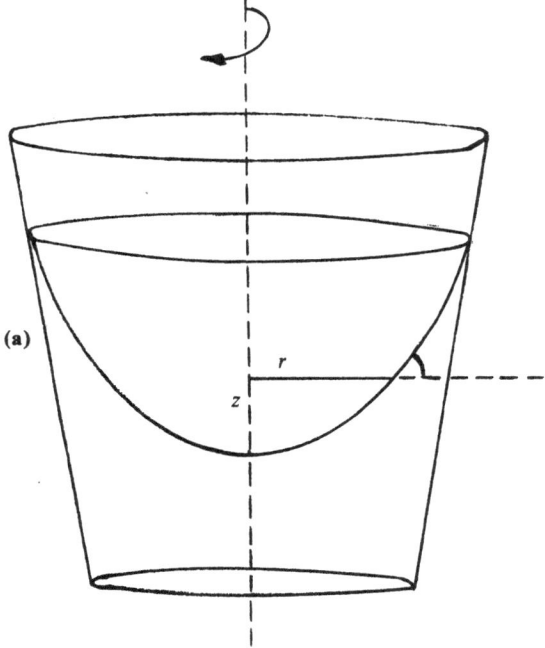

(a)

as if it were on a frictionless incline, which therefore exerts a normal force $\mathbf{N}$ on the drop. Newton's second law, $\mathbf{F}_{tot} = m\mathbf{a}$, gives

$$\mathbf{N} + m\mathbf{g} = m\mathbf{a}$$

where $m\mathbf{g}$ is the drop's weight and $\mathbf{a}$ is its centripetal acceleration. The vertical and centripetal components of this equation are

$$N \cos \theta - mg = 0$$
$$N \sin \theta = m\omega^2 r$$

(ω = angular velocity of the pail). Eliminating N, we find

$$\tan \theta = \frac{\omega^2 r}{g} \qquad (6.\text{N}.6)$$

which, incidentally, does not depend on m.

Next, the shape of the surface can be determined. Call z the altitude of a point on that surface (e.g., with respect to its lowest point); z will then be a function of r. From calculus we know that the slope of the surface is $dz/dr = \tan \theta$. Thus, from (6.N.6), we have

$$\frac{dz}{dr} = \frac{\omega^2 r}{g}$$

The most general function z satisfying this is

$$z = \frac{\omega^2 r^2}{2g} + \text{const} \qquad (6.\text{N}.7)$$

and we obtain the announced parabola. (The constant will only shift the reference level with respect to which one measures z; here the constant is chosen to be zero.)

This parabolic feature has suggested an intriguing application. Astronomers need large and perfect parabolic reflectors to serve as telescope mirrors; a rotating dish of mercury would seem ideal. Unfortunately, the ripples caused by the slightest vibration, as well as the fact that the dish cannot be steered, have so far prevented this kind of telescope from becoming practical.

Condensed Checklist

Angular variables:

$$\text{Angular position} = \theta = \frac{s}{r} \qquad (6.\text{A}.2)$$

$$\text{Angular velocity} = \frac{d\theta}{dt} = \omega = \frac{v_\theta}{r}$$

$$(6.\text{A}.15), (6.\text{A}.17)$$

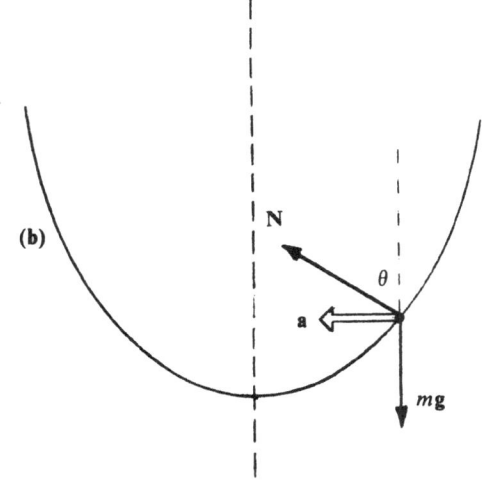

(b)

Figure 6.14. (a) The rotating pail: the water surface assumes the shape of a paraboloid. (b) The angle θ, shown in (a), between water surface and horizontal direction, is the same as that between the normal force $\mathbf{N}$ and the vertical.

Frequency and period:

$$f = \frac{1}{T}, \qquad \omega = 2\pi f = \frac{2\pi}{T}$$

(6.B.5), (6.B.7), (6.B.8)

Radial component of acceleration and force (uniform or nonuniform circular motion):

$$a_r = -\omega^2 r = \frac{v^2}{r}$$

(6.B.14), (6.C.1)

$$F_{\text{tot},r} = -m\omega^2 r = -\frac{mv^2}{r}$$

(6.B.15)

(Uniform circular motion: $\omega = $ const)

Derivative of a phase vector l:

$$\frac{dl}{dt} \text{ leads } l \text{ by } 90°.$$

(6.B.11)

$$\left| \frac{dl}{dt} \right| = \omega \, |l|$$

True or False

1. "Tangential velocity" is another name for "angular velocity."

2. If a particle is made to revolve uniformly around a *vertical* circle, the magnitude of the total force on the particle remains constant.

3. A racing car is being timed around a level, unbanked circular race track. If the car's speed is limited only by the risk of skidding, it achieves the largest number of circuits by staying as close as possible to the inner guard rail.

4. A cyclist, while rounding a corner, leans to the inside of the bend. In this way, part of the centripetal force which makes him turn is gravitational.

5. The fourth time derivative of a phase vector points in the same direction as the phase vector itself.

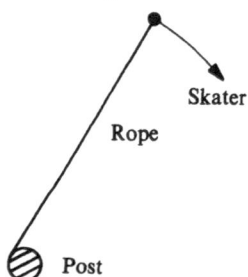

6. While a freely coasting skater holds the end of a rope, which winds itself around a fixed cylindrical post, as shown, the tension in the rope increases.

7. As a particle increases its speed along a circular orbit, the ratio of its centripetal acceleration to its kinetic energy remains constant.

Problems

Section A: Angular Variables

*6.1. A horse runs around a 100-meter circular track in 3 min. Find (a) its frequency of revolution in rev/sec; (b) its angular velocity in rad/sec and in deg/sec.

6.2. A phonograph record turns at $33\frac{1}{3}$ rpm. If a 50-cm length of groove passes under the needle during 1 sec, how far is the needle from the center of the record?

*6.3. The Moon circles the Earth in 27.3 days (this is the "true," or so-called sidereal period, in which the background stars, rather than the Sun, are used as angular reference marks). (a) Find the Moon's orbital frequency in rev/sec. (b) Find the Moon's orbital angular velocity ω in rad/sec. (c) If the radius of the Moon's (nearly circular) orbit is 3.8×10^8 meters, find the Moon's orbital speed v.

*6.4. The Earth circles the Sun in 365 days, 6 hours. (a) Find the Earth's orbital frequency in rev/sec. (b) Find the Earth's orbital angular velocity ω in rad/sec. (c) If the Earth–Sun distance (nearly constant) is 1.50×10^{11} meters, find the Earth's speed v.

6.5. A particle of mass m undergoes uniform circular motion with radius r, angular velocity ω, frequency f, and period T. (a) Express the kinetic energy in terms of m, ω, r; in terms of m, f, r; in terms of m, T, r. (b) If the Moon's mass is 7.3×10^{22} kg, find its orbital kinetic energy (see Problem 6.3). (c) If the Earth's mass is 6.0×10^{24} kg, find its orbital kinetic energy (see Problem 6.4).

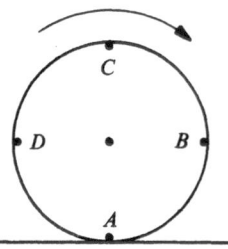

*6.6. A wheel (see figure) has four equally spaced dots A, B, C, D painted on its rim. It is rolling at a speed of 1 meter/sec on a straight horizontal track. When A is next to the ground, what are, in direction and magnitude, the velocities of A, B, C, D?

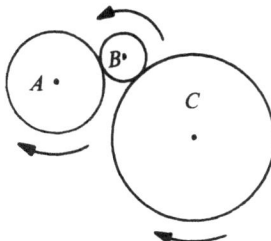

6.7. Three pivoted wheels A, B, C are in nonslipping contact, as shown in the adjoining figure. If their respective radii are given as r_A, r_B, r_C, find the ratio f_A/f_C, where f_A, f_C are the outer wheels' rotational frequencies.

Section B: Uniform Circular Motion

6.8. A 0.25-kg particle undergoes uniform circular motion with radius 0.5 meter and angular velocity 2 rad/sec. (a) What is the centripetal acceleration? (b) What is the centripetal force?

*6.9. A slingshot is uniformly whirling a stone of mass 0.15 kg in a horizontal circle of radius 0.20 meter, at a frequency of 5 revolutions per second. (a) What is the speed of the stone? (b) What is the centripetal force?

6.10. A 1.5 metric ton ($=1500$ kg) car, rounding a curve of radius 50 meters, is exerting a centrifugal force of 5000 newtons on the road surface. How fast is the car moving?

*6.11. A particle undergoes uniform circular motion with radius r and period T. (a) Express the centripetal acceleration a in terms of r and T. (b) From the data of Problem 6.3, calculate the centripetal acceleration of the Moon in units of meters/sec². (c) From the data of Problem 6.4, calculate the centripetal acceleration of the Earth around the Sun in units of meters/sec².

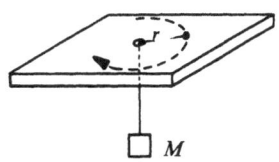

6.12. A particle of mass m is whirling, in uniform circular motion of radius r, on a horizontal frictionless plane; see the figure. It is tied to one end of a light string, which passes through a small frictionless hole and whose other end supports a stationary mass M. In terms of g, r, m, and M, find the frequency of revolution such that the situation can persist in a steady state.

6.13. A particle P undergoes uniform circular motion with angular velocity ω. If **r** is the radius vector of P, find the magnitude and direction of the derivatives $d\mathbf{a}/dt$, $d^2\mathbf{a}/dt^2$, where **a** is the acceleration vector. (Express the magnitudes in terms of ω and r; show the directions on a sketch together with **r**, **v**, and **a**.)

6.14. Review Sec. C iv of Chap. 2 and give a dimensional derivation of the formula v^2/r for the centripetal acceleration, up to a purely numerical factor.

*6.15. What speed is needed to launch an Earth satellite from the top of the atmosphere? (See Example 6.3; the low-altitude orbit needs the least launching speed.) How many times larger is this than the speed of sound in air (340 meters/sec)?

*6.16. What should be, in minutes, the length of day in order that there be zero effective gravity at the equator? (Make the unrealistic assumption that the Earth keeps its present shape.)

6.17. A set of n identical gold coins weigh at the north pole about what $n + 1$ such coins weigh at the equator. What is n? (Use $g_{pole} = 9.832$ meters/sec², $g_{equator} = 9.780$ meters/sec².)

6.18. In darkness, gravity determines that a certain type of seed should sprout vertically up. If such a seed is started in darkness in a pot mounted on a $33\frac{1}{3}$ rpm turntable at 25 cm from the center, at what angle to the vertical would you expect the sprout to grow? Inwards or outwards?

6.19. How fast, in km/h, should a hypothetical airplane circle the Earth in order to give its passengers an effective gravity equal to $g/2$?

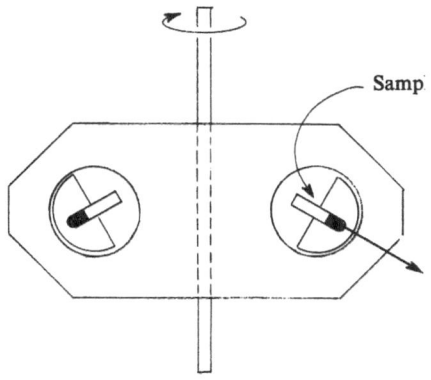

Samp

*6.20. A certain laboratory centrifuge (see figure) features a rotational frequency of 3000 rpm and an effective gravity of $1300g$. (a) How far is the sample from the axis of rotation? (b) What angle, in radians and in degrees, does the effective **g** vector make with the horizontal? (Use a small-angle approximation.)

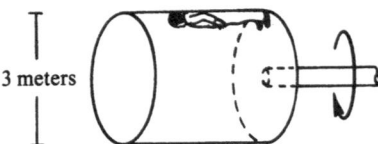

3 meters

6.21. A 3-meter-diameter amusement-park centrifuge spins at a constant rate about an axis that becomes horizontal while the customers are safely lying against its wall, as shown. (a) How many revolutions per minute are needed at the least? (b) What is the minimum gravity, in units of g, felt at the lowest point?

*6.22. A 2000-lb crate, freely suspended from a 50-ft cable, moves as a conical pendulum in a circle of radius 3 ft. What is the crate's period of revolution?

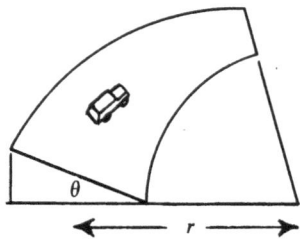

θ

r

*6.23. A curve in the road has approximate radius r. To prevent skidding, the road is banked (see the figure) at an angle θ to the horizontal, so that a car of any mass, moving at an officially approved speed v, exerts a purely normal force on the road surface. Express θ in terms of r, v, and g.

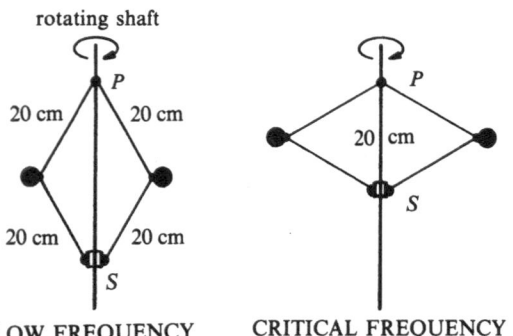

rotating shaft

P

20 cm | 20 cm

20 cm | 20 cm

S

LOW FREQUENCY

P

20 cm

S

CRITICAL FREQUENCY

6.24. (An engine governor). The speed of an engine can be regulated by means of a **governor**, as shown. At a critical frequency of revolution f_c, a light sleeve S begins to close a throttle (not shown). Find f_c if each of the four light articulated bars has length 20 cm, and if the critical position of S is 20 cm below the point of suspension P. Neglect all masses except the two small 0.5-kg spheres.

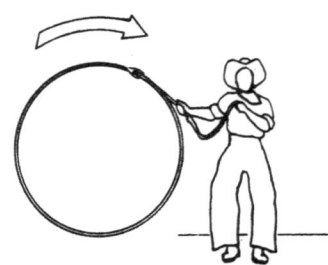

*6.25. As shown in the figure, a lasso is being whirled in a perfect steady circle, of diameter 1.0 meter and mass 2.0 kg, at a frequency of 4.0 rev/sec. (a) Consider one half of the loop. What is the centripetal acceleration, a_C, of its center of mass? (See Fig. 5.16 of Chap. 5.) (b) What is the net force on the half loop? (c) What is the tension in the rope? (Neglect gravity and the supporting end of rope; justify this neglect by comparing a_C with g.)

Section C: Nonuniform Circular Motion

6.26. (a) Use dimensional analysis, Sec. C iv of Chap. 2, to show that the answer to Example 6.8 cannot depend on r. (b) Modify Example 6.8 so that the particle is released at the top of the circle. (An infinitesimal nudge should be enough to make it fall from that position of unstable equilibrium.) How large is N when the particle is at B? (Express your answer in terms of m and g.)

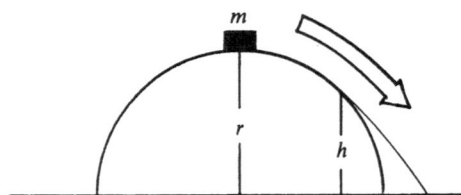

m

r

h

*6.27. A particle of mass m, resting in unstable equilibrium at the top of a frictionless hemispherical dome of height r (see the figure) is given a small nudge and slides to the ground. In terms of m, g, and r, find the height h at which the particle leaves the surface of the dome.

Answers to True or False

1. False.

2. True.

3. True (ω, not v, is what "gets it there"; for given ω, F decreases with r).

4. False ($\mathbf{g}$ has zero horizontal component; friction with the road supplies the centripetal force).

5. True.

6. True (kinetic energy is conserved, hence $v = $ const; but r decreases, hence v^2/r increases).

7. True [$(v^2/r)/\frac{1}{2}mv^2 = $ const].

Answers or Hints to Problems

6.1. (c) 2.0 deg/sec.

6.3. (a) See Fig. 6.10; (b) 2.66×10^{-6} rad/sec; (c) 1.01×10^3 meters/sec.

6.4. (a) See Fig. 6.10.

6.6. Hint: Consider velocities relative to the axle.

6.9. (a) 6.3 meters/sec; (b) 30 newtons.

6.11. (b) 2.69×10^{-3} meter/sec^2.

6.15. 8×10^3 meters/sec.

6.16. Same as (6.B.19).

6.20. (b) 4.4×10^{-2} deg.

6.22. See Example 6.5.

6.23. $\arctan(v^2/rg)$.

6.25. (a) 64π meters/sec^2; (b) 64π newtons; (c) 32π newtons.

6.27. $\frac{2}{3}r$. (Use conservation of energy to calculate v.)

CHAPTER 7

ROTATION AND TORQUE

Rotational motion, or spin, is visible everywhere; it is involved whenever an object changes its orientation. Our technical civilization relies on wheels and levers, and will continue to do so despite the modern engineer's justified abhorrence of "moving parts." In the natural world, rotation may be observed on levels from astronomical to subatomic. All astronomical entities, such as the Earth, the Galaxy, the stars, etc., rotate to some extent about their own axis. This rotation is quite independent of their orbital motion ("revolution"), as the familiar example of the Earth reminds us. On the molecular, atomic, and subatomic scale, particle spin is a key to many important and beautiful phenomena, some of which will be touched upon later in this book.

The concept of **torque** plays a central role in this chapter. Just as a force modifies the linear motion of a particle, so does a torque modify the rotational motion of an object. For example, a torque can impart some spin when none was originally present. It encounters an inertia as well; in rotational motion, the **moment of inertia** of an object is what diminishes the effect of a torque, just as, in linear motion, a particle's mass diminishes the effect of a force.

In the following we learn how a pivoted object responds to a torque, how to calculate torques, how to combine them, and how they can cancel to produce **rotational equilibrium**. We also define an "amount of spin," the **angular momentum**.

In the absence of a net torque, the angular momentum of a system turns out to be conserved. This piece of information is sometimes used to predict how an object will rotate; it is crucial, as well, in understanding both electronic and astronomical orbits.

A. Rotation About a Fixed Axis

Rotating objects have a wide variety of behaviors available to them. While spinning, the top illustrated in Fig. 7.1 continuously shifts the spatial direction of its axis. On the other hand, the coin of Fig. 7.2, while keeping a vertical axis of rotation, changes the direction of that axis with respect to its own (the coin's) geometry. Thus, *an axis of rotation may change its orientation in space, or it may change its orientation with respect to the spinning object itself*; these two processes may occur in combination.

Our own planet provides a stately example of these phenomena. The Earth's axis, when extended both ways into space, aims toward well identified points of the celestial sphere. The northern one is at present near the familiar star Polaris. Careful observations have shown this direction in space to be slowly varying. (It will approximately return to the present one in 26 000 years.)

Furthermore, the *geographical* location of the pole suffers a very small and rather irregular variation, confined inside a circular region no more than 26 meters in diameter. The pole returns (very roughly) to its original location every 14 months.

The complete motion of an arbitrary spinning object is notoriously difficult to visualize. Here we shall concentrate on cases where *the axis keeps a fixed direction*, both in space and within the object itself. This condition is most easily achieved through the use of a mechanical pivot or axle.

Introductory Definitions (A Preview)

For a smooth reading of this chapter, it is necessary to keep in mind from the outset where the

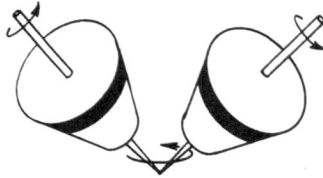

Figure 7.1. Rotation about a variable axis in space.

discussion is leading us. In this first section, the chapter's two main ideas are introduced, namely, the **moment of inertia** and the **torque**. As noted before, they are the analogues of *mass* and *force*. The moment of inertia, I, will be defined first, by the requirement that it yield a spinning object's kinetic energy in the form $\mathcal{K} = \frac{1}{2}I\omega^2$ (ω = angular velocity), modeled after a particle's $\mathcal{K} = \frac{1}{2}mv^2$. Next, the (total) torque will be defined as $\tau_{\text{tot}} = I\alpha$ (α = angular acceleration), similar to $\mathbf{F}_{\text{tot}} = m\mathbf{a}$. We begin with a brief review of ω and α.

i. Rotational Variables

Let us consider a **rigid body**, i.e., a system of particles whose mutual distances are constant. Figure 7.3 illustrates rigid-body rotation about a *fixed axis AB*. A typical particle P which belongs to the object undergoes *circular motion*, with some radius r, in a plane perpendicular to the axis, and about a center $\mathbb{O}$ which lies on that axis. Since A and B are fixed in space, the object's orientation is completely specified by just one variable, namely, the position of P on its circular orbit. Thus, *the description of fixed-axis rotation reduces to that of circular motion*, and can always be represented as such in an "axial view"; see Fig. 7.4.

Let θ be the angular coordinate of P. Then the **angular velocity** is, as in the preceding chapter,

$$\omega = \frac{d\theta}{dt} \qquad (7.\text{A}.1)$$

Figure 7.2. Rotation about a variable axis in the object.

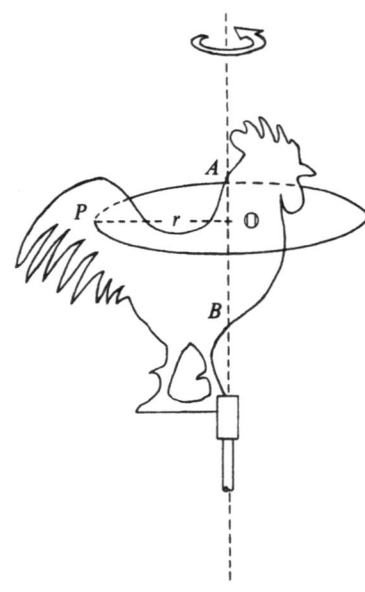

Figure 7.3. A weathervane as an example of a rigid body pivoted on a fixed axis.

We also define an **angular acceleration** α by

$$\alpha = \frac{d\omega}{dt} = \frac{d^2\theta}{dt^2} \qquad (7.\text{A}.2)$$

Since the object is rigid, all of its particles have the same ω and the same α; these variables apply to the spinning body as a whole.

ii. Moment of Inertia

Is there a rotational analogue of mass? More specifically, we seek to define, for a pivoted object,

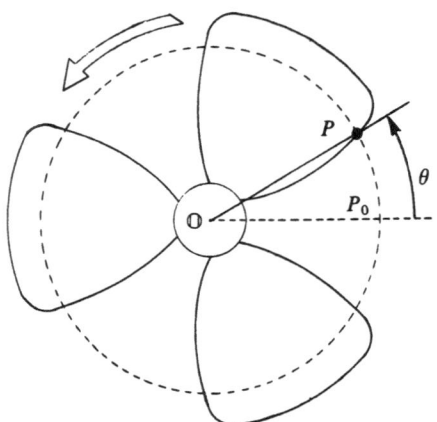

Figure 7.4. An axial view of fixed-axis rotation. The direction $\mathbb{O}P_0$ is fixed in space; the angular coordinate θ defines the object's orientation.

a quantity I (the **moment of inertia**), in terms of which the kinetic energy $\mathcal{K}$ of the whole rotating system reads

$$\mathcal{K} = \tfrac{1}{2}I\omega^2 \qquad (7.A.3)$$

in analogy with a particle's kinetic energy

$$\mathcal{K} = \tfrac{1}{2}mv^2 \qquad (7.A.4)$$

Equation (7.A.3) can be taken as defining I. We shall see presently that I depends on the shape and mass of the rigid body as well as on the axis about which it spins. In fact, the value of I can be computed just from the structure of an object, without the need for an energy measurement. This is done as follows.

Single Particle

We first calculate I for a single "off-axis" particle of mass m, for example, particle P in Fig. 7.4. (Here P happens to constitute a small portion of a larger rotating object.) Applying our knowledge of circular motion, we have, for that particle,

$$\mathcal{K} = \tfrac{1}{2}mv^2 = \tfrac{1}{2}m(r\omega)^2$$

or

$$\mathcal{K} = \tfrac{1}{2}(mr^2)\,\omega^2 \qquad (7.A.5)$$

This is nothing but the desired equation (7.A.3), in which

$$I = mr^2 \qquad \text{(single particle)} \qquad (7.A.6)$$

[Hence, moments of inertia are expressed in units of kg meter^2, although a more complete designation would be kg meter^2 rad^{-2}, in view of Eq. (7.A.3).]

Extended Object

To obtain the total moment of inertia I of a many-particle rigid system, we go back to definition (7.A.3). The total kinetic energy, $\mathcal{K}$, is the sum

$$\mathcal{K} = \tfrac{1}{2}I_1\omega^2 + \tfrac{1}{2}I_2\omega^2 + \cdots \qquad (7.A.7)$$

of the particles' individual kinetic energies; the

angular frequency ω is the same for all. By factoring out $\tfrac{1}{2}\omega^2$ we find

$$I = I_1 + I_2 + \cdots \qquad (7.A.8)$$

Thus the individual moments of inertia (all calculated about the same axis) are simply added. As an explicit recipe for calculating I, Eq. (7.A.8) reads

$$I = m_1 r_1^2 + m_2 r_2^2 + \cdots \qquad (7.A.9)$$
$$= \sum_i m_i r_i^2 \qquad (7.A.10)$$

where $m_1, m_2,...$ are the particles' masses, and $r_1, r_2,...$ their distances to the common axis of rotation.

Can a continuous object be treated in this manner? Equation (7.A.10) can be generalized to an infinite number of particles, each with an infinitesimal mass. The sum (7.A.10) then goes over into an integral. If a typical infinitesimal mass dm is at a distance r from the axis, then it contributes an infinitesimal moment of inertia

$$dI = (dm)(r^2) \qquad (7.A.11)$$

to the total amount I. Integrating both sides over the whole rigid body gives

$$\int dI = \int (dm)\, r^2$$

or

$$I = \int r^2\, dm \qquad (r = \text{distance to axis})$$

$$(7.A.12)$$

This is the continuous version of (7.A.10). Some examples follow.

Example 7.1. A Hoop. A thin, circular, uniformly made hoop of radius $r = 2.0$ meters and mass $M = 5.0$ kg lies in a fixed horizontal plane, and rotates about its center; see Fig. 7.5. What is the hoop's moment of inertia, I, about its vertical axis?

Figure 7.5. A rotating hoop, with a typical mass element m_1 singled out.

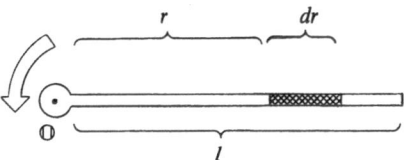

Figure 7.6. Calculating the moment of inertia of a thin uniform rod about one of its ends, O.

Consider a small portion of the hoop, having mass m_1. Its moment of inertia is

$$I_1 = m_1 r^2$$

If we consider in turn all such portions, then Eq. (7.A.9) yields

$$I = m_1 r^2 + m_2 r^2 + \cdots$$
$$= (m_1 + m_2 + \cdots)\, r^2 = Mr^2 \qquad (7.A.13)$$

In this case,

$$I = (5.0)(2.0)^2 \text{ kg meter}^2 = \underline{20 \text{ kg meter}^2}$$

(Owing to the fact that all particles are at the same distance r from the axis, this example turned out simple enough not to require a formal integration.)

Example 7.2. A Thin Rod. A thin uniform rod, of length l and mass M, is at right angles to an axis through one of its end points; see Fig. 7.6. What is its moment of inertia, I, about that axis?

Let r be the distance coordinate along the rod, starting from the axis. Consider an infinitesimal element of the rod, having length dr. Its mass is

$$dm = \left(\frac{dr}{l}\right) M \qquad (7.A.14)$$

$(dr)/l$ being the appropriate fraction of the total length. The corresponding infinitesimal moment of inertia is then

$$dI = (dm)\, r^2 = \frac{M}{l}\, r^2\, dr$$

with use of (7.A.14). Integrating both sides over the length of the rod yields

$$I = \int_0^l \frac{M}{l}\, r^2\, dr = \frac{M}{l}\int_0^l r^2\, dr = \frac{1}{3}\, Ml^2 \qquad (7.A.15)$$

(More complicated shapes may require the use of two- or three-dimensional integration. Figure 7.7

Figure 7.7. The moment of inertia, I, for various shapes. The objects' total mass in all cases is M. For additional results, see the method of parallel axes [formula (7.D.14)] and of perpendicular axes (Problem 7.40); see also Problems 7.41, 7.42. In the following all mass distributions are assumed uniform; all thin shapes are of constant thickness. (a) Single particle, off-axis; (b) thin rod about a perpendicular axis through one end; (c) thin rod about an oblique axis through one end; (d) thin rod parallel to the axis; (e) thin rod about a perpendicular axis through the middle; (f) flat rectangular vane about one of its medians; (g) flat rectangular vane about one side; (h) flat parallelogram-shaped vane about one side; (i) flat triangular vane about one side; (j) flat rectangular plate about a perpendicular axis through the center; (k) flat disk about its axis of rotational symmetry; (l) flat triangular plate about a perpendicular axis through a vertex; (m) rectangular parallelopiped about an axis through the center and parallel to a side; (n) solid right circular cylinder about its axis of rotational symmetry; (o) solid right circular cylinder about a central diameter; (p) thin circular ring about its axis of rotational symmetry; (q) thin right cylindrical shell about its axis of rotational symmetry; (r) thin right cylindrical shell about a central diameter; (s) thin circular ring about a diameter; (t) solid sphere about a diameter; (u) thin spherical shell about a diameter; (v) thin circular disk about a diameter; (w) solid right cone about its axis of rotational symmetry; (x) thin right conical shell of constant thickness about its axis of rotational symmetry.

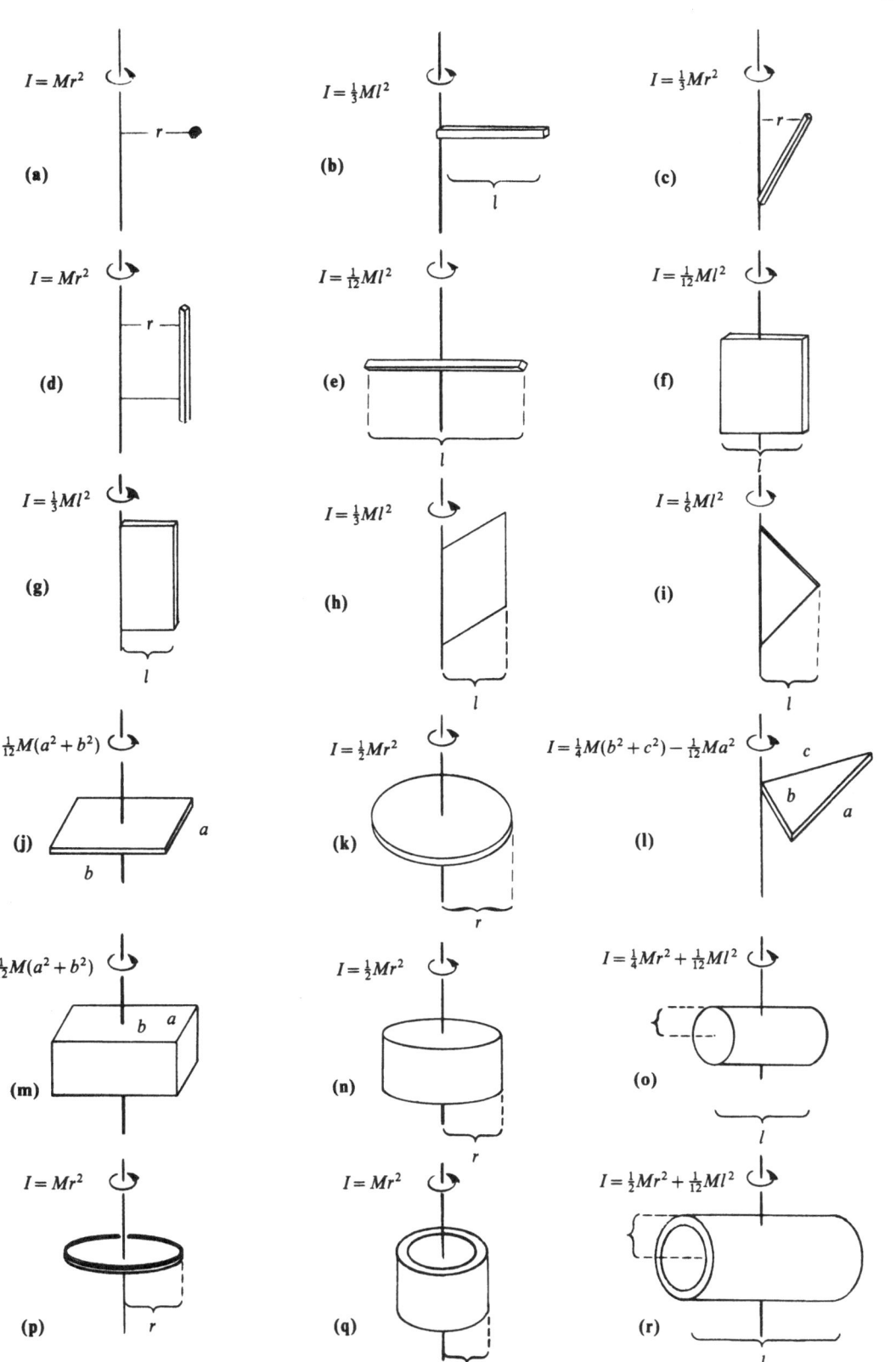

$I = Mr^2$ **(a)**

$I = \frac{1}{3}Ml^2$ **(b)**

$I = \frac{1}{3}Mr^2$ **(c)**

$I = Mr^2$ **(d)**

$I = \frac{1}{12}Ml^2$ **(e)**

$I = \frac{1}{12}Ml^2$ **(f)**

$I = \frac{1}{3}Ml^2$ **(g)**

$I = \frac{1}{3}Ml^2$ **(h)**

$I = \frac{1}{6}Ml^2$ **(i)**

$I = \frac{1}{12}M(a^2 + b^2)$ **(j)**

$I = \frac{1}{2}Mr^2$ **(k)**

$I = \frac{1}{4}M(b^2 + c^2) - \frac{1}{12}Ma^2$ **(l)**

$I = \frac{1}{12}M(a^2 + b^2)$ **(m)**

$I = \frac{1}{2}Mr^2$ **(n)**

$I = \frac{1}{4}Mr^2 + \frac{1}{12}Ml^2$ **(o)**

$I = Mr^2$ **(p)**

$I = Mr^2$ **(q)**

$I = \frac{1}{2}Mr^2 + \frac{1}{12}Ml^2$ **(r)**

(continued on next page)

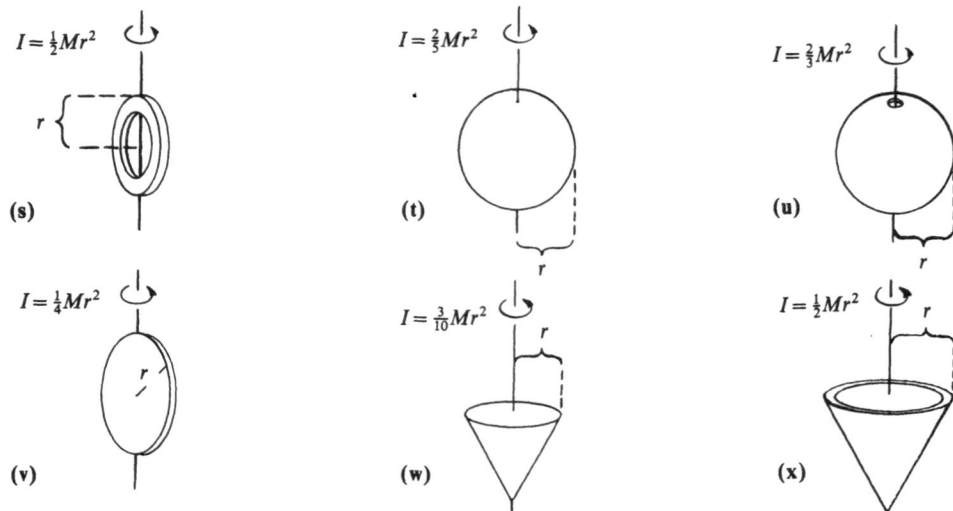

Figure 7.7 (*Continued*)

lists some moments of inertia as calculated for a variety of shapes. Some additional results can easily be obtained by application of the so-called parallel-axis theorem; see Sec. D iii further on.)

Example 7.3. A Flywheel Engine. An experimental streetcar, climbing a grade, derives its power from a solid uniform cylindrical flywheel, of mass $M = 2.0$ metric tons and radius $r = 0.50$ meter, initially spinning at 10 rev/sec. A suitable transmission connects the flywheel to the driving wheels; see Fig. 7.8. (a) Find the initial energy $\mathcal{K}$ stored in the flywheel. (b) If the car's total mass is $M' = 8.0$ metric tons, find the height h to which it can climb. (Assume it starts from rest; neglect friction.)

a. Energy. According to Fig. 7.7, we have for the flywheel's moment of inertia

$$I = \tfrac{1}{2}Mr^2$$

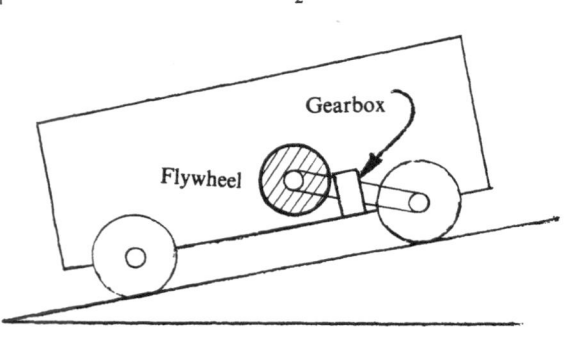

Figure 7.8. A flywheel-powered streetcar.

Therefore, from Eq. (7.A.3), we have

$$\mathcal{K} = \tfrac{1}{2}I\omega^2 = \tfrac{1}{2}(\tfrac{1}{2}Mr^2)(2\pi f)^2$$
$$= (\tfrac{1}{2})(\tfrac{1}{2})(2.0 \times 10^3)(0.50)^2 \left[(2\pi)(10)\right]^2 \text{ joules}$$
$$= \underline{4.9 \times 10^5 \text{ joules}}$$

b. Height. The kinetic energy used must equal the potential energy gained, or

$$\mathcal{K} = M'gh$$

($g = $ acceleration of gravity). Thus, we have

$$h = \frac{\mathcal{K}}{M'g} = \frac{4.9 \times 10^5}{(8 \times 10^3)(10)} \text{ meters} = \underline{6.1 \text{ meters}}$$

iii. Torque and Angular Acceleration

Having found the rotational analogue of a mass, we are now concerned with the analogue of a force. Newton's law for linear motion (say along the x axis),

$$F_{\text{tot},x} = ma_x \qquad (7.A.16)$$

will serve as a model.

Accordingly, let the **total torque** τ_{tot} (τ: lower case Greek tau), acting on a pivoted object, be defined through the following transcription of (7.A.16):

$$\boxed{\tau_{\text{tot}} = I\alpha} \qquad (7.A.17)$$

where I is the object's moment of inertia, and where α is its angular acceleration; see Eq. (7.A.2). *The torque τ_{tot} is due to the forces that tend to spin the object around its pivot*, as in the illustration of Fig. 7.9. Here τ_{tot} is an algebraic sum, $\tau_{tot} = \tau_1 + \tau_2$; each of these two terms is due to a separate force. We must learn to calculate a torque when the contributing forces are known.

iv. Torque and Work

As an intermediate step, let us consider how the total torque τ_{tot} affects an object's rotational kinetic energy $\mathscr{K}$. Again, linear motion provides a guide. We know that if a particle completes an interval dx under a force $F_{tot,x}$, its kinetic energy changes by an amount

$$d\mathscr{K} = F_{tot,x}\, dx \qquad (7.A.18)$$

(namely, the work done by $F_{tot,x}$). Similarly, if a pivoted object rotates through an angle $d\theta$ under a torque τ_{tot}, its kinetic energy changes, as we shall see, by

$$\boxed{d\mathscr{K} = \tau_{tot}\, d\theta} \qquad (7.A.19)$$

The right-hand side therefore represents the work done by a torque. Thus, a torque may be viewed as **work done per unit angle of rotation**, and is expressible in SI units of joule/rad. (The "rad" should never be omitted here; it makes clear that one is referring to a torque rather than to an energy.)

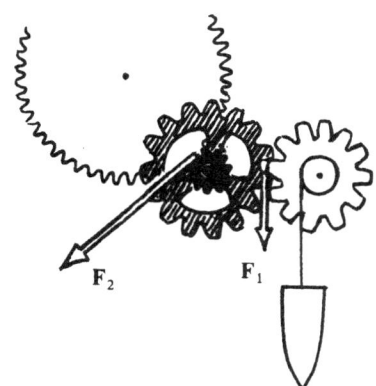

Figure 7.9. The forces $\mathbf{F}_1$, $\mathbf{F}_2$ contribute torques τ_1, τ_2 to the shaded gear with respect to its axis. Here τ_1 and τ_2 have opposite signs, corresponding to a clockwise versus counterclockwise action.

There is some practical interest in rewriting (7.A.19) in terms of the **power** supplied by τ_{tot}. Dividing both sides by dt, i.e., by the time corresponding to a rotation through $d\theta$, and recalling that $d\theta/dt = \omega$, we have

$$\boxed{\frac{d\mathscr{K}}{dt} = \tau_{tot}\,\omega} \qquad (7.A.20)$$

(The right-hand side is reminiscent of the expression $\mathbf{F}_{tot} \cdot \mathbf{v}$, representing the power supplied to a particle; see Chap. 4.)

Proof. Equation (7.A.19), or equivalently, (7.A.20), still needs to be proved. Starting from the expression for $\mathscr{K}$, and noting that $I = \text{const}$, we have

$$\frac{d\mathscr{K}}{dt} = \frac{d}{dt}\left(\frac{1}{2}I\omega^2\right) = I\frac{d\omega}{dt}\,\omega = \tau_{tot}\,\omega$$

The last step comes from definition (7.A.17) written in the form $\tau_{tot} = I\, d\omega/dt$; thus we obtain Eq. (7.A.20).

Contributing Torques

If the total torque on an object is the sum of several contributions,

$$\tau_{tot} = \tau_1 + \tau_2 + \tau_3 + \cdots \qquad (7.A.21)$$

then the total work also splits up in that manner: Equation (7.A.19) becomes

$$d\mathscr{K} = \tau_1\, d\theta + \tau_2\, d\theta + \tau_3\, d\theta + \cdots \qquad (7.A.22)$$

Isolating the contribution of, for example, τ_1, we have

$$\tau_1\, d\theta = d\mathscr{K} - \tau_2\, d\theta - \tau_3\, d\theta - \cdots \qquad (7.A.23)$$

This equation shows the fate of the work done by a torque τ_1 on a rotating object: that work goes partly into changing the rotational kinetic energy, and is partly expended against other torques. If $d\mathscr{K} = 0$ (steady rotation), then the work of τ_1 is done entirely against $-\tau_2$, $-\tau_3$, etc.

To Summarize

Any torque τ, acting on an object that rotates through an angle $d\theta$, does an amount of work

$$\boxed{d\mathcal{W} = \tau\,d\theta} \qquad (7.A.24)$$

Correspondingly, if the object has an angular frequency ω, the power transferred by τ is

$$\boxed{\frac{d\mathcal{W}}{dt} = \tau\omega} \qquad (7.A.25)$$

Example 7.4. A Motor. The shaft of a certain electric motor spins at a constant frequency $f = 6$ rev/sec. If the motor puts out 1000 watts of power, how much torque, τ, does it deliver?

The motor's shaft is the rotating object under consideration. (Since its kinetic energy does not change, the 1000 watts is being expended against an equal and opposite torque $-\tau$, perhaps due to a weight being lifted, or to friction.) Using positive quantities, we get

$$\text{Power} = \tau\omega = (\tau)(2\pi f)$$

so that

$$\tau = \frac{\text{power}}{2\pi f} = \frac{1000}{(2\pi)(6)}\,\frac{\text{joules}}{\text{rad}} = 27\,\frac{\text{joules}}{\text{rad}}$$

B. Torques in Terms of Forces

In the preceding section, torques were discussed in terms of their *effect* on the rotational motion of a pivoted object. Here we turn to *what causes a torque*. This means that we must first examine how the forces acting on an object are related, in space, to that object's axis of rotation.

i. Torque and Lever Arm

We *now study in detail how a force* **F** *produces a torque* τ. The method consists of comparing the work, $d\mathcal{W} = \tau\,d\theta$, done by τ, with that same work as calculated from **F**.

These concepts are conveniently visualized in terms of an ordinary door, grasped by the edge at

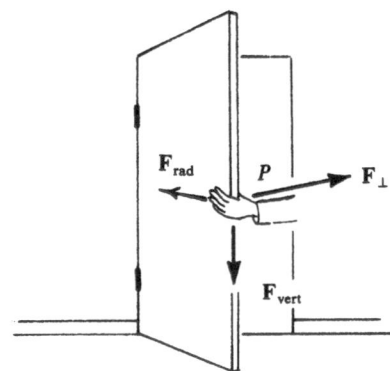

Figure 7.10. Three mutually perpendicular forces on a pivoted rigid body.

some point P. Figure 7.10 shows three kinds of force which may be exerted at that point. A vertical force such as $\mathbf{F}_{vert}$ cannot do any work since the hinges prevent vertical motion. A radial force such as $\mathbf{F}_{rad}$, which acts perpendicularly to the hinges, but along the plane of the door, is similarly "of no use." Of the three forces shown, only $\mathbf{F}_{\perp}$ (read "F perpendicular"), which is directed at right angles to the plane of the door, will transfer energy when the door rotates.

Figure 7.11 shows a top view, or axial view, of the situation. For a small rotation angle $d\theta$, the displacement of point P is the vector $d\mathbf{r}$, so that the work done by **F** is

$$d\mathcal{W} = \mathbf{F} \cdot d\mathbf{r} = F_{\perp}\,ds \qquad (7.B.1)$$

where the path length $ds = |d\mathbf{r}|$ is given by

$$ds = r\,d\theta \qquad (7.B.2)$$

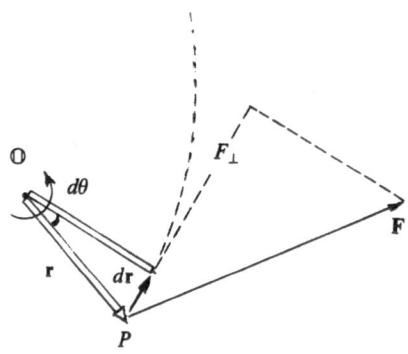

Figure 7.11. A top view ("axial view") of Fig. 7.10. The force **F**, which is not necessarily in the plane of the figure, is the resultant of all three forces, $\mathbf{F} = \mathbf{F}_{vert} + \mathbf{F}_{rad} + \mathbf{F}_{\perp}$.

r being the width of the door. The desired result for $d\mathcal{W}$ is then

$$d\mathcal{W} = rF_\perp \, d\theta \qquad (7.B.3)$$

This allows direct comparison with the effect of τ. We have, for the same $d\mathcal{W}$,

$$d\mathcal{W} = \tau \, d\theta \qquad (7.B.4)$$

so that

$$\boxed{\tau = rF_\perp} \qquad (7.B.5)$$

Let the vector $\mathbf{r}$ in Fig. 7.11 be called the **lever arm** associated with the force $\mathbf{F}$. Then (7.B.5) says that *the torque due to a force equals the length of the lever arm, multiplied by that component of the force that is perpendicular to the lever arm and to the axis of rotation.*

Static Torque

Formula (7.B.5) makes no reference to how fast a motion is taking place. Thus, the expression (7.B.5) for a torque is also valid in the limit *where no motion occurs at all.*

[*Units*: Equation (7.B.5) implies a torque unit of **newton meter**, equivalent to, and more often used than, the **joule/radian**. The fact that radians are conventionally mentioned in the latter designation and not in the former is a matter of custom.]

Algebraic Sum of Torques

Figure 7.12 shows how torques are combined algebraically according to the "counterclockwise =

positive, clockwise = negative" convention. The following example illustrates several of the ideas developed so far.

Example 7.5. Slamming a Door. A 16-kg, uniformly built, rectangular door, 0.75 meter wide, is originally at rest and open, at 45° to its closed position; see Fig. 7.13. Someone shuts it by exerting, near the edge, a horizontal force, maintained at 30° to the plane of the door; the force's magnitude is a constant 80 newton. (a) Find the applied torque about the hinges. (b) When the door slams, how much energy is dissipated in sound, heat, or breakage? (Assume frictionless hinges.) (c) How fast, in rad/sec, is the door rotating just before impact? (d) What is its angular acceleration? (e) How much time is needed to complete the motion?

---

a. Torque. In the notation of Fig. 7.11, we have

$$\tau = rF_\perp = rF \sin 30°$$
$$= (0.75)(80)(0.50) \text{ newton meters}$$
$$= \underline{\underline{30 \text{ newton meters}}}$$

b. Energy. The work done is

$$\mathcal{W} = \tau\theta \qquad (7.B.6)$$

[This is just Eq. (7.A.24) for constant τ.] Here $\theta = 45° = \pi/4$ rad, so that the kinetic energy is

$$\mathcal{K} = \mathcal{W} = (30)\left(\frac{\pi}{4}\right) \text{ joules} = \underline{\underline{24 \text{ joules}}}$$

c. Final angular velocity. From $\mathcal{K} = \frac{1}{2}I\omega^2$, we find

$$\omega = \left(\frac{2\mathcal{K}}{I}\right)^{1/2} \qquad (7.B.7)$$

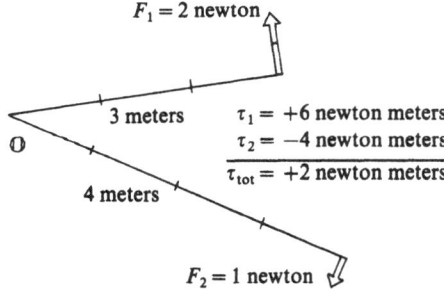

F_1 = 2 newton

3 meters $\tau_1 = +6$ newton meters
                $\tau_2 = -4$ newton meters
O $\overline{\tau_{\text{tot}} = +2 \text{ newton meters}}$

4 meters

$F_2 = 1$ newton

Figure 7.12. Here a positive torque τ_1, due to $\mathbf{F}_1$, and a negative one, τ_2, due to $\mathbf{F}_2$, add algebraically to give a net positive (i.e., counterclockwise) torque τ_{tot}.

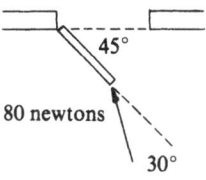

45°

80 newtons

30°

Figure 7.13. The door in Example 7.5.

The value of I can be looked up in Fig. 7.7g:

$$I = \tfrac{1}{3}Ml^2 \qquad (7.B.8)$$

where M, l are the door's mass and width. Equation (7.B.7) becomes

$$\omega = \frac{1}{l}\left(\frac{6\mathscr{H}}{M}\right)^{1/2}$$

$$= \frac{1}{0.75}\left[\frac{(6)(24)}{16}\right]^{1/2}\frac{\text{rad}}{\text{sec}} = \underline{4.0\,\frac{\text{rad}}{\text{sec}}}$$

d. Angular acceleration. The rotational equation of motion, $\tau = I\alpha$, gives

$$\alpha = \frac{\tau}{I} = \frac{3\tau}{Ml^2} \qquad (7.B.9)$$

see Eq. (7.B.8). Hence

$$\alpha = \frac{(3)(30)}{(16)(0.75)^2}\frac{\text{rad}}{\text{sec}^2} = \underline{10\,\frac{\text{rad}}{\text{sec}^2}}$$

e. Time. In Eq. (7.B.9), $\tau = \text{const}$ implies $\alpha = \text{const}$; we are dealing with **uniformly accelerated angular motion**, starting from rest. Therefore we can write

$$\theta = \tfrac{1}{2}\alpha t^2 \qquad (7.B.10)$$

in analogy with $x = \tfrac{1}{2}a_x t^2$, valid in uniformly accelerated linear motion. (The symbols are different, but the mathematics can be taken over.) Solving for t, we obtain

$$t = \left(\frac{2\theta}{\alpha}\right)^{1/2} = \left[\frac{(2)(\pi/4)}{10}\right]^{1/2}\text{sec} = \underline{0.40\ \text{sec}}$$

ii. Moment Arm

The torque contributed by a force, $\tau = rF_\perp$, can be calculated from an alternative, geometric construction, which is often used. Figure 7.14 shows an axial view of a force $\mathbf{F}$ acting at a point P. This figure exhibits two new concepts.

1. *The **line of action** of the force* $\mathbf{F}$. This is an infinite straight line, having the direction of $\mathbf{F}$, and passing through that force's point of application P.

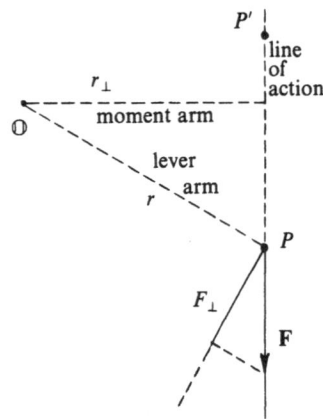

Figure 7.14. A force $\mathbf{F}$, applied to point P. If it were applied to any other point, say P', on the same line of action, its torque about O would still be the same, $r_\perp F$, in magnitude.

2. *The **moment arm** $r_\perp$ of* $\mathbf{F}$. This is the distance between the axis of rotation and the line of action; $r_\perp$ is therefore measured perpendicularly to both these lines, and is in general shorter than the lever arm r.

To demonstrate an alternative calculation, we assume for simplicity that $\mathbf{F}$, r, and $r_\perp$ are all in the plane of the figure. Then, by the similarity of two triangles, we see that

$$\frac{F_\perp}{r_\perp} = \frac{F}{r}$$

so that, in magnitude, we have

$$\boxed{|\tau| = rF_\perp = r_\perp F} \qquad \text{(F perpendicular to the axis)}$$

$$(7.B.11)$$

Thus, $|\tau|$ is also equal to *the moment arm times the force's magnitude.*

The new formula shows (see Fig. 7.14), that

> The torque due to $\mathbf{F}$ remains the same when that force's point of application is shifted along its line of action.

$$(7.B.12)$$

(This statement is true even if $\mathbf{F}$ is at an angle to the plane of the figure. Indeed, if $\mathbf{F}$ is shifted along its line of action, then the projection of $\mathbf{F}$ on the plane of the figure is similarly shifted along *its* line of action; the component of **F** perpendicular to the figure does not enter into the calculation of the torque.)

> **Example 7.6. A Wrench and Bolt.** A wrench of length $r = 0.25$ meter is at an angle of $30°$ to the vertical, see Fig. 7.15. A downward force **F** of magnitude 80 newton is exerted at the free end. What torque does **F** exert on the bolt, which is assumed horizontal and perpendicular to the wrench?
>
> _____
>
> Let us use, as in part (a) of the figure, the formula
>
> $$|\tau| = rF_\perp = (r)(F\sin 30°)$$
>
> $$= (0.25)(80)(0.50) \text{ newton meters}$$
>
> $$= \underline{\underline{10 \text{ newton meters}}}$$
>
> Alternatively, as in part (b) of the figure,
>
> $$|\tau| = r_\perp F = (r\sin 30°)(F)$$
>
> giving the same result. The sign of τ is negative according to our convention.

Forces Without Torque

We note that *a force whose line of action intersects an axis contributes zero torque about that axis.* To see this, we make $r_\perp = 0$ in Fig. 7.14. (As a limiting case, the force and the axis could be parallel.) This zero-torque condition would be true for any combination of the forces $\mathbf{F}_{vert}$ and $\mathbf{F}_{rad}$ on the door of Fig. 7.10 with respect to the hinges. Of course, even if a force is torqueless with respect to a certain axis, it will in general give nonzero torques about other axes.

iii. The Torque of Gravity

How much torque does gravity exert on a rigid body? The answer, of course, depends on the axis about which the torque is measured; but, no matter what the axis, *the torque has the value it would have if the whole object were concentrated at its center of mass*; in other words,

> The total weight vector of a rigid body may be considered as acting at the center of mass.

$$(7.B.13)$$

(In particular, gravity exerts zero torque about any axis through the center of mass.)

To understand this, we look back at Chap. 5. There we learned that the gravitational potential energy, $\mathscr{U}$, of an extended system is evaluated as if the whole system were concentrated at its center of mass $\mathbb{C}$:

$$\mathscr{U} = \mathscr{U}_\mathbb{C} \qquad (7.B.14)$$

We now compare two gravitational torques: τ, on some object shown in Fig. 7.16; and $\tau_\mathbb{C}$, on the same mass, concentrated at $\mathbb{C}$, and replacing the original object. Under the same small rotation $d\theta$, about the same $\mathbb{O}$, their potential energies decrease by $-d\mathscr{U} = -d\mathscr{U}_\mathbb{C}$; see (7.B.14). These are the amounts of work done by gravity, or, from (7.A.24), $\tau\, d\theta = \tau_\mathbb{C}\, d\theta$. Thus we have $\tau = \tau_\mathbb{C}$ as required by (7.B.13).

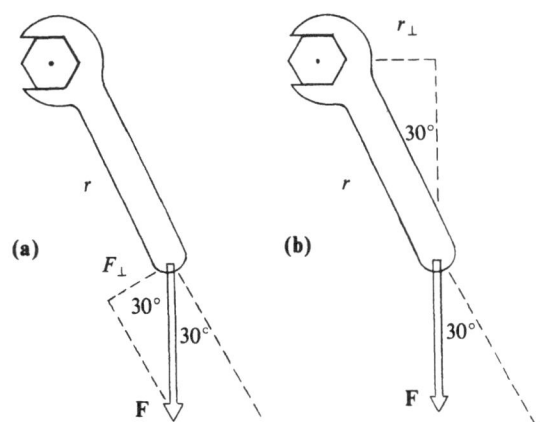

Figure 7.15. Two equivalent methods for calculating the torque exerted on a bolt through the intermediary of a wrench. (The weight of the wrench contributes its own additional torque, not calculated here.)

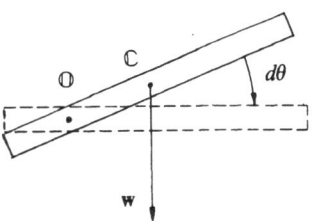

Figure 7.16. An object with center of mass $\mathbb{C}$ and weight **w** is allowed to rotate through $d\theta$ under the torque of gravity.

Example 7.7. A Horizontal Rod. A uniform horizontal flagpole has mass $M = 30$ kg and length $l = 4$ meters. What torque τ does it exert about its point of attachment P? (See Fig. 7.17.)

The whole weight vector $M\mathbf{g}$ may be considered to act at the midpoint $\mathbb{C}$ of the flagpole; see the figure. Therefore, the required torque has magnitude

$$|\tau| = \left(\frac{l}{2}\right)(Mg) = \left(\frac{4}{2}\right)(30)(10) \text{ newton meters}$$

$$= \underline{600 \text{ newton meters}}$$

Center of Gravity and Center of Mass: A Caution

Viewed on a planetary or astronomical scale, the acceleration of gravity is not uniform, but varies from place to place. The force of gravity then does not always act on a system's center of mass, but on a different point, the **center of gravity**. The latter is synonymous with "center of mass" only under a uniform vector $\mathbf{g}$.

C. Rotational Equilibrium

A whole new class of equilibrium problems become soluble through the application of the torque and center-of-mass concepts. These calculational methods have a venerable history, going as far back as Archimedes (third century B.C.), who successfully analyzed levers.

i. The Equilibrium Conditions

A rigid body is said to be in **rotational equilibrium** about a fixed axis if its angular acceleration

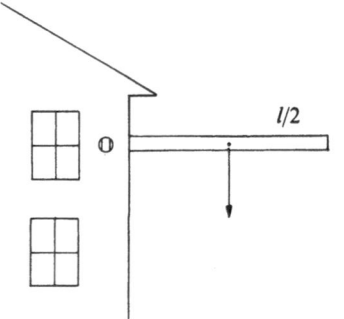

Figure 7.17. This horizontal uniform flagpole exerts a torque $\tau = -\frac{1}{2}lmg$ about $\mathbb{O}$.

α about that axis vanishes, i.e., if it is rotating at a constant angular velocity ω. We shall be mostly interested in the case where $\omega = 0$, i.e., where there is no motion whatsoever. Under this condition of **static equilibrium**, it is clear that *any axis is a fixed axis*. Therefore we can choose any desired location as the site of an imaginary pivot; a wise decision may greatly simplify our mathematics, as later examples will show.

The equilibrium conditions for a rigid body are

$$\boxed{\mathbf{F}_{tot} = 0 \qquad \text{("translational equilibrium")}}$$

$$(7.\text{C}.1)$$

and

$$\boxed{\tau_{tot} = 0 \qquad \text{("rotational equilibrium")}}$$

$$(7.\text{C}.2)$$

These are nothing but the equations of motion, $\mathbf{F}_{tot} = m\mathbf{a}$, and $\tau_{tot} = I\alpha$, taken with $\mathbf{a} = 0$, $\alpha = 0$. Equation (7.C.1) and its implications were studied at length in Chap. 3.

ii. Examples

The best-known elementary application of rigid-body equilibrium deals with the **lever**, which we have already treated by the energy method in Example 4.17 of Chap. 4. Its analysis in terms of torques is left to Problem 7.14.

The following examples illustrate the simultaneous use of conditions (7.C.1) and (7.C.2). We shall see that the success of an equilibrium analysis depends largely on (1) identifying the appropriate rigid body; (2) choosing a convenient axis of rotation.

Example 7.8. A Weighted Plank. A uniform plank of length $l = 2.0$ meters and weight $w = 100$ newtons is carried horizontally by workers A and B, one at each end; see Fig. 7.18. A can of paint, whose weight is $W = 150$ newtons, rests on the plank at a distance $a = 0.40$ meter from A. How much force, F_A or F_B, does each worker have to exert vertically for equilibrium to prevail?

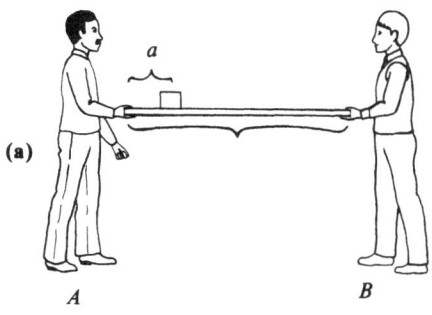

(a)

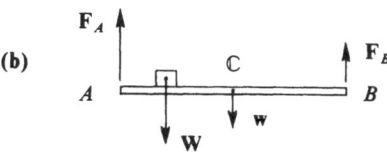

(b)

Figure 7.18. (a) Two workers, carrying a plank with a can of paint. Equilibrium is assumed. (b) The force diagram for the situation shown in (a).

Part (b) of the figure shows all the forces acting on the rigid body comprising the plank and the can; the plank's center of mass C is at its midpoint. Translational equilibrium, $F_{tot} = 0$, gives

$$F_A + F_B = w + W \qquad (7.C.3)$$

To enforce rotational equilibrium, any axis may be used; let us choose it through A. In this way, F_A will contribute zero to the resultant torque. The equilibrium condition $\tau_{tot} = 0$ then gives, in terms of the individual torques contributed by F_A, W, w, F_B:

$$(0)(F_A) - aW - \frac{l}{2}w + lF_B = 0$$

so that

$$F_B = \frac{a}{l}W + \frac{1}{2}w$$

$$= \left[\left(\frac{0.40}{2.0}\right)(150) + \left(\frac{1}{2}\right)(100)\right] \text{newtons}$$

$$= \underline{80 \text{ newtons}}$$

Inserting this result in (7.C.3) gives

$$F_A = (100 + 150 - 80) \text{ newtons} = \underline{170 \text{ newtons}}$$

The simplification resulting from choosing the axis on the line of action of a force, especially of an unknown force, should be carefully noted.

Example 7.9. An Equal-Arm Balance. The balance beam shown in Fig. 7.19 has an arm length l; it also has a depth s between the fulcrum P and the points of suspension A, B of the scales. If all the weights are negligible except w and W as shown, find the angle θ of the beam with the horizontal. Assume equilibrium.

Let the beam be the rigid body considered. In part (b) of the figure, let us choose an axis of rotation through some point ⓪, anywhere on the vertical line through the fulcrum. (There is no reason why ⓪ should be within the rigid body.) Translational equilibrium is already enforced in that figure, the fulcrum exerting an upward force $w + W$. Actually, this will not be relevant with our choice of rotation axis. We see from the figure that $\mathbf{W}$ has the moment arm

$$r_{\perp W} = l\cos\theta - s\sin\theta$$

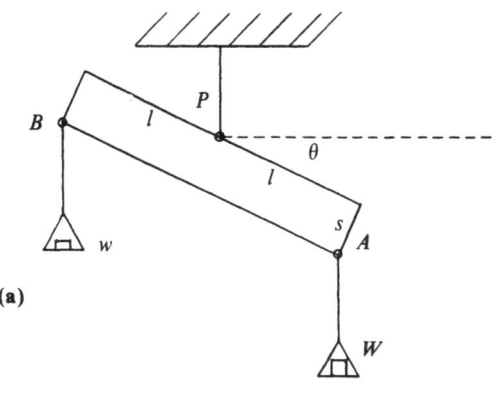

(a)

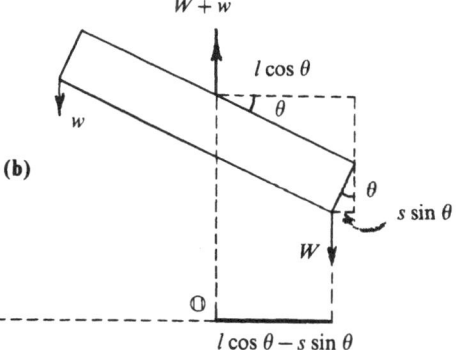

(b)

Figure 7.19. An equal-arm balance. The equilibrium conditions are being applied to the rectangular beam, whose own weight is neglected.

Similarly, **w** will have the moment arm (not shown)

$$r_{\perp w} = l \cos \theta + s \sin \theta$$

Rotational equilibrium about $\mathbb{O}$ gives, in terms of the torques contributed by **W**, **w**, and $-(\mathbf{W}+\mathbf{w})$,

$$-(l \cos \theta - s \sin \theta) W$$
$$+ (l \cos \theta + s \sin \theta) w + (0)(W+w) = 0$$

so that

$$\tan \theta = \left(\frac{l}{s}\right)\left(\frac{W-w}{W+w}\right) \qquad (7.C.4)$$

iii. Two and Three Forces in a Plane

Here we list a few geometrical constructions that are of use in simplifying force diagrams for rigid bodies.

1. Two Nonparallel Forces

Given a pair of forces $\mathbf{F}_1$, $\mathbf{F}_2$ with lines of action lying in a common plane, say the plane of Fig. 7.20, find the line of action of a single equivalent force

$\mathbf{F}_{tot} = \mathbf{F}_1 + \mathbf{F}_2$. (Here "equivalent" means not only in terms of the total force, but in terms of the total torque about any axis.)

Construction: See part (b) of the figure.

2. Two Parallel Forces

What happens to the preceding construction if $\mathbf{F}_1$ and $\mathbf{F}_2$ are parallel?

Construction: See Fig. 7.21, either part (a) or part (b) according to whether $\mathbf{F}_1$ and $\mathbf{F}_2$ point in the same or opposite directions. Let $\mathbf{F}_{tot}$ have its line of action at distances $r_{\perp 1}$, $r_{\perp 2}$ from the given lines of action. Then we have

$$\frac{r_{\perp 2}}{r_{\perp 1}} = \frac{F_1}{F_2} \qquad (7.C.5)$$

in both cases. (The resultant line of action is closer to the larger force.) The "inside" solution is used in (a), while the "outside" solution is used in (b). Verification of this result is left to the reader: the procedure is to choose an arbitrary axis perpendicular to the plane of the forces and then to compare the net torque from the pair $\mathbf{F}_1$, $\mathbf{F}_2$ with that from the properly situated $\mathbf{F}_{tot}$.

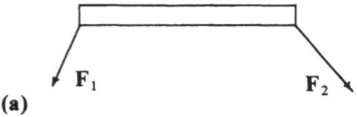

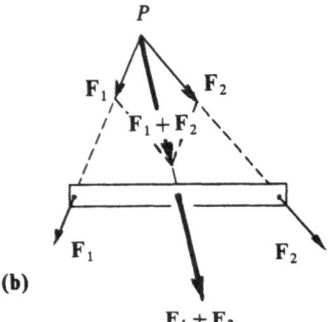

Figure 7.20. (a) A pair of forces $\mathbf{F}_1$, $\mathbf{F}_2$ in a plane. (b) Construction for finding the correct line of action of the resultant $\mathbf{F}_1 + \mathbf{F}_2$. We take advantage of the fact that all forces may be shifted along their line of action; $\mathbf{F}_1$, $\mathbf{F}_2$, and therefore their resultant, can all be assumed to act at the intersection P.

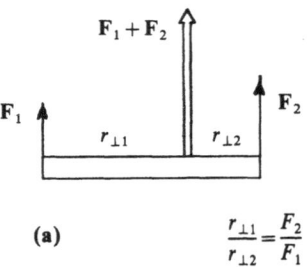

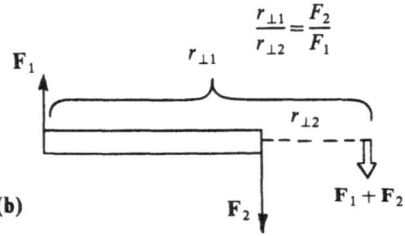

Figure 7.21. The correct line of action for the resultant $\mathbf{F}_1 + \mathbf{F}_2$ is found by the "leverlike" relation $r_{\perp 1}/r_{\perp 2} = F_2/F_1$. The geometry is quite different in the parallel (a) and antiparallel (b) cases.

3. Couple

The preceding construction breaks down immediately if $\mathbf{F}_1 = -\mathbf{F}_2$, because the "outside solution" of Fig. 7.21b cannot accommodate $r_{\perp 1} = r_{\perp 2}$:

> There exists no single force that is equivalent to a pair of equal and opposite forces with distinct lines of action.

$$(7.C.6)$$

Such a pair of forces is called a **couple**. It possesses another curious feature:

> A couple exerts the same torque
> $$\tau_{\text{tot}} = r_\perp F$$
> about all the axes perpendicular to the couple's plane.

$$(7.C.7)$$

See Fig. 7.22, where $\mathbf{F}$, $-\mathbf{F}$ are the individual forces of the couple, and where $r_\perp$ is the distance between their lines of action. This result is proved in the figure.

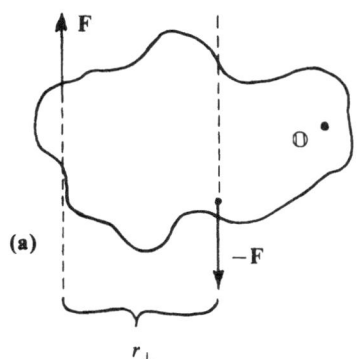

(a)

$r_\perp$

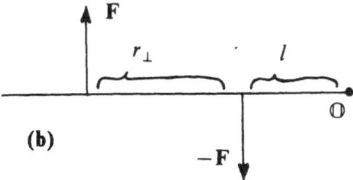

(b)

Figure 7.22. A couple, consisting of two equal and opposite forces with distinct lines of action. There is no equivalent resultant. The torque does not depend on the location O of the axis. As a proof, draw a line through O and perpendicular to $\mathbf{F}$ and $-\mathbf{F}$; see part (b) of the figure. The clockwise torque about O is $(F)(r_\perp + l) - (F)(l) = F_r$, independent of l.

From (7.C.7) we see, in particular, that equilibrium ($\tau_{\text{tot}} = 0$) requires $r_\perp = 0$ for a nonzero F. Thus:

> Equilibrium under two forces requires that they have the same line of action.

$$(7.C.8)$$

This is illustrated in Fig. 7.23.

4. Three Coplanar Forces

Under what conditions does one obtain equilibrium from three forces acting in the same plane? A partial but useful answer is as follows:

> Equilibrium under three coplanar forces requires that they be concurrent or parallel.

$$(7.C.9)$$

That is, the three lines of action must intersect at the same point (which is "at infinity" in the parallel case).

To see this, we return to Fig. 7.20. In order that a third force $\mathbf{F}_3$ (not shown in the figure) be in equilibrium with $\mathbf{F}_1$ and $\mathbf{F}_2$ (both shown), we need to have

$$\mathbf{F}_3 = -(\mathbf{F}_1 + \mathbf{F}_2)$$

(translational equilibrium), and, to avoid causing a couple, this force must have the same line of action as the force $\mathbf{F}_1 + \mathbf{F}_2$ shown in the figure. Thus, the line of action of $\mathbf{F}_3$ must go through P, and the three forces $\mathbf{F}_1$, $\mathbf{F}_2$, $\mathbf{F}_3$ are concurrent.

(The parallel case can be worked out on the basis of Fig. 7.21.)

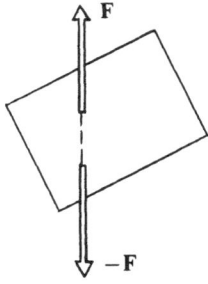

Figure 7.23. An object under translational *and* rotational equilibrium.

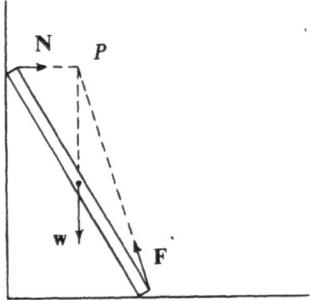

Figure 7.24. A leaning ladder as an illustration of the concurrence (in this case, at P) required for the equilibrium of three coplanar forces (here N, w, F). The wall is assumed frictionless, so that N must be normal to the wall.

Example 7.10. A Leaning Ladder. The convenience of condition (7.C.9) may be seen in Fig. 7.24, where a uniform ladder of weight w is leaning in equilibrium against a slippery wall. As a result of combined static friction and normal force, the floor exerts a total force **F** on the ladder. Does **F** point along the ladder, below the ladder, or above the ladder?

A simple construction gives the answer. First, draw **N**, the normal force exerted by the wall; then draw **w**, acting at the ladder's midpoint; finally, the concurrence of the three forces **N**, **w**, **F** (at point P) shows that

F points above the ladder.

D. Rotation and the Center of Mass

The center of mass $\mathbb{C}$ plays a privileged role in rigid-body rotation. For example, under a zero total force, point $\mathbb{C}$ undergoes uniform motion; in particular, it may be at rest; *such a free objects spins about its own center of mass.* In this section, some additional properties of $\mathbb{C}$ are discussed; they will prove useful in analyzing rigid-body rotation, especially for *rolling objects.*

i. Kinetic Energy and the Center of Mass

The *total kinetic energy* $\mathscr{K}_{\text{syst}}$ of an extended system may be conveniently split into two portions, **translational** and **internal**:

$$\mathscr{K}_{\text{syst}} = \mathscr{K}_{\text{transl}} + \mathscr{K}_{\text{int}} \qquad (7.D.1)$$

Let us discuss each of these two contributions.

The value of $\mathscr{K}_{\text{transl}}$ is obtained if one assumes that the whole mass m_{syst} of the system is concentrated at its center of mass $\mathbb{C}$. Thus, $\mathscr{K}_{\text{transl}}$ is defined by

$$\mathscr{K}_{\text{transl}} = \tfrac{1}{2} m_{\text{syst}} v_{\mathbb{C}}^2 \qquad (7.D.2)$$

$v_{\mathbb{C}}$ being the velocity of the center of mass. On the other hand, the amount $\mathscr{K}_{\text{int}}$ is what would be measured by an observer traveling along with the center of mass. For such an observer, the center of mass is stationary; in the case of a rigid body, $\mathscr{K}_{\text{int}}$ is the **rotational kinetic energy**. An algebraic proof of Eq. (7.D.1) is found in Note (i) at the end of this chapter.

ii. Rolling Objects

The application of these ideas to rolling objects is best illustrated by the following two examples.

Example 7.11. Kinetic Energy of a Rolling Hoop. A thin uniform circular hoop of mass M and radius r rolls without slipping on a flat surface, at a speed v. What is its total kinetic energy $\mathscr{K}_{\text{syst}}$?

In the decomposition (7.D.1), $\mathscr{K}_{\text{syst}} = \mathscr{K}_{\text{transl}} + \mathscr{K}_{\text{int}}$, we can set immediately

$$\mathscr{K}_{\text{transl}} = \tfrac{1}{2} M v^2 \qquad (7.D.3)$$

The second term (rotational energy relative to the hoop's center) is

$$\mathscr{K}_{\text{int}} = \tfrac{1}{2} I \omega^2 \qquad (7.D.4)$$

where I is the hoop's moment of inertia,

$$I = M r^2 \qquad (7.D.5)$$

and where ω is its angular velocity. Seen from the hoop's center $\mathbb{C}$, the flat surface moves backwards at speed v. Therefore the circular speed of the rim relative to $\mathbb{C}$ is v, and we have

$$\omega = \frac{v}{r} \qquad (7.D.6)$$

Equation (7.D.4) becomes

$$\mathscr{K}_{\text{int}} = \left(\frac{1}{2}\right)(Mr^2)\left(\frac{v}{r}\right)^2 = \frac{1}{2}Mv^2$$

(the rotational energy of a rolling hoop equals its translational energy); the conclusion is

$$\mathscr{K}_{\text{syst}} = \tfrac{1}{2}Mv^2 + \tfrac{1}{2}Mv^2 = \underline{\underline{Mv^2}} \quad (7.D.7)$$

One of the neatest and most familiar of gravitational experiments consists of clocking an object as it rolls down an incline. The following example shows one way of analyzing that situation on the basis of what we have just learned.

Example 7.12. Acceleration of a Rolling Hoop. What is the acceleration, a, of the hoop in Example 7.11 as it rolls down a plane whose angle with the horizontal is φ?

Absence of kinetic friction means conservation of mechanical energy,

$$\mathscr{U} + \mathscr{K} = \text{const} \quad (7.D.8)$$

Here, as we recall from Eq. (5.D.17) of Chap. 5, the potential energy is

$$\mathscr{U} = Mgz = Mgs\sin\varphi \quad (7.D.9)$$

(z = altitude of the hoop's center, $\mathbb{C}$, above a fixed level; s = coordinate of $\mathbb{C}$ parallel to the incline; see Fig. 7.25.) On the other hand, we have just seen that the kinetic energy is

$$\mathscr{K} = Mv^2 \quad (7.D.10)$$

(not $\tfrac{1}{2}Mv^2$). Thus, Eq. (7.D.8) reads

$$Mgs\sin\varphi + Mv^2 = \text{const}$$

or

$$gs\sin\varphi + v^2 = \text{const} \quad (7.D.11)$$

which describes a *uniformly accelerated* motion. Indeed, in Chap. 4 we saw that any constant acceleration implies

$$2as + v^2 = \text{const} \quad (7.D.12)$$

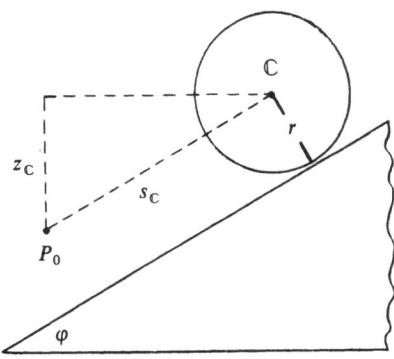

Figure 7.25. A hoop rolling down an incline. Point P_0 is a fixed chosen origin for $z_\mathbb{C}$ and $s_\mathbb{C}$.

Comparing (7.D.11) and (7.D.12), we find

$$a = \tfrac{1}{2}g\sin\varphi \quad (7.D.13)$$

which is half the acceleration of a particle sliding frictionlessly down the same plane. Ignoring the moment of inertia in an inclined-plane experiment can result in a grossly wrong measurement of g.

An alternative method for calculating a, based on torque rather than on energy, is examined in Problem 7.44.

iii. The Parallel-Axis Theorem

Separating the kinetic energy into a translational and a rotational part, as we have done, has yet another use. It provides a shortcut for deducing some moments of inertia from others that are already known. Suppose a rigid body of mass M has a known moment of inertia $I_\mathbb{C}$ about an axis through the center of mass $\mathbb{C}$. The problem is to find the same object's moment of inertia I_P about a new axis, say through point P, as in the eccentric disk of Fig. 7.26. The two axes are parallel and at a distance l from each other. The answer is

$$\boxed{I_P = I_\mathbb{C} + Ml^2} \quad (7.D.14)$$

That is to say, one only has to add the particlelike contribution Ml^2 to the original value. We see that, *of all axes in a fixed direction, the one through $\mathbb{C}$ is associated with the smallest moment of inertia.*

Formula (7.D.14) is readily established from a kinetic energy calculation. Let the rigid body rotate

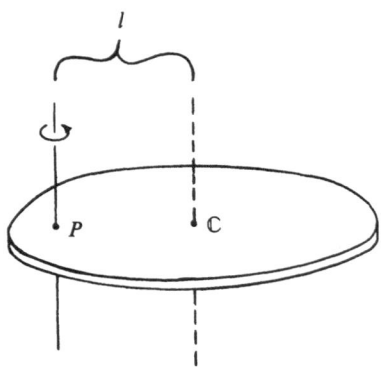

Figure 7.27. A braking car. The torque is clockwise because it must be referred to the car's center of mass $\mathbb{C}$, which is higher than the road.

Figure 7.26. The parallel-axis theorem relates the moment of inertia I_C (about an axis through the center of mass $\mathbb{C}$) to the moment of inertia I_P (about a parallel axis a distance l away).

---

(1) The center of mass, $\mathbb{C}$, behaves like a particle of mass M under the combined existing forces. This was explained in Chap. 5.

(7.D.18)

with angular velocity ω about P. According to the decomposition of Eq. (7.D.1), we have

$$\mathcal{K} = \mathcal{K}_{\text{transl}} + \mathcal{K}_{\text{int}} = \tfrac{1}{2}Mv_C^2 + \tfrac{1}{2}I_C\omega^2$$
$$= \tfrac{1}{2}M(\omega l)^2 + \tfrac{1}{2}I_C\omega^2 \qquad (7.D.15)$$

(We note that *the angular velocity ω is the same relatively to any parallel axis*, since the time needed for a complete turn is independent of which axis is considered motionless.) Another way of expressing the same quantity $\mathcal{K}$ is

$$\mathcal{K} = \tfrac{1}{2}I_P\omega^2 \qquad (7.D.16)$$

(fixed-axis rotation about P). When the right sides of (7.D.15) and (7.D.16) are equated, ω^2 drops out, with the desired relation (7.D.14) as the result.

Example 7.13. An Eccentric Disk. What is the moment of inertia of the disk in Fig. 7.26 if P is at the rim? (Assume the disk is uniform, with mass M and radius r).

The list of Fig. 7.7 gives $I_C = \tfrac{1}{2}Mr^2$, so that, from (7.D.14),

$$I_P = \tfrac{1}{2}Mr^2 + Mr^2 = \tfrac{3}{2}Mr^2 \qquad (7.D.17)$$

(This formula is equally valid if the disk has non-zero thickness, i.e., for a cylinder.)

iv. Angular Acceleration About the Center of Mass

The complete response of a rigid body (of mass M) to forces and torques can be summarized in two statements.

(2) The angular acceleration about any axis through $\mathbb{C}$ can be calculated from the existing torques about that axis, just as if $\mathbb{C}$ were at rest.

(7.D.19)

The latter statement can be understood without mathematics. An accelerated observer (i.e., moving with $\mathbb{C}$) sees all the same forces as a stationary observer, *plus* an effective uniform gravity pointing oppositely to his acceleration. (This was explained in Sec. F of Chap. 3.) Now we have seen, in connection with statement (7.B.13), that uniform gravity introduces zero torque of its own about a center-of-mass axis. Thus, *our accelerated observer sees the same torque as a stationary observer, provided the torque is measured about an axis through the center of mass.*

As an illustration, consider a braking car, Fig. 7.27. Its nose dips because the road exerts a *clockwise torque* about the car's center of mass $\mathbb{C}$.

E. Angular Momentum

In rotational motion we encounter a valuable conservation law, that of **angular momentum**. (Our collection of conserved quantities already includes mass, energy, and ordinary, or linear momentum.) Like the other conservation laws, this one can supply powerful conclusions with a minimum of calculation.

i. Angular Momentum in Fixed-Axis Rotation

In the beginning of this chapter we made use of linear-rotational analogies such as $M \leftrightarrow I$, $F_x \leftrightarrow \tau$, $v_x \leftrightarrow \omega$, etc., to analyze rotational motion.* Pursuing this approach further, we now introduce the rotational analogue of a particle's momentum, $p_x = mv_x$:

$$L = I\omega \qquad (7.E.1)$$

The quantity L is defined as the **angular momentum** of a rigid body whose moment of inertia is I and whose angular velocity is ω, all about a single fixed axis. (The SI unit of angular momentum is seen to be the kg meter^2/sec.)

For several objects, numbered 1, 2, etc., the **total angular momentum of the system** is defined additively:

$$L_{\text{syst}} = L_1 + L_2 + \cdots \qquad (7.E.2)$$

provided all these objects are pivoted on the same axis; they may, however, be spinning at different angular velocities. Formula (7.E.2) is an algebraic sum, i.e., two counterrotating bodies have opposite signs for their angular momenta; the direction of positive spin is chosen arbitrarily.

With this prescription in mind, we obtain the rotational analogue of momentum conservation. It turns out that *the combined angular momentum of a system does not change unless there is a net external torque on the system.* Note the word "external": a torque exerted by one part of the system on another part of the same system does not count. In short, we have

$$L_{\text{syst}} = \text{const} \quad \text{if} \quad \tau_{\text{tot}} = 0 \qquad (7.E.3)$$

(law of conservation of angular momentum). The following example illustrates the definition, and shows how to use the conservation law; a proof of (7.E.3) is given later.

* A summary of these analogies is to be found in the Condensed Checklist at the end of this chapter.

Example 7.14. Interacting Flywheels. Two flywheels, 1 and 2, are frictionlessly and independently mounted on the same axle, as illustrated in Fig. 7.28; they have moments of inertia $I_1 = 3$ kg meter^2, $I_2 = 5$ kg meter^2, and are spinning in opposite directions, both at 120 rpm. Next, a clutch presses the flywheels into frictional contact with each other. After some time they rotate as a unit. What is the final rotational frequency of the combination? (We have here the rotational analogue of an inelastic collision.)

Assuming that the clutch mechanism (not shown in the figure) exerts zero net torque of its own, we can use conservation of angular momentum. Let flywheel 1 have an initial angular velocity ω_i. Then 2 has an initial angular velocity $-\omega_i$. The total initial angular momentum is

$$L_{\text{syst},i} = I_1\omega_i - I_2\omega_i = (I_1 - I_2)\omega_i$$

If the final angular velocity of the combination is ω_f, then the final angular momentum is

$$L_{\text{syst},f} = (I_1 + I_2)\omega_f$$

Conservation gives $L_{\text{syst},f} = L_{\text{syst},i}$, or

$$(I_1 + I_2)\omega_f = (I_1 - I_2)\omega_i$$

Solving for ω_f, we obtain

$$\omega_f = \frac{I_1 - I_2}{I_1 + I_2}\omega_i$$

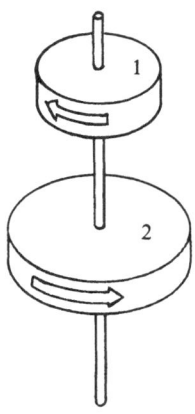

Figure 7.28. These counterrotating flywheels will be clamped against each other. Their resulting common angular velocity is calculated from the conservation of angular momentum.

Equivalently, in terms of frequencies ($f = \omega/2\pi$), we have

$$f_f = \frac{I_1 - I_2}{I_1 + I_2} f_i \qquad (7.E.4)$$

$$= \frac{3 - 5}{3 + 5} (120 \text{ rpm}) = \underline{\underline{-30 \text{ rpm}}}$$

The minus sign means that the final spin is in the opposite direction to the original spin of flywheel No. 1.

The inelasticity of the interaction implies that *the total kinetic energy is not conserved*; see Problem 7.47.

ii. Variable Moments of Inertia

Rigid bodies with fixed axes have constant moments of inertia. What happens when rigidity is given up and an object changes its moment of inertia, I, while it is rotating? Then we can no longer apply our definition $L = I\omega$, since a nonrigid body does not have a well-defined ω. However, it is still possible to define the angular momentum (this is done in the Notes at the end of this chapter) in such a way that it is conserved. If the rotating object begins and ends as a rigid body, then a conservation statement can still be made, relating the initial and final motions: $L_f = L_i$, or $I_f\omega_f = I_i\omega_i$, as previously.

A classic qualitative example is that of a skater who greatly increases her angular velocity by drawing herself up inside a narrow imaginary cylinder; see Fig. 7.29. The quadratic nature of the formula mr^2 (for each particle) makes her method very effective.

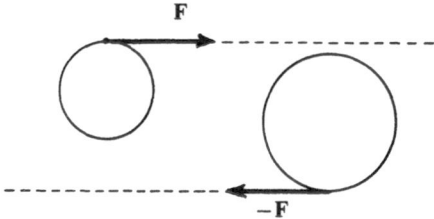

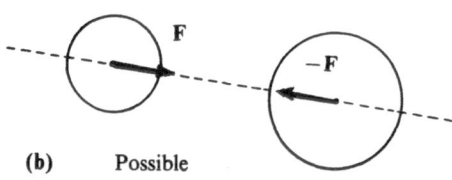

Figure 7.30. (a) A hypothetical attraction that violates principle (7.E.5). (b) A situation that agrees with (7.E.5).

iii. Why is Angular Momentum Conserved?

There remains to explain the constancy of L_{syst} in the absence of a net external torque. Let us consider fixed-axis rotation, leaving a more general treatment for the Notes at the end of this chapter.

Action and Reaction, Revisited

The starting point is a new look at Newton's third law (action and reaction). Observation leads to the following postulate:

> An action force, **F**, and its reaction, $-$**F**, act along the same line.

$$(7.E.5)$$

See Fig. 7.30. *This is the first new basic principle since Chap. 3 of this book*: it does not follow from anything we have seen before. Its immediate consequence, shown in Fig. 7.31, is that, *about any axis, the forces of action and reaction exert opposite*

Figure 7.29. A skater increases her angular velocity by decreasing her moment of inertia.

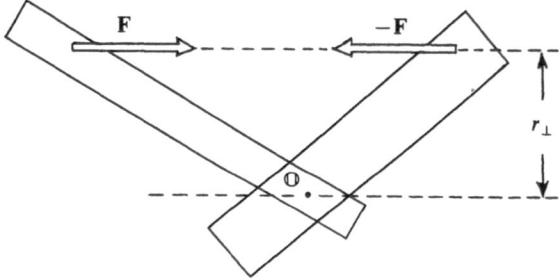

Figure 7.31. Because **F** and $-$**F** share their line of action, their torques about $\mathbb{O}$ are $+r_\perp F$, $-r_\perp F$, equal and opposite.

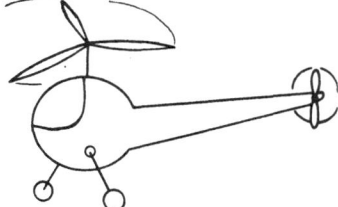

Figure 7.32. The small propeller is needed to prevent the helicopter from acquiring an angular velocity with respect to a vertical axis.

torques τ and $-\tau$. Thus, there is a law of action and reaction for torques.

The helicopter of Fig. 7.32 provides an illustration of that principle. If the engine exerts a torque τ on the large propeller, the latter necessarily exerts a torque $-\tau$ on the engine, and therefore on the vehicle as a whole, setting it into a spinning motion. To prevent this, a side propeller is made to exert the opposite torque, $+\tau$, on the helicopter, again about the large propeller axis.

Conservation

If several rigid bodies (Nos. 1, 2, etc.) share an axis of rotation, as did the flywheels of Example 7.14, then we have

$$\frac{dL_{\text{syst}}}{dt} = \frac{d}{dt}(I_1\omega_1 + I_2\omega_2 + \cdots)$$

$$= I_1\alpha_1 + I_2\alpha_2 + \cdots = \tau_1 + \tau_2 + \cdots \qquad (7.E.6)$$

If the torques are exerted only by the rigid bodies 1, 2,... on one another, then the right-hand side of (7.E.6) consists of mutually canceling pairs, perhaps $\tau_1 = -\tau_2$, etc.; hence we obtain $dL_{\text{syst}}/dt = 0$, as claimed.

iv. A Spectrum of Angular Momenta

To conclude this section, Fig. 7.33 presents a survey of angular momenta for representative spinning and revolving bodies.

Figure 7.33. A spectrum of angular momenta. The range of magnitudes is enormous, owing to the strong effect of size in the formula mr^2. The scale is logarithmic. Although arbitrarily large angular momenta appear to be possible, Nature has set a limit to the smallest nonzero angular momentum that can be realized. The electron spins about its axis with that particular value, 9.1×10^{-35} joule sec. An angular momentum can also be exactly zero, however. Chapter 27 (Atomic Structure) will touch upon this fundamental discontinuity, which cannot be understood on the basis of classical physics.

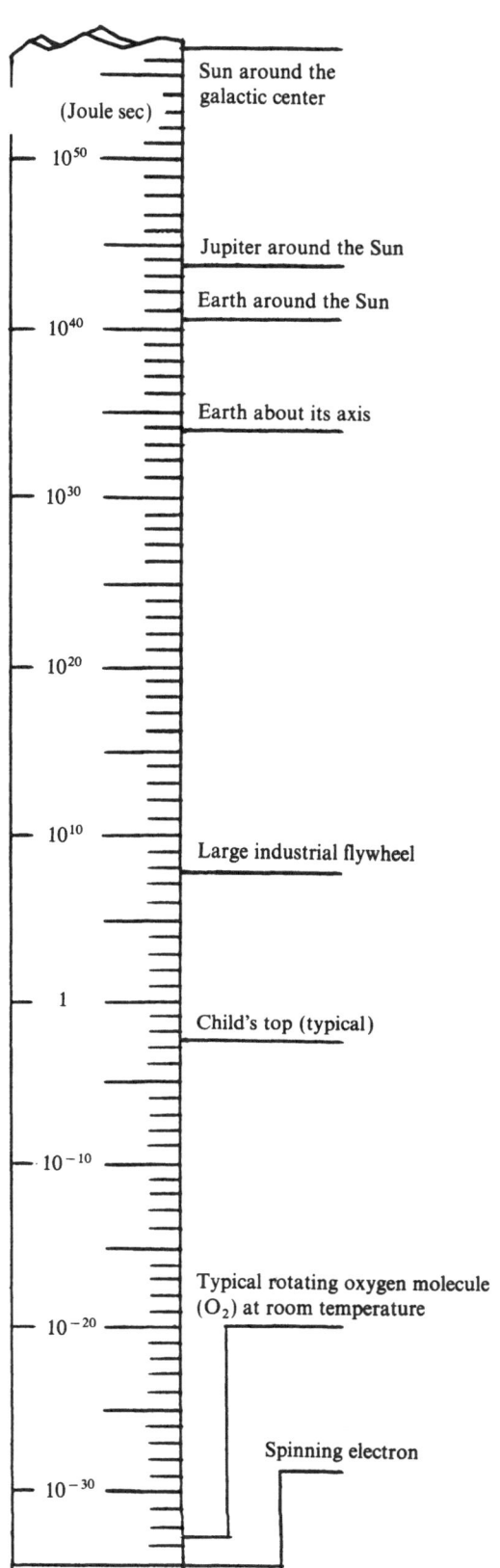

(Joule sec)

10^{50} — Sun around the galactic center

10^{40} — Jupiter around the Sun

Earth around the Sun

Earth about its axis

10^{30}

10^{20}

10^{10} — Large industrial flywheel

1 — Child's top (typical)

10^{-10}

10^{-20} — Typical rotating oxygen molecule (O_2) at room temperature

Spinning electron

10^{-30}

Notes

The first of these notes proves a statement, made in the text, about an object's internal (for example, rotational) kinetic energy; Notes (ii)–(iv) may be considered a mini-course on the angular momentum as a *vector*.

i. Translational and Internal Kinetic Energies

Equation (7.D.1) claims that a system's total kinetic energy is given by

$$\mathscr{K}_{\text{syst}} = \mathscr{K}_{\text{transl}} + \mathscr{K}_{\text{int}} \tag{7.N.1}$$

where

$$\mathscr{K}_{\text{transl}} = \tfrac{1}{2}m_{\text{syst}}v_{\text{C}}^2 \tag{7.N.2}$$

(v_{C} = velocity of the center of mass C), and where $\mathscr{K}_{\text{int}}$ is calculated as if C were at rest. The proof is as follows.

Let the system consist of particles 1, 2, etc., with velocities v_1, v_2, etc. In the center-of-mass system (that is, as seen by an observer traveling with C), a typical particle, say No. 1, is moving at a velocity $\mathbf{v}_1 - \mathbf{v}_{\text{C}}$. Thus, we have, by definition,

$$\mathscr{K}_{\text{int}} = \tfrac{1}{2}m_1 |\mathbf{v}_1 - \mathbf{v}_{\text{C}}|^2 + \tfrac{1}{2}m_2 |\mathbf{v}_2 - \mathbf{v}_{\text{C}}|^2 + \cdots \tag{7.N.3}$$

The first term involves

$$|\mathbf{v}_1 - \mathbf{v}_{\text{C}}|^2 = (\mathbf{v}_1 - \mathbf{v}_{\text{C}}) \cdot (\mathbf{v}_1 - \mathbf{v}_{\text{C}})$$
$$= v_1^2 + v_{\text{C}}^2 - 2\mathbf{v}_{\text{C}} \cdot \mathbf{v}_1$$

As applied to each term of (7.N.3), this results in

$$\mathscr{K}_{\text{int}} = (\tfrac{1}{2}m_1 v_1^2 + \tfrac{1}{2}m_2 v_2^2 + \cdots)$$
$$+ \tfrac{1}{2}(m_1 + m_2 + \cdots)v_{\text{C}}^2$$
$$- \mathbf{v}_{\text{C}} \cdot (m_1\mathbf{v}_1 + m_2\mathbf{v}_2 + \cdots) \tag{7.N.4}$$

We recognize the first term as the total kinetic energy $\mathscr{K}_{\text{syst}}$; in the second term we set $m_1 + m_2 + \cdots = m_{\text{syst}}$; and in the last term we recognize the total momentum of the system:

$$m_1\mathbf{v}_1 + m_2\mathbf{v}_2 + \cdots = m_{\text{syst}}\mathbf{v}_{\text{C}}$$

Therefore Eq. (7.N.4) reads

$$\mathscr{K}_{\text{int}} = \mathscr{K}_{\text{syst}} - \tfrac{1}{2}m_{\text{syst}}v_{\text{C}}^2$$

which amounts to (7.N.1).

ii. Torque and Angular Momentum as Vectors

To understand rigid-body rotation about a variable axis was a great challenge, not mastered until a 1765 publication by the mathematical genius Leonhard Euler. Our own ambition is limited to proving the conservation of angular momentum; for that purpose we must go back to fundamentals, and this means the consideration of single particles. The discussion is most easily conducted in terms of vectors.

The Torque Vector

Figure 7.34 shows how to define the torque vector τ due to a force $\mathbf{F}$ on a particle P. The definition is made *with respect to a chosen point O* rather than with respect to an axis.

Let $\mathbf{r}$ be the position vector of P, with O as the origin. Then the following three statements are valid:

First:
$$\tau \text{ is perpendicular to } \overrightarrow{OP} = \mathbf{r}, \text{ and to } \mathbf{F}. \tag{7.N.5a}$$

Second:

τ points *above* the plane containing $\mathbf{r}$ and $\mathbf{F}$, as drawn in the figure. This so-called **right-hand rule**

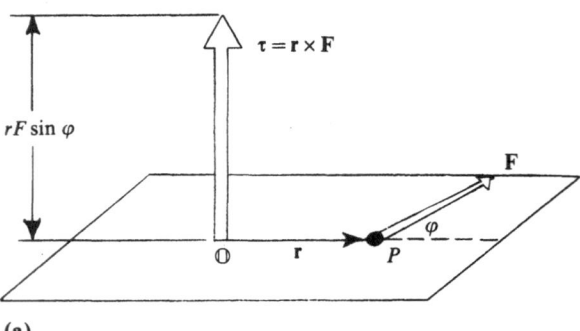

(a)

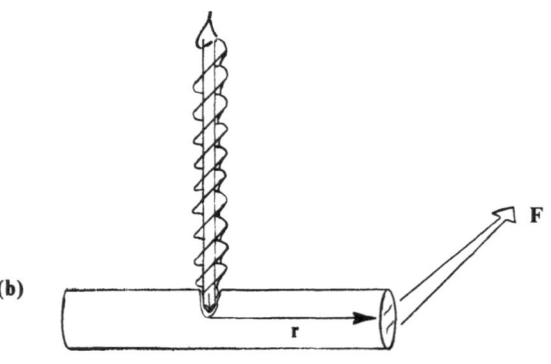

(b)

Figure 7.34. Construction of the torque vector.

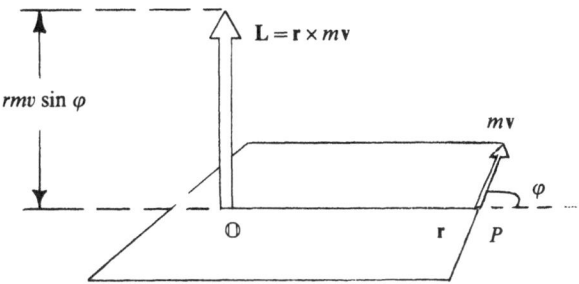

Figure 7.35. Construction of the angular momentum vector.

is memorized as follows. Imagine a corkscrew, with its handle along **r**, being turned by **F** [part (b) of the figure]. Then **τ**, by convention, points in the direction of the *advancing screw*. For example, in Fig. 7.12, the torques τ_1, τ_2, τ_{tot} about $\mathbb{O}$ are directed perpendicularly out of, into, and out of the page, respectively. (7.N.5b)

Third: The magnitude of **τ** is the ordinary numerical torque calculated about an axis along **τ**; thus, we have $\tau = r_\perp F = rF \sin \varphi$ in Fig. 7.34a. (7.N.5c)

This lengthy set of specifications is elegantly summarized by the **cross-product** notation.*

$$\tau = r \times F \qquad (7.N.6)$$

(also called a vector product). The innocent-looking cross-multiplication symbol, ×, is powerful indeed. It stands for the perpendicularity *and* the right-hand rule *and* the sin φ factor in (7.N.5).

In contrast to other kinds of multiplication, the order of factors in (7.N.6) is important. For any two vectors **A**, **B**, we have

$$A \times B = -(B \times A) \qquad (7.N.7)$$

This curious property may be checked in three steps. (1) Choose an arbitrary fixed pair of vectors **A**, **B**. (2) Apply rule (7.N.5b) with **r** = **A**, **F** = **B**. (3) Apply that same rule with **r** = **B** and **F** = **A**; the direction of **τ** will now be flipped over.

The Angular Momentum Vector

In analogy with torque, Fig. 7.35 displays the angular momentum vector **L** of a particle P relative to point $\mathbb{O}$. We have, by definition,

$$L = r \times mv \qquad (7.N.8)$$

where *m***v** is the momentum vector of P. In magnitude, we obtain in Fig. 7.35

$$L = rmv \sin \varphi \qquad (7.N.9)$$

Does this fit in with our previous definition of L?

---

* For some further details, see Appendix II, the Vector Cross Product.

Adapting (7.N.9) to circular motion about $\mathbb{O}$ ($\varphi = 90°$), we find

$$L = rmv = mr^2\omega = I\omega$$

in agreement with (7.A.6).

Additivity

A whole system's angular momentum is given by the vector sum

$$L_{syst} = L_1 + L_2 + \cdots \qquad (7.N.10)$$

of its individual particles' angular momenta. (In the illustration of Fig. 7.36, all contributions are along the axle, and so is L_{syst}.)

iii. The Rotational Equation of Motion

Just as a particle's equation of motion can be written as $d(m\mathbf{v})/dt = \mathbf{F}_{tot}$, so can the rotational equation of motion of a whole system about any chosen origin be written as

$$\frac{dL_{syst}}{dt} = \tau_{tot} \qquad (7.N.11)$$

To establish this, we require two preliminary mathematical facts: (1) How does one differentiate a cross product? The answer is, for any two vectors **A**, **B**,

$$\frac{d}{dt}(A \times B) = \frac{dA}{dt} \times B + A \times \frac{dB}{dt} \qquad (7.N.12)$$

as expected for a product; however, the order of the factors must be carefully preserved (here **A** before **B**). (2) The cross product of two vectors, say **A** and *m***A**, which point in the same direction, vanishes:

$$A \times mA = 0 \qquad (7.N.13)$$

(This may be seen from prescription (7.N.5c), applied to a force whose line of action goes through the origin.)

To verify Eq. (7.N.11), consider first a force **F** acting on a single particle, whose angular momentum is

$$L = r \times mv \qquad (7.N.14)$$

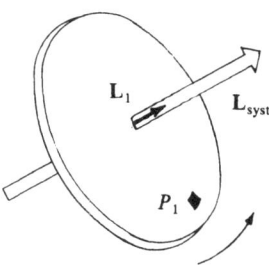

Figure 7.36. A fixed axis runs through the center of this wheel and is perpendicular to it. Particle P_1 contributes L_1 to the system's angular momentum L_{syst} about $\mathbb{O}$.

Differentiating, we have

$$\frac{d\mathbf{L}}{dt} = \frac{d\mathbf{r}}{dt} \times m\mathbf{v} + \mathbf{r} \times \frac{d(m\mathbf{v})}{dt}$$

$$= \mathbf{v} \times m\mathbf{v} + \mathbf{r} \times \mathbf{F}$$

or, since the first product vanishes,

$$\frac{d\mathbf{L}}{dt} = \tau$$

Adding such contributions for all particles gives the announced relation (7.N.11).

iv. Conservation of Angular Momentum

We have seen in Fig. 7.31 that a pair of action–reaction forces $\mathbf{F}$, $-\mathbf{F}$, results in a zero net torque with respect to any axis. Therefore *we can neglect all internal contributions* to the total torque τ_{tot} on a system, and Eq. (7.N.11) is equivalent to

$$\frac{d\mathbf{L}_{syst}}{dt} = \tau_{ext} \qquad (7.N.15)$$

where τ_{ext} is the vector combination of all externally applied torques. If the latter are absent, we have

$$\boxed{\frac{d\mathbf{L}_{syst}}{dt} = 0} \qquad (\tau_{ext} = 0) \qquad (7.N.16)$$

the anticipated conservation law.

Condensed Checklist   (Analogies between linear motion and fixed-axis rotation)

Linear		Rotational		
Distance coordinate x		**Angular coordinate** θ		Fig. 7.4
Velocity	$v_x = \dfrac{dx}{dt}$	**Angular velocity**	$\omega = \dfrac{d\theta}{dt}$	(7.A.1)
Acceleration	$a_x = \dfrac{dv_x}{dt}$	**Angular acceleration**	$\alpha = \dfrac{d\omega}{dt}$	(7.A.2)
Mass	m	**Moment of inertia**	$I = \sum_i m_i r_i^2$	(7.A.10)
			$(r = \text{distance to axis})$	
Momentum	$p_x = mv_x$	**Angular momentum**	$L = I\omega$	(7.E.1)
Force	F_x	**Torque**	$\tau = rF_\perp = r_\perp F$	(7.B.5), (7.B.11)
Kinetic energy	$\mathcal{K} = \tfrac{1}{2}mv_x^2$	**Kinetic energy**	$\mathcal{K} = \tfrac{1}{2}I\omega^2$	(7.A.3)
Work	$d\mathcal{W} = F_x\, dx$	**Work**	$d\mathcal{W} = \tau\, d\theta$	(7.A.24)
Power	$\dfrac{d\mathcal{W}}{dt} = F_x v_x$	**Power**	$\dfrac{d\mathcal{W}}{dt} = \tau\omega$	(7.A.25)

Laws of motion

$F_{tot,x} = ma_x$	$\tau_{tot} = I\alpha$	(7.A.17)
$\dfrac{dp_x}{dt} = F_{tot,x} = F_{ext,x}$	$\dfrac{dL}{dt} = \tau_{tot} = \tau_{ext}$	(7.E.6)
$(=0$ for isolated system$)$	$(=0$ for isolated system$)$	

Moments of inertia: Fig. 7.7

True or False

1. When asking for the moment of inertia of a sphere, one need not specify its axis of rotation.

2. A sphere has its largest moment of inertia about an axis tangent to its surface.

3. There is no upper limit to a given object's moment of inertia, since the axis can be chosen outside the object itself.

4. When a car travels fast enough, the rotational kinetic energy of its wheels exceeds the translational kinetic energy of the whole car.

5. The *angular velocity* of a pivoted rigid body can be calculated from its kinetic energy and its angular momentum.

6. The *moment of inertia* of a pivoted rigid body can be calculated from its kinetic energy and its angular momentum.

7. Just after it is being set in motion, a motor delivers negligible power.

8. Two equal and opposite forces must give a zero combined torque.

9. A bolt with vertical axis is being tightened, at a uniform rate of rotation, by a horizontally held wrench. About that axis, the torque exerted by the wrench on the bolt is greater than the torque exerted by the hand on the wrench.

10. When a boat's propeller reaches its maximum rate of rotation, the total torque on it reaches its maximum value.

11. A diver jumps off a diving board; if, while in the air, he brings his knees up to his chin, this decreases the rate at which he is tumbling.

12. If a spinning space station extends its solar collector panels away from its axis of rotation, the whole station's kinetic energy decreases.

Problems

Section A: Rotation About a Fixed Axis

*7.1. (a) How much kinetic energy is stored in a flywheel whose moment of inertia is 3×10^4 kg meter², if it rotates at 400 rpm? (b) How many watts of power could it deliver if its frequency were lowered to 300 rpm in 1 min?

*7.2. A homogeneous solid sphere of radius 0.10 meter and mass 16 kg rotates about one of its diameters at a frequency $f = 4$ rev/sec. Find its kinetic energy.

*7.3. The radius of gyration of an object about an axis is the distance from that axis at which the whole mass of the object could be concentrated without changing its moment of inertia. What is the radius of gyration of an object of mass M and moment of inertia I?

7.4. A crank is turned against a frictional torque of 0.05 newton meter. How many turns of the crank are needed to expend a work of 60 joules?

*7.5. An engine delivers 150 watts while exerting a torque of 3 newton meters. At how many revolutions per second is it rotating?

7.6. A given engine is rated at 150 horsepower when turning at 1000 rpm. How much torque, in SI units, should it exert?

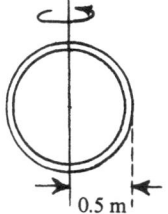

0.5 m

*7.7. A 4-kg uniform circular hoop of radius 0.5 meter is pivoted about a diameter, as shown. In order to spin it at 8 rad/sec (starting from rest) in 2 sec, what constant torque must be exerted about the pivot?

*7.8. A certain flywheel has a moment of inertia of 5 kg meter², and is subjected to a constant torque of 0.25 newton meter. (a) What is the flywheel's angular acceleration? (b) How much time does it need to reach a rotational frequency of 2 rev/sec, starting from rest?

7.9. It takes 6 sec for a certain motor to speed up from 1200 rpm to 1800 rpm. (a) Find its angular acceleration (assumed to be constant). (b) How many revolutions are made during this time interval?

7.10. Obtain by integration the expression for the moment of inertia of a cylinder, item (n) of Fig. 7.7. [You may start from the cylindrical shell, item (q) of that figure.]

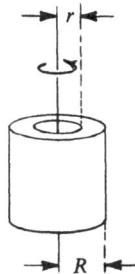

r

R

*7.11. A thick homogeneous cylindrical ring, of inside and outside radii r and R, respectively, has total mass M; see the figure. In terms of these quantities, find its moment of inertia about its cylindrical axis. (You may take a difference involving two solid cylinders, or you may evaluate an integral from r to R.)

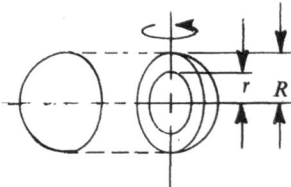

*7.12. A thick homogeneous spherical shell, of inside and outside radii r and R, respectively, has total mass M; see the figure. In terms of these quantities, find its moment of inertia about a diameter (cf. Problem 7.11).

Sections B: Torques in Terms of Forces; C: Rotational Equilibrium

7.13. To start a chain saw, one pulls a cord, which is wound around a pulley 10 cm in diameter. (a) Find the torque, in newton meters, resulting from the application of a 50-newton force. (b) What is the torque in ft lb?

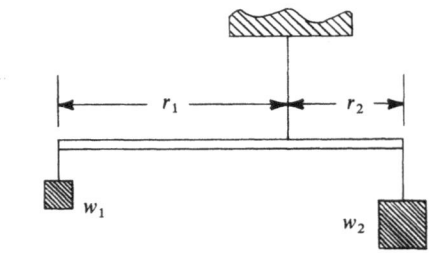

7.14. The horizontal level shown in the figure is in equilibrium, with weights w_1, w_2 acting at the ends. Neglecting the weight of the lever itself, derive the arm ratio $r_1/r_2 = w_2/w_1$.

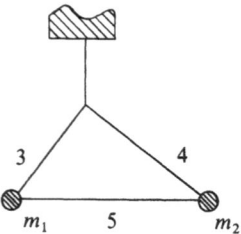

*7.15. A light rigid 3–4–5 triangle is suspended in equilibrium, with heavy masses m_1, m_2 attached as shown. What should be the mass ratio m_1/m_2 in order that the long side should be horizontal?

7.16. A helicopter, made as in Fig. 7.32, has an overhead engine torque of 4000 newton meters. The axles of the

two propellers are 8 meters apart. How much force must be exerted by the small propeller to maintain the orientation of the craft?

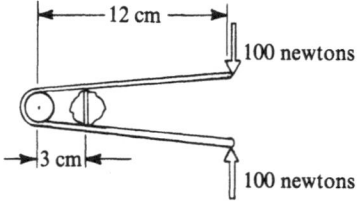

7.17. A nutcracker is 12 cm long. If a normal force of 100 newtons is exerted at the end of each half of the nutcracker, as shown, how much force is exerted on one side of a nut 3 cm from the hinge?

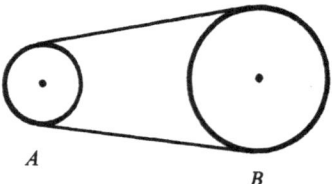

*7.18. (Transmission of a torque.) Pulleys A and B, with radii r_A and r_B, are connected by a belt as shown. If an external torque τ_A is exerted on pulley A by its axle, what torque τ_B will be exerted by the large pulley B on *its* axle? (Assume both pulleys are at equilibrium if the external torques are included; express your answer in terms of τ_A, r_A, r_B.)

7.19. A homogeneously made grindstone is a cylindrical wheel of radius 0.20 meter and mass 10 kg. With no power applied, the grindstone is initially coasting at 8 rev/sec. An axe blade is then pressed against its rim, with a resulting 3 newtons of friction. What is, in rev/sec, the rotational frequency after 2 sec?

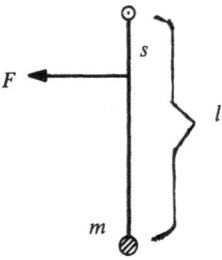

7.20. A small object of mass m is suspended vertically from a light rigid bar of length l, frictionlessly pivoted at the other end; see the figure. A horizontal force F is exerted at a distance s below the pivot. What horizontal acceleration does m achieve? (Express your answer in terms of l, s, F, m.)

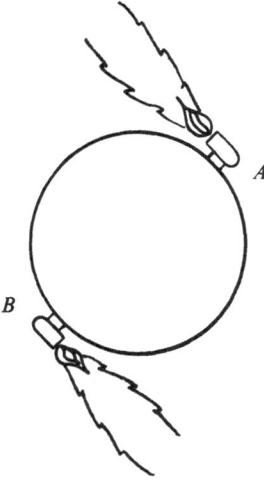

mass m and radius r. The cylinder is pivoted about its axis; see the figure. Find the downward acceleration of the suspended mass. (Express your answer in terms of m, M, g, r; neglect friction.)

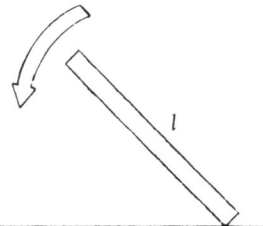

7.24. A uniform bar of mass m and length l falls with its lower end pivoted at ground level; see the figure. What is the vertical component of acceleration of the free end as it is about to strike the ground? (Express your answer in terms of m, l, g.)

*7.21. A certain space station may be approximated by a cylindrical shell of mass 30 metric tons (30 000 kg) and diameter 7 meters. It is originally spinning, at 1 revolution every 10 sec, about its cylindrical axis. To stop this, the small rockets (at A and B in the figure) are made to exert a tangential force of 20 newtons each. For how long must they be turned on? (Neglect loss of mass.)

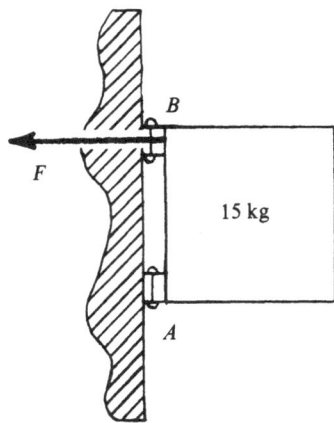

7.25. A uniform square gate (see the figure) has a mass of 15 kg, which is entirely supported by the lower hinge at A, so that the upper hinge, at B, exerts only a horizontal force F. Calculate F.

7.26. In Fig. 7.24, assume the ladder weighs 100 newtons and makes a $30°$ angle with the vertical. Find the forces N and F.

7.22. At the end of a so-called saccadic movement, a certain human eyeball undergoes an angular deceleration of 50 rad/sec^2. (The eye does not ordinarily turn smoothly.) How large a force F (in newtons and in dynes) must be exerted by the appropriate muscle, pulling as shown? (Approximate the eyeball by a homogeneous solid sphere of diameter 2.5 cm and mass 8 grams; assume it moves in its socket as if pivoted about a diameter, shown end-on as the point $\odot$.)

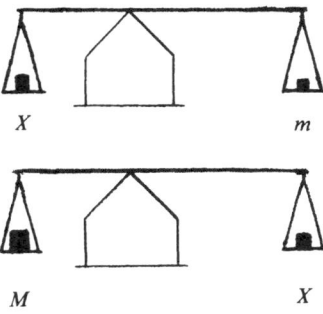

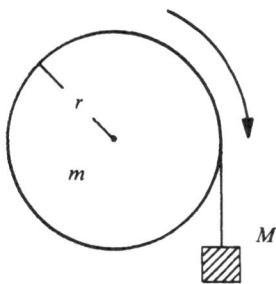

*7.23. A mass M falls vertically while attached to a light string which unwinds from a uniform solid cylinder of

*7.27. (An unequal-arm balance.) The figure shows a mass X being weighed successively on each arm of an unequal-arm balance. Equilibrium is achieved with the known masses m and M, respectively. Express X in terms of m and M; see also Problem 7.14.

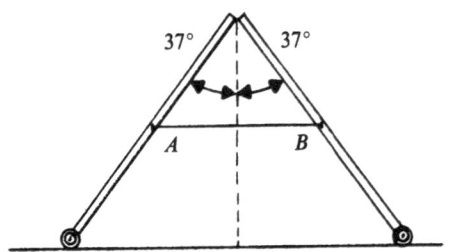

*7.28. The double ladder shown in the figure is frictionlessly hinged at the top. It is set up on small rollers and is prevented from collapsing only by the light chain AB attached to the ladders' midpoints. Each ladder is uniform, of mass 12 lb, and makes a 37° angle with the vertical. Find the tension in the chain.

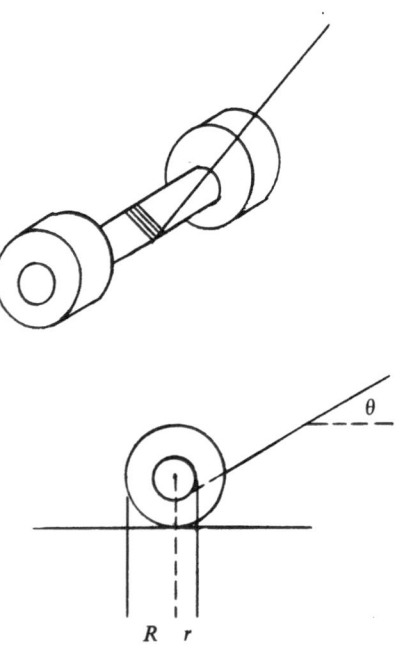

*7.29. ("Walking the dog.") A spool rolls without slipping on a horizontal plane (see the figure). The rolling ends have radius R; the inner cylinder has a smaller radius r. If the wound string is pulled at an angle θ to the horizontal, the spool will accelerate towards or away from the puller according to whether θ is smaller or larger than a critical value θ_0. (a) Find θ_0 in terms of R and r. (b) What happens in practice when a sufficient pull is exerted with $\theta = \theta_0$?

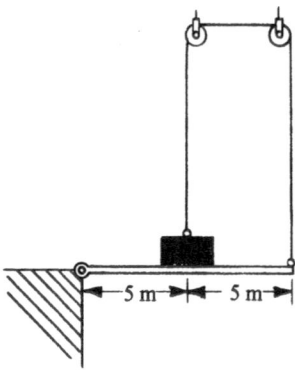

7.30. A 200-newton weight is supported on the middle of a horizontal light hinged board of length 10 meters, as shown. The light cord passes over two pulleys. What is the tension in the cord? (Neglect friction.)

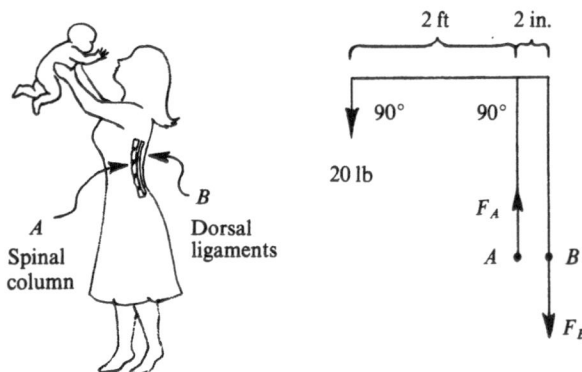

7.31. As shown in the figure, a woman lifts a 20-lb child at arms' length. In the simplified model, and assuming equilibrium, find the compressive force F_A on a typical vertebra at A, and the tension F_B in the dorsal ligaments at B. (Neglect all weights except the child's. Your result will then represent quantities over and above the already existing, usual loads.)

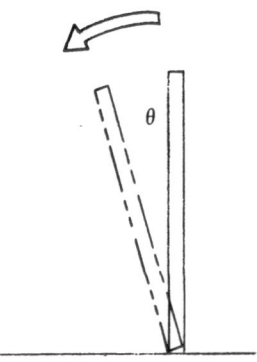

*7.32. A thin uniform bar of mass M and length l is vertically standing on one end, about which it is frictionlessly pivoted (see the figure); it is then given an infinitesimal nudge, and starts falling. Using the energy principle, calculate, as a function of θ, the angular velocity $\omega = d\theta/dt$, where θ is the bar's angle with the vertical. (Express your answer in terms of θ, l, m, g.)

7.33. (The perpendicular-axis theorem.) Consider a rectangular system of axes x, y, z. A certain particle of mass m has coordinates $(x, y, 0)$. Its moments of inertia are I_x, I_y, I_z about the three respective axes. (a) Show that $I_z = I_x + I_y$. (b) This relation is also true for a flat plate of arbitrary shape lying in the xy plane. Why? (c) Verify the consistency of this relation with the moments of inertia listed in Fig. 7.7, items (f), (j); (k), (v); (p), (s).

7.34. Demonstrate the relation (7.C.5), concerning the line of action of a single force equivalent to two parallel forces.

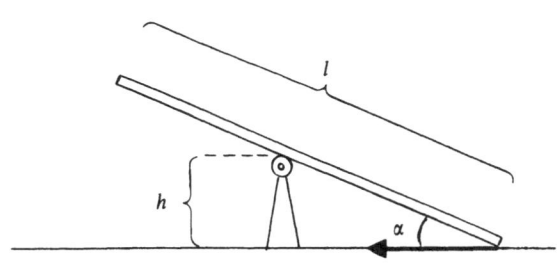

*7.35. In the figure, a uniform ladder of length l is being pulled over a roller by a light rope lying on the ground along the direction in which the ladder is leaning. If the roller is at a height h above ground, what angle α must be reached before the ladder tips over? (Assume h is small enough for the operation to be feasible.)

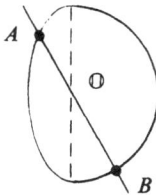

7.36. The figure shows a proposal for a perpetual-motion machine. A light lever is pivoted at O; equal weights slide at A and B on the intersections of the lever with a closed track. (The whole device is in a vertical plane.) The left part of the track is a half-ellipse; the right part is a semicircle. Since the left arm is always the shorter one, the machine should always be unbalanced clockwise, with kinetic energy to spare for carrying the lever past its "neutral" vertical position. Find qualitatively where the fallacy lies.

7.37. How should Eq. (7.C.4) be modified if the balance beam is a uniform rectangle of non-negligible weight w_0?

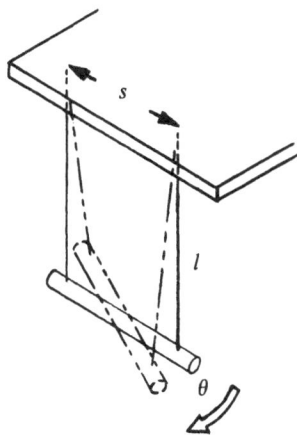

*7.38. (A bifiar torsion pendulum.) A horizontal bar is symmetrically suspended from two equal light parallel inextensible strings. Starting from the position where the strings are vertical, the bar is turned in the horizontal plane by a small angular deflection θ about its middle, as shown. (a) If the strings have length l and spacing s, by what height does the bar rise? (b) If the bar's weight is w, how much work has been done against gravity? (c) How much torque, about the center of the bar, corresponds to a deflection θ?

Section D: Rotation and the Center of Mass

*7.39. On an inclined plane making an angle θ with the horizontal, what is, in terms of g, the acceleration of the following rolling objects: (a) A solid uniform sphere? (b) A thin uniform spherical shell? (c) A uniform circular hoop?

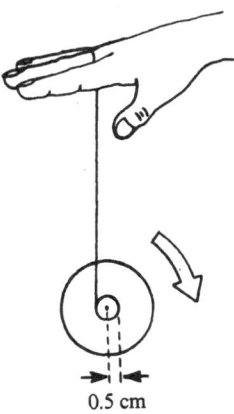

0.5 cm

7.40. A yo-yo of mass 0.10 kg, shown in the figure, descends at the end of a nearly vertical string whose other end is held fixed. If the axle from which the string unwinds is 0.5 cm in radius, and if the yo-yo's total moment of inertia about the axle is 1.5×10^{-4} kg meter^2, find its downward acceleration.

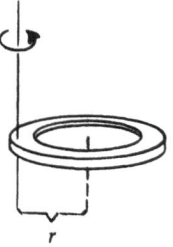

*7.41. By using the parallel-axis theorem. Sec. D iii, (a) verify the consistency of items (b) and (e) [or (f) and (g)] of Fig. 7.7; (b) calculate the moment of inertia for a thin uniform hoop of radius r and mass M about an axis through the rim and perpendicular to the plane of the hoop; see ιne adjoining figure.

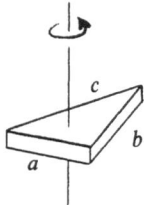

7.42. By using Fig. 5.16e of Chap. 5 and this chapter's Fig. 7.7 l, show from the parallel-axis theorem, Sec. D.iii, that the moment of inertia of a flat triangular plate about a perpendicular axis through its center of mass (see the adjoining figure) is $\frac{1}{36}M(a^2 + b^2 + c^2)$, where M is the mass and a, b, c the sides. [Note: the median towards a has length $= (b^2/2 + c^2/2 - a^2/4)^{1/2}$; the center of mass is at $1/3$ along that median from a.]

7.43. Consider Example 7.12, with the hoop replaced by a uniform solid disk, also of radius r and mass M. Find the acceleration a_C of its center of mass in terms of M, r, and the plane's angle φ with the horizontal.

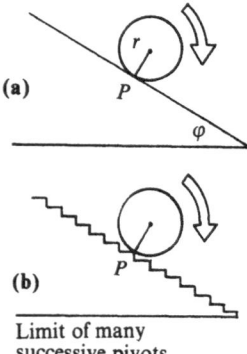

(a)

(b)

Limit of many successive pivots

*7.44. (Rolling down an incline, torque method.) A uniform solid right cylinder, with its axis horizontal, is

rolling without slipping down a plane whose angle with the horizontal is φ, as shown in (a). In terms of φ and the cylinder's mass M and radius r, (a) find the net torque on the cylinder about its point of contact P with the plane. (b) Using P as a *temporary* fixed pivot [see sketch (b) for justification], find the cylinder's angular acceleration α about P. (c) Hence find the acceleration a_C of the cylinder's center of mass. (See also Problem 7.43.)

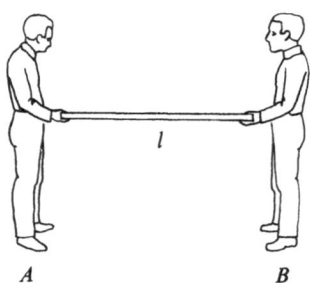

A   B

*7.45. As shown in the figure, a uniform plank of length l and mass m is supported horizontally, in equilibrium, by workers A and B. Each exerts a vertical force at his end. If B suddenly lets go while A holds fast, what is the new force momentarily exerted by A?

Section E: Angular Momentum

7.46. (a) What are the British and c.g.s. units of angular momentum? (b) How much are they worth in SI units?

7.47. What percentage of the initial kinetic energy is lost by the interacting flywheels of Example 7.14?

7.48. If the flywheels of Example 7.14 spin in opposite directions at frequencies of 20 rpm (for A) and 6 rpm (for B), what is the frequency of the final unit? Which flywheel keeps its original direction of spin?

*7.49. Two 1/2 metric ton polar bears sit on opposite points at the edge of a circular ice floe, of mass 20 metric tons and uniform thickness, which is slowly rotating in still water, with negligible friction. If both bears walk to the center of the floe, by what percentage is the latter's frequency of revolution increased when they meet?

*7.50. By combining data from Fig. 6.10 of Chap. 6 and this chapter's Fig. 7.33, estimate the following moments of inertia: (a) the Sun around the Galactic center; (b) the Earth around the Sun; (c) the Earth about its axis; (d) an oxygen molecule about its axis of rotation.

Answers to True or False

1. False (it could be eccentric).

2. False (see next item).

3. True.

4. False (these two energies keep a constant ratio).

5. True.

6. True.

7. True (power $= \tau\omega$, $\omega \approx 0$).

8. False (their lines of action may be different).

9. False (the wrench is in rotational equilibrium).

10. False (the rotational acceleration is what counts).

11. False (it increases the rate).

12. True (I increases, L stays constant, $\mathcal{K} = L^2/2I$).

Answers or Hints to Problems

7.1. (b) 2×10^5.

7.2. 20 joules.

7.3. $(I/M)^{1/2}$.

7.5. 8.

7.7. 2 newton meters.

7.8. (b) Apply $\omega = \alpha t$ (the analogue of $v_x = a_x t$).

7.11. $I = \frac{1}{2}m(R^2 + r^2)$.

7.12. $I = \frac{2}{5} m \dfrac{R^5 - r^5}{R^3 - r^3}$.

7.15. 16/9.

7.18. $(r_B/r_A)\,\tau_A$.

7.21. 1.6×10^3 sec.

7.23. Hint: Acceleration of M = acceleration of a point on the rim. Answer: $Mg/(M + \frac{1}{2}m)$.

7.27. $(mM)^{1/2}$.

7.28. Apply the equilibrium conditions to a single ladder; by symmetry, the force at the hinge must be horizontal.

7.29. (b) Slipping must occur.

7.32. $\omega = (6g/l)^{1/2} \sin(\theta/2)$.

7.35. $\sin \alpha = \sqrt{\frac{4}{3}} \cos \theta$, where $\cos 3\theta = -3\sqrt{3}\,h/l$.

7.38. (a) height $\approx l[1 - \cos(\theta s/2l)] \approx \theta^2 s^2/8l$; (c) use (7.A.24).

7.39. (a) $\frac{5}{7} g \sin \theta$.

7.41. (b) $2Mr^2$.

7.44. Hints: (a) Only gravity contributes; why? (b) Use $I_P = \frac{3}{2}Mr^2$; why? (c) Answer as in Problem 7.43.

7.45. Use a fixed axis through the end held by A.

7.49. 10%.

7.50. (c) 8×10^{37} kg meter^2.

CHAPTER 8

GRAVITY

Gravity is an obvious and all-pervading background to our lives. At first sight, it also appears to be quite simple. As long as we confine our attention to a single geographic neighborhood, the Earth's gravity can be represented by a uniform acceleration vector **g**. On an astronomical scale, however, gravity is not uniform, but obeys an **inverse-square-distance law**—the main new feature in what follows.

Why does gravity deserve further study? Astronomical relevance is only part of the answer. Gravity, which affects all matter, is perhaps the most fundamental of all physical phenomena; it is by no means completely understood as yet. Of more practical concern to us here is the fact that the inverse-square feature of gravity is shared by electricity and by magnetism; thus, *the present chapter is an essential prerequisite to our later study of these topics.* Furthermore, although few things could be as different as an atom and the solar system, a mathematical analogy will be found between them; the reason is the presence, in both, of an inverse-square attractive force.

The story of gravity is the recognition that free fall operates throughout the Universe as it does on Earth. Astronomical observation has, of course, played an essential role in this recognition. By their behavior, the Sun, the planets and their satellites, the comets, and, more recently, the man-made space vehicles, have allowed us to determine many basic aspects of gravity with great certainty and accuracy. Yet, on a cosmic scale, or for stars much more massive than the Sun, the functioning of gravity is still a subject of lively research and speculation, involving Einstein's theory of General Relativity as a tool. To the extent that gravity pertains to the microscopic world of fundamental particles, it is also, as of this writing, at the forefront of theoretical activity.

A. Astronomical Comments

Our present subject is gravity "in the large," which requires some astronomical data, especially distances and periods of time. Where do these data come from?

Over the millennia, they were obtained from a combination of naked-eye observation, primitive instruments, and sophisticated inference; the size of the Earth was already known to Eratosthenes around 240 B.C.

Today's blossoming of astronomical measurement, in terms of both scope and accuracy, is due to our advanced optics, to radio telescopes, and to manned and unmanned space travel. Table 8.1 hints at the quality of some of the data. That table is a reference for the entire chapter, including the problems.

Distances

Within the solar system, distances are most accurately obtained by radar and laser ranging, including triangulation with the help of a spacecraft. The Earth–Sun and Earth–Moon distances, r_{ES} and r_{EM}, are particularly well known; they vary somewhat with time because the relevant orbits are only approximately circular. The follow-

Table 8.1. Some Astronomical Data[a]

The Gravitational Constant	$G = 6.673 \times 10^{-11} \dfrac{\text{newton meters}^2}{\text{kg}^2}$

The Earth

Radius at sea level:	
Equatorial	6.378160×10^6 meters
Polar	6.356775×10^6 meters
Distance from the Sun:	
Mean[b,c]	1.4960×10^{11} meters $= 1$ AU
Minimum	1.4709×10^{11} meters
Maximum	1.5211×10^{11} meters
Period of rotation[d]:	8.6164090×10^4 sec ≈ 23.94 h
Period of revolution:	3.1558149×10^7 sec $= 1$ sidereal year
Mass (M_E):	5.976×10^{24} kg
$GM_E =$	$3.9861 \times 10^{14} \dfrac{\text{meters}^3}{\text{sec}^2}$

The Moon

Radius:	1.7382×10^6 meters
Distance from the Earth[e]:	
Mean	3.844×10^8 meters
Minimum	3.564×10^8 meters
Maximum	4.067×10^8 meters
Period of revolution ($=$period of rotation)[f]:	2.36059×10^6 sec $= 27.3217$ days
Mass (M_M):	7.350×10^{22} kg
$GM_M =$	4.9025×10^{12} meters^3/sec^2

The Sun

Radius:	6.960×10^8 meters
Period of rotation:	2.19×10^6 sec $= 25.4$ days
Mass (M_S):	1.989×10^{30} kg
$GM_S =$	1.3272×10^{20} meters^3/sec^2

The Planets

Planet	Mean distance from Sun (AU)	Mean distance from Sun (10^{11} meters)	Period of revolution (years)	Period of revolution (10^7 sec)	Equatorial radius (10^6 meters)	Mass/ Earth mass
Mercury	0.3871	0.579	0.2408	0.7599	2.44	0.056
Venus	0.7233	1.082	0.6152	1.9415	6.05	0.815
Earth	1	1.496	1	3.1558	6.38	1
Mars	1.5237	2.279	1.881	5.936	3.40	0.107
Jupiter	5.2028	7.783	11.86	37.43	71.9	317.8
Saturn	9.5388	14.270	29.46	93.0	60.3	95.15
Uranus	19.1819	28.696	84.014	265.13	25.6	14.54
Neptune	30.0611	44.971	164.79	520.05	24.8	17.23
Pluto	39.44	59.00	247.7	781.7	1.15	0.002

[a] All distances are center-to-center.
[b] The mean orbital radius of a planet equals (1/2)(maximum + minimum) distance to the Sun.
[c] This is the so-called **astronomical unit** (AU) of length.
[d] Shorter than the mean solar day because calculated with respect to the fixed stars rather than to the Sun.
[e] The Moon's orbital radius is shorter by 1.2% than the Earth–Moon distance, because both revolve about the center of mass of the pair.
[f] The rotation and revolution have the same period because the Moon always presents the same side to the Earth.

ing multiples help visualize the scale involved: if r_E is the Earth's radius, then

$$r_{EM} \approx 60 r_E, \qquad r_{ES} \approx 400 r_{EM} \qquad (8.A.1)$$

(More precisely, 60.3 and 389.)

At any one time, the Moon's distance from us is known to within an almost incredible ± 15 cm, thanks to laser ranging off reflectors that were placed on the lunar surface by Apollo astronauts; our precise knowledge of the speed of light is essential here, as the delay of a light signal is being timed.

Periods

Accurate timing, and a proper interpretation of visual data, are both involved in the determination of periods. The Moon's period of revolution is of special interest because it has played a large role in establishing Newton's theory of gravitation. The Moon circles the Earth with a slightly variable period whose average value is

$$T_M = 27.32 \text{ days} \qquad (8.A.2)$$

This may seem odd, since we are used to seeing the full Moon at intervals of about 29.5 days (its **synodic period**). However, the phases of the Moon result from *two* combined motions: Moon around Earth and Earth around Sun. The latter motion must, in a sense, be subtracted out to yield T_M, the Moon's **sidereal period**.

B. Universal Gravitation

Isaac Newton's historical synthesis of celestial and laboratory physics was published in 1687. This work incorporated the insights of Galileo concerning free fall and acceleration (our Chap. 2), and those of Kepler concerning the planetary orbits. (Kepler's rules will be described in the present chapter.) In order to develop his "system of the Universe," Newton had to create his theory of force, embodied in three celebrated laws (see our Chap. 3); in addition, he had to describe the properties of one special kind of force, that of gravity. This latter set of specifications is known as **Newton's law of universal gravitation**, and is no less famous than his other three laws. It forms the topic of the present section.

(Since 1915, when Einstein proposed his general theory of relativity, it has been known that corrections are needed to Newton's law of gravitation. These corrections, while they are of great fundamental importance, are typically of insignificant size except near certain very dense stars. They can, however, imply the formation of "black holes," a topic briefly discussed in a Note at the end of this chapter.)

i. Newton's Law of Attraction

On Earth, an object of mass m experiences a force $\mathbf{w} = m\mathbf{g}$, its weight, directed towards the Earth's center. Clearly the Earth must in some way be responsible for the presence of that force. According to the law of action and reaction, the object itself must exert a force $-\mathbf{w}$ on the Earth as a whole. Thus, the gravitational attraction is mutual and proportional to m.

More generally, consider any pair of particles (say with masses m, M); with Newton, we postulate the existence of action–reaction forces $\mathbf{F}$ (acting on m) and $-\mathbf{F}$ (acting on M), with the following properties:

1. $\mathbf{F}$ and $-\mathbf{F}$ have a common line of action, joining the two particles; see Fig. 8.1.
2. The common magnitude F is proportional to each of the masses,

$$F \propto m, \qquad F \propto M$$

3. F decreases as the inverse square of the distance, r, between the particles,

$$F \propto \frac{1}{r^2}$$

Property 1 belongs to the law of action and reaction; it was discussed in the preceding chapter and results in the conservation of angular momentum. Property 2 is directly motivated by the fact that weight is proportional to mass, $w \propto m$. Property 3 is a new insight, originally not based on any laboratory observation; it turns out to be essential

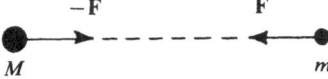

Figure 8.1. The pair of action–reaction forces due to the gravitational attraction between two particles.

in order to account for the observed behavior of astronomical systems.

The proportionalities for F can be condensed into a single formula,

$$F = G \frac{Mm}{r^2} \qquad (8.B.1)$$

G being a fixed number, the so-called **gravitational constant**, whose magnitude cannot be guessed by pure thought but must be measured or deduced from observation. By a method that will be explained subsequently, its value is found to be

$$G = 6.67 \times 10^{-11} \frac{\text{newton meters}^2}{\text{kg}^2} \qquad (8.B.2)$$

The reader is invited to check these units by writing out the dimensions in Eq. (8.B.1). Note the extraordinary smallness of G's numerical value. Our discussion will proceed for a while without using (8.B.2) explicitly.

The law of attraction is postulated in uncompromising generality. Any particle pair in the Universe, regardless of location, and regardless of motion, is assumed to satisfy it. Any amount of matter may be present between the two members of the pair. For example, Eq. (8.B.1) should be valid for the force between an apple in America and a pumpkin in Australia.

Calculations involving extended objects cannot be performed unless a **superposition principle** is assumed concerning the gravitational forces, as follows. Consider a system of particles A, B,... exerting a gravitational force on yet another particle P. Then

The combined gravitational
force on P is the vector
sum of the separate forces due $\qquad (8.B.3)$
to A, B,..., each taken
in isolation with P.

Each of these forces is calculated from Newton's law of gravitation for the appropriate pair of particles (A and P, or B and P, etc.).

Example 8.1. A Zero-Gravity Point. Two particles A, B, with masses $m_A = 4m$, $m_B = 9m$, are separated by a distance r. In relation to A and B, where should a third particle P be placed in such a way that the total gravitational force on it be zero? (We assume no other masses are present.)

Figure 8.2a shows qualitatively that, unless P is on the same line as A and B, it will always be acted on by a nonzero gravitational force. In part (b) of the figure we see how P can be positioned between A and B to effect a cancellation of forces. Let P have a mass m_P (which will turn out to be irrelevant, as will the gravitational constant G). Then, from the figure, the total force on P is, in the direction A to B,

$$F = -\frac{Gm_A m_P}{r_{AP}^2} + \frac{Gm_B m_P}{r_{BP}^2}$$

(note the signs). The requirement $F = 0$ yields

$$-\frac{m_A}{r_{AP}^2} + \frac{m_B}{r_{BP}^2} = 0$$

so that

$$\frac{r_{AP}}{r_{BP}} = \left(\frac{m_A}{m_B}\right)^{1/2} = \left(\frac{4}{9}\right)^{1/2} = \frac{2}{3}$$

Therefore

$$r_{AP} = \tfrac{2}{5}r, \qquad r_{BP} = \tfrac{3}{5}r$$

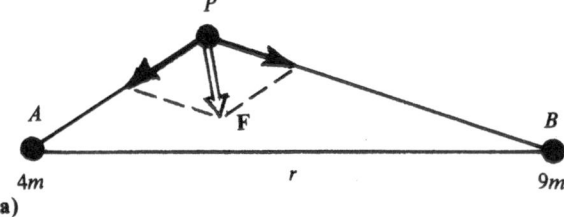

(a)

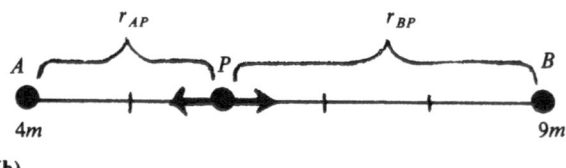

(b)

Figure 8.2. (a) If particle P is off the line joining A and B, there is always a nonzero force $\mathbf{F}$ on it. (b) P is located where the pulls of A and B cancel out.

is the "zero-gravity point." There is of course no way of obtaining zero gravity by placing P on the same line, but outside the interval AB, since then the individual forces would add instead of subtracting.

ii. Spherical Objects

The law of gravitation must in particular be applicable to the ordinary weight of an object on Earth. There should be no concern about treating a small object as a particle. The Earth, however, can surely not be considered as a particle in this case; see Fig. 8.3. If the Earth's mass distribution were known, its gravitational attraction for objects at its surface could be calculated from the superposition principle: one would have to integrate the vector forces arising from all the infinitesimal mass elements of which the Earth is made. Fortunately such a difficult calculation can be completely by-passed, owing to the following theorem.

> The gravitational attraction exerted on an outside object B, of any size and shape, by a spherically symmetric object A can be calculated as if A's whole mass were concentrated at its center.

(8.B.4)

The necessary tools will be available in Chap. 15 (Electric Charge and its Field) for a simple derivation of this statement. Here our study of the center of gravity (Chap. 7) will unfortunately not help —we are no longer dealing with a region of uniform gravity.

Statement (8.B.4) does not apply to the case where object B is *inside* object A; we postpone to Chap. 15 the consideration of this possibility as well.

A valuable aspect of statement (8.B.4) is the fact that the spherically symmetric object need not be of uniform density. It is sufficient that it be made of concentric spherical shells, each of which should be uniform. The Earth, as well as most other astronomical bodies, is to a good approximation made in this way, the deeper shells being the denser ones; see Fig. 8.4.

Attraction Between Two Spheres

The massive-sphere theorem (8.B.4) can be applied to a *pair* of spherically symmetric objects, as in Fig. 8.5. By a two-step process, explained in the figure, one concludes that

> The gravitational attraction between two spherically symmetric objects can be calculated as if each had its mass concentrated at its center.

(8.B.5)

iii. The Acceleration of Gravity, Revisited

How does the acceleration of an object P, under free fall toward Earth, depend on its distance r from the Earth's center? For arbitrary r, we denote that acceleration by $\mathbf{g}'$, keeping in mind that $\mathbf{g}' = \mathbf{g}$ at

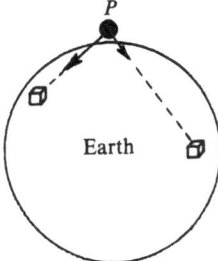

Figure 8.3. The weight of P is the vector sum of the forces of attraction due to all the particles of which the Earth is made. Two typical such mass elements are shown. It is not obvious from this figure that the Earth has the same effect as a single particle at its center. This can nevertheless be demonstrated. (If the Earth were, say, a cube, it would *not* be gravitationally equivalent to a point-mass at its center.)

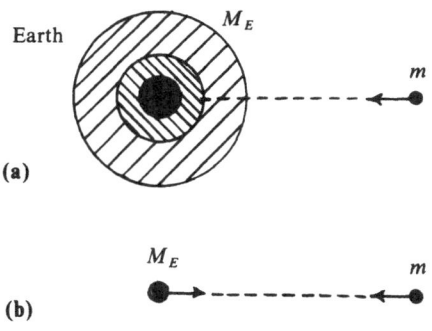

Figure 8.4. The nonuniform but nearly spherical Earth (mass M_E) and its effect on an outside particle (mass m). Situations (a) and (b) are equivalent. The total force on the Earth itself can also be calculated as if the Earth were a point mass; this follows from the law of action and reaction.

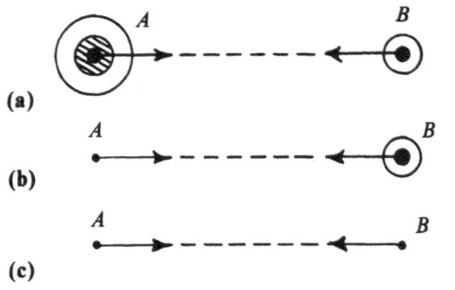

(a)

(b)

(c)

Figure 8.5. Attraction between two spheres. Situations (a) and (b) are equivalent by statement (8.B.4). Similarly, situations (b) and (c) are equivalent.

the Earth's surface. Like the force itself, $\mathbf{g}'$ points toward the center of the Earth and its magnitude decreases with distance in proportion to

$$\mathbf{g}' \propto \frac{1}{r^2} \tag{8.B.6}$$

The acceleration due to the Earth's gravity decreases as the inverse square of the distance to the Earth's center.

Let us supply the proportionality constant in (8.B.6). Suppose P has a mass m, while the mass of the Earth is M_E. The total force on the free-falling P is

$$F_{\text{tot}} = \frac{GM_E m}{r^2} \tag{8.B.7}$$

From Newton's second law we have

$$F_{\text{tot}} = mg'$$

We see from the two equations above that

$$\boxed{g' = \frac{GM_E}{r^2}} \tag{8.B.8}$$

Thus, the proportionality constant needed in Eq. (8.B.6) is GM_E, a combination of two rather unfamiliar factors. We observe in (8.B.8) that the mass m is not involved, so that the equivalence principle holds at all distances from the Earth (all objects have the same acceleration).

How can we obtain a numerical value for GM_E? For this purpose we take Eq. (8.B.8) at the surface

of the Earth, where $g' = g$ and $r = r_E$ (the Earth's radius). We have

$$g = \frac{GM_E}{r_E^2} \tag{8.B.9}$$

so that

$$GM_E = gr_E^2 = (9.8)(6.4 \times 10^6)^2 \frac{\text{meters}^3}{\text{sec}^2}$$

or

$$\boxed{GM_E = 4.0 \times 10^{14} \frac{\text{meters}^3}{\text{sec}^2}} \tag{8.B.10}$$

(For a more accurate value, based on data from artificial satellites, see Table 8.1.)

iv. The Falling Moon

Why doesn't the Moon fall into the Earth, since it is attracted by the latter? In our discussion of circular motion (Chap. 6) we have already recognized that the Moon in fact *does* fall, although it perpetually misses the Earth. The Moon's acceleration toward the center of the Earth is approximately

$$a_M = 2.69 \times 10^{-3} \text{ meter/sec}^2 \text{ (measured)} \tag{8.B.11}$$

(This result combines two pieces of data, the Moon's orbital radius and the Moon's sidereal period; see Problem 6.7.)

Is Newton's theory of gravitation capable of "predicting" a_M from other data? The answer is yes, as we have just seen: a_M should be the acceleration of free fall given by formula (8.B.8) for $r = r_{EM}$, as follows:

$$a_M = g' = \frac{GM_E}{r_{EM}^2} = \frac{4.0 \times 10^{14}}{(3.84 \times 10^8)^2} \frac{\text{meter}}{\text{sec}^2}$$

$$= 2.7 \times 10^{-3} \frac{\text{meter}}{\text{sec}^2} \tag{8.B.12}$$

The good agreement of this theoretical value with the measured one, Eq. (8.B.11), is one of the countless confirmations of Newton's inverse-square law.

A somewhat more intuitive, but completely equivalent, verification goes as follows. From the inverse-square behavior

$$a_M \propto \frac{1}{r_{EM}^2}, \qquad g \propto \frac{1}{r_E^2}$$

(with the same proportionality constant), we see that

$$\frac{g}{a_M} = \frac{r_{EM}^2}{r_E^2} = \left(\frac{r_{EM}}{r_E}\right)^2 = (60.3)^2 \qquad (8.B.13)$$

The left-hand side of that equation is $9.8/(2.69 \times 10^{-3}) = 3.64 \times 10^3$, in agreement with the right-hand side.

v. Weighing the Earth

The gravitational constant G is of basic significance, as it governs the behavior of all matter. Can its value be inferred from observed astronomical orbits? Equation (8.B.8) is typical in showing that the answer is no. The reason is that G occurs in combination with M, the mass of an astronomical object, which cannot be measured directly. Thus, either G or M must be determined before the other can be found. Free-fall experiments on Earth are of no help in this respect, as they are also described by (8.B.8). Therefore one must resort to a laboratory experiment for the measurement of G. This is done with two spheres of known masses M, m, and of known center-to-center distance r; one measures their mutual gravitational force F. Equation (8.B.1) can then be solved for G in terms of known quantities,

$$G = \frac{r^2 F}{mM} \qquad (8.B.14)$$

In practice, the experiment is quite delicate owing to the extreme weakness of the gravitational attraction between laboratory-sized objects. This experiment, first performed by Henry Cavendish (1798), is explained in Fig. 8.6. Although the details of the procedure have been improved, its basic principle has never been superseded. The constant G is at present the least accurately known of Nature's so-called fundamental constants; see Appendix V.

Once the value of G has been determined, one

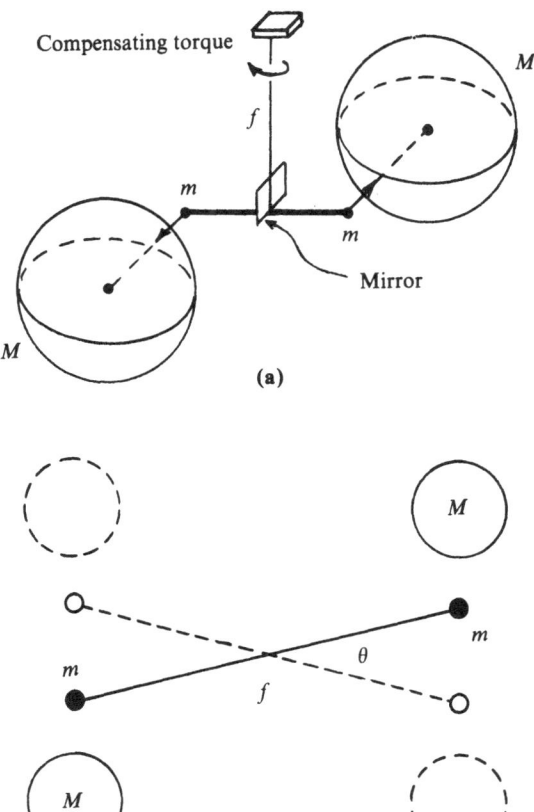

Figure 8.6. Cavendish's experiment. (a) The heart of this arrangement is a **torsion balance**, consisting of two small spheres m at the ends of a horizontal bar suspended in the middle from a thin fiber f. Each small sphere is attracted by a neighboring large sphere M. Consequently the fiber is slightly twisted, until, at equilibrium, it exerts on the bar a torque which compensates the gravitational torque due to the large spheres. Light reflected from the mirror provides a sensitive indication of the twist angle. (b) Top view. During the experiment, the large spheres are shifted in location (solid to dashed), causing the torsion to reverse itself. By measuring the angle θ, one can infer the change in torque, provided the fiber's elastic properties are known. Hence one obtains the gravitational forces involved, and, through use of Eq. (8.B.14), the value of G. (The fiber's properties are measured in a separate experiment, in which the balance's torsional oscillations are clocked; see Problem 13.27 further on.)

can immediately infer the mass of the Earth from the known value of GM_E, Eq. (8.B.10). The result is

$$M_E = \frac{GM_E}{G}$$

$$\approx \frac{4.0 \times 10^{14}}{6.7 \times 10^{-11}} \text{ kg}$$

$$\approx 6.0 \times 10^{24} \text{ kg} \qquad \text{(mass of the Earth)}$$

Cavendish, having measured G, had therefore the good fortune of being the man who "weighed the Earth."

The following example may help the reader gain an idea of the weakness of the gravitational interaction.

Example 8.2. Gravity from a Truck. A 70-kg passer-by walks next to a parked truck whose mass is 100 metric tons. What is the gravitational attraction on him, due to the truck? (Only a rough estimate is needed.)

For the purpose of obtaining an order-of-magnitude figure, let us assume a center-to-center distance of 4 meters between the passer-by and the truck; let us treat both unrealistically as spheres. If their respective masses are m and M we have, for the attractive force,

$$F \approx \frac{GmM}{r^2} \approx \frac{(7 \times 10^{-11})(70)(100 \times 10^3)}{(4)^2} \text{ newton}$$

$$\approx \underline{\underline{3 \times 10^{-5} \text{ newton}}}$$

comparable to the weight of a mosquito.

One moral of this example pertains to the numerous mechanical problems discussed in the previous chapters. Gravity, except that of the Earth, was of no relevance to these problems, and was correctly disregarded. This is true of all gravitational forces between laboratory-sized objects; Cavendish's experiment is the single exception to that rule.

C. Orbits

Newton's inverse-square law scores its most notable success when applied to planetary motion. Here we first examine the theory of **circular orbits**; these are at once very simple and a good approximation to reality. Next we describe without proof how the results obtained with circles can be generalized to **ellipses**, and how these generalized results fit **Kepler's observational rules** for the planets.

i. Circular Orbits

The planetary orbits are, to a good approximation, circular and concentric, with the Sun at the center; they all share nearly the same plane. This curiously simple configuration is also quite fortunate. The early stages of astronomy were greatly facilitated by this fact, and so will be our own analysis. (Incidentally, terrestrial life might not exist if the Earth's distance from the Sun kept changing drastically.)

Under the assumptions of a stationary Sun and circular orbits for the planets, two questions are discussed below.

1. How does a planet's period of revolution T around the Sun (i.e., its "year") depend on its distance r from the Sun?
2. What is the Sun's mass M_S?

1. Period for a Given Orbit

Let us apply Newton's second law,

$$F_{\text{tot}} = ma \qquad (8.\text{C}.1)$$

to a planet of mass m. The left-hand side of this equation is given by

$$F_{\text{tot}} = \frac{GM_S m}{r^2} \qquad (8.\text{C}.2)$$

The right-hand side involves the centripetal acceleration

$$a = \left(\frac{2\pi}{T}\right)^2 r \qquad (8.\text{C}.3)$$

as we recall from Chap. 6. Equation (8.C.1) now becomes

$$\frac{GM_S m}{r^2} = m\left(\frac{2\pi}{T}\right)^2 r \qquad (8.\text{C}.4)$$

The planet's mass m factors out and is therefore irrelevant. We are interested in relating T and r; we therefore rewrite (8.C.4) in the form

$$\frac{r^3}{T^2} = \frac{GM_S}{(2\pi)^2} \qquad (8.\text{C}.5)$$

The right-hand side of this equation, whatever its

value, is independent of the planet considered, and we thus arrive at the following rule:

$$\boxed{\dfrac{r^3}{T^2} \text{ is the same for all planets.}} \qquad (8.C.6)$$

That is to say:

$$\boxed{\begin{array}{c} \text{The square of a planet's} \\ \text{period of revolution is} \\ \text{proportional to the cube} \\ \text{of its distance from the Sun,} \\ T^2 \propto r^3. \end{array}} \qquad (8.C.7)$$

(This statement is also known as **Kepler's third rule**; see the next subsection.)

> **Example 8.3. Jupiter's Orbit.** Jupiter's period around the Sun is observed to be $T_J = 4330$ days. What is its distance r_{JS} from the Sun? (Express the result in terms of the Earth–Sun distance r_{ES}.)
>
> ────────────
>
> Using (8.C.6) to compare Jupiter and the Earth, we have
>
> $$\frac{r_{JS}^3}{T_J^2} = \frac{r_{ES}^3}{T_E^2} \qquad (8.C.8)$$
>
> where T_E is the terrestrial year. Solving for the desired ratio r_{JS}/r_{ES} gives
>
> $$\frac{r_{JS}}{r_{ES}} = \left(\frac{T_J}{T_E}\right)^{2/3} = \left(\frac{4330}{365}\right)^{2/3} = \underline{\underline{5.20}} \qquad (8.C.9)$$
>
> Jupiter is 5.20 times further away from the Sun than we are.

There is of course no reason why the period–distance relation (8.C.5) should be limited to planetary orbits. Rather than the Sun, the central body could, for example, be Jupiter itself, while its satellites would play the role of the planets. Similarly, the Earth could be the central body, and one would then compare parameters for the Moon and the artificial satellites. For this purpose, a useful number to keep for reference is the period

$$T_{\text{low}} = 84 \text{ min} \qquad (8.C.10)$$

of a satellite in low orbit; this number was obtained in Example 6.3 of Chap. 6.

As we consider wider and wider orbits, we see from (8.C.6) that T must increase more steeply than r. Therefore the period increases faster than the circumference of the orbit. We conclude that more remote planets not only have a smaller angular velocity around the Sun, but also a *smaller absolute speed*. Similarly, the Moon travels more slowly around the Earth than does a lower-orbiting satellite.

2. The Solar Mass

If the orbital radius r and period of revolution T of any one planet is known, then Eq. (8.C.5) can be solved to yield the Sun's mass,

$$M_S = \left(\frac{2\pi}{T}\right)^2 \frac{r^3}{G} \qquad (8.C.11)$$

Choosing the Earth to give us our parameters, and using the numerical shortcut

$$1 \text{ year} \approx \pi \times 10^7 \text{ sec} \qquad (8.C.12)$$

(to 0.5 % accuracy; this is just an amusing coincidence) we obtain from Eq. (8.C.11)

$$M_S \approx \left(\frac{2\pi}{\pi \times 10^7}\right)^2 \frac{(1.5 \times 10^{11})^3}{6.7 \times 10^{-11}} \text{ kg}$$

$$\approx 2.0 \times 10^{30} \text{ kg} \qquad (8.C.13)$$

In connection with this calculation, a few comments are appropriate.

1. It is clear that the mass of any astronomical body S (not necessarily the Sun) may be similarly inferred from the orbital radius and period of another body, E (not necessarily the Earth), which revolves around it. Classic examples are the mass of Jupiter (from any of its satellites), or the mass of the Moon (from a man-made lunar orbiter). The method may, in fact, be adapted to noncircular (elliptic) orbits.

2. Object E must be much less massive than object S in order for the method to work; otherwise the motion of S will be affected by E, and our assumption of a fixed S breaks down. A sufficiently light object E may be called a "test mass."

3. The mass of E, if not too large, is completely

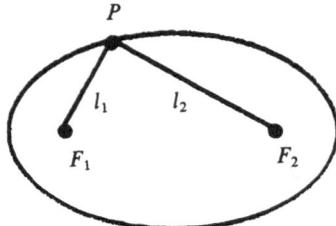

Figure 8.7. An ellipse. If l_1, l_2 are the distances between a point P and two fixed points F_1, F_2, then an ellipse may be defined as the locus of all points P, in a plane through F_1 and F_2, such that $l_1 + l_2 = \text{const.}$ (A thread fastened at F_1 and F_2, and stretched by a pencil point at P, will guide the latter along an ellipse.) Points F_1 and F_2 are called the ellipse's **foci.**

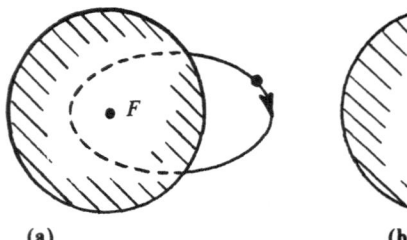

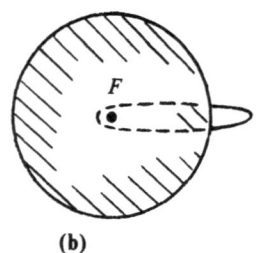

(a)   **(b)**

Figure 8.8. (a) A ballistic missile. Since, gravitationally, the Earth can be considered concentrated at its center, the missile's (unpowered) trajectory follows a portion of an ellipse with one focus, F, at the Earth's center. (b) A parabolic trajectory of free fall, such as was studied in Chap. 2, can be considered as the tip of an extremely elongated ellipse, with one focus, F, at the Earth's center.

irrelevant to the information we seek. This is an illustration of the **equivalence principle** on an astronomical scale. All test masses are subject to the same acceleration at a given location in space. Astronauts inside an unpowered, nonspinning space station detect no gravity because gravity accelerates them to the same extent as it accelerates the space station itself.

ii. Elliptic Orbits (Kepler's Rules)

Consider once more an object of mass m in orbit around a fixed object of much greater mass M. For the sake of our discussion we may think of a planet in orbit around the Sun.

We have seen that a circular orbit is always physically possible, since it allows $\mathbf{F}_{\text{tot}} = m\mathbf{a}$ to be satisfied. However, the circle is only a special case. Under the inverse-square law of force, the most general orbit is a **conic section**, of which the circle and the parabola are two familiar examples: all **closed orbits**, including the circle, are **ellipses**. Figure 8.7 shows a geometrical definition for the ellipse and its foci.*

We are now ready to state three famous observational rules, originally published in 1609 and 1619 by Johann Kepler. These rules were eventually shown by Newton to follow deductively from his own laws of motion, taken together with his inverse-square law. Kepler's rules for elliptic orbits are extensions of what we already know for circular orbits.

> **Kepler's first rule.** A planet's orbit is an ellipse, one of whose foci coincides with the center of the Sun.

(8.C.14)

(The other focus is empty and does not play any particular astronomical role.) We note that the rule allows the orbit to be a circle with the Sun at the center.

The same elliptic geometry applies just as well in a different context: the Earth, rather than the Sun, occupies the attractive focus; the orbiting bodies are satellites, or even **ballistic missiles**, coasting unpowered along a suborbital path; see Fig. 8.8.

> **Kepler's second rule.** The line segment joining the Sun and the planet sweeps out equal areas in equal times.

(8.C.15)

See Fig. 8.9. This rule, as it turns out, amounts to the conservation of angular momentum. In the cir-

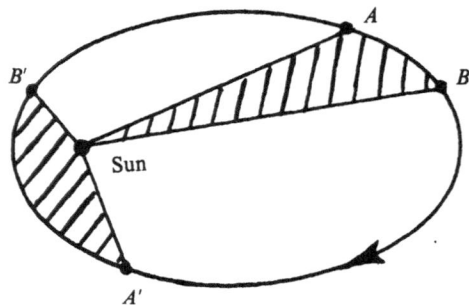

Figure 8.9. Kepler's second rule (rule of areas). If the portions AB and $A'B'$ of a planet's orbit are traveled in equal time intervals, then the shaded areas are equal. In other words, the ellipse's area is being swept out at a uniform rate, $\Delta(\text{area})/\Delta(\text{time}) = \text{const.}$

---

* Singular: Focus.

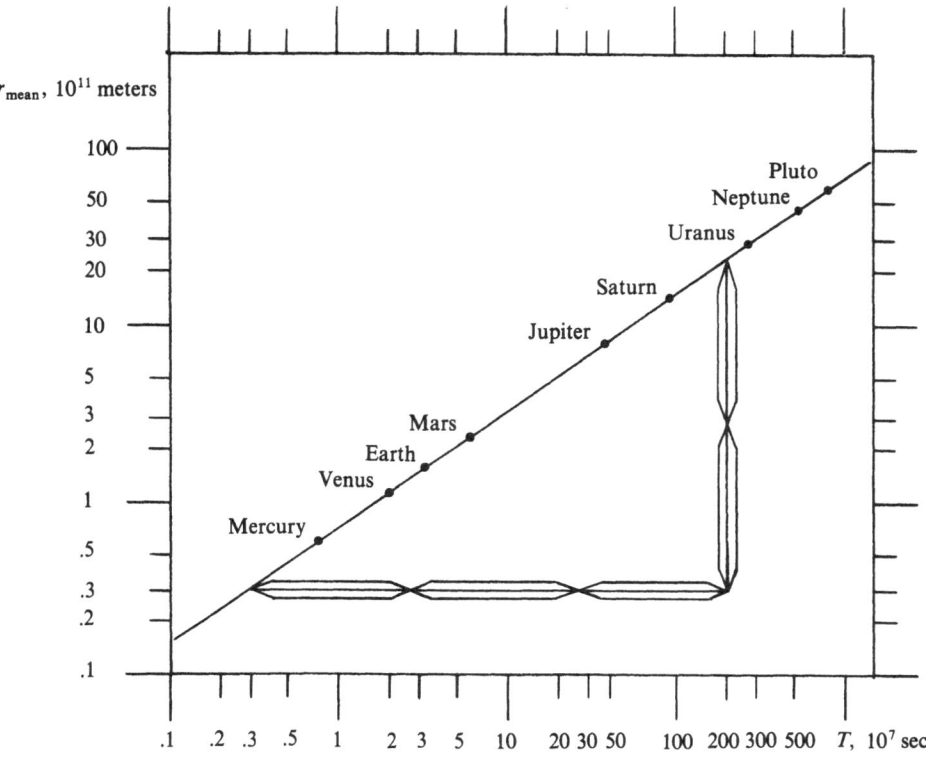

Figure 8.10. Kepler's third rule. The planetary data (period and mean orbital radius, see also Sec. E) are plotted on a log–log grid; they lie on a straight line whose geometrical slope is 2/3. This implies an equation of the form $r_{mean} = (\text{const}) \, T^{2/3}$. To see this, take the logarithm on both sides of the above, $\log r_{mean} = \log \text{const} + \frac{2}{3} \log T$, and note that $\log T$ and $\log r_{mean}$ are proportional to the distances actually laid out on paper. Some contemplation of this graph will show one reason why Newton's inverse-square law carries overwhelming conviction. (The error boxes that should be drawn are so small that they would fit inside the data points.)

cular case, (8.C.15) just reduces to the statement that the circular motion is uniform.

> **Kepler's third rule.** The square of a planet's period of revolution is proportional to the cube of its mean distance from the Sun,
> $$T^2 \propto (r_{mean})^3$$ (8.C.16)

[The "mean distance" equals half the sum of the longest and shortest distances; in the circular case it is simply the radius of the orbit, and we are back to (8.C.7).]

The excellent verification of Kepler's third rule is demonstrated in Fig. 8.10. Again: *All three rules are consequences of Newton's laws.*

D. Gravitational Potential Energy

Is mechanical energy conserved under the effect of gravitational forces? We have seen in Chap. 4

that these forces do indeed conserve energy in the laboratory (or in any region where the acceleration of gravity can be considered uniform). We have also seen how important this conservation law turned out to be to our understanding of complex systems. Here we proceed to demonstrate the same conservation law for the more general case where the gravitational forces vary with position, as they do on an astronomical scale.

i. Gravity and Work

Let us briefly recall which kind of force conserves the mechanical energy of a particle P. Suppose P undergoes a small displacement $d\mathbf{r}$ under a gravitational force $\mathbf{F}$. That force then does an amount of work $\mathbf{F} \cdot d\mathbf{r}$ on P. Conservation of energy means that a corresponding decrease, $-d\mathcal{U}$, must occur in the gravitational potential energy $\mathcal{U}$ that was stored. Thus, we have

$$\boxed{\mathbf{F} \cdot d\mathbf{r} = -d\mathcal{U}} \qquad (8.\text{D}.1)$$

$$\boxed{\begin{array}{c} \text{The work done by gravity must} \\ \text{equal the decrease in} \\ \text{gravitational potential energy.} \end{array}} \qquad (8.\text{D}.2)$$

If a function $\mathcal{U}$ of position exists such that (8.D.1) is true for any $d\mathbf{r}$, then we can say that gravity conserves mechanical energy, or that *the force of gravity is conservative*.

ii. The Potential Energy for an Inverse-Square Force

Our next task is to find such a function $\mathcal{U}$; because it is very simple, we first state the result and derive it later.

Consider the case of two particles (or spheres), of masses m and M, located at a distance r from each other. The correct expression for the gravitational potential energy $\mathcal{U}$ of this system is

$$\boxed{\mathcal{U} = -\frac{GMm}{r}} \qquad (8.\text{D}.3)$$

(note the sign). This inverse-distance function should be carefully distinguished from the inverse-square which expresses the force. Its dimensions are most easily checked as follows:

$$\mathcal{U} \sim \left(\frac{GMm}{r^2}\right)(r) \sim (\text{force})(\text{distance}) \sim \text{energy}$$

which is as it should be.

Figure 8.11 shows a qualitative graph of the function $\mathcal{U}$. The following observations may be of help in becoming acquainted with it. For definiteness, we may consider the mass M to be fixed at the origin, while the other mass, m, is allowed to vary its position.

1. We see from the graph that $\mathcal{U}$ increases with r. This reflects the fact that one needs positive work to lift an object against gravity.

2. A constant term may be added to $\mathcal{U}$ without

changing its physical significance: the modified expression

$$-\frac{GMm}{r} + \text{const}$$

would be just as correct as (8.D.3). Adding a constant just changes the reference level with respect to which the potential energy is defined.

3. Where is the reference level $\mathcal{U} = 0$ in the case of formula (8.D.3)? A look at that equation, or at the graph of Fig. 8.11, shows that *the reference "level" is infinitely far away from the center of attraction*. In other words, the potential energy of a particle at a distance r from the central mass is

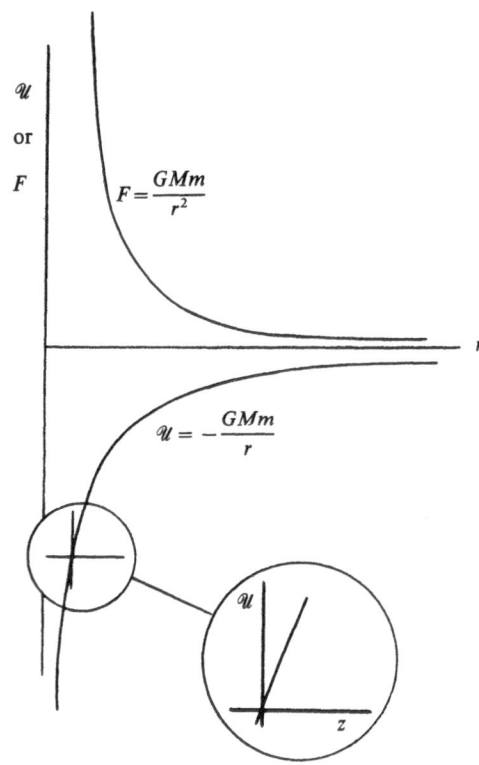

Figure 8.11. Bottom graph: The potential energy function $\mathcal{U}$. Top graph: The magnitude F of the gravitational force. Both are plotted as functions of the distance r from the attracting center. The graphs are drawn correctly to scale, but the units are arbitrary since GMm is not specified. (In practice, $\mathcal{U}$ does *not* approach $-\infty$ as $r \to 0$, contrary to what is suggested by the formula $1/r$ and by the graph. The reason is that actual attracting bodies, such as the Earth, have a nonzero radius, inside which the $1/r$ dependence is not valid.) Magnified inset: The gravitational potential energy $\mathcal{U}$ versus the altitude z in an Earth laboratory. Owing to the relatively small scale, we obtain the familiar linear dependence, $\mathcal{U} = mgz$.

defined relative to what that energy would be if the particle were much further away. How far is "much further" depends on the degree of accuracy one wishes to reach. In theoretical work, there is usually no problem in taking a rigorous limit $r \to \infty$.

4. We know from Chap. 4 that, on Earth, the gravitational potential energy is given by $\mathcal{U} = mgz + \text{const}$, where z can be taken as the altitude above sea level. In terms of a graph, one is saying that $\mathcal{U}$ is a *linear function* of altitude. Does this agree with our new result? In Fig. 8.11, the size of an earthly laboratory corresponds to a minuscule portion of the curve near $r = r_E$, the Earth's radius. Such a small segment can be considered straight, which accounts for the apparent linearity of $\mathcal{U}$ in the laboratory.

Example 8.4. The Speed of Escape. With what speed v_{esc} should a space vehicle be launched in order to escape completely from the gravitational pull of the Earth? (This means being able to reach arbitrarily large distances from the Earth; we assume unpowered flight after the initial speed has been acquired; atmospheric friction is neglected.)

Conservation of energy states that

$$\begin{pmatrix} \text{Total energy} \\ \text{at launch} \end{pmatrix} = \begin{pmatrix} \text{Total energy} \\ \text{when far away} \end{pmatrix}$$

or

$$\mathcal{U}_i + \mathcal{K}_i = \mathcal{U}_f + \mathcal{K}_f \qquad (8.D.4)$$

Initial values:

$$\mathcal{U}_i = -\frac{GM_E m}{r_E}, \qquad \mathcal{K}_i = \tfrac{1}{2}mv_{\text{esc}}^2$$

where M_E and r_E are the Earth's mass and radius, respectively.

Final values:

$$\mathcal{U}_f = -\frac{GM_E m}{r} \qquad \text{as} \quad r \to \infty$$

so that $\mathcal{U}_f = 0$. Also, with barely enough initial speed to escape, there will be no kinetic energy left as $r \to \infty$; thus, $\mathcal{K}_f = 0$. Hence Eq. (8.D.4) becomes

$$-\frac{GM_E m}{r_E} + \tfrac{1}{2}mv_{\text{esc}}^2 = 0 + 0$$

or

$$v_{\text{esc}} = \left[\frac{2(GM_E)}{r_E} \right]^{1/2} \qquad (8.D.5)$$

$$= \left[\frac{(2)(4.0 \times 10^{14})}{6.4 \times 10^6} \right]^{1/2} \frac{\text{meters}}{\text{sec}}$$

$$= \underline{\underline{11\,000 \, \frac{\text{meters}}{\text{sec}}}} \qquad (8.D.6)$$

(This is $\sqrt{2}$ times the launching speed needed to circle the Earth in low orbit; see Problem 6.15.)

iii. Proving the Inverse-Distance Formula

In order to verify the potential-energy formula

$$\mathcal{U} = -\frac{GMm}{r} \qquad (8.D.7)$$

we need to work out each side of the conservation equation (8.D.1),

$$\mathbf{F} \cdot d\mathbf{r} = -d\mathcal{U} \qquad (8.D.8)$$

for an arbitrary small displacement $d\mathbf{r}$. Is this equation satisfied?

Let us calculate the right-hand side first. Figure 8.12 shows two imagined spheres, of radii r and $r + dr$, drawn concentrically around a fixed particle (or sphere) of mass M. Along each of

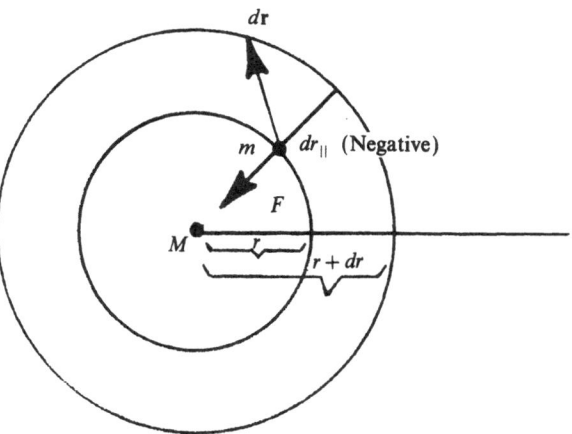

Figure 8.12. A particle of mass m moves along $d\mathbf{r}$ from a distance r to a distance $r + dr$ from the attracting center. The force $\mathbf{F}$ does a work $\mathbf{F} \cdot d\mathbf{r} = F \, dr_{\parallel}$, where $dr_{\parallel} = -dr$.

these spherical surfaces the potential energy $\mathscr{U}$, Eq. (8.D.7), is constant. For definiteness let us visualize an outward-moving particle of mass m, on which gravity is then expected to do negative work. As m moves from the inner sphere to the outer sphere, the change in potential energy is

$$d\mathscr{U} = d\left(-\frac{GMm}{r}\right) = \frac{GMm}{r^2}\,dr \quad (8.D.9)$$

We now turn to the left-hand side of (8.D.8). From Fig. 8.12 we see that

$$\mathbf{F} \cdot d\mathbf{r} = F\,dr_{\parallel} = -F\,dr = -\frac{GMm}{r^2}\,dr \quad (8.D.10)$$

according to the law of attraction. Comparison of results (8.D.9) and (8.D.10) completes the proof.

Integrating the $1/r^2$ Force to Get the $1/r$ Potential

We can motivate our "lucky" guess, Eq. (8.D.3), somewhat further. In Chap. 4 (Energy) we have developed a mechanism for calculating a potential energy $\mathscr{U}$ when given a **conservative force** $\mathbf{F}$. Note (ii) at the end of this chapter explains how to use that mechanism in the present case. Still, no matter what line of argument is chosen, care must be taken to verify the conservative nature of $\mathbf{F}$.

Notes

i. General Relativity and Black Holes

In regions of intense gravity, Newton's theory cannot be used. Rather, one has to apply the methods of general relativity, developed by Einstein. (See the introduction of Sec. B.) Within the solar system, gravity is not strong enough to make relativistic corrections important. Elsewhere, there probably exist certain small but very massive stars in whose vicinity Newton's law of attraction is a poor approximation. How concentrated in space does a given mass M have to be for Newtonian mechanics to break down completely? We do not have to understand general relativity in order to reach a good estimate. All we need to know is that, according to *special* relativity, the speed of light, $c = 3.0 \times 10^8$ meters/sec, is the largest that can ever be achieved by a material par-

ticle relative to any observer. Consider now a star of mass M and radius r. The speed of escape from that star is

$$v_{\mathrm{esc}} = \left(\frac{2GM}{r}\right)^{1/2} \quad (8.N.1)$$

according to Newton's theory; see Eq. (8.D.5). For a given M, how small must we make r in order that nothing be able to escape, even when launched at the speed of light? Setting $v_{\mathrm{esc}} = c$ in Eq. (8.N.1), and solving for r, we find

$$r = \frac{2GM}{c^2} \quad (8.N.2)$$

(Relativity, which ought to have been used here, gives the same result; $2GM/c^2$ is known as the **Schwarzschild radius** of the mass M.) If the no-escape assumption is taken seriously, we have here a **black hole**, so called because even "particles of light" cannot get out; it can be shown that, in an alternative model, *waves* of light would do no better.

As an example, setting $M = M_E$ (the mass of the Earth), we find $2GM_E/c^2 = 0.9$ cm (i.e., the Earth would have to be squeezed to the size of a marble in order to become a black hole). For the Sun, we find a Schwarzschild radius $2GM_S/c^2 = 3$ km, or about $1/250\,000$ of the Sun's actual radius. Although some stars do collapse drastically under their own gravity, the Sun is not believed to be a black hole candidate. The actual existence of any black hole has so far not been conclusively demonstrated, but only made probable.

ii. The $1/r$ Potential Energy by Integration

At the end of the chapter it was mentioned that the potential energy

$$\mathscr{U} = -\frac{GMm}{r} \quad (8.N.3)$$

could be obtained from the force

$$F = \frac{GMm}{r^2} \quad (8.N.4)$$

by integration rather than by guessing. Under the crucial assumption that $\mathbf{F}$ is conservative, we apply Eq. (4.C.5) of Chap. 4 to the case of a particle being lowered from A to B:

$$\mathscr{U}_A - \mathscr{U}_B = \int_A^B \mathbf{F} \cdot d\mathbf{r} \quad (8.N.5)$$

The path between A and B can be chosen to fit our convenience, as in Fig. 8.13. The integral now reads

$$\mathscr{U}_A - \mathscr{U}_B = \int_A^X \mathbf{F} \cdot d\mathbf{r} + \int_X^B \mathbf{F} \cdot d\mathbf{r}$$

$$= 0 - \int_{r_A}^{r_B} \frac{GMm}{r^2} dr \qquad (8.\text{N}.6)$$

The first integral vanishes because $\mathbf{F}$ is transverse ($\mathbf{F} \cdot d\mathbf{r} = 0$); in the second integral, both $\mathbf{F}$ and $d\mathbf{r}$ point toward M, so that we have $\mathbf{F} \cdot d\mathbf{r} = F|d\mathbf{r}| = (F)(-dr)$, noting that $-dr$ is positive. The result of (8.N.6) is

$$\mathscr{U}_A - \mathscr{U}_B = + \left[\frac{GMm}{r} \right]_{r_A}^{r_B} = \frac{GMm}{r_B} - \frac{GMm}{r_A} \qquad (8.\text{N}.7)$$

We now decide on a fixed reference point A ($\mathscr{U}_A = 0$) and an arbitrary variable point B ($r_B = r$, $\mathscr{U}_B = \mathscr{U}$); Eq. (8.N.7) gives

$$\mathscr{U} = -\frac{GMm}{r} + \frac{GMm}{r_A}$$

$$= -\frac{GMm}{r} \qquad (\text{if } r_A \to \infty)$$

This is the desired result.

Condensed Checklist

Universal attraction:

$$\boxed{F = G\frac{Mm}{r^2}} \qquad (8.\text{B}.1)$$

$$G = 6.67 \times 10^{-11} \text{ (SI units)} \qquad (8.\text{B}.2)$$

Orbits:

$$\frac{(r_{\text{mean}})^3}{T^2} = \frac{GM}{(2\pi)^2} \qquad \text{(fixed central mass)} \quad (8.\text{C}.5)$$

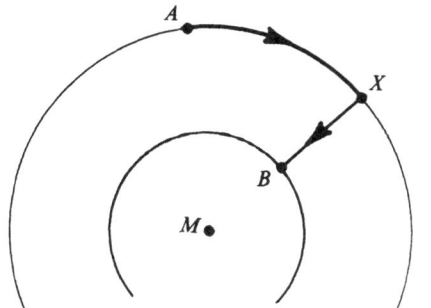

Figure 8.13. A convenient path of integration for calculating the potential drop $\mathscr{U}_A - \mathscr{U}_B$.

Potential energy:

$$\mathscr{U} = -\frac{GMm}{r} \qquad (8.\text{D}.3)$$

Speed of escape:

$$v_{\text{esc}} = \left(\frac{2GM}{r} \right)^{1/2} \qquad (8.\text{D}.5)$$

Astronomical data:
Table 8.1

True or False

1. To the extent that all planetary orbits are circular and coplanar, and if satellites are ignored, the solar system is well represented by a rigid rotating disk to which each planet is attached at the appropriate place.

2. Centrifugal force is what prevents the Earth from falling into the Sun.

3. To double the length of our year, the Earth would have to be twice as far from the Sun as it is now.

4. An interplanetary vehicle should depart eastward rather than westward in order to save fuel.

5. The gravitational constant G can be calculated entirely in terms of the acceleration of gravity at the Earth's surface, and the Earth's radius.

6. A satellite, orbiting the Earth at 3 Earth radii above the surface, has with respect to points at infinity, 1/4 the potential energy it had when on the ground.

7. If a satellite's orbit passes over the north pole, that satellite follows one of the Earth's meridians.

8. Cavendish's experiment amounts to a highly sensitive measurement of the Earth's attraction for a test mass.

Problems

Section B: Universal Gravitation

*8.1. Two spheres, 3 meters apart from center to center, attract each other with a force of 2×10^{-12} newton. At what distance from each other must they be placed to reduce their attraction to 2×10^{-14} newton?

8.2. Express the unit for G, Eq. (8.B.2), as a combination of kilograms, meters, and seconds.

*8.3. A space traveler weighs 150 lb on Earth. How much would he weigh on a planet half as massive as the Earth and with half its radius?

8.4. In one version of the Cavendish experiment, each large sphere is 10.0 cm in radius and has a mass of 47.5 kg. Each small sphere is 0.50 cm in radius and has a mass of 0.0101 kg. What is the gravitational force of attraction between a large and a small sphere when they are almost in contact?

*8.5. A carbon atom has a mass of 2×10^{-26} kg, and a radius of 0.8×10^{-10} meter. (a) What is the gravitational force between two such neighboring atoms if they are in contact? (b) What fraction is this of a carbon atom's weight on Earth? (c) Since a small piece of carbon is not appreciably deformed by its own weight, what can you conclude about the importance of interatomic gravity?

*8.6. Three particles, of masses $18M$, $16M$, and $8M$ (M being some unspecified mass), are located on the x axis at $x = 0$, l, $3l$ respectively. What is the magnitude of the total force on the $8M$ particle, due to the other two masses? (Express your answer in terms of G, M, l.)

8.7. In the preceding problem, what is the magnitude and direction of the total gravitational force on the $16M$ particle?

Section C: Orbits

8.8. Two artificial satellites, in circular orbits around the Earth, have periods of 8 h and 27 h, respectively. What is the ratio of their orbital radii?

8.9. Compare the ratios r^3/T^2 for the Moon and a low-orbit satellite of the Earth [see Eq. (8.C.10)]. Is Kepler's rule valid?

*8.10. A **geosynchronous satellite** is one which has a circular orbit above the Earth's equator, at such a distance r_{sync} from the Earth's center that the satellite remains forever above a fixed spot on Earth. Find r_{sync} in terms of the Earth's radius r_E. (Such satellites are used for microwave communication; the one circle in space that is available to them is sometimes called the "geosynchronous corridor.")

8.11. The asteroid (or minor planet) Ceres is at a mean distance of 2.77 AU from the Sun. What is Ceres' period, in seconds and in years? (1 AU = mean Earth–Sun distance = 1.50×10^{11} meters.)

*8.12. The asteroid Juno revolves around the Sun with a period of 4.36 years. Find, in astronomical units (cf. preceding problem), its mean distance from the Sun.

8.13. (a) In the graph of Fig. 8.10, what is, in terms of G and the Sun's mass M_S, the constant in $r_{mean} = $ (const) $T^{2/3}$? (b) Explain how to obtain the numerical value of that constant from the graph. (c) To the precision with which you can read the graph, what is that numerical value? (d) Plot the data for the first four

planets (Mercury to Mars) on a graph with uniform scales. Briefly compare the merits of both types of graph.

8.14. Derive the period–distance relation for circular orbits, Eq. (8.C.5), by dimensional analysis. (Search for exponents α, β, γ such that $r^3 \sim T^\alpha G^\beta M_S^\gamma$; this assumes the equivalence principle, according to which there should be no dependence on the planet's mass m; the factor $(2\pi)^{-2}$ cannot be obtained in this way; see also Sec. C iv of Chap. 2, and Problem 8.2.

Section D: Gravitational Potential Energy

*8.15. Calculate the escape speed from the surface of (a) the Moon; (b) Mars; (c) Jupiter; (d) the Sun.

8.16. (a) How far should one raise a 1-kg mass above the surface of the Earth in order to halve its potential energy relative to points at infinity? (Express your answer in Earth radii.) (b) By how many joules is its potential energy raised in that way?

8.17. Prove that the speed of escape from a planet, Eq. (8.D.5), is $\sqrt{2}$ times the speed needed for a low circular orbit around that planet.

*8.18. (a) An object falls to Earth from outer space with a speed v, measured just before interaction with the atmosphere. Show that, before being in the Earth's vicinity, it had a speed v_0 given by $v_0 = (v^2 - v_{esc}^2)^{1/2}$, where v_{esc} is the speed of escape from the Earth. (b) A certain meteorite is observed to fall with an upper-atmosphere speed of 14.2 km/sec. What would have been its speed without the Earth's gravitational pull?

*8.19. (a) From the data of Table 8.1, find the speed of the Earth in solar orbit. (b) A space vehicle is "parked" around the Sun in the same nearly circular orbit as the Earth, but has already escaped from the Earth itself. What initial speed does it need in order to escape from the solar system? (Assume unpowered flight after launching).

*8.20. (a) If a planet in circular orbit has a translational kinetic energy $\mathcal{K} = \frac{1}{2}mv^2$ and a gravitational potential energy $\mathcal{U} = -GmM_S/r$ ($r = $ distance from the Sun), show that $\mathcal{K} = -\frac{1}{2}\mathcal{U}$. (The total projectile energy is therefore still negative and rises for higher orbits). (b) How much energy must be given to an Earth satellite of mass 2 metric tons, already in low orbit, in order to lift it to many Earth radii away?

Answers to True or False

1. False (the outer planets lag behind the inner ones).
2. False (there is no centrifugal force; the Earth does accelerate toward the Sun).

3. False [the period is proportional not to the distance but to (distance)$^{3/2}$].

4. True (the Earth's rotation supplies an initial speed).

5. False (needs Cavendish's experiment).

6. True ($\mathscr{U}$ is inversely proportional to distance from the center).

7. False (the Earth rotates under the orbit).

8. False (that would just be weighing the test mass).

Answers or Hints to Problems

8.1. 30 meters.

8.3. 300 lb.

8.5. (a) 4×10^{-42} newton.

8.6. $48GM^2/l^2$.

8.10. A simple method: Use Kepler's third rule and Eq. (8.C.10). Answer: $6.7r_E$.

8.12. 2.67 AU.

8.15. (a) 2.4×10^3 meters/sec.

8.18. (b) 8.7 km/sec.

8.19. (a) $v = 2\pi r_{ES}/T = 30$ km/sec; (b) $(\sqrt{2})(30$ km/sec); see Problem 8.17.

8.20. (b) 6.3×10^{10} joules.

CHAPTER 9

FLUIDS

Anyone who has ever contemplated a smoking pipe, or a cup of coffee in which cream is being poured, must be impressed by the complexity of fluid behavior. Such complexity is not surprising, however. Each small visible portion of fluid is made of a huge number of molecules, moving to a considerable extent independently of each other. Yet this many-particle system has at least some simple properties, which form the topic of the present chapter.

A fluid ("that which can flow") can be either a liquid or a gas. A liquid is easily recognized by its equilibrium behavior. It settles to the bottom of its container and acquires a horizontal free upper surface. This behavior is caused by gravity, and minimizes the liquid's gravitational potential energy. A gas, on the other hand, will completely fill any container available to it.

There is no absolute distinction between solids, liquids, and gases. A substance commonly regarded as a solid, such as granite, may flow like a liquid under geologic forces and on a geologic time scale. Glass turns so gradually into a liquid when its temperature is raised, that the dividing line between its solid and liquid states can only be drawn arbitrarily. In a similar way, it is possible to turn a liquid into a gas, by heating it inside a rigid closed container, without any sharp transition being detectable.

The present chapter focuses on the most basic *mechanical property* of fluids, namely, the pressure that they exert.

A. Pressure and Density

We begin by reviewing two elementary concepts—**pressure** and **density**—which are of constant use in science as well as in daily life. These notions will then be applied to floating and submerged objects. In subsequent chapters, they will be found relevant to such diverse topics as sound waves and the utilization of heat.

i. The Absence of Shearing Forces in Static Fluids

Why is a fluid able to flow? Answering this question will give us an important clue as to the forces operating in a fluid. Actually, it is easier to examine first why a solid does *not* flow. Let us imagine a solid block of paraffin, Fig. 9.1, resting in equilibrium on a rough inclined metal plate. We recall that two forces, $F_{\parallel}$ (a "shearing force") and $F_{\perp}$ (a normal force) are then exerted by the plate on the block; $F_{\perp}$ prevents the block from sinking into the plate, while $F_{\parallel}$ prevents it from slipping.

Next, suppose the plate is slightly heated, thereby liquefying a thin adjoining layer of the block. The latter will still not sink into the plate; $F_{\perp}$ maintains its existence. But the block can no longer be kept from sliding down the plate, unless the latter is perfectly horizontal. Thus, in this case, static equilibrium implies horizontality; that is, it implies that *the plate exerts no shearing force on the liquid paraffin.* We now generalize our conclusion:

> Under static equilibrium, any contact force on a fluid surface must be exactly normal to that surface.

(9.A.1)

(This may serve as the technical definition of a fluid.) By the law of action and reaction, *a fluid in*

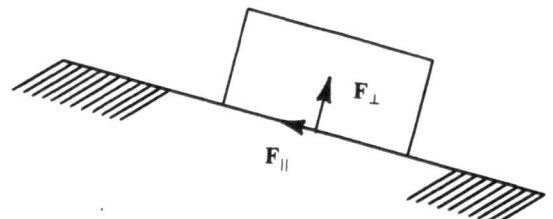

Figure 9.1. Static equilibrium of a block on an incline. The presence of a tangential force $\mathbf{F}_{\parallel}$ shows that there cannot be a fluid layer under the block. Caution: A uniformly sliding block (i.e., one in **nonstatic** equilibrium) can be supported by a fluid layer with nonzero $\mathbf{F}_{\parallel}$ (a "viscous force").

static equilibrium can therefore only exert a normal force on any surface in contact with it, be it fluid or solid. Such forces are called **hydrostatic**.

ii. Pressure

In order to examine the force exerted by a fluid surface, let us immerse, as in Fig. 9.2, a small solid block at some chosen location inside a fluid. What can we say about the contact force $\mathbf{F}$, due to the fluid, on one of the block's sides? (That side will be assumed flat and of area $\mathscr{A}$.) Part (b) of the figure shows how the block may, in principle, be set up as a force meter.

We already know that $\mathbf{F}$ is normal to the block's surface. Furthermore, it is directed into the block; it is a push, or compressive force.* The magnitude F is proportional to $\mathscr{A}$,

$$F = P\mathscr{A} \qquad (9.A.2)$$

The proportionality constant P is, by definition, the **pressure** that exists at the location of $\mathscr{A}$. Equivalently, we have

$$\boxed{P = \frac{F}{\mathscr{A}}} \qquad (9.A.3)$$

Pressure equals normal force per unit area.

The proportionality (9.A.2) can be understood from Fig. 9.3, where several identical force-measuring surfaces are set up in parallel. For the validity of (9.A.2) or (9.A.3), it is important that the area $\mathscr{A}$

---

* There has actually been some evidence for pulling forces in liquids; see "Negative Pressure" by A. T. Hayward, *American Scientist*, vol. 59, p. 434 (1971).

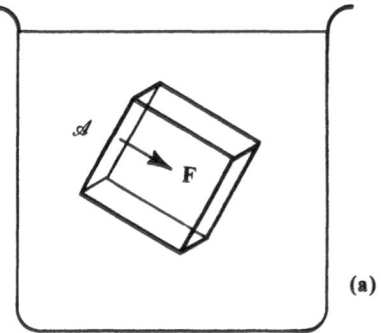

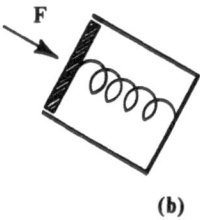

Figure 9.2. (a) A typical side, with area $\mathscr{A}$, of an immersed block is subjected to a normal force $\mathbf{F}$ due to the surrounding fluid. (b) An idealized box arrangement to measure $\mathbf{F}$. This involves a leakproof frictionless piston, a vacuum inside the box, and a calibrated spring. Any pressure-measuring device is called a **manometer**.

be flat, and also that P be constant over that area. (This will effectively be the case if $\mathscr{A}$ is small enough.)

The most striking property of P is that *its value does not depend on the direction that $\mathscr{A}$ is facing.* This is shown as follows. We isolate, in thought, a small body of fluid in the shape of a rectangular prism, with one vertical and one horizontal side; see Fig. 9.4. The whole fluid is assumed to be in static equilibrium, and therefore so is the prism.

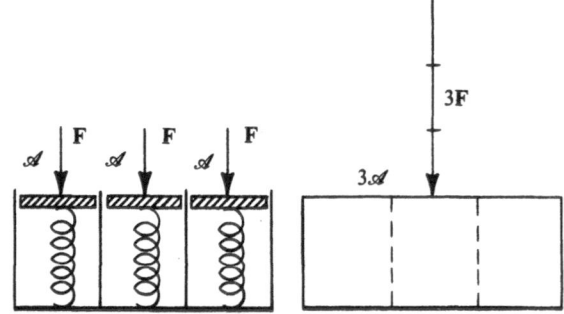

Figure 9.3. The proportionality of force and area is due to the fact that parallel forces just add numerically. The three forces shown are parallel because the three areas $\mathscr{A}$ are combined into a *flat* area $3\mathscr{A}$.

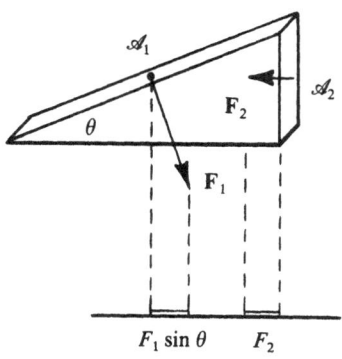

Figure 9.4. A prismatic portion of the fluid, constructed in order to demonstrate the nondirectional property of pressure. The horizontal equilibrium condition is examined.

In terms of horizontal components, the figure gives

$$F_2 = F_1 \sin \theta \qquad (9.A.4)$$

Geometry gives, for the corresponding areas,

$$\mathscr{A}_2 = \mathscr{A}_1 \sin \theta \qquad (9.A.5)$$

From (9.A.4) and (9.A.5) we see that

$$\frac{F_2}{\mathscr{A}_2} = \frac{F_1}{\mathscr{A}_1} \qquad (9.A.6)$$

that is to say, $\mathscr{A}_1$ and $\mathscr{A}_2$ feel the same pressure.

To summarize: the pressure P does not depend on the size, shape, or orientation of the small test area on which it is measured. Thus, in particular, there is no vector corresponding to P; *pressure is not a directional quantity*; it is a so-called **scalar**. We must keep in mind that the pressure does, in general, depend on the location of the test area (P is a function of position).

Units of Pressure

It follows from Eq. (9.A.3) that the SI unit of pressure is

$$1 \frac{\text{newton}}{\text{meter}^2} = 1 \text{ pascal} = 1 \text{ Pa} \qquad (9.A.7)$$

named after Blaise Pascal (1623–1662), who was among the first to recognize the nondirectional properties of pressure.

Other commonly used units include the lb/in^2 (pound per square inch, p.s.i.):

$$1 \frac{\text{lb}}{\text{in.}^2} \approx 6.89 \times 10^3 \text{ Pa} \qquad (9.A.8)$$

and the **atmosphere** (atm), defined exactly as

$$1 \text{ atm} = 1.01325 \times 10^5 \text{ Pa} \qquad (9.A.9)$$

This unit is not to be confused with the almost equivalent **bar**:

$$1 \text{ bar} = 10^5 \text{ Pa} \qquad (9.A.10)$$

exactly.

By design, 1 atm constitutes an accurate value for the average atmospheric pressure at sea level; we have

$$1 \text{ atm} \approx 14.70 \frac{\text{lb}}{\text{in}^2} \qquad (9.A.11)$$

A further unit, the millimeter of mercury, or **torr** (760 torr = 1 atm), will be motivated in Sec. B ii.

The atmospheric pressure is surprisingly unobtrusive in daily life. The 15-lb force exerted by the surrounding air on every square inch of our skin passes unnoticed. This is because we live in equilibrium with the atmosphere, and every square-inch element of our surface area is pushed outward with a compensating force of 15 lb, arising from our own internal pressure.

Example 9.1. An Inflated Tire. A certain car puts 600 lb of weight on each wheel. If the tires' gauge pressure is 30 lb/in^2, calculate the area of each tire which is in contact with the pavement. (The **gauge pressure** is the excess pressure above atmospheric.)

Let us ignore the stiffness of the rubber. The whole force F, exerted by the tire on the ground, is due to the pressure P inside; see Fig. 9.5. Let $\mathscr{A}$ be the required area. First imagine that a perfect vacuum prevails outside. Then we have

$$F = P\mathscr{A} \qquad (9.A.12)$$

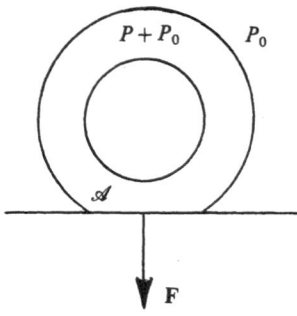

Figure 9.5. An inflated tire. ($P_0 = 1$ atm, $P =$ gauge pressure.) A force **F** is exerted by the tire on the road.

Now let us "turn on" the atmosphere, at a pressure $P_0 \approx 15$ lb/in². Simultaneously, let us also add an extra pressure P_0 *inside* the tire, in order to offset the outside pressure. Thus, the resulting net force on any part of the tire will remain the same as in the no-atmosphere case. (This is true even underneath the tire, as we cannot keep the atmosphere from seeping in between road and tire.) We see that the no-atmosphere solution remains correct in the actual situation, provided P is the *excess* pressure over atmospheric. Equation (9.A.12) gives

$$\mathscr{A} = \frac{F}{P} = \frac{600 \text{ lb}}{30 \text{ lb/in}^2} = \underline{\underline{20 \text{ in}^2}}$$

of contact area under each tire.

Example 9.2. Force on a Curved Surface. A cylindrical tank, of height h and diameter D, contains a gas at gauge pressure P. There is a welded seam S that bisects the tank, as shown in Fig. 9.6a. Find the force F (the total tension across the seam) that keeps both halves of the tank together.

First assume a vacuum outside the tank. Let us, equivalently, find the force (equal to F) exerted by the gas on one half of the enclosure. Since this surface (say of area $\mathscr{A}$) is curved, the formula $P = F/\mathscr{A}$ is not applicable. Instead, consider the portion G of the gas which is on one side of the seam, as in part (b) of the figure. Equilibrium for G means that, in magnitude,

(force on G due to wall)

= (force on G due to remaining gas)

or, since the right-hand side of this equation deals with a flat surface,

$$F = P\mathscr{A}'$$

where $\mathscr{A}' = Dh$ *is the area framed by the seam.* Thus, we obtain

$$F = \underline{\underline{DhP}} \qquad (9.A.13)$$

Next we increase the pressure, inside and outside, by 1 atm; this does not modify the result.

iii. A Spectrum of Pressures

Figure 9.7 shows some orders of magnitude that may be expected when one measures or calculates pressures.

iv. Density

For a sample of matter whose mass is m and whose volume is $\mathscr{V}$, the **density** ρ (lower-case Greek rho) is defined as

$$\boxed{\rho = \frac{m}{\mathscr{V}}} \qquad (9.A.14)$$

(*mass per unit volume*); its SI unit is therefore the kg/meter³.

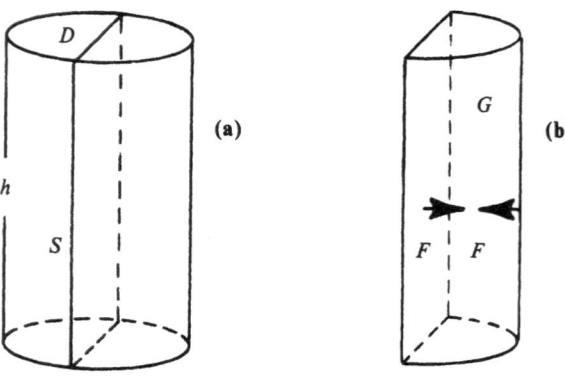

Figure 9.6. (a) Gas under pressure in a tank bisected by a seam S. (b) The action–reaction forces, of magnitude F, on a half-cylinder of gas. (The pictured forces are the resultants of infinitesimal forces distributed over the flat and curved surfaces, respectively.)

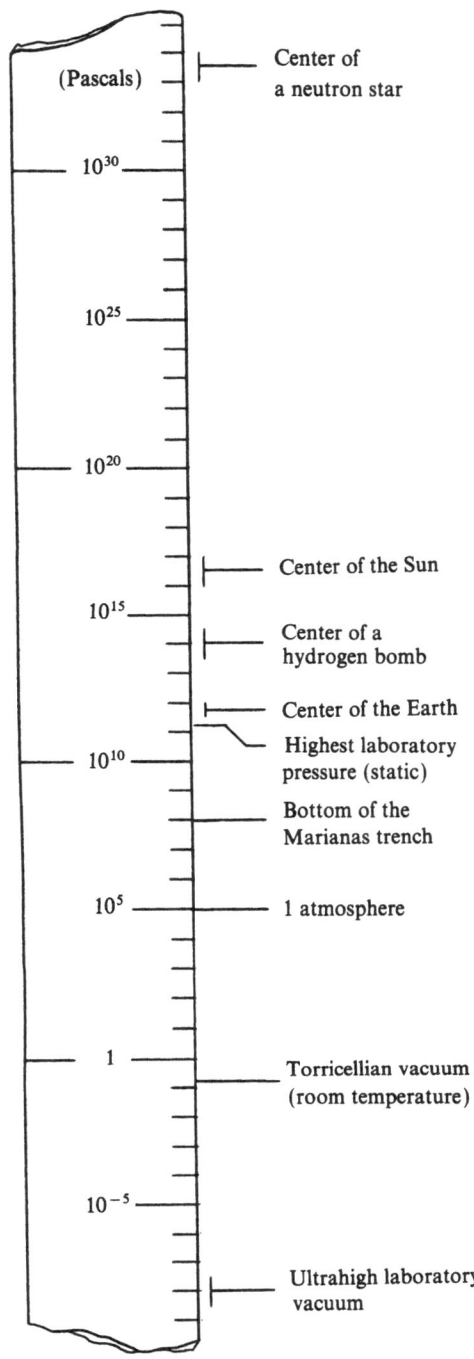

(Pascals)

Center of
a neutron star

10^{30}

10^{25}

10^{20}

Center of the Sun

10^{15}

Center of a
hydrogen bomb

Center of the Earth

Highest laboratory
pressure (static)

10^{10}

Bottom of the
Marianas trench

10^5 — 1 atmosphere

1

Torricellian vacuum
(room temperature)

10^{-5}

Ultrahigh laboratory
vacuum

Figure 9.7. Some pressures, shown on a logarithmic scale. Neutron stars (whose existence is almost certain) are so compressed by their own gravity that their density reaches that of an atomic nucleus. The highest laboratory pressure is obtained in a tiny press made of diamond, and is comparable to the pressure at the bottom of the Earth's mantle. A **Torricellian "vacuum"** is found over the barometer's mercury, and hence is limited only by the mercury's vapor pressure.

Pressure and density are mutually relevant, for three reasons:

1. At higher pressure, matter tends to occupy less space. (This is not a strict rule, because other effects, such as temperature, also come into play.)
2. The pressure of a fluid sample depends on how much weight it has to bear, and this in turn depends on the density of the fluid layers lying overhead.
3. The motion of a fluid is partly controlled by local differences in pressure and partly by the fluid's inertia, which depends on its density; this is important in the science of *fluid dynamics*.

Specific Gravity*

In practice it is often convenient to compare a substance's density ρ with that of water,

$$\rho_{\text{water}} = 1000 \, \frac{\text{kg}}{\text{meter}^3} \qquad (9.A.15)$$

almost exactly. For a substance with density ρ, we have by definition

$$\boxed{\text{Specific gravity} = \frac{\rho}{\rho_{\text{water}}}} \qquad (9.A.16)$$

This is a pure number and must be given without units.

Some actual densities are listed in Table 9.1.

The Near–Incompressibility of Liquids

How do liquids differ from gases at the molecular level? Each molecule has a small but rather well-defined volume; between the molecules is empty space. In a gas, that empty space accounts for nearly the whole volume of the container that holds the gas; the aggregate volume of the molecules themselves is almost negligible. The gas appears to fill the container because its molecules travel randomly and at a great speed in the available space. As a result, a gas is readily compressible into a smaller enclosure, for example, by

---

* A traditional, but inappropriate name: gravity is not involved.

Table 9.1. Some Densities ρ in kg/meter³

Gases,[a] at 1 atm, 0°C		Liquids, at 1 atm, 20°C	
Hydrogen	0.090	Gasoline	0.66–0.69
Helium	0.18	Ethyl alcohol	0.79
Nitrogen	1.25	Methyl alcohol	0.81
Air (dry)	1.29	Water[b]	1.00 $\quad \times 10^3$
Oxygen	1.43	Glycerine	1.27
Carbon dioxide	1.96	Sulfuric acid	1.83
		Mercury	13.5

Solids, at 1 atm, 20°C

Metals		Nonmetals	
Aluminum	2.70	Wood (seasoned)	0.11–1.3
Iron	7.87	Plexiglas	1.2
Copper	8.9	Graphite[c]	2.25
Lead	11.3 $\quad \times 10^3$	Glass	2.4–2.8 $\quad \times 10^3$
Uranium	19.1	Quartz	2.65
Gold	19.3	Granite	2.7
Iridium and osmium	22.5	Diamond[c]	3.5

[a] See also the method of Sec. C of Chap. 11 for predicting the density of many gases, simply and accurately.

[b] Other densities in kg/meter³ for water at 1 atm: minimum = 0.999973×10^3 (at 4°C); ice at 0°C: 0.92×10^3; dry steam (vapor) just above 100°C: 0.59; sea water: about 1.03×10^3.

[c] Diamond and graphite are two different forms of pure carbon.

a movable wall such as a piston. On the other hand, in a liquid the intermolecular space is comparable in extent to that occupied by the molecules themselves, just as happens in a solid. This is why solid enclosures and pistons cannot appreciably squeeze the fluid's molecules closer together without deforming their own structure and breaking down. In this chapter we consider liquids to be, for practical purposes, incompressible, i.e., *the density of a liquid is assumed to be independent of its pressure.*

The Range of Densities

Because of the atomicity of matter, all densities are, in some sense, averages taken over a region of space that contains many particles. With this in mind, one can speak of densities ranging from perhaps somewhat less then 7×10^{-33} kg/meter³, the average density of the visible Universe (with regions where the density must be far less), to more than $3 \times 10^{+17}$ kg/meter³ for nuclear interiors and probably for certain stars like neutron stars. Far denser stars (black holes) have even been postulated.

B. Archimedes' and Pascal's Principles

This section examines two aspects of fluids in **static equilibrium**: the buoyant force exerted on immersed or floating objects, and the variation in pressure with depth or altitude.

i. Buoyancy (Archimedes' Principle)

It is a familiar fact that an object's apparent weight (e.g., our own) decreases when it is immersed in water. A simple argument allows us to calculate the extent of that decrease.

Figure 9.8a shows an equilibrium situation where the apparent weight w' of an immersed object is being measured (by a spring balance for illustrative purposes).

In part (b) of the figure we consider a different equilibrium situation, where the object has been replaced by an identically shaped and identically positioned substitute *made of water.* (There is no spring balance.) The whole system, of course, is now nothing but a container of motionless water:

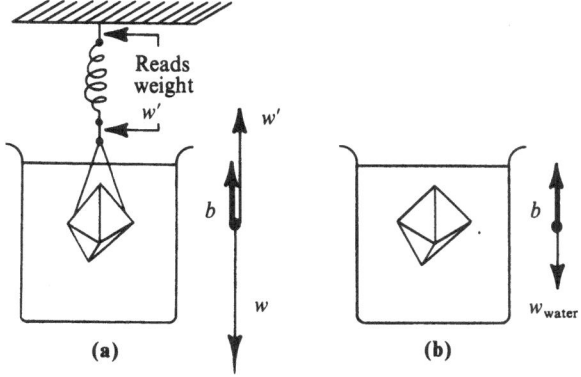

Figure·9.8. (a) Weighing an object immersed in water. The scale reads the **apparent weight** w'. The vector diagram shows the equilibrium of buoyancy, true weight, and reaction to apparent weight. (b) A body of water, exactly replacing the object of part (a). Here, buoyancy and true weight offset one another.

the substitute "object" is in equilibrium. Its weight, w_{water}, is evidently being canceled by the resultant of all pressure forces on its surface, the so-called **buoyancy b**; this vector, which is therefore vertical, is shown next to the figure. We have, in magnitude,

$$b = w_{water}$$ (9.B.1)

Incidentally, since rotational equilibrium also prevails in the water, **b** and w_{water} must share their line of action:

b goes through the center of mass of the water-made substitute.

(9.B.2)

Let us now go back to the original object. It presents to its surroundings the same surface as the substitute, and therefore the same buoyancy **b** must

be acting on it as on the substitute. This is shown next to part (a) of the figure. To summarize: If w is the true weight of an immersed object, the scale will read an apparent weight

$$w' = w - b$$ (9.B.3)

Here

The buoyancy b equals the weight of the displaced fluid.

(9.B.4)

(This is Archimedes' principle.) Equivalently, for an object of volume $\mathscr{V}$ in a fluid of density ρ, (9.B.4) reads

$$b = \rho \mathscr{V} g$$ (9.B.5)

($\rho \mathscr{V} =$ displaced mass.)

What happens if two density layers are present, with some fluid displaced in both? Then Archimedes' principle, (9.B.4), gives

$$b = (\rho_1 \mathscr{V}_1 + \rho_2 \mathscr{V}_2) g$$ (9.B.6)

See Fig. 9.9a. As a special case of this, the upper fluid might be air (whose density we neglect), so that

$$b = \rho \mathscr{V}_1 g$$ (9.B.7)

as in part (b) of the figure. Archimedes' principle is just as applicable in a fluid with many layers, or even with a continuous density variation.

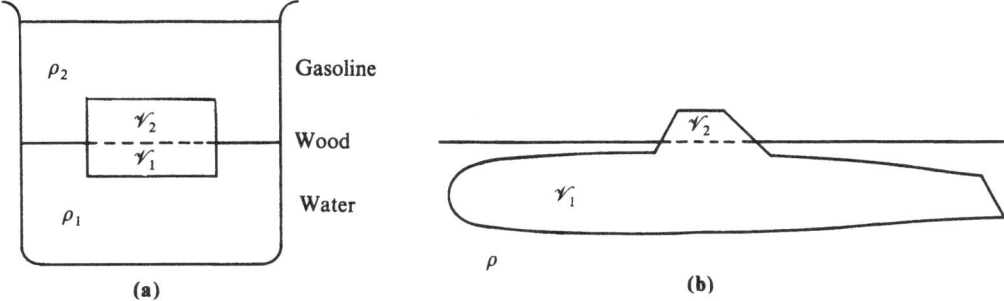

Figure 9.9. (a) Archimedes' principle with two density layers. The buoyancy still equals the total weight of fluid displaced by $\mathscr{V}_1$ and $\mathscr{V}_2$. (b) In this special case, the upper layer is air, whose density is effectively zero here.

Archimedes is said to have applied his principle to assay a gold crown by way of its density. His method was to compare the crown's weight in water and in air; see Problem 9.14.

Example 9.3. A Helium Balloon. Estimate the diameter of a spherical helium balloon that can lift a passenger of mass 75 kg, with an allowable mass of 25 kg for the balloon's envelope and attachments. Assume a temperature of 0°C; see Table 9.1.

Let $M = 100$ kg be the total load, let $\mathcal{V}$ be the required volume of helium, and let ρ_{He}, ρ_{air} be the densities of the helium and of the surrounding air. We have

True weight of the helium: $\qquad w_{He} = \rho_{He}\mathcal{V}g$

Buoyancy: $\qquad b = \rho_{air}\mathcal{V}g$

Load: $\qquad w_M = Mg$

At equilibrium, the total force on the balloon is

$$0 = w_{He} - b + w_M$$

or

$$0 = \rho_{He}\mathcal{V}g - \rho_{air}\mathcal{V}g + Mg$$

so that

$$\mathcal{V} = \frac{M}{\rho_{air} - \rho_{He}} = \frac{100}{1.29 - 0.18}\ \text{meters}^3$$

$$= 90\ \text{meters}^3$$

This is the volume of a sphere of diameter

$$2r = \left(\frac{6\mathcal{V}}{\pi}\right)^{1/3} = \underline{5.6\ \text{meters}}$$

The buoyancy on the passenger himself was neglected. Is this justified? (Consider the volume ratio of passenger to balloon.)

A Floating Stick

We conclude this discussion of Archimedes' principle with an illustration involving the buoyant force's line of action. Why does a stick float in a horizontal rather than in a vertical position?

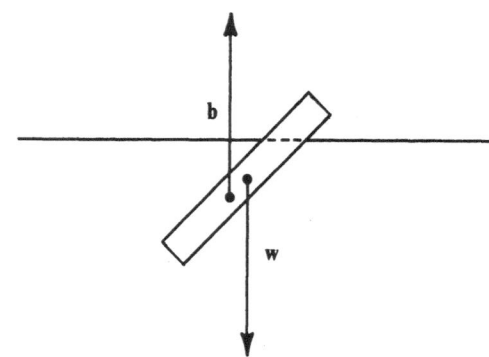

Figure 9.10. This object is in translational, but not in rotational equilibrium.

Figure 9.10 shows a uniform wooden stick, floating in water, in *translational* equilibrium. It is, however, not in *rotational* equilibrium, since the true weight **w** acts at the stick's midpoint, whereas the buoyancy **b** = −**w** acts at the midpoint of the submerged section. Therefore a vertical floating position is unstable: the slightest nudge would cause a net torque towards the horizontal position.

ii. Depth and Pressure (Pascal's Principle)

The preceding derivation of Archimedes' principle gives little indication of the mechanism that produces the buoyancy. The latter in fact arises from an excess pressure on the lower surfaces of the immersed object.

How does pressure increase with depth? To find out, let us isolate in thought a thin vertical prism in a uniform fluid; see Fig. 9.11. We are assuming

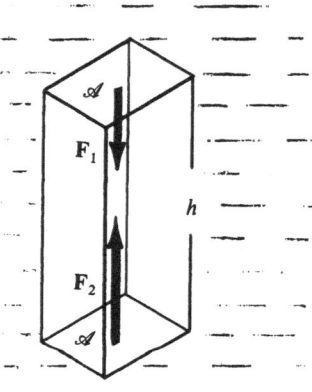

Figure 9.11. A prismatic portion of fluid, considered separately from its surroundings. The surface force $\mathbf{F}_1 + \mathbf{F}_2$ must offset the prism's true weight (not shown). The top and bottom surfaces are horizontal and have equal areas $\mathscr{A}$.

static equilibrium as before. Since all hydrostatic forces on the prism have to be normal to its surface, we obtain the vertical equilibrium condition

$$F_2 - F_1 = mg \qquad (9.B.8)$$

where mg is the true weight of the prism:

$$mg = \rho \mathcal{V} g = \rho h \mathcal{A} g$$

and where the hydrostatic forces

$$F_1 = P_1 \mathcal{A}, \qquad F_2 = P_2 \mathcal{A}$$

are the ones exerted on its top and bottom surfaces, respectively. Equation (9.B.8) becomes

$$(P_1 - P_2) \mathcal{A} = \rho h \mathcal{A} g$$

or

$$\boxed{P_2 - P_1 = \rho h g} \qquad (9.B.9)$$

(Pascal's principle). Thus, in a uniform fluid ($\rho = $ const), *pressure increases linearly with depth.*

There is, in fact, no need for one point to be vertically above the other. Formula (9.B.9) is valid for *any two points within a uniform static fluid.* (This follows from the fact that pressure does not change under horizontal displacements, a result that the reader may derive by considering a horizontal fluid prism.)

Example 9.4. The Mercury Barometer. Figure 9.12 shows some possible arrangements for determining the atmospheric pressure with a column of mercury in a glass tube. The enclosure above point A should contain no air. If a level difference $h = 760$ mm (typical at sea level) is observed, what is, in pascals, the corresponding atmospheric pressure P_0?

Neglecting the pressure of any mercury vapor above A (that pressure is of the order of 10^{-6} atm at room temperature), we have, at point A,

$$P_A = 0$$

At point B, we have

$$P_B = P_0$$

Therefore we obtain

$$P_0 = P_B - P_A = \underline{\rho_{Hg} h g} \qquad (9.B.10)$$

where ρ_{Hg} is the density of mercury. We find

$$P_0 = \left(13.6 \times 10^3 \, \frac{kg}{meter^3}\right)\left(9.8 \, \frac{meters}{sec^2}\right)$$
$$\times (0.760 \, meter)$$
$$= \underline{1.01 \times 10^5 \, Pa} \approx 1 \, atm$$

Recall definition (9.A.9) for that unit of pressure.

As Eq. (9.B.10) demonstrates, an accurate measurement with the mercury barometer requires an accurate knowledge of the local value of g and of the density of mercury. An auxiliary unit, 1 millimeter of mercury = 1 mm Hg = 1 torr (named after Torricelli, who invented the barometer), is defined such that

$$1 \, atm = 760 \, torr \qquad (9.B.11)$$

exactly. This gives

$$1 \, torr \approx 133 \, Pa \qquad (9.B.12)$$

From day to day, the actual barometric pressure varies by not more than 25 torr or so, on either side of the mean.

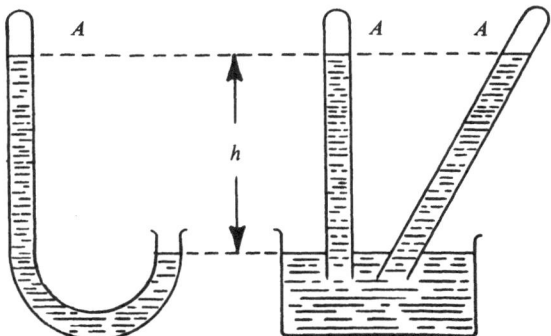

Figure 9.12. Three equivalent versions of a mercury barometer, an extraordinarily simple device invented in 1643 by Evangelista Torricelli. The level difference h is a measure of the atmospheric pressure. The enclosure above A contains only mercury vapor, whose pressure can be neglected in this case. (Note: The edges of a free mercury surface are ordinarily pulled down by an effect called **surface tension**. This effect can be neglected if the tube is wide enough, and if the position of the surface is defined as that of its flat portion.)

Pressure in a Moving Fluid

When accelerations are taking place in a fluid, Pascal's principle becomes inadequate; a note at the end of this chapter considers how it should be generalized.

iii. Application to the Atmosphere

In 1646, Pascal conducted the first measurements of atmospheric pressure as a function of altitude, using the newly invented mercury barometer. As he expected, the air pressure, P, decreases with rising altitude z. (The air density, ρ, decreases as well.) In order to find how steeply P decreases, we use Eq. (9.B.9) over an infinitesimal altitude rise. Denoting h by dz and $P_2 - P_1$ by $-dP$, we obtain

$$\boxed{dP = -\rho g \, dz} \qquad (9.B.13)$$

the infinitesimal version of Pascal's principle. The negative sign reminds us that pressure decreases with altitude.

The Lower Atmosphere

The air pressure P at an altitude z above sea level can very roughly (to about 10%, and up to the highest mountains) be approximated by an exponential formula,

$$P = P_0 e^{-z/\zeta} \qquad (9.B.14)$$

where P_0 is a characteristic pressure, ζ (lower-case Greek zeta) is a characteristic height, and $e \approx 2.72$ is the base of the natural logarithms. [Formula (9.B.14) would correctly describe an **isothermal atmosphere**, where the temperature is the same at all altitudes; see Problem 11.35 further on.] Given the density $\rho_0 = 1.3$ kg/meter^3 for air at sea level, let us determine P_0 and ζ.

(1) To determine P_0, we take $z = 0$ in Eq. (9.B.14). Then we have $P = 1$ atm, and the equation reads

$$\underline{\underline{1 \text{ atm} = P_0}} \qquad (9.B.15)$$

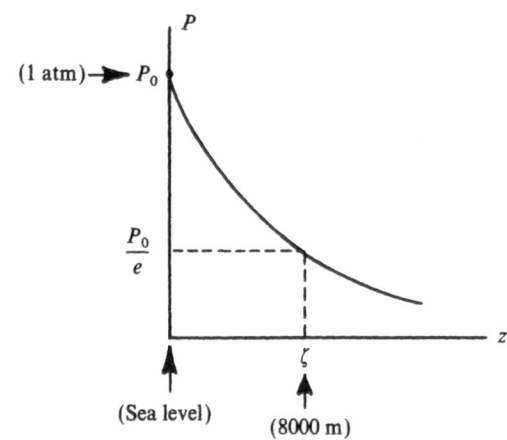

Figure 9.13. Atmospheric pressure against altitude (exponential approximation). The air density follows a similar curve.

(2) To determine ζ, we use Pascal's principle, (9.B.13). This may be written

$$-\frac{dP}{dz} = \rho g \qquad (9.B.16)$$

or, from (9.B.14),

$$\frac{P_0}{\zeta} e^{-z/\zeta} = \rho g \qquad (9.B.17)$$

Again taking $z = 0$, $\rho = \rho_0$, we obtain

$$\frac{P_0}{\zeta} = \rho_0 g \qquad (9.B.18)$$

$$\zeta = \frac{P_0}{\rho_0 g} = \frac{1 \text{ atm}}{\rho_0 g} \qquad (9.B.19)$$

$$= \frac{1.01 \times 10^5}{(1.3)(9.8)} \text{ meters} = \underline{\underline{8000 \text{ meters}}}$$

almost the height of Mount Everest. [The role of ζ in Eq. (9.B.14) is that at this altitude P is reduced to $1/e \approx 37\%$ of its value at sea level. In reality, at 8000 meters, the pressure is about 35% of 1 atm.] Figure 9.13 shows the qualitative behavior of formula (9.B.14). It is left to the reader to find how ρ depends on altitude.

Note

Bernoulli's Principle

How does pressure vary from place to place in a moving fluid? This question is vital to many people, including airplane designers and meteorologists. Unfortunately, owing to its complexity, the problem cannot be solved by a general formula.

The simplest cases are well understood, however. In the static limit we have Pascal's principle, (9.B.9), which relates pressure to depth. This note discusses the case of a *frictionless incompressible fluid under steady-state streamline flow.*

[A **steady-state flow** is one where all the fluid particles that succeed one another through a given location in space do so with the same velocity. Thus, each point in space is characterized by a fixed **local velocity** of the fluid. (As a given particle moves from point to point, it may of course change its velocity.) Under these conditions, a **streamline**—which is a fixed line in space—is defined as being the trajectory of a particle. Streamlines are, of course, infinite in number; when drawing them on paper one can only select a few representative ones, as is done in Fig. 9.14.]

Here our main result will be a comparison between the pressures P_A, P_B at two different points A, B along a single streamline; see Fig. 9.14. At first, we do not want to consider differences in altitude, and thus we assume the streamline to be horizontal. Our conclusion will turn out to be as follows:

> Higher speeds are associated with lower pressures.   (9.N.1)

Suppose v_A and v_B are the relevant speeds; in this illustration we have $v_B > v_A$, and consequently $P_B < P_A$. Quantitatively, we shall see that

$$P_A - P_B = \tfrac{1}{2}\rho(v_B^2 - v_A^2) \qquad (9.N.2)$$

where ρ is the (constant) density of the fluid; (9.N.2) is called **Bernoulli's principle** (Daniel Bernoulli, 1738) for a horizontal streamline.

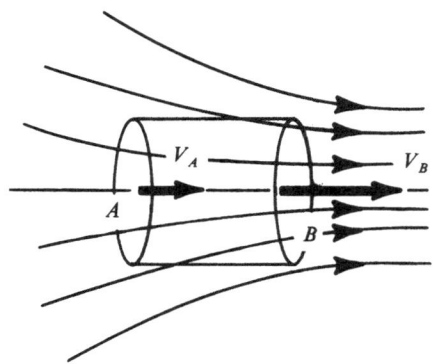

Figure 9.14. Top view of a set of horizontal streamlines. The local speed is assumed larger at B than at A. [The bunching of streamlines from A to B ensures that, in any given time, the same volume of fluid leaves (at right) as arrives (at left).] The cylinder is an imaginary construction used in deriving (9.N.2).

Derivation* of Eq. (9.N.2)

Consider, as in Fig. 9.14, two points A, B close enough together so that the streamline segment AB is essentially straight. An x axis is drawn through A and B.

We now apply Newton's law, $F_x = ma_x$, to a small right circular cylinder of mass m, made of the fluid itself, with base area $\mathscr{A}$ and axis AB. Since, in the figure, m accelerates to the right, P_A must exert a larger force than P_B, and rule (9.N.1) follows.

In more detail, we note that F_x is entirely due to pressure on the cylinder's flat ends. (Without friction, the pressure forces on the curved cylindrical surface are purely normal.) We therefore have

$$F_x = \mathscr{A}P_A - \mathscr{A}P_B = -\mathscr{A}\,\Delta P \qquad (9.N.3)$$

where $\Delta P = P_B - P_A$.

The cylinder's mass is given by

$$m = \rho\mathscr{A}\,\Delta x \qquad (9.N.4)$$

where Δx is the length of AB.

Finally, to estimate the cylinder's acceleration, we consider a fluid particle as it moves, over a time Δt, from A to B:

$$a_x = \frac{v_B - v_A}{\Delta t} = \frac{\Delta v}{\Delta t} \qquad (9.N.5)$$

With (9.N.3), (9.N.4), (9.N.5), the equation $F_x = ma_x$ now reads

$$-\mathscr{A}\,\Delta P = (\rho\mathscr{A}\,\Delta x)\left(\frac{\Delta v}{\Delta t}\right) \qquad (9.N.6)$$

The factor $\mathscr{A}$ can be dropped. In the limit, we set $\Delta v \to dv$, $\Delta t \to dt$, $\Delta x \to dx$, and obtain

$$-dP = \frac{\rho\,dx\,dv}{dt} \qquad (9.N.7)$$

or, with $dx/dt = v$,

$$-dP = \rho v\,dv \qquad (9.N.8)$$

Since ρ is a constant, we can integrate both sides of

---

* An elegant and often-used method of proof exploits the conservation of mechanical energy in a frictionless fluid; see, for example, Sec. 13-6 in P. A. Tipler's *Physics*, 2nd ed. (Worth, 1982). It is probably more instructive, however, to apply Newton's second law directly to a fluid particle—our approach in this Note.

(9.N.8) along a curved streamline between *distant* points A, B:

$$-\int_{P_A}^{P_B} dP = \rho \int_{v_A}^{v_B} v\, dv \qquad (9.N.9)$$

which yields the announced Eq. (9.N.2).

Points A, B at Different Levels

It is easy to add a gravitational term if AB is not horizontal; here we only quote the result,

$$\boxed{P_A - P_B = \tfrac{1}{2}\rho(v_B^2 - v_A^2) + \rho g(z_B - z_A)}$$

$$(9.N.10)$$

where z_A, z_B are the respective altitudes of A and B, as in Fig. 9.15. In this more general form of Bernoulli's principle, the contribution $\rho g(z_B - z_A)$ is familiar to us from Pascal's principle.

[As the reader will verify, (9.N.10) states that the quantity $P + \tfrac{1}{2}\rho v^2 + \rho g z$ has the same value at A and at B, or, for that matter, anywhere along the streamline.]

Example 9.5. The Pitot Tube. Figure 9.16 shows a wide tube in which water is flowing at speed v. A narrow tube (the Pitot tube), open at A and B, experiences a pressure difference in the two vertical branches, as can be measured in this example from a difference $h = 0.50$ cm between two mercury levels. Find v, assuming Bernoulli's principle to be applicable. [This device for measuring a speed of flow is due to Henri Pitot (1695–1771).]

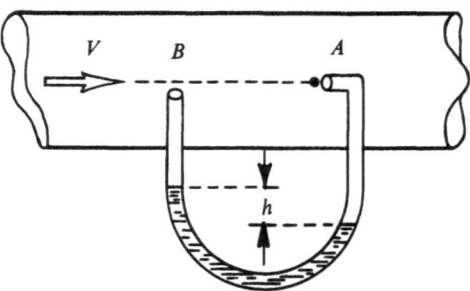

Figure 9.16. A Pitot tube.

Consider a special horizontal streamline (heavy dashed line in the figure). At B, the speed is v, while at A, it must be zero under steady-state conditions. (No water flows through the Pitot tube.) From (9.N.2) we have

$$P_A - P_B = \tfrac{1}{2}\rho_{\text{water}} v^2$$

On the other hand, from Pascal's principle,

$$P_A - P_B = (\rho_{\text{mercury}} - \rho_{\text{water}})\, gh$$

(The reader should verify this.) Comparing these two equations and solving for v, we find

$$v = \left[2\left(\frac{\rho_{\text{mercury}}}{\rho_{\text{water}}} - 1 \right) gh \right]^{1/2}$$

$$= [(2)(13.6 - 1)(10)(0.50 \times 10^{-2})]^{1/2}\, \frac{\text{meters}}{\text{sec}}$$

$$= \underline{\underline{1.12\, \frac{\text{meters}}{\text{sec}}}}$$

Caution: It is useful to know that, along a streamline, pressure tends to decrease with increasing speed. Nevertheless, Bernoulli's *quantitative* result, although of appealing simplicity and elegance, is of limited use. Most flows involve significant friction, and the equation, even when applicable, does not compare pressures on different streamlines. The student should beware of many fluid-dynamic "explanations" found in the literature, in which these restrictions are ignored.

Condensed Checklist

Pressure:

$$P = \frac{F}{\mathscr{A}} \qquad (9.A.3)$$

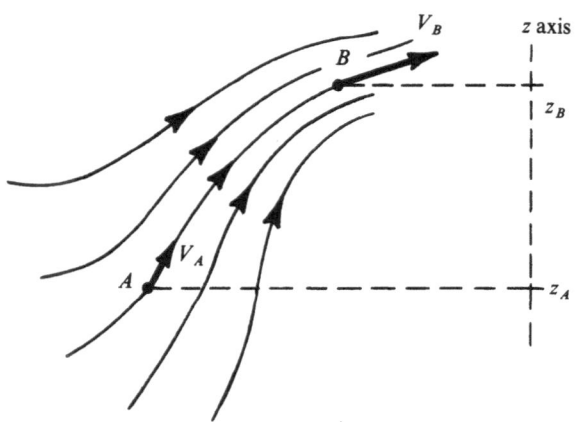

Figure 9.15. Side view of rising streamlines.

$$1 \text{ pascal} = 1 \text{ newton/meter}^2 \qquad (9.\text{A}.7)$$

$$1 \text{ atm} \approx 1.01 \times 10^5 \text{ pascals} \approx 14.7 \text{ lb/in}^2$$
$$(9.\text{A}.9), (9.\text{A}.11)$$

$$1 \text{ bar} = 10^5 \text{ pascals} \qquad (9.\text{A}.10)$$

$$1 \text{ atm} = 760 \text{ torr} = 760 \text{ mm Hg} \qquad (9.\text{B}.11)$$

Density:

$$\rho = \frac{m}{\mathscr{V}} \qquad (9.\text{A}.14)$$

$$\rho_{\text{water}} = 1000 \frac{\text{kg}}{\text{meter}^3} \qquad (9.\text{A}.15)$$

$$\text{Specific gravity} = \frac{\rho}{\rho_{\text{water}}} \qquad (9.\text{A}.16)$$

Table of densities: Table 9.1

Archimedes' principle:

$$w' = w - w_{\text{water}} \qquad (9.\text{B}.3)$$

Pascal's principle:

$$P_2 - P_1 = \rho g h \qquad (9.\text{B}.9)$$

$$dP = -\rho g \, dz \qquad (9.\text{B}.13)$$

True or False

1. Because of the air bubble (shown in the pipe), water levels A and B can be different under static equilibrium.

2. A glass of water, full to the brim, contains a floating ice cube; when the latter melts, some water must overflow.

3. A rowboat is floating in a pond; when some water is bailed out of the boat, the pond level remains unchanged.

4. If the rowboat contains a stone, and the latter is thrown into the water, the pond level remains unchanged.

5. If two different cities are at the same altitude they must have the same atmospheric pressure.

6. Pascal's altitude–pressure relation depends on static equilibrium being assumed.

7. Since a kite stays aloft in spite of its weight and of the string pulling it down, it must mean that the air pressure is considerably greater just underneath it than just above.

8. Pressure has the same units as energy per unit volume.

Problems

Section A: Pressure and Density

*9.1. According to Table 9.1, and under the pressure and temperature conditions listed there, what is the specific gravity of (a) hydrogen, (b) mercury, (c) iron?

9.2. A squirrel's bite can result in a pressure of $22\,000 \text{ lb/in}^2$ on a hard object. Assuming his tooth exerts a force of 5 lb, estimate the diameter of the contact area at the tooth's point.

*9.3. If the glider of Chap. 3, Fig. 3.1 is a disk of mass 15 grams and diameter 8 cm, what should be, in pascals, the gauge pressure of the underlying air cushion?

9.4. A 20 metric ton ground-effect vehicle (**hovercraft**), of base area 15 meters^2, is entirely supported, a few inches above ground, by an air cushion originating from internal propellers. Estimate, in atmospheres and in pascals, the gauge pressure of the underlying air.

*9.5. A 150-meter by 200-meter football stadium is covered by an air-supported dome made of flexible material, a typical square meter of which has a mass of 8 kg. Estimate the increase in pressure, in atmospheres and in millibars, experienced by a spectator entering the building through one of the airlock doors (neglect the curvature of the dome).

*9.6. The **mean density** $\bar{\rho}$ of a possibly inhomogeneous object is defined as its total mass divided by its total volume. What is $\bar{\rho}$ for a hollow aluminum sphere of outer radius 1.00 meter and inner radius 0.80 meter? (The cavity is considered part of the total volume.)

Section B: Archimedes' and Pascal's Principles

*9.7. A swimmer's true weight is 125 lb, while his apparent weight in water is only 5 lb. (a) What is his mean density, i.e., his density calculated as if it were uniform? (b) What is his mean specific gravity, defined in a similar way?

9.8. An ice cube (specific gravity 0.9) is floating in water. What fraction of the ice's volume is above the surface?

9.9. A disused 16 000-ft-deep oil well is full of water. What is the pressure, in atm, at the bottom?

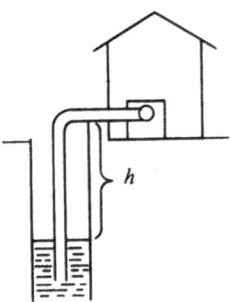

*9.10. (a) How high a column of water will exert a 1-atm gauge pressure at its base? (b) What maximum altitude difference h is permissible between a suction pump and the water level in a well (see the figure) in order that water may be drawn?

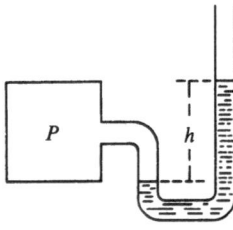

*9.11. (A **U-tube manometer**.) The figure shows gas enclosed at an unknown pressure P. Find an expression for P in terms of the atmospheric pressure P_0, the difference h between the two mercury levels, the density ρ of the mercury, and the acceleration g of gravity.

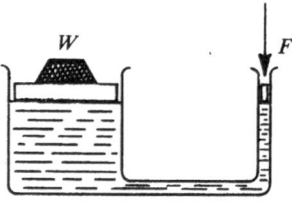

9.12. (A **hydraulic jack**.) The oil-filled cylinders in the figure have diameters 0.1 meter and 1.0 meter, respectively. What force F must be exerted on a piston in the small cylinder in order to lift a given weight W resting on a piston in the large cylinder? (Assume both pistons are at the same level; neglect their weight.)

9.13. Estimate by what fraction a person's weight is reduced by the buoyancy of the surrounding air. (Human specific gravity ≈ 1.)

*9.14. (a) A certain object's weight is found to be w in air and w' in water. In terms of w and w', what is the object's specific gravity? (b) A certain quartz crystal (specific gravity 2.65) weighs 0.12 newton in air and 0.052 newton in an unidentified liquid. Find the latter's specific gravity.

9.15. On Earth ($g = 10$), a certain rectangular piece of lead, floating in equilibrium on mercury, experiences a buoyancy of 25 newtons and keeps 1.0 cm of its height above the mercury's surface. If the experiment is repeated on the Moon ($g = 1.6$) with the same piece of lead, find (a) the buoyancy, (b) the height kept above the surface.

*9.16. Someone's eye has an internal gauge pressure of 18 mm of mercury ($= 0.024$ atm). What is the total hydrostatic force on one half of the eye's wall? (Consider the eyeball as a spherical enclosure 2.5 cm in diameter.)

9.17. Near sea level, by what percentage does the atmospheric pressure decrease (a) as one rises by 100 meters, (b) between the floor and the ceiling of a 4-meter-high laboratory?

*9.18. If you are on top of Mount Everest, what fraction of the atmosphere's total mass is below you?

*9.19. At approximately what altitude, $z_{1/2}$, is the atmospheric pressure equal to 0.5 atm?

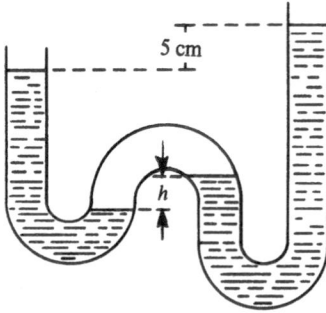

*9.20. The tube shown in the figure contains two sections of mercury, separated by an air bubble. If the end levels are 5 cm apart, find the difference h between the middle levels.

*9.21. Following Sec. B iii, derive a formula for the atmospheric air density ρ at altitude z in terms of the density ρ_0 at sea level and the characteristic height ζ.

9.22. The figure shows a proposal for a perpetual motion machine, where a wooden disk rotates in a vertical leakproof slot, cut in the wall of a water tank. The buoyancy of the immersed half disk should produce more than enough torque to overcome friction. Analyze the fallacy.

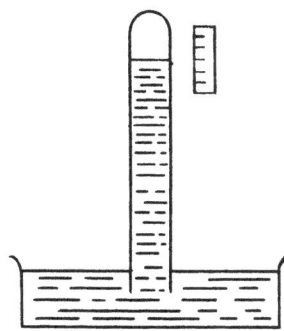

*9.23. The mercury barometer shown in the figure consists of a vertical tube with internal cross section 1 cm^2 and a cup in which the mercury has a free area of 10 cm^2. A scale next to the tube reads the atmospheric pressure in millimeters of mercury (torr). What is the spacing between two consecutive torr marks?

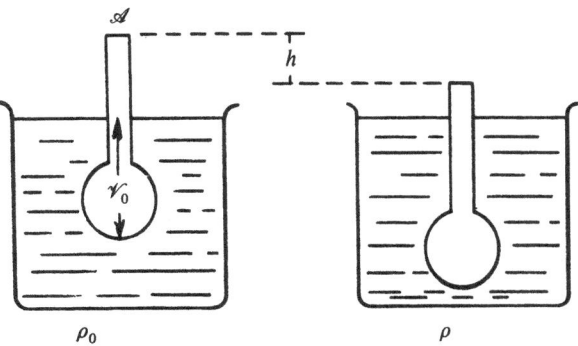

*9.24. (A **hydrometer**.) The density of a liquid may be measured by this device, shown in the figure. When floating in water (density ρ_0) its immersed volume is $\mathscr{V}_0$. Its vertical column has uniform cross section $\mathscr{A}$. When floating in a liquid of density ρ, it sinks by a further height h. Express ρ in terms of $\mathscr{V}_0$, ρ_0, $\mathscr{A}$, and h.

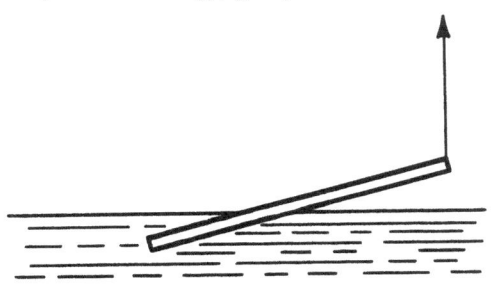

*9.25. A thin uniform wooden stick, of specific gravity 0.5, and floating in water, is lifted slightly by a string at one end; see the figure. Equilibrium is assumed. (a) Show that the string is vertical. (b) In terms of the stick's true weight w, calculate the tension in the string. (c) What fraction of the stick is under water?

Answers to True or False

(Note: Items 2, 3, and 4 are brain teasers that several eminent physicists have been known to answer wrongly at cocktail parties.)

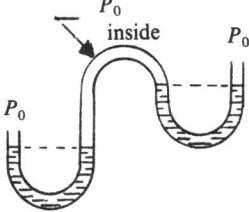

1. True (see the adjoining figure for a clear example).
2. False (the ice cube's weight is the same after melting, and hence the same amount of surrounding water is displaced: the level does not change.)
3. True (the weight of boat plus bailed-out water has not changed; hence the same amount of surrounding water is displaced).
4. False (the system of boat plus stone receives an added lift from the bottom of the pond as it supports the stone; hence less water is displaced, and the level sinks).
5. False (the atmosphere is not in static equilibrium).
6. True.
7. True (an effect of the wind).
8. True.

Answers or Hints to Problems

9.1. (a) 9.0×10^{-5}.

9.3. 7.3.

9.5. The stadium's size is irrelevant.

9.6. $1.32 \times 10^3 \text{ kg/meter}^3$.

9.7. (a) $1.04 \times 10^3 \text{ kg/meter}^3$; (b) 1.04.

9.10. (a), (b) 10.3 meters.

9.11. $P = P_0 + \rho h g$.

9.14. (a) $w/(w - w')$; (b) 1.50.

9.16. 1.2 newtons.

9.18. About 65%.

9.19. Hint: From (9.B.14), obtain $z_{1/2} = \zeta \ln 2$.

9.20. Hint: The middle levels are at the same pressure.

9.21. Use Eqs. (9.B.13) and (9.B.14).

9.23. (10/11) mm.

9.24. $\rho = \mathscr{V}_0 \rho_0 / (\mathscr{V}_0 + h \mathscr{A})$.

9.25. (b) $(\sqrt{2} - 1)w$; (c) $1 - 1/\sqrt{2}$.

CHAPTER 10

HEAT

This chapter introduces two concepts: **temperature** and **heat**. Both phenomena have always played a prominent role in man's daily life, everyone has a good intuitive grasp of them, and they are quite easy to measure. Yet they have been surprisingly slow in reaching precise quantitative status. (This happened, roughly, between 1800 and 1850.)

In this book, our understanding of temperature will be acquired in two steps, just as it was acquired historically. The present chapter defines **absolute temperature** through an experimental measuring procedure. Later, temperature is to be related to the disorderly motion of molecules.

Similarly for heat: First we present an experimental definition in terms of a temperature rise; this is followed by an account of Joule's measurement, which demonstrates that heat is nothing but an energy transfer. The relation between heat and molecular motion will be explored in the next two chapters. In this way, heat and temperature will be reduced to the earlier mechanical concepts of this book.

Temperature and heat are the basic ideas of **thermal physics**. Their scientific uses are endless, notably in chemistry and in the life sciences. In physics, their study reveals something of the molecular structure of matter, as we shall see in the next chapter. Alternatively, thermal concepts can be studied for their own sake, giving rise to the science of **thermodynamics** (see Chap. 12).

A. Temperature

The sensations of hot and cold are familiar ones. From our daily experience we are used to ranking objects along a one-dimensional scale (the temperature scale), stretching between "very cold" and "very hot." This observation must now be made quantitative.

Temperature is what one measures with a thermometer. This is not a circular definition, provided we describe the instrument, as well as the procedure for using it.

There are many kinds of thermometers; the reader is assumed to be familiar with at least some of them, such as the **mercury thermometer** or the **bimetallic strip** of Fig. 10.1. We shall see later how to calibrate the scale of a thermometer, so as to make all thermometers agree with one another. For the time being, any scale will do; our preliminary instrument is only required to detect *changes* in temperature.

i. Thermal Equilibrium

By definition, a thermometer measures *its own temperature*. Also by definition, it measures the temperature of any object with which it is in *thermal equilibrium*. The following idealized laboratory procedure will specify what is meant by that condition.

First we need to build an insulating, or **adiabatic** enclosure. This is a box such that, *inside it, any sample of matter shows zero response to a heating or a cooling of the box's outer surface*; see Fig. 10.2. By contrast, a nonzero response would consist of any quantity, such as the sample's pressure, its volume, its ratio of liquid mass to solid mass, etc., being modified from the outside. (In practice, a thermos bottle, or a box lined with a common insulator such

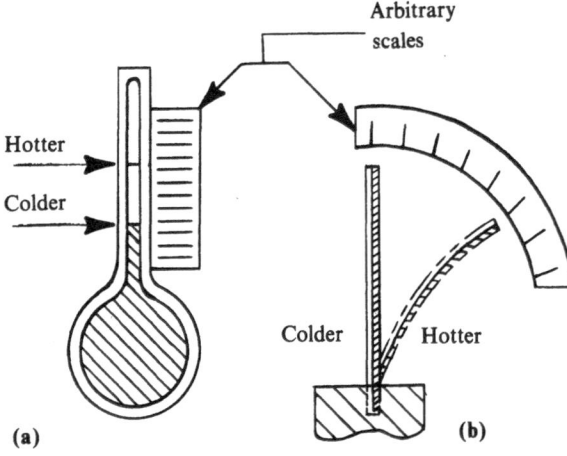

Figure 10.1. The mercury and bimetallic thermometers both exploit the difference in expansion between two substances. (a) When heated, mercury expands more than glass: the capacity of the bulb increases, but the mercury's volume increases even more; the overflow rises in the capillary tube. (b) Two strips of dissimilar metals, say copper (shaded) and aluminum (light) are welded or riveted together. Under heating, the aluminum side grows more than the copper side, and the combination bends.

as glass fiber, are good approximations to an adiabatic enclosure.)

We can now define thermal equilibrium:

> Two objects, left undisturbed and in mutual contact within a single adiabatic enclosure, eventually reach thermal equilibrium with each other.

(10.A.1)

How long should one wait? The answer is, until no more changes are observed to occur in either object.

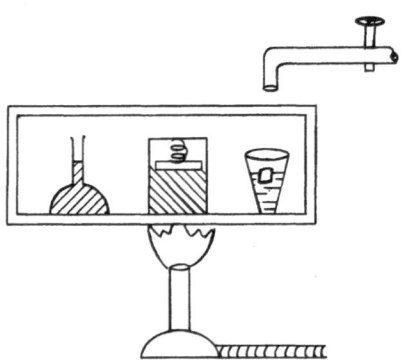

Figure 10.2. An adiabatic enclosure. External heating or cooling leaves the internal systems unchanged in every respect.

The following postulate* is made for any three systems A, B, C: *if A is in thermal equilibrium with B and with C, then B and C are in thermal equilibrium with each other.* This can be verified in practice through the application of procedure (10.A.1) to each of the three pairs; no changes in properties are ever observed to occur during these "confrontations." The following is a consequence, often taken for granted. Suppose A is a thermometer, while B and C are any objects; we see that, *if B and C are in thermal equilibrium with each other, they have the same temperature* (as read by A).

ii. Ideal Gases

Any device whose properties depend on temperature could, in principle, be used as a thermometer. For example, the number of times a cricket chirps in one minute is said to be fairly reliable in this respect. The cricket, of course, is a very complicated thermometer, and so is, to some extent, an expanding liquid such as mercury. In order to define a useful temperature scale for physicists a simple instrument is required, otherwise its connection to the basic natural laws is bound to be obscure. For this reason, the **ideal gases**, discussed below, are the substances whose behavior is used as an index of temperature, and all other thermometers are calibrated by reference to them.

As defined in practical terms,

> An ideal gas is any gas maintained under conditions of very low density.

(10.A.2)

What is the meaning of "very low density?" All gases, when sufficiently cooled and compressed, become liquid. Their molecules are then close together. The ideal-gas situation is at the other extreme:

"Very low density" means "very low compared to the density in the liquid state."   (10.A.3)

An ideal gas is therefore mostly vacuum, and therein lies its simplicity.

---

* Sometimes called the "zeroth law of thermodynamics."

The ideal gas, like the ideal spring and the ideal point-particle, is one of the great abstractions of physics. Although it cannot be realized exactly, it helps us make sense of the real world: it will continue to occupy center stage in our study of thermal phenomena.

For the purposes of this book many real gases can be considered ideal. This includes the best-known gases, at no more then a few atmospheres and when not much colder than room temperature; examples are: helium, neon, hydrogen, oxygen, nitrogen, fluorine, chlorine, methane, carbon dioxide, and even water vapor under conditions where it cannot condense (dry or unsaturated vapor).

iii. The Ideal-Gas Thermometer

Let us enclose an arbitrary amount of any ideal gas, such as air, in a cylinder that is closed off by a freely moving piston, as in Fig. 10.3. Under mechanical equilibrium, the pressure inside the cylinder equals the atmospheric pressure, which we assume to be constant during an experiment. It is observed that, *under heating, the volume of the enclosed gas increases*. It is therefore natural to define the temperature T of the gas sample as a multiple of its volume $\mathscr{V}$:

$$T = a\mathscr{V} \quad \text{(at constant pressure)} \quad (10.\text{A}.4)$$

where a is any convenient proportionality constant; the device may be called an **ideal-gas thermometer**.

To what extent is (10.A.4) an arbitrary definition? To find out, one could prepare many such thermometers, at different pressures, with different volumes, and using different ideal gases—say hydrogen, helium, etc.; their behaviors can then be compared. Figure 10.4 shows in principle how this is done for a pair of thermometers: (a) They start in thermal equilibrium with each other. (b) Heat is

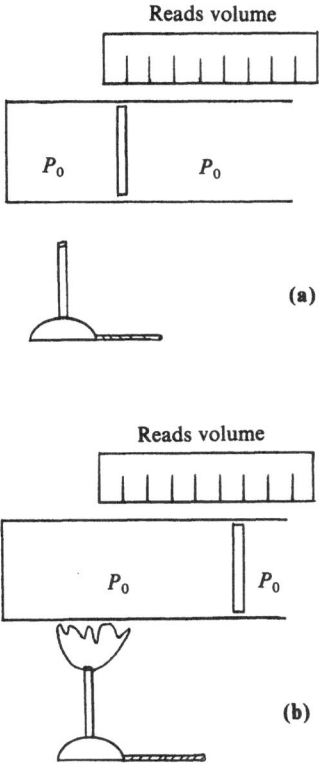

(a)

(b)

Figure 10.3. An ideal-gas thermometer. At a constant pressure P_0, the gas volume (measured by a scale as shown) is used to define the temperature. (Note: Any expansion of the container's walls is considered negligible compared to that of the gas.)

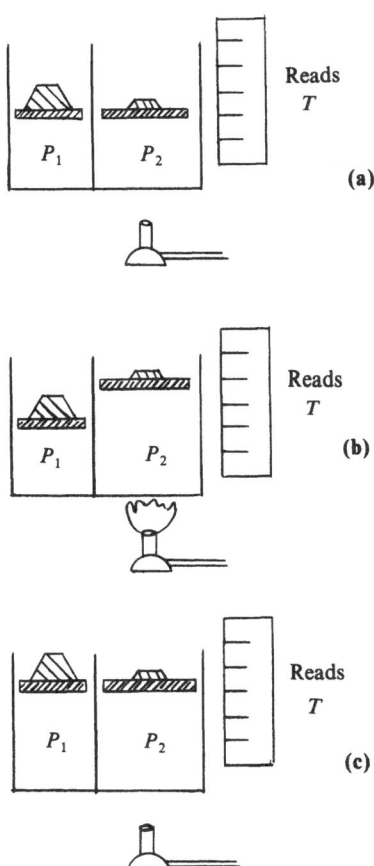

(a)

(b)

(c)

Figure 10.4. Universality of the ideal-gas thermometer. Thermometers No. 1 (left) and No. 2 (right) have different amounts of different gases at different pressures. Nevertheless, their temperatures always agree under thermal equilibrium.

applied to either or both thermometers. (c) They are allowed to reach a new thermal equilibrium.

In this way one discovers that the two gas volumes increase in the same proportion:

> If an ideal-gas sample is held at constant pressure while heated, its volume will increase by a factor that is independent of the size, composition, and pressure of the sample.   (10.A.5)

Therefore (10.A.4) appears to be a fundamental definition of temperature, since any type of ideal-gas sample will measure T in the same way, up to a proportionality constant.

Statement (10.A.5), together with the definition (10.A.4) of the temperature, may be written as a proportion involving any single ideal-gas sample at two temperatures T_1, T_2, with corresponding volumes $\mathscr{V}_1$, $\mathscr{V}_2$:

$$\boxed{\frac{\mathscr{V}_1}{T_1} = \frac{\mathscr{V}_2}{T_2}}$$ (any ideal gas at constant pressure)

(10.A.6)

This is often referred to as **Charles' law**. This law is partly a definition; but it also contains the important information that all ideal gases respond in the same way to a change in temperature.

A laboratory note: Gas thermometers, although valuable as primary standards, are to be avoided in practical temperature measurements. They are exceedingly slow and unwieldy compared to instruments such as those in Fig. 10.1.

An Absolute Zero of Temperature?

As a gas sample shrinks into a small volume, it loses its ideal nature. Therefore it is meaningless to define $T = 0$ as "the temperature at which an ideal gas occupies zero volume."

We shall introduce in Chap. 12 a new definition of T that is valid in an unrestricted range, while still agreeing with (10.A.4) under ideal-gas conditions. The absolute zero of temperature then has a well-defined meaning. However, although the absolute zero can be approached arbitrarily closely, it can never be reached. A satisfactory discussion of this intriguing question is unfortunately beyond the scope of the present book.

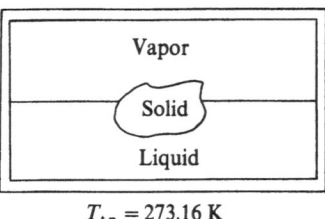

$T_{\text{t.p.}} = 273.16 \text{ K}$

Figure 10.5. A triple-point standard of temperature. As long as ice, liquid water, and *pure* water vapor are simultaneously present, the temperature inside this container is unique and equal to 273.16 K by definition.

Thermal Expansion of Liquids and Solids

At constant pressure most (but not all) substances increase their volume as their temperature is raised; liquids and solids expand by a much smaller factor than gases. A solid of uniform composition, when uniformly heated, will expand all its dimensions by the same factor. (Some exceptions are found among crystals.) For details, see Note (i) at the end of the chapter.

iv. Temperature Scales

Two thermometers are mutually consistent if they read the same temperature when in equilibrium with the same object (or with each other). Hence, if an object could be made whose temperature is not perturbed by contact with the environment, it could serve to calibrate all thermometers.

Such a reference is, in fact, quite simple to prepare. It consists of a sealed rigid vessel containing nothing but pure water in its three phases (liquid, solid, and vapor); see Fig. 10.5. If one attempts to raise the temperature of such a combination, for example by contact with a hot object, one discovers that the relative amounts of each phase will change (e.g., some ice will melt), but the temperature of the system will remain rigorously constant as long as some amount of each phase is still present. (Incidentally, the pressure of the vapor inside the vessel will also remain constant, at 6.0×10^{-3} atm.) One refers to this combination of pressure and temperature as the **triple point** of water. The SI unit of temperature is the **kelvin**,*

---

* Named after Lord Kelvin, and formerly known as the Kelvin degree (K°).

symbolized by K. Its value has been taken such that the triple point of water occurs at

$$T_{t.p.} = 273.16 \text{ K} \qquad (10.A.7)$$

exactly. This defines the **Kelvin scale, absolute scale,** or **thermodynamic temperature scale,** which is part of the International System. (The rationale for the number 273.16 is that it happens to yield a 100-kelvin interval between the freezing and boiling points of water at 1 atm.)

Example 10.1. A One-Liter Gas Thermometer. A certain hydrogen thermometer has a volume $\mathscr{V} = 1.0000$ liter $(= 1.0000 \times 10^{-3} \text{ meter}^3)$ when in equilibrium with triple-point water. What is the constant a of Eq. (10.A.4) for that thermometer?

We have, from (10.A.4),

$$a = \frac{T}{\mathscr{V}} = \frac{273.16 \text{ K}}{1.0000 \times 10^{-3} \text{ meter}^3}$$

$$= 273.16 \times 10^3 \frac{\text{K}}{\text{meter}^3}$$

in this particular case.

Summary: *Defining the Kelvin Temperature T*

Consider an ideal-gas sample kept at any constant pressure. Let its volume be $\mathscr{V}_{t.p.}$ when in equilibrium with a triple-point standard, and $\mathscr{V}$ when at temperature T. We then have

$$T = (273.16 \text{ K}) \frac{\mathscr{V}}{\mathscr{V}_{t.p.}} \qquad (10.A.8)$$

a relation that incorporates Charles' law, (10.A.6).

The Celsius Scale*

This is nothing but a shifted Kelvin scale. The Celsius degree (C°) is equal in size to the kelvin, but the zero of the scale is at the *ice point,* i.e., at the temperature of ice that is melting under 1 atm pressure. It then turns out that a temperature of

* Formerly centigrade scale.

100°C corresponds to the **boiling point** of water at 1 atm. [This is not a coincidence, as was pointed out in connection with (10.A.7).] On the Kelvin scale, the ice point is

$$T_{i.p.} \approx 273.15 \text{ K} \qquad (10.A.9)$$

very close to the triple point.

To summarize: Temperature *intervals* are numerically the same on the Kelvin and Celsius scales; if T_K kelvin and t_C°C are equivalent readings, they satisfy

$$t_C = T_K - 273.15 \qquad (10.A.10)$$

For most purposes one can take $T_{t.p.} = T_{i.p.} = 273$ K, corresponding to 0°C. It is convenient to remember that room temperature is roughly 300 K.

The Fahrenheit Scale

This scale, whose unit interval is the Fahrenheit degree (F°), is defined so that, in terms of temperature *readings,*

$$\begin{array}{c} 32°\text{F is equivalent to } 0°\text{C} \\ 212°\text{F is equivalent to } 100°\text{C} \end{array} \qquad (10.A.11)$$

both exactly. Thus, in terms of *intervals,*

$$1 \text{ F}° = \frac{100 - 0}{212 - 32} \text{ C}° = \frac{5}{9} \text{ C}° \qquad (10.A.12)$$

(The inversions F°, C° instead of °F, °C are customary to distinguish temperature *intervals* from actual temperatures.) Corresponding temperatures of t_F °F and t_C °C are therefore converted into one another, numerically, by

$$t_F = 32 + \tfrac{9}{5} t_C \qquad (10.A.13)$$
$$t_C = \tfrac{5}{9}(t_F - 32) \qquad (10.A.14)$$

The use of the Fahrenheit scale in scientific or technical work is to be discouraged.

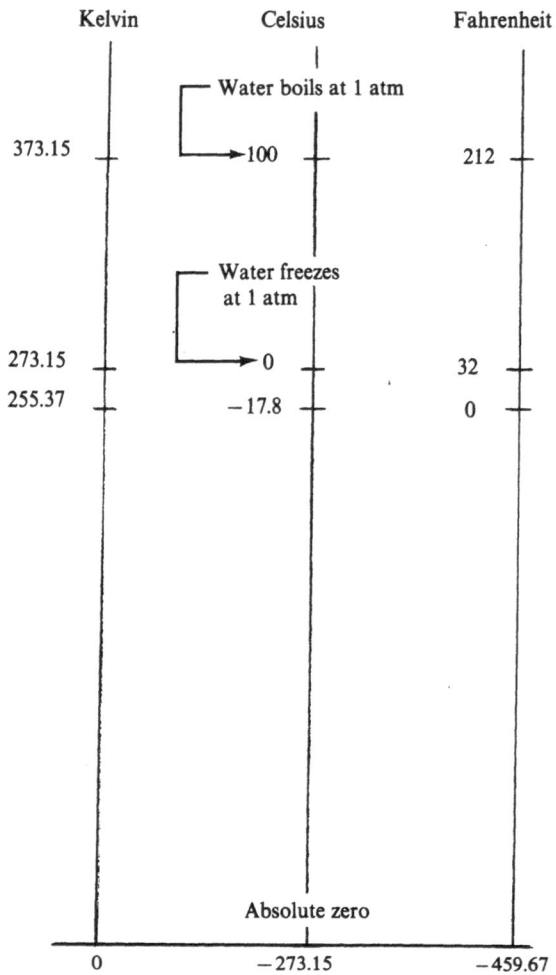

Figure 10.6. Some equivalent temperatures on three different scales.

Figure 10.6 illustrates the relation between the three scales.

v. A Spectrum of Temperatures

Figure 10.7 shows a range of existing temperatures, as observed or inferred.

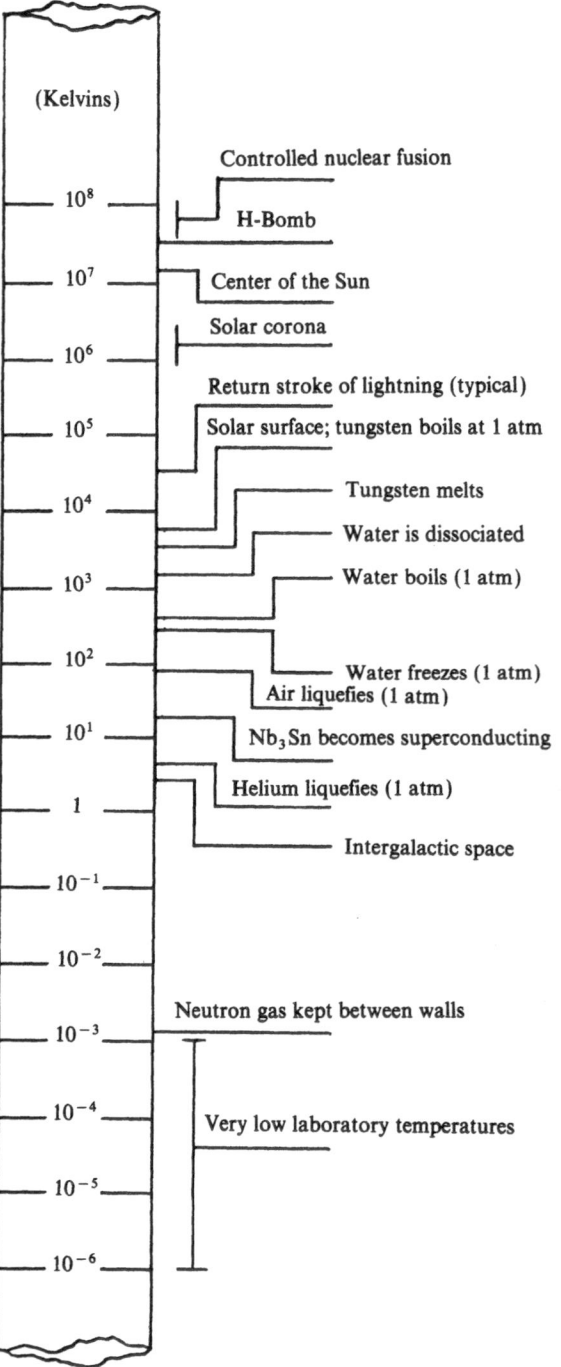

Figure 10.7. Some observed or calculated temperatures, shown on a logarithmic scale. Nuclear fusion reactors are briefly discussed in Chap. 28; temperatures in the range shown have actually been achieved in practice. Note the remarkable temperature gap between the tenuous corona, which surrounds the Sun, and the solar surface itself. Tungsten is a highly refractory metal used for bulb filaments. Dissociated water consists of its separated hydrogen and oxygen atoms. Niobium tin (Nb_3Sn) is an alloy often used as a superconductor; see Sec. D of Chap. 17. The temperature of intergalactic space is really that of the electromagnetic radiation that pervades it, and which is believed to originate from the (enormously hotter) "big bang" that started our Universe. The experimental low temperatures shown are obtained in small samples of matter by means of (in the coldest range) the interaction between magnetic fields and atomic nuclei. At about 10^{-3} K, material walls will reflect a neutron gas; neutrons ordinarily pass through matter almost unimpeded.

B. Heat as Transferred Energy

There are several ways of changing the temperature of a substance, notably through chemical reactions and through certain transfers of energy. Among the latter, **heat** bears a particularly close relation to temperature.

Anticipating somewhat on our story, we must caution the student that **heat is energy in transit**; as Chap. 12 ("Work from Heat") will emphasize, there is no such thing as the "total heat" contained in a system.

i. The Transfer of Heat

Let us bring two objects, initially at different temperatures, together inside a single adiabatic enclosure. As they approach a common temperature, the hotter object is said to **transfer** a certain amount of heat to the colder one. For the purpose of measurement, let some heat be transferred in that way into a 1-kg sample of pure water at atmospheric pressure and room temperature. Suppose the water's temperature increases by exactly 1 K. The amount of heat transferred is then defined to be 1 **kilocalorie** ($=1$ kcal):

---

1 kcal of heat raises the temperature
of 1 kg of water by 1 K.

$$(10.B.1)$$

(A much more fundamental, and slightly more precise, definition of the kilocalorie will be presented later in this section.) More generally, a water sample of mass m whose temperature is raised by a small amount ΔT requires a heat Q proportional to m and to ΔT:

$$Q = \left(\frac{m}{1 \text{ kg}}\right)\left(\frac{\Delta T}{1 \text{ K}}\right)(1 \text{ kcal}) \qquad \text{(for water)} \quad (10.B.2)$$

of which statement (10.B.1) is a special case. Thus, if the mass is measured in kilograms, and the temperature in kelvins, the heat transfer is given in kilocalories by this formula. We see that Q is negative when T decreases. Finally, we note that heat transfer is a two-sided phenomenon. If a cooler object B *receives*, by thermal contact, a positive heat Q from a hotter object A, the latter suffers a *negative* heat transfer $-Q$ (a heat loss Q). This should not be taken for granted, but can be tested, as in Fig. 10.8.

[**The Btu**. In engineering and economic practice, one still encounters the **British thermal unit** or Btu, a unit of energy based on the degree Fahrenheit and on a sample of water whose weight is 1 lb:

1 Btu raises the temperature
of 1 lb of water by 1 F°. $\qquad$ (10.B.3)

How many kilocalories is 1 Btu? For the above-mentioned sample, we have $m \approx 0.454$ kg, $\Delta T = (5/9)$C°. Thus, from Eq. (10.B.2),

$$Q \approx (0.454)(5/9) \text{ kcal}$$

or

$$1 \text{ Btu} \approx \underline{0.252 \text{ kcal}}\,] \qquad (10.B.4)$$

We shall see shortly that the kilocalorie, although convenient, is not the SI unit of heat transfer; the appropriate unit will turn out to be the joule.

How is Thermal Contact Established?

There are three principal ways of transferring heat from one object, A, to another, B.

1. **Conduction** occurs when A and B are in direct physical contact. Randomly moving molecules (and electrons) in A transmit some of their energy to those in B by a sequence of neighbor-to-neighbor forces. Thermal conduction is properly discussed in the context of electrical conduction, one of whose laws can be used as a model. For a brief mention of the topic, see Problem 17.8 in Chap. 17.

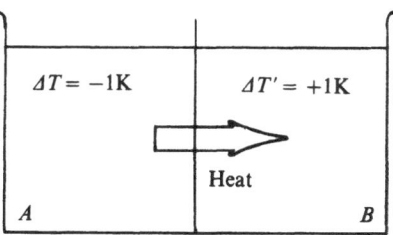

Figure 10.8. In this illustration, involving two equal samples of water, the heat lost by the sample at the left is independently measured in two ways: through its own temperature fall, and through the heat absorbed by the right-hand sample. Both agree.

2. **Convection** involves a third, traveling body, which takes up heat from A and delivers it to B. Moving fluids are important conveyors of heat, for example, in the atmosphere and oceans, as well as in most artificial heating and cooling systems. Convection heating is a complex and specialized subject which we shall not attempt to treat in these pages.

3. **Radiation**, which operates best (but not exclusively) through vacuum, involves the emission and absorption of electromagnetic waves. This is, for example, how the Sun heats the Earth. Electromagnetic waves are discussed in Chap. 22.

ii. Joule's Experiment

We have just observed that, in order to heat a sample of matter, one can bring it into thermal contact with an object at higher temperature; but this is not the only way. With some care, an entirely equivalent *mechanical* method can be devised. Figure 10.9 illustrates the two methods in the case of a water sample: (a) thermal contact, say with a flame; (b) stirring the water adiabatically by means of a propeller or paddle wheel, itself driven by a descending weight w. The frictional work $\mathscr{W}$ done by the stirrer can easily be measured from the altitude drop h of the weight: $\mathscr{W} = wh$. (We neglect the kinetic energy of the descending weight; with sufficient water friction, it never acquires significant speed.) This procedure was devised in the 1840s by James Prescott Joule. His qualitative and quantitative results are as follows.

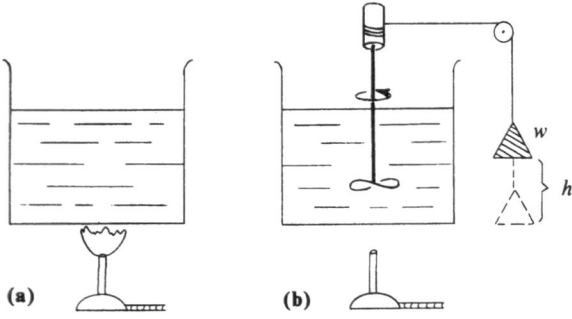

Figure 10.9. Joule's experiment is a comparison of thermal contact versus friction: (a) and (b) are equivalent heating procedures as far as the water sample is concerned.

First:
Stirring will raise the water's temperature.

Second:

> A 1-kg sample of pure water increases its temperature by 1 C° under a frictional work of about 4 200 joules.

(10.B.5)

Third:
When suitable precautions are taken to prevent heat transfers out of the water, the temperature rise depends *only* on the mechanical energy expended, and not on the construction of the paddle wheel, or on its rate of rotation, etc.

The Mechanical Equivalent of Heat

In short, a loss of 4200 joules in mechanical energy produces, in the water sample, an effect indistinguishable from a heat transfer of 1 kcal. This evidence is consistent with the idea that:

> Total energy is conserved;
> The observed 1 kcal is nothing
> but the missing 4200 joules.   (10.B.6)

Countless measurements, involving a wide variety of substances, have confirmed that *every heat transfer is an energy transfer*. The missing energy does not necessarily have to be converted through friction, but can be associated with inelastic impacts, chemical reactions (e.g., combustion), electric currents (as in heating coils and light bulbs), etc. In view of the overwhelming evidence, the kilocalorie has been officially redefined in terms of energy:

> 1 thermochemical kilocalorie
> $(= 1 \text{ kcal} = 1000 \text{ cal}) = 4184 \text{ joules}$ (10.B.7)

exactly.

The mechanical equivalent of heat is, to most of us, surprisingly large. On a human scale, 1 kcal is a very small amount of heat (a family's morning coffee typically requires 100 times that amount), while 4200 joules might be expended by an adult walking up one floor.

Generalized Definition of Heat

The evidence just presented leads to the following definition of a heat transfer Q into a system X.

> Q equals an energy transferred into X *by thermal contact*, from an object at a higher temperature than that of X.
>
> Q equals an energy transferred *by other means* (for example, the mechanical work of friction), but whose effect on X is indistinguishable from thermal contact.   (10.B.8)

(Caution: In Joule's experiment, the mechanical work $\mathcal{W}_{in}$ is *completely* equivalent to the heat transfer Q as far as its effect on the water sample is concerned. This complete equivalence is necessary for $\mathcal{W}_{in} = Q$ to be valid, and does not occur with most methods of doing work. To illustrate: when winding a clock, we perform a mechanical work $\mathcal{W}_{in}$ on the system; however, there exists no equivalent heat transfer to the clock that will produce the same result, and therefore the rise in temperature (if any) of the clock during winding is *not* a measure of $\mathcal{W}_{in}$.)

> **Example 10.2. Heating by Pouring?** Some water, initially at rest in a jug, is poured into a glass from a height $h = 1.0$ meter. Neglecting any heat transfer to the glass itself, calculate the water's rise in temperature.
>
> ———
>
> After the turbulence has died out, the water's gravitational potential energy loss has become a heat transfer Q. Let m be the mass of the water; we then have
>
> $$Q = mgh$$
>
> The temperature rise is, from Eq. (10.B.2),
>
> $$\Delta T = \left(\frac{Q}{m}\right)\left(\frac{\text{kg K}}{\text{kcal}}\right) = (gh)\left(\frac{\text{kg K}}{4200 \text{ joules}}\right)$$
>
> $$= \frac{(10)(1.0)}{4200}\text{ K} = \underline{0.0024 \text{ K}}$$
>
> which is hardly observable; note the conversion of kilocalories to joules.

Heat of Combustion and Dietary Calorie

Combustion is a chemical method of achieving a heat transfer. It is briefly discussed, as well as the dietary calorie, in Note (ii) at the end of the chapter.

C. Specific Heat

As heat is being transferred to a sample of matter, one of two types of change can be observed: (1) the sample's temperature may rise, or (2) part of the sample may change its phase (e.g., solid to liquid, or liquid to gas). Here we shall concentrate on the case where temperature rises; phase transitions will be discussed in the next section.

i. Definition of the Specific Heat

Let us return to the heating of a mass m of water through a small temperature interval ΔT. This, as we have seen in Eq. (10.B.2), requires a heat transfer

$$\boxed{Q = cm\,\Delta T}\qquad(10.C.1)$$

Here the proportionality constant c is just a shorthand for

$$c = 1.00\,\frac{\text{kcal}}{\text{kg K}}$$

$$\approx 4200\,\frac{\text{joules}}{\text{kg K}}\qquad\text{(for water)}\quad(10.C.2)$$

Other substances satisfy (10.C.1) with different values for c; this parameter is called the substance's **specific heat**, and is *defined* by Eq. (10.C.1). Some representative specific heats are listed in Table 10.1.

For gases (but not so much for solids and liquids) it makes a great deal of difference whether c is measured at constant pressure or under some other conditions such as constant volume. For definiteness all values quoted here are for constant pressure. Any temperature dependence of c may be ignored for our present purposes.

Other Units

If we compare the SI and British-system definitions (10.B.1) and (10.B.3) for the unit of heat, we

Table 10.1. Some Specific Heats, c (at constant pressure), in kcal/(kg K)

Metallic solids		Nonmetallic solids		Liquids		Gases	
Lead	0.030	Glass	0.12–0.20	Mercury	0.033	Carbon dioxide	0.20
Silver	0.050	Carbon: Diamond	0.12	Gasoline	0.5	Oxygen	0.22
Copper	0.092	Graphite	0.17	Ethyl alcohol	0.55	Air (dry)	0.24
Iron and steel	0.10	Stone, brick, concrete	0.15–0.22	Methyl alcohol	0.60	Nitrogen	0.25
Aluminum	0.22	Plexiglas	0.35	Water	1.00	Water vapor	0.48
		Ice (just below 0°C)	0.50	Ammonia	1.1	(just above 100°C)	
		Wood	0.6–0.7			Ammonia	0.51
		Paraffin	0.69			Helium	1.24
						Hydrogen	3.4

see that the specific heat of water can be expressed in equivalent ways as

$$1\frac{\text{Btu}}{(\text{lb-mass})\,(\text{F}°)} = 1\frac{\text{kcal}}{\text{kg K}}$$

$$= 1\frac{\text{kcal}}{\text{kg C}°}$$

$$= 1\frac{\text{cal}}{\text{gram C}°} \quad (10.\text{C}.3)$$

Example 10.3. An Inelastic Impact. A lead ball, of mass $m = 1$ kg, is dropped from a height $h = 20$ meters into a lead container, also of mass 1 kg; it never bounces out. If both objects were initially at the same temperature (say 20°C), estimate their temperature rise ΔT after they have come again to thermal equilibrium. (Neglect heat leaks to the outside; assume a specific heat $c = 0.03$ kcal/kg K for lead.)

Assuming the converted potential energy to be entirely equivalent to a heat transfer Q, we have

$$Q = mgh$$

On the other hand, for both objects together,

$$Q = c(2m)\,\Delta T$$

Combining these two equations, and noting that m cancels out, we have

$$\Delta T = \frac{gh}{2c} = \frac{(10)(20)}{(2)[(0.03)(4200)]}\text{ K}$$

$$= \underline{\underline{0.8\text{ K}}}$$

Note again the conversion of kilocalories to joules.

ii. Calorimetry and Heat Capacity

A controlled heat transfer is best obtained inside a **calorimeter**, that is, an adiabatic vessel equipped with a thermometer. This may be used to measure a substance's specific heat, as in the following example.

Example 10.4. Calorimetric Determination of a Specific Heat. A 50-gram piece of unknown metal, at 120°C, is deposited in a calorimeter containing 500 grams of water at 18.0°C. The final temperature reached is 20.0°C. Find the specific heat of the unknown. (Neglect any heat transfer to the calorimeter itself.)

Since no mechanical work is involved, we have some heat Q transferred out of the metal, and the same Q transferred into the water. Equality between these reads

$$(cm\,\Delta T)_{\text{water}} = [cm(-\Delta T)]_{\text{metal}} \quad (10.\text{C}.4)$$

Solving for c_{metal},

$$c_{\text{metal}} = \frac{m_{\text{water}}}{m_{\text{metal}}}\frac{(\Delta T)_{\text{water}}}{(-\Delta T)_{\text{metal}}}c_{\text{water}}$$

$$= \left(\frac{500}{50}\right)\left(\frac{20.0 - 18.0}{120 - 20.0}\right)\left(1.00\frac{\text{kcal}}{\text{kg K}}\right)$$

$$= 0.20\frac{\text{kcal}}{\text{kg K}}$$

Heat Capacity

It is often convenient to deal more directly with the heat transfer Q to an entire sample under a temperature change ΔT. For this purpose one defines the heat capacity C of an object by

$$\boxed{Q = C\,\Delta T} \qquad (10.C.5)$$

For a substance with specific heat c and total mass m, we recall that, equivalently,

$$Q = cm\,\Delta T \qquad (10.C.6)$$

so that

$$C = cm \qquad (10.C.7)$$

Example 10.5. Heat Capacity of a Lake. Estimate the heat capacity, in kcal/C°, of a lake 10 km² in area and of average depth 100 meters.

We have, according to Eq. (10.C.7),

$$C = c\rho \mathcal{V}$$

Here ρ is the density of water and $\mathcal{V}$ is the volume of the lake. Thus,

$$C = \left(1.00\,\frac{\text{kcal}}{\text{kg C}°}\right)\left(1000\,\frac{\text{kg}}{\text{meter}^3}\right)$$
$$\times\ (10\times 10^6\ \text{meters}^2)(100\ \text{meters})$$
$$= 1\times 10^{12}\,\frac{\text{kcal}}{\text{C}°}$$

(Land climates are strongly regulated by the enormous heat capacities of the oceans.)

D. Phase Transitions

Consider, once more, a substance at constant pressure; let it absorb some heat. Curiously enough, a *constant temperature* is sometimes observed during that process, but this phenomenon is possible only if some of the substance is changing its phase (e.g. solid **melting** to liquid, or liquid **vaporizing** to gas) at the same time. A solid-to-gas transition, or **sublimation**, can also occur, as well as its inverse process, but we must leave their discussion to more complete treatments.

i. Melting and Boiling Points

At 0°C and 1 atm, a piece of ice will remain forever in equilibrium with the surrounding water if one supplies no heat to the ice-plus-water combination. More generally, the following statements define the concepts of *melting and boiling points.*

At a given pressure, the melting point of a substance is that single temperature at which the solid and liquid phases may coexist in thermal equilibrium. (10.D.1)

Its boiling point is that single temperature at which the liquid and vapor phases may coexist in equilibrium. (10.D.2)

[Caution: The latter definition is valid only when the vapor phase is not mixed with another gas. For example, water can exist in equilibrium with a mixture of vapor and air (humid air) well below the boiling point. Truly boiling water releases bubbles that consist of pure water vapor. *Our whole discussion applies only to pure chemical substances.*]

ii. Latent Heat

To illustrate how and when a phase transition occurs, let us transfer heat to a 1-kg sample of ice at atmospheric pressure, going through melting, then through further heating of the water, then through boiling, and ending with steam. The resulting Celsius temperature t is plotted in Fig. 10.10 against the amount of heat, Q, so far transferred to the sample. We start, say, at $-60°C$, well below melting, and finish at $+160°C$, well above boiling.

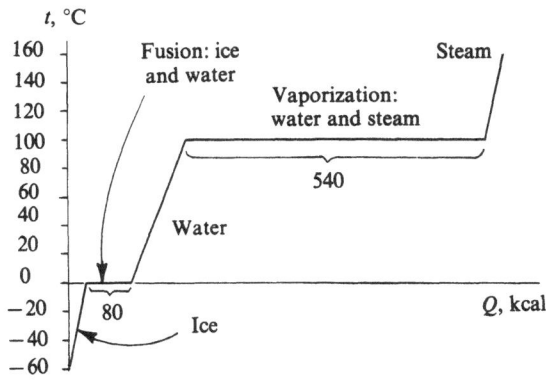

Figure 10.10 A 1-kg sample of ice, initially at $-60°C$, is gradually heated. The resulting temperature t is plotted against the total heat Q absorbed.

First, the temperature rises to 0°C. Since no pure water ice can exist above 0°C in a stable state, it follows that *all the ice must melt before the temperature can continue to increase.* The graph of Fig. 10.10 shows the experimental fact that 80 kcal must be absorbed during the total melting of 1 kg. This amount is called the (latent) **heat of fusion** for ice,

$$h_F = 80 \, \frac{\text{kcal}}{\text{kg}} \left(= 80 \, \frac{\text{cal}}{\text{gram}} \right) \qquad (10.D.3)$$

Similarly, above its boiling point at 100°C, water cannot stably exist in the liquid state, and therefore it must completely vaporize before the temperature rises above 100°C. During the vaporization of 1 kg of water, one measures the absorption of 540 kcal: the (latent) **heat of vaporization** for water is

$$h_V = 540 \, \frac{\text{kcal}}{\text{kg}} \left(= 540 \, \frac{\text{cal}}{\text{gram}} \right) \qquad (10.D.4)$$

One may, of course, start from steam and cool the sample, with attendant *yielding* rather than absorption of heat. The corresponding amounts may still be read from Fig. 10.10. Table 10.2 lists a few representative latent heats at 1 atm, together with the corresponding melting or boiling points. Note the unusually large latent heats of water compared to other materials, a circumstance that has a great effect on the Earth's climate.

In summary: To change the phase of a sample whose mass is m and whose latent heat is h, one must transfer to it a heat

$$\boxed{Q = \pm mh} \qquad (10.D.5)$$

($h = h_F$ or $h = h_V$); the positive sign applies to fusion and vaporization, while the negative sign applies to the inverse processes of *freezing* and *condensation.*

Example 10.6. "Timing" a Latent Heat. A liquid, with specific heat $c = 0.30$ kcal/kg K, is being heated electrically by a heater of constant output. During 15 min the temperature rises from 20°C to 80°C; it then remains constant for another 200 min until all the liquid has been boiled off. What is the heat of vaporization of the substance? (Assume a constant pressure and no heat leaks.)

Let Q be the heater's output during the first 15 min. We have

$$Q = cm \, \Delta T$$

where m is the sample's (unknown) mass and ΔT is the temperature rise. During the last 200 min, the output is

$$Q' = mh_V$$

The crux of the experiment is that the ratio

$$\frac{Q'}{Q} = \frac{h_V}{c \, \Delta T}$$

Table 10.2. Some Melting Points, Boiling Points, and Latent Heats, at 1 atm Pressure

Substance	Melting point T_{mp}, K	Heat of fusion $h_F, \frac{\text{kcal}}{\text{kg}}$	Boiling point T_{bp}, K	Heat of vaporization $h_V, \frac{\text{kcal}}{\text{kg}}$
Helium	(Does not solidify at 1 atm)		4.2	6.0
Hydrogen	14	15.0	20	107
Oxygen	55	3.3	90	51
Nitrogen	63	6.2	77	48
Ethyl alcohol	158	25	352	204
Ammonia	198	108	240	327
Mercury	234	2.7	630	71
Water	**273 (0°C)**	**79.7**	**373 (100°C)**	**539**
Lead	601	6.3	2013	222
Copper	1360	51	2840	1760
Iron	1810	65	3020	1620
Tungsten	3680	44	5930	1180

is known to equal 200 min/15 min. Solving for h_V, we have

$$h_V = \frac{Q'}{Q} c \, \Delta T = \left(\frac{200}{15}\right)(0.30)(80 - 20)\frac{kcal}{kg}$$

$$= 2.40 \frac{kcal}{kg}$$

Notes

i. Thermal Expansion

It was observed at the end of Sec. A iii that, at constant pressure, the volume of most substances increases with rising temperature. (A notable exception, water below 4°C, is illustrated in Fig. 10.11.) The amount of expansion is in general difficult to predict from theory; the measured result is conveniently expressed in terms of the **coefficient of volume expansion,**

$$\boxed{\alpha_{vol} = \frac{1}{\mathscr{V}}\frac{d\mathscr{V}}{dT}} \qquad (10.N.1)$$

Here $d\mathscr{V}$ is the change in the volume $\mathscr{V}$ caused by a change dT in the temperature. Thus, α_{vol} is the *fractional change in volume per unit change in temperature*; its unit is the K^{-1}, or equivalently, the $(C°)^{-1}$.

Example 10.7. The Ideal Gas Again. What is α_{vol} for an ideal gas at constant pressure in the vicinity of room temperature?

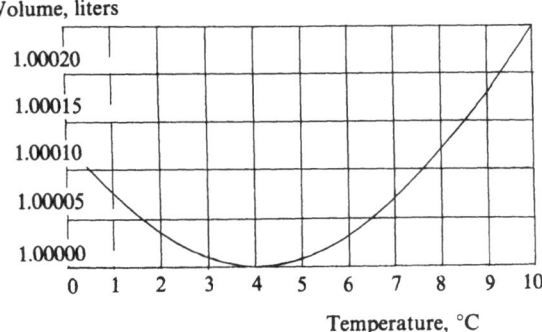

Volume, liters

Figure 10.11. The volume of 1 kg of water, showing its remarkable contraction from 0°C to 4°C. This behavior is attributed to the presence of different kinds of molecules, such as H_2O and $(H_2O)_2$, which change their relative abundances with temperature. Note, however, the minuscule *fractional* variations involved on the vertical axis.

Here the definition of temperature provides the answer. We have, from Eq. (10.A.4),

$$\mathscr{V} = \frac{T}{a}, \qquad \frac{d\mathscr{V}}{dT} = \frac{1}{a}$$

so that (10.N.1) becomes

$$\alpha_{vol} = \left(\frac{a}{T}\right)\left(\frac{1}{a}\right) = \frac{1}{T} \qquad (10.N.2)$$

$$\approx \frac{1}{300} K^{-1} = \frac{1}{300}(C°)^{-1}$$

at room temperature. This result illustrates the fact that α_{vol} is, in general, temperature dependent.

Linear Expansion

For a solid object of length l, the **coefficient of linear expansion**

$$\boxed{\alpha_{lin} = \frac{1}{l}\frac{dl}{dT}} \qquad (10.N.3)$$

expresses the fractional change in length per unit temperature change; α_{lin} again has units of K^{-1} or $(C°)^{-1}$. Occasionally, α_{lin} depends on whether l stands for length, width, or height; α_{lin} is then called **anisotropic**. This happens in some crystals (calcite may actually contract in one direction while expanding in another), or in fibrous materials such as wood. For most materials, however, α_{lin} is the same in all directions (α_{lin} is **isotropic**); the remainder of this note deals with isotropic materials.

There is a simple relation between the linear and volume coefficients. Consider a solid cube of side l. Its volume coefficient is given by

$$\alpha_{vol} = \frac{1}{\mathscr{V}}\frac{d\mathscr{V}}{dT} = \frac{1}{l^3}\frac{d(l^3)}{dT} = \frac{3}{l}\frac{dl}{dT}$$

Thus, we have

$$\boxed{\alpha_{vol} = 3\alpha_{lin}} \qquad (10.N.4)$$

Table 10.3 lists a few typical expansion coefficients. They are not very sensitive to pressure, nor, except in the case of water, to temperature.

ii. Heat of Combustion

Ordinary combustion, which we mentioned at the end of Sec. B, is a technological source of heat that goes back

Table 10.3. Some Thermal Expansion Coefficients, in $(C°)^{-1}$, at 1 atm and 20°C

α_{vol} (liquids), in $(C°)^{-1}$			α_{lin} (solids), in $(C°)^{-1}$	
Mercury	0.18		"7971 glass"[a]	<0.03
Water	0.21		Fused quartz	0.6
Glycerin	0.50	$\times 10^{-3}$	"Invar"[b]	$\lesssim 1$
Ethyl alcohol	1.1		Glass (ordinary)	8–11
Gasoline	1.15		Iron	12
Methyl alcohol	1.2		Copper	17
			Aluminum	25
α_{vol} (ideal gas): 3.3×10^{-3} $(C°)^{-1}$			Ice (just below 0°C)	50
see Eq. (10.N.2)			Nylon	80

The $\times 10^{-6}$ applies to the solids column.

[a] A Corning Glass Co. material intended for a space telescope.
[b] 64% iron, 36% nickel.

to human beginnings. The **heat of combustion** h_C of a fuel is defined as the *heat energy released to the surroundings per unit fuel mass* as the latter is completely burned; for definiteness one should assume a return to the original conditions of temperature and pressure after the heat has been transferred away. As an example, consider a simple reaction such as the burning of carbon in oxygen, $C + O_2 \rightarrow CO_2 +$ heat energy. To measure h_C, a mass m of carbon is enclosed, with excess oxygen, in a pressure and temperature resistant vessel of known heat capacity ("bomb calorimeter"). The sample is ignited electrically, thermal equilibrium is allowed to return, and the heat Q transferred to the calorimeter is measured. Then we have by definition (and after a small correction to Q if the original conditions have not fully returned)

$$h_C = \frac{Q}{m} \qquad (10.N.5)$$

The heat Q is attributed to a *decrease in the chemical energy* of the ingredients; many independent measurements, such as obtaining mechanical or electrical work,

rather than heat, from the same chemicals, have confirmed the conservation of energy in chemical phenomena. Table 10.4 lists the approximate heats of combustion for some fuels and foods.

The kilocalorie used in this chapter coincides with the common **dietary calorie** (the body's heat and work outputs are ultimately reducible to the heats of combustion characterizing the foodstuffs used as fuels; the oxygen is supplied through the lungs). We note the truly astounding mechanical energy equivalents of typical fuels [e.g., $(9000) \times (4200)$ joules $\approx 4 \times 10^7$ joules for 1 kg of ordinary fat; this would be enough to lift 100 people to the top of a skyscraper.]

Table 10.4. Some Heats of Combustion, h_C

Substance	$h_C \left(\dfrac{kcal}{kg} \right)$
Hydrogen	30 000
Gasoline	10 000
Carbon or coal	6 700
Alcohol (ethyl, methyl)	5 000
Dry wood	3 300
(Dehydrated foodstuffs:)	
Fats	9 000
Proteins	4 000
Carbohydrates	4 000

Condensed Checklist

Charles' law:

$$T = a\mathscr{V} \qquad (10.A.4)$$

(Any ideal gas; a depends on sample).

Temperature scales: Fig. 10.6

($T \leftrightarrow$ Kelvin, $t \leftrightarrow$ Celsius or Fahrenheit)

$$T = (273.16 \text{ K}) \frac{\mathscr{V}}{\mathscr{V}_{t.p.}} \qquad \text{(Ideal gas)} \quad (10.A.8)$$

$$t_C = T_K - 273.15 \qquad (10.A.10)$$

$$t_F = 32 + \tfrac{9}{5} t_C, \qquad t_C = \tfrac{5}{9}(t_F - 32) \quad (10.A.14)$$

The kilocalorie:

$$\Delta T = 1 \text{ K} \qquad (10.B.1)$$

Mass of water = 1 kg

Mechanical equivalent:

$$1 \text{ kcal} = 4184 \text{ joules} \qquad (10.\text{B}.7)$$

Specific heat and heat capacity:

$$Q = cm \, \Delta T = C \, \Delta T \quad (10.\text{C}.1), (10.\text{C}.5)$$

Table: Table 10.1

Latent heat:

$$Q = \pm mh \qquad (10.\text{D}.5)$$

Table (including transition temperatures):

Table 10.2

True or False

1. The density of an ideal gas, at given pressure, is inversely proportional to its absolute temperature.

2. The total amount of heat in a given object is proportional to its absolute temperature.

3. At a given temperature, the total amount of heat in an object is proportional to its heat capacity.

4. As a hollow steel sphere is uniformly heated, its outer radius increases, while its inner radius decreases.

5. Heat is always conserved.

6. As long as a sample of water is boiling at atmospheric pressure, no net heat is transferred from outside; rather, the only heat transfer occurs out of the liquid and into the vapor.

7. Suppose a number of objects at different temperatures are left together undisturbed inside an adiabatic enclosure of negligible heat capacity. Then, during any time interval, the *algebraic* sum of all their heat inputs is zero.

Problems

Section A: Temperature

*10.1. (a) What Celsius temperatures are equivalent to 68°F, 23°F, −22°F, 225 K, 518 K? (b) What Kelvin temperatures are equivalent to 3°C, 3.00°C, −272.12°C, 41°F, −31°F?

10.2. An oxygen sample is heated at a constant pressure of 0.5 atm, from 0°C to 100°C. By what factor is its density changed?

*10.3. If the Celsius temperature of a certain ideal gas is doubled, at constant pressure, its volume is increased by 50%. What is its original temperature, on the Celsius and Kelvin scales?

10.4. An 80-meter³ room remains in thermal and pressure equilibrium with the outdoors. Approximately what volume of air enters the room when the temperature drops from 20°C to 15°C?

*10.5. During a high-altitude flight, the outside temperature was announced in degrees, without the label "Celsius," "Fahrenheit," or "Kelvin." To a physics student among the passengers, that piece of data was sufficient to determine the actual temperature if he excluded other scales. What was the temperature?

10.6. A certain hot-air balloon can have its air heated to 70°C in a 20°C environment. What should be its heated volume in order to lift a total mass load (not counting the hot air itself) of 200 kg? At 20°C and 1 atm, $\rho_{\text{air}} \approx 1.3$ kg/meter³; see also Example 9.3 of Chap. 9.

Sections B: Heat as Transferred Energy; C: Specific Heat

*10.7. Heat in the amount of 360 kcal is transferred to 6 kg of water, initially at 20°C. What is the water's final temperature?

10.8. The disused Skylab space station, whose mass was 80 metric tons, had a speed of about 8000 meters/sec when it crashed into the atmosphere. How many kilocalories of thermal energy were shared between the air, the ground, and the station's debris? (1 metric ton = 1000 kg.)

*10.9. A 2.0-kg object, initially at 44°C, loses 15 kcal of heat, and its temperature goes down to 20°C. Calculate (a) its heat capacity, (b) its specific heat, assuming it is made of a single material.

10.10. How long would it take a 2000-watt electric heater to raise the temperature of an 80-meter³ room from 10°C to 20°C, assuming that only the air needs to be heated, and that there are no heat losses? ($c_{\text{air}} = 0.24$ kcal/kg C°, $\rho_{\text{air}} = 1.3$ kg/meter³.)

*10.11. A 1 metric ton automobile, initially going at 20 km/hr, brakes to a stop on level ground, without skidding. (a) How much heat, in kilocalories, is developed by friction? (b) If each of the four brake mechanisms has the heat capacity of 5 kg of iron, estimate their rise in temperature. ($c_{\text{iron}} \approx 0.10$ kcal/kg K, $\rho_{\text{iron}} \approx 8000$ kg/meter³; 1 metric ton = 1000 kg.)

10.12. By how many Celsius degrees would you expect the water's temperature just downstream from a 100-meter waterfall to exceed its temperature just upstream? (Neglect all effects other than gravitational.)

*10.13. A calorimeter of negligible heat capacity contains 0.20 kg of water at 0°C. If one adds 2.0 kg of mercury at 100°C, what is the final temperature, t, of the combination? ($c_{\text{mercury}} = 0.033$ kcal/kg C°.)

*10.14. A 1.0-kg mass of chocolate syrup is being stirred for 1.0 min by a blender whose beater rotates at 240 rpm under a torque of 0.020 newton meter. If the syrup's specific heat is 0.75 kcal/kg, (a) What is its heat capacity? (b) What is its temperature rise, in C°?

10.15. A 3.0-gram copper coin, initially at 80°C, is dropped in a 10-cm³ test tube full of water at 10°C. (a) What is the water's heat capacity? (b) What is the coin's heat capacity? (c) Assuming zero heat transfer to the test tube itself, and negligible impact energy, what is the water's final temperature? ($c_{\text{copper}} = 0.10$ kcal/kg C°.)

10.16. (An impossible dream; see Chap. 12). If the top 100 meters of all the oceans were cooled by 1 C°, and the extracted heat converted to useful energy, how many joules would be obtained? [For comparison, the United States consumes about 10^{20} joules per year; use the fact that about 3/4 of the Earth's surface is sea water; use c (sea water) $\approx c$ (fresh water).]

*10.17. A homeowner is thinking of installing heating wires under his 3 meter × 10 meter driveway in order to melt snow and ice. For a day when there is a 1-cm-thick ice cover at −10°C, make a lowest estimate of how many kW h would be needed. (For comparison, a typical family's monthly electric energy consumption is of order 1000 kW h.)

10.18. How many kilocalories of heat are generated in one minute just by the (totally inelastic) handclaps from 1000 people in a concert hall? Consider each hand, with the appropriate part of the forearm, as a 1-kg particle moving at 3 meters/sec just before impact; assume 2 claps per second (ironically, the sound energy generated is negligible in this context).

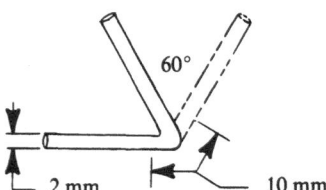

10.19. A sharply bent steel coathanger wire, of diameter 2 mm, is quickly and repeatedly straightened out and bent again, an angle of 60° being swept out 10 times, always *against* a torque of 1 newton meter; see the figure. The internal friction (between microcrystals) at the bend causes it to heat up. Assuming that no heat is transferred out of a 10-mm length of wire, estimate its

rise in temperature. ($c_{\text{steel}} \approx 0.10$ kcal/kg K; $\rho_{\text{steel}} \approx$ 8000 kg/meter³).

*10.20. Suppose, contrary to fact, that the Sun (mass = 2×10^{30} kg) has no internal source of energy and merely cools off through radiation. Assume its overall temperature is 1.5×10^7 K (this is believed to be its actual temperature at the center). If the Sun's specific heat were of the same order as that of hydrogen gas in the laboratory ($c \approx 3$ kcal/kg K), and if its rate of heat loss maintained itself at 4×10^{26} watts (its present value), after how many years would its temperature have dropped by half?

Section D: Phase Transitions

10.21. (a) Water is boiling in a 1000-watt electric kettle. What mass of water is boiled off in 1 min? (Neglect heat losses.) (b) If this kettle is regularly replenished with fresh water at 10°C, how much water, on the average, must be added every hour?

10.22. How much heat, in kilocalories, must be transferred to 8.0 kg of water, initially at 20°C, in order just to vaporize it at atmospheric pressure?

*10.23. Find the equilibrium temperature that results when 300 grams of ice at 0°C are added to 600 grams of water at 50°C.

*10.24. How much heat is given up when 20 kg of steam at 100°C is condensed at atmospheric pressure and cooled to 20°C?

10.25. A 50-kg mass of liquid nitrogen, initially at 77 K, is allowed to evaporate in a roomful (say 150 kg) of dry air, initially at 20°C. Estimate the room's equilibrium temperature, assuming that the process has expelled about 50 kg of the original air without mixing.

10.26. A 7.0-kg mass of molten lead, at 601 K, is poured into a certain mass m of water at 10°C. The lead solidifies, the whole water evaporates into the atmosphere, and the final temperature of the lead is 100°C. Find m.

Answers to True or False

1. True (density $\propto$ 1/volume).

2. False (there is no such thing as the total amount of heat).

3. False (see item 2).

4. False (both radii increase by the same factor).

5. False (see Joule's experiment).

6. False (the latent heat must be supplied).

7. True.

Answers or Hints to Problems

10.1. (b) 276 K, 276.15 K, 1.03 K, 278 K, 238 K.

10.3. 546 K.

10.5. Hint: Which temperature reads the same on the Celsius and Fahrenheit scales? Can a Kelvin temperature be negative?

10.7. 80°C.

10.9. (a) 0.62 kcal/C°; (b) 0.31 kcal/kg C°.

10.11. (a) 3.7 kcal; (b) 1.8 C°.

10.13. 25°C.

10.14. (b) 0.010 C°.

10.17. 27 kW h.

10.20. 1.5×10^7 years.

10.23. 6.7°C.

10.24. 1.24×10^4 kcal.

CHAPTER 11

MOLECULES AND GASES

This chapter is our first bridge between the study of ordinary objects and that of their microscopic constituents, the molecules. It is organized as follows.

After asking why we should believe in the existence of molecules, we explain how to gauge the number of molecules present in a bulk sample, and hence how to prepare a standard amount, namely, a **kilomole** (about 6×20^{26} molecules) of any substance.

Among all kinds of material, the **ideal gases** are the simplest to understand in molecular terms. Accordingly, they will be our main topic. We shall discuss them under two completely different aspects, which, however, support and complement one another.

First we outline the experimentally observed behavior of *bulk samples* of ideal gas; such a description involves their pressure, volume, and temperature. (Charles' law, formulated in the preceding chapter, will be one element of this comprehensive "ideal-gas law.")

Afterwards, these results will be interpreted in terms of molecules. To arrive at such an interpretation, we shall use elementary concepts of **kinetic theory**, which is the art of averaging the effects of moving molecules.

A. The Existence of Molecules

Up to this point we have studied *macroscopic* physics only, that is to say, the physics of objects whose size makes them directly visible.

If the use of a magnifying glass is not ruled out, "macroscopic" means "greater than 10^{-4} or 10^{-5} meter." On the other hand, the consideration of molecules opens the gate to **microscopic** physics, in which macroscopic equipment (we have no other) must be used to reveal things so small that a few generations ago they were expected to remain forever hypothetical. The atoms, in particular, have radii of the order of 10^{-10} meter, while the molecules roughly have the sizes one would expect from putting together the appropriate atomic sizes, as in the colored-ball models of chemistry lecture halls.

Chemical measurements are at the roots of modern atomic theory, whose earliest systematic form is found in a book by John Dalton (1808). The atomic concept itself goes back to Democritus (5th century B.C.)

i. Some Observational Facts

The available evidence for atoms and molecules is overwhelming. Aside from chemistry, here are four physical observations.

1. Condensation of Gases. The extreme difference in compressibility between a liquid and its vapor is evidence for hard molecules, of nonzero volume, that are close together in the liquid and (except for collisions) far apart in the vapor; "close" and "far" are to be understood in relation to the molecules' range of interaction. Unfortunately, this qualitative observation, taken by itself, gives us no clue as to the actual size of a molecule.

2. Crystals. Crystals have well-defined and recognizable angles between their cleavage planes. This is most naturally explained in terms of the orderly stacking of identical molecules or atoms, as

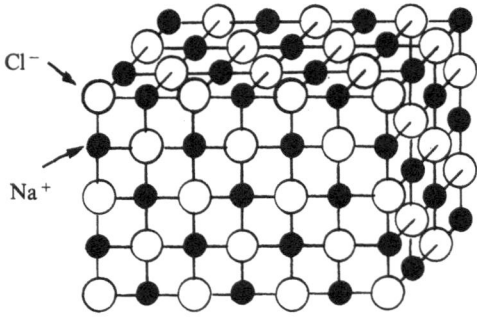

Figure 11.1. Cubic spatial arrangement of Na and Cl atoms (more accurately, Na^+ and Cl^- ions) in a crystal of common salt. A similar arrangement is found for the salts NaF, NaBr, and NaI.

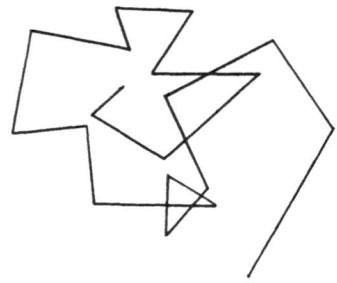

Figure 11.2. Typical "random walk" of a dust speck undergoing Brownian motion. The positions, recorded at equal time intervals, are joined by unrealistically straight segments: more detail would show random-walk behavior on a smaller and smaller scale, until the level of individual molecular collisions is reached.

in Fig. 11.1. The modern measurement of crystals by x-rays has revealed the size of interatomic spacings with amazing accuracy, as we shall explain in Chap. 24 (Wave Optics). In this way, the number of atoms in a given crystal becomes known, and by weighing an ordinary crystal one may find the mass of a single atom. Thus, *crystals are a primary source for our knowledge of atomic masses.*

*3. Brownian Motion.** Consider a sample of fluid, at uniform temperature, and well shielded from any possible cause of convection or vibration. If the fluid contains sufficiently small dust specks (e.g., in air they might be smoke particles), the latter are observed through an ordinary microscope to remain in perpetual, random motion; see Fig. 11.2. This behavior, with its sudden and irregular spurts, is explained in terms of a random and unbalanced bombardment of the particle from all sides by the molecules of the fluid. [The laws of chance do not guarantee balanced momentum transfers at all times, but only on the average, over a sufficiently long period. A statistical study shows that the side-to-side fluctuations should be more easily observed if the dust speck suffers less frequent impacts—hence it should be of small size. This is reasonable; in the extreme case where only one impact occurs during the time of observation, there cannot possibly be a balancing from all sides. A famous quantitative analysis by Einstein (1905) has established the molecular explanation of Brownian motion beyond any doubt.]

4. Direct Imaging. Few demonstrations carry as much conviction as the fluorescent-screen imaging of individual atoms (Fig. 11.3a). This technique, devised by E. W. Mueller, exploits the electric propulsion of ions away from the tip of a metal needle and onto a screen; see also Sec. E of Chap. 16.

More recently a very different device has appeared, the **scanning tunneling microscope**, which makes an accurate image of a wide variety of surface atoms and molecules. The ultrafine point of a sensing needle travels back and forth over the surface to be examined, all the while avoiding actual contact by a distance comparable to atomic dimensions. This incredibly precise maneuver, automatically performed and recorded, results in a relief map of the surface, atom by atom. One such scan is shown in Fig. 11.3b. (The word "tunneling" refers to the behavior of electrons at the tip of the sensing needle; the tunnel effect is discussed in textbooks on quantum mechanics.)

ii. The Permanence of Molecules

It is possible to modify the structure of a molecule drastically, for a example by a chemical reaction, by radiation, by heat, etc. However, one cannot *slightly damage* a molecule: it cannot be dented, scratched, or worn. Water may flow for millions of years without its chemical or physical properties being altered in the least; also, the remarkable permanence of certain living species after reproduction through thousands of generations is due to the fact that all genetic information is carried by molecules, which must change their structure by discrete steps or not at all.

* First noticed, with pollen grains in water, by Robert Brown in 1827.

(a)

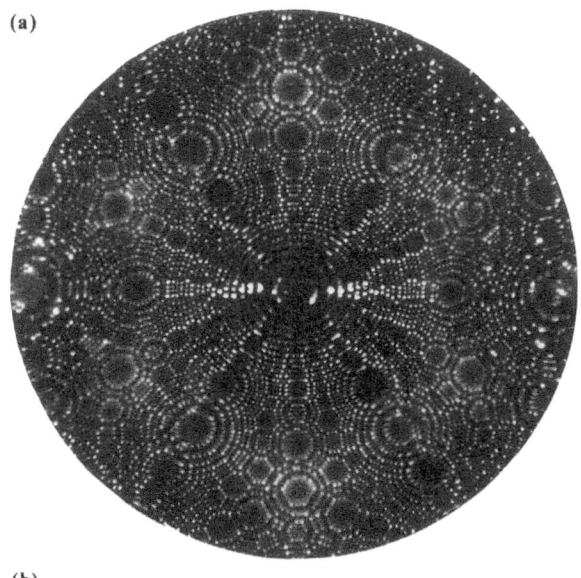

(b)

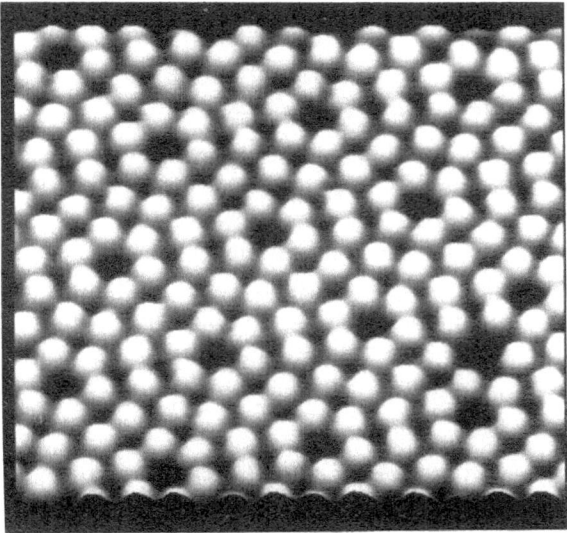

Figure 11.3. (a) The location of individual atoms (white spots) at the tip of a tungsten needle is exhibited on a fluorescent screen in Mueller's field ion microscope. (Reprinted by permission. Courtesy T. T. Tsong, Pennsylvania State University.) (b) The atoms of a silicon crystal surface as mapped by a scanning tunneling microscope. (Courtesy Ph. Avouris, IBM Thomas J. Watson Research Center.)

iii. The Size of Atoms

The picture of an atom as a smooth, hard sphere is not a realistic one, although it is a useful model for many purposes. The "outer coating" of an atom consists of the orbits of one or more electrons (identical, pointlike, electrically charged particles)

in rapid motion, and one achievement of modern physics has been to explain how such a fragile-seeming structure can have the properties described here. In a certain sense, nevertheless, an atom may be said to have a size. For example, an isolated hydrogen atom may be visualized as a sphere of radius 1.2×10^{-10} meter. This so-called **Van der Waals radius** determines the closest approach of two hydrogen atoms which collide without combining chemically; if they do combine, as in the usual molecule H_2, the two atoms share their electrons and interpenetrate significantly, with a resultant effective radius, for each, of only 0.38×10^{-10} meter, known as the **covalent radius** of hydrogen. No atom has a Van der Waals radius much above 2×10^{-10} meter.

B. Atomic Masses and the Kilomole

A near-chemical interlude is needed for our physical understanding of gases: we must learn to relate the properties of individual atoms and molecules to those of bulk matter. The necessary concepts, developed in the present section, deal with *individual atomic or molecular masses, numbers of atoms or molecules,* and *bulk masses.* They often elaborate a rather trivial theme: for a sample of identical particles, we have

$$\begin{pmatrix} \text{Mass of each} \\ \text{particle} \end{pmatrix} \times \begin{pmatrix} \text{number of} \\ \text{particles} \end{pmatrix} = \text{total mass}$$

i. Isotopes and the Atomic Mass Unit

Atoms are classified, on the basis of their chemical behavior, into a hundred-odd different kinds called the **chemical elements**; but even in a pure sample of a chemical element, not all atoms are identical. Owing to differences in their nuclei, two atoms of the same element may have slightly different masses; they are then called **isotopes** of one another.

It is, of course, correct to express the mass of an atom in kilograms; but a sometimes more convenient standard, the **atomic mass unit,*** u, has been defined as follows:

$$\boxed{\text{Mass of a single } ^{12}\text{C atom} = 12 \text{ u}} \qquad (11.B.1)$$

---

* Also called the **dalton**; the abbreviation amu is sometimes used.

exactly. Carbon 12 (^{12}C) is the most common isotope of carbon; its mass is defined as that of the neutral atom, with its standard complement of six electrons. *Atomic masses, by convention, always refer to the neutral atoms.*

By methods alluded to in the preceding section, the mass of a ^{12}C atom can be measured very accurately, with the following result for 1/12 of it:

$$1 \text{ u} \approx 1.66053 \times 10^{-27} \text{ kg} \qquad (11.B.2)$$

In terms of this unit, some other atomic masses are as follows: 13.003 u for ^{13}C, 1.008 u for ^{1}H, 15.995 u for ^{16}O, etc. [Note that the left superscripts 13, 1, 16 ("mass numbers") are integers which closely match the masses; they should not be confused with the subscripts used in chemical formulas.]

Isotopes are very unequally represented. To illustrate: Natural carbon is 98.9% ^{12}C and 1.1% ^{13}C, with a (usually negligible) trace of radioactive ^{14}C; in sea water, 15 out of 100 000 hydrogen atoms are ^{2}H (also called deuterium, D), the rest are ^{1}H.

ii. Average Atomic and Molecular Weights

For convenience, an atomic mass is turned into a *dimensionless number* through the omission of the atomic mass unit. Formally, we define

$$(\text{Atomic weight})(1 \text{ u}) = \text{atomic mass} \qquad (11.B.3)$$

For example, the atomic mass of ^{1}H is 1.008 u, its atomic weight is 1.008. (Gravity is of course not involved, but the word "weight" is unfortunately established by custom.)

For a natural mixture of isotopes, "atomic weight" means **average atomic weight**:

$$(\text{Atomic weight})(1 \text{ u}) = \left(\frac{\text{mass of a sample}}{\text{number of atoms}} \right)$$
$$(11.B.4)$$

Example 11.1. Atomic Weight of Chlorine. Natural chlorine consists of approximately 75% ^{35}Cl and 25% ^{37}Cl, with atomic weights close to 35 and 37, respectively. What is the (average) atomic weight of chlorine?

From (11.B.4) with a typical sample of 100 atoms, we obtain

$$\text{Atomic weight} = \frac{(75)(35) + (25)(37)}{100} = \underline{35.5}$$

Molecular Weight

In an entirely similar manner, we have, in the sense of an average,

$$(\text{Molecular weight})(1 \text{ u})$$
$$= \left(\frac{\text{mass of a sample}}{\text{number of molecules}} \right) \quad (11.B.5)$$

The molecular weight of a compound is just the sum of the appropriate atomic weights, as in the case of water, H_2O:

$$(\text{Molecular weight of } H_2O) \approx 1 + 1 + 16 = 18$$

Tabulated Atomic Weights

These are shown, for all stable and a few radioactive elements, in Table 11.1. The atomic number will not be discussed until we study modern physics; for the time being it may be considered as just a label that characterizes each element, and which could in fact serve as its chemical symbol.

iii. Avogadro's Number and the Kilomole

In 1811, Amedeo Avogadro made a fruitful suggestion: the size of a material sample should be described in terms of how many molecules it contains (a figure which, in his days, could not yet be estimated).

The SI standard for such large numbers is defined as follows. Let **Avogadro's number**, $\mathcal{N}_A$, be the number of atomic mass units in one kilogram:

$$\mathcal{N}_A = \frac{1 \text{ kg}}{1 \text{ u}} \qquad (11.B.6)$$

a dimensionless quantity. (Equivalently, $\mathcal{N}_A$ is *the number of atoms in 12 kg of ^{12}C*; this is a more often quoted definition.) It follows from (11.B.2) that

$$\mathcal{N}_A \approx \frac{1 \text{ kg}}{1.66053 \times 10^{-27} \text{ kg}}$$

Table 11.1. Average Atomic Weights[a]

Name	Symbol	Atomic number	Atomic weight	Name	Symbol	Atomic number	Atomic weight
Aluminum	Al	13	26.98	Nickel	Ni	28	58.71
Antimony	Sb	51	121.75	Niobium	Nb	41	92.91
Argon	Ar	18	39.95	Nitrogen	N	7	14.01
Arsenic	As	33	74.92	Osmium	Os	76	190.2
Barium	Ba	56	137.34	Oxygen	O	8	16.00
Beryllium	Be	4	9.01	Palladium	Pd	46	106.4
Bismuth	Bi	83	208.98	Phosphorus	P	15	30.97
Boron	B	5	10.81	Platinum	Pt	78	195.09
Bromine	Br	35	79.90	Plutonium[b]	Pu	94	239.05
Cadmium	Cd	48	112.40	Polonium[b]	Po	84	209.98
Calcium	Ca	20	40.08	Potassium	K	19	39.10
Carbon	C	6	12.01	Praseodymium	Pr	59	140.91
Cerium	Ce	58	140.12	Promethium[b]	Pm	61	147
Cesium	Cs	55	132.90	Protactinium[b]	Pa	91	231.04
Chlorine	Cl	17	35.45	Radium[b]	Ra	88	226.03
Chromium	Cr	24	52.00	Radon[b]	Rn	86	222.02
Cobalt	Co	27	58.93	Rhenium	Re	75	186.2
Copper	Cu	29	63.54	Rhodium	Rh	45	102.90
Dysprosium	Dy	66	162.50	Rubidium	Rb	37	85.47
Erbium	Er	68	167.26	Ruthenium	Ru	44	101.07
Europium	Eu	63	151.96	Samarium	Sm	62	150.35
Fluorine	F	9	19.00	Scandium	Sc	21	44.96
Gadolinium	Gd	64	157.25	Selenium	Se	34	78.96
Gallium	Ga	31	69.72	Silicon	Si	14	28.09
Germanium	Ge	32	72.59	Silver	Ag	47	107.87
Gold	Au	79	196.97	Sodium	Na	11	22.99
Hafnium	Hf	72	178.49	Strontium	Sr	38	87.62
Helium	He	2	4.00	Sulfur	S	16	32.06
Holmium	Ho	67	164.93	Tantalum	Ta	73	180.95
Hydrogen	H	1	1.01	Tellurium	Te	52	127.60
Indium	In	49	114.82	Terbium	Tb	65	158.92
Iodine	I	53	126.90	Thallium	Tl	81	204.37
Iridium	Ir	77	192.22	Thorium[b]	Th	90	232.04
Iron	Fe	26	55.85	Thulium	Tm	69	168.93
Krypton	Kr	36	83.80	Tin	Sn	50	118.69
Lanthanum	La	57	138.91	Titanium	Ti	22	47.90
Lead	Pb	82	207.19	Tungsten	W	74	183.85
Lithium	Li	3	6.94	Uranium[b]	U	92	238.05
Lutetium	Lu	71	174.97	Vanadium	V	23	50.94
Magnesium	Mg	12	24.30	Xenon	Xe	54	131.30
Manganese	Mn	25	54.94	Ytterbium	Yb	70	173.04
Mercury	Hg	80	200.59	Yttrium	Y	39	88.91
Molybdenum	Mo	42	95.94	Zinc	Zn	30	65.37
Neodymium	Nd	60	144.24	Zirconium	Zr	40	91.22
Neon	Ne	10	20.18				

[a] Further atomic and nuclear tables are found in Sec. D of Chap. 27 and Sec. A of Chap. 28.
[b] Radioactive; the given atomic weight pertains to the most common or most abundantly prepared isotope.

or

$$\mathcal{N}_A \approx 6.02217 \times 10^{26} \ (SI) \qquad (11.B.7)$$

The mention "SI" serves as a reminder that the definition is based on 1 kg of matter, in contrast to another often encountered definition based on the gram.

Avogadro's number is used to specify a certain amount of matter called the **kilomole** (kmole):

$$1 \text{ kmole} = \mathcal{N}_A \text{ molecules} \approx 6.02 \times 10^{26} \text{ molecules}$$

$$(11B.8)$$

We note that 1 kmole $= 10^3$ moles. The units "kmole" and "molecule" can both be treated as dimensionless.

The Generic Kilomole

By extension, definition (11.B.8) is adopted for any kind of object, not just molecules; but then the nature of the object must be mentioned for clarity. To illustrate:

$$1 \text{ kmole of } O_2 \text{ gas} = 1 \text{ kmole of } O_2 \text{ molecules}$$
$$= 2 \text{ kmole of O atoms}$$

Also:
1 kmole of neutral H atoms contains 1 kmole of electrons.

What is the Mass of a Kilomole?

The answer to this question follows directly from definition (11.B.5), in which we take the sample to be 1 kmole; this yields

$$\text{Molecular weight} = \frac{\text{mass of 1 kmole}}{(\mathcal{N}_A)(1 \text{ u})}$$

or, since the denominator is just 1 kg:

$$\boxed{\text{Mass of 1 kmole} = (\text{molecular weight})(1 \text{ kg})}$$

$$(11.B.9)$$

As we saw in (11.B.5), the molecular weight may be an average value. The following example illustrates that a kilomole may be $\mathcal{N}_A$ molecules of a mixture.

Example 11.2. One Kilomole of Air. Given that nitrogen (N) and oxygen (O) have atomic weights 14 and 16, respectively, find the mass of 1 kmole of dry air, which contains a ratio of 8 molecules of N_2 to 3 molecules of O_2.

The average molecular weight of air is

$$\text{Molecular weight} = \frac{(8)(14+14)+(3)(16+16)}{8+3}$$

$$= 29$$

From (11.B.9) we have

$$\text{Mass of 1 kmole of air} = \underline{\underline{29 \text{ kg}}}$$

iv. Avogadro's Law for Ideal Gases

Having just dealt with the question of a kilomole's mass, we may well ask: what is a kilomole's volume? This would at first seem to have no meaningful answer, since the volume of a given amount of matter depends on pressure, temperature, phase, etc., and most particularly on the kind of substance considered. It is therefore quite extraordinary that, as Avogadro hypothesized:

> At any given pressure and temperature,
> all *ideal* gases, including mixtures,
> have the same volume per kilomole.

$$(11.B.10)$$

Equivalently:

> A given volume of any ideal gas at given pressure and temperature always contains the same number of molecules.

$$(11.B.11)$$

[These statements are directly verifiable in the laboratory, since 1-kmole samples are easily prepared. For example, 1 kmole (12 kg) of carbon, C, makes 1 kmole of CO_2 gas; the oxygen O_2, needed for that preparation also amounts to 1 kmole; the relevant volumes of CO_2 and of O_2 can then be compared under ideal-gas conditions.]

Standard Temperature and Pressure

A 1-kmole sample of any ideal gas at 0°C and 1 atm ("standard temperature and pressure," abbreviated as "STP") provides a traditional reference point; its volume is measured to be

$$\mathscr{V}_{STP} \approx 22.4 \text{ meters}^3 \quad (1 \text{ kmole}) \qquad (11.B.12)$$

It follows that:

> At STP, 22.4 meters^3 of any ideal gas always contains $\mathscr{N}_A = 6.02 \times 10^{26}$ molecules.

$$(11.B.13)$$

We conclude this section with a pair of examples.

Example 11.3. Density of an Ideal Gas. Find the density of air at STP.

_____

Consider a 1-kmole sample. We have, according to Example 11.2,

$$\rho_{STP} = \frac{\text{mass}}{\text{volume}} = \frac{29 \text{ kg}}{22.4 \text{ meters}^3} = 1.3 \frac{\text{kg}}{\text{meter}^3}$$

More generally, the density of an ideal gas at STP is calculated as follows:

$$\rho_{STP} = \frac{(\text{av. mol. wt.})(1 \text{ kg})}{\mathscr{V}_{STP}}$$

$$= \left(\frac{\text{av. mol. wt.}}{22.4}\right)\left(\frac{\text{kg}}{\text{meter}^3}\right) \quad (11.B.14)$$

Example 11.4. Historical Molecules? In each breath, the reader of this book is likely to inhale $\mathscr{N}$ molecules from Avogadro's last breath. What is the order of magnitude of $\mathscr{N}$?

_____

This classic problem is to some extent an exercise in making rough assumptions. Let us say a human breath involves a volume $\mathscr{V} = 0.5$ liter $= 0.5 \times 10^{-3}$ meter^3 of air; we assume uniform mixing of Avogadro's last breath with the atmosphere. Let us simplify the latter into a spherical shell of uniform density and of thickness $h = 10\,000$ meters (see Fig. 9.13 of Chap. 9,

and Problem 9.46.) The atmosphere's volume is then

$$\mathscr{V}_{atm} = 4\pi r_{Earth}^2 h$$

$$= (12)(6.4 \times 10^6)^2 (10^4) \text{ meters}^3$$

$$\approx 5 \times 10^{18} \text{ meters}^3$$

Finally, let us assume standard temperature and pressure for all samples.

Avogadro's last breath contains a fraction $\mathscr{V}/\mathscr{V}_{STP}$ of 1 kmole, i.e., a number $(\mathscr{V}/\mathscr{V}_{STP})\mathscr{N}_A$ molecules; this is the number now distributed in the whole volume $\mathscr{V}_{atm}$. In every breath of volume $\mathscr{V}$ we catch a fraction $\mathscr{V}/\mathscr{V}_{atm}$ of these molecules, i.e.,

$$\mathscr{N} = \left(\frac{\mathscr{V}}{\mathscr{V}_{atm}}\right)\left(\frac{\mathscr{V}}{\mathscr{V}_{STP}}\right)\mathscr{N}_A$$

$$= \frac{(0.5 \times 10^{-3})^2 (6 \times 10^{26})}{(5 \times 10^{18})(22.4)} \approx \underline{\underline{1}}$$

an average of one molecule in every breath.

C. The Ideal-Gas Law

Here we continue our study of the ideal gases. The relation between their pressure, volume, and temperature will be described, with some help from the concept of kilomole and of Avogadro's law. The detailed role played by the molecules will not be studied until the next section.

Rather than first writing down a comprehensive law, and then examining its special cases, it will be more instructive—and nearly as simple—to approximate the historical process and put that equation together *from* its special cases.

i. Boyle's Law

Under increasing pressure, a gas sample will ordinarily shrink into a smaller volume, but not always; a sufficient rise in temperature will offset that tendency.

In order to disentangle the effects of pressure and temperature, let us equalize the initial and final temperatures of the gas, $T_1 = T_2$. Initial and final equilibrium with a constant environment is one way of achieving this condition.

With that procedure in mind, we change the pressure:

$$P_1 \to P_2$$

with a consequent change in volume

$$\mathscr{V}_1 \to \mathscr{V}_2$$

We then discover that *the volume has changed inversely with the pressure*; the product of the two is unchanged:

$$\boxed{P_1\mathscr{V}_1 = P_2\mathscr{V}_2} \qquad (T_1 = T_2) \quad (11.C.1)$$

This observation, first made by Robert Boyle (1627–1691), is strictly valid only for ideal gases; the importance of equal initial and final temperatures was not clear to Boyle; Edmé Mariotte pointed it out a few years later.

Boyle's law may be restated in a variety of useful ways, such as

$$P\mathscr{V} = \text{const} \qquad (T = \text{const}) \quad (11.C.2)$$

Figure 11.4 shows a plot of P against $\mathscr{V}$ (a so-called $P\mathscr{V}$ diagram) for an ideal gas under **isothermal** (that is, constant temperature) conditions.

Density and Pressure

Boyle's law enables us to relate the density of an ideal gas to its pressure. Let a certain gas have density ρ_1 at a pressure P_1. What is its density at the same temperature, but at a new pressure P_2?

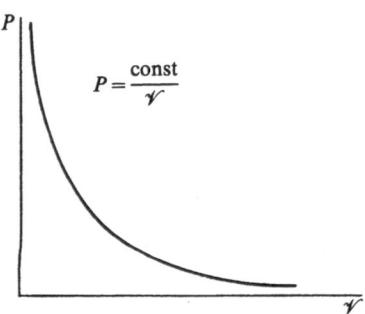

Figure 11.4. An isotherm, showing the pressure of an ideal-gas sample as a function of its volume at constant temperature. (Note: Any change that occurs at constant temperature is called an **isothermal process.**)

Consider a sample of mass M; its density is $\rho = M/\mathscr{V}$. According to (11.C.2) we then have

$$P = \frac{\text{const}}{\mathscr{V}} = (\text{const})(\rho) \qquad (11.C.3)$$

Thus, *isothermally*, *the density of an ideal gas is proportional to its pressure*; in the present case we have

$$\rho_2 = \frac{P_2}{P_1}\rho_1 \qquad (11.C.4)$$

ii. The Equation of State

The effect of both the pressure P and the absolute temperature T on the volume of an ideal-gas sample can now be summarized by the following proportionalities.

$$\text{Boyle's law:} \quad \mathscr{V} \propto \frac{1}{P} \qquad (\text{fixed } T) \quad (11.C.5)$$

$$\text{Charles' law:} \quad \mathscr{V} \propto T \qquad (\text{fixed } P) \quad (11.C.6)$$

as we saw in the preceding chapter.

In addition, for an N-kmole sample of any kind of ideal gas, we have

$$\text{Avogadro's law:} \quad \mathscr{V} \propto N \qquad (\text{fixed } P, T)$$
$$(11.C.7)$$

These three experimental results can be combined into a single formula,

$$\mathscr{V} \propto \frac{NT}{P} \qquad (\text{unrestricted ideal-gas sample})$$
$$(11.C.8)$$

The proportionality constant, still to be supplied, is, in view of Avogadro's law, independent of the nature of the gas; it is called the **universal gas constant** and is usually denoted by R. The proportionality (11.C.8) becomes

$$\mathscr{V} = R\frac{NT}{P} \qquad (11.C.9)$$

or

$$\boxed{P\mathscr{V} = NRT} \qquad (11.C.10)$$

This is the **ideal-gas law**, also known as the **equation of state** of an arbitrary sample of any ideal gas. The variables P, $\mathscr{V}$, T are said to characterize the **state** of the sample.

[A practical tip on the use of (11.C.10): in problem solving, it simplifies matters to consider 1 kmole samples ($N = 1$) whenever possible.]

Our collection of proportionalities can now be completed by

Gay-Lussac's law: $P \propto T$ (fixed $\mathscr{V}$)

$$(11.C.11)$$

This is one of the many consequences of (11.C.10).

Numerical Value of R

There remains to fix the value of R in the ideal-gas law. For that purpose, we examine Eq. (11.C.10) at standard temperature and pressure, and as applied to a 1-kmole sample. It then reads

$$(1 \text{ atm})(22.4 \text{ meters}^3) = (1 \text{ kmole})(R)(273 \text{ K})$$

$$(11.C.12)$$

from which we have

$$R = 0.082 \frac{\text{meter}^3 \text{ atm}}{\text{kmole K}} \qquad (11.C.13)$$

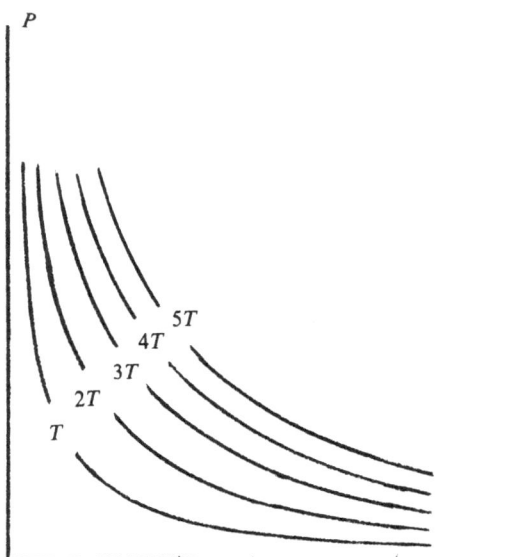

P

$5T$
$4T$
$3T$
$2T$
T

Figure 11.5. Ideal-gas isotherms for equally spaced temperatures.

In SI units ($1 \text{ atm} = 1.01 \times 10^5$ newtons/meter^2; also recall 1 newton meter = 1 joule) and, alternatively, in terms of kilocalories,

$$R = 8300 \frac{\text{joules}}{\text{kmole K}} = 2.00 \frac{\text{kcal}}{\text{kmole K}}$$

$$(11.C.14)$$

both values to about 1% accuracy.

The $P\mathscr{V}$ diagram of Fig. 11.4 can now be completed so as to exhibit the dependence on T as well. This is done in Fig. 11.5.

Example 11.5. Comparing Two Gas Samples. Bottles A and B, both containing expensive pure krypton gas, are being compared in price. Bottle A, in which the pressure is 2.0 atm and the temperature is 250 K, holds 0.50 meter^3. Bottle B, at 5.0 atm and 150 K, holds 0.25 meter^3. What is the ratio of the krypton masses, M_A and M_B, present in A and B?

For a given substance, the total mass is proportional to the number of kilomoles. Therefore we have

$$\frac{M_A}{M_B} = \frac{N_A}{N_B} = \frac{\left(\dfrac{P_A \mathscr{V}_A}{RT_A}\right)}{\left(\dfrac{P_B \mathscr{V}_B}{RT_B}\right)}$$

where the ideal-gas law (11.C.10) has been assumed. Hence we have

$$\frac{M_A}{M_B} = \frac{P_A \mathscr{V}_A T_B}{P_B \mathscr{V}_B T_A} = \frac{(2.0)(0.5)(150)}{(5.0)(0.25)(250)} = \underline{\underline{0.48}}$$

Note the convenience of canceling units in this type of calculation. Note that R also cancels out, as it always does when ratios between two gas samples are considered.

Example 11.6. Volume Expansion During Boiling. How many meters^3 of steam are made from each meter^3 of water, when boiled off at atmospheric pressure?

Let us pretend that steam is an ideal gas even in the vicinity of its condensation point; this is in fact a reasonably good approximation at 1 atm. Also we *consider* 1 *kilomole of substance*. As far as the volume $\mathscr{V}_{\mathrm{liq}}$ of 1 kmole of liquid water is concerned, we have

$$\mathscr{V}_{\mathrm{liq}} = \frac{\text{mass of 1 kmole}}{\text{density of water}}$$

$$= \frac{\left(\begin{array}{c}\text{molecular weight}\\ \text{of } H_2O\end{array}\right)(1 \text{ kg})}{1000 \text{ kg/meter}^3}$$

$$= \frac{18.0}{1000} \text{ meter}^3$$

Turning next to the volume $\mathscr{V}_{\mathrm{vap}}$ in the vapor phase, and using the gas law (11.C.10), we have

$$\mathscr{V}_{\mathrm{vap}} = \frac{RT}{P} = \frac{RT_{\text{boiling point}}}{1 \text{ atm}}$$

$$= \frac{(8300)(273 + 100)}{1.01 \times 10^5} \text{ meters}^3$$

$$= 30.6 \text{ meters}^3$$

The required ratio is

$$\frac{\mathscr{V}_{\mathrm{vap}}}{\mathscr{V}_{\mathrm{liq}}} = \frac{30.6}{(18.0/1000)} = \underline{\underline{1700}}$$

iii. Partial Pressures

The universality of the gas constant R in (11.C.10) implies a kind of **superposition principle** for the pressure due to several gases mixed in the same container. Consider a vessel of given volume $\mathscr{V}$, which we can alternatively fill with the following ideal gases, all at the same temperature T:

1. N_1 kmole of gas No. 1 (e.g., hydrogen);
2. N_2 kmole of gas No. 2 (e.g., helium);
3. The two above, added together.

According to Eq. (11.C.10), situations (1) and (2) yield pressures

$$P_1 = \frac{N_1 RT}{\mathscr{V}}, \qquad P_2 = \frac{N_2 RT}{\mathscr{V}} \qquad (11.\mathrm{C}.15)$$

respectively; in situation (3) we have $(N_1 + N_2)$ kmole of gas, with a consequent pressure

$$P = \frac{(N_1 + N_2) RT}{\mathscr{V}} \qquad (11.\mathrm{C}.16)$$

Thus,

$$P = P_1 + P_2 \qquad (11.\mathrm{C}.17)$$

More generally, in a container of given volume:

A mixture of ideal gases exerts a pressure that is the sum of the pressures each gas would exert if it were alone in that same container, at the same temperature.

$$(11.\mathrm{C}.18)$$

This is known as **Dalton's law** of partial pressures; it of course assumes gases that do not react chemically together.

Example 11.7. Removing Oxygen. A rigid vessel contains dry air (8 N_2 molecules to 3 O_2 molecules) at atmospheric pressure. A strip of magnesium is ignited inside and consumes all the oxygen to make solid magnesium oxide. After the remaining nitrogen has returned to its initial temperature, what is the pressure P_{final} in the container?

For given $\mathscr{V}$ and T, Eqs. (11.C.15), (11.C.16) imply a final pressure

$$P_{\mathrm{final}} = \frac{8}{8+3} P_{\mathrm{initial}} = \underline{\underline{\frac{8}{11} \text{ atm}}}$$

(Any increase in the solid's volume has been neglected.)

In a gas mixture, the *partial pressure* of one of the components is defined as the pressure it would exert if the other components were taken away. Thus, in dry air, the partial pressures of N_2 and O_2 are 8/11 atm and 3/11 atm, respectively.

D. Kinetic Theory

In the following, a molecular description of pressure and temperature will be developed. The model goes back to Daniel Bernoulli* (1700–1782), who anticipated by a long time the establishment of atomic theory. His idea consists of representing a gas by many pointlike particles bouncing elastically from the walls of a rigid box in which they are confined. [The extreme idealization involved in such a picture will be briefly pointed out in Note (i) at the end of this chapter.]

i. Pressure from Molecular Bombardment

Let us assume that the pressure of a gas on a wall of its container is due to the impact forces between molecules and wall. How does that pressure depend on the number of molecules and on their speed? The problem becomes manageable if we assume that *the multiple impacts are equivalent, on the average, to a steady force.*

We shall consider, at first, a model gas made of only one single molecule; the more realistic case of many molecules will follow simply.

A One-Molecule Gas

Figure 11.6 shows a single molecule of mass m and velocity v bouncing elastically between the frictionless walls of a cubic enclosure of side l. "Elastically" means that the molecule's kinetic energy,

$$\mathscr{K} = \tfrac{1}{2}mv^2 = \tfrac{1}{2}m(v_x^2 + v_y^2 + v_z^2)$$

is the same before and after impact; accordingly, the components v_x, v_y, v_z are affected as follows.

During a collision with a wall, say at A ($x=0$), lack of friction means zero tangential acceleration, so that v_y and v_z are unchanged. However, the normal component v_x is reversed, $v_x \rightarrow -v_x$. As a result, $\mathscr{K}$ is unchanged, as required by elasticity. We see that v_x is constant during all collisions except at the end walls $x=0$, $x=l$.

Effective Force

We consider one wall (shaded in Fig. 11.6), and ask for the effective force **F** it experiences, over a long time Δt, owing to the molecular impacts.

If $\Delta \mathbf{p}_{\text{trans}}$ denotes the momentum transferred

* Of fluid-dynamics fame; see the Note at the end of Chap. 9.

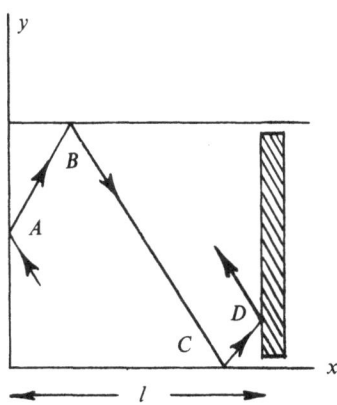

Figure 11.6. Trajectory of a single molecule in a box. Impacts at A and D reverse v_x; impacts at B and C have no effect on v_x. The time of travel between A and D is determined by v_x. (The z coordinate is not shown.) The right-hand wall is a massive piston that measures the effective pressure force through its change in momentum; its effect is that of just another fixed wall.

during Δt from molecule to wall, the answer is, from Newton's second law,

$$\mathbf{F} = \frac{\Delta \mathbf{p}_{\text{trans}}}{\Delta t} \qquad (11.\text{D}.1)$$

[The quantity $\Delta \mathbf{p}_{\text{trans}}$ is called an **impulse**. The wall can literally gain a momentum $\Delta \mathbf{p}_{\text{trans}}$, namely, if it is designed as a very massive piston; it gradually accelerates because of the many impacts, and thus functions as a force-measuring device. Yet, if sufficiently massive, it is seen by the gas as essentially a fixed wall. For the legitimacy of smoothing the impacts into a steady **F** as in (11.D.1), see Note (ii) at the end of this chapter.]

To arrive at a pressure, we now successively compute (1) the impulse to the wall from a single collision; (2) the impulse per unit time from a single molecule which keeps returning, and (3) the impulse per unit time (= effective force) from many molecules.

1. Impulse to the Wall

Let v_x (a positive number) be the x velocity just before impact with the right-hand wall. By conservation, the impulse to that wall equals the x momentum *lost* by the molecule,

$$(\Delta p_{\text{trans}})_x = mv_x - (-mv_x) = 2mv_x \qquad (11.\text{D}.2)$$

2. Impulse per Unit Time

To complete a round trip between two successive impacts with the same wall, a time

$$\Delta t = \frac{2l}{v_x} \qquad (11.D.3)$$

is needed. Hence the impulse per unit time, over as many collisions as we want, is

$$\frac{(\Delta p_{\text{trans}})_x}{\Delta t} = \frac{mv_x^2}{l} \qquad (11.D.4)$$

3. A Many-Molecule Gas

For many point molecules (labeled 1, 2,...), which never get in each other's way, the total impulse per unit time is just the sum

$$\frac{(\Delta p_{\text{trans}})_x}{\Delta t} = \frac{m_1 v_{1x}^2}{l} + \frac{m_2 v_{2x}^2}{l} + \cdots$$

which we write as

$$F_x = \frac{\sum m_i v_{ix}^2}{l} \qquad \begin{array}{l} \text{(sum over all molecules} \\ i = 1, 2, 3,...) \end{array}$$

$$(11.D.5)$$

F_x being the *normal force* on the right-hand wall. (Note that all molecules need not have the same mass.) Since the wall has area l^2, the pressure is $P = F_x/l^2$, or

$$P = \frac{\sum m_i v_{ix}^2}{l^3} = \frac{\sum m_i v_{ix}^2}{\mathscr{V}} \qquad (11.D.6)$$

where $\mathscr{V}$ is the volume of the box.

There is nothing special about the x direction, of course.* To get rid of the subscript x, we write

$$\sum m_i v_{ix}^2 = \sum m_i v_{iy}^2 = \sum m_i v_{iz}^2$$

$$= \tfrac{1}{3} \sum m_i (v_{ix}^2 + v_{iy}^2 + v_{iz}^2)$$

$$= \tfrac{1}{3} \sum m_i v_i^2 \qquad (11.D.7)$$

---

* The equivalence of all directions, as assumed here, is a property called **isotropy**.

($v_i =$ speed of the ith molecule). Equation (11.D.6) becomes

$$P = \frac{\sum m_i v_i^2}{3\mathscr{V}}$$

or, with some rearrangement of factors,

$$\boxed{P\mathscr{V} = \tfrac{2}{3} \sum \left(\tfrac{1}{2} m_i v_i^2\right)} \qquad (11.D.8)$$

Thus, $P\mathscr{V}$ equals 2/3 of the *total translational kinetic energy* of all the molecules. Equation (11.D.8) is as far as Bernoulli's model will take us. It foreshadows the ideal-gas law,

$$P\mathscr{V} = NRT \qquad (11.D.9)$$

ii. The Meaning of Temperature

Consider 1 kmole ($N = 1$) of an ideal gas. Comparison of Eqs. (11.D.8), (11.D.9) yields, for its total kinetic energy,

$$\sum \tfrac{1}{2} m_i v_i^2 = \tfrac{3}{2} RT \qquad (11.D.10)$$

This result is often expressed in terms of the mean kinetic energy *per molecule*. Dividing both sides of (11.D.10) by $\mathscr{N}_A$ (the number of molecules in the sample), we obtain

$$\boxed{\begin{pmatrix} \text{Mean translational} \\ \text{kinetic energy} \\ \text{per molecule} \end{pmatrix} = \tfrac{3}{2} kT} \qquad (11.D.11)$$

where the number

$$k = \frac{R}{\mathscr{N}_A} \qquad (11.D.12)$$

$$= \frac{8300}{6.02 \times 10^{26}} \frac{\text{joule}}{\text{K}}$$

$$= 1.4 \times 10^{-23} \frac{\text{joule}}{\text{K}} \qquad (11.D.13)$$

is called **Boltzmann's constant**. Equation (11.D.11) best summarizes the meaning of temperature in a gas. The word "translational" in (11.D.11) is intended to emphasize that any rotational or vibrational energy that the molecules may possess is not included in the $\frac{3}{2}kT$.

iii. The Root-Mean-Square Molecular Speed

How large is a typical molecular speed? To answer that question, we examine a pure gas, i.e., one where all molecules have the same mass m. With use of a bar to denote a mean, Eq. (11.D.11) reads

$$\tfrac{1}{2}m\overline{v^2} = \tfrac{3}{2}kT$$

so that we have

$$\overline{v^2} = 3\frac{k}{m}T \qquad (11.D.14)$$

essentially the required information. Each side of this result calls for a comment.

The left side, $\overline{v^2}$, known as the **mean square speed** (or mean square velocity), is defined, in terms of labeled molecules, by

$$\overline{v^2} = \frac{v_1^2 + v_2^2 + \cdots}{\mathcal{N}_A} \qquad (11.D.15)$$

assuming a 1-kmole sample. [Review how Eq. (11.D.11) was obtained.]

The right-hand side of (11.D.14) can be expressed in terms of bulk quantities; from Eq. (11.D.12) we have

$$\frac{k}{m} = \frac{R}{\mathcal{N}_A m} = \frac{R}{\text{mass of 1 kmole}} = \frac{R}{(\text{mol. wt.})(1\text{ kg})}$$

Hence the mean square speed, Eq. (11.D.14), is

$$\overline{v^2} = \frac{3RT}{(\text{mol. wt.})(1\text{ kg})} \qquad (11.D.16)$$

The square root of this quantity is called the **root-mean square speed**, or rms speed, or rms velocity,

$$\boxed{v_{\text{rms}} = \sqrt{\overline{v^2}} = \left[\frac{3RT}{(\text{mol. wt.})(1\text{ kg})}\right]^{1/2}} \qquad (11.D.17)$$

and is typical (but only typical) of the molecules' speed. It must be kept in mind that in a chaotic assembly such as a gas sample, essentially all speeds are likely to be represented. {A further caution: v_{rms} does not quite equal the mean speed $\bar{v}$. As a simple illustration, consider two numbers, say 1 and 7. Their mean value is $\frac{1}{2}(1+7) = 4$, while their rms value is $[\frac{1}{2}(1^2 + 7^2)]^{1/2} = 5$.}

It is worth noting that v_{rms}, Eq. (11.D.17), is not, in a technical sense, a microscopic quantity, since its evaluation involves macroscopic parameters only; the same can be said of $\overline{v^2}$.

Example 11.8. A Comparison with the Speed of Sound. Find the rms speed of a nitrogen molecule at room temperature (300 K).

Equation (11.D.17) gives

$$v_{\text{rms}} = \left[\frac{(3)(8300\text{ joule/K})(300\text{ K})}{28\text{ kg}}\right]^{1/2}$$

$$\approx 500\ \frac{\text{meters}}{\text{sec}}$$

At room temperature the speed of sound in air (which is largely nitrogen) is measured to be about 340 meters/sec. This appears reasonable; since a sound signal must be propagated by molecular collisions within the gas, that signal is not expected to travel faster than its messengers, the moving molecules themselves.

iv. The Equipartition of Energy

Consider two different gases, say pure hydrogen and pure helium, in separate containers but at the same temperature T. In each of these gases, the mean translational energy per molecule is $\frac{3}{2}kT$; see Eq. (11.D.11). This remarkable independence from the nature of the gas is due to the universality of k, which in turn follows from the universality of the gas constant R, an experimentally observed fact *amounting to Avogadro's law.*

More remarkably yet, this situation persists when the two gases are mixed. Each kind of molecule in the mixture has the same mean translational energy $\frac{3}{2}kT$. That same $\frac{3}{2}kT$ even characterizes *any single molecule* in the crowd; "mean" then stands for "averaged over time." For example, the particle of Fig. 11.2 may be viewed as a large molecule; its translational kinetic energy, although widely variable in time, is $\frac{3}{2}kT$ on the average, just as it is for each invisible gas molecule with which it collides. This is well verified experimentally, and provides a striking confirmation of the kinetic theory.

The universality of the value $\frac{3}{2}kT$ is a result known as the **equipartition of energy**. No general proof will be attempted here.

(The required formalism, known as *statistical mechanics*, is beyond the scope of this book. The reader should beware of "plausible proofs" of equipartition; these arguments often assume in disguised form the result that must be derived.)

Example 11.9. Atmospheric Helium. Our atmosphere contains about 5 atoms of helium to every 10^6 molecules of air. How much faster than a nitrogen molecule, N_2, does a He atom typically move?

Equipartition means that formula (11.D.17) for v_{rms} can be applied equally to both kinds of molecules in the mixture. For a given T, we have the proportionality

$$v_{rms} \propto \frac{1}{(\text{mol. wt.})^{1/2}} \qquad (11.D.18)$$

Therefore, we find the ratio

$$\frac{v_{rms,He}}{v_{rms,N_2}} = \left(\frac{\text{mol. wt. of } N_2}{\text{mol. wt. of He}} \right)^{1/2}$$

$$= \left[\frac{(2)(14)}{4} \right]^{1/2} = \underline{\underline{2.6}}$$

Thus, helium moves 2.6 times faster than nitrogen, irrespective of atmospheric temperature. The resulting speed at STP (see Example 11.8) is

$$v_{rms,He} \approx (2.6) \left(500 \, \frac{\text{meters}}{\text{sec}} \right) \approx 1300 \, \frac{\text{meters}}{\text{sec}}$$

(Some helium atoms, of course, go much faster.) In view of these speeds, permanent escape into outer space is significantly more probable for a helium atom than for a nitrogen molecule. For comparison, the universal speed of escape from Earth is about 11 000 meters/sec; see Eq. (8.D.6) of Chap. 8.

v. A Detailed Verification of the Kinetic Theory

An ingenious rotating-drum experiment, illustrated in Fig. 11.7a reveals how many gas molecules have any particular speed v. Since no two molecules have precisely the same speed, the practical question is: in a certain sample, how many molecules, $d\mathcal{N}$, have a speed between given values v and $v + dv$? The result obtained, called the **distribution of speeds**, is measured from the thickness of a deposition layer, part (b) of the figure, and agrees extremely well with the predictions of kinetic theory; in particular, the molecular root-mean-square speed is measured to be as given in Eq. (11.D.17).

vi. Kinetic Theory and Gravity

When considering molecules in a box we never gave a thought to gravity. Was this justified?

If gravity were important, the molecules would be slowed down appreciably while traveling from the bottom to the top of the box. This would imply, for a typical molecule, a potential energy difference $\Delta\mathcal{U}$ between these two levels that is comparable to the initial kinetic energy $\mathcal{K}$.

An estimate of the dimensionless ratio $\Delta\mathcal{U}/\mathcal{K}$ in a box of height l yields

$$\frac{\Delta\mathcal{U}}{\mathcal{K}} = \frac{mgl}{\frac{1}{2}m\overline{v^2}} = \frac{2gl}{\overline{v^2}} \qquad (11.D.19)$$

For definiteness, let us take a box of height 1 meter containing nitrogen at room temperature; see Example 11.8. We have then

$$\frac{\Delta\mathcal{U}}{\mathcal{K}} = \frac{(2)(10)(1)}{(500)^2} \approx 10^{-4}$$

so that gravity is indeed a minor influence over 1 meter. On the other hand, in practice we know that altitude differences of several hundred meters already result in appreciable gravitational effects, such as changes in pressure or density; see Sec. B iii of Chap. 9. At lower temperatures, $\overline{v^2}$ in Eq. (11.D.19) becomes smaller, and gravity should become more noticeable.

The lack of gravitational effects shows up in every quiet roomful of air, which *does not* separate into two layers—denser oxygen towards the floor,

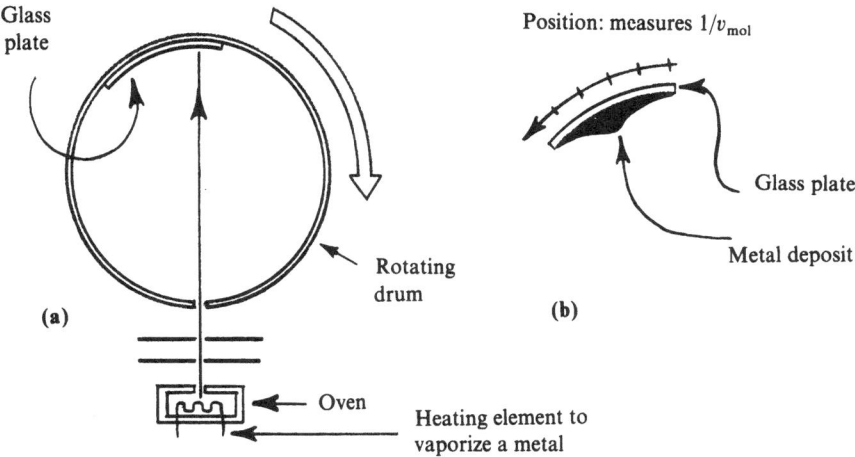

Figure 11.7. (a) The **rotating-drum experiment**. In this arrangement, whose principle was invented by O. Stern in 1920, an ideal monatomic gas (here a metal vapor) is allowed to escape from an oven through a well-aligned succession of small holes or slits, thus forming a narrow beam containing atoms of widely different speeds. A slit in a fast rotating drum admits the beam at regular intervals; the atoms strike a glass plate, to which they adhere. The fast atoms reach the leading section of the plate, while the slower ones deposit themselves further behind. (A high vacuum is needed to permit passage of the beam.) (b) The trailing distance along the glass depends *inversely* on the molecule's speed. The thickness of the metallic deposit is gauged by its absorption of light, as observed through the plate, and reveals the relative number of atoms having a given speed.

lighter nitrogen towards the ceiling. In the atmosphere as a whole, there is enough turbulence to keep O_2 and N_2 well mixed.

E. Internal Energy and Heat Capacity

One kind of molecular energy associated with temperature is of **translational** nature, amounting, as we have seen, to $\frac{3}{2}kT$ for the average molecule. This is the only kind of energy that appears in the ideal-gas formula (11.D.8) for the product $P\mathscr{V}$.

There exist other thermal energies, namely, **rotational** and **vibrational**, that do not affect that formula, and which arise from the molecules' spin and internal motion. These forms of energy do, however, contribute to the total energy budget of a gas sample.

We shall often consider the simplest gases, in which a molecule consists of just one atom (**monatomic** gases). Such molecules ordinarily show no evidence of rotation or vibration; their thermal energy is entirely translational. This is in contrast to **diatomic** or **polyatomic** molecules (consisting of two or more atoms), which exhibit those

additional forms of energy. It should be noted that *an ideal gas may belong to any of these types.*

(Monatomic gases are not rare. They include the noble gases, namely, helium, neon, etc., as well as many high-temperature gases such as metallic vapors. Other gases, such as diatomic hydrogen, will dissociate into isolated atoms at a sufficiently high temperature.)

To summarize, we have

$$\left(\begin{array}{c}\text{Mean thermal}\\ \text{energy per molecule}\end{array}\right)\left\{\begin{array}{l}= \frac{3}{2}kT \quad \text{(monatomic gases)}\\[4pt] > \frac{3}{2}kT \quad \text{(diatomic or polyatomic gases)}\end{array}\right.$$

(11.E.1)

i. Internal Energy

The internal energy U of a sample of matter is defined as the total thermal energy of all its molecules. This new concept must be carefully distinguished from that of heat, which is energy *being transferred* into or out of a system.

To find how U depends on temperature, we

simply recall the mathematical definition of a mean energy. For any bulk sample we have

Total thermal energy

= (number of molecules)

× (mean thermal energy per molecule)

$$(11.E.2)$$

For 1 kmole of a monatomic ideal gas, this reads

$$U = (N_A)(\tfrac{3}{2}kT)$$

or

$$\boxed{U = \tfrac{3}{2}RT} \qquad (11.E.3)$$

(1 kmole of monatomic gas)

For other kinds of ideal gas,

$$U > \tfrac{3}{2}RT \qquad (11.E.4)$$

(1 kmole of diatomic or polyatomic gas)

The irrelevance of pressure to these internal energies is remarkable and should be kept in mind:

> The internal energy of an ideal-gas sample depends only on its temperature.

$$(11.E.5)$$

ii. Heat Capacity at Constant Volume

If the internal energy U of a gas sample could be measured, its value would provide an important check of kinetic theory. However, there seems to be no direct way of doing so. Fortunately, we can observe something that is almost as revealing, namely, a *change* ΔU under a change ΔT in temperature.

Let us transfer some heat Q to 1 kmole of gas held in a rigid enclosure. Since zero work is done on the walls, and since energy is conserved, the whole amount Q must go into raising the gas' internal energy U,

$$Q = \Delta U \qquad \text{(constant volume)} \quad (11.E.6)$$

Dividing both sides by the consequent temperature rise ΔT, we obtain, by definition, the sample's **heat capacity at constant volume**,

$$C_\gamma = \frac{Q}{\Delta T} = \frac{\Delta U}{\Delta T} \qquad (11.E.7)$$

For a monatomic ideal gas, Eq. (11.E.3) predicts

$$C_\gamma = \frac{\Delta(\tfrac{3}{2}RT)}{\Delta T}$$

or

$$\boxed{\begin{array}{c} C_\gamma = \tfrac{3}{2}R \\[1mm] \approx 3.00 \, \dfrac{\text{kcal}}{\text{kmole K}} \end{array}} \qquad (11.E.8)$$

(monatomic ideal gas)

in excellent agreement with observation.

Example 11.10. Cooling a Sample of Argon. How much heat, Q_{out}, is transferred out of 5.0 kg of argon gas when it cools from 300°C to 100°C in a closed steel bottle?

A 50-kg mass of argon (atomic weight = 40) represents a number of kilomoles $N = 5.0/40$. From (11.E.6), (11.E.7), (11.E.8) we obtain

$$Q_{out} = N C_\gamma (-\Delta T) = \left(\frac{5.0}{40}\right)(3.0)(200) \text{ kcal}$$

$$= \underline{75 \text{ kcal}}$$

More generally, a gas may not be monatomic, and U may rise in a complicated way with T. In terms of an infinitesimal dT, Eq. (11.E.7) becomes

$$\boxed{C_\gamma = \frac{dU}{dT}} \qquad (11.E.9)$$

which is the general expression for a heat capacity at constant volume. Figure 11.8 compares C_γ as a function of T for any monatomic gas and for hydrogen (H_2). Some heat absorbed by the hydrogen serves to increase its molecules' rotational and vibrational energy. On the other hand, for any given rise dT, the translational energy

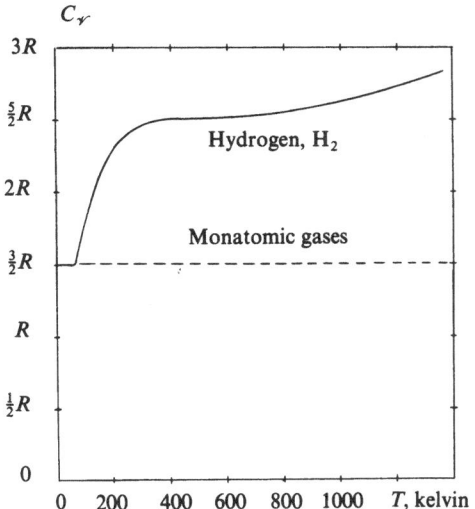

Figure 11.8. The measured kilomolar heat capacity $C_{\mathscr{V}}$ for H_2 gas (solid curve). For monatomic gases one observes the dashed curve, in agreement with theory. Note the curious fact that H_2 behaves monatomically at low temperature. Quantum theory is needed to explain that peculiarity. The $\frac{5}{2}R$ portion of the graph is taken as *evidence for the diatomic formula* H_2.

increases by the gas-independent amount $\frac{3}{2}R\,dT$. Therefore the total heat transfer is

$$dQ > \tfrac{3}{2}R\,dT$$

so that

$$C_{\mathscr{V}} > \tfrac{3}{2}R \qquad \text{(constant volume)} \qquad \text{(11.E.10)}$$

a general result *for 1 kmole of a diatomic or polyatomic gas.*

A rule of thumb, approximately valid for many gases at room temperature, gives their kilomolar heat capacity at constant volume as

$$C_{\mathscr{V}} \approx \begin{cases} \tfrac{5}{2}R & \text{(diatomic gases)} \\ 3R & \text{(polyatomic gases)} \end{cases} \qquad \text{(11.E.11)}$$

(This rule—which often breaks down—turns out to represent the equipartition of energy for rotational motion. Space is lacking to describe the argument in detail; let it only be said that its frequent failure has remained unexplained until the advent of quantum mechanics.)

Notes

i. Is the Kinetic Model Too Simple?

The assumption of point particles between static walls is so idealized that its success is almost undeserved. The following three complications have been ignored.

1. Even in an ideal gas, many collisions do occur between the molecules. In air at STP, for example, a molecule seldom travels much more than the minuscule distance of 10^{-7} meter, i.e., a few hundred times the molecular size, before its path is deflected in that manner; in this connection, see Problem 11.46.
2. When examined in detail, real walls are far from static. They are, themselves, made of vibrating molecules, which transfer kinetic energy to or from the gas molecules during collisions. Some gas molecules may even adhere to the wall for some time before being reemitted.
3. Translational and rotational energies are often converted into one another during collisions, at least for diatomic and polyatomic molecules.

Modern kinetic theory has risen to the challenge of these objections, and is able to incorporate the above features into its assumptions. What is more, such apparent complications are actually needed to explain how thermal equilibrium is reached within the gas, and between the gas and its container. The results of the analysis are then fortunately still consistent with what we have learned in this chapter.

ii. Averaging the Impacts

How legitimate is it to "smooth" the impacts from a particle bombardment, as we did in Eq. (11.D.1)? The answer is that many small impulses, over a long time, nearly amount to a smoothly behaved force. This can be understood if one imagines the bombarded wall to be massive, and mounted so as to accelerate under the collisions, as in Fig. 11.9. Then Fig. 11.10 shows that, as time progresses, the wall's momentum (and hence its velocity) is well approximated by what it would be under a continuous jet.

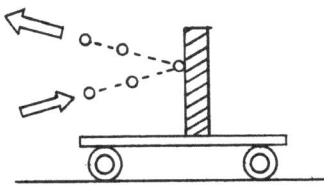

Figure 11.9. A mobile wall subjected to many repeated impacts due to projectiles.

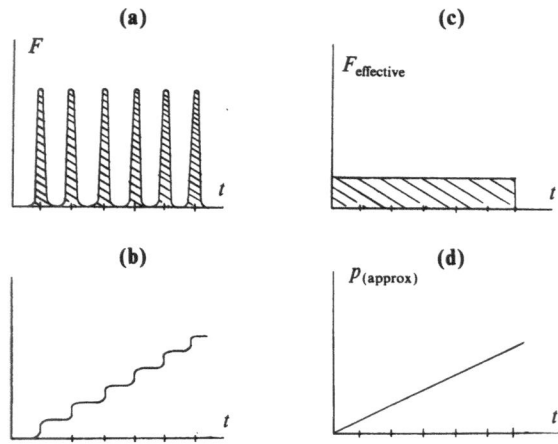

(a) F ... t

(b) ... t

(c) $F_{\text{effective}}$ ... t

(d) $p_{(\text{approx})}$ ... t

Figure 11.10. (a) Actual impact forces. (b) The target's momentum (initially zero). (c) Constant effective force. (d) The effective force results in a wall momentum approximating that in (b).

Condensed Checklist

Isotope notation:

$$\left.\begin{array}{l} ^A(\text{chemical symbol}) \\[4pt] A = \text{mass number} \quad (\text{an integer}) \\[4pt] (\text{Mass of 1 neutral atom}) = (\text{atomic mass}) \\[4pt] \approx (1\text{ u})(A) \quad (\text{not exact}) \end{array}\right\}$$

Below (11.B.2)

Atomic mass unit:

$$1\text{ u} = \tfrac{1}{12}(\text{atomic mass of } ^{12}\text{C}) \quad (\text{exact})$$

(11.B.1)

$$1\text{ u} \approx 1.66 \times 10^{-27}\text{ kg} \qquad (11.B.2)$$

$$\text{Atomic weight} = \frac{\text{atomic mass}}{1\text{ u}} \qquad (11.B.3)$$

Table of atomic weights: Table 11.1

Avogadro's number (SI):

$$\mathcal{N}_A = \frac{1}{1\text{ u}} \approx 6.02 \times 10^{26} \quad (11.B.6), (11.B.7)$$

Kilomole:

$$\boxed{1\text{ kmole (of molecules)} = \mathcal{N}_A \text{ molecules}} \quad (11.B.8)$$

$$(\text{mass of 1 kmole}) = (\text{mol. wt.}) \times (1\text{ kg}) \quad (11.B.9)$$

Composition of air:

$$3O_2 : 8N_2 \qquad \text{Examples 11.2, 11.7}$$

Molecular weight of air ≈ 29

Standard temperature and pressure:

"STP" means 0°C, 1 atm

Volume of 1 kmole ≈ 22.4 meters^3 (11.B.12)

(any ideal gas at STP)

Ideal-gas law:

$$\boxed{P\mathcal{V} = NRT} \qquad (N = \text{number of kmoles}) \qquad (11.C.10)$$

$$R \approx 8300 \frac{\text{joules}}{\text{kmole K}} \approx 2.00 \frac{\text{kcal}}{\text{kmole K}} \quad (11.C.14)$$

Mean translational energy:

$$\tfrac{1}{2}\overline{mv^2} = \tfrac{3}{2}kT \qquad (11.D.11)$$

$$k = \frac{R}{\mathcal{N}_A} \qquad (11.D.12)$$

$$= 1.4 \times 10^{-23} \frac{\text{joule}}{\text{K}} \qquad (\text{Boltzmann's constant})$$

(11.D.13)

Root-mean-square speed:

$$v_{\text{rms}} = \left[\frac{3RT}{(\text{mol. wt.})(1\text{ kg})}\right]^{1/2} \qquad (11.D.17)$$

Internal energy of 1 kmole:

$$\boxed{U \begin{cases} = \tfrac{3}{2}RT & (\text{monatomic ideal gas}) \\ > \tfrac{3}{2}RT & (\text{other ideal gases}) \end{cases}} \begin{array}{l} (11.E.3) \\ (11.E.4) \end{array}$$

Heat capacity of 1 kmole at constant volume:

$$\boxed{C_{\mathcal{V}} = \tfrac{3}{2}R \quad (\text{monatomic ideal gas})} \qquad (11.E.8)$$

$$C_{\mathcal{V}} > \tfrac{3}{2}R, \text{ in general nonconstant} \quad (11.E.11)$$

(other ideal gases)

True or False

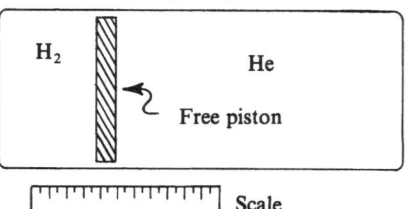

Closed glass cylinder

H$_2$ He   Free piston

Scale

1. Through the use of the ideal-gas law, the overall temperature of the illustrated device can be inferred from the piston's location if the total mass of each gas is known.

2. A given sample of ideal gas cannot have its pressure, temperature, and volume independently adjusted.

3. A given equilibrium sample, consisting of liquid, solid, and gaseous water, has an adjustable volume.

4. A given sample of ideal gas, at given temperature, has an adjustable internal energy.

5. When an initial sample of diatomic hydrogen H_2 is heated sufficiently to dissociate into H atoms, its volume will necessarily be doubled.

6. The mass of a single O_2 molecule can be calculated just from a knowledge of its molecular weight and of the value of the gas constant R.

7. In a crystal of known chemical composition, the volume per molecule can be calculated from the system's density and molecular weight. (Avogadro's number is assumed known.)

8. Atomic weight = (atomic mass)(g).

9. Two ideal gases can have the same temperature, yet different average speeds for their molecules.

10. In an equilibrium mixture of ^{4}He and ^{14}N, the mean kinetic energy of a He atom is 4/14 of that of a N atom.

11. In an ideal gas, the magnitude of a molecule's momentum typically increases as the square root of the gas' absolute temperature.

12. At STP, the internal energy is greater in 1 liter of oxygen (O_2) than in 1 liter of neon (Ne).

Problems

Section B : Atomic Masses and the Kilomole

11.1. Find, in atomic mass units and in kilograms, the mass of a water molecule, H_2O, and of a glucose molecule, $C_6H_{12}O_6$. (Use the chemical atomic weights.)

*11.2. Natural magnesium is 79 % ^{24}Mg (at. wt. = 24.0), 10 % ^{25}Mg (at. wt. = 25.0), and 15 % ^{26}Mg (at. wt. = 26.0). Calculate the chemical atomic weight of magnesium.

*11.3. If each atom of lead, Pb, contains 82 electrons, how many electrons are there in 1 gram of Pb?

*11.4. If hydrogen contains no neutrons, while each oxygen atom contains eight neutrons, how many neutrons are there in 1 metric ton (= 1000 kg) of water? (You do not need to know anything about neutrons.)

11.5. How many molecules are there in 1 milligram of methyl alcohol (CH_4O)?

11.6. If each carbon (C) atom contains six protons, while each sulfur (S) atom contains 16 protons, how many protons are there in 1.00 liter of carbon disulfide (CS_2, specific gravity = 1.26)?

*11.7. How many hydrogen molecules, H_2, are there per meter^3 in pure hydrogen at STP?

*11.8. Calculate the density, at STP, of (a) hydrogen, H_2; (b) helium, He; (c) carbon dioxide, CO_2.

*11.9. Natural copper is a mixture of ^{63}Cu (at. wt. = 62.93) and ^{65}Cu (at. wt. = 64.93). Given that the chemical atomic weight of copper is 63.54, what are the percent abundances of ^{63}Cu and ^{65}Cu?

11.10. Estimate the total number of air molecules in the Earth's atmosphere. (See Example 11.4.)

11.11. If 1 kmole of hydrogen (H_2) is mixed with 1 kmole of nitrogen (N_2), what is (a) the volume, (b) the average molecular weight, (c) the mass density of the mixture at STP?

*11.12. The Sun's surface composition is about 75 % hydrogen and 25 % helium by mass. What is the average atomic weight of that mixture?

Section C : The Ideal-Gas Law

11.13. Calculate the density of (monatomic) iron vapor at 1 atm and 2000°C. [(Atomic weight of Fe) = 56.]

11.14. What pressure is needed to hold 1.0 kmole of neon gas in a 1.0-meter^3 container at 0°C?

*11.15. A 3-gal volume of carbon dioxide gas (CO_2) at 200°C and 2 atm is compressed into 1 gal; the temperature rises but is then brought back to 200°C. What is the final pressure?

*11.16. A 1.0-gram sample of chlorine gas is kept at a constant volume while its temperature is lowered from 500°C to 100°C. If the initial pressure is 0.3 atm, what is the final pressure?

11.17. (A tire gauge). A 10-mm^3 air sample, initially inside a tire, is transferred to a pressure gauge, where it expands to a total volume of 30 mm^3 at atmospheric pressure. Find the gauge pressure in the tire, in atm. (Assume a constant temperature.)

11.18. In one cylinder of a certain gasoline engine, a gauge pressure of 18 atm and a temperature of 825°C are reached when the gas volume is 60 cm^3. How many kilomoles of gas mixture are present?

*11.19. A 1.00-liter cylinder contains 1.50 grams of methane gas (CH_4) at 150°C. What is the pressure, in pascals?

11.20. A 1.00-kg sample of O_2 gas is enclosed in a 1.00-meter^3 vessel, at a pressure of 1.00 atm. What is its temperature in °C?

11.21. A metal tube of volume 75 cm^3, originally open to the atmosphere, is sealed off while in an oven at 1000°C. How many kilomoles of air are left inside?

*11.22. Suppose that the gas reaction $2N_2O_3 \rightarrow 2N_2 + 3O_3$ proceeds in a closed rigid vessel; the mixture of products is allowed to come back to the temperature of the initial gas. By what factor has the pressure been changed?

*11.23. A 1.0-meter^3 volume of hydrogen gas (H_2) at STP is burned to yield 2.0 meters^3 of water vapor (H_2O) at atmospheric pressure, by the reaction $2H_2 + O_2 \rightarrow 2H_2O$. What is the kelvin temperature of the water vapor?

*11.24. At 27°C and 0.50 atm, a 1-meter^3 volume of a certain ideal gas has a mass of 0.77 kg. (a) How many kilomoles are involved? (b) What is the gas' molecular weight?

*11.25. Dry air at STP has a density of 1.29 kg/meter^3. Find its density at (a) 1.00 atm and 100°C; (b) 1.50 atm and 0°C; (c) 1.50 atm and 100°C.

11.26. Calculate the density of hydrogen sulfide gas, H_2S, at 100°C and 0.5 atm (Atomic weights: 1 for H, 32 for S.)

*11.27. Prove that the molecular weight of a gas sample of mass m, volume $\mathscr{V}$, pressure P, and temperature T is given by

$$\text{Molecular weight} = \frac{mRT}{(1 \text{ kg}) P\mathscr{V}}$$

*11.28. A 3-meter^3 volume of unknown gas, at 1 atm and 150°C, has a mass of 6.6 kg. What is its molecular weight? [See Problem 11.27.]

11.29. A certain hydrogen sample (H_2) has double the volume, double the pressure, and double the absolute temperature of a certain oxygen sample (O_2). What is the ratio between the total masses of the two samples?

*11.30. A 20-meter^3 volume of ethyl alcohol vapor, C_6H_6O, at 0.5 atm and 200°C, is compressed into 4 meters^3 while its temperature is raised to 300°C. What is its new pressure?

11.31. In a certain pump, a 0.015-liter volume of air, originally at 1 atm and 25°C, is compressed into 1/3 that volume, while its temperature rises to 45°C. What is the new pressure?

*11.32. At what pressure, in atmospheres and in pascals, does air at 1000 K have a concentration of 1 molecule per cm^3 (similar to interstellar conditions)?

11.33. The **compressibility** of a substance is defined as $-(1/\mathscr{V})(d\mathscr{V}/dP)$, i.e., as the fractional decrease in volume per unit pressure rise. Prove that, at constant temperature, the compressibility of an ideal gas equals $1/P$.

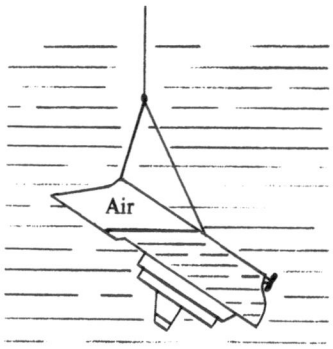

*11.34. A sunken ship, lying 300 meters under water, has an effective weight of 150 metric tons × g. An air bubble of volume 50 meters^3 is blown into it, and the ship is then hoisted up by cable; see the figure. Above what depth will it continue to rise without help from the cable? [Assume that no air escapes, and that the air temperature stays constant; neglect the atmospheric pressure, which will cause an error of only about 10 meters; see Eq. (9.B.9) of Chap. 9; 1 metric ton = 1000 kg.]

11.35. (The **isothermal atmosphere**.) Assume that temperature does not vary with altitude (an oversimplification). (a) Given the atmospheric pressure P_0 and density ρ_0 at sea level, determine ρ in terms of P at any altitude. (b) Inserting (a) into $dP = -\rho g \, dz$ [Eq. (9.B.13) of Chap. 9], prove that $P = P_0 e^{-z/\zeta}$, where $\zeta = P_0/\rho_0 g$ [Eqs. (9.B.14), (9.B.19) of Chap. 9.]

11.36. In a gaseous mixture of hydrogen and carbon dioxide at STP, hydrogen has a partial pressure of 0.3 atm. (a) What is the partial pressure of the carbon dioxide? (b) If the gases were originally unmixed at STP, what was the ratio of their volumes?

*11.37. In a certain room at 20°C, the partial pressure of water vapor, H_2O, is 0.03 atm. How many H_2O molecules are there per cm^3?

*11.38. A test tube at 150°C contains, at 1 atm, 50% methyl alcohol vapor, CH_4O, and 50% water vapor, H_2O, by mass. What are these substances' partial pressures, in atmospheres?

11.39. Two rigid enclosures, of volumes 1 liter and 3 liters, originally contain helium at 0.5 atm and hydrogen at 0.2 atm, respectively. What is the total pressure after the two enclosures have been connected by the opening of a stopcock between them? (Assume a single temperature for the whole process.)

11.40. At any given temperature below 100°C, the partial pressure of water vapor in the atmosphere cannot stably exceed a well-defined **saturation value** P_{sat}, above which condensation would occur. At 10°C, we have $P_{sat} = 9$ torr (1 atm = 760 torr). In terms of *mass*, what percentage of water vapor does saturated air contain at 1 atm and 10°C? The ideal gas law may be assumed even for the water vapor. (Note: In general, the air is not saturated; the **relative humidity** is defined as P/P_{sat}, where P is the actually prevailing partial pressure of water vapor.)

Section D: Kinetic Theory

11.41. Consider the molecule of Fig. 11.6. Assuming that its effective force F_x on the right wall depends only m, on v_x, and on l, derive $F_x \propto mv_x^2/l$ by dimensional analysis.

11.42. Find, for each of the following sets of numbers, the mean value and the rms value. (a) 2, 2, 5; (b) 35, 35, 35; (c) 1, 2, 3, 4.

*11.43. A certain hydrogen sample contains mostly H_2 molecules, but also some HD and some D_2 molecules. (Here H is taken to mean the isotope ^{1}H; D is ^{2}H; their atomic weights are 1.0 and 2.0, respectively.) At STP, what is the rms speed of the three kinds of molecule?

11.44. At its center, the Sun is a mixture of hydrogen nuclei ($m_H \approx 1$ u), helium nuclei ($m_{He} \approx 4$ u), and free electrons ($m_e \approx 1/2000$ u). If the temperature is 1.5×10^7 K, estimate the rms speed of each of these three types of particle.

*11.45. Supposing that a small helium balloon, of total mass 5 grams, and just supported by its buoyancy, could be kept at thermal equilibrium in a room of perfectly still air, estimate the rms speed of that balloon's Brownian motion. (Consider the balloon as a big molecule.)

11.46. (The **mean free path** in a real gas). The average distance $\bar{l}$ traveled by any one molecule between two consecutive collisions with other molecules is called its **mean free path**. Assuming that all molecules are identical spheres of diameter D, theory gives $\bar{l} = (\sqrt{2}\,\pi n D^2)^{-1}$, where n is the molecular concentration ($=$ number of molecules per unit volume of gas). (a) Explain in a few words why $\bar{l}$ should decrease as either n or D^2 increases. (b) Verify the formula's dimension. (c) Find $\bar{l}$ for nitrogen (N_2) at STP, treating the molecule as spherical (which it is not) and assuming $D \approx 2.9 \times 10^{-10}$ meter. (d) At what pressure, in atmospheres, does N_2 (at 0°C) have a mean free path of 1 cm?

*11.47. In a certain plasma (highly ionized gas), used for controlled-fusion experiments, a temperature of 0.5×10^7 K was achieved. If the plasma consisted of 2×10^{19} hydrogen ions per meter^3 (mass of 1 ion $\approx$ 1.7×10^{-27} kg) and an equal density of electrons (mass of 1 electron $\approx 1 \times 10^{-30}$ kg), estimate the pressure in pascals.

11.48. The temperature of a certain gaseous nebula is believed to be about 10 000 K. The gas is ionized and contains free electrons (mass $\approx 1 \times 10^{-30}$ kg). If equipartition is applicable, what is the rms speed of the electrons?

*11.49. Nuclear technology requires the (chemically impossible) separation of the isotopes ^{238}U and ^{235}U. One physical method utilizes the difference in the rms molecular speeds v_{rms} in gaseous uranium hexafluoride (UF$_6$). Most molecules of that gas contain ^{238}U, some contain ^{235}U, and all fluorine atoms are ^{19}F. At 200°C, what is the fractional difference, $(\Delta v_{rms})/v_{rms}$, between the values of v_{rms} for "light" and "heavy" UF$_6$ molecules? (Take the mass numbers 238, 235, 19 to be the approximate atomic weights.)

Section E: Internal Energy and Heat Capacity

11.50. What is, in kcal/kg K, the specific heat of hydrogen gas H_2 at constant volume and room temperature? Assume (11.E.11).

*11.51. A certain ideal monatomic gas has a specific heat $C_V = 56$ joules/kg K. What is its molecular weight?

*11.52. A steel bottle contains 1.5 kmole of helium gas at 20°C and 18 atm. How much heat, in kilocalories, must be given off by the helium if its temperature is to go down to -50°C?

*11.53. What is, in kcal/kg K, the specific heat of dry air at constant volume and room temperature? (Average molecular weight of air = 29; use $C_V = \frac{5}{2}R$, air being diatomic.)

11.54. How many kilocalories of heat are extracted in cooling each of the following gases from 100°C to 0°C: (a) 2 grams of oxygen, O_2; (b) 0.05 kmole of sulfur dioxide, SO_2; (c) 1 liter of methane, CH_4, at STP? [Assume (11.E.11).]

*11.55. How many kilocalories are needed to raise the temperature of 1 kg of hydrogen from 400°C to 600°C in a constant-volume vessel, starting at a pressure of 0.5 atm? (Refer to Fig. 11.8.)

11.56. A dirigible, filled with 500 meters^3 of helium at STP, is cruising at a speed of 25 km/h. Find the ratio of the helium's kinetic energy (calculated as if it were a single large object) to its internal energy. (Note. The gas' total kinetic energy is the sum of these two contributions, see Sec. D i of Chap. 7.)

*11.57. How many kilocalories are needed to heat

1 kmole of hydrogen at constant volume from 100 K to 300 K? (Refer to Fig. 11.8.)

*11.58. Helium at 300 K and neon at 500 K, both at the same pressure, are mixed and allowed to reach thermal equilibrium without heat loss to the outside. If the amounts are 2.0 kmole of helium to 3.0 kmole of neon, what is the final temperature? (He and Ne are monatomic and nonreactive gases.)

Answers to True or False

1. False (as T rises, both pressures rise in the same proportion, while the piston remains in the same place).
2. True (they are constrained by $P\mathscr{V} = NRT$).
3. True (transferring heat into the sample will change the relative amount of each phase).
4. False.
5. False (affected by pressure).
6. False (R and the molecular weight are macroscopic data).
7. True.
8. False.
9. True (the same kinetic energy can imply different speeds).
10. False (it is the same).
11. True [$mv \propto (\frac{1}{2}mv^2)^{1/2}$].
12. True (O_2 has rotational energy, Ne not).

Answers or Hints to Problems

11.2. 24.3.

11.3. 2.4×10^{23}.

11.4. 2.7×10^{29}.

11.7. 2.7×10^{25}.

11.8. See Table 9.1 of Chap. 9.

11.9. 69.5% ^{63}Cu, 30.5% ^{65}Cu.

11.12. First show that, by number of atoms, He:H = 1:12.

11.15. 6 atm.

11.16. 0.14 atm.

11.19. 3.3×10^5. [First determine how many kmole are involved: $N = $ (mass of sample)/(mol. wt.); use SI units for R.]

11.22. 5/2.

11.23. Hint: If the temperature did not change, we would have $\mathscr{V}_{H_2O} = \mathscr{V}_{H_2}$. (Why?) Answer: 546 K.

11.24. (b) 38.

11.25. (c) 1.42 kg/meter^3.

11.27. Hint: Molecular weight $= m/N$. (Why?)

11.28. 76.

11.30. Use $P_1\mathscr{V}_1/T_1 = P_2\mathscr{V}_2/T_2$. The chemical formula is not needed.

11.32. First calculate the volume occupied by $\mathscr{N}_A$ molecules.

11.34. 100 meters.

11.37. 7.5×10^{17}.

11.38. The temperature is not needed.

11.43. 1.5×10^3 meters/sec for HD.

11.45. $\sim 10^{-9}$ meter/sec. The mean translational kinetic energy of the whole balloon is $\frac{3}{2}kT$, $T \approx 300$ K.

11.47. The masses are irrelevant.

11.49. 0.5%. [Treat $\varDelta v_{rms} \to dv_{rms}$ as infinitesimal; use d(mol. wt.) $= 238 - 235 = 3$.]

11.51. 222.

11.52. The pressure is not needed.

11.53. 0.17 kcal/kg K.

11.55. 500 kcal; P does not enter the calculation.

11.57. Use $Q = \int_{T_1}^{T_2} C_{\mathscr{V}} \, dT$; i.e., find the area under the curve.

11.58. Consider the total internal energy.

CHAPTER 12

WORK FROM HEAT

The expenditure of work to produce heat, as in Joule's experiment (Chap. 10) is a commonplace achievement. It is also a familiar nuisance in many cases involving friction.

This chapter is more concerned with the reverse process, *changing heat into work*. Since 1769, when James Watt first patented his improvements on the steam engine, the problem of getting work from heat has been a major preoccupation in our industrial society, and still is to this day.

Heat engines are important quite independently of their practical aspects, however; their study has resulted in a great extension of our scientific understanding, as will be explained in these pages.

The main idea can be condensed as follows. *There exist two absolute limitations on the amount of work obtainable from a heat source*:

1. Total energy must be conserved, as we already know.
2. Even the best imaginable engine can convert only a portion of the available thermal energy into work.

How large a portion depends on the *temperature differences* that exist in the engine's environment.

This second limitation, which is new to us, is in some ways even more drastic than the first, as it prevents us from utilizing more than a few drops of the immense thermal energy that surrounds us. The two principles (1), (2) above are known as the **first and second laws of thermodynamics**.

The original statement of the first law (in its most powerful form, involving the interconversion between any kinds of energy) is attributed to Helmholtz (1847); his insight rested heavily on the work of Joule and several others.

Curiously enough, the second law, which is far subtler than the first, is of earlier origin. Its preliminary form, due to Carnot (1824), concerns the *efficiency of heat engines*, a question that we also address in this chapter. In the hands of Clausius and Kelvin, that practical rule became established as a basic law of Nature.

A. Heat and the Conservation of Energy (First Law of Thermodynamics)

Energy conservation is of great use in predicting the thermal behavior of matter. Here we shall apply that law to the simplest of all substances, the ideal gases. As a result, we shall obtain new predictions (going beyond those of the preceding chapter) involving the state variables P, V, T as well as the gases' heat capacity.

i. Work from Expansion

To obtain mechanical energy from heat, as in automobile engines, steam engines, etc., one utilizes a so-called **working substance**, usually a gas, which expands, displaces a piston, and thereby performs some work. A natural question, and clearly a practical one, is: "How much work does a given expansion yield?" Here we learn how to find the answer.

Figure 12.1 illustrates the simple case of a cylinder inside which the pressure P is uniform in space and constant in time. The piston has area $\mathscr{A}$, and is perpendicular to the cylinder's axis; it rises over a distance s. (Why it rises is not important to the

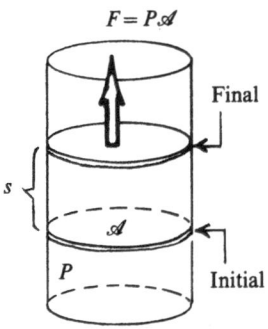

Figure 12.1. A fluid at pressure P is below the piston, which sweeps out a volume $\mathscr{A}\,\varDelta\mathscr{V}$.

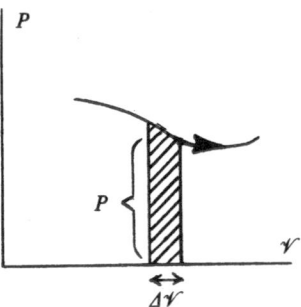

Figure 12.3. The shaded area equals the work done by a substance as its volume increases by $\varDelta\mathscr{V}$.

present discussion: the cause might be thermal expansion of the gas, or admission of more gas into the cylinder.) The work output of the gas is then

$$\mathscr{W} = Fs = P\mathscr{A}s$$

where $F = P\mathscr{A}$ is the force exerted by the gas on the piston. We recognize $\mathscr{A}s = \varDelta\mathscr{V}$ as the increase in the cylinder's volume, so that we have

$$\boxed{\mathscr{W} = P\,\varDelta\mathscr{V}} \qquad (P = \text{const}), \qquad (12.A.1)$$

the basic relation for a working substance. If $\varDelta\mathscr{V}$ is negative (a compression), then so is $\mathscr{W}$; there is a positive work, $-\mathscr{W}$, done *by* the piston *on* the gas. (*In this chapter, the symbol $\mathscr{W}$ is generally used to denote an output.*)

The simple cylindrical geometry of Fig. 12.1 is not needed for the validity of (12.A.1); expansion into an arbitrary shape results in the same formula, as is shown by the device of Fig. 12.2.

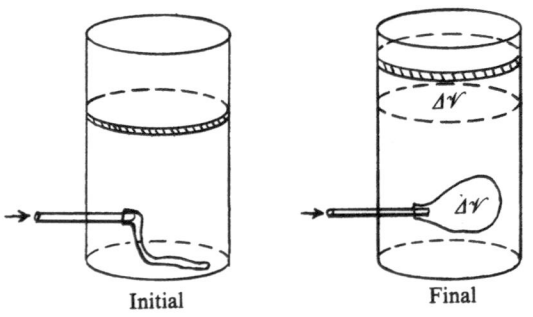

Initial   Final

Figure 12.2. The fluid entering the bag at pressure P does a work $P\,\varDelta\mathscr{V}$ on the bag even if the latter has an irregular shape. Proof: Assume an essentially weightless, incompressible fluid around the bag. The energy of that fluid does not change, hence all the work is transmitted to the piston, which sweeps out a volume $\varDelta\mathscr{V}$ equal to the gain $\varDelta\mathscr{V}$ in the bag's volume. Thus, we are back to Fig. 12.1.

More generally, suppose P varies during the expansion. Even so, P may be considered fixed during an infinitesimal expansion $\mathscr{V} \to \mathscr{V} + d\mathscr{V}$. In this case (12.A.1) must be written

$$\boxed{d\mathscr{W} = P\,d\mathscr{V}} \qquad (12.A.2)$$

a relation often used in what follows.

A graphical representation of Eq. (12.A.1) on a $P\mathscr{V}$ diagram is shown in Fig. 12.3 for the case when $\varDelta\mathscr{V}$ is very small. More generally, suppose we allow an arbitrary expansion, starting with volume $\mathscr{V}_i$ and ending with $\mathscr{V}_f$; this process may be considered as a succession of many small increases

$$\mathscr{V}_i \to \mathscr{V}_i + \varDelta\mathscr{V} \to \mathscr{V}_i + 2\varDelta\mathscr{V} \to \mathscr{V}_i + 3\varDelta\mathscr{V} \to \cdots$$

with corresponding values of the pressure during those increases,

$$P_1 \to P_2 \to P_3 \to \cdots$$

The total work done is (see Fig. 12.4),

$$\mathscr{W} = P_1\,\varDelta\mathscr{V} + P_2\,\varDelta\mathscr{V} + P_3\,\varDelta\mathscr{V} + \cdots \qquad (12.A.3)$$

The figure shows that

$$\boxed{\mathscr{W} = \text{area under the graph of } P \text{ against } \mathscr{V}}$$

$$(12.A.4)$$

To be precise, one should write Eq. (12.A.3) as an integral:

$$\mathscr{W} = \lim_{\varDelta\mathscr{V} \to 0} (P_1 + P_2 + \cdots)\,\varDelta\mathscr{V} \qquad (12.A.5)$$

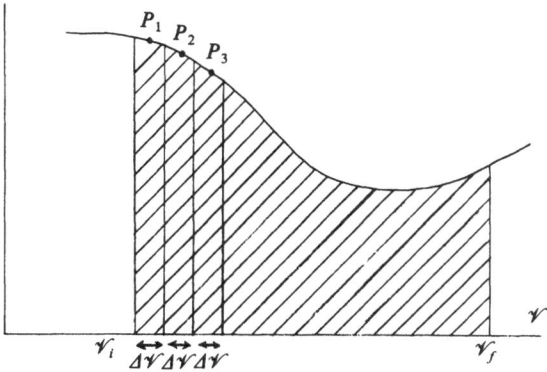

Figure 12.4. The shaded area equals the work done during the expansion $\mathcal{V}_i \to \mathcal{V}_f$.

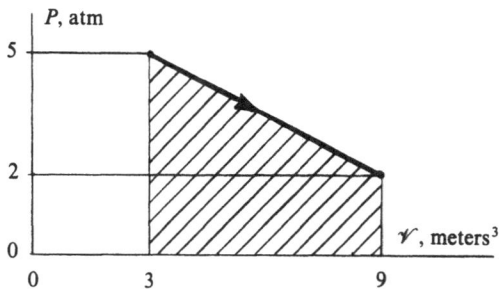

Figure 12.5. Example of an expansion. The shaded area equals the work done by the expanding gas.

or

$$\mathcal{W} = \int_{\mathcal{V}_i}^{\mathcal{V}_f} P\, d\mathcal{V} \qquad (12.\text{A}.6)$$

where $\mathcal{V}_i$ and $\mathcal{V}_f$ are the initial and final volumes.

Does this formula permit us to calculate $\mathcal{W}$? Not unless P is known as a function of $\mathcal{V}$; in other words, a curve must be specified on the $P\mathcal{V}$ diagram.

> **Example 12.1. Reading $\mathcal{W}$ from a Graph.** The $P\mathcal{V}$ diagram of Fig. 12.5 shows the expansion of a certain amount of gas against a piston. How much work is done?
>
> ---
>
> The answer amounts to the trapezoidal area on the diagram:
>
> $$\mathcal{W} = (9 - 3)\left(\frac{5+2}{2}\right)(\text{atm})(\text{meters}^3)$$
>
> $$= 21 \text{ atm meters}^3$$
>
> $$= (21)(1.0 \times 10^5 \text{ Pa})(\text{meters}^3)$$
>
> $$= \underline{\underline{2.1 \times 10^6 \text{ joules}}}$$

ii. The First Law of Thermodynamics

In drawing up the energy budget of a material sample, three kinds of contributions must be considered. They are (1) a possible **heat transfer** Q into the material; (2) a possible **work output**, $\mathcal{W}$, arising from the material's expansion; and (3) a stored supply of thermal energy, U, the **internal energy** of the material itself. (The reader should understand how to interpret a negative sign for Q or $\mathcal{W}$.)

Conservation of total energy can now be written down:

$$\boxed{Q = \Delta U + \mathcal{W}} \qquad (12.\text{A}.7)$$

This, the **first law of thermodynamics**, says that the fate of a heat input Q is, in general, to be distributed between a work output $\mathcal{W}$ and an increase in the stored internal energy U. (The symbol Q is consistently used for an *input*; the early industrial orientation of thermodynamics is at the origin of our "utilitarian" convention for the signs of $\mathcal{W}$ and Q.)

It is often useful to consider an infinitesimal heat input dQ, in which case Eq. (12.A.7) reads

$$dQ = dU + d\mathcal{W} \qquad (12.\text{A}.8)$$

or, more explicitly, see Eq. (12.A.2),

$$\boxed{dQ = dU + P\, d\mathcal{V}} \qquad (12.\text{A}.9)$$

(Again: We have $dQ > 0$ if heat is going *into* the sample, $d\mathcal{W} > 0$ if work is done *on* the environment.) This important formula will be applied in the next subsections to study the behavior of ideal gases under a variety of processes.

iii. Isothermal Expansion

The first law of thermodynamics is particularly simple for an ideal gas at constant temperature T. We recall that its internal energy U depends only on T, so that here we have

$$\Delta U = 0 \qquad (T = \text{const})$$

and Eq. (12.A.7) becomes

$$\boxed{Q = \mathscr{W}} \qquad (T = \text{const}) \qquad (12.A.10)$$

Thus, *heat is always absorbed in an isothermal expansion* (and released in an isothermal compression). Furthermore, we see that the work output accounts for 100% of the heat input. (Unfortunately, this promising feature will not persist in the operation of a heat engine, which must involve compression as well as expansion.)

Calculating the Work Output in an Isothermal Process

Let us expand N kilomoles of an ideal gas isothermally from an initial volume $\mathscr{V}_i$ to a final volume $\mathscr{V}_f$. How much work $\mathscr{W}$ does the gas then do on its container, say by displacing a piston? We can apply formula (12.A.6),

$$\mathscr{W} = \int_{\mathscr{V}_i}^{\mathscr{V}_f} P \, d\mathscr{V} = \int_{\mathscr{V}_i}^{\mathscr{V}_f} \frac{NRT}{\mathscr{V}} \, d\mathscr{V}$$

with the help of the ideal-gas law $P\mathscr{V} = NRT$. Since NRT is a constant, the integration yields

$$\boxed{\mathscr{W} = NRT(\ln \mathscr{V}_f - \ln \mathscr{V}_i) = NRT \ln \frac{\mathscr{V}_f}{\mathscr{V}_i}}$$

$$(T = \text{const}) \qquad (12.A.11)$$

Therefore also, for the *heat input*,

$$\boxed{Q = NRT \ln \frac{\mathscr{V}_f}{\mathscr{V}_i}} \qquad (T = \text{const}) \qquad (12.A.12)$$

in view of (12.A.10).

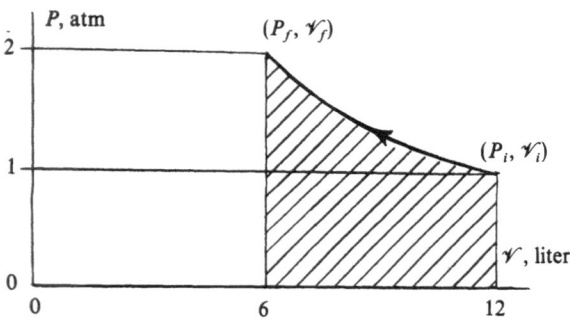

Figure 12.6. An isothermal compression.

Example 12.2. Heat from a 2:1 Compression.

Find the heat output from the isothermal compression of hydrogen, starting with 12.0 liters at STP and ending with 6.0 liters.

Figure 12.6 illustrates the process. From Eq. (12.A.12),

$$Q = NRT \ln \frac{\mathscr{V}_f}{\mathscr{V}_i} = P_i \mathscr{V}_i \ln \frac{\mathscr{V}_f}{\mathscr{V}_i}$$

$$= (1.00 \text{ atm})(12.0 \text{ liters}) \ln \frac{6.0}{12.0}$$

$$= [(1.00)(1.01 \times 10^5 \text{ Pa})]$$

$$\times (12.0 \times 10^{-3} \text{ meter}^3) \ln \frac{6.0}{12.0}$$

$$= -840 \text{ joules}$$

(The logarithm is negative.) Thus, the heat output is $-Q = +840$ joules. This also equals the work input, and hence represents the shaded area in Fig. 12.6.

In this example, we note that any ideal gas, not just hydrogen, would give rise to the same result.

iv. Two Kinds of Heat Capacity

When a gas is heated at constant volume, no work is done, and the heat input goes only into raising the internal energy. On the other hand, when heated *at constant pressure* (as in the gas thermometer of Chap. 10), the sample expands and does some work. This requires more heat for the same increase in temperature. We conclude that *the heat capacity of a gas sample is larger when measured at constant pressure than at constant volume*; in what follows we calculate how much larger.

Let us consider 1 kmole of ideal gas; its heat capacity is always defined as

$$C = \frac{dQ}{dT} \qquad (12.A.13)$$

Constant Volume

With $d\mathscr{V} = 0$ we obtain

$$dQ = dU \qquad (12.A.14)$$

see Eq. (12.A.9). Equation (12.A.13) then yields the heat capacity at constant volume,

$$C_v = \frac{dU}{dT} \qquad (12.A.15)$$

a quantity already studied in Sec. E of Chap. 11. We recall that

$$C_v \begin{cases} = \frac{3}{2}R & \text{(monatomic gas)} \\ > \frac{3}{2}R & \text{(other gases)} \end{cases} \qquad (12.A.16)$$

Constant Pressure

Here we have $dP = 0$, and therefore Eq. (12.A.9) may be written

$$dQ = dU + d(P\mathcal{V}) = dU + R\,dT$$

with help from the ideal-gas law $P\mathcal{V} = RT$. Equation (12.A.13) now gives the heat capacity at constant pressure:

$$C_P = \frac{dU}{dT} + R \qquad (12.A.17)$$

Comparison between (12.A.15) and (12.A.17) yields

$$C_P - C_v = R \qquad (12.A.18)$$

Thus, the difference between heat capacities is the same for all ideal gases, no matter whether monatomic, diatomic, etc.; this relation is in good agreement with observation. In particular, for a monatomic ideal gas [see Eq. (12.A.16)], there follows

$$C_P = \frac{3}{2}R + R = \frac{5}{2}R \qquad (12.A.19)$$

For other ideal gases, we have $C_P > \frac{5}{2}R$.

Example 12.3. A Specific Heat. Calculate, from theory, the specific heat (or heat capacity per unit mass) c_P at constant pressure for helium.

For 1 kmole we have

$$
\begin{aligned}
c_P &= \frac{C_P}{\text{mass of 1 kmole}} \\
&= \frac{(5/2)\,R}{4.00 \text{ kg}} = \frac{(5/2)(2.00)}{4.00} \frac{\text{kcal}}{\text{kg K}} \\
&= 1.25 \frac{\text{kcal}}{\text{kg K}}
\end{aligned}
$$

v. Adiabatic Expansion

We have seen two kinds of expansion that require a heat input; this is the case at constant temperature and at constant pressure. Consider next a different kind of experiment, where *heat transfers are blocked*,

$$Q = 0 \qquad (12.A.20)$$

during the expansion (or compression) of an ideal-gas sample. Such a process is called **adiabatic**. In practice, this condition may result from container walls that are themselves adiabatic; alternatively, a rapid change in volume may allow insufficient time for an appreciable heat transfer. (In Chap. 10 an adiabatic partition was defined as one across which changes in *temperature* are not transmitted; but changing a temperature by thermal contact means a heat flow. Therefore an adiabatic partition is also one across which a *heat transfer* is impossible.) We now examine the effect of $Q = 0$ on the variables $P, \mathcal{V}, T$ during the expansion of say, 1 kmole of ideal gas; the relation

$$P\mathcal{V} = RT \qquad (12.A.21)$$

is already known to hold.

Temperature–Volume Relation

Adiabatic expansion is a highly effective method for *cooling a gas*. (This is why the air is cold at high altitudes; see Problem 12.26.) The explanation is that some energy must supply the work of expansion. None comes from outside ($Q = 0$); hence the internal energy has to be depleted, and as U decreases, so does T. We now calculate the effect.

Consider a volume expansion $\mathcal{V} \to \mathcal{V} + d\mathcal{V}$, with a corresponding temperature change $T \to$

$T + dT$. Adiabatically, the first law (12.A.9) may be written

$$P \, d\mathscr{V} = -dU \qquad (12.A.22)$$

We note that

$$dU = C_v \, dT \qquad (12.A.23)$$

see Eq. (12.A.15). This relation, as the presence of C_v reminds us, was derived from constant-volume heating. It is, however, valid under all circumstances, including adiabatic, because U depends only on T. Thus, (12.A.22) reads

$$P \, d\mathscr{V} = -C_v \, dT \qquad (12.A.24)$$

To eliminate P, divide by corresponding sides of (12.A.21):

$$\frac{d\mathscr{V}}{\mathscr{V}} = -\frac{C_v}{R} \frac{dT}{T} \qquad (12.A.25)$$

If C_v is temperature independent, this yields, by integration,

$$\ln \mathscr{V} = -\frac{C_v}{R} \ln T + \text{const}$$

or, taking the exponential on both sides and combining factors,

$$\boxed{\mathscr{V} T^{C_v/R} = \text{const}} \qquad (12.A.26)$$

We see that T falls as $\mathscr{V}$ rises, and vice versa. In the case of a monatomic gas, (12.A.26) reads

$$\mathscr{V} T^{3/2} = \text{const} \quad (\text{monatomic}) \qquad (12.A.27)$$

Pressure–Volume Relation

To eliminate T, we raise both sides of (12.A.26) to the power R/C_v, and then multiply by corresponding sides of $P\mathscr{V}/T = \text{const}$. The result is

$$\boxed{P\mathscr{V}^{\gamma} = \text{const}} \qquad (12.A.28)$$

where the exponent is given by

$$\gamma = 1 + \frac{R}{C_v} \qquad (12.A.29)$$

(γ is the lower-case Greek gamma), or

$$\boxed{\gamma = \frac{C_P}{C_v}} \qquad (12.A.30)$$

[Caution: this ratio, or equivalently C_v, must be temperature independent for our new "adiabatic laws" (12.A.26), (12.A.28) to be valid; that assumption is what allowed us to integrate (12.A.25) as we did. As an illustration, Fig. 11.8 of Chap. 11 shows an "inapplicable range" for H_2 gas, extending approximately from 60 K to 250 K. For *infinitesimal expansions* we do not have to worry about that question; the reader may differentiate (12.A.21) to obtain $dP/P + d\mathscr{V}/\mathscr{V} = dT/T$, and combine this result with (12.A.25) to obtain

$$\frac{dP}{P} = -\gamma \frac{d\mathscr{V}}{\mathscr{V}} \qquad (12.A.31)$$

no matter whether γ is constant or not.]

Adiabatic versus Isothermal Curves

From Eq. (12.A.29) we note that $\gamma > 1$ in all cases; for example, in a monatomic gas we have

$$\gamma = \frac{\frac{5}{2}R}{\frac{3}{2}R} = \frac{5}{3} \quad (\text{monatomic}) \quad (12.A.32)$$

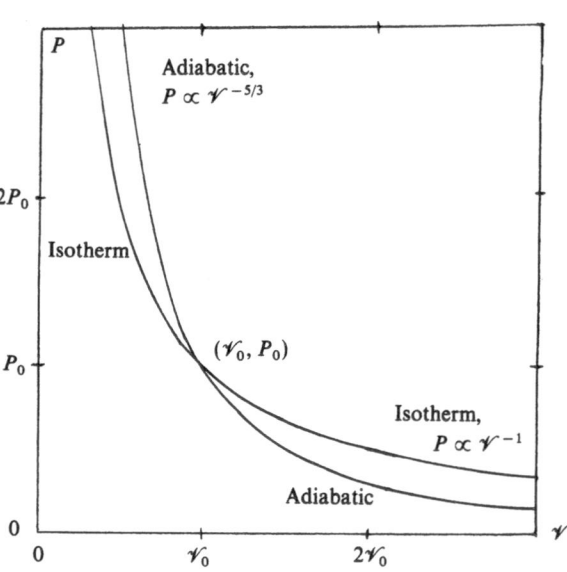

Figure 12.7. An isotherm and an adiabatic, drawn to scale for an ideal monatomic gas ($\gamma = 5/3$, the case in which they differ most). The same state $(\mathscr{V}_0, P_0)$ of the same sample is modified isothermally in one case, adiabatically in the other.

see Eq. (12.A.19); for hydrogen above 250 K, Fig. 11.8 of Chap. 11 shows that

$$\gamma \approx \frac{(\frac{5}{2}+1)\,R}{\frac{5}{2}R} = \frac{7}{5}$$

as is the case for many other diatomic gases near room temperature.

When plotted as a $P\mathscr{V}$ diagram, the curve for $P\mathscr{V}^{\gamma} = \text{const}$ is called an **adiabatic**. In the vicinity of any given P and $\mathscr{V}$, *the adiabatic is steeper than the isotherm* for the same gas, see Fig. 12.7. This is because, under compression, the temperature rises in the adiabatic case, whereas of course it does not in the isothermal case. Therefore, in the adiabatic case, the pressure rise in $P = RT/\mathscr{V}$ gets a contribution from T as well as from $1/\mathscr{V}$.

Example 12.4. A 1:2 Expansion. Some neon gas is expanded adiabatically to twice its original volume. If it starts at $\mathscr{V}_i = 1.00$ liter, $P_i = 1.00$ atm, what is its final pressure, P_f?

Neon is monatomic, so that Eq. (12.A.28) gives

$$P\mathscr{V}^{5/3} = \text{const}$$

or

$$P_f \mathscr{V}_f^{5/3} = P_i \mathscr{V}_i^{5/3}$$

Solving for P_f,

$$P_f = \left(\frac{\mathscr{V}_i}{\mathscr{V}_f}\right)^{5/3} P_i$$

$$= \left(\frac{1}{2}\right)^{5/3} (1 \text{ atm}) = \underline{0.315 \text{ atm}}$$

(If the given expansion were isothermal, the answer would be 0.500 atm.)

vi. Free Expansion

It would be comfortable to believe that every gas process can be represented by a $P\mathscr{V}$ curve. Such is not the case, however. Some processes have a starting point and an end point on a diagram, but *no continuous line joining them*. Consider an ideal gas originally confined in a volume $\mathscr{V}_i$ at a pressure P_i

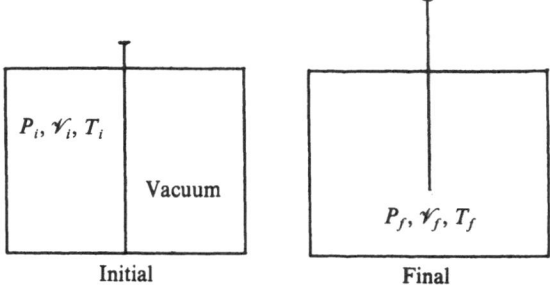

Figure 12.8. Free expansion of a gas into a vacuum. If the partition slides out, no work is done.

and temperature T_i; see Fig. 12.8. A so-called **free expansion** occurs when a baffle is slid out or punctured, and the gas spreads into the (previously empty) remainder of the vessel, the total volume now being $\mathscr{V}_f$. No heat is transferred to or from the outside. When thermal and pressure equilibrium is finally reached, what are the temperature T_f and pressure P_f? The key observation is that the gas has done *zero net work*, since no piston has been moved. Therefore Eq. (12.A.7), $Q = \Delta U + \mathscr{W}$, reads

$$0 = \Delta U + 0 \qquad (12.A.33)$$

We see that U has remained unchanged, and hence also T. Thus,

$$T_f = T_i \qquad (12.A.34)$$

Boyle's law can now be applied:

$$P_f = \frac{\mathscr{V}_i}{\mathscr{V}_f} P_i \qquad (12.A.35)$$

In this case the $P\mathscr{V}$ diagram just consists of the two points shown in Fig. 12.9. During expansion the

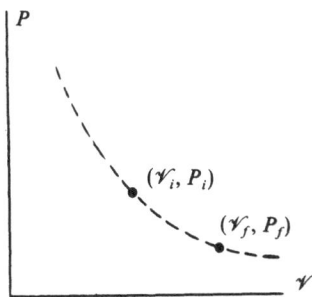

Figure 12.9. In the free expansion of an ideal gas, the initial and final states are on the same isotherm. However, the process cannot be described by a $P\mathscr{V}$ curve between these two points.

pressure is very nonuniform in space; a $P\mathscr{V}$ diagram is therefore inadequate to describe these turbulent intermediate states.

vii. Summary

There are, of course, infinitely many ways of tampering with the variables P, $\mathscr{V}$, T of a sample. Each of these ways, involving its characteristic heat transfer Q, is called a **process**. We have selected some processes for discussion, namely, those that are either particularly simple, or of particular interest for our upcoming study of heat engines. Those cases may be listed as follows:

1. Isothermal ($T = \text{const}$);
2. Constant-pressure or constant-volume (for a comparison of heat capacities);
3. Adiabatic ($Q = 0$); and
4. Free-expansion.

The relevant formulas are collected in the end-of-chapter checklist.

B. Heat Engines and the Second Law of Thermodynamics

A device for transforming *heat into work* is called a **heat engine**. In the following we shall study an idealized class of such engines. When operated in reverse, heat engines become **refrigerators**. The relation between these two modes of operation will provide us with an extraordinary limitation on the efficiency of any heat engine whatsoever. This, in turn, will give us a more fundamental understanding of the absolute temperature scale.

The key to such sweeping statements will be a reasonable-looking postulate which, roughly, says that "heat flows spontaneously only from higher temperature to lower temperature." This principle, when carefully worded, is known as the **second law of thermodynamics**.

i. Cyclic Processes

In order to visualize a heat engine, consider a cylinder, equipped with a piston, and containing a given working substance. After a series of compressions and expansions, combined with heat transfers (positive and negative), the substance has done a net work $\mathscr{W}_{tot}$ on the piston, and has absorbed a total heat Q_{tot} from its environment.

The energy budget of such an operation is greatly simplified if the working substance has returned exactly to its original state, as illustrated in Fig. 12.10; one then says that the engine has gone through a **cycle**. Since the internal energy U returns to its earlier value, one has

$$\varDelta U = 0 \qquad (12.\text{B}.1)$$

and therefore the first law of thermodynamics, Eq. (12.A.7), reads

$$Q_{tot} = \mathscr{W}_{tot} \qquad \text{(cyclic process)} \quad (12.\text{B}.2)$$

Thus, over a complete cycle, *the internal energy of a cyclic engine plays no part in its work output.*

Real-life heat engines are more or less well approximated as cyclic. The classic case, which Carnot and his contemporaries had in mind, was the steam engine, where the working substance (water) is indeed put many times through the same vaporization and condensation process. On the other hand, the internal-combustion engine, which powers the automobile, is not truly cyclic. The exhaust gas is not fed back to the air intake; there would be no cycle even if it were, since the exhaust's composition would be different from that of the intake. (The fuel burns inside the cylinder.) Nevertheless, the automobile engine is often approximated in theory as being cyclic.

ii. A Simple Heat Engine

Although quite primitive and not very practical, the following design, shown in Fig. 12.11, illustrates

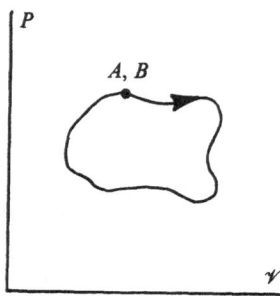

Figure 12.10. A cyclic process, where the initial and final states *A, B* are identical. It is left to the reader to show that the area inside the curve equals the total work output $\mathscr{W}_{tot}$ over the cycle; see Fig. 12.4.

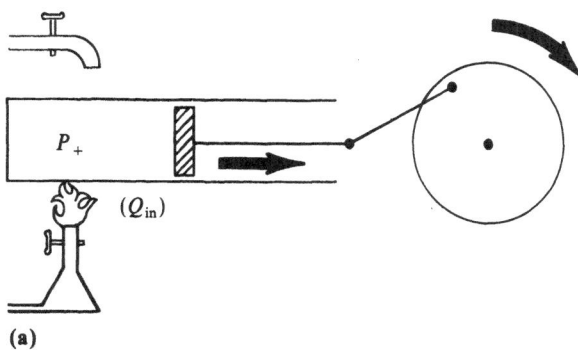

(a)

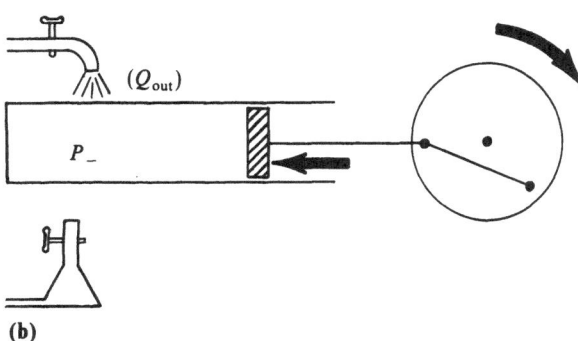

(b)

Figure 12.11. A simple heat engine. (a) Expansion stroke. (b) Compression stroke.

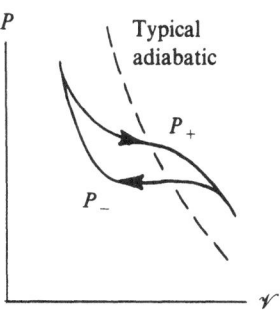

Figure 12.12. $P\mathscr{V}$ diagram for the engine of Fig. 12.11.

Application of the first law gives

$$\boxed{\mathscr{W}_{\text{tot}} = Q_{\text{in}} - Q_{\text{out}}} \qquad (12.\text{B}.3)$$

iii. The Performance of a Heat Engine

Source Temperatures

Suppose Q_{in} comes from thermal contact with a so-called heat reservoir ("hot source") at a constant temperature T_+, the flame in Fig. 12.11; Q_{out} is transferred into another heat reservoir ("cold source") at a lower constant temperature T_-, the running water in the figure. One says that the engine *operates between the temperatures T_+ and T_-*.

Flow Diagram

Figure 12.13 is of a type often used to visualize a heat engine's performance. It exhibits in an intuitive way T_+, T_-, Q_{in}, Q_{out}, $\mathscr{W}_{\text{tot}}$, and the fact that $\mathscr{W}_{\text{tot}} = Q_{\text{in}} - Q_{\text{out}}$.

the basic features of a heat engine. The working substance in the cylinder alternately expands at a high pressure P_+ while absorbing a large amount of heat Q_{in} from a flame [part (a) of the figure], and is compressed at a low pressure P_- while releasing a small amount of heat Q_{out} into some cold water [part (b) of the figure]. Let us neglect the atmospheric pressure at first.

During expansion, P_+ does a large amount of work on a mechanical connecting rod; during compression, the connecting rod does a small amount of work against P_-. At the end of the cycle, the whole engine is by definition back in its initial state, but a net positive amount of work $\mathscr{W}_{\text{tot}}$ has been obtained, and a net amount of heat $Q_{\text{tot}} = Q_{\text{in}} - Q_{\text{out}}$ has been extracted from the environment. The $P\mathscr{V}$ cycle for this engine might look like the one in Fig. 12.12.

Although the atmospheric pressure P_0 outside the cylinder affects the detailed operation of the engine, it does not change the overall heat–work balance: since P_0 is constant, the work it does during compression cancels the work done on it during expansion. Thus $\mathscr{W}_{\text{tot}}$ is unaffected.

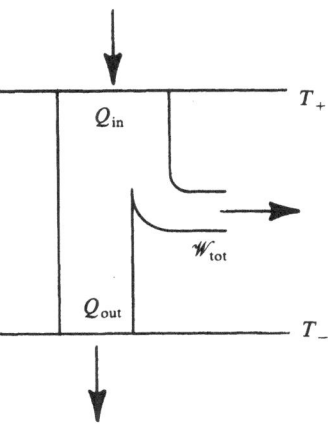

Figure 12.13. Flow diagram for a heat engine. The widths of the fictitious "pipes" are drawn proportional to the quantities Q_{in}, Q_{out}, $\mathscr{W}_{\text{tot}}$ so as to exhibit conservation.

Efficiency

The total conversion of heat into work is the amateur engineer's dream; it would mean $Q_{in} = \mathscr{W}_{tot}$, $Q_{out} = 0$. As we shall see, however, this can never be realized. A measure of closeness to that goal is provided by the **efficiency** (more accurately, the **thermodynamic efficiency**) of the engine, defined as

$$\varepsilon = \frac{\mathscr{W}_{tot}}{Q_{in}} \qquad (12.B.4)$$

which must be a number between zero and 100%. That number does not, however, quite reflect the engine's **practical efficiency**, since part of $\mathscr{W}_{tot}$ may be expended against friction, an effect neglected here. As we shall see, an engine with $\varepsilon = 100\%$ is a basic impossibility, even without friction.

At this point a digression is called for, away from the topic of engines and into something far more general and important. This will then lead us back to the question of an engine's efficiency.

iv. The Second Law of Thermodynamics

Conservation of energy should not prevent heat from flowing spontaneously out of a cold object (a heat output Q_{out}) into a hot one (a heat input Q'_{in}). On the basis of that conservation law, the only requirement is $Q_{out} = Q'_{in}$; *and yet such a phenomenon has never been observed*. If a red hot poker is dipped in cold water, we do not expect the poker to become white hot and the water to freeze. Such common observations have been promoted by Clausius (1850) to a universal law:

> There exists no process whose net result is a heat transfer from a cold to a hot object, without any other energy transfer.

$$(12.B.5)$$

This is one of the many equivalent ways of stating the **second law of thermodynamics**.

Do we have here an entirely new principle, unrelated to the laws of mechanics? The answer is no. The science of **statistical mechanics** utilizes the fact that heat is related to random molecular energy in order to *derive* statement (12.B.5) purely from Newtonian mechanics and probability theory. It is then found that the "nonexistent" processes in question are in fact, *overwhelmingly improbable* although strictly speaking possible. We shall continue to treat them as nonexistent.

v. Perfect Efficiency Is Unattainable

Applying the second law of thermodynamics can involve more formal logic than is usual in the physical sciences; that logic often exploits a **thought experiment**, not actually performed. To illustrate: The second law, (12.B.5), implies that

> Heat transferred out of a system can never be converted entirely into mechanical energy.

$$(12.B.6)$$

This we shall show by denying (12.B.6) and thereby eventually contradicting the second law (12.B.5).

The argument is shown schematically in Fig. 12.14. An initial situation, (a), involves a cold and a hot object, at temperatures T_- and T_+.

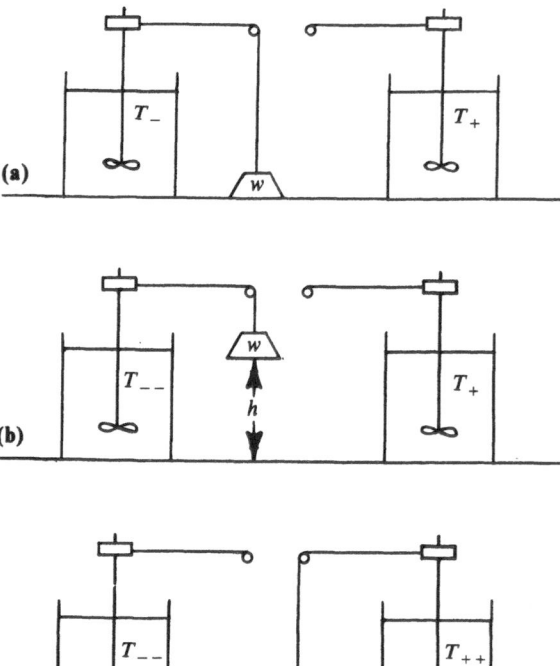

Figure 12.14. A sequence of events that would violate the second law, (12.B.5).

There is also a suspended weight w for storing potential energy. Suppose that, in violation of (12.B.6), the cold object lowers its temperature to T_{--} by losing a heat Q_{out}, which is entirely converted into a potential energy $wh = Q_{out}$, see part (b) of the figure. Next, the weight is shifted to another pulley and lowered, thereby transferring to the hot object a heat $Q'_{in} = wh$: this can always be done, as in Joule's experiment. The result, shown in part (c) of the figure, contradicts (12.B.5), since, compared to (a), the only change is the "uphill" transfer of heat. Statement (12.B.6) again says that *there are no 100% efficient engines*. A much more informative version of this result is derived later in this section.

From (12.B.6) it follows, for example, that Joule's experiment, Sec. B ii of Chap. 10, cannot be reversed: one cannot lift a weight by *cooling* water with a stirrer. More generally, kinetic friction cannot be reversed: one cannot propel an object simply by cooling a rough surface on which it lies.

vi. Reversible Heat Engines

In the light of the preceding discussion one might wonder if any work-to-heat conversion can be reversed at all. The answer is that such a reversibility is never exact, but that it can be approximated arbitrarily closely, as follows.

Consider a working substance that undergoes a cyclic process; at every instant of time, the values of P, $\mathscr{V}$, and T are assumed to be well defined; the $P\mathscr{V}$ diagram might, for example, be as shown in Fig. 12.12. By suitably controlling the volume of the sample, as well as the incoming and outgoing heat transfers, one can always retrace the cycle backwards, i.e., counterclockwise on the $P\mathscr{V}$ diagram. As a result, all the energy transfers suffer a change in sign; in the engine of Fig. 12.11 the piston now feeds a net positive work $-\mathscr{W}'_{tot}$ *into* the working substance; during compression, a positive heat Q'_{out} is transferred *into* the hot source; during expansion, a heat Q'_{in} is *removed* from the cold source. The heat engine now transfers heat from cold to hot, and has become a **refrigerator**. This does not manifestly violate the second law (12.B.5), because the net result of a cycle also involves a consumption of mechanical energy, $-\mathscr{W}'_{tot}$.

All the preceding observations can be condensed in the flow diagrams of Fig. 12.15, where we also see that

$$-\mathscr{W}'_{tot} = \mathscr{W}_{tot}$$

$$Q'_{in} = Q_{out} \qquad (12.B.7)$$

$$Q'_{out} = Q_{in}$$

These relations summarize the reversed operation of an engine as far as its energy budget is concerned. There remains to discuss how the source temperatures are related to that of the working substance.

Matching the Temperatures

Let us run our engine between two sources with given temperatures T_+, T_-. An exact reversal requires that the temperature T of the working substance retrace its values in reverse order. Consider the part of the cycle where a heat Q_{in} is transferred from the hot source into the engine. For any heat to flow at all, T must be *somewhat lower* than T_+.

Next consider the corresponding part of the reversed cycle ("refrigerator mode"), where heat flows out of the engine into the hot source. This now requires that T be *somewhat higher* than T_+.

We conclude that an exact reversal cannot be achieved. It can, however, be approximated if we make T very close to T_+ during the "hot" part of the cycle; as a consequence, T will not have to be modified appreciably when the cycle is reversed. In

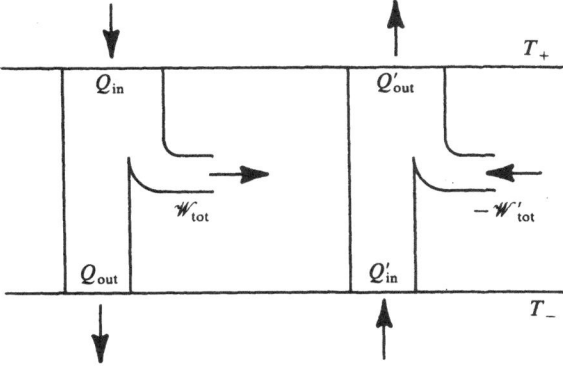

Figure 12.15. Exactly reversed operation of a heat engine. In this diagram, all quantities Q_{in}, Q_{out}, $\mathscr{W}_{tot}$, Q'_{in}, Q'_{out}, $-\mathscr{W}'_{tot}$ are positive.

a similar way, T must closely match T_- during the "cold" part of the cycle. Thus, ideally,

$$\left.\begin{array}{l}\text{During a heat transfer,}\\ \text{the working substance}\\ \text{and the source must be}\\ \text{at the same temperature.}\end{array}\right\} \text{(reversible engine)}$$

(12.B.8)

(Note: A reversible or nearly reversible engine would be of very little practical use. It turns out that the rate of heat transfer approaches zero as the relevant temperature difference approaches zero. Hence a reversible engine works "infinitely slowly." Why, then, should the concept be of such interest to us? The answer is that it will allow us to derive a genuine limitation for real engines.)

Our discussion has strict implications for an engine operated reversibly between two fixed-temperature sources at T_+ and T_-, as in Fig. 12.15:

1. One portion of the cycle should proceed at a constant working-substance temperature T_+.
2. Another portion should similarly be at constant T_-.
3. No heat is to be transferred during other portions of the cycle.

We recognize (1) and (2) as *isothermal* processes, and (3) as *adiabatic*. Thus,

A reversible engine operating between
two fixed temperatures has a cycle
consisting only of isotherms and adiabatics.

(12.B.9)

This is known as a **Carnot cycle**; its $P\mathcal{V}$ diagram is illustrated in Fig. 12.16 for an ideal gas.

Engines based on the Carnot cycle have a remarkable property: when operating between given temperatures, *they all have the same efficiency* (less than 100%, of course); furthermore,

When operating between two given
temperatures, no heat engine can be more
efficient than a Carnot-cycle engine.

(12.B.10)

This will now be demonstrated.

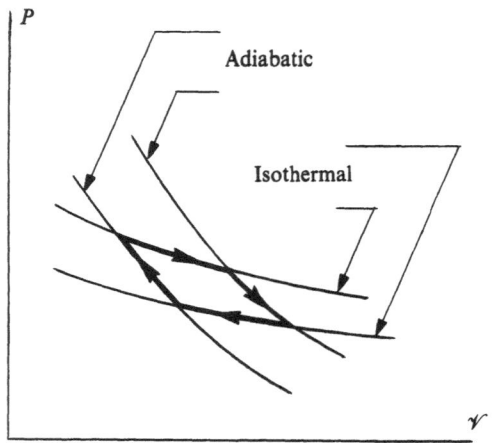

Figure 12.16. A Carnot cycle, made of isotherms and adiabatics. (Not to scale.)

vii. "Reversible" Means "Most Efficient" (Proof)

Consider two heat engines: a reversible one, R, and another, hypothetical one, X, whose efficiency is greater than R's. It is our task to show that X cannot exist.

In Fig. 12.17a, both engines are compared; it is assumed that each is run over an appropriate number of cycles so as to yield the same total work $\mathcal{W}_{\text{tot}}$. The assumed efficiencies compare as follows:

$$\left(\frac{\mathcal{W}_{\text{tot}}}{Q_{\text{in}}}\right)_X > \left(\frac{\mathcal{W}_{\text{tot}}}{Q_{\text{in}}}\right)_R$$

so that

$$(Q_{\text{in}})_X < (Q_{\text{in}})_R$$

as shown on the diagram.

Next, let us operate R in reverse (as a refrigerator), while X is operated normally; the work output $\mathcal{W}_{\text{tot}}$ from X is used as an *input* to operate R; see Fig. 12.17b. The diagram shows that a net positive heat is transferred to the hot source. This is in violation of the second law (12.B.5), since the only other net transfer is a heat loss from the cold source. Thus, X cannot exist.

A further conclusion is that *any two reversible engines* R, R' *must have the same efficiency* between given heat sources.

Indeed, we have just seen that R' *cannot be more efficient than* R. If in this argument, the roles of R

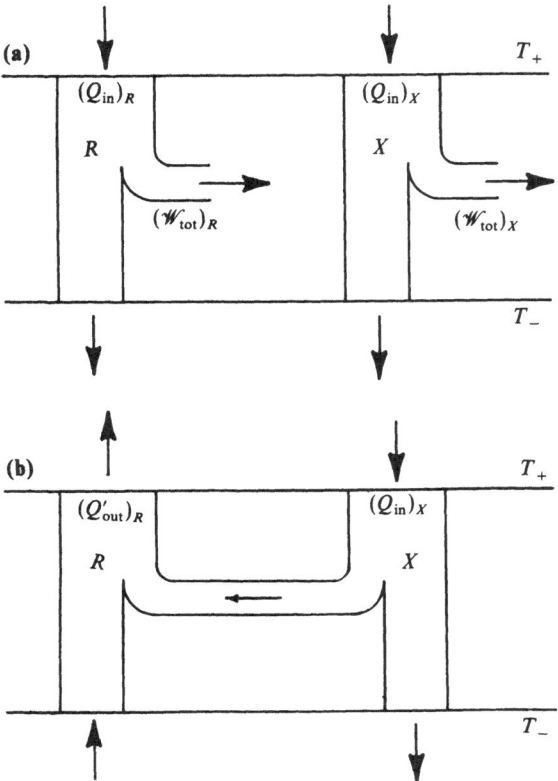

Figure 12.17. (a) Engine X is assumed to be more efficient than the reversible engine R. This means that for the same $\mathcal{W}_{tot}$, X has a smaller heat input than R. (b) The work from X is used to operate R in reverse. The net heat transfer from T_- to T_+ violates the second law.

and R' are interchanged, we see also that R *cannot be more efficient than* R'. Hence their efficiencies must be the same.

viii. What Is the Maximum Efficiency?

Since the efficiency ε_{rev} of a reversible heat engine depends only on T_+ and T_-, but not on its construction, there is no loss of generality in taking a monatomic ideal gas for the working substance. We know how it behaves, and hence it is only a matter of calculation to find the following relation between Q_{in} and Q_{out}:

$$\boxed{\frac{Q_{in}}{T_+} = \frac{Q_{out}}{T_-}}\qquad \text{(any reversible engine)}$$

$$(12.B.11)$$

(So as not to disrupt our present flow of ideas, we postpone this calculation to the end of the section.)

The efficiency is, by definition,

$$\varepsilon_{rev} = \frac{\mathcal{W}_{tot}}{Q_{in}} = \frac{Q_{in} - Q_{out}}{Q_{in}} = 1 - \frac{Q_{out}}{Q_{in}}$$

where the first law has been used in the numerator. With Eq. (12.B.11), this gives directly

$$\boxed{\varepsilon_{rev} = 1 - \frac{T_-}{T_+} = \frac{T_+ - T_-}{T_+}}$$

$$\text{(any reversible engine)}\quad (12.B.12)$$

This is the desired result, which clearly can never reach 100%.

Equations (12.B.11) and (12.B.12) are equivalent statements, and hence both are true independently of the engine's constructional details, working substance, etc.

[If a cold source could be maintained at absolute zero, $T_- = 0$, then $\varepsilon_{rev} = 100\%$ would be implied by (12.B.12). But this cannot happen; the absolute zero of temperature is unreachable even for a short period of time. This inaccessibility is referred to as the "third law of thermodynamics" because it is not logically contained in the first two laws.]

Example 12.5. A Steam Engine. What maximum efficiency ε_{max} can be expected from a steam engine operating between 20°C and 120°C?

The answer, from Eq. (12.B.12), is

$$\varepsilon_{max} = \varepsilon_{rev} = \frac{120 - 20}{120 + 273} = \underline{\underline{25\%}}$$

Note that, in the numerator, which is a difference, Celsius or Kelvin temperatures may indifferently be used.

In the **refrigerator mode** (energy flows are reversed but temperatures are unchanged), Eq. (12.B.11) becomes

$$\frac{Q_{out}}{T_+} = \frac{Q_{in}}{T_-}\qquad \text{(reversible refrigerator)}$$

$$(12.B.13)$$

A refrigerator is called a **heat pump** when one is

interested in utilizing Q_{out}, for example in order to heat a house while (involuntarily) refrigerating the outdoors. This is further discussed in problems 12.39 and 12.43.

ix. Fundamental Nature of the Kelvin Scale

As already noted, the basic reversibility equation, (12.B.11), is remarkable for not depending on the working substance or other details of the engine; a measurement of Q_{in}/Q_{out} universally yields T_+/T_-. In other words, *a reversible engine is an absolute thermometer*: if T_- is regulated by some standard device—perhaps a triple-point cell—then the engine is an instrument for measuring T_+. In this way, *all absolute temperatures, except one, can be determined without reference to any particular substance*; even an ideal gas is not required.

In practice, of course, such a procedure would be —at least in most cases—absurdly cumbersome and inaccurate; it is mentioned only to confirm the importance and uniqueness of the Kelvin scale.

x. Proof of the Reversibility Relation

Equation (12.B.11) still must be derived. We shall make use of Fig. 12.18, which shows a Carnot cycle; our working substance is 1 kmole of an ideal monatomic gas. The portions BC and DA are adiabatic, while AB and CD are isothermal, at temperatures T_+ and T_-, respectively.

We begin by examining the isothermal portions. Formula (12.A.12) tells how much heat, Q_{in}, is absorbed by the sample under expansion AB:

$$Q_{in} = RT_+ \ln \frac{\mathcal{V}_B}{\mathcal{V}_A} \qquad (12.B.14)$$

Similarly, under compression CD, we have

$$Q_{out} = RT_- \ln \frac{\mathcal{V}_C}{\mathcal{V}_D} \qquad (12.B.15)$$

From these two equations we form the expression that we must show to be zero:

$$\frac{Q_{in}}{T_+} - \frac{Q_{out}}{T_-} = R\left(\ln \frac{\mathcal{V}_B}{\mathcal{V}_A} - \ln \frac{\mathcal{V}_C}{\mathcal{V}_D} \right)$$

$$= R \ln \frac{\mathcal{V}_B \mathcal{V}_D}{\mathcal{V}_A \mathcal{V}_C} \qquad (12.B.16)$$

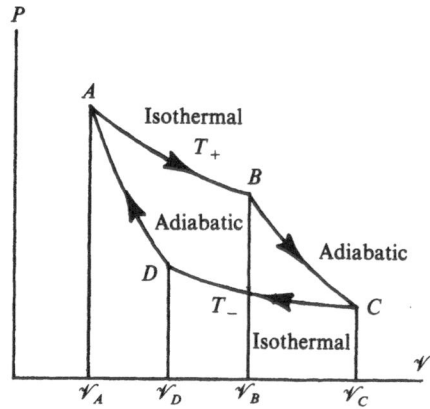

Figure 12.18. Carnot cycle (not to scale) for a monatomic ideal gas used as the working substance of a heat engine.

In order to evaluate this, we now make use of the adiabatic portions of the cycle. Along BC, we have

$$\mathcal{V} T^{3/2} = \text{const} \qquad (12.B.17)$$

see Eq. (12.A.27). This means

$$\mathcal{V}_B T_+^{3/2} = \mathcal{V}_C T_-^{3/2} \qquad (12.B.18)$$

or

$$\frac{\mathcal{V}_B}{\mathcal{V}_C} = \left(\frac{T_-}{T_+} \right)^{3/2} \qquad (12.B.19)$$

Similarly, for process DA,

$$\frac{\mathcal{V}_D}{\mathcal{V}_A} = \left(\frac{T_+}{T_-} \right)^{3/2} \qquad (12.B.20)$$

Multiplying the last two formulas together,

$$\frac{\mathcal{V}_B \mathcal{V}_D}{\mathcal{V}_A \mathcal{V}_C} = 1 \qquad (12.B.21)$$

Therefore the right side of (12.B.16) vanishes, as announced.

C. Entropy and the Waste of Energy*

The word "waste" ordinarily involves a value judgment. Here, however, that word will be used in a more technical sense: "waste" will denote a per-

* This section can be left out without loss of continuity.

manently lost opportunity to obtain some work. A lavish waste of energy occurs all the time in natural and man-made processes.

One illustration of such a waste is the free expansion of a gas (Sec. A vi); that expansion could have been used to drive a piston. Here an opportunity is lost forever because some form of work would be required if one wished to squeeze the gas back into its original volume.

Another case of wasted energy is the conduction of heat from a hotter object to a colder one: some work could have been obtained from a portion of the transferred heat, as we have seen in the preceding section. Here again, the irreparable character of the waste arises from the fact that the heat cannot be made to flow back into the hotter object unless a work-consuming device (such as a refrigerator) is used. In short, *a wasteful process is an irreversible one.*

In such a process, it is not possible to calculate precisely how much energy is wasted. This is because the wasteful process would have to be compared to another one that is not wasteful, and there is a variety of the latter, yielding different amounts of work.

As it turns out, a quantity exists (not an energy) that does measure the degree of waste, or irreversibility. The **entropy**, S, of a system can be defined in such a way that, whenever S is constant, there is no waste; any increase in S measures the seriousness of the waste; S never decreases.

Here we must limit ourselves to a definition of entropy, with some illustrations where it increases. For an appreciation of its great practical usefulness, we refer the student to the more advanced textbooks.

i. Defining an Entropy Change (Reversible Processes)

The reversible engine of the preceding section gives us the necessary hint on how to define S. As we know, the reversibility relation (12.B.11),

$$\frac{Q_{in}}{T_+} - \frac{Q_{out}}{T_-} = 0 \qquad (12.C.1)$$

must hold over a Carnot cycle, not only for an ideal-gas sample, but *for any system whatsoever.* That equation has the appearance of a conserva-

tion law, valid in all reversible processes. Let us assume, subject to confirmation, that there exists a physical parameter S (the **entropy**), which characterizes a sample of working substance. Let

$$\frac{Q_{in}}{T_+} = (\Delta S)_+ \qquad (12.C.2)$$

be defined as the sample's **increase in entropy** due to the heat input Q_{in}, and similarly, let

$$\frac{Q_{out}}{T_-} = -(\Delta S)_- \qquad (12.C.3)$$

be the **decrease in entropy** due to Q_{out}. Then (12.C.1) reads

$$(\Delta S)_+ + (\Delta S)_- = 0 \qquad (12.C.4)$$

Thus, *any system's entropy is conserved over a complete reversible cycle* (a Carnot cycle as far as this proof goes).

This conclusion suggests (correctly as it turns out) that, no matter how the state of a system is changed, its entropy S always returns to its initial value whenever the system returns to its initial state. Any state is therefore labeled by its own unique value of S; in short, entropy (like internal energy, pressure, etc., but *unlike heat*) is a "state function." (For a rigorous proof, not confined to Carnot cycles, consult a thermodynamics textbook.)

The overall argument can now be summarized as follows. Starting from the second law of thermodynamics ("heat does not flow uphill by itself"), we ultimately conclude that every system, in every state, has a certain entropy which increases by Q/T under a reversible process with heat input Q under a system temperature T.

A comparison with the concept of energy is appropriate. We recall that the total energy of a system is a somewhat arbitrary quantity (it depends on a reference level for the potential energy); but an energy *change* has direct physical meaning. Similarly, we make no commitment as to the value of an object's entropy S, but its change ΔS is defined as a measurable amount.

Infinitesimal Change in Entropy

Definitions (12.C.2), (12.C.3) pertain only to isothermal changes. Much more generally, under a small heat input dQ, the change dS in the entropy of a sample is *defined* as

$$\boxed{dS = \frac{dQ}{T}} \qquad \text{(any reversible process)}$$

$$\text{(12.C.5)}$$

where T is the sample's absolute temperature. It is understood that, while dQ is being absorbed, *the state of the sample changes in a reversible manner*; for example, (12.C.5) is not meant to apply to a free expansion.

We next look at some special cases. In an isothermal process ($T = \text{const}$), Eq. (12.C.5) can be integrated to yield

$$\Delta S = \frac{Q}{T} \qquad \text{(reversible isothermal process)}$$

$$\text{(12.C.6)}$$

just as in Eqs. (12.C.2), (12.C.3).

In an adiabatic process ($dQ = 0$), Eq. (12.C.5) becomes $dS = 0$. Hence we have

$$\boxed{\Delta S = 0} \qquad \text{(reversible adiabatic process)}$$

$$\text{(12.C.7)}$$

The following example illustrates the calculation of ΔS in a more general situation.

Example 12.6. Heating a Kilogram of Water. A 1.00-kg sample of liquid water is heated from 0°C to 100°C at atmospheric pressure. Find, in kcal/K, the increase ΔS in its entropy.

We have a reversible process as far as the water is concerned, hence (12.C.5) is applicable:

$$dS = \frac{dQ}{T} = \frac{C\,dT}{T} \qquad \text{(12.C.8)}$$

where $C = 1.00$ kcal/K is the heat capacity of the sample. Integration of (12.C.8) gives

$$\Delta S = C \int_{273\,\text{K}}^{373\,\text{K}} \frac{dT}{T}$$

$$= \left(1\,\frac{\text{kcal}}{\text{K}}\right) \ln\frac{373}{273} = \underline{\underline{0.312\,\frac{\text{kcal}}{\text{K}}}}$$

ii. Entropy and Irreversibility

In the preceding example, the heating was reversible *as far as the water sample was concerned.* However, the system consisting of the water and its environment (the "world") may have suffered an irreversible change, in which heat has flowed "downhill."

Here we demonstrate that any process will generate some *net positive entropy*, unless that process is exactly reversible *when all affected systems are included.* (In this reversible case the total existing entropy is conserved.)

Let us consider the world as seen by a heat engine. That world is made of three parts: the hot source (h.s.), the cold source (c.s.), and the working substance (w.s.). We define entropy (like energy) as *additive*, so that the "entropy of the world" is the sum

$$S_{\text{tot}} = S_{\text{h.s.}} + S_{\text{c.s.}} + S_{\text{w.s.}} \qquad \text{(12.C.9)}$$

Reversible Process

Referring once more to Fig. 12.18, suppose engine goes through only part of its cycle, say the process AB. We now claim that, during that process, *the total entropy is conserved*, $\Delta S_{\text{tot}} = 0$.

As a proof, we just take the whole world to be our system; there are no external heat transfers (the only heat transfers are between different parts of the overall system); furthermore, the process is reversible. For such an overall-adiabatic reversible process Eq. (12.C.7) applies, and we obtain $\Delta S_{\text{tot}} = 0$ as announced.

Irreversible Process

Next we assume that a "malfunction" occurs during part of the cycle, say during AB in Fig. 12.18; that process is now no longer reversible. This is indicated schematically in Fig. 12.19. As a result, the efficiency is less than before, when we had

$$\frac{Q_{in}}{T_+} = \frac{Q_{out}}{T_-} \qquad (12.C.10)$$

For a given Q_{in}, we get less work, and hence a greater Q_{out}. In this case, therefore,

$$\frac{Q_{in}}{T_+} < \frac{Q_{out}}{T_-} \qquad (12.C.11)$$

over a complete cycle.

How does this affect the entropy of the world? We have, for its three pieces,

$$\Delta S_{h.s.} = \frac{-Q_{in}}{T_+} \qquad (12.C.12)$$

(since $-Q_{in}$ is the heat transfer to the hot source),

$$\Delta S_{c.s.} = \frac{+Q_{out}}{T_-} \qquad (12.C.13)$$

and

$$\Delta S_{w.s.} = 0 \qquad (12.C.14)$$

(since the initial and final states are identical). In total,

$$\boxed{\Delta S_{tot} = -\frac{Q_{in}}{T_+} + \frac{Q_{out}}{T_-} > 0} \qquad (12.C.15)$$

by (12.C.11). Thus, positive entropy has been generated during the cycle. Since this could not have happened in any of its reversible portions, it

must have occurred in stage AB. Therefore, as announced,

> An irreversible process increases the total existing entropy. (12.C.16)

An increase in the total entropy is in fact used as a quantitative test of irreversibility.

All natural and man-made phenomena generate entropy to some extent. This happens most prominently in a direct heat transfer from a hot to a cold object, but also in free expansion; in friction; in fluid turbulence; in electric currents through resistors; and in the mixing of originally pure substances. Statistical mechanics teaches that any breakdown of organization in favor of randomness is also associated with an increase in total entropy. It must be kept in mind that such processes *can* be reversed in selected regions, but at the expense of an additional increase in entropy elsewhere. As an illustration, living systems can increase their numbers and their level of organization at the expense of a gigantic irreversible heat transfer out of the Sun.

The Second Law Once More

In order to encompass any kind of physical process, reversible or irreversible, we can extend (12.C.16) to read as follows:

> No process whatsoever can decrease the total existing entropy.

(12.C.17)

As it turns out, (12.C.17) is a way of formulating *the second law of thermodynamics*; statements (12.C.17) and (12.B.5) are entirely equivalent.

iii. Examples

Let us calculate the increase in total entropy resulting from two typical irreversible processes.

Example 12.7. Conduction of Heat. Two reservoirs, at 80°C and 30°C, are briefly brought into contact, the resulting heat transfer $Q = 3.0$ joules causing negligible temperature changes. Find ΔS_{tot} for the two-reservoir system.

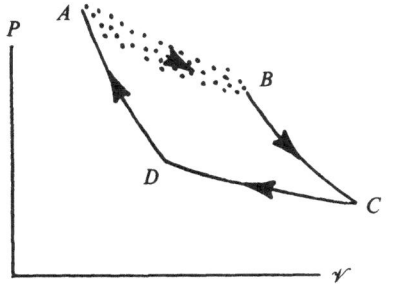

Figure 12.19. Process AB is irreversible; the rest of the cycle is reversible.

We have

$$\Delta S_{\text{tot}} = \frac{-Q}{T_+} + \frac{+Q}{T_-} = Q\left(\frac{1}{T_-} - \frac{1}{T_+}\right) \quad (12.\text{C}.18)$$

$$= (3.0)\left(\frac{1.}{273+30} - \frac{1}{273+80}\right)\frac{\text{joule}}{\text{K}}$$

$$= +1.4 \times 10^{-3}\frac{\text{joule}}{\text{K}}$$

Example 12.8. Free Expansion. As discussed in Sec. A vi, suppose 1 kmole of nitrogen, originally at standard temperature and pressure, is allowed to double its volume by leaking through a small hole into a previously evacuated enclosure. Calculate the total entropy change, assuming the system is adiabatically enclosed.

Here we are dealing with an irreversible process, which cannot be used for calculating ΔS_{tot}. An equivalent *reversible* method consists of expanding the gas *isothermally* to twice its original volume against a piston. [Equation (12.A.34) tells us that the correct final temperature is reached in this way.]

From Eq. (12.A.12) we obtain the equivalent heat input,

$$Q_{\text{in}} = RT\ln\frac{\mathcal{V}_f}{\mathcal{V}_i}$$

so that

$$\Delta S_{\text{tot}} = \frac{Q_{\text{in}}}{T} = R\ln\frac{\mathcal{V}_f}{\mathcal{V}_i} \quad (12.\text{C}.19)$$

$$= (2.00)(\ln 2)\frac{\text{kcal}}{\text{K}} = +1.39\frac{\text{kcal}}{\text{K}}$$

We see from (12.C.19) that, remarkably enough, for a given amount of freely expanding ideal gas, *the entropy change depends only on the final-to-initial volume ratio*.

D. The Logic of this Chapter

Thermodynamics is more abstract than most other topics in this book. To facilitate a review of the chapter we present here its logical flow. An

arrow indicates a logical deduction, or a contribution to such a deduction. The first law of thermodynamics is already tacitly assumed in this diagram.

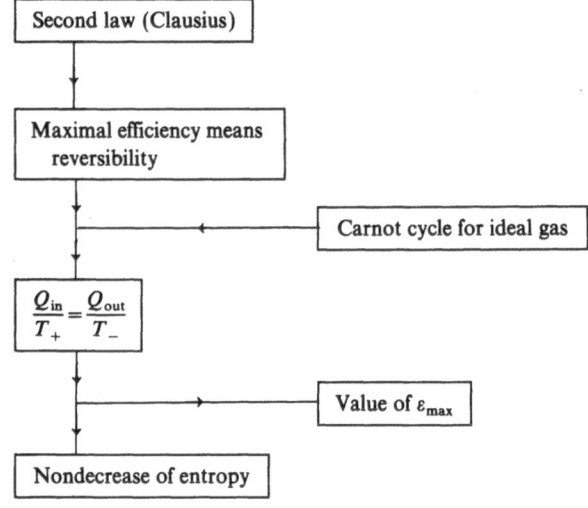

Condensed Checklist

Work output:

$$d\mathcal{W} = P\,d\mathcal{V} \quad (12.\text{A}.2)$$

$$\mathcal{W} = \int_{\mathcal{V}_i}^{\mathcal{V}_f} P\,d\mathcal{V} \quad (12.\text{A}.6)$$

$$\mathcal{W} = \text{area under the } P\mathcal{V} \text{ curve} \quad (12.\text{A}.4)$$

First law:

$$Q = \Delta U + \mathcal{W} \quad (12.\text{A}.7)$$

$$dQ = dU + P\,d\mathcal{V} \quad (12.\text{A}.9)$$

(The following items, through "free expansion," are for *ideal gases*.)

Heat capacities:

$$C = \frac{dQ}{dT} \quad (12.\text{A}.13)$$

$$dU = C_v\,dT \quad (12.\text{A}.15)$$

(any process where T is well-defined; not just constant volume)

$$C_P = C_v + R \quad (12.\text{A}.18)$$

$$C_P = \tfrac{5}{2}R \quad (\text{monatomic}) \quad (12.\text{A}.19)$$

$$R \approx 8300 \text{ J/(kmole K)} \approx 2.00 \text{ kcal/(kmole K)}$$

Chap. 11, Eq. (11.C.14)

Isothermal expansion or contraction:

$$\boxed{\Delta U = 0,} \quad \begin{cases} Q > 0 & \text{for} \quad \text{expansion} \\ Q < 0 & \text{for} \quad \text{contraction} \end{cases} \quad \text{Sec. A iii}$$

$$P_V = \text{const} \qquad \text{Chap. 11, (11.E.5), (11.C.2)}$$

$$Q = \mathcal{W} = NRT \ln(\mathcal{V}_f / \mathcal{V}_i) \qquad \text{(12.A.11), (12.A.12)}$$

Adiabatic expansion or contraction ($Q = 0$):

$$\left. \begin{array}{l} \mathcal{V} T^{C_v/R} = \text{const} \\ P\mathcal{V}^\gamma = \text{const} \end{array} \right\} \text{(if } C_v \text{ independent of } T) \quad \begin{array}{l} \text{(12.A.26)} \\ \text{(12.A.28)} \end{array}$$

$$\gamma = \frac{C_P}{C_v} = 1 + \frac{R}{C_v} \qquad \text{(12.A.29), (12.A.30)}$$

Free expansion:

$$\Delta U = 0, \qquad Q = 0 \qquad \text{(12.A.33)}$$

$$P_f \mathcal{V}_f = P_i \mathcal{V}_i \qquad \text{(12.A.35)}$$

(The remaining items involve any substances.)

Second law:

$$\boxed{\begin{array}{c} \text{No "uphill" flow of heat without} \\ \text{some other energy transfer.} \end{array}} \qquad \text{(12.B.5)}$$

First law for cyclic process:

$$\boxed{\mathcal{W}_{\text{tot}} = Q_{\text{in}} - Q_{\text{out}}} \qquad \text{(12.B.3)}$$

Second law for cyclic process:

$$\boxed{\frac{Q_{\text{out}}}{T_-} = \frac{Q_{\text{in}}}{T_+}} \qquad \text{(reversible engine)} \qquad \text{(12.B.11)}$$

$$\frac{Q_{\text{in}}}{T_-} = \frac{Q_{\text{out}}}{T_+} \qquad \text{(reversible refrigerator or heat pump)}$$

$$\text{(12.B.13)}$$

Efficiency:

$$\varepsilon = \frac{\mathcal{W}_{\text{tot}}}{Q_{\text{in}}} \qquad \text{(12.B.4)}$$

Reversible engine:

$$\boxed{\varepsilon_{\text{rev}} = \frac{T_+ - T_-}{T_+}} \qquad \text{(12.B.12)}$$

All reversible engines working between the same two temperatures have the same efficiency,

$$\boxed{\varepsilon_{\text{rev}} = \varepsilon_{\text{max}}} \qquad \text{Sec. B vii}$$

Entropy*:

$$dS = \frac{dQ}{T} \qquad \text{(reversible process)} \qquad \text{(12.C.5)}$$

$$\Delta S_{\text{tot}} = 0 \qquad \text{(reversible process)} \qquad \text{Sec. C ii}$$

$$\boxed{\Delta S_{\text{tot}} > 0} \qquad \text{(irreversible process)} \qquad \text{(12.C.16)}$$

True or False

1. If $P\mathcal{V} = NRT$ is valid (ideal gas), it means that the heat capacity C_v is independent of T.

2. A gas sample, while expanding, can do an amount of work that is larger than the heat it absorbs.

3. During an isothermal compression, positive heat must be released.

4. While it is being compressed into a smaller volume, an ideal gas sample may absorb positive heat.

5. For an ideal gas, the value of C_P may sometimes be used to disprove the hypothesis that it is monatomic.

6. P, $\mathcal{V}$, T, and U all change during the adiabatic compression of an ideal gas.

7.* A system increases its entropy only if it absorbs heat.

8. In an ideal gas, all processes are reversible.

9.* After a sample has gone through a cyclic process, its internal energy is unchanged but its entropy has increased.

10.* The entropy of a gas sample always increases during a reversible adiabatic expansion.

11.* The entropy of a gas sample decreases during an isothermal compression.

* Leave out if Sec. C has been left out.

Problems

Section A: Heat and the Conservation of Energy (First Law of Thermodynamics)

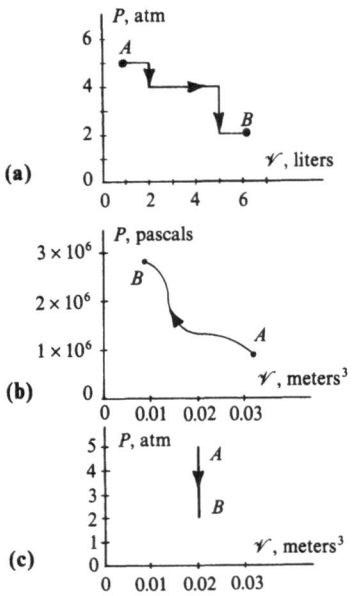

(a)

(b)

(c)

*12.1. Estimate, in joules, the work done by the substances in the PV processes AB of the adjoining Figs. (a), (b), (c).

12.2. How much work is done by a gas sample while it expands from a volume of 3×10^{-3} meter^3 to a volume of 5×10^{-3} meter^3 at constant atmospheric pressure?

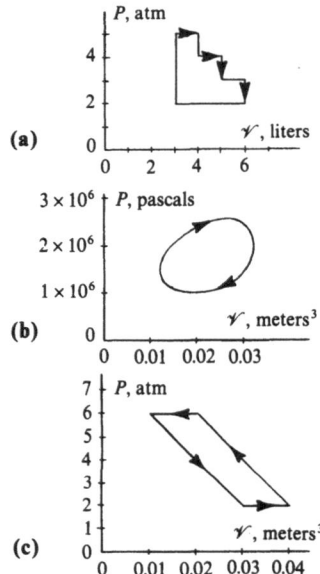

(a)

(b)

(c)

*12.3. In a cyclic process, the total work output W_{out} equals the area inside the closed PV curve; we have

$W_{out} > 0$ or $W_{out} < 0$ according to whether the process follows the curve clockwise or counterclockwise. Estimate, in joules, the work done by the substances of the adjoining Figs. (a), (b), (c) during one cycle.

12.4. A small piece of dry ice (solid CO_2) is dropped to the bottom of a test tube of volume 10 cm^3 and gently sublimates (vaporizes) until the air in the test tube is expelled and replaced by gaseous CO_2. How much work, in joules, has the gas done against the atmosphere? (Neglect the initial volume of the dry ice.)

12.5. A 0.12 meter^3 volume of oil is admitted into the cylinder of a hydraulic jack while it is raising a 1.0 metric-ton car by 2.0 meters. What is the oil's (constant) pressure, in pascals? (1 metric ton = 1000 kg.)

*12.6. How much heat, in kcal, is needed to heat 1 kmole of helium, at constant pressure, from 0°C to 100°C?

*12.7. When 25 kcal of heat is transferred at constant pressure to 5 kmole of (monatomic) krypton gas (atomic weight = 84), initially at 0°C, what is that sample's final temperature?

*12.8. At 20°C, some gases' kilomolar heat capacities C_v at constant volume, expressed in kcal/(kmole K) are as follows: carbon dioxide CO_2, 6.9; oxygen O_2, 5.0; nitrogen N_2, 5.0; ammonia NH_3, 6.7. Assuming them to be ideal gases, calculate (a) their kilomolar heat capacities C_P at constant pressure; (b) their specific heats c_P at constant pressure. [Atomic weights: Table 11.1 of Chap. 11.]

12.9. A sample of ideal gas, originally at STP, with a volume of 1 meter^3, is expanded isothermally to twice its volume, then heated at constant volume back to its original pressure, then cooled at constant pressure back to its original volume. Draw the PV diagram for that history.

*12.10. How much more heat, in kcal, is obtained from cooling 1/2 kmole of an ideal gas from 400 K to 300 K at constant pressure than at constant volume?

*12.11. When 50 joules of heat is transferred to 0.10 kg of (monatomic) neon gas (atomic weight = 20) at constant pressure, what is its temperature rise?

12.12. A certain substance contracts at a constant pressure of 6.0×10^6 Pa between initial and final volumes of 0.075 meter^3 and 0.065 meter^3. It also releases 8.0×10^4 joules of heat. By how much has its internal energy decreased?

*12.13. At constant atmospheric pressure, a gas doubles its volume, starting at 1.0 meter^3. It is also supplied with 40 kcal of heat. What is, in joules, the change ΔU in its internal energy?

12.14. A certain gas sample is expanded at a constant

pressure of 1 atm from a volume of 3.0 meters^3 to one of 5.0 meters^3. At the same time, it absorbs 1.0×10^3 kcal of heat. What is its change in internal energy?

12.15. Derive Eq. (12.A.31), which states that $dP/P = -\gamma \, d\mathcal{V}/\mathcal{V}$ for an infinitesimal adiabatic process in an ideal gas. Do not assume that $\gamma = $ const.

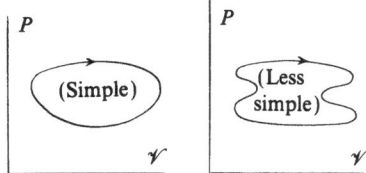

*12.16. Prove that a closed $P\mathcal{V}$ curve implies a work output $\mathcal{W}_{tot}$ (for each cycle) that equals the area *inside* the curve. (For simplicity, assume a convex shape as shown.)

12.17. While a certain tire is being inflated, its volume increases by 0.05 liter, its gauge pressure remaining about constant at 2 atm. If it helps support a car of mass 1 metric ton, by what height has the car's center of mass been raised? (Assume no net energy is transferred to the tire's rubber; 1 metric ton = 1000 kg.)

*12.18. The volume $\mathcal{V}$ of a certain cylinder is being decreased by a piston, while gas is let out so that the pressure in the cylinder is $P = P_0 - a\mathcal{V}$, where $P_0 = 15$ atm, $a = 2400$ atm/meter^3. How much work, in joules, must the piston do to reduce $\mathcal{V}$ from 0.01 meter^3 to zero?

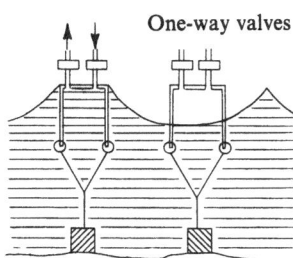

One-way valves

*12.19. The figure shows a proposal for harnessing the sea's wave energy. As the water level changes, air is expelled from, or drawn into a floating open-bottomed, anchored container. The air drives a turbine to generate electric power. What approximate container volume $\mathcal{V}$, in meters^3, is needed if a useful work of 10^7 joules is desired every time the air is expelled at a constant gauge pressure of 0.2 atm? (Work against the atmospheric pressure is not useful.)

12.20. While a certain bubble of real gas rises in water, its pressure, as a function of time, is $P = P_0 - at$; its volume is $\mathcal{V} = \mathcal{V}_0 + bt$. ($P_0$, $\mathcal{V}_0$, a, b are constants.) Between an initial time $-\tau$ and time zero, how much work does the expanding gas do against the surrounding water? (Express your answer in terms of P_0, $\mathcal{V}_0$, a, b, τ.)

12.21. A 1-kmole sample of helium gas is originally at STP. (a) Indicate, on a $P\mathcal{V}$ diagram, the point corresponding to that condition. (b) Through that point, draw an isotherm as a dashed line. (c) Draw the following cycle as a solid line on the same diagram: the sample is compressed adiabatically to twice its original pressure, then cooled at constant volume back to its original pressure, then heated at constant pressure to its original volume. (Use Fig. 12.7 as a guide; perfect accuracy should not be attempted.)

*12.22. Consider N kilomoles of an ideal gas. Prove that the work required for an isothermal compression from a volume $\mathcal{V}_i$ to a volume $\mathcal{V}_f$ is $\mathcal{W}_{in} = NRT \ln(\mathcal{V}_i/\mathcal{V}_f) = P_i \mathcal{V}_i \ln(\mathcal{V}_i/\mathcal{V}_f)$, where P_i is the initial pressure.

12.23. How much work, in joules, is required to compress 1 meter^3 of air isothermally into a 1/2-meter^3 volume? Assume an initial pressure of 1 atm. (See Problem 12.22.)

*12.24. How much heat, in kilocalories, is required to expand isothermally 1 kmole of hydrogen, H_2, initially at STP, to twice its volume? (See Problem 12.22.)

12.25. Consider an ideal gas with a constant value of γ. Use $P\mathcal{V} = NRT$ and $P\mathcal{V}^\gamma = $ const to derive (a) an adiabatic relation between T and $\mathcal{V}$; (b) an adiabatic relation between T and P. (c) For air ($\gamma = 7/5$) show that, adiabatically, $T = (\text{const}) \times P^{2/7}$.

*12.26. Because of the wind, some amount (a "parcel") of atmospheric air rises by 100 meters, fast enough to undergo adiabatic expansion. What is its temperature drop? [Assume that pressure decreases by 1% for every 100 meters of altitude; use $T = (\text{const}) \times P^{7/2}$; see Problem 12.25; assume an initial temperature of 300 K.] The observed altitude dependence of the temperature can be fairly well predicted in this way.

12.27. In an adiabatic process, the temperature–volume relation for air is $\mathcal{V} T^{5/2} = $ const; see also Problem 12.25. If the 20°C air in a bicycle pump is compressed to 1/3 of its volume, what is its final Celsius temperature? (Assume the compression is fast enough to be adiabatic.)

12.28. In Problem 12.27, how much gauge pressure (i.e., over the original atmospheric one) must be reached in the pump?

12.29. Show that, when an ideal gas contracts at constant pressure, it obeys $Q_{out} = C_P \mathcal{W}_{in}/R$. (Assume C_P is constant.)

Section B: Heat Engines and the Second Law of Thermodynamics

12.30. What is the thermodynamic efficiency of a Carnot engine operating between 30°C and 300°C?

*12.31. In order to liquefy helium (at about 4 kelvin) someone plans to extract 0.001 kcal from the sample; the rejected heat goes into liquid hydrogen (at about 20 kelvin). At least how many kcal of work input are required? [The least amount of work is expended in the reversible case.]

*12.32. With concave mirrors, the solar radiation can be concentrated so as to make the temperature of a small spot on Earth comparable to that of the Sun's surface itself (6000 kelvin). With a more modest 3000 kelvin hot source and a 20°C coolant, what ideal efficiency could be reached by a heat engine?

12.33. If a river at 10°C is available, at what minimum temperature should a nuclear reactor be run so that 50% of its generated heat may be converted to electric power (by way of mechanical work)?

12.34. A furnace at 300°C transfers 1000 kcal to a steam engine which is cooled by water at 10°C. (a) Set an upper limit on the amount of mechanical work, in joules, that can be obtained. (Assume the engine has gone through a whole number of cycles.) (b) How much heat, in kcal, would then be transferred to the coolant?

*12.35. A certain gasoline engine has a maximum temperature, in its cylinders, of 500°C; the coolant fluid is at 80°C. (a) Set an upper limit on the engine's efficiency. (b) If the engine puts out 20 hp of mechanical power, at least how many kcal of heat must the coolant remove per minute? (1 hp = 735 watt.)

*12.36. A food refrigerator releases 3 kcal of heat to its surroundings for every kcal of heat removed from its interior. (a) What is its coefficient of performance, defined as heat removed per work input? (b) If the interior temperature is 2°C, set a limit on the exterior temperature. Is this a lower or an upper limit?

12.37. A heat pump supplies 100 kcal of heat per hour to the inside of a house, which is at 15°C, while the outdoor temperature is −5°C. Set a lower limit, in watts, on the mechanical (or electric) power consumption.

12.38. What minimum work, in joules, does a refrigerator need in order to extract 100 kcal from a compartment at 0°C when the environment is at 20°C? [The work is minimum when the operation is reversible.]

*12.39. In thermodynamics, electric energy counts as mechanical energy, because each can be completely transformed into the other. Suppose a home is heated by direct conversion of an electric energy $\mathcal{W}$ into heat. (a) To obtain that same amount of heat, how much electric energy $\mathcal{W}'$ would be needed if used to operate a heat pump between the outdoor and indoor temperatures, −10°C and +20°C respectively? (b) What percentage of $\mathcal{W}$ would be saved? (c) What does this percentage

become if the outdoor temperature is +10°C? [For optimum conditions, assume a reversible heat pump.]

12.40. A heat engine is needed to produce electrical power while operating between 300°C and 100°C, the hot source yielding 5 kcal every second. Four proposals are submitted: No. 1 promises almost 21 000 watts of electric power output; No. 2, at least 7300 watts; No. 3, at least 5000 watts; and No. 4, almost 500 watts. On the basis of this information alone, which proposal would you consider first, and why?

12.41. Prove that the reversibility equation (12.B.11), $Q_{in}/T_+ = Q_{out}/T_-$, follows from Eq. (12.B.12) for the maximum efficiency, $\varepsilon_{rev} = 1 - T_-/T_+$.

12.42. The **coefficient of performance** of a refrigerator is defined as $\varepsilon' = Q_{in}/\mathcal{W}_{tot,in}$, i.e., the ratio, over one cycle, of the heat extracted from the cold reservoir to the total work input. (a) Express ε' in terms of T_+ and T_- for a reversible refrigerator. This will yield the maximum ε'. (b) If T_+ and T_- are allowed to vary, between what limits can ε' vary? (c) What relation between T_+ and T_- (very close or very distant) promotes good performance?

*12.43. A **heat pump** is a refrigerator used as a further heater of its already hot source; in the winter it might cool the outdoors while heating the house. Its coefficient of performance is defined as $\varepsilon = Q_{out}/W_{tot,in}$; cf. Problem 12.42. (a) Express ε in terms of T_+ and T_- for a reversible heat pump; ε will then be maximum. (b) If T_+ and T_- can vary, what is the range of possible ε? (c) What relation between T_+ and T_- promotes good performance?

12.44. A power plant is proposed that would extract heat from the warm layers (25°C) of a tropical ocean and reject heat into the deeper cold layers (10°C). If 10^9 watts of electric power is desired (about the capacity of a typical large utility), (a) at least how much heat, in kcal, must be extracted per second from the warm water? (b) Make a rough estimate of how many meters^3 of water must be processed per second. [Take the specific heat of sea water to be 1 kcal/(kg K).]

Section C: Entropy and the Waste of Energy

*12.45. How much more entropy does 1 kg of 100°C steam at atmospheric pressure possess than the boiling water which it initially was? (Latent heat of vaporization = 540 kcal/kg; assume boiling can be done reversibly.)

12.46. As a 1-kg sample of water freezes at atmospheric pressure, by how much does its entropy decrease? (Latent heat of fusion = 80 kcal/kg; assume freezing can be done reversibly.)

*12.47. Two 1-kg samples of water, at 0°C and 100°C, are mixed together. Find the increase in the total entropy.

12.48. A 20-g sample of hydrogen gas is heated at atmospheric pressure from 100°C to 300°C. Assuming a kilomolar heat capacity $C_P = \frac{5}{2}R$, find the sample's entropy change in kcal/K.

*12.49. A 1-kg block of iron, initially at 200°C, is cooled by being dipped into a very large amount of water at 10°C. Find, in kcal/K, the total entropy change. (Specific heat of iron = 0.10 kcal/kg K.)

12.50. Into how large an evacuated volume should 5 kmole of air, originally occupying 100 meters³, be freely released, in order that its entropy should rise by 10 kcal/K?

12.51. Give an alternative version of the argument leading to statement (12.C.16) concerning the entropy increase in an irreversible process; follow the text, but consider the engine to be acting in the refrigerator mode.

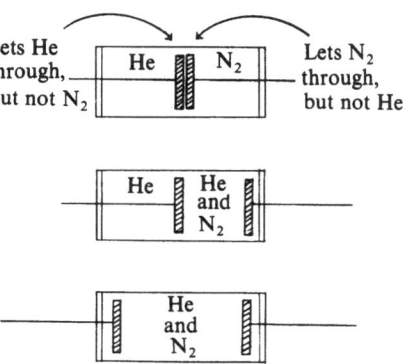

*12.52. Two 1-kmole samples, one of helium, the other one of nitrogen, are each in a 22.4-meters³ half of a cylinder at 1 atm and 0°C. They are initially separated by **semipermeable** pistons as shown in the figure, and then subjected to a two-step process. (1) The helium expands isothermally into the whole cylinder by doing work against one piston. (2) The nitrogen similarly expands against the other piston. Find the system's entropy change during each step, and add them. You have just calculated an **entropy of mixing**. (The semipermeable pistons are fictitious; they would be difficult to realize in practice.)

Answers to True or False

1. False (true for a monatomic ideal gas).
2. True (it can cool and give up some internal energy).
3. True.
4. True (if its temperature is raised sufficiently).
5. True (if $C_P > \frac{5}{2}R$).
6. True.
7. False (true for reversible processes but not in general).
8. False.
9. False.
10. False (it stays constant).
11. True (it can put out heat reversibly).

Answers or Hints to Problems

12.1. (b) -3.8×10^4; (c) zero.

12.3. (c) -4×10^3 joules.

12.6. 500 kcal.

12.7. The atomic weight is not needed.

12.8. (b) See Table 10.1 of Chap. 10.

12.10. 100 kcal.

12.11. 0.5 K.

12.13. 6.6×10^4.

12.16. Put a leftmost point A and a rightmost point B on the loop; divide the cycle into AB and BA; subtract one area from another.

12.18. 3×10^3 joules.

12.19. Useful work = (gauge pressure) × (volume change).

12.22. $\mathcal{W}_{\text{in}} = -\int_{\mathcal{V}_i}^{\mathcal{V}_f} P \, d\mathcal{V} = -NRT \int_{\mathcal{V}_i}^{\mathcal{V}_f} d\mathcal{V}/\mathcal{V}$.

12.24. 3.8×10^2 kcal.

12.26. First show that $dT/T = \frac{7}{2} \, dP/P$.

12.31. Use Eq. (12.B.13) to find Q_+; then $\mathcal{W}_{\text{in}} = Q_+ - Q_-$.

12.32. 90%.

12.35. (b) 180 kcal.

12.36. Use Eq. (12.B.13).

12.39. (c) 96.6%.

12.43. (a) $T_+/(T_+ - T_-)$; (b) $1 < \varepsilon < \infty$; (c) very close.

12.45. 1.45 kcal/K.

12.47. Mixing is irreversible. Replace by thermal contact.

12.49. $(0.10)[(190/283) - \ln(473/283)] = 0.016$.

12.52. The semipermeable pistons allow you to ignore the effects of one gas while dealing with the other.

CHAPTER 13

VIBRATIONS

The preceding four chapters were a digression into the bulk properties of matter. We now return to single-particle mechanics in order to explore vibrational (or oscillatory) motion. The simplest kind of vibration, exemplified by a pure musical tone, is known as a **harmonic oscillation**. It is found in Nature, if one knows how to look (or listen), as an ingredient of nearly all vibrations—mechanical, electrical, or other. The next chapter will demonstrate its close connection with waves.

Newton's laws of motion will again be our principal tool here. The essentials of uniform circular motion (Chap. 6) should be reviewed, as they are relevant to the present material.

Forces responsible for vibrations (**elastic forces**) are easiest to visualize in terms of **springs**.

A. Hooke's Law and the Ideal Spring

The spring is a modest man-made device. Yet its behavior serves as a simple model for some of the most fundamental and interesting of natural phenomena: those involving vibrations and waves. By analyzing the spring we are preparing ourselves to deal with a vast area of experience, including sound, alternating electric currents, radio waves, and light.

When studying the concept of force in Chap. 3, we made great use of the **ideal spring**, a device which, at a standard extension, provided us with a standard force. Never did we ask how that force would vary if the spring's extension were changed. Here we do ask that question; its answer will be our key to understanding vibrations. This answer—the ideal-spring law—was first formulated, from obser-

vations on real springs, by Robert Hooke (1635–1703), a student of Boyle (of ideal-gas fame) and a contemporary of Newton.

i. Force versus Stretch or Compression

Let one end of a spring be attached to a fixed point, and let the position of the other end, P, be denoted by x; see Fig. 13.1. By definition, one has $x = 0$ when zero force is applied to the spring; the point $x = 0$ is called the **equilibrium position** (or neutral position) of P; x is called the **displacement** of P, or the **elongation** (extension) of the spring. A negative value of x indicates a compression.

We can now state the spring's basic feature, known as **Hooke's law**:

> The spring exerts a restoring force proportional to its elongation or compression. (13.A.1)

That is to say,

$$F_x = -kx \qquad (13.A.2)$$

where the minus sign indicates a direction opposite to the displacement; this is the "restoring" feature of the force. The (positive) proportionality constant k, a measure of stiffness, is called the **spring constant**; its SI unit is the newton/meter; F_x is called a **linear restoring force** because its graph as a function of x is a straight line (Fig. 13.2).

To measure a spring constant, it is not necessary to know the equilibrium position of the spring's end

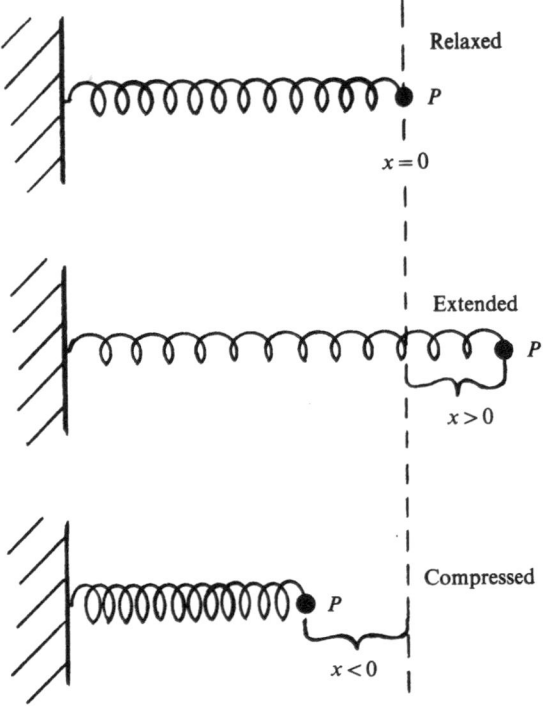

Figure 13.1. The displacement, x, as defined for the end of a spring. (An alternative definition, equally good, would involve the opposite sign: $x < 0$ for extension, $x > 0$ for compression.)

point. Graphically, the *slope* on Fig. 13.2 is sufficient information:

$$k = -\frac{\Delta F_x}{\Delta x} \qquad (13.A.3)$$

Thus, physically, only *changes* in force and displacement need be observed to get k. If such

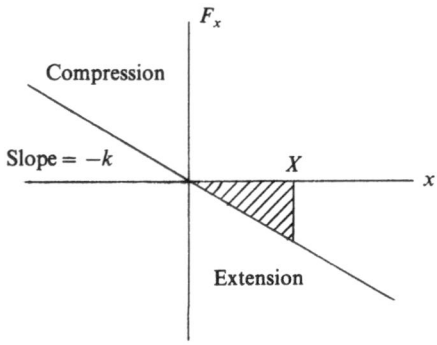

Figure 13.2. The force exerted by the spring, plotted against displacement. The negative slope indicates the restoring property. The shaded area and point X need not concern us now, but will be used later in deriving the potential energy of the spring.

changes obey (13.A.3) for a fixed k, we get a straight-line graph and hence Hooke's law, although, of course, the equilibrium point is not specified by Eq. (13.A.3).

Many systems exhibit such a springlike behavior, that is, one characterized by Hooke's law.

Example 13.1. A Diving Board. An 80-kg man stands at the end of a diving board. If he were carrying an additional 3-kg load, his point of support would be 0.10 meter lower. Assuming Hooke's law, what is the spring constant k of the diving board's endpoint?

First we note the restoring character of the force exerted *by* the board *on* the man's feet: the additional force, needed to support the 3 kg, is upwards, while the additional displacement is downwards. Thus, if x is measured upwards, we have (using $g = 10$ meters/sec^2),

$$\Delta F_x = +(3 \text{ kg}) \left(10 \, \frac{\text{meters}}{\text{sec}^2} \right)$$

$$\Delta x = -0.10 \text{ meters}$$

and from (13.A.3)

$$k = \frac{(3)(10)}{0.10} \frac{\text{newtons}}{\text{meter}} = 300 \, \frac{\text{newtons}}{\text{meter}}$$

Example 13.2. A Vertical Loaded Spring. Figure 13.3 shows a suspended spring with constant $k = 200$ newtons/meter. First it is relaxed (a), and later (b) it carries a load of mass $m = 0.5$ kg. By what distance l has the spring been elongated?

Assuming equilibrium for the load, $F_{\text{tot}} = 0$, we have

$$(\text{weight}) + (\text{restoring force}) = 0$$

$$mg - kl = 0 \qquad (13.A.4)$$

or

$$l = \frac{mg}{k} \qquad (13.A.5)$$

$$l = \frac{(0.5)(10)}{200} \text{ meter} = \underline{0.025 \text{ meter}}$$

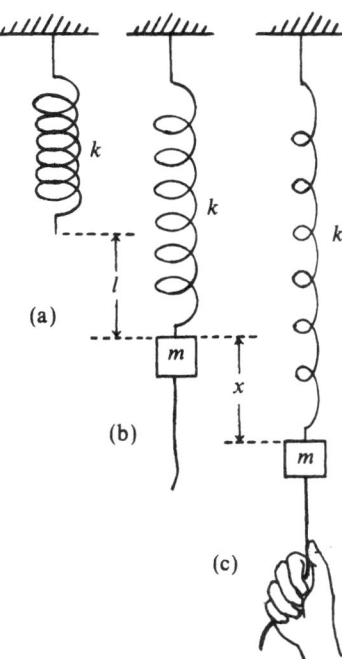

Figure 13.3. The vertical spring, shown relaxed in (a), acquires a new equilibrium point when loaded, as shown in (b); if further displaced as in (c), it still obeys Hooke's law, with the original k, but about the new equilibrium point.

Example 13.3. The Vertical Loaded Spring, continued. The load of the preceding example is pulled down by hand a further distance x; see Fig. 13.3c. What is then the combined downward force, $(F_{comb})_x$, of spring and gravity on the load? (The hand's force is not included in this question.)

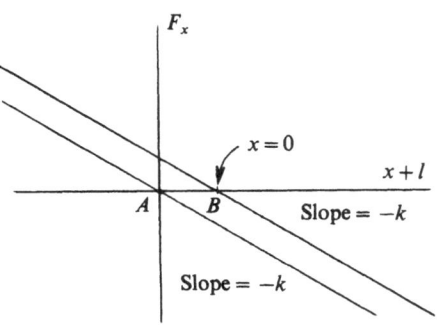

Figure 13.4. Addition of a constant to the original restoring force (line through A) produces the new restoring force (line through B). This is again Hooke's law, but with a new equilibrium point at B. The horizontally plotted variable is $x + l$.

The total elongation is now $x + l$. Therefore we have

$$(F_{comb})_x = mg - k(l + x)$$

or, with (13.A.4),

$$\underline{(F_{comb})_x = -kx}$$

Thus, the combined force on the load again obeys Hooke's law, with the same force constant as for the unloaded spring. The equilibrium position $x = 0$ is, however, shifted by the amount l calculated in the preceding example. Noting that the weight mg is just a constant force, we conclude that

> Hooke's law is preserved when a constant force is added; the equilibrium point is shifted, but the spring constant remains the same. (13.A.6)

Such a conclusion is also apparent at a glance from Fig. 13.4.

Real Springs

We have just studied the behavior of an *ideal spring*. Real springs can be made very close to the specifications of Hooke's law, but not perfectly so. Any actual spring has an **elastic limit**, i.e., an elongation beyond which it becomes permanently deformed; even before that limit is reached deviations from direct proportionality will occur. In general, *a good fit to Hooke's law requires that the displacement be small*. In what follows, ideal springs will, in addition, be assumed to have negligible mass. This assumption allows us to ascribe all gravitational and inertial effects to the attached loads rather than to the spring itself.

ii. The Potential Energy of a Spring

In order to pull or push the end of a spring away from its equilibrium point, work must be expended. That work is stored by the spring as **elastic potential energy**, and may be recovered by allowing the spring to return to its equilibrium point while doing work. We now demonstrate this property quantitatively by calculating the potential energy $\mathcal{U}$ stored in a spring with constant k under an elongation x.

In Fig. 13.5, the end P is shown moving from x to $x + dx$. The work done *on* the spring is

$$\text{(Applied force)(displacement)} = (kx)(dx)$$

Equivalently, the additional stored energy is

$$d\mathcal{U} = kx\,dx$$

Integration gives the desired result:

$$\mathcal{U} = \int kx\,dx = \tfrac{1}{2}kx^2 + \text{const} \qquad (13.A.7)$$

As always, the constant term in the potential energy is arbitrary, since only the changes $\Delta\mathcal{U}$ are physically meaningful. Conventionally,

> One takes the elastic potential energy to be zero at the neutral point:
> $\mathcal{U} = 0$ when $x = 0$.

$$(13.A.8)$$

Equation (13.A.7) then becomes

$$\boxed{\mathcal{U} = \tfrac{1}{2}kx^2} \qquad (13.A.9)$$

An Alternative Derivation

A graphical derivation of (13.A.9) from Fig. 13.2 consists of asking for the (negative) work done by

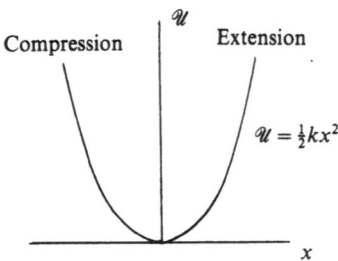

Figure 13.6. A spring's potential energy relative to the equilibrium point.

the spring as x goes from zero to a maximum value X. We have

$$\mathcal{U}(0) - \mathcal{U}(X) = \int_0^X F_x\,dx$$

$$= \text{shaded area in Fig. 13.2}$$

$$= \tfrac{1}{2}(X)(kX) = -\tfrac{1}{2}kX^2$$

We now return to our result, (13.A.9), whose graph is a parabola (Fig. 13.6). Note, in that figure, the symmetry of $\mathcal{U}$ about the equilibrium point, and the fact that $\mathcal{U}$ is positive for compression as well as for extension.

The existence of a potential energy function $\mathcal{U}$ means that a linear restoring force is *conservative*, like a gravitational force, but unlike a frictional force.

The spring's potential energy has a location in space. It must be viewed as *distributed over the spring's own material*. In reality, it is the sum of a very large number of interatomic potential-energy contributions.

Example 13.4. A Spring Gun. A vertically aiming gun uses a light spring whose constant is $k = 4000$ newtons/meter (see Fig. 13.7), origi-

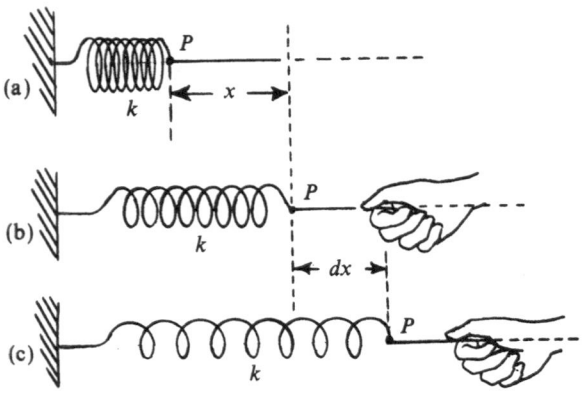

Figure 13.5. Doing work on a spring. (a) The equilibrium point is shown for reference. (b), (c): The spring does negative work during a displacement dx.

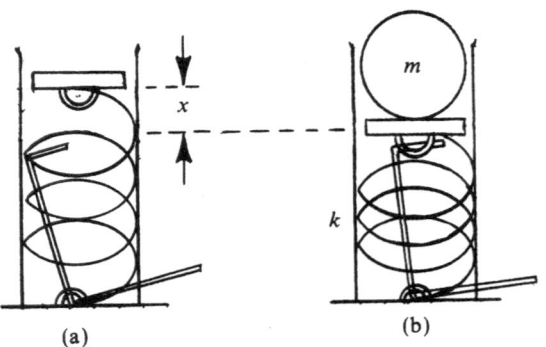

Figure 13.7. The spring gun (a) relaxed, (b) ready to go off.

nally compressed a distance $x = 2$ cm below its unloaded equilibrium point, and supporting a steel ball of mass $m = 8$ grams. To what height h above its initial position does the ball rise after the spring is released? Neglect friction.

<hr />

Let $\mathscr{U}_{spr}$ and $\mathscr{U}_{grav}$ be the spring's elastic and the ball's gravitational potential energies, respectively; take the ball's initial level as a reference. Conservation of energy from the initial to the final situations gives, with zero final projectile speed (no kinetic energy terms),

$$(\mathscr{U}_{spr} + \mathscr{U}_{grav})_f = (\mathscr{U}_{spr} + \mathscr{U}_{grav})_i \qquad (13.\text{A}.10)$$

(The spring, if light enough, has negligible kinetic and gravitational energies; the final projectile speed is zero.) Equation (13.A.10) becomes

$$0 + mgh = \tfrac{1}{2}kx^2 + 0$$

so that

$$h = \frac{kx^2}{2mg}$$

$$= \frac{(4000)(0.02)^2}{(2)(0.008)(10)} \text{ meters} = \underline{\underline{10 \text{ meters}}}$$

iii. Angular Displacement and the Simple Pendulum

So far we have analyzed linear restoring forces. The same methods are applicable to linear restoring *torques*, such as exist in the systems of Fig. 13.8. The angular analogue of Hooke's law, (13.A.2), is

$$\boxed{\tau_\theta = -\kappa\theta} \qquad (13.\text{A}.11)$$

where τ_θ is the torque exerted on the system's mass, θ is its **angular displacement** away from the equilibrium orientation in which zero torque is exerted, and κ is a constant of proportionality (**angular spring constant**) with SI units of newton meters/rad. (Note: κ is the lower-case Greek kappa.)

The Simple Pendulum

This familiar device, which uses no spring, is the best known illustration of Hooke's law in angular

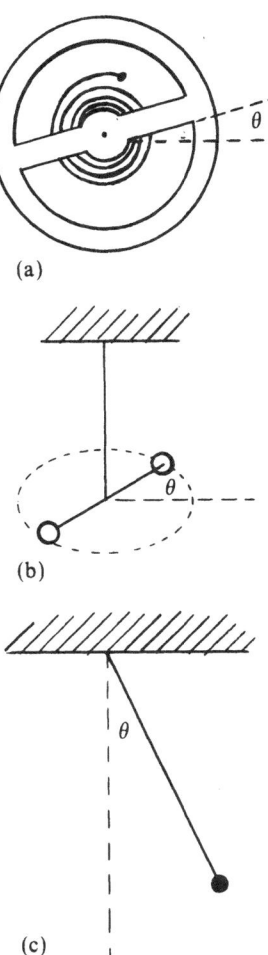

(a)

(b)

(c)

Figure 13.8. Examples of a restoring torque, i.e., one exerted in the direction of minus θ. (a) The spiral spring of a watch. (b) A torsion pendulum; see also Problem 13.27. (c) A simple pendulum; see also Fig. 13.9 further on.

form. It consists of a particle with mass m (the pendulum's bob) at one end of a string (or light rod) whose other end is held fixed. In Fig. 13.9, it is shown pulled off the vertical by an angle θ. What is the total torque τ_θ on the particle (due to string and gravity) about the point of suspension, $\mathbb{O}$?

The force exerted by the string has its line of action through $\mathbb{O}$, and therefore it exerts zero torque about that point. Hence the torque is due only to gravity. We recall from Chap. 7 that

$$\tau = (\text{force}) \times (\text{moment arm})$$

or, in this case,

$$\tau_\theta = -(mg)(l \sin \theta) \qquad (13.\text{A}.12)$$

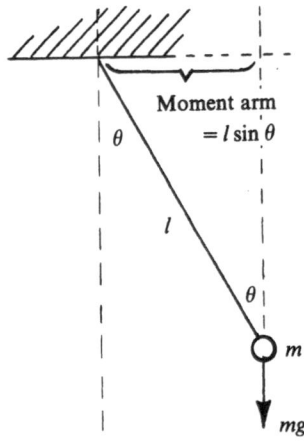

Figure 13.9. A simple pendulum.

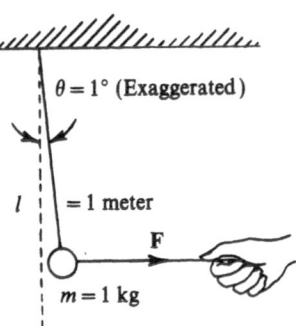

Figure 13.10. An external force **F** is applied to a pendulum.

from the geometry of Fig. 13.9. The minus sign is needed to express the fact that τ_θ is a restoring torque. If θ is small, we can set $\sin\theta \approx \theta$ provided θ is expressed in radians. With that approximation in mind we rewrite (13.A.12) as

$$\tau_\theta \approx (mgl)\theta \qquad (13.A.13)$$

Reference to Eq. (13.A.11) then yields the angular spring constant,

$$\boxed{\kappa = mgl} \qquad (13.A.14)$$

To conclude: the simple pendulum obeys Hooke's law, (13.A.11), for small displacements from the vertical, but it deviates from that law for larger angles.

Example 13.5. An External Force on a Pendulum. What horizontal external force F (see Fig. 13.10) must be applied to the bob in order to hold the pendulum off the vertical by 1 degree of angle if the bob's mass is 1 kg and the string is 1 meter long?

Rotational equilibrium gives, in terms of magnitudes,

External torque = restoring torque

or

$$Fl = mgl\theta$$

where, in the left side, we have used the fact that **F** is nearly orthogonal to the string; the right side comes from Eq. (13.A.13). Therefore, with $1° \approx 0.02$ rad, we obtain

$$F = mg\theta = (1)(10)(0.02) \text{ newton} = \underline{0.2 \text{ newton}}$$

Note: This problem can also be solved without the angular form of Hooke's law; look back, for example, at Problem 3.26.

B. Harmonic Motion

How does a particle move when attached to the end of an ideal spring? If pulled and released, it will move to the equilibrium point; inertia will bring it past that point until compression of the spring reverses the motion. In the absence of friction, this results in a special kind of periodic back-and-forth behavior called **harmonic motion**, the simplest example of a vibration.

This section has two principal aims: (1) to describe harmonic motion, and (2) to relate its properties to that of the spring.

Besides *linear* oscillators like the loaded spring, there exist many other kinds, such as *rotational* ones, of which a swinging pendulum is an example. We shall discuss these briefly.

i. Harmonic Motion versus Uniform Circular Motion

Harmonic motion will first be defined without reference to a spring; later, we shall see that it originates in Hooke's law.

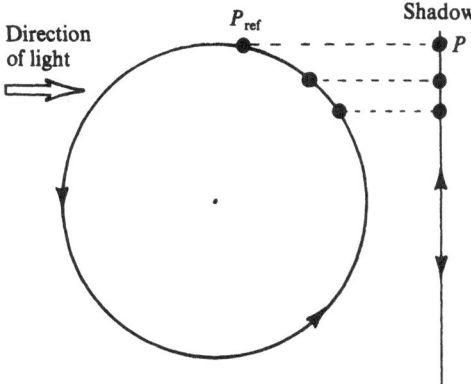

Figure 13.11. In this arrangement, P_{ref} revolves uniformly, and its shadow P undergoes harmonic motion. The three positions are separated by equal time intervals.

As informally defined:

> Harmonic motion is uniform
> circular motion seen edge-on.   (13.B.1)

Mathematically speaking, it is the *projection* of uniform circular motion on an axis in the plane of that motion. For example, in Fig. 13.11, harmonic motion is observed for the shadow, P, of the revolving particle P_{ref} (the **reference particle**). The position of P is entirely determined by the position of P_{ref} on the so-called **reference circle**. In Fig. 13.12, the position x of P is plotted as a function of time; this is accomplished by looking at successive positions of P_{ref}. Conventionally, motion along the reference circle is counterclockwise.

It is important to become familiar with the nomenclature of harmonic motion. The list below should be studied in connection with Fig. 13.12.

One **cycle** denotes the motion of P during one complete revolution of P_{ref}. This might, for example, be one back-and-forth trip of P, starting and ending at the topmost point.

The **period**, T, is the time needed for one cycle.

The **frequency**, f, is the number of cycles per unit time. It is related to T by

$$f = \frac{1}{T} \qquad (13.B.2)$$

and is expressed in SI units of hertz (Hz), equivalent to cycles per second or cps.

The **angular frequency**, ω, equals by definition the angular velocity of P_{ref} along its circle; it is related to f and T by

$$\omega = 2\pi f = \frac{2\pi}{T} \qquad (13.B.3)$$

and is expressed in rad/sec.

The **amplitude** (or displacement amplitude), A, is the maximum displacement of P. The coordinate x of P oscillates between the extremes $+A$, $-A$; the amplitude is always defined positive; it equals the radius of the reference circle.

The **phase angle** (or phase), φ, is the angular position, measured from point $\mathbb{O}$ on the reference circle. For simplicity, we choose the origin of time, $t = 0$, at the moment when P_{ref} is at $\mathbb{O}$. Then the phase is given by

$$\varphi = \omega t \qquad (13.B.4)$$

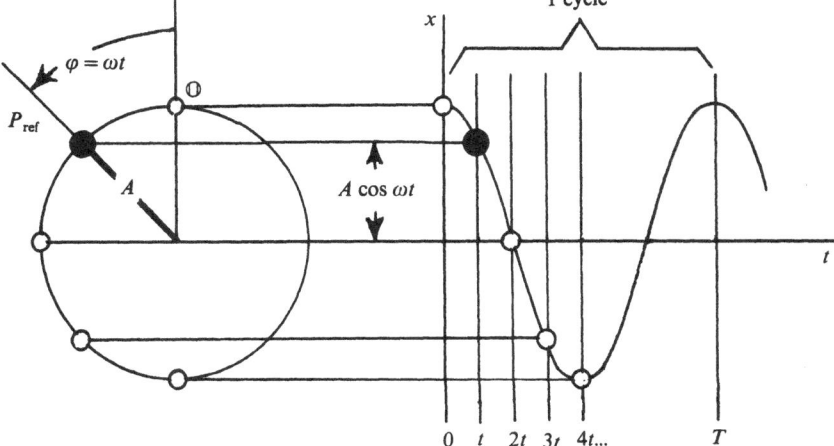

Figure 13.12. How to plot the position of P as a function of time. Here that position (or, equivalently, the vertical coordinate of P_{ref}) is recorded at equal time intervals. The symbols are explained in the text.

Harmonic Motion as a Cosine Function

In Fig. 13.12 we see that

$$x = A \cos \varphi \qquad (13.B.5)$$

or

$$\boxed{x = A \cos \omega t} \qquad (13.B.6)$$

The graph is a *cosine curve*. [Note: By starting P_{ref} at a different point of the reference circle (which point?), one can express the motion as $x = A \sin \omega t$, or, even more generally, as $x = A \cos(\omega t + \alpha)$, α being an adjustable **initial phase**.]

This brings us to a more precise definition, as compared with the informal one, (13.B.1): *harmonic motion is a sinusoidally time-dependent motion.*

ii. Harmonic Motion and Hooke's Law

Harmonic motion has now been defined and described. Our next task is to find out what kind of force makes it happen; the answer, as we shall demonstrate, is "a linear restoring force." The reader ought to be reasonably conversant with **phase vectors** and their time derivatives (see especially Figs. 6.5 and 6.6 of Chap. 6) in order to appreciate the argument used below.

Let the position of P_{ref} in Fig. 13.13 be represented by a phase vector $\mathbf{r}$. Then the successive time derivatives $\mathbf{v} = d\mathbf{r}/dt$, $\mathbf{a} = d^2\mathbf{r}/dt^2$ are phase vectors as well. The figure reminds us that

$$\mathbf{a} = -\omega^2 \mathbf{r} \qquad (13.B.7)$$

which is the familiar centripetal acceleration of Chap. 6. Looking at the x components of $\mathbf{r}$, $\mathbf{v}$, and $\mathbf{a}$, we find x, v_x, a_x, respectively; these are the displacement, velocity, and acceleration of the actual particle P under harmonic motion. We can now forget about P_{ref} and, instead, concentrate on P itself. From Eq. (13.B.7) it follows that

$$\boxed{a_x = -\omega^2 x} \qquad (13.B.8)$$

Multiplying both sides by the mass m of the particle, we have equivalently

$$F_{\text{tot},x} = -m\omega^2 x \qquad (13.B.9)$$

This proportionality, $F_{\text{tot},x} \propto -x$, is precisely Hooke's law. We conclude that

$$\boxed{\text{Hooke's law results in harmonic motion.}}$$

$$(13.B.10)$$

How is the spring constant, k, related to the vibration? A comparison of (13.B.9) with Hooke's law (assumed responsible for the total force)

$$F_{\text{tot},x} = -kx \qquad (13.B.11)$$

produces the result

$$k = m\omega^2$$

or

$$\boxed{\omega = \left(\frac{k}{m}\right)^{1/2}} \qquad (13.B.12)$$

This yields the so-called

$$\boxed{\textbf{natural frequency,} \quad f = \frac{1}{2\pi}\left(\frac{k}{m}\right)^{1/2}} \qquad (13.B.13)$$

$$\boxed{\text{and } \textbf{natural period,} \quad T = 2\pi\left(\frac{m}{k}\right)^{1/2}} \qquad (13.B.14)$$

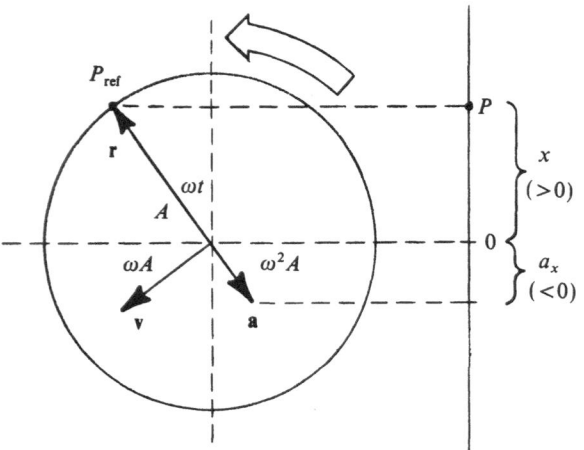

Figure 13.13. The position and acceleration of P, on the axis at right, are the projections of the corresponding phase vectors.

of a loaded spring. We see that *the natural frequency increases with the stiffness of the spring, and decreases with the mass of the load*; also, remarkably,

> The frequency is independent of the amplitude.

$$(13.B.15)$$

(Caution: This is true only when Hooke's law holds exactly.)

Comparing dimensions on both sides of (13.B.12) is a small but rewarding exercise: in case of doubt, it offers a quick way of securing the proper combination of k and m.

iii. Velocity and Acceleration under Harmonic Motion

Figure 13.13 can be put to some further use. The values of the x components x, v_x, a_x can be read off as follows:

$$x = A \cos \omega t \qquad (13.B.16)$$

$$v_x = -\omega A \sin \omega t \qquad (13.B.17)$$

$$a_x = -\omega^2 A \cos \omega t \qquad (13.B.18)$$

[Equation (13.B.16) is not new; see (13.B.6).]

We now have all the details of the motion. In particular, let us note that $\cos \omega t$ and $\sin \omega t$ have maximum value 1. Therefore the maximum values for $x, v_x,$ and a_x are

$$x_{max} = A \qquad (13.B.19)$$

$$v_{max} = \omega A \qquad (13.B.20)$$

$$a_{max} = \omega^2 A \qquad (13.B.21)$$

Here x_{max} is the displacement amplitude, as we have seen before; v_{max} may be called the **velocity amplitude**.

It helps to visualize the timing of the motion's zeroes and maxima. With the help of Fig. 13.13, we can assemble Table 13.1. [The reader should, at this point, run a few simple consistency checks. (1) Differentiate x, Eq. (13.B.16), twice in succession with respect to t. Do you obtain Eqs. (13.B.17),

Table 13.1. Special Points

Quantity	Magnitude at equilibrium point	Magnitude at end points
x	0	max
v_x	max	0
a_x	0	max

(13.B.18)? (2) Insert $\omega = (k/m)^{1/2}$, Eq. (13.B.12), into the expression (13.B.21) for a_x. Is the absolute value consistent with $|F_x| = |kx|$ at the full-amplitude points?]

Example 13.6. Duration of a Round Trip. A 3-kg load is suspended from an ideal spring whose constant is 150 newtons/meter. The load is lifted 5 cm above its equilibrium point and is then released from rest. In how much time will it return to its original height?

When the load has zero speed, its absolute displacement is largest. Hence the required time is the natural period T; see Fig. 13.14. We have according to Eq. (13.B.14),

$$T = 2\pi \left(\frac{m}{k}\right)^{1/2} = 2\pi \left(\frac{3}{150}\right)^{1/2} \text{sec} = \underline{0.89 \text{ sec}}$$

The 5-cm amplitude is irrelevant.

Example 13.7. Speed at the Equilibrium Point. How fast is the load of the preceding example moving when it passes through its equilibrium point?

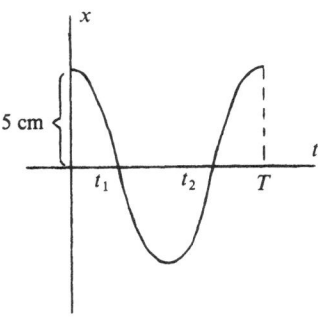

Figure 13.14. The displacement x of a load which starts at rest and returns to its starting point. The total time required is T.

According to Eq. (13.B.20), we have, for the required speed,

$$v_{\max} = \omega A = \left(\frac{k}{m}\right)^{1/2} A$$

$$= \left(\frac{150}{3}\right)^{1/2} (0.05) \frac{\text{meter}}{\text{sec}} = 0.35 \frac{\text{meter}}{\text{sec}}$$

This is the speed corresponding to times t_1 or t_2 in Fig. 13.14.

iv. Rotational Oscillation and the Period of a Pendulum

We conclude our discussion of natural periods by considering rotational (or angular) oscillation; see Fig. 13.15. Here a rigid body, with moment of inertia I, and pivoted about a fixed axis, has an **angular displacement** θ that oscillates harmonically about an equilibrium orientation $\theta = 0$. What is the natural period T of the pivoted object?

There is no need for a fresh derivation of T, as the linear-rotational analogy yields the answer. (The symbols are different; the mathematics is the same.) Recalling the definition of an angular spring constant κ from

$$\tau_\theta = -\kappa\theta$$

[see Eq. (13.A.11)], and making the replacements $k \to \kappa$, $m \to I$, we obtain for the period

$$\boxed{T = 2\pi \left(\frac{I}{\kappa}\right)^{1/2}} \qquad \text{(13.B.22)}$$

in analogy to $T = 2\pi(m/k)^{1/2}$. Comparing units on both sides of (13.B.22) provides an additional check.

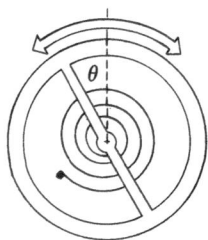

Figure 13.15. A rotational oscillator (balance wheel of a watch). The radial spoke is at an angle θ from its equilibrium orientation (dashed).

Example 13.8. Designing a Balance Wheel. When using a spiral spring ("hairspring") with angular constant $\kappa = 3.0 \times 10^{-5}$ newton meter/rad, what moment of inertia I does one need for the balance wheel of a watch (see Fig. 13.15) if its period of oscillation is to be $T = 0.20$ sec? (Neglect the moment of inertia of the spring itself.)

Formula (13.B.22) gives

$$I = \kappa \left(\frac{T}{2\pi}\right)^2$$

$$= (3.0 \times 10^{-5}) \left(\frac{0.20}{2\pi}\right)^2 \text{ kg meter}^2$$

$$= 3.0 \times 10^{-8} \text{ kg meter}^2$$

Period of a Simple Pendulum

The pendulum is surely the best known of all rotational oscillators. What is its natural period if its length is l and its bob has mass m?

We have already determined the angular spring constant, $\kappa = mgl$ [see Eq. (13.A.14)]; the pendulum's moment of inertia is $I = ml^2$. With these results formula (13.B.22) reads

$$T = 2\pi \left(\frac{ml^2}{mgl}\right)^{1/2}$$

or

$$\boxed{T = 2\pi \left(\frac{l}{g}\right)^{1/2}} \qquad \text{(13.B.23)}$$

The mass does not enter.

Example 13.9. A One-Second Pendulum. What is the length of a "one-second pendulum" (1 second for a one-way swing) on Earth?

From Eq. (13.B.23), and using $T/2 = 1$ sec, we have

$$l = \left(\frac{T}{2\pi}\right)^2 g = \left(\frac{2}{2\pi}\right)^2 (9.8) \text{ sec} = \underline{0.993 \text{ meter}}$$

fortuitously* close to 1 meter. (Note: Measuring the period and the length of a simple pendulum provides one of the best determinations of the local value of g.)

Large Oscillations

The validity of (13.B.23) requires the amplitude of oscillation to be small. This, we recall, is what gave us Hooke's law for a simple pendulum. What happens to the period if the amplitude is not small? This is a difficult problem, and only the result is presented, Fig. 13.16.

Real Pendulums

If the bob is an extended object ("physical pendulum"), its moment of inertia and gravitational torque are only slightly more complicated to determine; see Problem 13.29.

Can the bob be an extended object and yet behave like a particle? The answer is yes, if the **bifilar** arrangement of Fig. 13.17 is used. Since the bob has only translational motion but no spin of its own, its equations of motion are the same as if the whole object were concentrated at its center of mass $\mathbb{C}$, and we again have a simple pendulum.

v. Total Energy of a Harmonic Oscillator

An oscillating spring with constant k and load m has at any time a total energy

$$\mathscr{E} = \mathscr{U} + \mathscr{K} = \tfrac{1}{2}kx^2 + \tfrac{1}{2}mv^2 \quad (13.B.24)$$

where Eq. (13.A.9) is used for $\mathscr{U}$. Since a linear restoring force is conservative, we know that $\mathscr{E}$ is constant; for an explicit verification of this fact, insert (13.B.16), (13.B.17) in the above. As the system oscillates, its energy goes over from entirely potential (at the extreme positions) to entirely kinetic (at the equilibrium point, where the speed must therefore be maximum). Thus we have

$$\mathscr{E} = \tfrac{1}{2}kA^2 = \tfrac{1}{2}mv_{equil}^2 \quad (13.B.25)$$

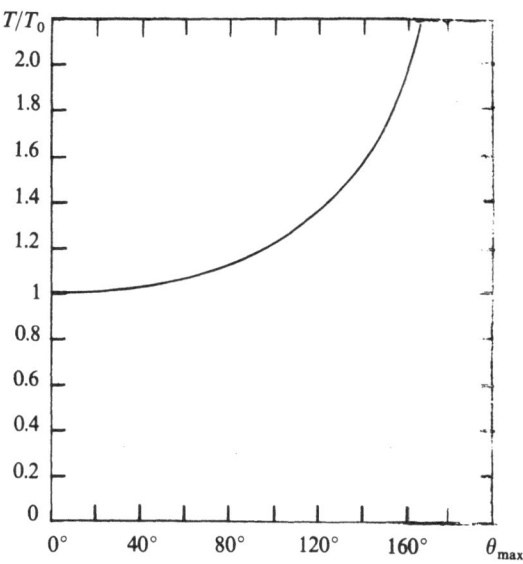

Figure 13.16. If a simple pendulum's maximum angle θ_{max} from the vertical is not small, its period T deviates from $T_0 = 2\pi(l/g)^{1/2}$; the ratio T/T_0 is plotted here against θ_{max}. The string should be replaced by a light rod to permit swinging over amplitudes larger than 90°; T approaches infinity as θ_{max} approaches 180°.

Example 13.10. Falling on a Spring. Figure 13.18 shows a particle of mass m, originally at rest, falling from a height h on top of a relaxed ideal spring whose constant is k. (a) By what amount l must the particle compress the spring before reaching its maximum speed, v_{max}? (b) Calculate v_{max}.

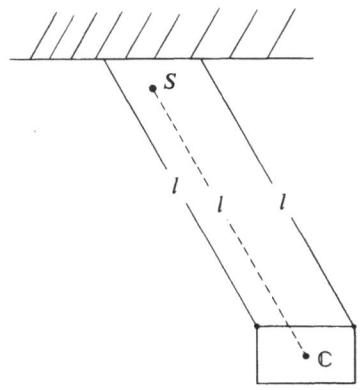

Figure 13.17. The bifilar pendulum oscillates like a simple pendulum: the center of mass $\mathbb{C}$ describes a circle about a fixed point S, and with a radius equal to the string's length l.

---

* Perhaps not quite fortuitously: the one-second pendulum was an early candidate (1790) for the standard meter.

a. *Amount of compression*: While supported by the spring, the particle undergoes harmonic motion. Therefore the maximum speed occurs at the equilibrium point of the loaded spring, with a compression

$$l = \frac{mg}{k} \qquad (13.B.26)$$

see Eq. (13.A.5). Further compression still occurs thereafter, but at decreasing speed.

b. *Maximum speed*: Conservation of energy from the start of free fall to the equilibrium point gives

$$\mathcal{U}_f + \mathcal{K}_f = \mathcal{U}_i + \mathcal{K}_i$$

(f = final, i = initial). If the gravitational potential energy is calculated *with respect to the loaded spring's equilibrium point*, then only the spring contributes to $\mathcal{U}_f$, according to $\mathcal{U}_f = \frac{1}{2}kl^2$; see Eq. (13.A.9). Also, the particle is dropped from rest, so that $\mathcal{K}_i = 0$. The conservation equation becomes

$$\tfrac{1}{2}kl^2 + \tfrac{1}{2}mv_{max}^2 = mg(h+l) + 0$$

or

$$v_{max} = \left[2g(h+l) - \frac{kl^2}{m} \right]^{1/2}$$
$$= \left[g\left(2h + \frac{mg}{k} \right) \right]^{1/2} \qquad (13.B.27)$$

where (13.B.26) was used in the last step.

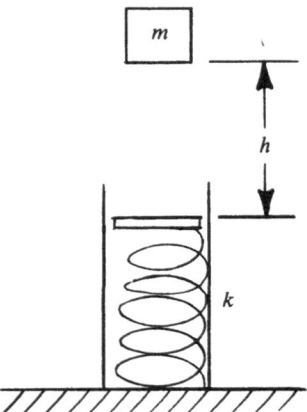

Figure 13.18. A mass m falls on an initially relaxed spring.

Figure 13.19. This device illustrates the superposition (top piston) of three motions (lower pistons). The liquid is assumed incompressible.

C. Superposition and Sound

In the real world, vibrations are apt to be very complicated, and harmonic motion is the exception rather than the rule. Nevertheless, scientific and engineering analyses rely heavily on harmonic motion in order to deal with more intricate cases. The present section will give some indication as to why this is so.

i. Vibrations Can Be Superposed

The oscillations found in Nature, even when not harmonic, can be considered as *superpositions* of harmonic motions. The meaning of a superposition is illustrated by the device of Fig. 13.19, in which all pipes have the same cross section. If the three lower

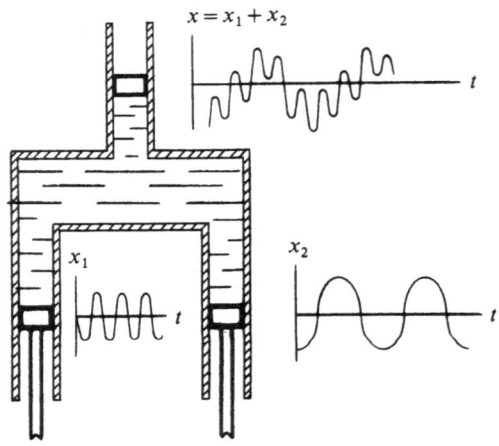

Figure 13.20. The lower pistons' harmonic motions x_1, x_2 and their superposition x are shown as functions of time.

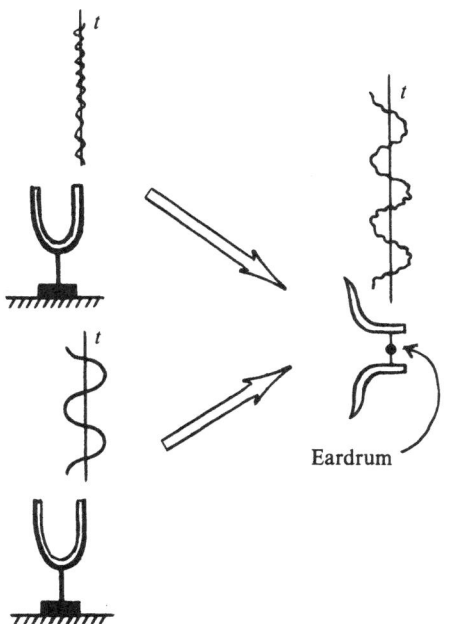

Figure 13.21. The eardrum's motion is a superposition, arising from two tuning forks.

pistons move upwards by distances x_1, x_2, x_3, then the upper piston moves upwards by a distance

$$x = x_1 + x_2 + x_3$$

Its motion is said to be a **superposition** of the other pistons' motions. In the example of Fig. 13.20, two lower pistons, each with the same area as the top piston, are made to undergo harmonic motion with very different amplitudes and frequencies. A far

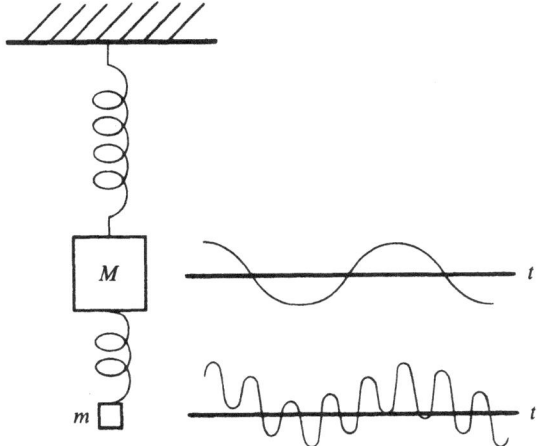

Figure 13.22. The large mass has a motion that is essentially unaffected by that of the small one. If the latter oscillates harmonically with respect to the large one, its total motion is the superposition shown.

more common form of this phenomenon occurs without the help of pipes. As an illustration, Fig. 13.21 shows two harmonic motions, simultaneously transmitted through the air to someone's eardrum. A typical point on the eardrum then undergoes a motion that is essentially a superposition of the original motions. As another example, the system of Fig. 13.22 will exhibit, for the small load, a motion quite similar to the one plotted for x in Fig. 13.20.

To summarize: Superposing several vibrations means, mathematically, taking the sum of their displacements. Superposition is achieved physically if vibrations are transmitted from several sources to the same object, as in Figs. 13.19–13.21, or if a single *system* is allowed to vibrate without external influence, as in Fig. 13.22, but with appropriate initial conditions.

Here we are interested in the **superposition of harmonic motions**. We shall look first at superpositions that are themselves periodic.

ii. Harmonics

Even a complicated vibration may still be periodic, i.e., it may repeat itself precisely at regular time intervals; an illustration is given in Fig. 13.23.

Since we now understand harmonic motion, let us use it to synthesize more intricate—but still periodic—vibrations. As an illustration, we combine several harmonic motions in such a way that the resulting superposition repeats itself, say, every second. First we ask what harmonic motion, taken by itself, has that one-second property. Figure 13.24 shows that there are many candidates,

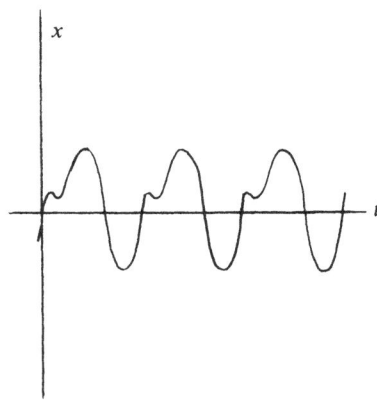

Figure 13.23. Example of a periodic but nonharmonic motion. This might be the vibration of a point on a piano string.

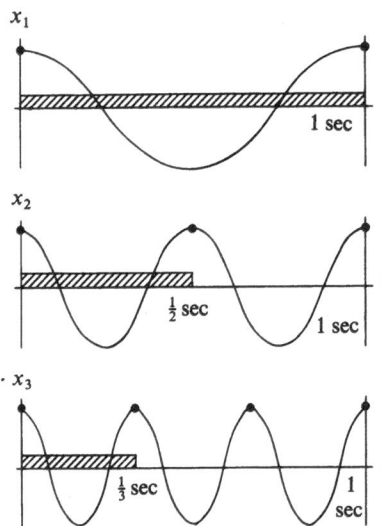

Figure 13.24. These three harmonic motions, as well as infinitely many others, repeat themselves every second.

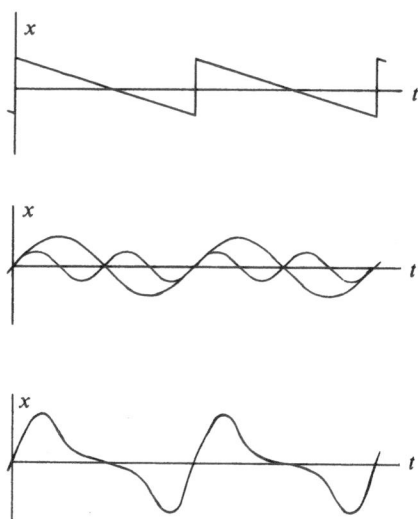

Figure 13.25. The top "saw tooth" curve is a good approximation to the stick-slip motion of a bowed violin string. It follows the bow during the downstroke and snaps back almost instantaneously during the upstroke; the bow is assumed not to change its motion. The middle set of curves show the first and second harmonics, whose superposition (bottom curve) is already a remarkably good reproduction of the sawtooth. Inclusion of a few more harmonics would reconstruct the sawtooth almost exactly.

namely, the harmonic motions with periods 1 sec, 1/2 sec, 1/3 sec, etc. Since their sum also exhibits the 1-sec periodicity, we conclude that a *superposition of harmonic motions with periods* 1 sec, 1/2 sec, 1/3 sec, etc., *is itself a periodic vibration with period* 1 sec. The frequencies involved are 1 Hz, 2 Hz, 3 Hz, etc.; the overall frequency is 1 Hz. More generally:

> A superposition of harmonic motions with frequencies f, $2f$, $3f$, etc. is itself a *periodic* (but not harmonic) vibration with period f.

(13.C.1)

In such a superposition, f is called the **fundamental frequency**; the harmonic motion with frequency f is called the **first** or **fundamental harmonic**, the one with frequency $2f$ is the **second harmonic**, and so on.

(The first few harmonics are heard as musical tones. For example, if $f = 440$ Hz is heard as an A, then $2f = 880$ Hz is the next higher A, and $3f = 1320$ Hz is the next E.)

Remarkably, the overall vibration can be *anything* with frequency f. It was shown mathematically by Fourier in 1807 that

> Any periodic vibration whatsoever is a superposition of harmonic motions.

(13.C.2)

An illustration of that fact is provided by the vibration of any single point on a violin string, Fig. 13.25.

iii. Pitch and the Acoustic Spectrum

A harmonic oscillation of the eardrum is heard as a so-called **pure musical tone**, or tone of definite **pitch**. Pitch is a subjective property related to the frequency; high frequency means high pitch, and conversely. As a frequency is doubled ($f \rightarrow 2f$), the corresponding pitch is said in musical terminology to be raised by **one octave**. The human ear is, at its best, sensitive to frequencies between about 20 Hz and 20 000 Hz (amounting to approximately how many octaves?). The range narrows considerably as the subject gets older. Figure 13.26, a spectrum of mechanical frequencies, is meant to put these data in perspective.

iv. Beats

In the preceding discussion we have combined frequencies that are related like simple integers. Another quite different and interesting case of superposition involves two frequencies that are

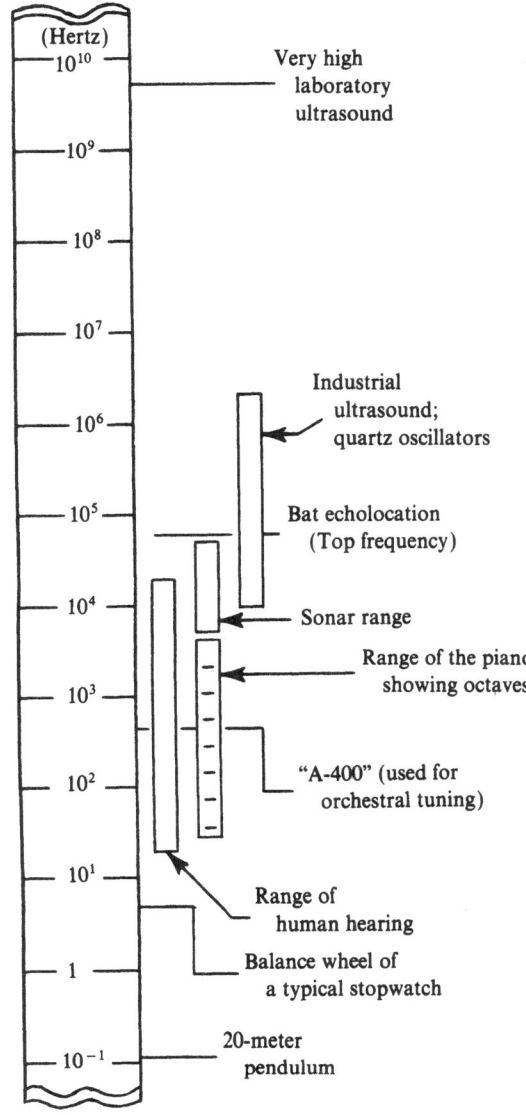

Figure 13.26. A spectrum of mechanical frequencies (mostly acoustic, i.e., for sound) shown on a logarithmic scale.

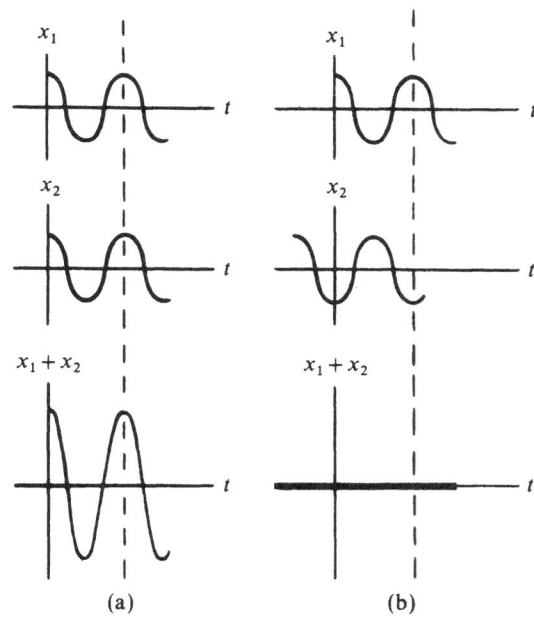

Figure 13.27. The top two motions are superposed to yield the bottom one. (a) Constructive interference. (b) Destructive interference.

A laboratory illustration of this seeming paradox is shown in Fig. 13.29.

Next, suppose the two frequencies are slightly different, say $f_1 = 6$ Hz and $f_2 = 7$ Hz as in Fig. 13.30. The oscillations may start in phase, but they eventually become out of phase by 180°; later they are in phase again. The corresponding reinforcements and cancellations are shown in the figure; each rise in overall amplitude is called a **beat**. If the two amplitudes are not the same, one still observes the beats, but with incomplete cancellation, as in Fig. 13.31. A so-called **moiré pattern**, Fig. 13.32, provides an intriguing way of visualizing the phenomenon.

For any f_1 and f_2 (f_2 only slightly larger than f_1), what is the beat frequency f_b? To solve this, we

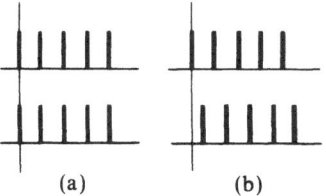

Figure 13.28. The top two curves of Fig. 13.27 are symbolized here by their peaks only; more cycles are shown. (a) When the peaks coincide, constructive interference occurs. (b) When the peaks alternate, the interference is destructive.

nearly equal. To see what happens, we first let both frequencies be exactly the same, as well as the amplitudes. The resulting superposition then only depends on the **relative phase**, as shown in Fig. 13.27. The two oscillations reinforce each other maximally if in phase and cancel each other completely if *out of phase by 180°*. They are said to **interfere constructively** or **destructively** in these respective cases. Figure 13.28 shows more schematically what happens in terms of the displacement peaks of each harmonic motion.

Thus, we are led to the curious but true statement that adding two sounds may produce silence.

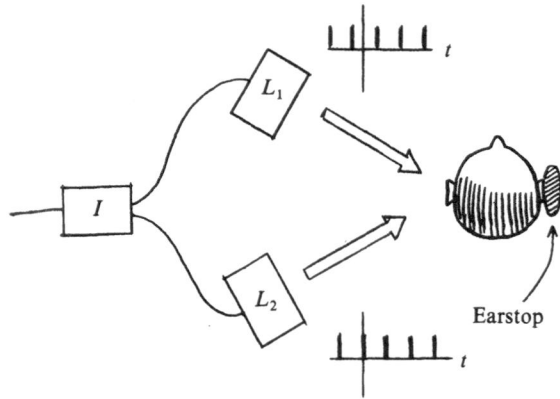

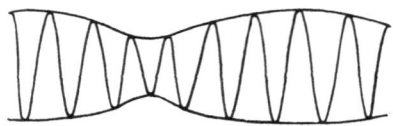

Figure 13.31. Beats between two harmonic motions with nearly equal frequencies, but different amplitudes.

Figure 13.29. Sound + sound → silence (destructive interference). In this experiment, the two loudspeakers L_1, L_2 are controlled by a common input I which makes them vibrate at the same amplitude and frequency, but in opposite phases. If the open ear is equally distant from L_1 and L_2, the received vibrations will still be of equal amplitude, and out of phase by 180°; very little will be heard. Disconnecting one speaker will greatly bring up the sound level.

Example 13.11. Tuning a Violin. The A string of a violin, which is slightly out of tune, is bowed while an A (440 Hz) tuning fork is struck. If 2 beats per second are heard, what is the actual frequency of the string?

We have, from (13.C.3), either

$$f_1 = 440 \text{ Hz}, \qquad f_2 = 440 \text{ Hz} + 2 \text{ Hz} = \underline{\underline{442 \text{ Hz}}}$$

or

$$f_2 = 440 \text{ Hz}, \qquad f_1 = 440 \text{ Hz} - 2 \text{ Hz} = \underline{\underline{438 \text{ Hz}}}$$

There is no information to indicate whether the string or the tuning fork is higher.

use the diagram of Fig. 13.33, starting at A with the two oscillations in phase, and terminating at B as soon as they are in phase again. The individual periods are $1/f_1$ and $1/f_2$. We see that n cycles of one oscillation match $n + 1$ cycles of the other (in the figure, $n = 6$; in general, n is not an integer). Since the time interval AB equals the beat period $1/f_b$, we obtain

$$\frac{1}{f_b} = \frac{n}{f_1}$$

$$\frac{1}{f_b} = \frac{n+1}{f_2}$$

Eliminating n gives

$$\boxed{f_b = f_2 - f_1} \qquad (13.\text{C}.3)$$

The beat frequency equals the frequency difference.

D. Resonance

An oscillator can respond dramatically to an external force, especially to a force that itself oscillates periodically. One sometimes hears the story of a soprano shattering a glass by her unaided voice. It does not matter to us here whether this is legend or truth, for the phenomenon can be reliably generated in the laboratory; see Fig. 13.34. The glass is the oscillator, endowed with a natural fre-

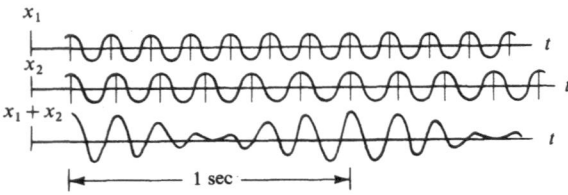

Figure 13.30. The superposition of the top two motions is shown at the bottom. Interference alternates between constructive and destructive. In this illustration the beat period equals 1 sec.

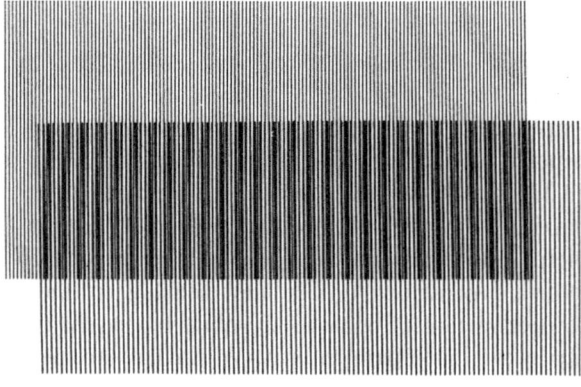

Figure 13.32. A moiré pattern (two superposed patterns whose line spacings are slightly different) for visualizing beats.

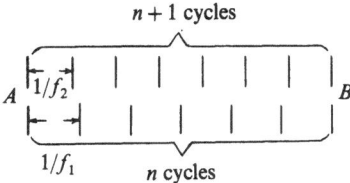

Figure 13.33. A single beat, arising from frequencies f_1, f_2.

quency f_0; that frequency is clearly evident as the very pure tone produced when the glass is lightly rubbed or struck in a suitable way. (The glass has many natural frequencies, but only one need concern us here.) The external oscillating force, or input, of frequency f_{in}, is exerted by the air that touches the glass, and is transmitted from a loudspeaker. The breaking occurs when the loudspeaker's frequency matches very accurately that of the glass:

$$\boxed{f_{in} = f_0} \qquad (13.D.1)$$

This is the basic equation for all resonance

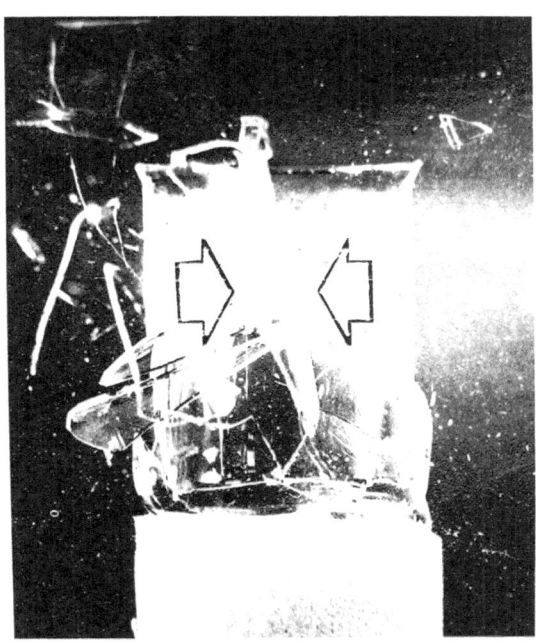

Figure 13.34. A glass beaker can break when driven to vibrate at its natural frequency. The loudspeaker is behind the beaker, between the arrows. [Reprinted by permission from the *Science of Hi-Fidelity*, by K. W. Johnson, W. C. Walker, and J. D. Cutnell (Kendall/Hunt, 1981); photographed by W. C. Walker.]

phenomena. In short, *resonance is a peak response which occurs when a driving frequency matches a natural one.*

i. Features Common to All Resonances

Many other examples of resonance come to mind: a child's swing under slight but well-timed pushes; a car with defective shock absorbers being driven at just the "right" speed on a wavy road surface; the legendary bridge collapsing under the rhythmic step of an army; the actual collapse of the Tacoma Narrows Bridge responding to a periodic wind force (see Fig. 13.35); a piece of sheet metal in a household appliance, or in a car, rattling in response to the matching frequency of a motor; the air column in a musical wind instrument, responding to periodic vibrations at the mouthpiece.

All these cases are mechanical, but electrical and atomic systems are equally prone to resonate. This is, for example, how a radio tuner responds to a sharply defined radio frequency, or how a color filter blocks light of a certain frequency. Finally, consider an application from advanced nuclear technology: an electron in a ^{235}U atom is made to resonate when subjected to laser radiation of a special frequency. The corresponding electron in ^{238}U has a slightly different natural frequency, and therefore does not respond. One day a cheap and efficient separation process may exploit this difference in order to extract the energetically precious ^{235}U from among the much more abun-

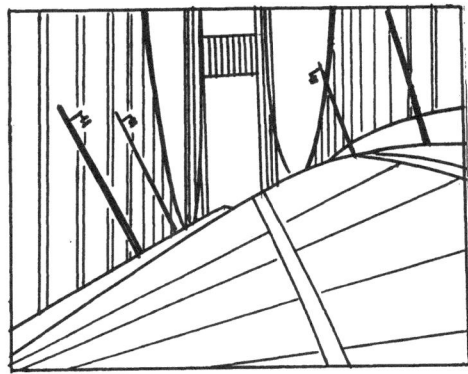

Figure 13.35. On 7 November 1940, wind eddies around this suspension bridge matched the latter's natural frequency of 0.23 Hz for torsional waves, with disastrous results (the wind speed was only 45 miles/hr). As in a flute, *steady* blowing can generate *periodic* eddies through a so-called "nonlinear interaction." The sketch is after a contemporary photograph.

dant but less useful ^{238}U atoms. (Chemistry is unable to discriminate between them.)

All cases of resonance share the following characteristics:

---

1. A modest external periodic force, if applied during a sufficient number of cycles, produces a large or catastrophic amplitude of vibration.

2. An accurate matching of natural and input frequencies is needed. (Hence the natural frequency is also called the **resonant frequency**; the input frequency is also called the **driving frequency**.)

3. Friction or other energy losses reduce the obtainable response but also reduce the necessity for sharp tuning.

(13.D.2)

ii. Resonance in a Loaded Spring

This chapter's favorite device, the loaded spring, demonstrates in a simple way how and why resonance occurs. Consider, as in Fig. 13.36, a spring with constant k, one end of which is loaded with a mass m, the other end being held in a clamp that moves harmonically with a displacement amplitude A_{in} (the input amplitude). The load is supported on a frictionless horizontal plane. In the simplest situation, a **steady state** prevails, in which the load and

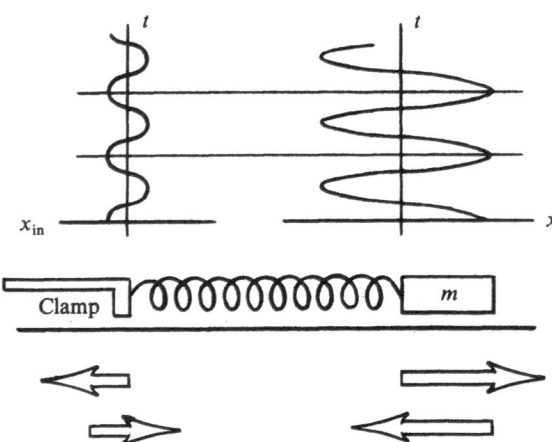

Figure 13.36. Resonance in a loaded spring. A small oscillation of the clamp at left causes a large oscillation of the load. The positions of clamp and load are shown plotted against time; note that x is positive to the right, x_{in} to the left.

the clamp both oscillate at the same angular frequency ω_{in}. Given the natural angular frequency

$$\omega_0 = \left(\frac{k}{m}\right)^{1/2} \qquad (13.D.3)$$

of the loaded spring, as well as ω_{in} and A_{in}, what is the response amplitude, A, of the load?

For definiteness, suppose $\omega_{in} > \omega_0$. To make the system oscillate at this larger-than-natural frequency, Eq. (13.D.3) shows that one must effectively increase the value of k "felt" by the load. This means increasing the restoring force, which in turn means increasing the spring's elongation at the clamp's end. In order to achieve this, let the harmonic oscillations of load and clamp be synchronized as in Fig. 13.36: the clamp's displacement x_{in} is at any time proportional to the load's displacement x. In other words, x_{in}/x is a constant throughout the motion:

$$\frac{x_{in}}{x} = \frac{A_{in}}{A}$$

giving

$$x_{in} = \frac{A_{in}}{A} x \qquad (13.D.4)$$

Hooke's law for the load reads

$$F_x = -k(x + x_{in})$$

or, with (13.D.4),

$$F_x = -k\left(1 + \frac{A_{in}}{A}\right) x$$

Thus k has effectively been increased according to

$$k \to k\left(1 + \frac{A_{in}}{A}\right)$$

In view of the proportionality $\omega \propto \sqrt{k}$, the frequency has been increased as follows:

$$\omega_0 \to \omega_0\left(1 + \frac{A_{in}}{A}\right)^{1/2}$$

or

$$\omega_{in} = \omega_0 \left(1 + \frac{A_{in}}{A}\right)^{1/2} \qquad (13.D.5)$$

Solving this for A, we obtain

$$A = \frac{A_{in}}{(\omega_{in}/\omega_0)^2 - 1} \qquad (\omega_{in} > \omega_0)$$

A similar argument holds for the case $\omega_{in} < \omega_0$, but here the clamp must move in the same direction as the load, instead of against it. Both cases are covered by a formula involving the absolute value:

$$A = \frac{A_{in}}{|(\omega_{in}/\omega_0)^2 - 1|} \qquad (13.D.6)$$

This is plotted in Fig. 13.37. We see that, for a given input amplitude, *the response amplitude becomes arbitrarily large as ω_{in} approaches ω_0*, a clear case of resonance.

> **Example 13.12. Doubling the Input Amplitude.**
> Suppose you are holding one end of a spring, with a load suspended from the other end; see Fig. 13.38. With what frequency f_{in} should you move your hand in order to give the load twice the amplitude of your hand? Assume a natural frequency of 1.00 Hz; also assume $f_{in} > 1.00$ Hz.
> _____

Equation (13.D.5) gives

$$f_{in} = f_0 \left(1 + \frac{A_{in}}{A}\right)^{1/2}$$

$$= (1.00 \text{ Hz})(1 + 1/2)^{1/2} = \underline{1.22 \text{ Hz}}$$

(There is also another possible frequency, smaller than 1.00 Hz; see Problem 13.38.)

"Hitting the Resonance"

On the basis of Fig. 13.37, what happens at $\omega_{in} = \omega_0$? Here A is infinite, but of course no actual system can have an infinite amplitude; some damage may occur instead. Alternatively, if frictional losses (also called **damping losses**) are present, it can be shown that the following occurs:

1. The actual amplitude is finite but still maximal at $\omega_{in} \approx \omega_0$; see the dashed line in Fig. 13.37.
2. At that frequency, the input clamp neither moves against the load, nor in the same direction, but rather *leads* the motion of the load in time by about 1/4 cycle.

How large an amplitude can be achieved at resonance? The answer depends on how much damping is present; the less damping, the higher the maximum amplitude. In addition, the resonance behavior depends on the *kind* of friction (dry, viscous, etc.) that prevails. This is why a quantitative discussion of damping in mechanical systems is quite involved; it will not be attempted here.

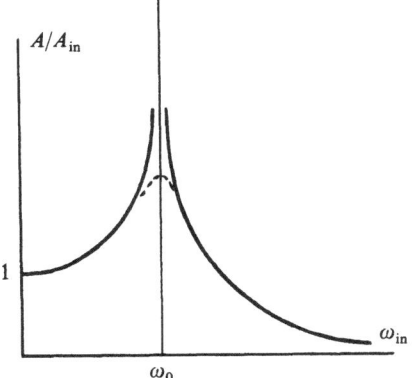

Figure 13.37. The resonance curve for the phenomenon of Fig. 13.36. The response amplitude A is plotted in units of the (fixed) input amplitude, for a range of input frequencies ω_{in}. The dashed curve corresponds to the presence of some friction.

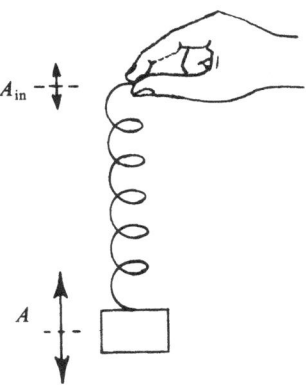

Figure 13.38. At a certain frequency, the load responds to the hand with a factor of 2 in the amplitude.

Condensed Checklist

Hooke's law results in harmonic motion.   (13.B.10)

Hooke's law:

$$F_x = -kx$$ (13.A.2), (13.A.6)

also for *vertical* loaded spring.

Harmonic motion:

$$x = A \cos \omega t \qquad (13.B.6)$$

Natural angular frequency:

$$\omega = \left(\frac{k}{m}\right)^{1/2} \qquad (13.B.12)$$

Harmonic motion checklist:

Amplitude A;   phase $\varphi = \omega t$

Angular frequency ω;   Frequency $f = \dfrac{\omega}{2\pi}$;

Period $T = \dfrac{2\pi}{\omega}$

$x_{max} = A$; $v_{max} = \omega A$; $a_{max} = \omega^2 A$ Fig. 13.12

Total energy:

$$\mathscr{E} = \tfrac{1}{2}kA^2 = \tfrac{1}{2}mv_{equil}^2 \qquad \text{Example 13.10}$$

Linear versus rotational oscillator (correspondences):

$$x \to \theta, \qquad F_x \to \tau_\theta, \qquad m \to I, \qquad k \to \kappa \qquad (13.B.22)$$

Simple pendulum:

$$T = 2\pi \left(\frac{l}{g}\right)^{1/2} \qquad (13.B.23)$$

Periodic **vibration of frequency f:**

The harmonics have frequencies $f, 2f, 3f$, etc.

(13.C.1)

Octave:

Frequency ratio $= 2$ Sec. C iii

Constructive interference: $0°$ phase difference.

Fig. 13.27

Destructive interference: $180°$ phase difference.

Beat frequency:

$$f_{beat} = f_{higher} - f_{lower} \qquad (13.C.3)$$

Resonance:

$$f_{in} = f_0 \qquad (13.D.1)$$

$$\text{Resonance factor} = \frac{1}{|(\omega_{in}/\omega_0)^2 - 1|} \qquad (13.D.6)$$

True or False

1. The restoring force exerted by a given spring, when extended, is proportional to its length.

2. If a pingpong ball keeps bouncing vertically (and elastically) on a table, it is performing harmonic motion.

3. A vertically suspended loaded spring has an amplitude of vibration that is proportional to the load.

4. As a vertically suspended loaded spring is further extended by an external force, its combined elastic *and gravitational* potential energy increases.

5. The time-averaged energy of vibration in a harmonic oscillator equals its peak kinetic energy.

6. The total vibrational energy of a loaded spring can be calculated from a knowledge of the vibrating mass, the frequency, and the amplitude.

7. In the above, knowledge of the frequency is superfluous.

8. The maximum deflection of a given harmonic oscillator from equilibrium is proportional to its speed at the equilibrium point.

9. If harmonic vibrations of two different frequencies are superposed, the result is a harmonic vibration of intermediate frequency.

10. If two tones, one octave apart, are sounded together, they can in some cases be made to produce complete silence, provided their amplitudes are the same.

11. As the frequencies of two tones approach one another during a tuning operation, the beats between them become faster and faster until they are no longer noticed.

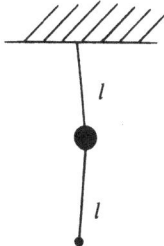

12. Suppose a simple pendulum of length l has a very massive bob, from which is suspended another simple pendulum, with the same length l and a very light bob, as shown. After the massive bob is given a small nudge, the light bob will exhibit a resonant behavior.

Problems

Section A : Hooke's Law and the Ideal Spring

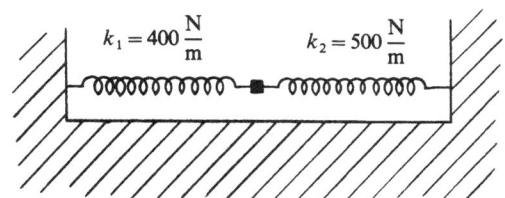

*13.1. In the figure, both springs are relaxed (unstretched). Find the total spring constant k for particle P under a horizontal displacement. (The spring's outer ends are held fixed.)

13.2. A spring scale (see the figure) supports, at rest, two blocks of masses 5 kg and 7 kg. After the 5-kg block is removed, the tray reaches a new equilibrium position, 0.04 meter higher. Assuming Hooke's law, what is the spring constant of the scale?

*13.3. Suspending a 5.0-kg load lowers the equilibrium point of a certain spring by 0.080 meter. What is the spring constant?

*13.4. By how much should a spring with constant 100 newtons/meter be elongated in order to store 2 joules of potential energy?

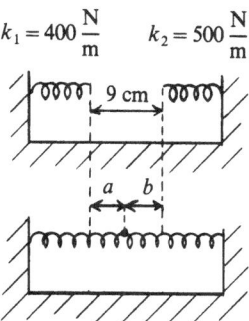

*13.5. In the figure, both springs are shown first relaxed, and then at equilibrium when joined. What are the displacements a, b in the latter situation?

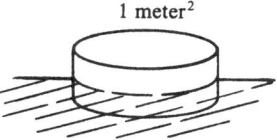

*13.6. A wooden cylinder of basis area 1 meter^2 floats on water, as shown. What is its spring constant, as measured by the application of a static external force to sink it by a given amount?

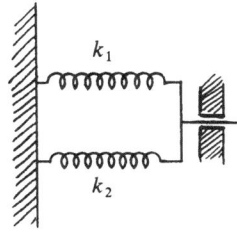

13.7. Two springs, of given force constants k_1, k_2 and equal equilibrium lengths, are tied in parallel as shown. What is the overall spring constant, k, for the free end of the combination?

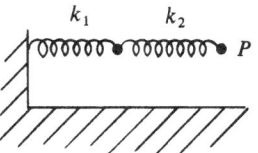

*13.8. Two springs, with constants k_1, k_2 are tied in series as shown. What is the overall spring constant k for the free end P?

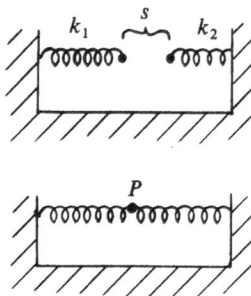

13.9. Two horizontal springs, with given constants k_1 and k_2, are tied in a stretched condition, as shown. The gap between them was a given amount s when both were relaxed. What is the overall spring constant for point P under horizontal displacements?

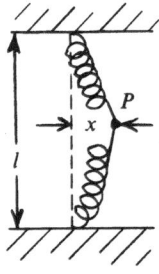

*13.10. Two identical stretched springs, each exerting a force of magnitude F, are joined at point P, as shown. Their outer ends are fixed, and are a distance l apart. What is the spring constant for small lateral displacements x of P?

Section B : Harmonic Motion

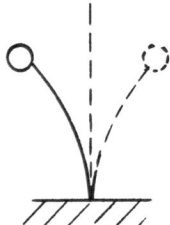

13.11. A 0.5-kg lead ball is attached to the top of a long flat spring, of negligible mass, and whose lower end is clamped, as shown. A static 15-newton force will displace the ball 3 cm to the side. Find the natural frequency of oscillation, f.

13.12. By what factor must the load of a harmonic oscillator be multiplied to triple its frequency?

*13.13. If the amplitude of a given harmonic oscillator is doubled, by what factor does this multiply (a) the frequency, (b) the total energy, (c) the maximum speed?

13.14. Each prong of a certain A (440-Hz) tuning fork vibrates with a 1.0-mm amplitude at the tip. (a) What is the latter's maximum speed? (b) What is its maximum acceleration in units of $1g$?

*13.15. A 1.5-kg block is suspended from a spring with constant 80 newtons/meter. Find (a) the shift in the equilibrium point, and (b) the natural frequency in Hz.

13.16. From Eq. (13.B.24), verify explicitly that the energy of a harmonic oscillator is conserved.

*13.17. A pendulum clock is brought to the Moon, where the acceleration of gravity is 1/6 that of Earth. Will the clock be slow or fast, and by what factor?

13.18. A 1-kg load, when suspended from a certain spring, displays a natural period of 1 sec. What is the spring constant?

*13.19. (a) A certain light spring becomes elongated by 0.25 meter when a certain load is suspended at equilibrium, as in Fig. 13.3. What is the natural period of vertical oscillation? (Yes, the information is sufficient.) (b) What is the length of a simple pendulum with the same natural period?

*13.20. A load of mass 0.10 kg hangs from a long, light spring. When pulled down 0.20 meter below its equilibrium position and released from rest, it vibrates with a period of 2 sec. (a) What is its speed as it passes through the equilibrium position? (b) By how much will the spring shorten when the load is removed?

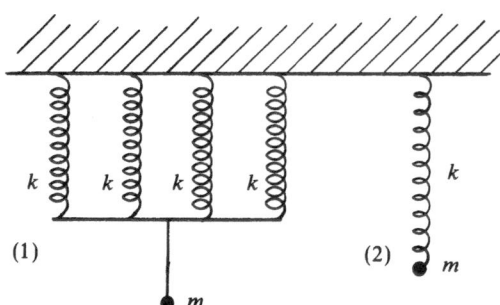

13.21. What is the ratio of natural periods, T_1/T_2, for the two oscillators shown in the figure? (All springs are of identical construction; consider vertical motion only.)

*13.22. A 0.5-kg load, hanging from the end of a spring, oscillates with maximum speed 0.75 meter/sec and amplitude 0.25 meter. What is the spring constant?

*13.23. A harmonic oscillator with given load m has a given frequency f and a given amplitude A. To what maximum force is the load subjected?

13.24. A certain automobile engine is operating at 300 rpm. If one of the pistons has a 5-cm length of stroke and a 0.25-kg mass, find (a) its maximum speed; (b) its maximum acceleration; (c) the maximum force to which it is subjected. (Assume its motion is harmonic; see also Problem 13.23.)

13.25. A harmonic oscillator has amplitude 25 cm and frequency 3 Hz. When its displacement is 5 cm, what is (a) its speed, (b) its acceleration?

*13.26. A harmonic oscillator is started from rest at a displacement of 25 cm. If its frequency is 3 Hz, find the earliest time at which the displacement is 5 cm.

13.27. The figure shows a **torsion pendulum** such as was used by Cavendish to measure the gravitational constant (Fig. 8.6 of Chap. 8). If the pendulum oscillates with a period of 2 minutes, and consists of a light 30-cm-long rod suspended at the middle, with a 1-gram mass at each end, find (a) the angular spring constant κ; (b) the torque τ needed for a static 1° deflection from equilibrium; and (c) the perpendicular force F needed at each end for that deflection.

13.28. (The pendulum as a nearly linear oscillator.) Consider a simple pendulum of length l and mass m. Over small oscillations, the bob's trajectory is almost straight. Let s_{max} be the bob's maximum length of arc away from equilibrium, i.e., its approximate displacement amplitude. Then Eq. (13.B.20) reads $v_{max} = \omega s_{max}$, or $v_{max} = (2\pi/T)s_{max}$. (a) How does this formula read in terms of l and g rather than T? (b) A certain grandfather clock has a simple pendulum of length 0.75 meter and mass 0.15 kg. If the bob swings 8.0 cm on either side of equilibrium, what is its speed as it goes through equilibrium?

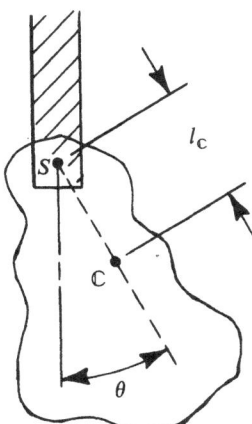

*13.29. The figure shows a real (or physical) pendulum, of total mass m, with its center of mass $\mathbb{C}$ at a distance $l_{\mathbb{C}}$

from its point of suspension S. If its radius of gyration (Problem 7.3) is l_{gyr}, find (a) the gravitational restoring torque about S when the angular displacement is θ; (b) the angular spring constant for small θ; (c) the natural period for small θ; (d) the length of the simple pendulum having the same period. (e) Calculate the natural period of a uniform rod of length l suspended from one end.

Sections C : Superposition and Sound ; D : Resonance

*13.30. From Fig. 13.26, over how many octaves does human hearing extend?

13.31. (a) Similarly to Fig. 13.30, plot the maxima of a 5-Hz and a 6-Hz oscillator over a 2-sec interval, starting in phase. (b) At what times do the beat maxima occur? Indicate their positions on the diagram.

*13.32. Two sources of sound, having periods of 7.500×10^{-3} sec and 7.510×10^{-3} sec, are heard at the same time. What is the beat period?

13.33. How much time elapses between two beats when tones of frequencies 261.6 Hz and 261.0 Hz are heard together?

*13.34. In Fig. 13.36, the driving clamp moves with an amplitude $A_{in} = 5$ cm. Find the load's steady-state amplitude A when the clamp's frequency is 10 Hz while the loaded spring's natural frequency is 11 Hz. (Assume no friction.)

13.35. In Fig. 13.37, and with $\omega_0 = 7$ rad/sec, what should $|\omega_{in} - \omega_0|$ be in order that $A = 100 A_{in}$? (Assume no friction.)

*13.36. Two pendulums, of lengths 2.01 meters and 2.03 meters, are oscillating in phase. How long must one wait to see them oscillate out of phase by 180°?

*13.37. In the experiment of Fig. 13.38, with natural frequency $f_0 = 1.00$ Hz, what should be the driving frequency f_{in} in order that the load have the same amplitude as the hand?

13.38. What is the alternative frequency, $f_{in} < 1.00$ Hz, in Example 13.12?

Answers to True or False

1. False (length is not the same as deflection from equilibrium).

2. False (only periodic).

3. False (unrelated to the load).

4. True (the increase in the former more than makes up for the decrease in the latter; total energy must increase if net external work is done).

5. True (they are both equal to the total energy).

6. True.

7. False.

8. True.

9. False (not a harmonic vibration at all).

10. False.

11. False (slower and slower).

12. True.

Answers or Hints to Problems

13.1. 900 newtons/meter.

13.3. 625 newtons/meter.

13.4. 0.2 meter.

13.5. Use $k_1 a = k_2 b$ (action and reaction), $a + b = 9$ cm.

13.6. Use buoyancy = (displaced volume) × (density of water) × (g).

13.8. Assume a static external force on P, require (net force on spring No. 2) = 0; introduce displacements x_1, x_2 for each spring. Answer: $k = k_1 k_2 / (k_1 + k_2)$.

13.10. Use $F \approx$ const. Answer: $4F/l$.

13.13. (a) 1; (b) 4; (c) 2.

13.15. (b) 1.16 Hz.

13.17. See also Problem 2.71.

13.19. (a) 1.0 sec; (b) 0.25 meter.

13.20. The mass information is not needed.

13.22. $k = 4.5$ newtons/meter.

13.23. $(2\pi f)^2 mA$.

13.26. First show that $t = (1/2\pi f) \cos^{-1}(x/A)$.

13.29. (b) mgl_C; (c) $2\pi l_{gyr}/\sqrt{gl_C}$; (d) l_{gyr}^2/l_C.

13.30. About 10 octaves.

13.32. 5.6 sec.

13.34. 29 cm.

13.36. 300 sec.

13.37. About zero, or $(\sqrt{2})(1.00$ Hz$)$.

CHAPTER 14

WAVES

Having examined how a single object vibrates, we are now ready to explore how its vibration may be transmitted some distance away, to other objects; this occurs through the intermediary of **waves**.

Nature provides a constant display of many kinds of waves; all of them share to a considerable extent the features discussed in this chapter. Here we shall deal mostly with **mechanical waves**, and in particular with those that propagate sound.

A later chapter takes up the topic of **electromagnetic waves**, of which light and radio signals are familiar cases. (Gravitational waves are also believed to exist, and may some day be caught by a man-made detector.)

The concept of waves finds its most revolutionary application in the science of **wave mechanics**, according to which a particularly subtle kind of wave determines the outcome of many measurements involving particles; we shall briefly touch upon that subject when we discuss atomic physics.

Wave motion is a large topic. Accordingly this book approaches it in two installments. The present chapter, being an introduction to the subject, considers *one-dimensional propagation* only; that is, our waves will be assumed to travel (possibly back and forth) along a single direction in space.

Much later (in Chap. 24, Wave Optics), we ask what happens when waves come from several directions at once. There the context will be electromagnetic rather than mechanical, but this is incidental; most wave properties apply equally to both fields of study.

A. Traveling Waves in Strings

A string under tension exhibits wave propagation in a simple way, which is nevertheless typical of all wave motions. Our discussion therefore greatly transcends the narrow topic of strings; it should be carefully followed for later reference.

To understand the plan of this section, it helps to recall our approach to harmonic motion. First came a mere *description* (in mathematical terms); afterwards, the oscillation was related to a force via Newton's law of acceleration.

Similarly, here we first learn how to describe a traveling wave mathematically; special attention will be paid to an important shape, the sinusoidal one. Later in the section, we shall see how wave propagation follows from Newton's laws as applied to a string.

i. Waves as Delayed Signals

Consider a uniform horizontal stretched string, as in Fig. 14.1. When in static equilibrium, it is a straight line, along which a coordinate axis, x, can be laid out. (Effects such as bending under gravity will be neglected; this is legitimate when the tension is sufficient.)

Let us start shaking the string's left end, P_0, *in a vertical direction*, while keeping the tension constant. The vertical distance y between P_0 and its equilibrium position is given by some function of time,

$$y = \mathcal{F}(t) \qquad (14.A.1)$$

The function $\mathcal{F}$ is controlled "by hand."

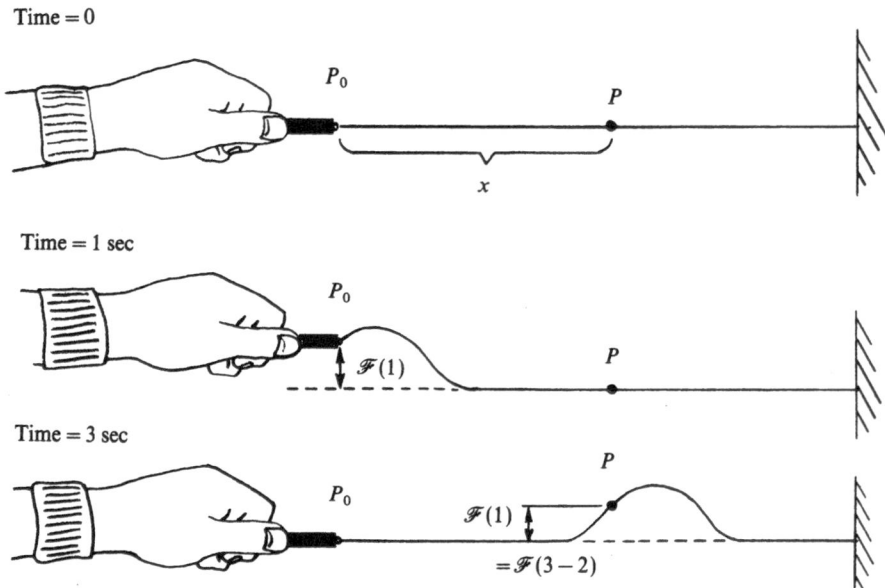

Figure 14.1. Shaking one end of a string under tension results in a traveling wave. In this illustration, the string is shown at times 0 sec, 1 sec, ad 3 sec. Point P repeats the motion $\mathscr{F}(t)$ of point P_0, but with a 2-sec delay. The particular wave form illustrated here, a single traveling "hump," is called a **pulse**.

Transferring our attention to another point P, located further along the string at a horizontal coordinate x, we observe the following experimental fact: P repeats the motion of P_0 with a time lag that is proportional to x, its distance from P_0:

$$\text{time lag} = \frac{x}{V} \qquad (14.A.2)$$

The constant V is the so-called **wave velocity**, or **speed of propagation** of waves in the string. It follows from Eq. (14.A.1) that P, at time t, has a vertical coordinate

$$y = \mathscr{F}\left(t - \frac{x}{V}\right) \qquad (14.A.3)$$

(That is to say, the deflection of P at time t is the same as that of P_0 at time $t - x/V$.)

The Moving Wave Form

The delayed effect just described can be obtained artificially as follows. We enclose the string in a rigid tube with a suitable shape (see Fig. 14.2), and we pull the tube to the right at speed V while preventing the string itself from moving to the right. Consideration of the figure shows that any

point P at x then duly repeats the motion of P_0 after a delay x/V.

In actual fact, this is precisely what happens, without the aid of a tube. We conclude that, as far as **horizontal motion** is concerned,

> The shape of the string (the **wave form**) travels uniformly while the string itself stays in place.

$$(14.A.4)$$

This is an important characteristic of traveling waves: *the medium itself* (in this case the string) *does not travel with the wave*.

For a given time behavior $\mathscr{F}(t)$, what is the shape of the traveling wave? Let us take a "snapshot" of the whole string at some special time t_1. Then the vertical position of each point is given

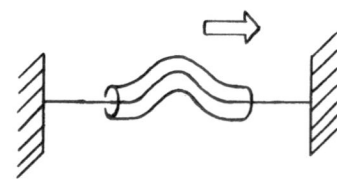

Figure 14.2. The shape travels, but the string does not.

by (14.A.3): $y = \mathscr{F}(t_1 - x/V)$, where, however, t_1 is a given number, and x is considered variable.

It is important to become familiar with both interpretations of (14.A.3): (1) at given x, it specifies the vertical position of a point on the string as a function of time; (2) at given t, it specifies the wave form as a function of position.

That the wave form $\mathscr{F}(t - x/V)$ travels to the right is seen as follows: x must increase with t in order to keep $t - x/V$, and hence the *number* $\mathscr{F}$, unchanged. Similarly, a form $\mathscr{F}(t + x/V)$ would exhibit leftward travel.

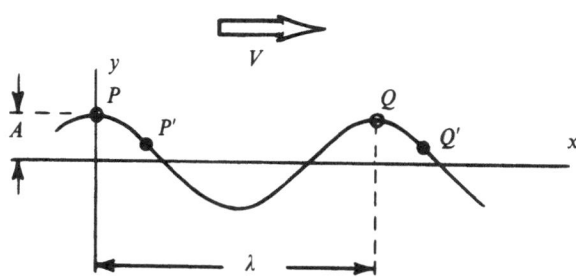

Figure 14.3. A traveling sinusoidal wave: A = amplitude, λ = wavelength; if points P', Q' are a distance λ apart, their vertical displacement is the same.

ii. Sinusoidal* Waves

Figure 14.3 shows a wave form that, at $t = 0$, is proportional to a cosine curve; it is assumed to travel in the $+x$ direction at speed V. **The wavelength** λ (lower case Greek lambda) *is defined as the distance PQ between two consecutive maxima, or "crests," of the wave.* More generally, the string has the same deflection at any two points (such as P' and Q' in the figure) which are a distance λ apart in the x direction.

As the wave progresses to the right, point Q in Fig. 14.3 will go down and then up again to its original position, while the crest that was originally at Q is replaced by the one that was at P. We recognize this down-and-up motion of Q as *one cycle of Q's oscillation.* The time required to complete one cycle is the same for all points of the string, and is defined as the **period** T of the wave.

Period–Wavelength Relation

During time T, the wave crest at P shifts to Q. Hence the whole wave must advance by a distance λ. If the speed of propagation is V, this means

$$\boxed{\lambda = VT} \qquad (14.A.5)$$

a fundamental relation for *periodic waves* of any kind. It may be written in a variety of ways; in terms of the frequency $f = 1/T$, we have

$$\boxed{\lambda f = V} \qquad (14.A.6)$$

---

* A sine curve is just a displaced cosine curve; in both cases the shape is referred to as **sinusoidal**.

Example 14.1. The Speed of Radio Waves. How crucial is the fact that we are dealing with a string? The answer is, not at all. To illustrate —without proof—the universality of (14.A.6) let us anticipate our discussion of radio waves, knowing nothing about them other than the fact that they are traveling waves. Given an FM radio set whose frequency–wavelength dial is marked both in MHz (1 MHz = 10^6 Hz) and in meters, and observing that the 100-MHz mark also reads 3.00 meters, what is the speed of radio waves?

Setting $f = 100 \times 10^6$ Hz, $\lambda = 3.00$ meters, we get from Eq. (14.A.6)

$$V = (3.00)(100 \times 10^6) \text{ meters/sec}$$

$$= \underline{3.00 \times 10^8 \text{ meters/sec}}$$

which, not by chance, is also the speed of light.

Harmonic Motion and Sinusoidal Waves

Let us see what wave develops if a typical point of the string, say at $x = 0$, is made to oscillate harmonically:

$$y = A \cos \omega t \qquad (14.A.7)$$

This is a special case of Eq. (14.A.1), and, correspondingly, Eq. (14.A.3) gives for any other point on the string (at x) the delayed harmonic motion

$$y = A \cos \omega \left(t - \frac{x}{V} \right) \qquad (14.A.8)$$

The wave form can be examined at any time, for example at $t = 0$:

$$y = A \cos\left(-\frac{\omega x}{V}\right) = A \cos\frac{\omega x}{V} \quad (14.A.9)$$

precisely a cosine curve in space. (At an appropriate time other than zero, we might have a sine curve.) The conclusion is that

> A traveling sinusoidal wave corresponds to harmonic motion for the individual points on the string.

$$(14.A.10)$$

The maximum deflection A in Eq. (14.A.8) and in Fig. 14.3 is called the **amplitude** of the wave, and ω is its **angular frequency**.

There are many ways of writing the wave formula, (14.A.8), according to one's preferred parameters among ω, f, T, V, λ, etc. Here we just list an illustration involving T and λ. With

$$\omega = \frac{2\pi}{T} \quad (14.A.11)$$

Eq. (14.A.8) becomes

$$y = A \cos 2\pi\left(\frac{t}{T} - \frac{x}{VT}\right)$$

or

$$\boxed{y = A \cos 2\pi\left(\frac{t}{T} - \frac{x}{\lambda}\right)} \quad (14.A.12)$$

The formally inclined reader will notice that λ plays in space the role that T plays in time. Observe the factor of 2π: the argument of the cosine is expressed in radians.

iii. The Speed of a Wave

When not of excessive amplitude or frequency, the waves in a string display a remarkable property (which is shared by sound waves in a fluid and by light waves in vacuum): all wave forms travel at the same speed V. In particular, *all sinusoidal waves travel at the same speed V regardless of their*

frequency and amplitude. (This is true in a given string, kept at a given tension; V will, however, vary with tension, and from string to string.)

This property should not be taken for granted. In the case of sound signals in air, our perception of the world would be quite bizarre if the speed of propagation were frequency dependent. For example, music would become unrecognizable to people sitting at the back of a concert hall, because, among other effects, the note sequence would be jumbled: a high and low note, if played simultaneously, would get to the listener at different times.

[There do exist waves whose speed is frequency dependent (a property called **dispersion**), for example, certain sound waves in stiff plates or bars. When such objects are struck they "sing" in a characteristic fashion, especially if large: this is a real version of the imaginary concert-hall effect mentioned higher. Other cases of dispersion are found in water surface waves, and in light waves traveling through glass or other matter.]

Dimensional Calculation of V

A simple dimensional argument yields V, up to a dimensionless factor, in terms of the string properties. This does not go without some judicious assumptions, as follows:

1. The speed V is independent of the wave form.
2. The string never deviates much from its straight equilibrium shape, so that the tension F is constant and equal to its value at equilibrium. Hence *we ignore the elastic properties of the string's material*; these would be important only under changes in F.
3. The relevant parameters are the string's linear mass density μ (defined as mass per unit length), and its tension F; both μ and F are taken to be constant along the string. Dimensionally, we have

$$\mu \sim \frac{\text{kg}}{\text{meter}}, \qquad F \sim \text{newton} \sim \frac{\text{kg meter}}{\text{sec}^2}$$

$$(14.A.13)$$

We require

$$V \sim \frac{\text{meter}}{\text{sec}} \quad (14.A.14)$$

Let us try an expression of the form

$$V \propto \mu^\alpha F^\beta \qquad (14.\text{A}.15)$$

with unknown α, β. Dimensionally, this reads

$$\frac{\text{meter}}{\text{sec}} \sim \left(\frac{\text{kg}}{\text{meter}}\right)^\alpha \left(\frac{\text{kg meter}}{\text{sec}^2}\right)^\beta \qquad (14.\text{A}.16)$$

Successively comparing powers of second, kilogram, and meter, we have

$$-1 = -2\beta, \qquad 0 = \alpha + \beta, \qquad 1 = -\alpha + \beta$$

The first two conditions give $\beta = 1/2$, $\alpha = -1/2$, making the third condition automatic. In conclusion,

$$V \propto \left(\frac{F}{\mu}\right)^{1/2}$$

The complete result turns out to be, in fact,

$$\boxed{V = \left(\frac{F}{\mu}\right)^{1/2}} \qquad (14.\text{A}.17)$$

Example 14.2. Traveling Pulse in a Piano String. A segment AB of a piano string has a length $l = 2.0$ meters and a mass $m = 0.030$ kg; it is under a tension $F = 1500$ newtons. How much time, Δt, does it take for a hammer blow at A to shake the bridge on which the string rests at B?

We have

$$\Delta t = \frac{l}{V}$$

or, with (14.A.17), and noting that $\mu = m/l$,

$$\Delta t = l\left(\frac{m/l}{F}\right)^{1/2} = \left(\frac{ml}{F}\right)^{1/2}$$

$$= \left[\frac{(0.030)(2.0)}{1500}\right]^{1/2} \text{sec} = \underline{6.3 \times 10^{-3} \text{sec}}$$

iv. Wave Motion from Newton's Laws*

The existence of wave behavior as previously described, including Eq. (14.A.17) for the speed, will now be derived from Newtonian mechanics.

* This subsection, as well as the explanatory remark (14.D.2) further on, may be omitted without loss of continuity.

We assume, as before, a string of uniform linear mass density μ, under a constant tension F. For the sake of simple mathematics, this derivation is limited to waves of small amplitude. (More precisely, the *slope* of the string should be small everywhere. We note that waves with large slopes can exist as well, but they obey modified equations.)

Let us denote by **F** the *vector force* (whose magnitude is the tension) that is exerted, across a point at x, by the right portion of string on the left portion. Figure 14.4 shows its y component to be

$$F_y = F \sin \theta \approx F \tan \theta \qquad (\text{small } \theta)$$

or, in this approximation,

$$\boxed{F_y = F\frac{dy}{dx}} \qquad (14.\text{A}.18)$$

since $\tan \theta = dy/dx$, the slope of the string.

Next consider the *net* force on a small segment of string, of length dx. Figure 14.5 shows that force to be $d\mathbf{F}$, the change in **F** between one end of the segment and the other. Newton's second law for that small segment is therefore

$$d\mathbf{F} = (\mu \, dx)\, \mathbf{a}$$

(mass $= \mu \, dx$, acceleration $= \mathbf{a}$). Dividing by dx and taking the y component, we obtain

$$\frac{dF_y}{dx} = \mu a_y$$

On the left-hand side, Eq. (14.A.18) for F_y can be

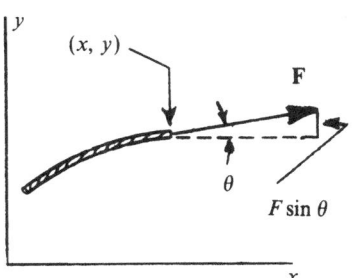

Figure 14.4. That portion of the string (not shown) which is to the right of (x, y) exerts a force **F** on the portion shown: the y component of **F** is $F \sin \theta \approx F \tan \theta$ if the string is nearly parallel to the x axis.

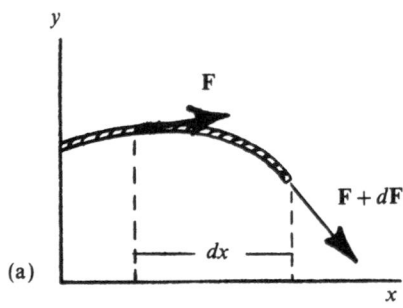

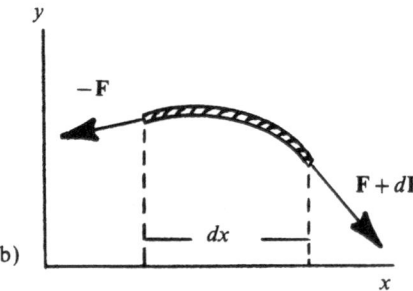

Figure 14.5. (a) Showing how **F**, see Fig. 14.4, changes as the x coordinate changes by dx. (b) Action and reaction on the left end of segment dx shows that $d\mathbf{F}$ is the net force on that segment.

used; on the right side, we set $a_y = d^2y/dt^2$. The result is a **wave equation**,

$$F\frac{d^2y}{dx^2} = \mu \frac{d^2y}{dt^2}$$

more usually written

$$\boxed{F\frac{\partial^2 y}{\partial x^2} = \mu \frac{\partial^2 y}{\partial t^2}} \qquad (14.A.19)$$

The special notation $\partial^2 y/\partial x^2$ (a so-called **partial differentiation**) is only a reminder that the differentiation d^2/dx^2 is to be carried out on a "snapshot" of the string; that is to say, the time t is treated as a constant. Similarly, $\partial^2 y/\partial t^2$ operates on one particular point of the string; thus, x is held fixed while d^2/dt^2 is performed.

Let us see what (14.A.19) has to say about the wave form (14.A.8),

$$y = A \cos \omega \left(t - \frac{x}{V} \right) \qquad (14.A.20)$$

Insertion in (14.A.19) gives

$$-FA\frac{\omega^2}{V^2} \cos \omega \left(t - \frac{x}{V} \right)$$

$$= -\mu A\omega^2 \cos \omega \left(t - \frac{x}{V} \right)$$

This is valid *for all x and all t*, provided we take

$$\frac{F}{V^2} = \mu \qquad (14.A.21)$$

as claimed in Eq. (14.A.17). To summarize:

> 1. The wave equation (14.A.19) expresses Newton's law of acceleration for each particle of the string.
> 2. The wave equation has a solution, (14.A.20), which is a traveling sinusoidal wave whose speed of propagation is $(F/\mu)^{1/2}$, independent of A and ω; there is no restriction on the values of A or ω.

$$(14.A.22)$$

[It is worth noting what happens to the wave formula, (14.A.20), when x is changed to $-x$:

$$y = A \cos \omega \left(t + \frac{x}{V} \right) \qquad (14.A.23)$$

This represents a *leftward* traveling wave, which is just as good a solution of Eq. (14.A.19) as the rightward wave is.]

Arbitrary Wave Forms

Any function of the form $y = \mathscr{F}(t \pm x/V)$ satisfies the wave equation, provided Eq. (14.A.21) for V holds; this confirms that any wave form (not just sinusoidal) can indeed travel as described at the beginning of this section. In order to check this, the reader may subject $\mathscr{F}(t - x/V)$ to the same operations as $A \cos \omega(t - x/V)$ above.

Power in a Wave

A traveling wave transmits power; this is true for sound and electromagnetic waves as well. Note (i) at the end of this chapter explains, in the case of a string, how this comes about.

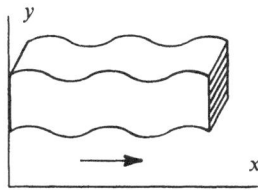

Figure 14.6. Shear waves (transverse mechanical waves) in a solid.

B. Classifying Waves

The material whose motion constitutes a wave is called the wave's **medium of propagation**; this might be a piece of string, as in the preceding section. The wave itself travels, typically at a uniform velocity, while the medium vibrates but does not travel. (A wave can exist without any medium at all; electromagnetic waves can propagate in a perfect vacuum, as we shall see later.) In all cases one must distinguish the *direction of propagation* from the *direction of vibration*; by doing so we are naturally led to classify waves into different kinds.

i. Transverse Waves

In the rectangular coordinate system we have used for the string, the propagation occurred in the x direction, while the vibration was in the y direction. The resulting waves are called **transverse**, i.e., the vibration is at right angles to the propagation.

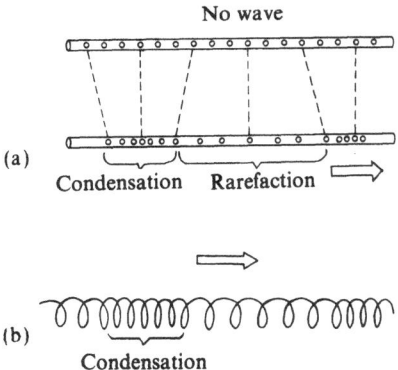

(a)

No wave

Condensation   Rarefaction

(b)

Condensation

Figure 14.7. (a) Longitudinal wave in an extensible string. The situations of static equilibrium and of a traveling wave are compared. The dots can be thought of as selected particles of the string. The string as a whole does not travel. (b) Longitudinal wave in a helical spring. Here, "condensation" means a region of tighter coils. (On a small scale, where part of a single coil is examined, one observes a slight bending of the wire, and hence from this point of view the wave is transverse.)

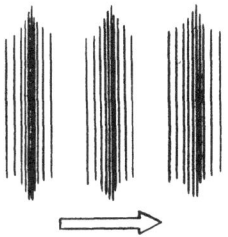

Figure 14.8. A longitudinal wave in a solid or fluid. The darker regions are denser; note the analogous features in Fig. 14.7a.

Other examples are **shear waves** in a solid, a traveling deformation illustrated in Fig. 14.6, and electromagnetic waves.

ii. Longitudinal Waves

A string, if extensible, is also capable of vibrating as shown in Fig. 14.7a. Here the directions of vibration and of travel are parallel, and the wave is called **longitudinal**. Another illustration, involving a helical spring, is shown in part (b) of the figure. Longitudinal waves commonly occur as *sound waves* in solids and fluids; see Fig. 14.8. *A fluid does not carry transverse sound waves*: we have seen in Chap. 9 that it cannot sustain shearing stresses. (This is, incidentally, used as an important clue to the fact that the Earth's outer core, Fig. 14.9, is liquid rather than solid.)

iii. Mixed Waves

A surface wave in water, Fig. 14.10, is a case where the vibrating particles move both transversely and in parallel to the wave's propagation.

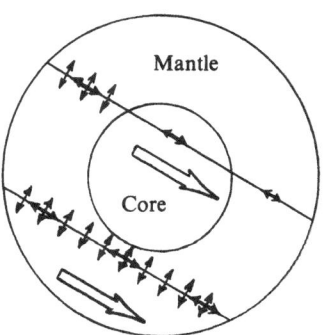

Mantle

Core

Figure 14.9. Longitudinal vibrations due to earthquakes are transmitted through the liquid core; transverse ones are not. In the solid mantle, both kinds of wave are transmitted. A small central portion of the core may be solid. (In this figure we ignore any refraction of the waves, i.e., a change in their direction of propagation due to the Earth's nonuniform composition.)

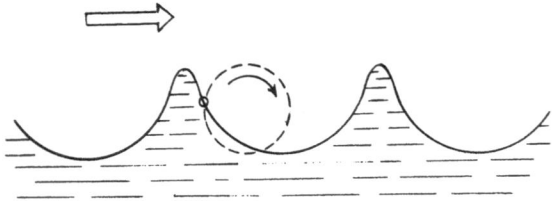

Figure 14.10. Over deep water, in a traveling wave of single frequency, a surface particle describes a circle, and thus mixes transverse with longitudinal motion. (The wide arrow indicates the wave velocity.)

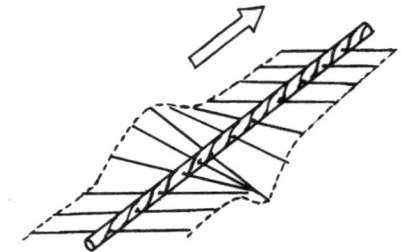

Figure 14.11. Torsional wave in a "herringbone" consisting of transverse bars mounted on a torsionally elastic "spine." Figure 13.35 of Chap. 13 showed a torsional wave in a suspension bridge.

iv. Torsional Waves

Figure 14.11 shows an example of waves involving torsion: the motion of each particle is transverse, but in addition there exists a *rotation axis* along the direction of propagation.

C. Sound in Gases

Air is undoubtedly the most familiar medium of sound propagation. Here we briefly examine how a gas (viewed as a bulk substance—no individual molecules considered) transmits a mechanical wave. This is one of the many aspects of **acoustics**, the science of sound. Lack of space prevents us from doing justice to this beautiful and important topic; our main objectives are to point out the mathematical resemblance with a vibrating string, and to quote a formula for predicting the speed of sound in a gas. The section ends with a brief discus-

sion of the Doppler effect—a frequency shift due to a relative motion between the listener and the source of sound.

i. Plane Waves

Figure 14.12a shows a volume of gas in static equilibrium, cut into equal imaginary parallel slices. If a piston on the left begins to set up alternate compressions and expansions, a *longitudinal oscillation* will propagate to the right, as shown in part (b) of the figure. Let the equilibrium coordinate of a typical cut (separating two slices) be x, and let y be the cut's time-dependent displacement from equilibrium ($x + y =$ coordinate of the cut during vibration). Then, at any time, y may be plotted against x (i.e., we consider cuts that are further and further to the right); see (c) in the figure, which shows the special case of a **sinusoidal**

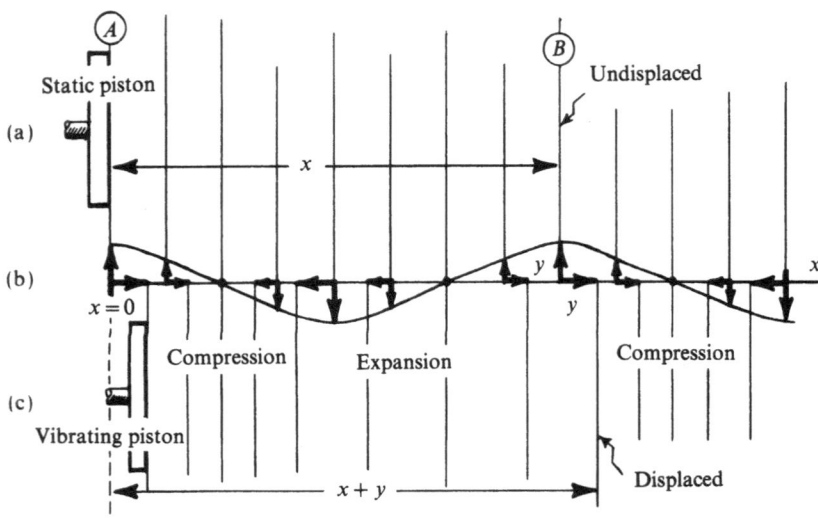

Figure 14.12. How to plot the displacement y in a longitudinal wave which propagates in the x direction. (a) Static equilibrium for a set of equal parallel slices. (b) During vibration some slices are expanded, others are compressed; a typical cut is displaced over a distance y. (c) The curve shows y, plotted *transversely* at the location of the undisturbed cut. Planes such as A, B, located at the crests of y, are called **wave fronts**.

wave. As the graph leads us to suspect, the formula for such a traveling sound wave turns out to be just like its counterpart in a string, Eq. (14.A.8):

$$y = A \cos \omega \left(t - \frac{x}{V} \right) \qquad (14.C.1)$$

where V is now the speed of sound.

The gas slices considered here are enclosed between plane cuts, each of which vibrates as a whole: we are dealing with **plane waves**. It is often convenient to represent a sound wave by drawing just the surfaces of maximum y (for example planes A, B in Fig. 14.12a). Such surfaces define the "crests" of the wave, and are called **wave fronts**. Approximately plane waves are not difficult to obtain in practice; see Fig. 14.13.

Pressure in a Sound Wave

Sound turns out to be a *pressure wave* as much as a *displacement wave*. At any point, the pressure P swings alternately above and below its average value P_0. (In daily life, $P_0 \approx 1$ atm.) The relevant quantity, $P - P_0$, behaves like y: sinusoidal in time at any one location, sinusoidal in space at any one time. This behavior of $P - P_0$ is due to the successive compressions and expansions of each gas slice. A careful look at Figs. 14.12b and 14.12c indicates that at points of greatest compression the displacement is zero; this is also the case at points of greatest rarefaction. We conclude that (1) at any location, the pressure peaks and bottoms in alternation with the displacement; (2) in a "snapshot" representation of the wave, the high and low pressure points alternate with the high and low displacement points. The details are pursued in Problem 14.14.

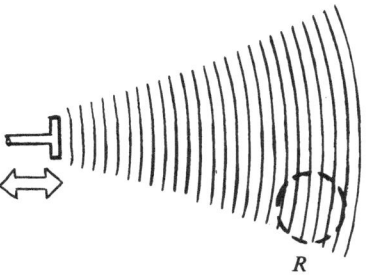

Figure 14.13. Approximately plane waves occur immediately in front of a plane loudspeaker diaphragm, or in a limited region such as R, far from the source of sound.

ii. The Speed of Sound

We recall that, in a string, the wave speed is $V = (F/\mu)^{1/2}$, Eq. (14.A.17). The corresponding result in an ideal gas is

$$V = \left(\frac{\gamma P_0}{\rho} \right)^{1/2} \qquad (14.C.2)$$

(Laplace's formula), where P_0 is the equilibrium pressure of the gas, ρ is its equilibrium density, and γ is the ratio of its heat capacities:

$$\gamma = \frac{C_P}{C_V} = \begin{cases} \frac{5}{3} & \text{(monatomic gases)} \\ \frac{7}{5} & \text{(air and many other diatomic} \\ & \text{gases at ordinary temperatures)} \end{cases}$$

$$(14.C.3)$$

See, for example, Eq. (12.A.32) of Chap. 12. If it were not for the curious dimensionless factor of γ, formula (14.C.2), which we shall not derive, would be the perfect analogue of $V = (F/\mu)^{1/2}$. (Newton himself missed that factor in his calculation of V, an omission due to the imperfect state of gas theory in his day. Laplace's correction came more than a century later.)

Formula (14.C.2) for V is often more useful in terms of the gas' molecular weight than in terms of its density. Consider 1 kmole, with a volume $\mathcal{V}$ and a mass necessarily equal to the molecular weight times 1 kg. We have

$$\rho = \frac{(\text{mol. wt.})(1 \text{ kg})}{\mathcal{V}} \qquad (14.C.4)$$

and Eq. (14.C.2) becomes

$$V = \left[\frac{\gamma P \mathcal{V}}{(\text{mol. wt.})(1 \text{ kg})} \right]^{1/2}$$

or

$$V = \left[\frac{\gamma RT}{(\text{mol. wt})(1 \text{ kg})} \right]^{1/2} \qquad (14.C.5)$$

We see that, at given temperature, in an ideal gas,

$$\boxed{\text{The speed of sound is independent of pressure.}} \qquad (14.C.6)$$

[Why does this not contradict (14.C.2)?] The speed does, however, increase with temperature.

At this stage the reader should take a moment to compare (14.C.5) with our result for the root-mean-square speed of a molecule, Eq. (11.D.17) of Chap. 11.

Example 14.3. Speed of Sound in Air. Find the speed of sound in dry air at STP, given that air has a mean molecular weight of 29.

Formula (14.C.5) gives directly

$$V = \left[\frac{\left(\frac{7}{5}\right)\left(8320 \frac{\text{joules}}{\text{K}}\right)(273 \text{ K})}{(29)(1 \text{ kg})} \right]^{1/2}$$

$$= 331 \frac{\text{meters}}{\text{sec}}$$

in good agreement with observation. (Recall that the root-mean-square molecular speed is about 500 meters/sec; see Example 11.8 of Chap. 11.)

Actually, one of the best uses of formula (14.C.5) consists of "turning it around" to obtain γ for a gas, from a measurement of V. This is often more convenient and more accurate than a calorimetric measurement of C_γ or C_P.

For comparison, Table 14.1 lists the speed of sound in selected solids and liquids; the temperature dependence is much less than in gases.

Power in a Sound Wave

Most users of audio equipment are well aware of power levels in sound transmission. Sound energy propagates in a fluid much as it does in a vibrating string, and for a similar reason: here the work is done by pressure forces. The **intensity level** of a sound wave is related to its power, and is expressed in **decibels**; that terminology is explained in a Note at the end of the chapter.

iii. The Doppler Effect

This may be defined as a change in the observed frequency of a wave due to a relative motion between the observer and the source of the wave. To illustrate: While a car is approaching, its horn sounds higher pitched than while it is receding. In the analysis of this familiar effect, the speed of the car will have to be compared with the *speed of propagation of the sound waves*.

In general, we need to consider at least two types of situation: a traveling source, and a traveling listener. The two motions may, of course, occur in combination.

Suppose the source (S) vibrates at a given frequency f; then the observer ($\mathbb{O}$) receives a frequency f', which is related to f as follows:

$$f' > f \quad \text{if} \quad S \text{ and } \mathbb{O} \text{ approach one another}$$

$$f' < f \quad \text{if} \quad S \text{ and } \mathbb{O} \text{ recede from one another}$$

$$(14.C.7)$$

The amount $f' - f$ is called the **Doppler shift** in the frequency, after Christian Johann Doppler, who, in 1842, found how to calculate its value. We next turn to Doppler's formulas.

[This discussion avoids certain complications by assuming that (1) the speeds of S and $\mathbb{O}$ are constant and smaller than the speed of sound in air; (2) S and $\mathbb{O}$ move along the same straight line.]

Table 14.1. Speed of Sound in Various Substances[a]

Solids[b]	Speed (meters/sec)	Liquids	Speed (meters/sec)
Glass (typical)	5550	Sea water	1500
Iron	5130	Water	1461
Aluminum	5104	Mercury	1407
Nickel	4973	Methyl alcohol	1143
Copper	3560	Carbon disulfide	1060
Lead	1322	Ether	1032

[a] At about standard temperature and pressure.
[b] Longitudinal waves in thin rods.

A Traveling Observer

Let S be at rest, in stationary air, while $\mathbb{O}$ approaches S at speed v, as illustrated in Fig. 14.14; $\mathbb{O}$ then hears a frequency

$$\boxed{f' = \left(1 + \frac{v}{V}\right) f} \qquad \text{(traveling observer)}$$

(14.C.8)

where V is the speed of sound in air.

As a proof, we note that $\mathbb{O}$ receives consecutive wave fronts which are spaced by the same wavelength λ as seen by an observer at rest; but the speed of the wave front relative to $\mathbb{O}$ is

$$V' = V + v \qquad (14.C.9)$$

Therefore $\mathbb{O}$ measures a frequency

$$f' = \frac{V'}{\lambda} = \frac{V + v}{V/f} \qquad (14.C.10)$$

leading to (14.C.8).

A Traveling Source

Let S have a speed v of approach towards a stationary $\mathbb{O}$, again in still air. The frequency received by $\mathbb{O}$ is now

$$\boxed{f' = \frac{f}{1 - v/V}} \qquad \text{(traveling source)}$$

(14.C.11)

So as not to overload this section, we leave a detailed explanation of (14.C.11) to Note (iii) at the end of the chapter.

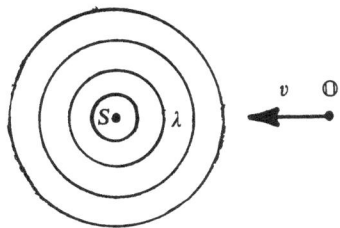

Figure 14.14. A stationary S and moving $\mathbb{O}$.

Small-Speed Approximation

Formulas (14.C.8) and (14.C.11) are more similar than they appear: when v is small compared to V, they become identical. To see this, we can use in (14.C.11) the approximation $(1 - v/V)^{-1} \approx 1 + v/V$, valid when $v/V \ll 1$.

Receding Motions

In the above, v is positive (it is a speed); but under the substitution $v \to -v$, the discussion becomes applicable to speeds v of separation rather than of approach.

Example 14.4. Two Flying Bats. A leading bat, A, is being followed by another bat, B, who is trying to catch up. Bat B emits a regular series of echolocation signals (short sound pulses) at a repetition rate $f = 155.00$ pulses/sec. If the bats' respective speeds are $v_A = 1.8$ meters/sec and $v_B = 2.3$ meters/sec, find the rate f' at which A hears the signals from B. Use $V = 340$ meters/sec for the speed of sound in still air.

(We first note that the Doppler formulas must hold just as well for the repetition rate of a sound signal as for vibrational frequencies.)

If A were stationary, it would detect a rate

$$\tilde{f} = \frac{f}{1 - v_B/V} \qquad (14.C.12)$$

Being in motion, it actually receives

$$f' = \left(1 - \frac{v_A}{V}\right) \tilde{f} \qquad (14.C.13)$$

(Watch the sign.) Eliminating $\tilde{f}$, we have

$$f' = \frac{1 - v_A/V}{1 - v_B/V} f \qquad (14.C.14)$$

$$= \left(\frac{1 - 1.8/340}{1 - 2.3/340}\right)(155.00) \frac{\text{pulses}}{\text{sec}}$$

$$= \underline{\underline{155.23 \frac{\text{pulses}}{\text{sec}}}}$$

[Note that (14.C.14) would read $f' = f$ if $v_A = v_B$.]

Electromagnetic Waves

The Doppler shift of a radar beam, after reflection from a car, reveals its speed to the police. Astronomers infer the speed of a star from the Doppler shift of its emitted light.

As long as the moving object has a speed v that is small compared to the speed of light c, Doppler's formula (14.C.8) [or, equivalently, (14.C.11)] is applicable here, with V replaced by c. On the other hand, a relativistic formula is needed when v is comparable to c. That formula will be one of the topics of Chap. 25 (Relativity).

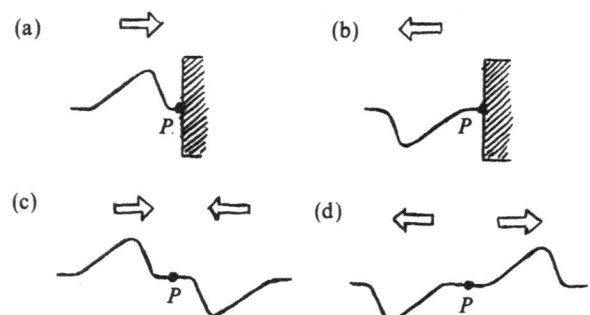

Figure 14.16. Reflection from a fixed point, (a) and (b), can be simulated by the superposition of two oppositely traveling waves, (c) and (d).

D. Standing Waves

So far we have looked at traveling waves only. Clearly some waves do not fall in that category. For example, if two traveling waves are sent along a string, one from each end, what is the result when they meet? As we shall see, the answer to that question leads to an interesting prediction: a vibrating string has a large set of **natural frequencies**, closely related to the **harmonics** of the preceding chapter; the same is true for air volumes and other vibrating objects.

i. Superposition

Figure 14.15 shows a typical encounter between two traveling waves in a string. What happens is remarkable: *the vertical coordinates of the two waves simply add*; after the waves have separated, each is seen to proceed as if the other had never

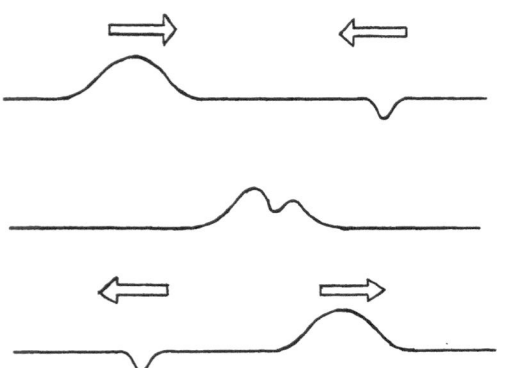

Figure 14.15. Encounter between two oppositely traveling waves in a string.

existed. The rule for combining any two wave forms is as follows.

> If each of two wave forms $\mathscr{F}_1(t, x)$ and $\mathscr{F}_2(t, x)$ can individually occur in the string, then so can their sum (or superposition), $\mathscr{F}_1(t, x) + \mathscr{F}_2(t, x)$.

$$(14.D.1)$$

(Note that $\mathscr{F}_1, \mathscr{F}_2$ can travel in the same or in opposite directions.) This should remind us of the preceding chapter, where such superpositions were discussed as functions of the time only.

[The reason for (14.D.1) goes back* to the wave equation (14.A.19),

$$F \frac{\partial^2 y}{\partial x^2} = \mu \frac{\partial^2 y}{\partial t^2} \qquad (14.D.2)$$

If y_1 and y_2 are two different wave forms, each obeying (14.D.2), it will be seen that $y_1 + y_2$ obeys it also; we express this property of (14.D.2) by saying that it is a **linear** wave equation.]

ii. Reflection

Suppose a traveling wave (Fig. 14.16a), encounters the end of the string, which is tied to a fixed point P on the wall. No mechanical energy can be transferred to a motionless object; friction, which could transfer some heat, is assumed absent; where, then, does the wave energy go? It can only be sent

---

* Omit this explanation if Sec. A iv has been omitted.

back, i.e., *the wave must be reflected*, as in part (b) of the figure.

What is the shape of the reflected wave? The following artifice, which makes use of the superposition principle, provides the answer: *the reflected wave has the same shape as the incoming wave, but is flipped top to bottom, and right to left.* To see this, consider a new arrangement, (c) of the figure, where the fixed end at P has been done away with; the string extends further to the right. Also, an additional wave has been added, which travels oppositely to the first. It has the "flipped" shape, and both waves are equally far from P. When they both reach P, the observed result is their sum, which shows an exact cancellation at P (destructive interference). *This particular point is motionless as required.* Afterwards, the "new" wave is seen as if it were the "old" one, reflected from P; compare parts (b) and (d) of the figure. This concludes the argument, the basis of which is that it does not matter to the string's behavior through what stratagem P is held motionless.

iii. Making a Standing Wave

If a string is stretched between two rigid supports, its wave energy is trapped; the wave travels back and forth, interfering with itself. (In practice, friction would eventually damp that motion to a stop.) Much can be learned from the simplest case, where a sinusoidal traveling wave and its reflection are superposed. We do not have to worry about flipping the sinusoidal shape: it would remain sinusoidal. Also, let us not worry at first about where the fixed points are located.

Observed Result

Consider the two waves

$$y_1 = A \cos \omega \left(t - \frac{x}{V} \right)$$

$$y_2 = A \cos \omega \left(t + \frac{x}{V} \right) \tag{14.D.3}$$

which differ only in their direction of propagation. The resulting combination, $y_{tot} = y_1 + y_2$, taken at various times, is illustrated in Fig. 14.17. The whole traveling aspect has disappeared, and we have a so-called **standing wave**: *any two points of the string oscillate with the same (or opposite) phase.* The amplitude of oscillation of a single point on the string is zero at certain points (called **nodes**), and has a maximum value $2A$ at other points (**antinodes**, or **loops**), located halfway between the nodes. We shall see that

Consecutive nodes are a distance $\lambda/2$ apart.

$$\tag{14.D.4}$$

Here λ is the wavelength of the underlying traveling waves (14.D.3).

Mathematical Derivation

All these features of a standing wave result from the well-known identity

$$\cos a + \cos b = 2 \cos \frac{a-b}{2} \cos \frac{a+b}{2} \tag{14.D.5}$$

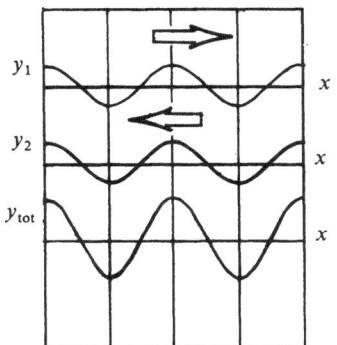
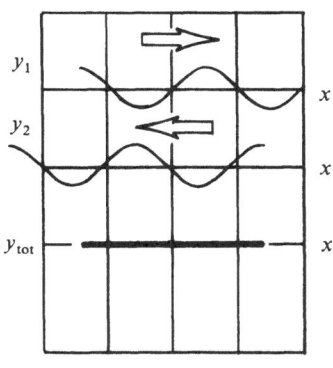
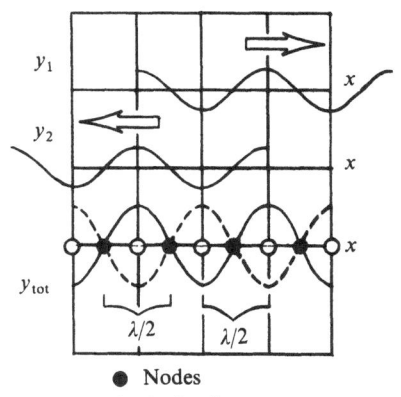

Figure 14.17. The bottom curve is a standing wave obtained by superposing the two top curves, $y_{tot} = y_1 + y_2$. The figure shows three successive "snapshots." The dashed curve shows the earlier shape for comparison with the new one.

As applied to the superposition $y_{tot} = y_1 + y_2$ [see Eq. (14.D.3)], it yields

$$y_{tot} = 2A \cos \frac{\omega x}{V} \cos \omega t$$

or

$$y_{tot} = \left(2A \cos \frac{2\pi x}{\lambda}\right) \cos \omega t \quad (14.D.6)$$

For any given point x, we therefore have a harmonic motion with amplitude $2A |\cos(2\pi x/\lambda)|$, illustrated by the last curve of Fig. 14.17, which has all the higher-mentioned properties.

The Natural Frequencies of a String

Let a string, for example a guitar string, be held under a given tension between two fixed points a distance L apart. Suppose all points vibrate at the same frequency ω, thus collectively producing a pure tone. As we have seen, this can be achieved by a standing wave; with a bit more rigor in the mathematics we could have shown that the standing wave is the only way of obtaining a pure tone.

Since the end points are fixed, see Fig. 14.18, the string's length must be a whole number n of half-wavelengths:

$$L = n\frac{\lambda}{2} \quad (n = 1, 2, 3, ...) \quad (14.D.7)$$

Thus, only certain wavelengths are allowed,

$$\lambda = \frac{2L}{n} \quad (n = 1, 2, 3, ...) \quad (14.D.8)$$

corresponding to the frequencies ($f = V/\lambda$)

$$\boxed{f = \frac{nV}{2L}} \quad (n = 1, 2, 3, ...) \quad (14.D.9)$$

This remarkable formula yields all the natural, or resonant, frequencies of a string with fixed ends. We recognize them as the **harmonics** (Sec. C ii of Chap. 13) corresponding to a **fundamental frequency**

$$\boxed{f_1 = \frac{V}{2L}} \quad (14.D.10)$$

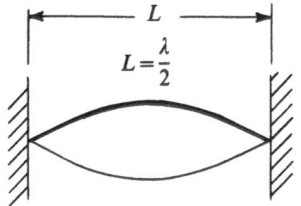

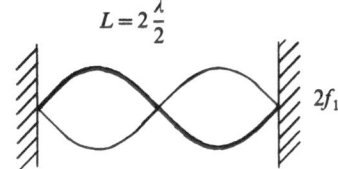

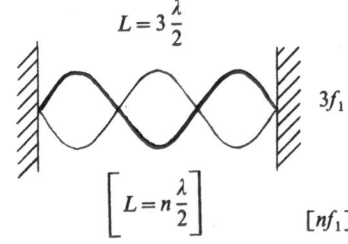

Figure 14.18. Examples of standing waves in a string held at both ends. The string is shown at two times when its displacement from equilibrium has maximum magnitude. (The heavy and light curves are "snapshots" taken half a period apart.) The allowed frequencies are of the form $f_1, 2f_1, 3f_1, ...$.

in which V of course depends on the string's tension and linear density.

Example 14.5. Information from Two Natural Frequencies. A 3.0-meter-long string has two consecutive resonant frequencies at 50 Hz and 55 Hz. Find (a) the lowest resonant frequency f_1, and (b) the speed of wave propagation along the string.

a. Lowest frequency: The resonant frequencies are $f_1, 2f_1, 3f_1, ...$, so that two consecutive frequencies differ by precisely f_1:

$$f_1 = 55 \text{ Hz} - 50 \text{ Hz} = \underline{5 \text{ Hz}}$$

(50 Hz is the ninth harmonic).

b. From Eq. (14.D.10), we have

$$V = 2Lf_1 = (2)(3.0)(5) \text{ meters/sec}$$

$$= \underline{30 \text{ meters/sec}}$$

Superposition Again

A less special disturbance generally gives the *sum* of several standing waves. The resulting vibration is itself no longer a standing wave, nor is it a traveling wave. Here we confine ourselves to an attempt, shown in Fig. 14.19, at visualizing a superposition of the fundamental and the second harmonic.

iv. Standing Waves in Air

The various kinds of waves mentioned in Sec. B can occur as standing waves. A striking example is provided by the torsional standing waves in the Tacoma Narrows bridge, Fig. 13.35 of Chap. 13. In the following, however, we focus our attention on sound.

Standing waves in a gas are very much like standing waves in a string; it is not important that they are longitudinal rather than transverse. The best known application is to wind instruments, such as flutes, oboes, organs, etc., where the vibration is that of a confined air column. Here we depend on the *reflection of a wave at the end of a pipe*. If that end is closed off, it must correspond to a node of the (longitudinal) displacement. Some resulting standing wave patterns are illustrated in Fig. 14.20.

An alternative way, often used to produce a reflection in a pipe, is to keep one or both of its ends open to the atmosphere. (This works only for pipes of diameter much smaller than the wavelength involved.) It may seem absurd that sound energy should find its exit blocked by an open end, but the explanation is that the pressure is nearly constant in time at that point, and equal to the atmospheric pressure, instead of oscillating as it does inside the pipe. A constant pressure means a zero *average* power transfer across the opening: the air vibrates in and out, but does as much work on the atmosphere as is done on it by the atmosphere. Figure 14.21 shows standing waves in a pipe with

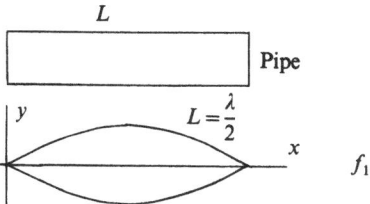
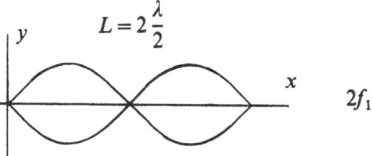
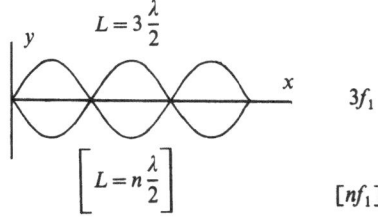

Figure 14.20. Some standing sound waves in a cylindrical pipe closed at both ends. The air's displacement is being plotted here. This situation is analogous to that in a string held at both ends, Fig. 14.18.

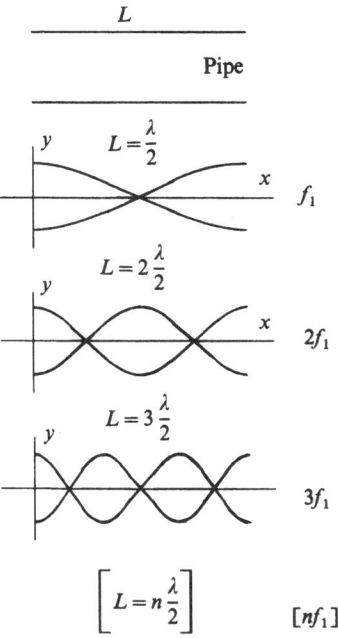

Figure 14.21. Standing waves in a pipe with both ends open. The allowed frequencies are just as in Figs. 14.20 and 14.18.

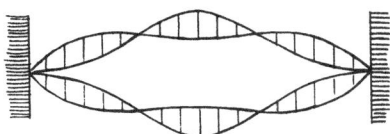

Figure 14.19. Example of a second harmonic superposed on the fundamental. The shaded vibration occurs at 3 times the frequency of the wider shape's vibration.

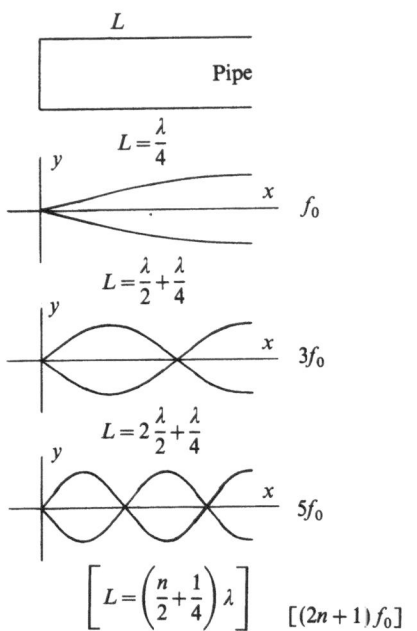

Figure 14.22. Standing waves in a pipe closed at one end. The allowed frequencies are $f = (n/2 + 1/4) V/L$ ($n = 0, 1, 2,...$). They amount to an incomplete harmonic series $f_0, 3f_0, 5f_0,...$, where the fundamental is $f_0 = V/4L$.

both ends open, while in Fig. 14.22 one end is closed and the other is open. Note the new formulas replacing (14.D.7) and (14.D.9).

To summarize: In a pipe, a closed end corresponds to a **displacement node**; an open end corresponds (as can be shown) to a **displacement antinode**.

Notes

i. Power in a Wave (Vibrating String)

At the end of Sec. A it was mentioned that a traveling wave can be used to transfer energy from one place to another; this is illustrated in Fig. 14.23. Let us calculate,

in the case of a sinusoidal wave, how much energy is transferred per unit time. Specifically, in Fig. 14.24, how much power *enters* the wave at P owing to the harmonically vibrating clamp?

The force exerted by the clamp has a vertical component of magnitude $|Fy|$. According to Eq. (14.A.18) we have, in terms of the string's slope at P,

$$F_y = F\frac{dy}{dx} \qquad (14.\text{N}.1)$$

In the figure, P happens to be moving down, as the reader should verify. Also, F_y is upward. Hence, positive work is being done by the clamp on P, namely at the following rate:

$$\text{Power} = |(\text{force on } P)_y \, (\text{velocity of } P)_y|$$

$$= \left(F\frac{dy}{dx}\right)\left(-\frac{dy}{dt}\right) \qquad (14.\text{N}.2)$$

(minus sign needed for positivity). We note that only the y components are relevant, because the velocity of P is purely vertical. Switching to the partial-derivative notation, we obtain

$$\text{Power} = -F\frac{\partial y}{\partial x}\frac{\partial y}{\partial t} \qquad (14.\text{N}.3)$$

Insertion of the cosine formula (14.A.8) gives

$$\text{Power} = (-F)\left[\frac{A\omega}{V}\sin\omega\left(t-\frac{x}{V}\right)\right]$$

$$\times\left[-A\omega\sin\omega\left(t-\frac{x}{V}\right)\right]$$

$$= +\frac{A^2F\omega^2}{V}\sin^2\omega\left(t-\frac{x}{V}\right)$$

$$= \frac{A^2F\omega^2}{V}\sin^2\omega t \qquad (14.\text{N}.4)$$

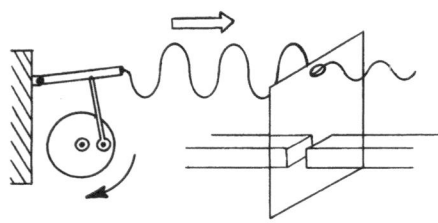

Figure 14.23. A fanciful way of transferring energy by a wave. The plate and bars on the right are heated by their mutual friction.

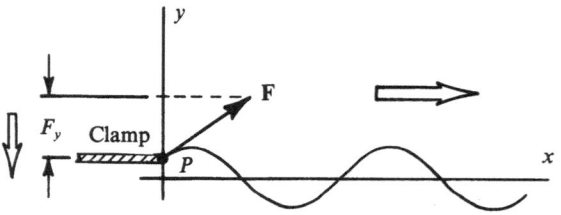

Figure 14.24. The vertically oscillating clamp on the left feeds power into the string. The string exerts a force **F** on the clamp. Since, in this illustration, the clamp is moving down (against F_y), it is doing positive work. The wide horizontal arrow indicates the direction of wave propagation.

the last step because P is at $x = 0$. This result is plotted in Fig. 14.25 as a function of time. We see that the power fluctuates symmetrically above and below the value $\frac{1}{2}A^2F\omega^2/V$ (dashed line), which therefore represents the *time-averaged* power:

$$\boxed{\text{Average power} = \frac{1}{2}\frac{A^2F\omega^2}{V}} \qquad (14.\text{N}.5)$$

(When speaking of the power transmitted by a sinusoidal traveling wave, one usually has this average power in mind.) Our result shows that, in such a wave,

> The power increases quadratically with the amplitude, and quadratically with the frequency.

$$(14.\text{N}.6)$$

[The *quadratic amplitude dependence* of the power is a very general result, applicable to other kinds of waves. It can be traced to the two factors in Eq. (14.N.2), each of which is proportional to the amplitude. It is also reminiscent of the A^2 dependence of the *stored energy* in a harmonic oscillator.]

ii. Decibels

The strength of a sound signal—referred to at the end of Sec. C ii—may be gauged by the power that reaches a given area, perhaps the cross-section of someone's ear canal. Since that power fluctuates strongly during each cycle of a vibration, it needs to be averaged over several such fluctuations to yield a meaningful figure. Accordingly, one defines the **intensity** of a sound wave by

$$\boxed{\text{Intensity} = \frac{\text{time-averaged power}}{\text{area of detector}}}$$

$$(14.\text{N}.7)$$

(It is assumed that the area in question is facing the incoming wave.) We see from (14.N.7) that the intensity is to be expressed in watts/meter^2.

In practice one defines the **intensity level** of a sound wave by comparing its intensity to a standard intensity,

$$\boxed{\text{Standard intensity} = 10^{-12}\,\frac{\text{watt}}{\text{meter}^2}}$$

a value chosen because it approximates the human threshold of hearing at a frequency of 1000 Hz, to which we are very sensitive. The comparison is made in terms of *intensity ratios*. By definition, we have

$$\boxed{\begin{aligned}\text{Intensity level} &= \log_{10}\left(\frac{\text{actual intensity}}{\text{standard intensity}}\right) \\ &\quad \times (10 \text{ decibels})\end{aligned}}$$

$$(14.\text{N}.8)$$

The designation 10 decibels = 10 db is traditionally preferred to its equivalent, 1 **bel**, a dimensionless unit named after Alexander Graham Bell (1847–1922), the inventor of the telephone.

Example 14.6. Extreme Intensity Levels. What is the intensity level (a) of a 10^{-12}-watt/meter^2 sound wave (hearing threshold), (b) of a 1-watt/meter^2 wave (pain threshold)?

From recipe (14.N.8) we obtain

$$\text{Intensity level}$$
$$= \begin{cases} \log_{10}\dfrac{10^{-12}}{10^{-12}}(10\text{ db}) = \underline{0\text{ db}} & (a) \\[2ex] \log_{10}\dfrac{1}{10^{-12}}(10\text{ db}) = \underline{120\text{ db}} & (b) \end{cases}$$

A rough correlation of familiar sounds with their intensity levels is shown in Table 14.2.

Some useful rules of thumb, which the reader may wish to deduce from (14.N.8), are as follows.

1. Repeatedly *adding* the same number of decibels to the intensity level amounts to repeatedly *multiplying* the intensity by a fixed factor.
2. Adding 10 db corresponds to multiplying the intensity by a factor of 10.

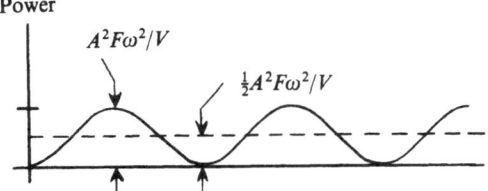

Power

$A^2F\omega^2/V$

$\frac{1}{2}A^2F\omega^2/V$

Figure 14.25. The power supplied by the clamp in Fig. 14.24, plotted as a function of time. Zero power coincides with maximum absolute displacements of the clamp. The dashed line is the **average power**.

Table 14.2. Some Intensity Levels

Kind of sound[a]	Intensity level (db)	Intensity (watts/meter^2)
Threshold of hearing	0	1×10^{-12}
Breathing	10	1×10^{-11}
Whispering	20	1×10^{-10}
Page turning	35	3×10^{-9}
Conversation	60	1×10^{-6}
Street traffic	70	1×10^{-5}
Jackhammer	95	3×10^{-3}
Threshold of pain	120	1

[a] With the source at a "typical" distance.

3. Adding 3 db corresponds to multiplying the intensity by almost exactly a factor of 2 (Note: $\log_{10} 2 \approx 0.301$.)

iii. Doppler Shift from a Moving Source

We want to derive formula (14.C.11),

$$f' = \frac{f}{1 - v/V} \qquad (14.N.9)$$

It is instructive to reason in two stages.

1. A Steady Wind

As a preliminary, we consider a wind of speed v, blowing from a *stationary* observer $\mathbb{O}_0$ toward a *stationary* source S, as shown in Fig. 14.26. In this case, $\mathbb{O}_0$ hears the same frequency f_1 as is emitted by S; the *wind has no effect on the frequency*.* Indeed, since every wave front takes the same time to travel from S to $\mathbb{O}_0$, the vibration at $\mathbb{O}_0$ is identical to that at S except for a constant time delay.

On the other hand, the wave fronts do travel at a reduced speed

$$V' = V - v \qquad (14.N.10)$$

with a corresponding reduced wavelength

$$\lambda' = \frac{V'}{f} = \frac{V - v}{f} \qquad (14.N.11)$$

---

* Caution: If S and $\mathbb{O}_0$ were not at rest relative to one another, the wind would, in general, affect the frequency.

As Fig. 14.26 indicates, the wave fronts crowd together in front of S.

2. Back to the Moving Source

In Fig. 14.26, we may consider the air to be at rest, while S travels to the right at speed v. The speed of the wave fronts is then restored as V. A new observer, $\mathbb{O}$, at rest with respect to the air, hears a frequency

$$f' = \frac{V}{\lambda'} = \frac{V}{(V - v)/f} \qquad (14.N.12)$$

[with use of (14.N.11)]; this is just formula (14.N.9).

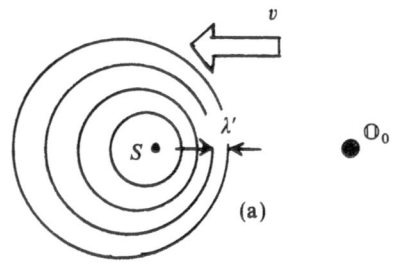

(a)

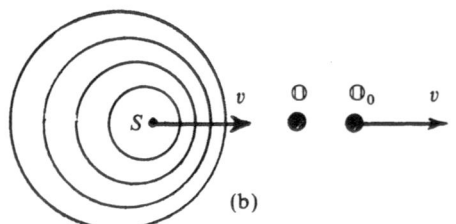

(b)

Figure 14.26. (a) A stationary S, a stationary $\mathbb{O}_0$, and moving air. (b) Equivalently, the air is considered stationary while S and $\mathbb{O}_0$ have speed v. We ask for the frequency heard by a stationary $\mathbb{O}$.

Condensed Checklist

Sinusoidal wave traveling in $\pm x$ direction:

$$y = A \cos \omega(t \mp x/V) \qquad (14.A.8)$$

$$= A \cos 2\pi \left(\frac{t}{T} \mp \frac{x}{\lambda} \right) \qquad (14.A.12)$$

$$\boxed{\lambda f = V \qquad \text{or} \quad \lambda = VT}$$

(any traveling sinusoidal wave)   (14.A.6), (14.A.5)

Wave velocity:

$$\boxed{V = \left(\frac{F}{\mu} \right)^{1/2}} \qquad \text{(transverse wave in a string)}$$

$$(14.A.17)$$

$$V = \left(\frac{\gamma P_0}{\rho} \right)^{1/2} = \left[\frac{\gamma RT}{(\text{mol. wt.})(1 \text{ kg})} \right]^{1/2} \text{(sound in a gas)}$$

$$= 331 \frac{\text{meters}}{\text{sec}} \text{ (dry air at STP)}$$

$$(14.C.2), (14.C.5), \text{Example } 14.3$$

Wave equation*:

$$F \frac{\partial^2 y}{\partial x^2} = \mu \frac{\partial^2 y}{\partial t^2} \qquad \text{(transverse wave in a string)}$$

$$(14.A.19)$$

Doppler shift:

$$\boxed{f' = \left(1 + \frac{v}{V} \right) f} \qquad \text{(moving observer)} \quad (14.C.8)$$

$$f' = \frac{f}{1 - v/V} \qquad \text{(moving source)} \quad (14.C.11)$$

Standing wave with antinode at $x = 0$:

$$y \propto \cos \frac{2\pi x}{\lambda} \cos \omega t \qquad (14.D.6)$$

Distance between nodes: $\boxed{\lambda/2}$ (14.D.4)

* Disregard if Sec. A iv has been omitted.

Standing wave in segment of length L:

$$\text{Fig. } 14.18, (14.D.7), \text{Figs. } 14.20, 14.21$$

$$\boxed{L = \frac{n\lambda}{2}}, \qquad f = \frac{nV}{2L} \quad (n = 1, 2, 3, \ldots) \quad (14.D.9)$$

(nodes at both ends, or antinodes at both ends)

$$\boxed{L = \left(\frac{n}{2} + \frac{1}{4} \right) \lambda},$$

$$\text{Fig. } 14.22$$

$$f = \left(\frac{n}{2} + \frac{1}{4} \right) \frac{V}{L} \qquad (n = 0, 1, 2, \ldots)$$

(node at one end, antinode at the other)

True or False

1. In a given gas at room temperature and atmospheric pressure, a sound wave propagates faster when its frequency is higher.

2. The amplitude of a wave in a string may be determined from the frequency, the wavelength, and the linear mass density.

3. A vibration that is a sinusoidal function of time propagates in a uniform string as a wave that is a sinusoidal function of the distance.

4. To state that the size of a musical wind instrument is related to the wavelength of the tones it emits is a valid explanation for the higher-pitched tones heard from smaller instruments.

5. Someone who inhales helium before speaking should have a lower-pitched voice than usual.

6. In order to obtain the lowest possible tone from an organ pipe of given length, one should make it with one end closed and the other end open.

7. If the speed of sound in air is lower on a certain mountain top than in the valley, this can be attributed to the pressure difference.

8. Although deuterium (D_2) and helium (He) gases have the same molecular weight, they exhibit different speeds of propagation for sound at equal temperatures.

9. When emitted forwards in still air by a traveling source, acoustical wave fronts propagate at a speed that is increased by an amount equal to the speed of the source.

Problems

(Unless otherwise stated, take the speed of sound in air to 340 meters/sec, its room-temperature value.)

Sections A: Traveling Waves in Strings; B: Classifying Waves

*14.1. In a certain string, the speed of the waves is 25 meters/sec. What is the wavelength corresponding to a 16-Hz vibration?

14.2. What frequency is required to set up a wave of wavelength 0.32 meter in a string where the speed of propagation if 75 meters/sec?

14.3. In a certain guitar string, a 560-Hz vibration corresponds to a 35-cm wavelength. (a) At what speed does a wave propagate along the string? (b) If its linear density is 6.5×10^{-3} kg/meter, what is the string's tension?

*14.4. A certain red light is transmitted in space with a wavelength of 6.6×10^{-7} meter. If the speed of all light waves is 3.0×10^{8} meters/sec, what is the period of oscillation for red light?

*14.5. Boats A and B are moored 14 meters apart on a lake; A is being rocked, and B bobs up and down once every second, owing to the wave spreading from A. On a photograph of the scene, 10 wave crests are counted from A to B (including one at A and one at B). How fast did the wave propagate?

14.6. The free portion of a certain piano string is 1.5 meters long, has a mass of 0.250 kg, and is under a tension of 5400 newtons. In what time does a pulse travel from one of its ends to the other?

*14.7. A certain telephone cable is under a tension of 2000 newtons. A branch falls on the cable, and a bird, perched 30 meters away, feels the impact after 1.0 sec. What is, in kg/meter, the cable's linear density?

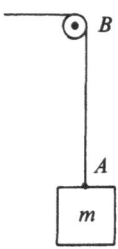

14.8. A construction worker wishes to weigh a load m hanging from a cable of length $AB = 15$ meters, as shown. He strikes the cable at A; the pulse, reflected back and forth between A and B, is counted to have covered the distance AB eight times in 2.0 sec. If the cable has a mass

of 50 kg, find m in kilograms. (Assume an approximately constant tension throughout the cable; is this assumption consistent with your result?)·

*14.9. A 36-meter length of telephone cable is stretched almost horizontally between two poles in such a way that its tension equals 10 times its weight. Assuming the tension to be nearly uniform, at what speed does a transverse mechanical wave propagate along the cable?

14.10.[†] For waves in a string, the lack of dispersion and the superposition principle can be proved if the string's slope is small everywhere (Sec. A iv). Show that, in the case of a sinusoidal wave, this assumption amounts to $A \ll \lambda$ (amplitude very small compared to wavelength).

Section C: Sound in Gases

14.11. What is the wavelength, in air, of sound corresponding to the musical note A (440 Hz)?

*14.12. Referring to Fig. 13.26 of Chap. 13, what is the wavelength, in air, of (a) a very high ultrasound, (b) the two extreme frequencies audible by man, (c) the highest-pitched sound emitted by a bat? Is this the longest or the shortest wavelength of which the bat is capable?

*14.13. Referring to Table 14.1, what frequency is required to set up a longitudinal sound wave of wavelength 1.00 cm in iron?

14.14. From the compressions and expansions shown in Fig. 14.12, plot *qualitatively* your expectation of how the excess pressure, $P - P_0$, varies with x. On the same graph, copy the curve for y versus x. If the wave travels to the right, does $P - P_0$ lead or lag relative to y? By about what fraction of a cycle?

*14.15. (a) Give an explicit dimensional verification of formula (14.C.2) for V in a gas. (b) For smooth **water surface waves** (small slope everywhere), the speed of propagation, V, is independent of the amplitude. It does, however, depend on the wavelength λ. Apply dimensional analysis to express V in terms of λ, ρ (the water's density), and g (the acceleration of gravity). An overall dimensionless factor will remain undetermined.

14.16. An ideal pure monatomic gas (helium, neon, or argon) at 400 K propagates sound at 527 meters/sec. Identify the gas in Table 11.1 of Chap. 11.

*14.17. A certain ideal gas at STP is found to have a density of 2.5 kg/meter^3. Sound propagates in it at 234 meters/sec. Find the gas' kilomolar heat capacity C_{γ}. (Express your answer in terms of the gas constant R.)

[†] Omit if Sec. A iv has been omitted.

14.18. An ideal gas, with a given value of $\gamma = C_P/C_V$ and a given temperature, propagates sound waves at speed V. In terms of V and γ, what is the rms molecular speed in that gas at that temperature? [See Eq. (11.D.17) of Chap. 11.]

14.19.[†] Prove that if $y = \mathscr{F}(T \pm x/V)$, for an arbitrary function $\mathscr{F}$, then the wave equation, (14.A.19), is satisfied for $V = (F/\mu)^{1/2}$.

(Problems 14.20–14.31 below deal with the Doppler effect.)

*14.20. What frequency is heard by a stationary listener when a 235.4-Hz horn is moving away from him at 18.76 meters/sec?

14.21. What frequency is heard while one is driving at 35 meters/sec towards a stationary 4755-Hz whistle in stationary air?

14.22. A jet airplane, whose motors are emitting a 8400-Hz whistle, has just taken off; at the airport, the whistle is still being heard with a frequency of 6300 Hz. How fast is the plane flying?

*14.23. What is the frequency change perceived by a stationary policeman for the horn of a car as it approaches him at 30 meters/sec and then recedes from him at that same speed? The horn's own frequency is 359 Hz.

14.24. What frequency drop is heard by the occupants of the car in Problem 14.23 if the policeman is blowing a 2550-Hz whistle?

*14.25. A ship, sailing at 12 meters/sec towards a shore installation, launches a series of regularly spaced torpedoes in the forward direction at a rate of 2.0 torpedoes per minute; their speed relative to the water is 36 meters/sec. How many torpedoes per minute hit the target? (Assume sea currents are absent.)

14.26. In terms of the speed of sound, V, (a) how fast should one approach a stationary source of sound in order to hear its pitch an octave higher than emitted? (b) How fast should a source approach the listener in order to be heard an octave higher than it vibrates?

*14.27. Two 120.00-Hz buzzers, activated by the same electric signal, are several meters apart. (a) Calculate the frequency received from each buzzer by someone running at 5 meters/sec from one to the other. (b) What beat period does he hear? (See also Problem 14.40 further on.)

14.28. A cyclist is whistling at 440 Hz while riding away from a wall at 10 meters/sec. Given that the wall reflects sound at the same frequency it receives, (a) what reflected frequency does the cyclist hear? (b) What beat

frequency does he hear between his own and the reflected signals?

*14.29. In Example 14.4, find the repetition rate heard by bat B after its signals have been reflected off bat A. (Use the fact that an object receives and reflects at the same frequency.)

*14.30. Simplify formula (14.C.14) by using the fact that v_A and v_B are small compared to V.

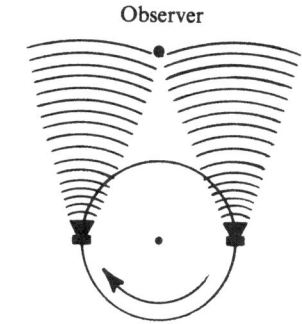

Observer

14.31. Two small loudspeakers are mounted, at the ends of a diameter, on a platform of radius 1.5 meters, as shown. Their vibration is synchronized at 3450 Hz. If the platform rotates at 20 rpm, what is the beat frequency heard by a fixed, far distant observer as both speakers face him from the same distance?

Section D: Standing Waves

14.32. What are the frequencies obtainable from a 5-meter-long organ pipe, assuming both ends are open?

*14.33. The human ear shows a slight increase in sensitivity at a frequency where the ear canal (an approximately cylindrical tube closed at one end) has its fundamental resonance. For an ear canal of length 2.5 cm, estimate that frequency.

*14.34. A steel wire, of mass 4 gram, is stretched between two points 75 cm apart. Its fundamental frequency (for transverse vibration) is 300 Hz. What is, in newtons, the tension in the wire?

14.35. Consider a string of given mass, length, and tension, clamped at both ends. Show that the time needed for a pulse to make one round trip between the two ends equals the string's fundamental period of vibration.

*14.36. By what percentage does a musical wind instrument change its frequency when warmed up from room temperature, say 20°C, to nearly body temperature, say 30°C? (For comparison, one musical semitone means a change of about 6%.)

14.37. A flute, designed to play its lowest note (C) at 262 Hz at room temperature, is filled with helium, also at

[†] Omit if Sec. A iv has been omitted.

room temperature, before being played. Estimate its lowest frequency while this situation lasts.

*14.38. The **Kundt tube**. The speed of sound in a gas is conveniently and accurately measured if one sets up a standing wave of known frequency in a closed pipe of known length, see the adjoining figure. Any light powder will collect at the displacement nodes, and thus makes their locations visible. In the arrangement shown, one observes four such nodes. (a) If the tube is 56 cm in length, and the frequency is 1250 Hz, what is the speed of sound? (b) What is the next frequency at which standing waves are observed?

14.39. A cylindrical tube, one of whose ends is closed by a movable piston, has its open end exposed to a vibrating tuning fork of unknown frequency f. For certain settings of the piston (tuning fork frequency = resonant frequency of the air column), a louder sound is heard. If this happens at two consecutive tube lengths of 18 cm and 24 cm, calculate f.

14.40. Solve part (b) of Problem 14.27 by considering the distance between nodes in a standing wave.

Answers to True or False

1. False (speed is independent of frequency).
2. False (amplitude is adjustable independently of these quantities).
3. True.
4. True (smaller λ means larger f).
5. False (a higher-pitched voice because the speed increases as ρ decreases).
6. True.

7. False (at given temperature, the speed of sound is independent of pressure).
8. True (their densities are equal).
9. False (the speed is independent of the motion of the source).

Answers or Hints to Problems

14.1. 1.56 meters.

14.4. 2.2×10^{-15} sec.

14.5. 14/9 meters/sec.

14.7. 2.2 kg/meter.

14.9. 59 meters/sec.

14.12. (b) About 17 meters and 1.7 cm.

14.13. 5.13×10^5 Hz.

14.15. (b) The complete result is $V = (g\lambda/2\pi)^{1/2}$.

14.17. Use $\gamma = i + R/C_\gamma$. Answer: $C_\gamma = 2.8R$.

14.20. 223 Hz.

14.23. -64 Hz.

14.25. 3.

14.27. (b) 0.28 sec.

14.29. The formula is

$$\frac{\left(1 - \frac{v_A}{V}\right)\left(1 + \frac{v_B}{V}\right) f}{\left(1 - \frac{v_B}{V}\right)\left(1 + \frac{v_A}{V}\right)}$$

14.30. Your result should involve the relative velocity $v_B - v_A$.

14.33. 3.4×10^3 Hz.

14.34. 1080 newtons.

14.36. $+1.7\%$.

14.38. (b) 1.6×10^3 Hz.

CHAPTER 15

ELECTRIC CHARGE AND ITS FIELD

We are now reaching an important milestone—the beginning of electrical science. In these pages, the subject will be developed, as it was historically, from an initial study of *electric charges at rest*; this forms the topic of **electrostatics**. (Moving charges exhibit an additional range of phenomena, **magnetism**, about which more later.)

The basic electrostatic experiments and their interpretation are due to many pioneers, notably Benjamin Franklin (1706–1790) and Charles Augustin de Coulomb (1736–1806).

From their results one concludes that electric charges exert, on each other, forces that have *an inverse-square dependence on distance*, just like the gravitational forces; and, furthermore, that *charge is conserved*. Charge and mass are, however, very different from each other in some of their properties. Unlike mass, charge can occur with positive or negative sign; a repulsion is observed between two charges of like sign, an attraction between charges of opposite sign.

This chapter does two things. First, it describes the above-mentioned features in detail. Next, it introduces what is, in terms of our earlier chapters, a drastically new way of thinking about forces, based on the concept of a vector **field**. According to this view, an electric charge fills all space around it with invisible **field lines**. By imagining the presence of these lines, we greatly help our intuition, as well as our quantitative thinking, in a way that will carry over into the study of magnetism.

The importance of electricity in our understanding of the world is almost self-evident. What needs to be stressed is electricity's wide range of relevance; it extends to the phenomena of optics, chemistry, atomic physics, nuclear physics, and relativity, to name only some representative disciplines; *electricity is a key to modern physics.*

A. The Electric Force

Let us define an electric charge as *what exerts an electric force on another charge* (or, by action and reaction, as *what experiences an electric force*). Is this a circular definition? Not if we possess an independent way of recognizing electric forces. For that purpose, all we need is a single laboratory example to serve as our initial standard of comparison.

i. Recognizing Static Electricity

In a time-honored procedure, a glass rod and a piece of silk are rubbed together. Both objects then experience a mutual *attraction*, observable over separations of up to several centimeters. Two pieces of glass, each rubbed with silk, repel each other, and so do two pieces of silk, rubbed with glass. These objects are said to be **electrically charged**, and the observed forces are electric ones.

It is now a simple matter to recognize other charged objects; they are attracted or repelled by the charged glass or silk.

Electric forces are unique in the complete set of properties that they exhibit. A brief survey of the other known forces will document that fact.

Electric versus Gravitational Forces. As mentioned earlier, the phenomenon of repulsion (glass–

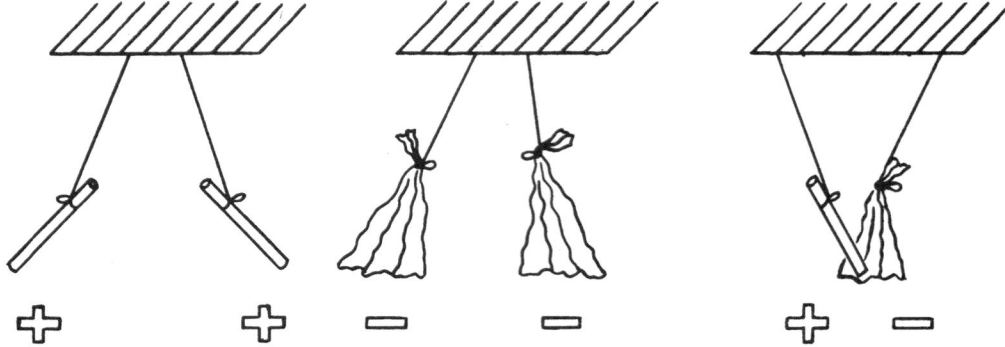

Figure 15.1. Glass and silk, after having been rubbed together, illustrate the fact that like charges repel while opposite charges attract.

glass or silk–silk) has no gravitational counterpart. Electric and gravitational forces also differ enormously in their typical magnitudes. Between two laboratory objects, electric forces can easily be set up that are billions of times stronger than these objects' mutual gravitational attraction. Astronomical bodies are quite a different story, of course, owing to their huge mass and small total charge.

Electric versus Magnetic Forces. Although magnets also act at a distance, magnetic and electric forces can easily be distinguished; the remarkable preference of magnetic forces for *a narrow class of substances related to iron* has no electric counterpart. For example, the force from a silk-rubbed piece of glass does not single out iron objects.

Deeper differences between electric and magnetic forces will emerge later in this book, as will the fact that they are, nevertheless, closely related.

Electric versus Nuclear and Subnuclear Forces. The other forces of Nature—nuclear forces, and those connected with radioactivity—are hard to confuse with electric forces. Those so-called **strong** and **weak** forces have a range of action so short as to be kept well within the atom; in fact, their range does not extend appreciably beyond the nuclear surface itself; hence these forces are never experienced by visible objects in the laboratory.

Macroscopic Contact Forces. The contact forces of mechanics (pressure-related, frictional, elastic) originate, in fact, from electric and magnetic forces between neighboring atoms.

ii. The Sign of a Charge

Franklin himself introduced the convention, still in use today, that defines the sign of a charge. In the glass-and-silk procedure *the glass is said to be positively charged*, while the silk is negative. Whatever the convention, *charges of like sign* (+ + or − −) *repel, while those of unlike sign* (+ −) *attract*, a principle illustrated in Fig. 15.1.

There is nothing special about glass and silk, of course. Many pairs of different substances can be rubbed together with a similar effect; the sign of their charges (always opposite for the two members of the pair) can be tested by whether they are attracted or repelled by a glass standard. It appears that all substances can be ordered in a sequence A, B, C, ..., such that A becomes positive when rubbed against B, B becomes positive when rubbed against C, etc. This sequence is called the **triboelectric series**.*

iii. The Measurement and Transfer of Charge

Just as an arbitrary mass can be manufactured through the piling up of standard masses, so can an electric charge be made from *a combination of standard charges*. What makes this procedure feasible is that charge is readily transferred from

---

* For details, see *Contact and Frictional Electricity*, by W. R. Harper (Oxford, 1967). A ready-made standard more convenient than glass and silk is provided by a roll of 3M Scotch Magic Transparent Tape. During unrolling, the adhesive side becomes positive, while the previously underlying plastic becomes negative.

one object to another, especially in the case of objects made from conducting substances, or **conductors**. Simple contact is then sufficient to transfer a charge; within a conductor, charges will spontaneously move under the effect of their mutual attraction or repulsion.

Metals, graphite, ionized gases, and aqueous solutions of salts, acids, or bases are examples of conductors. Nonconducting substances are called **insulators** or **dielectrics**; they include vacuum, dry air, glass, rubber, sulfur, most fabrics and plastics, paper, and dry wood. Humid air is a sufficient conductor to permit charges to "leak away" in frictional-electricity experiments but, fortunately, is an insulator for most other purposes such as the use of batteries or power outlets.

Using a Standard Charge as a Measuring Unit

We obtain a recognizable amount of charge, to be used as a standard, by means of an **electroscope**. Figure 15.2 shows a simple example of that device. An arbitrary setting of the electroscope's leaves may be chosen to define the standard of charge.

Next, just as a measuring cup is used to gauge a volume of water, an electroscope can be repeatedly "filled" to its standard level, and then "emptied," in order to count the standard units in a given

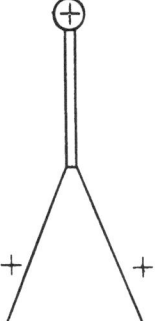

Figure 15.2. This electroscope consists of a metal knob and stem, from which two light flexible metal leaves are suspended. A charge Q, if put on the knob, will spread itself over this conducting system; by mutual repulsion, the leaves will form an angle that increases with Q. By choosing a given angle we define a standard, Q_{st}. (The relation between Q and the angle is not simple to predict, but this does not matter if we are just interested in defining a single Q_{st}.) The apparatus involves a suitable dielectric support (not shown). One must avoid extraneous charges in the vicinity, as they will affect the charges' distribution in the electroscope and hence the angle between the leaves.

charged object. An ingenious, more or less automatic procedure for doing so is shown in Fig. 15.3. It is left to the reader to devise a method for measuring fractions of the standard charge, and for empirically calibrating an electroscope so that the angle of its leaves indicates how many standard charges it contains. (The law of repulsion between the leaves, as a function of charge, need *not* be known in advance for this purpose; the law will emerge from the calibration procedure.)

Verifying the Algebraic Addition of Charges

Various experiments can be performed, in which charges of opposite signs are combined; the simplest case is shown in Fig. 15.4. In this way we can check by observation that charges do add algebraically; two charges Q and Q' combine into a total charge

$$Q_{tot} = Q + Q' \qquad (15.A.1)$$

no matter what the signs of Q and Q'. For example, Q_{tot} is measured to be zero if $Q' = -Q$.

Conservation of Charge

If leakage is prevented by careful experimental procedure, such as the use of good dielectrics, *all experiments performed with a system of initial charge Q will keep the system's total charge equal to*

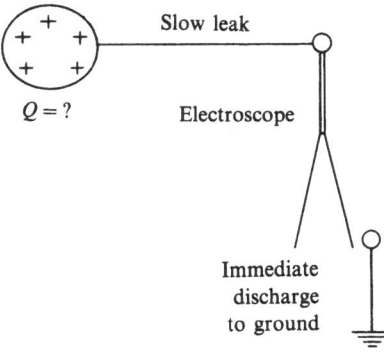

Figure 15.3. An electroscope that automatically discharges itself when it holds one standard charge. The discharge is by contact with a suitably positioned conducting knob, which leaks the charge away to ground. The charge Q being measured is very slowly conducted to the electroscope, for example by a cotton thread. The measurement consists of counting how many times the leaves spread out. Ideally, an estimate should also be made of the small residue of charge left at the end.

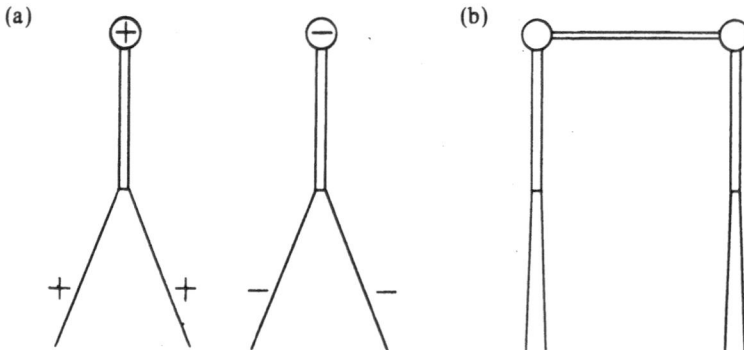

Figure 15.4. Equal and opposite charges (a) combine to zero net charge via a conducting bridge (b).

Q. In particular, an isolated system with zero charge can never develop a net charge. If a charge Q is produced somewhere in such a system, an opposite charge $-Q$ must also arise. (In the glass–silk system, a charge Q on the glass implies a charge $-Q$ on the silk.)

The conservation of total charge is one of the best-established laws of physics, on a par with conservation of energy and of momentum. So far there has been no indication that this law will ever have to be revised.

(Note: If a net charge did vanish or appear during the electroscope procedure for measuring charges, that procedure would not even be self-consistent. Thus, our quantitative definition of charge would become meaningless without the conservation property.)

Sign of a Transferred Charge

Figure 15.5 shows two alternative ways to transfer a positive charge Q from object A to object B, both being originally neutral.

> As seen from outside the objects,
> a transfer $+Q$ from A to B is
> equivalent to a transfer $-Q$ from
> B to A.   (15.A.2)

In practice it is known that the latter is what really happens, at least in solid conductors: *the (negative) electrons are the moving charge carriers.* However, we shall deliberately slip into the habit of saying that a charge $+Q$ was transferred from A to B, even if, actually, $-Q$ went from B to A. This is a widespread convention in electrical science, and, as long as we are aware of its two possible interpretations, it will not lead us into errors or contradictions. Later, when we study individual electrons

or ions, we shall no longer be able to afford such a casual attitude.

In short: Until further notice, *all charge transfers will be discussed as if arising from a flow of positive charge.*

iv. Coulomb's Law

In 1785, Coulomb demonstrated by means of a torsion balance* that two charges, each confined to a very small object ("point charges") exert on one another a force that, except for its sign, behaves mathematically just like Newton's gravitational force. Coulomb's result, for two positive charges q and Q separated by a distance r, is a repulsive force **F** of magnitude

$$F = \frac{KQq}{r^2}$$

with a proportionality constant K reminiscent of G in Newton's law of gravitational attraction.

Let us view the force **F** as *acting on q*; the reaction on Q is then $-\mathbf{F}$. In order to incorporate the direction of **F** into the formula, we introduce a unit vector **u** ($|\mathbf{u}| = 1$), which points along the direction from Q to q. The force *vector* **F** acting on q is then given by

$$\mathbf{F} = \frac{KQq}{r^2}\mathbf{u} \qquad (15.A.3)$$

see Fig. 15.6. One additional virtue of (15.A.3) is that it remains valid for either sign of Q or q,

---

* Cavendish's gravitational experiment was performed somewhat later (in 1798); both men appear to have hit independently on the use of the torsion balance for that purpose.

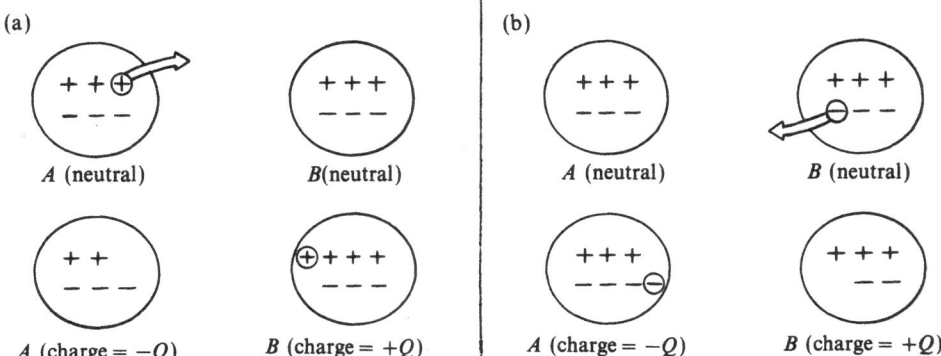

Figure 15.5. The net result of transferring a positive charge, (a), or of transferring the corresponding negative charge the other way, (b), is the same.

correctly indicating an attraction when their signs are opposite. Equation (15.A.3) is known as **Coulomb's law**.

(The $1/r^2$ feature of Coulomb's law had been anticipated as early as 1767 by Joseph Priestley, and was soon recognized, by Cavendish among others, as being very precisely true, at least over laboratory distances. Under the assumption of a $1/r^n$ law, the value of the exponent n has, in recent times,* been found equal to 2 within 1 part in 10^{10}. For an indication of how this was done, see Example 15.4 further on.)

v. The Unit of Charge from Coulomb's Law

Let us perform a measurement of the type summarized by Eq. (15.A.3). The numbers F and r are then determined in terms of newtons and meters, and the combination KQq may be calculated. Once the unit of charge is chosen (say a standard charge collected in the laboratory from an electroscope), then Q and q are known individually, and from KQq we obtain a value for K. In short, *the value of K depends on the choice for a unit of charge.*

Most of us would probably choose units such that K comes out equal to 1, and indeed, this idea is the basis of the once-popular "electrostatic system" of units. However, engineering tradition, as congealed into the International System, has resulted in the choice

$$K \approx 9.00 \times 10^9 \text{ SI units} \qquad (15.A.4)$$

* Plimpton and Lawton, 1936.

[To be exact, $K = (2.99792458)^2 \times 10^9$ SI units; this arbitrary-looking number will receive some motivation in Note (i) of Chap. 18 and in Sec. A iii of Chap. 22.] The reader should be aware that, in the literature, K is usually denoted by the expression $1/(4\pi\varepsilon_0)$.

The unit of charge corresponding to the choice (15.A.4) is called the **coulomb** ($= 1$ C). Some feel for that quantity may be obtained from an estimate of the force between two 1-coulomb charges, placed 1 meter apart; Eq. (15.A.3) gives, in magnitude,

$$F = \frac{(9 \times 10^9)(1)^2}{(1)^2} \text{ newtons} \qquad (15.A.5)$$

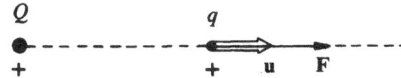

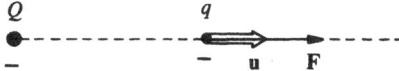

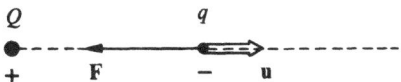

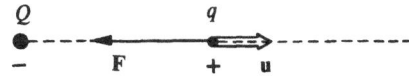

Figure 15.6. The electric force **F** exerted by Q on q for all possible signs of charge. The reaction force on Q is not shown; **u** is the unit vector defined in the text.

This huge force is about the weight of a dozen air-craft carriers. Clearly, therefore, no one will ever isolate whole coulombs on a laboratory scale; we note, however, that the total charge stored in a thundercloud may easily be of the order of 10 coulombs.

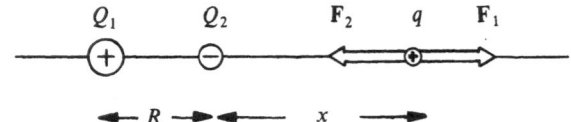

Figure 15.7. Where should q be placed in order to experience zero net electric force from Q_1 and Q_2?

vi. The Quantum of Charge

There exists an amount of charge, denoted by e, that is so small that it cannot be further subdivided.* By methods such as are discussed in later chapters, e is measured to be

$$e \approx 1.6 \times 10^{-19} \text{ coulombs} \qquad (15.A.6)$$

An object can be exactly neutral (have zero total charge), or it can have a charge $+e$, $-e$, $+2e$, $-2e$, etc.; e is a basic building block for charges; one says that *charge is quantized*. The atomic and nuclear particles are characterized by charges $-e$ for the electron, $+e$ for the proton, and zero for the neutron.

What is ordinarily called *electrically neutral matter*, for example a glass object before it has been rubbed with silk, only appears neutral because the positive nuclei and negative electrons are very uniformly interlaced. When glass is charged by friction, large numbers of electrons leave its surface to become attached to the silk, thus leaving an excess of positive charge on the glass. An ordinary hand-held glass rod can easily lose 10^{13} electrons or more in this way.

vii. The Superposition Principle

Just as in the gravitational case, the force exerted on a given charge q by several other charges Q_1, Q_2,... (with any assortment of signs) equals the *vector sum* of the forces that would be exerted on q by each of Q_1, Q_2,... separately. This principle, deduced from observation, is of great practical importance, because it provides us with a recipe

(vector addition) for predicting the effect of all kinds of charge configurations.

Example 15.1. A Zero-Force Point. Figure 15.7 shows two point charges, $Q_1 = +9 \times 10^{-6}$ coulomb and $Q_2 = -4 \times 10^{-6}$ coulomb, located a distance $R = 2$ cm apart. Where will a test charge $q = +5 \times 10^{-6}$ coulomb experience zero electric force?

We know that q must lie precisely on the horizontal line in the figure, for otherwise it would experience two nonparallel forces (from Q_1 and Q_2); two nonparallel vectors never cancel. (Why?)

Let us place q at some distance x beyond Q_2. (Why not between Q_1 and Q_2, or to the left of Q_1?) From Coulomb's law, the forces from Q_1 and Q_2 on q are, in magnitude,

$$F_1 = \frac{K|Q_1 q|}{(R+x)^2}, \qquad F_2 = \frac{K|Q_2 q|}{x^2}$$

They are directed as shown in the figure. Their vector sum vanishes by assumption: $F_1 = F_2$, or

$$\frac{|Q_1|}{(R+x)^2} = \frac{|Q_2|}{x^2}$$

(The values of K and q drop out.) We find

$$\frac{R+x}{x} = +\left(\frac{|Q_1|}{|Q_2|}\right)^{1/2} = \frac{3}{2}$$

The negative root cannot be used, since $x > 0$, $R + x > 0$. Therefore we have

$$x = 2R = \underline{4 \text{ cm}}$$

(It is instructive to look back at Example 8.1 of Chap. 8, involving gravitation, for a case amounting to Q_1 and Q_2 having the same sign.)

---

* Since about 1964 a search has been under way for particles ("quarks") with charge $\pm e/3$, $\pm 2e/3$, etc.; the motivation comes from basic theory. At present it is rather generally believed that no such fractional charges can ever be isolated, although they undoubtedly exist; they are thought always to combine among themselves in amounts 0, $\pm e$, $\pm 2e$, etc., to form the observed neutrons, protons, and other well-known particles.

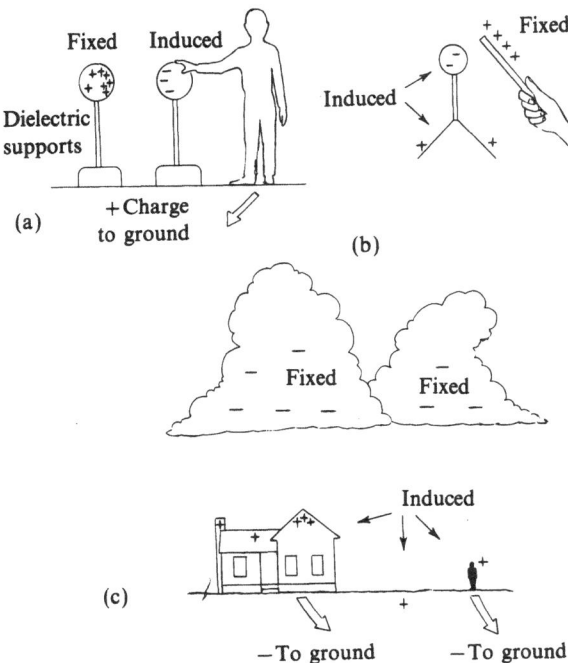

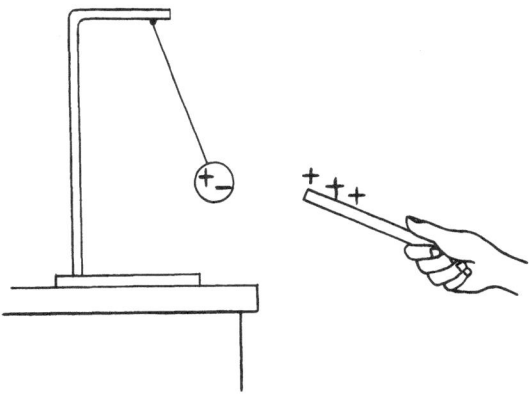

Figure 15.9. The suspended sphere (traditionally, a pith ball at the end of a silk thread) has zero net charge, but is attracted to the rod because the negative induced charge is closer.

ix. Mass versus Charge (Concluding Remarks)

The possibility of two signs, and the typical interaction strength, are not the only distinctions between mass and charge. Equally important is the fact that we have three ways of determining the

Figure 15.8. Examples of electrostatic induction due to a fixed charge. (a) Positive charge is repelled away through the experimenter's body. (b) The over-all neutral electroscope still shows a deflection, due to the proximity of a charged glass rod. (c) The negatively charged thunderclouds induce a positive charge under them, especially in elevated objects. (In real life, thunderclouds contain some positive as well as negative regions.)

viii. Electrostatic Induction

Conducting objects can be charged by action at a distance, as shown in Fig. 15.8. If an originally neutral object finds itself in the vicinity of a fixed charge (positive in the illustrations), migration of charge will occur; positive charge will be repelled away, leaving an excess of *electrostatically induced* negative charge.

In this way, an initially neutral object can experience a net attraction towards a fixed charge, as illustrated in Fig. 15.9. (Even such poor conductors as bits of paper will exhibit this phenomenon in the vicinity of a strongly charged piece of glass or plastic.)

An ingenious mechanized way of exploiting the induction principle is illustrated in Fig. 15.10. Considerable charges can be rapidly built up in such **electrostatic generators**, of which a wide variety have been invented.

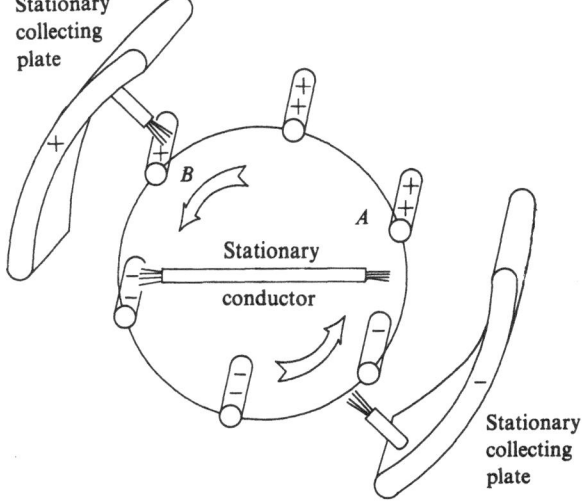

Figure 15.10. Simplified diagram of the Dirod, invented by Professor A. D. Moore of the University of Michigan. The Dirod is a portable-sized machine that accumulates large opposite charges on its two metal collecting plates. The small cylinders represent individually insulated metal rods; the whole set of rods is mounted on an insulating wheel and rotates like a squirrel cage. (A motor is used for that purpose.) Initially, a small charge is always accidentally present on one of the collecting plates (say a negative charge on the lower one). Here the situation is shown just after rod *A*, while passing by that charge, has been made positive by induction; its minus charge has been conducted into the opposite rod through the horizontal bar. Another rod (*B*) has just shared its similarly acquired plus charge with the upper plate. The contacts are made through flexible "brushes."

amount of a mass, whereas we only have two ways for the amount of a charge.

1. The most "primitive" way of making multiples of a standard mass is to collect n identical objects and put them together. This can also be done with charge (electroscope method).
2. The amount of mass shows up as *inertia* in $\mathbf{F}_{tot} = m\mathbf{a}$. This is how a unit of mass is defined: the proportionality constant in Newton's law of motion $\mathbf{F}_{tot} = (\text{const})(m)(\mathbf{a})$ is made to be 1. *A similar phenomenon is not available for charges.*
3. The amount of a mass shows up a third time in Newton's law of gravitation. Since the unit of mass has been fixed by (2), we have no control over the proportionality constant G, which must therefore be measured. In the electric case, Coulomb's law has a constant K (the analogue of G). However, K can still be adjusted as long as the unit of charge is not yet fixed.

B. The Electric Field

Coulomb's law of force involves two point charges. In this section we focus our attention on one of the two charges (it does not matter which one), and we postulate that it causes an **electric field** to exist around itself. The electric field, in turn, acts on the second charge. This field is therefore a kind of intermediary, which transmits the electrical force from one charge to the other.

The electric field is an indispensable idea, which must become thoroughly familiar to the student. It will serve as a model when we introduce its nonidentical twin, the **magnetic field**, in a later chapter. Together, these two fields will eventually permit us to analyze the phenomenon of **radiation**: the fields will turn out to have an existence of their own, quite apart from the charges that produce them.

(Could a gravitational field have been defined in Chap. 8? Most definitely; and it would have been useful as well, had we pursued the subject somewhat further.)

i. Coulomb's Law in Two Steps

Consider a given point-charge Q at a fixed location, as in Fig. 15.11. We explore the distant effects

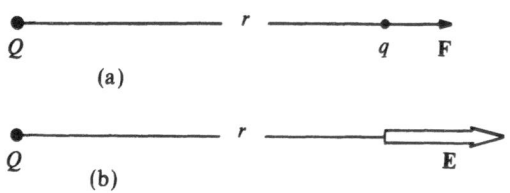

Figure 15.11. (a) The test charge q experiences a force $\mathbf{F}$. (b) q is gone, but Q still produces a *field* $\mathbf{E}$.

of Q by placing another point-charge q (a "test charge") at an arbitrary distance r from Q; by Coulomb's law, q will be subjected to a force

$$\mathbf{F} = q\left(\frac{KQ}{r^2}\mathbf{u}\right) \tag{15.B.1}$$

Here the second factor, to be denoted by $\mathbf{E}$, is a vector that depends only on Q and on the *location* of q; the first factor is just the amount of test charge. In short, we write

$$\mathbf{F} = q\mathbf{E} \tag{15.B.2}$$

The philosophy underlying this new notation is as follows. Even before q was brought in, Q had already prepared the space around itself in a manner summarized by the factor

$$\boxed{\mathbf{E} = \frac{KQ}{r^2}\mathbf{u}} \tag{15.B.3}$$

The test charge, when introduced, merely contributes to the force through a factor of q. In the above, $\mathbf{E}$ is called the **electric field vector** (or just the **electric field**) due to Q; *it is a property of the space around Q and does not depend on whether there is a test charge available to measure it.* Every point in space has its local $\mathbf{E}$ vector. [Equation (15.B.3) is sometimes called Coulomb's law in preference to (15.A.3).]

It is clear from (15.B.2) that

$\mathbf{F}$ and $\mathbf{E}$ have the same or opposite directions
according to the sign of the test charge
($\mathbf{F}$ is the affected vector; $\mathbf{E}$ is not).   (15.B.4)

General Definition of the Electric Field Vector

Equation (15.B.3) represents a special case of $\mathbf{E}$ in which it is produced by a single point charge Q.

More generally, many charges may be present. In order to cover all cases one defines **E** by "turning Eq. (15.B.2) around":

$$E = \frac{F}{q} \qquad (15.B.5)$$

This is a complete prescription for measuring **E**. In words:

> The electric field at any point equals the electric force that *would* act on a test charge at that point, divided by the amount of the hypothetical test charge. (15.B.6)

A convenient slogan is "the electric field is the force per unit test charge," although, of course, the word "unit" should be taken with a grain of salt; a 1-coulomb test charge is a highly unrealistic concept. In real life, test charges are made as weak as possible so as not to disturb the situation one wishes to probe: *the test charge should cause no significant induction on nearby conductors.*

Unit of Electric Field

Equation (15.B.5) shows that **E** is measured in **newtons/coulomb**. (For reasons explained in the next chapter, that unit is also called the **volt/meter**.)

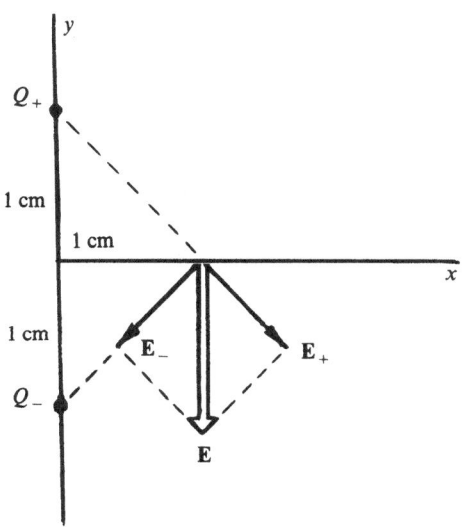

Figure 15.12. A calculation of the electric field, **E**, due to Q_+ and Q_-.

Superposition of Fields

Because electric forces obey the superposition principle, so do electric fields. Example 15.1 can be interpreted in terms of the superposed electric fields of two charges Q_1, Q_2; the resulting field was shown to be zero at a special point.

Will **E** always be independent of the test charge q, even when many charges Q_1, Q_2,... are present? The superposition principle assures us that this is so: the total force on q is made of many terms, all of which contain the same factor q. That factor is removed by prescription (15.B.5).

Example 15.2. Field from a Dipole. Two point charges, $Q_+ = 5 \times 10^{-8}$ coulomb and $Q_- = -5 \times 10^{-8}$ coulomb, are located at coordinates $(0, 1)$ and $(0, -1)$ in the xy plane (readings are in centimeters). What is, in direction and magnitude,[*] the electric field at point $(1, 0)$? (A combination $+Q$, $-Q$ of two equal and opposite point charges is called an **electric dipole**; see also Fig. 15.18b further on.)

Figure 15.12 shows the individual electric field vectors E_+, E_-, and their sum $E = E_+ + E_-$. We see that

1. The vector **E** points along the $-y$ direction.
2. Its magnitude E (the diagonal of the square in the figure) is given by

$$E = \sqrt{2} \, E_+ = \sqrt{2} \left(\frac{KQ_+}{(\sqrt{2}\ \text{cm})^2} \right)$$

$$= \frac{(9 \times 10^9)(5 \times 10^{-8})}{(1.41)(10^{-2})^2} \frac{\text{newtons}}{\text{coulomb}}$$

$$= 3.2 \times 10^6 \frac{\text{newtons}}{\text{coulomb}}$$

ii. A Spectrum of Electric Fields

Figure 15.13 shows the orders of magnitude to be expected from typically encountered electric fields.

iii. Field Lines of a Single Point Charge

If time and patience were infinite, one could measure the field of a point charge Q at all loca-

---

[*] The magnitude $|E| = E$ is often called the **field intensity** or **field strength**.

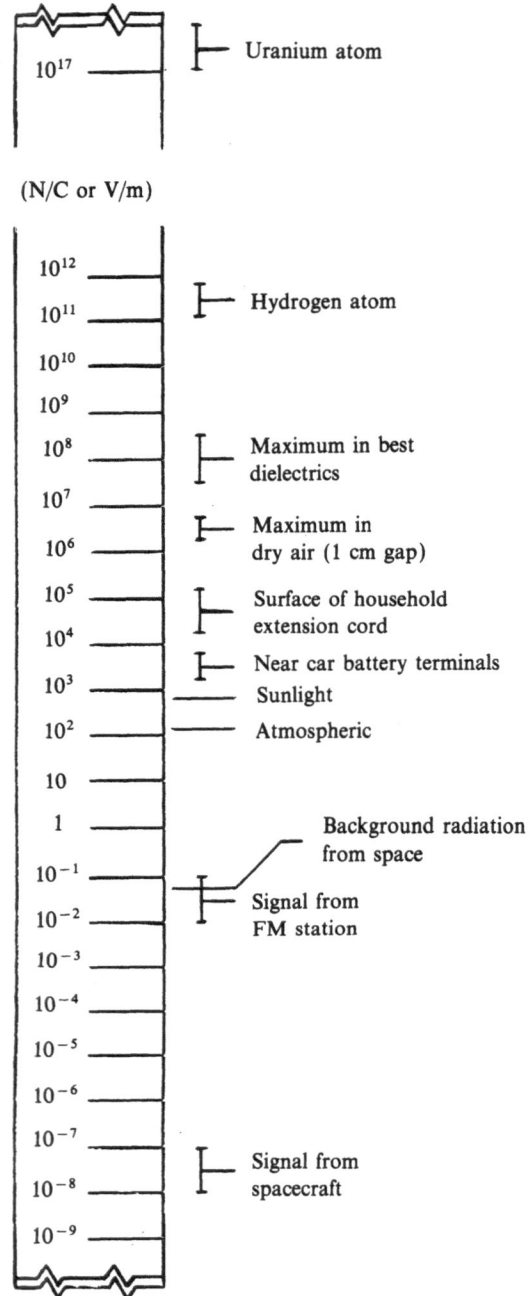

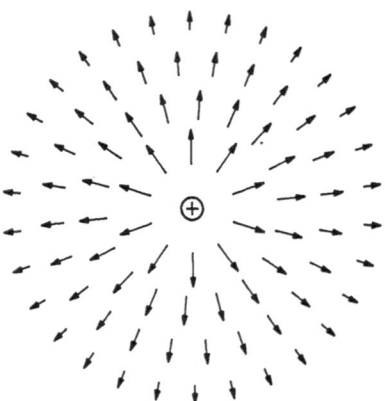

Figure 15.14. A map of **E** at representative points in the vicinity of a positive point charge.

tions around it. The overall situation could then be represented by a map, somewhat like Fig. 15.14.

Michael Faraday, the great intuitive genius of electromagnetism, discovered in the 1820s that such a map may be improved by the introduction of **field lines**,* drawn, in the case of a single point charge Q, according to the following prescription, illustrated in Fig. 15.15.

1. Each field line is a straight line which extends from the point charge out to infinity.
2. Each field line is marked by an arrow which points away from Q if Q is positive, and towards Q if it is negative. (The lines "start," or "end" on Q.)
3. The lines are as uniformly distributed as possible, so as not to favor any special orientation in space.
4. The total number $\mathcal{N}$ of lines extending from a charge is proportional to Q itself. By convention,

$$\boxed{\mathcal{N} = 4\pi K Q}$$ (15.B.7)

or

$$\mathcal{N} \approx 1.1 \times 10^{11} Q \qquad \text{(SI units)}$$

Figure 15.13. Some electric field strengths, on a logarithmic scale. Top, the field typically experienced by an electron deep in a uranium atom, or in a hydrogen atom, and whose source is the nucleus occupying the atom's center. Next, the highest fields that can be sustained in good dielectrics (such as plastics and glasses) without causing a spark discharge ("dielectric breakdown"); in dry air at atmospheric pressure the figure is about 3×10^6 newtons/coulomb across a 1-cm gap; these breakdown fields are therefore the highest that can be sustained in ordinary laboratory experiments; in the atmosphere, higher fields lead to lightning. Further down the scale are the fields to be expected at the surface of the insulation around a typical working 120-volt

* Also called **flux lines** or, misleadingly, "lines of force."

extension cord, or near a car battery. The "atmospheric" field (120 newtons/coulomb) is a fairly constant downward field which prevails at most geographic locations under fair weather conditions, and to which we are quite insensitive. The other listings are not electrostatic, but pertain to the rapidly fluctuating fields found in radiation, say in full sunlight on Earth, in dark interstellar space, and in radio signals.

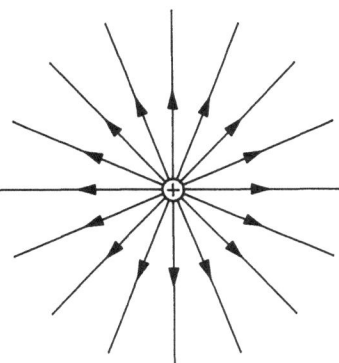

Figure 15.15. A better map: representative field lines for a positive point charge.

(This number is not, in general, an integer; also, in most cases, there are either too many lines to draw individually, or else too few to spread uniformly. In practice, one ignores these annoyances, and one draws as many representative lines as one wishes, pretending that their number is $4\pi KQ$. Caution: One still must draw twice as many lines emanating from a charge $+2Q$ as from a charge $+Q$.)

Two properties of the field lines make them a good map of the field $\mathbf{E}$:

---

1. The *direction* of $\mathbf{E}$ at any point is the direction of the field line through that point.   (15.B.8)

2. The *magnitude* of $\mathbf{E}$ at any point equals the density of field lines in the vicinity of that point.   (15.B.9)

Property 1 should be clear from Fig. 15.15. To understand property 2, we must define the **density of lines**.

Consider an imaginary surface of area $\mathscr{A}_\perp$, drawn perpendicularly to the lines. The density of lines equals the number of lines, $\mathscr{N}$, crossing $\mathscr{A}_\perp$, divided by $\mathscr{A}_\perp$.

$$\text{Line density} = \frac{\mathscr{N}}{\mathscr{A}_\perp} \qquad (15.B.10)$$

This concept is illustrated in Fig. 15.16.

We can now verify the truth of statement (15.B.9), which says that

$$E = \frac{\mathscr{N}}{\mathscr{A}_\perp} \qquad (15.B.11)$$

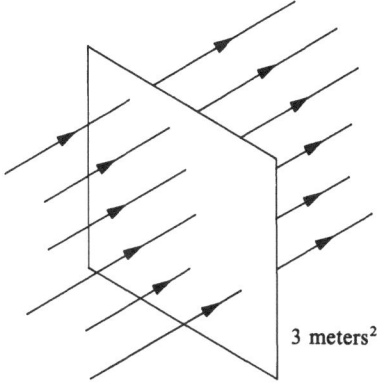

Figure 15.16. In this illustration, six lines intersect an area of 3 meters^2 at right angles. The density of lines is (6 lines)/(3 meters^2) = 2 lines/meter^2.

Draw, as in Fig. 15.17, a spherical surface of radius r and centered around Q. The surface is everywhere perpendicular to the field lines, and therefore, the right side of (15.B.11) gives

$$\frac{\mathscr{N}}{\mathscr{A}_\perp} = \frac{4\pi KQ}{4\pi r^2} = \frac{KQ}{r^2}$$

But this, by Coulomb's law, is just E, the left side of (15.B.11). That relation is now proved; the factor $4\pi K$ in (15.B.7) was expressly designed with that result in mind.

A Field Unit Based on Line Counting

If a line may be thought of as a physical unit, then Eq. (15.B.11) shows that an alternative unit for E is the line/meter^2:

$$1\,\frac{\text{newton}}{\text{coulomb}} = 1\,\frac{\text{line}}{\text{meter}^2} \qquad (15.B.12)$$

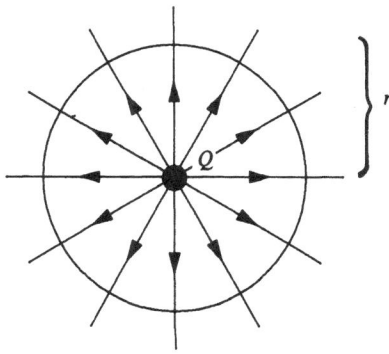

Figure 15.17. "E equals the density of lines." This is proved here for a point charge.

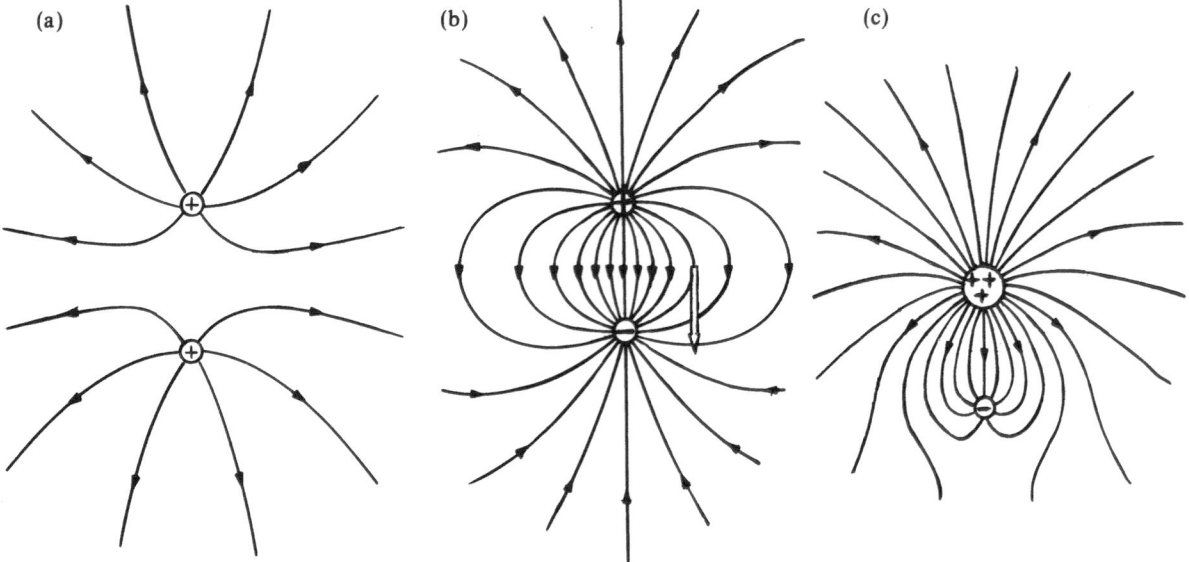

Figure 15.18. Representative field lines for systems of two charges. (a) $+Q$, $+Q$ (both positive). (b) $+Q$, $-Q$ (a **dipole**); the emphasized vector is a reminder of Fig. 15.12. (c) $+3Q$, $-Q$.

iv. Field Lines When Several Charges Are Present

Properties (15.B.8), (15.B.9) provide a general recipe for mapping an electric field **E**, even if more than one point-charge is involved. What must be done in all cases is to "sprinkle" the space with field lines in such a way that, everywhere,

1. The lines are in the direction of **E**;
2. The line density equals E.   (15.B.13)

Figure 15.18 shows some examples involving two charges. Just as in the single-charge case, *all lines are uninterrupted*, except where they begin (on a positive charge) or end (on a negative charge). We shall see later why this simple feature must always be present.

In sketching field lines, it is also useful to know that *two field lines never intersect one another*. If they did, the intersection would be a point at which **E** has two directions; that is to say, the force on a test charge would have two directions—a nonsensical statement.

C. The Electric Flux and Gauss' Law

This section will exploit a property that was built into the field lines: counting the number of lines

reveals how much charge is "sending them out." In the laboratory, field lines cannot be counted, of course, but the counting procedure can be restated in terms of fields, which are indeed measurable.

By thinking in this manner, we shall obtain, at the end of the section, the resulting field of variously shaped charges; this will be achieved through completely elementary calculations.

i. Electric Flux as a Line Count

Although field lines are a mathematical fiction and cannot be seen, it is useful to define an amount of charge Q in terms of the number $\mathcal{N}$ of lines emanating from it, as in Eq. (15.B.7). For the purpose of counting field lines, we draw an imaginary closed surface S around Q (see Fig. 15.19), and define the **electric flux** through S as

$$\text{Flux} = \mathcal{N} = \text{number of lines crossing } S \quad (15.\text{C}.1)$$

There is a sign convention to be observed: a line crossing *from inside to outside* is counted positive, while from outside to inside it is counted negative. This has the welcome consequence that the flux from a charge has the sign of that charge (see Fig. 15.19); also, the shape of S can be such that it has incoming as well as outgoing lines (see Fig. 15.20) without the total flux being affected.

(a)

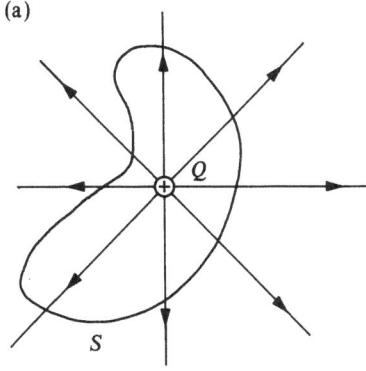

(b)

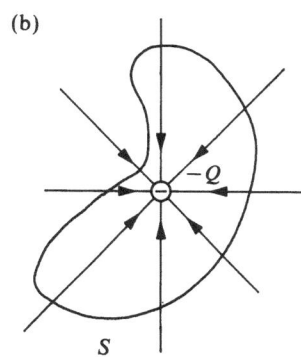

Figure 15.19. (a) The charge Q is "measured" by a count of the lines crossing S from inside to outside. (b) By convention, this number is negative when lines enter the surface.

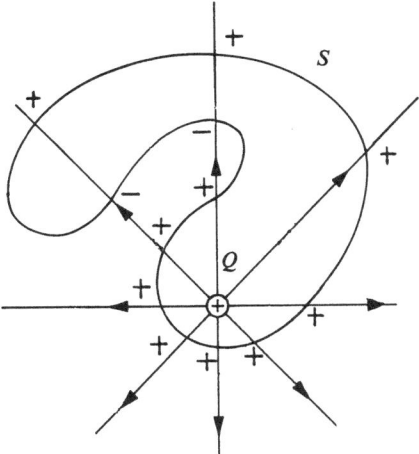

Figure 15.20. Each entry $(-)$ is canceled by an exit $(+)$. In this figure the net number of exits is still 8, in spite of the complicated shape of S.

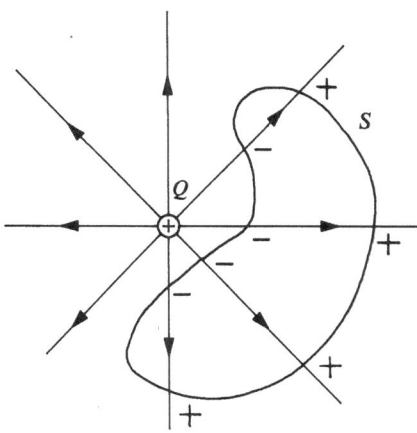

Figure 15.21. Each entry $(-)$ is again canceled by an exit $(+)$, and this time the *net* number of exits (the flux) is zero. The reason is that the charge Q is outside the surface S.

What if the point-charge lies outside S, as in Fig. 15.21? Then we have as many incoming as outgoing lines, and, according to the sign convention, the total flux through S vanishes.

To summarize: the total flux, $\mathcal{N}$, out of a closed surface S is given by [see Eq. (15.B.7)]

$$\mathcal{N} = \begin{cases} 4\pi KQ & (Q \text{ inside } S) \\ 0 & (Q \text{ outside } S) \end{cases} \qquad (15.C.2)$$

ii. Electric Flux in Terms of the Field

In the real world, measurements must be conducted in terms of the field rather than in terms of field lines. Thus we must ask: What is, in terms of $\mathbf{E}$, the flux $\mathcal{N}$ through a surface?

Rather than a closed surface we consider first a plane area $\mathcal{A}$, Fig. 15.22. We also assume a *uniform field* (meaning $\mathbf{E} = $ constant vector). We recall that the line density is given by $E = \mathcal{N}/\mathcal{A}_\perp$, where $\mathcal{A}_\perp$ is

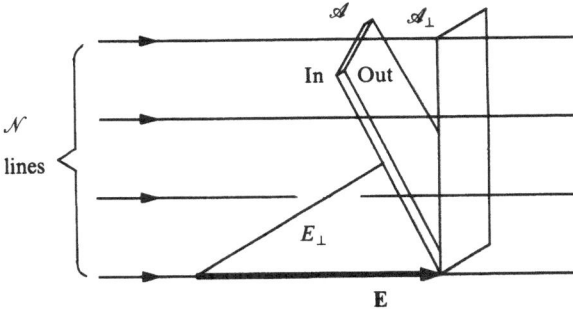

Figure 15.22. The flux, $\mathcal{N} = E\mathcal{A}_\perp$, can also be written $\mathcal{N} = E_\perp \mathcal{A}$. This follows from the similar-triangle relation $E/\mathcal{A} = E_\perp/\mathcal{A}_\perp$.

the *perpendicular* area crossed by the $\mathcal{N}$ lines. We also make an arbitrary agreement on what to call the *outward* and *inward* sides of $\mathcal{A}$. Solving for the flux $\mathcal{N}$, we get

$$\mathcal{N} = E\mathcal{A}_\perp \qquad (15.C.3)$$

or, since $E\mathcal{A}_\perp = E_\perp \mathcal{A}$ from similar triangles in the figure,

$$\boxed{\mathcal{N} = E_\perp \mathcal{A}} \qquad \text{(plane } \mathcal{A}, \text{ uniform } \mathbf{E)} \quad (15.C.4)$$

This expression is defined as a *positive* number if $\mathbf{E}$ is directed *outwards*, and negative in the opposite case; $E_\perp$ is the component of $\mathbf{E}$ normal to the surface $\mathcal{A}$. Thus,

> The flux through an area equals the
> normal component of the field
> times the area. (15.C.5)

iii. Gauss' Law

Single Point Charge

Equation (15.C.4) is the "dictionary" we need to translate line language (left side of the equation) into field language (right side). Let us apply that dictionary to the charge-measuring prescription (15.C.2),

$$\mathcal{N} = 4\pi KQ$$

for a point charge Q inside an imaginary enclosure S; see Fig. 15.23. For simplicity, we pretend that S is composed of many plane areas $\mathcal{A}$, each of which

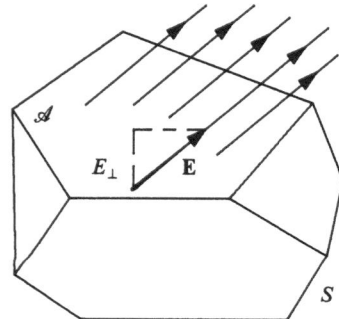

Figure 15.23. Integrating the flux coming out of a closed faceted surface S. We must add all the $E_\perp \mathcal{A}$ contributions (shown only for one facet of S; the enclosed charge is not shown). In the smooth limit the sum becomes an integral.

is in a region of uniform field $\mathbf{E}$. Then the total flux $\mathcal{N}$ is [see Eq. (15.C.4)]

$$\mathcal{N} = \sum E_\perp \mathcal{A} \qquad (15.C.6)$$

a sum ranging over all those plane areas, each with its own value of $E_\perp$.

In reality, S might be curved, and $\mathbf{E}$ might be nonuniform; but if each region $\mathcal{A}$ is small enough, Eq. (15.C.6) becomes as exact as we please: in the limit it will involve infinitely many small contributions. We obtain a so-called **closed-surface integral** over S. (Such an integration is denoted by the symbol $\oiint$, in which the two integral signs remind us that the surface is two-dimensional, while the loop means that the surface is closed.) The element of area, $\mathcal{A}$, must now be written as an infinitesimal, $d\mathcal{A}$. Equation (15.C.6) reads

$$\mathcal{N} = \oiint E_\perp \, d\mathcal{A} \qquad (15.C.7)$$

The flux of a point charge, Eq. (15.C.2), rewritten in field language, is

$$\oiint E_\perp \, d\mathcal{A} = \begin{cases} 4\pi KQ & (Q \text{ inside } S) \\ 0 & (Q \text{ outside } S) \end{cases} \qquad (15.C.8)$$

Many Point Charges

The above remains true if many point-charges $Q_1, Q_2, Q_3,\ldots$ are present. This "grand conclusion" is called **Gauss' law** (Karl Friedrich Gauss, 1839):

$$\boxed{\oiint E_\perp \, d\mathcal{A} = 4\pi KQ_{\text{enclosed}}} \qquad (15.C.9)$$

> The electric flux out of any closed
> surface S equals $4\pi K$ times the
> net charge enclosed by S. There may
> also be charges outside S, but they
> do not contribute to the net flux. (15.C.10)

It cannot be overemphasized that, although charges and fields are well defined in each situation, the closed surface ("Gaussian surface") is only a construction of the mind, and that its shape and position may be chosen to fit one's convenience.

[The proof of Gauss' law in the presence of

several charges applies the **superposition principle** to (15.C.8). In a three-charge illustration, Fig. 15.24, suppose that Q_1 and Q_2 are inside S, while Q_3 is outside. Each by itself, Q_1, Q_2, Q_3 would have fields $\mathbf{E}_1$, $\mathbf{E}_2$, $\mathbf{E}_3$; the actual field is $\mathbf{E} = \mathbf{E}_1 + \mathbf{E}_2 + \mathbf{E}_3$. Separately, according to (15.C.8), these field obey

$$\oiint (E_1)_\perp \, d\mathscr{A} = 4\pi K Q_1$$

$$\oiint (E_2)_\perp \, d\mathscr{A} = 4\pi K Q_2$$

$$\oiint (E_3)_\perp \, d\mathscr{A} = 0$$

Adding the corresponding sides gives

$$\oiint (E_1 + E_2 + E_3)_\perp \, d\mathscr{A} = 4\pi K (Q_1 + Q_2)$$

which is precisely (15.C.9).]

For our purposes, Gauss' law is more easily stated, memorized, and applied in terms of field lines than in terms of fields. Equation (15.C.9) gives *the flux out of a closed surface S*:

Flux $= \mathscr{N} =$ (total number of lines out of S)

$$= 4\pi K Q_{\text{enclosed}} \qquad (15.\text{C}.11)$$

[The thoughtful reader may have realized that the superposition principle is hard to formulate in the language of field lines. This is why an intermediate statement in terms of the field itself, (15.C.8) or (15.C.9), is an essential part of the argument leading to (15.C.11).]

Application : The Continuity of Field Lines

It is now easily demonstrated that

Field lines cannot begin or end except on charges.

$$(15.\text{C}.12)$$

If they could originate in empty space, as in Fig. 15.25, a contradiction would exist. Indeed, draw a Gaussian surface S, as shown; since new lines originate within S, the net flux out of S is nonzero. This implies a nonzero charge in S, contrary to the assumption.

A rough but helpful slogan to memorize in connection with (15.C.12) is that "every positive coulomb sends out $4\pi K$ lines; every negative coulomb takes in $4\pi K$ lines."

iv. Field of a Charged Object

A charge usually occupies an extended region of space, such as a glass or metal surface, rather than a single point. If the charge distribution is given, how do we predict its electric field $\mathbf{E}$? Gauss' law provides an elementary method for solving that problem. To be successful, however, *we require the direction* of $\mathbf{E}$ to be already known; symmetry often yields the necessary clue in this regard.

How to Use Gauss's Law

Before turning to specific cases, we summarize the recipe for calculating $\mathbf{E}$ at any desired point P.

1. Sketch the field lines as dictated by symmetry.
2. Through P, draw a well-chosen Gaussian surface S that intersects the field lines at right angles, or, in some places, runs parallel to them.
3. Verify, from symmetry, that the line density is uniform wherever any lines intersect S.

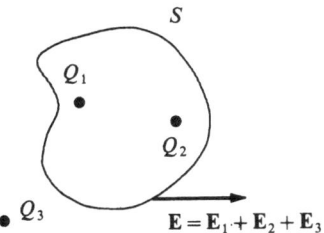

Figure 15.24. The total field vector at a typical point of the Gaussian surface; each charge contributes something; however, the integrated contributions of Q_3 will cancel.

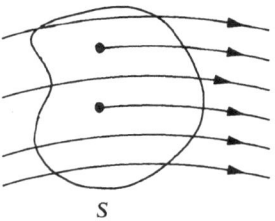

Figure 15.25. Here two field lines originate in empty space—an impossibility.

4. Determine the number of lines, $\mathcal{N}$, originating within S ($4\pi K$ per unit charge).

5. Find the area $\mathcal{A}$ over which lines intersect S. (Some parts of S may not be intersected by lines.)

6. The desired result is

$$E = \frac{\mathcal{N}}{\mathcal{A}_\perp} \qquad (15.C.13)$$

giving the magnitude of E at any point on $\mathcal{A}_\perp$; see Eq. (15.B.11).

(It should be clear that this is not a "graphical method," i.e., no quantities need ever be measured off the sketch; a scale drawing in not required.)

Example 15.3. A Spherical Charge. Given: A spherically symmetric charge, of total amount Q. Find the field $\mathbf{E}$ *outside the charge*, at a distance r from the center; see Fig. 15.26.

Let us assume a positive Q for definiteness.

1. Spherical symmetry leaves the radial direction as the only possible one for $\mathbf{E}$.

2. Take S to be a sphere, concentric with the charge, and of radius r.

3. Spherical symmetry implies a uniform line density over S.

4. $$\mathcal{N} = 4\pi K Q$$

5. $$\mathcal{A} = 4\pi r^2$$

6.
$$E = \frac{4\pi K Q}{4\pi r^2} = \underline{\underline{\frac{KQ}{r^2}}} \qquad (15.C.14)$$

The field outside a spherical charge is the same as that of an equal point charge at the center.

(15.C.15)

Here we could just as well be dealing with the *gravitational effect of a spherical mass*; we now have the long-delayed proof of this result. (Newton had obtained it the hard way, after several years of intense preoccupation. Gauss' law was not available to him.)

Example 15.4. A Hollow Spherical Shell. Figure 15.27 shows a hollow, spherically symmetric charge distribution. It is left to the reader to prove the curious result that

Everywhere inside a hollow spherical shell of charge, its field vanishes exactly, $\mathbf{E} = 0$. (15.C.16)

That conclusion follows from Gauss' law, which in turn depends crucially on the $1/r^2$ nature of the electrostatic force. Therefore the measurement of $\mathbf{E}$ inside a hollow charged sphere provides a test of the $1/r^2$ law; being a null experiment ($\mathbf{E} = 0$), it can be made extremely sensitive. (Null experiments do not wreck highly sensitive instruments, or even push them off scale.) Some of our best information is obtained, essentially, in that manner, with a precision quite out of the reach of torsion-balance measurements such as Coulomb's.

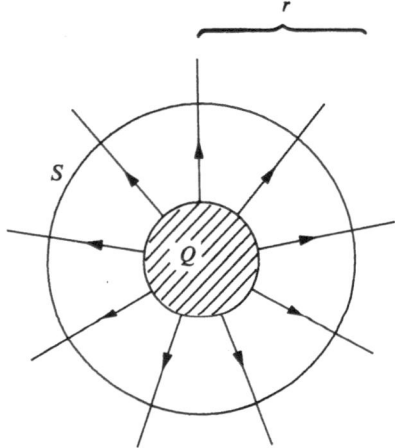

Figure 15.26. Calculation of E for an extended spherical charge.

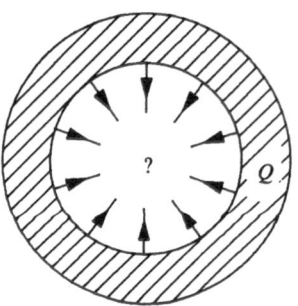

Figure 15.27. To prove: The field vanishes everywhere inside a spherical shell of charge.

Example 15.5. An Infinite Straight Wire.
Consider a very long straight wire (infinite for
practical purposes) with uniform linear charge
density λ (i.e., there is a charge λ per unit length
of the wire); see Fig. 15.28. What is the field $\mathbf{E}$ at
a distance r from the wire?

We may think of a positive λ for definiteness.

1. "Infinite wire" implies that all points along
   the wire are equivalent, or, as is often
   stated, that there are no observable **end
   effects**. Cylindrical symmetry thus means
   that $\mathbf{E}$ must be along a perpendicular
   drawn through the wire.
2. Let S be a cylinder of finite length l, radius
   r, and chosen coaxial with the wire.
3. By cylindrical symmetry, there is a uniform
   line density over the cylindrical surface; the
   two flat ends of S are not intersected by any
   field lines, and hence will contribute zero
   flux.
4. The total charge inside S is the amount
   residing on a length l of wire: $Q_{\text{enclosed}} = l\lambda$.
   Therefore we have

$$\mathcal{N} = (4\pi K)(l\lambda)$$

5.
$$\mathcal{A}_{\perp} = 2\pi rl$$

($\mathcal{A}_{\perp}$ is the cylindrical area, not including
the flat ends.)

6.
$$E = \frac{4\pi K l\lambda}{2\pi rl}$$

or

$$E = \frac{2K\lambda}{r} \qquad (15.\text{C}.17)$$

The length l drops out as it should, being related
to a purely fictitious surface S. Note the r^{-1}
dependence of the field, as opposed to the r^{-2} of
Coulomb's law. [For an **infinite sheet** of charge,
the r dependence drops out altogether; see
Problem 15.22.]

Caution: The magic of Gauss' law is limited
to special cases. If a problem is not sufficiently
symmetric—*and even though Gauss' law still applies*

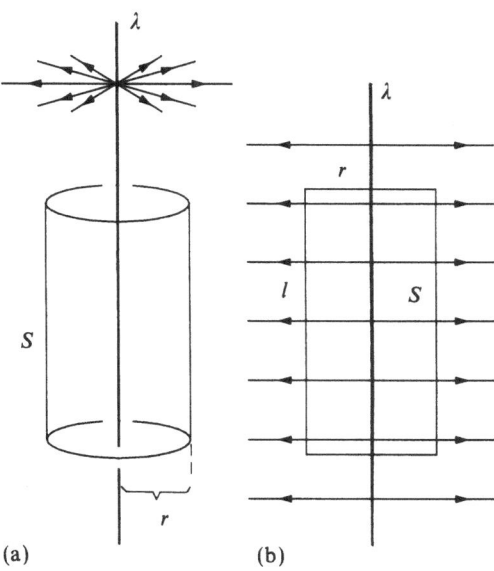

Figure 15.28. (a) Some field lines for the infinite straight wire;
perspective drawing of the Gaussian surface S. ("Infinite," in
practice, means that the ends of the wire are very far from where
$\mathbf{E}$ is measured; "very far" means many times the distance r.)
(b) Side view of S and representative field lines.

—there is no avoiding a laborious use of the superposition principle, i.e., one must integrate contributions to $\mathbf{E}$ from all parts of the charge distribution.
When Gauss' law is effective, it is by far the simpler
method; by contrast, see the Note at the end of this
chapter dealing with the infinite wire through direct
integration.

Note

The Straight Charged Wire (Integration Method)

Let us do Example 15.5 without the benefit of Gauss'
law. What is the field magnitude, E, at a distance r from
an infinite, straight, uniform charged wire? (Linear
charge density $= \lambda$, say positive.)

The following "brute force" method is applicable
whenever we know the location of all charges. It consists
of (1) subdividing the overall charge into many pointlike
elements dQ; (2) finding, by Coulomb's law, the con-
tribution $d\mathbf{E}$ of each dQ; (3) taking the vector sum, $\int d\mathbf{E}$,
of all the $d\mathbf{E}$.

(Only the r component of $\mathbf{E}$ needs to be calculated
here: Fig. 15.29 shows that, in collecting the contributions of all charge elements, we can pair these symmetrically. Their resultant field vector then points directly
away from the charged wire. This confirms our earlier,
more intuitive symmetry argument.)

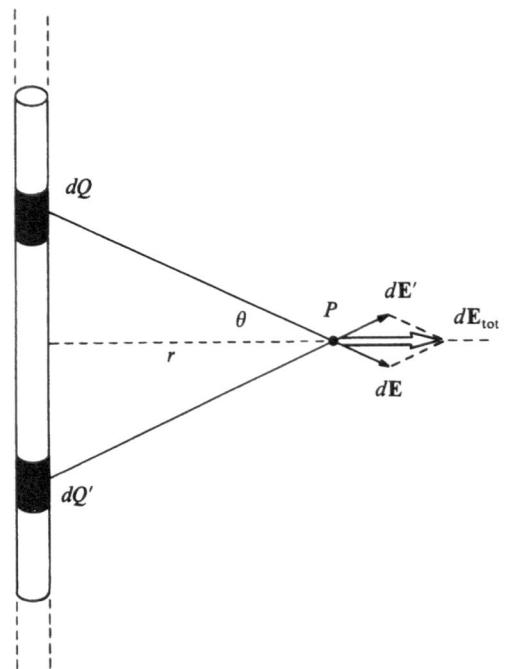

Figure 15.29. Symmetry argument to show that the resultant field is in the *r* direction. We take $dQ' = dQ$, equidistant from the point of observation, *P*. All elements can be paired in this way.

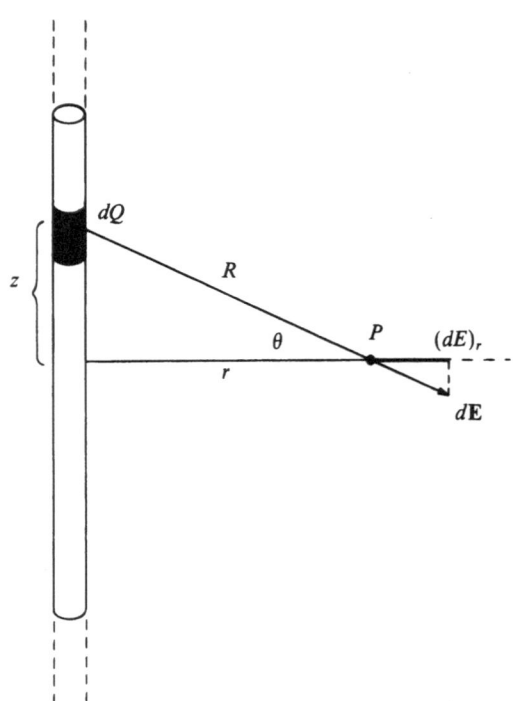

Figure 15.30. Variables z, R, θ for the calculation of E.

1. *Location of dQ*

Figure 15.30 introduces, besides r, the coordinate z, the distance R, and the angle θ; r is a fixed, chosen distance, but z, R, and θ vary as we consider different charge elements dQ along the wire. Point P, at which E is measured, is considered fixed.

All variables can be expressed in terms of θ. [The reason for doing so will be seen in (3) below.] From the figure, we have

$$R = \frac{r}{\cos \theta}, \qquad z = r \tan \theta \qquad (15.N.1)$$

An infinitesimal segment dz carries a "point charge"

$$dQ = \lambda \, dz = \frac{\lambda r \, d\theta}{\cos^2 \theta} \qquad (15.N.2)$$

2. *Field Contribution of dQ*

At point P we find, in magnitude,

$$|d\mathbf{E}| = \frac{K \, dQ}{R^2} \qquad (15.N.3)$$

[from Coulomb's law in the form (15.B.3)], or, using both Eqs. (15.N.1),

$$|d\mathbf{E}| = \frac{K\lambda}{r} \, d\theta \qquad (15.N.4)$$

We just want the r component,

$$(d\mathbf{E})_r = |d\mathbf{E}| \cos \theta = \frac{K\lambda}{r} \cos \theta \, d\theta \qquad (15.N.5)$$

3. *Total Field Contribution*

If E satisfies the superposition principle, so does each of its components. Hence we integrate over the whole wire:

$$E = \int (dE)_r = \frac{K\lambda}{r} \int_{-\pi/2}^{+\pi/2} \cos \theta \, d\theta$$

$$= \frac{2K\lambda}{r}$$

in agreement with Eq. (15.C.17). We now see that z, R, and θ had to be reduced to a single variable, namely, the variable of integration. Choosing θ was an arbitrary decision, but it turned out to make the integral easy to carry out.

Condensed Checklist

Conservation of charge in a closed system:

$$\boxed{\Delta Q_{tot} = 0}$$

Below (15.A.1)

Coulomb's law:

$$\boxed{\mathbf{F} = \frac{KQq}{r^2}\mathbf{u}}$$

(15.A.3)

$$\left.\begin{array}{l} Q, q \text{ in coulombs} \\[2mm] K = \dfrac{1}{4\pi\varepsilon_0} \approx 9.00 \times 10^9 \text{ SI units} \end{array}\right\}$$ (15.A.4)

Elementary charge $= e \approx 1.6 \times 10^{-19}$ coulomb

(15.A.6)

Electric field vector (defined):

$$\mathbf{E} = \frac{\mathbf{F}}{q} \qquad (q = \text{any test charge})$$

(15.B.5)

Single point-charge:

$$\mathbf{E} = \frac{KQ}{r^2}\mathbf{u}$$

(15.B.3)

Superposition of fields:

$$\mathbf{E}_{total} = \mathbf{E}_1 + \mathbf{E}_2 + \cdots$$
$$(\mathbf{E}_1 \text{ from } Q_1, \mathbf{E}_2 \text{ from } Q_2, \text{ etc.})$$

Stated above
Example 15.2

Field line properties:

1. Direction along $\mathbf{E}$ (15.B.8)

2. Density $= E$ (15.B.9)

3. Density $= \dfrac{\mathcal{N}}{\mathcal{A}_\perp}$ (definition)   (15.B.10)

4. Begin only on $+$charges, end only on $-$charges

Below (15.B.13); (15.C.12)

Electric flux through $\mathcal{A}$ (line definition):

1. Define outward side

2. Flux $= \mathcal{N} =$ number of lines *out* of $\mathcal{A}$ (lines *into* $\mathcal{A}$ are counted negative)

(15.C.1)

Electric flux through $\mathcal{A}$ (field definition):

$$\boxed{\mathcal{N} = E_\perp \mathcal{A}} \qquad (\text{constant } E, \text{ plane } \mathcal{A}) \quad (15.C.4)$$

Gauss' law:

The flux out of a closed surface is

$$\boxed{\oiint E_\perp \, d\mathcal{A} = 4\pi K Q_{enclosed}}$$

(15.C.9), (15.C.11)

Special shapes:

Sphere, $E = \dfrac{KQ}{r^2}$ (outside)	(15.C.14)
Spherical shell, $E = 0$ (inside)	(15.C.16)
Solid sphere (inside)	Problem 15.21
Infinite straight wire, $E = \dfrac{2K\lambda}{r}$	(15.C.17)
Infinite cylindrical shell,	Problem 15.25
Flat plate,	Problem 15.22
Hemispheric shell,	Problem 15.26
Two parallel plates,	Problem 15.23 and next chapter, Sec. C.

True or False

1. An electric field is a force.

2. "Electric flux" is synonymous with "electric field."

3. The vector $\mathbf{E}$ is used to denote an electric flux.

4. Halfway between two point charges of equal magnitudes and opposite signs, the electric field vanishes. (Assume there are no other sources of field.)

5. The electric field of a solid uniform sphere of charge decreases in strength inside the sphere as one approaches its center.

6. If all electric charges in the Universe had their signs reversed, the direction of all electrostatic forces would be reversed.

7. If an electric field is probed by means of a test charge, the sign of the field is reversed when the test charge is replaced by one of opposite sign.

8. The field lines of an infinite uniform flat plate of positive charge run parallel to the plate.

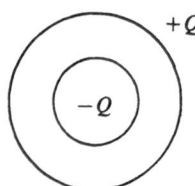

9. Consider two oppositely charged concentric spherical shells, as shown. The two charges are uniformly distributed and of equal magnitude. Then their field vanishes everywhere except between the two shells.

10. Gauss' law is valid only for charge distributions having a suitable symmetry.

11. The trajectory of a charged particle always coincides with an electric field line. (Neglect all forces other than electric.)

12. If a charged particle is released from rest, its trajectory always coincides with an electric field line. (Neglect all forces other than electric.)

13. About the preceding two items: under only electric forces, a charged particle cannot follow a curved field line because the transverse force needed to curve the trajectory would be missing.

Problems

Section A : The Electric Force

*15.1. What force is exerted by a field of 6000 newtons/coulomb on a 2×10^{-4} gram mass carrying 5×10^{-6} coulomb of charge?

*15.2. What is the SI unit of K [see Eq. (15.A.3)]? (Express your answer in terms of coulombs, meters, kilograms, seconds.)

15.3. Two equal point charges repel each other with a force of 0.50 newton when separated by a certain distance. What will be the force of repulsion if each charge is doubled, while the distance between them is halved?

15.4. In a hot plasma (ionized gas) used for nuclear fusion experiments, two deuterons (charge on each $= +e = 1.6 \times 10^{-19}$ coulomb) come within 1.0×10^{-12} meter of each other. What is the force of repulsion between them?

*15.5. The hydrogen atom consists of a proton (charge $= +e$) and an electron (charge $= -e$), separated by a fluctuating distance which we take as $r \approx 5 \times 10^{-11}$ meter for definiteness. (a) What is the ratio of the electric to the gravitational attraction between these two particles? (b) What is the ratio of the electric force, exerted by the electron on the proton, to the latter's weight on Earth? (c) What do you conclude about the importance of gravity in atomic physics? (proton mass $\approx 2 \times 10^{-27}$ kg, electron mass $m \approx 10^{-30}$ kg, $e \approx 2 \times 10^{-19}$ coulomb, $G \approx 7 \times 10^{-11}$ SI units, $g \approx 10$ meters/sec^2.)

Section B : The Electric Field

*15.6. (a) What would be, in Example 15.2, the direction and magnitude of the electric force on a $-0.20\ \mu$C (microcoulomb) test charge located at $(1, 0)$? (b) How many field lines terminate on a charge of -2×10^{-6} coulomb?

*15.7. A proton, of charge 1.6×10^{-19} coulomb and mass 1.7×10^{-27} kg, is in an electric field of strength 250 newtons/coulomb. (a) Find its acceleration, assuming the electric force is the only one present. (b) Compare the result in (a) with the acceleration of gravity, $g = 10$ meters/sec^2. Is gravity negligible?

15.8. If a single electron (mass $= 9 \times 10^{-31}$ kg, charge $= -1.6 \times 10^{-19}$ coulomb), starting from rest, stays in a constant uniform E field of only 5 newtons/coulomb for a period of only 3×10^{-6} sec, what speed will it attain? (Neglect gravity.)

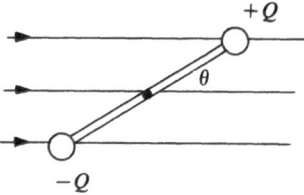

*15.9. Consider a test **dipole**, consisting of point charges $\pm Q$ ($Q > 0$) at the ends of a rigid insulating bar of length l. The dipole is pivoted about its center, as shown, and makes an angle θ with an externally produced uniform E field (that is, the vector E is constant over all space). (a) Sketch the force vectors acting on $+Q$, $-Q$, and show that the total electric force on the dipole vanishes. (b) In terms of Q, E, θ, l, find the torque τ that E exerts about the pivot. Is τ clockwise or counterclockwise? What is the stable orientation of the dipole? What would be the answer for τ if the pivot were at $+Q$ or at $-Q$?

15.10. A microscopic oil droplet has a mass of 2.0×10^{-18} kg, and is charged by just one excess electron (charge $= -1.6 \times 10^{-19}$ coulomb). (a) What electric field

E is needed to keep the droplet suspended against gravity? (b) Does **E** point up or down?

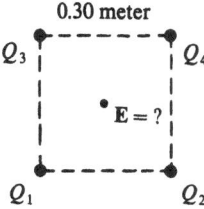

*15.11. Four point charges Q_1, Q_2, Q_3, Q_4 occupy the corners of a square of side 0.30 meter, as shown. If $Q_1 = Q_2 = 9 \times 10^{-6}$ coulomb, and $Q_3 = Q_4 = -18 \times 10^{-6}$ coulomb, what is, in direction and magnitude, the electric field at the center of the square?

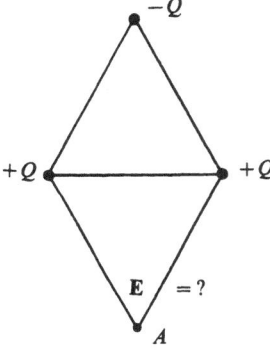

*15.12. Find the electric field vector **E** at point A if given charges $+Q$, $+Q$, $-Q$ occupy the other three points as shown. (The two triangles are equilateral, of side a, and are in the same plane.)

15.13. Sketch your idea of representative field lines in the case of three equal negative point-charges at the corners of an equilateral triangle.

Section C : The Electric Flux and Gauss' Law

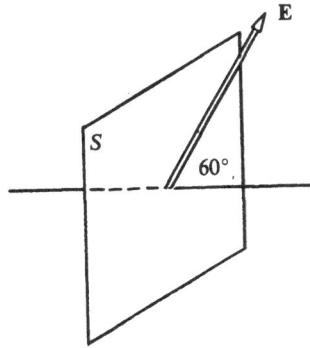

15.14. A uniform electric field of 5000 newtons/coulomb is at 60° to the normal of a plane surface S, as shown. What is the electric flux (number of field lines) through 1 cm² of that surface?

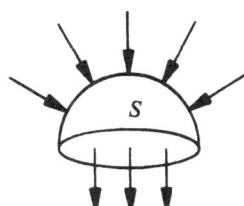

*15.15. A hemispherical box S, of radius 3 meters, has, on its surface, a field **E** that is everywhere normal to the surface; its magnitude is everywhere $E = 500$ newtons/coulomb. On the curved part of the box, **E** points inwards, while on the flat part it points outwards; see the figure. (a) What is the total flux out of S? (b) How much charge is in S?

15.16. The unit of electric flux, "1 field line," is defined such that E has the units of (field lines)/meter²; see Eq. (15.B.12). (a) Express the "field line" in terms of coulombs, meters, kilograms, and seconds. (b) Find the unit of K (in Coulomb's law) in terms of field lines, kilograms, meters, and seconds.

15.17. Show explicitly that the units are the same on both sides of Eq. (15.C.17), $E = 2K\lambda/r$. (See also Problem 15.2.)

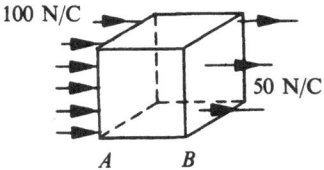

*15.18. When the electric field **E** is measured over the surface of a certain cube (side = 2 meters), it is found that **E** is everywhere parallel to an edge AB, as shown. On the left face, $E = 100$ newtons/coulomb; on the right face, $E = 50$ newtons/coulomb. How much charge is inside the cube?

15.19. Prove that the interior field of a hollow spherical shell vanishes; see Fig. 15.27.

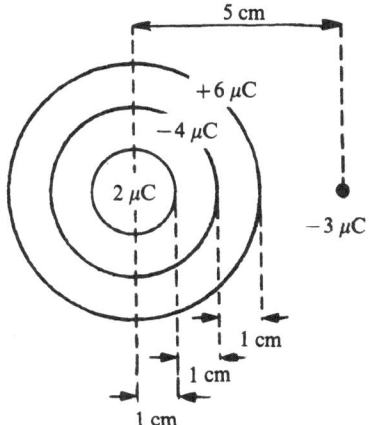

*15.20. Three uniformly charged, concentric spherical shells contain the charges shown (Note: 1 μC = 1 microcoulomb = 10^{-6} coulomb). (a) What force, in newtons, does this structure exert on a $-3\ \mu C$ test charge, 5 cm from the center? Which way does the force point? (b) What is **E** at that location (magnitude and direction)?

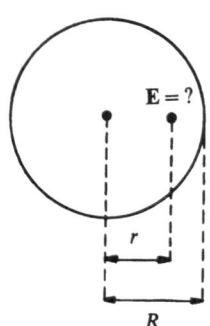

E = ?

r

R

*15.21. A total positive charge Q is spread uniformly over the *volume* of a sphere of radius R, as shown. Find (a) the direction, and (b) the magnitude of **E** at an interior point, a distance r from the center $(r < R)$.

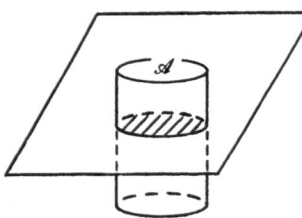

$\mathscr{A}$

*15.22. A plane surface of infinite extent is uniformly charged; its surface charge density is $\sigma > 0$ (σ = charge/ per unit area). (a) By symmetry, what is the direction of **E** at any distance from the surface? Sketch the field lines. (b) By Gauss' law, calculate E at a distance z from the surface. (Use the Gaussian surface shown, with arbitrary flat area A.)

*15.23. Two infinite parallel plates, separated by a distance s, have respective surface charge densities $\pm\sigma$ ($\sigma > 0$). (a) Show that if **E** = **0** on one side of the *pair* of plates, **E** = **0** also on the other side. (b) Under those conditions, find the direction and magnitude of **E** anywhere between the two plates. (c) Sketch the field lines. (Gaussian surfaces: in analogy with the preceding problem.)

15.24. Use Gauss's law to show that if, inside a region, there is a uniform electric field (**E** constant in magnitude and direction), then the region contains no charges.

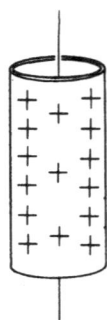

15.25. An infinitely long cylindrical shell, see the figure, is uniformly charged over its whole surface. (a) Show that its outside field is the same as if the charge were concentrated on a wire along the axis. (b) Show that **E** = **0** everywhere inside the shell.

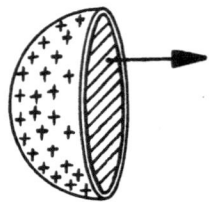

*15.26. The adjoining figure shows a uniformly charged, open hemispheric shell. By completing the sphere and using the superposition principle, show that, in the single-hemisphere case, **E** anywhere on the flat shaded surface is perpendicular to that surface.

Answers to True or False

1. False (a force per unit charge).
2. False (field is flux per unit area).
3. False (**E** is a field; flux is not a vector).
4. False (true if both signs are the same).
5. True.
6. False [all $q\mathbf{E}$ would become $(-q)(-\mathbf{E})$].
7. False (**E** is independent of q).
8. False (they must originate in the plate).
9. True (by Gauss's law, or by superposition of the two shells).
10. False (always valid, but more convenient when symmetry exists).
11. False (the direction of the acceleration does).
12. False (the starting direction is that of **E**, however).
13. True.

Answers or Hints to Problems

15.1. 0.03 newton.

15.2. $\dfrac{\text{kg meter}^3}{\text{sec}^2\,\text{coulomb}^2}$.

15.5. (a) r is irrelevant; (c) gravity is utterly negligible in atomic physics.

15.6. (a) 0.64 newton in the $+y$ direction. (b) 2×10^5.

15.7. (b) Gravity is overwhelmed by a factor of 4×10^{-10}.

15.9. $\tau = QlE \sin \theta$.

15.11. $54\sqrt{2} \times 10^5$ newtons/coulomb, vertically up in the figure.

15.12. $(\sqrt{3} - 1/3)\,KQ/a^2$, down.

15.15. (b) $\dfrac{(3)^2\,(500)}{(4)(9 \times 10^9)}$ coulomb.

15.18. First calculate the net outgoing flux, $[-(100)(2)^2 + (50)(2)^2]$ lines. Then use Gauss's law.

15.20. Only the total charge $+4\,\mu C$ matters here; (b) 1.4×10^7 newtons/coulomb, radially outwards.

15.21. (a) Radially out. (b) Subdivide the charge into a solid sphere of radius r and a shell between r and R; you may then ignore the shell; the sphere of radius r contains a fraction $(\frac{4\pi}{3}r^3)/(\frac{4\pi}{3}R^3)$ of the total charge.

15.22. Side view:

E does not depend on z. (b) The curved side of the cylindrical surface does not contribute: both flat ends contribute the same by symmetry (total flux $= 2\mathscr{A}E$). Answer: $E = 2\pi K\sigma$.

15.23. See Sec. C i of the next chapter.

15.26. First assume that $\mathbf{E}$ is *not* perpendicular, but use symmetry to draw $\mathbf{E}$ with an off-perpendicular component that is purely radial. What interior field results from joining two hemispheres together?

CHAPTER 16

THE ELECTRIC POTENTIAL

We have just seen how a force motivates the idea of an electric field (the electric force per unit test charge).

This chapter promotes a new but similar concept. Here again, we consider a test charge, but we focus our attention on its electric **potential energy** rather than on any force; we remember from mechanics how the energy point of view improves our understanding and manipulative power.

Just as before, the test charge is only a measuring device—a means to an end. We do not want to be bothered with choosing its amount, and therefore we prefer to speak of the *electric potential energy per unit test charge*, a quantity that is called, for short, the **electric potential**, or simply the **potential**, at the location of the test charge.

Field and potential have an important feature in common. Both are properties of space, unaffected by the value of the test charge. They do, however, depend completely on their sources, that is to say, on the charges that produce them.

In another respect, field and potential are quite dissimilar. The field has a direction in space; it is a vector, describable by three numbers, its components. The potential, on the other hand, is a single number (a *scalar*, as it is usually called). This distinction is the same as exists between force and potential energy. Because of its scalar property, the potential is easier to work with than the field.

Of even more practical importance than potentials are **potential differences** ("voltages"); their role is best understood from an analogy. When water is piped from one place to another, one must, in designing and operating the pumps, consider the altitude difference between the two locations; the reason lies in the gravitational work to be done on the water. Similarly, when electric charge is "piped" (in a conducting wire) between two terminals of a circuit, the electric potential difference between those terminals is an essential parameter. Potential is analogous to altitude in this respect.

The highlights of this chapter can be listed as follows. After defining the potential, we shall learn how it and the **E** field *may be calculated from one another*. We shall see how these ideas bear on the *electrostatics of conducting objects*. A special circuit element, the **capacitor**, will be singled out for discussion; aside from being useful in electrical practice, it teaches us a basic fact: the electrostatic field, even in empty space, stores energy.

Later chapters will consider the all-important role of potential differences in controlling electric currents.

A. Introducing the Potential

In what follows we define the electric potential that exists at any point in space, and we examine in some detail the relation between **field** and **potential**. We then show how a potential is mapped graphically by means of **equipotential surfaces**. [These ideas go back to the work of George Green (1828).]

i. Potential near a Point Charge

Figure 16.1 shows a thought experiment designed to probe the effect of a fixed point-charge Q. A test charge q, initially at point A, a distance r

Figure 16.1. A charge q moves out to infinity; the starting point, A, determines the total electric work done. (This procedure measures the potential energy at A.)

from Q, is carried to a much more distant point ("out to infinity") along some arbitrary path. How much work does the electric force, originating from Q, do on q? (The answer will also tell us how much work must be done against that force in order to bring q back from infinity to A.)

This problem has been solved already in connection with gravity. (A brief review of Chap. 8, Sec. D, might help at this stage.) If, instead of the fixed charge Q, we had a mass M, while q were replaced by a test mass m, the desired work (which is independent of path shape) would be the gravitational potential energy of m relative to points at infinity,

$$\mathcal{U}_{\text{grav}} = -\frac{GMm}{r}. \qquad (16.A.1)$$

(We note once more that $1/r$ is a behavior quite distinct from $1/r^2$.)

In the present electric case we obtain, correspondingly, the **electrostatic potential energy** of q relative to points at infinity,

$$\mathcal{U} = +\frac{KQq}{r} \qquad (16.A.2)$$

Note the plus sign, which indicates a repulsion if Q and q have the same sign. (The work done by the field in pushing q outwards is then positive.) We owe formula (16.A.2) to the fact that *the electrostatic force*, like the gravitational one, *is conservative*.

As in the preceding chapter, we are interested in the effect of Q and not in the amount of test charge. We therefore rewrite (16.A.2) as

$$\mathcal{U} = \Phi q \qquad (16.A.3)$$

where

$$\boxed{\Phi = \frac{KQ}{r}} \qquad (16.A.4)$$

The quantity Φ (capital Greek phi) is defined as the **electric potential** relative to points at infinity, where Φ is set equal to zero. That potential is due entirely to Q, and is measured at a distance r from where Q is located.

ii. General Definition of the Potential

We note that (16.A.4) is a special case, involving a single charge Q. We can do much better, and define the potential due to an arbitrary configuration of charges. For that purpose we rewrite Eq. (16.A.3):

$$\boxed{\Phi = \frac{\mathcal{U}}{q}} \qquad (16.A.5)$$

As a slogan: "Potential is potential energy per unit test charge." More carefully worded: *The electric potential at any point equals the electric potential energy which would belong to a test charge at that point, divided by the amount of the hypothetical test charge.* The existence and value of Φ does not, however, depend on any test charge being there at all.

(At this stage the reader should review the exactly parallel discussion of the **E** field in Sec. B i of the preceding chapter.)

Units

From Eq. (16.A.5) we see that the potential is measured in units of joules/coulomb; the shorthand for this is

$$1 \frac{\text{joule}}{\text{coulomb}} = 1 \text{ volt} = 1 \text{ V} \qquad (16.A.6)$$

in honor of Alessandro Volta (1745–1827), the inventor of the electric cell.

Superposition

Figure 16.2 illustrates a situation where one asks for the potential Φ due to an arbitrary set of fixed charges Q_1, Q_2, etc. The answer amounts to a superposition of potentials:

$$\Phi = \frac{KQ_1}{r_1} + \frac{KQ_2}{r_2} + \cdots \qquad (16.A.7)$$

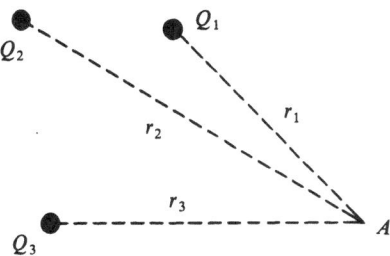

Figure 16.2. The total potential at A is a sum given by Eq. (16.A.7).

that is to say, one takes the *algebraic sum* of the individual potentials, each calculated as if the charge that produces it were there all by itself. The reason for this simple rule will appear in the following subsection.

Example 16.1. A Potential Difference. Figure 16.3 shows two fixed charges, $Q_1 = 2\ \mu C$, $Q_2 = -4\ \mu C$, located at opposite corners of a rectangle of sides $a = 6$ meters, $b = 3$ meters. How much net work, $\mathcal{W}_{AB}$, is done by the electric field on a charge $q = -0.5\ \mu C$, moving between the other two corners (from A to B) along the path shown?

Let Φ_A, Φ_B be the potentials at A and B. Since the required work is the potential energy drop from A to B, it is given by

$$\mathcal{W}_{AB} = q\Phi_A - q\Phi_B = q(\Phi_A - \Phi_B) \qquad (16.A.8)$$

irrespective of the path's shape.
We find, from the figure,

$$\Phi_A = \frac{KQ_1}{b} + \frac{KQ_2}{a}$$

$$\Phi_B = \frac{KQ_1}{a} + \frac{KQ_2}{b}$$

yielding, for formula (16.A.8),

$$\mathcal{W}_{AB} = Kq\left(\frac{Q_1}{b} + \frac{Q_2}{a} - \frac{Q_1}{a} - \frac{Q_2}{b}\right)$$

$$= Kq(Q_1 - Q_2)\left(\frac{1}{b} - \frac{1}{a}\right)$$

The last step is not really necessary, but makes the formula more pleasing to work with. The result is

$$\mathcal{W}_{AB} = (9 \times 10^9)(-0.5 \times 10^{-6})$$
$$\times (2 \times 10^{-6} + 4 \times 10^{-6})(\tfrac{1}{3} - \tfrac{1}{6})\ \text{joule}$$
$$= -4.5 \times 10^{-3}\ \text{joule}$$

(The negative sign means that work is actually done by q *against* the field.) This problem would have been just about insoluble without help from the potential, or at least from the potential-energy concept.

iii. Field versus Potential

It will now become apparent that the same physical information is conveyed whether one specifies the values of Φ or of $\mathbf{E}$ throughout a region of space; however, Φ is simpler and more economical because it amounts to a single function of position, whereas $\mathbf{E}$ amounts to three functions E_x, E_y, E_z.

To document this equivalence we first show how a potential *difference* between two locations may be obtained from a knowledge of $\mathbf{E}$. Later we turn to the value of the potential itself.

Suppose $\mathbf{E}$ is known everywhere in space. Then the force $\mathbf{F} = q\mathbf{E}$ on a test charge is known as well, and therefore also the work $\mathbf{F} \cdot d\mathbf{r}$ done by that force over a small displacement $d\mathbf{r}$ of the test charge. Over a whole path (A to B in Fig. 16.4) this is integrated to give the potential energy drop from A to B:

$$\mathcal{U}_A - \mathcal{U}_B = \int_A^B q\mathbf{E} \cdot d\mathbf{r} \qquad (16.A.9)$$

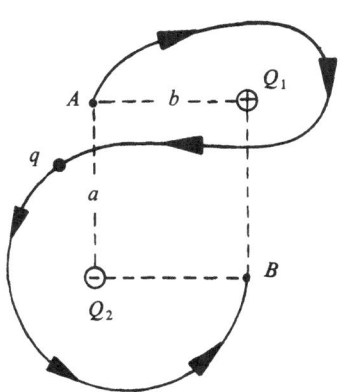

Figure 16.3. A charge q moves from A to B along the curved path.

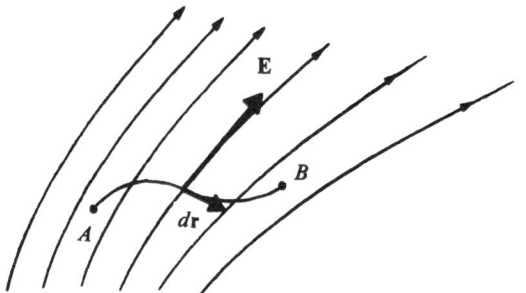

Figure 16.4. Calculation of the potential difference between A and B.

Dividing both sides by q yields the corresponding potential drop:

$$\boxed{\Phi_A - \Phi_B = \int_A^B \mathbf{E} \cdot d\mathbf{r}} \qquad (16.A.10)$$

This may be a difficult integral in practice, but in principle we can always think of it as being carried out. As we have learned in the gravitational case, the force is conservative and the shape of path is unimportant; it may be chosen in any convenient manner.

Equation (16.A.10) is most simply realized in a uniform field $\mathbf{E} = \text{const}$, with the path chosen as a straight displacement $\overrightarrow{AB}$. Equation (16.A.10) now reads

$$\Phi_A - \Phi_B = \mathbf{E} \cdot \overrightarrow{AB} \qquad (16.A.11)$$

or

$$\Phi_A - \Phi_B = (E_{AB}) \, |\overrightarrow{AB}| \qquad (16.A.12)$$

where E_{AB} is the component of $\mathbf{E}$ in the direction from A to B, and where $|\overrightarrow{AB}|$ is the path length.

Units Again

We observe that, dimensionally, (16.A.12) reads

$$\text{volt} \sim (\text{electric field})(\text{meter})$$

This yields the most widely used name for the SI unit of electric field, namely, the volt/meter:

$$1 \frac{\text{volt}}{\text{meter}} = 1 \frac{\text{newton}}{\text{coulomb}} = 1 \frac{\text{field line}}{\text{meter}^2}$$

$$(16.A.13)$$

The designation "field line/meter^2" is not in general use, but helps our intuition.

Example 16.2. Potential of a Thundercloud. The electric field under a certain cloud is measured to be a uniform, upward-pointing $E = 5 \times 10^4$ volts/meter. If the base of the cloud is at a height $h = 800$ meters above ground, find its potential relative to ground.

Let us apply (16.A.12) along the path AB shown in Fig. 16.5. Since $\mathbf{E}$ is constant and parallel to $\overrightarrow{AB}$, we obtain

$$\Phi_{\text{ground}} - \Phi_{\text{cloud}} = Eh$$
$$= (5 \times 10^4)(800) \text{ volts}$$
$$= \underline{4 \times 10^7 \text{ volts}}$$

The cloud base is at -4×10^7 volts relative to ground.

Potential Relative to Points at Infinity

The potential-difference equation (16.A.10) can be made to yield Φ for some point A relative to points at infinity. We just take B very far out, so that $\Phi_B \to 0$; Eq. (16.A.10) now becomes

$$\Phi_A = \int_A^\infty \mathbf{E} \cdot d\mathbf{r} \qquad (\text{any path}) \quad (16.A.14)$$

If both sides of this equation are multiplied by the value of a test charge q, the result summarizes precisely the thought experiment which opened this section. The result (16.A.14) is an explicit demonstration that Φ at any point may be calculated from $\mathbf{E}$; however, $\mathbf{E}$ must be known over a whole range of points.

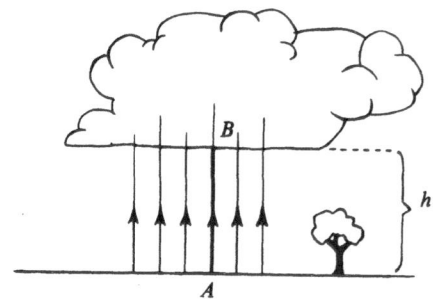

Figure 16.5. Negative thundercloud over positive ground.

Example 16.3. Potential of a Spherical Shell. A positive amount of charge, Q, is uniformly spread over a spherical surface of radius R. Find (a) the radial electric field E_r, and (b) the potential Φ, at a distance r from the center.

a. Electric field: From the preceding chapter we recall that, everywhere inside the sphere, we have $\mathbf{E} = 0$; outside, $\mathbf{E}$ is like that of a point-charge Q at the center. In short,

$$E_r = \begin{cases} 0 & (r < R) \\ \dfrac{KQ}{r^2} & (r > R) \end{cases} \qquad (16.A.15)$$

This is plotted in Fig. 16.6a.

b. Potential: This time let us carry a test charge q *inward* from $r = \infty$ (where $\Phi = 0$) to $r = 0$. As long as we are still outside the sphere ($r > R$), Q might as well be a point charge, giving a potential energy KQq/r, and hence $\Phi = KQ/r$. However, inside the sphere, where $\mathbf{E} = 0$, no work is done, and the potential energy of q no longer changes within that region. Thus, Φ itself is constant for $r < R$, and retains its surface value KQ/R. We conclude that

$$\Phi = \begin{cases} \dfrac{KQ}{R} \ (=\text{const}) & (r < R) \\ \dfrac{KQ}{r} & (r > R) \end{cases} \qquad (16.A.16)$$

The graph is shown in Fig. 16.6b.

Proof of Superposition

Rule (16.A.7), according to which individual potentials are simply added up, follows directly from the relation (16.A.14) between Φ and $\mathbf{E}$. Setting, in that formula, $\mathbf{E} = \mathbf{E}_1 + \mathbf{E}_2 + \cdots$, where $\mathbf{E}_1, \mathbf{E}_2,\ldots$ are the fields due to the individual charges $Q_1, Q_2,\ldots$, we obtain directly

$$\Phi_A = (\Phi_1)_A + (\Phi_2)_A + \cdots$$

where $(\Phi_1)_A$ is the potential that Q_1 produces at A, etc. In short: Superposition for Φ follows from superposition for $\mathbf{E}$.

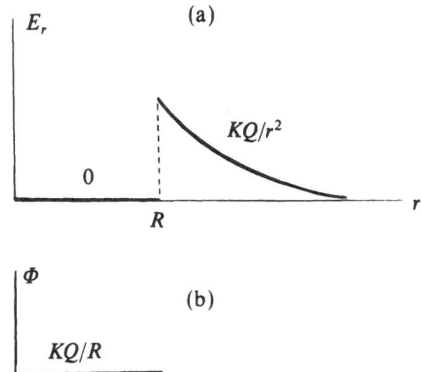

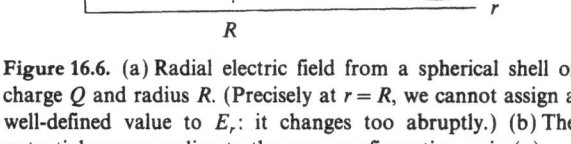

Figure 16.6. (a) Radial electric field from a spherical shell of charge Q and radius R. (Precisely at $r = R$, we cannot assign a well-defined value to E_r: it changes too abruptly.) (b) The potential corresponding to the same configuration as in (a).

Obtaining the Field from the Potential

There remains a logical loose end in our claim that Φ and $\mathbf{E}$ carry essentially the same information. From Eq. (16.A.14) we see how Φ is obtained from $\mathbf{E}$; but how is $\mathbf{E}$ obtained from Φ?

We already have the answer in the case of a uniform $\mathbf{E}$; Eq. (16.A.12) yields any desired component of $\mathbf{E}$ from

$$E_{AB} = \frac{\Phi_A - \Phi_B}{|\overrightarrow{AB}|} \qquad (16.A.17)$$

If $\overrightarrow{AB}$ is chosen parallel to the x axis, we obtain E_x, and similarly for the other two components.

(If $\mathbf{E}$ varies in space, we simply take points A and B so close together that $\mathbf{E}$ is effectively constant; Example 16.3 then yields the well-defined *local* value of E_{AB} after we let $|\overrightarrow{AB}| \to 0$ in the style of a derivative.)

iv. Equipotentials

Just as the field lines are a graphical representation of the electric field, **equipotentials** can now be introduced to help visualize the potential in space.

> An equipotential is defined as the locus of all points at which the value of the potential is the same.

$$(16.A.18)$$

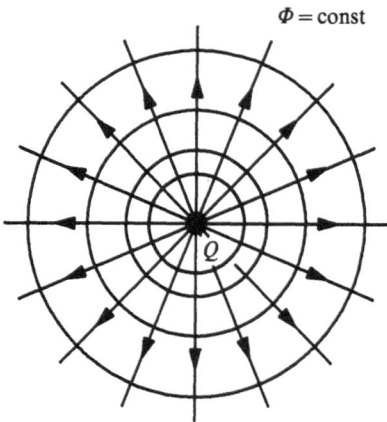

Figure 16.7. Around a point-charge, the equipotentials are concentric spheres.

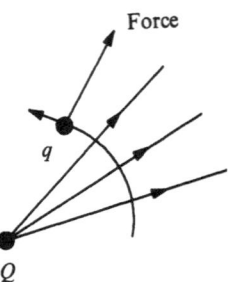

Figure 16.8. An argument showing that the field lines are at right angles to the equipotentials.

To illustrate: what are the equipotentials in the vicinity of a single point charge Q?

From $\Phi = KQ/r$ we see that all points with the same value of r (a spherical surface) have the same value of Φ. Hence, for a point charge, *each equipotential is a spherical surface centered on the charge*; see Fig. 16.7.

In that figure, the field lines always intersect the equipotentials at right angles. This, as a simple argument will now show, is a general feature of all electric field configurations, no matter how complicated.

Suppose we move a test charge q at right angles to the field lines. Then the motion is at right angles

to the force, and zero work is done; see Fig. 16.8. Therefore, as claimed, the potential energy $q\Phi$ of the test charge does not change.

To conclude:

> The field lines always meet the equipotential surfaces at right angles.

(16.A.19)

Two other illustrations of that fact are shown in Fig. 16.9.

Analogy to Contour Lines

On a topographical map, the contour lines play the role of equipotentials. A contour line is the locus of all points whose altitude is the same; see

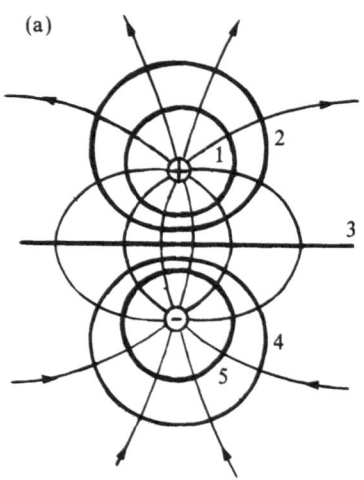

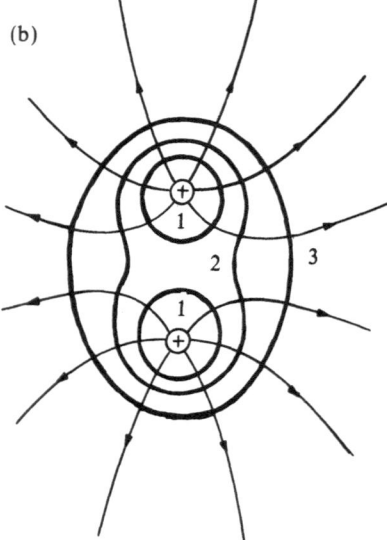

Figure 16.9. (a) Equipotential surfaces (heavy lines) and field lines for a pair of opposite charges. Surfaces 1–5 are in order of decreasing potential. (b) Equipotentials and field lines for a pair of equal charges; surfaces 1, 2, 3 are at decreasing potentials.

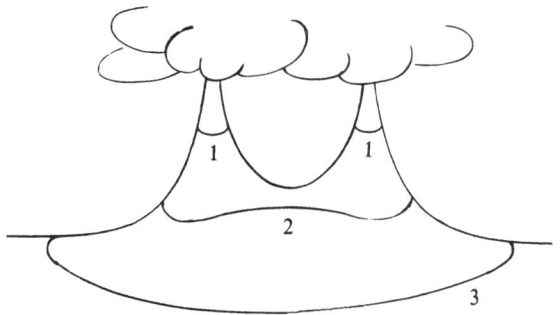

Figure 16.10. Contour lines on twin mountain peaks. Seen from above they should appear as in Fig. 16.9b.

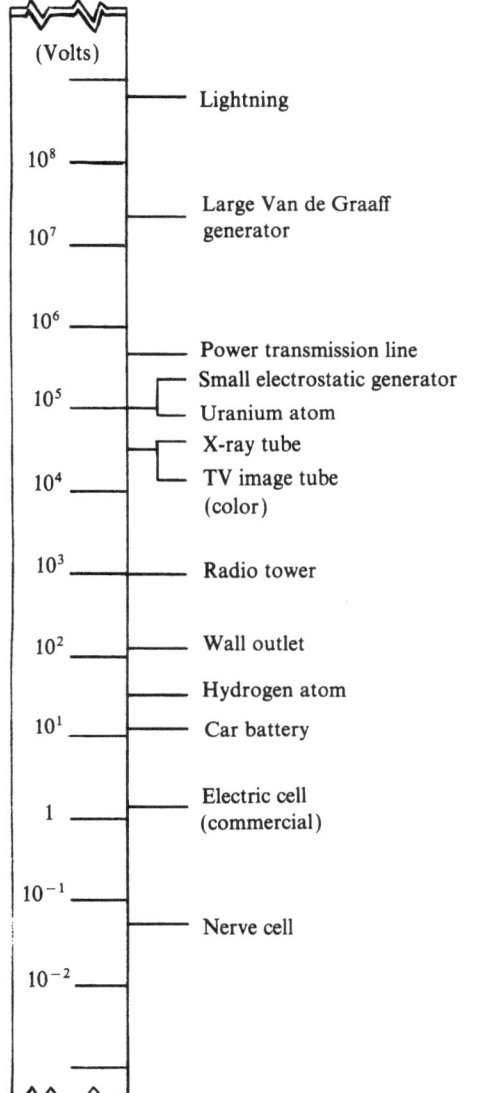

Figure 16.11. Some typical potential differences (relative to ground, or between two terminals), on a logarithmic scale. The values shown should not be thought of as precise, but may vary

Fig. 16.10. Therefore an explorer walking along a contour line (and playing the role of a test mass) does not change his potential energy. If a reference level is agreed upon, we can speak, for example, of a 100-meter altitude contour, just as we speak of a 100-volt equipotential. The direction of steepest downhill slope on the geodesic map is analogous to the direction of a field line, i.e., to the direction of **E**.

v. A Spectrum of Potentials

Figure 16.11 exhibits some potential differences as realized in Nature or in the laboratory.

B. Single Conductors in Equilibrium

In this chapter, for definiteness, let a conductor be thought of as a piece of metal. In such an object, each atom may contribute one (or a few) of its outer electrons to a common pool of so-called **conduction electrons**, which are free to wander throughout the conductor, very much in the manner of gas molecules inside a container. Here we shall not look at individual electrons; for simplicity, we pretend that, in a conductor, all charges are smoothly spread out rather than "grainy," as they really are. This will not affect the validity of our results except on a microscopic scale.

The electrical behavior of a conductor is simplest when electrostatic *equilibrium* prevails. This means that there is no motion of charge, and that the electric potential (and hence also the field) is constant in time, although of course it may vary from place to place.

The purpose of this section is to derive a few simple rules that must be obeyed by the charges and the field, in and around a conductor at equilibrium.*

* More precisely, we are dealing with conductors whose composition and temperature are the same at all points. However, we shall not worry about that restriction, which is unimportant in most experiments.

over wide ranges. The Van de Graaff generator is mentioned further on; see Fig. 16.20. The uranium and hydrogen potentials are typical of those experienced within the atom by the electrons. The x-ray and image tube potentials are used to give kinetic energy to electron beams. The nerve cell potential occurs across the cell membrane. Several potentials quoted are not really electrostatic, but fluctuate rapidly, notably the potential difference between the top and bottom of an emitting radio tower.

i. Some Equilibrium Rules

We first note that

> 1. Inside a conducting material
>   at equilibrium, the electric
>   field vanishes everywhere:
>
> $$\mathbf{E} = 0$$

(16.B.1)

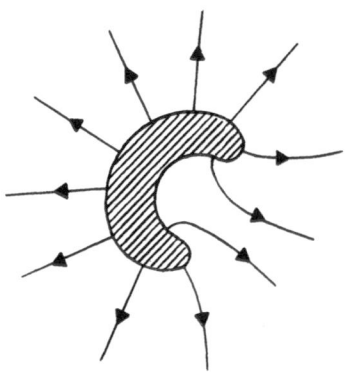

Figure 16.13. The field lines meet the conductor at right angles.

This may be understood as follows. If, by some external means, an electric field $\mathbf{E}$ were set up in a metal, the conduction electrons, being negative, would respond by migrating against the direction of $\mathbf{E}$; there would be a lack of (static) equilibrium. Therefore equilibrium implies $\mathbf{E} = 0$.

How does a piece of metal achieve this $\mathbf{E} = 0$ condition? It rearranges its conduction electrons so as to cancel the original field; an illustration of this phenomenon is shown in Fig. 16.12.

Next, consider the *potential* inside the conductor.

> 2. In a conducting material at
>   equilibrium the electric potential
>   has a constant value,
>
> $$\Phi = \text{const}$$

(16.B.2)

This follows directly from Eq. (16.A.10). For any path AB inside the conductor, we have

$$\Phi_A - \Phi_B = \int_A^B \mathbf{E} \cdot d\mathbf{r} = 0$$

since $\mathbf{E} = 0$, so that Φ has the same value at any two points.

We see that the whole conducting object is *an equipotential region of space*. In particular, *the surface of the conductor is an equipotential surface*. It follows immediately, by statement (16.A.19), that

> 3. Just outside a conductor at
>   equilibrium, the field lines, and
>   hence $\mathbf{E}$, are perpendicular to
>   the conductor's surface.

(16.B.3)

see Fig. 16.13.

(The field strength $|\mathbf{E}|$ just outside the conductor is directly related to the surface charge density on the conductor in that vicinity; see Problem 16.15.)

Our next inference concerns the presence of charged regions inside a conducting substance—in fact, there are none:

> 4. The interior of a conducting substance
>   at equilibrium is everywhere neutral.

(16.B.4)

To prove this we first assume the opposite. Suppose, as in Fig. 16.14, that some charge $Q \neq 0$

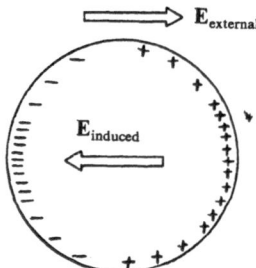

Figure 16.12. This conducting sphere achieves zero net field in its interior. The field of the induced charges exactly cancels the exterior one: $\mathbf{E}_{\text{external}} + \mathbf{E}_{\text{induced}} = 0$.

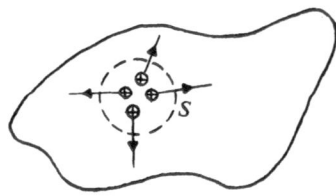

Figure 16.14. Impossibility of charges being present inside the conductor.

is present somewhere in the conductor. Let us draw a Gaussian surface S around Q, *keeping S still entirely within the conductor*. We know that a net flux $4\pi KQ \neq 0$ comes out of S. Since the flux is the surface integral of $E_\perp$, it follows that $\mathbf{E} \neq \mathbf{0}$ at least somewhere on S; but this contradicts our previous conclusion (16.B.1).

Statement (16.B.4) leads at once to the remarkable deduction that, if there are any charges at all on a conductor at equilibrium:

> 5. All charges are confined to the conductor's surface.

$$(16.B.5)$$

[Why does the proof of (16.B.4) still allow surface charges?]

Example 16.4. Designing a High-Potential Electrode. A large electrostatic generator, used in a nuclear physics laboratory, is designed to raise the potential of a metal globe to $\Phi = 3 \times 10^6$ volts relative to ground ("ground" = "points at infinity" = the floor and walls of the room). If we assume $E_{max} = 2 \times 10^6$ volts/meter as the breakdown field in air, what should be the globe's radius, R?

For definiteness, assume a positively charged conducting sphere. Then statement (16.B.5), together with spherical symmetry, tells us that we are dealing with a uniformly charged shell of radius R. Sparking, when it happens, will begin where the field is strongest, namely, just outside the metal, with

$$E = E_{max} = \frac{KQ}{R^2} \qquad (16.B.6)$$

and

$$\Phi = \frac{KQ}{R} \qquad (16.B.7)$$

Taking the ratio, we have

$$\frac{\Phi}{E_{max}} = R \qquad (16.B.8)$$

We see that the obtainable potential is directly proportional to the radius of the electrode. In this case,

$$R = \frac{3 \times 10^6 \text{ volts}}{2 \times 10^6 \text{ volts/meter}} = \underline{\underline{1.5 \text{ meters}}}$$

ii. Hollow Conductors

A region of space surrounded by conducting material, as in Fig. 16.15, has some curious properties of its own, which we now enumerate.

> 1. In an empty cavity surrounded by a conductor at equilibrium, the field vanishes.

$$(16.B.9)$$

("Empty" means that no separate charged objects have been introduced into the cavity.) As a proof, assume once more the opposite: $\mathbf{E} \neq \mathbf{0}$, i.e., there are some field lines.

A given line cannot bend back on itself; otherwise it would provide a closed path around which a test charge could be endlessly carried. In this way we could obtain an infinite amount of work and the world's energy problems would be solved!

Could a line terminate on the conductor's inner surface? It cannot, for a similar reason; both impossibilities are illustrated in Fig. 16.16. The only alternative is $\mathbf{E} = \mathbf{0}$.

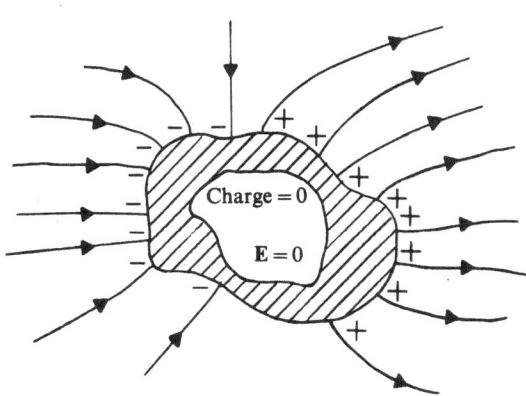

Figure 16.15. A cavity inside a conductor has zero field and zero surface charge at all points. In this illustration the inside is completely shielded from the external field.

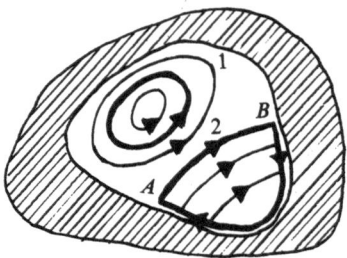

Figure 16.16. Both line 1 (a closed loop) and line 2 (from *A* to *B*) would allow nonconservation of energy. A test charge carried along the indicated paths (heavy lines) would keep receiving net work from the field. (From *B* to *A*, no work is done because here the path is perpendicular to the field lines.)

Shielding

We see that a conducting enclosure *shields* its interior from any electrostatic field. An approximated enclosure such as a wire cage ("Faraday cage") often suffices to protect an experimenter, or some delicate apparatus, from dangerous electric fields; similarly, inside a car one is fairly safe from the effects of lightning. (Why is a similar gravitational shield impossible?)

We have seen that charges may collect on a conductor's surface. Nevertheless, at equilibrium, and if the cavity is empty of charge:

> 2. The inner wall of a hollow conductor is neutral everywhere.

(16.B.10)

The proof (by means of a small Gaussian surface straddling a small portion of the inner wall) is left to the reader.

As an illustration (Fig. 16.17), let us touch the inner surface of a nearly closed metal vessel with a small conducting object carrying a net charge Q. When contact is made, the whole structure becomes a single hollow conductor, and Q migrates to the outer surface—the small object is completely neutralized. By doing so repeatedly, one can in principle accumulate a huge charge on the outer surface of the vessel, if it is insulated. (Faraday's original demonstration of this principle made use of an ice pail as the cavity.) When suitably mechanized, this operation becomes the principle of Van de Graaff's electrostatic machine; see further on.

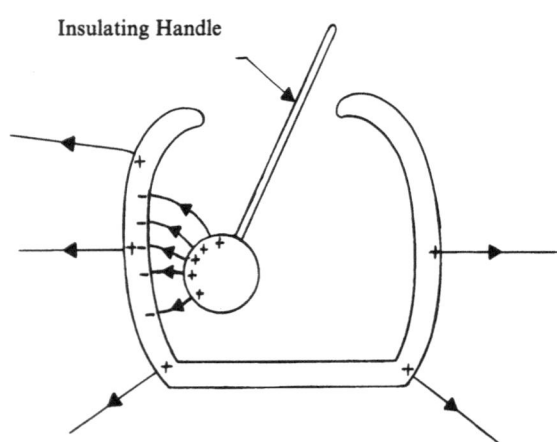

Insulating Handle

Figure 16.17. This metal "ice pail" is a good approximation to an enclosed cavity. When contact is made, the positive charge will leave the small positive ball and neutralize the induced negative charge. The pail's total charge is thereby increased.

To end this collection of "conductor facts," consider a hollow conductor in whose cavity a charged object (total charge $= Q$) has been enclosed. Then, in contrast to the preceding statement, there must be, at equilibrium, a net charge on the inner wall of the enclosure:

> 3. If a cavity in a conductor contains a net charge Q (not touching the wall), then the wall itself has a surface charge of net amount $-Q$

(16.B.11)

see Fig. 16.18. Thus, the charge $+Q$ *induces* a charge $-Q$. The proof makes use of the fact that

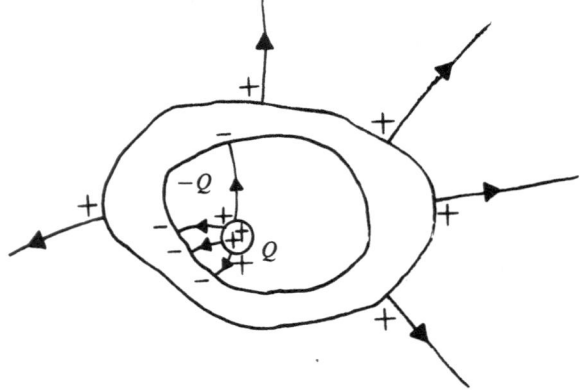

Figure 16.18. The charge Q induces a total charge $-Q$ on the inner surface. This was approximately true also in Fig. 16.17.

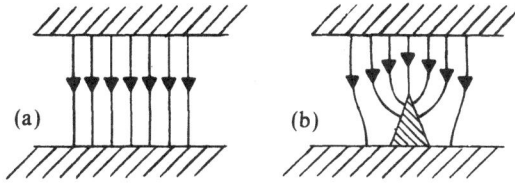

Figure 16.19. (a) A field configuration without a sharp point. (b) Perpendicularity at the point draws the lines together.

each field line out of Q must terminate on the wall, since it can go no further. Each line connects a positive charge to an equal-sized negative charge.

iii. The Effect of Sharp Points

A charged metallic point can produce an extremely intense electric field in its immediate vicinity. This may result in a breakdown, into ions and electrons, of the surrounding air, which then becomes conducting and will discharge the point. In electrostatic machines, rounded conductors are used where charge leakage is to be prevented. In other parts of such machines, sharp points are advantageously used to transfer charge through the air; the lightning rod, invented by Franklin, is another example of that use.

Why do sharp points act in that manner? The explanation lies in the perpendicularity of the field lines to the conductor's surface [statement (16.B.3)]. Figure 16.19 indicates that the result is a high density of lines at the tip. The figure should not be considered a proof, of course.

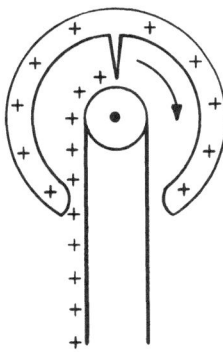

Figure 16.20. The hollow metal collector of a Van de Graaff generator. Charge has been sprayed (further down) on the insulating conveyor belt, and leaks off into the sharp point: it then migrates to the outer surface, and no longer produces any field in the cavity.

The Van de Graaff Generator

This machine combines the principles of Faraday's ice pail and of sharp points in order to transfer charge from a conveyor belt to a high-potential electrode; see Fig. 16.20. Such generators are used in nuclear physics laboratories to impart a large kinetic energy to charged particles.

C. Capacitors

The capacitor, one of the simplest electric devices, consists of *two conductors separated by a dielectric*. It can be used to store impressive amounts of charge with only a modest potential difference. It was invented, or rather discovered, in 1746, when Pieter van Musschenbroek, in Leyden (Holland), jolted himself with a capacitor he had inadvertently assembled. He had been charging the water in a glass jar held in his hand; the "conductors" were his hand and the water, and the dielectric was the glass. The myriad capacitors which, today, are indispensable components of even the most sophisticated electronic circuits are nothing but modified versions of that "Leyden jar." How a capacitor actually performs in a circuit will be studied in detail later in the book.

i. The Parallel-Plate Capacitor

Let us bring a pair of flat, parallel metal plates in close proximity to each other, but without mutual contact; see Fig. 16.21; we have made a **parallel-plate capacitor**. Suppose the plates have equal and opposite charges, $\pm Q$. We know from the preceding section that these charges distribute themselves

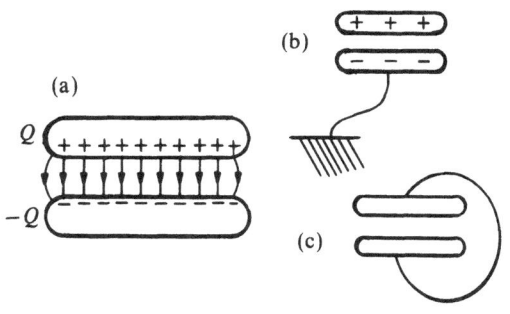

Figure 16.21. (a) A parallel-plate capacitor. (b) The opposite charges keep facing one another even if one of the plates is connected elsewhere (say to ground). (c) Discharge has occurred after both plates were connected together *even from the back*.

along the surfaces of the plates. More specifically, *the charges stay on the inner surfaces*, facing each other; also, *they spread themselves uniformly* over these inner surfaces.

Since the gap between the plates is narrow, every point in that region is very close to both conducting surfaces, and therefore, in the gap, **E** is everywhere perpendicular to the plates [recall statement (16.B.3)]. Thus, all the field lines are parallel, and hence their density remains constant ($E = $ const) as one moves along the field lines from one plate to the other; see Fig. 16.21. (An exception occurs at the edge, where the flat geometry comes to an end. There the field lines bend out, an effect called **fringing**.)

Is E also constant as one moves *across* the field lines, i.e., along the gap? The answer is yes, as can be seen from the fact that both plates are equipotentials with potentials Φ_+ (higher) and Φ_- (lower), respectively. Along any short field line AB connecting the two plates, as in Fig. 16.22, the field strength can be calculated from

$$E = \frac{\Phi_+ - \Phi_-}{s} \qquad (16.C.1)$$

where $s = |\overrightarrow{AB}|$ is the distance between the plates; recall Eq. (16.A.17). This calculation clearly yields the same number everywhere along the plates.

In summary:

> Throughout the capacitor's gap, the field **E** is uniform and perpendicular to the plates.

$$(16.C.2)$$

Let the **potential difference** between the plates be denoted by

$$V = \Phi_+ - \Phi_- \qquad (16.C.3)$$

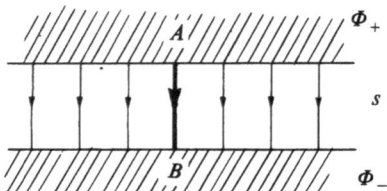

Figure 16.22. The field strength is constant from A to B, as well as along the direction of the plates' surface.

(This is correctly called the **voltage** across the capacitor, if expressed in volts.) Then Eq. (16.C.1) gives

$$E = \frac{V}{s} \qquad (16.C.4)$$

The direction of **E** is of course "downhill," i.e., from high to low potential.

Why do the charges stay on the inner surfaces? It is not enough to say that they attract one another, as the discharging behavior of Fig. 16.21c demonstrates. The answer is that the charges remain on the inner sides "in order that" no electric field occur within the metal of the plates. The superposition principle makes this clear. Assume, contrary to what we want to prove, that some of the charge spreads to the outer surfaces. Then the situation would be as shown in Fig. 16.23d. The situation is viewed as a superposition of single layers of charge, as built up in parts (a)–(d) of the figure. That description implies $\mathbf{E} \neq \mathbf{0}$ inside the metal, an unacceptable conclusion. Part (c) of

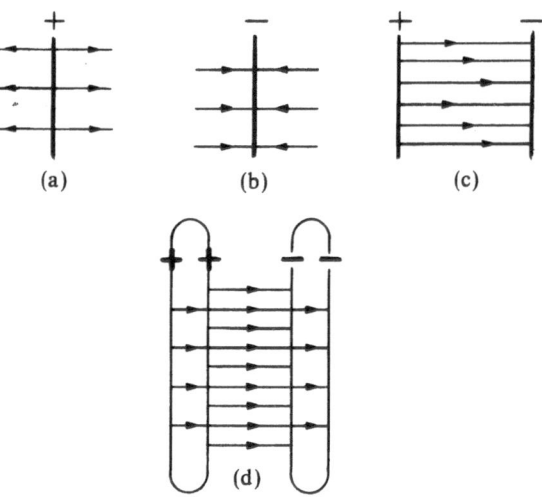

Figure 16.23. An imaginary superposition is used to build up the surface charges and their fields. (Fringing and end effects not shown.) (a) A single plane of charge; the field must be symmetric; see also Problem 15.22 in the preceding chapter; (b) The charge on (a) is reversed. (c) Superposing (a) and (b) to show a double-strength field between the plates, zero field outside; this demonstrates the confinement of **E** between the plates. (d) Superposing two configurations of type (c), but with different surface spacings. We find a nonzero **E** inside the metal, which is impossible. Therefore no charge can reside on the outer surfaces.

the figure also demonstrates that *the capacitor's field is*, in fact, *confined between the plates.*

ii. Capacitance

How much charge, $\pm Q$, is stored on either plate of the capacitor? The answer is proportional to the potential difference V, as we shall now see.

Consider the positive plate, and draw an imaginary Gaussian box around Q, as in Fig. 16.24; Q is then proportional to the emerging flux $\mathcal{N}$,

$$Q = \frac{\mathcal{N}}{4\pi K} = \frac{E\mathcal{A}}{4\pi K} \qquad (16.C.5)$$

where $\mathcal{A}$ is the plate area. Setting $E = V/s$ [Eq. (16.C.4)], we obtain

$$Q = \frac{V\mathcal{A}}{4\pi Ks} \qquad (16.C.6)$$

or

$$\boxed{Q = CV} \qquad (16.C.7)$$

where

$$C = \frac{\mathcal{A}}{4\pi Ks} \qquad (16.C.8)$$

is called the **capacitance** of the capacitor; it depends only on the area $\mathcal{A}$ and the spacing s, and is therefore basically a *geometric property*. The formula for C is simplified by the introduction of a new constant, "epsilon zero,"

$$\varepsilon_0 = \frac{1}{4\pi K} \approx 8.85 \times 10^{-12} \text{ SI units} \quad (16.C.9)$$

Equation (16.C.8) becomes

$$\boxed{C = \varepsilon_0 \frac{\mathcal{A}}{s}} \qquad (16.C.10)$$

Formula (16.C.7), or $C = Q/V$, leads to the slogan "capacitance equals charge stored per unit voltage." The SI unit of capacitance is therefore the coulomb/volt, or, equivalently,

$$1 \text{ farad} = 1 \text{ F} = 1 \frac{\text{coulomb}}{\text{volt}} \quad (16.C.11)$$

in honor of Michael Faraday.

> **Example 16.5. Large Size of the Farad.** In order to make a 1-farad capacitor from two plates separated by $s = 1$ mm of air, how large an area $\mathcal{A}$ should one plan for?
>
> ---
>
> Setting $\varepsilon_0 \approx 10^{-11}$ SI units, we have from Eq. (16.C.10)
>
> $$\mathcal{A} = \frac{(1)(10^{-3})}{10^{-11}} \text{ meters}^2 = \underline{\underline{10^8 \text{ meters}^2}}$$
>
> which would cover a small town.

This result is not surprising; the farad is huge because the coulomb is huge. Commercially obtainable capacitances are usually comparable to a microfarad; the capacitors are more often rolled up than flat, but this feature does not invalidate our discussion. Compared to the tiny spacing s, the surfaces involved still have very large radii of curvature; hence they can be considered flat in spite of appearances.

Exceptionally large capacitances [up to thousands of μF (microfarads)], resulting from a very small spacing s, are found in **electrolytic capacitors**. Here one metal plate dips into a conducting salt solution, which serves as the other "plate." The dielectric separation is an oxide layer, a few molecules thick, coating the metal. This layer turns out to be self-healing against charge leaks, provided the metal is kept at a higher potential than the solution.

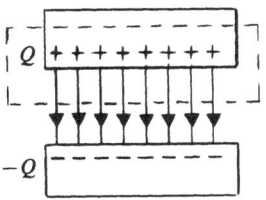

Figure 16.24. Using Gauss' law to calculate the charge Q from the field **E**. Fringing is neglected; the gap shown here is exaggerated for clarity.

Dielectric Constant

Practical capacitors usually incorporate a solid dielectric, such as paper or plastic, to keep the two conductors apart, and to reduce the possibility of sparking. Since our analysis assumed an air gap (actually, a vacuum, which is equivalent for this purpose), formula (16.C.10) needs to be amended. This is done through an empirical factor, κ (lower-case Greek kappa), the **dielectric constant** of the insulator. We still have $Q = CV$, but C is given by

$$C = \kappa \varepsilon_0 \frac{\mathscr{A}}{s} \qquad (16.\text{C}.12)$$

($\kappa = 1$ for vacuum); the effect is always an *increase* in C. Most commercial dielectrics have $2 \lesssim \kappa \lesssim 3$; among the highest commonly used values is $\kappa \approx 7.0$, for mica. In (16.C.12), C is no longer a purely geometric quantity. Time and space do not permit us to go into the theory of the dielectric constant; it is related to the charge distribution in the dielectric's molecules.

Only the Potential Difference is Relevant

We summarize here two important aspects of capacitors:

> The charges on the two plates are always equal and opposite ($\pm Q$).

$$(16.\text{C}.13)$$

> They depend only on the potential difference $V = \Phi_+ - \Phi_-$,

$$(16.\text{C}.14)$$

rather than on the individual Φ_+, Φ_-. Why are capacitors so simple? Their basic feature is that:

> The spacing s is very small compared to the overall dimensions of the device.

$$(16.\text{C}.15)$$

A note at the end of the chapter indicates why the small spacing is important for obtaining (16.C.13), (16.C.14).

iii. Absolute Measurement of a Potential

The potential has been defined in terms of the work done by a field on a test charge. As a laboratory procedure, however, this recipe is at once too clumsy and too abstract.

There does exist a practical instrument, the **electrometer**, which allows an electrostatic measurement of potential differences. (More convenient methods, based on electric currents, will be described later; they depend on a comparison with an already known potential difference. The electrostatic method does not, and is therefore called **absolute**.)

The simplest type of electrometer, illustrated in Fig. 16.25, is nothing but a capacitor in which the electrostatic force of attraction between the plates is being measured. If w is the weight needed to replace the effect of the field, the potential difference across the plates is given by

$$V = s \left(\frac{2w}{\varepsilon_0 \mathscr{A}} \right)^{1/2} \qquad (16.\text{C}.16)$$

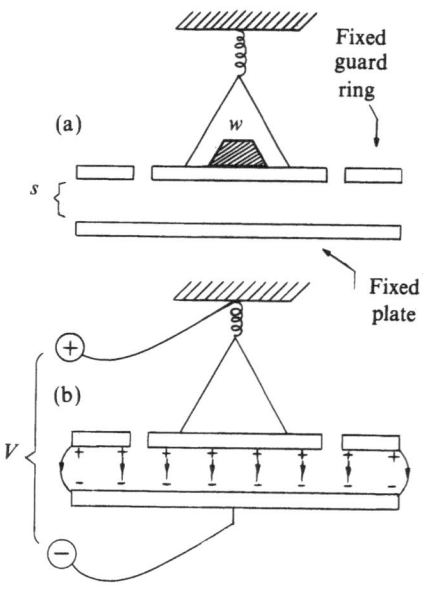

Figure 16.25. An absolute electrometer. (a) With zero potential difference; the central plate is kept lined up with its continuation (the "guard ring") by means of a weight w. (b) With a potential difference V; the weight is no longer required. The guard ring prevents fringing at the edge of the movable plate. (Fringing would invalidate the force calculation.) This design is not very sensitive and needs at least 1000 volts or so to give a reliable reading. More sophisticated versions can be very sensitive and precise, however. [Lord Kelvin invented the guard ring (about 1865) and many of the later improvements.]

where s is the spacing and $\mathscr{A}$ is the area of the movable plate. The derivation of that result is left to Problem 16.36.

iv. Capacitors in Combination

Several capacitors, when connected by their terminals, as in Fig. 16.26, may be considered as a single new capacitor, whose capacitance we now calculate. Flatness of the plates need not be assumed, nor does the dielectric have to be air or vacuum.

Parallel Combination

The overall capacitance in part (a) of the figure is defined in terms of the total charge supplied to one of the terminals:

$$C = \frac{Q_1 + Q_2 + \cdots}{V} \qquad (16.C.17)$$

where V is the potential difference between those

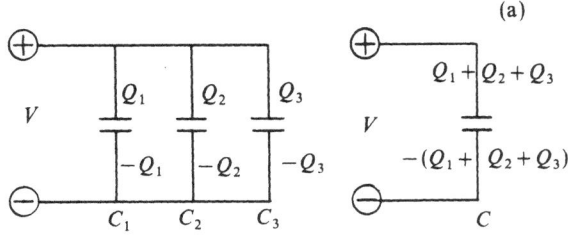

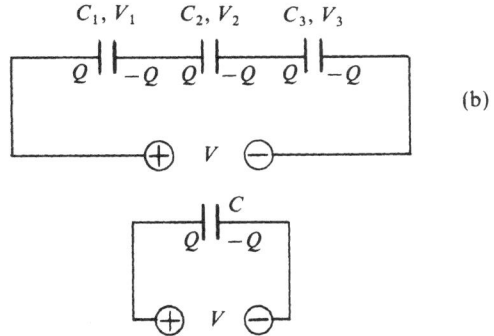

Figure 16.26. (a) A parallel combination C_1, C_2, C_3, and its equivalent, C. (By definition, in a parallel combination, the $\oplus$ terminals of all capacitors are directly connected with one another; the same is true for all the $\ominus$ terminals.) (b) A series combination; a charge $-Q$ is always connected to $+Q$ on the next capacitor, because no net charge is created in the connections. (By definition, in a series combination, single capacitors alternate with single wire connections.)

terminals. We note that each capacitor is connected *across the same potential difference V*, so that $Q_1 = C_1 V$, $Q_2 = C_2 V$, etc. Insertion in (16.C.17) gives

$$\boxed{C = C_1 + C_2 + \cdots} \qquad \text{(parallel)}$$

$$(16.C.18)$$

In a parallel combination one must add the capacitances.

Series Combination

In part (b) of Fig. 16.26, and as explained in its caption, we see that the charges are the same, $\pm Q$, on all pairs of plates. The overall potential difference is

$$V = V_1 + V_2 + \cdots \qquad (16.C.19)$$

because V_1, V_2,... represent successive steps through which the potential Φ rises. The overall capacitance, C, is defined as before in terms of V and of the charge Q supplied to a *single* terminal, in such a way that $V = Q/C$; likewise, for each circuit element we have $V_1 = Q/C_1$, $V_2 = Q/C_2$, etc. Therefore Eq. (16.C.19) reads

$$\boxed{\frac{1}{C} = \frac{1}{C_1} + \frac{1}{C_2} + \cdots} \qquad \text{(series)}$$

$$(16.C.20)$$

In a series combination one must add the inverse capacitances.

Example 16.6. Series – Parallel Combination. With a potential difference of 100 volts across the combination of Fig. 16.27a how much charge is stored in the 3 μF capacitor?

First we find the overall capacitance. The encircled parallel combination has a capacitance $1 \mu F + 2 \mu F = 3 \mu F$; the new and equivalent arrangement, part (b) of the figure, has a capacitance C that obeys

$$\frac{1}{C} = \frac{1}{3\ \mu F} + \frac{1}{3\ \mu F} = \frac{2}{3}\ (\mu F)^{-1}$$

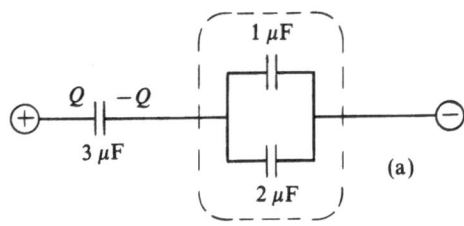

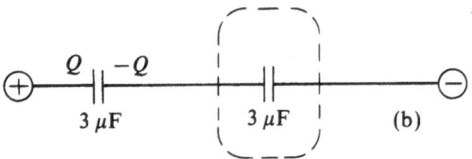

Figure 16.27. A series-parallel combination; (a) is equivalent to (b).

which gives

$$C = \tfrac{3}{2}\mu\text{F}$$

The charge Q on the leftmost plate is

$$Q = CV = (\tfrac{3}{2}\mu\text{F})(100 \text{ volts}) = \underline{\underline{150 \ \mu\text{C}}}$$

Can all cases be analyzed into series and parallel combinations? The answer is that some cases cannot, as in the illustration of Fig. 16.28. The solution is nevertheless found by straightforward (if lengthy) algebra; one must deal with all charges and potential differences in the system.

D. Energy of the Electric Field

Is the electric field only a mathematical abstraction? Here we shall find that *an electric field in vacuum stores energy*, and therefore that the field has indeed a degree of physical reality. This conclusion will be reached through our knowledge of the capacitor.

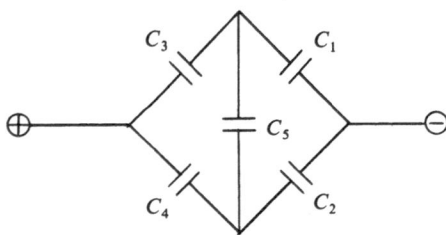

Figure 16.28. This circuit has no series or parallel subcircuits.

i. Work Needed to Charge a Capacitor

Figure 16.29 illustrates a hypothetical procedure for building up charges $\pm Q$, with potential difference V, on a parallel-plate capacitor with capacitance C. Small amounts of charge are carried mechanically from the negative to the positive plate, into which they are released. How much work must be invested?

At an intermediate stage (denoted by primes), where the charges are $\pm Q'$ and the potential difference V', the transfer of a small charge dQ' costs an amount of work $d\mathscr{W}'$ given by the increase in the potential energy of dQ':

$$d\mathscr{W}' = V' \, dQ' = \frac{Q'}{C} \, dQ' \qquad (16.\text{D}.1)$$

Integrating from $Q' = 0$ to $Q' = Q$ yields

$$\mathscr{W} = \int_0^Q \frac{Q' \, dQ'}{C} = \frac{1}{2}\frac{Q^2}{C} \qquad (16.\text{D}.2)$$

Thus we obtain

$$\boxed{\text{Stored energy} = \frac{1}{2}\frac{Q^2}{C} = \frac{1}{2}CV^2 = \frac{1}{2}QV.}$$

$$(16.\text{D}.3)$$

(These alternative expressions result from $Q = CV$.) The factor $1/2$ can be understood intuitively as resulting from an average between charging a zero-voltage capacitor (the initial situation), and a capacitor with voltage V (the final situation): $\tfrac{1}{2}[(Q)(0) + QV] = \tfrac{1}{2}QV$. This remark is not meant as a derivation, of course.

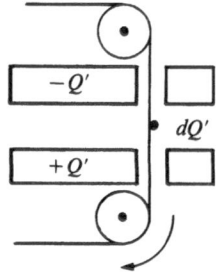

Figure 16.29. A conveyor belt carries a charge dQ' from the top plate to the bottom one.

Example 16.7. Smallness of the Stored Energy. A potential difference of 12 volts is set up between the plates of a typical capacitor (say 1 μF) when the latter is connected across a typical car battery. (a) Find the energy stored in the capacitor. (b) How does it compare with the energy stored in the battery, typically 10^6 joules?

a. Formula (16.D.3) gives, for the energy,

$$\tfrac{1}{2}CV^2 = \tfrac{1}{2}(10^{-6})(12)^2 \text{ joule} \approx \underline{10^{-4} \text{ joule}}$$

b. Relation to battery energy:

$$\frac{\text{Capacitor energy}}{\text{Battery energy}} = \frac{10^{-4}}{10^6} = \underline{\underline{10^{-10}}}$$

For storing energy in large amounts, a typical capacitor is clearly the wrong device. But it can store or release its energy much faster than a battery, and therein lies a possible practical advantage.

ii. Field Energy

Let us see how the stored energy, Eq. (16.D.3), is related to the field between the plates of the capacitor. From (16.D.3) we take the formula

$$\text{Stored energy} = \tfrac{1}{2}CV^2 \qquad (16.\text{D}.4)$$

and set

$$V = Es \qquad (16.\text{D}.5)$$

[recall Eq. (16.C.4)]; also, assuming a vacuum between the plates, we have

$$C = \varepsilon_0 \frac{\mathscr{A}}{s} \qquad (16.\text{D}.6)$$

see (16.C.10). The result for (16.D.4) is

$$\text{Stored energy} = \frac{\varepsilon_0}{2} \mathscr{A} s E^2 \qquad (16.\text{D}.7)$$

The combination $\mathscr{A}s$ is very suggestive: it is nothing but the volume of the capacitor's gap. Dividing both sides of (16.D.7) by $\mathscr{A}s$ yields the energy stored per unit volume of the gap, or **energy density**:

$$\boxed{\text{Energy density} = \frac{\varepsilon_0}{2} E^2} \qquad (16.\text{D}.8)$$

This must be interpreted as *the energy density of the electric field*. We conclude that the energy stored in the capacitor is in fact located in the field itself.

In the context of particle mechanics, as developed in the first half of this book, this concept is completely new and revolutionary. In fact, it was not until the work of Einstein (1905) that the idea of an energy-carrying field occupying a vacuum became fully accepted; even Maxwell, the brilliant architect of electromagnetic theory, had wrongly considered it necessary to postulate an invisible space-filling substance, the "ether."

Both in the subjects of magnetism (Chap. 19) and of electromagnetic waves (Chap. 22), further confirmation for the field-energy concept will emerge.

E. Particles Controlled by Electric Fields

Charged particles, moving in free space under the effect of an electric field, are at the heart of many ingenious devices. Here are some instances that illustrate electrostatic principles, and that are useful in their own right as well.

i. The Field-Ion Microscope

Figure 16.30 is a simplified diagram of a **field-ion microscope**, invented by Erwin W. Muller in 1955. On the screen of this instrument, the image of single atoms can be displayed.

A fine metal needle, whose tip is only a few hundred atoms wide, is given a positive potential (1000 volts or more) with respect to another conductor of arbitrary shape. The surrounding space is evacuated except for a small residual pressure (a few times 10^{-6} atm) of some gas such as hydrogen. The field lines near the needle illustrate the effect of points; see part (a) of the figure. At the tip, the field is so intense as to tear electrons out of the nearby gas atoms. The resulting positive ions are propelled radially outwards along nearly straight trajectories; see part (b) of the figure. (This is because almost their whole momentum is acquired in the intense radial field near the needle.) Eventually the ions strike a special translucent coating (such as zinc sulfide) on the glass enclosure; the coating emits light under the ions' impacts.

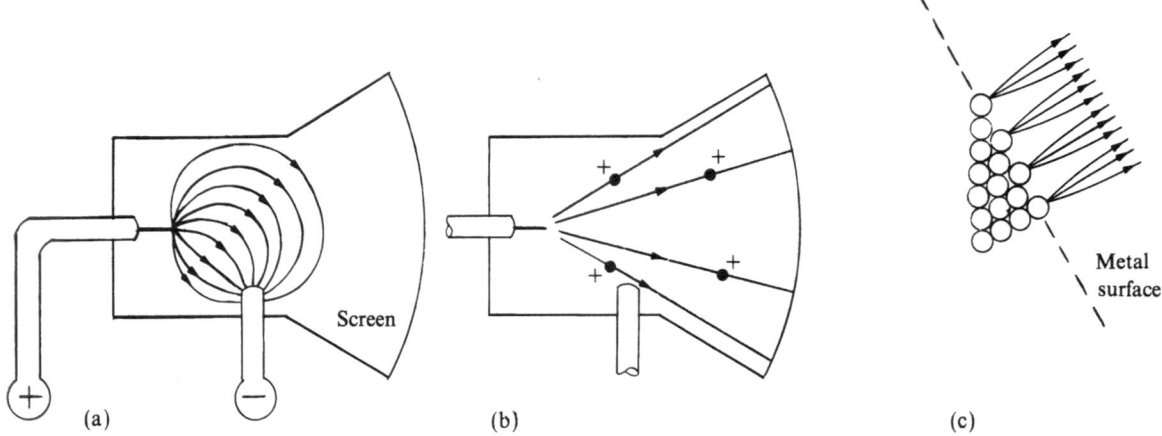

Figure 16.30. The field-ion microscope. (a) Field lines. (b) Ion trajectories. (c) Stronger field near protruding metal atoms.

To understand why at least some individual atoms on the needle are mapped into bright spots on the screen, we refer once more to the effect of points, part (c) of the figure; the more protruding atoms give a more intense field, and hence create more ions. One image can be seen in Chap. 11, Fig. 11.3a.

ii. The Electron Gun

Beams of electrons, traveling in vacuum or in a very dilute gas, are versatile tools for the scientist. Not only do they reveal some properties of the electrons themselves, but they are the main component of electron microscopes, electronic image tubes, x-ray tubes, etc.

Figure 16.31 shows a simple **electron gun**, designed to produce such a beam. The electrons start at the cathode, with approximately zero velocity. After traveling to the anode they come out with a kinetic energy $\mathcal{K} = -\Delta\mathcal{U}$, which is their potential energy drop from cathode to anode. Thus, each electron in the beam emerges with energy

$$\boxed{\mathcal{K} = eV} \qquad (16.E.1)$$

where V is the potential rise between the electrodes, and $-e \approx -1.6 \times 10^{-19}$ coulomb is the electron's charge.

Here, let us introduce the **electron-volt** as a con-

venient and often-used unit of energy (not SI). We have

$$1 \text{ electron-volt} = 1 \text{ eV} = (e)(1 \text{ volt})$$
$$\approx 1.6 \times 10^{-19} \text{ joule} \quad (16.E.2)$$

That is to say,

> 1 electron-volt equals the kinetic energy gained by an electron that is accelerated through a potential difference of 1 volt.

$$(16.E.3)$$

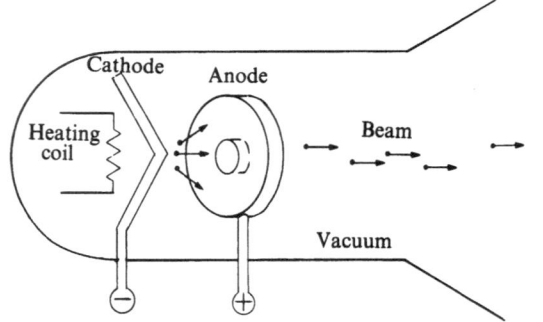

Figure 16.31. An electron gun (simplified design). The two conductors are called **electrodes**; the negative one is the **cathode** and the positive one is the **anode**. The cathode, which is heated by an electric coil, is coated with a material, such as barium oxide, which readily allows electrons to "evaporate" into the evacuated tube. Most electrons strike the anode, but some pass through a fine hole and emerge as a beam. The electrodes form a kind of capacitor; the field does not extend beyond the entrance of the hole and therefore the electrons do not change their kinetic energy once they have reached the anode potential.

We shall tend to avoid the standard abbreviation "eV" because the beginner frequently confuses it with the product eV which occurs in (16.E.1); the latter can of course equal an arbitrary number of electron volts. In terms of this unit, (16.E.1) reads

$$(\mathscr{K} \text{ expressed in electron-volts}) = \frac{\mathscr{K}}{1 \text{ electron–volt}}$$

$$= \frac{eV}{(e)(1 \text{ volt})} = \frac{V}{1 \text{ volt}}$$

$$(16.E.4)$$

The energy, in electron volts, is numerically equal to the voltage.

The units keV $(=10^3 \text{ electron–volts})$, MeV $(=10^6 \text{ electron–volts})$, GeV $(=10^9 \text{ electron–volts})$ are also of frequent use.

Example 16.8. Speed of an Electron. Calculate the speed of a 0.030-keV electron.

We have, for the kinetic energy,

$$\mathscr{K} = \tfrac{1}{2}mv^2$$

where $m \approx 9.1 \times 10^{-31}$ kg is the electron's mass. Solving for v, we obtain

$$v = \left(\frac{2\mathscr{K}}{m}\right)^{1/2}$$

$$= \left[\frac{(2)(0.030 \times 10^3)(1 \text{ electron–volt})}{m}\right]^{1/2}$$

$$= \left[\frac{(2)(0.030 \times 10^3)(1.6 \times 10^{-19})}{9.1 \times 10^{-31}}\right]^{1/2} \frac{\text{meters}}{\text{sec}}$$

$$= 3.2 \times 10^6 \frac{\text{meters}}{\text{sec}}.$$

Notice the conversion of electron-volts to joules.

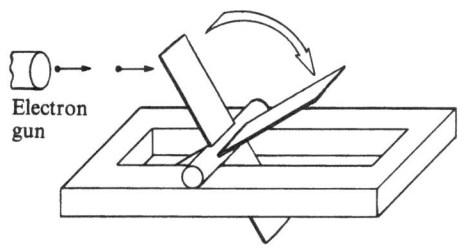

Electron gun

Figure 16.32. The paddle wheel turns under the electrons' impact.

Demonstrating the Electron's Momentum

A classic experiment, illustrated in Fig. 16.32, shows qualitatively the force of an electron bombardment on a small paddle wheel, and therefore demonstrates that electrons possess momentum.

iii. The Cathode-Ray* Tube

In this instrument, an electron beam can be aimed instantaneously with great precision. As it passes between the plates of a capacitor, it suffers a deflection whose angle is controlled by the capacitor's voltage; see Fig. 16.33. The beam's deflection is, in fact, often used to measure that voltage. The electrons' trajectory is practically unaffected by gravity, see Problem 16.48.

Cathode-ray tubes are found in virtually all laboratories; they have been for generations the workhorses among scientific measuring instruments. In one form, they are known as television image tubes.

Example 16.9. Calculating a Deflection. Assume, in Fig. 16.33, potential differences of $V = 400$ volts for the gun and $V' = 2$ volts for the capacitor; the latter has a length $l = 5$ cm in the beam direction and a spacing $s = 0.5$ cm. By what angle θ is the beam deflected?

In terms of the final velocity components v_x, v_y, we have

$$\tan \theta = \frac{v_y}{v_x} \qquad (16.E.5)$$

As in Example 16.8, we have, for the electron gun,

$$\tfrac{1}{2}mv_x^2 = eV \qquad (16.E.6)$$

In the capacitor, where the field is V'/s, the electron experiences a constant acceleration

$$a_y = \frac{\text{force}}{m} = \frac{(e)(V'/s)}{m} \qquad (16.E.7)$$

* The electron beam was thought to be a "ray" before its true nature was understood. That name occasionally persists.

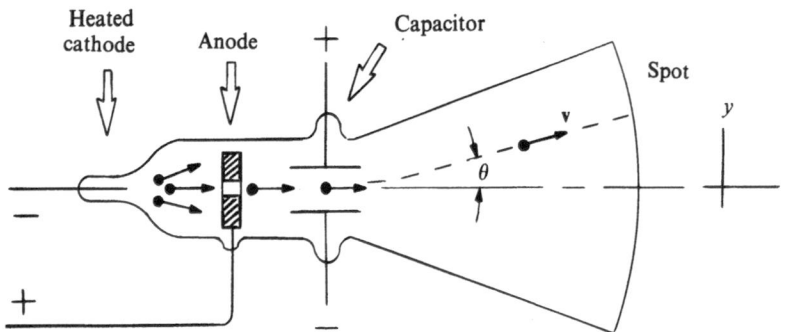

Heated cathode

Anode

Capacitor

Spot

Figure 16.33. A cathode-ray tube, comprising an electron gun and a deflecting capacitor. (A second capacitor is usually included, elsewhere along the beam, in order to deflect the beam horizontally as well.) The electrons are launched parallel to the deflecting plates. A bright spot on the translucent screen shows the beam's position at any time.

The electron remains between the plates for a time $t = l/v_x$; therefore its final y velocity is

$$v_y = a_y t = \frac{eV'l}{msv_x} \tag{16.E.8}$$

which can be written

$$v_x v_y = \frac{eV'l}{ms} \tag{16.E.9}$$

The ratio of (16.E.9) to (16.E.6) yields the required result:

$$\frac{v_y}{v_x} = \frac{1}{2}\left(\frac{V'}{V}\right)\left(\frac{l}{s}\right) \tag{16.E.10}$$

In this case, (16.E.5) becomes

$$\tan\theta = \frac{1}{2}\left(\frac{2}{400}\right)\left(\frac{5}{0.5}\right) = 0.025$$

or

$$\theta = \underline{\underline{1.5°}}$$

(Refer back to Problem 2.77 for an additional comment on the electrons' trajectory.)

iv. Millikan's Experiment

Robert A. Millikan, in 1911, systematically observed tiny charged droplets of oil that he had injected in the gap of a capacitor. These particles, much larger and heavier than ions, were still so small as to bridge the distinction between microscopic and macroscopic; Millikan used them to obtain the first direct and convincing measurement of e, the quantum of charge.

The simplest version of this experiment is shown in Fig. 16.34. The droplet, whose weight w has been inferred from separate observations, is held in equilibrium against gravity by the capacitor's known field E. If q is the drop's charge, then we have

$$w = qE \tag{16.E.11}$$

from which q is calculated,

$$q = \frac{w}{E} \tag{16.E.12}$$

Millikan found that, on the various drops, the observed value of q always took the form

$$\boxed{q = ne} \tag{16.E.13}$$

where n is a positive or negative integer, and where

$$\boxed{e \approx 1.6 \times 10^{-19} \text{ coulomb}} \tag{16.E.14}$$

Sometimes a droplet acquired an elementary charge from the air while being observed.

(Millikan's method has now been superseded by far more accurate, but indirect, ways of evaluating e.)

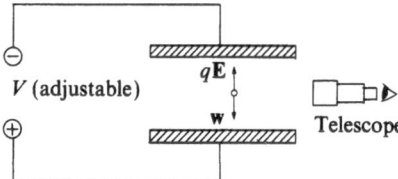

V (adjustable)

$q\mathbf{E}$

w

Telescope

Figure 16.34. Millikan's experiment. The capacitor voltage V can be continually adjusted by hand in such a way that the field **E** balances the oil droplet against gravity. In practice, V is of the order of several kV. The droplet's altitude is measured against cross-hairs in the small telescope at right.

Note

What Makes a Capacitor so Simple?

In Sec. C ii we saw that only the potential difference $V = \Phi_+ - \Phi_-$ determines the capacitor's charge. Why not the individual Φ_+, Φ_-? The answer lies in the initial assumption of Sec. C: the charges on the plates are $+Q$ and $-Q$. Therefore the real question is, why can we always assume equal magnitudes for these two charges?

For an explanation, we look to the special model of Fig. 16.35, in which the "plates" are two concentric metal shells. In general, the charges on both shells are different. We denote them by $-Q$ (inner shell) and $Q+q$ (outer shell). The corresponding potentials Φ_-, Φ_+ are assumed given.

We now prove that, when the spacing s between the shells is very small compared to the shells' radii, then Q is very large compared to q, and consequently both shells have charges of nearly equal magnitude,

$$Q \approx Q + q \qquad (16.N.1)$$

The charges are distributed among surfaces 1, 2, 3, 4, as shown in Fig. 16.35. On surface 1, the charge must be zero (inner surface of an empty hollow conductor); surfaces 2 and 3 have charges $-Q$, $+Q$ (all field lines out of surface 2 must terminate on surface 3); finally surface 4 has charge q.

We now calculate Q and q in terms of Φ_+, Φ_-. Apply-

ing the derivation of Sec. C ii to the thin gap between the two shells, we have Eq. (16.C.6),

$$Q = \frac{V \mathscr{A}}{4\pi K s} \qquad (16.N.2)$$

Here $\mathscr{A}$ is the area of surface 2 or 3. (These areas are practically the same; also, the gap is treated as flat because it is very thin.)

To find q, we set $R_4 =$ radius of surface 4. From that radius on, the potential is that of an equivalent point charge at the center. That charge (the total charge) is just q, and we find, for the potential at R_4,

$$\Phi_+ = \frac{Kq}{R_4}$$

from which

$$q = \frac{R_4 \Phi_+}{K} \qquad (16.N.3)$$

How does q compare with Q? From (16.N.2), (16.N.3) we have

$$\frac{q}{Q} = 4\pi \frac{R_4}{\mathscr{A}} \frac{\Phi_+}{V} s \qquad (16.N.4)$$

which indeed approaches zero as $s \to 0$ (R_4, $\mathscr{A}$, Φ_+, and V are fixed). This concludes the argument. *The small gap is responsible for the near-equality of the charges' absolute values*, Eq. (16.N.1). That conclusion can, as it turns out, be extended to all capacitor shapes.

Condensed Checklist

Potential:

$$\boxed{\Phi = \frac{\mathscr{U}}{q}} \qquad (\mathscr{U} = \text{potential energy of test charge } q)$$

$$(16.A.5)$$

$$\boxed{\Phi_A - \Phi_B = \int_A^B \mathbf{E} \, d\mathbf{r}} \qquad (\text{any shape of path})$$

$$(16.A.10)$$

$$\frac{\Phi_A - \Phi_B}{|\overrightarrow{AB}|} = E_{AB} \qquad (E_{AB} = \text{component of } \mathbf{E} \text{ along } \overrightarrow{AB};$$
$$\mathbf{E} \text{ assumed uniform})$$

$$(16.A.17)$$

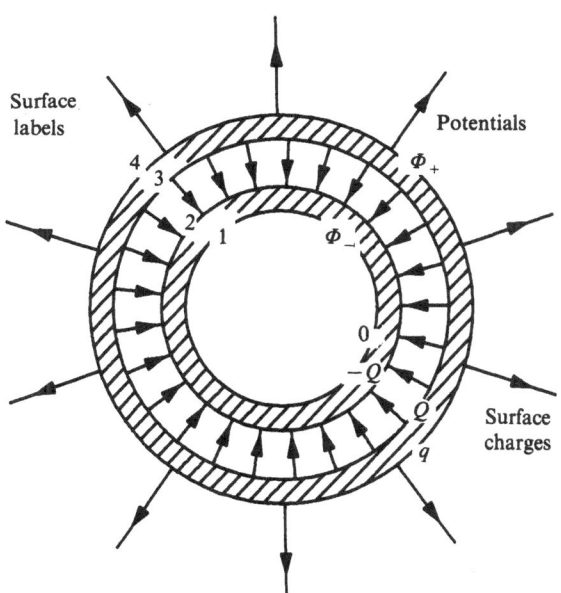

Figure 16.35. A spherical capacitor. The charges Q, q are taken positive for definiteness.

$$\boxed{\Phi = \frac{KQ}{r}}$$ (outside spherical or point charge Q)

(16.A.4), (16.A.16)

$\Phi = \text{const}$ (inside spherical charged shell)

(16.A.16)

Equipotentials and conductors: qualitative facts   Sec. B

Capacitors:

$$\boxed{E = \frac{V}{s}}$$ (16.C.4)

$$\boxed{Q = CV}$$ (16.C.7)

$$C = \kappa\varepsilon_0 \frac{\mathcal{A}}{s}$$ (16.C.12)

$\varepsilon_0 \approx 8.85 \times 10^{-12}$ SI units   (16.C.9)

$$\kappa \begin{cases} = 1 & \text{(vacuum)} & \text{(16.C.10)} \\ > 1 & \text{(other dielectrics)} & \text{(16.C.12)} \end{cases}$$

$$\boxed{\begin{array}{l} C = C_1 + C_2 + \cdots \quad \text{(parallel) (16.C.18)} \\ \dfrac{1}{C} = \dfrac{1}{C_1} + \dfrac{1}{C_2} + \cdots \quad \text{(series) \ (16.C.20)} \end{array}}$$

Stored energy $= \dfrac{1}{2}QV = \dfrac{1}{2}CV^2 = \dfrac{1}{2}\dfrac{Q^2}{C}$ (16.D.3)

Field energy density:

$$\boxed{\frac{\text{Energy}}{\text{Volume}} = \frac{\varepsilon_0}{2}E^2}$$ (16.D.8)

Electrometer:

$$V = s\left(\frac{2w}{\varepsilon_0 \mathcal{A}}\right)^{1/2}$$ (16.C.16)

Units:

$1 \text{ volt} = 1 \dfrac{\text{joule}}{\text{coulomb}}$ (potential)

(16.A.6)

$1 \dfrac{\text{newton}}{\text{coulomb}} = 1 \dfrac{\text{volt}}{\text{meter}}$ (field)

(16.A.13)

$1 \text{ farad} = 1 \dfrac{\text{coulomb}}{\text{volt}}$ (capacitance)

(16.C.11)

$1 \text{ electron volt} = \dfrac{e}{1 \text{ coulomb}}$ (1 joule)

$= 1.6 \times 10^{-19}$ joule   (energy)

(16.E.2)

Cathode-ray tube deflection:

$$\tan\theta = \frac{1}{2}\left(\frac{V'}{V}\right)\left(\frac{l}{s}\right)$$ (16.E.10)

Millikan's experiment:

$$w = qE$$ (16.E.11)

$$q = ne$$ (16.E.13)

$$\boxed{e \approx 1.6 \times 10^{-19} \text{ coulomb}}$$ (16.E.14)

True or False

1. The sign of a potential depends on the sign of the test charge with which it is measured.

2. A single electron, accelerated over 1 volt, changes its energy by 1 joule.

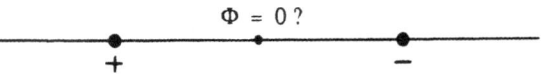
$\Phi = 0$?

3. The halfway point between two point charges of opposite sign and equal magnitude is at zero potential relative to points at infinity; see the figure.

4. The halfway point between two identical point charges is at zero potential relative to points at infinity.

5. The potential always decreases in the direction of **E**

6. Negative test charges are attracted towards high potentials.

7. If the field strength at one single point A is known the potential at A (relative to points at infinity) can calculated.

8. Straight parallel field lines imply flat parallel equipotentials.

9. Straight field lines imply flat equipotentials.

10. The potential inside a uniform *solid* sphere of positive charge increases towards the center.

11. An arbitrarily high potential can be found if one comes close enough to a given positive point charge.

12. When a capacitor's plates are pushed closer together, their potential difference can remain constant, provided their charge is increased in magnitude.

13. If a capacitor's gap width is increased while the charges are kept constant, the potential difference between the plates decreases.

14. A charged particle, projected into a parallel-plate capacitor gap, follows a parabolic trajectory.

15. The speed of the electrons emerging from an electron gun is proportional to the applied voltage.

16. If some electrons in the cathode-ray tube of Example 16.9 had an unusual mass, their deflection would be different from the others'.

17. Millikan's experiment involves balancing an electron's weight against an upward electric force.

Problems

Sections A: Introducing the Potential;
B: Single Conductors in
Equilibrium

*16.1. How much work, in joules, is needed to bring two 5-μC point charges, initially far apart, within 1.00 mm of one another?

16.2. If a uranium nucleus has a charge of $+92e = (92)$ $(1.6 \times 10^{-19}$ coulomb), estimate at what distance from it the potential is as quoted in Fig. 16.11. (Ignore the electrons which surround the nucleus.)

*16.3. A certain stroke of lightning releases 1.5×10^8 joules of energy, mostly as heat. If this event is due to 20 coulombs of charge being transferred from ground to cloud, estimate the potential difference V between these. (Assume $V = \text{const}$ during the discharge; this is an oversimplification.)

*16.4. In a certain laboratory, the electrostatic potential is 500 volts on the west wall and -300 volts on the east wall; the distance between those walls is 8 meters. What are the direction and strength of the electric field in the laboratory?

16.5. On a fine day, the downward field, outdoors, is about 100 volts/meter. If a 1.80-meter person did not dis-

turb that field by his presence, what would be the potential of his head relative to his feet? (Actually, the local field is reduced almost to zero because people are reasonably good conductors.)

*16.6. In analogy with the electric potential $V = \mathscr{U}/q$, define a **gravitational potential**, $V_{\text{grav}} = \mathscr{U}_{\text{grav}}/m$, where m is a test mass. (a) Specify the SI units of V_{grav}. (b) Estimate numerically the gravitational potential of the ceiling relative to the floor in a 3-meter-high laboratory. (c) Estimate numerically the gravitational potential at the surface of the Earth with respect to points at infinity, and neglecting other bodies such as the Sun.[†]

16.7. A square, 12 cm on the side, has three of its corners occupied by equal 0.5-μC charges. Find the electrostatic potential (relative to points at infinity) (a) at the unoccupied corner, (b) at the center of the square.

*16.8. What potential difference is needed to accelerate a proton (charge $= +e$, mass $= 1.7 \times 10^{-27}$ kg) to a speed of 10^6 meters/sec?

16.9. Two point charges, 10 cm apart, have values of $+2 \, \mu$C and $-3 \, \mu$C, respectively. Find a point, located on the same straight line as the charges, where the potential is the same as infinitely far away.

16.10. (a) Sketch your idea of some representative equipotentials in the vicinity of three equal negative charges occupying the corners of an equilateral triangle. (b) Check for right-angle intersections with the lines of Problem 15.13.

16.11. As in Problem 16.10, but use two pairs of opposite charges, at the corners of a square; see the figure.

*16.12. Using Eq. (16.B.7), find how much charge is put on the metal sphere in Example 16.4.

16.13. Do Problem 16.12 by applying Gauss' law.

16.14. Prove Statement (16.B.10), about the inner wall of a hollow conductor.

[†] Neglecting the Sun means that "points at infinity" are defined to be many Earth radii away, but still close by compared to the Sun.

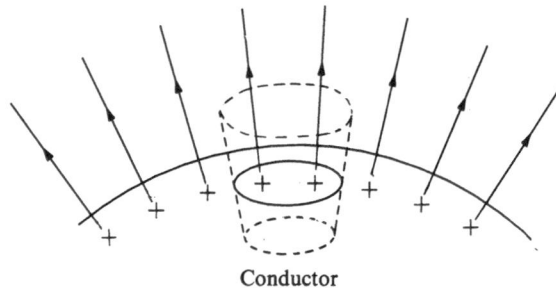

Conductor

*16.15. The figure shows the charged surface of a conductor, and a suggested choice of Gaussian surface (dashed) enclosing an infinitesimal area. If, in that region, the surface charge density is σ (=charge per unit area), find the field strength E just outside the conductor. See the preceding chapter for Gauss' law.

Sections C: Capacitors;
D: Energy of the Electric Field

*16.16. (a) Verify the value of ε_0, Eq. (16.C.9). (b) What are the SI units of ε_0, in terms of kilograms, meters, seconds, and coulombs?

*16.17. In terms of the charge Q, the area $\mathscr{A}$, and the spacing s, find the field E in a parallel-plate air capacitor.

16.18. What maximum potential is permitted between capacitor plates that are 1 cm apart in air if sparking occurs at a field of 3×10^6 volts/meter?

*16.19. Find the capacitance of a capacitor made of two parallel disks, of radius 0.25 meter, and separated by a 0.5-mm plate of mica, whose dielectric constant equals 7.

16.20. Two square copper plates, 8 cm on the side, are kept 1.0 mm apart in air. How much charge is needed on either plate to produce a potential difference of 500 volts?

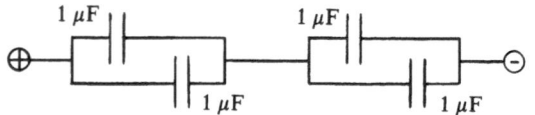

*16.21. Find the total capacitance of the four-element combination shown.

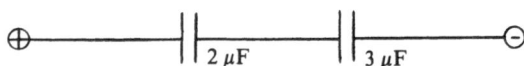

16.22. Find the total capacitance of the two-element combination shown.

16.23. A capacitor has a charge of 8 μC and a voltage of 500 volts. Find (a) its capacitance, (b) its total energy.

*16.24. A certain electrolytic capacitor is rated at 10 μF, and its plate area is 100 cm². If the molecules of its oxide layer are 6×10^{-10} meter in diameter, find how many molecules thick it is. (Assume a dielectric constant $\kappa = 2$.)

16.25. The capacitance C of a *single* conductor (rather than of a capacitor) may be defined through $Q = C\Phi$, where Φ is its potential relative to points at infinity. (a) What is C for an isolated metal sphere of radius R? (Express your answer in terms of ε_0 and R.) (b) How large should R be in order to yield $C = 1$ farad?

16.26. Three different capacitors, labeled 1, 2, 3, are available. How many different capacitances can you make by connecting some or all of 1, 2, 3, in different ways? (List the diagrams but do not calculate anything; ignore the fact that for special values of 1, 2, 3 some combinations may look different but have the same capacitance.)

*16.27. In Example 16.6, find the charges on the 1-μF and 2-μF capacitors.

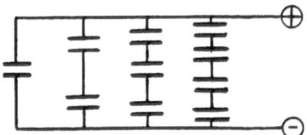

*16.28. Find the total capacitance of the ten-element combination shown, if each capacitor has a 1-μF capacitance.

16.29. Find the total capacitance of the three-element combination shown.

16.30. A certain air capacitor has plates of area 5 cm², spaced 1 mm apart. What field strength between these plates would be needed to store 1 joule of electrostatic energy, if we assume that sparks between the plates could somehow be prevented?

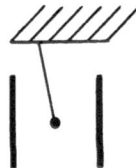

*16.31. A pith ball of weight w and charge q is suspended from a silk thread inside a vertical capacitor of spacing s and area $\mathscr{A}$, as shown. If the capacitor's voltage is V, find the angle θ that the thread makes with the vertical. (Express your answer in terms of q, V, s, $\mathscr{A}$, ε_0, w.)

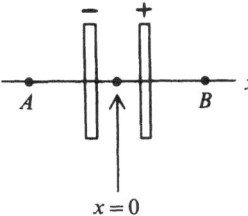

$x = 0$

*16.32. A certain parallel-plate capacitor has a field $E = 5000$ volts/meter between its plates, whose separation is 1.0 cm. Plot the potential Φ as a function of the coordinate x along a perpendicular path from A to B, as shown. Take $\Phi = 0$ at $x = 0$. (Note: because of fringing, E is not exactly zero outside the plates. If your graph were continued to remote distances, Φ would approach zero as $x \to \pm\infty$.)

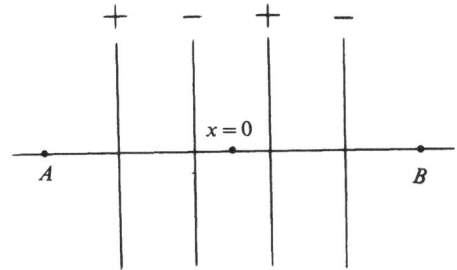

$x = 0$

16.33. Four plane parallel uniform layers of charge, each amounting to $\pm 5\,\mu C$, are arranged as shown; they are square, 10 cm on the side, and separated by equal distances of 0.20 cm. Make a graph of the potential Φ as a function of the coordinate x along a perpendicular path from A to B through the layers' centers. Take $\Phi = 0$ at $x = 0$ (See also Problem 16.32.)

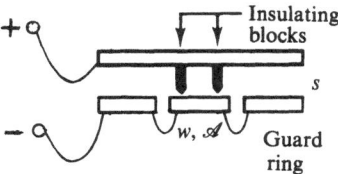

*16.34. In the horizontal-plate arrangement shown, the lower plate, whose weight is w, is barely held up against gravity by electrostatic attraction to the top plate. In terms of w, the spacing s, the plate area $\mathscr{A}$, and ε_0, what are (a) the potential difference V between the top and bottom plates? (b) the electrostatic energy between the movable and top plates? (c) the ratio of this energy to the work needed to lift the bottom plate to the level of the upper one against gravity alone?

*16.35 (Force between capacitor plates). (a) How much energy is stored in a parallel-plate air capacitor with charge Q? (Express your answer in terms of ε_0, s, $\mathscr{A}$, and Q.) (b) How much additional energy is stored if the plates are moved apart a further small distance Δs? (Assume no charge is allowed to enter or leave the plates.) (c) Since the result of (b) must equal the mechanical work needed to pull one (movable) plate away from the other (fixed) plate, what is the force between the two plates?

16.36. Use the result of Problem 16.35 (c) to derive the electrometer formula (16.C.16).

*16.37. (a) What is the energy density of the electric field in the vicinity of a uniformly charged spherical surface of radius R and charge Q? (Express your answer in terms of ε_0, Q, R, and the distance r from the center.) (b) What is, by integration, the total energy of the sphere's field?

16.38. (a) A metal sphere of total surface charge Q' and radius R is given an additional charge dQ'. How much work does this require if dQ' is brought in from far away? (b) By integrating the answer to (a) from $Q' = 0$ to $Q' = Q$, find the energy stored in a shell of charge Q. Your answer should agree with that of Problem 16.37(b).

16.39. Verify the last two energy formulas (16.D.3).

Section E: Particles Controlled by Electric Fields

(Note: electron mass $= 9.11 \times 10^{-31}$ kg.)

16.40. In a certain field ion microscope, the needle tip has a radius of 3×10^{-7} meter; it is at the center of a glass bulb (serving as the screen), of radius of 0.10 meter. What magnification factor can we expect?

*16.41. Find, in SI units, the momentum of a 0.1-keV electron.

*16.42. In a certain cathode-ray tube, a 500 electron volt beam suffers a 1.0 degree deflection after traveling for 5.0 cm between a pair of plates which are 1.0 cm apart. Find the deflection voltage.

16.43. Suppose the voltage V of an electron gun is known. If, somehow, the speed v of the electron beam could be measured, the electron's charge-to-mass ratio e/m could be determined. Express e/m in terms of V and v. (See the next chapter for practical ways of making that measurement.)

16.44. In the context of Millikan's experiment, express the charge q in terms of the droplet's weight w, the capacitor voltage V, and the plate separation s.

*16.45. In a certain field ion microscope, the needle tip has a radius of 3×10^{-7} meter. If a surface field of 2.5×10^{10} volts/meter is required, find the applied voltage. (Approximate the needle tip as an isolated sphere;

the voltage is then computed relatively to points at infinity.)

16.46. In a 0.5-kV electron gun, (a) what is (in joules) the kinetic energy of an electron when it reaches the anode? (b) What is its speed at that time?

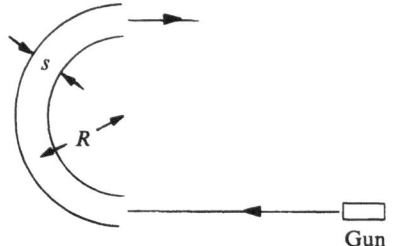

Gun

16.47. An electron gun, of voltage V, sends a beam into a semicylindrical capacitor gap, as shown. If the gap width is s and the radius is R, find the voltage V' needed between the plates in order to send the beam back as shown. (Assume the beam is in a plane perpendicular to the cylinder axis.)

*16.48. Consider a 0.030-keV, initially horizontal electron beam, as in Example 16.8. Over what height is it pulled down by gravity after traveling 0.50 meter?

16.49. Suppose the droplet in Fig. 16.34 were made of water, had a diameter of 0.05 mm, and carried a charge of $-10e$. Find the potential difference between the plates (assumed 1.0 cm apart) that would be needed to support the droplet.

*16.50. The following is an approximate description of a nuclear experiment done by Rutherford (about 1908). A helium nucleus (alpha particle), of charge $+2e$ and mass $m_\alpha = 6.7 \times 10^{-27}$ kg, is shot straight at a gold nucleus (much more massive and considered to be fixed in space), of charge $+79e$. If the α particle has a kinetic energy $\mathscr{K}$ when still far from its target, (a) find the distance l of closest approach between the two particles (in terms of $\mathscr{K}$, m_α, e, and the constant K of Coulomb's law). (b) Find l in meters if $\mathscr{K} = 5.5$ MeV. (Consider both particles as point charges.)

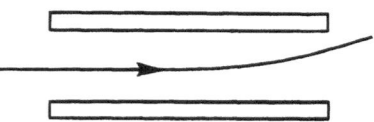

*16.51. In the cathode-ray tube of Example 16.9, suppose the beam enters the deflecting gap halfway between the plates, as shown. At what ratio V'/V does the deflected beam begin to graze the edge of a plate? (Neglect fringing of the field; find the algebraic and numerical answers.)

*16.52 (Interstellar propulsion by an ion gun). It can be shown that the ultimate speed v_f reached by a rocket star-

ting from rest in interstellar space is $v_f = v_0 \ln(M_+/M_-)$, where v_0, the "specific impulse," is the exhaust speed of the rocket engine, and where $M_\pm$ is the rocket's mass with and without its initial fuel supply, respectively. Let $v_0 = 1500$ meters/sec for a certain gas combustion engine. For the same M_+, M_-, how much faster would v_f be if the engine were an ion gun, accelerating hydrogen ions (charge $= +e$, mass $= 1.7 \times 10^{-27}$ kg) over a potential difference of 10 000 volts? (This speed would, however, be very slow to acquire; note also that electrons would have to be ejected to keep the space ship neutral; what would happen if it became ever more negative?)

Answers to True or False

1. False.
2. False (by 1 electron volt).
3. True.
4. False (both contributions have the same sign).
5. True.
6. True.
7. False (we need the field along a path of integration).
8. True (the equipotentials are perpendicular to the lines).
9. False (e.g., around a point charge).
10. True (because **E** is outward, as shown by Gauss' law).
11. True.
12. True ($V = Q/C$, where C increases).
13. False (it increases because E remains constant and $V = Es$).
14. True.
15. False (true for the kinetic energy).
16. False (the deflection depends only on geometry and on the ratio of two voltages).
17. False (the weight of an oil drop).

Answers or Hints to Problems

16.1. 225 joules.

16.3. 7.5×10^6 volts.

16.4. 100 volts/meter eastwards.

16.6. (c) -6.3×10^7 meters^2/sec^2.

16.8. 5.3×10^3 volts.

16.12. 5×10^{-4} coulomb,

16.15. $E = 4\pi K\sigma$.

16.16. (b) coulomb^2 sec^2 kg^{-1} meter^{-3}.

16.17. See Eq. (16.C.5).

16.19. 2.4×10^{-8} farad.

16.21. $1 \ \mu F$.

16.24. 30.

16.27. First find the voltage across the $3 \ \mu F$, then across the $1 \ \mu F$ or $2 \ \mu F$.

16.28. $(25/12) \ \mu F$.

16.31. $\mathscr{A}$ is not needed.

16.32. Hint: The schematic appearance is

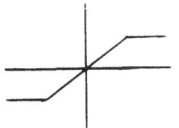

16.34. (a) Formula (16.C.16); (c) 1.

16.35. (c) $Q/(2\varepsilon_0 \mathscr{A})$.

16.37. (b) Hint: The field energy in a shell of radius r and thickness dr is $(Q^2 \ dr)/(8\pi\varepsilon_0 r^2)$.

16.41. 5.4×10^{-24} kg meter/sec.

16.42. 3.5 volts.

16.45. See Eq. (16.B.8).

16.48. 1.1×10^{-13} meter.

16.50. (a) m_α is not needed. (b) 4.1×10^{-14} meter; see also Note ii at the end of Chap. 28.

16.51. Use $s/2 = \frac{1}{2}a_y t^2$, $l = v_x t$, eliminate t.

16.52. See Example 16.8.

CHAPTER 17

DIRECT CURRENTS

We have seen that a conducting object tolerates no electric field in its interior *as long as static equilibrium prevails*. Suppose now that an electric field is nevertheless maintained there by some means. Then static equilibrium cannot exist; some motion of charge will occur. For example, a copper wire connected across the terminals of a battery does indeed have an electric field in it. Conduction electrons then flow along the wire, amounting to an **electric current**.

In 1826 Georg Ohm formulated his famous law, which relates the current in the wire to the applied potential difference. These two quantities are *simply proportional to one another*. (Ohm's law did not become well established, however, until good current-measuring devices became available in the late 1830s.)

Any circuit element to which Ohm's law can be applied is called a **resistor**. Resistors, and batteries as well, are the chief topics of this chapter. At first they are studied only in terms of how they affect a circuit's voltages and currents; later, a modest amount of space is devoted to their inner workings.

This chapter concentrates on **direct currents** ("dc"), which never reverse their direction. More specifically, the currents are assumed to be constant in time (with one exception, in the case of a discharging capacitor). A later chapter will introduce alternating currents ("ac"), to which certain important dc results no longer apply.

A. Ohm's Law

The flow of charge in a wire is in some ways analogous to the flow of water in a pipe. Figure 17.1

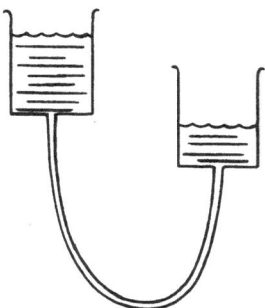

Figure 17.1. A water current in a capillary tube is the analogue of an electric current in a wire. If the tube is sufficiently narrow, there is negligible inertial (oscillatory) behavior for the two levels; the current is proportional to the level difference (hydraulic "Ohm's law"); water volume is conserved, as is the charge. In an electric wire, a potential difference plays the same role as the level difference does here.

shows a hydraulic model that can be kept in mind. It is not to be taken literally, of course; electricity is not a liquid, and wires are not pipes.

i. The Battery as a Black Box

In order to study electric currents, or to apply them in practice, electrostatic generators are almost useless. Because their voltages are so high (thousands to millions of volts), and because they store so little energy, the currents they produce are mainly short-lived sparks.

By contrast, batteries provide an extremely constant potential difference (typically a few volts) between their terminals; a steady flow of charge can be obtained for many hours. A (partial) explanation of these features is postponed to the notes at the end of the chapter. For the time being we just view the device as a "black box" with two terminals. We use the symbol of Fig. 17.2 to represent an "ideal battery," i.e., one that supplies a constant voltage.

(A typical wall outlet is very different from a battery, in that it is an ac source, whose voltage varies sinusoidally with time.)

ii. The Electric Current

The net amount of charge flowing per unit time through a given cross section of a wire is defined as the **electric current**, I, through that cross section; see Fig. 17.3. Thus, we have

$$I = \frac{dQ}{dt} \qquad (17.A.1)$$

where dQ is the net charge that passes during a small time dt. For a constant current, we can equivalently write

$$I = \frac{\Delta Q}{\Delta t} \qquad (I = \text{const}) \qquad (17.A.2)$$

The SI unit of current is clearly the coulomb/sec, or **ampere**:

$$1 \text{ ampere} = 1 \text{ A} = 1 \frac{\text{coulomb}}{\text{sec}} \qquad (17.A.3)$$

[André-Marie Ampère (1775–1836) is famous for his studies on the magnetic effects of currents; see Chap. 19 further on.]

In household appliances, typical currents are of the order of 1–10 A. Relatively modest currents through the human body are well known to be lethal, particularly in the range from 0.1 to 0.2 amperes.

The following example illustrates the importance of counting only the net charge when evaluating currents, and reminds us of the sign convention for charges.

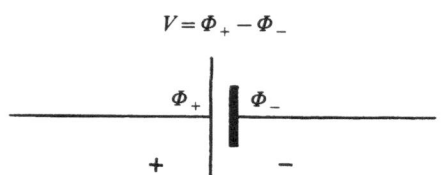

$$V = \Phi_+ - \Phi_-$$

Figure 17.2. An ideal battery with terminal potentials Φ_+ and Φ_-.

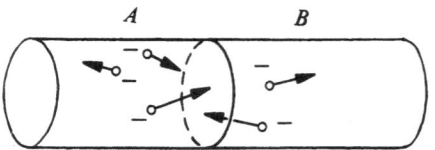

Figure 17.3. A wire in which electrons move in various directions. The net charge crossing from A to B per unit time is defined as the net current in the AB direction.

Example 17.1. Current from a Two-Way Electron Traffic. In Fig. 17.3, let 2.0×10^{18} electrons cross over from A to B every second, while, during the same time, 1.5×10^{18} electrons cross over from B to A. What is the current I in the AB direction?

During $\Delta t = 1$ sec we have

$$\Delta Q = (+2.0 \times 10^{18})(-e) + (-1.5 \times 10^{18})(-e)$$

so that

$$I = \frac{\Delta Q}{\Delta t}$$
$$= \frac{(2.0 \times 10^{18} - 1.5 \times 10^{18})(-1.6 \times 10^{-19} \text{coulomb})}{1 \text{ sec}}$$
$$= -0.08 \text{ ampere}$$

(Equivalently, the current is $+0.08$ ampere in the BA direction.)

We note that a current, if constant in time, *must also be the same at all cross sections of the wire.* Otherwise some locations would experience ever-increasing accumulations of charge, with the result that huge forces of repulsion would soon be set up.

iii. Ohm's Law and Resistance

Let us connect a metal wire across the terminals of a voltage supply, as in Fig. 17.4. As a result of the potential difference V, an electric field exists in the wire, and a current I flows from high to low potential. Ohm's law states that I is proportional to V,

$$I = \frac{V}{R} \qquad (17.A.4)$$

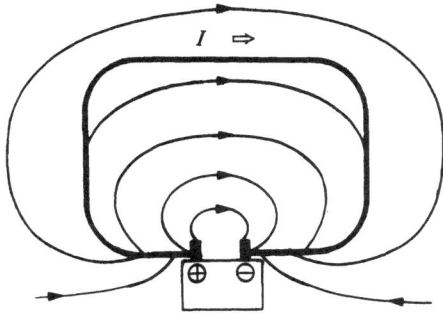

Figure 17.4. A battery with a metal wire across its terminals. Inside the wire, the electric field lines are parallel to the wire itself. The overall field (sample lines shown) is only slightly distorted by the presence of the wire.

with a proportionality constant $1/R$. The quantity R is called the **resistance** of the wire; we see from (17.A.4) that the SI unit of R is the volt/ampere, or **ohm** (Ω, capital Greek omega):

$$1\ \Omega = 1\ \text{ohm} = 1\ \frac{\text{volt}}{\text{ampere}} \qquad (17.\text{A}.5)$$

[It should be kept in mind that Ohm's law is purely empirical, and while it is valid for a metallic conductor, *kept at a fixed temperature*, it is wrong for some other conductors. If a common fluorescent lamp tube, which contains gases at low pressure, is connected (at fixed temperature) across a voltage V which is gradually increased, the current I will not rise in proportion to V, but in a much more complicated manner.]

Example 17.2. Total Charge over a Given Time. How much charge, ΔQ, passes through a bulb filament of resistance $R = 20\ \Omega$ if it is connected across a potential difference $V = 12$ volts for a time $\Delta t = 1$ h?

From definition (17.A.2) we have

$$\Delta Q = I\ \Delta t \qquad (17.\text{A}.6)$$

or, with Ohm's law, (17.A.4),

$$\Delta Q = \frac{V\ \Delta t}{R} \qquad (17.\text{A}.7)$$

$$= \frac{(12)(3600)}{20}\ \text{coulombs}$$

$$= \underline{2200\ \text{coulombs}}$$

Figure 17.5. The symbol for a resistor. The current I flows from high to low potential.

This would be enormous as an isolated charge; actually, of course, all parts of the circuit remain neutral.

Resistors

A piece of conductor (say a wire) that obeys Ohm's law is called a **resistor** when it is used as a circuit element. It is then symbolized by the diagram of Fig. 17.5. Resistors are the simplest and most common of all electrical devices. They serve to control voltages by means of currents, or vice-versa; in addition, they play a well-known role as heating elements.

Verification of Ohm's Law

The proportionality $I \propto V$ is simple enough, but how can it be verified in the laboratory? We need a procedure for measuring a voltage V, and another one for measuring a current I. Then we can check that, with a given metal wire, doubling V will result in doubling I, etc.

In practice, V and I are read off the dials of so-called voltmeters and ammeters. However, these instruments make use of magnetism, and it is important to convince ourselves that the required measurements, and the verification of Ohm's law, can in principle be conducted without any reference to magnetism.

In the preceding chapter we have seen that a voltage can be measured with an electrometer. The question is how to measure a current. One method, described in the next section, exploits the conservation of energy and the dissipation of heat in resistors.

B. Electric Power

Whenever a voltage supply produces a current, it also *delivers energy*, which may appear in various forms—heat, mechanical work, etc. How much energy is supplied per unit time? This is an important practical question, which we now address.

i. Power Delivered by a Voltage Supply

Figure 17.6 shows a current I traversing an unspecified system (perhaps a heating coil, or a motor) connected across a voltage V. For the sake of giving that system a name, one calls it a "load." During a time interval dt, a charge dQ leaves the positive terminal of the voltage supply (at potential Φ_+); during the same time, an equal amount of charge dQ enters the negative terminal (potential Φ_-). As far as the voltage supply is concerned, the net effect is that of a charge dQ flowing "downhill" between potentials Φ_+ and Φ_-. The loss of electric potential energy is

$$-d\mathcal{U} = \Phi_+ \, dQ - \Phi_- \, dQ$$
$$= (\Phi_+ - \Phi_-) \, dQ = V \, dQ \quad (17.B.1)$$

This must also be the amount of energy delivered to the load:

$$d(\text{energy}) = V \, dQ \quad (17.B.2)$$

Per unit time, we have

$$\frac{d(\text{energy})}{dt} = V \frac{dQ}{dt}$$

or

$$\boxed{\text{Power} = VI} \quad (17.B.3)$$

This important relation tells us that any load, or appliance, receives *a power that is the product of the current through it times the voltage across its terminals*; what it does with that power depends on

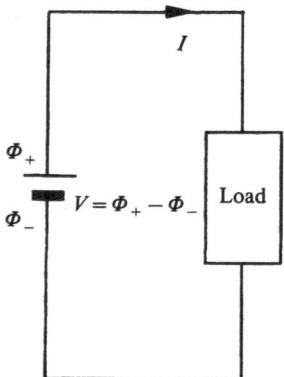

Figure 17.6. A voltage supply sends a current through a load.

what kind of appliance it is. Note the generality of formula (17.B.3); it does not depend on Ohm's law being valid anywhere in the system. Note also the often-used dimensional consequence of (17.B.3):

$$1 \text{ watt} = (1 \text{ A})(1 \text{ volt}) \quad (17.B.4)$$

ii. Heat Dissipation in a Resistor

It is an easily observed fact that resistors evolve heat when traversed by an electric current; heating coils (of which incandescent lamp filaments are examples) are familiar to all of us. The load in Fig. 17.6 might be such a resistor.

Assuming that the electric energy is entirely converted to heat, we get from Eq. (17.B.3)

$$\text{Power} = \frac{d(\text{heat})}{dt} = VI \quad (17.B.5)$$

(independently of Ohm's law), or

$$\boxed{\begin{array}{c}\text{Power} = \dfrac{d(\text{heat})}{dt} = RI^2 \quad (17.B.6) \\[2mm] = \dfrac{V^2}{R} \quad (17.B.7)\end{array}}$$

after using Ohm's law, Eq. (17.A.4).

Example 17.3. Calorimetric Measurement of a Current. Figure 17.7 shows an experimental arrangement in which a voltage supply measured at $V = 12$ volts is found to raise the temperature of 0.10 kg of water by 5C° after 15 min. Find the current I.

According to (17.B.5), we have, for a constant current,

$$I = \frac{1}{V} \frac{\Delta(\text{heat})}{\Delta t} \quad (17.B.8)$$

where

$$\Delta(\text{heat}) = (0.10 \text{ kg})(5\text{C}°) \frac{1 \text{ kcal}}{\text{kg C}°} = 0.50 \text{ kcal}$$
$$= (0.50)(4200 \text{ joules}) = 2100 \text{ joules}$$

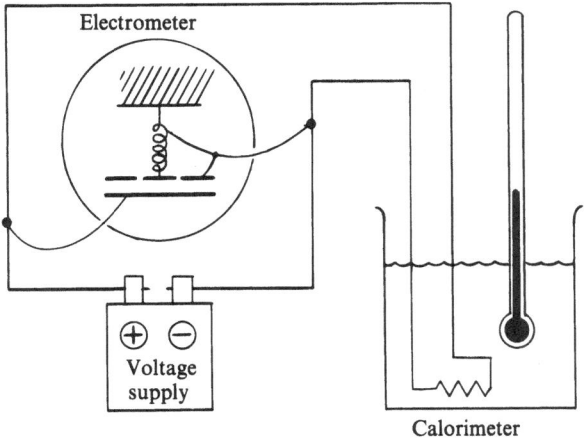

Figure 17.7. The water in the calorimeter is heated by a resistor.

Hence, from (17.B.8),

$$I = \frac{2100 \text{ joules}}{(12 \text{ volts})(15)(60 \text{ sec})} = \underline{\underline{0.19 \text{ A}}}$$

(Strictly speaking, this is an average over the duration of the measurement. Indeed, a minor change of resistance can be expected during the 5 C° rise.)

Thus, a current can be measured from "first principles." It should be noted that this application repeats Joule's water-stirring experiment with the heating coil as a replacement for the stirrer. However, the purpose is *not* to confirm the value 1 kcal ≈ 4200 joules, or to test the conservation of energy. Rather, that value, and the principle of energy conservation, are put to use in order to measure the current.

The heat evolved in a resistor is called **Joule heat**. Like the heat dissipated in Joule's water stirring experiment, or in the capillary tube of Fig. 17.1, it is *the result of a thermodynamically irreversible process.*

Example 17.4. Current and Resistance in a Light Bulb. A certain incandescent light bulb is rated at 100 watts and 120 volts under normal use. (a) How much current does it draw? (b) What is its resistance? (Assume a dc supply.)

(a) Current: According to Eq. (17.B.3) the current, across $V = 120$ volts, is

$$I = \frac{\text{power}}{V} = \frac{100 \text{ watts}}{120 \text{ volts}} \approx \underline{\underline{0.83 \text{ A}}}$$

(b) Resistance: From Ohm's law, (17.A.4), we have

$$R = \frac{V}{I} \qquad (17.B.9)$$

$$= \frac{120 \text{ volts}}{0.83 \text{ A}} \approx \underline{\underline{145 \ \Omega}}$$

Alternatively, without using the result of (a), we have from (17.B.7)

$$R = \frac{V^2}{\text{power}} \qquad (17.B.10)$$

$$= \frac{(120)^2}{100} \approx \underline{\underline{144 \ \Omega}}$$

Note that a bulb filament is a resistor, in which the electric power is completely converted to heat. Much of the latter is then converted to light.

C. Circuits

Circuit theory is of practical use to every experimental scientist. It is the basis of all electrical technology, including much laboratory instrumentation. The circuits discussed in this chapter are made of resistors and batteries, except in one case where a capacitor will be included. Commercial resistors are often made of carbon rather than metal, but Ohm's law still applies to them.

Having seen how voltages and currents are measured in terms of basic units (coulombs, kilograms, meters, and seconds), we now assume that commercially available **voltmeters** and **ammeters** are used for those measurements whenever needed. These (properly calibrated) instruments must, until the next chapter, remain unexplained "black boxes," equipped with two terminals and a dial. They are used as shown in Fig. 17.8.

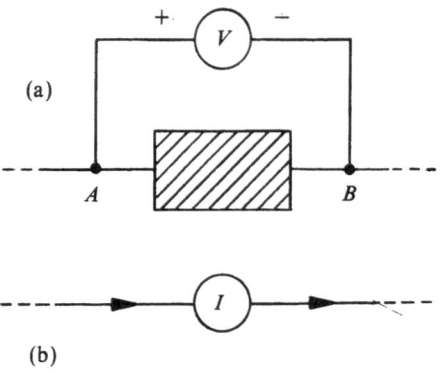

(a)

(b)

Figure 17.8. (a) A voltmeter, which reads the voltage between points A and B. This instrument draws some current and therefore perturbs the circuit, but it is built with a high resistance to minimize this effect. (b) An ammeter. This is a low-resistance instrument, which reads the current passing through it.

i. Series and Parallel Combinations

Just as we did earlier with capacitors, we may connect resistors in series (Fig. 17.9) or in parallel (Fig. 17.10). In this way a great variety of resistances may be created from just a few available values. The combination possesses two terminals, between which there exists a voltage V; into one terminal, and out of the other, flows a total current I. The combination amounts to a single resistor of overall resistance R such that

$$I = \frac{V}{R} \qquad (17.C.1)$$

What is R in terms of the individual R_1, R_2,...?

Series Combination

Adding the potential steps in Fig. 17.9, we have

$$V = V_1 + V_2 + \cdots = IR_1 + IR_2 + \cdots$$
$$= I(R_1 + R_2 + \cdots).$$

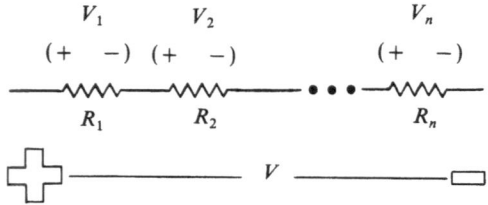

Figure 17.9. Series connection.

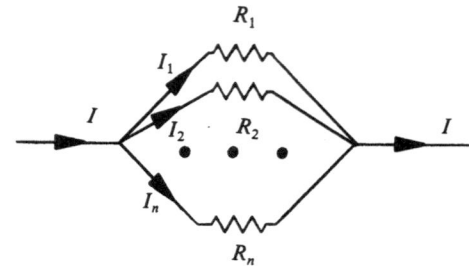

Figure 17.10. Parallel connection.

where we have used the fact that all resistors have the same current I. Comparing this result with (17.C.1) in the form

$$V = IR \qquad (17.C.2)$$

we see that

$$\boxed{R = R_1 + R_2 + \cdots} \qquad \text{(series)} \quad (17.C.3)$$

One consequence is that a piece of wire with uniform composition and thickness has *a resistance proportional to its length*,

$$R \propto l \qquad \text{(uniform wire)} \qquad (17.C.4)$$

as illustrated in Fig. 17.11.

Parallel Combination

In Fig. 17.10, conservation of charge implies that

$$I = I_1 + I_2 + \cdots = \frac{V}{R_1} + \frac{V}{R_2} + \cdots$$
$$= V\left(\frac{1}{R_1} + \frac{1}{R_2} + \cdots\right)$$

since each resistor is connected across the same

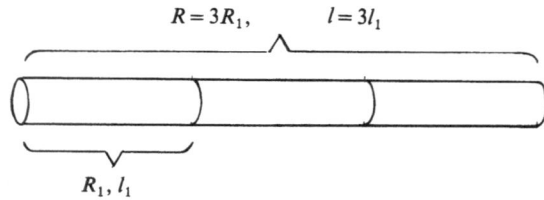

Figure 17.11. These three equal segments of wire amount to a series combination. We see that $R/R_1 = l/l_1$.

potential difference V. Comparison with (17.C.1) shows that

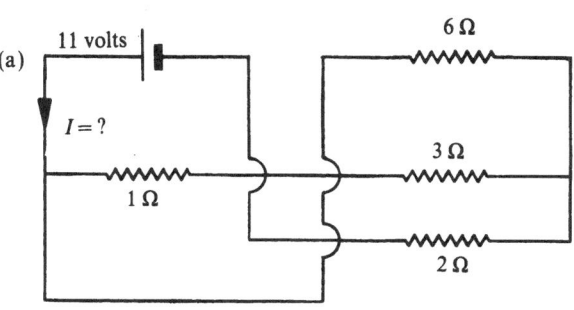

$$\frac{1}{R} = \frac{1}{R_1} + \frac{1}{R_2} + \cdots \quad \text{(parallel)} \quad (17.C.5)$$

In Summary: *We must add the resistances in a series combination, and the inverse resistances* in a parallel combination.*

Example 17.5. Analyzing a Circuit. In the circuit of Fig. 17.12a find the current I supplied by the battery. (The small bridges indicate wires crossing over without contact.)

A rearrangement of the wires gives the simpler-looking diagram shown in part (b) of the figure. (The reader ought to check this carefully.)

By formula (17.C.3), the 1-Ω, 3-Ω combination has an overall resistance

$$1\,\Omega + 3\,\Omega = 4\,\Omega$$

see part (c) of the figure. Portion AB of the circuit then has an inverse resistance given by (17.C.5),

$$\frac{1}{R_{AB}} = \frac{1}{4\,\Omega} + \frac{1}{6\,\Omega}$$

or

$$R_{AB} = 2.4\,\Omega$$

see part (d) of the figure. With inclusion of the 2-Ω resistor, (17.C.3) gives a total resistance

$$R = 2.4\,\Omega + 2\,\Omega = 4.4\,\Omega$$

The required current is, therefore,

$$I = \frac{11\ \text{volts}}{4.4\,\Omega} = \underline{\underline{2.5\ \text{amperes}}}$$

by Ohm's law.

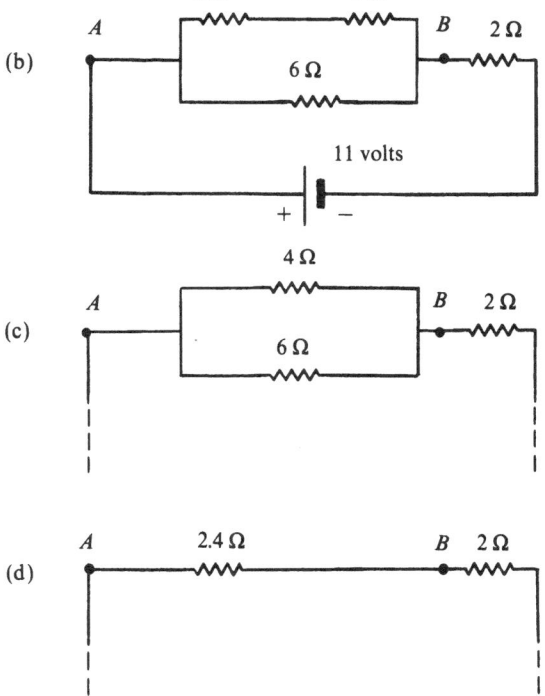

Figure 17.12. Circuit for Example 17.5.

The Effect of Thickness

Let us assume that a current is uniformly distributed through the cross section of a wire of uniform composition (rather than flowing, for example, just along its surface). Then (17.C.5) can be used to demonstrate that *the wire's resistance is inversely proportional to its cross-sectional area* $\mathscr{A}$:

$$R \propto 1/\mathscr{A} \quad \text{(uniform wire)} \quad (17.C.6)$$

This relation is indeed observed with a dc current, and is a valuable confirmation of its uniform distribution. The verification of (17.C.6) in a simple case is left to Problem 17.37.

---

* An inverse resistance is also called a **conductance**.

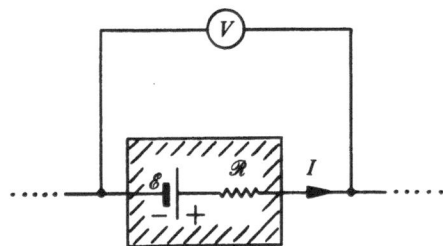

Figure 17.13. The box represents a real battery; it behaves as if containing the two elements shown.

ii. Real Batteries

Even the best batteries show a decreased voltage while a current is being drawn from them. When the battery is then disconnected, the observed voltage comes back to its former value. This behavior is to be expected quite aside from the battery's inner details, as the following simple argument shows.

A good conductor across the battery's terminals puts them effectively in contact with one another ("shorts them"); they are then at the same potential. In a less extreme case, the potential difference ought to be at least decreased, and this is indeed what happens.

When their current is not too large, batteries are found to behave like the model shown in Fig. 17.13, which consists of an ideal battery (with constant voltage $\mathscr{E}$) in series with a constant resistance $\mathscr{R}$. The (positive) quantity $\mathscr{E}$ is called the real battery's **electromotive force*** (abbreviated as **emf**), and $\mathscr{R}$ is its **internal resistance**. (Ordinarily, $\mathscr{E}$ is quoted on the battery's label as its "voltage.")

How is the actual voltage, V, related to $\mathscr{E}$, $\mathscr{R}$, and I? The answer can be read directly off Fig. 17.13, where the voltage V consists of two steps,

$$\boxed{V = \mathscr{E} - I\mathscr{R}} \qquad (17.C.7)$$

(Note the sign of the second term.) In particular, we see that when $I = 0$ one has

$$V = \mathscr{E} \qquad \text{(zero current)} \qquad (17.C.8)$$

and that a shorted battery ($V = 0$) implies

$$I = \frac{\mathscr{E}}{\mathscr{R}} \qquad (17.C.9)$$

the largest current attainable without additional batteries.

---

* A voltage is clearly not a force, but the name is traditional.

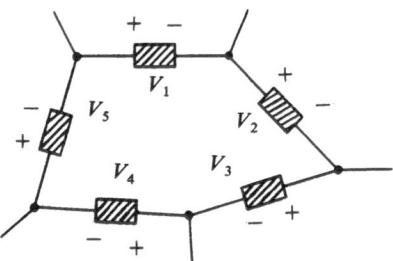

Figure 17.14. Kirchhoff's voltage rule. The $(+-)$ assignments are arbitrary; at least one of these directions must be wrong, but no matter—the corresponding V will come out negative. To prove the rule, write each V as a potential difference between two terminals, say $V_1 = \Phi_1 - \Phi_2$, $V_2 = \Phi_2 - \Phi_3$, etc.; common terminals have a common potential. We then have $V_1 + V_2 + \cdots = (\Phi_1 - \Phi_2) + (\Phi_2 - \Phi_3) + \cdots + (\Phi_5 - \Phi_1) = 0$. Note for future reference that a circuit element (black box) can be any device with a well-defined terminal potential difference, for example a capacitor; the voltage then need not even be constant in time.

iii. Kirchhoff's Rules

Not all circuits can be analyzed into series and parallel connections. For some help with more difficult cases one invokes a pair of common-sense principles called **Kirchhoff's rules**.

Voltage Rule*

The voltages, all measured in the same direction around any single loop of a circuit, add up to zero:

$$\boxed{V_1 + V_2 = \cdots = 0} \qquad (17.C.10)$$

A detailed explanation will be found in the caption of Fig. 17.14. (Because the existence of a potential means conservation of energy for a test charge, one commonly says that "Kirchhoff's voltage rule expresses the conservation of energy.")

Current Rule†

All currents flowing towards a common junction have zero sum:

$$\boxed{I_1 + I_2 + \cdots = 0} \qquad (17.C.11)$$

see Fig. 17.15. ("Kirchhoff's current rule expresses the conservation of charge.") By convention, a

---

* Also known as "loop rule."
† Also known as "node rule" or "junction rule."

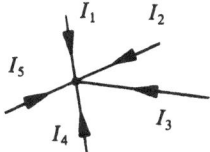

Figure 17.15. Kirchhoff's current rule. Here at least one of the I's will turn out negative. The rule means that zero net charge accumulates at the junction; it is not necessary to assume constant currents.

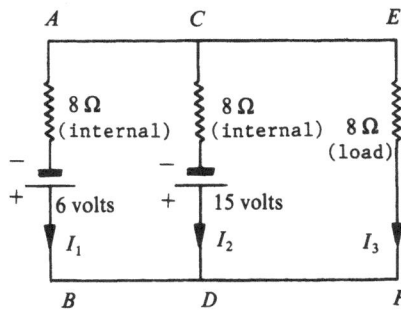

Figure 17.16. Circuit for Example 17.6.

positive current flowing away from a junction is treated as a negative current towards the junction.

Example 17.6. Two Real Batteries and a Resistor. The batteries in Fig. 17.16 have emf's of 6 volts and 15 volts, respectively; their internal resistances are both 8 Ω, as is the external load. Find the current in the 6-volt battery.

First we label with arrows the currents I_1, I_2, I_3 on the figure. *It does not matter if the wrong direction is guessed*; this will only show up as a negative sign in the result.

Current rule: At junction D, we have

$$I_1 + I_2 + I_3 = 0 \qquad (17.C.12)$$

(Hence at least one of the currents flows oppositely to the marked arrow.) Application of the current rule at junction C yields $(-I_1) + (-I_2) + (-I_3) = 0$, which is nothing new.

Voltage rule: There are three loops to be found: $ABFEA$, $CDFEC$, and $ABDCA$. We first add the consecutive increases in potential around $ABFEA$, assuming the currents to be flowing as shown. (Careful attention must be paid to the signs in what follows.)

$$-(8\,\Omega)\,I_1 + 6\text{ volts} + (8\,\Omega)\,I_3 = 0 \quad (17.C.13)$$

Similarly, around $CDFEC$,

$$-(8\,\Omega)\,I_2 + 15\text{ volts} + (8\,\Omega)\,I_3 = 0 \quad (17.C.14)$$

Solving the equations: Eqs. (17.C.12), (17.C.13), and (17.C.14) are sufficient to extract the three unknowns I_1, I_2, I_3. Therefore there is no need to apply Kirchhoff's rule to the third loop.

Let us eliminate I_3 by means of (17.C.12): $I_3 = -I_1 - I_2$. Insertion in (17.C.13), (17.C.14) yields a system of two equations for I_1 and I_2,

$$16I_1 + 8I_2 = 6\text{ amperes} \qquad (17.C.15)$$

$$8I_1 + 16I_2 = 15\text{ amperes} \qquad (17.C.16)$$

Solving for I_1, we find

$$I_1 = -\tfrac{1}{8}\text{ ampere}$$

Thus, the current flows *backwards* into the 6-volt battery. A look at Sec. B i should convince the reader that the battery now *receives*, rather than supplies, net power. If suitably designed, the battery can in fact be recharged in this way (storage battery).

iv. The Potentiometer

Once a stable potential difference, say that of a battery, has been accurately measured, for example by an electrometer, it can be used as a standard for the purpose of measuring other potential differences.

Comparing two potential differences is a far easier—and usually more accurate—proposition than making an absolute measurement. The **potentiometer** is a device that finds *the ratio of two potential differences* in terms of *the ratio of two lengths*. One possible design is shown in Fig. 17.17, in which the emf's $\mathscr{E}_1$ and $\mathscr{E}_2$ of two batteries are being compared. A uniform wire AB carries a constant current I which originates from a third, "working" battery. (Neither the value of I, nor the working battery's emf need to be measured.) The emf's $\mathscr{E}_1$ and $\mathscr{E}_2$ are each in series with a **galvanometer** G_1

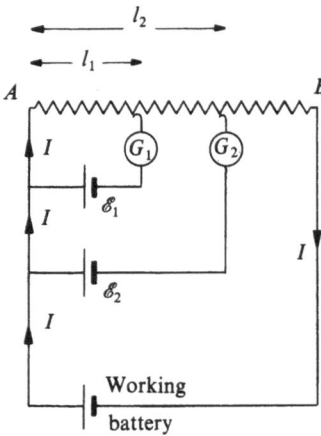

Figure 17.17. A potentiometer circuit.

or G_2. These are sensitive detectors of electric current; their construction will be discussed in Chap. 18. The procedure, a **null measurement**, consists of shifting the sliding contacts until both galvanometers read zero current. At this setting one has

$$\frac{\mathscr{E}_1}{\mathscr{E}_2} = \frac{l_1}{l_2} \qquad (17.\text{C}.17)$$

where l_1 and l_2 are the wire lengths indicated in the figure. As a proof, we note that the null readings in G_1 and G_2 imply that the current has the same value I in all segments of AB; furthermore, neither the internal resistances of batteries 1 and 2, nor those of the galvanometers, play any role in the measurement. Applying Kirchhoff's voltage rule around loops $A, G_1, \mathscr{E}_1, A$ and $A, G_2, \mathscr{E}_2, A$, we find

$$-IR_1 + \mathscr{E}_1 = 0, \qquad -IR_2 + \mathscr{E}_2 = 0 \quad (17.\text{C}.18)$$

where R_1 and R_2 are the resistances of the wire segments l_1 and l_2. Eliminating I, we have

$$\frac{\mathscr{E}_1}{\mathscr{E}_2} = \frac{R_1}{R_2} = \frac{l_1}{l_2} \qquad (17.\text{C}.19)$$

as announced. [Recall proportionality (17.C.4).]

Commercially available potentiometers easily reach accuracies of 1 part in 10^5. (High accuracy is typical of null measurements because the measuring instrument—here a galvanometer—can be, essentially, as delicate as one pleases.)

Why is using a potentiometer more reliable than simply connecting a voltmeter across the terminals of a battery to read its emf? The answer is that a voltmeter draws some current; therefore, according to Eq. (17.C.7), the reading will always be lower than the actual emf (by an unknown amount, if the battery's internal resistance is unknown).

v. A Resistor–Capacitor (RC) Loop

The discharge of a capacitor (capacitance $= C$) through a resistor (resistance $= R$) provides an interesting application of Ohm's law; Fig. 17.18 illustrates such a process. How long does it last?

The model of Fig. 17.A.1 helps visualize what goes on; each water tank plays the role of a capacitor plate. The rate of flow decreases to zero as the two levels become equalized; in electric language, the current I approaches zero as the capacitor voltage V approaches zero. The discharge is substantially completed (never totally) in a characteristic time, the so-called **time constant** of the circuit, which we now calculate. The precise

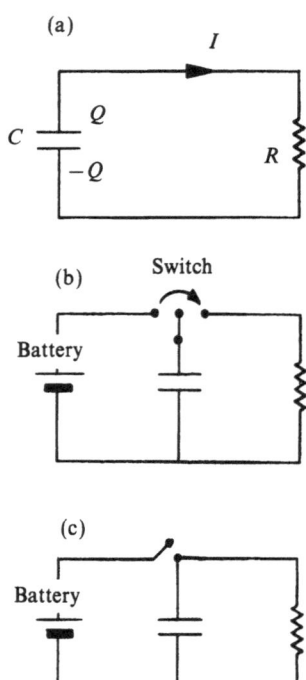

Figure 17.18. (a) An RC loop. The capacitor is initially charged; it discharges through the resistor. (b) A switching arrangement to effect this. The switch is in the left and right positions for $t < 0$ and $t > 0$, respectively. (c) An alternative arrangement. The switch is closed and open for $t < 0$ and $t > 0$, respectively.

question is: "If the capacitor voltage V equals V_0 (given) at time zero, what is its value at any later time t?"

From Fig. (17.18) we see that I is the rate of *decrease* of the charge Q,

$$I = -\frac{dQ}{dt} = -\frac{d}{dt}(CV) \qquad (17.C.20)$$

The last step follows from the definition of capacitance.

Across the resistor, Ohm's law gives, for the same current,

$$I = \frac{V}{R} \qquad (17.C.21)$$

Comparing (17.C.20) and (17.C.21), we find

$$\boxed{\frac{V}{R} = -C\frac{dV}{dt}} \qquad (17.C.22)$$

a differential equation for V with solution

$$V = V_0 e^{-t/RC} \qquad (17.C.23)$$

(V_0 = potential difference at time zero, $e = 2.718....$) In order to test whether this is indeed the solution, one can substitute (17.C.23) in both sides of (17.C.22). Any given value of V_0 is seen to be acceptable. [It is shown in differential equation texts that (17.C.23) is the most general solution; see also our Problem 17.28.]

It is left to the reader to demonstrate that the charge and the current also exhibit that behavior:

$$Q = Q_0 e^{-t/RC} \qquad (17.C.24)$$
$$I = I_0 e^{-t/RC} \qquad (17.C.25)$$

The same **time constant**

$$\boxed{\tau = RC} \qquad (17.C.26)$$

governs all these decays; its meaning is illustrated in Fig. 17.19.

Example 17.7. Designing a Time Circuit. Consider once more the circuit of Fig. 17.18, with $C = 3.0\ \mu\text{F}$. Estimate the resistance R needed to obtain a 90% discharge in a time $t = 2.0$ msec.

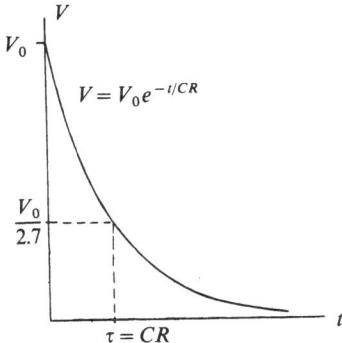

Figure 17.19. Exponential decay of the capacitor's voltage. After a time τ, V is reduced to $1/e \approx 37\%$ of its initial value. (Fluid mechanics teaches us that the system of Fig. 17.1 follows a similar exponential law.)

The charge must decrease to 0.10 times its original value. We require

$$\frac{Q}{Q_0} = e^{-t/RC} = 0.10$$

$$-\frac{t}{RC} = \ln 0.10 = -2.3$$

from which

$$R = \frac{t}{(2.3)(C)} = \frac{2.0 \times 10^{-3}}{(2.3)(3.0 \times 10^{-6})}\ \Omega = \underline{\underline{290\ \Omega}}$$

D. Conduction

Why is Ohm's law valid in metals? A complete explanation is difficult and will not be attempted, as it involves advanced concepts in the quantum theory of solids. In a simplified scenario, the conduction electrons suffer random collisions with the metal's atoms and cause the observed Joule heat.

Here we address two simple questions. First, can we predict the resistance of a wire from its geometrical dimensions and from its composition? Second, at what *average* velocity must the electrons be migrating to produce a given current?

i. Resistivity

The resistance R of a wire depends on its dimensions and on its material. How do we separate these two effects?

Table 17.1. Some Resistivities at 20°C

Silver	1.63	
Copper	1.72	
Aluminum	2.83	
Tungsten	5.51	
Iron	9.7	
Lead	22	$\times 10^{-8}\ \Omega$ meter
Mercury	97	
Carbon (graphite)	1400	
Glass (typical)	10^{13}	
Silicon oxide	10^{21}	
Aluminum oxide	10^{22}	

We already know [Eqs. (17.C.4), (17.C.6)] that R is proportional to the wire's length, $R \propto l$, and inversely to the wire's cross section $\mathscr{A}$, $R \propto \mathscr{A}^{-1}$.

To summarize these two results, we set

$$R = r\frac{l}{\mathscr{A}} \qquad (17.D.1)$$

where *the proportionality constant r no longer depends on the wire's geometry*; r is called the electrical **resistivity*** of the material of which the wire is made; its SI unit is the ohm meter.

The range of observed resistivities is enormous, see Table 17.1. A material's resistivity depends considerably on the presence of impurities; the table assumes pure substances. At room temperature, silver is the best known conductor; note the huge resistivity of glass or ceramic substances such as silicon oxide and aluminum oxide; they are among the best insulators.

Example 17.8. A 1-Ω, 1-cm³ Resistor. To what length l should a volume $\mathscr{V} = 1.0$ cm³ of silver ($r = 1.6 \times 10^{-8}\ \Omega$ meter) be drawn in order to have a resistance $R = 1.0\ \Omega$?

Formula (17.D.1) states that

$$R = r\frac{l}{\mathscr{A}} \qquad (17.D.2)$$

On the other hand, the wire's volume is

$$\mathscr{V} = l\mathscr{A} \qquad (17.D.3)$$

* Its inverse, $1/r$, is called the **conductivity**.

Eliminating $\mathscr{A}$ from (17.D.2) and (17.D.3) gives

$$l = \left(\frac{\mathscr{V}R}{r}\right)^{1/2} = \left[\frac{(1.0\times10^{-6})(1.0)}{1.6\times10^{-8}}\right]^{1/2} \text{ meters}$$

$$= \underline{7.9 \text{ meters}}$$

In metals, r increases with temperature. (In other materials, such as graphite, it usually decreases.) As a homely illustration, incandescent bulbs typically fail just as the light is switched on; the still-cold metal filament carries a larger current, which fatally overheats a small constriction before it heats the remainder of the filament.

At very low temperature, a large number of metals and alloys exhibit a spectacular, *total lack of resistance*: they have $r = 0$. This condition is known as **superconductivity**. A superconducting ring will carry a current for indefinite periods of time without a power supply; no heat is dissipated. However, a superconductor does not tolerate an arbitrarily large current without reverting to its normal (resistive) state. Until recently, the highest temperatures at which superconductivity had been observed were in the vicinity of 20 K. Lately, however, superconductors have been shown to exist at astonishingly "high" temperatures such as 90 K or higher. The eventual discovery of room-temperature superconductivity might revolutionize electrical technology; in particular, a vehicle might be made to float in a stable manner over a magnetic field. Superconducting coils for electromagnets, and superconducting transmission lines, have already demonstrated their usefulness. (Why currents should have any relation at all to magnetism is examined in great detail in the next few chapters.)

ii. The Electrons' Drift Velocity

Consider a copper wire in which a known current is flowing. How fast, in meters/sec, do the conduction electrons migrate along the wire to give rise to this current? A rough estimate can be given under the assumption that *each copper atom contributes one of its* 29 *electrons to the pool of conduction electrons*. (That single electron is the one that is least tightly bound to the atom. Solid-state physicists have determined that the assumption of 1 conduction electron per atom is fairly accurate for copper, silver, and some other metals; it is a poor estimate in many other cases.)

We shall denote by n the number of conduction electrons per unit volume, or equivalently, the number of copper atoms per unit volume.

The velocity estimate—on the basis of a known current—goes as follows. A segment of the wire, with length Δl and cross section $\mathscr{A}$, is shown in Fig. 17.20. For simplicity we pretend in this calculation that all conduction electrons move to the right with the same velocity of migration, called the **drift velocity** and denoted by v_d. Their combined charge is given, in absolute magnitude, by

$$\Delta Q = (e)(\text{number of electrons})$$

$$= (e)(n)(\text{volume}) = en\mathscr{A}\,\Delta l \quad (17.D.4)$$

If ΔQ takes a time Δt to move completely through the right end of the wire segment, then the current is given by

$$I = \frac{\Delta Q}{\Delta t} = en\mathscr{A}\frac{\Delta l}{\Delta t}$$

or

$$\boxed{I = en\mathscr{A}v_d} \quad (17.D.5)$$

Solving for v_d, we have

$$v_d = \frac{I}{en\mathscr{A}} \quad (17.D.6)$$

(The combination $I/\mathscr{A}$ is called the **current density**.)

For definiteness, let $I = 1$ ampere and $\mathscr{A} = 1 \text{ mm}^2$; these are typical "household" values. To calculate n, it is easiest to consider 1 kmole of copper atoms. We have

$$n = \frac{\text{number of atoms in 1 kmole}}{\text{volume of 1 kmole}} \quad (17.D.7)$$

$$= \frac{\text{Avogadro's number}}{(\text{mass of 1 kmole})/(\text{mass density of copper})}$$

$$= \frac{(\text{Avogadro's number})(\text{mass density})}{(\text{atomic weight})(1 \text{ kg})} \quad (17.D.8)$$

$$= \frac{(6 \times 10^{26})(9 \times 10^3)}{63} \frac{\text{atoms}}{\text{meter}^3}$$

$$= 8.5 \times 10^{28} \frac{\text{atoms}}{\text{meter}^3} \quad (17.D.9)$$

Equation (17.D.6) yields

$$v_d = \frac{1}{(1.6 \times 10^{-19})(8.5 \times 10^{28})(1 \times 10^{-6})} \frac{\text{meter}}{\text{sec}}$$

$$\approx 7 \times 10^{-5} \frac{\text{meter}}{\text{sec}} \quad (17.D.10)$$

less than a snail's pace. (By contrast, an electron, accelerated across only 1 volt *in vacuum*, acquires a speed of more than 10^5 meters/sec.) The explanation for the low value lies in the fact that v_d is only the average velocity of an electron. Its instantaneous velocity, which is much larger, and which is due to thermal motion, changes randomly because of collisions with the copper ions; Fig. 17.21 gives a more realistic picture than

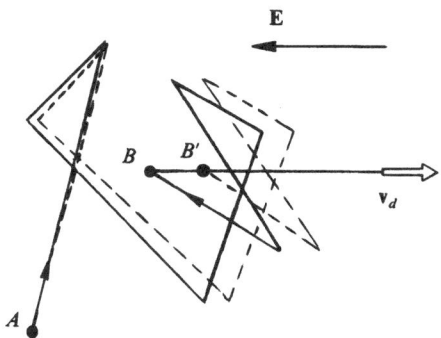

Figure 17.21. Solid line: Hypothetical trajectory of a conduction electron as it collides with the metal's positive ions. Zero electric field is assumed. Progress from A to B is due to chance, and is equally likely to occur in another direction. Dashed line: The same trajectory, modified by an electric field $\mathbf{E}$ to the left. (A rightward acceleration occurs between collisions.) The additional displacement BB' is due to the field and results in a net drift velocity $\mathbf{v}_d$.

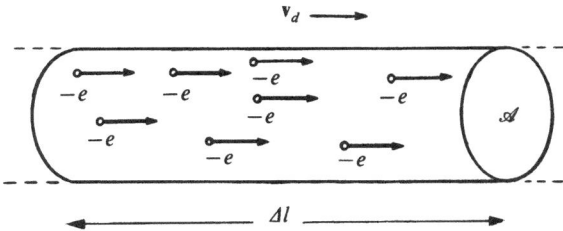

Figure 17.20. A model where all electrons travel at the same velocity $\mathbf{v}_d$.

Fig. 17.20. The wind provides a rough analogy: its speed, corresponding to v_d, is very small compared to the thermal speed of the individual air molecules. [We recall from Chap. 11 that an air molecule typically travels faster than sound. An electron, being much lighter, should move even faster, since the root-mean-square speed increases as $(\text{mass})^{-1/2}$. Actually, electrons cannot be treated as classical particles. When quantum mechanics is correctly taken into account, the electrons' thermal speed is found to be much higher yet.]

Caution: *Speed of Electric Signals*

The slow speed v_d bears no relation to the high speed (comparable to that of light) at which electrical effects are propagated along a wire. The closing of a switch, for example, is a disturbance whose messenger is an electric field rather than traveling electrons. (In a similar way, the opening of a water valve sets distant water in motion almost instantaneously because the messenger is a pressure wave, propagating at the speed of sound.)

Notes

Ordinary electric batteries consist of one or several **voltaic cells** connected in series and/or in parallel. Each cell is made of a pair of dissimilar electrodes dipping in an **electrolyte**, or ionic solution; see Fig. 17.22. The cell's inventor, Alessandro Volta (1800), used silver, zinc, and salt water, but since his time a huge number of combinations have been found useful. Here we discuss a zinc–copper combination introduced by John Frederic

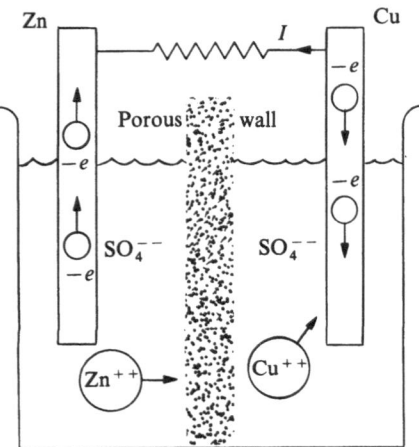

Figure 17.22. A zinc–copper cell (Daniell cell).

Daniell in 1836. Its main virtue, as far as we are concerned, is to provide a particularly simple illustration of the cell's principle. Later, we see how a cell yields an intriguing recipe for measuring Avogadro's number in terms of the electron's charge.

i. What Happens in a Cell

As we see in Fig. 17.22, the zinc (Zn) dips in an aqueous solution of zinc sulfate (in the ionized form $Zn^{++}SO_4^{--}$); the copper (Cu) dips in a copper sulfate solution ($Cu^{++}SO_4^{--}$). The two solutions are in contact and diffuse into one another through a porous wall.

Suppose the two electrodes are connected by an external resistor. Then the following processes occur simultaneously.

1. At the zinc electrode, some Zn atoms go into solution as Zn^{++}, each leaving two electrons behind in the metal:

$$Zn \to Zn^{++} + 2e^- \qquad (17.N.1)$$

2. At the copper electrode, an equal number of Cu^{++} ions deposit themselves as metallic copper; this requires each of them to extract two electrons from the electrode:

$$Cu^{++} + 2e^- \to Cu \qquad (17.N.2)$$

3. Nothing happens to the SO_4^{--} ions. They play the purely passive role of keeping the solution neutral.
4. The above-mentioned charges $-2e$ leave the zinc electrode to enter the resistor, while the same amount of charge leaves the resistor to enter the copper electrode; Joule heat is evolved in the resistor.
5. The Zn^{++} region of the electrolyte expands through the porous wall; the Cu^{++} region shrinks.
6. Some Joule heat is evolved also in the electrolyte, corresponding to the cell's internal resistance.
7. The chemical energy lost in the net reaction

$$Zn + Cu^{++} \to Zn^{++} + Cu \qquad (17.N.3)$$

equals the total corresponding Joule heat. This amounts to about 1 electron-volt for each electron transferred. Hence it can be demonstrated that the cell's emf is about 1 volt. Because this emf depends only on the existence of reaction (17.N.3), *it is independent of the size or geometry of the cell.*

The process just described can continue until all the Cu^{++} ions are depleted, or until the zinc electrode is completely dissolved, whichever comes first.

ii. Avogadro's Number and the Electron's Charge

For every two electrons that leave the top of the zinc electrode, one zinc atom goes into solution. Hence if we could count the electrons, we would thereby count the atoms too.

Let 1 kilomole of electrons flow out of the zinc electrode. Then 1/2 kilomole of zinc atoms, or 32 kg of the metal, will be dissolved. (In practice, these amounts would be scaled down for convenience.) The total charge is therefore

$$\Delta Q = \mathcal{N}_A e \qquad (17.N.4)$$

($\mathcal{N}_A$ = Avogadro's number).

Experimentally, ΔQ is found from

$$\Delta Q = I \, \Delta t \qquad (17.N.5)$$

where I is the current, assumed constant for simplicity, and Δt is the time needed to dissolve the 32 kg of zinc. When the left side of (17.N.4) is measured in this way it is found to be

$$\Delta Q \approx 9.65 \times 10^7 \text{ coulombs} \qquad (17.N.6)$$

(One thousandth of this amount is called 1 **faraday**.) Equation (17.N.4) now reads

$$\boxed{9.65 \times 10^7 \text{ coulombs} = \mathcal{N}_A e \quad \text{(SI)}} \qquad (17.N.7)$$

providing us with an important relation between e and $\mathcal{N}_A$.

Clearly, this method does not depend on the special choice of zinc; however, we need to be given an ionization number by the chemists. Assuming Zn^+ rather than Zn^{++} would give us the wrong result by a factor of 2. (Actually, the method can be "turned around" to determine the ionization number if it is already known for one element.)

Condensed Checklist

Current:

$$\boxed{I = \frac{dQ}{dt}} \qquad (17.A.1)$$

$$1 \text{ A} = 1 \frac{\text{coulomb}}{\text{sec}} \qquad (17.A.3)$$

Ohm's law:

$$\boxed{I = \frac{V}{R}} \qquad (17.A.4)$$

$$1 \, \Omega = 1 \frac{\text{volt}}{\text{A}} \qquad (17.A.5)$$

Power:

$$\boxed{\text{Power} = \begin{cases} VI & \text{(general)} \\ RI^2 = \dfrac{V^2}{R} & \text{(resistor)} \end{cases}} \qquad (17.B.3)$$

$$(17.B.6), (17.B.7)$$

$$\text{Watt} = \text{volt ampere} \qquad (17.B.4)$$

Circuit connections:

$$\boxed{\begin{array}{l} R = R_1 + R_2 + \cdots \\ \dfrac{1}{R} = \dfrac{1}{R_1} + \dfrac{1}{R_2} + \cdots \end{array}} \quad \begin{array}{l} \text{(series)} \qquad (17.C.3) \\ \text{(parallel)} \quad (17.C.5) \end{array}$$

Real battery:

$$V = \mathscr{E} - I\mathscr{R} \qquad (17.C.7)$$

Kirchhoff's rules:

(Voltage) $V_1 + V_2 + \cdots = 0$ (17.C.10)

(Current) $I_1 + I_2 + \cdots = 0$ (17.C.11)

RC loop:

$$V = V_0 e^{-t/RC} \qquad (17.C.23)$$

Current density:

$$\frac{I}{\mathscr{A}} = -env_d \qquad (17.D.5)$$

Resistivity r:

$$\boxed{R = r \frac{l}{\mathscr{A}}} \qquad (17.D.1)$$

Table: 17.1

True or False

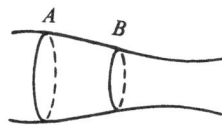

1. If, from A to B in the figure, a wire's cross section decreases by half, the (steady) current at B must have twice the value it has as A.

2. "Electromotive force" is the technical name for a battery's voltage.

3. A real battery, rated at 12 volts, delivers to a given load twice the power that would be obtained from a 6-volt battery.

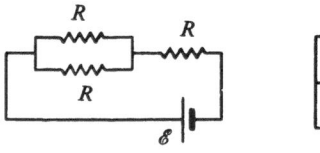 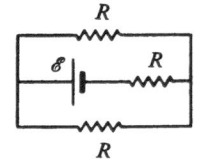

4. The adjoining figure shows two representations of the same circuit.

5. If connected in series, a battery and a resistor may be exchanged without any effect on the current through them.

6. A potentiometer gives the same information as an electrometer, although the accuracy may be different.

7. A potentiometer gives the same information as a voltmeter, although the accuracy may be different.

8. A given RC loop takes twice as long to discharge itself by 50% if its original charge is doubled.

9. If a wire of given mass is drawn to twice its length, its resistance is multiplied by 4. (Assume an unchanged resistivity.)

10. The drift velocity of electrons in a conductor is limited mainly by the *mutual* collisions between conduction electrons.

11. If the applied electric field is doubled, each conduction electron's instantaneous velocity is doubled.

Problems

Section A: Ohm's Law

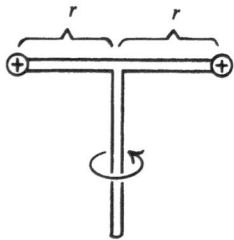

*17.1. A certain electrostatic machine involves a pair of metal spheres, whirling at the end of an insulating bar, as shown. If each sphere carries 5×10^{-8} coulomb of charge, and the rotation proceeds at 30 rev/sec, what effective current exists along the spheres' trajectory?

17.2. A certain storage battery delivers a 1-A current for 50 hours. To how much total charge, flowing out of the positive terminal, does this correspond?

*17.3. The dielectric belt in a certain Van de Graaff generator carries 3.0×10^{-8} coulomb of charge per meter length; its speed is 8.0 meters/sec. How much current does it deliver?

17.4. Express the ohm in terms of kilograms, meters, seconds, and coulombs.

17.5. (a) What is the resistance of a bulb filament rated at 120 volts, 60 watts (dc)? (b) How much current does it use?

*17.6. In a simplified view of the hydrogen atom, the electron is considered to be in circular orbit about the proton. Taking $r = 0.5 \times 10^{-10}$ meter for the orbit's radius, and $v = 2 \times 10^6$ meters/sec for the electron's speed, calculate an average electric current that may be said to circulate along the orbit.

17.7. The dielectric belt in a Van de Graaff generator carries a surface charge density (charge per unit area) σ; the width of the belt is w and its speed is v. Find the current. (See also Problem 17.3.)

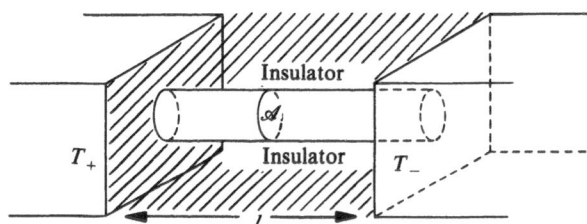

*17.8. (Heat conduction.) The figure shows a rod of length l, cross section $\mathscr{A}$, and uniform material, in contact with two steady temperatures $T_+ > T_-$. If $T_+ - T_-$ is not too large, the heat transfer per unit time along the rod is given by $dQ/dt = (T_+ - T_-)/R_{th}$, a law similar to Ohm's. Here R_{th} is a "thermal resistance," given by $R_{th} = r_{th} l/\mathscr{A}$ [cf. Eq. (17.D.1)]; the "thermal resistivity" r_{th} is a property of the material. Tables commonly list **thermal conductivities** $1/r_{th}$. What should be the side of a copper cube that must transmit 200 kcal in 30 min between two opposite faces at temperatures of 100°C and 350°C? Use a thermal conductivity of 9.4×10^{-2} kcal sec^{-1} meter^{-1} (C°)$^{-1}$.

Sections B: Electric Power; C: Circuits

17.9. Find the heating coil's resistance in Example 17.3.

*17.10. List the resistances you can make if three 4 Ω resistors are available. (Show the diagrams.)

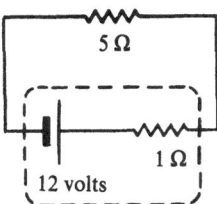

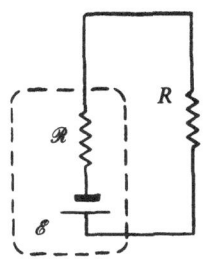

*17.11. (a) Find the current drawn from a 12-volt battery, with a 1-Ω internal resistance, in series with a 5-Ω resistor, as shown. (b) How much power is dissipated in the 5-Ω resistor?

17.12. A 50-volt supply sends a 10-A current into a winch motor. Assuming no heat losses, find at what speed a 1000-kg mechanical load can be lifted.

*17.19. A battery, whose emf is $\mathscr{E} = 6.0$ volts and whose internal resistance is $\mathscr{R} = 1.0\ \Omega$ is connected as shown to a load of resistance R. (a) Find the current I in R when its value is 0 Ω, 1 Ω, 2 Ω, 3 Ω; plot I against R. (b) Find the power dissipated in R; plot it as in (a). (c) At what value of R is that power a maximum?

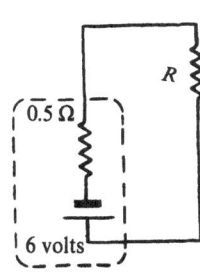

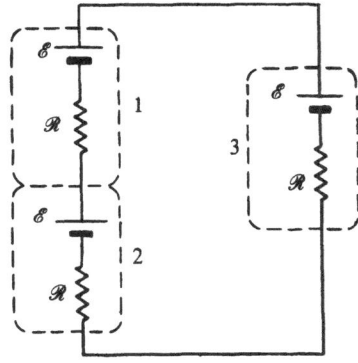

*17.13. A 6-volt battery, of internal resistance 0.52 Ω, is connected as shown to a load of resistance R, which draws a current I. (a) If R is varied, what are the smallest and largest possible values of I? (b) Plot the battery's terminal voltage, V, against R.

17.14. A 5.0-volt battery, with internal resistance 8 Ω, is connected across a resistance R. The rate of heat dissipation in R is 0.50 watt. (a) Find two possible values of the current. (b) What are the two corresponding values of R?

*17.15. An electric car of mass 500 kg, going up a ramp, rises in altitude at a rate of 0.1 meter/sec. If the car's battery has an emf of 50 volts and an internal resistance of 0.8 Ω, estimate two possible values of the current. (Which one is realized depends on the detailed construction of the motor.)

17.16. List the values of the resistances which you can make if two 1-Ω resistors and one 2-Ω resistor are available. (Show the diagrams.)

17.17. List the values of the resistances that you can make if three resistors, of 1 Ω, 2 Ω, and 3 Ω, are available. (Show the diagrams.)

*17.18. In Example 17.5, find (a) the voltage across the 2-Ω resistors; (b) the current in the 6-Ω resistor; (c) the voltage across the 3-Ω resistor.

17.20. Three storage batteries, with equal emf's and equal internal resistances $\mathscr{R}$, are connected as shown. Assume that battery No. 3 is being charged. (a) How long does it take to store a chemical energy $\mathscr{U}$ in it? (b) During the same time, how much energy is dissipated in heat in all three batteries?

17.21. Calculate the currents I_2 and I_3 in Example 17.6.

*17.22. In Example 17.6, calculate I_1 if the 15-volt battery is replaced by a 12-volt battery, without change in its internal resistance.

17.23. In Example 17.6, verify that Kirchhoff's voltage rule is satisfied around loop $ABDCA$.

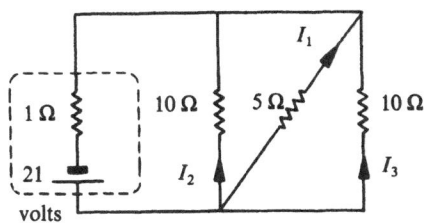

*17.24. Find currents I_1, I_2, I_3 in the circuit shown. (The dashed box holds a battery with its internal resistance.)

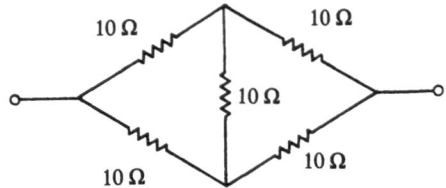

*17.25. What is the overall value of the resistance between the terminals in the figure?

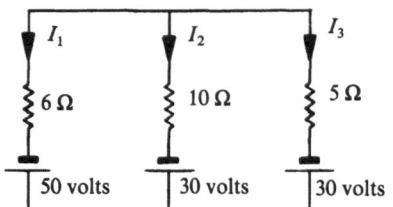

17.26. Find currents I_1, I_2, I_3 in the circuit shown. (The batteries' internal resistances are neglected.)

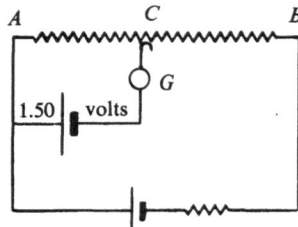

*17.27. The figure shows a potentiometer with only one battery (emf $= 1.50$ volts) connected, in addition to the working battery. The slide wire AB is 2.00 meters long and $0.100 \, \text{mm}^2$ in cross section; its resistivity is $5.00 \times 10^{-6} \, \Omega$ meter. A null current through G corresponds to a distance $AC = 0.750$ meter. What is the current in the slide wire?

17.28. Show how to integrate Eq. (17.C.22) for the voltage of a discharging capacitor. (First divide both sides by V and multiply by dt.) You should obtain (17.C.23).

17.29. Consider the discharging capacitor of Fig. 17.18. (a) In terms of V and I, write an equation stating that "the rate of decrease of electrostatic energy in the capacitor equals the rate of heat evolution in the resistor" (conservation of energy). (b) Show that this amounts to Eq. (17.C.22).

17.30. (a) Derive Eqs. (17.C.24), (17.C.25) for Q and I in an RC circuit. (b) Express I_0 in terms of Q_0, R, C.

17.31. Verify that the product RC has units of time.

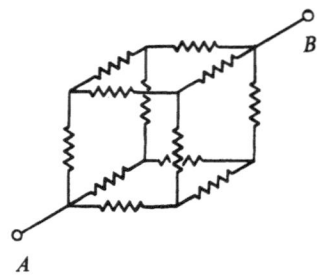

*17.32. Twelve equal resistors (each resistance $= R$) are connected as the edges of a cube; see the figure. Find the overall resistance between points A and B.

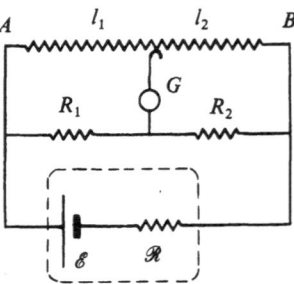

*17.33. (The **Wheatstone bridge**). In the circuit shown, resistor AB is a uniform smooth wire. The sliding contact is shifted until galvanometer G reads zero current. Derive a relation between resistances R_1, R_2 in terms of the lengths l_1, l_2. (This is a highly accurate and useful null method of comparing two resistances, similar to the potentiometer for comparing two emf's.)

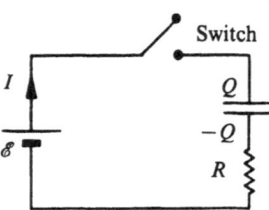

*17.34. In the figure, $\mathscr{E}$ represents a constant source of potential difference. A switch is closed at time $t = 0$, and a current I begins to charge the capacitor. (a) Find Q, I, and the potential difference V across the capacitor as functions of time. (Assume $\mathscr{E}$, R, C are given.) (b) Plot V and I against t. (c) With $\mathscr{E} = 100$ volts, $R = 1000 \, \Omega$, $C = 500 \, \mu\text{F}$, find the time needed to reach 90% of the ultimate charge on C.

Section D: Conduction

17.35. Find the resistance, at room temperature, of a 1.0-meter column of mercury with a 1.0-mm^2 cross section.

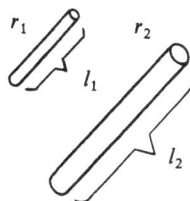

*17.36. Two cylindrical metal bars, of resistivities r_1 and r_2, are exact scale models of one another, as shown, but have the same resistance. (a) What is the ratio of their lengths, l_1/l_2? (b) What is l_1/l_2 numerically in the case of a mercury column and an iron bar, both at room temperature?

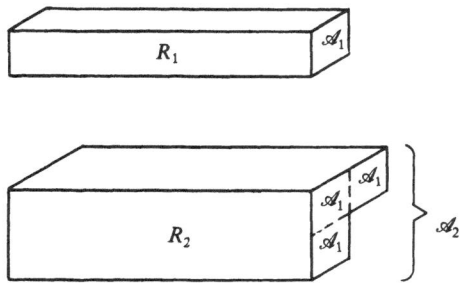

17.37. Two metal bars of equal length and material have cross sections $\mathscr{A}_1$, $\mathscr{A}_2$ shaped as shown, with $\mathscr{A}_2 = 3\mathscr{A}_1$. Assuming a uniform distribution of current throughout the bars, show that the corresponding resistances are $R_2 = R_1/3$, illustrating the relation $R \propto 1/\mathscr{A}$. (Use the parallel-combination rule.)

17.38. A strip of aluminum, 0.10 meter long and 1 mm wide, is deposited on glass by evaporation. How thick should it be in order to have a resistance of $10^{-2}\,\Omega$ between its ends?

17.39. A 5.0-meter-long bar of square cross section, 3.0 cm on the side, and of uniform composition, is found to carry a 0.90-A current under a potential difference of 70 mV. Which substance listed in Table 17.1 *could* it be made of?

*17.40. A wire, of cross section $\mathscr{A}$ and resistivity r, contains a uniform longitudinal field **E** which causes a current I. Express the current density $I/\mathscr{A}$ in terms of E and r.

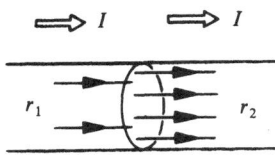

*17.41. The figure shows a junction between two wires with the same diameter but different resistivities r_1 and

r_2; the electric field is indicated by lines. (a) Which is larger, r_1 or r_2? (b) Show that some net charge Q must be present on the interface; what is its sign? (c) Evaluate Q in terms of r_1, r_2, and the current I. (See also the preceding problem.)

*17.42. (a) In a wire of resistivity r, how does the conduction electrons' drift velocity v_d depend on the applied field E? (Express v_d in terms of E, r, the density n of conduction electrons, and the electron charge e.) (b) In copper, how large a field E is needed to produce $v_d = 1$ cm/sec?

17.43. How much heat per unit volume and per unit time is evolved as the result of an electric field E in a wire of resistivity r?

Answers to True or False

1. False (more concentrated at B, but equal in coulombs/sec to that at A).
2. False (true at zero current).
3. False (depends on current as well).
4. True.
5. True.
6. False (only compares two voltages).
7. False (the potentiometer gives a zero-current reading; there is a current through the voltmeter).
8. False (same time).
9. True.
10. False (collisions with atoms).
11. False (true for the drift velocity).

Answers or Hints to Problems

17.1. 3×10^{-6} A.

17.3. Hint: Consider the charge and the time involved as 1 meter goes past the collector.

17.6. 1.0×10^{-3} A.

17.8. 4.7 mm.

17.10. 7 different values.

17.11. (b) 20 watts.

17.13. (a) largest: 12 A.

17.15. Larger value: 50 A.

17.18. (c) 4.5 volts.

17.19. (c) 1 Ω.

17.22. Zero.

17.24. $I_1 = 3$ A.

17.25. The middle resistor might as well be removed: use symmetry to compare potentials at its two ends.

17.27. 0.04 A.

17.32. As a current flows through AB, note which points have the same potential by symmetry. They may be joined. Answer: $5R/6$.

17.33. $R_1/R_2 = l_1/l_2$.

17.34. Hint: try $Q =$ decaying exponential + constant.

17.36. (b) Hg 10 times longer than Fe.

17.40. Hint: Apply Ohm's law to an arbitrary length l of the wire; then use $V = El$.

17.41. Hint: Use Gauss' law.

17.42. (a) Use Eq. (17.D.6).

CHAPTER 18

MAGNETIC FORCES

The age of electromagnetism was launched in 1819, when Hans Christian Oersted noticed, during one of his lecture demonstrations, that an electric current deflects a magnetized needle: *magnetism is a manifestation of electricity*. After a few months of intense international experimentation, it became clear that (1) steady electric currents produce steady magnetic forces, and (2) these forces are felt by the currents themselves. Most implications, including some of the deeper theoretical ones, were already worked out by 1825, largely through the investigations of Ampère.

Here we start with some traditional information on the performance of permanent **magnets**; this leads to the introduction of the **magnetic field* B** in analogy to the already familiar electric field **E**. Then (no longer adhering to the exact historical development) we study the effect of **B** on **currents** or **moving charges**. A discussion of Oersted's discovery, namely, the production of **B** by moving charges, is postponed to the next chapter.

It must be kept in mind that the present material is not an end in itself. A good working knowledge of it is essential to the understanding of later topics, including alternating currents and electromagnetic waves.

A. Magnetic Poles and Fields

A magnetic force, like an electric force, should be defined *through a demonstration* rather than in the abstract. For that purpose we can use any commercially obtainable **permanent magnets**, typically made of some iron-based alloy. A magnetic force is exerted, at a distance, by such a magnet on another one. Other and more fundamental cases of magnetic force, involving electric currents rather than permanent magnets, will emerge later on.

i. The Ideal Bar Magnet

We now add a new item to our already extensive collection of ideal devices. Let an **ideal bar magnet**, Fig. 18.1, be a thin rigid rod whose two ends are, respectively, designated as the **positive** or **north-seeking** ("*N*") **pole**, and the **negative** or **south-seeking** ("*S*") **pole**. We postulate the following properties (approximately realized in actual magnets such as compass needles).

1. As in Fig. 18.1, let the magnet be suspended in equilibrium from its center of gravity. Then its *N* pole points northward (with, in the northern hemisphere, a downward **inclination** as well; in the southern hemisphere, the *N* pole tilts up.) No identifiable disturbances, such as other magnets or current-carrying wires, are assumed to be near.

2. When two bar magnets are in one another's vicinity, they exhibit a *force of repulsion* between their like poles (*N* and *N*, or *S* and *S*), and a force of attraction between their unlike poles (*N* and *S*).

3. The force between two poles is an action–reaction pair, directed along the line that joins the poles; it decreases in strength as the inverse square, r^{-2}, of the distance r between the poles.

---

* The official name is "magnetic induction," but "magnetic field" is common parlance among physicists. Following E. M. Purcell (*Electricity and Magnetism*, McGraw-Hill, New York, 1985), we deliberately adopt the latter nomenclature as less confusing. Alternatively, we often just speak of the "**B** field."

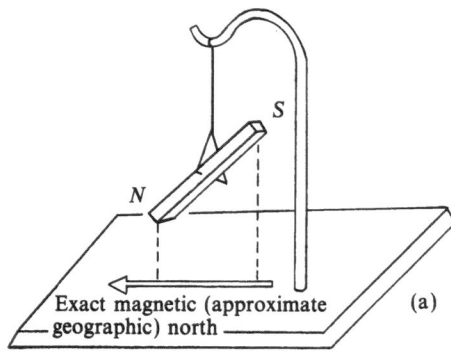

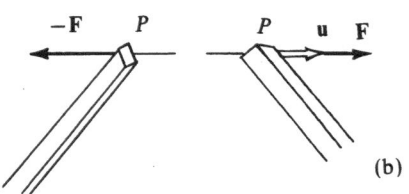

Figure 18.1. (a) Equilibrium orientation of a freely suspended bar magnet. (b) Coulomb's law holds between two poles; to confirm this by measurement, one may keep the other two poles (not shown) far away in relation to r; their force contributions are then negligible. In this illustration, both poles are of the same sign (NN or SS); **u** is a unit vector indicating the direction of **F**.

4. An ideal bar magnet is assumed to be *permanent*, a feature that will be explained presently.

As the reader has surely recognized, these properties are largely a repetition of what is observed in the case of electric charges. In analogy with the amount of charge, they allow us to measure an amount of magnetic pole, or **pole strength** P. This is done, as in Fig. 18.1b, in terms of the magnetic force **F** exerted by P on another pole p located a distance r away in vacuum (or, for practical purposes, in air):

$$\boxed{\mathbf{F} = K' \frac{Pp}{r^2} \mathbf{u}} \qquad (18.A.1)$$

According to its overall sign, this formula is interpreted as a repulsion ($+$) or an attraction ($-$). The constant K', like K in Coulomb's law for electric charges, is determined by convention. (This amounts to choosing a unit of magnetic pole.) For

better or for worse, the International System has settled on*

$$K' = 10^{-7} \qquad \text{(SI units)} \qquad (18.A.2)$$

exactly. To get a feel for the corresponding unit pole, we examine (18.A.1) in the case where $P = p = 1$ SI unit; let $r = 1$ meter. We see that

Two unit poles, 1 meter apart, in vacuum, repel one another with a force of 10^{-7} newtons   (18.A.3)

This may be considered the SI definition of the unit pole.[†] (A compass needle has poles typically of order 1 SI unit.)

We shall always assume an ideal magnet to be *permanent*; that is, its pole strength is unaffected by the proximity of other magnets, and also remains constant in time. Real magnets are never quite permanent.

Equation (18.A.1) is sometimes called **Coulomb's law** for magnetic poles, although it was first announced by John Michell in 1750; Coulomb did verify it on his torsion balance. For comparison we recall Coulomb's *electrostatic* law

$$\mathbf{F} = K \frac{Qq}{r^2} \mathbf{u} \qquad (18.A.4)$$

for charges Q, q a distance r apart, with

$$K \approx 9.00 \times 10^9 \text{ (SI)} \qquad (18.A.5)$$

Magnetic Peculiarities

The following are special features of magnetism, without counterpart among electric charges. First, the two poles P_N, P_S of any given magnet are always opposite in sign and equal in magnitude:

$$P_S = -P_N \qquad (18.A.6)$$

A second feature, logically related to the first, is

---

* The notation most often seen in the literature is $K' = \mu_0/4\pi$. Note that μ_0 ("mu zero") is in the numerator, unlike ε_0 in $K = 1/4\pi\varepsilon_0$.
† Note to the instructor: we depart from the classical literature in presenting poles as *effective* sources of **B** rather than as *actual* sources of **H**.

that no one has ever been able to separate P_N from P_S; see Fig. 18.2. In contrast to charges, *isolated magnetic poles have not been observed, and perhaps do not exist.**

Example 18.1. A Pair of Real Magnets. Two identical store-bought magnets repel one another with a force of 0.5 newton when in the configuration of Fig. 18.3a. By idealizing them as in Fig. 18.3b, estimate roughly the strength of their poles.

We first calculate the force on one pole, $+P$, of the top magnet. The contributions are $\mathbf{F}_+$, due to the lower $+P$, and $\mathbf{F}_-$, due to the lower $-P$. Their magnitudes are

$$F_+ = \frac{K'P^2}{r^2}, \qquad F_- = \frac{K'P^2}{(\sqrt{2}\,r)^2} \qquad (18.A.7)$$

By left–right symmetry, only the vertical force component F_y need be considered, as the reader ought to check carefully for himself. From the figure,

$$F_y = F_+ - (\cos 45°)\,F_- = F_+ - \frac{F_-}{\sqrt{2}}$$

$$= \frac{K'P^2}{r^2}\left(1 - \frac{1}{2\sqrt{2}}\right) \qquad (18.A.8)$$

Both ends of the top magnet contribute equally to the total force F_{tot} on it:

$$F_{\text{tot}} = 2F_y = \frac{2K'P^2}{r^2}\left(1 - \frac{1}{2\sqrt{2}}\right) \qquad (18.A.9)$$

Solving for P, we obtain

$$P = r\left[\frac{F_{\text{tot}}}{2K'\left(1 - \dfrac{1}{2\sqrt{2}}\right)}\right]^{1/2} \qquad (18.A.10)$$

$$= (2 \times 10^{-2})\left[\frac{0.5}{(2)(10^{-7})(0.65)}\right]^{1/2} \text{ (SI)}$$

$$= \underline{\underline{39 \text{ SI unit poles}}}$$

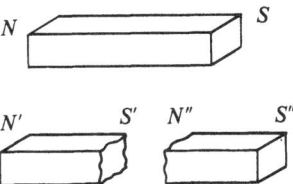

Figure 18.2. As a magnet is broken, new poles develop at the break; N' and S' are again of equal strength, and so are N'' and S''.

ii. The Magnetic Field

The use of test charges (Chap. 15) has a magnetic analogue. Let us use a **test pole** of known strength p in order to explore the magnetic effects in a region of space. Suppose a magnetic force $\mathbf{F}$ is found to act on p when the latter is at some given location. A **magnetic field B** is then said to exist at that point; $\mathbf{B}$ is a vector defined by

$$\boxed{\mathbf{F} = p\mathbf{B}} \qquad (18.A.11)$$

The field $\mathbf{B}$ is independent of p, and exists even when no test pole is used at all; $\mathbf{B}$ does, however, depend on the location of the measurement. (Caution: p always has a partner, $-p$, at the other end of the test magnet. In general, the force on $-p$ must be taken into account if the total force on the magnet is measured. With a sufficiently long magnet, however, $-p$ can be kept in a region of zero magnetic field, and ignored.)

The reader is assumed to have already acquired

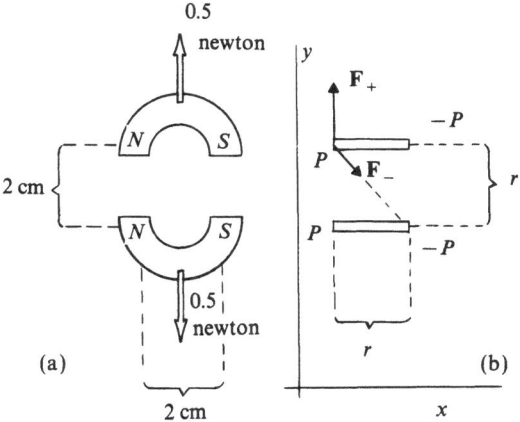

Figure 18.3. (a) Overall repulsion between two magnets. (b) Force contributions on a single pole (idealized). Forces between two poles on the same magnet are ignored. (Why?)

---

* These so-called **magnetic monopoles** are, however, being searched for in several laboratories, as they are a necessary ingredient in some theories of elementary particles; they might be microscopic particles from outer space.

a good grasp of the electric analogue to Eq. (18.A.11),

$$\mathbf{F} = q\mathbf{E} \qquad (18.A.12)$$

and therefore (18.A.11) will receive rather little discussion of its own.

Field Lines

Magnetic field lines are defined in analogy with electric field lines. They have the following properties, established in Chap. 15 in connection with the **E** field.

1. At any point in space, the *direction* of the field line is the direction of **B**.
2. The *line density*, or number of field lines per perpendicular unit area, $\mathcal{N}/\mathcal{A}_\perp$, equals the magnitude of **B**:

$$\boxed{B = \frac{\mathcal{N}}{\mathcal{A}_\perp}} \qquad (18.A.13)$$

see Fig. 18.4.
3. A line never begins or stops in free space.
4. A magnetic pole of strength P "sends out" $4\pi K'P$ lines into free space; near the pole, the lines go out radially and are uniformly sprinkled in all directions. (If the pole is negative it "takes in" the lines, rather than sending them out.)
5. The interior of the bar magnet is not accessible to probing with a test pole. Hence we cannot yet say anything as to the nature of **B** or of its field lines inside that thin, but very special region. This is illustrated in Fig. 18.5, which shows the field lines of a bar magnet.

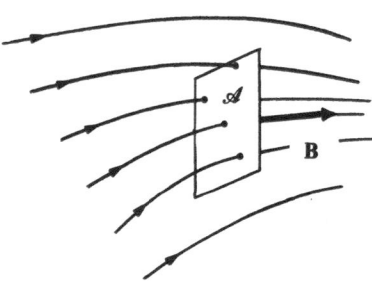

Figure 18.4. The field as a line density. In this illustration we have $B = (4 \text{ lines})/\mathcal{A}$. The quantity $\mathcal{N} = 4$ lines is called the **magnetic flux** through $\mathcal{A}$.

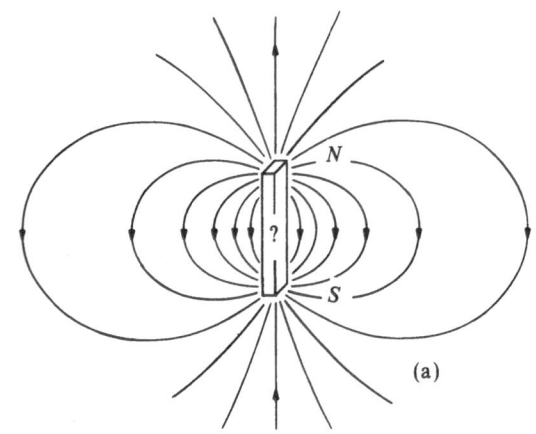

(a)

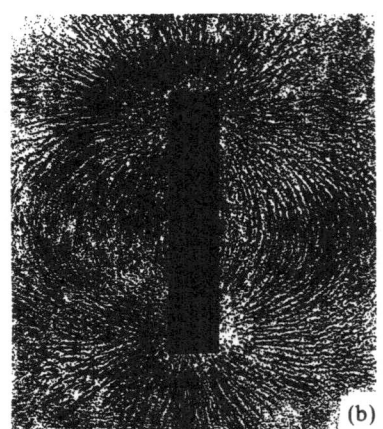

(b)

Figure 18.5. (a) Field of a bar magnet; compare with the electric-dipole field of Fig. 15.18b in Chap. 15. In the next chapter we shall see what happens to the field lines inside the magnet. (b) Iron filings, sprinkled on a sheet of paper over a bar magnet, make the field lines visible. Each bit of iron becomes a tiny bar magnet and aligns itself with the field.

Electric charges, by contrast, have no such zones of mystery, since they can in principle be isolated in space and inspected from all sides.

Units of B

The SI unit of B is the **tesla** (the standard unit abbreviation is T), after Nikola Tesla (1856–1943), who perfected the use of time-varying magnetic fields in motors. From Eq. (18.A.11) we see that

$$1 \text{ tesla} = 1 \frac{\text{newton}}{\text{SI unit pole}} \qquad (18.A.14)$$

Those of us who like to think in terms of field lines may equivalently write [see Eq. (18.A.13)]

$$1 \text{ tesla} = 1 \frac{\text{line}}{\text{meter}^2} = 1 \frac{\text{weber}}{\text{meter}^2} \quad (18.A.15)$$

When being counted, magnetic field lines are referred to as "webers" in the SI nomenclature. (Wilhelm Eduard Weber is known for his investigations of terrestrial magnetism, in collaboration with Gauss, in the 1830s.) We have

$$1 \text{ weber} = 1 \text{ Wb} = 1 \text{ magnetic field line} \quad (18.A.16)$$

Finally, we must mention a once prevalent, and still often used, unit of B, the **gauss** (G):

$$1 \text{ tesla} = 10^4 \text{ gauss} \quad (18.A.17)$$

exactly.

Field Contributed by Each Pole

If a single magnetic pole P existed by itself, it would contribute, at a distance r from itself, a field

$$\mathbf{B} = \frac{K'P}{r^2} \mathbf{u} \quad (18.A.18)$$

whose direction is radially away from P if $P > 0$. This relation is analogous to $\mathbf{E} = (KQ/r^2)\mathbf{u}$ and follows from Coulomb's magnetic law (18.A.1), together with $\mathbf{B} = \mathbf{F}/p$.

More realistically, we must deal with several magnetic poles; they contribute the *vector sum* of such terms. This is illustrated in Fig. 18.6, and exemplifies the *superposition principle*, which is valid for the **B** field just as for the **E** field.

The Earth's Magnetic Field

The most familiar **B** field is perhaps that of the Earth. Its lines, as plotted above ground, seem to be issuing from a bar magnet hidden inside the planet; see Fig. 18.7. No one believes there really is such a magnet; the cause of the terrestrial **B** field, although still somewhat mysterious, is thought to be electric currents that exist deep within the Earth, and which are somehow related to its rotation. Geological evidence from magnetic minerals shows that the Earth's "magnet" has exchanged its N and S poles many times in the past, although the Earth's spin has remained fairly steady. The Earth's field varies between about 3×10^{-5} tesla (0.3 G) at the equator and 7×10^{-5} tesla (0.7 G) at the polar latitudes.

iii. A Spectrum of Magnetic Fields

The Universe exhibits a wide range of magnetic field strengths; some typical ones are displayed in Fig. 18.8.

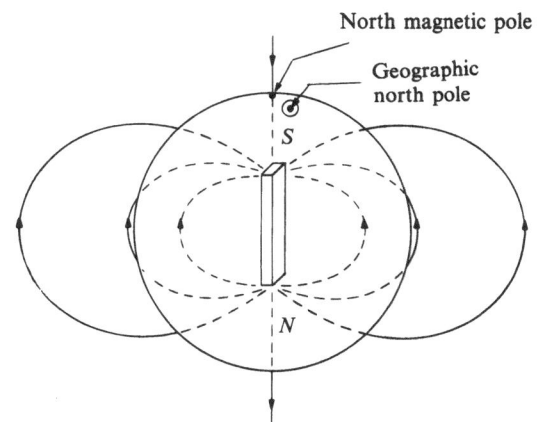

Figure 18.7. Magnetic field of the Earth. The north magnetic pole migrates slowly over the years. At present, it is near Bathurst Island, in northern Canada, and is not identical with the geographic north pole, which lies on the Earth's axis of rotation, 11.5 degrees of arc away. Since it attracts the N pole of a compass needle, the north magnetic pole is an S pole. In this figure, the bar magnet, as well as the dashed part of the field lines, are purely fictitious. Also, at higher altitudes than shown here, the lines are drastically distorted by solar effects.

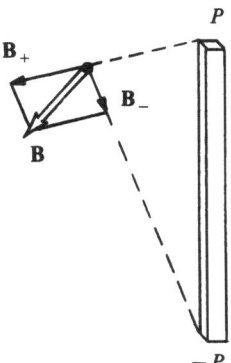

Figure 18.6. The fields $\mathbf{B}_+$, $\mathbf{B}_-$, contributed by the two poles, add vectorially to give the total field $\mathbf{B} = \mathbf{B}_+ + \mathbf{B}_-$. Figure 18.5 can be obtained from this principle.

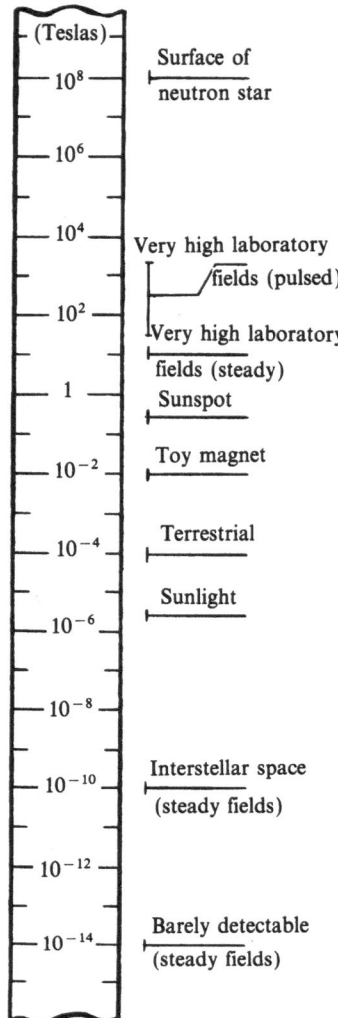

Figure 18.8. Some magnetic fields, on a logarithmic scale. Some very dense and small stars are believed to consist of close-packed neutrons; these stars have surface spots with the high fields shown. Sunspots have much lower fields. The highest laboratory fields can be made only in short-lived pulses, by explosive techniques. Light consists of vibrating electric and magnetic fields; shown here is a typical value of B contributed by sunlight, as received on Earth. The lowest fields shown are detected by small superconducting devices.

iv. Torque on a Dipole

The term "**magnetic dipole**" is often applied to an ideal bar magnet—two equal and opposite poles at a given distance from one another. (With two equal and opposite charges we would have an **electric dipole**; see Problem 15.9 in Chap. 15.) Taking a compass needle as an example, let us evaluate (1) the force, and (2) the torque to which it is subjected in a **uniform magnetic field** (B constant in

direction and in magnitude). We note incidentally that, at any given geographical location, the terrestrial B field may be considered uniform. The situation is shown in Fig. 18.9, where a dipole of length l and poles $\pm p$ is at an angle α to the field.

1. Force

The two contributions, shown in Fig. 18.9, are $\mathbf{F}_+$ and $\mathbf{F}_-$, opposite in direction and equal in magnitude:

$$\mathbf{F}_+ = p\mathbf{B}, \qquad \mathbf{F}_- = -p\mathbf{B}$$

The total magnetic force is

$$\mathbf{F}_+ + \mathbf{F}_- = 0 \qquad (18.A.19)$$

A dipole experiences zero net force in a uniform field.

2. Torque

With reference to Fig. 18.9 we have, about the center, a torque τ with counterclockwise contributions only:

$$\tau = \sum (\text{force})(\text{moment arm})$$

$$= (F_+)\left(\frac{l}{2}\sin\alpha\right) + (F_-)\left(\frac{l}{2}\sin\alpha\right)$$

$$= (2)(pB)\left(\frac{l}{2}\sin\alpha\right)$$

or

$$\tau = plB\sin\alpha \qquad (18.A.20)$$

The torque is directed so that it tends to align the

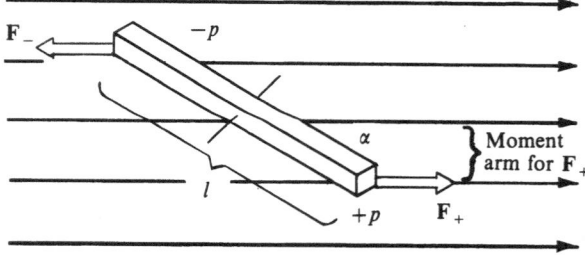

Figure 18.9. Magnetic dipole in a uniform B field.

dipole with the field. The combination *pl* is called the **magnetic moment** of the compass needle:

> Magnetic moment
> = (pole strength)(dipole length)

(18.A.21)

This is what a magnetic torque measures, rather than the individual factors *p* or *l*; Eq. (18.A.20) reads

> $\tau = $ (magnetic moment)$(B \sin \alpha)$

(18.A.22)

[There is a shortcut to the torque formula (18.A.20). We recall from mechanics that a **couple** (two equal and opposite forces, as we have here) exerts a torque that remains the same when calculated about any axis parallel to the original one. Consider, then, the axis through $-p$; the contribution from $\mathbf{F}_-$ vanishes, but now the moment arm of $\mathbf{F}_+$ is $l \sin \alpha$, and we have

$$\tau = (pB)(l \sin \alpha) + 0 \qquad (18.A.23)$$

the same result as (18.A.20).]

v. What are Magnetic Poles Made of?

A brief caution to the reader: poles are *not* simply the magnetic equivalent of charges. Although electricity consists of elementary charges, $\pm e$, *there are no elementary magnetic poles*, or, if they exist, they are exceedingly rare on Earth.

Instead, *magnetic poles are made of electric currents*, hidden not just at the tips, but throughout the volume of a magnet, and caused by spinning electrons. The curious reader may wish to look ahead, at Fig. 19.26 in the next chapter, to see how poles are explained in terms of field lines.

The true origin of magnetic poles is expressed in their SI unit:

> 1 SI unit of magnetic pole = 1 meter ampere

(18.A.24)

see Problem 18.58 after reading Sec. C further on.

B. Magnetic Force on a Moving Charge

Naive experiments, attempting to connect electricity and magnetism, are apt to be disappointing: *an electric charge, at rest in a steady* **B** *field, experiences zero force* (unless, of course, an **E** field also happens to be present.) Similarly, *an electric charge at rest generates zero magnetic field.* In both types of experiment, a nonzero result is found *only when the charge under consideration is moving.* The force on a moving charge is described in what follows; the production of **B** *by* moving charges is taken up in the next chapter.

i. The Right-Hand Force Rule

Consider a particle with electric charge *q*, moving at velocity **v** in a magnetic field **B**. What is the magnetic force **F** on that particle? The situation is shown in Fig. 18.10.

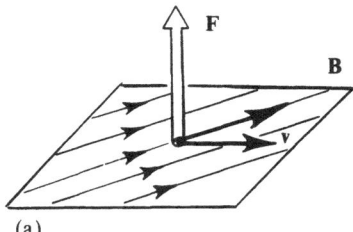

(a)

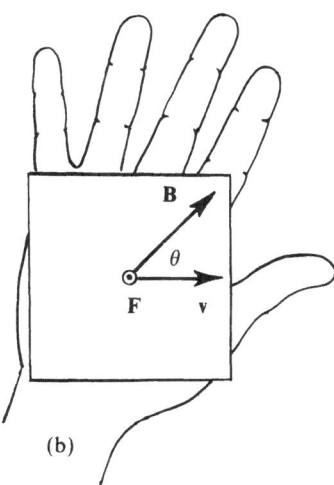

(b)

Figure 18.10. (a) A positive particle with velocity **v**, in a magnetic field **B**, experiences a force **F**, perpendicular to the plane containing **B** and **v**. (b) In this top view, the vector **F**, indicated by a dot, points toward the reader. According to the right-hand rule, the thumb points along $+\mathbf{v}$, the other fingers along $+\mathbf{B}$; the palm then "pushes" in the direction of $+\mathbf{F}$. For a negative particle, the direction of **F** must be reversed.

First we state the rule for calculating **F**; that rule is experimentally based, and some of its manifestations will be discussed later.

Magnitude

The force **F** has magnitude

$$F = qvB \sin \theta \qquad (18.B.1)$$

where θ is the angle between the vectors **v** and **B**. We see that $v = 0$ means $F = 0$, as mentioned before.

Direction

The vector **F** is at right angles to both the velocity **v** and the field **B**.   (18.B.2)

This still leaves the possibility of two directions, opposite to one another. The correct choice is found by application of the **right-hand rule**, explained in Fig. 18.10b.

The factor $\sin \theta$ in Eq. (18.B.1) deserves special comment. It means a zero force whenever $\theta = 0$ or $\theta = 180°$, that is to say, whenever **v** and **B** are in the same (or opposite) direction.

A charged particle moving along a line of **B** experiences zero magnetic force.   (18.B.3)

Furthermore, we see that, for a given speed v, the force is a maximum when $\theta = 90°$:

At given speed, the maximum force is encountered when the motion is at right angles to the magnetic field lines.   (18.B.4)

Novelty of the Magnetic Force Law

Are there other kinds of velocity-dependent forces? The answer is yes; the frictional force on an object traveling in a fluid turns out to increase, unlike dry friction, with the object's velocity. However, fluid friction points along $-\mathbf{v}$; it is a longitudinal force. Here, in the magnetic case, **F** points at right angles to **v**; we are dealing with a *transverse force*. Frictionless planes or tracks, as well as strings with one end nailed down, provide examples of transverse forces; but they are not proportional to the velocity. Here we are indeed discovering a new type of force law.

The Vector Cross Product

The following convention is used to represent the vector **F**, with its full information of magnitude and direction:

$$\boxed{\mathbf{F} = q\mathbf{v} \times \mathbf{B}} \qquad (18.B.5)$$

This compact formula, known as a **cross product**, stands for the whole set of instructions (18.B.1), (18.B.2), together with the right-hand rule of Fig. 18.10b. As written in this form, the law is due to Oliver Heaviside (1889); we shall take it as a basic postulate of electromagnetism, that is to say, not as a consequence of anything more elementary.*

A Mathematical Comment

In Eq. (18.B.5) the notation $q\mathbf{v} \times \mathbf{B}$ designates the vector $\mathbf{v} \times \mathbf{B}$, multiplied by the number q. All the following are equivalent:

$$q\mathbf{v} \times \mathbf{B} = q(\mathbf{v} \times \mathbf{B}) = (q\mathbf{v}) \times \mathbf{B} = \mathbf{v} \times (q\mathbf{B})$$
$$(18.B.6)$$

This is what we expect of a product. However, the factors **v** and **B** are not treated equally; it matters which is written first. The reader may verify that

$$\mathbf{v} \times \mathbf{B} = -\mathbf{B} \times \mathbf{v} \qquad (18.B.7)$$

To check this, reverse the roles of **v** and **B**, pretending that **B** is the velocity and **v** the magnetic field. In a three-dimensional sketch, $\mathbf{B} \times \mathbf{v}$ will be seen to point oppositely to $\mathbf{v} \times \mathbf{B}$.

For more details about the cross product, see Appendix II.

---

* Equation (18.B.5) is worth contemplating for some time. It unifies electricity and magnetism. With some care, a definition of **B** can be extracted from the formula. Indeed, in today's physics, **B** is commonly defined via (18.B.5) rather than from a force on a magnetic pole, as in (18.A.11). Both ways are valid.

Example 18.2. A Magnetic Lift. Near the equator, a certain negative nitrogen ion, N^-, of mass $m = 2.3 \times 10^{-26}$ kg and charge $-e = -1.6 \times 10^{-19}$ C, is subjected to a northward terrestrial magnetic field of magnitude $B = 3.0 \times 10^{-5}$ tesla. The ion is traveling to the southwest at such a speed v that its weight is just offset by the magnetic force on it. Find v.

First we verify that the magnetic force points straight up. Figure 18.11 shows this to be the case.

Next we require the magnetic force to equal the gravitational force in magnitude,

$$evB \sin \theta = mg$$

$$v = \frac{mg}{eB \sin \theta} \qquad (18.B.8)$$

$$= \frac{(2.3 \times 10^{-26})(10)}{(1.6 \times 10^{-19})(3.0 \times 10^{-5})(\sin 45°)} \frac{\text{meter}}{\text{sec}}$$

$$= \underline{\underline{6.8 \times 10^{-2} \frac{\text{meter}}{\text{sec}}}}$$

approximately the walking speed of an ant; compare this with the typical thermal speed of a nitrogen ion at room temperature, which is faster than sound. At such more usual speeds, the ion's weight would be overwhelmed by the magnetic force.

Electric and Magnetic Units

The reader who wonders why the magnetic formula $\mathbf{F} = q\mathbf{v} \times \mathbf{B}$ contains no special proportionality constant in addition to the factors q, $\mathbf{v}$, $\mathbf{B}$ is referred to End Note (i) of this chapter, which explains the relation between electric and magnetic

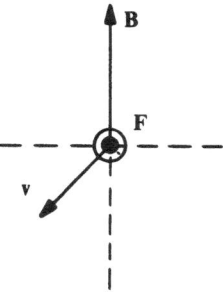

Figure 18.11. Top view of the N^- ion traveling southwest.

units in the International System; it also brings up an intriguing dimensional hint as to how electromagnetism predicts the speed of light.

ii. Particle Motion in E and B Fields

Before examining some simple trajectories in detail, we must make a few sweeping (and useful) statements about the motion of test charges.

Consider a particle of charge q that has been launched in a region of space where a magnetic field **B** exists, and possibly an electric field **E** as well. No forces other than electric and magnetic will be assumed; gravitational forces can usually be neglected in connection with ions or electrons. The electric force $q\mathbf{E}$, and the magnetic one $q\mathbf{v} \times \mathbf{B}$, give rise to a combined total force

$$\boxed{\mathbf{F}_{\text{tot}} = q\mathbf{E} + q\mathbf{v} \times \mathbf{B}} \qquad (18.B.9)$$

or, in terms of the particle's acceleration **a**,

$$m\mathbf{a} = q\mathbf{E} + q\mathbf{v} \times \mathbf{B} \qquad (18.B.10)$$

This is the particle's equation of motion, as formulated by Hendrik Lorentz in 1892.

Initial Conditions

In practice, Eq. (18.B.10) could be prohibitively complicated to deal with; but in principle, it may be solved to yield the particle's complete motion, that is to say, its position **r** as a function of time.

How much does one need to know about initial conditions? What we learned early in mechanics can be extended to the present case, and will be stated without proof:

> If **E** and **B** are known at all points of space, and if the initial position $\mathbf{r}_0$ and the initial velocity $\mathbf{v}_0$ are given, then the motion is determined at all later times. (18.B.11)

This piece of knowledge will help us predict the behavior of a charged particle in given **E** and **B** fields. All we need is to discover some motion *consistent* with the force (18.B.9). If that motion fits the assumed initial conditions, then we have our solution; it is unique.

Charge-to-Mass Ratio

If **E**, **B**, and the initial conditions are all given, the particle's motion can depend only on its charge q and mass m. Here a simplification occurs: the particle's behavior depends just on the *ratio q/m* rather than on q and m individually. To see this, we just divide both sides of (18.B.10) by m:

$$\mathbf{a} = \frac{q}{m}\,(\mathbf{E} + \mathbf{v} \times \mathbf{B}) \qquad (18.\text{B}.12)$$

Thus, the acceleration depends just on q/m, and therefore, with a fixed initial position and velocity, so does the whole motion.

Example 18.3. A "Black-Box" Mass Spectrometer. A mass spectrometer is an instrument in which a mixture of charged particles is split into individual trajectories (**beams**), according to their mass. Figure 18.12 shows such a device, whose mode of operation is unspecified except for the fact that it makes use of **E** and **B** fields. Suppose the starting mixture just contains singly ionized helium, He^+ [mass ≈ 4.00 atomic mass units (u), charge $= +e$], doubly ionized helium, He^{2+} (mass ≈ 4.00 u, charge $= +2e$), and ionized deuterium D^+ (mass ≈ 2.00 u, charge $= +e$), all practically at rest. Only two beams are seen to emerge. What is their composition?

Particles with equal charge-to-mass ratio follow identical trajectories since their initial conditions are the same. Those ratios are

$$He^+: \frac{e}{4\,u}$$

$$\left.\begin{array}{l} He^{2+}: \dfrac{2e}{4\,u} \\[2ex] D^+: \dfrac{e}{2\,u} \end{array}\right\} \text{(equal)}$$

Hence, <u>one beam contains He^+, the other a mixture of He^{2+} and D^+</u>. The apparatus ought to be called a q/m spectrometer rather than a mass spectrometer.

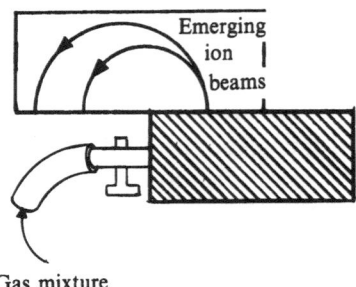

Figure 18.12. A mass spectrometer without the details; for a closer look, see Fig. 18.17 further on.

Lack of Work from B

Another simplification occurs when the magnetic force is the only one present. In this case,

> A charged particle maintains a constant speed v (pure **B** field)

$$(18.\text{B}.13)$$

although the direction of **v** may change. To prove this, we examine the force $\mathbf{F} = q\mathbf{v} \times \mathbf{B}$: (1) it is transverse (**F** perpendicular to **v**, or $\mathbf{F} \cdot \mathbf{v} = 0$), and (2) it amounts to the total force acting on the particle. The work-energy theorem of mechanics then tells us that the particle's kinetic energy $\frac{1}{2}mv^2$ changes at a rate

$$\frac{d(\frac{1}{2}mv^2)}{dt} = \mathbf{F} \cdot \mathbf{v} = 0 \qquad (18.\text{B}.14)$$

Since $\frac{1}{2}mv^2$ is constant, so is v. In summary: like all transverse forces, *the magnetic force on a charged particle does zero work.*

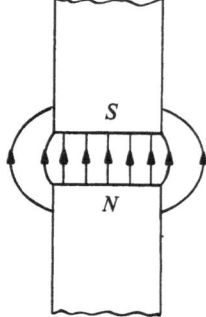

Figure 18.13. A magnet whose pole pieces are flat, parallel, and close together, often turns out to produce a **B** field that is uniform, apart from some fringing, in analogy with a parallel-plate capacitor.

We are now ready to study some special cases; *uniform fields* are assumed, that is to say, **B** (or **E**) has a constant magnitude and direction everywhere in the region where we study the particle's motion. Furthermore, we assume in this chapter that all fields are constant in time. Not only do these assumptions make the physics simple, but they can be fairly well realized in practice. A parallel-plate capacitor will do the job for the **E** field; magnets with flat pole pieces, as in Fig. 18.13, are used for the **B** field. Alternatively, current-carrying coils of the appropriate shape may be used, as the next chapter will explain.

iii. Measuring Velocities and Charge-to-Mass Ratios

Rather than opposing a magnetic force with gravity, as in Example 18.2, let us play an electric and a magnetic force against one another, as in Fig. 18.14. Suppose a beam of charged particles, all moving at velocity **v**, traverses a region where a given horizontal magnetic field **B** has been set up; a vertical electric field **E** is adjusted in strength until the beam suffers no net deflection (it emerges in the same direction as when **E** = 0, **B** = 0). This implies a zero net vertical force,

$$\mathbf{F}_{\text{electric}} = -\mathbf{F}_{\text{magnetic}} \qquad (18.B.15)$$

or

$$q\mathbf{E} = -q\mathbf{v} \times \mathbf{B} \qquad (18.B.16)$$

[The reader should verify that both sides of (18.B.16) are vertical vectors.] In magnitude, and with q factored out, we have

$$E = vB \sin 90° = vB \qquad (18.B.17)$$

Solving for v, we obtain

$$\boxed{v = \frac{E}{B}} \qquad (18.B.18)$$

independently of the particles' mass and charge, including its sign.

Application : Determining e/m for Electrons

Let the arrangement of Fig. 18.14 be set up in a cathode-ray tube, whose electron gun has an accelerating voltage V. If the crossed fields **E** and **B** are adjusted for zero beam deflection, one can express the electron's charge-to-mass ratio, e/m, in terms of V, E, and B. (This historic determination was first achieved by J. J. Thomson in 1897.) The calculation is as follows.

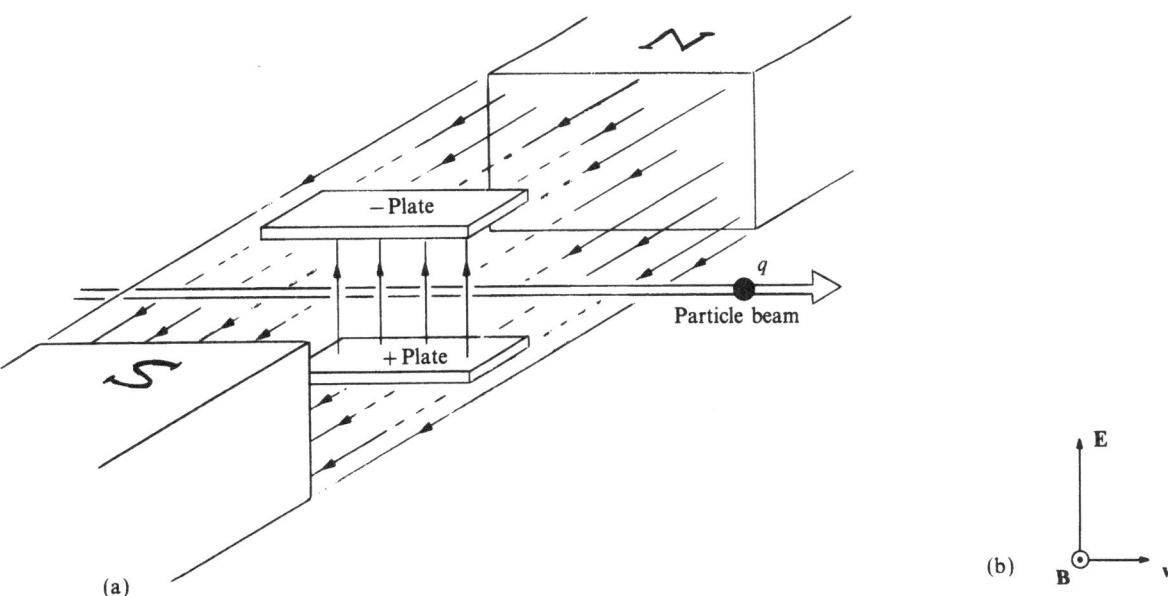

Figure 18.14. (a) Setting up crossed **E** (vertical) and **B** (horizontal) fields. (b) Front view of the fields and of the particles' velocity.

First we note the electrons' speed, given by Eq. (18.B.18). Another way of finding v is from the electron-gun data; for each electron we have

Emerging kinetic energy

= potential energy drop in the gun

or

$$\tfrac{1}{2}mv^2 = eV \qquad (18.\text{B}.19)$$

The two calculated values of v must agree; inserting (18.B.18) in (18.B.19) gives

$$\frac{1}{2} m \left(\frac{E}{B}\right)^2 = eV$$

or

$$\frac{e}{m} = \frac{1}{2V}\left(\frac{E}{B}\right)^2 \qquad (18.\text{B}.20)$$

involving only measurable quantities on the right. The numerical result turns out to be

$$\boxed{\frac{e}{m} \approx 1.76 \times 10^{11} \ \frac{\text{coulombs}}{\text{kg}}} \qquad (18.\text{B}.21)$$

(Improved modern procedures yield 1.7588028×10^{11} C/kg, making e/m one of the most accurately known fundamental constants.) Any determination of e, such as Millikan's, taken together with (18.B.21), yields the value of the electron's mass, $m \approx 9.1 \times 10^{-31}$ kg. The fact that the same result (18.B.21) is obtained under a wide range of experimental parameters argues powerfully for the particle nature of electrons.

A beam of protons (hydrogen nuclei) can be studied just as well as an electron beam. Knowledge of the ratios $e/(\text{proton mass})$ and $e/(\text{electron mass})$ then leads to a value of the proton-to-electron mass ratio:

$$\boxed{\frac{\text{Proton mass}}{\text{Electron mass}} \approx 1836} \qquad (18.\text{B}.22)$$

This demonstrates that most of an atom's mass is contained in its nucleus.

Example 18.4. The Deflecting Potential in J. J. Thomson's Experiment. In the procedure just described, suppose $B = 2.0 \times 10^{-4}$ tesla; let the deflecting capacitor have a gap width $s = 0.50$ cm; and assume an electron gun with $V = 6.0$ kvolts. Find the voltage V', which, applied between the deflecting plates, results in a null deflection of the beam.

The deflecting electric field is

$$E = \frac{V'}{s} \qquad (18.\text{B}.23)$$

Formula (18.B.20) then gives

$$\frac{e}{m} = \frac{1}{2V}\left(\frac{V'}{sB}\right)^2 \qquad (18.\text{B}.24)$$

or, solving for V',

$$V' = sB\left[(2V)\left(\frac{e}{m}\right)\right]^{1/2} \qquad (18.\text{B}.25)$$

$$= (0.50 \times 10^{-2})(2.0 \times 10^{-4})$$

$$\times [(2)(6.0 \times 10^3)(1.76 \times 10^{11})]^{1/2} \text{ volts}$$

$$= \underline{46 \text{ volts}}$$

iv. Motion in a Pure B Field

How does a charged particle behave in a uniform **B** field when no forces other than magnetic are present?

Motion Parallel to **B**

Let us launch the particle in a direction parallel to **B**.

The result will be uniform motion
along a field line.   (18.B.26)

As a proof, recall that a velocity **v** parallel to **B** implies a zero magnetic force; see statement (18.B.3). Zero force, in turn, implies uniform motion. Finally we note that the initial condition, $\mathbf{v}_0$ parallel to **B**, is satisfied.

Motion Perpendicular to B

It is far more interesting to start the particle perpendicularly to the field lines. Here *the speed,* $v = |\mathbf{v}|$, *is still constant,* see statement (18.B.13). The resulting behavior, illustrated in Fig. 18.15, is *uniform circular motion in a plane perpendicular to the field lines.* Let us now make sure that the equation of motion

$$\mathbf{a} = \frac{q}{m}\mathbf{v} \times \mathbf{B} \qquad (18.B.27)$$

is satisfied ($\mathbf{a}$ is the acceleration vector).

1. Direction of $\mathbf{a}$. The right-hand rule, applied to $\mathbf{v}$ and $\mathbf{B}$ in Fig. 18.15, predicts a centripetal direction for $\mathbf{a}$, as required.

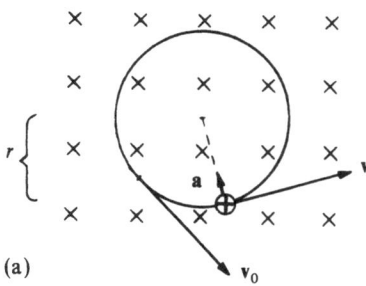

(a)

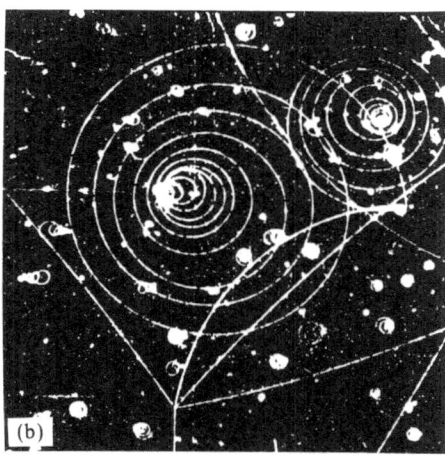

(b)

Figure 18.15. (a) Uniform circular motion of a positive particle, perpendicular to a uniform $\mathbf{B}$. The field lines, shown as crosses, point into the page. (b) Bubble-chamber trajectory of a positron (large spiral) and electron (somewhat smaller spiral) in a magnetic field. Why are we seeing spirals rather than perfect circles? The particles' speed gradually decreases, owing to collisions with the liquid medium of the chamber; as we shall see, the curvature of the track is affected by speed. Other, nearly straight, tracks were made by much faster or much more massive particles. (By permission, Nahmin Horwitz, Syracuse University.)

2. Magnitude of $\mathbf{a}$. From Eq. (18.B.27) we have

$$a = \frac{q}{m}vB\sin 90° = \frac{q}{m}vB \qquad (18.B.28)$$

This is a new piece of information, from which we shall presently extract the radius of the orbit as well as its frequency.

Finally, we see in Fig. 18.15 that the initial condition ($\mathbf{v}_0$ perpendicular to $\mathbf{B}$) is satisfied.

v. Orbital Radius r in a Uniform B

Recalling the centripetal acceleration formula of mechanics,

$$a = \frac{v^2}{r} \qquad (18.B.29)$$

(for uniform circular motion), we have, from (18.B.28),

$$\frac{v^2}{r} = \frac{q}{m}vB$$

or

$$\boxed{r = \frac{mv}{qB}} \qquad (18.B.30)$$

This result is sometimes expressed in terms of the **curvature**, $1/r$, of the trajectory; for a given charge, and in a given magnetic field, *the curvature depends inversely on the momentum*; the faster the particle, the straighter the trajectory. Two examples follow.

Example 18.5. Another Determination of e/m. An almost evacuated glass vessel contains an electron gun, which is aimed perpendicularly to a uniform $\mathbf{B}$ field; see Fig. 18.16. The (circular) electron beam is made visible by the glow of the residual gas. Express the electron's charge-to-mass ratio e/m in terms of the gun voltage V, the magnetic field B, and the orbital radius r.

The electron's speed, v, must obey the two relations

$$r = \frac{mv}{eB} \qquad (18.B.31)$$

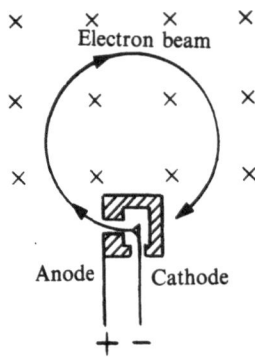

Figure 18.16. An electron gun is used to realize the situation of Fig. 18.15. Note the reverse bending of the orbit (the traveling charges are negative). The electrons are emitted to the left, and eventually strike the back of the gun.

[see Eq. (18.B.30)], and

$$\tfrac{1}{2}mv^2 = eV \qquad (18.B.32)$$

[see Eq. (18.B.19)]. Eliminating v, we obtain

$$\frac{e}{m} = \frac{2V}{r^2 B^2} \qquad (18.B.33)$$

the desired expression.

Example 18.6. A Mass Spectrometer. Figure 18.17 shows schematically the arrangement of a mass spectrometer. Ions, having gone through an accelerating voltage V, emerge from a slit and complete a semicircular orbit of radius r, perpendicular to a uniform magnetic field **B**. They eventually blacken a spot (actually a small line parallel to the slit) on a photographic film, thereby recording the size of their orbit, which in turn reveals their charge-to-mass ratio. With $V = 1.000$ kvolt and $B = 5.000 \times 10^{-2}$ tesla, how far from the slit are singly ionized atoms of the carbon isotope ^{14}C detected? [Use a mass of 14.00 atomic units $= (14.00)(1.661 \times 10^{-27}$ kg).]

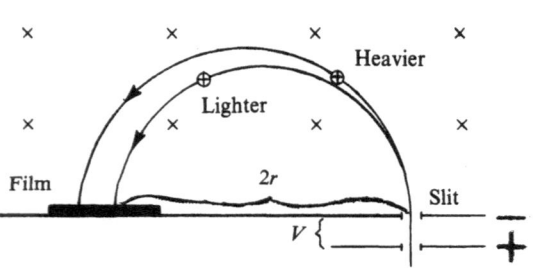

Figure 18.17. Principle of the mass spectrometer. Before entering the accelerating gap with voltage V, the ions may be considered at rest.

The orbit's radius satisfies (18.B.33), where m is now the ion's mass, rather than that of an electron; the charge is $+e$. Solving for r, we have

$$r = \frac{1}{B}\left(\frac{2Vm}{e}\right)^{1/2} \qquad (18.B.34)$$

$$= \frac{1}{5.000 \times 10^{-2}}$$

$$\times \left[\frac{(2)(1.000 \times 10^3)(14.00)(1.661 \times 10^{-27})}{1.602 \times 10^{-19}}\right]^{1/2}$$

$$\times \text{ meter}$$

$$= \underline{0.3408 \text{ meter}}$$

vi. Frequency of Revolution in a Uniform B

We just saw how the equation of motion (18.B.28) predicts the radius of a circular orbit; the same equation also yields the particle's angular velocity ω if we use the circular motion relations

$$a = \omega^2 r, \qquad v = \omega r \qquad (18.B.35)$$

Equation (18.B.28) then becomes

$$\omega^2 r = \frac{q}{m}\,\omega r B \qquad (18.B.36)$$

or

$$\boxed{\omega = \frac{qB}{m}} \qquad (18.B.37)$$

[Alternatively, we can solve (18.B.30) for v/r.] The striking feature of this result is that *it does not depend on the orbiting particle's speed, nor* (therefore) *on its orbital radius*; $\omega/2\pi$ is called the particle's **cyclotron frequency** in a field B, for reasons that will appear in the next subsection.

Caution

In the laboratory and in astronomical environments, speeds comparable to that of light,

3×10^8 meters/sec, are not rare for electrons, protons, and other ions. Our simple analysis then needs to be corrected for relativistic effects. In particular, ω [Eq. (18.B.37)] becomes speed dependent. *The formulas in this section are restricted to speeds small compared to that of light.*

Example 18.7. Two Nearly Equal Frequencies. A deuterium ion D^+ and a doubly charged helium ion He^{2+} are in concentric orbits at right angles to a magnetic field of strength $B = 0.10$ tesla; see Fig. 18.18. The D^+ particle has charge $+e$ and mass $m_D = 2.0135$ atomic mass units (u), while the He^{2+} has charge $+2e$ and mass $m_{He} = 4.0015$ u. (Use $1\,u \approx 1.66 \times 10^{-27}$ kg.) (a) Find these particles' frequencies of revolution. (You may round off the masses for this purpose.) (b) How many times per second does the He^{2+} overtake the D^+?

a. Frequency: The rounded charge-to-mass ratios are equal; recall Example 18.3. The (almost) common frequency of revolution, f, is

$$f = \frac{\omega}{2\pi} = \left(\frac{1}{2\pi}\right)\left(\frac{q}{m}\right) B \qquad (18.B.38)$$

by Eq. (18.B.37). Thus, we have

$$f = \frac{1}{2\pi}\left(\frac{e}{2.0\ u}\right) B \qquad (18.B.39)$$

$$= \frac{1}{2\pi}\left[\frac{1.6 \times 10^{-19}}{(2.0)(1.66 \times 10^{-27})}\right](0.10)\ \frac{rev}{sec}$$

$$= 7.7 \times 10^5\ \frac{rev}{sec} \qquad (18.B.40)$$

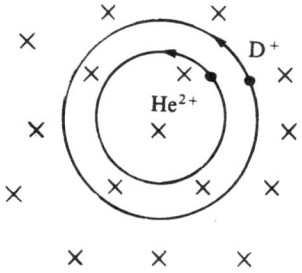

b. Passing frequency: Since $(q/m)_{He}$ is somewhat larger than $(q/m)_D$, the frequency of revolution is somewhat larger for He^{2+} than for D^+, and He^{2+} regularly overtakes D^+. Projecting the two circular motions on a single line, as was done in Sec. B-i of Chap. 13 in order to describe harmonic motion, we see that "overtaking" means that both harmonic motions are in phase. The required frequency is therefore the **beat frequency** f_{beat}, given by formula (13.C.3) of Chap. 13. Here

$$f_{beat} = f_{He} - f_D \qquad (18.B.41)$$

$$= \frac{1}{2\pi}\left(\frac{q}{m}\right)_{He} B - \frac{1}{2\pi}\left(\frac{q}{m}\right)_D B$$

$$= \frac{1}{2\pi}\left(\frac{2e}{4.0015\ u}\right) B - \frac{1}{2\pi}\left(\frac{e}{2.0135\ u}\right) B$$

$$= \left[\frac{eB}{(2\pi)(1\ u)}\right]\left(\frac{2}{4.0015} - \frac{1}{2.0135}\right) \quad (18.B.42)$$

The factor in square brackets happens to be 2 times expression (18.B.39). Hence we obtain

$$f_{beat} = (2)(7.7 \times 10^5)(3.16 \times 10^{-3})\ sec^{-1}$$

$$= 4.9 \times 10^3\ sec^{-1}$$

Overtaking occurs 4900 times per second.

vii. Some Further Illustrations

The Cyclotron

This celebrated instrument was developed in 1931 by Ernest O. Lawrence and M. Stanley Livingston. Its function is to launch a beam of protons, which are meant to collide with target nuclei; the protons' kinetic energy must be made high enough to overcome the Coulomb repulsion, which tends to keep the projectile and the target apart. The proton–nucleus encounter can result in a so-called **nuclear reaction**. The cyclotron takes advantage of two features just discussed—the circular orbits in a uniform **B** field, and their constant frequency of revolution. Figure 18.19 shows how the protons, released nearly from rest, are accelerated in many small stages and spiral outwards until they smash into a stationary target made of any chosen material. Protons can acquire

Figure 18.18. Two different ions in a uniform **B** field. Their velocities are very different, but their angular velocities are almost identical in this special case.

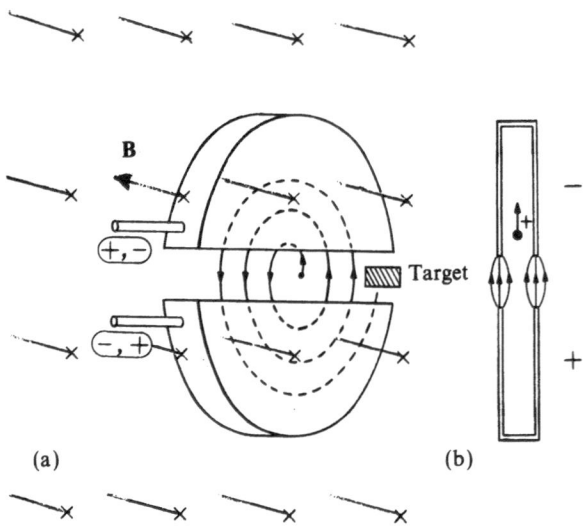

(a)   (b)

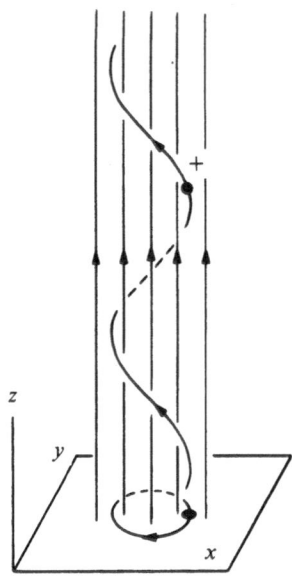

Figure 18.19. (a) Principle of the cyclotron. A flat cylindrical box is cut along a diameter to form a pair of Ds, or **dees**. A potential difference exists between them, and is made to alternate in sign at a constant frequency f. Meanwhile, a steady uniform **B** field is maintained in a direction perpendicular to the plane of the dees. A proton, released between the dees, is attracted inside the negative one, where it becomes shielded from all electric fields—the dee acts as a Faraday cage. The proton always feels the **B** field, however, and describes, inside the dee, a semicircular trajectory. The basic idea is to match the frequency of the potential difference to that of the orbit, $f = eB/2\pi m$. The proton finds the potential difference reversed when it emerges from the dee, and is therefore accelerated once more as it crosses the gap; every crossing increases its kinetic energy in this way. Hence the semicircular paths gradually increase in radius. New protons are being fed continuously at the center, resulting in a high-energy beam at the periphery. (b) Cross section through the dees, showing the electric field in the gap, and a passing proton.

up to about 20 MeV of kinetic energy in this manner. More advanced machines, which take relativistic effects into account, are able to reach much higher energies.

Helical Orbits

We have seen what happens to a charged particle injected in a direction parallel or perpendicular to a uniform **B**. We now describe, without calculations, the case of a general angle between **B** and **v**. The orbit is then a **helix**, shown in Fig. 18.20. As one might expect, it combines features of both the parallel and perpendicular cases.

1. In the figure, the z motion is uniform (the z component of the velocity is constant, $v_z = \text{const}$).

Figure 18.20. Helical trajectory of a positive particle in a uniform **B** field.

2. As projected on a plane perpendicular to **B**, the motion is uniform circular, *with the same angular velocity ω* as given by Eq. (18.B.37).

Magnetic Bottles

Real-life magnetic fields are seldom uniform. Orbits in a *nonuniform* **B** are illustrated in Fig. 18.21. They loop themselves around a "tube" of field lines, i.e., *each turn of the orbit revisits the same field lines as the previous turn*. (While this description is only an approximation, it is in most cases a very good one.) Furthermore, where the orbit reaches a sufficiently strong **B** field, it migrates back. In part (b) of the figure, the particle is moving back and forth along a confined orbit; it is trapped in a "magnetic bottle." [Part (c) shows a type of magnetic bottle where the orbit's migration is not reversed.] Research in controlled nuclear fusion utilizes magnetic bottles to confine hot ionized gases (plasmas). If a material wall were used instead, its contact would cool the plasma to unacceptably low temperatures.

A magnetic bottle on a grand scale is provided by the Earth's **B** field high above the atmosphere; see Fig. 18.22. High-speed protons and electrons, whose direct or indirect origin lies in outer space, spiral as shown in the figure, and cause auroras to appear where they leak into the upper atmosphere.

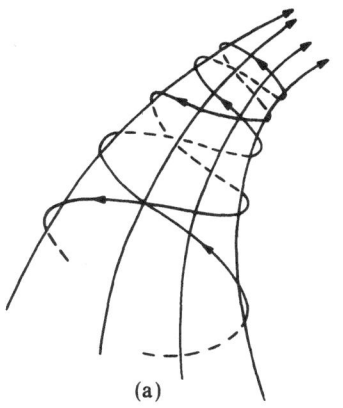 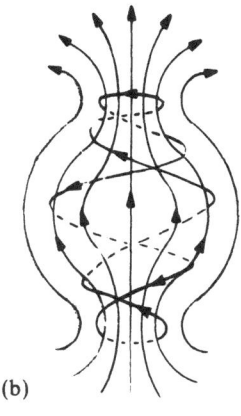 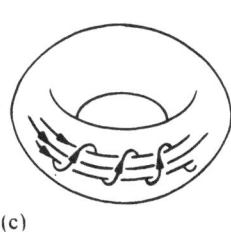

Figure 18.21. (a) Trajectory in a nonuniform **B** field. (Wrapping a tape around a cone will help visualize the effect.) (b) A magnetic bottle with two turning points. (c) A magnetic bottle without turning points.

Magnetic Lenses

In an electron microscope, the specimen is "illuminated" by a beam of electrons rather than by a light beam. Electron beams are focused as they pass through the **B** fields of current-carrying coils, much as light beams are focused by lenses. How a coil produces a magnetic field is explained in the next chapter.

C. Magnetic Force on a Current

It is far easier to set up and control currents in wires than particle beams in evacuated tubes. In both cases we are dealing with moving charges, and magnetic forces are observed. Here we learn to predict the force exerted by a given field on a given current.

The demonstration in Fig. 18.23 shows that the magnetic force on a current is actually transmitted to the conductor that carries the current, even when the latter is not confined to a wire. The magnetic forces discussed in this section are, accordingly, *forces on conducting materials.*

i. Current Elements

Currents in wires consist of electrons that move largely at random. We shall need to idealize that situation in two ways. First, we consider the sign of the moving charges. The magnetic force, $q\mathbf{v} \times \mathbf{B}$ on

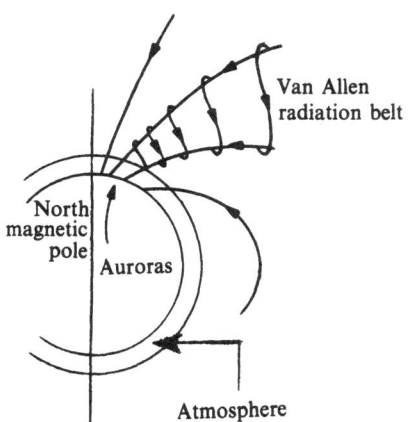

Figure 18.22. Motion of a proton in the Earth's magnetic field. *Regions exist where many high-energy protons or electrons are trapped; they form the "radiation belts" discovered by James Van Allen through the use of artificial satellites.*

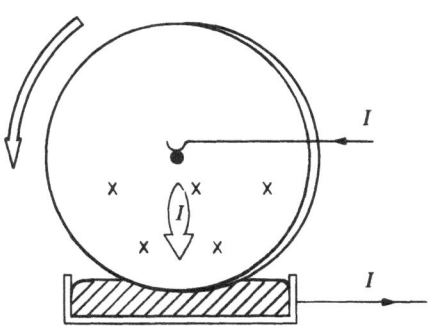

Figure 18.23. A copper wheel conducts a current between an axle contact and a rim (mercury) contact. A magnetic field (lines into the page, and indicated by crosses) produces a transverse force and makes the wheel turn. This so-called **homopolar motor** is due to Peter Barlow (1823); present-day electric motors are usually of a very different design.

a single particle, changes sign if q does, $q \to -q$, and also if $\mathbf{v}$ does, $\mathbf{v} \to -\mathbf{v}$. Therefore a positive charge, q, moving in a direction $\mathbf{v}$, experiences the same magnetic force as a negative charge, $-q$, moving in the opposite direction, $-\mathbf{v}$. In what follows we accordingly continue to *represent a current of (negative) electrons by an opposite current of positive charges*; this is common practice.

A second idealization consists of pretending that all charges move at the same velocity $\mathbf{v}$ (equal to the **average** or **drift velocity**) in the direction of the wire. This is so unrealistic that some justification is needed; the concerned reader is referred to Note (ii) at the end of the chapter.

We now consider the situation of Fig. 18.24, where charges q succeed one another at distance intervals l and time intervals t along the wire, like marbles in a pipe. The resulting current I is related to q by

$$q = It \qquad (18.C.1)$$

and the charges' speed is

$$v = \frac{l}{t} \qquad (18.C.2)$$

We are interested in the force on the wire when it is placed in a uniform $\mathbf{B}$ field. In particular, the force $\mathbf{F}$ on a single segment l amounts to the force on a single charge q. We have

$$\mathbf{F} = q\mathbf{v} \times \mathbf{B} \qquad (18.C.3)$$

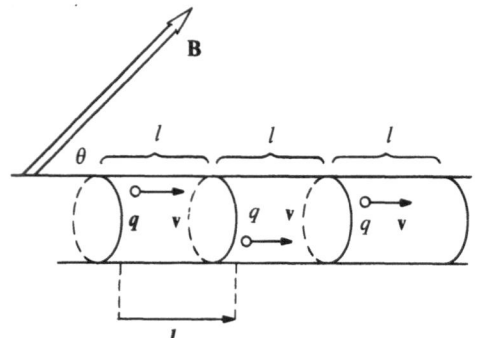

Figure 18.24. Charges moving in a wire. The vector l serves to indicate the length and direction of a wire segment; l points along $+\mathbf{v}$. The $\mathbf{B}$ field makes an arbitrary angle with the wire.

With (18.C.1) and (18.C.2), the magnitude qv in (18.C.3) can be rewritten

$$qv = (It)\left(\frac{l}{t}\right) = Il \qquad (18.C.4)$$

Using vector notation to incorporate the direction of the segment l, as shown in Fig. 18.24, we have

$$\boxed{q\mathbf{v} = I\mathbf{l}} \qquad (18.C.5)$$

and (18.C.3) becomes

$$\boxed{\mathbf{F} = I\mathbf{l} \times \mathbf{B}} \qquad (18.C.6)$$

the magnetic force on the wire element l. This cross product yields a magnitude

$$F = IlB \sin \theta \qquad (18.C.7)$$

where θ is the angle between the current element and the $\mathbf{B}$ field; in the right-hand rule, the thumb points in the direction of the current.

Longer Wire Segments

Formula (18.C.6) is valid, not only for an infinitesimal element of wire, but also for any straight wire segment, of length and direction l, placed in a uniform $\mathbf{B}$ field. We leave the proof of this statement to the reader. (Hint: break up the segment into many small elements, and add all forces.)

Example 18.8. Lifting a Conducting Rod. A small metal rod, of mass $m = 1.0$ gram, is laid across a pair of parallel conducting bars which themselves are lying on a horizontal table, as shown in Fig. 18.25. The parallel bars, which are a distance $s = 1.0$ cm apart, are connected to a DC supply, which sends a current I through the small rod. A field of 0.05 T is maintained in a direction parallel to the bars. If the angle between rod and bars is 30°, how much current is needed to lift the rod (and temporarily break the circuit)?

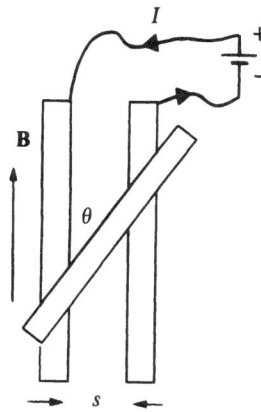

Figure 18.25. Top view of a small rod lying on two bars and carrying a current I.

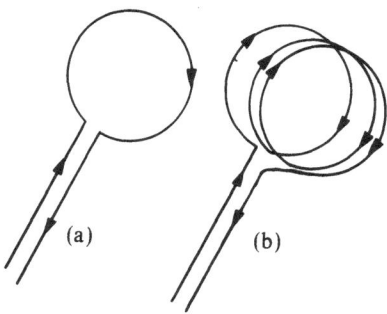

Figure 18.26. (a) A loop of current. The two leads contribute zero magnetic force and torque because of canceling direction and nearly identical location. (b) A coil, i.e., several loops in series. An N-turn coil with current I amounts to a single loop with current NI.

The right-hand rule gives an upward magnetic force **F** when I is directed as shown in the figure. Balancing against gravity means

$$F = mg \qquad (18.C.8)$$

or, from Eq. (18.C.7),

$$IlB \sin \theta = mg \qquad (18.C.9)$$

where l is the length of rod that carries a current. From the figure we see that $l = s/\sin \theta$, and therefore (18.C.9) becomes

$$IsB = mg$$

so that

$$I = \frac{mg}{sB} = \frac{(1.0 \times 10^{-3})(10)}{(1.0 \times 10^{-2})(0.05)} \text{ A} = \underline{\underline{20 \text{ A}}}$$

Note the irrelevance of $\theta = 30°$.

ii. Current Loops

Magnetic forces are of particular interest when acting on coils or loops; see Fig. 18.26. Such objects serve a huge variety of purposes; we may list some of them as follows:

1. Measuring electric currents, for example in a device called the **galvanometer**.
2. Generating magnetic torques, as in **electric motors**.
3. Generating magnetic fields; this is one of the next chapter's topics.

4. Generating currents and voltages, as in a **dynamo**; this topic is left for Chap. 20 (Magnetic Induction).
5. Most fundamentally, the behavior of coils provides the clue needed to understand the existence of magnets.

Rectangular Loop in a Uniform **B**

Let us place a rectangular loop of wire, with sides a and b, in a uniform horizontal **B** field, as illustrated in Fig. 18.27; side a is horizontal and perpendicular to **B**, while b makes an angle α with the vertical. Assuming a current I in the loop, we

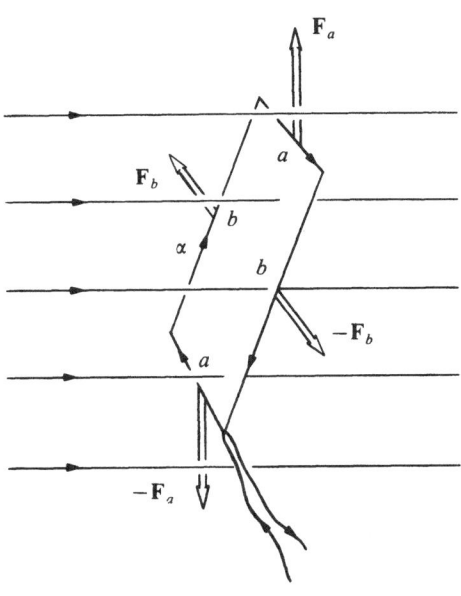

Figure 18.27. Rectangular current loop in a uniform magnetic field.

now calculate, first, the total magnetic force, and second, the torque about the lower side a.

Force

Figure 18.27 shows the force acting on each side of the rectangle, as obtained from the right-hand rule.

The forces $\pm \mathbf{F}_a$ on sides a are vertical, while the forces $\pm \mathbf{F}_b$ on sides b are parallel to side a. By symmetry, we must consider each force $\pm \mathbf{F}_a$, $\pm \mathbf{F}_b$ to be acting at the center of its respective segment; this is shown in the figure. The total force is

$$\mathbf{F} = \mathbf{F}_a - \mathbf{F}_a + \mathbf{F}_b - \mathbf{F}_b = 0 \quad (18.C.10)$$

The loop experiences a zero net force when in a uniform $\mathbf{B}$ field.

Torque

A side view corresponding to Fig. 18.27 is shown in Fig. 18.28. The forces $\pm \mathbf{F}_b$ cancel one another because they have a common line of action; they are omitted from the figure. The forces $\pm \mathbf{F}_a$ constitute a couple (whose torque is the same about any axis parallel to sides a). Calculating with respect to the lower side a, we find a counterclockwise torque whose single contribution is

$$\tau = (\text{force})(\text{moment arm}) \quad (18.C.11)$$

$$= (F_a)(b \sin \alpha) \quad (18.C.12)$$

$$= (IaB \sin 90°)(b \sin \alpha) \quad (18.C.13)$$

where the last step makes use of (18.C.7). Thus we obtain

$$\tau = IabB \sin \alpha \quad (18.C.14)$$

or

$$\boxed{\tau = (I)(\text{area of loop})(B \sin \alpha)} \quad (18.C.15)$$

Here let us state, without proof, that *this formula is valid for any flat loop—not just a rectangle.*

iii. A Current Loop as a Magnetic Dipole

If the above reminds us of the magnetic dipole (Sec. A-iv), it is no coincidence. The construction given in Fig. 18.29 shows that *the loop simulates a dipole.* The latter is to be drawn perpendicularly through the plane of the loop, in such a way that the current, viewed from the S pole, circulates clockwise. This fictitious dipole has a length l and a pole strength p such that its magnetic moment, pl, is given by

$$\boxed{\text{Magnetic moment} = pl = (\text{area of loop})(I)}$$

$$(18.C.16)$$

(The individual factors p, l may be varied at will as long as their product is given by the above.)

In support of this claim ("a loop is equivalent to a dipole") we examine Eq. (18.C.15). It now reads

$$\tau = (\text{magnetic moment})(B \sin \alpha) \quad (18.C.17)$$

If our fictitious dipole were real, this is precisely the torque it would experience; recall (18.A.22). The reader should verify that the geometry is exactly that of Fig. 18.9.

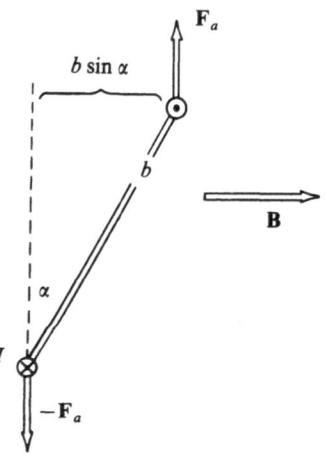

Figure 18.28. Side view of Fig. 18.27.

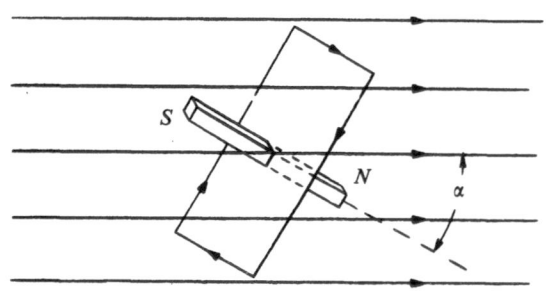

Figure 18.29. Magnetic dipole, simulating the loop of Fig. 18.27.

To Summarize :

As far as force and torque are concerned, a current loop, in a uniform **B**, is indistinguishable from a magnetic dipole perpendicular to the plane of the loop, and with a magnetic moment given by (18.C.16). The quantity (area)(I) is, in fact, known as *the loop's* magnetic moment.

We are now in possession of a first hint that actual magnets are, in fact, nothing but "black boxes" that contain hidden loops of current. This view will be confirmed and discussed in the next chapter.

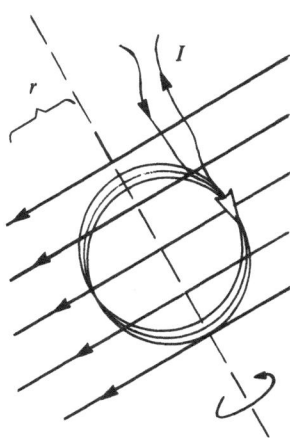

Figure 18.30. Torque on a coil in the Earth's **B** field.

iv. The Galvanometer

This sensitive and accurate measuring device is due, in its modern form, to Arsène d'Arsonval (1882); it is illustrated in Fig. 18.31. A permanent magnetic field **B** exerts a torque τ on a rectangular coil of wire, which is suspended from a fiber in the

Example 18.9. A "Compass Coil." Figure 18.30 shows a circular 10-turn coil, of radius $r = 0.10$ meter, and carrying a current $I = 2.0$ A; the coil is pivoted about one of its diameters. (a) What is the torque τ on the coil when its plane is lined up with a terrestrial field $B = 6.0 \times 10^{-5}$ tesla? (b) A compass needle is built so as to experience, when perpendicular to **B**, the same torque as the coil in (a). If the needle's length, l, equals the diameter of the coil, what should be the pole strength p?

a. Torque: The torque on an N-turn coil is N times the torque on a single loop. With this factor, we obtain for (18.C.17) and (18.C.16) a magnitude

$$\tau = (N)(\text{area})(I)(B \sin 90°) = N\pi r^2 I B \quad (18.C.18)$$

$$= (10)(\pi)(0.10)^2 (2.0)(6.0 \times 10^{-5}) \text{ newton meter}$$

$$= 3.7 \times 10^{-5} \text{ newton meter} \quad (18.C.19)$$

b. Pole strength of the equivalent needle: The magnetic moments of the N combined loops and of the needle must be equal:

$$pl = (N)(\text{area})(I) \quad (18.C.20)$$

or, with $l = 2r$,

$$p = \frac{(N)(\pi r^2)(I)}{2r} = \frac{\pi}{2} NrI \quad (18.C.21)$$

$$= \frac{\pi}{2} (10)(0.10)(2.0) \text{ SI units}$$

$$= \underline{3.14 \text{ SI units}}$$

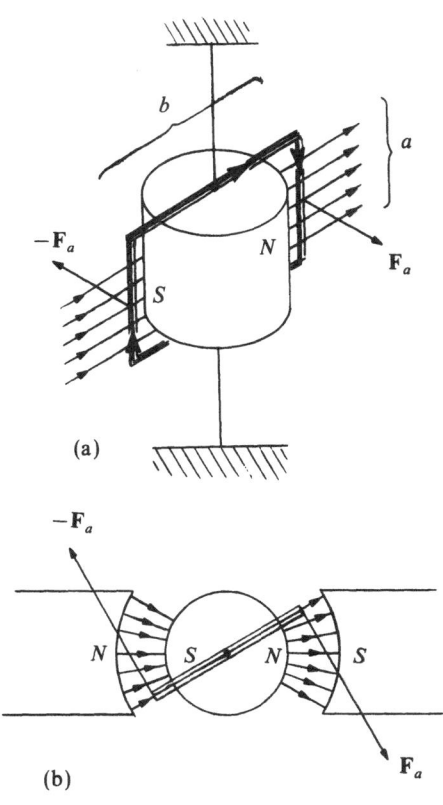

(a)

(b)

Figure 18.31. (a) The galvanometer's rectangular coil, with sides a and b, pivots around a fixed cylindrical magnet. (b) Top view of (a), showing the external pole pieces needed to produce radial field lines.

manner of Coulomb's or Cavendish's torsion balance. The restoring torque τ', due to the fiber, exactly balances τ when equilibrium is reached:

$$\tau + \tau' = 0 \qquad (18.C.22)$$

From the figure we see that

$$\tau = (N)(ab)(I)(B \sin 90°) \qquad (18.C.23)$$
$$= (\text{const})I \qquad (18.C.24)$$

The angular form of Hooke's law reads here

$$\tau' = -\kappa\theta = (\text{const})\theta \qquad (18.C.25)$$

for some angular spring constant κ. Equation (18.C.22) then ensures that

$$\boxed{\theta = (\text{const})I} \qquad (18.C.26)$$

The equilibrium deflection is proportional to the current in the coil. That deflection is often measured by optical means, as we described in connection with Cavendish's gravitational experiment.

The proportionality constant in (18.C.26) could, in principle, be determined from an accurate measurement of a, b, B, and κ, but in practice the instrument is simply calibrated with a known current. That calibrating current must have been separately measured by absolute methods.

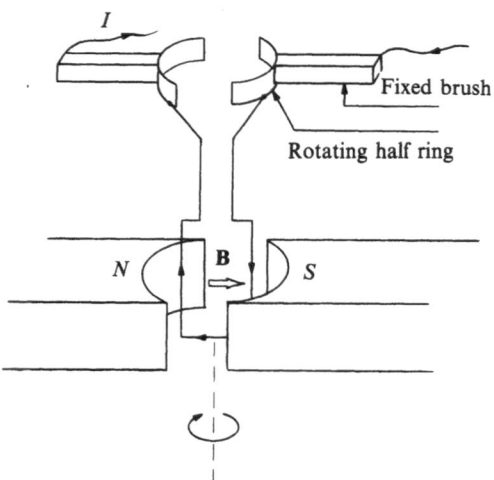

Figure 18.32. Principle of the dc motor. In practice, the rotating coil has many turns and is wound on a soft-iron core, which turns out to enhance the magnetic field. This motor was developed in 1831 by Joseph Henry.

With minor changes aiming at ruggedness and portability, the galvanometer as described here forms the heart of most ammeters and voltmeters.

v. The dc Motor

The galvanometer is almost a motor; a slightly different construction permits the coil to continue its rotation indefinitely. The basic scheme of a motor is shown in Fig. 18.32. The rotating coil is mounted on an ordinary axle, and a special switching arrangement, the **commutator**, ensures that the current in the coil is reversed at every half turn; this prevents the torque from reversing its direction.

Notes

i. Relations between Electric and Magnetic Units

The force formula $\mathbf{F} = q\mathbf{v} \times \mathbf{B}$ involves a "proportionality constant" which is exactly unity:

$$\mathbf{F} = (1)\, q\mathbf{v} \times \mathbf{B} \qquad (18.N.1)$$

Is this luck or good management? For comparison, we recall the electric and magnetic forms of Coulomb's law,

$$\mathbf{F}_{\text{elec}} = K \frac{Qq}{r^2} \mathbf{u} \qquad (18.N.2)$$

$$\mathbf{F}_{\text{mag}} = K' \frac{Pp}{r^2} \mathbf{u} \qquad (18.N.3)$$

with proportionality constants

$$\begin{aligned} K &\approx 9.00 \times 10^9 \text{ (SI)} \\ K' &= 10^{-7} \text{ (SI)} \end{aligned} \qquad (18.N.4)$$

The answer is "good management," and the logic is as follows:

1. $K' = 10^{-7}$ in (18.N.3) is truly arbitrary; it fixes the size of the unit pole and of the tesla.
2. In (18.N.1), where the unit of $\mathbf{B}$ is now fixed, the choice of 1 as a proportionality constant fixes the size of the coulomb.
3. The constant K in (18.N.2) can no longer be adjusted, because the units of q and Q are already fixed; the units of F and r have been assumed all along to be those of the metric system. In order for (18.N.2) to account numerically for Coulomb's law as observed, K is then constrained to be $K \approx 9.00 \times 10^9$ SI units.

The Constant K and the Speed of Light

In Chapter 22 (Electromagnetic Waves) we shall learn that K is intimately connected with c, the speed of light in vacuum, through a famous relation,

$$\boxed{\left(\frac{K}{K'}\right)^{1/2} = c} \qquad (18.\text{N}.5)$$

In rough numerical form, this reads

$$\left(\frac{9 \times 10^9}{10^{-7}}\right)^{1/2} \frac{\text{meters}}{\text{sec}} = 3 \times 10^8 \frac{\text{meters}}{\text{sec}} \qquad (18.\text{N}.6)$$

Let us check that, indeed, $(K/K')^{1/2}$ *has the dimensions of a speed*; the argument is in three steps.

1. Dimensional Relation between E and B.
The total force on a charge q moving in a combined $\mathbf{E}$ and $\mathbf{B}$ field is

$$\mathbf{F}_{\text{tot}} = q\mathbf{E} + q\mathbf{v} \times \mathbf{B} \qquad (18.\text{N}.7)$$

Both terms must have the same dimensions; in terms of units,

$$E \sim (\text{speed})(B)$$

or

$$\boxed{\frac{E}{B} \sim \text{speed}} \qquad (18.\text{N}.8)$$

2. Dimensional Relation between Charge and Pole Strength.
The forces on a test charge q and on a test pole p are

$$F_{\text{elec}} = qE, \qquad F_{\text{mag}} = pB \qquad (18.\text{N}.9)$$

Their ratio (force/force ~ 1) gives

$$1 \sim \frac{qE}{pB}$$

or, with (18.N.8),

$$\boxed{\frac{p}{q} \sim \text{speed}} \qquad (18.\text{N}.10)$$

[Does this agree with Eq. (18.A.24)?]

3. Dimensional Relation between K and K'.
Comparing in an analogous way Coulomb's laws (18.N.2) and (18.N.3), we obtain

$$1 \sim \frac{K}{K'} \frac{q^2}{p^2} \qquad (18.\text{N}.11)$$

(Here we have used the fact that q and Q are expressed in the same units, and similarly for p and P.) This yields

$$\frac{K}{K'} \sim \frac{p^2}{q^2} \sim (\text{speed})^2 \qquad (18.\text{N}.12)$$

with the help of (18.N.10). Therefore

$$\left(\frac{K}{K'}\right)^{1/2} \sim \text{speed} \qquad (18.\text{N}.13)$$

as announced.

ii. Average Magnetic Force on Conduction Electrons

In Sec. C-i we calculated the magnetic force on a current by pretending that all electrons have the same velocity $\mathbf{v}$, their drift velocity along the wire. This gives the correct result because we observe only the *aggregate* force $\mathbf{F}$ on all the electrons in a wire segment. The details of the argument are as follows.

In terms of the individual electrons (No. 1, No. 2, etc.), we have

$$\mathbf{F} = \mathbf{F}_1 + \mathbf{F}_2 + \cdots \qquad (18.\text{N}.14)$$

$$= -e\mathbf{v}_1 \times \mathbf{B} - e\mathbf{v}_2 \times \mathbf{B} - \cdots \qquad (18.\text{N}.15)$$

If the segment contains $\mathcal{N}$ electrons, we rewrite the above as

$$\mathbf{F} = -\mathcal{N}e\left(\frac{\mathbf{v}_1 + \mathbf{v}_2 + \cdots}{\mathcal{N}}\right) \times \mathbf{B} \qquad (18.\text{N}.16)$$

$$= (\text{total charge})(\textbf{average velocity}) \times \mathbf{B} \qquad (18.\text{N}.17)$$

the desired result. [We have relied on the **distributive property**

$$\mathbf{v}_1 \times \mathbf{B} + \mathbf{v}_2 \times \mathbf{B} + \cdots = (\mathbf{v}_1 + \mathbf{v}_2 + \cdots) \times \mathbf{B} \qquad (18.\text{N}.18)$$

See also Appendix II.]

Condensed Checklist

Coulomb's magnetic law:

$$\boxed{\mathbf{F} = K'\frac{Pp}{r^2}\mathbf{u}} \qquad (18.\text{A}.1)$$

$$K' = 10^{-7} \text{ (SI units)} \qquad (18.\text{A}.2)$$

Magnetic field:

$$\boxed{B = \frac{F}{p}} \qquad \text{(18.A.11)}$$

$$B = \frac{\mathcal{N}}{\mathcal{A}_\perp} \qquad \text{(line density)} \qquad \text{(18.A.13)}$$

$$B = K' \frac{P}{r^2} \mathbf{u} \qquad \text{(from fictitious single pole } P\text{)} \qquad \text{(18.A.18)}$$

$$B = \mathbf{B}_1 + \mathbf{B}_2 + \cdots \qquad \text{(vector sum of contributions)} \qquad \text{Fig. 18.6}$$

Units:

$$1 \text{ line} = 1 \text{ weber} \qquad \text{(18.A.16)}$$

$$1 \text{ tesla} = 1\,\frac{\text{weber}}{\text{meter}^2} = 1\,\frac{\text{newton}}{\text{SI unit pole}} \qquad \text{(18.A.15), (18.A.14)}$$

$$1 \text{ tesla} = 10^4 \text{ gauss} \qquad \text{(18.A.17)}$$

Dipole:

$$\text{Magnetic moment} = pl \qquad \text{(18.A.21)}$$

$$\text{Torque} = (\text{magnetic moment})(B \sin \alpha) \qquad \text{(18.A.22), (18.C.17), (18.C.15)}$$

Force on a moving charge:

$$\boxed{F = q\mathbf{v} \times \mathbf{B}} \qquad \text{(18.B.5)}$$

$$F = qvB \sin \theta \qquad \text{(18.B.1)}$$

$$\text{Right-hand rule} \qquad \text{Fig. 18.10}$$

Trajectory: depends on $\mathbf{E}$, $\mathbf{B}$, q/m, initial position, and initial $\mathbf{v}$.   Sec. B ii

$$\mathbf{a} = \frac{q}{m}(\mathbf{E} + \mathbf{v} \times \mathbf{B}) \qquad \text{(Lorentz's equation of motion)} \qquad \text{(18.B.12)}$$

$$\text{In pure } \mathbf{B}: \quad |\mathbf{v}| = \text{const} \qquad \text{(18.B.13)}$$

Electron data:

$$\frac{e}{m} \approx 1.76 \times 10^{11} \frac{\text{coulombs}}{\text{kg}} \qquad \text{(18.B.21)}$$

$$\frac{\text{proton mass}}{\text{electron mass}} \approx 1836 \qquad \text{(18.B.22)}$$

Mutually perpendicular E, B, v:

$$\text{No deflection means } v = \frac{E}{B} \qquad \text{(18.B.18)}$$

Circular motion (pure $\mathbf{B}$):

$$r = \frac{mv}{qB} \qquad \text{(18.B.30)}$$

$$\omega = \frac{qB}{m} \qquad \text{(18.B.37)}$$

Current element (moving charges):

$$q\mathbf{v} = I\mathbf{l} \qquad \text{(18.C.5)}$$

Force on a current element:

$$\boxed{F = I\mathbf{l} \times \mathbf{B}} \qquad \text{(18.C.6)}$$

$$F = IlB \sin \theta \qquad \text{(18.C.7)}$$

Flat coil or loop (generalized to N turns):

$$\text{Magnetic moment} = (N)(\text{area})(I) \qquad \text{(18.C.16)}$$

$$\text{Direction of equivalent dipole} \qquad \text{Fig. 18.29}$$

$$\text{Torque} = (\text{magnetic moment})(B \sin \alpha) \qquad \text{(18.C.17), (18.C.15), (18.A.22)}$$

Galvanometer:

$$\theta \propto I \qquad \text{(18.C.26)}$$

True or False

1. The magnetic force on a magnetic pole obeys the right-hand rule.

2. At a given distance, the force between two SI units of magnetic pole equals the force between two SI units of electric charge.

3. "Weber" is synonymous with "tesla/meter^2."

4. At the Earth's south pole, the $\mathbf{B}$ field points almost vertically up.

5. Since a dipole tends to line itself up *with* the field, the torque on it must be greatest when it is lined up *against* the field.

6. By suitably combining two identical magnetic dipoles, each of magnetic moment pl, one can construct a dipole moment $2pl$.

7. The torque *vector* on a magnetic dipole is perpendicular to **B** and to the dipole itself.

8. The **B** field propels an electric charge along the magnetic field lines.

9. Electric charges are constrained to move perpendicularly to the magnetic field lines.

10. Electric·charges always accelerate perpendicularly to the magnetic field lines.

11. Item 10 is true under purely magnetic forces.

12. Under a purely magnetic force, and when viewed along the direction of **B**, the trajectory of an electron loops counterclockwise around the field lines.

13. If, in a given uniform **B** field, and in the absence of nonmagnetic forces, an electron has a constant velocity **v**, then its motion must be parallel to **B**.

14. The magnetic force on a *negative* charge is described by the rule of Fig. 18.10b, with the left hand substituted for the right.

15. The existence of undeflected trajectories in crossed **E** and **B** fields demonstrates that a **B** field can be completely cancelled by a suitable **E** field.

16. An electron beam, split in two by a given set of crossed **E**, **B** fields, must consist of electrons with two different initial kinetic energies.

17. In Fig. 18.23, the direction of rotation is obtainable from the right-hand rule.

18. If, in Fig. 18.23, the direction of I were reversed, with **B** still pointing into the page, the wheel would still rotate in the direction shown.

19. Torques and forces, in a uniform **B**, cannot distinguish a horizontal loop (with the current clockwise as seen from above), from a bar magnet with its N pole up.

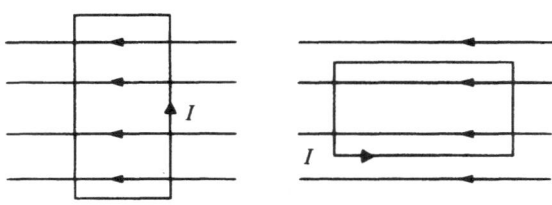

20. A flat current loop, whose plane is lined up with a uniform **B** field, experiences the same (nonzero) torque, about a vertical axis, in both orientations shown.

Problems

Section A : Magnetic Poles and Fields

*18.1. One end of a magnetized needle is an N pole of strength 0.25 SI units. In what field strength does it experience a magnetic force of 0.50 newton?

18.2. A magnetic field of 3.0 teslas exerts a force of 0.18 newton on one end of a certain compass needle. What is its pole strength?

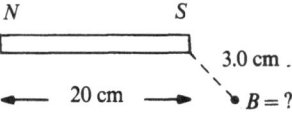

*18.3. What is the field strength B at 3.0 cm distance from the S pole of a 20-cm-long bar magnet whose pole strength is 2.0 SI units? (See the figure; neglect contributions from the N pole.)

18.4. Verify that, in Fig. 18.3, the total x component of force on each magnet vanishes.

18.5. From Coulomb's magnetic law, Eq. (18.A.1), determine the SI units of K' in terms of SI pole units, kilograms, meters, and seconds. (See also Problem 18.44.)

*18.6. A uniform **B** field has lines that intersect a plane surface of area 0.01 m^2 at an angle of 45°. If $B = 0.5$ tesla, how many lines intersect the surface? (A fractional result shows that "line counting" must not be taken literally.)

*18.7. A compass needle of length 2.0 cm experiences a torque of 1.0×10^{-7} newton meter when at right angles to a terrestrial field of 0.50×10^{-4} tesla. What is the needle's pole strength?

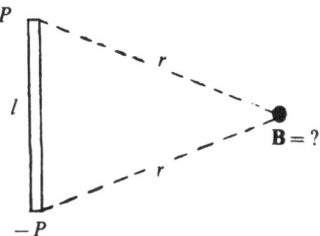

*18.8. Find, in magnitude and direction, the **B** field at a point equally distant (distance $= r$) from both poles of a bar magnet whose length is l, as shown. (Express your answer in terms of l, r, and the pole strength P; take $P > 0$.)

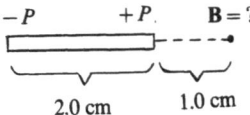

*18.9. A certain bar magnet has a length $l = 2.0$ cm and a pole strength $P = 8.0$ SI units. Find the field **B** at a point $r = 1.0$ cm beyond the positive pole along the magnet's direction, as shown.

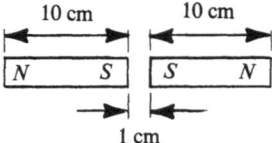

18.10. Two identical bar magnets, when arranged as shown, repel one another with a force of 5×10^{-3} newton. (Neglect the effect of the N poles and of the Earth's field.) Then consider one of the magnets by itself. What force does a terrestrial field of 6×10^{-5} tesla exert on one of the poles?

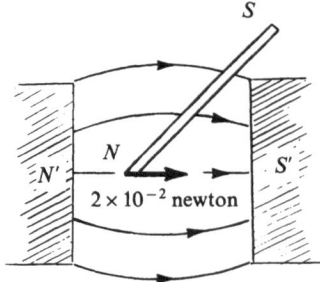

18.11. One of the magnets in Problem 18.10 is subsequently positioned, as shown, between the pole pieces of a larger magnet. A force of 2×10^{-2} newton is observed in the direction shown. What is the strength of **B** between the pole pieces?

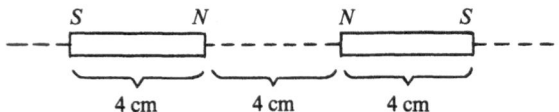

*18.12. Two identical magnets, of length 4.0 cm and pole strength 1.0 (SI), are positioned along a line with both N poles facing one another at 4.0 cm distance, as shown. (a) Find, in magnitude and direction, the total force on the right-hand magnet. Does your result conflict with statement (18.A.19)? (b) Find, in magnitude and direction, the contribution of the leftmost S pole to that force. (c) Find, in percent of the total force, the contribution of the NN force.

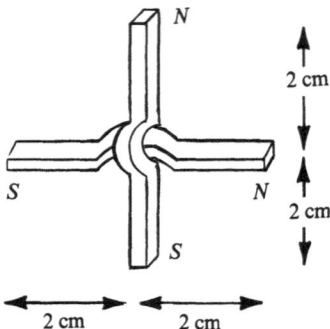

18.13. Two identical magnets, of length 4.0 cm and pole strength 1.0 (SI) are arranged at right angles as shown. Find (a) the total force on the vertical magnet; (b) the torque about its center; (c) the torque about its S pole.

18.14. Sketch your idea of representative field lines in the arrangement of Problem 18.12.

*18.15. At the Earth's north pole, where $B = 7 \times 10^{-5}$ tesla, what is the maximum magnetic torque on a compass needle of magnetic moment equal to 0.1 SI unit? What angle does the torque axis make with the vertical?

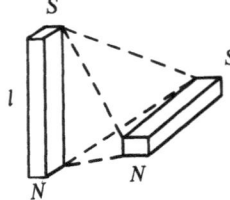

*18.16. Two identical bar magnets, of length l and pole strength p, occupy opposite edges of a regular tetrahedron, as shown. Calculate the magnetic (a) force and (b) torque on one magnet, due to the other. In which direction is the torque axis? Use the result of Problem 18.8; express your answer in terms of p and l.

18.17. A compass needle is freely suspended at its center, about which its moment of inertia is 1.0×10^{-8} kg meter². After being disturbed, it oscillates at a frequency of 2.0 Hz in a terrestrial field of 0.6×10^{-4} tesla. (a) Find the needle's magnetic moment. (b) If the needle's length is 3.0 cm, what is its pole strength?

Section B : Magnetic Force on a Moving Charge

18.18. A charge of -5×10^{-7} C travels due west, at a speed of 3×10^6 m/s, in a vertically upward **B** field of 0.8 tesla. Find the direction and magnitude of the magnetic force on the charge.

*18.19. On the basis of Eq. (18.B.5), express the tesla in terms of (a) coulombs, newtons, meters/second; (b) coulombs, kilograms, meters, seconds.

18.20. What would be the ratio of magnetic force to weight for the N^- ion of Example 18.2 if its speed were that of sound (its direction of motion being unchanged)?

*18.21. What is the radius of a circular electron orbit perpendicular to a field of 5.0×10^{-3} tesla, if the electron's speed is 3.0×10^7 meters/sec?

*18.22. (a) What alternating-voltage frequency, in MHz, is needed in a cyclotron made to accelerate protons, and in which the **B** field is 1.50 tesla? (b) If the orbit reaches a radius of 0.30 meter, what energy, in MeV, do the protons achieve?

18.23. Two identical charged particles, of kinetic energies $\mathcal{K}$ and $2\mathcal{K}$, are moving perpendicularly to the same uniform **B** field. What is the ratio of their orbits' radii?

*18.24. When a lithium ion (Li^+), of mass 6 atomic mass units, has a kinetic energy of 20 MeV in a cyclotron's **B** field of 1.5 teslas, what is the radius of its orbit?

18.25. How is the answer to Problem 18.18 modified if the charge travels to the southwest, all else remaining the same?

18.26. How is the answer to Problem 18.18 modified if the **B** field points (horizontally) to the northwest, all else remaining the same?

*18.27. An electron rises along a helical path about a vertical axis. Each turn of the helix takes 2×10^{-6} sec to complete. What is the direction and magnitude of the existing **B** field? (Assume the force is purely magnetic.)

18.28. What is, in meter^{-1}, the track curvature of a 1.0-keV electron moving perpendicularly to a field of 1.0 tesla?

*18.29. A region of space contains uniform **E** and **B** fields, pointing upwards and to the southeast, respectively, as well as a *straight* electron beam traveling due east. What is the electrons' speed v, in terms of E and B?

18.30. A deuteron (mass $= 2$ atomic mass units, charge $= +e$) moves in a circular orbit of radius 0.30 meter in a field of 0.08 tesla. Find the deuteron's (a) momentum; (b) kinetic energy in electron volts; (c) frequency of revolution.

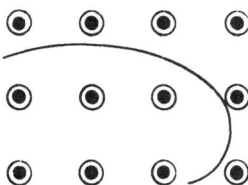

18.31. The figure shows a charged-particle track drawn to natural size. A magnetic field of 0.25 tesla points per-

pendicularly out of the page (circled dots). (a) Did the particle travel clockwise or counterclockwise? (Assume energy was gradually lost.) (b) Of what sign is the charge? (c) Assuming a charge of magnitude e, estimate the particle's momentum near the left end of the track.

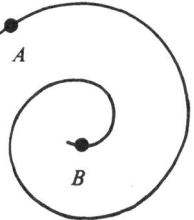

18.32. The figure, drawn to natural size, shows the track of an electron that loses energy by collisions with the molecules of some medium. If the field (perpendicular to the page) is $B = 1.0 \times 10^{-2}$ tesla, (a) make a rough estimate, in keV ($= 10^3$ electron volts), of the electron's energy at A and at B. (The track is in the plane of the figure.) (b) Does **B** point into or out of the page?

*18.33. Suppose that, in J. J. Thomson's cathode-ray tube arrangement, an undeflected beam is obtained; the electron-gun voltage V is then doubled, while the magnetic field B is kept unchanged. By what factor must be deflecting-plate voltage V' be multiplied to yield an undeflected beam once more?

18.34. In one version of J. J. Thomson's e/m experiment, a gun voltage of 4000 volts is used; the magnetic field is 0.050 tesla. If the distance between the deflecting plates is 3 mm, find their potential difference V' for zero deflection.

*18.35. A cyclotron works only when the alternating-voltage frequency is carefully matched to Eq. (18.B.37). Hence a similarly built device can be used for high-precision measurements of q/m. (It is then called an **omegatron**.) If high-energy particles emerge when the frequency is $f = 2321.1$ Hz under a field $B = 2.2790 \times 10^{-4}$ tesla, find their q/m.

*18.36. A good quality mass spectrometer can yield separate beams for He^{2+} and D^+, in contrast to what happens in Example 18.3, Section B-ii. How far apart are the He^{2+} and D^+ lines (assumed distinguishable) on the mass spectrometer of Example 18.6, Section B-v? See Example 18.7, Section B-vi, for the mass values.

18.37. Show that the helical motion of Fig. 18.20, when projected on the xy plane, has the angular velocity $\omega = qB/m$.

*18.38. In Fig. 18.20, let the particle's velocity vector be written as $v = v_{\parallel} + v_{\perp}$, where $v_{\parallel}$ points along **B**, while $v_{\perp}$ is parallel to the xy plane. Show that the radius r of the trajectory's xy projection is $r = mv_{\perp}/qB$.

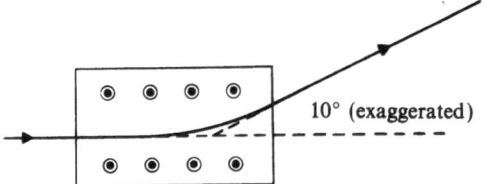

18.39. A beam of 2.0-keV electrons passes perpendicularly through a 5.0-cm-wide region of approximately uniform **B** field, as shown in the figure. (a) If the emerging beam makes an angle of 10° with its original direction, calculate *B*. (Consider 10° to be a small angle, and the magnetic force vector to be constant in magnitude and direction.) (b) Find the vertical **E** field that could replace the **B** field so as to yield a *downward* 10° deflection.

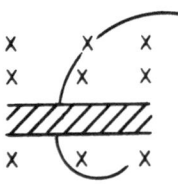

18.40. The figure represents (to scale but not natural size) the track of a charged particle that penetrated a metal plate and emerged on the other side (this is possible at sufficiently high energy); the track is seen as a trail of condensation droplets in a so-called cloud chamber. The track is in the plane of the figure; a uniform **B** points perpendicularly into the page. (a) If kinetic energy was lost in the metal, did the particle travel upwards or downwards? (b) Is the particle positive or negative? (c) Estimate the ratio of the speeds before and after passage through the plate. [This problem is based on Carl D. Anderson's discovery (1930) of the **positron**, a positive particle otherwise identical to the electron.]

Section C : Magnetic Force on a Current

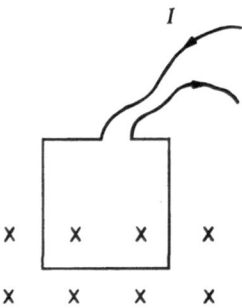

18.41. A rectangular loop, in a vertical plane, and with horizontal sides of 2.0 cm, is partially within a perpendicular 0.025-tesla *B* field, as shown. Find the current

I needed in the loop to give it a lift of 0.30 newton. Discuss the effect of the vertical sides.

*18.42. A horizontal bar, 2.0 meters long, carries a current of 300 A. If the bar's mass is 5.0 kg, what is the weakest *B* field which can support its weight?

18.43. On the basis of Eq. (18.C.16), what is the SI unit of magnetic moment, (a) in terms of amperes, meters, kilograms, seconds? (b) In terms of coulombs, meters, kilograms, seconds? (See also Problem 18.44.)

*18.44. (a) On the basis of Eq. (18.C.16), what is the unit of magnetic pole in terms of amperes, kilograms, meters, seconds? (b) Consequently, with Eq. (18.A.1), what is the unit of *K'* in terms of coulombs, kilograms, meters, seconds? (See also Problem 18.43.)

*18.45. What should be the linear density (mass per unit length) of a wire, carrying a current of 100 A, which would circle the Earth at the equator and be just held up against gravity by the terrestrial **B** field?

18.46. In Problem 18.41, the loop is subsequently pivoted about a vertical axis until its plane is at an angle θ to the **B** field. Find θ such that the lift becomes 0.15 newton.

18.47. In Example 18.9, Section C-iii, obtain the result of part (b) from the result of part (a).

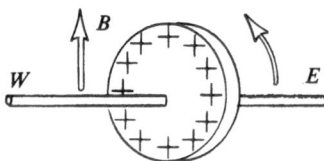

*18.48. The figure shows an insulating wheel, of radius 0.20 m; its plane is parallel to a **B** field, and it rotates at 25 rev/sec. A total charge of $+3.0 \times 10^{-6}$ coulomb is uniformly spread along its rim. (a) What is the equivalent loop current? (b) What is, in direction and magnitude, the magnetic torque on the wheel in a vertical **B** field of 0.50 tesla? (Refer to the axle's direction as east-west.)

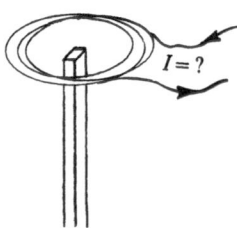

18.49. One pole (strength = 5.0 SI units) of a long vertical magnet lies at the center of a 30-turn circular coil of radius 10 cm as shown. If the whole coil experiences a 1.8×10^{-4}-newton vertical force, find the current in the wire. (Neglect the more distant pole.)

18.50. A certain galvanometer has a 2 cm × 2 cm coil. The **B** field has a strength of 0.06 tesla; it yields a torque of 1.5×10^{-3} newton meter when the current is 0.05 A. How many turns of wire are used?

*18.51. Suppose a galvanometer is redesigned with all geometric dimensions doubled (including wire thickness), while its **B** field is unchanged. (a) For a given current, by what factor is the magnetic torque multiplied? (b) By what factor is the coil's resistance multiplied?

*18.52. A DC motor is redesigned with all geometric dimensions doubled, so as to accommodate more turns of wire, whose thickness is unchanged. If the **B** field is also unchanged, (a) by what factor is the torque multiplied for a given current? (b) By what factor is the coil's resistance multiplied?

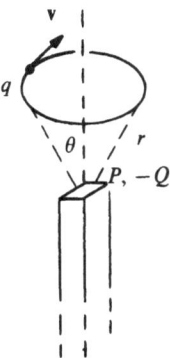

*18.53. A long magnet, whose S pole is far enough away to be disregarded, has a negative point charge $-Q$ added to its N pole, whose strength is P. A particle of positive charge q revolves at constant speed v in a circular orbit, a distance r away from the pole, as shown. Find the cone half-angle θ such that the combined electric and magnetic force on q is centripetal. (Express your answer in terms of Q, P, q, v, r.) Note, incidentally, that only a special value of the particle's mass will result in uniform circular motion if no external mechanical forces are applied.

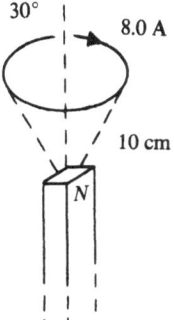

18.54. A long vertical magnet, of pole strength 5.0 SI units, lies on the axis of a horizontal circular loop with

current 8.0 A, as shown. If the N pole is 10 cm from the wire, which in turn is 30° away from the axis as seen from the pole, find, in direction and magnitude, the total magnetic force on the loop. (Assume the S pole is far enough away to be neglected.)

*18.55. In Fig. 18.23, assume the current in the wheel goes straight down along the radial line (this is a slight oversimplification). What torque does a uniform field of 0.030 tesla exert on the wheel if $I = 2.0$ A, and the radius equals 0.10 meter?

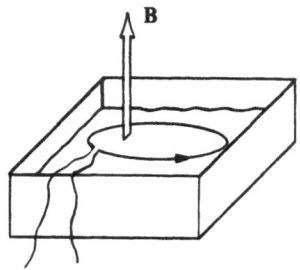

*18.56. In the arrangement shown, a circular loop of insulated wire, with current I and radius r, is floating on mercury in a uniform vertical **B** field. Find the tension in the wire in terms of I, B, and r.

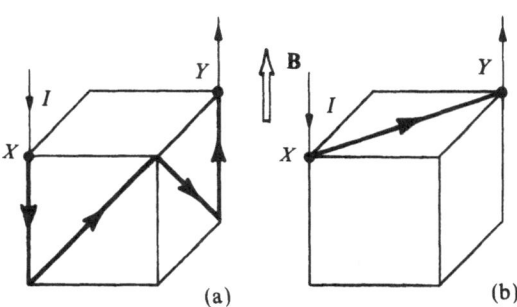

(a)   (b)

*18.57. A cube of side l, with one pair of faces perpendicular to a uniform **B** field, as shown in (a), has a current I flowing from X to Y along the indicated segments. (a) In terms of I, l, and B, find the total force vector on the cube. (b) Repeat the calculation for the more direct path shown in (b). (c) Prove that any path from X to Y gives the same result for given I.

*18.58. (a) From Eq. (18.A.11), express the SI unit of magnetic pole in terms of the newton and the tesla. (b) From Eq. (18.C.6), express the tesla in terms of the ampere, the meter, and the newton. (c) Combine (a) and (b) to write the SI unit of magnetic pole without involving the tesla.

Answers to True or False

1. False (it is like the electric force on a charge).
2. False ($K' \neq K$).
3. False (tesla meter^2).
4. True (it is a magnetic N pole).
5. False (at right angles).
6. True.
7. True.
8. False.
9. False.
10. False (see next item).
11. True.
12. False (clockwise).
13. True.
14. True.
15. False (most trajectories *are* deflected).
16. True.
17. True.
18. False (in the opposite direction).
19. False (N pole down).
20. True (same size and direction of the magnetic moment).

Answers or Hints to Problems

18.1. 2.0 teslas.

18.3. 2.2×10^{-4} (SI).

18.6. 3.5×10^{-3} lines.

18.7. 0.10 (SI).

18.8. $K'pl/r^3$.

18.9. 7.1×10^{-3} tesla.

18.12. (a) No conflict because here **B** is not uniform.

18.15. Hint: the needle must be horizontal; why?

18.16. (a) 0; (b) $K'p^2/l$ about any axis perpendicular to both magnets.

18.19. (b) kg/(sec coulomb).

18.21. 3.4×10^{-2} meter.

18.22. (b) 9.5 MeV.

18.24. 1.1 meters.

18.27. Hint: the period is the same as for a circular orbit.

18.29. $\sqrt{2} \, E/B$.

18.33. Use Eq. (18.B.24).

18.35. 6.3993×10^7 coulombs/kg.

18.36. Denote the small differences in r and m/q by dr and $d(m/q)$; differentiate Eq. (18.B.34), where m/e is replaced by m/q.

18.38. Hint: use $\mathbf{v} \times \mathbf{B} = \mathbf{v}_{\parallel} \times \mathbf{B} + \mathbf{v}_{\perp} \times \mathbf{B} = \mathbf{v}_{\perp} \times \mathbf{B}$, $\mathbf{F}_{\perp} = m\mathbf{a}_{\perp}$, etc.

18.42. **B** can be made weakest when perpendicular to the bar; why?

18.44. (b) $\dfrac{\text{kg meters}}{\text{coulomb}}$.

18.45. Consider, e.g., a 1-meter length of wire.

18.48. (b) 4.7×10^{-6} newton meter, clockwise about the wheel's horizontal diameter.

18.51. Use formula (18.C.23).

18.52. Hint: there are 4 times as many turns.

18.53. $\theta = \tan^{-1} \dfrac{KQ}{K'Pv}$.

18.55. Integrate the current elements' contributions over the radius. Answer: 3×10^{-4} newton meter.

18.56. Integrate the rightward force components over the right half-loop; the resultant is twice the tension. (Why?) Answer: IBr.

18.57. (c) Decompose the path in vector pieces $\mathbf{l}_1, \mathbf{l}_2, \ldots$; use $\mathbf{l}_1 \times \mathbf{B} + \mathbf{l}_2 \times \mathbf{B} + \cdots = (\mathbf{l}_1 + \mathbf{l}_2 + \cdots) \times \mathbf{B}$.

18.58. See Eq. (18.A.24).

CHAPTER 19

MAGNETIC FIELDS FROM CURRENTS

We have just learned how a **B** field can be *detected* through its effect on an electric current. Conversely, we next learn how a **B** field is *produced* by an electric current.

From the preceding chapter, we infer that a current must produce a magnetic field. The argument involves Newton's law of action and reaction, and goes as follows.

A magnet exerts, at a distance, a force **F** on a current-carrying wire. Therefore the wire must exert a reaction force $-\mathbf{F}$ on the magnet. This, in turn, implies that the wire *produces a magnetic field* at the location of the magnet. This simple argument is so powerful that it will yield, later in this chapter, an exact quantitative expression for the **B** field of a current.

In what follows, we work out in detail the **B** field for variously shaped currents; in the process, we arrive at an efficient short cut (Ampère's law) for calculating **B** in symmetric situations. We shall see, also, that a **B** field can exert, on conductors, what amounts to a pressure. Finally, we review some of the evidence, already touched upon in the preceding chapter, that a magnet consists of nothing but circulating charge, hidden inside the magnet's material; *magnetic poles are fictitious*, as Ampère had suggested as early as 1820.

A. The Law of Biot and Savart

We shall first describe without proof the magnetic field produced by a current. This descrip-tion was first given in 1820 by Jean-Baptiste Biot and Félix Savart, less than a year after Oersted discovered that a compass needle is deflected by a current.

Circular and straight wires provide the simplest illustrations of the Biot–Savart law, and we shall examine their **B** field in some detail.

When using the formulas of this chapter, we keep in mind the experimental fact that *the superposition principle holds for magnetic fields*, at least in vacuum and in air. Hence, we can predict the field $\mathbf{B}_{\text{total}}$ due to several currents in combination if we know $\mathbf{B}_1$, $\mathbf{B}_2$,... from each. At any location, we have a *vector sum* $\mathbf{B}_{\text{tot}} = \mathbf{B}_1 + \mathbf{B}_2 + \cdots$. (That principle was already applied, in the preceding chapter, to fields from permanent magnets.)

i. Magnetic Field from a Current in a Wire

Consider, as in Fig. 19.1a, a wire of arbitrary shape, carrying a current I. In order to specify its **B** field, we first mentally cut the wire into small pieces, or **current elements**, whose length and direction we indicate by the vectors l_1, l_2, etc. Let us now construct **B** at a given location X. The figure shows a typical contribution, B_n, arising from element l_n, located at point A on the wire. The observed field **B** at point X is the vector sum of all such partial fields:

$$\mathbf{B} = \mathbf{B}_1 + \mathbf{B}_2 + \cdots \qquad (19.A.1)$$

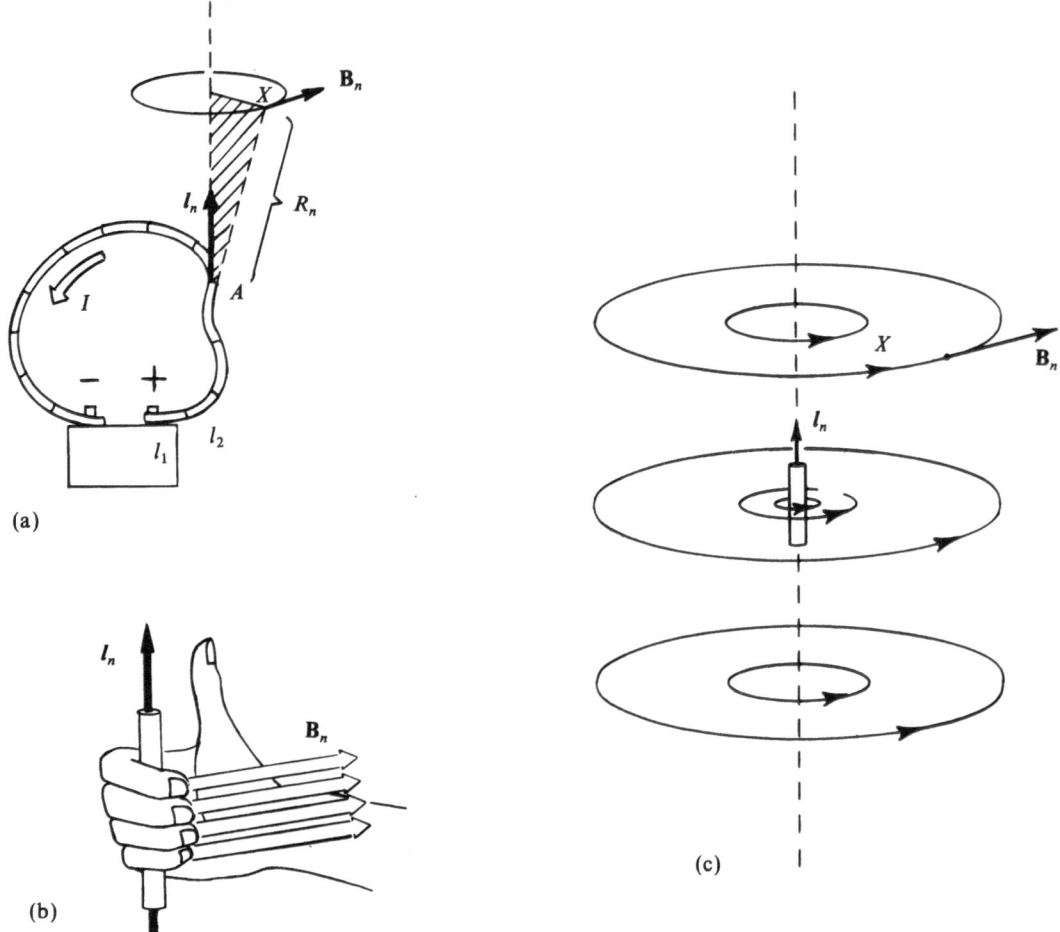

Figure 19.1. (a) The unobservable contribution $\mathbf{B}_n$ from a current element Il_n, according to Biot and Savart. The observed $\mathbf{B}$ is the sum of all such contributions. (b) The right-hand rule. (c) These are false field lines; however, they show better than (a) the direction of $\mathbf{B}_n$ at various points. The real field lines would indicate $\mathbf{B}$, not $\mathbf{B}_n$.

(Caution: $\mathbf{B}$ is physically measurable, but the individual $\mathbf{B}_n$ are not—*they are only mathematical constructions.*)

Magnitude of Each $\mathbf{B}_n$

The recipe of Biot and Savart calls for a magnitude

$$\boxed{B_n = K' \frac{Il_n}{R_n^2} \sin \varphi_n} \qquad (19.A.2)$$

Here R_n is the distance from A to X; φ_n is the angle between l_n and the direction of segment AX; the constant $K' = 10^{-7}$ (SI units) is the same as in Coulomb's law for *magnetic poles*.

Formula (19.A.2) is not quite unexpected if examined factor by factor. Even though a current element cannot be isolated experimentally (the moving charge must come from somewhere, and it must go somewhere), Eq. (19.A.2) incorporates something like a superposition principle. The proportionality $B_n \propto I$ agrees with the idea that two identical current elements *side by side* amount to a single one with current $2I$; accordingly, B_n becomes $2B_n$. Similarly for $B_n \propto l_n$, if we consider two identical elements *end to end*. The factor $1/R_n^2$ gives the familiar inverse square dependence already encountered in gravity, in electrostatics, and in magnetism from permanent magnets. The directional factor $\sin \varphi_n$ is truly the unusual feature in (19.A.2).

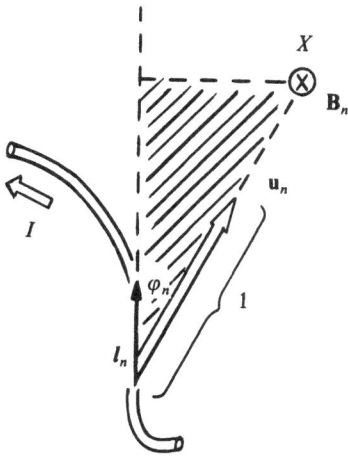

Figure 19.2. Detail from Fig. 19.1.

(The reader ought to check in detail that the magnitude of $\mathbf{B}_n$ comes out as in Eq. (19.A.2), and its direction as in Fig. 19.1 or 19.2.)

We stress once more that (19.A.3) is a mathematical prescription but not a physical law; it acquires physical meaning when inserted into the complete sum (19.A.1).

Summary

The $\mathbf{B}$ field, observed at any selected point X, and due to a wire with current I, is given by

$$\mathbf{B} = K' \frac{I l_1 \times \mathbf{u}_1}{R_1^2} + K' \frac{I l_2 \times \mathbf{u}_2}{R_2^2} + \cdots \quad (19.A.4)$$

where all elements l_1, l_2,... of the wire contribute to the sum. In the limit where l_1, l_2,... become infinitesimally small and infinitely numerous, formula (19.A.4) becomes exact. Representing the typical element l_n by dl, and denoting the sum by an integral sign, we symbolize the exact limit of (19.A.4) by

$$\mathbf{B} = \int_{\text{wire}} \frac{K' I \, dl \times \mathbf{u}}{R^2} \quad (19.A.5)$$

This is the law of Biot and Savart.

Direction of Each $\mathbf{B}_n$

Figure 19.1a shows that $\mathbf{B}_n$ is at right angles to the (shaded) plane which contains l_n and the point X. To which side of the plane does $\mathbf{B}_n$ point? The correct direction is selected by a new **right-hand rule**, illustrated in part (b) of the figure. (1) Grasp the current element in such a manner that your thumb indicates the direction of I; (2) your other fingers then curl around the wire so as to indicate the direction of $\mathbf{B}_n$. (Note how this prescription is patterned after the previously learned right-hand rule.)

Cross-Product Notation*

As similarly happened in the preceding chapter, these instructions are neatly summarized in a cross product of two vectors. One of these two is l_n, and the other, $\mathbf{u}_n$, not shown in Fig. 19.1, is defined as a vector of unit magnitude ($|\mathbf{u}_n| = 1$), which points in the direction of the segment AX; $\mathbf{u}$ is illustrated in Fig. 19.2. (The plane of that figure is the shaded plane in Fig. 19.1.)

The complete specifications for $\mathbf{B}_n$ now take the form

$$\mathbf{B}_n = \frac{K' I l_n \times \mathbf{u}_n}{R_n^2} \quad (19.A.3)$$

Reversal of the Current

If the direction of I is reversed, the wire's $\mathbf{B}$ field also reverses its direction everywhere. This can be seen from (19.A.4) or (19.A.5), where we replace each l_n by $-l_n$ (dl by $-dl$), or, alternatively, change I into $-I$.

ii. A Circular Loop

Field at the Center

The simplest application of the law of Biot and Savart consists of determining $\mathbf{B}$ at the center X of a circular loop of given radius r, carrying a given current I; the geometry is shown in Fig. 19.3. A typical l_n is represented by a cross (vector pointing into the page). The contribution $\mathbf{B}_n$ points along the axis of the loop. (Check for agreement with the

---

* See Appendix II for a review of the definition and properties of the cross product.

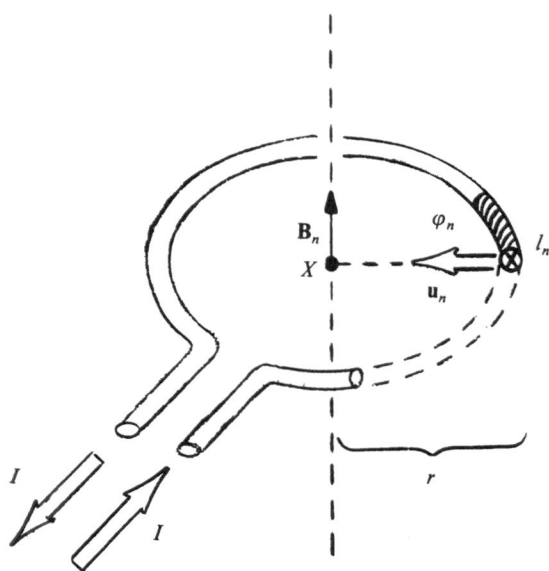

Figure 19.3. Contribution $\mathbf{B}_n$ at the center of a circular loop. The two leads, which connect the loop to a source of current, contribute negligibly to $\mathbf{B}$ because they follow nearly identical paths, with opposite currents.

right-hand rule.) In this case we have $\varphi_n = 90°$, $R_n = r$. From (19.A.2) we obtain

$$B_n = \frac{K'Il_n}{r^2} \qquad (19.\text{A}.6)$$

The resultant, $\mathbf{B} = \mathbf{B}_1 + \mathbf{B}_2 + \cdots$, also points along the axis; here the magnitudes simply add:

$$B = \frac{K'Il_1}{r^2} + \frac{K'Il_2}{r^2} + \cdots \qquad (19.\text{A}.7)$$

$$= \frac{K'I}{r^2}(l_1 + l_2 + \cdots) \qquad (19.\text{A}.8)$$

$$= \frac{K'I}{r^2}(2\pi r) \qquad (19.\text{A}.9)$$

the last step because $l_1 + l_2 + \cdots$ is just the circumference of the loop. In conclusion, we have

$$\boxed{B = \frac{2\pi K'I}{r}} \qquad (19.\text{A}.10)$$

Field Elsewhere on the Axis

In a slight extension of the above, we consider the same loop, but calculate $\mathbf{B}$ at an arbitrary point

(X in Fig. 19.4) on the loop's axis. We still have $\varphi_n = 90°$; $R_n = R$ has the same value for all n. The contribution $\mathbf{B}_n$ no longer points along the loop's axis; its magnitude, from Eq. (19.A.2), is

$$B_n = \frac{K'Il}{R^2} \qquad (19.\text{A}.11)$$

As Fig. 19.4 shows, each $\mathbf{B}_n$ can be paired with $\mathbf{B}_{n'}$ arising from the oppositely located element on the loop. Their components parallel to the plane of the loop cancel, while their components along the axis add. Hence we only need to calculate that component, $(B_n)_{\text{axis}}$. Let θ be the angle under which the loop's radius r is seen from X. The figure yields

$$(B_n)_{\text{axis}} = B_n \sin\theta = \frac{K'Il_n \sin\theta}{R^2} \qquad (19.\text{A}.12)$$

The resultant, $\mathbf{B} = \mathbf{B}_1 + \mathbf{B}_2 + \cdots$, must point along the axis, and its magnitude is

$$B = B_{\text{axis}} = (B_1)_{\text{axis}} + (B_2)_{\text{axis}} + \cdots$$

$$= \frac{K'I \sin\theta}{R^2}(l_1 + l_2 + \cdots) \qquad (19.\text{A}.13)$$

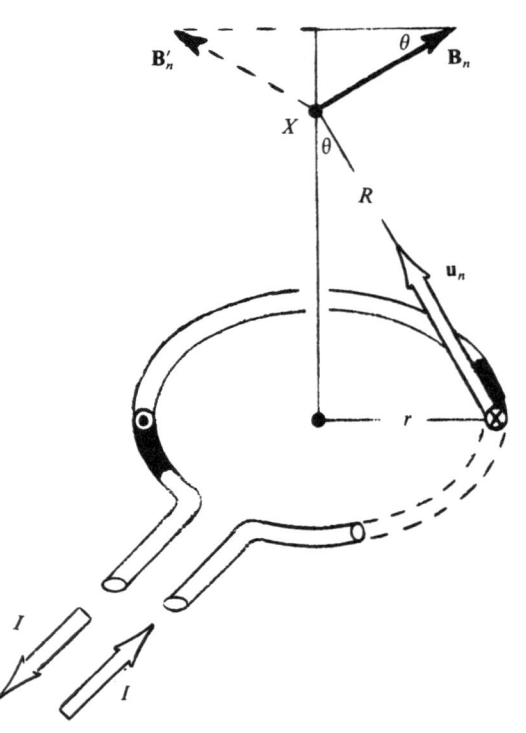

Figure 19.4. Contribution $\mathbf{B}_n$ on the axis of a circular loop. (Why are the two angles θ equal?)

from (19.A.12). Equivalently,

$$B = \frac{K'I \sin \theta}{R^2} (2\pi r)$$

or, since $\sin \theta = r/R$,

$$\boxed{B = \frac{2\pi K' I r^2}{R^3}} \qquad (19.A.14)$$

In the special case where X is the center of the loop we have $R = r$, and (19.A.14) reduces to (19.A.10) as it must. It is interesting that B, in Eq. (19.A.14), should obey an **inverse-cube** distance law relative to the loop as a whole.

The general appearance of a circular loop's **B** field is illustrated in Fig. 19.5. Note its resemblance, some distance away from the loop, to the field of a bar magnet.

Example 19.1. A Coil and a Bar Magnet. A circular horizontal 10-turn coil of radius $r = 5.0$ cm carries a current of 2.0 A. A bar magnet of length 5.0 cm and pole strength $p = 3.0$ SI units is positioned as shown in Fig. 19.6, along the coil's vertical axis, with its N pole at the coil's center. What is the total force **F** exerted by the coil on the bar magnet?

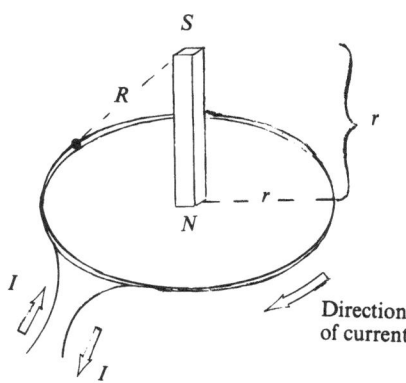

Figure 19.6. A bar magnet interacting with a flat coil.

Let the *coil's* magnetic field be $\mathbf{B}_N$ or $\mathbf{B}_S$ at N or S, respectively. Then the required force vector is

$$\mathbf{F} = p\mathbf{B}_N + (-p)\mathbf{B}_S \qquad (19.A.15)$$

Both $\mathbf{B}_N$ and $\mathbf{B}_S$ point downwards along the axis of the coil; their magnitude is

$$B_N = \frac{2\pi K'(10I)}{r} \qquad (19.A.16)$$

$$B_S = \frac{2\pi K'(10I)r^2}{R^3} = \frac{2\pi K'(10I)r^2}{(\sqrt{2}r)^3} \qquad (19.A.17)$$

(from the figure, we have $R = \sqrt{2}r$). By Eq. (19.A.15), where $\mathbf{B}_N$ is the larger vector,

$$\underline{\mathbf{F} \text{ points vertically down}}$$

and has a magnitude

$$\begin{aligned}
F &= p\frac{2\pi K'(10I)}{r} + (-p)\frac{2\pi K'(10I)r^2}{(\sqrt{2}r)^3} \\
&= \frac{2\pi K'(10I)p}{r}\left(1 - \frac{1}{2\sqrt{2}}\right) \\
&= \frac{(2\pi)(10^{-7})(10)(2.0)(3.0)}{0.050}\left(1 - \frac{1}{2\sqrt{2}}\right) \text{ newton} \\
& \qquad\qquad\qquad\qquad\qquad\qquad\qquad (19.A.18)
\end{aligned}$$

$$= \underline{4.9 \times 10^{-4} \text{ newton}}$$

iii. A Long Straight Wire

To illustrate the Biot–Savart law in its integral form (19.A.5), consider a long straight wire with

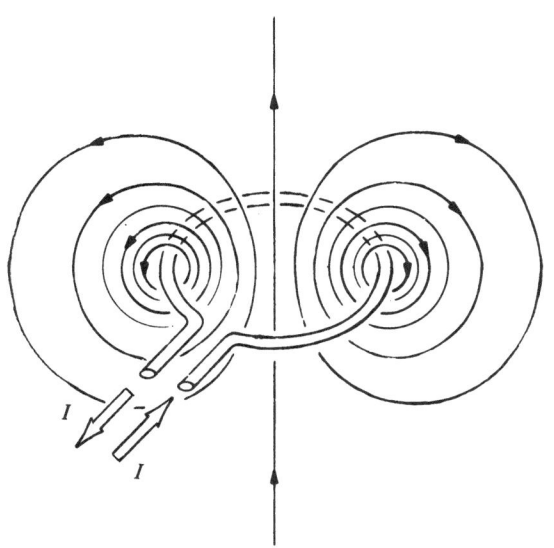

Figure 19.5. Representative magnetic field lines for a circular loop. Each line is in a plane which contains the loop's axis.

current I; see Fig. 19.7. What is its **B** field at point X? We assume the wire extends infinitely far in both directions. "Infinitely far" means many times the distance r at which we observe the field, so that end effects may be ignored; the return wires needed to complete the circuit and maintain the current I are likewise far away and negligible in their effects. In more rigorous treatments than given here, these assumptions can be justified, and we shall not worry about them further.

We use the wire as a z axis, with the origin ⓪ chosen as being the perpendicular projection of point X. We first calculate the contribution $d\mathbf{B}$ of a current element with length $dl = dz$. The right-hand rule yields a vector $d\mathbf{B}$ (indicated by a cross) which points perpendicularly into the plane of the figure; its magnitude is

$$|d\mathbf{B}| = \frac{KI\,dz}{R^2}\sin\varphi \qquad (19.A.19)$$

see Eq. (19.A.2). On the right of this equation, all variables can be expressed in terms of the angle φ and the fixed distance r shown in the figure. We have (noting that z happens to be negative here)

$$-z = r\cot\varphi, \qquad dz = \frac{r\,d\varphi}{\sin^2\varphi}, \qquad R = \frac{r}{\sin\varphi}$$
$$(19.A.20)$$

Equation (19.A.19) now reads

$$|d\mathbf{B}| = \frac{K'I(r\,d\varphi/\sin^2\varphi)}{(r/\sin\varphi)^2}\sin\varphi$$

$$= \frac{K'I}{r}\sin\varphi\,d\varphi \qquad (19.A.21)$$

Next we integrate all contributions $d\mathbf{B}$ along the wire; thus φ ranges between zero and $+180°$. Since all vectors $d\mathbf{B}$ point in the same direction, their magnitudes simply add, and we obtain

$$B = \int_{\varphi=0}^{\varphi=\pi}|d\mathbf{B}| = \frac{K'I}{r}\int_0^\pi\sin\varphi\,d\varphi \qquad (19.A.22)$$

or

$$\boxed{B = \frac{2K'I}{r}} \qquad (19.A.23)$$

the desired result. The direction of **B** is that of $d\mathbf{B}$ in the figure. The complete field lines are concentric circles in planes perpendicular to the wire, as illustrated in Fig. 19.8. The **inverse-distance behavior** exhibited by (19.A.23) is analogous to that of the **E** field due to a long charged wire.

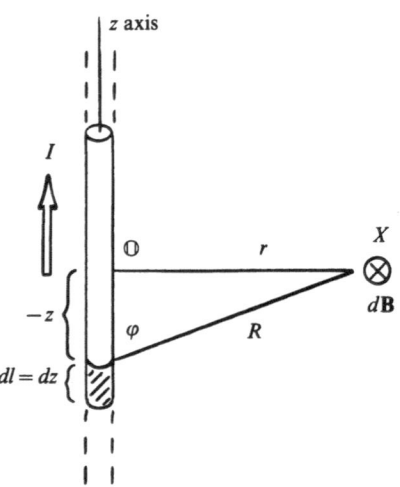

Figure 19.7. Contribution $d\mathbf{B}$ from the shaded element of an infinite straight wire. The z coordinate increases towards the top of the figure.

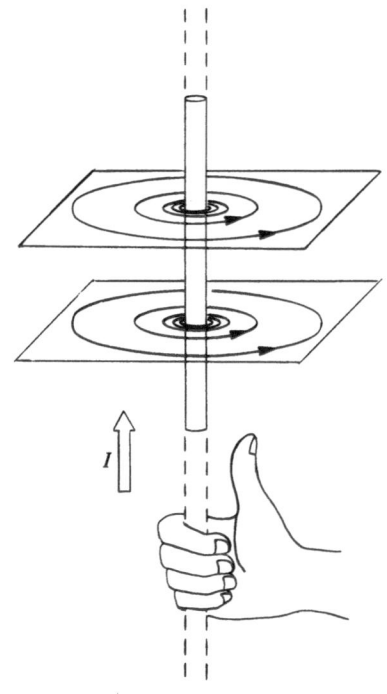

Figure 19.8. Typical magnetic field lines of an infinite straight wire.

Note the similarity of formulas (19.A.23) and (19.A.10), which differ only by a factor of π. The reason is not very deep: K', I, and r can be combined in only one manner, namely, $K'I/r$, to yield the dimensions of B.

The Same Result Without Integration

An action–reaction method, which avoids the setting up and evaluation of an integral, also yields formula (19.A.23). It is presented as a note at the end of this chapter.

Field Very near a Curved Wire (Fig. 19.9)

> A curved wire can be considered straight and infinite if its **B** field is measured at close range.　(19.A.24)

That is to say, our straight-wire results hold with arbitrarily high percentage accuracy if **B** is measured at a small enough distance r from any wire. [The proof is based on considering, in Fig. 19.9, a segment s which is small, but of fixed, chosen size. In the Biot–Savart sum it contributes, at X, a field that becomes arbitrarily large (as $1/r$) when r becomes sufficiently small. The whole remainder of the wire contributes an amount that remains finite and hence can be neglected in comparison. The shape of that remainder therefore is unimportant when r approaches zero.]

Example 19.2. Force between Parallel Currents.
Figure 19.10 shows two parallel wires: an infinite one, with current $I_1 = 2.0$ A, and a segment XY of length $l = 20$ cm, with current $I_2 = 3.0$ A. If the distance between the wires is $r = 1.0$ cm, find the magnetic force between them.

The problem is solved in two steps. We calculate (1) the **B** field which I_1 produces at the location of I_2; (2) the force on I_2 due to that field.

1. Field at I_2. Equation (19.A.23) gives

$$B = \frac{2K'I_1}{r} \qquad (19.A.25)$$

The vector **B** is directed into the plane of Fig. 19.10.

2. Force on I_2. From the preceding chapter we recall that

$$F = I_2 lB \sin 90° \qquad (19.A.26)$$

90° being the angle between segment XY and the local **B** field; the right-hand force rule then gives a vector **F** directed perpendicularly towards the infinite wire, i.e., *a force of attraction* between the wires. Inserting (19.A.25) into (19.A.26), we obtain

$$F = I_2 l \left(\frac{2K'I_1}{r} \right) (1) \qquad (19.A.27)$$

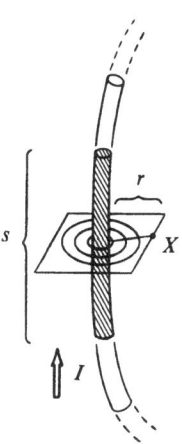

Figure 19.9. Typical magnetic field lines close to a curved wire.

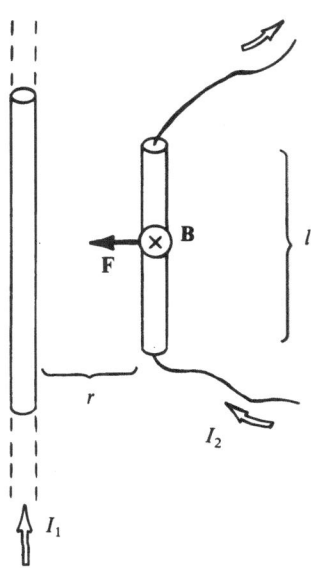

Figure 19.10. Interaction between parallel straight currents.

or

$$F=\left(\frac{2K'I_1I_2}{r}\right)l \qquad (19.A.28)$$

In the present case,

$$F=\frac{(2)(10^{-7})(2.0)(3.0)(20\times10^{-2})}{1.0\times10^{-2}}\text{ newton}$$

$$=\underline{\underline{2.4\times10^{-5}\text{ newton}}}$$

Note the smallness of this number in spite of the closeness of the wires and the respectable-sized currents.

Some Comments on Eq. (19.A.28)

We see that the force is proportional to each of the two currents and inversely proportional to their distance. If the direction of I_2 is reversed, attraction turns into repulsion: *parallel currents attract, antiparallel currents repel.* Equation (19.A.28) has inspired the *official SI definition of the ampere* as the current that, when present in two parallel wires, one of which is 1 meter long and the other infinite, gives rise to a mutual force of 2×10^{-7} newton, exactly, if the wires are one meter apart. (Note: When both wires are infinite, the mutual force is infinite; that is, it can be made arbitrarily large with sufficiently long wires.)

The Pinch Effect

An actual wire has nonzero thickness, and can be thought of as carrying infinitely many parallel currents which all attract one another. Hence, if the wire were compressible, it would contract to a thinner cross section. (We recall that a force on a current is a force on the current-carrying matter.) The parallel currents would then come closer together, and their force of attraction would increase. The result ought to be a catastrophic collapse if the wire were perfectly compressible. A metal is far from meeting that condition, but an ionized gas (plasma) is a conductor, and is compressible enough to collapse significantly under an intense current.

In many experiments on controlled nuclear fusion, one attempts to compress a high-temperature plasma without contact with material walls. In the 1950s, when that research program was launched, the pinch effect seemed ready-made for the purpose. Unfortunately, although the pinch did occur, the electric discharge had a wildly unstable path and could not be kept away from the walls of the reaction chamber. As of this writing, one is only beginning to master the pinch effect.

B. Proof by Action and Reaction

Why must exactly the same factor $K'=10^{-7}$ (SI) be used in both the Biot–Savart and magnetic Coulomb laws? By addressing this question, we also teach ourselves that two laws are really a single one: the transverse force of Chap. 18, and the transverse field of the present chapter. Our tool will be Newton's third law of motion.

Consider the arrangement of Fig. 19.11, involving a magnetic pole P and an arbitrary loop with current I. The argument is in three steps. (1) We calculate the force exerted by P on I; (2) by Newton's law of action and reaction, this must equal minus the force exerted by I on P; (3) considering P as a test pole, we obtain the **B** field produced by I at the location of P.

1. Force Exerted by P on I. Let l be an arbitrary loop element, located at A. We want the

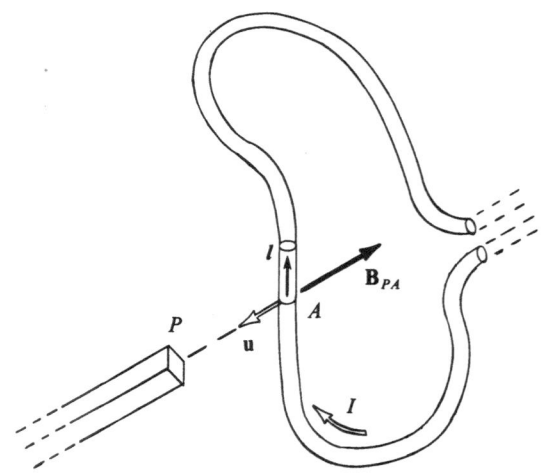

Figure 19.11. The field, $\mathbf{B}_{PA}$, of a positive magnetic pole P, seen at point A of a current loop. If the unit vector $\mathbf{u}$ is defined as shown, then $\mathbf{B}_{PA}$ points in the $-\mathbf{u}$ direction. The other pole, $-P$, is far enough away to be ignored.

force on I due to P, and thus we need the **B** field, $\mathbf{B}_{PA}$, produced by P at A. We have

$$\mathbf{B}_{PA} = -\frac{K'P}{R^2}\mathbf{u} \qquad (19.B.1)$$

where **u** is the unit vector shown in Fig. 19.11, and R is the distance from P to A. The resulting force exerted by P on I is

$$\mathbf{F} = I\mathbf{l} \times \mathbf{B}_{PA} \qquad (19.B.2)$$

$$= -I\mathbf{l} \times \left(\frac{K'P}{R^2}\mathbf{u}\right) = -\frac{K'PI}{R^2}\mathbf{l} \times \mathbf{u} \qquad (19.B.3)$$

The force exerted by P on the whole loop is therefore

$$\mathbf{F}_{PI} = -\left(\frac{K'PI}{R_1^2}\mathbf{l}_1 \times \mathbf{u}_1 + \cdots\right) \qquad (19.B.4)$$

a sum ranging over all the current elements. [Both Eqs. (19.B.1) and (19.B.2) are taken from the preceding chapter.]

2. Force Exerted by I on P. That force is just the reaction

$$\mathbf{F}_{IP} = -\mathbf{F}_{PI} = +\frac{K'PI}{R_1^2}\mathbf{l}_1 \times \mathbf{u}_1 + \cdots \qquad (19.B.5)$$

3. Field at P, due to I. As a test pole, P detects a field

$$\mathbf{B} = \frac{\mathbf{F}_{IP}}{P} \qquad (19.B.6)$$

or, from (19.B.5),

$$\mathbf{B} = \frac{K'I}{R_1^2}\mathbf{l}_1 \times \mathbf{u}_1 + \cdots \qquad (19.B.7)$$

as announced in (19.A.4).

C. Ampère's Law

The law of Biot and Savart permits us, *in principle*, to calculate the magnetic field of any system of steady currents, no matter how complicated. In practice, however, only the simplest situations can be worked out. Fortunately, in some cases a valuable shortcut is available, based on Ampère's law, which we now formulate and apply.

i. Statement of Ampère's Law

Let us move a magnetic test pole p over a small path $d\mathbf{r}$ in a region where the magnetic field is **B**; see Fig. 19.12. Since **B** exerts a force $p\mathbf{B}$ on p, the work done is $d\mathcal{W} = p\mathbf{B} \cdot d\mathbf{r}$. By taking p around a closed path, as in the figure, we obtain a total work

$$\mathcal{W} = p \oint \mathbf{B} \cdot d\mathbf{r} \qquad (19.C.1)$$

where $\oint$ symbolizes the closed-path integral.

After such a round trip, neither p nor **B** have changed. (We assume tentatively that only permanent magnets are involved in the production of **B**.) Can we gain a positive energy $\mathcal{W}$ in this way? This would make a perpetual-motion machine possible, and solve the world's energy problem.

As it happens, both theory and experiment tell us that conservation of energy cannot be violated, even with magnets. Thus, we set $\mathcal{W} = 0$, or

$$\oint \mathbf{B} \cdot d\mathbf{r} = 0 \qquad (19.C.2)$$

This is **Ampère's law in the absence of currents**.

Next suppose, as in Fig. 19.13, that an electric current I threads the closed path along which we would like to carry the pole p. Now we can no longer express the conservation of energy in the form (19.C.2). One reason is that p has a partner,

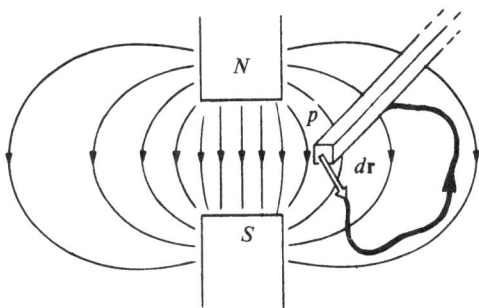

Figure 19.12. A magnetic test pole, p, describing a closed loop in a permanent magnetic field. The other pole, $-p$, is far away in a region of zero magnetic field.

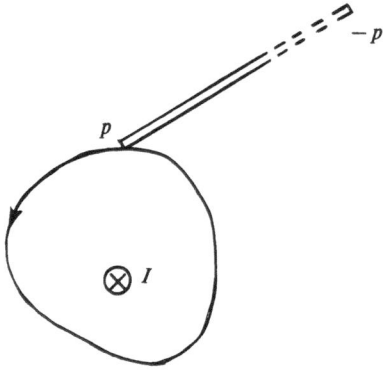

Figure 19.13. A current I (directed into the figure), prevents the pole p from completing its intended circuital path.

$-p$, at the other end of the bar magnet. As the figure shows, we cannot take p completely around I without doing one of two things:

1. Intersect the wire with the magnet. Not only is this technically difficult, but, more importantly, such an operation would add an unknown energy term to (19.C.1).
2. Take $-p$ around the wire as well. This might make another energy contribution to (19.C.1), because although $-p$ is far away, its path would be very long.

Another reason is that the current comes from a battery or generator, which *could* absorb or release energy.

In conclusion: Eq. (19.C.2) breaks down when the path of p loops around a current. In the present case, the revised version of (19.C.2) turns out to be (as we shall verify later)

$$\oint \mathbf{B} \cdot d\mathbf{r} = 4\pi K' I_{\text{enclosed}} \qquad (19.C.3)$$

Any closed path integral of $\mathbf{B}$ *is proportional to the total current around which it loops.* This is the general form of **Ampère's law*** in the presence of steady currents. (That equation involves no mention of a magnetic pole; in what follows the help of test poles will no longer be required.)

In general, the current I_{enclosed} may have positive and negative contributions, as Fig. 19.14 illustrates.

* Also known as Ampère's **circuital law**.

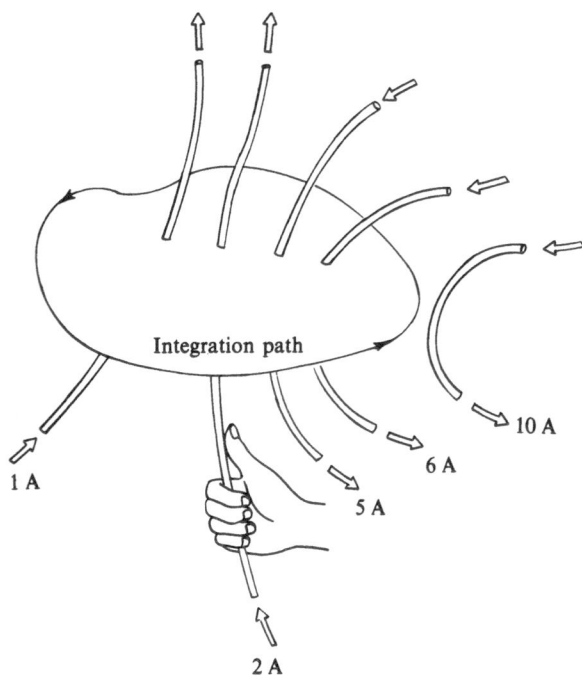

Figure 19.14. How to apply Ampère's law in the presence of many currents. The path follows the general direction of the fingers curled, for example, around the 2-A current. That current, and all others threading the loop in the same direction, are then counted positive; the others are counted negative. "Threading the loop" means intersecting a surface which is bounded by the loop. Here the total enclosed current is $I_{\text{enclosed}} = (+1 + 2 - 5 - 6)\,\text{A} = -8\,\text{A}$; the 10 A current makes no contribution because it does not thread the loop.

Recalling Gauss' Law

It is interesting, and a good review of concepts, to compare Ampère's and Gauss' laws; they are similar in some respects, as Table 19.1 demonstrates.

In symmetric situations, both laws provide elegant calculational shortcuts.

ii. A Limited Verification

Let us check the validity of (19.C.3) in a simple case, that of the infinite straight wire. As the

Table 19.1. Comparison of Ampère's and Gauss' Laws

Ampère's law	Gauss' law of Chap. 15
Deals with $\mathbf{B}$.	Deals with $\mathbf{E}$.
Uses an imaginary closed path of arbitrary shape.	Uses an imaginary closed surface of arbitrary shape.
Involves the total enclosed current.	Involves the total enclosed charge.

integration path, we choose one of the (circular) field lines; see Fig. 19.15. We move in the direction of **B**, and set $|d\mathbf{r}| = ds =$ element of path length. Since **B** is parallel to $d\mathbf{r}$ we obtain, for the left side of (19.C.3),

$$\oint \mathbf{B} \cdot d\mathbf{r} = \oint B \, ds = (B)(2\pi r) \qquad (19.C.4)$$

where we have used **B** = const (at constant r), and $\oint ds = 2\pi r$. Inserting B as taken from Eq. (19.A.23), we find from (19.C.4)

$$\oint \mathbf{B} \cdot d\mathbf{r} = \left(\frac{2K'I}{r}\right)(2\pi r) = 4\pi K'I$$

in agreement with (19.C.3). We see, incidentally, that the direction in which we loop around the wire must be guided by the right-hand rule in order that both sides of (19.C.3) should agree in sign.

General Verification

When the integration path is arbitrary, and furthermore winds around an arbitrarily shaped current, the verification of Ampère's law is somewhat more involved. A note at the end of this chapter shows how to carry it out.

iii. Thick Wires

Before we apply Ampère's law to any situation, we must examine the shape of the field lines. For practice, we consider a solid vertical cylindrical wire; see Fig. 19.16. The (vertical) current is

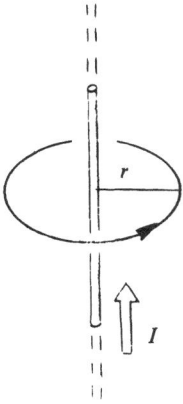

Figure 19.15. A special case of Ampère's law.

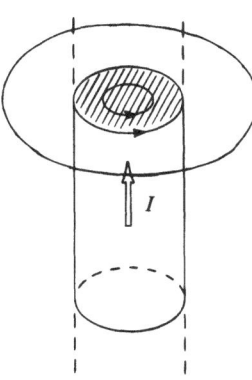

Figure 19.16. A thick wire with two representative magnetic field lines.

uniformly distributed over the cross section. We demonstrate in a note at the end of the chapter that the field lines are circles, concentric with the wire, and in horizontal planes, *just as for a thin wire*; the magnitude B is constant along each field line. These features hold true inside the wire as well as outside it. Furthermore, the **B** field, *outside the wire*, does not depend in any way on the wire's thickness, but only on its total current; the derivation of this result, as well as a calculation of B inside the wire, is left to Problems 19.38 and 19.39.

The following example shows how to use our knowledge of the field line's shape in order to apply Ampère's law; it also shows how to treat a two-way current.

Example 19.3. Field inside a Coaxial Cable. A long straight cylindrical cable consists of a thin insulated wire at the axis, with current $I = 5.0$ A flowing into the plane of Fig. 19.17, and a surrounding cylinder of radius $a = 3.0$ cm, with an equal return current of 5.0 A, uniformly distributed over its cross section. Find the value of B at a distance of $r = 2.0$ cm from the thin wire.

In Ampère's law,

$$\oint \mathbf{B} \cdot d\mathbf{r} = 4\pi K' I_{\text{enclosed}} \qquad (19.C.5)$$

we select an integration path which coincides with a field line of radius r. The left side of (19.C.5) is

$$\oint \mathbf{B} \cdot d\mathbf{r} = \oint B \, ds = (B)(2\pi r) \qquad (19.C.6)$$

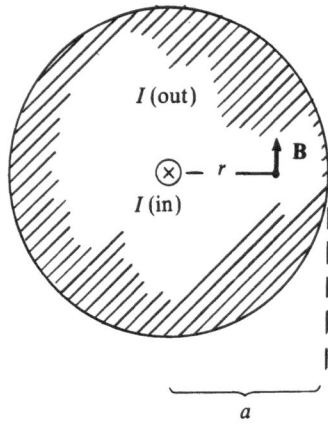

Figure 19.17. Cross section of a coaxial cable.

iv. Solenoids

A solenoid, shown in Fig. 19.18, is a more interesting configuration than a straight wire. As we shall see, it does for the magnetic field what the parallel-plate capacitor does for the electric field: provide a small chamber in which a uniform field of controlled strength may be set up. It is therefore much used in the laboratory and in electric instrumentation. Several electromagnetic principles can be learned from it as well.

Let us calculate the **B** field produced by a perfectly smooth and infinitely long solenoid; this will approximate the field of a real solenoid whose end effects are negligible.

To evaluate the right side of (19.C.5) we note that the contour encloses an amount of return current of magnitude I' which is proportional to the cross section it occupies; thus,

$$I' = \frac{\pi r^2}{\pi a^2} I = \frac{r^2}{a^2} I \qquad (19.C.7)$$

Since I' is opposite in direction to I, it must be subtracted:

$$I_{\text{enclosed}} = I - I' = I - \frac{r^2}{a^2} I \qquad (19.C.8)$$

Equation (19.C.5) now reads

$$(B)(2\pi r) = 4\pi K' \left(1 - \frac{r^2}{a^2}\right) I$$

so that

$$B = \frac{2K' \left(1 - \dfrac{r^2}{a^2}\right) I}{r} \qquad (19.C.9)$$

$$= \frac{(2)(10^{-7})[1 - (2.0/3.0)^2](5.0)}{2.0} \text{ tesla}$$

$$= \underline{\underline{2.8 \times 10^{-7} \text{ tesla}}}$$

We note that $B = 0$ *outside* the coaxial cable. (Why?)

Direction of B

We claim that **B** is everywhere parallel to the axis. To see this, we first rule out a radial component of **B** by the same symmetry argument as for the cylindrical wire of Fig. 19.16; that argument is set out in note (iii) at the end of the chapter. Next we consider circles in the same plane as, and concentric with, one of the solenoid's turns. The **B** field has zero component tangential to these circles. As a proof, we apply Ampère's law, using one of the circles as an integration path; see Fig. 19.19. We find

$$(B_{\text{tangential}})(\text{circumference}) = 0$$

because zero net current threads the path. (Each turn of wire is considered a closed circle.)

Magnitude of B

We first note that **B** = **0** *everywhere outside the solenoid*. This may be seen from Ampère's law applied to the path of Fig. 19.20, together with the reasonable assumption that **B** vanishes at an infinite distance from the solenoid.

Next, we use Ampère's law with the path of Fig. 19.21, in order to calculate the magnitude B inside the solenoid. We find that $\mathbf{B} \cdot d\mathbf{r}$ contributes along segment XY only. Indeed, we have **B** = **0** outside, while, along $X'X$ and YY', **B** is perpendicular to the path ($\mathbf{B} \cdot d\mathbf{r} = 0$).

Let s be the length of XY; the enclosed current is then $I_{\text{enclosed}} = NI$, where N is the number of wire

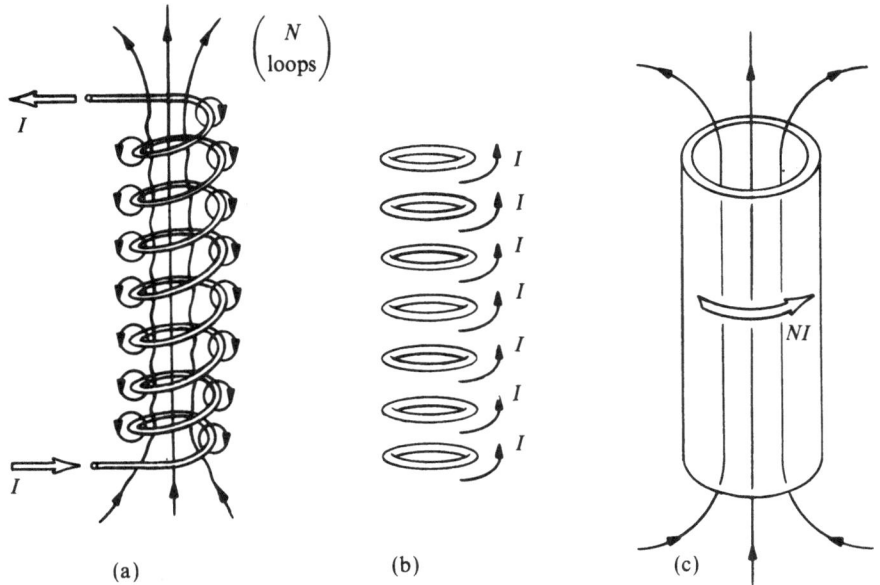

Figure 19.18. (a) A real solenoid, with some of its magnetic field lines. (b) An almost equivalent stack of loops (field lines omitted). (c) An almost equivalent uniform cylindrical shell of current, with some field lines. (Unless otherwise mentioned, a solenoid is assumed to have the shape of a right cylinder.)

turns within s, and I is the current in each turn. Thus, Ampère's law reads

$$Bs = 4\pi K' NI \qquad (19.C.10)$$

or

$$\boxed{B = \frac{4\pi K' NI}{s}} \qquad (19.C.11)$$

There is no dependence on position. To summarize: *Inside a long solenoid, its* **B** *field is uniform and*

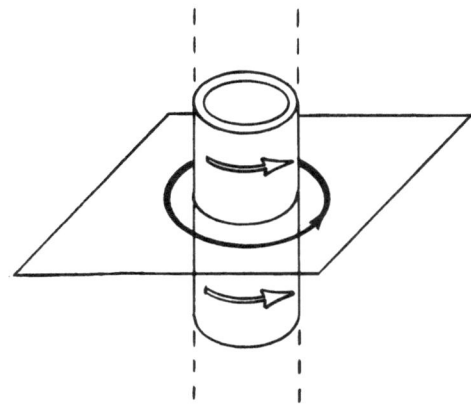

Figure 19.19. The integration path (heavy line) is threaded by zero net current.

proportional to the current per unit length of solenoid; outside the solenoid, its **B** *field vanishes (end effects are neglected).*

Example 19.4. A Current Balance. [The absolute determination of a current is carried out with great precision by **current balances**, of which the following type is due to Joseph Pellat (1887).] A horizontal balance beam (Fig. 19.22) has a flat horizontal coil rigidly attached to one end; a tray is hanging from the other end at

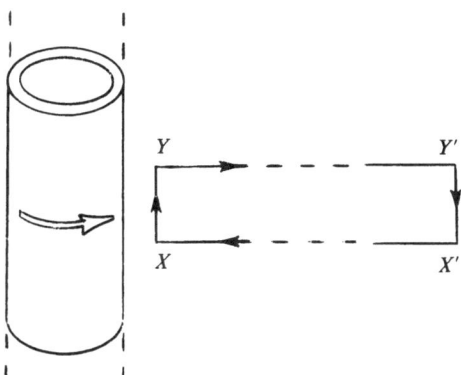

Figure 19.20. Because **B** is vertical, segments YY' and $X'X$ do not contribute to $\oint \mathbf{B} \cdot d\mathbf{r}$. Segment $Y'X'$ contributes negligibly. Thus, we have $\oint \mathbf{B} \cdot d\mathbf{r} = (B)|\overrightarrow{XY}|$, where B is evaluated at the segment XY, and where $|\overrightarrow{XY}|$ is the length of that segment. Since, also, $\oint \mathbf{B} \cdot d\mathbf{r} = 0$ (why?), we see that $B = 0$ at segment XY.

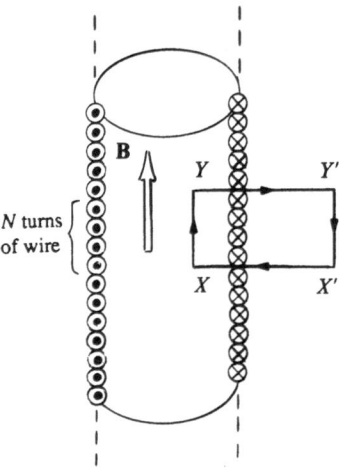

Figure 19.21. Using Ampère's law along path $XYY'X'X$ in order to evaluate B inside the solenoid.

a distance $x = 15$ cm from the fulcrum. The coil is surrounded by a long solenoid (length = $s = 20$ cm) whose axis coincides with the beam. The solenoid has $N = 400$ turns of wire, while the coil has $N' = 8$ turns, of area $\mathscr{A} = 2.0$ cm^2; both carry the same unknown current I (they are connected in series by flexible contacts). Equilibrium exists with $I = 0$ and an empty tray; a mass $m = 0.030$ gram is needed in the tray to restore equilibrium after I is turned on. Find the magnitude of I.

As usual when calculating magnetic forces or torques, we assume the coil "sees" the solenoid's field, but not its own. Here each turn of the coil is parallel to the solenoid's **B** field whose strength is

$$B = \frac{4\pi K' N I}{s} \qquad (19.C.12)$$

The total torque on the N' turns is therefore [see Eq. (18.C.15) of the preceding chapter]

$$\tau = N' \mathscr{A} I B \qquad (19.C.13)$$

$$= \frac{4\pi K' N N' \mathscr{A} I^2}{s} \qquad (19.C.14)$$

after insertion of (19.C.12). As we recall, this torque arises from a couple, whose axis is perpendicular to the plane of Fig. 19.22 but is otherwise arbitrarily located.

Compensation of torques occurs when the added weight mg and the corresponding additional force from the fulcrum combine into a couple whose torque has a magnitude equal to the coil's,

$$mgx = \tau \qquad (19.C.15)$$

Comparing (19.C.14) and (19.C.15), and solving for I, we have

$$I = \left(\frac{mgxs}{4\pi K' N N' \mathscr{A}}\right)^{1/2} \qquad (19.C.16)$$

$$= \left[\frac{(0.03 \times 10^{-3})(10)(0.15)(0.20)}{(4\pi)(10^{-7})(400)(8)(2.0 \times 10^{-4})}\right]^{1/2} \text{ amperes}$$

$$= \underline{3.3 \text{ amperes}}$$

Other Long Solenoids

A long straight solenoid *may have a base of arbitrary shape*, as illustrated in Fig. 19.23; what we have learned for the circular-base solenoid remains valid. Neglecting end effects, we state (without proof) that the **B** field is confined inside; it is uniform and parallel to the wall; its strength is still given by formula (19.C.11).

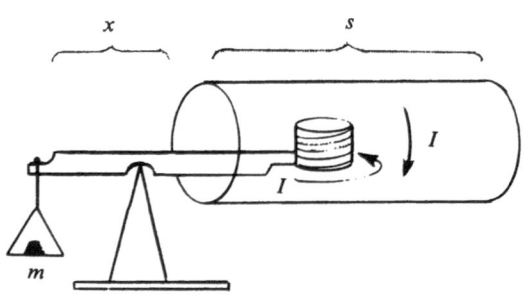

Figure 19.22. A Pellat current balance. In this illustration, the interior field of the long solenoid is horizontal and to the right, as the reader should verify.

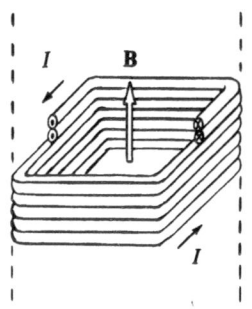

Figure 19.23. A solenoid with a noncircular base.

Real Solenoids

A solenoid of finite length allows field lines to escape from its ends, and, thus, has some amount of **B** field outside its turns. That field, which we discuss in the next section, is of special interest because it makes the solenoid act like a bar magnet.

v. Magnetic Pressure

The wire of a solenoid experiences an outward push under the solenoid's own **B** field. This curious effect is illustrated in Fig. 19.24, which shows the end view of a solenoid. Mechanically, the effect of the **B** field on the solenoid's wall is like that of a pressurized gas on the wall of a pipe. Faraday, the inventor of the field-line concept, visualized the lines of **B** somewhat as a compressed bunch of thick rubber bands. While this picture should not be taken literally, it does give a feel for what goes on.

To verify the phenomenon on the basis of Fig. 19.24, we first use the right-hand rules to check (1) the direction of **B** as related to that of I, and (2) the direction of the forces as related to I and **B**. (Important: Going through these checks with a reversed I still yields outward forces.)

Let us now estimate the force **F** on a small length l of wire. The inward-facing side of the wire is in a field of magnitude B, while the outward side sees zero field. As a guess, we take an effective field B_{eff} that is a compromise between these two values,

$$B_{eff} = \tfrac{1}{2}(B+0) = \tfrac{1}{2}B \qquad (19.C.17)$$

The outward magnetic force on l is then

$$F = IlB_{eff} \sin 90° = \frac{IlB}{2} \qquad (19.C.18)$$

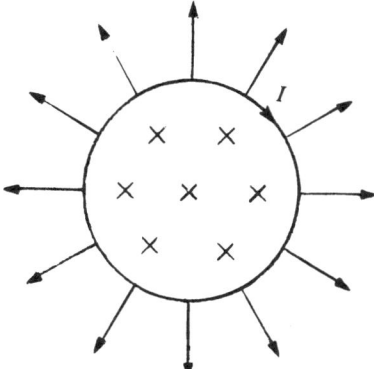

Figure 19.24. A solenoid, with field lines (crosses) pointing into the page; the radial arrows show the magnetic forces on the wire.

Since I is related to B, we can express F in terms of B alone. Solving Eq. (19.C.11) for I gives

$$I = \frac{sB}{4\pi K'N} \qquad (19.C.19)$$

where N is the number of turns in a length s of solenoid; (19.C.18) now reads

$$F = \frac{IsB^2}{8\pi K'N} \qquad (19.C.20)$$

To find the effective pressure, we consider an element $\mathscr{A} = ls$ of the solenoid's wall area (l measured along the wire, s across the wires). This area contains N elements of wire, and therefore undergoes a force

$$F_{\mathscr{A}} = NF = \frac{lsB^2}{8\pi K'}$$

The corresponding pressure is $F_{\mathscr{A}}/\mathscr{A} = F_{\mathscr{A}}/ls$, or

$$\boxed{\text{Magnetic pressure} = \frac{B^2}{8\pi K'}} \qquad (19.C.21)$$

In spite of our rough averaging in (19.C.17), the above turns out to be the exact result. (Caution: Unlike a hydrostatic pressure, this one is not direction-independent; here it strictly acts *transversely to the field lines*.)

Equation (19.C.21) points to an important limitation on the setting up of strong **B** fields, namely, the mechanical strength of materials.

Example 19.5. A Maximum B Field. A solenoid, whose wire has negligible mechanical strength, is wound inside a thin brass pipe. If this kind of pipe bursts at a gauge pressure of 3.0 atm (1 atm $\approx 10^5$ newtons/meter^2), how strong a **B** field can the solenoid be allowed to produce?

Equation (19.C.21) yields

$$B = [(8\pi K')(\text{pressure})]^{1/2} \qquad (19.C.22)$$

$$= [(8\pi)(10^{-7})(3 \times 10^5)]^{1/2} \text{ tesla} = \underline{0.87 \text{ tesla}}$$

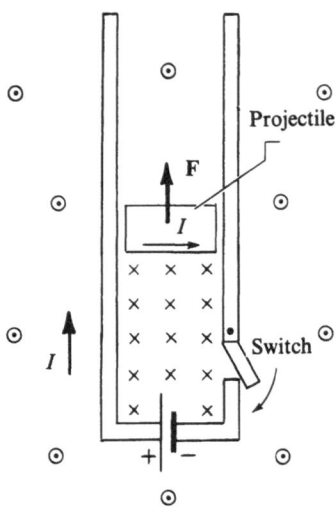

Figure 19.25. A rail gun. A powerful current source is connected to two vertical conducting rails. The circuit is completed by a conducting projectile and a switch. The field lines (crosses) through the circuit are denser than the return lines (circled dots), and hence the net magnetic pressure acts outwards.

The Rail Gun

Figure 19.25 shows how to launch a projectile by the explosive force of compressed field lines. The method works well in practice. (An alternative point of view: The side and bottom segments of the gun are Biot–Savart sources, contributing to **B** at the projectile's location. The right-hand force rule yields the direction of the force **F**. Omitting the Biot–Savart contribution of the projectile itself does not seriously invalidate this argument.)

D. Magnetic Poles : An Illusion

The end of a bar magnet behaves as a magnetic pole only when seen from outside the magnet's material. A microscopic observer, traveling among the atoms of the magnet in order to get an "inside view," would discover that the pole is nothing but an illusion, created by *the spinning motion of the atomic electrons about their own axis.* The resulting tiny current loops, which exist not only at the ends of the bar but throughout its substance, are what produces the observed field.*

The evidence against true poles can be summarized as follows.

---

* Caution: This is a figurative description. The correct one employs quantum mechanics.

1. Single poles cannot be isolated.
2. Magnetic theory has no need for poles: the Biot–Savart law, which only involves currents, suffices to explain any observed magnetic field; similarly, the magnetic forces on hidden currents suffice to explain all forces (and torques) acting on magnets. This economy of concepts is very appealing to theoreticians.
3. Modern physics has given us a good understanding of electron spin and of its relation to magnetic fields; we have a detailed mechanism for the existence of magnets.

The following material is mostly illustrative and makes no attempt at rigor or generality; it features *the solenoid as a model for the bar magnet.*

In order to understand a magnetic substance, we must look at it microscopically. Unfortunately, here we can give no more than a brief description of electron spin as the cause of magnetism in ironlike materials.

i. Gauss' Law for the B Field

If the field lines issuing from the ends of a magnet are not due to true poles, where do they come from? Figure 19.26a shows an idealized model of the actual situation: *the lines continue through the magnet without interruption*; this is how the "mystery zone" of Fig. 18.5, Chap. 18, must be visualized. Part (b) of Fig. 19.26 shows a hidden current which *could* be responsible for the lines in part (a).

Here we recall a feature of electrostatics. In the absence of charges, the net electric flux coming out of any closed surface is zero; equivalently, *the field lines are uninterrupted in the absence of charges.* This is one aspect of Gauss' law.

In Fig. 19.26 we have, likewise, drawn uninterrupted field lines, and we have assumed the absence of true magnetic poles. In order to establish the correctness of that model, therefore, we must show that Gauss' law is valid for the **B** field, *as produced by currents.* This is done in analogy with electrostatics; however, we must start from a current element rather than from a point-charge.

Figure 19.27 shows that the field $d\mathbf{B}$ contributed by a current element $I\, dl$ obeys Gauss' law: the field lines for $d\mathbf{B}$ are uninterrupted circles. The next step is to look at the full field **B**. When the contributions

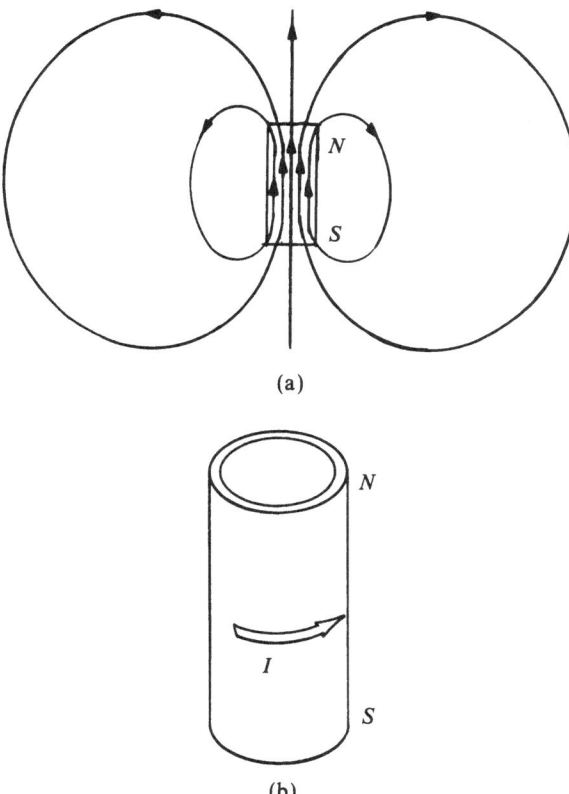

(a)

(b)

Figure 19.26. (a) Magnetic field lines inside a bar magnet. (b) A possible solenoid-type current, occupying the outer surface of the bar magnet. Here the solenoid has finite length; its end effects are essential, since they give rise to the outwardly observed field.

$d\mathbf{B}$ from all elements are added together, as the Biot–Savart law prescribes, the result, $\mathbf{B}$, again necessarily obeys Gauss' law. The proof involves the superposition principle just as in Chap. 15, and we shall not repeat it here; it must use field language rather than line language. The conclusion is

$$\begin{pmatrix} \text{Net magnetic flux coming} \\ \text{out of any closed surface} \end{pmatrix} = 0 \qquad (19.\text{D}.1)$$

or

$$\oiint B_\perp \, d\mathscr{A} = 0 \qquad (19.\text{D}.2)$$

Here $B_\perp$ represents the component of $\mathbf{B}$ normal to the chosen surface, on which $d\mathscr{A}$ is the element of area. "Flux" may be interpreted as "total number of lines." The zero right side of (19.D.1) or (19.D.2) tells us that magnetism is formulated without true magnetic poles.

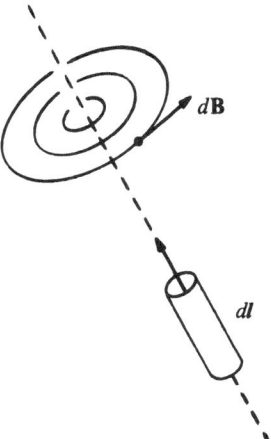

Figure 19.27. The (unobservable) field lines of a single current element; symmetry around its axis means that the line density is constant as one follows a field line. Hence the field lines are uninterrupted.

ii. Spinning Electrons as the Hidden Current Loops

Most electrons in a magnetic sample undergo spinning and orbital motions whose magnetic effects cancel one another. In certain chemical elements, however, each atom contains one or a few electrons (2 out of 26 in an iron atom) which "conspire" with their counterparts in neighboring atoms so as to spin about parallel axes, thereby combining their magnetic moments into a larger magnetic moment; see Fig. 19.28. Such substances

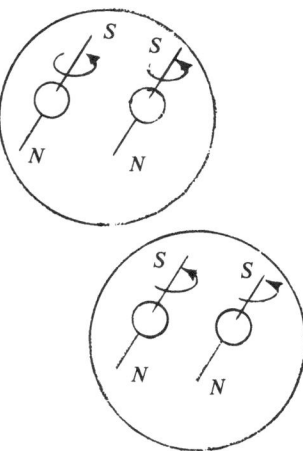

Figure 19.28. Schematic view of two neighboring atoms (large circles) in a piece of magnetized iron. Only two specially oriented spinning electrons (small circles) are shown in each atom. (Are the N and S poles incorrectly reversed? No, the electron's charge is negative.)

are called **ferromagnetic**, and can be made into magnets. Each spinning electron can be considered as a tiny loop of current, whose magnetic moment is

$$\mathcal{M}_{electron} \approx 9 \times 10^{-24} \text{ ampere meter}^2 \quad (19.D.3)$$

Ferromagnetism is found primarily in iron, nickel, and cobalt, as well as in many of their alloys and compounds.

An external **B** field can be used to promote the above-mentioned alignment of electrons' magnetic moments throughout a ferromagnetic object. That object then becomes a magnet, with its N pole pointing in the direction of the external **B**.

When the external field is removed, the magnet may persist (these "permanent" magnets may nevertheless be demagnetized by a strong enough opposing **B** field); alternatively, the sample's magnetic moment may vanish when the external field does. Which of these behaviors (called "hard" or "soft," respectively) occurs depends on the alloy's composition and on its metallurgical preparation.

iii. A Solenoid as a Model for a Bar Magnet

Figure 19.29a shows very schematically the end view of a bar magnet, with the electron-spin currents responsible for its magnetic field. In parts (b) and (c), the picture is progressively made still more schematic, leading to an interpretation in terms of current loops girdling the magnet in solenoid fashion. We see that the "equivalent-solenoid" picture is rather rough, but it is a simple and useful model for actual magnets.

We therefore ask: Given an N-turn solenoid of length s, base area $\mathcal{A}$, and current I, what pole strength, $\pm P$, does each end simulate?

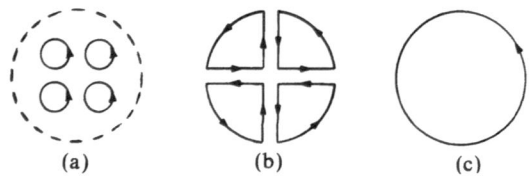

Figure 19.29. (a) End view of four aligned spinning electrons; the small loops represent their effective currents. (b) A distorted view, which is meant to show that the touching currents cancel one another. (c) A loop equivalent to (b).

The key to predicting the strength P is to note that as many field lines emerge from the solenoid as are inside it (Gauss' law). In terms of the inner field B, we have a total number of lines (i.e., a magnetic flux)

$$\mathcal{N} = B\mathcal{A} = \left(\frac{4\pi K' NI}{s}\right)\mathcal{A} \quad (19.D.4)$$

see Eq. (19.C.11); also, we recall from our study of magnets that a pole P sends out a number of lines

$$\mathcal{N} = 4\pi K' P \quad (19.D.5)$$

Comparing (19.D.4) and (19.D.5),

$$\boxed{P = \frac{NI}{s}\mathcal{A}} \quad (19.D.6)$$

the desired result. *The solenoid's end mimicks a pole whose strength is proportional to the current per unit solenoid length, and proportional to the solenoid's base area.* Formula (19.D.6) reminds us, also, that the SI unit of magnetic pole is the meter · ampere; see Eq. (18.A.24) of the preceding chapter.

Example 19.6. Force between a Solenoid and a Loop. A long solenoid, with $N/s = 100$ turns per centimeter, has one of its ends at the center of a circular loop, as in Fig. 19.30. The solenoid has a base area $\mathcal{A} = 0.50 \text{ cm}^2$ and a current $I = 4.0$ amperes, while the loop has a radius $r = 8.0$ cm and a current $I' = 20$ amperes. Find the force **F** exerted by the solenoid on the loop. (The solenoid's other end is far enough to be neglected.)

_____

The solenoid's end acts like a positive pole

$$P = \frac{N}{s} I\mathcal{A} \quad (19.D.7)$$

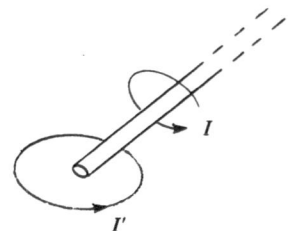

Figure 19.30. Solenoid and loop.

The loop's field, at that location, is [see Eq. (19.A.10)]

$$B_{\text{loop}} = \frac{2\pi K' I'}{r} \qquad (19.D.8)$$

vertically up. Hence the force on the solenoid is a vertically upward one, of magnitude

$$F = P B_{\text{loop}} = \left(\frac{N}{s} I \mathscr{A}\right)\left(\frac{2\pi K' I'}{r}\right) \qquad (19.D.9)$$

$$= \left(\frac{100}{10^{-2}}\right)(4.0)(0.50 \times 10^{-4})$$

$$\times \frac{(2\pi \times 10^{-7})(20)}{8.0 \times 10^{-2}} \text{ newton}$$

$$= \underline{3.14 \times 10^{-4} \text{ newton}}$$

The required force is the reaction to this: it also has this magnitude and points <u>vertically down</u>.

iv. Electromagnets

An electromagnet, illustrated in Fig. 19.31, is one of the most effective sources of magnetic field. It was invented in 1823 by William Sturgeon and further developed by Joseph Henry, who could, by 1831, lift more than a ton of iron with a current from an ordinary battery.

The device demonstrates the huge contribution which a **soft-iron core** can make to the magnetic field of a coil. We state, without theory, some features of that core, which are due to the lining up of some of its electron spin directions.

1. Wherever its continuity is broken by a gap, as at X and Y in the figure, the core exhibits strong pairs of opposite and equal poles.
2. These poles change sign when the current in the coil is reversed.
3. The coil's **B** field is enormously intensified in the core; a factor of 100,000 can be realized over what would be observed in air with the same current.
4. The core confines within itself almost all the magnetic field lines produced by the coils.
5. For full effect, an unbroken iron "circuit" is essential; the slightest gaps, such as those in the figure, severely weaken the field everywhere in the core.

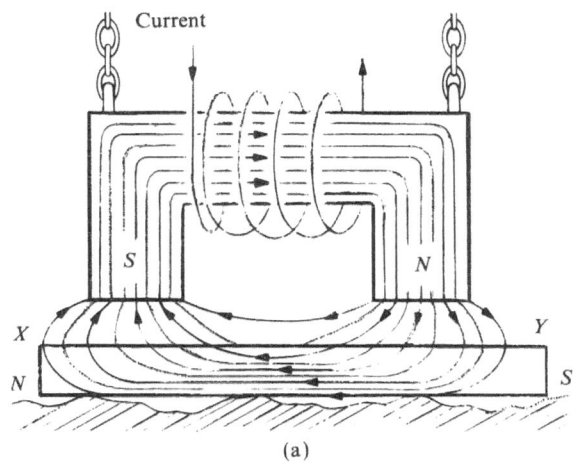

(a)

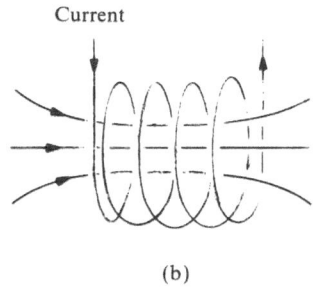

(b)

Figure 19.31. (a) An electromagnet, about to lift an iron rail. The latter will serve to complete the core "circuit." (b) **B** field of the same coil, with the same current, but without the core.

Notes

i. The Long-Wire Formula from Action and Reaction

Here we turn to a "trick" derivation of formula (19.A.23), $B = 2K'I/r$. The idea is to consider the mutual force between a flat row of bar magnets and a long straight loop of wire with current I, as illustrated in Fig. 19.32. The north poles (of total amount P) are in the plane of the loop and divide it lengthwise in half; the south poles are far enough to be neglected. Except for its shorter sides, which we ignore, the loop amounts to a pair of long straight wires. In what follows we (1) calculate the force exerted by the magnets on the loop; (2) require this to equal minus the force exerted by the loop on the magnets; and (3) consider the magnets as providing a row of test poles which measure the magnetic field of the two long wires. In short, the argument follows the pattern of this chapter's Sec. B.

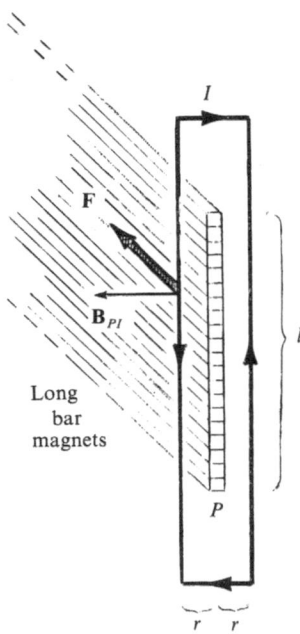

Figure 19.32. A row of north poles in a long rectangular loop. With *l* large enough, any effects in the vicinity of the two shorter sides ("end effects") can be ignored.

1. Force on the Loop. We first ask for the field $\mathbf{B}_{PI}$ due to the poles, and felt at any point on the loop's wire. This magnetic field looks mathematically just like the *electrostatic* field $\mathbf{E}$ of a line of charges. In Chap. 15, and with the help of Gauss' law, we found

$$E = \frac{2K(Q/l)}{r} \qquad (19.\text{N}.1)$$

where Q/l is the charge per unit length, and r is the distance from the charges. In the present (magnetic) case we must make the replacements $E \to B_{PI}$, $K \to K'$, $Q \to P$; r is half the loop's width. We obtain

$$B_{PI} = \frac{2K'(P/l)}{r} \qquad (19.\text{N}.2)$$

Therefore the magnet's force on either wire has magnitude

$$F = IlB_{PI} \sin 90°$$

$$= (Il)\left(\frac{2K'P}{lr}\right)(1) = \frac{2K'PI}{r} \qquad (19.\text{N}.3)$$

$\mathbf{F}$ points the same way for both wires; the force exerted by the magnets on the loop is $2F$.

2. Force of Loop on Magnets. By Newton's third law, this must also have magnitude $2F$.

3. Field of the Loop. The long wires produce a field whose direction is assumed known: at the row of poles, both sides contribute the same field **B** perpendicular to the loop's plane. Thus, the actual field seen by the magnetic poles has magnitude $2B$. In terms of the row of test poles P, we have

$$2F = (P)(2B) \qquad (19.\text{N}.4)$$

so that

$$B = \frac{F}{P} = \frac{2K'I}{r} \qquad (19.\text{N}.5)$$

the desired formula.

ii. General Verification of Ampère's Law

In Sec. C we found it very easy to check Ampère's law,

$$\oint \mathbf{B} \cdot d\mathbf{r} = 4\pi K'I \qquad (19.\text{N}.6)$$

by considering a circular contour around a straight infinite wire. We now show that (19.N.6) remains true when I is carried by a wire of arbitrary shape, and when the left side of (19.N.6) is evaluated along an arbitrary closed path surrounding I, as in Fig. 19.33.

First we note that *the result of the integration*, $\oint \mathbf{B} \cdot d\mathbf{r}$, does not change when the path is shrunken at will, provided the current remains enclosed in it. To prove this remarkable assertion, we consider three ways of traveling around the current for the calculation of $\oint \mathbf{B} \cdot d\mathbf{r}$. Part (a) of the figure shows the original path, perhaps beginning and ending at Y, and giving a result $[\oint \mathbf{B} \cdot d\mathbf{r}]_{\text{original}}$. In part (b), we decide to end with a shortcut from X to Y, but, on passing X, we forget our intention and keep following the original path. When we reach Y, we suddenly remember; we backtrack to X, take the shortcut, and complete the roundtrip. This time we have completed a *double loop*—the original one, plus a smaller

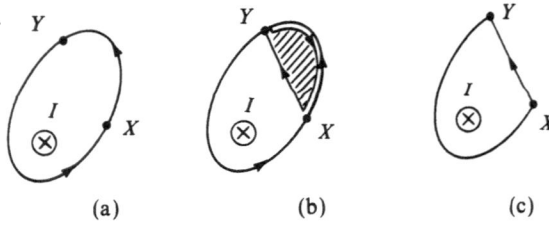

Figure 19.33. Equivalent paths for Ampère's law.

one which just encompasses the shaded region. Now the complete loop integral is

$$\left[\oint \mathbf{B} \cdot d\mathbf{r}\right]_{\text{double loop}} = \left[\oint \mathbf{B} \cdot d\mathbf{r}\right]_{\text{original}} + \left[\oint \mathbf{B} \cdot d\mathbf{r}\right]_{\text{shaded}}$$

$$(19.\text{N}.7)$$

but we recall from (19.C.2) that

$$\left[\oint \mathbf{B} \cdot d\mathbf{r}\right]_{\text{shaded}} = 0 \qquad (19.\text{N}.8)$$

since no current is enclosed. Thus we have not changed the value of the path integral:

$$\left[\oint \mathbf{B} \cdot d\mathbf{r}\right]_{\text{double loop}} = \left[\oint \mathbf{B} \cdot d\mathbf{r}\right]_{\text{original}} \quad (19.\text{N}.9)$$

Finally, in part (c), we attempt the trip once more. This time we do remember the shortcut, but again there is no change in the path integral. The reason is that, in (b), we have traveled the arc XY twice in opposite directions; these two contributions to $\oint \mathbf{B} \cdot d\mathbf{r}$ cancel out because every $d\mathbf{r}$ changes sign on the return trip. Thus, we have

$$\left[\oint \mathbf{B} \cdot d\mathbf{r}\right]_{\text{shrunken path}} = \left[\oint \mathbf{B} \cdot d\mathbf{r}\right]_{\text{double loop}}$$

$$(19.\text{N}.10)$$

Summarizing (19.N.9) and (19.N.10), we see that the shrunken path contributes the same integral as the original one; this was to be shown.

We now calculate $\oint \mathbf{B} \cdot d\mathbf{r}$ in Fig. 19.33 by shrinking the integration path into a very small circle centered on the current I. No matter what the shape of the wire, it can then be considered straight and infinite; we have verified this case in Sec. C.

For completeness, we must ask what happens when the path is shrunken around several currents. The answer is provided by Fig. 19.34.

iii. The Field Lines of a Thick Wire

How can we predict the concentric circles of Fig. 19.16? First, as in Fig. 19.35a, we imagine the thick wire to be made of many thin ones. As we know, each of these *by itself* would produce a horizontal $\mathbf{B}$, and hence the superposition of these fields is likewise horizontal. In short: The total $\mathbf{B}$ field has zero vertical component.

The following symmetry argument shows that $\mathbf{B}$ has zero radial component ($B_r = 0$) as well. [The meaning of B_r is illustrated in part (b) of the figure.] Suppose, contrary to what must be shown, that B_r is positive (outward).

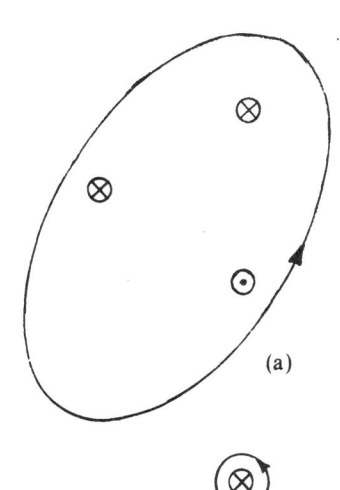

(a)

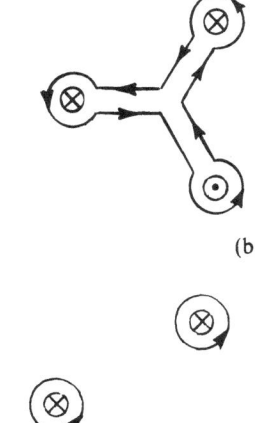

(b)

(c)

Figure 19.34. Shrinking a path when three currents are enclosed. (a) The original path. (b) The shrunken path; the inner connections cancel out. (c) We get the sum of three path integrals.

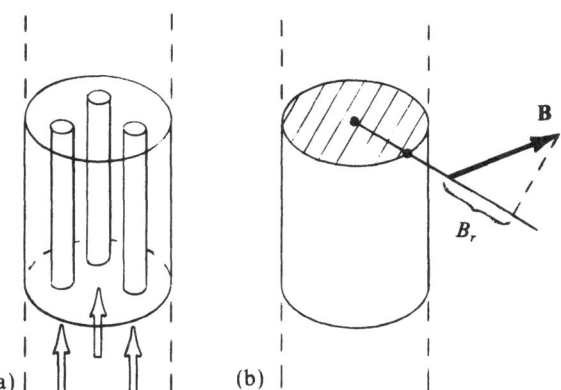

(a)   (b)

Figure 19.35. (a) A thick wire is equivalent to a bunch of thin wires (only three representative thin wires are shown). (b) An incorrect assumption: the radial component B_r should in fact be zero.

Let us reverse the current in the wire; since **B** is thereby reversed, B_r becomes negative (inward). Next let us, without changing the current, turn the wire top to bottom by 180° until the current is effectively reversed once more. During this operation, **B** has not changed its orientation *relative to the wire*; hence it still tilts inwards.

However, the new physical arrangement is completely equivalent to the original one, where **B** tilted outwards. This is absurd unless $B_r = 0$ exactly.

Only one component of **B** remains: it is tangent to one of the earlier-described concentric circles. There are no privileged horizontal directions in Fig. 19.35, and hence **B** must have the same value everywhere on its circle.

Condensed Checklist

(In this chapter, perhaps more than in any other, it is essential to have the correct geometric pictures in mind when the checklist is being reviewed.)

Law of Biot and Savart:

$$\boxed{\mathbf{B} = \int d\mathbf{B} = \int_{\text{wire}} \frac{K'I\, dl \times \mathbf{u}}{R^2}} \qquad (19.\text{A}.5)$$

$$|d\mathbf{B}| = \frac{K'I \sin\varphi}{R^2}\, dl, \qquad K' = 10^{-7} \quad (19.\text{A}.2)$$

Circular loop of radius r:

$$B = \frac{2\pi K'I}{r} \qquad \text{(center)} \qquad\qquad (19.\text{A}.10)$$

$$B = \frac{2\pi K'Ir^2}{R^3} \qquad \text{(anywhere on axis)} \qquad (19.\text{A}.14)$$

Long straight wire:

$$B = \frac{2K'I}{r} \qquad (19.\text{A}.23)$$

Parallel currents:

$$\frac{F}{l} = \frac{2K'I_1 I_2}{r} \qquad (19.\text{A}.28)$$

Ampère's law:

$$\boxed{\oint \mathbf{B} \cdot d\mathbf{r} = 4\pi K' I_{\text{enclosed}}} \qquad (19.\text{C}.3)$$

Field in a thick wire (radius $= a$):

$$B = \frac{2K'Ir}{a^2} \qquad \text{Problem 19.39}$$

Field in a long solenoid:

$$\boxed{B = \frac{4\pi K'NI}{s}} \qquad (19.\text{C}.11)$$

"Pole" at the end of a solenoid:

$$P = \frac{N}{s} I\mathscr{A} \qquad (19.\text{D}.6)$$

Magnetic pressure:

$$\text{Magnetic pressure} = \frac{B^2}{8\pi K'} \qquad (19.\text{C}.21)$$

Gauss' law (nonexistence of true poles):

$$\boxed{\oiint \mathbf{B}_\perp\, d\mathscr{A} = 0} \qquad (19.\text{D}.2)$$

Aligned electron spin (single electron):

$$\text{Magnetic moment} = 9 \times 10^{-24} \text{ ampere meter}^2$$
$$(19.\text{D}.3)$$

True or False

1. According to Biot and Savart, a current element contributes zero field at points along the straight-line extension of that element.

2. Magnetic field lines begin or end on currents.

3. The magnetic field lines of a circular loop current are circles parallel to the loop.

4. The magnetic field lines of a long straight wire are parallel to the wire.

5. If two current loops attract one another, and both currents are then reversed, the attraction becomes a repulsion.

6. Over and above their magnetic interaction, two wires carrying parallel currents repel one another electrostatically because the currents consist of moving electrons with the same (negative) sign.

7. Ampère's law is valid only in cases with cylindrical symmetry.

8. The path integral $\oint \mathbf{B} \cdot d\mathbf{r}$, evaluated over a (circular) field line around an infinite straight wire, becomes smaller when we choose a wider circle.

9. A thick cylindrical wire, carrying a uniformly distributed current in the axial direction, has its maximum magnetic field strength on the axis itself.

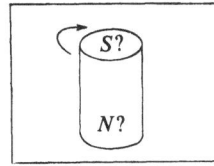

10. If, as in the adjoining figure, a black box contains a solenoid whose current circulates clockwise as seen from above, then external measurements cannot distinguish it from a bar magnet with its N pole down.

11. There is zero magnetic field outside the coaxial cable of Example 19.3.

12. If two solenoids, with the same **B** field, have different diameters, then the wider solenoid experiences the greater magnetic pressure.

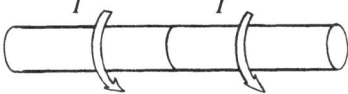

13. When two identical long solenoids, with equal currents, are put end to end with their currents in the same direction, as shown, the resulting double-length solenoid has, inside it, a **B** field twice as strong as before.

Problems

Sections A : The Law of Biot and Savart;
B : Proof by Action and Reaction

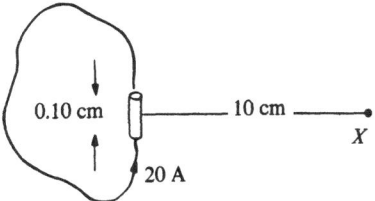

*19.1. Find, in direction and magnitude, the Biot–Savart contribution to **B**, at point X on the figure, of the indicated current element (a 0.10-cm wire segment carrying 20 A); X is at 10 cm from the element and in a direction at right angles to it.

*19.2. The B field at the center of a circular loop of radius 8.0 cm is measured to be 0.20×10^{-4} tesla. If that field is just due to the loop itself, what is the current in that loop?

*19.3. A small object of charge 6×10^{-7} coulomb is being whirled at 200 revolutions per second around a circle of radius 1.0 cm. Estimate what magnitude of B field this produces at the center of the circle.

19.4. In a simplified picture of the hydrogen atom, an electron (charge $= -e = -1.6 \times 10^{-19}$ coulomb) undergoes uniform circular motion around a fixed proton. If the radius of the orbit is taken as 5.3×10^{-10} meter, and the electron's speed as 2.2×10^6 meters/sec, find (a) the effective current when the orbit is viewed as a circular loop; (b) the strength of the **B** field at the proton's location.

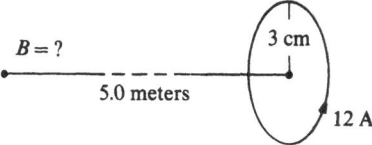

*19.5. What field B does a circular loop of current produce at a point on its axis, 5.0 meters from its center, as shown? The current is 12 A and the loop's radius is 3.0 cm.

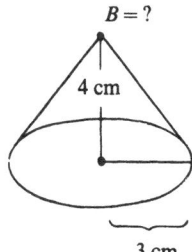

19.6. A circular current loop, with radius 3.0 cm and current 2.0 A, occupies the base of a right cone of altitude 4.0 cm, as shown. Find the magnetic field B at the apex of the cone.

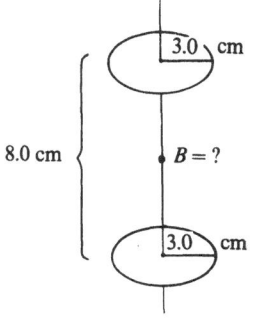

19.7. Two circular loops, each of radius 3.0 cm, have the same current (10 A) as shown. The loops are in two horizontal planes, separated by a distance of 8.0 cm, and their centers are on the same vertical line. Find B halfway between the two centers.

19.8. What is the magnitude of the **B** field at a distance of 1 meter from a long straight wire with current 1 A?

***19.9** (Magnetic units). On the basis of the force equation $\mathbf{F} = I l \times \mathbf{B}$, what is the meaning of the tesla in terms of coulombs, kilograms, meters, seconds? (b) Using the result of (a) as well as Eq. (19.A.22), find the dimensions of K' in terms of coulombs, kilograms, meters, seconds. (See also Problem 18.10.)

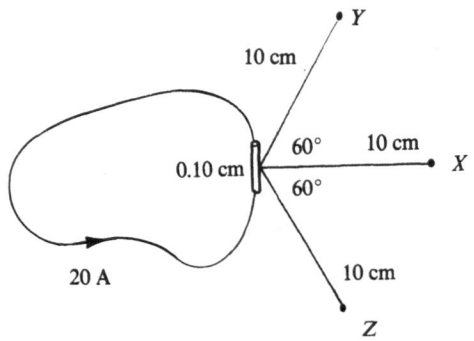

19.10. Repeat Problem 19.1 for points Y and Z lying at 10 cm from the element 60° forwards and backwards relative to X, as shown.

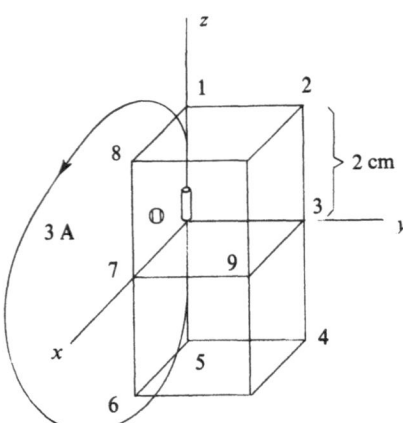

19.11. A wire with current 3 A has a small element of length 0.001 meter which is located at the origin $\mathbb{O}$ and points in the z direction, as shown. Find, in magnitude and direction, its Biot–Savart contribution to the **B** field at corners 1–9 of the two cubes in the figure. The cubes have 2-cm sides. (To indicate directions, make a sketch, with appropriately labeled angles.)

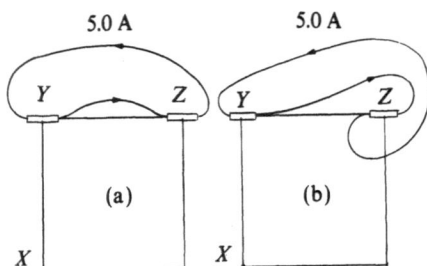

***19.12.** From a wire with current 5.0 A, we select two small elements, each of length 0.0010 meter, which lie as shown at corners Y, Z of a square with side 1.0×10^{-2} meter. Find, in direction and magnitude, the two elements' combined Biot–Savart contribution to the **B** field at point X, in the illustrated cases (a) and (b).

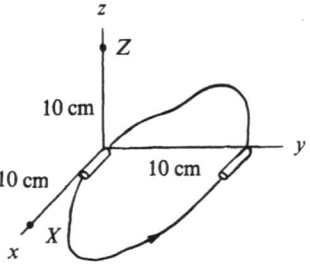

19.13. A loop with current 5.0 A lies in the xy plane, as shown, with two parallel elements, each 0.10 cm long, selected for our attention. What is their joint Biot–Savart contribution (in direction and magnitude) to **B**, (a) at X, (b) at Z? (The right-hand element, as well as points X and Z, are at 10 cm from the origin; the elements' length is exaggerated in the sketch.)

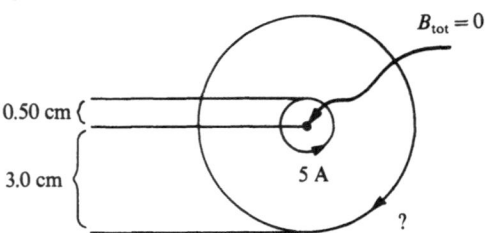

***19.14.** Two concentric circular loops, with radii 0.50 cm and 3.0 cm, as shown, cancel one another's **B** field at their center. If the smaller loop has a current of 5.0 A, what must be the current in the larger loop? (The sketch is not to scale.)

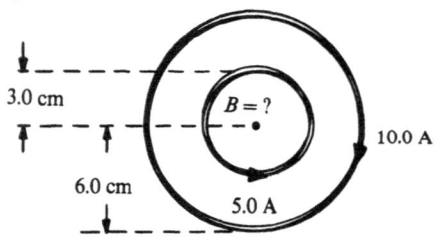

19.15. Two circular flat coils lie concentrically in the same plane, as shown. Coil No. 1 has a wire current of

5.0 A, 20 turns of wire, and a 3.0-cm radius. For coil No. 2 the corresponding figures are 10.0 A, 40 turns, and 6.0 cm. Find the magnitude and direction of **B** at the coil's center.

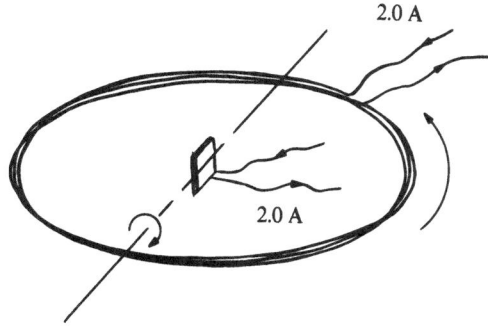

*19.16. A square 10-turn coil of side 0.50 cm and wire current 2.0 A is pivoted as shown at the center of a circular 50-turn coil of radius 12 cm and wire current 2.0 A. When the two coils are in perpendicular planes, as shown, what is, in magnitude and direction, the torque exerted by the small coil on the large one?

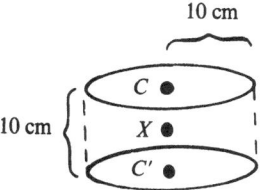

19.17. Two circular coils, each with 100 turns, an 8-A wire current, and a 10-cm radius, occupy the ends of a right cylinder of altitude 10 cm, as shown. The currents are in the same direction. Calculate B (a) at point X, the center of the cylinder; (b) at points C, C', the centers of the loops. (In the vicinity of X, the coils give their most uniform **B** field when spaced in this way; they are then known as **Helmholtz coils**.)

*19.18. Consider a simplistic model in which the Earth's magnetic field is due to a circular loop of current flowing under the equator at 1/4 of a terrestrial radius below the surface. (a) How much current is needed to produce the 7×10^{-5} tesla field at the Earth's poles? (Earth's radius $\approx 6.4 \times 10^6$ meters.) (b) Does the current flow from west to east or from east to west? (c) How much *surface* current, at the equator, would give the same effect at the poles?

19.19. Two long straight parallel wires, 6 cm apart, have currents in the same direction. Wire No. 1 carries 20 A, while wire No. 2 carries 40 A. At what distance from each wire, and in the same plane as both, is a point where their combined **B** field vanishes?

19.20. Two long wires, lying the the xy plane, are parallel to the y axis, at $x = 2$ cm and $x = 4$ cm. Their respective currents are 5.0 A in the $+y$ direction and 3.0 A in the $-y$ direction. Find the resulting **B** field (direction and magnitude) on the x axis at $x = 0$, 1 cm, 3 cm, 5 cm, 10 cm, 20 cm.

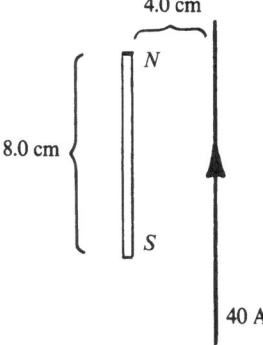

*19.21. An 8.0-cm-long bar magnet with pole strength ± 2.0 SI units is parallel to, and at 4.0 cm from, a long straight wire with current 40 A, as shown. Find, in direction and magnitude, (a) the force, (b) the torque, exerted by the wire on the magnet.

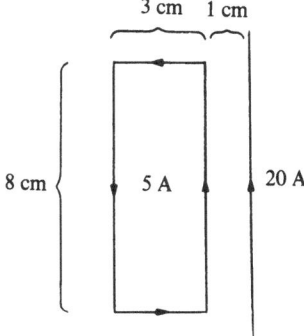

19.22. Find the direction and magnitude of the force exerted by the long straight wire on the rectangular loop in the figure. (Both objects lie in the same plane.)

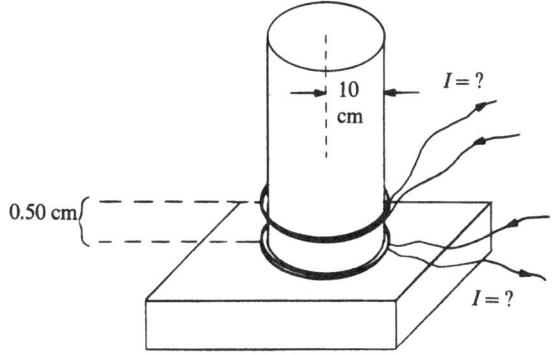

*19.23. Two 30-turn circular coils, each of radius 10 cm and carrying the same wire current, are loosely arranged, as shown, around a vertical wooden column. The upper coil, of mass 1.0 kg, is kept magnetically 0.50 cm above the bottom one. Find the wire current in the coils. (Approximate each coil as a tight bunch of long straight wires.)

19.24. Do Problem 19.17 for points 10 and 11.

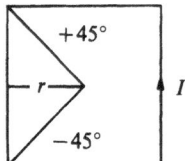

*19.25 (A square loop). (a) In the straight-wire configuration of Fig. 19.7, find the contribution to integral (19.A.22) made by the portion of wire between $\theta = \pm 45°$ [instead of $\pm 90°$ as in (19.A.22).] (b) Use this result to find B at the center of a square loop with current I and side $2r$, as shown. (c) What fraction is this of B at the center of a circular loop with current I and radius r?

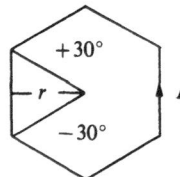

*19.26 (A hexagonal loop). (a) By considering, instead of (19.A.22), the integral $\int_{-30°}^{+30°} \cos \theta \, d\theta$, adapt Problem 19.25 to calculating B at the center of a regular hexagonal loop with current I and radius r for the inscribed circle; see the figure. (b) What fraction is this of B at the center of a circular loop with current I and radius r?

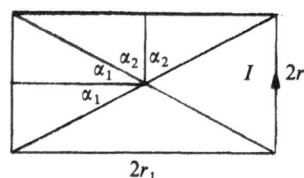

*19.27 (A rectangular loop). (a) By considering, instead of (19.A.22), the integral $\int_{-\alpha}^{+\alpha} \cos \theta \, d\theta$, where α is some fixed angle, adapt Problem 19.25 to calculating B at the center of a rectangular loop with current I, half-sides r_1, r_2, and corresponding angles α_1, α_2; see the figure. (Express your result in terms of r_1, r_2, α_1, α_2, I, K'. (b) Eliminate α_1, and α_2 by setting $\sin \alpha_1 = r_2/(r_1^2 + r_2^2)^{1/2}$, $\sin \alpha_2 = r_1/(r_1^2 + r_2^2)^{1/2}$. (c) Does your result reduce to that of Problem 19.25 as $r_1 = r_2 = r$?

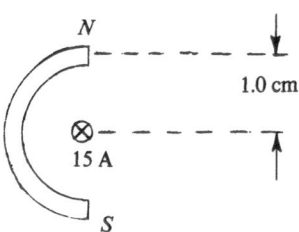

*19.28. A semicircular horseshoe magnet, of radius 1.0 cm, and characterized by two point poles of strength ± 5.0 SI units, lies in the plane of the figure, perpendicularly to, and concentrically with a long straight wire with current 15 A, directed into that plane. Find the direction and magnitude of the force exerted by the magnet on the wire.

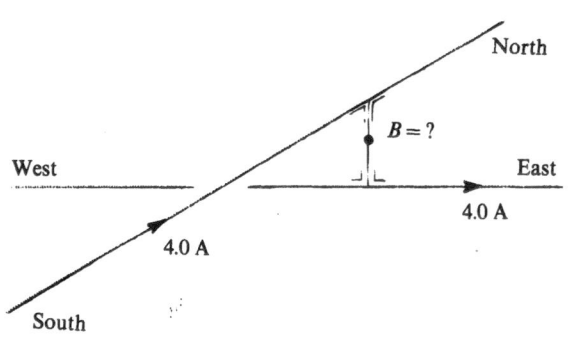

19.29. Two long straight horizontal wires, shown in the figure, have the same 4.0-A current. One wire goes from west to east; the other is 2.0 cm higher and goes from south to north. Find, in direction and magnitude, the **B** field at the midpoint of their common (vertical) perpendicular. (Ignore the Earth's field.)

19.30. Use $d\mathbf{F} = I \, d\mathbf{l} \times \mathbf{B}$ to calculate the force exerted by the bar magnet on the coil in Example 19.1. (Your result should agree with the law of action and reaction.)

Sections C : Ampère's Law; D : Magnetic Poles : An Illusion

*19.31. A 5.0-A current in a certain long straight cylindrical bar is found to cause a **B** field of strength 6.0×10^{-5} tesla at the bar's surface. What is the bar's radius?

19.32. Calculate the magnitude of the **B** field in a solenoid with length 25 cm, radius 1.5 cm, and 1000 turns of wire carrying a 0.75-A current.

*19.33. (a) How much current is needed in a 10-cm-long, 100-turn solenoid, with radius 0.50 cm, in order to produce a 0.02-tesla field inside? (b) What is the resulting magnetic pressure on the wires?

19.34. What is the effective pole strength at the end of a 100-turn solenoid of radius 3.0 mm, length 9.0 cm, and wire current 10 A?

*19.35. A bar magnet, of length 6.0 cm, cross section 0.80 cm^2, and pole strength 2.0 SI units, is being replaced by an equivalent 1200-turn solenoid with the same dimensions. What current is needed in the solenoid?

*19.36. Two identical 10,000-turn solenoids, with length 1.5 meters and base area 1.0 cm^2, are arranged parallel to one another, with an 8.0-cm separation, as shown. If each has a wire current of 0.50 A (directions as shown) what is the total force they exert on one another? (Neglect the interaction between top and bottom ends.)

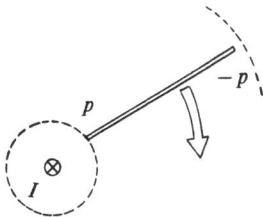

19.37. The figure shows a radially directed bar magnet with poles p, $-p$, which revolves in a plane perpendicular to a long, straight, constant current I. Prove that when each pole has described a complete circle (with the wire at the center), the wire's magnetic field has done zero net work on the magnet. [Your result is relevant to item (b) in the discussion of Fig. 19.13.]

19.38. (a) On the basis of Fig. 19.16, give a formula for B outside the wire, at a distance r from its axis; compare your result with Eq. (19.A.23). (b) If $I = 100$ A, and if the wire has a radius $a = 1.0$ cm, find B at its surface.

*19.39. (a) On the basis of Fig. 19.16, give a formula for B inside the wire at a distance r from its axis; assume the wire has radius a. (b) Combining this result with that of Problem 19.38, plot B against r from $r = 0$ to r much larger than a; in your graph, assume a total current $I = 100$ A and a radius $a = 1.0$ cm.

19.40. In the arrangement of Example 19.3, plot the B field against distance from the cable's axis, from r very small to r larger than 3.0 cm.

*19.41. A long solenoid, with 20 turns per centimeter, has a wire current of 12 A and a radius of 0.30 cm. Find the tension in the wire contributed by the magnetic pressure.

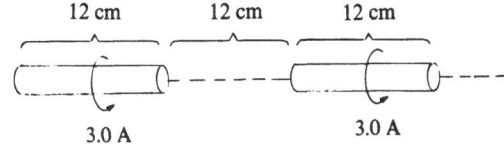

*19.42. Two identical 1000-turn solenoids, with length 12 cm and base area 2.0 cm^2, each carry a 3.0-A wire current. When positioned as shown, what force do they exert on one another?

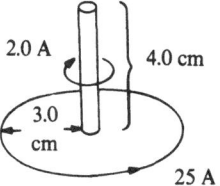

*19.43. A horizontal circular loop with current 25 A and radius 3.0 cm is placed with its center at one end of a 100-turn vertical solenoid of length 4.0 cm and base area 0.20 cm^2 as shown. If the solenoid's wire has a current of 2.0 A, find the force exerted by the solenoid on the loop.

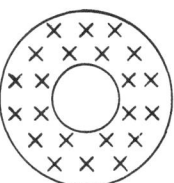

19.44. The figure shows a hollow cylindrical wire in cross section. Assuming that a uniformly distributed current flows into the page as shown (crosses), prove that its **B** field vanishes everywhere inside the wire's cavity.

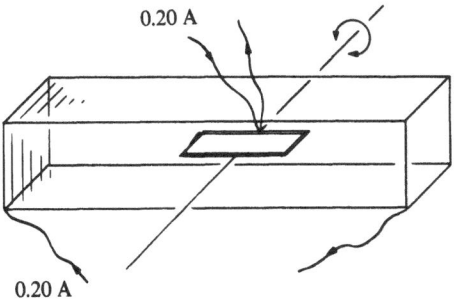

*19.45. A square coil, 1.0 cm on the side, with a current of 0.20 A and 100 turns of wire, is pivoted as shown inside a parallelopiped-shaped solenoid. The solenoid is 30 cm long and uniformly wound with 2000 turns, also carrying a 0.20-A current. Find the magnitude of torque exerted by the coil on the solenoid, when the plane of the coil is directed along the length of the solenoid.

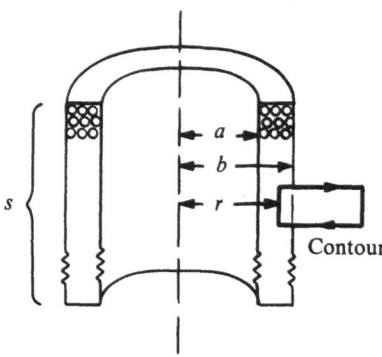

Contour

*19.46. A thick-walled solenoid of length s, inner radius a, and outer radius b contains N turns of wire which fill the wall uniformly, as shown, and which carry a current I. The length s is much larger than a or b. Given that the solenoid's **B** field is parallel to its axis, and that it vanishes outside the wall, (a) use Ampère's theorem with the contour shown in order to find a formula for B *in the wall* at a distance r from the axis. (b) Plot B against r, from $r = 0$ to $r > b$; in your graph, use the values $a = 1$ cm, $b = 2$ cm, $s = 10$ cm, $N = 1000$, $I = 1$ A. (In this problem, we are smoothing the field irregularities caused by the individual turns of wire.)

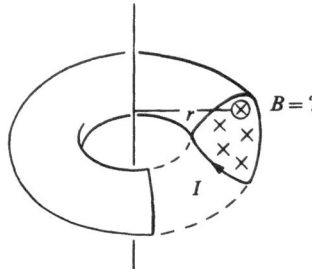

$B = ?$

19.47 (Toroidal coil). A coil, of arbitrarily shaped cross section, has rotational symmetry about some fixed axis, as shown. There are N turns of wire, carrying a current I. The magnetic field lines (shown as crosses, and confined inside the coil) are circles, coaxial with the coil. Find the field strength B at a distance r from the axis. (Use Ampère's law; express your answer in terms of K', I, N, r; compare your result with that for a solenoid of length $2\pi r$.)

Answers to True or False

1. True.
2. False (they loop around currents).
3. False (see item 2).
4. False (see item 2).

5. False (stays the same).
6. False (there are stationary positive ions as well).
7. False (always valid).
8. False (path independent).
9. False ($\mathbf{B} = 0$ on axis).
10. True.
11. True.
12. False (pressure depends only on **B**).
13. False (no change).

Answers or Hints to Problems

19.1. 2.0×10^{-7} tesla.

19.2. 2.5 A.

19.3. Use an effective loop current of $(6 \times 10^{-7}$ coulomb$) (200 \text{ sec}^{-1})$.

19.5. In formula (19.A.14), take $R \approx 5.0$ meters.

19.9. (b) kg meters coulomb^{-2}

19.12. (a) 6.8×10^{-6} tesla; (b) 3.2×10^{-6} tesla; perpendicularly into the figure in both cases.

19.14. 30 A.

19.16. Approximate the field of the large coil by its value at the center; calculate the torque on the small coil; the reaction torque is opposite.

19.18. (a) 8×10^{8} A.

19.21. (a) zero; (b) 3.2×10^{-5} newton meter.

19.23. 21 A.

19.25. (b) $4\sqrt{2}K'I/r$; (c) 90%.

19.26. (a) $6K'I/r$; (b) 95.5%.

19.27. (b) $4K'I(r_1^2 + r_2^2)^{1/2}/r_1 r_2$.

19.28. Find the reaction force first.

19.31. 1.7×10^{-2} meter.

19.33. (a) 16 A.

19.35. 1.25 A.

19.36. First calculate the effective poles.

19.39. (a) $2K'Ir/a^2$.

19.41. See Example 9.2 of Chap. 9; note that (force on 1/2 turn) = (2) (tension), as in Problem 18.56.

19.42. See Problem 18.12.

19.43. See Example 19.1.

19.45. Use action–reaction for torques.

19.46. (a) $\dfrac{4\pi K'NI}{s}\left(\dfrac{b-r}{b-a}\right)$.

CHAPTER 20

MAGNETIC INDUCTION

A current produces a magnetic field. Conversely, does a magnetic field produce a current? Such an effect was eagerly looked for (by Faraday, among others) in the days following Oersted's experiment. There were dreams of generating currents without the help of batteries, by means of wire loops cleverly located in the vicinity of permanent magnets. Those efforts were completely fruitless until Faraday discovered that *the loops and magnets have to move with respect to one another* in order to generate electric currents. His experiments and those of Joseph Henry, independently performed in the 1830s, established the principle of **magnetic induction*** : *a time-varying magnetic field generates an electric field; this electric field can produce a current in a closed wire loop.*

Faraday could not, in general, calculate the strength or direction of the generated E field, but he discovered how to predict its cumulative effect around a complete circuit. That effect may be measured in volts, and is known as an **electromotive force** (emf), just as in a battery. According to Faraday, the emf around a circuit equals the rate at which *magnetic* field lines get removed from the total line count ("magnetic flux") enclosed by the circuit.

That discovery—the **law of induction**—has revolutionized technology as well as science. Dynamos, transformers, and oscillating (alternating-current) circuits are the most immediate applications; they will be discussed in this chapter and the next.

When, further in this book, we reach the topic of electromagnetic waves, we shall see that Faraday's

induction principle is one key factor in understanding why such waves must exist.

It will help to keep in mind the plan of this chapter. First, we demonstrate that any piece of equipment that moves in a static **B** field experiences an additional **E** field due to its motion. Second, we use that result to show the existence of an emf in a wire loop that leaves or enters a static **B** field; we calculate that emf—that is to say, we derive Faraday's law—from the conservation of energy. Eventually, we shift our point of view: the loop is considered at rest, while the **B** field varies in time.

A. Electric Field Seen by a Moving Object

Consider an idealized instrument, an "E meter," which serves to measure the magnitude and direction of the electric field at any location. Such a meter is illustrated in Fig. 20.1.

In what follows, we shall be dealing with fields as seen by observers in motion. The **E** meter of Fig. 20.2 is one such "observer." What it reads is *defined* as the **E** field in a reference frame which moves along with the meter.

To put it differently, the meter "does not care" whether or not it is moving; it always "considers itself" to be at rest when giving a reading. For consistency, we must then also assume that any other piece of equipment (e.g., a piece of conductor), traveling along with the **E** meter, experiences the same **E** field as measured by the meter.

This way of thinking is applied in Fig. 20.2, in

---

* Unrelated to the electrostatic induction of Chap. 15.

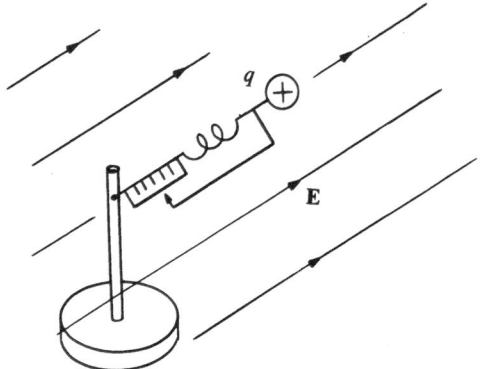

Figure 20.1. A fanciful "**E** meter," consisting of a positive charge q attached to a directional force meter. Gravity is assumed negligible. If the force on the charge is **F**, the scale is calibrated to read $E = F/q$.

which we assume zero electric field, $\mathbf{E} = \mathbf{0}$, but in which the meter is being moved at a speed **v** perpendicular to a local **B** field. From Chap. 18 we know that the test charge q undergoes a force

$$\mathbf{F} = q\mathbf{v} \times \mathbf{B} \qquad (20.\text{A}.1)$$

That force is shown in the figure.

Let us now describe the measurement as it is seen by an observer who moves along with the **E** meter. To him, the test charge is at rest, and therefore *it cannot experience any magnetic force.* Thus, that observer considers **F** to be purely electric:

$$\mathbf{F} = q\mathbf{E}' \qquad (20.\text{A}.2)$$

Here **E**′ is the electric field which the observer measures with his "stationary" **E** meter; **E**′ is called

the **motional electric field**. Comparison with (20.A.1) yields

$$\boxed{\mathbf{E}' = \mathbf{v} \times \mathbf{B}} \qquad (20.\text{A}.3)$$

In summary: *An observer who moves across a pure, static magnetic field* (no electric field present) *sees, in addition to that magnetic field, a transverse electric field* as well. That electric field is given by the formula for the magnetic force on a unit charge.

At this stage, our outlook has deepened in an important way. An electric field is no longer an "absolute" phenomenon to us. Rather, we see that *the observed electric field depends on the observer's motion*, at least when a magnetic field is present. That conclusion foreshadows the special theory of relativity.

Example 20.1. A Magnetohydrodynamic (MHD) Generator. A horizontal stream of mercury passes in an insulating pipe at a speed $v = 5.0$ meters/sec perpendicularly to a horizontal field $B = 0.30$ tesla; see Fig. 20.3. Calculate the motional electric field **E**′ that the mercury experiences. (In a completed circuit, involving a stationary electrode pair and a wire, **E**′ will produce an electric current I, as illustrated in the figure; the flowing mercury becomes a **generator**.)

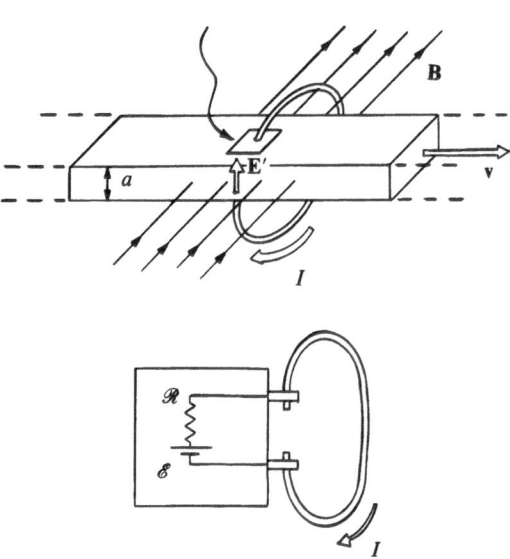

Figure 20.3. An MHD generator. Only one of the two electrodes is visible. The mercury flows to the right. The generator performs like a battery (illustrated).

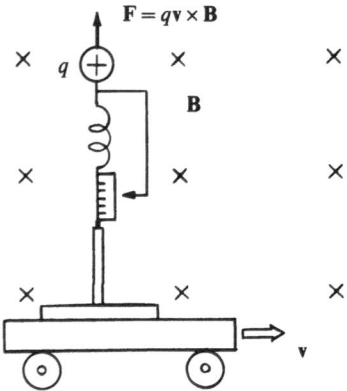

Figure 20.2. An **E** meter traveling in a static **B** field (**B** perpendicularly into the page).

The right-hand rule yields the direction of E'; that direction, vertically upwards, is shown in Fig. 20.3. In magnitude, we have

$$E' = vB \sin 90° \qquad (20.A.4)$$

$$= (5.0)(0.30)(1)\frac{\text{volts}}{\text{meter}} = 1.5\,\frac{\text{volts}}{\text{meter}}$$

$$(20.A.5)$$

Electromotive Force from Motion in a B Field

Should we expect the MHD generator to exhibit an emf $\mathscr{E}$, like a battery? The answer is yes. Between the two electrodes in Fig. 20.3 there exists, just as in a battery, a potential difference $V = \mathscr{E} - \mathscr{R}I$; here $\mathscr{R}$ is the resistance of the mercury path and I is the produced current. In the following we estimate $\mathscr{E}$ by a dimensional argument.

Example 20.2. Estimating the emf. Assuming that $\mathscr{E}$ depends only on E' and on the width, say $a = 2.0$ cm, of the mercury stream in Fig. 20.3, find a formula for $\mathscr{E}$ and estimate its numerical value.

The relevant units are

$$\mathscr{E} \sim \text{volt}, \qquad E' \sim \frac{\text{volt}}{\text{meter}}, \qquad a \sim \text{meter}$$

$$(20.A.6)$$

Only one combination is possible, namely,

$$\mathscr{E} \propto E'a$$

with an unspecified proportionality constant; this is as far as our dimensional estimate will take us. The full answer is

$$\mathscr{E} = E'a \qquad (20.A.7)$$

$$= (1.5)(2.0 \times 10^{-2}) \text{ volt} = 3.0 \times 10^{-2} \text{ volt}$$

(Using $E' = vB$ [see Eq. (20.A.4)], we note that (20.A.7) equivalently reads

$$\mathscr{E} = vBa \qquad (20.A.8)$$

a formula that will reappear in the next section.}

B. Moving Circuits

We can generate voltages and currents in circuits just by displacing them in a magnetic field. Here we calculate the batterylike **electromotive force** to which that effect is equivalent. For definiteness, we specialize to a simple geometry; but our result will be exact, rather than a dimensional estimate as in the preceding section. Our tool will be the conservation of energy.

i. Pulling a Loop out of a B Field

Figure 20.4 shows the moving-loop effect in a particularly simple case. A rectangular conducting wire loop, with sides a and b, is being slid out, along its own plane, from between the pole pieces of a magnet. The magnetic field lines, which are perpendicular to the loop, are shown in Fig. 20.5 as crosses. As we know from the preceding section, the trailing side of the loop sees an electric field $E' = vB \sin 90°$, or

$$E' = vB \qquad (20.B.1)$$

where v is the loop's speed (**v** is parallel to sides b), and where B is the magnet's field; the direction of $\mathbf{E}'$ is shown in the figure. The loop's leading side is already out of the **B** field, and therefore sees zero electric field.

The net effect of $\mathbf{E}'$ is to create (or **induce**) a current around the loop, as if a battery had been temporarily inserted into the circuit [part (b) of the figure]. Thus, *a loop of wire, while being pulled out of a* **B** *field, acquires a temporary electromotive force $\mathscr{E}$* ("induced emf").

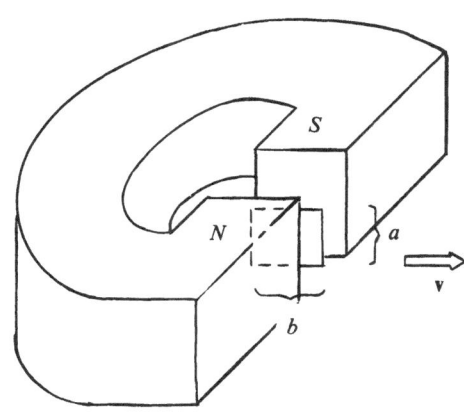

Figure 20.4. Conducting loop being pulled out of a **B** field.

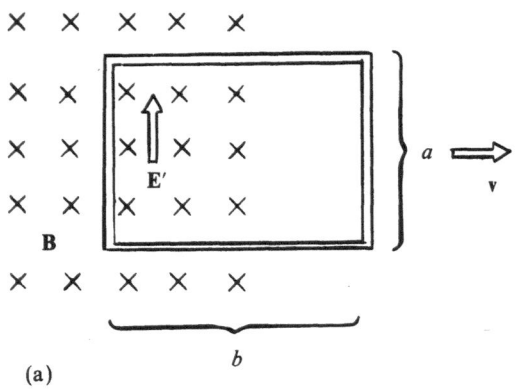

(a)

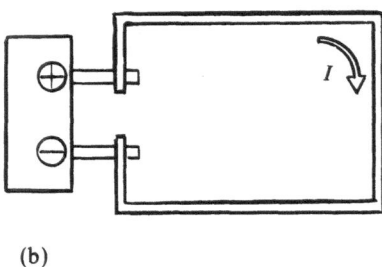

(b)

Figure 20.5. (a) The motional electric field, **E'**, for Fig. 20.4. (b) A battery, simulating the effect of **E'**.

ii. Calculating the Induced emf

Given the geometry of the loop (sides a, b), the strength of the **B** field, and the speed v at which the loop is being pulled, *we now predict the value of $\mathcal{E}$ from the energy principle.* Suppose the induced current supplies energy to an electric load, namely, a resistance, a motor, etc. Where does that energy come from? The answer is that *someone must do mechanical work in order to pull the loop out of the magnetic field.* The idea of the following derivation is to equate these two energies.

In Fig. 20.6 the pulling-out process is shown once more, with inclusion of an unspecified load; the loop carries a (temporarily steady) clockwise current I. For simplicity we assume that the loop

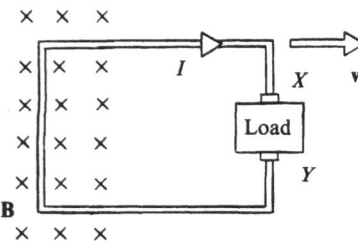

Figure 20.6. The loop of Fig. 20.5a with a load inserted.

itself has negligible resistance. As in the case of a battery, the emf, $\mathcal{E}$, can then be identified with the potential drop, V, from terminal X to terminal Y. We recall from Eq. (17.B.3) of Chap. 17 that the power supplied to the load is

$$\text{Power to load} = VI = \mathcal{E}I \quad (20.B.2)$$

What is the mechanical power needed to pull the loop to the right at a steady speed v? The net force F exerted by **B** on the loop acts entirely on the left side a (the horizontal sides experience mutually canceling forces). According to the right-hand rule of Chap. 18, **F** points to the left, *oppositely to the loop's motion*; its magnitude is

$$F = IaB \sin 90° = IaB \quad (20.B.3)$$

The mechanical power used in overcoming **F** is

$$\text{Mechanical power} = vF = vIaB \quad (20.B.4)$$

Conservation of energy requires

$$\text{Power to load} = \text{mechanical power} \quad (20.B.5)$$

or

$$\mathcal{E}I = vIaB$$

or

$$\boxed{\mathcal{E} = vBa} \quad (20.B.6)$$

as anticipated by (20.A.8). Note that we never had to calculate the current I; the value of b is likewise of no relevance; finally, we did not need E' as given in (20.B.1). The argument was powerful indeed.

Summary

The loop of Fig. 20.4, while being pulled to the right, behaves as if containing a source of clockwise emf, given by (20.B.6). *That small formula is our main result, and will grow into a general principle in the next section.*

Resistance of the Loop

In the above we neglected the loop's resistance. To incorporate it correctly, we use the model of

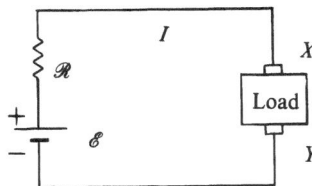

Figure 20.7. Equivalent circuit to Fig. 20.6.

Fig. 20.7. The "temporary battery" still has an emf given by $\mathscr{E} = vBa$. In addition, its internal resistance $\mathscr{R}$ equals the loop's total, ordinary resistance. This combination yields the experimentally observed current and voltage between the terminals X and Y.

iii. Induced Current and Magnetic Drag

A closer look at the induced current I and at the force **F** is best achieved through a pair of examples.

Example 20.3. Induced Current. In Fig. 20.4, let the loop have sides $a = 2.0$ cm, $b = 4.0$ cm; let its resistance be $\mathscr{R} = 0.05\ \Omega$; assume a magnetic field $B = 0.30$ tesla. Find the current I in the loop while it is being pulled at a steady 6.0 meters/sec.

From Ohm's law, applied to the model of Fig. 20.7 (in which terminals X, Y must be connected together, without a load), we have

$$I = \frac{\mathscr{E}}{\mathscr{R}} = \frac{vBa}{\mathscr{R}} \tag{20.B.7}$$

$$= \frac{(6.0)(0.30)(0.020)}{0.05}\ \text{A} = \underline{0.72\ \text{A}}$$

$$\tag{20.B.8}$$

The value of b has not been used.

Example 20.4. Magnetic Drag. We have seen that the loop experiences a force or "drag" **F** that opposes its motion. Calculate F, using the preceding example's data.

According to (20.B.3), we have

$$F = IaB = \left(\frac{vBa}{\mathscr{R}}\right) aB$$

with use of (20.B.7). Thus,

$$F = \frac{a^2 B^2}{\mathscr{R}} v \tag{20.B.9}$$

Numerically, using (20.B.3) and (20.B.8) for convenience, we have

$$F = IaB = (0.72)(0.020)(0.30)\ \text{newton}$$

$$= \underline{4.3 \times 10^{-3}\ \text{newton}}$$

Remarks on the Magnetic Drag

The magnetic force that opposes the motion of a closed loop is proportional to the loop's speed, according to Eq. (20.B.9). (If the loop were cut open, that force would be zero; why? Incidentally, the reader should also consider the case of a closed loop *entering* the **B** field; the result is, again, a force that *opposes* the motion.) In short, the loop behaves as if the **B** field were a thick (or "viscous") liquid.

Note that the *overall* **B** field in Fig. 20.4 is non-uniform because it is nonzero over only part of the loop. If the loop were moving entirely within the uniform portion of the field, it would have zero current and zero drag: the leading and trailing sides would contribute oppositely to the total emf.

[Caution: Would it become impossible to move the loop if its resistance $\mathscr{R}$ were negligible? Formula (20.B.9), with $\mathscr{R} \to 0$, might lead one to think so. However, the answer is no. Our analysis breaks down for a very small $\mathscr{R}$; the reason is that we never examined the effect of the loop's own magnetic field, arising from its induced current I. The required correction is complicated and we shall not pursue it; as a rule of thumb, formulas (20.B.7) and (20.B.9) are all right as long as *the loop's own magnetic field remains negligible in comparison with the external field*. We always assume this to be the case unless otherwise mentioned, or unless, as in the next chapter, there is no external magnetic field at all.]

Eddy Currents

All conducting objects display the phenomenon we have just discussed. In general, *a conductor that moves in a nonuniform **B** field is subjected to a viscouslike drag.*

Suppose, for example, that Fig. 20.4 illustrates a solid rectangular conducting plate of sides a, b, rather than a wire. Induced currents, possibly quite intense because of the low resistance, will circulate clockwise throughout the plate; they are known as

eddy currents. Such currents can also be shown to arise whenever a solid conductor at rest is subjected to a time-varying **B** field. Joule heat is always generated in the process; induction melting of metals is one industrial application of this phenomenon.

In electric motors and transformers, eddy-current heating in the magnetic iron cores is a nuisance. A good preventive measure is *lamination*, the insertion of nonconducting gaps throughout the material.

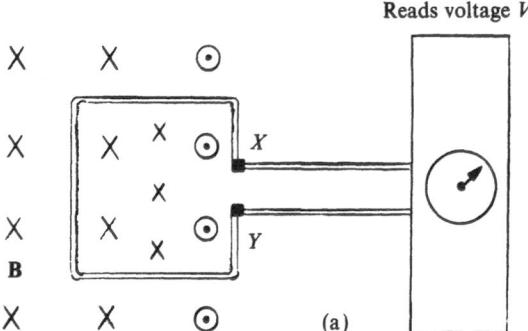

C. Faraday's Law of Induction

There are many shapes of loops, and many ways of moving them in an inhomogeneous **B** field. What determines a loop's emf in each case? According to Faraday, the single relevant factor is the *rate of decrease of the magnetic flux through the loop*. Most importantly, this is true even for *a motionless loop in a time-varying **B** field*.

i. Statement of Faraday's Law

Figure 20.8 summarizes the phenomenon. Part (a) shows the actual situation: a loop of resistance $\mathscr{R}$ which encloses a changing magnetic flux $\mathscr{N}$. Part (b) shows an equivalent circuit: its terminals X, Y act precisely like their counterparts in (a). Thus, the actual loop behaves like a battery with (temporary) emf

$$\mathscr{E} = -\frac{d\mathscr{N}}{dt} \qquad \text{(Faraday's law)}$$

(20.C.1)

and internal resistance $\mathscr{R}$. This is true irrespective of the current, or of the external equipment connected to the terminals. *Faraday's law is the central result of this chapter.*

Let us briefly review the concepts assembled in Eq. (20.C.1), with special attention paid to their most difficult features—directions and signs. It will be important to keep in mind that *a current can equally well flow against or along an induced emf*, just as in a battery under charging or discharging conditions; an additional (external) source of emf may have to be enlisted for the purpose.

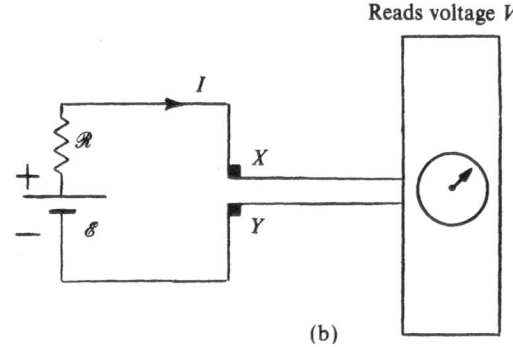

Figure 20.8. (a) Wire loop around a magnetic flux. Here some field lines go into the page, some come out. The loop is cut open at XY, and connected to some "black-box" equipment. We assume the gap XY to be very small, because Faraday's law is strictly valid only for complete loops. (b) No matter where, around its circumference, the loop is being tapped, this equivalent circuit is what the black box in (a) responds to.

ii. The Electromotive Force, $\mathscr{E}$, Reviewed

We refer to Fig. 20.8. For definiteness, let terminal X be at a higher potential than terminal Y, by an amount V (= "terminal voltage"). Let the loop have total resistance $\mathscr{R}$ and a clockwise current I. Then the clockwise electromotive force $\mathscr{E}$ is related to these quantities by

$$V = \mathscr{E} - \mathscr{R}I$$

(20.C.2)

just as in a battery. Equation (20.C.2) may be considered the *definition* of a loop's emf, in terms of easily measured quantities V, $\mathscr{R}$, I.

In particular, the emf equals the terminal voltage, $\mathscr{E} = V$, when the current or the internal resistance is negligible.

A more fundamental expression for $\mathscr{E}$ will be given in Sec. E further on.

iii. The Magnetic Flux, $\mathcal{N}$, Reviewed

We recall that

> The magnetic flux through a loop is the number $\mathcal{N}$ of magnetic field lines that thread the loop.   (20.C.3)

The procedure for counting $\mathcal{N}$ is as follows:

1. Let the loop be viewed (Fig. 20.8) with an emf defined as positive when clockwise and negative when counterclockwise.
2. Let the loop serve as a frame, or boundary, for an imagined surface S. It is convenient, but not necessary, that S be flat.
3. Count the number of field lines that intersect S; a line going into the plane of the figure is counted positive; if coming out it is counted negative. Thus, in Fig. 20.8, and if each cross and dot is literally taken as a single line, the flux through the loop is $\mathcal{N} = (+5 \ -2)\,\text{Wb} = +3\,\text{Wb}$.

From Line Language to Field Language

In electrostatics we spent some time discussing the concept of **electric flux**; its definition was precisely analogous to the present one. We had to acknowledge that field lines are fictitious and cannot, in practice, be counted; we therefore reformulated the flux in terms of a *surface integral of the field*, which is indeed measurable. We urge the reader to review the details of that argument; there is no point in reproducing it here. In the present (magnetic) case the flux is

$$\mathcal{N} = \iint_S B_\perp \, d\mathcal{A}$$ (20.C.4)

At any single point of the surface S, $B_\perp$ is the component of **B** in the direction perpendicular to S. If **B** points away from the viewer, $B_\perp$ is positive; if towards him, $B_\perp$ is negative.

If S is flat and of total area $\mathcal{A}$, and if in addition **B** is a uniform field, then (20.C.4) simply reads

$$\mathcal{N} = B_\perp \mathcal{A}$$ (20.C.5)

Note one important difference between the kind of surface discussed here and in the electrostatic case. In electrostatics we were mostly interested in closed surfaces, whereas here we focus on a "bounded" surface, i.e., a piece of surface whose edge is the loop. Hence our use of the integration symbol $\iint$ rather than $\oiint$.

iv. SI Units, Reviewed

Each magnetic field line is a unit of flux, namely,

$$1 \text{ line} = 1 \text{ weber}$$ (20.C.6)

Thus, $\mathcal{N}$ *is expressed in webers*; correspondingly, **B** has units of webers/meter^2 ($=$ teslas).

In terms of units, Eq. (20.C.1) gives the interesting relation

$$1 \text{ volt} = 1 \, \frac{\text{weber}}{\text{second}}$$ (20.C.7)

v. Lenz's Sign Rule

The minus sign in Faraday's law, Eq. (20.C.1), has everything to do with the "clockwise emf" convention of Fig. 20.8. Fortunately, we are dispensed from puzzling out the signs thanks to a rule pointed out in 1834 by Heinrich Lenz:

> The current induced by a change in flux opposes that change.

    (20.C.8)

This rule tells us the *direction* of the induced emf whenever we are given the time-dependent behavior of the flux. The meaning of (20.C.8) is as follows: Assume that the induced emf is allowed to produce a current, which then has a magnetic field of its own. This magnetic field, in turn, contributes to the total flux. According to Lenz, the direction of that contribution is such as to offset (partly) the change in flux. The next subsection will illustrate the principle in detail.

vi. Verification of Faraday's Law

Does the familiar rectangular-loop case of Fig. 20.4 fit Faraday's law? We reproduce the relevant geometry in Fig. 20.9. Let us examine (1) the emf, (2) the flux, (3) the flux's rate of decrease, (4) the validity of Lenz's rule.

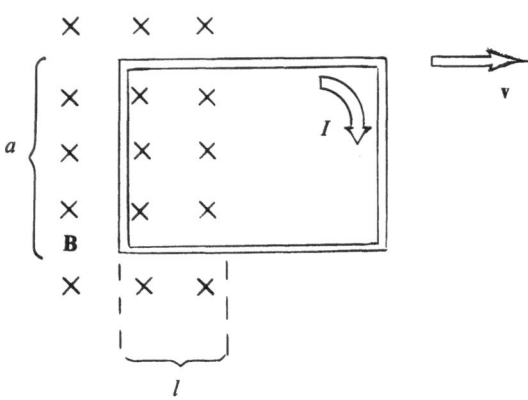

Figure 20.9. Verification of Faraday's law.

1. emf

This we have seen to be

$$\mathcal{E} = Bva \qquad (20.C.9)$$

clockwise and hence positive.

2. Flux

If the left-hand side of the loop is at a distance l from the edge of the field, as shown, we have

$$\mathcal{N} = \int B_\perp \, d\mathcal{A} = (B_\perp)(la) = (+B)(la) \qquad (20.C.10)$$

positive since **B** points into the figure.

3. Rate of Decrease of the Flux

The loop's motion shortens l at a rate $-dl/dt = v$; therefore we have from (20.C.10) (B and a are time independent)

$$-\frac{d\mathcal{N}}{dt} = B\left(-\frac{dl}{dt}\right)a = Bva \qquad (20.C.11)$$

Comparison of (20.C.9) and (20.C.11) yields

$$\mathcal{E} = -\frac{d\mathcal{N}}{dt}$$

as required by Faraday's law.

4. Lenz's Rule

The induced current goes clockwise in Fig. 20.9. According to the right-hand rule, that current produces a magnetic field that *increases* the field already present. Thus, it *opposes the decrease* in flux ("tries" to maintain the constancy of the flux), in conformity with Lenz's rule.

D. Application to Coils

Faraday's law applies not only to loops that move in a **B** field, but also to fixed loops in time-varying **B** fields. Both kinds of arrangement are illustrated in what follows; for additional generality, we note that several loops may be combined into a coil. The dynamo and the transformer are perhaps the two best-known uses of Faraday's law.

i. Induction in a Coil

Consider a conducting N-turn coil in a time-varying **B** field. Each turn of wire is an almost-closed loop and exhibits an induced emf, $\mathcal{E}_0$. If the magnetic flux is the same in each turn, the total emf generated in the coil is

$$\boxed{\mathcal{E} = N \mathcal{E}_0} \qquad (20.D.1)$$

because the N turns are connected in series (we are reminded of N batteries, connected in series). The cases of a coil that is so tightly wound as to look almost like a single turn, or that of a solenoid that is placed in a uniform **B** field, are illustrations where each turn has the same flux, so that Eq. (20.D.1) applies.

ii. The Dynamo

In various modified forms, the dynamo is the most common industrial generator of electric power. As idealized in Fig. 20.10, it is just an N-turn coil ($N = 1$ in the figure), rotating at a frequency f about an axis perpendicular to a permanent magnetic field **B**, which is constant in time. Each of the coil's terminals is in sliding contact with a "brush." If **B** is uniform in space, and the coil frames an area $\mathcal{A}$, how does the terminal voltage V (X relative to Y in the figure) depend on time? (We assume open terminals, i.e., zero current.)

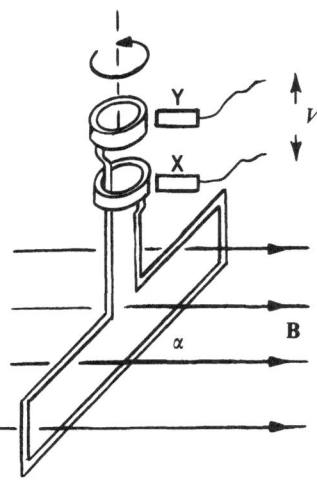

Figure 20.10. Principle of the dynamo. In practice, the rotating coil, or **armature**, has many turns and a soft-iron core to increase the magnetic flux.

At time t, let the plane of the coil make an angle

$$\alpha = \omega t \qquad (\omega = 2\pi f) \qquad (20.D.2)$$

with the magnetic field; see Fig. 20.10. The magnetic flux through the coil is given by Eq. (20.C.5),

$$\mathcal{N} = B_\perp \mathcal{A} = B\mathcal{A} \sin \alpha = B\mathcal{A} \sin \omega t$$

$$(20.D.3)$$

Each turn develops an emf

$$\mathcal{E}_0 = -\frac{d\mathcal{N}}{dt} = -\omega B\mathcal{A} \cos \omega t \quad (20.D.4)$$

The terminal voltage equals the total emf,

$$\boxed{V = \mathcal{E} = N\mathcal{E}_0 = -N\omega B\mathcal{A} \cos \omega t}$$

$$(20.D.5)$$

This is an **alternating voltage**, which depends sinusoidally on time. It can be prevented from reversing itself (it can be "rectified") through the use of a split-ring commutator, as was illustrated in Fig. 18.32 of Chap. 18. The time behavior of V is plotted in Fig. 20.11.

As we can see, the rectified dynamo and the DC motor share their basic design; in fact, *the very same piece of equipment can be used either as a motor or as a generator.*

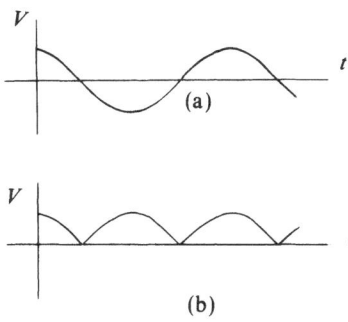

Figure 20.11. (a) Voltage output of the dynamo. (b) Rectified voltage.

[Note: the sign in (20.D.5) is not very meaningful until the polarity has been puzzled out in Fig. 20.10.]

iii. Mutual Induction between Two Fixed Coils

All our previous illustrations have dealt with moving circuits; but fixed circuits, placed in a time-varying **B** field, provide some of the main applications of the induction principle.

Consider, as in Fig. 20.12, two neighboring coils. Let us send a time-dependent current I_1 in coil 1, while the terminals of coil 2 remain open.

A chain of effects now takes place. Coil 1 sets up a time-dependent magnetic field **B**; some of the field

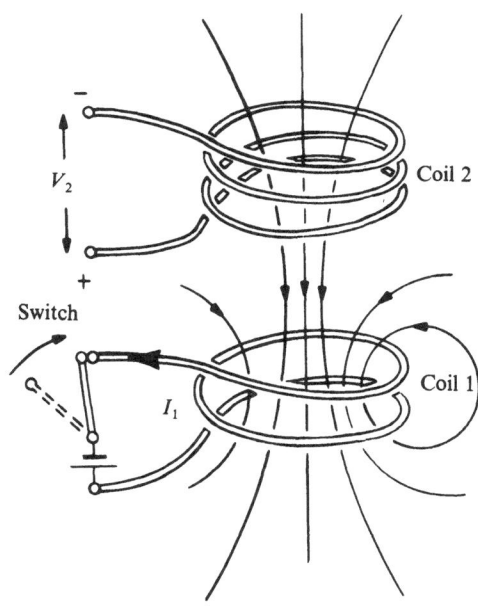

Figure 20.12. Time-dependent magnetic flux entirely produced by coil 1 (with the help of battery and switch) and partly shared by coil 2.

lines go through coil 2 and provide it with a time-dependent magnetic flux $\mathcal{N}_2$; that time dependence creates an electromotive force, $\mathscr{E}_2$, and hence a terminal voltage, $V_2 = \mathscr{E}_2$, in coil 2. In summary: *A time-dependent current I_1 in coil 1 induces an emf $\mathscr{E}_2$ in coil 2.*

For any two-coil configuration, there exists a numerical coefficient, the **mutual inductance**, which serves to gauge the strength of the effect. That coefficient is discussed in a Note at the end of this chapter.

Example 20.5. Solenoid and Loop. In Fig. 20.13, coil 1 is a solenoid with $N = 500$ turns, length $s = 10$ cm, and cross section $\mathscr{A} = 2.0$ cm^2. It is surrounded by a single open loop of irregular shape. The solenoid's current, I, is uniformly increased, in a time $T = 3.0 \times 10^{-6}$ sec, from zero to $I_{max} = 0.80$ amperes; it is then uniformly decreased to zero, at the same rate. Find, as a function of time, the terminal voltage V of the open loop.

_____

If the loop's emf is $\mathscr{E}$, we have

$$V = \mathscr{E} = -\frac{d\mathcal{N}}{dt} \qquad (20.D.6)$$

where $\mathcal{N}$ is the flux inside the solenoid. (There is negligible flux outside the solenoid, between it and the loop.)

At any time, the solenoid's flux is

$$\mathcal{N} = B\mathscr{A} = \left(\frac{4\pi K' NI}{s}\right)\mathscr{A} \qquad (20.D.7)$$

where we have used Eq. (19.C.11) of Chap. 19. Hence we get

$$\frac{d\mathcal{N}}{dt} = \frac{4\pi K' N\mathscr{A}}{s}\frac{dI}{dt} \qquad (20.D.8)$$

Let us conduct the experiment between times $-T$ and $+T$. We then have, as in Fig. 20.14a,

$$\frac{dI}{dt} = \pm\frac{I_{max}}{T} \qquad (20.D.9)$$

according to whether $-T < t < 0$ or $0 < t < T$, respectively. Using this in Eq. (20.D.8), and then in (20.D.6), we obtain, for $-T < t < 0$,

$$V = -\frac{4\pi K' N\mathscr{A} I_{max}}{sT} \qquad (20.D.10)$$

$$= -\frac{(4\pi)(10^{-7})(500)(2.0 \times 10^{-4})(0.80)}{(0.10)(3.0 \times 10^{-6})} \text{ volt}$$

$$= \underline{-0.34 \text{ volt}} \qquad (20.D.11)$$

During the next interval, $0 < t < T$, the opposite voltage exists; the result is plotted in Fig. 20.14b. It is left to the reader to determine, from Lenz's rule, which terminal, X or Y, starts at a higher potential than the other.

iv. The Iron-Core Transformer

This remarkable application of mutual induction is found in many industrial, scientific, and

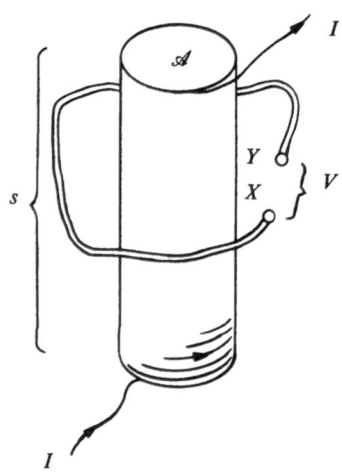

Figure 20.13. An irregular wire loop around a solenoid.

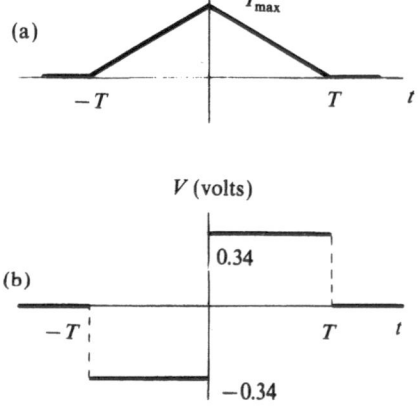

Figure 20.14. (a) Solenoid current in Fig. 20.13. (b) Voltage output of the loop in Fig. 20.13.

household appliances. It has four terminals X_1, Y_1, X_2, Y_2 (Fig. 20.15) with the basic feature that a voltage V_1 between X_1 and Y_1 automatically results in a proportional voltage $V_2 = \alpha V_1$ between X_2 and Y_2; α is a constant multiplier. *The voltage V_1 (and hence also V_2) must oscillate in time* for the appliance to be effective; if V_1 is predominantly in one direction, excessive currents build up, and resistive effects can no longer be neglected.

As the figure shows, the transformer consists of two coils linked by a closed soft-iron core about which they are tightly wound. Such a core *confines within its own substance nearly every magnetic field line* generated by any loops in either of the coils; recall the electromagnet of the preceding chapter's Sec. D iv.

Let the total magnetic flux in the core, produced by all the turns together, be $\mathcal{N}$. Each turn then develops the same emf,

$$\mathscr{E}_0 = -\frac{d\mathcal{N}}{dt} \qquad (20.D.12)$$

If the coils have N_1 and N_2 turns, respectively, their emf's are

$$\mathscr{E}_1 = N_1 \mathscr{E}_0, \qquad \mathscr{E}_2 = N_2 \mathscr{E}_0 \qquad (20.D.13)$$

so that

$$\boxed{\frac{\mathscr{E}_1}{\mathscr{E}_2} = \frac{N_1}{N_2}} \qquad (20.D.14)$$

Neglecting the coils' resistances, we have

$$\frac{V_1}{V_2} = \frac{N_1}{N_2} \qquad (20.D.15)$$

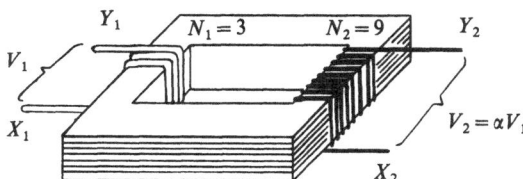

Figure 20.15. An iron-core transformer. An input voltage V_1, set up across the left coil ("primary winding"), causes an output voltage αV_1 ($= 3V_1$ in this case) to appear across the right coil ("secondary winding"). Which winding is used as primary and which as secondary is a matter of choice. (According to Lenz's rule, do XY and $X'Y'$ have the same or opposite polarities?) The core is **laminated**, i.e., made of mutually insulated plates in order to prevent eddy currents in the iron; such currents would dissipate energy and produce unwanted heating.

The multiplier in Fig. 20.15 is $\alpha = N_2/N_1$. In that figure, an instantaneous voltage $V_1 = 20$ volts (X_1 relative to Y_1) results in a simultaneous voltage $V_2 = \frac{9}{3}V_1 = 60$ volts (X_2 relative to Y_2), since the coils have 3 and 9 windings, respectively.

E. The Induced Electric Field

We have learned that in electrostatics, the **E** field is conservative; a test charge, carried around a closed path, does not gain or lose any energy.

In a time-varying **B** field, on the other hand, *the **E** field is no longer conservative.* A circulating charge q can gain (or lose) a net energy, $q\mathscr{E}$, corresponding to an emf, $\mathscr{E}$, induced over the length of its closed path.

To put it differently, the induced emf is a manifestation of an **induced electric field**. In our introductory moving-loop illustration, that field, **E′**, was due to the motion; but an induced electric field also exists in static-loop cases, such as that of Fig. 20.13. Here the **B** field is almost totally confined to the inside of the solenoid. How, then, is it able to affect the loop? The answer is that the changing magnetic flux surrounds itself with an electric field; the latter extends outside the solenoid.

The point of this section is that *the induced electric field exists in space even if no wire loop is available to detect it.* Accordingly, in what follows we express the law of induction in terms of the induced electric field **E**, rather than in terms of the emf, $\mathscr{E}$. By omitting the prime (we shall now talk about **E** rather than **E′**) we indicate that the observer is considered at rest, while **B** varies in time.

i. The emf as a Loop Integral

The moving loop of Sec. B was found to develop an emf $\mathscr{E} = vBa$, where the scenario is that of Fig. 20.5a.

We also saw that $vB = E'$ is the motional electric field seen by the trailing side a of the loop. Therefore $\mathscr{E}$ reduces to

$$\mathscr{E} = E'a$$

or, with our new point of view (loop considered at rest while **B** is being removed),

$$\mathscr{E} = Ea \qquad (20.E.1)$$

leaving out the prime.

Many shapes of loops and fields are possible. Is there a general form of (20.E.1) that applies to all cases? There is indeed: the formula turns out to be an integral over the loop,

$$\mathscr{E} = \oint \mathbf{E} \cdot d\mathbf{r} = \oint E_{\parallel} \, ds \qquad (20.E.2)$$

The symbol $\oint$ reminds us to integrate over the complete loop; $d\mathbf{r}$ is the typical loop element (of length $ds = |d\mathbf{r}|$); $E_{\parallel}$ is the component of $\mathbf{E}$ along $d\mathbf{r}$; see Fig. 20.16. Thus, *the emf equals the loop integral of the electric field's longitudinal component.*

Verifying the Integral Formula (20.E.2)

Let us go back to the moving loop (Fig. 20.5) for an evaluation of that integral; E' is now called E. Starting with the left-hand side of the loop and going clockwise, we have the four sides' contributions,

$$\mathscr{E} = \oint E_{\parallel} \, ds = Ea + 0 + 0 + 0 \qquad (20.E.3)$$

(since $E_{\parallel} = 0$ along both horizontal sides, and since there is zero field on the right). Thus, our result, (20.E.3), agrees with (20.E.1); we have here a simple check, of course, not a proof.

ii. Faraday's Law in Empty Space

Using the loop integral (20.E.2), we now rewrite Faraday's law, $\mathscr{E} = -d\mathcal{N}/dt$, as

$$\oint \mathbf{E} \cdot d\mathbf{r} = -\frac{d\mathcal{N}}{dt} \qquad (20.E.4)$$

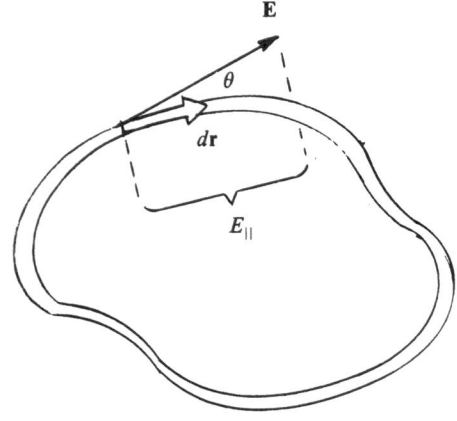

Figure 20.16. Loop integration of $\mathbf{E} \cdot d\mathbf{r} = \mathbf{E} |d\mathbf{r}| \cos \theta = E_{\parallel} \, ds$.

The wire loop can now be done away with; it becomes an imaginary closed path of arbitrary shape, still used to define the flux $\mathcal{N}$ as well as the path integral $\oint \mathbf{E} \cdot d\mathbf{r}$. (The abstractness of such a picture becomes less severe if one visualizes a wire loop whose resistance progressively increases, until the loop no longer conducts and can be dispensed with.) In applying (20.E.4), we shall allow ourselves one simplifying feature: *the closed loop will always be chosen stationary and of fixed shape.*

Faraday's law, in the form (20.E.4), is one of the fundamental building blocks of electromagnetic theory.

The following example involves the induced $\mathbf{E}$ field in empty space.

Example 20.6. The Solenoid. Consider once more the solenoid of Example 20.5, but without the surrounding loop of wire. Find, as a function of time, the electric field just outside the solenoid's cylindrical surface. Assume (as can be shown from symmetry) that the induced *electric* field lines are circles, concentric with the solenoid's turns; see Fig. 20.17.

Cylindrical symmetry implies a constant magnitude, $E = E_{\parallel}$, along a field line. Using that line as an integration loop, we obtain from (20.E.4)

$$\mathscr{E} = \oint E_{\parallel} \, ds = (E)(\text{circumference}) = 2\pi r E \qquad (20.E.5)$$

where r is the solenoid's radius. Thus, we find

$$E = \frac{\mathscr{E}}{2\pi r} \qquad (20.E.6)$$

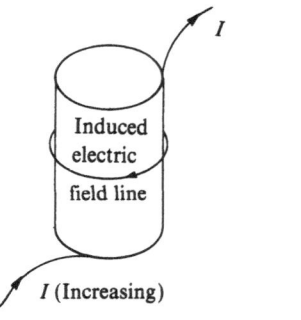

Figure 20.17. One typical induced electric field line just outside a solenoid. (Check for Lenz's rule.)

Here $\mathcal{E}$ is the same as for the irregular loop in Example 20.5, because the same flux is enclosed. This means [Eq. (20.D.11)] that

$$\mathcal{E} = \pm 0.34 \text{ volt} \qquad (20.E.7)$$

during rising (falling) flux. The cross section $\mathcal{A} = 2.0 \text{ cm}^2$ gives $r = 0.80 \text{ cm}$, so that Eq. (20.E.6) finally gives

$$E = \pm \frac{0.34}{(2\pi)(0.80 \times 10^{-2})} \frac{\text{volts}}{\text{meter}}$$

$$= \underline{\pm 6.8 \frac{\text{volts}}{\text{meter}}} \qquad (20.E.8)$$

according to whether $-3.0 \times 10^{-6} \text{ sec} < t < 0$ or $0 < t < 3.0 \times 10^{-6} \text{ sec}$. Lenz's sign rule confirms that **E** points in a direction opposite to the solenoid's current while the latter is increasing, and in the same direction while it is decreasing.

iii. The Betatron

This electron accelerator, designed in 1941 by Donald Kerst, yields high-energy electrons for impact experiments with nuclear targets. (The cyclotron, Sec. B vii of Chap. 18, yields protons for a similar purpose.) It is now superseded, but still provides one of the most beautiful demonstrations of the reality of an induced electric field.

Figure 20.18 shows the principle of the instru-

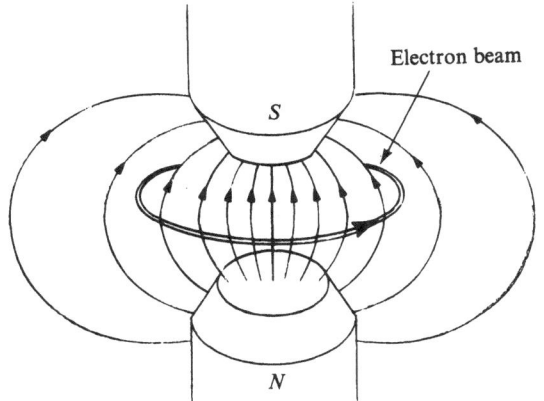

Figure 20.18. Principle of the betatron. The (laminated) soft iron pole pieces are shaped so as to yield a specially reduced **B** field at the electrons' orbit; see Problem 20.27. The coils of the electromagnet are not shown. The rotational symmetry of this geometry about a vertical axis is what allows us to infer the induced *E* from the induced emf.

ment. It consists of an electromagnet whose **B** field varies with time; the edge of that field holds an electron beam in a *fixed* circular orbit (cf. Problem 20.33) inside an evacuated enclosure with nonconducting walls (not shown).

The right-hand rule for magnetic forces determines which way the electrons circulate under (say) an upward **B**; the direction is shown in the figure. Then Lenz's rule tells us that the electrons are indeed speeded up, not slowed down, by an increasing flux.

Example 20.7. Final Speed and Energy in a Betatron. We discuss a small, early model of the instrument, in which relativistic considerations may be ignored (the electrons' speed remains much smaller than $c = 3 \times 10^8$ meters/sec). Suppose the orbit's circumference is $l = 0.20$ meter. Let the magnetic flux, $\mathcal{N}$, enclosed by the orbit, increase within a time interval of 1.5×10^{-2} sec from nearly zero to a final value $\mathcal{N} = 3.0 \times 10^{-5}$ Wb; $\mathcal{N}$ does not necessarily grow at a constant rate. If the electrons are initially nearly at rest, what is (a) their final speed v_f, (b) their final kinetic energy $\mathcal{K}_f$ in electron volts? [The induced electric field lines are circles, one of which coincides with the electron beam; ignore any electrostatic field, so that $\mathbf{E} = \mathbf{E}'$ as in Example 20.6.]

Since all vector directions are worked out in Fig. 20.18, we no longer worry about sign conventions. Let E be the induced electric field strength at the beam's location. For an increasing $\mathcal{N}$, Faraday's law (20.E.4) becomes

$$El = \frac{d\mathcal{N}}{dt} \qquad (20.E.9)$$

(a) Speed: Consider the electron's equation of motion, $\mathbf{F}_{\text{tot}} = m\mathbf{a}$. Its component along the orbit reads

$$eE = m \frac{dv}{dt} \qquad (20.E.10)$$

Eliminating E from (20.E.9) and (20.E.10),

$$\frac{dv}{dt} = \frac{e}{ml} \frac{d\mathcal{N}}{dt} \qquad (20.E.11)$$

where we note that $e/(ml)$ is just a constant factor.

Integrating over time, with negligible initial values of v and $\mathcal{N}$, we have, at any time,

$$v = \frac{e}{ml}\mathcal{N} \qquad (20.E.12)$$

The speed increases as the flux; the detailed time dependence of $\mathcal{N}$ was not needed in the calculation. At the end of the process we have

$$v_f = \frac{(1.6 \times 10^{-19})(3.0 \times 10^{-5})}{(9.1 \times 10^{-31})(0.20)} \frac{\text{meters}}{\text{sec}}$$

$$= 2.6 \times 10^7 \frac{\text{meters}}{\text{sec}} \qquad (20.E.13)$$

(b) Kinetic energy: In terms of electron volts,

$$\frac{\mathcal{K}_f}{1 \text{ electron volt}} = \frac{\frac{1}{2}mv_f^2}{1 \text{ electron volt}}$$

$$= \frac{(1/2)(9.1 \times 10^{-31})(2.6 \times 10^7)^2}{1.6 \times 10^{-19}}$$

$$= 2.0 \times 10^3 \qquad (20.E.14)$$

This betatron is yielding 2.0-keV electrons. (At best, a betatron achieves somewhat more than 300 MeV.)

Note

Mutual Inductance

As illustrated in Figs. 20.12 and 20.13, a time-varying current I_1 in one coil induced an emf $\mathcal{E}_2$ in a second coil. This phenomenon (**mutual induction**) involves a coefficient (the **mutual inductance** between the two coils) which we now motivate and define.

First we must convince ourselves that $\mathcal{E}_2$ is proportional to the rate of change of the inducing current dI_1/dt. (This is strictly true only when the coils have no magnetic core in them.)

From the law of Biot and Savart (Chap. 19), we can in principle determine **B** at any point in the vicinity of coil 2, using coil 1 as the source. Here we need not bother with the detailed integral, but we just note the proportionalities

$$I_1 \propto \mathbf{B} \propto \mathcal{N}_2 \qquad (20.N.1)$$

For example, doubling I_1 means doubling **B** at every location; this, in turn, doubles $\mathcal{N}_2$. The proportionality constant M_{12} between I_1 and $\mathcal{N}_2$ is the announced **mutual inductance** of the two coils:

$$\boxed{\mathcal{N}_2 = M_{12}I_1} \qquad (20.N.2)$$

Faraday's law,

$$\mathcal{E}_2 = -\frac{d\mathcal{N}_2}{dt} \qquad (20.N.3)$$

becomes

$$\boxed{\mathcal{E}_2 = -M_{12}\frac{dI_1}{dt}} \qquad (20.N.4)$$

We note that M_{12} can be positive or negative, depending on which direction around the coils is defined as positive for the purpose of specifying I_1 and $\mathcal{E}_2$.

Equation (20.N.2) implies that the mutual inductance of two coils (in air) depends just on their geometry and relative position; but M_{12} is usually far easier to measure from (20.N.4) than to calculate from the geometry. In general, M_{12} increases as the coils are placed closer together.

Units

The mutual inductance may be defined either by (20.N.2) or by (20.N.4):

$$M_{12} = \frac{\mathcal{N}_2}{I_1} = -\frac{\mathcal{E}_2}{dI_1/dt} \qquad (20.N.5)$$

Accordingly, its SI unit, called the **henry**, is given by

$$\boxed{1 \text{ henry} = 1 \text{ H} = 1\frac{\text{weber}}{\text{ampere}} = 1\frac{\text{volt second}}{\text{ampere}}}$$

$$(20.N.6)$$

[Joseph Henry (1797–1878) is the discoverer of self-induction, a phenomenon to be discussed in the next chapter.]

Example 20.8. Solenoid and Loop. What is the mutual inductance of the pair of coils described in Example 20.5?

Let the solenoid be No. 1 and the loop No. 2. We found that $\mathcal{N}_2$ equals $\mathcal{N}$, the flux in the solenoid itself; from Eq. (20.D.7), we have

$$\mathcal{N}_2 = \left(\frac{4\pi K'NI_1}{s}\right)\mathcal{A} \qquad (20.N.7)$$

so that, from (20.N.2), and with the data of Example 20.5, we obtain

$$M_{12} = \frac{\mathcal{N}_2}{I_1} = \frac{4\pi K'N\mathcal{A}}{s} \qquad (20.N.8)$$

$$= \frac{(4\pi)(10^{-7})(500)(2.0 \times 10^{-4})}{0.10} \text{ henry}$$

$$= \underline{1.3 \times 10^{-6} \text{ henry}}$$

(A magnetic core in the solenoid might boost this number by a factor of many thousands.)

Reciprocity

Going back to the solenoid-and-loop arrangement of Fig. 20.13, suppose we must find the emf induced in the solenoid by a variable current in the loop. This might seem to involve the difficult task of calculating the magnetic flux, due to the irregular loop, through every single turn of the solenoid. Here a **reciprocity theorem**, which we quote without proof, comes to the rescue: the mutual inductance between two coils is reciprocal, i.e., its value does not depend on which coil has the inducing current and which the induced emf.

More specifically, suppose we know M_{12} in

$$\mathcal{E}_2 = -M_{12}\frac{dI_1}{dt} \qquad (20.N.9)$$

and we require M_{21} in

$$\mathcal{E}_1 = -M_{21}\frac{dI_2}{dt} \qquad (20.N.10)$$

Reciprocity assures us that

$$\boxed{M_{21} = M_{12}} \qquad (20.N.11)$$

Condensed Checklist

Motional electric field:

$$\mathbf{E}' = \mathbf{v} \times \mathbf{B} \qquad (20.A.3)$$

Electromotive force from segment (all right angles):

$$\mathcal{E} = vBa \qquad (20.B.6)$$

Magnetic drag (rectangular loop):

$$F = \frac{a^2 B^2 v}{\mathcal{R}} \qquad (20.B.9)$$

Faraday's law for the emf (single turn):

$$\boxed{\mathcal{E} = -\frac{d\mathcal{N}}{dt}} \qquad (20.C.1)$$

or

$$\oint \mathbf{E} \cdot d\mathbf{r} = -\frac{d\mathcal{N}}{dt} \qquad (20.E.4)$$

where

$$\mathcal{N} = \iint B_{\perp}\, d\mathcal{A} \qquad (20.C.4)$$

Terminal voltage:

$$V = \mathcal{E} - \mathcal{R}I \qquad \text{(defines } \mathcal{E}) \qquad (20.C.2)$$

Lenz's rule:

An induced current opposes the flux change.

$$(20.C.8)$$

Dynamo's alternating emf (open circuit):

$$\mathcal{E} = -N\omega B\mathcal{A} \cos \omega t \qquad (20.D.5)$$

Iron-core transformer:

$$\frac{\mathcal{E}_1}{\mathcal{E}_2} = \frac{N_1}{N_2} \qquad (20.D.14)$$

Speed in a betatron (nonrelativistic):

$$v = \frac{e}{ml}\mathcal{N} \qquad (20.E.12)$$

True or False

1. If an originally neutral, vertical metal rod moves eastwards in a northward **B** field, the rod's upper end becomes positively charged.

2. If a stationary observer sees a pure static **B** field (no **E** field), a moving observer usually sees a combined **B** and **E** field.

3. Item 2 is true if the words "stationary" and "moving" are interchanged.

4. If observer X sees a pure static **B** field (no **E** field), observer Y, who moves with respect to X in a direction parallel to **B**, still sees zero electric field.

5. The motional **E** field points in the direction of the static **B** field that causes it.

6. The motional **E** field points in the direction of motion of the observer.

7. A magnetic flux enclosed by a circuit induces in that circuit an emf proportional to the flux.

8. "Volt" is synonymous with "weber second."

9. At the instant a time-varying magnetic flux equals zero, it induces zero emf.

10. At the instant a time-varying magnetic flux reaches a maximum, it induces zero emf.

11. According to Lenz, a current induced in a circuit tends to decrease the flux already enclosed.

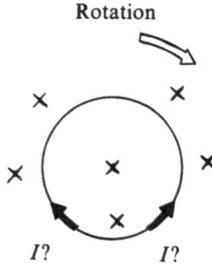

Rotation

$I?$ $I?$

12. Consider a circular loop, whose plane is perpendicular to a uniform **B** field. If the loop rotates in its own plane, as shown, it develops a nonzero electric current opposite to its sense of rotation.

13. The last words of item 12 should be "along its sense of rotation."

14. In Fig. 20.13, an increasing I causes X to be at a higher potential than Y.

15. A galvanometer, sent by mail, has less chance of arriving with a broken coil suspension if its terminals have been shorted than if they have not.

Problems

Sections A: Electric Field Seen by a Moving Object; B: Moving Circuits

*20.1. If you sweep your hand, as shown, at a speed of 5 meters/sec, between the pole pieces of a large 1-tesla electromagnet, approximately what voltage do you temporarily create between your wrist and finger tips, which are, say, 20 cm apart?

20.2. A certain 3-meter-wide whale is swimming horizontally at 10 meters/sec near the north magnetic pole, where the field is of order 10^{-4} tesla. What emf does the whale develop across its body?

*20.3. A vertical metal rod, 0.30 meter long, moves eastwards at 10 meters/sec in a uniform northward **B** field of 0.020 tesla. (a) Which end of the rod becomes positively charged? (b) As seen by someone traveling with the rod, what is the total electric field in the rod?

20.4. Suppose the moving loop in Fig. 20.4 is a square of side 3.0 cm, and that the **B** field between the pole pieces has a strength of 1.5 tesla. If the loop's resistance is 2.0 Ω, (a) at what speed must it be pulled to develop a current of 5.0 A? (b) What is the magnetic drag?

*20.5. An E meter is moving at 5.0 meters/sec in a direction 30° east of north in a uniform northward **B** field of strength 0.015 tesla. Find the direction and strength of the electric field measured.

20.6. In connection with Fig. 15.13 of Chap. 15 it was mentioned that a typical downward **E** field of 120 volts/meter exists in the Earth's lower atmosphere. Perhaps some of this is due to the Earth's rotation in its own **B** field (consider the latter as stationary). For simplicity, assume you are at the equator, where **B** is horizontal ($B \approx 0.3 \times 10^{-4}$ tesla). (a) Is the motional contribution upward or downward? (b) How large is it?

20.7. Electric power may be obtained directly from hot combustion gases in an MHD generator like that of Fig. 20.3. The ionized (hence conducting) gas moves with velocity **v** as shown. (a) If $v = 1000$ meters/sec, $B = 0.20$ tesla, and the spacing between electrodes is 0.50 meter, find the induced emf, $\mathscr{E}$. (b) If the electrodes are connected to a load, and draw a current $I = 10$ A, find the retarding force F on the gas. (Use the $Il \times \mathbf{B}$ force law.) (c) From $\mathbf{v}$ and $\mathbf{F}$, find the mechanical power depletion of the gas. (d) Check that this equals the electric power supplied to the circuit.

*20.8. If, in Problem 20.2, we assume a 1-Ω resistance for the whale together with the return circuit through sea water, estimate the order of magnitude of the magnetic drag on the animal. Do you believe it is important compared to the hydrodynamic drag?

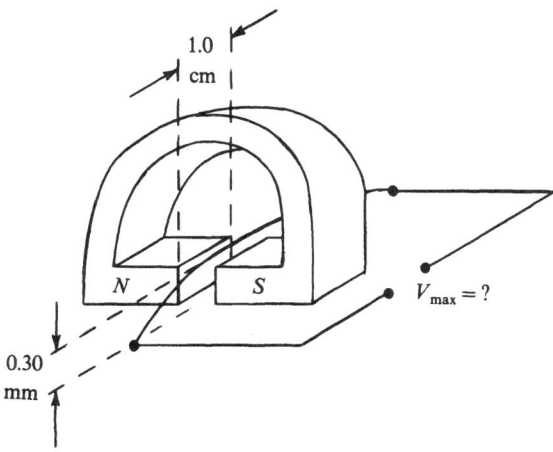

*20.9. A conducting wire is vibrating harmonically in a vertical plane, at a frequency of 3000 Hz, transversely to a horizontal 0.08-tesla field, as shown. Within the field, which is 1.0 cm wide, the amplitude of vibration is 0.30 mm. Find the amplitude of the alternating voltage measured between the ends of the wire.

*20.10 (Faraday's wheel). Consider Barlow's wheel motor, Fig. 18.23 of Chap. 18. Faraday showed it to be a generator ("homopolar generator") as well. Assuming a disk radius of 0.10 meter, a frequency of 5.0 rev/sec, and a uniform $\mathbf{B}$ field of 0.20 tesla, (a) find the voltage between axle and rim when the external circuit is open. (b) With all directions as shown in the figure, which is the positive terminal? (Note: This generator is occasionally used when high currents and low voltages are desired.)

Section C: Faraday's Law of Induction

20.11. A circular wire loop of radius 2.0 meters is in a uniform field of 0.25 tesla that points at 45° to the loop's plane. Find the magnetic flux through the loop.

*20.12. A square, closed loop of wire of side 8.0 cm has a resistance of 20 Ω. Its plane is intersected at 45° by a uniform $\mathbf{B}$ field whose strength, as a function of the time

t, is given by $B = (5.0 \times 10^{-5} \text{ tesla}) (e^{-t/3.0 \ \mu\text{sec}})$. Find the current in the loop as a function of time. (Ignore the flux due to the induced current.)

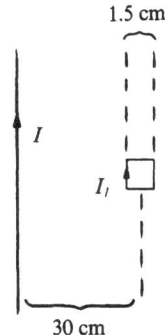

*20.13. The current I in a long straight wire has the time behavior $I = (2.0 \text{ A}) \cos[(2\pi)(60 \text{ Hz})t]$. A small, square, closed wire loop with side 1.5 cm and resistance 8.0 Ω is in the same plane as the straight wire, at a distance of 30 cm from it, as shown. (a) Find, as a function of t, the current I_l in the loop. (b) Plot both currents so as to show their relative phase; use the same time scale, but you may use different current scales. (Assume I_l contributes negligibly to the total $\mathbf{B}$ field.)

*20.14. The plane of a rectangular 50-turn coil, 20 cm × 30 cm, is perpendicular to a uniform $\mathbf{B}$ field, which changes uniformly in time from 0.400 tesla to 0.380 tesla in a 0.6-μsec interval. (a) Find the emf in the coil. (b) If the coil is a closed circuit with total resistance 800 Ω, estimate the induced current.

20.15. A large square loop of wire, of side 1.0 km, is laid on the ground for the purpose of detecting changes in the Earth's vertical component of magnetic field. If the latter changes by 2×10^{-8} tesla in 1 minute (which may happen because of solar wind effects), estimate the voltage measured at that time across a cut in the loop.

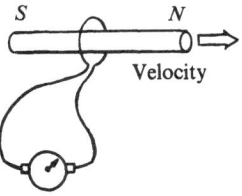

*20.16. A permanent bar magnet, moving lengthwise at constant velocity, is made to pass through a wire loop of small diameter, as shown. Make a qualitative graph, with correct sign, of the induced current I as a function of the time, before, during, and after passage. The small arrow alongside the wire indicates the positive-current convention for this problem.

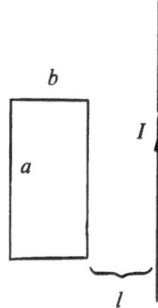

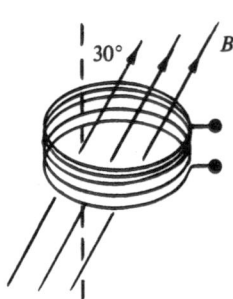

*20.17. Consider a rectangular loop with sides a, b. In the same plane and parallel to sides a, there is a long straight wire with current I, at a distance l from the nearest side a, as shown. Obtain, by integration, the flux through the loop (whose own current is zero). Express your answer in terms of K', I, a, b, l.

20.18. Find the emf induced in the rectangular loop of Problem 19.22 when the current in the long straight wire decreases uniformly from 20 A to 18 A in 3 μsec. (Use the result of Problem 20.17.)

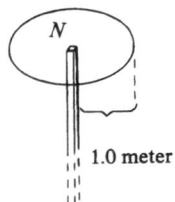

*20.19. A long bar magnet has its N pole, of strength 1.0 SI unit, at the center of a horizontal circular loop of radius 1.0 meter, as shown. Find the upward flux through the loop.

20.20. Let the magnetic pole of Problem 18.49 move through the plane of the coil at a downward vertical velocity of 20 meters/sec. At that time, (a) what emf is induced in the coil? (b) Clockwise or counterclockwise as seen from above? (See also Problem 20.19.)

Section D: Application to Coils

*20.21. A 100-turn, 20-cm-long solenoid, with base area 0.50 cm^2 and open terminals, is placed in a uniform **B** field which is parallel to the solenoid's axis and increases uniformly from zero to 0.030 tesla in 2.0 × 10^{-3} sec. What is, during that time, the solenoid's terminal voltage?

20.22. A circular coil of area 20 cm^2, made of 5 equal turns of wire, lies in a uniform **B** field oriented at 30° to the axis of the coil; see the figure. At what rate, in teslas/sec, must B change in order to generate a 0.50-volt emf in the coil?

*20.23. A certain transformer is designed to accept a 150-volt maximum across its primary winding, which has 25 turns. How many turns are needed in the secondary if a maximum of 150 000 volts is desired as the output? Neglect the coils' resistances.

20.24. A circular 20-turn coil, with radius 0.50 cm and open terminals, rotates about one of its diameters at a constant frequency of 10 rev/sec. The axis of rotation is perpendicular to a uniform **B** field of strength 0.030 tesla. Plot the coil's voltage as a function of time.

*20.25. A 100-turn circular coil, of area 60 cm^2 and resistance 9.0 Ω, rotates about one of its diameters, which is perpendicular to a 0.15-tesla magnetic field. What rotational frequency will give a peak current of 3.0 A in the shorted coil?

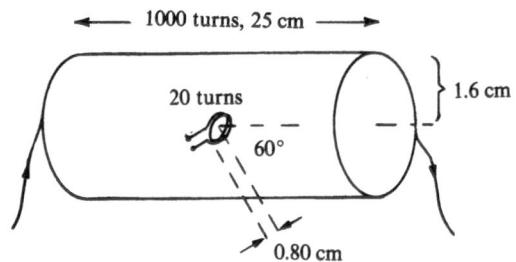

*20.26. A 1000-turn cylindrical solenoid, of length 25 cm and radius 1.6 cm, surrounds a 20-turn circular coil of radius 0.80 cm, with open terminals, as shown; the coil and the solenoid have their axes tilted 60° relative to one another. Find the voltage between the small coil's terminals while the solenoid increases its current at the rate of 3000 A/sec.

20.27. In the arrangement of Fig. 20.12, suppose each ampere in coil 1 produces 3.0 Wb through each turn of coil 2. At what rate, in A/sec, must we change current I_1 in order to obtain a potential difference $V_2 = 5$ volts across coil 2?

*20.28 (A search coil, or snatch coil). A single closed wire loop, of total resistance $\mathscr{R}$, is initially at rest in a static **B** field, and encloses an initial flux $\mathscr{N}_0$. It is then removed to a distant place, where the **B** field vanishes. (a) Express the loop current, at any time t, in terms of the rate of flux decrease, $-d\mathscr{N}/dt$; assume I produces negligible **B** of its own. (b) Setting $I = dQ/dt$, where Q is the total flow of charge up to time t, find an expression for Q in terms of $\mathscr{N} - \mathscr{N}_0$. (c) What is, in terms of $\mathscr{N}_0$, the charge Q_{tot} which flows during the complete removal the loop from the **B** field? (d) What does this result become if the loop is replaced by a tight N-turn coil? (Note: Q_{tot} can be measured by a so-called "ballistic galvanometer." This permits a practical estimate of the initial **B** field.)

*20.29 (Power from a dynamo). Consider an alternating-voltage dynamo, similar to that of Fig. 20.10, but with an N-turn coil. Suppose the terminals are connected together through a circuit of total resistance R, in which the coil's own resistance is included for purposes of calculation. (a) Find the current I in the circuit as a function of time. (Assume that R is large enough so that I negligibly affects the existing **B** field.) (b) Find the absolute value of the resulting magnetic torque τ on the coil, in terms of I, $\mathscr{A}$, B, N, and the angle $\alpha = \omega t$ of Fig. 20.10. (c) Is this torque a retarding or accelerating one? Explain with a sketch. (d) Eliminate I, i.e., express τ in terms of your answer to (a). (e) At time t, how much mechanical power is being expended to overcome the torque τ? (f) Verify that your answer to (e) equals the rate of Joule heat dissipation in the circuit. (g) For constant ω, what is the average mechanical power used by the dynamo?

20.30 (Back emf). Consider the dynamo of Fig. 20.10, but used as a dc motor (with a split-ring commutator); assume an N-turn coil carrying a current I small enough so that it negligibly affects the existing **B** field. The rotation occurs against a mechanical load, such as friction. (a) If the rotation is in the direction shown, in which direction must the current flow? (Show on a sketch.) Is this the same direction as for the current produced in dynamo fashion? (b) In terms of N, ω, B, $\mathscr{A}$, what is the emf developed in the coil? Explain why it is called a back emf. (c) Just as for a battery being charged, a current flowing against an emf supplies energy into the source of emf. Neglecting the coil's resistance, calculate the power ($=$ current $\times$ emf) supplied at time t by the current I. (Use $\alpha = \omega t$ and express your answer in terms of I, ω, t, $\mathscr{A}$, B, N.) (d) For given ω, calculate the torque τ delivered by the motor. (e) What is the mechanical power output as a function of t? Compare with (c).

Section E: The Induced Electric Field

*20.31. Find the electric field strength induced *inside* the solenoid of Example 20.6, at 0.40 cm from its axis.

20.32. Consider the solenoid of Example 20.6 during its rising-current stage. Plot the induced field strength, both inside and outside the solenoid, as a function of the distance r from its axis.

*20.33 (Betatron condition). Consider the betatron of Example 20.7. (a) Given the speed v, the flux $\mathscr{N}$, and the orbit's circumference l, determine the value of B needed at the orbit to enforce circular motion. (Consider the magnetic centripetal force; express B in terms of e, m, v, and l.) (b) Define the average B field by $B_{av} = \mathscr{N}/\mathscr{A}$, where $\mathscr{A}$ is the area enclosed by the orbit. Use Eq. (20.E.12) to express B_{av} in terms of e, m, v, and l. (c) How is B related to B_{av}? (Note: A suitable shape for the electromagnet's pole pieces will achieve this condition to a good approximation; the fine tuning is, amazingly, done by the orbit itself as it "chooses" the exactly appropriate radius.)

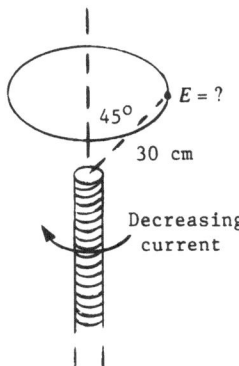

20.34. A long thin solenoid, with 20 turns/cm and cross section 1.0 cm^2, has a wire current of 1.5 A, which reduces uniformly to zero in $1.0 \ \mu\text{sec}$. During that interval, what is the induced electric field in space, 30 cm away from the solenoid's end, at 45° from its axis, as shown?

Answers to True or False

1. True.
2. True.
3. True.
4. True (because $\mathbf{v} \times \mathbf{B} = \mathbf{0}$).
5. False (at right angles).
6. False (at right angles).
7. False (proportional to the rate of decrease).
8. False (webers/second).

9. False (the flux's rate of decrease is what counts).

10. True.

11. False (tends to decrease the flux's rate of change).

12. False.

13. False (in neither of these two cases is the flux changing).

14. False (lower potential).

15. True (because swings are damped by the magnetic drag only if an induced current is permitted).

Answers or Hints to Problems

20.1. 1 volt [use (20.A.8)].

20.3. (b) Zero, because zero current must mean zero field; the field from the charge cancels the motional field.

20.5. Vertically up, 0.038 volt/meter.

20.8. 10^{-6} newton.

20.9. First review in Chap. 13 how to obtain the wire's velocity.

20.10. Integrate the motional emf's of all the small segments along the radius.

20.12. $(3.8 \times 10^{-3}\,\text{A})\,(e^{-t/3.0\,\mu\text{sec}})$.

20.13. Consider **B** as uniform in the loop.

20.14. (b) 125 A.

20.16. Use Fig. 19.26 of Chap. 19 for the field lines inside the magnet.

20.17. $2K'Ia \ln\left(1 + \dfrac{b}{l}\right)$.

20.19. Span the loop with a domed surface that avoids the N pole.

20.21. 0.075 volt; the 20-cm length is irrelevant, except to indicate a "long" solenoid.

20.23. 2.5×10^4 turns.

20.25. 48 rev/sec.

20.26. 0.030 volt; the 1.6-cm radius is irrelevant.

20.28. (c) $Q_{\text{tot}} = \mathcal{N}_0/\mathcal{R}$.

20.29. (g) $\frac{1}{2}(N\omega B\mathcal{A})^2/R$.

20.31. ± 3.4 volts/meter.

20.33. (c) $B = \frac{1}{2}B_{\text{av}}$.

CHAPTER 21

INDUCTANCE AND ALTERNATING CURRENTS

An alternating current ("ac") is one that reverses itself periodically in time; the typical dependence is sinusoidal, as in the adjoining figure. This chapter discusses basic aspects of ac circuits.

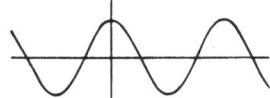

(An alternating current plotted as a function of time.)

We are surrounded by ac devices. The power supplied to homes and industries by electric companies is ac power; the usual frequency is 60 Hz in America, 50 Hz in Europe. The practical advantages of ac over dc were made clear around the turn of this century by Nikola Tesla and others (against the wrongheaded objections of Thomas Edison, a champion of dc). Some of these useful features are as follows.

1. The simplest rotating generators have emf's that are sinusoidal in time, as we recall from the preceding chapter.
2. An alternating voltage lends itself to easy multiplication, with minor power loss, by a transformer. Very high alternating voltages (hundreds of kilovolts) are used industrially for transmission by long-distance power lines. Only weak currents are then needed (because the transmitted power is proportional to current times voltage), and Joule dissipation in the wires is minimized. At the user's end, the voltage is reduced to a safe value by another transformer.

3. Many electric motors incorporate a rotating **B** field, produced without moving parts by alternating currents. (Details in Problem 21.19). Such motors can entirely avoid the use of brush collectors.
4. By carefully regulating the rate of rotation of their own generators, power companies control the rate of the synchronous motors in their customers' clocks as well.

Turning now to a completely different range of applications, we note that much of ac technology is related to the transmission, not of power, but rather of information. Just as a vibrating mechanical object can radiate sound waves in matter, an oscillating current can radiate radio waves, in vacuum as well as in matter. This is a topic for the next chapter, but it provides some of the motivation for our present study of ac circuits.

In what follows we learn how to make an electric oscillator. For that purpose, we introduce an additional circuit element, the **inductor** (which generally takes the form of a coil), and we combine it with a capacitor. These two objects store magnetic and electric field energy, respectively, much as a mass stores kinetic energy and a spring stores elastic potential energy. Just as in the mechanical case, an electric oscillation involves a back-and-forth transfer between the two forms of energy. Completing the analogy, resistance acts like friction, dissipating the electromagnetic energy into heat. The phenomenon of resonance (in response to an externally imposed electric vibration) exists here as it does in the mechanical case.

A. Self-Induction

We have learned that a coil develops an emf when it is linked by a time-dependent magnetic flux. So far, we have considered that flux as originating from outside; the coil under investigation was assumed to contribute negligibly. (The only exception was our discussion of the transformer, where we made no commitment as to which coil makes the magnetic flux; both coils do, in general.) Here we look at the opposite situation—a coil that is completely responsible for its flux. Faraday's law still operates, and the coil's emf equals the rate of decrease of the coil's own magnetic flux. This phenomenon, discovered by Joseph Henry in 1830, is called **self-induction**, and the coil, when used as a circuit element (like a resistor or a capacitor), is known as an **inductor**.

i. Inductance

We claim that a coil's self-induced emf, $\mathscr{E}$, is proportional to the rate of decrease, $-dI/dt$, of the *current* in that coil:

$$\mathscr{E} \propto -\frac{dI}{dt} \qquad (21.A.1)$$

The proportionality constant, denoted by L, is called the coil's **inductance**:

$$\boxed{\mathscr{E} = -L\frac{dI}{dt}} \qquad (21.A.2)$$

To understand (21.A.1), consider the coil of Fig. 21.1. Let a magnetic field $\mathbf{B}_1$ be produced *by the whole coil* somewhere inside the first turn. We shall not take the trouble to apply the Biot–Savart

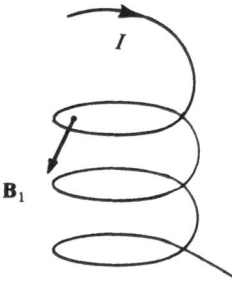

Figure 21.1. The field $\mathbf{B}_1$ is the integrated contribution of the whole coil at this particular point.

law in order to calculate $\mathbf{B}_1$, but, if we did so, we would find a result proportional to I:

$$\mathbf{B}_1 \propto I$$

The reason is that each current element contributes in proportion to I. For the whole flux through turn 1, as well as through turns 2, 3, etc., we therefore also have

$$\mathcal{N}_1 \propto I, \qquad \mathcal{N}_2 \propto I, \qquad \text{etc.} \qquad (21.A.3)$$

We are interested in the coil's emf. In the present case Faraday's law gives

$$\mathscr{E} = -\frac{d\mathcal{N}_1}{dt} - \frac{d\mathcal{N}_2}{dt} - \cdots$$

$$= -\frac{d}{dt}(\mathcal{N}_1 + \mathcal{N}_2 + \cdots) \qquad (21.A.4)$$

Since $\mathcal{N}_1 + \mathcal{N}_2 + \cdots$ is proportional to I, we set

$$\boxed{\mathcal{N}_1 + \mathcal{N}_2 + \cdots = LI} \qquad (21.A.5)$$

for some fixed number L. Equation (21.A.4) then reads

$$\mathscr{E} = -\frac{d}{dt}(LI) = -L\frac{dI}{dt} \qquad (21.A.6)$$

as announced in (21.A.2).

To summarize, there are two equivalent definitions, (21.A.2) and (21.A.5), for the inductance L of a coil:

1. *L is the self-induced emf per unit rate of decrease in current;*
2. *L is the sum of fluxes through all turns, per unit current.*

Lenz's Rule

An inductance must be positive. This fact goes together with the minus sign in Eq. (21.A.2), and means that

> The self-induced emf is in such a direction as to oppose a change in current.

$$(21.A.7)$$

Equivalently, the emf tends to oppose a change in the flux; statement (21.A.7) is Lenz's rule as applied to self-induction. We see that the number L expresses an inertialike property, somewhat as mass "tends to oppose" a change in velocity. For an explanation of (21.A.7) in terms of energy conservation, see Problem 21.9.

SI Units of L

Inductance is measured in **henries**, in honor of Joseph Henry. From definition (21.A.2), $L = -\mathscr{E}/(dI/dt)$, we have

$$1 \text{ henry} = 1 \text{ H} = 1 \frac{\text{volt second}}{\text{ampere}} = 1 \text{ ohm second} \tag{21.A.8}$$

Equivalently, from (21.A.5), we have

$$1 \text{ henry} = 1 \frac{\text{weber}}{\text{ampere}} \tag{21.A.9}$$

[Commercial inductors have values for L typically ranging from μH ($= 10^{-6}$ henry) to mH ($= 10^{-3}$ henry). By winding the coil around a soft-iron core that itself forms a closed ring (like a transformer core), one may achieve an inductance of several henries, owing to the intensification of the flux; however, the proportionality of $\mathscr{E}$ and $-dI/dt$ then breaks down for large currents; L is no longer strictly a constant.]

Inductance of an Air-Core Solenoid

For most coils the inductance is easier to measure than to calculate, but a long solenoid is simple enough to analyze. We assume a solenoid with N turns of wire, a length s, and a base of cross section $\mathscr{A}$. Using $\mathscr{N}_1 + \mathscr{N}_2 + \cdots = N\mathscr{N}_1$ in Eq. (21.A.5), and solving for L, we have

$$L = \frac{N\mathscr{N}_1}{I} = \frac{(N)(\mathscr{A}B)}{I} \tag{21.A.10}$$

Setting

$$B = \frac{4\pi K'NI}{s} \tag{21.A.11}$$

as in formula (19.C.11) of Chap. 19, we find

$$L = \frac{4\pi K'N^2\mathscr{A}}{s} \tag{21.A.12}$$

Note that I has dropped out, as it must; note also the dependence on the *square* of the number of turns.

Example 21.1. Large Size of the Henry. Someone hopes to design an air-core inductor of inductance $L = 1.0$ henry as a pocket-size solenoid of length $s = 10$ cm and base area $\mathscr{A} = 1.0$ cm^2. What should be (a) the number turns, N; (b) the spacing between turns?

(a) From (21.A.12) we find

$$N = \left(\frac{sL}{4\pi K'\mathscr{A}}\right)^{1/2}$$

$$= \left[\frac{(10 \times 10^{-2})(1.0)}{(4\pi)(10^{-7})(1.0 \times 10^{-4})}\right]^{1/2} = \underline{\underline{2.8 \times 10^4}}$$

(b) This results in a spacing

$$\frac{s}{N} = \frac{10 \times 10^{-2}}{2.8 \times 10^4} \text{ meter} = \underline{\underline{3.6 \times 10^{-6} \text{ meter}}}$$

a microscopic distance.

Inductors in Circuits

Figure 21.2 shows the conventional symbol for an inductor, together with Lenz's rule for the direction of its emf. If the inductor's internal resistance $\mathscr{R}$ is negligible, its self-induced emf is equal to its terminal voltage, $\mathscr{E} = V$. If $\mathscr{R}$ is not negligible, we use the model shown in Fig. 21.3; here the emf still equals the terminal voltage of the symbolic inductor. Thus, *we only need to think in terms of terminal voltages*, even when analyzing real circuits.

In a circuit made of inductors, the *inductances combine just as resistances would*; see, for example, Fig. 21.4 and Problem 21.20.

ii. Magnetic Energy

An *electric* field **E** in vacuum has an energy density $\frac{1}{2}\varepsilon_0 E^2$ ($= E^2/8\pi K$). We came to that conclusion, in Sec. D ii of Chap. 16, after asking for the

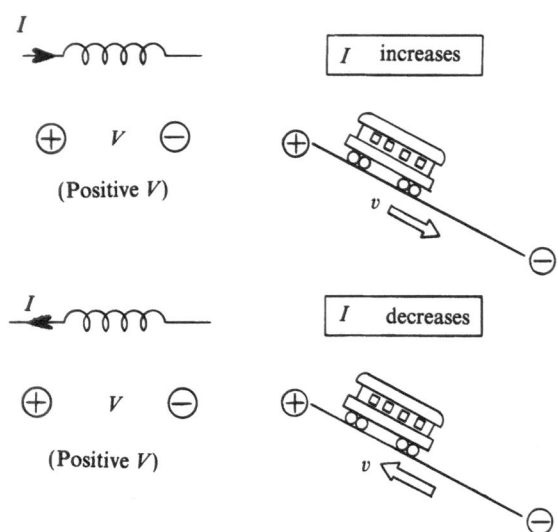

Figure 21.2. Lenz's rule in an inductor. The left end is at the higher potential. The train car coasting on a slope helps memorize that rule for positive V and I. Higher potential is represented by higher altitude, and the speed v increases or decreases as I does.

energy needed to charge a capacitor. Analogously, for a magnetic field **B** in vacuum, we have

$$\boxed{\text{Energy density} = \frac{B^2}{2\mu_0} = \frac{B^2}{8\pi K'}} \quad (21.A.13)$$

[Note: μ_0 (mu-zero) has the standard meaning

$$\mu_0 = 4\pi K' = 4\pi \times 10^{-7} \quad (\text{SI}) \quad (21.A.14)$$

and is known as the "permeability of the vacuum."] To derive (21.A.13) we compute the energy $\mathcal{W}$ needed to set up a current I in a coil of inductance L.

Figure 21.5a shows the inductor, in which the current is gradually increased from zero to a final value I; at an intermediate time t, the current has some value I'. Part (b) views the inductor as a load, whose terminal voltage has magnitude

$$V = L \frac{dI'}{dt} \quad (21.A.15)$$

see Eq. (21.A.2).

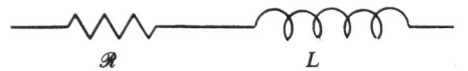

Figure 21.3. How to analyze a coil with inductance L and resistance $\mathcal{R}$.

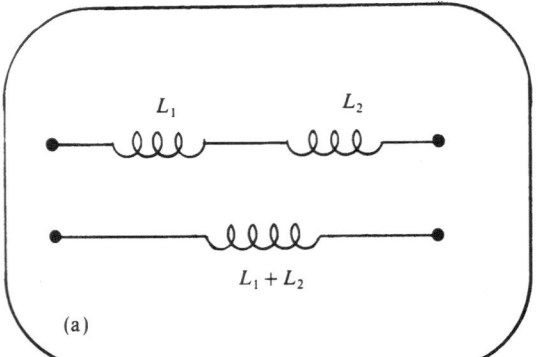

(a)

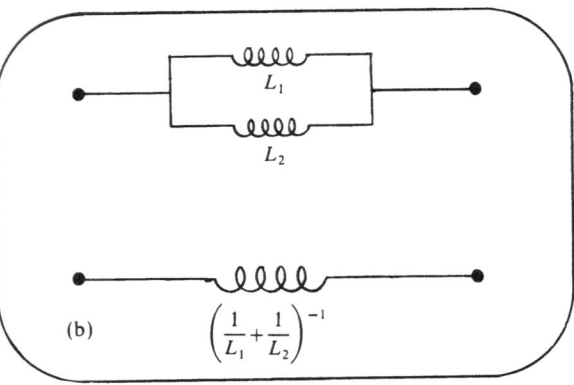

(b)

Figure 21.4. The top and bottom circuits are equivalent in each case. (a) Inductors in series: $L_{\text{tot}} = L_1 + L_2 + \cdots$. (b) Inductors in parallel: $1/L_{\text{tot}} = 1/L_1 + 1/L_2 + \cdots$.

As we recall from Chap. 17, Eq. (17.B.3), the power supplied to the load is

$$\frac{d\mathcal{W}}{dt} = VI' = LI' \frac{dI}{dt} \quad (21.A.16)$$

Hence, during dt, we supply an energy

$$d\mathcal{W} = LI' \, dI' \quad (21.A.17)$$

Integrating from $I' = 0$ to $I' = I$,

$$\mathcal{W} = \int_0^I LI' \, dI' \quad (21.A.18)$$

or

$$\boxed{\mathcal{W} = \tfrac{1}{2}LI^2} \quad \begin{array}{l}\text{(magnetic energy} \quad (21.A.19)\\ \text{stored in a coil)}\end{array}$$

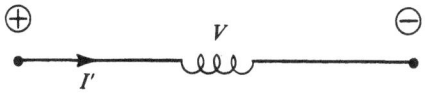

(a)

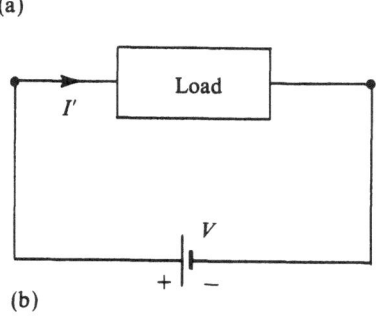

(b)

Figure 21.5. (a) An increasing current I', through an inductor. (b) The inductor as a load, and the power supply.

Recall the analogue, for a capacitance C with plate charges $\pm Q$,

$$\mathcal{W} = \frac{1}{2}\frac{Q^2}{C} \qquad (21.\text{A}.20)$$

As we can see, (21.A.19) must apply to any kind of coil, even one that includes a magnetic core, provided L is constant over the range of currents considered.

Special Case: A Long Solenoid

This situation is rather unique in that we know L explicitly [Eq. (21.A.12)] as well as the relation between I and B [Eq. (21.A.11)]. Thus we can evaluate $\mathcal{W}$ [Eq. (21.A.19)] in terms of B. We obtain

$$\mathcal{W} = \frac{1}{2}LI^2 = \frac{1}{2}\left(\frac{4\pi K'N^2\mathcal{A}}{s}\right)\left(\frac{sB}{4\pi K'N}\right)^2$$

$$= \frac{B^2}{2(4\pi K')}\mathcal{A}s \qquad (21.\text{A}.21)$$

Since $\mathcal{A}s$ is the volume occupied by the field, the above finally gives

$$\text{Energy density} = \frac{\mathcal{W}}{\mathcal{A}s} = \frac{B^2}{2(4\pi K')}$$

as announced in (21.A.13). We have justifiably

neglected the nonuniformity of **B** at the solenoid's ends, as well as the weak field that exists outside the solenoid; see also Problem 21.21.

iii. The *RL* Loop

Let us look in detail at a simple case where magnetic energy dissipates into heat. The circuit of Fig. 21.6 begins by carrying a steady current I_0, maintained by a battery. At time zero, through the use of a switch, the battery is replaced by a wire of negligible resistance. The current I then decreases gradually. Heat is evolved in the resistor R, while the magnetic energy stored in the inductor L is being depleted at the same rate. The time behavior of I is shown in Fig. 21.7.

We now compute I as a function of time, in order to show that this scenario really takes place, and in order to estimate the characteristic time ("time constant") needed for the current to decay. Figure 21.6b indicates the voltage drops across R and L. (At this stage, Kirchhoff's rules, Figs. 17.14 and 17.15 of Chap. 17, should be reviewed.)

Voltage rule:

$$V_L + V_R = 0 \qquad (21.\text{A}.22)$$

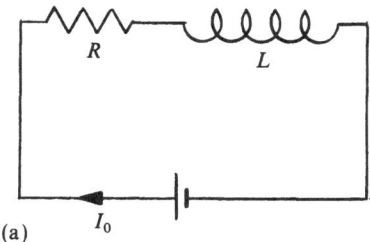

(a)

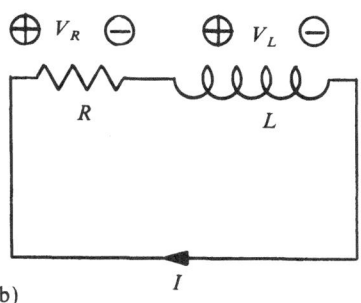

(b)

Figure 21.6. (a) Steady current through an *RL* circuit. (b) A switch (not shown) shorts out, and then disconnects, the voltage supply. The $(+-)$ assignments for V_R and V_L are arbitrary; in fact, V_L will turn out negative.

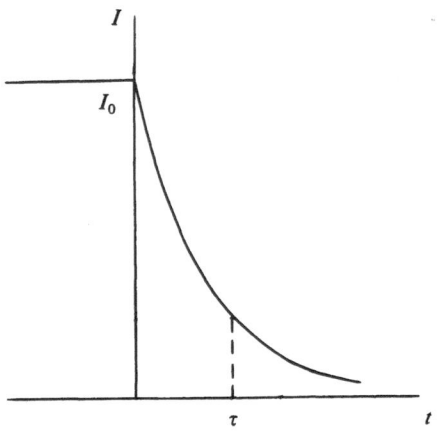

Figure 21.7. Decay of I in the RL circuit. The time constant τ is discussed later.

Ohm's law (or definition of resistance):

$$V_R = +RI \qquad (21.A.23)$$

Faraday's law (or definition of inductance):

$$V_L = +L\frac{dI}{dt} \qquad (21.A.24)$$

Inserting (21.A.23) and (21.A.24) in (21.A.22) gives

$$L\frac{dI}{dt} + RI = 0 \qquad (21.A.25)$$

This is the basic differential equation for the current in an RL loop. In order to solve it, we multiply by dt and divide by I, thus separating the I and t variables:

$$L\frac{dI}{I} + R\,dt = 0 \qquad (21.A.26)$$

Integrating from $t=0$, $I=I_0$ to arbitrary final values t, I, we have

$$L(\ln I - \ln I_0) + R(t-0) = 0 \quad (21.A.27)$$

Solving for I,

$$\ln I = \ln I_0 - \frac{R}{L}t \qquad (21.A.28)$$

or

$$I = I_0 e^{-(R/L)t} \qquad (21.A.29)$$

The constant L/R has the dimensions of time, as is clear from Eq. (21.A.26). In terms of SI units,

$$\frac{1\text{ henry}}{1\text{ ohm}} = 1\text{ second} \qquad (21.A.30)$$

see also Eq. (21.A.8). Thus, with

$$\boxed{\frac{L}{R} = \tau = \text{time constant}} \qquad (21.A.31)$$

Eq. (21.A.29) reads

$$I = I_0 e^{-t/\tau} \qquad (21.A.32)$$

After a time τ, the current has decayed to $1/e \approx 37\%$ of whatever its initial value was. The energy budget is the subject of Problem 21.18.

Exponential Decay: Seen Before

We have already encountered exponential decay, namely, for the voltage in an RC loop (time constant $= RC$). That case is governed by a differential equation identical to (21.A.25) except for the symbols. A brief review (Sec. C v of Chap. 17) is recommended at this stage. In the RC case we "solved" the differential equation by just stating the answer and trying it out. The same could have been done here, but instead we took the chance to improve our mathematical "bag of tricks."

Example 21.2. Time Constant of a Toroidal Coil. A certain toroidal coil, shown in Fig. 21.8, has an average circumference $s = 20$ cm, a number $N = 400$ of equally spaced turns of wire, and a total resistance $\mathscr{R} = 2.5\ \Omega$; each turn is a circle of area $\mathscr{A} = 0.50$ cm^2. There are no terminals; the wire forms a closed circuit. At time zero there exists in the wire a current $I_0 = 3.0$ mA (previously started by induction with the help of a primary coil, now disconnected). (a) Find the time constant τ of the toroidal coil. (b) At what time t has the current decayed to a value $I = 0.03$ mA? [Note: A thin toroidal coil yields the same inductance, formula (21.A.12), as a solenoid of equal length s; see also Problem 19.47.]

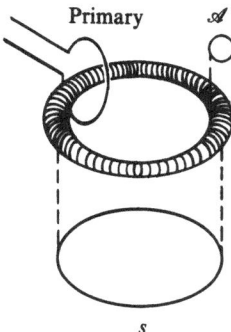

Figure 21.8. A toroidal coil in which a current, started by a primary, is then allowed to decay.

(a) Time constant: From (21.A.12) we have

$$L = \frac{4\pi K' N^2 \mathscr{A}}{s}$$

$$= \frac{(4\pi \times 10^{-7})(400)^2 (0.50 \times 10^{-4})}{0.20} \text{ henry}$$

$$= 5.0 \times 10^{-5} \text{ henry}$$

Idealizing the actual coil as an inductance L in series with a resistance $\mathscr{R}$, we obtain

$$\tau = \frac{L}{\mathscr{R}} = \frac{5.0 \times 10^{-5}}{2.5} \text{ sec} = \underline{\underline{2.0 \times 10^{-5} \text{ sec}}}$$

(b) Decay to 0.03 mA: Equation (21.A.32), $I = I_0 e^{-t/\tau}$, yields

$$t = \left(\ln \frac{I_0}{I} \right) \tau \qquad (21.A.33)$$

$$= (\ln 100)\tau = 4.6\tau$$

Here we have

$$t = (4.6)(2.0 \times 10^{-5} \text{ sec}) = \underline{\underline{9.2 \times 10^{-5} \text{ sec}}}$$

B. Natural Oscillations in a Circuit

The simplest oscillating circuits are those that can be put together from just an inductor and a capacitor. We first look at the idealized, resistance-free case. Later, more realistically, we take resistance into account.

i. The LC Loop (Basic Equation)

Figure 21.9 shows an LC loop, initially with capacitor plate charges $\pm Q_0$ and zero current. What are, at a later time t, the charge Q and current I? We proceed as in Eqs. (21.A.22), (21.A.23), (21.A.24).

Voltage rule:

$$V_L + V_C = 0 \qquad (21.B.1)$$

Inductance:

$$V_L = L \frac{dI}{dt} \qquad (21.B.2)$$

Capacitance:

$$V_C = \frac{Q}{C} \qquad (21.B.3)$$

Inserting (21.B.2), (21.B.3) in (21.B.1), we obtain

$$L \frac{dI}{dt} + \frac{Q}{C} = 0 \qquad (21.B.4)$$

We now have two functions Q, I to determine, in contrast to the RL case where we only had I. However, Q and I are related by

$$I = \frac{dQ}{dt} \qquad (21.B.5)$$

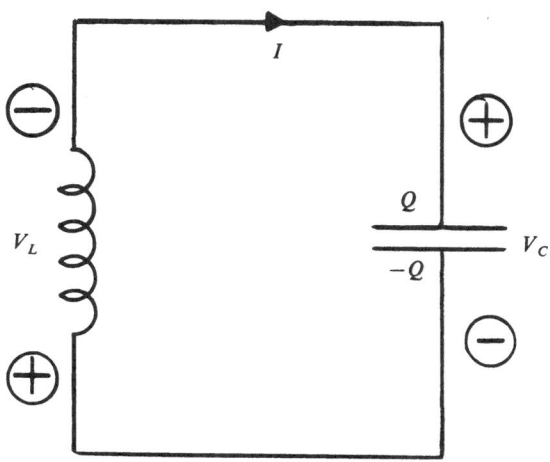

Figure 21.9. A capacitor is discharged through an inductor. The directions of I and V_L are assigned arbitrarily.

since the current, as defined in Fig. 21.9, is the rate at which charge increases on the top plate of the capacitor. Equation (21.B.4) now reads

$$L\frac{d\left(\dfrac{dQ}{dt}\right)}{dt}+\frac{Q}{C}=0$$

or

$$\boxed{L\frac{d^2Q}{dt^2}+\frac{Q}{C}=0}\qquad(21.B.6)$$

This basic differential equation must be solved for Q as a function of t.

ii. The *LC* Loop (Solution)

No new calculation is needed. We just recall [Chap. 13, Eq. (13.B.11)] the equation of motion of a particle of mass m vibrating at the end of a spring of constant k, namely, $F_{tot,x}+kx=0$, or

$$m\frac{d^2x}{dt^2}+kx=0\qquad(21.B.7)$$

whose general solution was found to be harmonic motion with angular frequency

$$\omega=\left(\frac{k}{m}\right)^{1/2}\qquad(21.B.8)$$

We shall now discover a mathematical analogy: *the inductor is like a mass, the capacitor is like a spring.*

Frequency in the LC Loop

In going back from (21.B.7) to (21.B.6), we make the substitutions $x\to Q$, $m\to L$, $k\to 1/C$. Hence Q oscillates sinusoidally in time with angular frequency $(k/m)^{1/2}\to[(1/C)/L]^{1/2}$, or

$$\boxed{\omega=\frac{1}{(LC)^{1/2}}}\qquad(21B.9)$$

Thus,

$$f=\frac{\omega}{2\pi}=\frac{1}{2\pi(LC)^{1/2}}\qquad(21.B.10)$$

is the natural frequency of the LC loop.

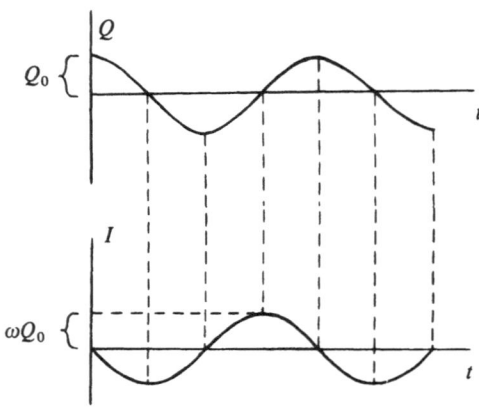

Figure 21.10. Time behavior of Q and I in the *LC* circuit of Fig. 21.9. The charge and current amplitudes are Q_0 and ωQ_0, respectively. Vertical dashed lines are 1/4 cycle apart.

Once the behavior of Q is known, I can be found from (21.B.5), $I=dQ/dt$; the current is the analogue of the velocity $v_x=dx/dt$ of the vibrating mass.

Initial Conditions

A positive Q_0 and a zero initial current correspond, mechanically, to a positive x_0 and zero initial velocity. The harmonic motion is then

$$x=x_0\cos\omega t\qquad(21.B.11)$$

($x_0=$ displacement amplitude). In the *LC* case, (21.B.11) becomes

$$Q=Q_0\cos\omega t\qquad(21.B.12)$$

with ω given by (21.B.9). From (21.B.5) we obtain the current,

$$I=\frac{d(Q_0\cos\omega t)}{dt}$$

or

$$I=-\omega Q_0\sin\omega t$$

We now have our desired result for Q and I; their behavior is plotted in Fig. 21.10. We see that *I leads* (anticipates) Q by 1/4 cycle in time; that is, a peak in I occurs 1/4 period before a peak in Q, etc.

Example 21.3. Tuning a Circuit. An *LC* loop, with $L=1.0\ \mu H$, has an adjustable capacitance C. Find C if the desired natural frequency is to be $f=800$ kHz (a radio frequency).

Solving (21.B.10) for C, we have

$$C = \frac{1}{(2\pi f)^2 L}$$

$$= \frac{1}{[(2\pi)(800 \times 10^3)]^2 (1.0 \times 10^{-6})} \text{ farad}$$

$$= \underline{\underline{0.040 \ \mu F}}$$

iii. The LC Loop (Energy Budget)

Let us see how the energy is apportioned, in the course of time, between magnetic and electrostatic.

At any time t, the magnetic energy stored in the inductor is, according to formula (21.A.19),

$$\mathcal{W}_L = \frac{1}{2} LI^2 = \frac{L}{2} \omega^2 Q_0^2 \sin^2 \omega t \quad (21.B.13)$$

Setting $\omega^2 = (LC)^{-1}$ [Eq. (21.B.9)], we have more simply

$$\mathcal{W}_L = \frac{Q_0^2}{2C} \sin^2 \omega t \qquad (21.B.14)$$

Similarly, the **electrostatic energy** in the capacitor is

$$\mathcal{W}_C = \frac{Q^2}{2C}$$

as we learned in electrostatics. With (21.B.12) this becomes

$$\mathcal{W}_C = \frac{Q_0^2}{2C} \cos^2 \omega t \qquad (21.B.15)$$

The two stored energies $\mathcal{W}_L$, $\mathcal{W}_C$ are plotted in Fig. 21.11. In summary,

1. The total energy,

$$\mathcal{W}_{tot} = \mathcal{W}_L + \mathcal{W}_C = \frac{Q_0^2}{2C} (\sin^2 \omega t + \cos^2 \omega t)$$

$$= \frac{Q_0^2}{2C} = \text{const} \qquad (21.B.16)$$

is conserved.

2. The energy is converted back and forth between electric and magnetic.

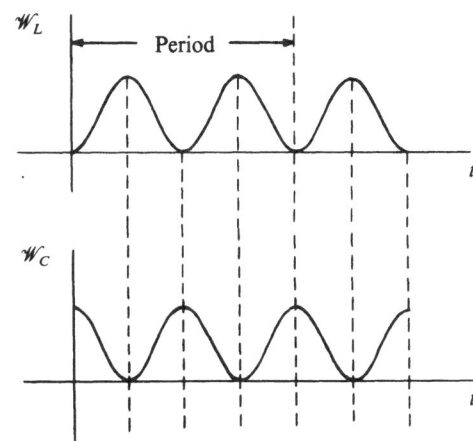

Figure 21.11. Time behavior of the magnetic and electric energies in the LC circuit. Vertical dashed lines are 1/4 cycle apart.

3. Since the time averages of $\sin^2 \omega t$ and of $\cos^2 \omega t$ are the same (namely, 1/2), it follows that, on the average, the same energy, $Q_0^2/4C$, is stored in the inductor and in the capacitor.

iv. The RLC Loop (Lightly Damped)

Mechanical oscillations die out if they receive no outside energy; the cause is friction. A similar decay, or **damping**, occurs in electric oscillations, and is due to any resistance that the circuit may include.

We consider the RLC loop of Fig. 21.12a, assuming a small R. In this way, the pure LC behavior, with which we are already familiar, will not be drastically changed. More specifically, we assume that, *over one cycle of the oscillation, the energy dissipated is small compared to the total energy stored.* Thus, the oscillation lasts many cycles without much damping, as shown in part (b) of the figure. Furthermore, it is legitimate to approximate the angular frequency of oscillation by

$$\omega = \frac{1}{(LC)^{1/2}} \qquad (21.B.17)$$

see (21.B.9).

The extent of damping in a given time is described in a note at the end of this chapter.

[Light damping requires a small R; but small compared to what? The circuit's only other rele-

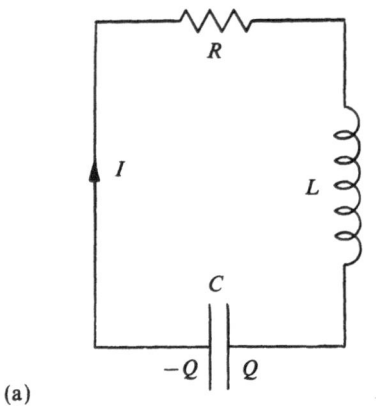

(a)

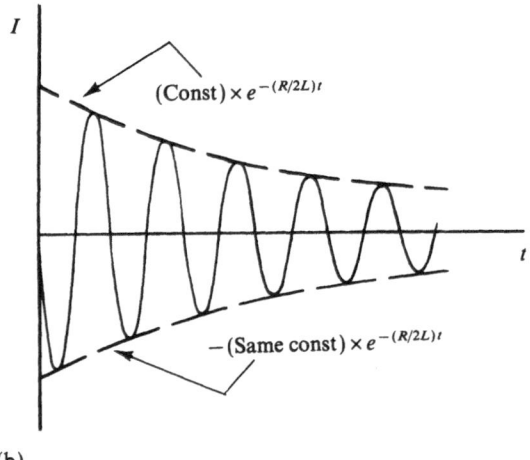

(b)

Figure 21.12. (a) An *RLC* loop. (b) Light damping of the current in an *RLC* loop with small *R*. The decaying amplitude (dashed lines) is discussed in Sec. B iv and in an end note.

vant parameters are L and C, out of which only the combination $(L/C)^{1/2}$ has dimensions of resistance,

$$\left(\frac{L}{C}\right)^{1/2} \sim \text{ohm} \qquad (21.\text{B}.18)$$

as the reader should verify; see Problem 21.35. In conclusion, whenever R is small relative to $(L/C)^{1/2}$,

$$\boxed{R \ll \left(\frac{L}{C}\right)^{1/2}} \qquad (21.\text{B}.19)$$

the damping is an effect of secondary importance; the amplitude during two consecutive cycles is about the same.]

C. Alternating Currents

Here we shall be concerned with the question, "how does a given circuit respond to a given alternating voltage?" We consider, as in Fig. 21.13, a pair of terminals—perhaps a household wall outlet—with a potential difference

$$V = V_{\text{max}} \cos \omega t \qquad (21.\text{C}.1)$$

(V_{max}, a positive constant, is the **voltage amplitude**). A circuit, connected as shown, carries an alternating current I in response to V; we want to describe I as a function of time.*

i. Phase Shift and Impedance

The response of a circuit is unpredictable unless we know something of its construction. Here we assume that the box in Fig. 21.13 contains nothing more than resistors, inductors, or capacitors (no switches, batteries, dynamos, transistors, etc.) Furthermore, those elements have constant values for their respective R, L, or C. Then, as it turns out, I oscillates harmonically with the same frequency as V:

$$I = I_{\text{max}} \cos(\omega t + \delta) \qquad (21.\text{C}.2)$$

Here I_{max} is the (positively defined) **current amplitude**, and δ is the (constant) **phase shift**, which tells us by what fraction of a cycle I leads ($\delta > 0$) or trails ($\delta < 0$) relative to V.

[**Switches and transients**. The no-switch assumption in the circuit of Fig. 21.13 must be taken seriously. It implies that the alternating voltage has been connected forever (in practice, for a time longer than the natural decay time of the circuit). An extra, decaying term, known as a **transient**, is needed in (21.C.2) if the power has recently been turned on. A transient might look somewhat like Fig. 21.12b; in this section we always assume that it is absent, i.e., that the circuit is undergoing a so-called **steady-state** oscillation.]

Three simple illustrations follow.

---

* To the instructor: In contrast to Chap. 13, this one avoids phase vectors. For ac theory, phase vectors are second best to complex numbers, which all continuing students of this material should eventually use.

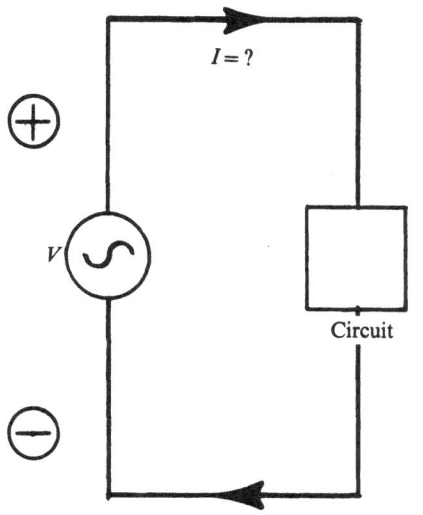

Figure 21.13. An alternating voltage is connected to an unspecified circuit.

Phase in a Resistor

In Fig. 21.14 we just use Ohm's law,

$$I = \frac{V}{R} = \frac{V_{max}}{R} \cos \omega t \qquad (21.C.3)$$

Comparison with (21.C.2) shows that

$$\boxed{I_{max} = \frac{V_{max}}{R}, \qquad \delta = 0} \qquad (21.C.4)$$

In a resistor, current and voltage are in phase.

Let us note at this point that a phase shift δ cannot physically be told apart from $\delta \pm 360°$, or $\delta \pm 2 \times 360°$, etc. All our statements about δ will carry this ambiguity, *which is of no consequence.*

Phase Shift in an Inductor

The circuit of Fig. 21.15 obeys $V_L = V$, or

$$L \frac{dI}{dt} = V \qquad (21.C.5)$$

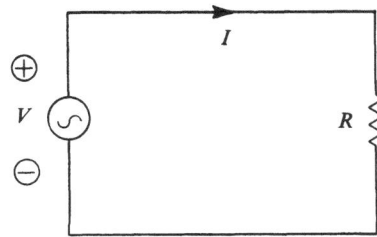

Figure 21.14. Resistor across an alternating voltage.

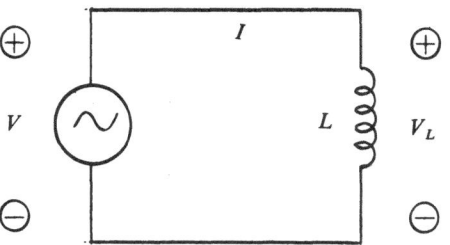

Figure 21.15. Inductor across an alternating voltage.

Inserting (21.C.2), we obtain

$$-\omega L I_{max} \sin(\omega t + \delta) = V_{max} \cos \omega t \qquad (21.C.6)$$

The two sides must have the same amplitude *and* the same phase:

$$\omega L I_{max} = V_{max}$$
$$-\sin(\omega t + \delta) = \cos \omega t \qquad (21.C.7)$$

from which we obtain (taking, for example, $\omega t = 0$)

$$\boxed{I_{max} = \frac{V_{max}}{\omega L}, \qquad \delta = -90°} \qquad (21.C.8)$$

A maximum of V, such as occurs at $\omega t = 0$, also occurs in I at $\omega t = +90°$. Thus,

$$\boxed{\text{In an inductor, the current lags by } 90° \\ \text{(a quarter cycle) relative to the voltage.}}$$

$$(21.C.9)$$

Phase Shift in a Capacitor

In the circuit of Fig. 21.16 the relation $V_C = V$ reads

$$\frac{Q}{C} = V \qquad (21.C.10)$$

To involve the current, we take the time derivative on both sides of (21.C.10), recalling that $dQ/dt = I$:

$$\frac{I}{C} = \frac{dV}{dt} \qquad (21.C.11)$$

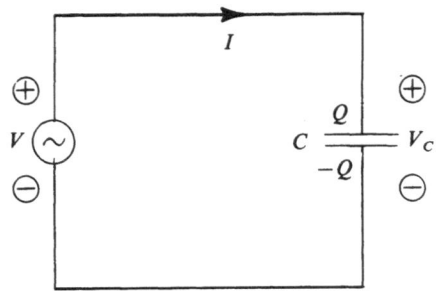

Figure 21.16. Capacitor across an alternating voltage.

or

$$\frac{I_{max}}{C} \cos(\omega t + \delta) = -\omega V_{max} \sin \omega t \quad (21.C.12)$$

Comparing amplitude and phase on both sides (consider in particular the time when $\omega t = -90°$) yields

$$\boxed{I_{max} = \omega C V_{max}, \qquad \delta = +90°} \quad (21.C.13)$$

> In a capacitor, the current leads the voltage by 90°.

$$(21.C.14)$$

Impedance

The resistance of a dc circuit element was defined as $R = V/I$; in a similar way, the **impedance*** Z of an ac circuit element is defined as

$$\boxed{Z = \frac{V_{max}}{I_{max}}} \quad (21.C.15)$$

The special cases we have just seen are

$$Z = \begin{cases} R & \text{(purely resistive circuit)} \\ \omega L & \text{(purely inductive circuit)} \\ \dfrac{1}{\omega C} & \text{(purely capacitive circuit)} \end{cases}$$

$$(21.C.16)$$

* More precisely, the **absolute impedance**, which, in more complete treatments, is the magnitude of a complex-number impedance.

Further, we state without proof that a series *RLC* combination, Fig. 21.17, has an impedance

$$Z = \left[R^2 + \left(\frac{1}{\omega C} - \omega L \right)^2 \right]^{1/2} \quad (21.C.17)$$

The naive result, $R + 1/\omega C + \omega L$, suggested by dc circuitry, is wrong. The reason is that, although the maximum voltages are indeed proportional to R, ωL, and $1/\omega C$, they are not realized simultaneously, but lead or lag one another. Hence they cannot simply be added around the loop in Kirchhoff's voltage rule.

ii. Average ac Power

In Fig. 21.13, how much power, $\mathscr{P}$, is being transferred into the box? At any instant of time, we can take over the conclusion of Eq. (17.B.3), Chap. 17, and write

$$\mathscr{P} = VI \quad (21.C.18)$$

$$= (V_{max} \cos \omega t)[I_{max} \cos(\omega t + \delta)] \quad (21.C.19)$$

Thus, $\mathscr{P}$ oscillates in a rather involved way. While $\mathscr{R}$ is positive, net power is supplied by the voltage source; while $\mathscr{P}$ is negative, the box feeds power back into the source. The explanation is that the circuit stores time-varying amounts of energy in its inductors and capacitors. Nevertheless, if the box contains any resistor at all, net power must, on the average, be dissipated as Joule heat. This average is what we now calculate.

Using the identity

$$\cos \alpha \cos \beta = \tfrac{1}{2}\cos(\alpha + \beta) + \tfrac{1}{2}\cos(\alpha - \beta)$$

$$(21.C.20)$$

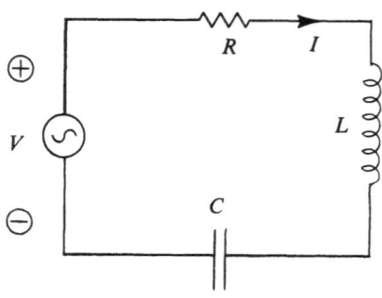

Figure 21.17. A series *RLC* circuit across an alternating voltage. This arrangement will be the next section's topic.

we rewrite (21.C.19) as

$$\mathscr{P} = \tfrac{1}{2}V_{max}I_{max}[\cos(2\omega t + \delta) + \cos\delta] \quad (21.C.21)$$

The first cosine oscillates in time and averages to zero; the second cosine is time independent. Hence the time average of $\mathscr{P}$ is

$$\mathscr{P}_{av} = \tfrac{1}{2}V_{max}I_{max}\cos\delta \quad (21.C.22)$$

In this formula, $\cos\delta$ is known as the **power factor**. It is largest when V and I are in phase ($\delta = 0$), as in a plain resistor. In this special case the factor of 1/2 is most directly understood from the fact that *the time average of $\cos^2\omega t$ is 1/2.*

One immediate consequence of $\mathscr{P}_{av} \geq 0$ is that the phase shift δ always satisfies

$$-90° \leq \delta \leq 90° \quad (21.C.23)$$

How does the power formula, (21.C.22), apply to the separate R, L, and C cases? From Eqs. (21.C.4), for a pure resistance, we get

$$\mathscr{P}_{av} = \frac{V_{max}^2\cos 0°}{2R} = \frac{V_{max}^2}{2R} \quad (\text{pure } R) \quad (21.C.24)$$

For pure L or pure C we have $\delta = \pm 90°$, $\cos\delta = 0$, and hence

$$\mathscr{P}_{av} = 0 \quad (\text{pure } L \text{ or } C) \quad (21.C.25)$$

This last result is reasonable since there can be no heat dissipation without a resistance.

Root-Mean-Square Quantities

Consider a resistance R in series with a voltage source V. The power consumption is

$$\mathscr{P} = \frac{V^2}{R} \quad (\text{dc supply}) \quad (21.C.26)$$

$$\mathscr{P}_{av} = \frac{1}{2}\frac{V_{max}^2}{R} \quad (\text{ac supply}) \quad (21.C.27)$$

With an improved nomenclature, the factor 1/2 can be dropped, and *the ac and dc formulas will look alike.* We define the **root-mean-square voltage** V_{rms} and the **root-mean-square current** I_{rms} as

$$V_{rms} = \frac{V_{max}}{\sqrt{2}}, \qquad I_{rms} = \frac{I_{max}}{\sqrt{2}} \quad (21.C.28)$$

(Caution: This definition is restricted to sinusoidal V or I.) Equation (21.C.27) now reads

$$\mathscr{P}_{av} = \frac{V_{rms}^2}{R} \quad (\text{pure } R) \quad (21.C.29)$$

as desired. More generally, from (21.C.22),

$$\mathscr{P}_{av} = V_{rms}I_{rms}\cos\delta \quad (21.C.30)$$

{The name "root-mean-square" comes from the fact that the time average of V^2 is

$$(V^2)_{av} = V_{max}^2(\cos^2\omega t)_{av} = (V_{max}^2)(\tfrac{1}{2}) = V_{rms}^2$$

so that

$$[(V^2)_{av}]^{1/2} = V_{rms} \quad (21.C.31)$$

and similarly for I_{rms}.}

In practice, ac power supplies and appliances are rated in terms of V_{rms} and I_{rms}; in power ratings one quotes $\mathscr{P}_{av}$.

Example 21.4. A Light Bulb. What is the peak current, I_{max}, in a 200-watt incandescent bulb connected to a 120-volt ac supply?

Using (21.C.28) and (21.C.29), we find

$$I_{max} = \sqrt{2}\,I_{rms} = \sqrt{2}\,\frac{\mathscr{P}_{av}}{V_{rms}}$$

$$= \frac{(\sqrt{2})(200)}{120}\text{ A} = \underline{2.36\text{ A}}$$

iii. Resonance in a Lightly Damped RLC Loop

If a circuit possesses a natural angular frequency ω_0, it will **resonate** in response to an external ac source whose angular frequency ω matches ω_0; we are reminded of the **loaded spring**.

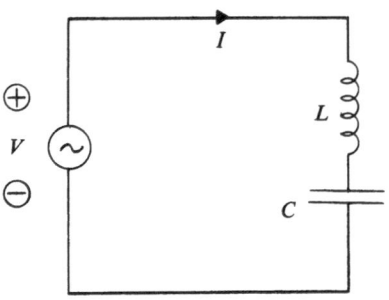

Figure 21.18. A series LC circuit across an alternating voltage.

We now examine how the circuit's behavior changes as ω approaches ω_0.

As in Fig. 21.18, the external voltage

$$V = V_{max} \cos \omega t$$

is in series with L and C; at first we assume no resistance. The circuit's response to V is, as always, of the form

$$I = I_{max} \cos(\omega t + \delta) \qquad (21.C.32)$$

see Eq. (21.C.2); the current amplitude I_{max} depends strongly on ω.

In this case, I_{max} is found to increase near resonance according to

$$\boxed{I_{max} \approx \frac{V_{max}}{2L\,|\omega - \omega_0|}} \qquad (21.C.33)$$

This formula is derived in Note (ii) at the end of the chapter. As we can see, I_{max} "blows up"

when ω approaches ω_0, a behavior characteristic of **undamped resonances** ("undamped" means "without energy dissipation"). Formula (21.C.33) is plotted qualitatively in Fig. 21.19, for a fixed V_{max}.

Damped Resonance

The infinite current predicted by (21.C.33) at $\omega = \omega_0$ is of course a physical impossibility. Any amount of resistance, R, will damp it to a finite value, as displayed by the heavy curve in Fig. 21.19. That more realistic resonance curve may be obtained as a plot of formula (21.C.17); but for a physical understanding, we argue as follows.

Let the circuit, with R included, be as in Fig. 21.20; let us examine its response at the exact

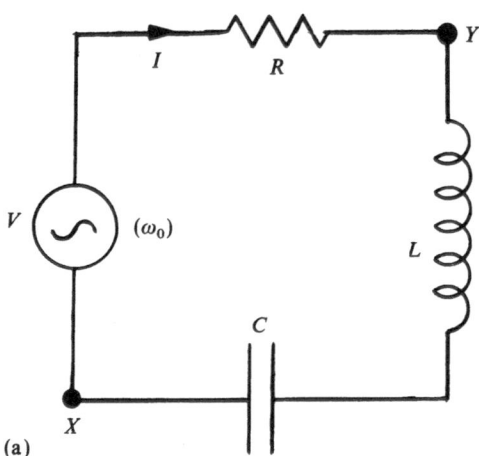

(a)

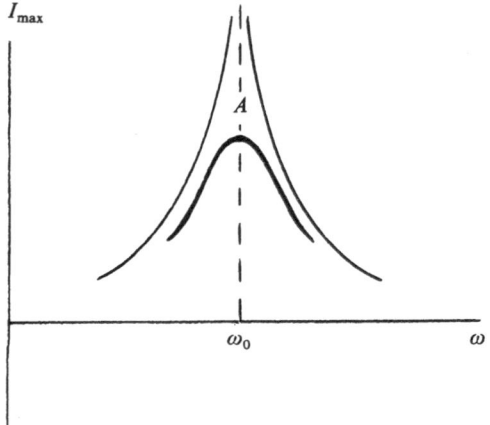

Figure 21.19. Thin curve: Undamped response, as a function of applied angular frequency. Heavy curve: Damped response.

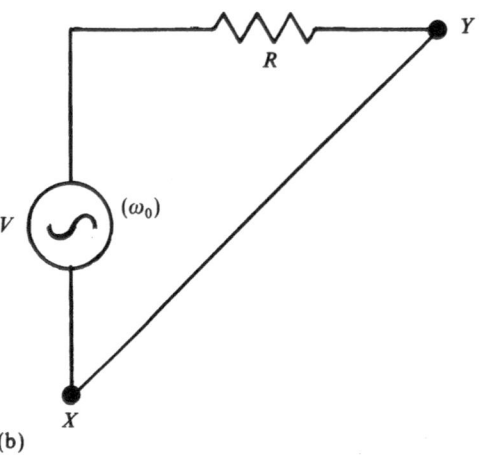

(b)

Figure 21.20. (a) The resonating circuit. (b) The LC part is shunted out.

natural angular frequency $\omega = \omega_0$, where that resonance should be close to a maximum.

No calculation is needed. Indeed, points X and Y in the figure are at the same potential; as in Fig. 21.9b, the voltages across C and L must cancel at resonance. Hence, as far as the response in R is concerned, the LC portion of the circuit may be shorted out; see part (b) of Fig. 21.20. The current through R then has amplitude

$$I_{\max} = \frac{V_{\max}}{R} \qquad \text{(at resonance)} \qquad (21.\text{C}.34)$$

This must also be the current through the original RLC circuit. The response (21.C.34) is indicated by point A in Fig. 21.19. In short,

> At resonance, the series RLC circuit behaves like a pure resistance R.

$$(21.\text{C}.35)$$

Notes

i. The RLC Loop (Time Behavior)

Precisely how does the oscillation shown in Fig. 21.12 decay with time? Our clue is that energy is dissipated at a rate $\mathscr{P} = RI^2$, depleting the energies $LI^2/2$ and $Q^2/2C$ stored in the inductor and capacitor. [For these formulas, see Eqs. (17.B.6) of Chap. 17 and (21.A.19), (21.A.20) of the present chapter.]

A calculation, not shown here, can be performed, based on that observation. For the charge on the capacitor, the result is

$$Q \approx Q_0 e^{-(R/2L)t} \cos \omega t \qquad (21.\text{N}.1)$$

and correspondingly, for the current,

$$I \approx -\omega Q_0 e^{-(R/2L)t} \cos \omega t \qquad (21.\text{N}.2)$$

These are approximations, valid for light damping only. They differ from the undamped formulas by the **exponential decay factor** $e^{-(R/2L)t}$, plotted as dashed lines in Fig. 21.12b. The characteristic time is proportional to L/R, and thus becomes infinite for $R=0$, as expected when damping is absent.

ii. Deriving the Resonance Formula

Formula (21.C.33),

$$I_{\max} \approx \frac{V_{\max}}{2L \, |\omega - \omega_0|} \qquad (21.\text{N}.3)$$

describes the response of an LC loop to an external voltage of amplitude $V_{\max}$ and angular frequency ω. The relevant circuit is shown in Fig. 21.18. If the source V were shorted out, we would be left with a pure LC loop, whose natural angular frequency is

$$\omega_0 = \frac{1}{(LC)^{1/2}} \qquad (21.\text{N}.4)$$

First let, for definiteness, the external frequency be lower than the natural one,

$$\omega < \omega_0 \qquad (21.\text{N}.5)$$

In order to obtain (21.N.3), we make a hypothetical experiment, *which preserves the frequency ω that was being imposed.* As in Fig. 21.21, we tamper with the left portion of the circuit in such a way that the right part keeps its former behavior. We

1. Replace the source V by an appropriate inductor L', and
2. Allow the new circuit to oscillate at *its* natural angular frequency

$$\omega = \frac{1}{[(L + L')C]^{1/2}} \qquad (21.\text{N}.6)$$

which is less than ω_0, as intended.

That oscillation can have any amplitude. Suppose the voltage amplitude across L' happens to be $V_{\max}$, simulating the original source. Because L' has impedance $\omega L'$ [see Eq. (21.C.16)], the current amplitude through L' is

$$I_{\max} = \frac{V_{\max}}{\omega L'} \qquad (21.\text{N}.7)$$

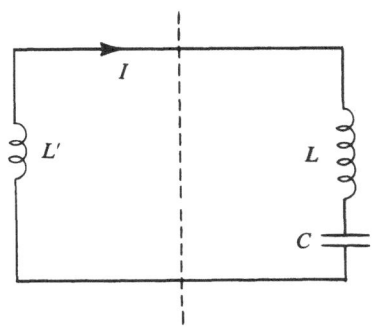

Figure 21.21. In Fig. 21.18, the source of V has been replaced by L'.

(Since I lags 90° behind V, we conclude that the current in an LC loop lags by that amount behind the driving voltage when $\omega < \omega_0$.)

We are not really interested in the auxiliary quantity L'. To eliminate it we solve (21.N.6) for L':

$$L' = \frac{1}{C}\left(\frac{1}{\omega^2} - LC\right) = \frac{1}{C}\left(\frac{1}{\omega^2} - \frac{1}{\omega_0^2}\right) \quad (21.N.8)$$

with the help of (21.N.4). Equation (21.N.7) then reads

$$I_{max} = \frac{V_{max}}{\dfrac{\omega}{C}\left(\dfrac{1}{\omega^2} - \dfrac{1}{\omega_0^2}\right)} = \frac{C\omega\omega_0^2 V_{max}}{\omega_0^2 - \omega^2} \quad (21.N.9)$$

This expression becomes much simpler if we assume that ω is only a few percent away from ω_0. Then we can set, in the numerator, $\omega \approx \omega_0$, and in the denominator

$$\omega_0^2 - \omega^2 = (\omega_0 + \omega)(\omega_0 - \omega) \approx 2\omega_0 |\omega - \omega_0| \quad (21.N.10)$$

where $|\omega - \omega_0|$ is the discrepancy between the applied ω and the natural ω_0. Equation (21.N.9) now becomes

$$I_{max} \approx \frac{C\omega_0^2 V_{max}}{2|\omega - \omega_0|} \quad (21.N.11)$$

which leads to (21.N.3).

The same expression is valid for $\omega > \omega_0$, as the reader can derive for himself by using an auxiliary capacitance, C', instead of L'. (For $\omega > \omega_0$ the current will be seen to lead the voltage by 90°.)

Condensed Checklist

Inductance:

$$\boxed{\mathscr{E} = -L\frac{dI}{dt}} \quad (= V \text{ if } \mathscr{R}=0) \quad (21.A.2)$$

$$\mathscr{N}_1 + \mathscr{N}_2 + \cdots = LI \quad (21.A.5)$$

$$1 \text{ henry} = 1 \frac{\text{volt second}}{\text{ampere}} = 1 \text{ ohm second} = 1 \frac{\text{weber}}{\text{ampere}}$$
$$(21.A.8), (21.A.9)$$

$$L = \frac{4\pi K' N^2 \mathscr{A}}{s} \quad \text{(long solenoid)} \quad (21.A.12)$$

Combinations:

$$L_{tot} = L_1 + L_2 + \cdots \quad \text{(series)}$$
$$\frac{1}{L_{tot}} = \frac{1}{L_1} + \frac{1}{L_2} + \cdots \quad \text{(parallel)}$$

Fig. 21.4

Magnetic energy:

$$\boxed{\mathscr{W} = \tfrac{1}{2}LI^2} \quad (21.A.19)$$

$$\text{Energy density} = \frac{B^2}{8\pi K'} \quad (21.A.13)$$

The RL loop:

$$L\frac{dI}{dt} + RI = 0 \quad (21.A.25)$$

$$\boxed{I = I_0 e^{-t/\tau}, \qquad \tau = \frac{L}{R}}$$
$$(21.A.32), (21.A.31)$$

The LC loop:

$$L\frac{d^2Q}{dt^2} + \frac{Q}{C} = 0 \quad (21.B.6)$$

$$\boxed{\omega_0 = \frac{1}{(LC)^{1/2}}} \quad (21.B.9)$$

$$\mathscr{W}_{tot} = \frac{1}{2}LI^2 + \frac{1}{2}\frac{Q^2}{C} = \text{const}$$

$$\left(\frac{1}{2}LI^2\right)_{av} = \left(\frac{1}{2}\frac{Q^2}{C}\right)_{av} = \frac{1}{2}\mathscr{W}_{tot}$$

Sec. B.iii

The RLC loop (low damping):

$$\omega_0 = \frac{1}{(LC)^{1/2}} \quad (21.B.17)$$

Resonating series RLC circuit (low resistance, ω near ω_0):

$$I_{max} \approx \frac{V_{max}}{2L|\omega - \omega_0|} \quad \text{(neglect } R) \quad (21.C.33)$$

$$I_{max} \approx \frac{V_{max}}{R} \quad (\omega = \omega_0)$$

Average ac power:

$$\mathscr{P}_{av} = \tfrac{1}{2}V_{max}I_{max}\cos\delta = V_{rms}I_{rms}\cos\delta$$
$$(21.C.22), (21.C.30)$$

$$\mathscr{P}_{av} = V_{rms}I_{rms} \quad \text{(pure resistance)}$$

$$\boxed{V_{rms} = V_{max}/\sqrt{2}, \qquad I_{rms} = I_{max}/\sqrt{2}} \quad (21.C.28)$$

Impedance:

$$Z = \frac{V_{max}}{I_{max}} \qquad (21.C.15)$$

$Z = R$,	V and I in phase (resistor)	(21.C.16);
$Z = \omega L$,	I lags 90° behind V (inductor)	(21.C.4), (21.C.9),
$Z = \dfrac{1}{\omega C}$,	I leads V by 90° (capacitor)	(21.C.14)

True or False

1. The inductance of a coil increases with the frequency of the current it carries.

2. The impedance of a coil is independent of the frequency of the current it carries.

3. The impedance of a capacitor increases with the frequency of its applied voltage.

4. The impedance of a resistor is frequency independent.

5. Impedance is measured in ohms.

6. To double the magnetic energy stored in a coil, we must double its current.

7. Low damping in an RLC loop means that R is small compared to $(L/C)^{1/2}$

8. All three combinations, $(LC)^{1/2}$, L/R, RC, have dimensions of time.

9. If both L and C are doubled in an LC loop, its natural frequency remains unchanged.

10. A given light bulb draws the same average power, whether it is connected to a 120-volt ac or dc outlet.

11. A 2-A, ac appliance is designed to draw an average (or mean) alternating current of 2 A.

12. The impedance of a series RLC circuit is about maximum at resonance.

13. A series RLC circuit dissipates maximum power when driven at its natural frequency.

14. If an RLC loop of natural frequency ω_0 is driven at an external frequency ω, the resulting steady-state frequency is intermediate between ω_0 and ω.

15. Alternating voltages and currents have zero time averages: $V_{av} = 0$, $I_{av} = 0$.

Problems

Section A: Self-Induction

*21.1. Express the henry in terms of kilograms, meters, seconds, and coulombs.

*21.2. A 0.60-mH inductor is connected across a constant 2.0-volt potential difference. If the initial current is zero, what is it after 1.0 sec? (Assume zero resistance.)

21.3. Obtain Eq. (21.A.21) by taking the ratio of the corresponding sides of Eqs. (21.A.10) and (21.A.11).

*21.4. A certain electromagnet stores 20 joules of magnetic energy when the current in its winding is 50 A. What is the inductance of that winding?

21.5. Show, by direct substitution, that (21.A.29) is a solution of (21.A.25).

21.6. Verify Eq. (21.A.33) from (21.A.32).

21.7. At time $t = 0$, the potential difference across the resistor in an RL loop is $V_R = 2.5$ volts; at time $t = 50$ μsec that voltage is 0.50 volt. Find V_R at time 150 μsec.

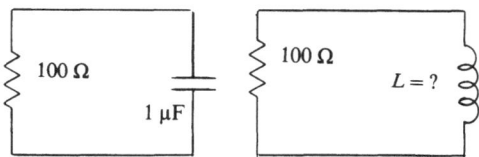

21.8. The two loops shown in the figure have the same time constant. Find the inductance L.

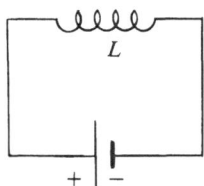

21.9. Suppose you have discovered an "anti-inductor," characterized by a negative inductance L. Connect it as shown across the terminals of a hypothetical rechargeable battery whose terminal voltage is constant. Explain how you can obtain unlimited amounts of energy from your discovery.

21.10. What constant potential difference between the terminals of an 8.0×10^{-6} henry coil is needed in order to increase its current by 0.25 A in 4.0×10^{-5} sec?

*21.11. A certain coil exhibits a current that decreases steadily from 3.0 A at time zero to 1.0 A at time 5.0 μsec. If the coil's inductance is 800 μH, (a) what is the potential difference across the inductor? (Assume zero resistance.) (b) Show in a sketch how the directions of current and voltage are related.

21.12. A certain coil increases its current from 3.0 A to 5.0 A in 2.0×10^{-3} sec while the potential difference between its terminals is 12.0 volts. Assuming zero resistance, find the coil's inductance.

*21.13. A certain inductor stores 0.0050 joule of magnetic energy when its current is 4.0 A. How much more energy is stored when the current is increased to 4.1 A?

*21.14. A certain N-turn solenoid, of given base area $\mathscr{A}$ and length s, has a given magnetic energy $\mathscr{W}$. Find the magnetic pressure (Chap. 19) acting against its cylindrical wall.

21.15. An RL loop has $R = 600\,\Omega$, $L = 0.75$ mH. How much time is needed for a current in the loop to decay to 10% of its value?

*21.16. (a) How much time is needed for the magnetic energy stored in the inductor of Problem 21.15 to decrease to 10% of its value? (b) By what factor does the rate of power dissipation in the resistor decrease in that time?

21.17. An RL loop has $R = 300\,\Omega$; at a certain time, the loop's current is 0.090 A, and 1.00 μsec later it has decreased to 0.040 A. What is the inductance L?

*21.18. In the RL loop of Fig. 21.6, (a) find, in terms of I_0, R, L, the initial energy stored in the inductor. (b) How much power is being dissipated in the resistor at time t? (c) Integrate your answer to (b) from $t = 0$ to $t = \infty$, and compare the result with (a); comment.

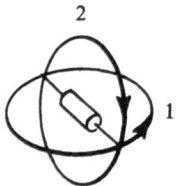

*21.19 (Rotating **B** field). Two mutually perpendicular, circular wire loops, of radius r, share a diameter, as shown; they carry alternating currents of equal amplitude. $I_1 = I_{max} \cos \omega t$, $I_2 = I_{max} \sin \omega t$ (I_2 lags by 90° relative to I_1). (a) Determine the field vectors $\mathbf{B}_1$, $\mathbf{B}_2$ at the center of the loops, as produced only by I_1 or I_2, respectively. (b) Combine $\mathbf{B}_1$, $\mathbf{B}_2$ into the actual field $\mathbf{B} = \mathbf{B}_1 + \mathbf{B}_2$; find the magnitude and direction of $\mathbf{B}$ as a function of time. (Your result should be a rotating $\mathbf{B}$ vector; a conducting cylinder, pivoted as shown, will be dragged into rotation by the field lines; we then have a simplified ac motor.)

21.20. Prove (a) the series formula, (b) the parallel formula for inductors, Fig. 21.4. (Review how it was done for resistors in Chap. 17.)

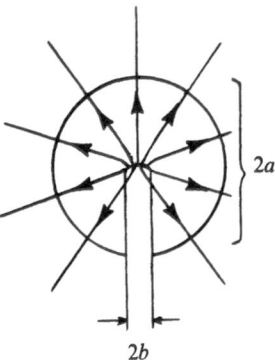

*21.21. A certain long solenoid (or magnet) has a flux $\pm \mathscr{N}$ issuing from each end. Consider one end as a point magnetic pole, and assume the other end is far enough to be neglected. (a) Calculate, by integration, the magnetic field energy $\mathscr{W}_a$ outside a sphere of radius a centered on the solenoid's end, as shown. (Use the $1/r^2$ field only; do not include the uniform inner field.) (b) Assuming a radius b and length s for that same solenoid, find the magnetic energy $\mathscr{W}$ inside it. (c) For a rough estimate of end effects, take $a = b$ and calculate the ratio $2\,\mathscr{W}_a/\mathscr{W}$. Under what condition can end effects be neglected, as was done in Sec. A ii?

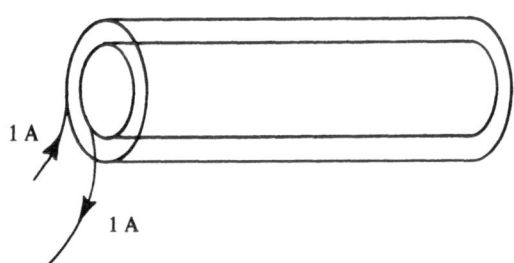

*21.22. Two 1000-turn, 20-cm-long solenoids, of radii 2.0 cm and 3.0 cm, respectively, are arranged coaxially, as shown. Their windings are connected in series and are opposite in direction. (a) Find the total magnetic energy stored when the current equals 1 A. (b) Use Formula (21.A.19) to find the inductance of the combined coils. Repeat (a), (b) for windings having the same direction.

21.23. In the RL loop of Fig. 21.6, find, as a function of time, the voltage V_L across the inductor (express your answer in terms of I_0, R, L, t); make a qualitative graph of V_L against t. (b) Suppose that the battery in part (a) of the figure has a 20-volt potential difference, and that $R = 100\,\Omega$, $L = 3.0$ mH. What is V_L just after the switch is thrown?

*21.24. In Fig. 21.6, start the circuit with zero current in arrangement (b); then, at $t = 0$, flip the switch to arrangement (a); the dc voltage, V_0, is assumed known. (a) Calculate I as a function of time. (Express your result in terms of V_0, L, R, t.) (b) Make a qualitative graph of your result. (c) What value does I approach as $t \to \infty$?

Sections B : Natural Oscillations in a Circuit; C : Alternating Currents

21.25. From the definition of the henry and the farad, prove that $(LC)^{1/2}$ has dimensions of time, as implied by Eqs. (21.B.9), (21.B.10).

*21.26. An LC loop has $L = 0.50\,\mu H$, $C = 40\,\mu F$. (a) What is its natural frequency f? (b) How many times per second is its magnetic energy a maximum?

21.27. The current in a certain LC loop oscillates at a frequency of 350 kHz. The capacitance is $C = 2.5 \times 10^{-3}\,\mu F$. Find the inductance L.

21.28. Calculate (a) the inductance, (b) the capacitance, that present a 15-Ω impedance at a 60-Hz frequency.

21.29. Show how Eqs. (21.C.16) give us (a) the henry, (b) the farad, in terms of the ohm and the second.

*21.30. An inductor must be found that will have a voltage amplitude of 0.030 volt between its terminals when it carries an alternating current of amplitude 20 A and frequency 60 Hz. Find the necessary (a) impedance, (b) inductance.

21.31. A capacitor with $C = 25\,\mu F$ is connected across a 120-volt (rms), 60-Hz supply. (a) What is the capacitor's impedance? (b) What is its rms current? (c) What is its peak current?

*21.32. A coil of inductance $0.80\,\mu H$ and negligible resistance is connected across an alternating voltage of amplitude 3.0 mV and frequency 5000 Hz. What is (a) the coil's impedance, (b) the maximum current, (c) the maximum energy in the coil?

*21.33. What are (a) the current amplitude, (b) the rms current, (c) the average current, (d) the average dissipated power in a 300-Ω resistor when it is connected across an alternating voltage of amplitude 80 volts?

21.34. When connected across a 240-volt (rms), 60-Hz supply, a certain circuit is found to carry an rms current of 40 A, which leads the voltage by a 60° phase shift. What is the average power drawn by that circuit?

21.35. (a) Find exponents α, β such that the combination $L^\alpha C^\beta$ has dimensions of resistance. [Use henry $\sim$ volt sec A^{-1}, farad $\sim$ coulomb volt^{-1}, ohm $\sim$ volt A^{-1}. Argue deductively; the unique answer is (21.B.18).] (b) Verify the dimensions of (21.B.18) on the basis of the impedance formulas (21.C.16).

*21.36. The current in a certain LC loop oscillates naturally at a frequency of 4.0 MHz. If the maximum capacitor charge is $3.0 \times 10^{-5}\,\mu C$, what is the maximum current?

21.37. A certain series LC circuit, with $L = 75\,\mu H$, has a natural frequency of 45.0 kHz. Find the current amplitude through the circuit when it is connected across an alternating voltage of amplitude 0.030 volt and frequency 44.8 kHz.

21.38. (a) Verify that $Q = Q_0 \cos \omega t$, Eq. (21.B.12), is a solution of Eq. (21.B.6). (b) Show that $Q = Q_0 \cos \omega t + Q_0' \sin \omega t$, for $Q_0' = $ const, is a more general solution. (c) In Fig. 21.9, the initial current is zero. Why does this assumption imply $Q_0' = 0$ in (b)? (d) Show that a time-shifted solution, $Q = Q_0'' \cos(\omega t + \delta)$ ($\delta = $ const, $Q_0'' = $ const) is of the form suggested in (b).

*21.39. What is the rms current through an incandescent bulb rated at 1000 watts and 240 volts, but connected to an alternating 120-volt (rms) supply?

*21.40. A certain circuit draws an average power of 18 kW when connected across an alternating voltage of amplitude 340 volts. The current lags relative to that voltage by a certain phase shift. If the current amplitude is 150 A, find the phase shift in degrees of angle.

21.41. A series LC circuit, with $L = 0.2500$ mH and $C = 80.00\,\mu F$, is in series with an alternating voltage of amplitude 15 mV; the current amplitude is 5.0 A. What is the applied frequency (assumed higher than the natural one)?

*21.42. A certain series RLC circuit resonates at 60 Hz. At that frequency, it dissipates an average power of 200 watts when connected across 120 volts (rms). Find the resistance R of that circuit.

*21.43. The peak current in a certain (naturally oscillating) LC loop is 8.0 mA, while the peak capacitor charge is $0.020\,\mu C$. Find the oscillation frequency.

21.44. In Problem 21.49, assume that the peak magnetic energy is 5.0×10^{-10} joule. Find the inductance L and the capacitance C.

21.45. Consider the series RLC circuit of Fig. 21.17. (a) Using the fact that all the power dissipation occurs in the resistor, write an expression for $\mathscr{P}_{av}$ in terms of just I_{max} and R. (b) Combining this result with Eqs. (21.C.22), (21.C.15), prove that the circuit's phase shift obeys $\cos \delta = R/Z$. (c) Hence show that, in the series RLC circuit, the average power is $\mathscr{P}_{av} = RV_{max}^2/(2Z^2)$.

21.46. Consider the exact result (21.C.17) for Z in a series RLC circuit. (a) Calculate, in terms of R, L, and C, the value of Z at $\omega = 1/(LC)^{1/2}$. (b) Show that this value is a minimum of Z as a function of ω.

*21.47. A certain transformer has a primary input consisting of an rms current I, with phase lag δ_1 relative to the voltage; the corresponding output is I_2, δ_2 at the secondary. If the number of primary and secondary turns is N_1, N_2, and if negligible power is dissipated in the transformer, (a) find I_2 in terms of the other data. (Use conservation of energy.) (b) If $\delta_1 = \delta_2 = 0$, $N_1 : N_2 = 1:100$, $I_1 = 0.50$ A, find I_2.

Answers to True or False

1. False.

2. False.

3. False (it decreases).

4. True.

5. True.

6. False.

7. True (means R small compared to the impedance $\omega_0 L$ or $1/\omega_0 C$).

8. True.

9. False (it is halved).

10. True.

11. False (the average alternating current always vanishes; the statement would be true for the rms current).

12. False (it is about minimum).

13. True.

14. False (it equals ω).

15. True (they oscillate symmetrically between positive and negative values).

Answers or Hints to Problems

(Note: Forgetting the conversion $\omega = 2\pi f$ is a common source of mistakes.)

21.1. kg meter^2 coulomb^{-2}.

21.2. 3.3×10^3 A.

21.4. 0.016 henry.

21.11. (a) 320 volts.

21.13. Hint: Use Eq. (21.A.17); set $dI' = 0.1$ A.

21.14. $\mathcal{W}/(s\mathcal{A})$, because energy density = magnetic pressure.

21.16. Hint: (a) Energy $\propto I^2$, hence energy $\propto e^{-2Rt/L}$; (b) same fraction as in (a); why?

21.18. (c) Conservation of energy requires (a), (c) to be the same.

21.19. $|\mathbf{B}| = 2\pi K' I_{\max}/r$.

21.21. (a) $\mathcal{N}^2/(8\pi\mu_0 a)$; (b) $\mathcal{N}^2 s/(2\pi\mu_0 b^2)$; (c) $b \ll 2s$.

21.22. (a) 5×10^{-3} joule; (b) 1×10^{-2} henry.

21.24. (c) Hint: We reach a dc situation.

21.26. (b) 7.1×10^4.

21.30. 4.0 μH.

21.32. (c) 5.7×10^{-9} joule.

21.33. (a) 0.27 A; (b) 0.19 A; (c) 0; (d) 10.7 watts.

21.36. Hint: $I_{\max} = \omega Q_{\max}$; see, e.g., Fig. 21.10.

21.39. Hint: First find R from the rating. Answer: 2.1 A.

21.40. Hint: An intermediate result is $\cos \delta = 36\,000/(150)(340)$.

21.42. Hint: Treat the circuit as a pure R.

21.43. Use Fig. 21.10.

21.47. (a) $I_2 = (N_1/N_2)(\cos \delta_1/\cos \delta_2) I_1$.

CHAPTER 22

ELECTROMAGNETIC WAVES

Electromagnetic radiation (and especially light) occupies center stage among physical phenomena. Life exists thanks to the huge flow of such radiation from the Sun. We receive most of our conscious information in the form of light signals. Other important forms of electromagnetic waves are radio, television, and radar waves, as well as x rays and gamma rays. In the laboratory, electromagnetic radiation is studied not only for its own sake; it provides the most versatile of all research tools, revealing the inner structure of atoms, or permitting the chemical analysis of stars.

Electromagnetic waves became an established concept when James Clerk Maxwell, in the years between 1864 and 1873, deduced their existence and properties in a celebrated theory. Maxwell's prediction was firmly rooted in the "waveless" experimentation of Coulomb, Ampère, and Faraday. The theory's first direct confirmation was reported as late as 1888 by Heinrich Hertz, who measured radio waves emitted by an oscillating circuit.

Maxwell's calculated speed for electromagnetic waves coincides with the observed speed of light signals. This indicated—but did not prove—that light is electromagnetic. In Maxwell's days, the electron was still undiscovered, and no mechanism suggested itself that could account for the emission of light by atoms.

In this chapter we learn why electromagnetic waves must exist. More importantly for a practical understanding of such waves, we learn how they vary with space and time, as well as how much energy and momentum they carry. Finally, we make an inventory of the kinds of electromagnetic waves found in Nature.

A. Theory of Electromagnetic Waves

Electromagnetism, as we have learned it, is slightly incomplete. We must first explain why this is so, and then we must supply the missing ingredient, called **Maxwell's displacement current**. The completed theory then predicts the following.

1. Electromagnetic waves exist; they consist of oscillating **E** and **B** fields.
2. These waves are transverse; that is to say, the **E** and **B** vectors point at right angles to the direction in which the wave travels; **E** and **B** are also perpendicular to one another.
3. The waves propagate in vacuum at a speed c, calculable from the electric and magnetic proportionality constants $K \approx 9 \times 10^9$ (SI), $K' = 10^{-7}$ (SI):

$$c = \left(\frac{K}{K'}\right)^{1/2} \approx 3 \times 10^8 \frac{\text{meters}}{\text{sec}} \qquad (22.\text{A}.1)$$

This is equal to the measured speed of light signals.
4. Electromagnetic radiation carries energy and momentum, in amounts given by simple formulas involving **E** and **B**.

i. The Displacement Current

In the past few chapters, we have focused our attention on circuits; that is, we have looked primarily at what goes on in wires. In contrast, we shall now be looking at wave propagation, which is a vacuum phenomenon. As a transition between the two topics, we ask: can a "piece of vacuum" be considered part of a circuit? More specifically, we want to know if the empty space between the parallel capacitor plates of Fig. 22.1 can be considered "connected in series" with the long straight wire and the plates themselves. The answer is, it *must* be so considered if the **B** field around such an arrangement is to be correctly predicted.

In order to understand this, let us pass for a limited period of time a steady current I through the wire, as shown; the charge $\pm Q$ on each plate then increases steadily in magnitude, and therefore so does the **E** field between the plates.

The electric current I can of course not pass through the gap. But in Maxwell's conception, a kind of surrogate current I_D does exist there, contributed by the rising **E** field. Since this I_D is not a true current, it goes under a somewhat different name, the **displacement current**. Because it serves as a surrogate for I, it is postulated in Fig. 22.1 to have the same value as I:

$$I_D(\text{in the gap}) = I(\text{in the wire})$$

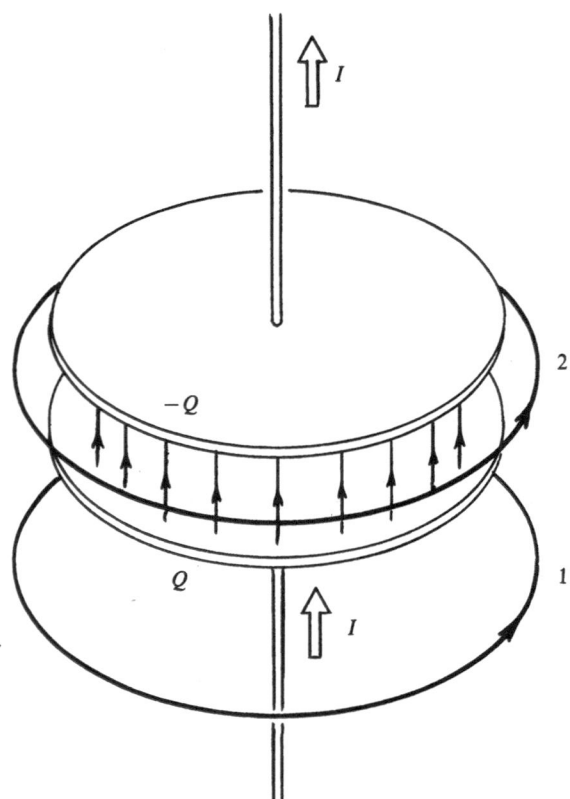

Figure 22.1. According to Maxwell, a capacitor in which the electric flux is changing generates a magnetic field around its gap (e.g., line 2), just as the wire does (e.g., line 1).

while, as we know,

$$I(\text{in the gap}) = 0$$

Does I_D really exist or is it pure invention? The answer, just as Maxwell predicted, is that

> I_D produces the same **B** field around the gap as I produces around the wire.   (22.A.2)

In other words, the physical existence of I_D is made manifest by the **B** field that it produces. This is illustrated by magnetic field line 2 in Fig. 22.1. Thus, according to Maxwell,

> A time-dependent **E** field surrounds itself with a **B** field.   (22.A.3)

Around the gap, we now must write Ampère's law in the form

$$\oint \mathbf{B} \cdot d\mathbf{r} = 4\pi K' I_D \qquad (22.A.4)$$

(Naively, the right-hand side would have been set equal to zero.) The displacement current is expressed in amperes just as though it were a true current.

Let us find out how I_D depends on the changing **E** that produces it. For this purpose we shall see that the **electric flux** $\mathcal{N}_E$ in the gap (i.e., the total number of electric field lines coming out of the bottom plate) is a useful number. As we recall from the beginning of electrostatics, we have

$$\mathcal{N}_E = 4\pi K Q$$

and therefore

$$\frac{d\mathcal{N}_E}{dt} = 4\pi K \frac{dQ}{dt} = 4\pi K I$$

by conservation of charge. But we have defined I_D equal to I, and so

$$I_D = \frac{1}{4\pi K} \frac{d\mathcal{N}_E}{dt} \qquad (22.A.5)$$

This formula summarizes Maxwell's definition of I_D, valid in all situations:

> The displacement current through any surface equals $1/4\pi K$ times the rate of increase of electric flux across that surface.

(The surface in question might be the circular area spanning field line 2 in Fig. 21.1.)

The Displacement Current as a Surface Integral

Around a loop that encloses only vacuum with a changing electric flux, but zero ordinary electric current, Eq. (22.A.4) reads

$$\oint \mathbf{B} \cdot d\mathbf{r} = \frac{K'}{K} \frac{d}{dt} \iint E_\perp \, d\mathscr{A} \qquad (22.A.6)$$

where we have used the "field-language" definition of $\mathscr{N}_E$; here $d\mathscr{A}$ is an element of the surface spanning the loop, perhaps again No. 2 in Fig. 22.1. Numerically, in (22.A.6), we have

$$\frac{K'}{K} = \frac{1}{c^2} \qquad (22.A.7)$$

where

$$c = 3.00 \times 10^8 \text{ meters/sec} \qquad (22.A.8)$$

is the speed of light; note the extreme smallness of K'/K. The explanation for (22.A.7) will emerge presently.

Example 22.1. The B Field around a Capacitor Gap. In Fig. 22.1, assume plates of radius $r = 3.0 \times 10^{-2}$ meter, and a positive upward E field which (momentarily) grows at a rate $dE/dt = 5.0 \times 10^{14}$ volts/(meter sec). Assuming axial symmetry around the capacitor, find the B field at the edge of the gap.

Let us integrate the left side of (22.A.6) along line 2 in the figure. Assuming a radius about equal to r, and using (22.A.7), we have

$$(B)(2\pi r) = \frac{1}{c^2} \left(\frac{dE}{dt}\right)(\pi r^2) \qquad (22.A.9)$$

so that

$$B = \frac{1}{c^2} \left(\frac{dE}{dt}\right)\left(\frac{r}{2}\right) \qquad (22.A.10)$$

$$= \frac{1}{(3.0 \times 10^8)^2} (5.0 \times 10^{14}) \left(\frac{3.0 \times 10^{-2}}{2}\right) \text{ tesla}$$

$$= \underline{8.3 \times 10^{-5} \text{ tesla}}$$

ii. Summary of Electrodynamics

To us, the displacement current will not be of interest for its own sake. Rather, we must understand what functions it serves.

1. It is the final ingredient needed to make electromagnetism a consistent theory when time-dependent E fields are involved. [As we see from (22.A.5), I_D vanishes when E is constant in time; this is why we had no need for I_D in the steady-current situations of Chap. 19.]
2. In vacuum, and far away from true charge currents, the displacement current I_D is the only available source for producing B fields; I_D therefore plays an essential role in the vacuum propagation of electromagnetic waves.

Maxwell's Theory

The basic equations of electromagnetism are presented in Table 22.1. Together, these equations are only the bare skeleton, or symbolic reminder, of the complete physical theory; but mathematically, they form a self-consistent and beautiful whole. The concepts they involve are reviewed in Table 22.2.

(Why are Coulomb's and Biot–Savart's inverse square laws not included in Table 22.1? The reason is to be found in more advanced mathematical treatments of the theory. There it is shown that these inverse square laws can be *derived* from those in Table 22.1. Thus, in a sense they *are* present in the table after all.)

To conclude this summary, we state (without proof) one of the most important qualitative predictions of electromagnetic theory:

> Any accelerating charge necessarily emits an electromagnetic wave. (22.A.11)

Table 22.1. Overview of Maxwell's Theory[a]

Law	Formula	This book's reference
Gauss' Law for **E**	$\oiint E_\perp \, d\mathscr{A} = 4\pi K Q_{\text{encl}}$	Chap. 15, Eq. (15.C.8)
Gauss' Law for **B**	$\oiint B_\perp \, d\mathscr{A} = 0$	Chap. 19, Eq. (19.D.2)
Faraday's Law of Induction	$\oint \mathbf{E} \cdot d\mathbf{r} = -\left(\dfrac{d\mathscr{N}_B}{dt}\right)_{\text{encl}}$	Chap. 20, Eq. (20.E.4)
Ampère's Circuital Law, revised by Maxwell[b]	$\oint \mathbf{B} \cdot d\mathbf{r} = \begin{cases} 4\pi K' I_{\text{encl}} \\[2mm] \dfrac{K'}{K}\left(\dfrac{d\mathscr{N}_E}{dt}\right)_{\text{encl}} \end{cases}$	Chap. 19, Eq. (19.C.3) Chap. 22, Eq. (22.A.6)

[a] The subscript "encl" means "enclosed within the arbitrary fixed surface or path over which the equation's left side is evaluated."
[b] If true and if displacement currents exist in the same region, then both right-hand contributions must be added together.

For example, radio waves come from the oscillating charges in ac circuits; thermal radiation comes from the disorderly motion of electrons in matter.

iii. Structure and Speed of a Wave

There are many shapes of electromagnetic waves, of which the plane one is the simplest. This will now be our topic; we present it in three steps:

1. We recapitulate Maxwell's theory for **E** and **B** in vacuum.
2. We describe the features of a plane wave.
3. We show that such a wave does indeed obey Maxwell's equations. That verification will yield Maxwell's celebrated prediction for the wave's speed of propagation.

1. Field Equations in Vacuum

Let us go back to the four laws of Table 22.1. In the absence of any charges and currents ($Q = 0$, $I = 0$), they read

$$\oiint E_\perp \, d\mathscr{A} = 0 \tag{22.A.12}$$

$$\oiint B_\perp \, d\mathscr{A} = 0 \tag{22.A.13}$$

$$\oint \mathbf{E} \cdot d\mathbf{r} = -\left(\frac{d\mathscr{N}_B}{dt}\right)_{\text{encl}} \tag{22.A.14}$$

$$\oint \mathbf{B} \cdot d\mathbf{r} = \frac{K'}{K}\left(\frac{d\mathscr{N}_E}{dt}\right)_{\text{encl}} \tag{22.A.15}$$

Note how **E** and **B** play very similar roles.

Table 22.2. Concepts in Maxwell's Theory

Concept	Meaning	This book's reference
Charge Q	Defined additively by electroscope measurement; unit fixed by Coulomb's (static) law.	Chap. 15, Sec. A
Current I	Charge flowing per unit time.	Chap. 17, Eq. (17.A.1)
Electric and magnetic fields **E**, **B**	What exerts the Lorentz force on a test charge.	Chap. 18, Eq. (18.B.9)
Electric and magnetic fluxes $\mathscr{N}_E$, $\mathscr{N}_B$	The number of field lines through a piece of surface.	Chap. 15, Eq. (15.C.7) Chap. 20, Eq. (20.C.4)

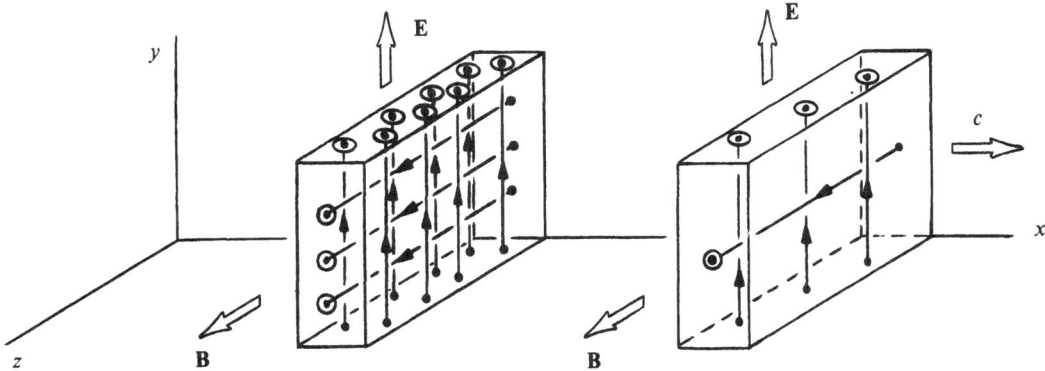

Figure 22.2. Two sample slices of space, showing the field lines in an intense and less intense part of a wave. The slices and their field lines must be imagined as moving to the right at speed c. The present figure is a "snapshot." The actual lines continue indefinitely in the y and z directions.

2. A Plane Wave

Figures 22.2 and 22.3 show what is meant by a traveling plane electromagnetic wave (not sinusoidal in this illustration). Its qualitative features are as follows:

- Lines of **E** are straight and parallel, perpendicular to the direction of propagation.
- Similarly for the lines of **B**.
- **E** and **B** are mutually perpendicular.
- The line densities (i.e., the magnitudes E and B) vary only in the direction of propagation.
- E and B are everywhere proportional to one another,

$$E \propto B \qquad (22.A.16)$$

For example, we see that E and B vanish, or are maximal, together.

- The whole pattern of field lines moves, as a rigid unit, at a constant speed c. (Note: The symbol c is in general use to denote the speed of an electromagnetic wave in vacuum.)
- There exists a right-hand rule, symbolized by the corkscrew in Fig. 22.4. When the handle is twisted from the **E** direction to the **B** direction, the screw advances in the same direction as the wave.

3. Are the Field Equations Obeyed?

In the situation of Fig. 22.2, Gauss' laws (for **E** and **B**) are manifestly satisfied. Indeed, the field lines are uninterrupted, and hence no net flux can leave or enter any closed surface. (See Chap. 15 for a review.)

Equations (22.A.14) and (22.A.15) are more interesting. We first test Faraday's law,

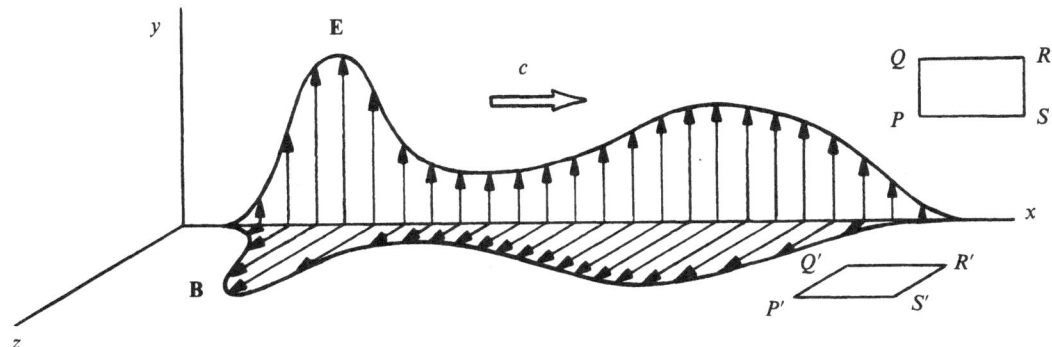

Figure 22.3. A "complete" record of **E** and **B** as a function of x, at one single instant of time. (The small rectangles will serve later as contours for Faraday's or Ampère's law; they lie in the xy and xz planes, respectively.)

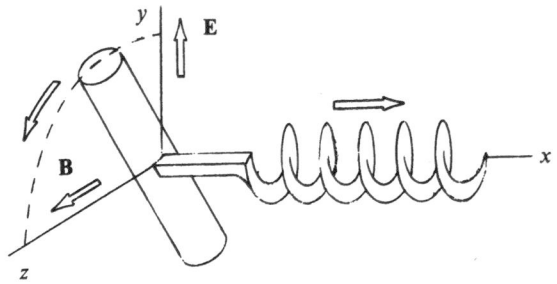

Figure 22.4. The right-hand rule for traveling electromagnetic waves.

Eq. (22.A.14), around the motionless rectangular loop $PQRSP$ in Fig. 22.3. The pattern of field lines moves past the loop with speed c. In the figure, the right end of the loop still experiences zero field.

How does Eq. (22.A.14) read in this case? The equation's left side receives a contribution only from segment PQ. That left side becomes

$$\oint \mathbf{E} \cdot d\mathbf{r} = Ea \qquad (22.A.17)$$

where a is the length of PQ.

The right-hand side of (22.A.14) was already evaluated in a slightly different situation (moving loop, static field) in Eq. (20.C.11) of Chap. 20. The reader should have no difficulty in adapting the argument to the present circumstances; the speed v must only be replaced by c. We find

$$-\left(\frac{d\mathcal{N}_B}{dt}\right)_{\text{encl}} = Bca \qquad (22.A.18)$$

Equating (22.A.17) and (22.A.18), we obtain

$$\boxed{E = cB} \qquad (22.A.19)$$

at the level of segment PQ in Fig. 22.3. Since that segment is arbitrarily situated, we have here the announced proportionality (22.A.16), valid at all points. Thus, *in terms of SI units*,

> The E field is c times larger than
> the B field (in a plane wave).   (22.A.20)

(Other systems of units may have factors different from c; for example, $E = B$ is true in the so-called Gaussian system.)

If we next decide to test Eq. (22.A.15) around the same loop $PQRSP$ in Fig. 22.3, we learn nothing new; we just obtain the identity $0 = 0$. (Why?) In order to test (22.A.15) we need a contour with a new orientation, such as rectangle $P'Q'R'S'P'$ in Fig. 22.3. This yields, analogously to (22.A.19),

$$B = \frac{K'}{K} Ec \qquad (22.A.21)$$

Again, rectangle $P'Q'R'S'P'$ gives $0 = 0$ for Eq. (22.A.14).

(One last check before we proceed: Lenz's rule must be verified around $PQRSP$, while the counterpart of that rule, namely, the right-hand rule for **B**, with the increasing **E** flux acting as a current, applies to $P'Q'R'S'P'$, as it does in Fig. 22.1.)

The Speed c

Are Eqs. (22.A.19) and (22.A.20) compatible? Multiplying their corresponding sides together, we find the consistency requirement

$$\boxed{c = \sqrt{\frac{K}{K'}} \approx 3.00 \times 10^8 \, \frac{\text{meters}}{\text{sec}}} \qquad (22.A.22)$$

a long-promised and important result. This is the wave's speed in vacuum, and for most practical purposes it is the same in air.

Experimentally, it turns out that c is measured far more accurately for microwaves than K can ever be determined from static experiments. [For details, see Problem 22.8 further on, and Note (i) at the end of Chap. 18.] The general trust in Maxwell's theory is such that the measured speed c is used, through (22.A.22), in order to fix $K = (10^{-7} \text{ SI units})(c^2)$.

Polarization

In Fig. 22.3, the **E** vectors, plotted in the xyz coordinate system, single out a plane known as the **plane of vibration** of the wave, here the xy plane. More simply, we may refer to the "plane of **E**"; the xz plane is the "plane of **B**."

(Caution: The term "plane wave" has nothing to do with the polarization. Rather, it refers to the fact that the surfaces of constant **E** and **B** are planes, in this case parallel to the yz plane.)

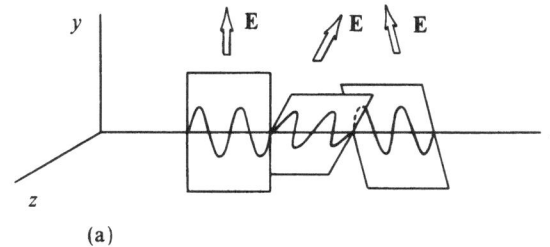

(a)

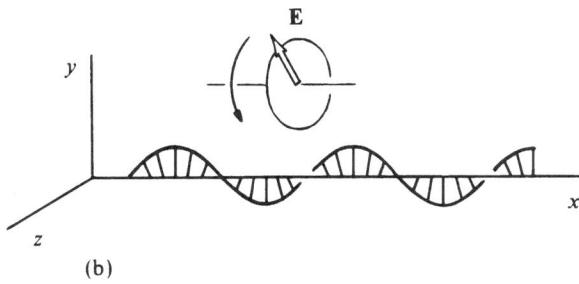

(b)

Figure 22.5. Waves that are not linearly polarized. (a) The plane of vibration is different for different sections of the wave. (b) The **E** vector rotates uniformly in the yz plane; this wave is called **circularly polarized**.

When a plane of vibration exists, as it does in Fig. 22.3, the wave is said to be **linearly polarized**. Not all plane electromagnetic waves are linearly polarized; Fig. 22.5 shows waves which are not.

B. Energy and Momentum in a Wave

Much earlier in this book we learned that a *sound* wave carries energy; the amount of that energy is proportional to the square of the vibration amplitude. A similar statement holds for electromagnetic waves; here lies the explanation for the heat we receive from the Sun, and for countless other instances of radiative energy transfer found in Nature.

Electromagnetic momentum also exists, but it is of much less practical importance—it is usually below the level at which it might be detected. Theoretically, however, the momentum of a wave is a valuable concept, because it, together with electromagnetic energy, is among the clues that will lead us to both relativity and quantum mechanics.

i. Energy

To find the electromagnetic energy, we just put together the earlier, static formulas, which gave us

the energy density of the **E** and **B** fields separately. The total energy density is the sum of both:

$$\text{Energy density} = \frac{E^2}{8\pi K} + \frac{B^2}{8\pi K'} \quad (22.\text{B}.1)$$

This can be written more simply. In a plane traveling wave we have $B = E/c$, according to (22.A.19). The above becomes

$$\text{Energy density} = \frac{E^2}{8\pi K} + \frac{E^2/c^2}{8\pi K'} \quad (22.\text{B}.2)$$

With the help of $c^2 = K/K'$ [Eq. (22.A.22)] we see that both energy terms are equal: in vacuum, *the wave carries the same amount of electric and magnetic energy*. Thus, (22.B.2) becomes

$$\text{Energy density} = \frac{E^2}{4\pi K}$$

(plane wave traveling in vacuum) (22.B.3)

Of more practical interest is the question: how much energy can be collected per unit time and per unit area by a surface facing an incoming wave? (We might be dealing with the eye's pupil facing the incoming light, as in Fig. 22.6. The desired quantity is called the **energy current density**, and is usually denoted by S. Thus, we have

$$S = \frac{\left(\begin{array}{c}\text{energy crossing a perpendicular}\\ \text{area } \mathcal{A} \text{ during } dt\end{array}\right)}{\mathcal{A}\, dt}$$

(22.B.4)

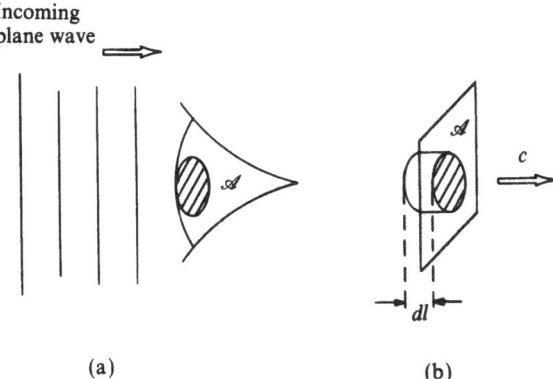

(a)   (b)

Figure 22.6. (a) The area $\mathcal{A}$ faces the wave and intercepts a certain amount of energy. (b) Showing that the energy current density equals c times the energy density.

S is an *instantaneous power per unit area*. (What is the SI unit for this quantity?) In (22.B.4), "perpendicular" means "perpendicular to the wave's direction of propagation." Part (b) of Fig. 22.6 shows how to calculate S. The small cylinder, which travels with the wave, carries a total energy

$$d\mathcal{W} = (\text{energy density}) (\text{volume}) \quad (22.B.5)$$

$$= (\text{energy density}) (\mathcal{A}\, dl) \quad (22.B.6)$$

The whole cylinderful of energy crosses $\mathcal{A}$ in a time $dt = dl/c$, and we obtain

$$S = \frac{d\mathcal{W}}{\mathcal{A}\, dt} = \frac{(\text{energy density})(\mathcal{A}\, dl)}{\mathcal{A}(dl/c)} \quad (22.B.7)$$

or

$$S = (c)(\text{energy density}) \quad (22.B.8)$$

$$S = \frac{cE^2}{4\pi K} \quad (22.B.9)$$

from Eq. (22.B.3). Expressed symmetrically in E and B, the above becomes

$$S = \frac{EB}{4\pi K'} \quad (22.B.10)$$

as the reader should verify for himself.

ii. Poynting Vector

A suitably defined cross product,

$$\mathbf{S} = \frac{\mathbf{E} \times \mathbf{B}}{4\pi K'} \quad (22.B.11)$$

known as the **Poynting vector**, shows the direction of energy transfer as well as its magnitude. As a check, we go back to Fig. 22.3, and verify that the direction of $\mathbf{S}$ is that of the wave propagation, while the magnitude $|\mathbf{S}|$ satisfies (22.B.10). (See Appendix II for a review of the cross product.)

The Poynting vector enjoys a remarkable property, *not* shared by (22.B.10). It correctly specifies the local electromagnetic energy flow in all cases—not just that of a plane traveling wave. If $\mathbf{E}$ and $\mathbf{B}$ enclose an angle θ (not necessarily $90°$), Eq. (22.B.11) shows that the right-hand side of (22.B.10) must be corrected by a factor of $\sin \theta$. We

shall not prove the general validity of (22.B.11); that result is due to John Henry Poynting (1884).

Example 22.2. Poynting Vector for Static Fields. A certain long straight wire, of uniform resistivity, carries a steady current I. Consider an imaginary cylindrical box (circumference $= l$, altitude $= h$), coaxial with the wire, as shown in Fig. 22.7. (a) At what net rate does electromagnetic energy, if any, enter the box? (Assume a uniform $\mathbf{E}$ field everywhere parallel to the wire.) (b) From circuit theory, how much power is dissipated as heat in the wire segment enclosed by the box? (c) Discuss the relation between (a) and (b).

———

(a) Electromagnetic power: *There are no waves in this example*; nevertheless, Eq. (22.B.11) must hold. The figure shows the direction of $\mathbf{B}$; its magnitude is (see Chap. 19)

$$B = \frac{2K'I}{r} \quad (22.B.12)$$

($r =$ distance from the wire's axis). The Poynting vector $\mathbf{S}$ is radially inward everywhere (even inside the wire; why?) and therefore energy enters the box only through the curved surface—not through the flat top or bottom. Thus, a steady stream of electromagnetic energy keeps converging on the wire, and must be dissipated there as Joule heat, as we shall see in (b).

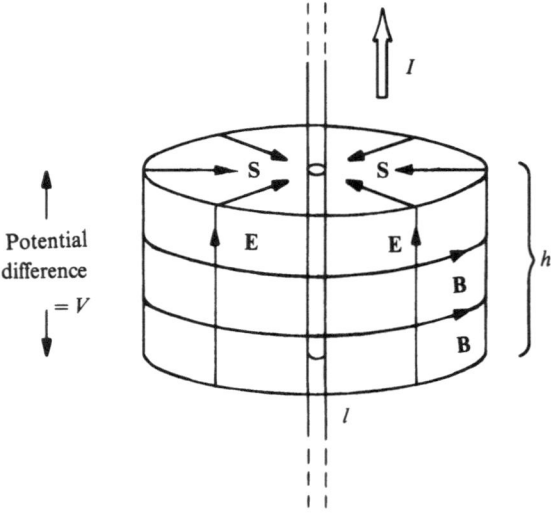

Figure 22.7. The Poynting vectors converge on the wire.

The magnitude of **S** is

$$S = \frac{EB \sin 90°}{4\pi K'} = \frac{(E)\left(\dfrac{2K'I}{r}\right)(1)}{4\pi K'}$$

$$= \frac{EI}{2\pi r} = \frac{EI}{l} \qquad (22.B.13)$$

at the curved surface. The rate of energy transfer is then

Electromagnetic power $= (S)$ (curved area)

$$= \left(\frac{EI}{l}\right)(lh) = Ehl$$

$$(22.B.14)$$

Now we recognize Eh as the potential drop

$$V = Eh \qquad (22.B.15)$$

between the ends of the wire segment. Thus, (22.B.14) reads

Electromagnetic power $= \underline{\underline{VI}} \qquad (22.B.16)$

(b) Joule heat: From Chap. 17 we get directly

Power dissipated $= \underline{\underline{VI}} \qquad (22.B.17)$

(c) Comparison: The equality of (a) and (b) points to two alternative ways of looking at a single process. In (b) the conduction electrons are considered to be losing potential energy. In (a), our bookkeeping ignores potential energy, and lists field energy instead. These two views were already present in Sec. D of Chap. 16, where the energy of a capacitor was ascribed to the charges' potential energy *or* to the field energy. If we included both we would be guilty of double counting.

iii. Intensity

Let us return to Fig. 22.6, which shows wave energy being collected through an opening. In general, that energy arrives in a lumpy manner. *As a function of time*, the **E** field in a plane wave might be such as in Fig. 22.8a; then the rate at which energy is collected behaves as in part (b). In

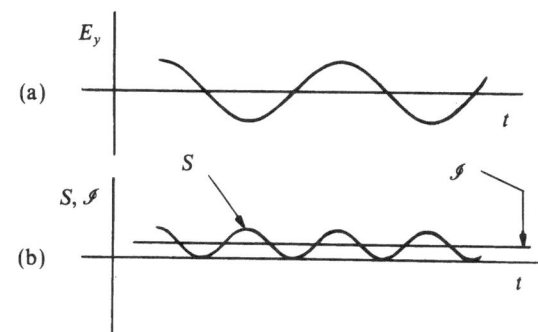

Figure 22.8. (a) A wave that oscillates in time (at a single chosen location). (b) The energy current density, S, which involves E^2, occurs in lumps; the intensity, $\mathscr{I}$, is the smoothed-out version of S.

practice, we are not interested in the individual "lumps." The eye might receive 10^{16} cycles of radiation in a single second; a radio, 10^6 cycles in a second. Clearly, a "smoothed out" rate is what must be defined. That concept is called the **intensity** of the wave:

> At any point in space, the intensity, $\mathscr{I}$, of a wave is its time-averaged energy current density S.

$$(22.B.18)$$

In other words, $\mathscr{I}$ is a *time-averaged power per unit area*; see part (b) of Fig. 22.8. That average must be taken over a large number of lumps, say between times t_1 and t_2. Mathematically, the definition is

$$\mathscr{I} = S_{av} = \frac{\left(\begin{array}{c}\text{energy collected per unit}\\ \text{perpendicular area between } t_1 \text{ and } t_2\end{array}\right)}{t_2 - t_1}$$

$$(22.B.19)$$

$$= \frac{\int_{t_1}^{t_2} S\, dt}{t_2 - t_1} \qquad (22.B.20)$$

Example 22.3. Sinusoidal Time Dependence. At a collecting surface, let $E = E_{max} \cos \omega t$, where $E_{max} = 20$ volts/meter and $\omega = 5$ kHz. (a) Assuming a plane traveling wave, find its intensity. (b) If the collecting surface is an opening of area $\mathscr{A} = 2$ meters^2 facing the incoming radiation, how much energy does it receive in a time $\Delta t = 1$ hour? [Use $(\cos^2 \omega t)_{av} = 1/2$.]

(a) Intensity: We have, by (22.B.9),

$$S = \frac{c(E_{max} \cos \omega t)^2}{4\pi K} \qquad (22.B.21)$$

Therefore

$$\mathcal{I} = \frac{cE_{max}^2}{4\pi K}(\cos^2 \omega t)_{av} = \frac{cE_{max}^2}{8\pi K}$$

$$= \frac{(3 \times 10^8)(20)^2}{(8\pi)(9 \times 10^9)} \frac{watt}{meter^2} = \underline{0.5 \frac{watt}{meter^2}}$$

$$(22.B.22)$$

Note the irrelevance of ω.

(b) Energy: By (22.B.19), the total energy received is

$$\mathcal{W} = \mathcal{I} \mathcal{A} \, \Delta t$$

$$= (0.5)(2)(3600) \text{ joules} = \underline{\underline{3600 \text{ joules}}}$$

$$(22.B.23)$$

Radiation from a Center

Far enough from a source of radiation, such as a light bulb or a star, a portion of the wave can be considered plane, as in Fig. 22.9. If the source emits a given, fixed power, radiated equally in all directions, then the intensity at a distance r from the center is

$$\mathcal{I} = \frac{power}{spherical \; area} = \frac{power}{4\pi r^2} \quad (22.B.24)$$

Intensity decreases as the inverse square of the

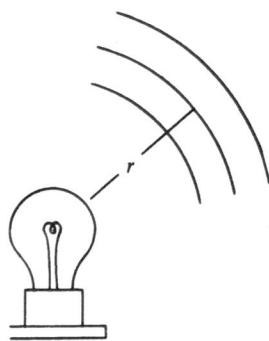

Figure 22.9. The intensity weakens as r^{-2} away from a small source.

observer's distance from the radiating center. This feature is required by the conservation of energy; see Problem 22.23.

iv. Momentum

In contrast to the energy of an electromagnetic wave, its momentum (which points in the direction of propagation), is very small and hard to detect. A mirror, when reflecting light, should experience a pressure (**radiation pressure**) due to a momentum transfer from the light. Indeed, after reflection, the light must have its own momentum reversed, somewhat like particles elastically bombarding a wall.

Radiation pressure does in fact exist, and has been observed in delicate experiments, some of which involve intense laser beams shining on microscopic beads.

The pressure of sunlight is responsible for turning the tails of most comets away from the Sun. There have been serious proposals for harnessing solar radiation pressure by means of huge, ultralight reflecting "sails" for the purpose of long-range space travel.

Here we ask: How much momentum does a given electromagnetic wave contain per unit volume? The answer is as follows: *any portion of the wave stores a momentum equal to $1/c$ times the energy stored there.* Thus,

$$\boxed{\text{Momentum density} = \frac{1}{c}(\text{energy density})}$$

$$(\text{traveling wave}) \quad (22.B.25)$$

The smallness of $1/c$, in practical units, accounts for the difficulty of observing electromagnetic momentum.

Equation (22.B.25) can be derived most simply from an idealized radiation-pressure experiment, illustrated in Fig. 22.10. Here a uniform metal plate is facing the incoming wave. The plate is taken so thin as to be almost transparent to the wave. (This is somewhat unrealistic: such a thin metal plate or foil would be difficult to manufacture.) The plate's role is to remove some energy from the wave; we shall see that it necessarily removes some momentum as well. The scenario is as follows.

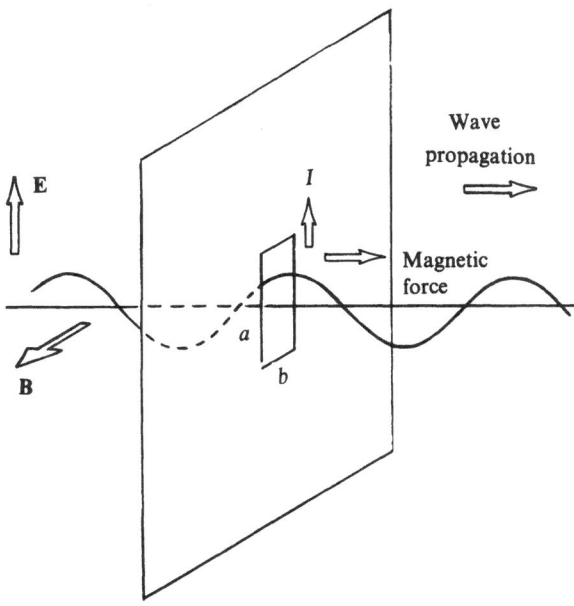

Figure 22.10. Wave passing through a thin conducting plate; the latter experiences a radiation pressure whose origin is magnetic.

The incoming **E** field sets up a current in the plate. Joule dissipation then accounts for the energy loss. On the other hand, the incoming **B** field exerts a force on the current, and therefore transfers some momentum to the plate. This accounts for the momentum removed from the wave.

In Fig. 22.10, we assume that the wave is transmitted through the plate with very little change; we also assume that negligible wave energy is reflected back. [The reader who is worried about this last statement is referred to Note (i) at the end of this chapter.]

Now for the quantitative argument. We focus our attention on a portion of the plate, namely, a rectangular strip of sides a, b; these sides are parallel to **E** and **B**, respectively. In the direction of **E**, the strip has a resistance R, carries a current I, and is subjected to a voltage $V = aE$.

Current

Ohm's law gives

$$I = \frac{V}{R} = \frac{aE}{R} \qquad (22.B.26)$$

Energy Absorbed

During a short time dt, the energy $d\mathcal{W}$ dissipated in the strip obeys

$$\frac{d\mathcal{W}}{dt} = I^2 R = \frac{a^2 E^2}{R} \qquad (22.B.27)$$

Force on the Strip

The force, **F**, is magnetic, and has absolute value

$$F = IaB = \frac{a^2 EB}{R} \qquad (22.B.28)$$

with use of (22.B.26); **F** points in the direction of propagation of the wave. Since **F** equals the rate of momentum transfer from the wave, $d\mathbf{p}/dt$, Eq. (22.B.28) reads

$$\frac{dp}{dt} = \frac{a^2 EB}{R} \qquad (22.B.29)$$

Comparing Energy and Momentum

Taking the ratio of the corresponding sides of (22.B.29) and (22.B.27), we find

$$\frac{dp}{d\mathcal{W}} = \frac{B}{E} = \frac{1}{c} \qquad (22.B.30)$$

the last step with the help of (22.A.19). Thus, a procedure that removes an energy $d\mathcal{W}$ from the wave also removes a momentum $dp = d\mathcal{W}/c$. Integrating both sides of this equality, we conclude that any portion of the wave has an energy $\mathcal{W}$ and a momentum p that satisfy

$$p = \frac{\mathcal{W}}{c} + \text{const} \qquad (22.B.31)$$

But the constant must vanish, because a nonexistent wave ($\mathcal{W} = 0$) also has zero momentum, $p = 0$. Equation (22.B.31) then reads

$$p = \frac{\mathcal{W}}{c} \qquad (22.B.32)$$

as announced in (22.B.25). Furthermore, the direction of **F** in Fig. 22.10 shows that **p** is indeed along the wave's direction of propagation.

C. The Electromagnetic Spectrum

In order to describe an electromagnetic wave experimentally, we must address at least one of two questions. (1) How is the wave produced? (2) How is it detected? From both points of view, the simplest waves (as in the case of sound) are those that *oscillate sinusoidally in time*. The reason is that emitters and detectors often consist of harmonically oscillating charges and currents. These may vary in size from many meters (radio towers) to less than a billionth of a meter (single molecules or atoms).

i. Sinusoidal Plane Waves

For definiteness, imagine a plane wave traveling in vacuum. Here, just as with sound, a harmonic vibration in time means a sinusoidal variation in space as well.

Figure 22.11 is a "snapshot" of a plane, traveling, linearly polarized, electromagnetic wave with definite wavelength λ. By definition,

$$\lambda = \begin{pmatrix} \text{distance (measured along the direction} \\ \text{of propagation) between two successive} \\ \text{maxima of any } \mathbf{E} \text{ field (or } \mathbf{B} \text{ field)} \\ \text{component} \end{pmatrix}$$

$$(22.\text{C}.1)$$

(It does not matter which component of $\mathbf{E}$, or of $\mathbf{B}$, we are referring to: λ has the same value for all.) As in the case of sound, *a definite wavelength λ is*

associated with *a definite frequency f*: the two are related by the important equation

$$\boxed{\lambda f = c} \quad \text{(in vacuum)} \qquad (22.\text{C}.2)$$

The reader should briefly review Sec. A of Chap. 14 for a discussion of this point.

Parameters of the Wave

Consider a plane traveling wave like that of Fig. 22.11; we assume linear polarization. We must keep in mind the following list of features, which need to be independently specified for a complete description of the wave.

1. The direction of propagation (here the $+x$ direction).
2. The plane of $\mathbf{E}$ (here the xy plane).
3. The wavelength λ.
4. The amplitude $E_{\max}$ (i.e., the maximum value of $|\mathbf{E}|$).
5. The phase, which indicates the position of the moving wave pattern at some chosen reference time like $t = 0$. (We shall not be concerned with phases in this chapter.)

If λ is specified, there is no need to give f; by (22.C.2), λ and f are equivalent pieces of information (because c is known once and for all). In fact, we shall indifferently label radiation by its frequency or by its wavelength.

Similarly, the amplitude $E_{\max}$ and the intensity $\mathscr{I}$ yield the same information; recall Eq. (22.B.22), which states that $\mathscr{I} = cE_{\max}^2/8\pi K$. We shall see that

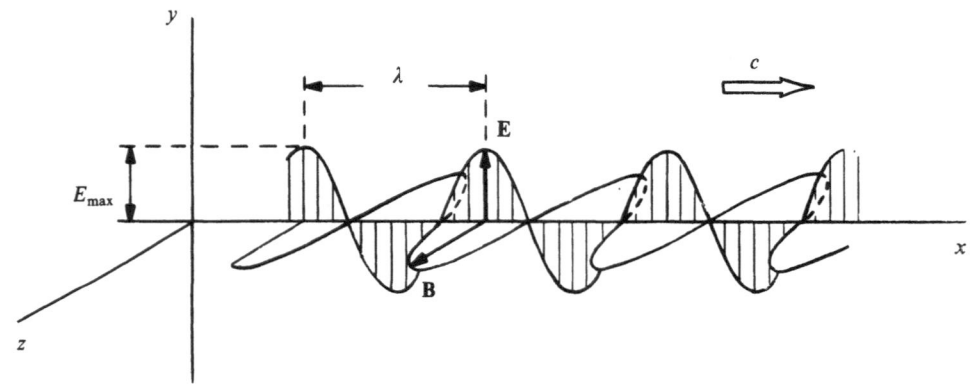

Figure 22.11. A sinusoidal wave, shown at a single time. (Compare with Fig. 22.3.)

waves *are more conveniently characterized by their intensity than by their amplitude.*

ii. The Spectrum

Maxwell's theory predicts that any accelerating charge must emit an electromagnetic wave. If the radiating source is a charge or current distribution that oscillates harmonically (i.e., sinusoidally in time), two simple rules govern the production of the wave.

1. The frequency of the wave equals the frequency of the source.
2. For a given source geometry, the amplitude of the wave is proportional to the amplitude of the source (i.e., to the total amount of vibrating charge, or to the maximum current.)

Figure 22.12 illustrates the broadcasting of a radio wave by the oscillating current and charge in an antenna, as well as its detection by two kinds of receiving antennas. As a rule, emitting and receiving structures *interact preferentially with wavelengths of size comparable to themselves.* Figure 22.13 classifies all electromagnetic waves by their frequency f or, equivalently, by their wavelength $\lambda = c/f$. The indicated production and detection modes are only typical; many more mechanisms are available than mentioned here.

Figure 22.14 gives more details of the visible spectrum than could be fitted into Fig. 22.13. (Note: It is customary to label visible and near-visible radiation by wavelength, rather than by frequency.)

As it passes through matter, radiation suffers many kinds of change; its direction of travel may be affected, and in most cases its intensity is reduced. However, it is a rule of thumb that *the wave's original frequency is left intact* during and after passage through a material medium at rest. We shall always take this rule for granted.

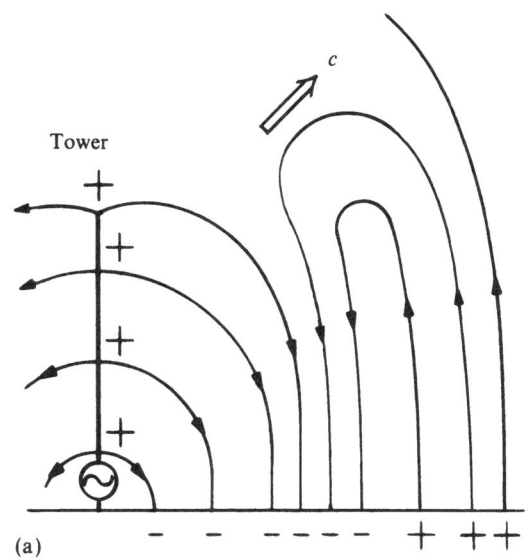
(a)

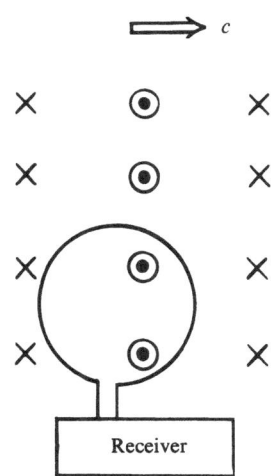
(b)

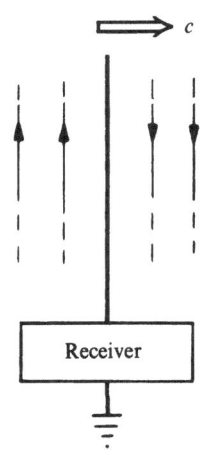
(c)

Figure 22.12. (a) Electric field lines around a radio tower. The spreading wave is linearly polarized, with a vertical **E** plane. *Zero energy is radiated vertically up from the tower.* (b) A loop antenna detects the wave through a magnetically induced current. The loop's best orientation coincides with the **E** plane; the dots and crosses are the lines of **B**. (c) A "dipole antenna" for receiving the wave. The antenna is vertical, to fit the **E** plane. Some lines of **E** are shown.

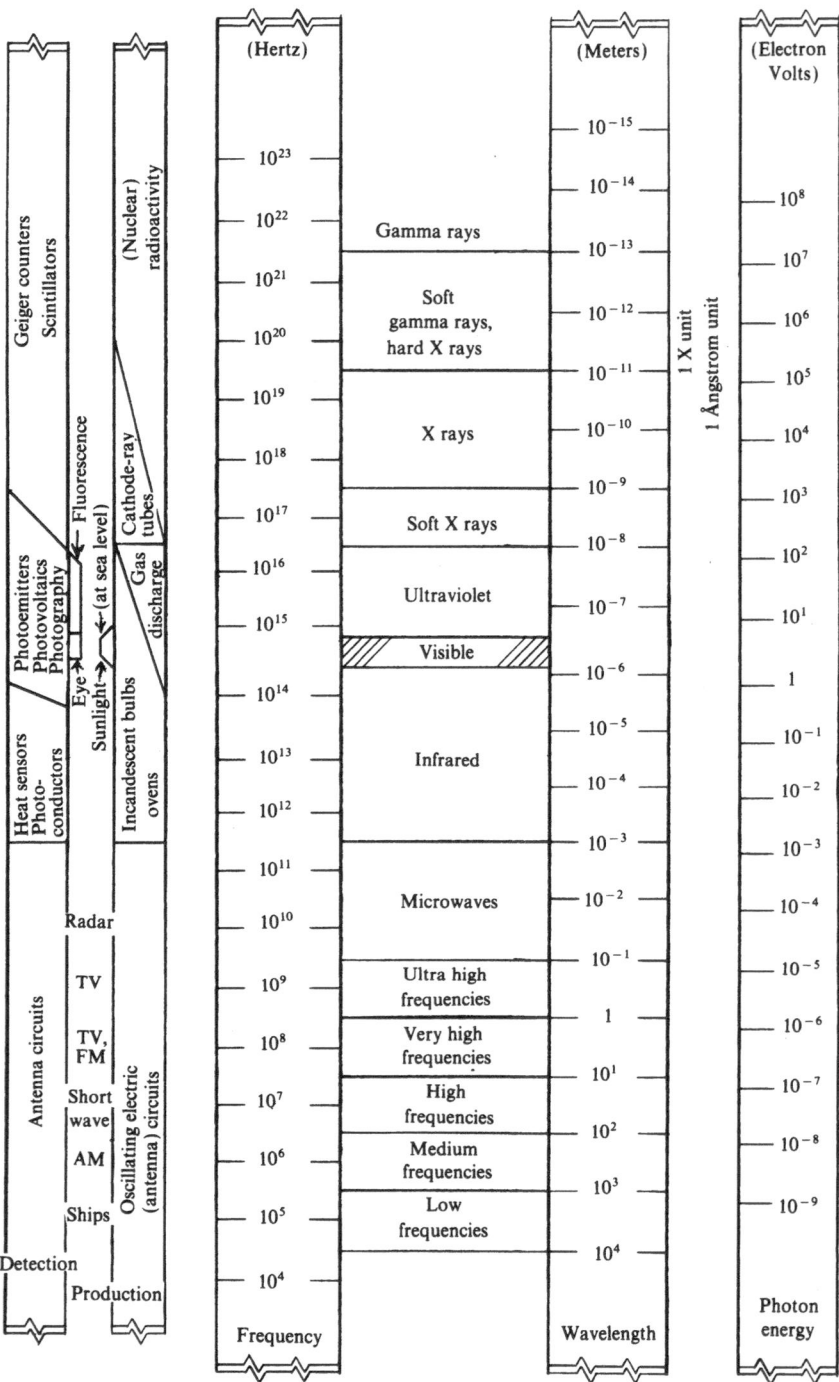

Figure 22.13. The electromagnetic spectrum. (Caution: The artificially sharp boundaries must not be taken literally.) Middle section, the top down: Gamma rays are emitted by radioactive nuclei; x rays result from the impact of fast electrons with stationary targets; in addition to the visible solar light, we receive some ultraviolet, and considerable infrared, at the Earth's surface. Millimetric and longer wavelengths are relevant to radar and communications; we list some frequency bands under their official names. Among the means of detection listed, some are self-explanatory. Photoconductors are substances whose conductivity improves dramatically under infrared radiation. When exposed to light, photoemitters release electrons in vacuum; photovoltaic cells develop a potential difference. Invisible ultraviolet light is absorbed by fluorescent substances, which then convert the energy to visible light. Scintillators work similarly with gamma and x rays. Ionization chambers, of which the Geiger counter is an example, detect gamma rays by the resulting ions created in a gas; the latter conducts a pulse of current. Discussion of the right-hand scale ("photon energy") must be postponed until Chap. 26 (Waves Versus Particles).

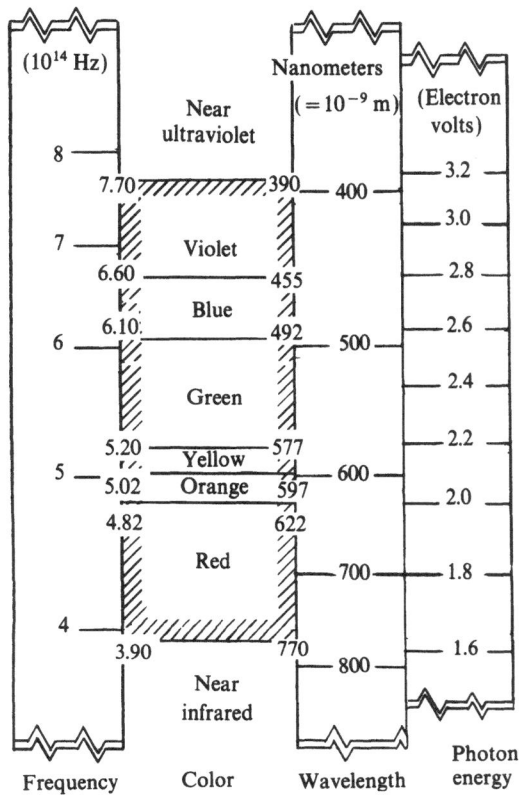

Figure 22.14. The visible band (shaded in Fig. 22.13) shown in more detail. **White light** is a mixture of all those frequencies.

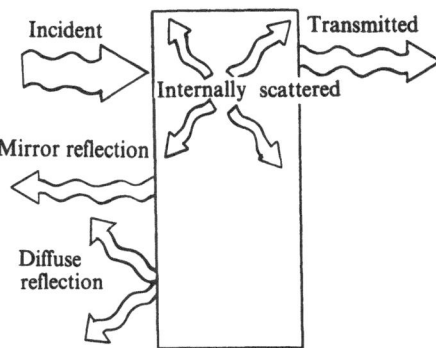

Figure 22.15. The fate of radiation incident on matter (absorption not shown).

Transmission, Reflection, Scattering, and Absorption

Figure 22.15 symbolizes how radiation may be affected by a layer of matter.

1. **Transmitted** radiation emerges on the other side, in a well-defined direction related to the original direction of propagation. (The possible changes of direction will be studied later, under the headings of refraction and diffraction.)

2. **Reflected** radiation is sent back in a well-defined direction (mirror reflection), or in all directions (diffuse reflection, as from this sheet of paper.) Reflection takes place at the surface of incidence.

3. Some **internally scattered** radiation is spread diffusely into all directions *within the medium*, like a headlight beam in the fog. Much of the diffusely scattered light may emerge from the substance and be observed from the outside.

4. Whatever part of the incident radiation does not appear in (1), (2), or (3) is the **absorbed energy**. It goes into heating the medium, or modifying it chemically, or into a variety of other effects.

As an illustration, we note that, in practice, a metal plate partly absorbs and partly reflects radio or radar waves; there is no transmission and no internal scattering in this case. As another illustra-

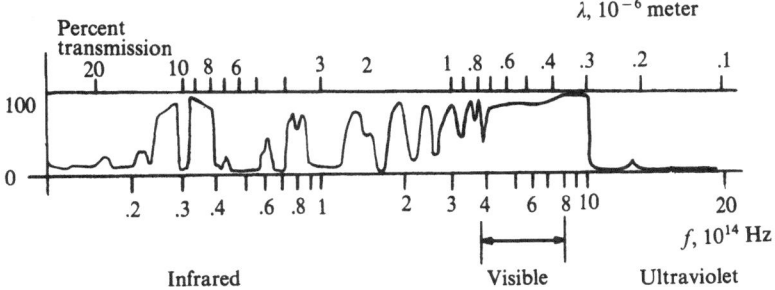

Figure 22.16. Transmission spectrum of the Earth's atmosphere. Note the excellent transparency in the visible and the near ultraviolet. Various gases are responsible for blocking the radiation at other frequencies. Ozone (O_3) causes the ultraviolet cutoff; water vapor (H_2O), as well as carbon dioxide (CO_2), absorb at many wavelengths in the infrared. (Data from A. Miller and J. Thompson, *Elements of Meteorology*, 2nd ed., C. E. Merrill Co., 1975.)

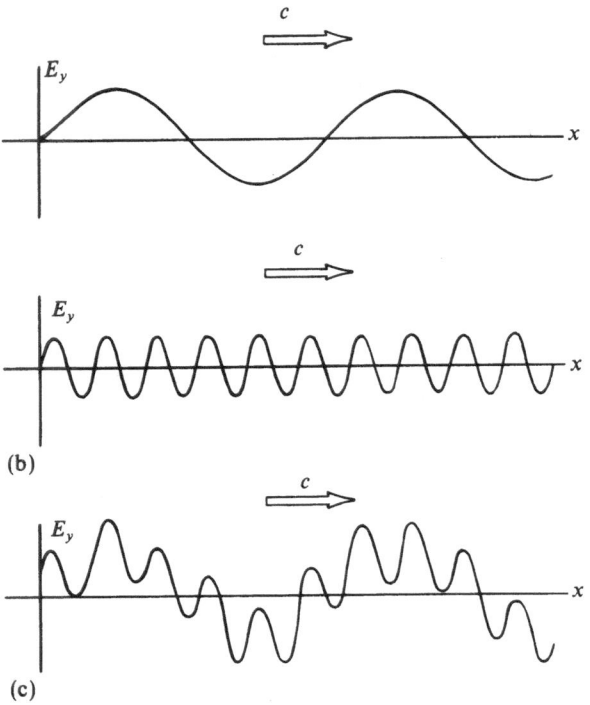

Figure 22.17. Part (c) displays a superposition of the two frequencies shown in (a) and (b).

tion, involving our atmosphere, Fig. 22.16 shows how drastically the extent of transmission may depend on the incident frequency.

In most cases, *transparency requires that the medium be a dielectric.*

iii. Mixed Frequencies

So far we have dealt with waves having a single, well-defined frequency. Such a wave is called **monochromatic.*** On the other hand natural radiation is a superposition of waves with different frequencies, and hence with different wavelengths. This fact is symbolized in Fig. 22.17, which shows only two superposed waves ("component waves"). The overall field is not sinusoidal, in time or in space, and hence possesses no single amplitude. However, the wave has a definite intensity, namely, its average energy current density, as in Eqs. (22.B.19), (22.B.20).

It now turns out that *intensities are additive.* To see the meaning of "additive," consider a device that separates the component waves from one

* Meaning, literally, "of a single color."

another. The intensity $\mathscr{I}_1$ or $\mathscr{I}_2$ of each can be measured individually. It is then a matter of experimental verification that the intensity of the mixed wave is given by $\mathscr{I} = \mathscr{I}_1 + \mathscr{I}_2$. Figure 22.18 is an idealized illustration of that relation. In the case of arbitrarily many frequencies we have

$$\boxed{\mathscr{I} = \mathscr{I}_1 + \mathscr{I}_2 + \mathscr{I}_3 + \cdots} \qquad (22.C.3)$$

The additivity of intensities can be shown to follow, *as a time average,* from the additivity of energy; each component wave conveys its own energy.

In order to separate a mixed wave's frequency components one utilizes a great variety of devices: photographic color filters, the color-sensitive pigments in the retina of one's eyes, prisms (next chapter), diffraction gratings (Chap. 24), and, in the radio range, tuned circuits, as in Chap. 21, Sec. B. Discrimination between frequencies (or wavelengths) is never perfectly sharp, and varies a good deal in quality from method to method. Any apparatus for exhibiting (or recording) the intensity of a wave's frequency components may be called a **spectroscope** (or **spectrometer**).

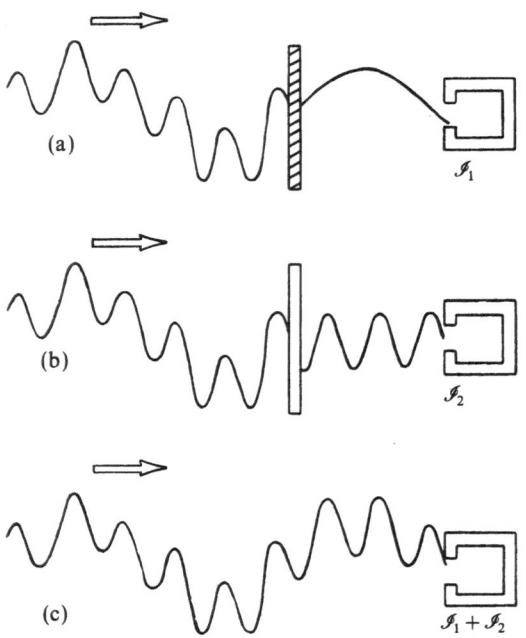

Figure 22.18. Adding intensities. A mixture of two frequencies comes in; the filters in (a) and (b) pass only the low and high frequencies, respectively. In (c), the superposition is measured to have the sum of the intensities measured in (a) and (b).

Note

Negligible Reflection from a Thin Plate

Section B iv described the passage of a wave through a thin conducting plate. The transmitted wave was assumed to look like the incoming wave, only a bit weaker. In actual fact, some radiation is reflected as well, but its energy is negligible compared to what is absorbed.

To see this, we note that the incoming wave creates an alternating current I in the plate; that current, if considered by itself, emits in both directions a wave illustrated in Fig. 22.19. Here we are interested in the reflected energy, collected at P_1. More specifically, we ask for the *ratio of energy reflected to energy absorbed*.

As we have seen in Eq. (22.B.27), the power absorbed in a rectangular strip is

$$\text{Absorbed power} = I^2 R \qquad (22.\text{N}.1)$$

(R = resistance of the strip).

On the other hand, the power reflected by the strip is carried by the **E** and **B** fields which I radiates. These fields can in principle be computed from a knowledge of I; if I is given, the resistance R need not be specified in that calculation, and we have, according to circuit theory,

$$\text{Reflected power} = (\text{const}) \, I^2 \qquad (22.\text{N}.2)$$

where the proportionality constant is independent of I and, especially, independent of R. (We omit the detailed argument; the wave is assumed sinusoidal, as in Fig. 22.19; strictly speaking, I should be interpreted as the root-mean-square current, the conducting strip being an ac circuit.) Thus, we find a ratio

$$\frac{\text{Reflected power}}{\text{Absorbed power}} = \frac{(\text{const}) \, I^2}{I^2 R} \qquad (22.\text{N}.3)$$

$$= \frac{\text{const}}{R} \qquad (22.\text{N}.4)$$

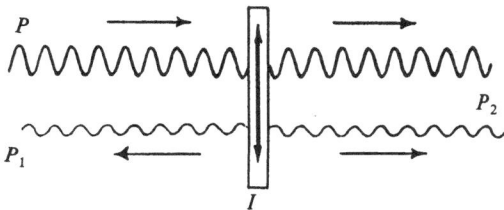

Figure 22.19. Reflection versus absorption in a thin plate. The wave comes from P; the reflected and transmitted signals are detected at P_1 and P_2. (Destructive interference causes the intensity reduction at P_2.)

As $R \to \infty$ (very thin plate), that ratio approaches zero, as claimed.

Condensed Checklist

Displacement current:

$$\boxed{I_D = \frac{1}{4\pi K}\frac{d\mathcal{N}_E}{dt}} \qquad [K \approx 9.00 \times 10^9 \text{ (SI)}] \quad (22.\text{A}.5)$$

$$\oint \mathbf{B} \cdot d\mathbf{r} = 4\pi K' I_D \qquad (22.\text{A}.4)$$

Maxwell's theory: Tables 22.1 and 22.2 of Sec. A

Maxwell's theory without charges or currents:

$$\oiint E_\perp \, d\mathscr{A} = 0, \qquad \oiint B_\perp \, d\mathscr{A} = 0$$

$$(22.\text{A}.12), (22.\text{A}.13)$$

$$\oint \mathbf{E} \cdot d\mathbf{r} = -\frac{d}{dt}\iint B_\perp \, d\mathscr{A},$$

$$\oint \mathbf{B} \cdot d\mathbf{r} = +\frac{K'}{K}\frac{d}{dt}\iint E_\perp \, d\mathscr{A}$$

$$(22.\text{A}.14), (22.\text{A}.15)$$

Speed of electromagnetic waves in vacuum:

$$\boxed{c = \left(\frac{K}{K'}\right)^{1/2}}$$

$$\approx 3.00 \times 10^8 \, \frac{\text{meters}}{\text{sec}} \qquad (22.\text{A}.22)$$

Plane traveling wave in vacuum:

$$\boxed{\begin{array}{c} E = cB \\ \textbf{E, B orthogonal, transverse} \end{array}} \qquad (22.\text{A}.19)$$

$$\text{\textbf{E, B} orthogonal, transverse} \qquad \text{Figs. 22.2 and 22.3}$$

Energy (any electromagnetic field in vacuum):

$$\text{Energy density} = \frac{E^2}{8\pi K} + \frac{B^2}{8\pi K'}$$

$$(22.\text{B}.1), (22.\text{B}.2), (22.\text{B}.3)$$

(both tems are equal)

$$\boxed{\text{Poynting vector} = \mathbf{S} = \frac{\mathbf{E} \times \mathbf{B}}{4\pi K'}} \qquad (22.\text{B}.11)$$

Energy current density $= S$

$$(22.\text{B}.4), (22.\text{B}.10)$$

Energy and momentum (traveling wave):

$$S = (c)(\text{energy density}) \quad (22.\text{B}.8)$$

$$\text{Momentum density} = \frac{1}{c}(\text{energy density}) \quad (22.\text{B}.25)$$

$$\text{Intensity} = \mathscr{I} = S_{\text{av}} \quad (22.\text{B}.19)$$

Monochromatic traveling wave:

$$\lambda f = c \quad (22.\text{C}.2)$$

$$\mathscr{I} = \frac{c}{8\pi K} E_{\text{max}}^2 \quad (22.\text{B}.22)$$

Mixed frequencies:

$$\mathscr{I}_{\text{tot}} = \mathscr{I}_1 + \mathscr{I}_2 + \cdots \quad (22.\text{C}.3)$$

True or False

1. The displacement current has SI units of amperes.

2. The displacement current through an area is proportional to the electric flux through that area.

3. Faraday's law, revised by Maxwell, involves the displacement current enclosed by the integration path.

4. A changing **E** field produces a **B** field.

5. In a traveling wave, the maxima of **E** and of **B** alternate in space.

6. In a traveling wave, the **E** vector vibrates transversely to the direction of propagation, while the **B** field vibrates parallel to it.

7. Linear polarization implies that **E** and **B** vibrate in parallel directions (but perpendicular to the direction of propagation).

8. Energy current density is measured in watts/meter^3.

9. Intensity is measured in watts/meter^2.

10. In order to double the intensity of an electromagnetic wave, one may double its frequency while leaving its amplitude the same.

11. In order to double the intensity of an electromagnetic wave, one must quadruple its amplitude.

12. In a traveling plane wave, the amplitude of **E** can be adjusted independently of the amplitude of **B**.

13. The Poynting vector is parallel to **E**.

14. The amplitude of a spherically spreading electromagnetic wave decreases as the inverse distance from the radiating source.

15. The ratio of energy density to momentum density has the dimensions of speed.

16. For an electromagnetic wave in vacuum, higher frequency always means smaller wavelength.

17. X rays are typically emitted by antennas.

18. X rays have a wavelength comparable to the size of an atom.

19. Radio waves in the low-frequency band have a wavelength that might fit into a room.

Problems

Section A : Theory of Electromagnetic Waves

*22.1. A certain air capacitor, each of whose parallel plates has an area of 0.25 meter^2, uniformly increases its electric field from 2.0×10^4 volts/meter to 3.5×10^4 volts/meter in 0.050 sec. What is, during that time, the displacement current between the plates?

*22.2. A parallel-plate air capacitor, of area 0.15 meter^2 for each plate, and gap width 4.0×10^{-4} meter, decreases its voltage uniformly from $+3.0$ volts to -2.0 volts in 0.0025 sec. (a) What is, during that time, the displacement current between the plates? (b) What is the electric current through one of its terminals?

*22.3. An air capacitor, whose parallel circular plates have a radius of 1.5 cm, and whose gap is 0.50 mm wide, has a voltage momentarily given by $V = (2.5 \times 10^5 \text{ volts/sec})(t)$. Find (a) the displacement current between the plates; (b) the **B** field at the edge of the gap.

22.4. In a certain plane wave that propagates upwards, the magnetic field vector, at one particular place and time, has a strength of 2.0×10^{-12} tesla; it points due west. What is, in magnitude and direction, the electric field at that same place and time? (Assume no other fields than those of the wave.)

*22.5. An electric power line, carrying a 60-Hz alternating current, is emitting an electromagnetic wave. What is its wavelength?

22.6. A certain radio wave and a certain sound wave have the same wavelength in air. By what factor is the radio wave's frequency larger than the sound wave's frequency? (Use $c \approx 3 \times 10^8$ meters/sec, speed of sound ≈ 300 meters/sec).

*22.7. From certain standing wave (resonance) experiments, similar to those performed with sound, the speed of light in vacuum was found to be $c = 2.9979250 \times 10^8$ meters/sec. Calculate K in Coulomb's law to that precision.

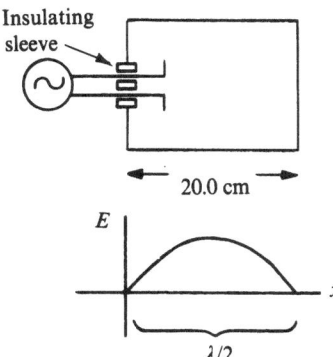

$\lambda/2$

*22.8 (Standing wave). A small antenna sets up an electromagnetic wave inside a metal box of length 20.0 cm (see figure). The wave is reflected back and forth at both ends (like sound in a closed pipe), and the electromagnetic energy in the box is observed to be a maximum at certain sharp frequencies, of which the lowest is f. If the length of the box is then half the wavelength of a traveling wave, find f.

22.9. The antenna of a certain radio receiver is a metal rod whose length l is adjusted for reception of a 100.0 MHz station. Find l if its value is desired to be a quarter wavelength.

*22.10. If, in Fig. 22.1, the ordinary wire current is given by $I = 2.0$ A, estimate the **B** field between the plates (whose radius is 0.020 meter), halfway along the plate radius; assume axial symmetry.

22.11. Verify, in Example 22.1, that the calculated **B** field is the same as it would be for an infinite straight wire without the capacitor. Procedure: from dE/dt, calculate dQ/dt, where Q = charge on a plate; this gives I in the wire. (The result depends on having assumed axial symmetry in the capacitor's vicinity.)

22.12. Show that the Ampère–Maxwell law (22.A.15), when applied to loop $PQRSP$ in Fig. 22.3, just yields the identity $0 = 0$.

22.13. Show in detail how the Ampère–Maxwell law (22.A.15) gives Eq. (22.A.21).

22.14 (Rate of magnetic flux change). Review the derivation of Eq. (20.C.11) in Chap. 20, and give an adapted version of it to obtain Eq. (22.A.18) of the present chapter.

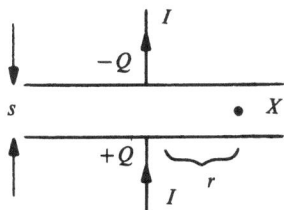

*22.15. A long straight wire is interrupted by an air capacitor with parallel circular plates of radius R and

small gap s. A steady current I is flowing in the wire, as shown. (a) Find, in magnitude and direction, the magnetic field **B** at a point X halfway between the plates, at a distance r from the central axis, at a time when the plates have charges $\pm Q$. (Express your answer in terms of I, Q, s, r, R, c, or some of these.) (b) How is the answer modified if X is moved a distance x closer to the top plate?

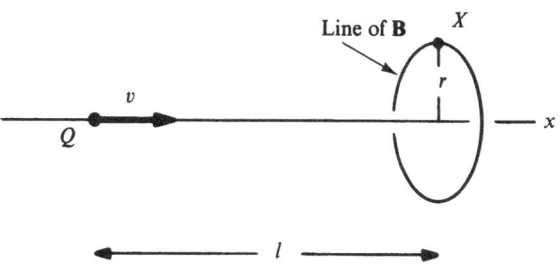

*22.16 (**B** field of a moving point charge; see also the next problem). A point charge Q is moving at speed v along the x axis, as shown. (a) What is the electric flux, (b) the displacement current, through a fixed imaginary disc of radius r, coaxial with the x axis at a distance l from the moving charge? (Assume l is large compared to r.) (c) Find the **B** field at a point X on the rim of the disk, above the x axis. (Express your answers in terms of Q, v, K, K', c, l, r, or some of these.)

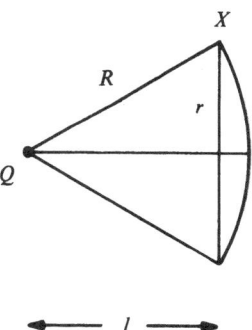

*22.17 (**B** field of a moving point charge, continued). Repeat the preceding problem with l not necessarily much larger than r. In (a) you may use the geometrical theorem which states that the spherical cap (heavy line in the figure) has an area equal to $2\pi R(R-l)$, R being the distance between the moving charge and point X. To calculate (b), first express the flux in terms of r and l through $R = (r^2 + l^2)^{1/2}$; note $v = -dl/dt$, $r = $ const. In the end, compare your answer to (c) with the Biot–Savart contribution (at X) of a current element at the location of Q.

Section B: Energy and Momentum in a Wave

*22.18. In order that an eye pupil, of radius 2.0 mm, should receive 8.0×10^{-6} watt of radiative power when facing a light bulb 4.0 meters away, what should be the power of the bulb? (Assume no radiation is absorbed anywhere along the way.)

*22.19. A certain radar antenna emits a plane wave of 6.0 kW with a beam cross section of 4.0 meters^2. Find (a) the wave's intensity, (b) its average energy density, (c) the electric field amplitude E_{max}, (d) the magnetic field amplitude B_{max}.

*22.20. Assuming that a light intensity $\mathscr{I} = 1000$ watts/meter^2 reaches us from the Sun, calculate the root-mean-square electric field $E_{rms} = [(E^2)_{av}]^{1/2}$ in that radiation.

22.21. A certain small antenna radiates with an amplitude $E_{max} = 2000$ volts/meter, as measured 10 meters away. Assuming an equal output in all directions (always an oversimplification for antennas), find (a) the electric field amplitude and (b) the intensity, 5 km away.

*22.22. A certain microwave generator, designed for use in a kitchen oven, is required (in a laboratory test) to produce 2.0 calories/sec over a beam cross section of 150 cm^2 (1 cal = 4.2 joules). Assuming a monochromatic plane traveling wave for simplicity, estimate the required amplitude E_{max}.

22.23. Using conservation of energy, prove the inverse-square relation (22.B.24).

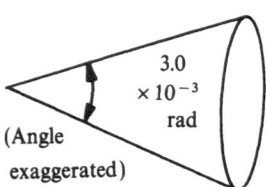

(Angle exaggerated)

3.0 $\times 10^{-3}$ rad

22.24. A certain spacecraft emits 20 watts of radio power. (a) What intensity would be received at 1.0 astronomical unit ($= 1$ AU = Earth–Sun distance $\approx 1.5 \times 10^{11}$ meters) if power were sent equally in all directions? (b) Assume that the whole power is concentrated in a cone of opening angle 3.0×10^{-3} radian by a reflecting dish at the spacecraft; see the figure. What intensity is received inside that cone, 1 AU away? [A cone of opening α cuts out an area $4\pi R^2 \sin^2(\alpha/4)$ from a sphere of radius R.] (c) How much of this power could be collected by a receiving dish of radius 1.2 meters?

22.25. (a) By considering, somewhat as in Fig. 22.6b, a cylinder of radiation incident at right angles on a completely absorbing plate of area $\mathscr{A}$, prove that $\mathscr{A}$ experiences a radiation force given by $F = $ (momentum density)$(c)(\mathscr{A})$. (b) What happens to this formula if the radiation suffers a mirrorlike reflection from $\mathscr{A}$?

Answers to True or False

1. True.
2. False (rather, to its rate of change).
3. False (should be Ampère's law).
4. True.
5. False (they come together).
6. False.
7. False.
8. False (should be like intensity).
9. True.
10. False.
11. False.
12. False.
13. False.
14. True (because $E_{max} \propto \sqrt{\mathscr{I}}$).
15. True.
16. True.
17. False.
18. True (see Fig. 1.1 of Chap. 1).
19. False.

Answers or Hints to Problems

22.1. 6.6×10^{-7} A.

22.2. (b) same as (a) without further calculation.

22.3. (a) 3.1×10^{-6} A.

22.5. 5.0×10^6 meters.

22.7. 8.9875543×10^9 (SI).

22.8. 7.5×10^8 Hz.

22.10. Hint: A circular contour at $r/2$ encompasses 1/4 the total displacement current.

22.15. Hint: A circle through X and coaxial with the capacitor encompasses r^2/R^2 times the total displacement current; (a) independent of Q, S; (b) not modified.

22.16. (a) Hint: If the disk were slightly convex, it would be part of a sphere of radius l and centered on Q; the total flux through that sphere is $4\pi KQ$. See also the following note.

22.17. Note: We take for granted that the **E** field from Q is given by Coulomb's law, even though Q is moving. Actually this is valid only if v is small compared to c.

22.18. 130 watts.

22.19. (b) 5.0×10^{-6} joule/meter^3; (c) 1.1×10^3 volts/meter.

22.20. Hint: Use Eq. (22.B.9), and $\mathscr{I} = S_{av}$. Answer: 6.1×10^2 volts/meter.

22.22. (a) 6.5×10^2 volts/meter.

CHAPTER 23

GEOMETRICAL OPTICS

From the infinite spectrum of electromagnetic waves, we now single out a narrow range—about one "octave"—for closer examination, namely, visible light. Here the detector of choice is of course the human eye.

The science of light—**optics**—is an extensive one, with obvious importance in scientific research and in daily life. It is conventionally divided into three portions: **geometrical optics**, to which the present chapter is an introduction; wave optics, discussed in the next chapter; and quantum optics (Chap. 26).

Geometrical optics is, basically, the science of **lenses, prisms**, and **mirrors**. It addresses the question of *how light changes its direction* of propagation in response to a transparent or reflecting body. The answers were well known long before the wave nature of light became established; the laws of geometrical optics may be formulated in terms of **rays** and without any reference to waves. For at least a century following Newton, rays were, in fact, believed to be the trajectories of particles. Using our modern hindsight, we take waves as the starting point of our discussion; but eventually, we abandon waves in favor of rays. Doing this, however, amounts to making an approximation. For reasons that will become clearer in the next chapter, ray optics is accurate only when the wavelength of light, λ, is infinitesimal compared to all dimensions of the equipment (mirror, lens, prism, diaphragm). This is commonly the case. For blue-green light and a 1-cm-diameter lens, we have

$$\frac{\lambda}{\text{Lens diameter}} \approx \frac{500 \text{ nm}}{1 \text{ cm}} = \frac{1}{20\,000}$$

A. Wave Fronts and Rays

We first need to relate the concept of a ray to that of a wave. Consider a steadily produced, linearly polarized, monochromatic light wave, traveling through space. This is the only kind of wave we need to discuss in the present chapter; actual light waves, although usually more complicated than this, are nothing but superpositions, or mixtures, of such simple waves.

Unless otherwise noted, we assume the medium of propagation to be *perfectly transparent*; the wave does not lose any energy as it propagates.

Suppose that, at a specific instant of time, we "freeze" the wave in a kind of snapshot, and explore its **E** field. We then find certain surfaces (heavy lines in Fig. 23.1) along which the field strength $|\mathbf{E}|$ is a maximum; these surfaces are called **wave fronts**. If we next "unfreeze" the wave and allow it to propagate as it physically must, we notice that, somewhat later, the wave fronts have traveled to the position of the dashed lines; still later, wave front X has moved to the position previously occupied by wave front Y, etc.

We now define a **ray** as *any fixed curve which intersects all wave fronts at right angles*; examples are AA' and BB' in the figure.

Wave fronts, and therefore also rays, can exist inside any transparent medium. Although the wave fronts may have a variety of shapes,

> The rays are always straight
> within a uniform medium.   (23.A.1)

In a nonuniform medium, the rays may bend ("be refracted") gradually or abruptly, according

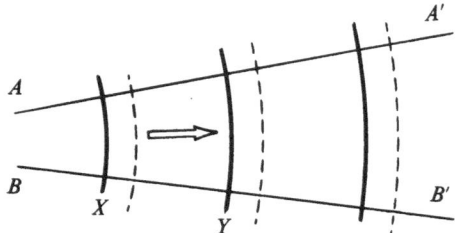

Figure 23.1. Rays in relation to wave fronts.

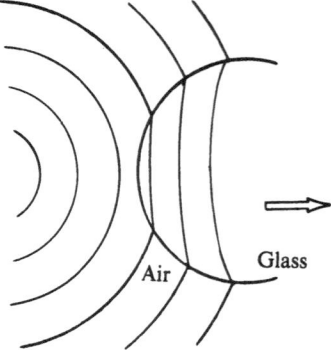

Figure 23.2. Continuity of the wave fronts across an interface.

to whether the medium changes its properties gradually or abruptly.

Furthermore, within a transparent region,

> Each wave front is a continuous,
> uninterrupted surface,   (23.A.2)

even at an abrupt change in medium, as in Fig. 23.2.

[Statements (23.A.1) and (23.A.2) are presented here without proof. They can be deduced from Maxwell's theory of electromagnetic waves.]

What is the distance between two successive wave fronts? In the preceding Chapter's Fig. 22.11 we explained that points of maximum $|\mathbf{E}|$ are found at half-wavelength intervals along a ray. (Incidentally, a maximum of $|\mathbf{E}|$ always coincides with a maximum of $|\mathbf{B}|$.) Our convention* may be said to define "half-cycle wave fronts."

We observe in Fig. 23.2 that the wave fronts are more closely spaced (the wavelength is shorter) in glass than in air; this is due to a slower propagation in glass. If, in any two media A and B, the wave fronts have speeds v_A and v_B, then the ratio of wavelengths is

$$\frac{\lambda_A}{\lambda_B} = \frac{v_A T}{v_B T} \qquad (23.A.3)$$

where T is the period, necessarily the same in both media. Therefore we have

$$\boxed{\frac{\lambda_A}{\lambda_B} = \frac{v_A}{v_B}} \qquad (23.A.4)$$

In each substance, *the wavelength is proportional to the speed of propagation.* (Why is T the same

everywhere? We recall that a wave is a delayed signal. At each point in the medium, the vibration repeats the incident time dependence.)

Pictorially, rays have some advantages over wave fronts, besides being straight; they show the direction of propagation; they are stationary, while wave fronts are moving; the same rays may serve for different wavelengths, as well as for a mixture of wavelengths. We shall think in terms of either rays or wave fronts as convenience dictates.

Light Beams

A ray is a geometrical abstraction, but in practice it may be approximated by a thin beam, as in the pinhole arrangement of Fig. 23.3. Flashlights and projectors, but especially lasers, can emit nearly cylindrical, or **collimated** beams. (Caution: We shall see in the next chapter that, even with the help of pinholes, a light beam cannot be made arbitrarily thin.)

B. Refraction of a Plane Wave

Next we examine, in terms of rays, how a plane wave changes course as it enters a new medium.

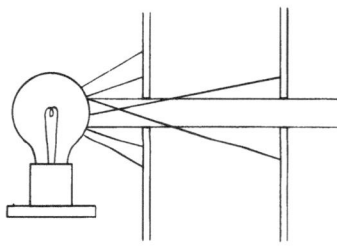

Figure 23.3. Making an approximately collimated beam with two successive pinholes.

---

* In the literature, a wave front is often defined as any surface in space on which, as a function of time, $\mathbf{E}$ has the same phase at all points. Our definition is a special case where that phase is 0° or 180°. A more common definition takes all points with 0° phase only; these "full-cycle wave fronts" are a distance λ rather than $\lambda/2$ apart.

Light propagates through certain materials —gases, glass, water, many plastics—qualitatively pretty much as it does through a vacuum, although with certain quantitative differences, mainly a lower speed. These transparent materials are predominantly **dielectrics** (electric insulators).

Dielectrics are transparent because they lack resistive, or Joule dissipation; this means that negligible energy is absorbed from the electromagnetic wave. (Transparency also requires a smooth composition of the medium, which would otherwise be cloudy, i.e., the light would be diffusely scattered even if it is not absorbed.)

Nevertheless, transparent substances do interact with light waves. Within each molecule, the electrons respond to the wave's oscillating **E** field with a vibration of their own; each electron acts as a small receiving and emitting antenna. The emitted electromagnetic waves combine with the incoming wave to produce the observed total wave.

If the mechanism is so complicated, why, then, does a dielectric act merely as a kind of modified vacuum? A partial answer is that, as seen by a light wave, the dielectric appears as a continuous (rather than atomic) medium. In order to document this, we must know the wavelength λ of the light and the distance l between neighboring atoms of the medium. If l is very small compared to λ, we are prepared to accept the smoothness statement.

In the visible range, we conservatively have $\lambda \approx 400$ nm (violet light). All atoms have diameters of order $l \approx 10^{-10}$ meter, and in solid or liquid dielectrics they touch one another. Thus we have $\lambda/l \approx 400 \text{ nm}/10^{-10} = 4000$. This should be large enough to reassure us about taking optical media as continuous.

Correspondingly, we ignore atomic-sized irregularities in the **E** and **B** fields, caused by individual electrons or nuclei; **E** and **B** shall designate *smooth averages* taken over a region containing several atoms, but still much smaller than a wavelength. Just as in a vacuum, the surfaces of maximum $|\mathbf{E}|$ (and $|\mathbf{B}|$) are then the wave fronts.

i. Snell's Law

Figure 23.4 shows two uniform transparent media, A and B, with a plane interface XY. A beam of parallel rays (the **incident rays**) originate in A *and enter* B (where they are called the **refracted rays**). The incident and refracted portions of one

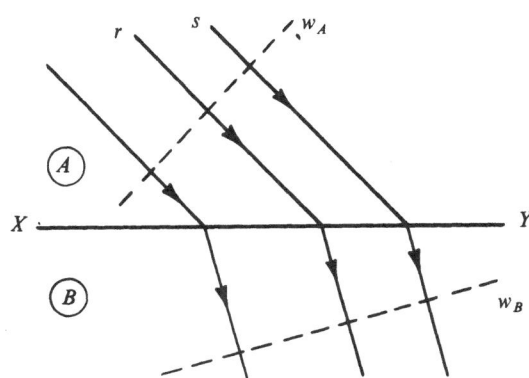

Figure 23.4. Refraction of parallel rays; w_A and w_B are wave fronts.

complete ray define the **plane of incidence**; in Fig. 23.4 it is the plane of the figure. We can now state that

The plane of incidence is normal to the interface.

(23.B.1)

This fact is to be understood from symmetry: we draw the plane that contains the incident ray and that is also normal to the interface; this can always be done. In Fig. 23.4, there cannot be a preference for refraction into the page or out of it. Hence the ray must stay within the plane of the figure.

Since the incident rays are mutually parallel, the refracted rays must be mutually parallel also. Indeed, the environment does not change under a shift in the XY direction. In Fig. 23.4, under such a shift, incident ray r can be made to replace incident ray s exactly, while the interface has not changed. Hence r must suffer the same refraction as s did. (This argument is said to invoke **translational symmetry** in the XY direction.)

Drawing the wave fronts (for example w_A and w_B), we see that

A plane wave, incident on a plane interface, gives a plane refracted wave. (23.B.2)

Next we ask by what angle the ray's direction is changed. Figure 23.5 names the relevant angles. Traditionally, one draws a line n (the **normal**) at right angles to surface XY. Its angle θ_A (θ_B) with the incident (refracted) ray is called the **angle of**

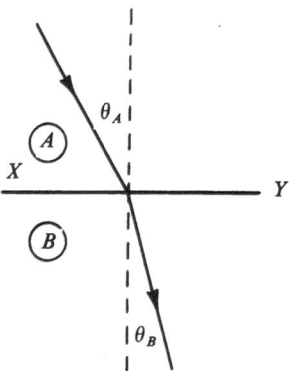

Figure 23.5. Refraction of a single ray; n, θ_A, θ_B are defined in the text.

incidence (angle of refraction): θ_A and θ_B are related by

$$\frac{\sin \theta_A}{\sin \theta_B} = \frac{v_A}{v_B} \qquad (23.B.3)$$

where v_A, v_B are the speeds of propagation of the wave fronts in media A and B.

Equation (23.B.3) is known as **Snell's law**, after Willebrord Snell van Royen, who noted around 1620 that the ratio $\sin \theta_A / \sin \theta_B$ is the same for all angles of incidence. (He probably did not suspect that speeds of propagation played any role at all.)

Proof of Snell's Law

Figure 23.6 shows a wave front w, which, after a time interval t, has migrated to a new position w'. It also shows two rays, r and s, drawn where w and w' intersect XY. During time t, the traveling wave front covers distances $v_A t$ and $v_B t$ in the two media,

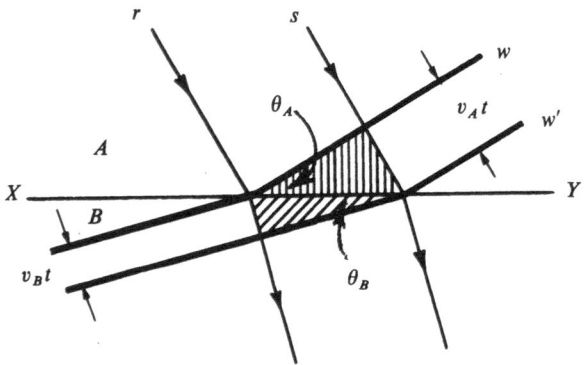

Figure 23.6. Proof of Snell's law. Note the closer-spaced wave fronts in medium B.

as indicated. We now examine the two shaded right triangles in the figure. Their common hypotenuse has length

$$\frac{v_A t}{\sin \theta_A} = \frac{v_B t}{\sin \theta_B} \qquad (23.B.4)$$

Equation (23.B.3) follows directly.

Remarks on Snell's Law

Except in certain crystals and long-molecule plastics, the speed of light v is independent of the direction in which the light travels. To a fair approximation, v is also independent of the light's frequency, and therefore the ratio v_A/v_B depends principally on the choice of substances A and B.

An important aspect of Snell's law is its **reversibility**: the derivation applies equally to light traveling from A to B and from B to A in Fig. 23.7. Hence, the directions of propagation may be reversed while all angles are kept the same; see Fig. 23.7.

ii. The Index of Refraction

Refraction depends on geometry as well as on material properties. To disentangle these two factors, one labels each substance by its **index of refraction** (or refractive index), defined as

$$\boxed{n = \frac{c}{v}} \qquad (23.B.5)$$

This dimensionless ratio compares the speed of light in vacuum, c, with its speed v in that substance. (We must keep in mind that smaller speeds

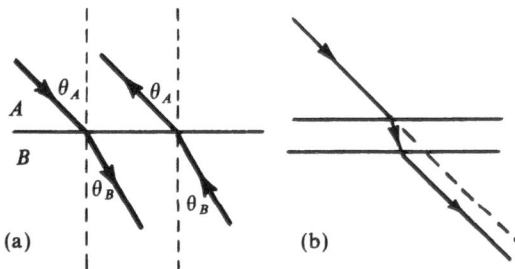

Figure 23.7. (a) Reversibility of a ray. (b) As an application, the ray's direction is unchanged after it has crossed the uniform glass plate. (What is the detailed argument?)

Table 23.1. Some Indices of Refraction for Yellow Light
($\lambda = 589$ nm in vacuum)

Solids at 20°C		Liquids at 20°C		Gases at 1 atm and 0°C	
Clear plastics	1.46–1.54	Water	1.33	Helium	1.000036
Fused quartz	1.46	Ethyl alcohol	1.36	Hydrogen	1.00013
Crown[a] glass	1.52	Glycerol	1.47	Oxygen	1.00027
Canada balsam[b]	1.53	Sucrose[d] solution	1.49	Dry air	1.00029
Sodium chloride	1.54	(80% by weight in water)		Nitrogen	1.00030
Flint[c] glass	1.58–1.89	Carbon disulfide	1.63	Methane	1.00044
Diamond	2.42				
Rutile (TiO$_2$)	2.6				

[a] I.e., without lead.
[b] A resin for cementing optical glasses.
[c] I.e., containing lead oxide.
[d] For example, cane sugar.

go with larger indices of refraction.) In terms of n, Snell's law, (23.B.3), becomes

$$\boxed{\frac{\sin \theta_A}{\sin \theta_B} = \frac{n_B}{n_A}} \qquad (23.B.6)$$

Note the exchange of A and B on the right. The vacuum, of course, has $n = 1$, and gases have n only very slightly larger than 1. Unless otherwise specified, we shall take $n = 1$, exactly, in air. The value of n for some common optical media is listed in Table 23.1; typically, n is larger than unity. Temperature and frequency dependences are slight.

or

$$\theta_B = 32.1°$$

Therefore we have

$$x = (10\,\text{cm}) \tan \theta_B = (10\,\text{cm})(\tan 32.1°) = \underline{6.27\,\text{cm}}$$

iii. Dispersion

Light rays of different frequencies are refracted through somewhat different angles. This phenomenon, which we have neglected so far, goes under the name of **dispersion**. It can alternatively be described as a slight *frequency dependence of the*

Example 23.1. A Water Surface. A cylindrical beaker, of radius 10.0 cm and height 20.0 cm, is half full of water ($n = 1.33$), as shown in Fig. 23.8. A light beam is directed to the center C of the water surface, and grazes the edge of the beaker. Where is the illuminated spot S on the bottom?

———

Let x be the horizontal distance between S and the center. From Snell's law,

$$\frac{\sin \theta_A}{\sin \theta_B} = \frac{n}{1}$$

we find

$$\sin \theta_B = \frac{\sin \theta_A}{n} = \frac{\sin 45°}{1.33} = 0.532$$

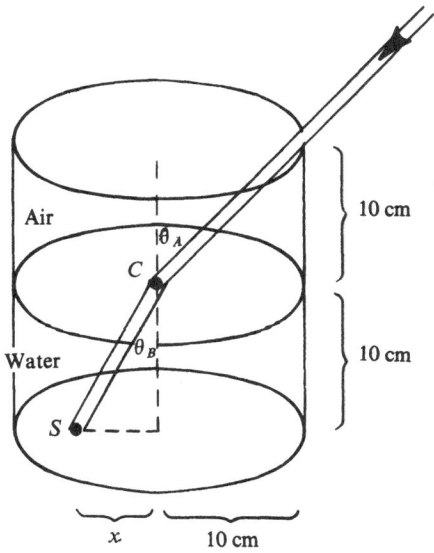

Figure 23.8. A thin light beam refracted by water.

Vacuum wavelength, λ, nm

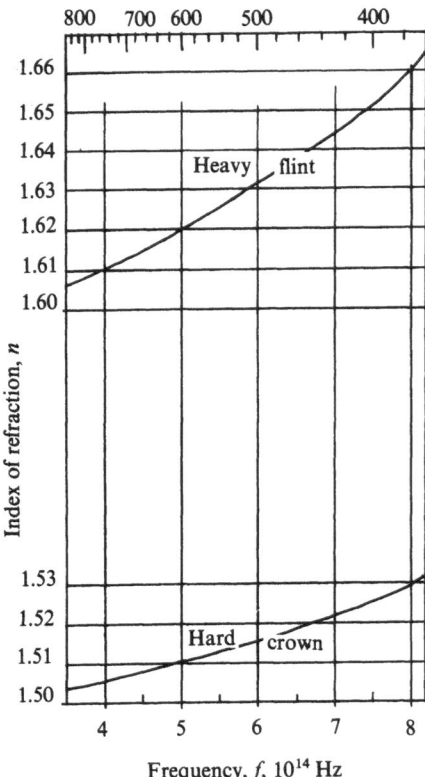

Figure 23.9. Dispersion in two sample glasses. The origin of coordinates (not shown) is far below. The index of refraction typically increases with frequency in completely transparent substances.

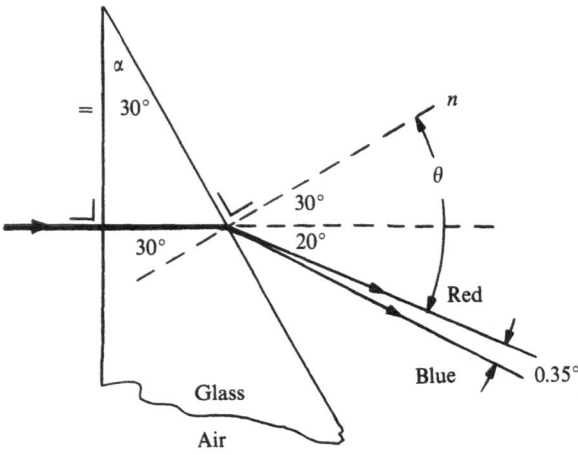

Figure 23.10. A thin light beam (perhaps purple in color) is dispersed by a prism.

index of refraction. (For an illustration involving glass, see Fig. 23.9.) Yet another equivalent description states that the wave fronts have a *slightly different speed of propagation at different frequencies.* All transparent materials exhibit dispersion to some degree; only in vacuum does light of every wavelength propagate at the same speed c.

By taking advantage of dispersion, we can separate a light beam with mixed frequencies into its spectral colors. Figure 23.10 shows this phenomenon in a glass prism. Another familiar instance, the rainbow, is caused by dispersion of the sunlight in raindrops.

Example 23.2. Dispersion in a Prism. Let the glass prism of Fig. 23.10 have an angle $\alpha = 30°$. A light beam, incident at right angles on one of the prism's surfaces, consists of mixed blue and red light. Both beams emerge at about 20° from the original direction; the blue ray is refracted 0.35° more than the red. (a) What is the glass' index of refraction n? (b) By what amount Δn does n for the blue light exceed n for the red?

(a) Index of refraction: The incident ray continues in a straight line (why?) until it meets the second glass–air interface. There its angle of incidence is 30° (why?); let its angle of refraction be θ, as shown; we first consider both emerging rays as being a single one. From the figure we see that

$$\theta = 30° + 20° = 50°$$

Applying Snell's law,

$$\frac{n}{1} = \frac{\sin \theta}{\sin 30°} \qquad (23.B.7)$$

we obtain

$$n = \frac{\sin 50°}{\sin 30°} \approx \underline{\underline{1.53}}$$

(b) Dispersion: We next consider, in Eq. (23.B.7), one set of values θ, n for the blue light and another set for the red. The small differences are conveniently represented by differentials: $\Delta\theta \approx d\theta$, $\Delta n \approx dn$. Differentiating both sides of (23.B.7), we have

$$dn = \frac{\cos \theta \, d\theta}{\sin 30°} \qquad (23.B.8)$$

[Caution: $d\theta$ must be in radians for this differentiation formula to be valid. Here we have

$$d\theta = 0.35° = \frac{(0.35°)(2\pi)}{360°} \text{ rad} = 0.0061 \text{ rad}$$

$$= 0.0061.]$$

From (23.B.8), the desired result is

$$dn = \frac{(\cos 50°)(0.0061)}{\sin 30°} = \underline{\underline{0.0078}}$$

C. Reflection of a Plane Wave

Reflection is an even more familiar effect than refraction. Light generally undergoes mirror reflection when it encounters *a smooth interface between two transparent media*. (But there is no reflection if the two indices of refraction are equal.) Alternatively, mirror reflection occurs at a *polished metal surface*; the metal's electrical conductivity is responsible, as can be shown from electromagnetic theory. In this section we shall be concerned with *plane* reflecting surfaces only.

i. The Laws of Reflection

Figure 23.11 depicts the phenomenon, and illustrates the fact that reflection often occurs together with refraction or absorption.

As with refraction,

> A plane wave, incident on a plane surface, gives a plane reflected wave.   (23.C.1)

With reference to Fig. 23.11,

> The incident and reflected portions of a ray are in a plane (the **plane of incidence**) that is normal to the reflecting surface.
> (23.C.2)

Finally, *the angles of reflection and of incidence are equal,*

$$\boxed{\theta' = \theta} \qquad (23.C.3)$$

The symmetry arguments leading to (23.C.1) and

(23.C.2) are as in Sec. B and will not be repeated here. A simple proof of (23.C.3) goes as follows.

Figure 23.12 singles out two of the wave fronts shown in Fig. 23.11 and also shows portions of two rays. The distance $\lambda/2$ between the two wave fronts w, w' must be unchanged after reflection, i.e., the

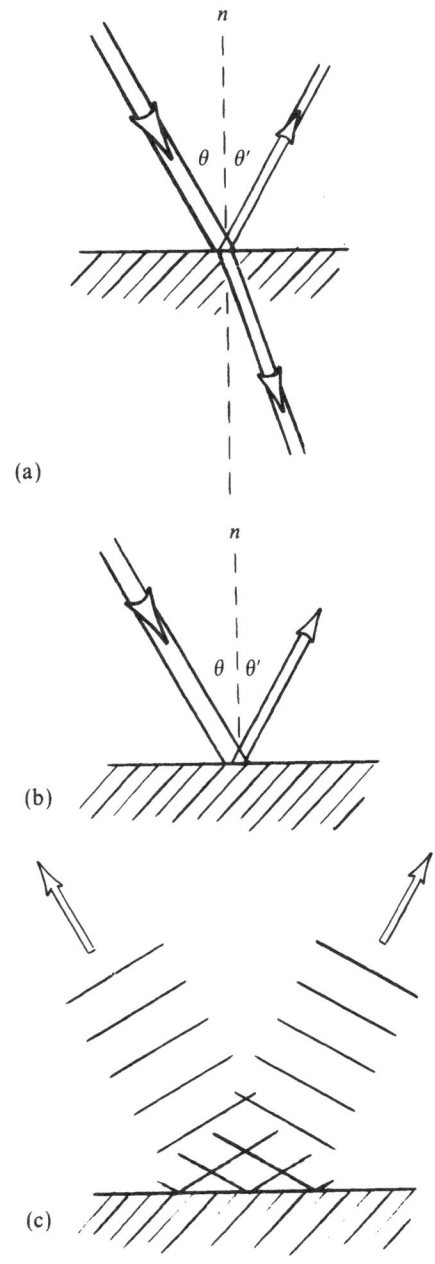

(a)

(b)

(c)

Figure 23.11. (a) At a smooth interface between two transparent media, an incident beam is partly reflected, partly refracted. The width of a ray symbolizes intensity. (b) A case where some absorption occurs instead of refraction. (This could be a metal surface.) (c) Continuity of the wave fronts upon reflection.

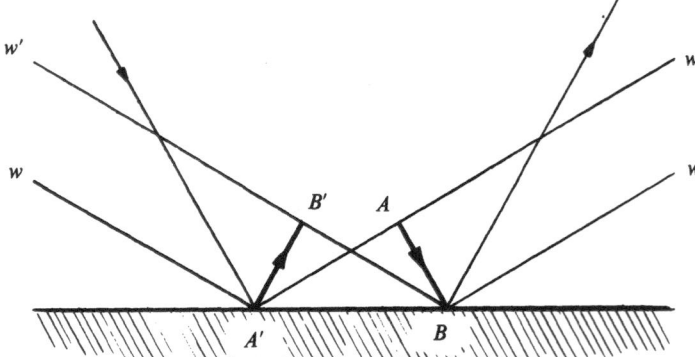

Figure 23.12. Proof of the angular equality (23.C.3).

ray segments AB and $A'B'$ have the same length. Hence the two right triangles ABA' and $B'A'B$ are congruent, and angle $ABA' =$ angle $B'A'B$.

That is, the incident and reflected rays make the same angle with the reflecting surface. Figure 23.11 then shows that they also make the same angle with the normal.

ii. Total Internal Reflection

Even the best metallized mirror absorbs some of the incident light. Total reflection can, however, be achieved in an unlikely manner, namely, at the interface between two transparent media. For definiteness we consider, in Fig. 23.13, a ray incident, from the glass side, on a glass–air interface (index of refraction of glass $= n$). Snell's law, Eq. (23.B.6), gives, for the angle of refraction θ',

$$\sin \theta' = \frac{n}{1} \sin \theta \qquad (23.C.4)$$

where θ is the angle of incidence.

If θ increases, so does θ', until $\theta' = 90°$, that is to say, until the refracted ray grazes the glass surface. This happens when $\sin \theta' = 1$, or, from (23.C.4), when $\sin \theta = 1/n$. If θ is made larger still, Eq. (23.C.4) gives

$$\sin \theta' > 1$$

a mathematical impossibility for a real angle θ'. Therefore *no refraction can occur* when $\sin \theta > 1/n$.

Assuming negligible absorption in air or glass, we conclude that *all the incident energy is reflected back* into the glass.

More generally, when two media of indices $n_{larger} > n_{smaller}$ are involved, the conditions for total internal reflection can be summarized as follows:

1. It occurs on the larger-index side of the interface.
2. It occurs only for angles of incidence θ larger than a **critical value** θ_c, obeying

$$\boxed{\sin \theta_c = \frac{n_{smaller}}{n_{larger}}} \qquad (23.C.5)$$

Two applications are shown in Fig. 23.14.

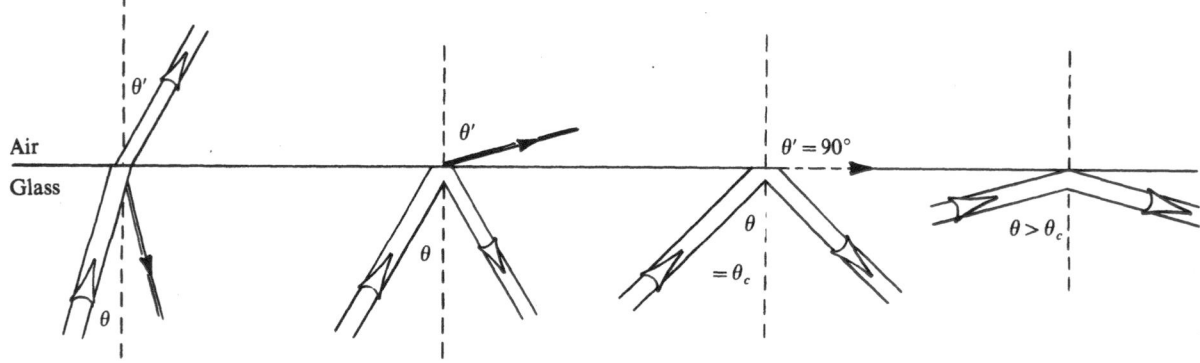

Figure 23.13. The angle of incidence, θ, is gradually increased; reflection improves, and becomes total from $\theta = \theta_c$ on.

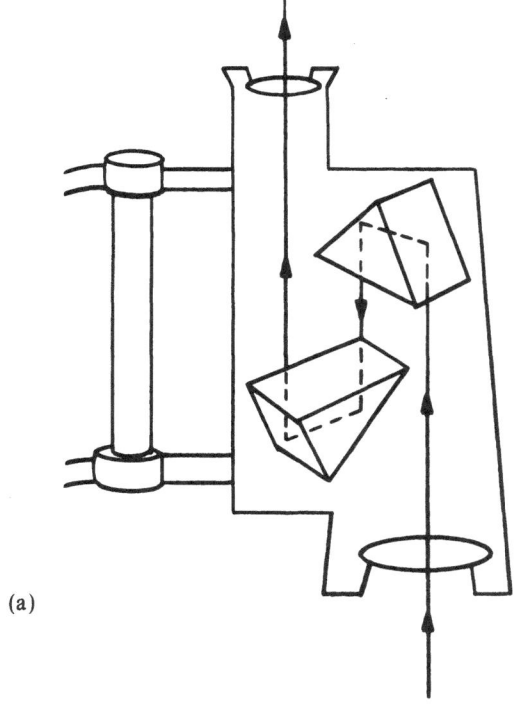

(a)

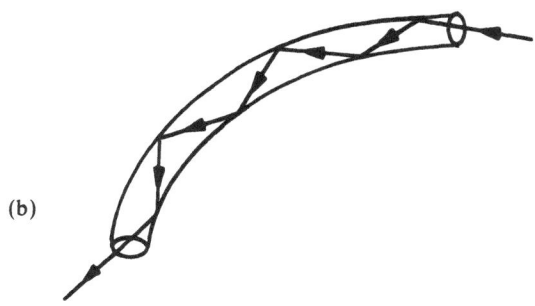

(b)

Figure 23.14. (a) Totally reflecting prisms in **binoculars**; the path of one light ray is indicated. (b) **Optic fiber.** This acts as a "light pipe," leading the incident light around bends by means of multiple total internal reflections. A bundle of many fibers can transmit an image (as a mosaic of lighted spots). Fiber optics is particularly useful in medicine as it permits the physician to light and view internal organs.

Example 23.3. A Cone of Invisibility. A circular leaf, of radius $r = 10$ cm, is floating on still water, as shown in Fig. 23.15. Consider an imaginary cone of altitude h, whose apex P is vertically below the leaf's center. Anywhere within that cone, a tadpole is invisible from above the water. Given that fact, determine h.

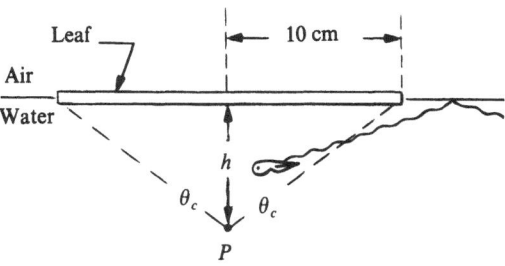

Figure 23.15. A tadpole hiding under a circular leaf.

Since the path of a ray is reversible, we know that, from within the cone, the tadpole cannot see above the water. (The surface beyond the leaf's edge looks to the tadpole like a perfect mirror.) Thus, a ray through P, coming from just beyond the leaf's edge, is just barely totally reflected. The cone half-angle θ_c equals the critical angle of incidence (why?) and obeys

$$\sin \theta_c = \frac{1}{n} = \frac{1}{1.33}, \qquad \theta_c = 48.8°$$

Therefore we have, from the figure,

$$h = \frac{r}{\tan \theta_c} = \frac{10 \text{ cm}}{\tan 48.8°} = \underline{\underline{8.8 \text{ cm}}}$$

D. Image Formation by Lenses

How are images formed? Our visual system—an elaborate teaming-up of eyes and brain—interprets incoming light signals whenever possible in terms of objects having a definite location in the environment. The basic principle used in this reconstruction is quite independent of any details in the visual system; it applies to cameras just as well, and is illustrated in Fig. 23.16.

We consider the object as a collection of points. If several rays, all originating from a single point P, enter the visual system, their information is processed in such a way that, in effect, they are extrapolated back to where they intersect, thus revealing the location of P. *This extrapolation is always done on the presumption of straight rays.* This is why the eye is readily deceived—or helped—by optical instruments, as in Fig. 23.16, where a mirror makes P appear to be at Q, or where a similar trick is performed by a lens.

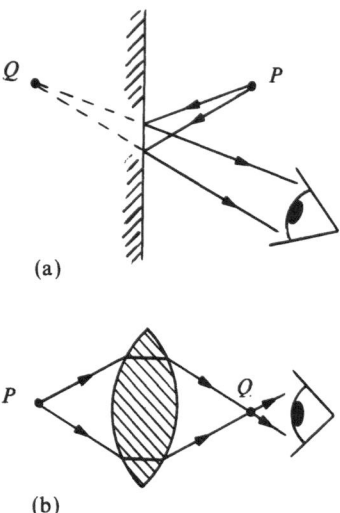

(a)

(b)

Figure 23.16. An actual object (point P) is seen at Q, from where the rays appear to originate. (a) Mirror reflection (Q is a virtual image). (b) Focusing by a lens. (Point Q is a real image; a sheet of paper or a photographic film at Q could capture it as a projected image.)

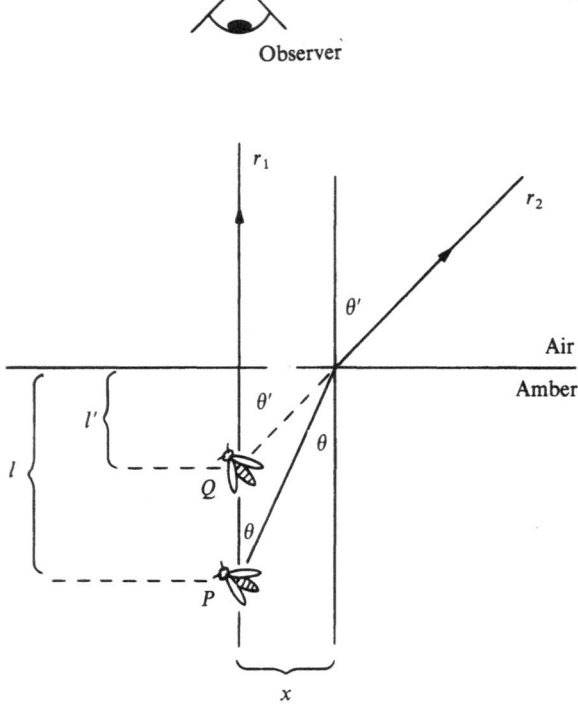

Figure 23.17. Point P is seen at Q, closer to the block's surface than it really is.

We shall ordinarily pay attention to one single light-emitting point P at a time. The nomenclature is as follows. In Fig. 23.16, P is the **object point**, Q is the **image point**. The image point is said to be **virtual**, as in part (a) of the figure, when the actual rays do not meet at Q; it is **real**, as in part (b), when the rays actually intersect there.

i. Flat Transparent Block

To illustrate quantitatively the concept of a virtual image, we consider an example.

Example 23.4. An Insect in Amber. A small insect P is imbedded in amber (a fossil resin), whose index of refraction is, let us say, $n = 1.50$. If P is at a distance $l = 2.0$ cm below the amber's flat polished surface, at what distance l below the surface does it *appear* to be, i.e., where is its virtual image Q? Assume we are sighting normally to the surface.

Figure 23.17 shows two rays r_1, r_2 originating from P. For convenience, r_1 is taken to follow the normal through P. For r_2, the angles of incidence and refraction are θ and θ'; r_2 emerges from the amber at a distance x from r_1, as shown.

Snell's law gives

$$\frac{\sin \theta}{\sin \theta'} = \frac{1}{n} \qquad (23.\text{D}.1)$$

From the appropriate right triangles we see that

$$\tan \theta = \frac{x}{l}, \qquad \tan \theta' = \frac{x}{l'} \qquad (23.\text{D}.2)$$

Eliminating x, we have

$$\frac{\tan \theta}{\tan \theta'} = \frac{l'}{l} \qquad (23.\text{D}.3)$$

The left-hand sides of (23.D.1) and (23.D.3) are practically the same, because if our eye is just above P it collects rays for which θ is very small, so that $\tan \theta \approx \sin \theta$, $\tan \theta' \approx \sin \theta'$. Comparison of (23.D.1) and (23.D.3) then gives

$$\frac{l'}{l} = \frac{1}{n} \qquad (23.\text{D}.4)$$

or

$$l' = \frac{l}{n} = \frac{2.0 \text{ cm}}{1.5} \approx \underline{\underline{1.3 \text{ cm}}}$$

The insect also appears vertically compressed by that same factor $1/n = 1/1.5$. (Why?)

ii. Lenses

For definiteness we shall think of a lens as being a piece of glass surrounded by air, although in general it could be any transparent medium immersed in another.

Two equivalent descriptions of a lens' action are illustrated in Fig. 23.18; that figure also shows the distinction between a converging and a diverging lens.

Ray Picture

Each ray is refracted twice, namely, at the two glass–air interfaces. If the incident rays originate from a common point P, then there exists another point Q at which the rays (or their straight-line extensions) intersect again. A **converging lens**, (Fig. 23.18a) is thicker in the middle, and bends the rays inwards; part (b) of the figure shows a **diverging lens**, which is thicker at the edge, and which bends the rays outwards.

Wave Picture

The converging or diverging behavior of the rays is more easily understood on the basis of parts (c) and (d) of the figure. In a nutshell, the principle is that a wave front travels more slowly in glass than in air. In part (c) we see the central portion of the wave front being held back in relation to the peripheral portions, because it must cross a greater thickness of glass. The wave front's curvature is thereby changed in the right way to produce convergence; for quantitative details, see Note (i) at the end of this chapter. Part (d) of Fig. 23.18 similarly illustrates a diverging process.

Conventions and Nomenclature

An assortment of lenses is shown in Fig. 23.19. For simplicity we assume each lens surface to be part of a sphere, whether **convex** (bulging) or **concave** (hollow). High-quality lens surfaces are purposely made to deviate slightly from perfect spheres, however.

The **center of curvature** of a lens' surface is the center of the sphere which fits that surface; the **radius of curvature** is just the radius of that sphere. (In the same sense, one may speak of the center and radius of curvature of a spherical wave front; the center of curvature of a plane is considered to be at infinity in the normal direction.) The lens' **axis** is the straight line through both its centers of curvature; it crosses the lens through its thickest (or thinnest) part and meets both surfaces at right angles.

Let us now look at a symmetric lens, Fig. 23.20. The lens' **center** $\textcircled{1}$ (not to be confused with the centers of curvature of its surfaces) is halfway

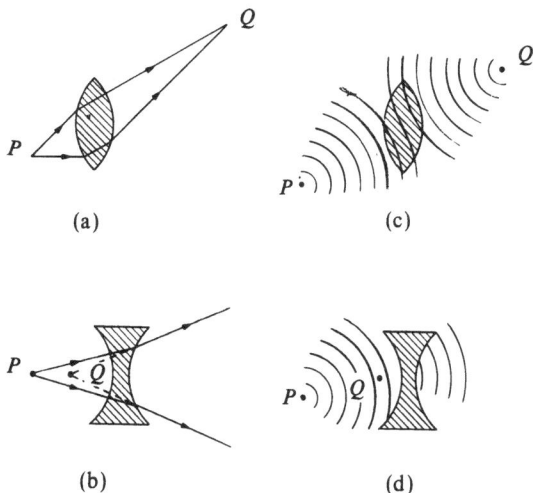

Figure 23.18. Image formation by a lens. (a), (b), ray picture; (c), (d), wave picture.

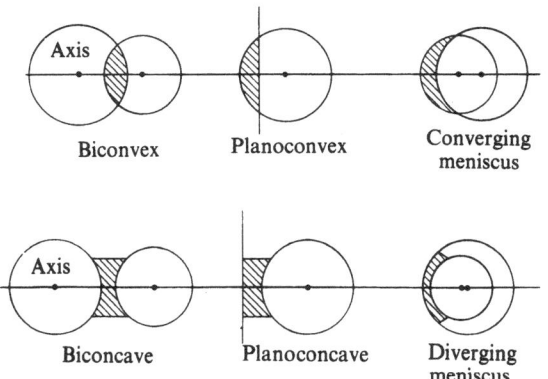

Figure 23.19. A variety of lenses.

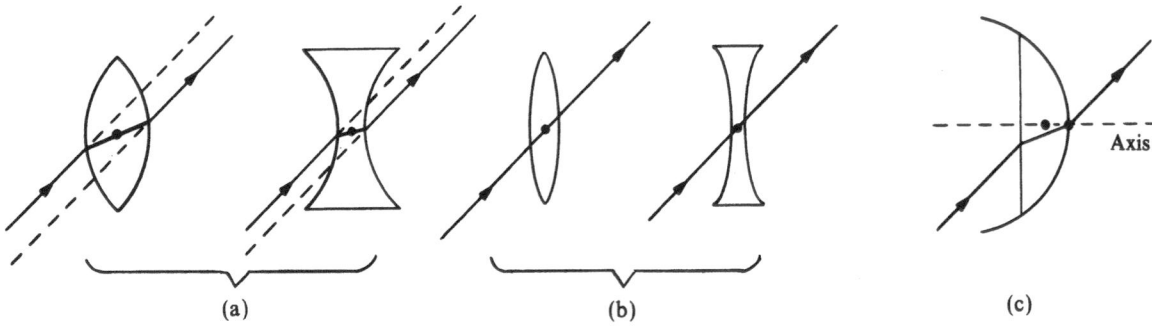

Figure 23.20. (a) Symmetric lenses: a ray through O is undeviated (why?) (b) Thin-lens approximation: the ray is considered unshifted as well. (c) Plano-convex illustration: the undeviated ray passes close to O.

between the two surfaces, on the axis, inside the glass. It is denoted by a dot in Fig. 23.20. A ray through O is undeviated in direction and, if the lens is thin, almost unshifted as well. We approximate such a ray by a straight line; see part (b) of the figure. If the lens is not symmetric, but very thin, we still make that approximation because the undeviated rays miss O by very little; see part (c) of the figure.

iii. The Thin-Lens Equation

We first consider the converging lens of Fig. 23.21, with an object point P and a (real) image point Q. The distances from P and Q to the lens' center are denoted by p and q, and are called the **object distance** and **image distance**, respectively. For a given lens, each value of p implies a well-defined value of q. The two are related through the **lens equation**,

$$\boxed{\frac{1}{p}+\frac{1}{q}=\frac{1}{f}} \qquad (23.D.5)$$

where the constant distance f depends only on the lens' shape and material; f is called the **focal length** (or focal distance) of the lens. Several features of Eq. (23.D.5) should be noted:

1. Since the equation is symmetric in p and q, object and image points are interchangeable; that is to say, the direction of light propagation along the rays may be reversed. This idea also implies that both sides of the lens are equivalent; its action remains unchanged when it is flipped over.

2. If P approaches the lens, Q recedes on the other side: to verify this we may examine Eq. (23.D.5) for decreasing values of p. Thus, P and Q move in the same direction. (However, their mutual distance does not turn out to be constant.)

3. When P is very far away (p infinite, $1/p = 0$), Eq. (23.D.5) gives $1/q = 1/f$, or $q = f$. Thus, *an object point "at infinity" is imaged at a distance f from the lens.* (What converse statement describes the case where Q is very far away?)

4. In later work we shall see that q or f can meaningfully be negative.

5. We see that the inverse focal length, $1/f$, is a convenient quantity; it is known as the lens' **power.** Its unit, the inverse meter, is called the **diopter** in optics:

$$\frac{1}{\text{meter}} = 1 \text{ diopter} \qquad (23.D.6)$$

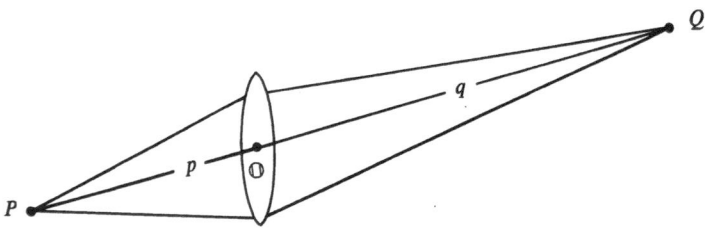

Figure 23.21. Object and image distances (p, q, respectively), defined for a thin converging lens.

6. The lens equation (23.D.5) is a good approximation to reality only *if the lens is thin compared to its focal length.*

Example 23.5. Relating Two Image Points. A certain thin converging lens, when held 20 cm from a sheet of paper, makes a sharp image of the Sun (and even burns a hole in the paper). When that same lens is held at a distance $p = 50$ cm from a candle flame, the flame's image is found, again with a sheet of paper, to be at a distance q from the lens. Find q.

The Sun is "at infinity," and therefore its image is at a distance

$$f = 20 \text{ cm}$$

from the lens. According to (23.D.5), the image of the flame obeys

$$\frac{1}{q} = \frac{1}{f} - \frac{1}{p} = \frac{1}{20 \text{ cm}} - \frac{1}{50 \text{ cm}} = 0.03 \text{ cm}^{-1}$$

$$q = \frac{1}{0.03} \text{ cm} = \underline{\underline{33 \text{ cm}}}$$

Caution: Small-Angle Requirement

The lens equation is really an approximation, good only when all the rays are nearly parallel to the lens' axis (such rays are called **paraxial rays**). We shall always assume situations in which Eq. (23.D.5) can be considered exact.

Derivation of the Lens Equation

Equation (23.D.5) is derived in Note (i) at the end of this chapter. As a by-product of that derivation we obtain the so-called **lensmaker's formula**, which predicts f in terms of the lens' curvatures and index of refraction.

iv. Ray Diagrams

Given a complete object (rather than just an object point), where is its image? Is its orientation reversed? Is it enlarged? A graphical method helps find the answer to these questions; Fig. 23.22 shows how it works.

The object is symbolized by just two of its points, P_0 and P, whose images are Q_0 and Q. Thus, the wide arrows $\overrightarrow{P_0P}$ and $\overrightarrow{Q_0Q}$ in the figure stand for object and image; P_0 and Q_0 are chosen to lie on the lens axis. Three basic facts are used:

1. If $\overrightarrow{P_0P}$ is perpendicular to the axis, so is $\overrightarrow{Q_0Q}$. This is because an off-axis tilting—from P_0Q_0 to PQ—involves essentially no change in the object distance ($\overline{P_0\mathbb{O}} \approx \overline{P\mathbb{O}}$), and hence also no change in the image distance ($\overline{Q_0\mathbb{O}} \approx \overline{Q\mathbb{O}}$). Object and image are assumed small, but on diagrams their size may be exaggerated for convenience.
2. A ray such as PQ, which goes through the center of the lens, is undeflected.
3. An incident ray such as PP', which is parallel to the lens' axis, must be deflected into the lens' **focal point** F'. (A lens has two focal points, F and F', lying symmetrically on either

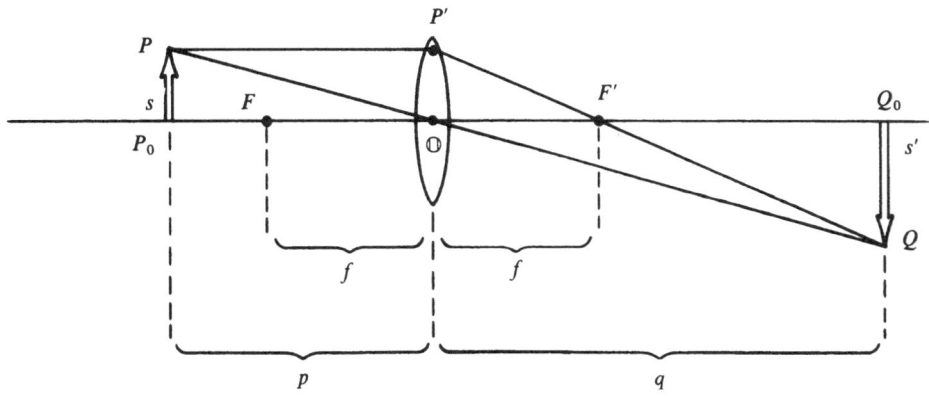

Figure 23.22. Ray diagram for a thin converging lens.

side. By definition, *F* and *F'* are on the lens' axis, each at a distance *f* from the lens' center $\mathbb{O}$.) To see why *PP'F'* is the proper path, we may think of both *PP'* and $P_0\mathbb{O}$ as coming from a single object point located at left infinity. According to the lens equation, they meet at a distance *f* on the other side of the (converging) lens.

In Fig. 23.22, the image is **inverted**; comparison of the similar triangles $P_0 P \mathbb{O}$ and $Q_0 Q \mathbb{O}$ shows that image and object sizes (*s'*, *s*, respectively) are related by a **magnification factor***

$$\text{Magnification} = \frac{s'}{s} = \left|\frac{q}{p}\right| \qquad (23.\text{D}.7)$$

This factor may often be less than unity (a reduction). We conclude this section with some examples.

Example 23.6. A Camera. In Fig. 23.22 (not drawn to scale), suppose the object is a painting of width $s = 12$ cm; assume the camera has a thin lens of focal length 3.0 cm. At what distance *p* from the painting should the lens be held in order for the photographic image to be $s' = 20$ mm wide on the film?

Let *q* be the adjustable distance between lens and film (image distance). The "magnification" is (with *p* and *q* positive)

$$\frac{q}{p} = \frac{s'}{s} \qquad (23.\text{D}.8)$$

The lens equation, (23.D.5), is

$$\frac{1}{p} + \frac{1}{q} = \frac{1}{f}$$

Solving this pair of relations for *p* gives

$$p = \left(1 + \frac{s}{s'}\right) f$$

$$= \left(1 + \frac{0.12}{0.020}\right)(0.030) \text{ meter} = \underline{0.21 \text{ meter}}$$

---

* *Also called **lateral magnification**, to distinguish it from a magnification in the direction of the lens' axis. In the literature, inversion is often denoted by a negative magnification. We shall, however, content ourselves with positively defined magnifications, while relying on ray diagrams to test for inversion—a safer and more instructive procedure; see also Table 23.2 further on.

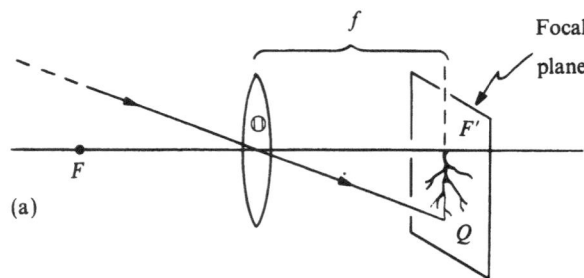

(a)

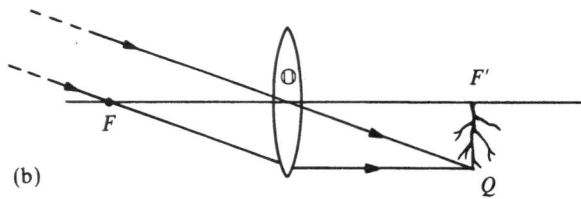

(b)

Figure 23.23. Object (not shown) at left infinity. (a) Single-ray diagram. (b) Two-ray diagram. Ray *FQ'* is refracted parallel to the axis. (Proof: Reverse direction, letting *Q* be the object.)

Object at Infinity

Suppose the object is a tree, hundreds of meters away. Figure 23.23 needs only a single ray (through $\mathbb{O}$), because we already know that the image will be at a distance *f* behind the lens. Segment *PP'* in Fig. 23.22 is now inaccessible anyway. If a second ray is desired, it can be traced as in Part (b) of Fig. 23.23; for further details see Problem 23.32.

(Note: The lens has two **focal planes**, defined as perpendicular to the axis, one through each focal point. In Fig. 23.23 the image lies in the focal plane through *F'*.)

The following example involves a **virtual image** and shows how to interpret a **negative image distance**.

Example 23.7. A Magnifying Glass. A thin converging lens, of focal length $f = 10$ cm, is held at a distance $p < f$ in front of a gem stone whose diameter is 3.0 mm. (a) Show that the image "distance," *q*, is negative. (b) Suppose that, seen through the lens, the stone appears to be 25 cm behind the lens; find *p*. (c) Find the magnification.

(a) Image distance: Figure 23.24 shows the relevant ray diagram. The image is **virtual** and

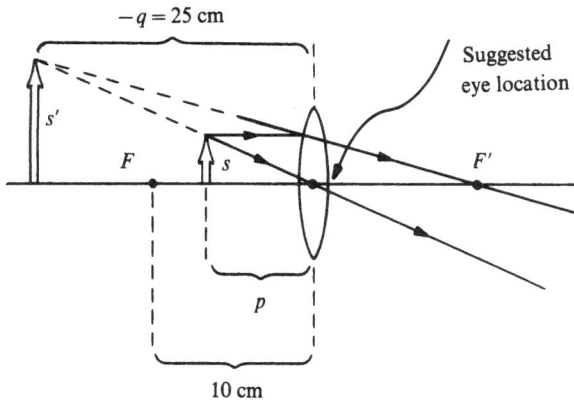

Figure 23.24. Any converging lens can be used as a magnifier; the object (*s*) must be between *F* and the lens; the eye is just behind the lens, or further away if convenient.

erect (i.e., oriented like the object.) From the lens equation we have, for the image distance *q*,

$$\frac{1}{q} = \frac{1}{f} - \frac{1}{p} \tag{23.D.9}$$

a negative number since $1/p > 1/f$. As the figure suggests, *a negative q implies an image on the same side of the lens as the object.*

(b) Object distance: Here we set $q = -0.25$ meter; *the minus sign is essential.* Equation (23.D.9) gives

$$\frac{1}{p} = \frac{1}{f} - \frac{1}{q} \tag{23.D.10}$$

$$= \left(\frac{1}{0.10} - \frac{1}{(-0.25)}\right) \text{ diopters}$$

$$= +14 \text{ diopters}$$

so that

$$p = \frac{1}{14} \text{ meter} = \underline{0.071 \text{ meter}}$$

(c) Magnification: From the appropriate similar triangles in the figure we find

$$\frac{s'}{s} = \left|\frac{q}{p}\right| \tag{23.D.11}$$

$$= \frac{0.25}{0.071} = \underline{\underline{3.5}}$$

(See also Problem 23.34.)

The Eye's "Near Distance"

The 25-cm figure in the preceding example is not quite arbitrary: it is the closest comfortable viewing distance, or **near distance**, for a typical adult human eye. Thus, in the example, the eye could comfortably be placed very close to the lens, as shown in Fig. 23.24.

The following example illustrates the **negative focal length** of a diverging lens.

Example 23.8. A Diverging Lens. Seen through a diverging spectacle lens (focal length $f < 0$), a person of height $s = 1.80$ meters, standing at a distance $p = 3.0$ meters, seems only 2.5 meters away, and of smaller stature. Find (a) the power of the lens; (b) the image size s'.

Figure 23.25 shows the geometry. Note the virtual nature of the image, and the negative sign of $q = -2.5$ meters. Just as in the converging lens, an incident ray PP', parallel to the axis, is deflected through a focal point F; however, F is on the same side as the image, and lies on the ray's extension.

(a) Power: From the lens equation we have

$$\frac{1}{f} = \frac{1}{p} + \frac{1}{q} = \left(\frac{1}{3.0} + \frac{1}{-2.5}\right) \text{ diopter}$$

$$= \underline{\underline{-0.067 \text{ diopter}}}$$

(The minus sign is a general result for the power of a diverging lens, and hence also for its focal length.)

(b) Image size: Using

$$\text{Magnification} = \left|\frac{q}{p}\right| = \frac{2.5}{3.0} = 0.83$$

we get

$$s' = (\text{magnification})(s) = (0.83)(1.80) \text{ meters}$$

$$= \underline{1.5 \text{ meters}}$$

Close Combinations

Several thin coaxial lenses, stacked close together as shown in Fig. 23.26, are equivalent to a

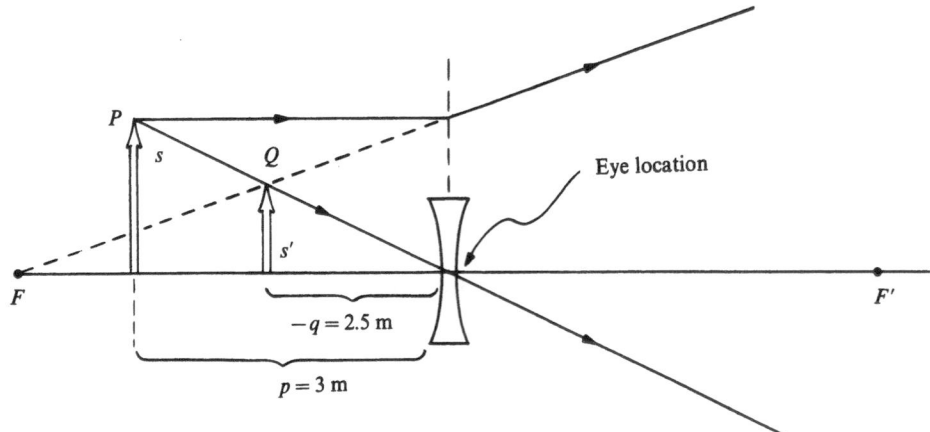

Figure 23.25. A lens for **near-sightedness** "brings the object closer"; see also Problem 23.36. Note how the plane of the lens may be extended as far as convenient (dashed line) for sketching purposes; to scale, the actual lens would be minuscule.

single new lens. If the combination is still thin, *its power is the algebraic sum of the individual powers*:

$$\frac{1}{f} = \frac{1}{f_1} + \frac{1}{f_2} + \cdots \qquad (23.D.12)$$

This is due to the fact that the power of a thin lens depends essentially on the total thickness of glass traversed by each paraxial ray; note (i) at the end of this chapter illustrates this in detail.

v. Summary (Thin Lens)

Our approach to lenses has been mainly through examples, and therefore we need a systematic case listing. This is presented below. In connection with the lens equation (23.D.5), we must keep the following points in mind:

1. The object distance, p, is always assumed positive.
2. The image distance, q, is positive if object and image are on opposite sides of the lens, negative if on the same side.
3. The focal length, f, is positive for a converging lens, negative for a diverging lens.
4. The (positively defined) magnification always equals $|q/p|$.
5. In Fig. 23.27, we imagine the object as gradually moving from left infinity to the lens itself; the corresponding image behavior is described in Table 23.2. The object and image always move in the same direction along the axis, except when the object passes through F in the converging case; the image then jumps discontinuously from right infinity to left infinity.

Such a concerted motion means that, as the object is transformed into its image, no right–left flip occurs along the axial direction. This conclusion is illustrated in Fig. 23.28: the image of a right hand is still a right hand.

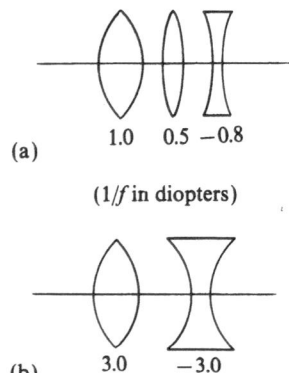

(a)

1.0   0.5 −0.8

(1/f in diopters)

(b)

3.0   −3.0

Figure 23.26. Close combinations of lenses: the powers add. (a) Total power = $(1.0 + 0.5 − 0.8)$ diopter = 0.7 diopter; (b) Total power = $(3.0 − 3.0)$ diopter = 0.0 diopter. (The lenses cancel one another.)

Figure 23.27. (For use with Table 23.2.) Possible locations of object and image on the axis; the lens is at O.

Table 23.2. (To be Used with Fig. 23.27.)
Image Formed by a Thin Spherical Lens[a]

Object location	Image location	Nature of image
Converging lens		
Left ∞	F'	$R \Downarrow$
Region a	a'	$R \Downarrow$
F	Left ∞	$V \Uparrow$
Region b	a, then b	$V \Uparrow$
O (i.e., touching the lens)	O	$V \Uparrow$
Diverging lens		
Left ∞	F	$V \Uparrow$
a, f, b	b	$V \Uparrow$
O	O	$V \Uparrow$

A diverging lens is incapable of forming a real image unless aided by another lens or curved mirror.

[a] R, real; V, virtual; $\Uparrow$, erect; $\Downarrow$, inverted.

vi. The Compound Microscope

The use of magnifying lenses goes back to Antiquity, and spectacles were worn in the Middle Ages. The idea of combining two lenses on the same axis, thus obtaining a microscope or a telescope, originated around 1600; Galileo was among the first users of both inventions. (In a present-day optical instrument, a light beam may traverse as many as a dozen lenses, made of different glasses; several lenses may be cemented surface to surface.)

This section has space for only one illustration —the two-lens microscope. The key to its design is a two-stage magnification, depicted in Fig. 23.29: the real image s', made by lens 1, serves as an object for lens 2.

In the figure, lens 1 (the **objective**) is shown with its focal points F_1, F_1'; lens 2 (the **eyepiece**) has focal points F_2, F_2'. The object s is barely to the left of F_1. This pushes the real image s' far to the right. Taken as an object for lens 2, s' is barely to the right

of F_2, and as a result its virtual image s'' is far out to the left, say at a comfortable 25 cm from the eyepiece; the observer's eye is located essentially at the eyepiece. Both focal lengths f_1, f_2 are designed to be short compared to the distance between the lenses, i.e., short compared to the microscope tube. (For clarity the figure is not drawn to scale in that respect.)

Overall Magnification

We calculate this quantity, defined as s''/s, from the product of two magnifications:

$$\frac{s''}{s} = \left(\frac{s'}{s}\right)\left(\frac{s''}{s'}\right) \tag{23.D.13}$$

With reference to lens 1 only, we have a magnification

$$\frac{s'}{s} = \left|\frac{\text{image distance}}{\text{object distance}}\right| \approx \frac{l}{f_1} \tag{23.D.14}$$

where l is the distance between F_2 and lens 1, or

$$l \approx \text{microscope length} \tag{23.D.15}$$

Similarly, using lens 2,

$$\frac{s''}{s'} \approx \frac{0.25 \text{ meter}}{f_2} \tag{23.D.16}$$

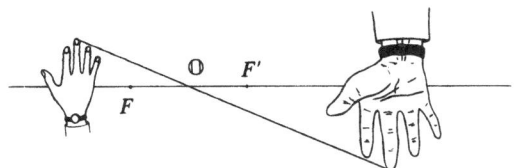

Figure 23.28. A lens does not exchange right and left. Note that the hand's width is more magnified than its length. The complete hand image is shown for clarity only; actually, the object stands in the way of some of its rays.

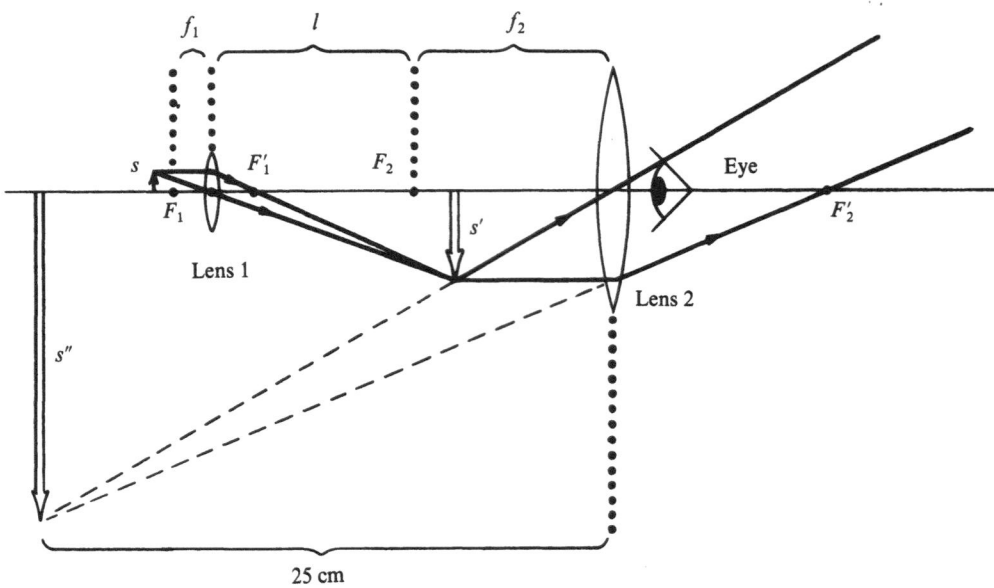

Figure 23.29. The **compound microscope**: a two-lens arrangement. The lens sizes are exaggerated for convenience. The rays (heavy lines) are not really broken at s'; we simply switch to a new set of rays as we transfer our attention from lens 1 to lens 2. In a scale drawing, these new rays could be extrapolated back to originate from lens 1.

Inserting (23.D.14) and (23.D.16) in (23.D.13), we get

$$\boxed{\frac{s''}{s} \approx \frac{(0.25 \text{ meter})l}{f_1 f_2}} \qquad (23.D.17)$$

This formula explains why a large l and, especially, small f_1, f_2 are conducive to a high magnification.

An obvious question arises: can we magnify without limit, for example, by increasing l sufficiently? The answer is yes, but at the price of such a blurring of the image that, eventually, no advantage is gained. This limitation is partly technical: a higher magnification puts greater demands on lens quality. Much more fundamental is the fact that geometrical optics is not valid on a scale comparable to the wavelength of light λ. The result is that *an object feature that is smaller than λ cannot be sharply imaged.* This will be brought up again in the next chapter.

E. Image Formation by Mirrors

Reflection is simpler than refraction, and therefore mirrors are simpler than lenses. (We are, furthermore, substituting the action of a single surface to that of a pair.) In what follows we briefly discuss the familiar plane mirror, as well as the concave mirror, which is of some scientific and technological importance as a focusing device.

i. The Plane Mirror

Figure 23.30a shows an object point P and its image Q produced by a plane mirror; we shall see that P and Q lie on the same normal and at the same distance $l = l'$ on either side of the mirror; Q *is necessarily virtual,* but may of course appear strikingly real in a good mirror.

We now verify on the figure that Q is where stated; we require that all reflected rays from P extrapolate back to Q.

First, the normal through P is itself one of those rays, so that P and Q are indeed on the same normal. Second, if PR is an arbitrary ray, the following angles are equal: $\alpha = \beta$ (the incident ray intersects two parallel lines); $\beta = \gamma$ (law of reflection); $\gamma = \delta$ (intersection with two parallel lines again). Thus, we have $\alpha = \delta$; the two right triangles have one rectangular side in common, and are congruent; therefore $l = l'$, as announced.

Right–Left Transformation

An extended object, in this case a right hand, is shown in part (b) of Fig. 23.30. Its image looks like

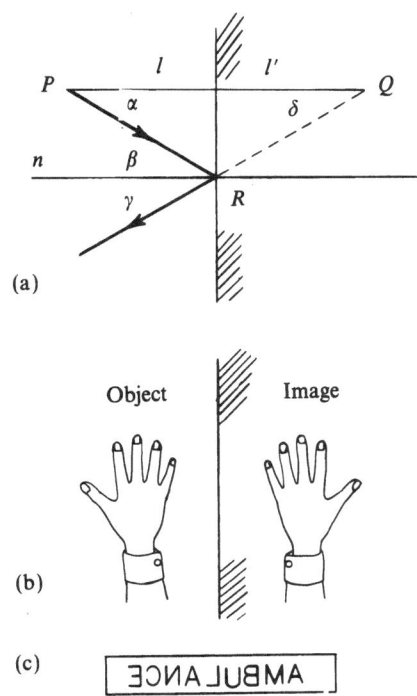

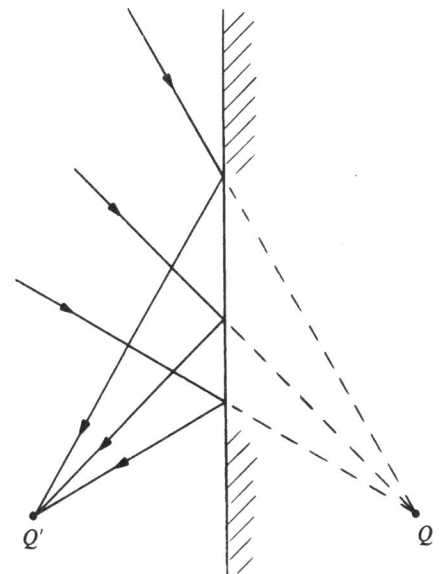

Figure 23.31. A "virtual object" Q, and its real image Q'.

Figure 23.30. (a) Reflection by a plane mirror. (b) Right and left are exchanged. (c) The word "AMBULANCE" is meant to be read in a rear-view mirror.

a left hand. This represents a different kind of inversion than is observed with lenses; compare with Fig. 23.28. *In the presence of several mirrors the right-to-left transformation occurs only after an odd number of reflections*, whether from plane, concave, or convex mirrors.

A Real Image from a Plane Mirror?

Figure 23.31 shows this to be possible if the incident rays are convergent—they might be coming from a lens. Before having a chance to make a real image at Q, they are intercepted by the mirror; the real image is at Q' instead. Points Q and Q' are symmetrically located on either side of the mirror, as shown. To demonstrate that Fig. 23.31 is correct, we only need to reverse the direction of light propagation; Q' becomes the object, and Q is its correctly located virtual image. Sometimes Q is called a "virtual object," of which Q' is the real image.

ii. Spherical Mirrors

Consider a concave, or **converging** mirror, shaped like a shallow spherical dish. Figure 23.32

shows the mirror's axis, its center ⊙, its center of curvature C, and its single focal point F; the focal length f is the distance ⊙F. *An incident ray parallel to the axis is always reflected through F; this is F's defining property.* Furthermore, *a ray incident at ⊙ is reflected symmetrically to the other side of the axis*, as shown. These two features enable us to

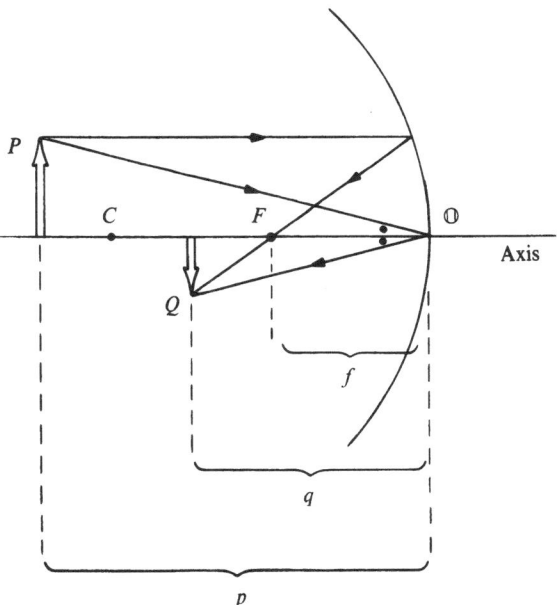

Figure 23.32. Image formation by a converging mirror. The axis is normal to the mirror's surface at ⊙, and hence bisects the angle between incident and reflected rays. The image is inverted (as shown) as well as having its right and left interchanged (not shown).

draw ray diagrams for mirrors, just as we did for lenses.

There exists a **mirror equation**, relating object and image distances p, q; it reads

$$\boxed{\frac{1}{p}+\frac{1}{q}=\frac{1}{f}} \qquad (23.E.1)$$

precisely as the lens equation (23.D.5); Fig. 23.32 is our "reference diagram," in which p, q, and f are all positive. In other cases, p is still assumed positive, but q and f may be negative, as we explain later on. A spherical-mirror image may be real or virtual, and its magnification is always

$$\boxed{\text{Magnification}=\left|\frac{q}{p}\right|} \qquad (23.E.2)$$

[Equation (23.E.1) is, once more, an approximation, which depends on the mirror being shallow (relative to f), and on the rays being paraxial.] Figure 23.33 shows f in relation to the mirror's radius of curvature R:

$$\boxed{f=\frac{R}{2}} \qquad (23.E.3)$$

This relation follows directly from (23.E.1) if we consider an object point located at the sphere's center of curvature C; as we see from Fig. 23.33, the incident rays are focused back into C because they

strike the mirror at right angles. Thus we take, in Eq. (23.E.1), $p=q=R$; (23.E.3) then follows.

Derivation of the Mirror Equation

See note (ii) at the end of this chapter.

Example 23.9. An Astronomical Telescope. Figure 23.34 shows a converging mirror, of focal length $f=2.0$ meters, being used to photograph the Moon. In this so-called "prime-focus" arrangement, the film is placed at the mirror's focal point. If the Moon subtends an angular diameter $\alpha=31$ minutes of arc in the sky, how large, in centimeters, is the image's diameter s' on the film?

———————

The Moon is effectively at infinity; therefore the image distance is $q=f$. Let s be the Moon's (unspecified) diameter, while p is its (large but unspecified) distance from us. According to (23.E.2), we have a "magnification" (in fact, a tremendous reduction)

$$\frac{s'}{s}=\frac{f}{p} \qquad (23.E.4)$$

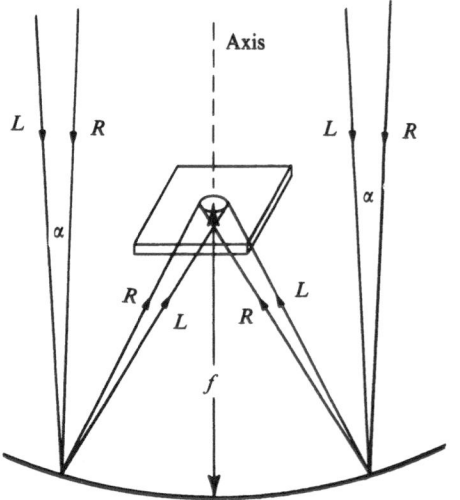

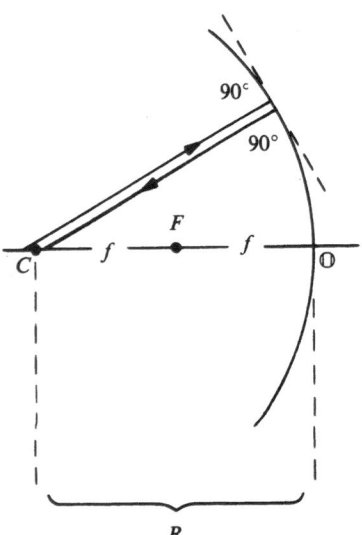

Figure 23.33. The focal point F is halfway between $\mathbb{O}$ and C.

Figure 23.34. (Not to scale. L, from left edge of the Moon; R, from right edge of the Moon.) The Moon's disk is projected, in perfect focus, on a photographic plate mounted at the reflector's focal point. Two typical bundles of rays are shown. The relatively small plate does not block image formation, even though it stands directly in the line of sight. It does prevent a small rectangle of the mirror from being used, however. (The reader should make a sketch showing the Moon as a sphere, and the location of the telescope as a point; identify the quantities s, p, α; and convince himself that $\alpha \approx s/p$.)

so that

$$s' = \frac{fs}{p} = (f)\left(\frac{s}{p}\right) = f\alpha \qquad (23.E.5)$$

The last step, $s/p = \alpha$, is legitimate for a small angle α expressed in radians; here

$$\alpha = \left(\frac{31}{60}\right)\left(\frac{2\pi}{360}\right) \text{rad}$$

$$= 0.0090 \text{ rad} = 0.0090$$

so that, from (23.E.5),

$$s' = (2.0)(0.0090) \text{ meter}$$

$$= 0.018 \text{ meter} = \underline{1.8 \text{ cm}}$$

Angular Magnification

When looking through a telescope, we observe an often spectacular magnification. This "angular magnification" is the subject of Problems 23.45 and 23.54.

Newtonian Eyepiece for a Reflector

In the preceding example, the image cannot be readily examined by eye, since one would have to stick one's head in front of the mirror. The arrange-

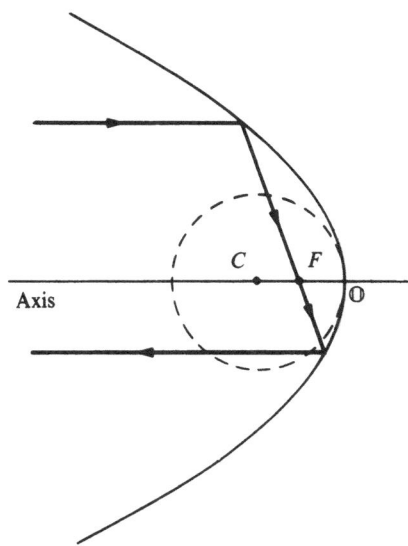

Figure 23.36. A parabolic mirror. This is a surface of revolution about OC. The focal point F coincides with the geometric focus of the parabola, halfway between C and O; point C is the center of that sphere which most closely fits the paraboloid at O. We see that a ray *issuing* from F is reflected parallel to the axis; a light or radio source at F is reflected into a collimated cylindrical beam.

ment of Fig. 23.35, first used by Newton in a reflecting telescope made by himself, is one of several existing ways to resolve the difficulty.

Parabolic Mirrors

Optimum focusing requires a slightly non-spherical mirror. A parabolic cross section, as in Fig. 23.36, will ensure that all rays that are incident

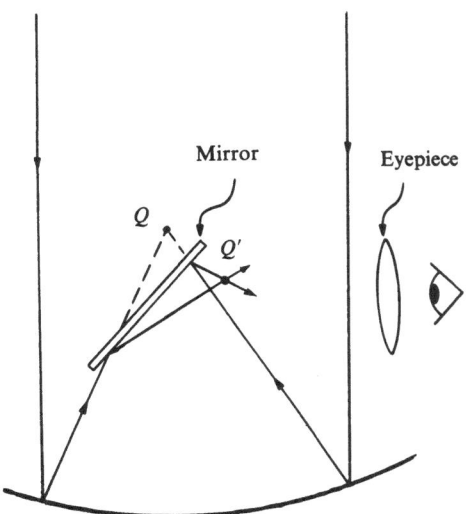

Figure 23.35. In Newton's arrangement, the image, which would otherwise be at Q, is formed at Q' by a small auxiliary plane mirror; Q' is examined through a magnifier (the eyepiece). Point Q is the "virtual object" for Q' as in Fig. 23.31. Compare this ray diagram with that of Fig. 23.34: here we see two rays from a single point on the Moon; there we saw two points being imaged, with two rays each.

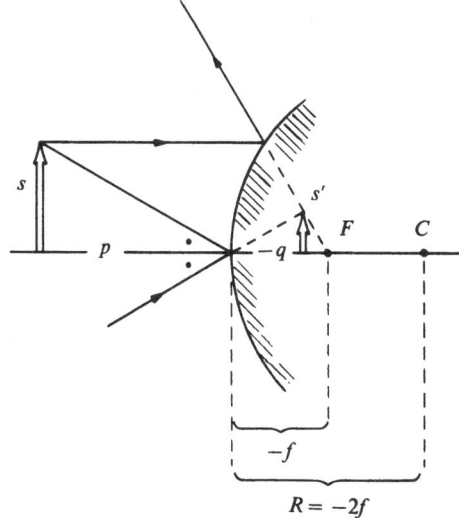

Figure 23.37. A diverging mirror; object s has an erect virtual image s'.

(a)

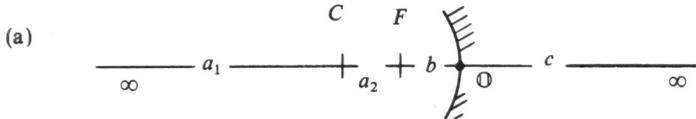

(b)

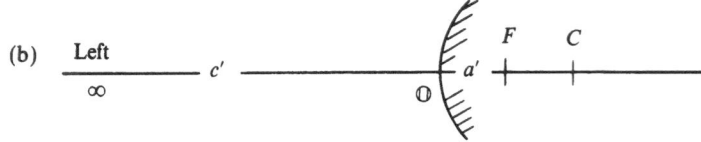

Figure 23.38. (For use with Table 23.3.) Possible object and image locations on the axis of a shallow mirror.

parallel to the axis are reflected into a single point F. For lack of space, we must omit a geometrical proof of this statement. (We still keep in mind that, even with a parabolic mirror, perfect sharpness is only a geometrical-optics idealization: we neglect the size of the radiation's wavelength.) Short-wave radio and radar "dishes," as well as high-quality astronomical reflectors, are usually parabolic. Our preceding discussion of spherical mirrors, being an approximation, is applicable to parabolic mirrors as well.

Diverging Mirror

Figure 23.37 depicts the operation of a convex, or diverging mirror. The mirror equation (23.E.1) is still applicable, *with negative f*; the value of q is negative as well, implying that object and (virtual) image are on opposite sides of the mirror.

Summary

Figure 23.38 shows the possible object locations in front of a mirror; Table 23.3 describes the resulting image. The following features, analogous to those of lenses, should be noted:

1. The object distance, p, is assumed positive.
2. The image distance, q, is positive if object and image are on the same side of the mirror, negative if on opposite sides.
3. The focal length f is positive for a converging mirror, negative for a diverging mirror.
4. The (positively defined) magnification equals $|q/p|$.
5. Object and image always move in opposite directions along the axis, except for discontinuous jumps between right and left infinity.

Table 23.3. (To be Used with Fig. 23.38.)
Image Formed by a Spherical Mirror[a]

Object location	Image location	Nature of image
Converging mirror, Fig. 23.38a		
Left ∞	F	$R\Downarrow$
Region a_1	a_2	$R\Downarrow$
C	C	$R\Downarrow$
Region a_2	a_1	$R\Downarrow$
F	Right ∞	$V\Uparrow$
Region b	c	$V\Uparrow$
$\mathbb{O}$	$\mathbb{O}$	$V\Uparrow$
Diverging mirror, Fig. 23.38b		
Left ∞	F	$V\Uparrow$
c'	a'	$V\Uparrow$
$\mathbb{O}$	$\mathbb{O}$	$V\Uparrow$

A diverging mirror cannot form a real image unless aided by another mirror or lens.

[a] R, real; V, virtual; $\Uparrow$, erect; $\Downarrow$, inverted.

Notes

The lens equation, $1/p + 1/q = 1/f$, follows from Snell's law with the help of some geometry. But the argument is rather dry and carries little physical insight; it has helped give a bad name to geometrical optics. The derivation below is based on waves; a similar one applies to mirrors.

i. Derivation of the Lens Equation

By examining the progress of a wave front, we obtain two results: (1) we derive Eq. (23.D.5); (2) we express f in terms of the lens' properties. First, however, we need a geometrical digression.

The Sagittal Theorem

This theorem deals with the following question: given a sphere of radius R, what is the thickness ($\overline{AB}$, or s in Fig. 23.39) of a "cap" of radius $\overline{AA'} = h$? The answer, for a *shallow* cap, is

$$s \approx \left(\frac{h^2}{2}\right)\left(\frac{1}{R}\right) \qquad (23.\text{N}.1)$$

To prove this we recall from Euclid that, for two lines BB' and $A'A''$ intersecting at A, we have

$$(\overline{AA'})(\overline{AA''}) = (\overline{AB})(\overline{AB'})$$

or

$$h^2 = (2R - s)(s) \approx (2R)(s) \qquad (23.\text{N}.2)$$

Solving for s gives Eq. (23.N.1).

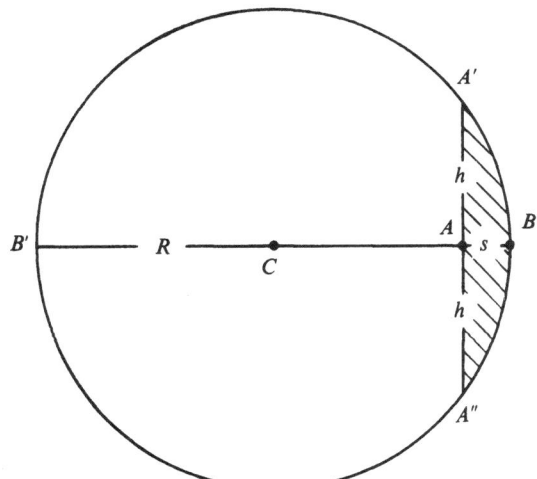

Figure 23.39. Estimating the thickness s of a spherical cap (shaded); C is the center of the sphere. (AB is the **sagitta**, or arrow, of the bow $A'A''$.)

In optical work it is often convenient to speak of a spherical surface's **curvature**, defined as its inverse radius of curvature,

$$\boxed{\text{Curvature} = \frac{1}{R}} \qquad (23.\text{N}.3)$$

(In particular, a flat surface has infinite R and hence zero curvature, as expected.) Formula (23.N.1) is proportional to the curvature of the cap.

Application to Image Formation

Figure 23.40 shows a wave front (arc AB) which originates from P and has just reached the lens at A. Somewhat later, the wave front has progressed through glass and air to the position of arc $A'B'$; it then converges on Q.

Two approximations will be made. First, we measure the image and object distances p, q, from the lens' surfaces rather than from its center; this is all right for a thin lens. Second, we assume a typical point B of the wave front to move along a line BB' that is parallel to the lens' axis; this is the "paraxial" approximation.

The key to our derivation is that it takes B the same time to reach B' as it does A to reach A'. Equal paths in glass (dotted segments) automatically take equal times, and need not be examined. There remain only the heavy line segments to be compared.

The lengths of these segments are shown in Fig. 23.40 on the basis of the sagittal theorem; the relevant curvatures are $1/p$ (initial wave front), $1/R_1$, $1/R_2$ (left and right lens surfaces), and $1/q$ (final wave front). The time of transit is given by (time) = (distance)/(speed), where the speed is c in air and c/n in glass. Thus the condtion

$$\text{Time for } \overline{BB'} = \text{time for } \overline{AA'}$$

becomes

$$\frac{1}{c}\frac{h^2}{2}\left(\frac{1}{p} + \frac{1}{R_1} + \frac{1}{R_2} + \frac{1}{q}\right) = \frac{n}{c}\frac{h^2}{2}\left(\frac{1}{R_1} + \frac{1}{R_2}\right)$$
$$(23.\text{N}.4)$$

Here $h^2/2$, which occurs everywhere, has been factored. Collecting terms, we get

$$\frac{1}{p} + \frac{1}{q} = (n-1)\left(\frac{1}{R_1} + \frac{1}{R_2}\right) \qquad (23.\text{N}.5)$$

The independence of this result from h^2 is important, and shows that all parts of the wave front cooperate to form the image point.

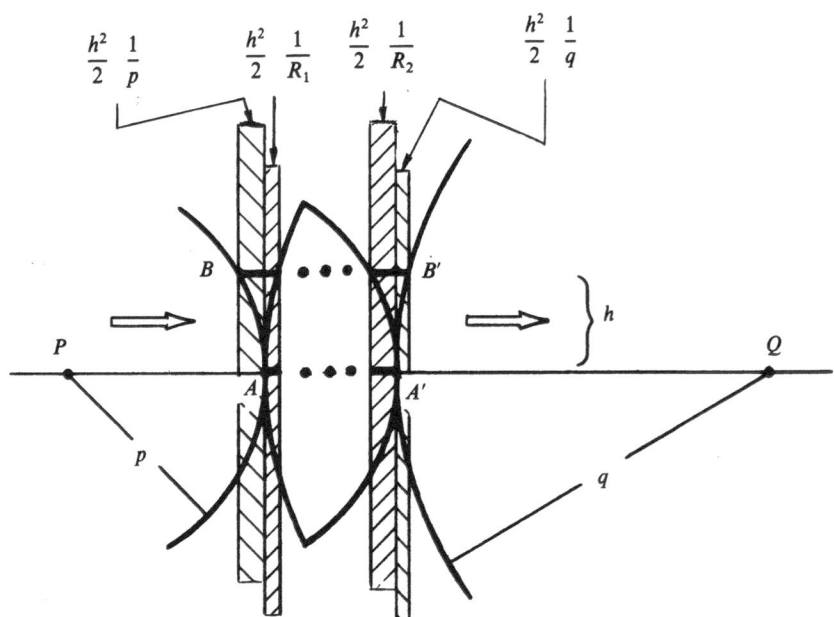

$$\frac{h^2}{2}\frac{1}{p} \qquad \frac{h^2}{2}\frac{1}{R_1} \qquad \frac{h^2}{2}\frac{1}{R_2} \qquad \frac{h^2}{2}\frac{1}{q}$$

Figure 23.40. Progress of a wave front through a lens. (The shaded strips serve to indicate the size of small path segments.)

In (23.N.5) we recognize the lens equation, (23.D.5), with an additional piece of information,

$$\frac{1}{f} = (n-1)\left(\frac{1}{R_1} + \frac{1}{R_2}\right) \qquad (23.N.6)$$

the so-called "lensmaker's formula."

[Incidentally, this result confirms our expectation that a piece of window glass, for which $1/R_1 = 1/R_2 = 0$, is a lens of zero power; also that a lens of any shape, but made of vacuum or air ($n = 1$), has zero power as well.]

A Concave Lens Surface Has Negative Curvature

Suppose at least one lens surface is concave, with radius of curvature R. The preceding argument still yields formula (23.N.6), provided the concave surface contributes a **negative curvature** term, $-1/R$ instead of $+1/R$. A diverging lens then turns out to have a **negative power**, a fact already used in Sec. D.

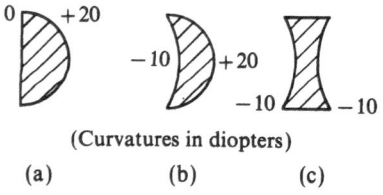

(Curvatures in diopters)

(a)　　　　(b)　　　　(c)

Figure 23.41. Three lenses, shown with their curvatures.

Example 23.10. Three Typical Lenses. Calculate the power and focal length of each of the three lenses shown in Fig. 23.41; the surfaces are labeled by their curvature in diopters; assume $n = 1.50$.

According to (23.N.6), the powers are

$$\frac{1}{f} = (n-1)\left(\frac{1}{R_1} + \frac{1}{R_2}\right)$$

$$= (1.5 - 1) \times \left\{ \begin{array}{l} 0 + 20 \\ -10 + 20 \\ -10 - 10 \end{array} \right\} \text{ diopters}$$

$$= \left\{ \begin{array}{l} 10 \\ 5 \\ -10 \end{array} \right\} \text{ diopter} \quad \begin{array}{l} (a) \\ (b) \\ (c) \end{array}$$

The corresponding focal lengths are

$$f = \left\{ \begin{array}{l} 0.10 \\ 0.20 \\ -0.10 \end{array} \right\} \text{ meter} \quad \begin{array}{l} (a) \\ (b) \\ (c) \end{array}$$

Off-Axis Geometry

As already illustrated in Fig. 23.21, P and Q do not have to be on the lens' axis for the lens equation to be valid. Does the preceding derivation have anything to say

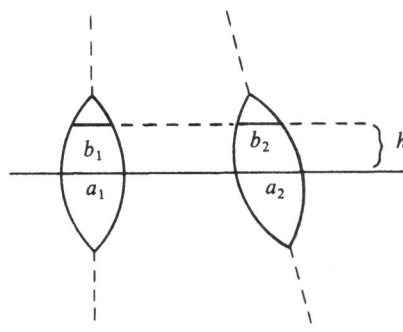

Figure 23.42. A slight tilt of the lens keeps $a_1 \approx a_2$, $b_1 \approx b_2$, etc.; the image is hardly affected.

about off-axis positions? To find out, we tilt the lens very slightly, as in Fig. 23.42. We recall that the only relevant aspect of the lens was the glass thickness traversed at any distance h from the line PQ. That thickness is negligibly changed by a small tilt, and therefore the lens equation remains valid, provided P and Q are not too far from the lens' axis (the paraxial condition again).

ii. Derivation of the Mirror Equation

Since this derivation parallels that of the lens equation, we shall be content with a sketch of the situation —Fig. 23.43—and a brief outline of the argument. The incoming and reflected wave fronts (arcs AB and $A'B'$) are shown just before and after reflection from a converging mirror with center $\mathbb{O}$. We require the travel time from

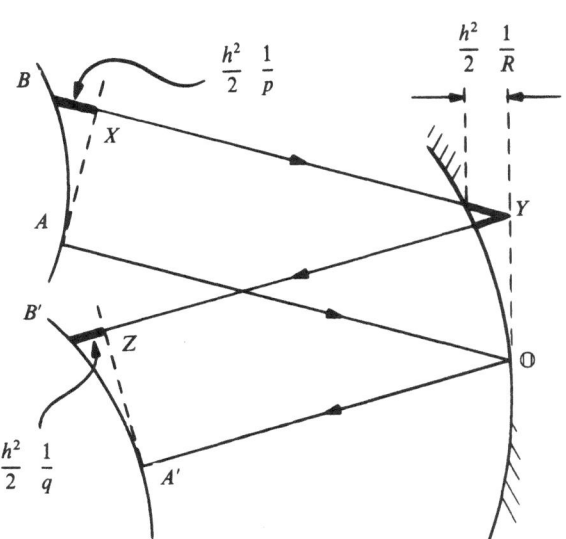

Figure 23.43. Derivation of the mirror equation. Distances and angles are exaggerated for clarity. To check equality of paths XYZ and $A\mathbb{O}A'$, we may think of $AXZA'$ as a rectangular strip "folded" about $\mathbb{O}Y$.

B to B' to equal that from A to A'. If the wave fronts and mirror are replaced by plane versions (dashed lines) tangent at A, A', and $\mathbb{O}$, then equal path lengths XYZ and $A\mathbb{O}A'$ are obtained. (The reflection of a plane wave is another plane wave.) In actuality, to path XYZ we must add the heavy segments at X and Z, while we must subtract from it those at Y. These segments' lengths are indicated in terms of $h = \overline{\mathbb{O}Y}$ and the curvatures. Zero path change means

$$\frac{h^2}{2}\left(\frac{1}{p} + \frac{1}{q} - \frac{2}{R}\right) = 0 \qquad (23.\text{N}.7)$$

leading to the required equation, (23.E.1), with

$$f = \frac{R}{2} \qquad (23.\text{N}.8)$$

Condensed Checklist

Index of refraction:

$$\boxed{n = \frac{c}{v}} \qquad (23.\text{B}.5)$$

Table: Table 23.1 of Sec. B

Wave fronts:

$$\text{Spacing} = \frac{\lambda}{2} \quad \text{Fig. 22.11 of Chap. 22}$$

Two media:

$$\frac{\lambda_A}{\lambda_B} = \frac{v_A}{v_B} = \frac{n_B}{n_A} \qquad (23.\text{A}.4), (23.\text{B}.5)$$

Snell's law:

$$\boxed{\frac{\sin\theta_A}{\sin\theta_B} = \frac{v_A}{v_B} = \frac{n_B}{n_A}}$$

$$(23.\text{B}.3), (23.\text{B}.6)$$

Dispersion:

n depends on f Sec. B iii

Reflection:

$$\boxed{\theta' = \theta} \qquad (23.C.3)$$

Total internal reflection:

$$\sin \theta_c = \frac{n_{\text{smaller}}}{n_{\text{larger}}} \qquad (23.C.5)$$

Virtual image in flat block:

$$\frac{l'}{l} = \frac{1}{n} \qquad (23.D.4)$$

Lens or mirror equation:

$$\boxed{\frac{1}{p} + \frac{1}{q} = \frac{1}{f}} \qquad (= \text{power})$$

$$(23.D.5), (23.E.1)$$

Unit: meter^{-1} = diopter; $\qquad (23.D.6)$

Mirror's focal length: $\boxed{f = \dfrac{R}{2}} \qquad (23.E.3)$

Magnification:

$$\frac{s'}{s} = \left| \frac{q}{p} \right| \qquad (23.D.7), (23.E.2)$$

Closely spaced lenses:

$$\frac{1}{f} = \frac{1}{f_1} + \frac{1}{f_2} + \cdots \qquad (23.D.12)$$

Summary of cases:

Lens   Table 23.2 of Sec. D

Mirror   Table 23.3 of Sec. E

Magnifying glass:

$$\frac{s'}{s} \approx \frac{0.25 \text{ meter}}{f} \qquad \text{Problem 23.34}$$

Compound microscope:

$$\frac{s''}{s} \approx \frac{(0.25 \text{ meter})l}{f_1 f_2} \qquad (23.D.17)$$

Telescope:

$$s' = f\alpha \qquad (23.E.5)$$

Angular magnification: Problems 23.45, 23.54

True or False

1. A ray always intersects a wave front at right angles.

2. Rays are sometimes curved.

3. Two successive wave fronts are everywhere equidistant.

4. Wave fronts are closer together where the index of refraction is smaller.

5. Monochromatic light cannot be dispersed by a prism.

6. In glass, the index of refraction decreases with increasing frequency.

7. A completely transparent prism bends a green light beam less than a yellow one.

8. The refracted ray is always closer to the interface normal than the incident ray.

9. The angle of incidence is measured between the incident ray and the interface.

10. One condition for total internal reflection is that the angle of incidence be small enough.

11. For a given angle of incidence, one condition for total internal reflection is that the index of refraction be small enough in the substance where the light beam travels.

12. A perfectly transparent object should be invisible.

13. A virtual image cannot be photographed as if it were an object.

14. If a ray is incident, parallel to the axis, on a thin lens, it does not get deflected.

15. A lens can focus rays from one of its focal points into the other.

16. A one-lens slide projector, used with a distant screen, should have the slide at the lens' focal point.

17. In a mirror, your right and left hands are interchanged, and yet your head and feet are not. Hence a mirror can tell vertical from horizontal.

18. A spherical mirror's center of curvature is also its focal point.

19. Your image in a reflecting Christmas tree globe is inverted.

20. Distant trees seen in a reflecting garden globe appear to be at the globe's center of curvature.

Problems

Sections A: Wave Fronts and Rays;
B: Refraction of a Plane Wave;
C: Reflection of a Plane Wave

*23.1. Using Table 23.1, calculate the speed of light waves in (a) water, (b) carbon disulfide, (c) diamond. (Use $c = 3.00 \times 10^8$ meters/sec.)

23.2. In a certain plastic, light waves propagate at 2.02×10^8 meters/sec. Fing the plastic's index of refraction.

*23.3. A light beam enters a plane transparent piece of material at an angle of incidence of 50°; its angle of refraction is 31°. Find the material's index of refraction.

23.4. What is the angle of refraction for a ray incident at 70° on a flat surface of a sodium chloride crystal ($n = 1.54$)?

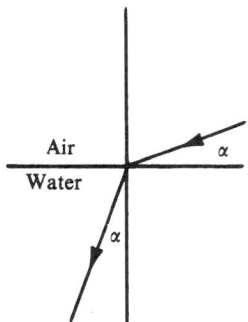

23.5. The angles α (see figure) are equal for a certain ray being refracted from air into water ($n = 4/3$). Calculate α.

*23.6. A certain material, with index of refraction n, exhibits total internal reflection, with critical angle $\theta_c = 75°$, when immersed in glycerol (index of refraction = 1.47). Find n.

23.7. Obtain Eq. (23.B.6) from Eq. (23.B.3).

*23.8. When a certain ray is refracted from glass to water (indices of refraction 1.60 and 1.33, respectively), the angle of refraction equals twice the angle of incidence. Calculate the angle of incidence.

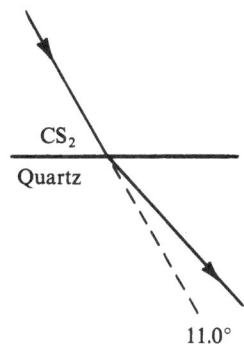

23.9. At a plane interface between carbon disulfide and fused quartz (indices of refraction 1.63 and 1.46, respectively) a certain ray, incident from the CS$_2$ side, is deflected by 11.0° from its original direction, as shown. What is its angle of incidence?

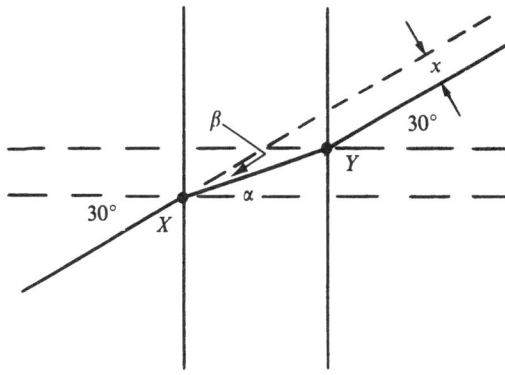

*23.10. The figure shows a thin light beam passing through a flat uniform glass plate of thickness 2.5 cm and index of refraction 1.50. If the angle of incidence is 30°, calculate (a) the first angle of refraction, α in the figure; (b) the length of segment XY; (c) angle β; (d) the distance x by which the beam is shifted laterally.

23.11. Consider the prism of Fig. 23.10, but with an arbitrary angle α. Find, in terms of α and n, the deflection δ of the beam from its original direction. (In Fig. 23.10, $\delta = 20°$.)

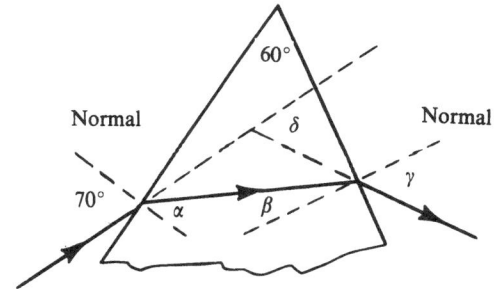

*23.12. Find the angle of deviation, δ, for a light beam incident at a 70° angle on a 60° glass prism, as shown. (Take $n = 1.60$.)

*23.13. A collimated beam of white light (containing all visible frequencies) is incident, from inside a piece of heavy flint glass, on its interface with air. All frequencies above 6.0×10^{14} Hz are totally reflected, while all those below that frequency are only partly reflected. What is the beam's angle of incidence, to three significant figures? (Use Fig. 23.9.)

23.14. In Example 23.2, let the blue and red lights have vacuum wavelengths of 486 nm and 656 nm, respectively. Suppose the incident beam also contains yellow light, which is refracted between the other two at $0.21°$ from the blue. By using a linear interpolation, estimate the vacuum wavelength of the yellow component.

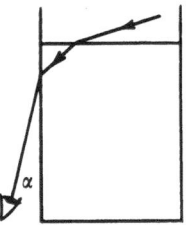

23.15. At what maximum angle α to the (vertical) side of an aquarium should one look up, as shown, in order to see beyond the water surface? (Take $n = 1.33$ for water.)

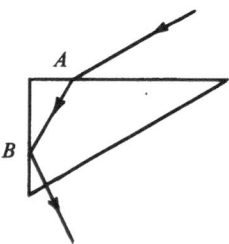

*23.16. Consider a rectangular prism, as shown. (a) What is its minimum index of refraction such that *any* ray entering one rectangular side (at A) will be totally reflected if it reaches the other (at B)? Assume the ray is in the plane of the figure (other rays will be even more surely totally reflected). (b) Will glass perform in that manner?

23.17. A monochromatic light wave, of frequency 5.5×10^{14} Hz, travels down a tube of length 2.5 meters, containing air at standard temperature and pressure (index of refraction 1.0003). If the air is now pumped out, how many more (or fewer) half-cycle wave fronts fit in the tube?

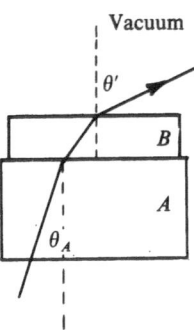

*23.18. A certain transparent plate is made of two horizontal uniform layers A, B with indices of refraction n_A and n_B, as shown. An upward beam makes an angle θ_A with the vertical in A, and emerges at an angle θ' to the vertical. (a) Calculate θ' in terms of Q_A, n_A, n_B, and the thickness of both layers. (b) Calculate θ' in terms of θ_A, n_A, and the thickness of layer A, assuming layer B to be absent. (c) What do you conclude about the effect on θ' of piling up more transparent plates on top of layer A? (d) To see a certain star, one must aim a telescope at $60.000°$ to the vertical. At what angle to the vertical is the star's true direction? (Consider the atmosphere as a set of horizontal layers; use $n = 1.0003$ for the air at the telescope's altitude.)

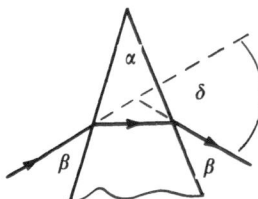

*23.19. The minimum deflection of a ray by a prism can be shown to occur in the symmetric situation shown. (The angles β are equal.) Find the ray's net deflection, δ, in terms of the prism's angle α and its index of refraction n.

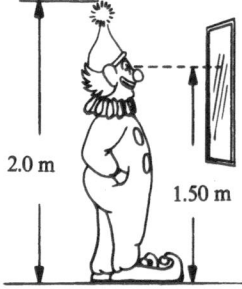

*23.20. A clown wants to see himself full length (2.0 meters with his hat) in a vertical mirror, as shown. If his eye is 1.50 meters from the floor, (a) what minimum length of mirror does he need? (b) How high above the floor is the mirror's bottom edge?

Section D: Image Formation by Lenses

*23.21. What focal length is needed for a converging lens to produce a real image 60 cm away from the object when the lens is halfway between the two? (Sketch a ray diagram.)

23.22. What focal length is needed for a converging lens to produce a real image 35 cm away from the object, when the lens is at 12 cm from the object?

*23.23. What power is needed for a diverging lens to produce a virtual image 12 meters away from itself, when the lens is 20 meters away from the object?

23.24. Do the algebra that solves Eqs. (23.D.8) and (23.D.5) for p.

23.25. Draw a version of Fig. 23.25 in which P is further from the lens than F.

23.26. Sketch how a point source of light at the focal point of a converging lens produces a collimated light beam. Can this be done with a diverging lens?

*23.27. A 3.0-cm-wide object is located near the axis of a converging lens with focal length 18 cm. If the object is at 30 cm from the lens, (a) where, and (b) how wide is the image? Sketch a ray diagram.

23.28. Do Problem 23.27 for a diverging lens (focal length $= -18$ cm).

*23.29. An object is successively placed at distances of 50, 25, 12, 5, and 3 cm from a converging lens of focal length 10 cm. Find the corresponding image distances. Sketch ray diagrams for the 25-cm and 5-cm cases.

23.30. Do Problem 23.29 for a diverging lens (focal length $= -10$ cm).

*23.31. A certain slide projector uses a single lens. If a 35-mm-wide slide must give a 1.5-meter-wide image on a screen at 8.0 meters from the lens, find the distance between slide and lens. (Sketch a ray diagram.)

23.32. On Fig. 23.22, trace a ray starting out as PF. Justify your construction by considering Q as the object, P as the image.

*23.33. What is the minimum distance between object and *real* image in the case of a converging lens of focal length f?

*23.34. (a) With $q = -25$ cm and f arbitrary, express the magnification s'/s of a lens or mirror in terms of f. (b) Simplify your result, assuming that f is very small compared to 25 cm. (This is the case for strong magnifiers.)

23.35. Suppose one requires a magnifying glass with magnification 5.0 when the image is 25 cm behind the lens. What should be the lens' power?

*23.36. A certain near-sighted eye sees clearly objects that

are at most 2.5 meters away. Prescribe the (minimum) power of a correcting lens that will permit seeing to infinity. Sketch a ray diagram; you do not need to use the optics of the eye itself.

*23.37. A certain far-sighted eye sees clearly objects that are at least 0.60 meter away. What is the power of the weakest lens that will permit sharp seeing of objects 0.25 meter away? Sketch a ray diagram; you do not need to use the optics of the eye itself.

23.38. (A one-lens refracting telescope). For taking photographs, a refracting telescope needs only its (converging) objective lens; the film is positioned at the focal point. (The telescope is just a camera adjusted for infinity.) How large an image of the Moon is formed if the lens has a power of 1.5 diopters? (The Moon's diameter subtends 31 minutes of arc in the sky.)

23.39. A certain compound microscope has focal lengths of 0.5 cm and 4.0 cm for objective and eyepiece, respectively. The lenses are 15 cm apart. Find the magnification in the case of an image at 25 cm from the eyepiece.

*23.40. In Problem 23.39, suppose the final image is 3 cm wide. Find (a) the size of the intermediate image (s' in Fig. 23.29); (b) the size of the object; (c) the distance between object and the objective lens' closest focal point.

23.41. Figure 23.22 shows a thin lens, approximately occupying a plane, and such that (1) every incident ray parallel to the axis is focused at F, and (2) every ray through the center is undeflected. From triangle similarities, derive the lens equation (23.D.5).

*23.42. Where should a small object be positioned in order that a converging lens of power $1/f$ give a virtual image at one of its focal points? Support your answer with a ray diagram.

23.43. A magnifying glass, of power 20 diopters, is held 10 cm above a postage stamp which is lying on a table. (a) Where is the image? Is it real or virtual? (b) If the lens is now lowered at a speed of 1 cm/sec, how fast and in what direction is the stamp's image moving?

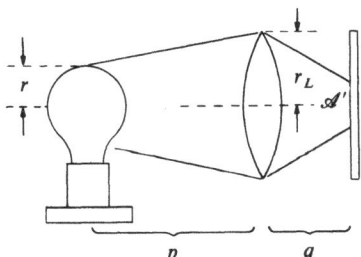

*23.44 (Image intensity[†]). A frosted light bulb of radius r radiates a power $\mathscr{P}$, isotropically in all directions, in the

---

† Related to, but not identical to illumination; this latter concept is based on the sensitivity of the eye.

form of visible light; see the figure. A converging lens of radius r_L focuses the bulb's image on a screen or film. Object and image distances are p and q. (a) Find $\mathscr{I}$, the power emitted per unit area of the bulb. (b) Find $\mathscr{P}'$, the power reaching the lens and therefore reaching the image. (Assume bulb and lens see one another under small angles.) (c) Find the area $\mathscr{A}'$ of the image. (d) Find the average intensity $\mathscr{I}' = \mathscr{P}'/\mathscr{A}'$ at the image. (e) Find the ratio $\mathscr{I}'/\mathscr{I}$, and show that it depends only on r_L and q. (Express your answers in terms of $\mathscr{P}$, r, r_L, p, q, or some of these.) (f) Estimate what fraction $\mathscr{I}'/\mathscr{I}$ of the Sun's actual surface intensity, ignoring atmospheric effects, is projected (in focus) on a piece of paper by a lens of radius $r_L = 4$ cm and focal length 12 cm.

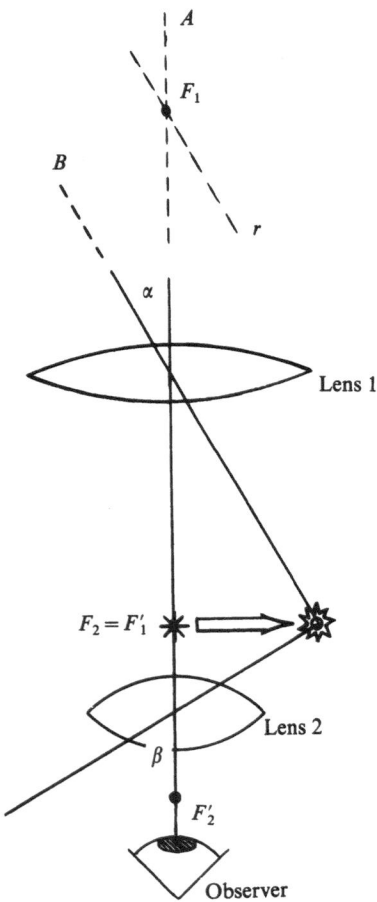

23.45 (The **refracting telescope**). The two converging lenses (see figure) share one focal point ($F_1' = F_2$). Lenses 1 and 2 have focal lengths f_1 and f_2, respectively. The angular distance α between two stars (A, B) is seen as β by the observer. (One of the stars is taken on the telescope's axis for convenience.) (a) Trace the complete path of another ray, r, coming from B and going through F_1. (Extend the lenses' width as needed.) (b) Show that the image is inverted and at infinity. (c) Calculate, in terms of f_1 and f_2, the **angular magnification** β/α for small α. (See also Problem 23.54.)

Section E: Image Formation by Mirrors

23.46. A coin, 1.2 cm in diameter, is facing a concave mirror whose radius of curvature is 30 cm. If the coin is at 20 cm from the mirror, (a) where is its image? (b) Describe the image's orientation relative to the object. (c) What is the image's diameter? (Sketch a ray diagram.)

*23.47. Do Problem 23.46 with the same coin and the same mirror, but 10 cm apart.

*23.48. Prove that, although a lens or mirror magnifies a small object by a factor $|q/p|$ laterally (i.e., in a direction transverse to the axis), the magnification along the axis is $|q/p|^2$.

23.49. A small soap bubble, of diameter 3 mm, is floating 15 cm away from a mirror of focal length 10 cm, on the mirror's axis. (a) Where is its image? Is it real or virtual? (b) What is the image's diameter, at right angles to the axis? (c) How deep is the image along the axis? (See the preceding problem; sketch a ray diagram.)

23.50. Explain how Eq. (23.E.1) is applicable to a plane mirror.

*23.51. Someone's face is 20 cm away from a concave shaving mirror and 50 cm away from its image. Find (a) the magnification; (b) the mirror's focal length. (c) Is the image real or virtual? Erect or inverted? (Sketch a ray diagram.)

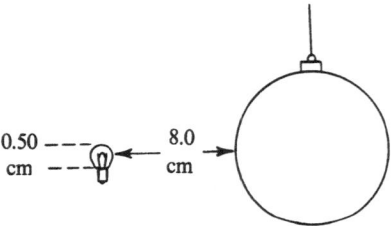

23.52. A small light bulb, 0.50 cm in diameter, is 8.0 cm away from a Christmas tree globe, in which its image is 0.10 cm in diameter; see the figure. Find the radius of the globe.

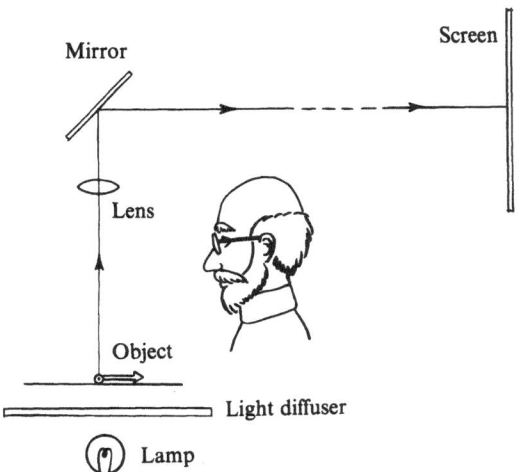

23.53. In the adjoining sketch of an overhead projector, the object (say a text on a transparency) is symbolized by *two* arrows, one of which looks like a point because it is directed out of the page. By appropriately completing the diagram, demonstrate that the audience sees the text in the same up-down and right-left orientation as does the lecturer.

*23.54 (Angular magnification of a reflecting telescope). Suppose that, in Fig. 23.35, the intermediate image Q' is at the focal point of the eyepiece. Let the mirror and eyepiece have focal lengths f_1, f_2, respectively. (a) Where is the new image seen through the eyepiece? (b) What is its angular size β if it corresponds to a piece of the sky with angular size α? (c) What is the ratio β/α (the telescope's angular magnification)? Assume α, β are small.

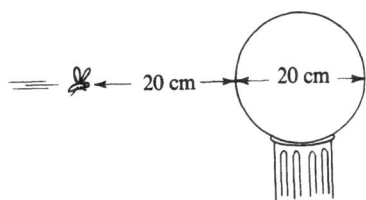

23.55. A bee, flying at 0.5 meter/sec towards the center of a reflecting garden globe of diameter 20 cm, as shown, sees her image flying towards her. When the bee is 20 cm away from the glass surface, how fast is her image moving (a) relative to the globe, (b) relative to the bee?

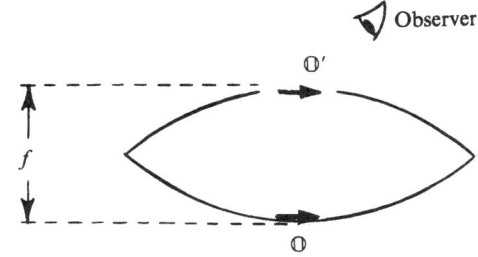

23.56. A certain illusion device consists of two equal concave mirrors, of focal length f, facing one another with

their centers $\mathbb{O}, \mathbb{O}'$ a distance f apart, as shown. Prove that a small object lying at $\mathbb{O}$ has a natural-sized image at $\mathbb{O}'$ (where an opening is provided for viewing from the top).

Answers to True or False

1. True.
2. True (in a gradually changing medium).
3. False (see Fig. 23.2).
4. False (the opposite).
5. True (it can be refracted).
6. False (the opposite).
7. False (the opposite).
8. False (it depends on which medium has the larger n).
9. False.
10. False (it should be large enough).
11. False (large enough).
12. False (refraction and reflection make it visible).
13. False.
14. False.
15. False.
16. True.
17. False (your left hand only becomes a right hand).
18. False.
19. False.
20. False (halfway closer to the surface).

Answers or Hints to Problems

23.1. (c) 1.24×10^8 meters/sec.

23.3. 1.49.

23.6. 1.52.

23.8. The identity $\sin 2\alpha = 2 \sin \alpha \cos \alpha$ may help.

23.10. (b) $\overline{XY} = 2.5/\cos \alpha$; (c) use $\beta = 30° - \alpha$; (d) $x = \overline{XY} \sin \beta = 0.48$ cm.

23.12. Hint: $\alpha + \beta = 60°$, and $\delta = 70° + \gamma - 60°$ in the figure. Answer: $51°$.

23.13. $37.8°$.

23.16. Hint: Take the extreme case where A is entered at grazing incidence; then show that $\theta_c \leqslant 45°$ is needed. (b) Yes.

23.18. (a) $\theta' = \sin^{-1}(n_A \sin \theta_A)$, no dependence on n_B or thicknesses.

23.19. Hints: Split the prism into a pair of prisms with angles $\alpha/2$; go back to Problem 23.11; use $\delta = 180° - 2\beta - \alpha$. Answer: $\delta = 2 \sin^{-1}[n \sin(\alpha/2)] - \alpha$.

23.20. (a) Mirror length independent of eye height.

23.21. 15 cm.

23.23. -0.033 diopter.

23.27. (b) 4.5 cm.

23.29. $q = -10$ cm for $p = 5$ cm.

23.31. 19 cm.

23.33. $4f$; you must show that this gives a true minimum.

23.34. (a) $\dfrac{s'}{s} = \dfrac{25\ \text{cm}}{f} + 1$; (b) $\dfrac{s'}{s} \approx \dfrac{25\ \text{cm}}{f}$.

23.36. Hint: An object at infinity must give a virtual image at 2.5 meters.

23.37. Hint: An object at 0.25 meter must give a virtual image at 0.60 meter.

23.40. (a) Use (23.D.16); (b) use (23.D.14).

23.42. Halfway between the focal point and the lens.

23.44. (d) $(\mathscr{P}r_L^2)/(4\pi q^2 r^2)$; (f) $1/q$.

23.47. (c) 3.6 cm.

23.48. Hint: Call the *longitudinal* object and image sizes $|dp|$ and $|dq|$; differentiate the lens equation.

23.51. (a) 1.5; (b) 60 cm.

23.54. First do without the auxiliary mirror; (b) is not affected. (c) f_1/f_2.

CHAPTER 24

WAVE OPTICS

If a light ray were the trajectory of some kind of particle, then geometrical optics *might* be exactly valid, and the concept of wave fronts would not be needed. Such a particle theory was favored by Newton, while his contemporary, Christian Huygens, held out for waves.

We now know that geometrical optics is inadequate to explain some important properties of light. Three of these properties will be discussed in this chapter. They are (1) interference, (2) diffraction, and (3) polarization. They cannot be explained satisfactorily except in terms of waves.

1. **Interference** is observed when several electromagnetic vibrations are superposed; it also exists as a mechanical phenomenon in sound waves. We recall from the mechanical discussion the essential role played by the **phase** of the individual vibrations.

How can the interference of light be observed? The answer is that bright or dark regions, called **fringes,** * are seen according to whether the interference is constructive or destructive. An apparatus specially made to exhibit fringes is called an **interferometer**; we shall study one famous design due to Michelson.

2. **Diffraction** is observed behind slits, pinholes, or partial obstructions to a light beam. An obstructing edge casts a shadow, as we all know; but in addition, light will, to some extent, bend (**diffract**) around a corner so as to penetrate the region where ray optics tells us there should be total shadow. Thus, near the edge of an obstruction, light fails to propagate in a straight line; this effect cannot be

attributed to reflection or refraction. Diffraction of light usually passes unnoticed, because most objects that are visible to the naked eye do not bend light beams appreciably. However, diffraction becomes conspicuous in the presence of slits or pinholes *whose size is comparable to the light's wavelength*. (We are, of course, used to *hearing* around corners. The diffraction of sound waves is sometimes entirely responsible for this, although usually reflections from nearby objects or walls also contribute.)

As an application, we shall examine the **diffraction grating**, a system of closely spaced slits that takes advantage of diffraction to separate wavelengths in the manner of a spectroscope.

Not all light sources are equally suitable for interference and diffraction experiments; the wave must meet a certain simplicity requirement known as **coherence**. The uniquely clean and simple light generated by a **laser** possesses that property to a high degree. The introduction of the laser in the early 1960s has greatly expanded the scope of interference and diffraction studies.

3. **Polarization** follows from the transverse nature of the electromagnetic wave; the concept was briefly discussed in Chap. 22, Sec. A. Here we shall see how polarization may be obtained and detected by suitable filters that work in the visible range.

What Did Geometrical Optics Overlook?

The conclusions of geometrical optics are imperfect because they are backed up by an imperfect argument. The laws of refraction, for example, were derived, in Sec. B i of the preceding chapter, partly

* For this chapter's computer-generated fringe illustrations the author is indebted to Daniel and Pierre Wellner.

from symmetry considerations. But that symmetry involved a plane wave and a plane interface, both without edges. Therefore, highly curved optical surfaces or narrow light beams may be expected to give rise to new phenomena, and they do.

For a glimpse at the modern synthesis of particle and wave theories, see Chap. 26 further on.

A. Interference by Reflection

When studying mechanical vibrations (or sound) in Chap. 13, we saw that a single object can *simultaneously execute two* (or more) *superposed harmonic oscillations*. That is to say, the object's displacement x from equilibrium is given by a sum such as

$$x = A \cos \omega t + A \cos(\omega t + \phi) \quad (24.A.1)$$

In this illustration, both amplitudes A are equal, and so are both angular frequencies ω. However, the oscillations have a **phase difference** ϕ. With $\phi = 180°$, Eq. (24.A.1) reads $x = 0$ (for all times); we have **destructive interference**. On the other hand, if both oscillations are in phase ($\phi = 0$), we get

$x = 2A \cos \omega t$; the resultant amplitude $2A$ is the largest obtainable from an adjustment of ϕ; we have **constructive interference**.

i. Parallel Reflecting Surfaces

Even a perfectly transparent substance reflects some light from its surface. If viewed against a dark background, such reflections are easy to observe; they permit many experiments on the wave nature of light. [What determines the amount of reflection? Note (i) at the end of this chapter briefly touches upon that question.]

Figure 24.1 depicts a simple case of wave interference. Here a plane monochromatic wave w is incident normally on a uniform transparent plate. An observer above the plate sees the superposition of two waves. One of these, a, has been reflected from the top surface; the other one, b, from the bottom surface. We state without proof that their amplitudes may be considered equal. According to their relative phases, a and b interfere constructively, destructively, or in some intermediate way. We now ask: given the index of refraction n of the plate and the wavelength λ_0 of the light in vacuum, what should be the plate thickness s so that a and

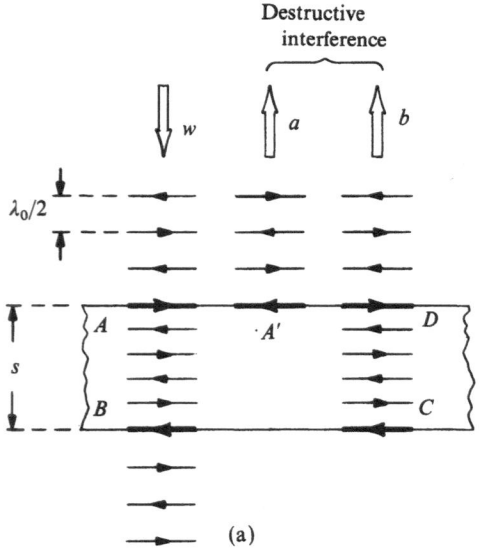
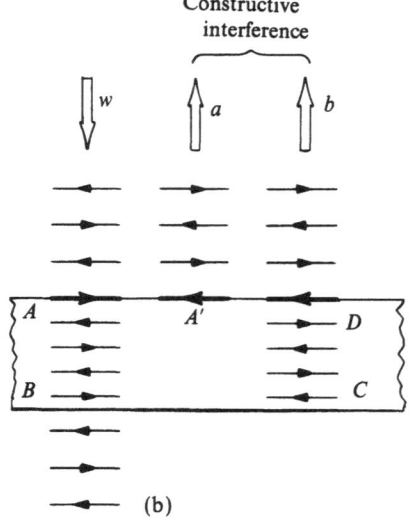

Figure 24.1. Reflection interference from an unmetallized transparent film, for given wavelength λ_0 and two different film thicknesses, $s = 5\lambda/2$ in (a), $s = 4\lambda/2 + \lambda/4$ in (b). Waves w, a, b are really superposed but are drawn separate for clarity. Multiple reflections are not shown. Arrows show the direction of **E** at the wave fronts. Note the reversal of **E** upon reflection at A, A'. This **phase reversal** always accompanies external reflection; it is predicted by electromagnetic theory. In contrast, note how the phase is *preserved* during internal reflection, at points B, C, in part (a); in part (b), the bottom surface does not coincide with a wave front but is a quarter-wave down from a wave front, at the moment the "snapshot" is taken.

b interfere *destructively* (overall reflection is suppressed)? Alternatively, how thick a plate gives *constructive* interference?

In Fig. 24.1, the wave fronts of a and b must coincide for either of the above to happen. The figure is a "snapshot," taken as one incident wave front (heavy bar) coincides with the top surface; this wave front is also present in a. We now require b to have a wave front at that location as well. Therefore the roundtrip $\overline{AB} + \overline{CD} = 2s$ must equal a whole number of wave front spacings $\lambda/2$,

$$2s = k\frac{\lambda}{2} \qquad (24.A.2)$$

where k is a nonnegative integer and λ is the wavelength in the dielectric. Alternate wave fronts have opposite phases, and therefore, according to whether k is odd or even, interference is constructive or destructive; but which goes with which? One special case suggests the right correspondence. An infinitely thin plate ($k = 0$) should amount to no plate at all, and therefore should give zero reflection. Hence, for $k = 0, 2, 4, 6,...$, the interference is destructive, while for $k = 1, 3, 5,...$ it is constructive.

Finally, we recall that λ is *shorter than its vacuum version λ_0* because, in the plate, the wave fronts move more slowly by a factor $1/n$:

$$\boxed{\lambda = \frac{\lambda_0}{n}} \qquad (24.A.3)$$

[The reader might wish to review Eqs. (23.A.4) and (23.B.5) of the preceding chapter.]

Equation (24.A.2) may be written

$$s = \frac{k\lambda}{4} \qquad (24.A.4)$$

or, with (24.A.3),

$$\boxed{s = \frac{k\lambda_0}{4n}} \qquad \left(\begin{array}{l} k \text{ even: destructive} \\ k \text{ odd: constructive.} \end{array}\right) \qquad (24.A.5)$$

In particular, a half-wave film ($s = \lambda/2$) is nonreflective.

Example 24.1. A Quarter-Wave Soap Film. A certain film of soap water was viewed in green light ($\lambda_0 = 550$ nm) while being stretched; it became shiny one last time before bursting. How thick was it when maximally reflecting? Assume the least possible thickness; plane waves; normal incidence; and an index of refraction equal to that of water ($n = 1.33$).

In formula (24.A.5), we must take $k = 1$. Hence the film's thickness was

$$s = \frac{(1)(550 \text{ nm})}{(4)(1.33)} = \underline{\underline{103 \text{ nm}}}$$

(If we estimate the size of a water molecule at 0.2 nm, the film was about $103/0.2 = 500$ molecules thick.)

When is a film too thin to be seen? The preceding example shows, via formula (24.A.5), that a transparent film appreciably thinner than $\lambda/4$ reflects no light; in particular, *if less than a few hundred atoms thick, it cannot be detected under visible light.*

Antireflective Coatings

Zero reflected light requires the *amplitudes* reflected from both interfaces to be equal. This, as it turns out, requires comparable jumps ("discontinuities") in the index of refraction at both interfaces. In the preceding discussion, both discontinuities were of equal magnitude. [See Note (i) at the end of this chapter for further details.]

Figure 24.2 illustrates an alternative route to equal reflected amplitudes. A layer of thickness s and index n_1 coats a glass of larger index n_2; the layer's index has the *intermediate* value. As s varies,

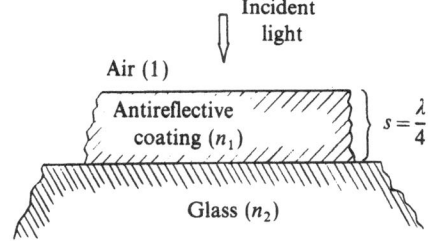

Figure 24.2. The incident light encounters two successive *increases* $1 \to n_1 \to n_2$ in the index of refraction. If the transparent coating has thickness $\lambda/4$, perpendicular reflection is suppressed. (Totally so, as can be shown, if $n_1 = \sqrt{n_2}$.)

we again get enhanced or suppressed reflection; formula (24.A.4) is still valid. However,

> An even k now goes with constructive,
> an odd k with destructive interference.
>
> (24.A.6)

(As a check, the zero-thickness limit, $k = 0$, now means full reflection.) For example, a quarter-wave (rather than half-wave) layer suppresses reflection. High-quality camera lenses are often coated in this manner and look purplish because $s = \lambda/4$ makes use of a λ chosen near the middle of the visible spectrum; red and violet are more reflected than yellow or green. The coating substance is typically magnesium fluoride (MgF_2).

Summary

Interference by reflection occurs under the following conditions.

1. The two beams come from the splitting of a single original beam by partial reflection.
2. In addition, the light is accurately monochromatic (implying regularly spaced wave fronts).
3. For maximum interference (constructive or destructive), the path difference between the two beams must amount to a whole number, k, of intervals $\lambda/2$.
4. Changing the path difference by $\lambda/2$ will interchange "constructive" with "destructive"; the case of zero path difference tells us whether even k goes with constructive or destructive.

ii. Interference Fringes

Figure 24.3 shows a transparent wedge of gradually varying thickness s, lighted from above and seen by the light it reflects. If s changes gently enough, we can apply the preceding arguments. Thus, regions where $s = k\lambda/4$ (k even) are nonreflective, and therefore seen as dark bands, or **dark fringes**. Where k is odd, we see a bright fringe. From one dark fringe to the next, k becomes $k + 2$; the thickness changes by an amount

$$\Delta s = \frac{(k+2)\lambda}{4} - \frac{k\lambda}{4} \quad \text{or} \quad \boxed{\Delta s = \frac{\lambda}{2}} \quad (24.A.7)$$

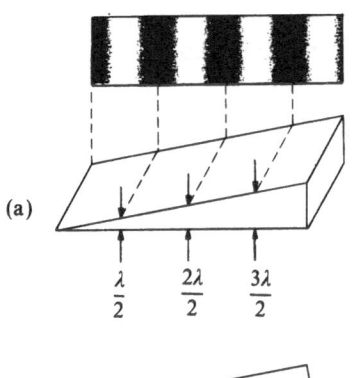

(a)

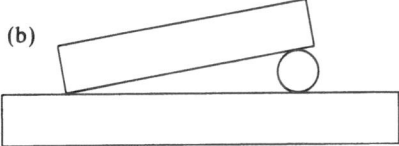

$$\frac{\lambda}{2} \qquad \frac{2\lambda}{2} \qquad \frac{3\lambda}{2}$$

(b)

Figure 24.3. (a) Fringes in a transparent wedge. (b) A thin fiber between two flat plates creates a transparent wedge of air, showing the same fringe pattern as in (a).

The dark fringes are contour lines whose intervals represent steps of $\lambda/2$ in thickness; along a fringe, thickness is constant; the less parallel the two surfaces, the more crowded the fringes.

The following example involves a fringe pattern first observed by Newton. (For some reason best understood by historians of science, Newton was not thereby converted to the wave theory of light.)

Example 24.2. Newton's Rings. A spherical lens surface, resting on a flat horizontal glass plate, gives rise to circular fringes when seen from above by reflected light; see Fig. 24.4. If the third dark fringe (not counting the dark spot at the center) has radius $r = 2.0$ mm, find the curvature $1/R$ of the lens' lower surface. Assume light of wavelength $\lambda = 590$ nm. Use the following geometrical result [the "sagittal theorem," see also Note (i) at the end of the preceding chapter]: a shallow cap of radius r, cut from a sphere of radius R, has a thickness

$$s = \frac{r^2}{2R} \qquad (24.A.8)$$

The thickness s at the third fringe is

$$s = (3)\left(\frac{\lambda}{2}\right) \qquad (24.A.9)$$

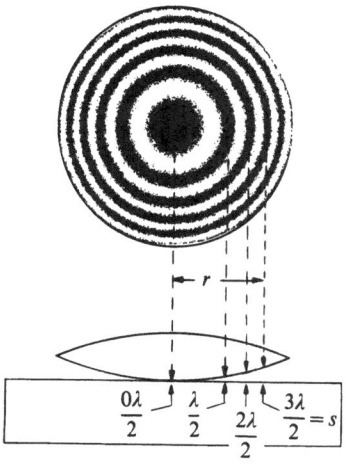

Figure 24.4. Newton's rings. Here we have a curved air wedge.

see Eq. (24.A.7). Solving (24.A.8), (24.A.9) for $1/R$, we have

$$\frac{1}{R} = \frac{3\lambda}{r^2}$$

$$= \frac{(3)(590 \times 10^{-9})}{(2.0 \times 10^{-3})^2} \text{ diopter} = \underline{\underline{0.44 \text{ diopter}}}$$

iii. Some Remarks on Interference

Viewing by Reflection

In order to see fringes in the film of Fig. 24.3, the eye should receive a reflected signal from more than one point on the film, i.e., from more than one direction. *An extended source*, such as a diffusely lighted ceiling, is necessary.

Reflection from transparent substances is weak. Hence a dark background should be provided under the film to prevent swamping of the signal by extraneous light.

Poor Viewing by Transmission

Since energy is conserved, a bright reflected fringe corresponds to a "dark" transmitted fringe; but since little energy is subtracted from the incident beam, the transmitted fringe has little contrast and is not easily visible. (A device called the Fabry–Pérot interferometer uses two partly silvered surfaces, which result in enhanced fringes as seen by transmission.)

Oblique Viewing of a Plate

Consider once more a uniform transparent plate of thickness s; see Fig. 24.5. Suppose a dark spot is seen at normal incidence, as explained before. The path difference between the two reflected normal rays is $2s$, but, any other angle of incidence θ, the path difference is no longer $2s$. As the viewing angle θ is increased, the path difference is alternately right for constructive and destructive interference. Each fringe corresponds to a given value of θ, and hence forms a ring. The general appearance is similar to that of Newton's rings.

Thin Films

Unless carefully prepared equipment is used, interference is seen ordinarily only from very thin films. There are several reasons for this:

1. Thick plates usually have surfaces whose parallelism is so low as to produce very crowded fringes, to such an extent that they cannot be told apart.
2. The obliqueness effect of Fig. 24.5 is much more pronounced for thick plates than for thin ones. Hence, again, an extreme crowding of the fringes, even for an ideally flat uniform plate.
3. Over great thicknesses (several centimeters), most real waves can no longer be represented by exact sine curves; we shall see why further on in Sec. C ("Noncoherent Light").

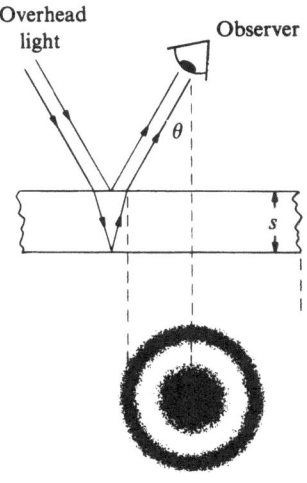

Figure 24.5. Circular fringes in a flat plate.

Color Effects

The location of a bright fringe depends on the wavelength of its light. Therefore, white light separates into fringes of different colors. This effect is familiar in soap bubbles, in gasoline films on water puddles, etc. The color separation is not perfect, however, because of overlap between fringes.

iv. The Michelson Interferometer

One important application of interference is in the accurate measurement of length; we have already seen that the counting of fringes may yield the thickness of a transparent film.

In 1881, Albert Michelson constructed his celebrated **interferometer**. This instrument is, figuratively speaking, a ruler whose markings are interference fringes. It measures distances in terms of the wavelength of light; conversely, it can measure a wavelength in terms of a known distance. One basic design, shown in Fig. 24.6a, involves a monochromatic light source collimated at A (details not shown) into a parallel beam, and three mirrors M_1, M_2, M_3, all of which are perpendicular to the plane of the figure; M_1 and M_2 are perpendicular to one another, while M_3, which is semireflecting (a lightly metallized, or "half-silvered," glass plate), is at 45° to M_1 and M_2.

The figure shows two possible paths for a ray from A. Part of the light intensity issuing from A is first transmitted through M_3, but is reflected by M_3 on the way back; its path is $AM_2M_3\mathbb{O}$ ($\mathbb{O}$ is the observer's eye). Another part, first reflected by M_3 and later transmitted, follows $AM_3M_1\mathbb{O}$. (Some light is reflected twice or transmitted twice by M_3, but it need not be considered here.)

Interference between both signals occurs at $\mathbb{O}$, just as in the thin-film arrangements discussed earlier. In the idealized case of Fig. 24.6, where M_3 is just a single semireflecting surface, the interference is constructive when both path lengths are equal, that is, when M_1 and M_2 are equidistant from M_3.

In practice, M_2 may be mounted on a micrometer screw and shifted forwards or backwards over precisely controlled distances; see part (b) of Fig. 24.6. When M_2 moves over a distance Δs, the signal's roundtrip distance to M_2 increases by $2\Delta s$. Just as in the thin-layer case, interference

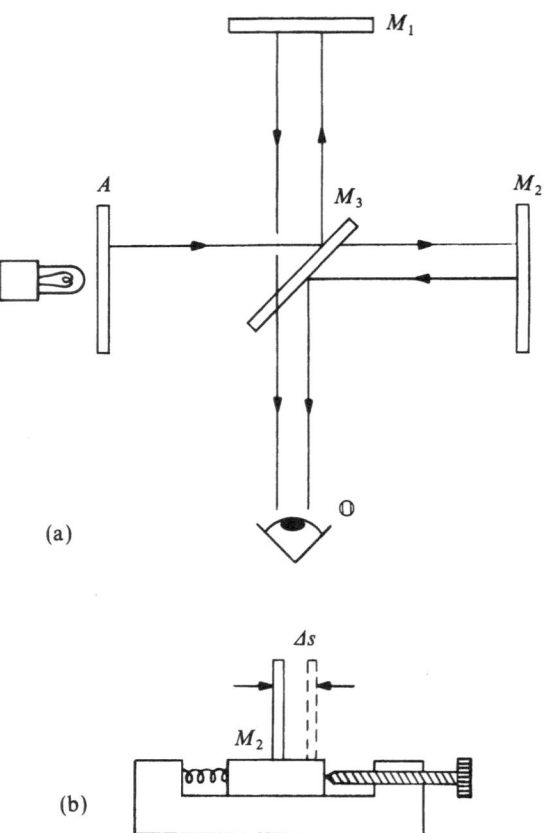

Figure 24.6. The Michelson interferometer. (a) Basic design; the instrument is of table-top size. Mirror M_3 is called the **beam splitter**. Collimating and focusing lenses at A and $\mathbb{O}$ are not shown. The source at A could be a laser. (b) Mirror M_2 may be shifted.

will turn from constructive to destructive, to constructive once more when

$$2\Delta s = \lambda$$

Thus, in this case, Δs is given by

$$\Delta s = \begin{pmatrix} \text{mirror displacement} \\ \text{between successive} \\ \text{bright signals} \end{pmatrix} = \frac{\lambda}{2} \quad (24.A.10)$$

In order to see this alternation of brightnesses while M_2 is being moved, $\mathbb{O}$ must look at M_1 in the normal direction, as Fig. 24.6 indicates.

Fringes in the Michelson Interferometer

In Fig. 24.6, observer $\mathbb{O}$ sees two images of the source at A; one image is a reflection from M_1, the other from M_2. These two images are superposed and interfere with one another; the interference pattern consists of *fringes*, as illustrated in Fig. 24.7a. To understand this, we first ask how $\bigcirc$ sees the mirrors M_1 and M_2. Mirror M_1 is seen

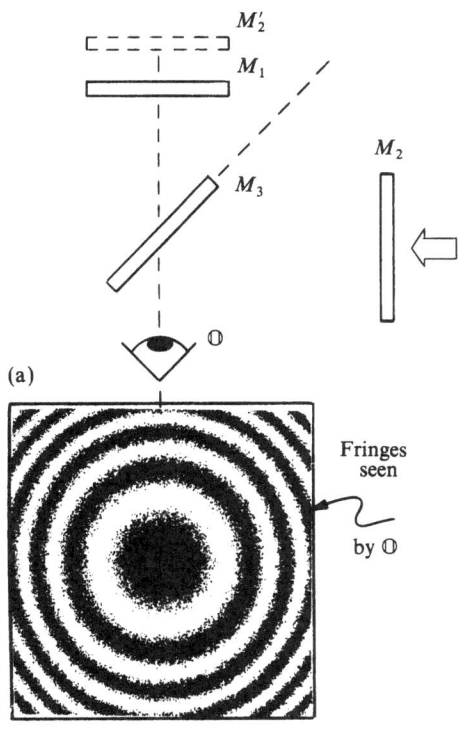

(a)

directly (through M_3). On the other hand, M_2 is seen as a reflected image in M_3; that image is shown as M_2'. Together, M_1 and M_2' act like the front and back of a transparent plate.

Suppose M_1 and M_2' coincide at first. Later, as they separate, $\bigcirc$ sees the circular fringes characteristic of a uniform plate [recall Fig. 24.5]; the fringes become tighter as the plate thickens; the central spot alternates between bright and dark; it corresponds to the normal incidence path of Fig. 24.6. In practice, M_2 is often purposely tilted very sightly; M_1 and M_2' then form a wedge, with fringes as shown in part (b) of Fig. 24.7. As M_2 moves, the wedge changes in thickness; the effect is that of the wedge—and fringe pattern—of Fig. 24.3, moving to the left or right.

To Summarize

As M_2 is moved by any amount Δs, a number N of bright fringes (or bright central spots) succeed one another before the observer; Δs is given by

$$\boxed{\Delta s = \frac{N\lambda}{2}} \qquad (24.A.11)$$

[We have multiplied (24.A.7) or (24.A.10) by N.]

Example 24.3. Motion of the Fringes. Under red light of wavelength $\lambda = 644$ nm in air, a slightly tilted Michelson interferometer mirror is advanced at a speed v. The fringe pattern is seen to shift laterally at a rate of 12 bright fringes per second. Find v.

In a time Δt, the mirror advances by

$$v = \frac{\Delta s}{\Delta t} = \frac{N\lambda}{2\Delta t} \qquad (24.A.12)$$

See (24.A.11). Setting $N/\Delta t = 12$ sec^{-1}, we have

$$v = \frac{(12)(644 \times 10^{-9})}{2} \frac{\text{meter}}{\text{sec}} = \underline{\underline{3.8 \times 10^{-6} \frac{\text{meter}}{\text{sec}}}}$$

Note that the angle of tilt affects the tightness of the fringes, but not their rate of passage $N/\Delta t$.

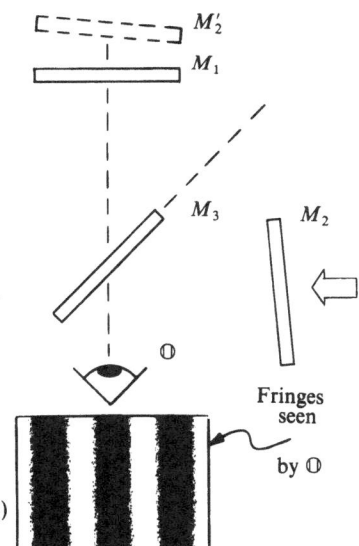

(b)

Figure 24.7. Fringes in the Michelson interferometer. (The arrow shows the direction of the incoming light.) (a) Perpendicular mirrors. (b) Tilted mirrors.

The Precise Measurement of Length

By shifting the micrometer carriage of M_2 over a chosen segment of a meterstick, and by counting fringes, one can express the meter as a number of wavelengths. A highly monochromatic and reproducible source of light is needed, of course. For many years the meter has, in fact, been defined officially by such a fringe count, until 1983, when a new definition, in terms of the speed of light, was substituted.

A Role in Relativity

In the next chapter we shall see how the Michelson interferometer provides experimental evidence for the theory of special relativity.

v. X Rays in Crystals

Interference by the atomic layers in crystals gives an elegant and precise way of *measuring how atoms are spaced*. This wonderful piece of information allows us, for example, to count the atoms in a crystal and hence to obtain Avogadro's number; recall Chap. 11, Sec. A. The technique, launched in 1912 by Max von Laue and by the father–son team of Sir William H. and Sir William L. Bragg, is called **x-ray diffraction**. We need not worry about that terminology; our point of view will just be based on thin-film interference.

A crystal can be conceived as a stack of identical layers, each of which is one molecule (or one atom) deep. Unfortunately, visible light, which has a wavelength much larger than the atomic spacing, is unsuited to interference experiments in this case; recall the comments following the soap film example 24.1. How small a wavelength is needed? Rough estimates of atomic diameters (and spacings) are in the vicinity of 10^{-10} meter.

Hence the appropriate wavelength is of order $\lambda \sim 10^{-10}$ meter. This puts the required radiation squarely in the x-ray range.

Consider now a plane wave of x-radiation (in practice, a thin **collimated beam** is adequate) incident on a crystal, as illustrated in Fig. 24.8. In part (b) of the figure, we idealize each atomic layer (called a lattice plane) as a smooth reflecting surface, perpendicular to the plane of the figure; the rays are in the plane of the figure.

Allowed Reflection Angles

We first ask: What is the angle of incidence, θ, that makes the reflections r_1, r_2 from the *top two* lattice planes interfere constructively?

In Fig. 24.8 we see that both rays suffer identical reflections; however, their paths differ by a length $\overline{AB} + \overline{BC}$. If the **lattice spacing** (that is to say, the distance between consecutive lattice planes) is s, we have

$$\overline{AB} + \overline{BC} = 2s \cos \theta \qquad (24.A.13)$$

where θ is the beam's angle of incidence. (*We ignore any refraction of the beam* because all earthly substances have an index of refraction $n \approx 1$ for x rays.)

Constructive interference requires a whole number k of wavelengths for the path difference:

$$\overline{AB} + \overline{BC} = k\lambda \qquad (24.A.14)$$

Comparison with (24.A.13) gives **Bragg's condition**,

$$\boxed{\cos \theta = \frac{k\lambda}{2s} \qquad (k = 1, 2, 3,...)} \qquad (24.A.15)$$

A peak intensity is reflected at each of these angles, the **Bragg angles**. (Note: k ranges only over a

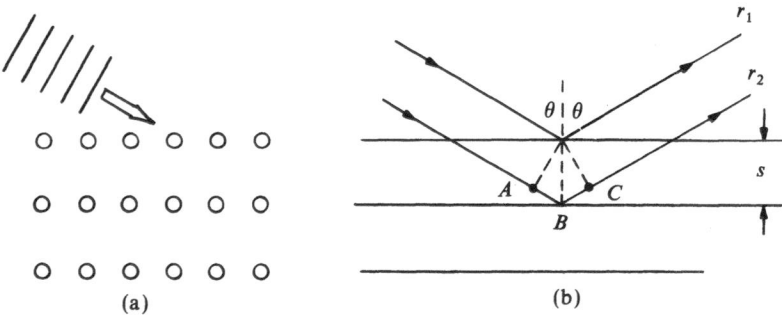

Figure 24.8. (a) Plane wave incident on a crystal. (b) Idealized model of (a).

limited set of values because $\cos \theta > 1$ is mathematically impossible.)

A Greater Intensity from Many Layers

In practice, the x ray beam penetrates the crystal, almost undiminished in intensity, through thousands of atomic layers; each layer supplies its own weak reflected wave, which joins the other reflected waves. When the Bragg condition (24.A.15) is met, all those waves—not just the first two reflections—interfere constructively. (Proof: The first is in phase with the second, the second with the third, etc.) Therefore *the whole crystal reflects maximally at angles that obey Bragg's condition* (24.A.15).

A Greater Sharpness from Many Layers

How much light is reflected if θ is very slightly off the exact values (24.A.15)? The answer is, *a negligible amount*. Thus, Bragg reflection occurs only at very sharply defined angles.

In this respect, crystals differ greatly from thin films, where the reflected intensity varies smoothly between constructive and destructive interference. In a crystal, *we have destructive interference at all angles except the Bragg angles*; the great number of layers is responsible for that fact. Note (ii) at the end of this chapter attempts to indicate why.

Photographing a Bragg Pattern

Figure 24.9 illustrates an arrangement for recording the Bragg angles on photographic emulsion. A monochromatic x-ray beam of wavelength λ is sent through a sample of a pure crystalline substance. This commonly used method involves, not a single

crystal, but a conglomerate, or powder, made of *many small crystals*. Those crystals which, by chance, happen to be at the proper angle to the incident beam will reflect it, as part (b) of the figure illustrates. The reflected ray strikes the photographic film at a distance

$$r = l \tan(180° - 2\theta) \qquad (24.A.16)$$

from the base A of the direct beam; θ is the relevant Bragg angle, and l is the distance between sample and film.

The many spots that come from the small crystals form a ring of radius r about A. For every allowed value of θ one observes one such ring, provided θ is not too large. In the geometry of Fig. 24.9, rays with $\theta < 45°$ miss the film entirely.

Figure 24.10 shows an actual Bragg pattern.

Many Kinds of Lattice Planes

Why does Fig. 24.10 show so many rings? One reason is the existence of various families of lattice planes in each crystal; Fig. 24.11 gives a side view of three such families in NaCl.

Example 24.4. Several Rings from One Set of Planes. Suppose an x-ray beam of wavelength $\lambda = 0.100$ nm is incident on a powdered sample in which one of the lattice spacings is $s = 0.210$ nm. (a) Find the relevant Bragg angles. (b) In the arrangement of Fig. 24.9, how many rings are photographed, corresponding to these angles?

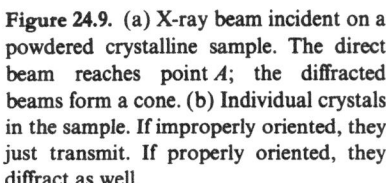

Figure 24.9. (a) X-ray beam incident on a powdered crystalline sample. The direct beam reaches point A; the diffracted beams form a cone. (b) Individual crystals in the sample. If improperly oriented, they just transmit. If properly oriented, they diffract as well.

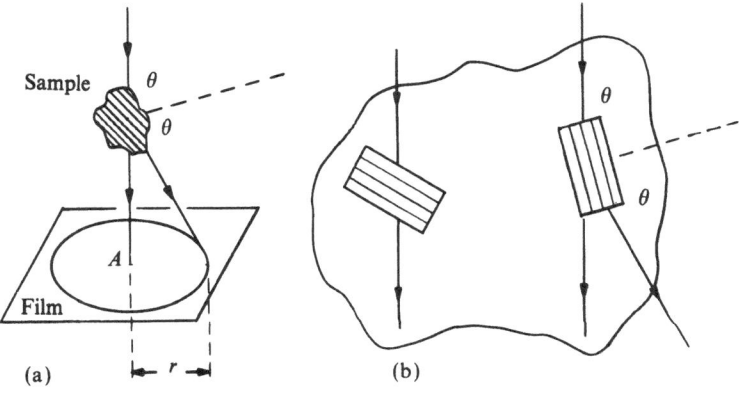

(a)   (b)

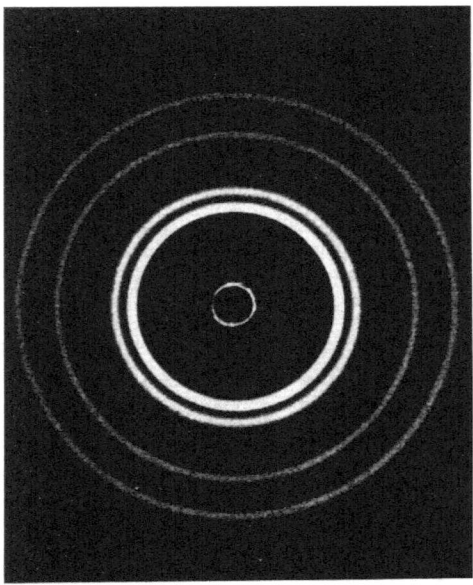

Figure 24.10. An actual Bragg pattern, obtained as in Fig. 24.9, for x rays passing through aluminum foil. The small circle at the center is the partially blocked direct beam. (Courtesy Education Development Center, Inc., Newton, MA.)

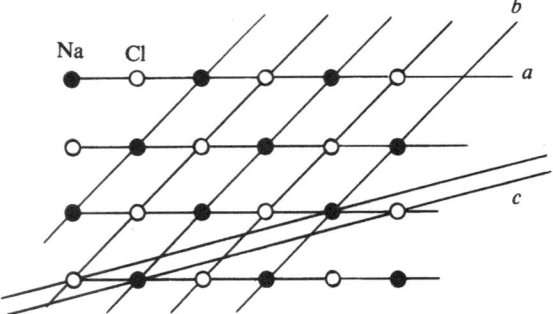

Figure 24.11. Some families, *a*, *b*, *c*, of lattice planes in rock salt, NaCl, a cubic crystal. Only a limited number of families give prominent diffraction rings; planes *c*, for example, do not. They contain very few atoms and are weak reflectors. See also Problem 24.14.

lengths, down to some minimum wavelength (maximum frequency). By reflection from a single crystal, through a selected Bragg angle, a single wavelength can be picked out of the tube's mixed output.

How is the Wavelength Measured?

The critical reader must have wondered how, in the laboratory, λ is known in the first place; that information is needed to measure a crystal lattice. The answer will be found in the next section.

Once a crystal lattice spacing has been measured, it, in turn, can serve as a standard for measuring x-ray wavelengths.

(a) Bragg angles: Bragg's condition, (24.A.15), is

$$\cos \theta = \frac{k\lambda}{2s}$$

$$= \frac{0.100}{(2)(0.210)} k = 0.238k$$

Since the right-hand side cannot exceed 1, the largest k is 4. For the allowed values of k we find

k	1	2	3	4
θ	76.2°	61.6°	44.4°	17.8°

(b) Number of rings: Not missing the film (which is assumed as large as needed) means $\theta > 45°$. Thus, only $\theta = 76.2°$ and $\theta = 61.6°$ qualify. We capture two rings.

How Are Monochromatic X Rays Obtained?

An x-ray tube does not produce monochromatic rays; rather, it produces a distribution of wave-

B. Diffraction from Slits

Light bends around corners—it is **diffracted**. Few people would have believed such a statement until, in 1801, Thomas Young, using a pair of parallel slits, clearly demonstrated the phenomenon; he thereby also proved that light consists of waves, as Christian Huygens had already postulated in 1678.

Here we shall study Young's two-slit experiment, as well as variations on that theme—not surprisingly, a single slit and a set of many slits, or **grating**.

i. Huygens' Construction

In order to follow the diffraction of a monochromatic plane wave around a corner, let us first restrict the width of the wave, as in Fig. 24.12, by means of an opening in an opaque screen. What shape of wave front emerges? A prescription due to Huygens correctly predicts it, as follows.

1. Subdivide the incoming wave front—the section framed by the opening—into many small surface elements. "Small" means small compared to the wavelength.
2. Consider each element as a **point-source** that sends out its own wave, indicated in the figure by a spherical wave front ("wavelet") expanding like a bubble in the forward direction.
3. At any time t after their creation, all the wavelets have the same radius, ct, where c is the speed of wave propagation.
4. Draw the new overall wave front, taken at time t, as a surface tangent to all the wavelets —in geometrical language, their **envelope**. That envelope is our desired result.

ii. Interference Between Wavelets

How intense is the diffracted light? Does the edge throw a shadow, as expected? Huygens' wavelets help us answer that question.

In order to achieve maximum simplicity, we restrict ourselves in several ways. We superpose at first only two wavelets, ignoring all the others, as in Fig. 24.13; we just examine the cases of constructive

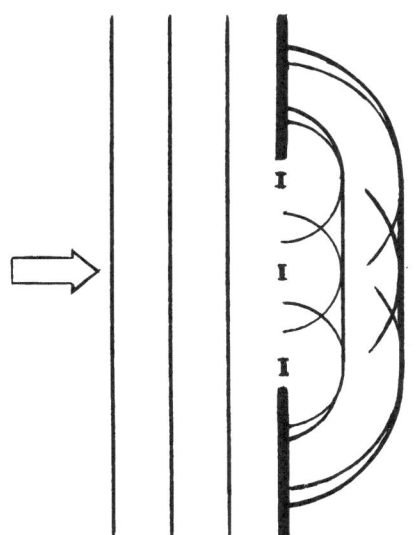

Figure 24.12. Huygens's construction. Three representative segments of the opening (small vertical bars) are shown giving rise to a set of spherical wavelets. Each emerging wave front (two are shown) is an envelope of the wavelets.

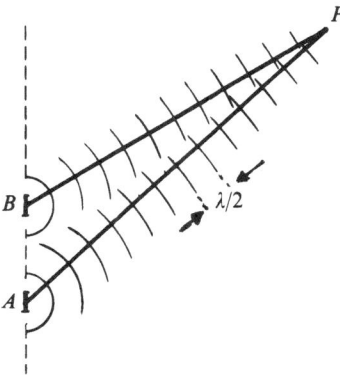

Figure 24.13. Constructive interference of two wavelets. A plane wave is assumed to have reached A and B from the left with equal amplitudes at A and B.

or destructive interference, but nothing intermediate; finally, we ask for the emerging intensity at a point P that is far from the opening (that is, far in terms of the wavelength and in terms of the opening size). The following rules then result:

1. Equal surface elements A, B of the incoming wave front contribute equal amplitudes at point P.
2. Therefore, if P is simultaneously located on a wavelet from A and on one from B, the two contributions interfere constructively (when in phase) or destructively (when 180° out of phase).
3. Quantitatively stated: Consider distances $\overline{AP}$, $\overline{BP}$, differing by a whole number k of half wavelengths $\lambda/2$,

$$|\overline{AP} - \overline{BP}| = k\frac{\lambda}{2} \qquad (24.B.1)$$

Interference is then

$$\left.\begin{array}{l} \text{Constructive} \\ \text{Destructive} \end{array}\right\} \quad \text{for} \quad k \left\{\begin{array}{l} \text{even} \\ \text{odd} \end{array}\right. \qquad (24.B.2)$$

iii. Young's Two-Slit Experiment

Figure 24.14a shows a modern version of Young's experiment; the incoming monochromatic plane wave originates from a laser; it falls on a pair of equal, parallel, closely-spaced slits. At a distance l beyond these, on a screen, one finds *a set of many parallel bright fringes*; the central fringe, located equally far from both slits, is the brightest. Note the

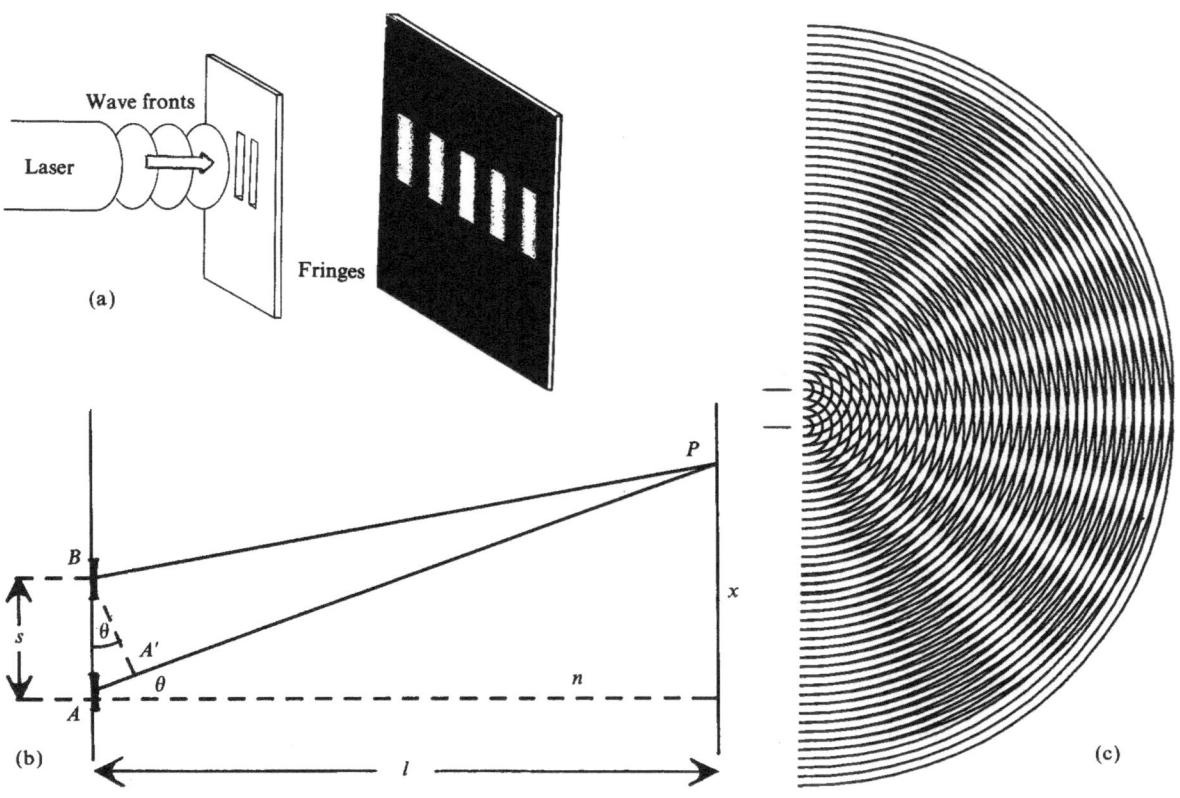

Figure 24.14. Young's two-slit experiment (not drawn to scale). The slits are perhaps a fraction of a millimeter apart. (a) Experimental setup; see Fig. 24.15, further on, for an intensity profile of the fringes. (b) Geometry for interference at P. In practice, s is measured *between the centers* of the slits. (c) By following the intersections of the wavelets, we can visualize the zones of constructive or destructive interference. The slits' location is indicated by two dashes.

breakdown of naive ray optics, which incorrectly predicts two bright lines on the screen—one for each slit.

Given the wavelength λ, the distance l, and the spacing s between the slits, where are the fringes? We can find out from conditions (24.B.1) and (24.B.2). (Each slit is a line rather than a point source, and hence emits a cylindrical—rather than spherical—wavelet; but we are still *approximately* correct if we use Huygens' prescription.)

In part (b) of Fig. 24.14 we note the relevant path difference to a point P on the screen:

$$|\overline{AP} - \overline{BP}| = \overline{AA'} = k\frac{\lambda}{2} \qquad (24.B.3)$$

Since P is far away, angle $AA'B$ is essentially a right angle, and (24.B.3) reads

$$s \sin \theta = k\frac{\lambda}{2} \qquad (24.B.4)$$

where θ is the **diffraction angle**, that is, the angle between the forward beam direction n (normal to the plane of the slits) and the direction from the slits towards P.

Thus, from Eq. (24.B.4), a *maximum intensity* (k even) results from diffraction at angles θ such that

$$\sin \theta = 0, \pm \frac{\lambda}{s}, \pm 2\frac{\lambda}{s}, \pm 3\frac{\lambda}{s}, \text{etc.} \qquad (24.B.5)$$

These diffraction angles are said to be of **zeroth, first, second,... order.**

On the other hand, a *vanishing wave intensity* is found at

$$\sin \theta = \pm \frac{\lambda}{2s}, \pm \frac{3\lambda}{2s}, \pm \frac{5\lambda}{2s}, \text{etc.} \qquad (24.B.6)$$

In terms of the transverse distance x on the screen (Fig. 24.14b), we have

$$x = l \tan \theta \quad \text{or} \quad \boxed{x \approx l \sin \theta} \quad (24.B.7)$$

the last step because, in practice, the relevant values of θ are small; the fringes fade out rapidly with θ. The value of $\sin \theta$ in (24.B.7) is taken from (24.B.5) or (24.B.6) according to whether a bright or dark location is involved.

Part (c) of Fig. 24.14 shows an overview of the cylindrical wavelets from both slits.

Example 24.5. Closeness of the Slits. A 1-mm spacing (between Young's bright fringes) is desired on a screen at 1 meter from a pair of slits whose spacing is s. If light of wavelength $\lambda = 600$ nm is available, estimate the needed s.

From Eqs. (24.B.7) and (24.B.5), we have a bright fringe on the screen at each of the positions

$$x = 0, \qquad \pm \frac{l\lambda}{s}, \qquad \text{etc.}$$

The spacing is

$$\Delta x = \frac{l\lambda}{s} \qquad (24.B.8)$$

so that

$$s = \frac{l\lambda}{\Delta x} \qquad (24.B.9)$$

$$= (1) \frac{(600 \times 10^{-9})}{1 \times 10^{-3}} \text{ meter}$$

$$= 6 \times 10^{-4} \text{ meter} = \underline{0.6 \text{ mm}}$$

This implies a rather delicate experiment, and helps us understand why diffraction fringes are not readily seen in daily life. A *wider spacing Δx of the fringes implies an even closer spacing s of the slits.*

Fraunhofer versus Fresnel Diffraction*

Young's fringe pattern, illuminating a distant screen, bears little resemblance to the slits them-selves. By contrast, on a screen close to the slits, and with wider and further-separated slits, we would see an ordinary projection, namely, two bright lines with the slits' own spacing. Thus, there is competition between two effects: diffraction and geometrical projection. The following nomenclature is commonly used:

1. **Fraunhofer diffraction.** This is a pattern observed far enough away from the slits (or other obstacle) so that resemblance to the geometrical projection is lost; here we have "pure" diffraction. Young's fringes is one illustration of this; *the present chapter concentrates on Fraunhofer diffraction.* (Through an appropriate use of lenses or mirrors, a Fraunhofer pattern may be brought to any convenient distance from the diffracting obstacle.)

2. **Fresnel diffraction.** This exhibits complicated diffraction fringes, often combined with a recognizable geometrical shadow—a compromise between the two competing effects. It is observed at small distances, and is more difficult to analyze than the Fraunhofer case. When squinting in a strong light, we sometimes see Fresnel fringes around the shadow cast on our retina by small impurities floating in our eye's internal fluid.

iv. The Diffraction Grating

Let us replace Young's two-slit system by a set of many parallel, equidistant slits of equal width as illustrated in Fig. 24.15. Such a device is known as a **grating**. A plane wave of wavelength λ, incident normally to the plane of the grating, is diffracted on the other side into a set of sharply defined directions given by the bright-fringe formula (24.B.5). Thus, the result of increasing the number of slits is to sharpen Young's fringes, without affecting their location. The fringes' brightness is greatly improved as well, and they become easily observable.

What we have learned about thin-film interference helps us understand what happens here. *A grating is related to a double slit somewhat as a crystal* (in x-ray diffraction) *is related to a thin film.*

The grating's main features are explained as follows.

Same Fringe Location as for the Double Slit

Let the slits at the right of Fig. 24.15 be numbered 1, 2, 3, etc., starting from the bottom.

---

* Joseph von Fraunhofer (1787–1826): Prolific researcher in optics; discovered dark lines in the solar spectrum; invented the diffraction grating. Augustin Fresnel (1788–1827): Champion and theoretician of the wave hypothesis; used experiments in diffraction and polarization to prove his point.

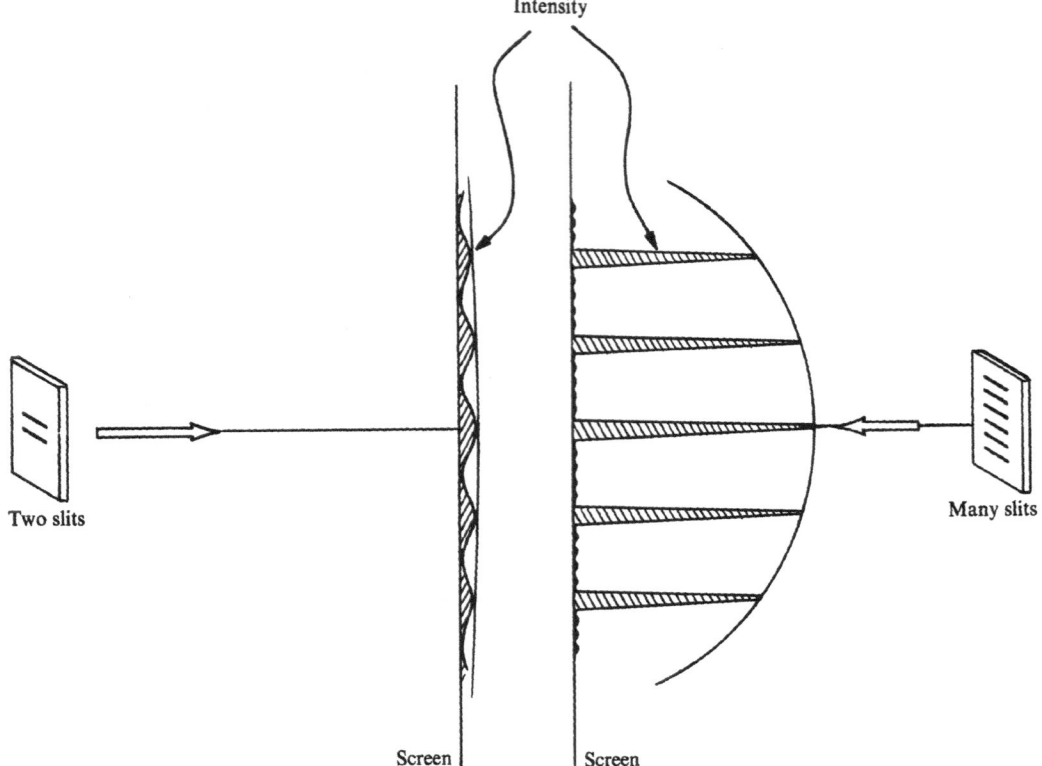

Figure 24.15. Comparison between a grating and a pair of slits, both with the same spacing, and equally distant from a projection screen; the incident wavelength is the same in both cases. The grating enhances intensity and sharpness. Actual gratings may have as many as 20 000 lines (slits) per centimeter.

Consider those directions in which the wavelets from slits 1 and 2 are in phase; they give Young's fringes, the "double slit" being the set 1, 2. Now the wavelets from 2 and 3 are automatically also in phase, and so are those from 3 and 4, etc.; we have overall constructive interference.

Greater Brightness

The total transmitted energy is proportional to the number of slits (the slit width is assumed given).

Greater Sharpness

Suppose there are N slits. The peak amplitude (at the center of a bright fringe) is proportional to N because contributions from all slits are superposed and in phase with one another. Therefore the *intensity* at that point is proportional to N^2. On the other hand, the total energy shared by all the fringes is proportional to the energy admitted by the N slits, that is to say, the total energy is proportional to N. We conclude that the *fraction* of the total energy that is concentrated into the center of each fringe goes as $N^2/N = N$, and so rises with N; for large N, all the energy must be bunched into extremely sharp fringes. (This interesting argument, peculiar to wave theory, has made use of the distinction between amplitude and intensity.)

How Are Gratings Made?

Carefully engineered "ruling engines" are able to engrave, with a diamond point, many thousands of rigorously parallel and equidistant lines on a single centimeter of a silvered glass surface. An inexpensive replica of such a grating can be made from a plastic coating applied to the ruled surface and then peeled off. Thus, a grating consists of grooves rather than slits; but the diffractive effect turns out to be the same.

The Grating Spectrometer

Figure 24.16 illustrates one arrangement for recording diffraction angles photographically. Part (a) of the figure shows the **zeroth-order** fringe S_0 (a real, undiffracted image of the illuminated slit S). Part (b) shows the formation of S_1, the **first-order** diffracted image of S. The diffraction process takes place at the grating, between the two lenses, in the manner discussed above. The left lens (collimating lens) serves to produce a plane wave; the right lens (imaging lens) focuses the parallel diffracted rays; the photographic film constitutes its focal plane.

The diffraction angle θ_1 is found geometrically from

$$\tan \theta_1 = \frac{x_1}{l} \qquad (24.B.10)$$

where the distances x and l are shown on the figure; x is measured on the film, while l is the focal length of the (thin) lens. (Verify on the figure that θ_1 indeed equals the diffraction angle for the parallel rays emerging from the grating.)

Monochromatic light reveals its wavelength λ through

$$\lambda = s \sin \theta_1 \qquad (24.B.11)$$

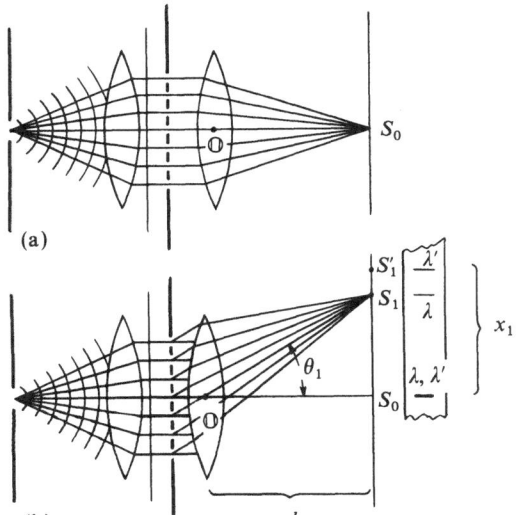

(a)

(b)

Figure 24.16. A grating spectrometer. Diagrams (a) and (b) must be thought of as superposed. The lines shown at the right are images of the slit S. If the light from S is a mixture, say of wavelengths λ, λ', then the corresponding lines might be S_0, S_1 (for λ) and S_0, S_1' (for λ').

see Eq. (24.B.5) in the first-order case, $\theta = \theta_1$; θ_1 is known from (24.B.10).

[The reader who is worried about whether the wavelets from all slits arrive in phase (i.e., combine constructively) at S_0, S_1, etc., in spite of the lens, is referred to Note (i) at the end of the preceding chapter, which shows that the wavelets would be retarded by their travel through the glass in just such a way as to arrive simultaneously at the image point.]

The Wavelength of X Rays

X rays are difficult to diffract through an artificial grating because of their high penetrating power and small wavelength. However, in 1922, Arthur H. Compton discovered that x rays are totally reflected from smooth surfaces at angles of incidence close to 90° (grazing incidence). By using a finely ruled **reflection grating** (Fig. 24.17) he was able to deduce the value of λ for x rays. The grating therefore became a standard of length, yielding first λ, then (from λ) the lattice spacings of many crystals. Thereafter it became more convenient to use these crystals as standards in their own right; several related fundamental constants were obtained in terms of lattice spacings, notably Avogadro's number and the charge of the electron. Available space does not permit us to examine reflection gratings in detail; their principle is almost identical to that of the transmission gratings we have studied.

Conclusion

The diffraction grating is a basic laboratory tool, because it accurately yields the wavelength of light in terms of directly measured distances. Notice from (24.B.11) that the longer the wavelength, the larger the diffraction angle. This is the reverse of what occurs with a prism; compare with Fig. 23.10 of the preceding chapter.

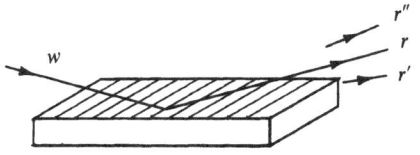

Figure 24.17. A reflection grating: r is the reflected ray as in geometrical optics; r' and r'' are first-order diffracted.

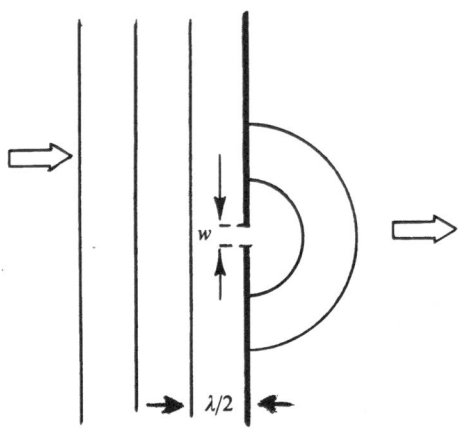

Figure 24.18. Diffraction by a single narrow slit.

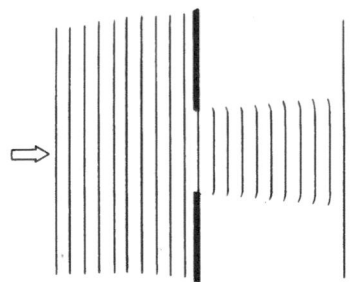

Figure 24.19. Near-absence of diffraction from a wide slit. (Figure not drawn to scale.) A faint structure of narrow fringes (not shown) exists near the shadow's edge.

Example 24.6. Measuring a Wavelength. Suppose lines S_0 and S_1 in Fig. 24.16 are $x_1 = 28.5$ cm apart, while the photographic film is at $l = 50.0$ cm from the center of the imaging lens. If the grating has precisely 10 000 lines per cm, find the wavelength λ corresponding to S_1.

From (24.B.10) we have

$$\theta_1 = \tan^{-1}\frac{x_1}{l} = \tan^{-1}\frac{28.5}{50.0} = 29.7°$$

so that, from (24.B.11), we have

$$\lambda = s \sin\theta_1 = \frac{0.01\ \text{meter}}{10\ 000}\sin 29.7°$$

$$= 4.95 \times 10^{-7}\ \text{meter} = \underline{495\ \text{nm}}$$

v. The Single Slit

Consider a monochromatic plane wave incident normally on a single slit. If the latter is narrow relative to the wavelength, as in Fig. 24.18, it radiates a cylindrical wavelet. All directions parallel to the plane of the figure and to the right of the slit are illuminated. (The forward direction is slightly favored, but the decrease of intensity towards the sides is gradual and not very important.)

At the other extreme, our daily experience suggests that a slit which is many wavelengths wide projects itself sharply as a collimated light beam having the slit's width; see Fig. 24.19.

The intermediate situation is our main topic here; the slit is typically several wavelengths wide. The resulting illumination on a *distant* screen is shown in Fig. 24.20. We observe a central bright fringe that subtends a total angle $\Delta\theta$ (as seen from the slit) such that

$$\boxed{\sin\frac{\Delta\theta}{2} = \frac{\lambda}{w}} \qquad (24.B.12)$$

where w is the slit width. We note that *the fringe becomes wider as the slit becomes narrower.* Most of the energy transmitted by the slit turns up in that

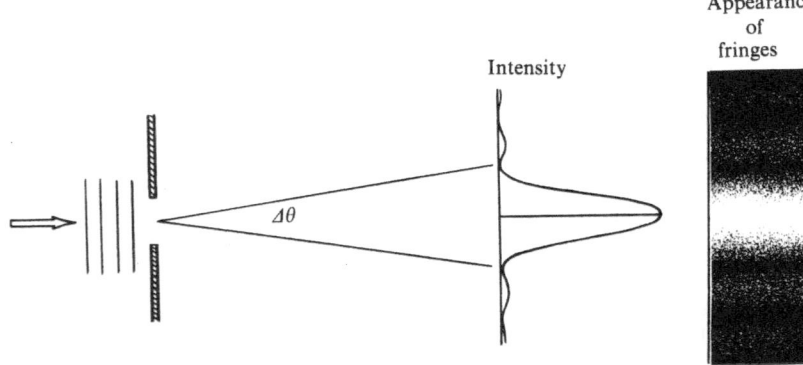

Figure 24.20. Diffraction from a single slit (general case; figure not drawn to scale.)

central fringe. On either side, we find a succession of much weaker fringes, each only about half as wide as the central one.

Predicting the Fringe Widths

The bright fringes are, of course, separated by dark fringes. It is easier to calculate the positions of the latter. A construction illustrated in Fig. 24.21 yields an angle θ in whose direction the net radiated amplitude is zero.

We divide the slit into a large number of equal small zones 1, 2, etc., each of which radiates a cylindrical wavelet. Next, we single out two such zones, 1 and 1', whose wavelets experience a path difference $\lambda/2$ (heavy segment) on their way to P; they therefore interfere destructively at that point.

As in the double-slit case, we write

$$\overline{AA'} = \overline{AB} \sin \theta = \frac{\lambda}{2} \qquad (24.B.13)$$

All subsequent pairs, 22', 33', etc., also must have canceling contributions. The two equal segments $\overline{AB} = \overline{BC}$ consist entirely of canceling pairs; there-

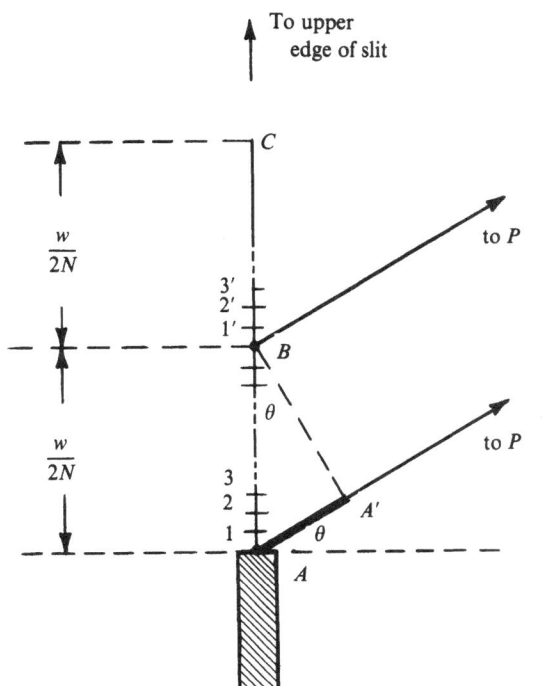

Figure 24.21. Geometry for obtaining the directions of zero intensity.

fore the overall segment AC sends zero intensity in direction θ.

Suppose now that $\overline{AC}$ is contained an integral number of times, N, in the total slit width. Then the cancellation repeats itself N times, with the conclusion that the slit *as a whole* does not radiate in direction θ. Since we have $\overline{AB} = w/2N$ (w = slit width), Eq. (24.B.13) reads

$$\frac{w}{2N} \sin \theta = \frac{\lambda}{2} \qquad (24.B.14)$$

$$\sin \theta = \frac{N\lambda}{w} \qquad (N = 1, 2, 3,...) \quad (24.B.15)$$

In summary, the dark fringes are in directions given by

$$\sin \theta = \pm \frac{\lambda}{w}, \ \pm 2 \frac{\lambda}{w}, \ \pm 3 \frac{\lambda}{w}, ... \qquad (24.B.16)$$

where the extra $\pm$ is inserted to account for the symmetrically flipped directions where θ is negative. From (24.B.16) we see, in particular, that the central bright fringe is enclosed between $\sin \theta = \pm \lambda/w$, as announced in (24.B.12).

Note the absence of $N = 0$ in (24.B.15). In the forward direction ($\theta = 0$), all wavelets are in phase, and the intensity is maximum.

[Has our construction missed any dark fringes? No. A more thorough argument, which we must skip, establishes that (24.B.16) is a complete list.]

The following negative example serves to emphasize that diffraction is inconspicuous in daily life.

Example 24.7. A Very Wide Slit. An imperfectly closed door leaves a slit of width $w = 1$ cm, which throws a line of sunshine on a wall at a distance $l = 5$ meters. The line appears quite blurred. Is this due to diffraction?

A typical wavelength is $\lambda = 500$ nm. Diffraction produces a spread $\Delta\theta$ given by

$$\Delta\theta \approx \frac{2\lambda}{w} \qquad (24.B.17)$$

[small-angle approximation to (24.B.12)]. The bright line on the wall would have a corresponding width,

$$\Delta x = l\,\Delta\theta = \frac{2l\lambda}{w} \qquad (24.B.18)$$

$$= \frac{(2)(5)(500 \times 10^{-9})}{10^{-2}} \text{ meter} = \underline{\underline{0.5 \text{ mm}}}$$

a minuscule effect. [In actual fact, a considerable blurring comes from the Sun's nonzero size. An angular diameter $(\Delta\theta)_{Sun} \approx 0.54° \approx 0.010$ rad for the Sun implies $\Delta x \approx l(\Delta\theta)_{Sun} = (5)(0.010)$ meter $= 5$ cm, presumably the observed effect.]

C. Noncoherent Light

Suppose we replace the illuminated slits in Young's experiment by two straight, parallel, glowing lamp filaments, with a monochromatic filter if necessary. This naive procedure would result in *no fringes at all being observed*. Why? We shall explain in what follows that the two filaments are a **noncoherent** pair of sources, while the two slits are **mutually coherent** sources. The outcome of a wave experiment depends in an essential way on the coherence of the source.

i. A Single Source

Figure 24.22a shows a plane monochromatic wave. The **E** field, viewed at a single time, is plotted along the direction of travel.

Lasers and radio transmitters can generate an excellent approximation to such a perfect sine wave; most other electromagnetic sources—natural or artificial—cannot.

Part (b) of the figure is a simplified model of an *approximately* monochromatic wave, such as may be obtained from a gas-discharge tube. (Outside the laboratory, examples of such tubes are neon advertising signs, sodium or mercury street lamps, etc.) In such cases the radiation comes in **wave trains**, or **wave packets**, two of which are shown in the figure; they are produced as separate bursts of radiation by individual atoms. *The phases of two packets are, in general, shifted relative to one another*; when an atom radiates a new wave packet,

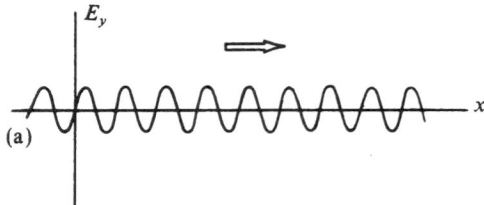

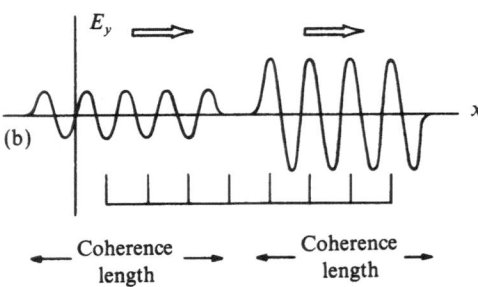

Figure 24.22. (a) Phase-coherent wave. (b) Partly noncoherent wave. The equidistant reference marks help exhibit the phase shift between the two wave packets.

it does not "remember" what it—or any of its neighbors—has radiated before.

The wave in (a) is said to exhibit **phase coherence**, while in (b) it has some degree of **phase noncoherence**. The **coherence length** of a wave is the distance interval over which it is indistinguishable from an exact sine wave. A good laboratory gas-discharge lamp gives a coherence length of the order of at most a few centimeters, but laser light may have a coherence length measured in kilometers.

"Thick" Film Interference Requires Good Coherence

Figure 24.23 demonstrates why no interference can be observed from a film *whose thickness appreciably exceeds half the coherence length*. In the figure's scenario, the wave packets arrive with long intervals between them. We therefore only need to consider a single packet; two packets seldom overlap. This assumption is realistic for weak light sources, and is made here for simplicity.

A simulated thick film can be seen in the Michelson interferometer, Fig. 24.7; the "film" surfaces are M_1 and M_2'. Therefore, the coherence length of the light sets a limit to the difference between the instruments' arm lengths. The fringes

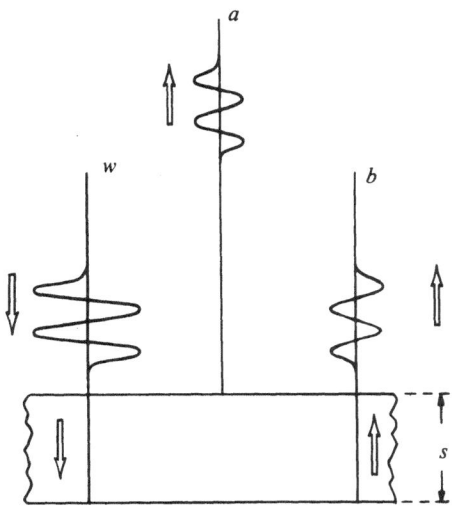

Figure 24.23. The incoming wave packet w gives rise to waves a and b, which cannot interfere because their path difference makes them nonoverlapping. Wave packet a is a distance $2s$ ahead of wave packet b. Hence there can be no interference unless the coherence length equals at least $2s$.

and B cannot synchronize themselves, their wave packets reach P at unrelated times, and do not interfere. Both sources are said to be **mutually noncoherent**. Taken together, they form a single extended source, which is called **spatially noncoherent**.

A laser amounts to a **spatially coherent** source (as well as being phase coherent); this is demonstrated, as in Fig. 24.14, by the fact that interference fringes are observed.

iii. Intensities from Noncoherent Sources Must Be Added

Figure 24.24 shows that the average energy reaching P over a long period of time is simply the sum of the average energies sent by both sources; in short, the intensity $\mathscr{I}$ observed at P equals the sum of the intensities $\mathscr{I}_A$, $\mathscr{I}_B$ which would be observed from either source alone:

are detected most clearly when the arms are about equal.

(Mutually noncoherent wave packets are not always widely separated in time; but even if they overlap, they give rise to no observable interference. The detailed proof, which we shall not give here, is based on the random phase shift which exists between the wave packets, as in Fig. 24.22b. Interference then fluctuates rapidly between constructive and destructive, and all fringes are "washed out.")

$$\boxed{\mathscr{I} = \mathscr{I}_A + \mathscr{I}_B} \qquad \text{(mutually noncoherent sources)}$$

$$(24.\text{C}.1)$$

In contrast, mutually coherent sources, say with $\mathscr{I}_A = \mathscr{I}_B$, give an intensity at P ranging from zero (destructive interference) to $4\mathscr{I}_A$ ($2 \times$ amplitude implies $4 \times$ intensity).

Similarly, the intensity reflected from a *thick* film equals the sum of both surfaces' noncoherent contributions; (24.C.1) is valid here also.

Equation (24.C.1) fits our daily experience with mutually noncoherent sources. Two reading lamps produce, on a book's page, the sum of both lamps' intensities.

ii. Two Sources

Figure 24.24 returns to our introductory question. Can two lamp filaments A and B give rise to interference at P? Since the radiating atoms in A

iv. Two Different Frequencies Do Not Interfere

Consider two superposed vibrations (mechanical or electromagnetic):

$$x = A \cos \omega_A t + B \cos(\omega_B t + \phi) \quad (24.\text{C}.2)$$

If $\omega_A = \omega_B$, interference occurs, the resulting intensity depending on the phase difference ϕ. If $\omega_A \neq \omega_B$, the intensity (which is a time average) is just *the sum of the two intensities* corresponding to A and B; thus, Eq. (24.C.1) is valid for a superposi-

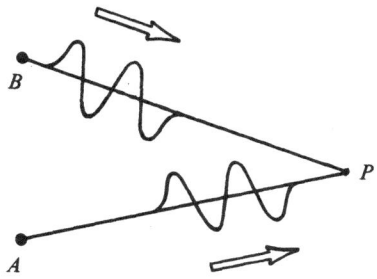

Figure 24.24. Wave packets from mutually noncoherent sources A, B do not interfere.

tion of different colors; there is no interference. Verification is left to Problem 24.43.

[In Chap. 13 (Vibrations), we learned that nearly equal ω_A and ω_B result in amplitude beats. Ordinarily, in the case of light, the beats are too fast to be observed and are averaged out by the eye or other detector.]

D. Limits of Resolution

Our ability to see fine detail is unfortunately limited by the wave nature of light. This section attempts to indicate why and to what extent this is the case.

i. The Pinhole Camera

Let us go back to Fig. 24.18, which shows light being diffracted through a single slit. Suppose, instead, that the figure illustrates a circular opening (or **circular aperture**) of diameter D, instead of the width w. The resulting diffraction pattern, Fig. 24.25, will then itself be circular. The central bright spot, framed by the first dark fringe, is the **Airy disk**, after Sir George Airy, who first calculated its diameter is 1835. As viewed from the opening, that diameter subtends an angle $\Delta\theta$ which, as usually happens in the laboratory, we assume to be quite small compared to 1 radian. It is then given by

$$\boxed{\Delta\theta \approx 2.44\,\frac{\lambda}{D}} \qquad \text{(from a circular aperture)}$$

$$(24.\text{D}.1)$$

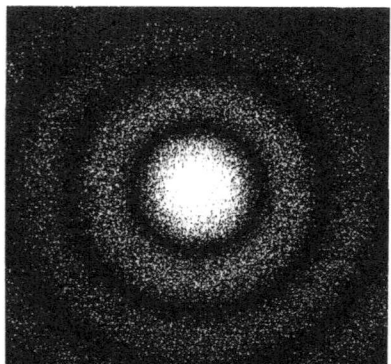

Figure 24.25. Diffraction fringes from a circular opening.

This formula, in which λ is assumed small compared to D $(\lambda \ll D)$ is analogous to

$$\Delta\theta \approx 2\,\frac{\lambda}{w} \qquad \text{(from a slit)}$$

recall (24.B.17); the two formulas describe the same basic phenomenon in different geometries. The factor 2.44 comes out of Airy's rather difficult calculation, which we shall not reproduce here. The bright fringes that surround the central spot are weak; most of the energy is emitted inside the cone given by (24.D.1).

As an application, we ask for the diameter Δx of the Airy disk in a pinhole camera, Fig. 24.26, when the object is a distant point; the pinhole has diameter D and is at a distance l in front of the film; the camera is aimed at the object.

In monochromatic light of wavelength λ, we have, from (24.D.1),

$$\Delta x \approx l\,\Delta\theta \approx 2.44\,\frac{l\lambda}{D} \qquad (24.\text{D}.2)$$

The remarkable conclusion is that *making the pinhole smaller makes the image "point" more blurred* (Δx increases with decreasing D); in other words, *a single ray cannot be isolated.* As before, formula (24.D.2) is valid for

$$D \gg \lambda \qquad (24.\text{D}.3)$$

Neglecting the Geometric Projection of the Pinhole

Geometrically, the pinhole is expected to project a spot of its own size. We neglect this fact by assuming the size Δx of the Airy disk to be large compared to that of the pinhole:

$$\Delta x \gg D \qquad (24.\text{D}.4)$$

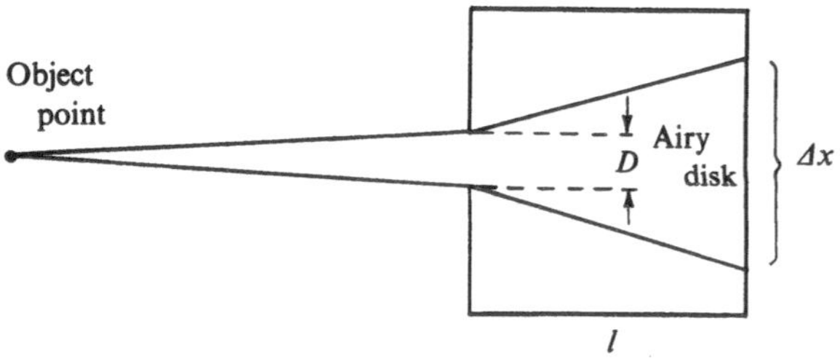

Figure 24.26. Making a pinhole smaller widens the Airy disk.

or, from (24.D.2),

$$2.44 \frac{l\lambda}{D} \gg D \qquad (24.D.5)$$

or

$$\boxed{l \gg \frac{D^2}{\lambda}} \qquad (24.D.6)$$

where we have left out the 2.44 since we are only dealing with orders of magnitude.

If (24.D.6) is valid, then *diffraction is entirely responsible* for blurring the picture. This is Fraunhofer diffraction, as explained at the end of Sec. B iii.

Example 24.8. Optimum Size of a Pinhole. Suppose that, in Fig. 24.26, we have $\lambda = 500$ nm (representative of visible light) and $l = 10$ cm. Paying no attention to (24.D.6), how large should we take D in order to obtain the sharpest possible picture?

There are two opposing factors. A very small D blurs the picture, but so does a very large D because of its geometrical projection; for a

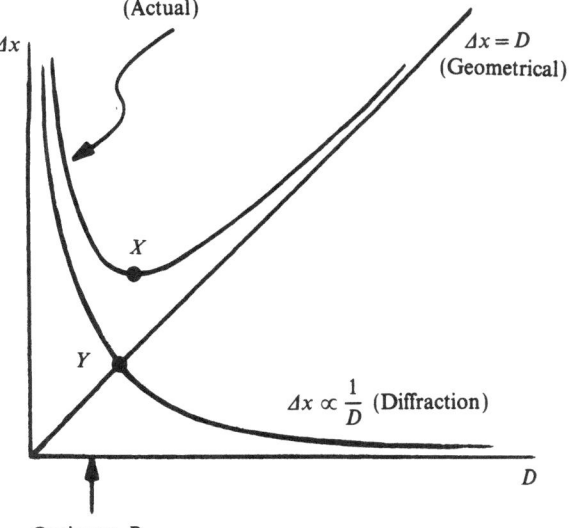

Figure 24.27. The width Δx of the image, plotted against the pinhole diameter D. The falling curve is valid for small D, the rising straight line for large D. The value of D at their intersection Y gives an estimate for the optimum D. (Minimum Δx occurs at X.)

qualitative graph, see Fig. 24.27. The best we can do is to decrease D, thus reducing the geometrical spot, until the diffraction effect (24.D.2) begins to take over. Roughly, the geometrical blurring should equal the diffractive blurring,

$$D \approx \Delta x \approx 2.44 \frac{l\lambda}{D} \qquad (24.D.7)$$

Solving for D, we obtain

$$D \approx (2.44 l\lambda)^{1/2} \approx (l\lambda)^{1/2} \qquad (24.D.8)$$

leaving out the factor 2.44 because our imprecise argument can only result in an order of magnitude estimate. Numerically, we have

$$D \approx [(0.10)(500 \times 10^{-9})]^{1/2} \text{ meter}$$

$$\approx \underline{2 \times 10^{-4} \text{ meter}}$$

about the diameter of an actual pin. The amount of light reaching the film may of course be miserably inadequate.

Angular Resolution of a Pinhole Camera

How close together can two object points be before their images fuse together, as in Fig. 24.28? This somewhat vague question becomes a definite one with the help of a few assumptions:

1. Only diffraction (not geometrical shadowing) is responsible for blurring the image.

2. When the images are just barely distinct, they consist of two Airy disks that overlap by one radius, as indicated in Fig. 24.28. (This criterion was proposed by Lord Rayleigh in 1896.)

3. We are asking for the angular separation α shown in the figure. The answer is

$$\alpha \approx \frac{\Delta\theta}{2} \approx 1.22 \frac{\lambda}{D}$$

from Eq. (24.D.1); α is the **angle of resolution** of the camera. It is sufficiently accurate to write

$$\alpha \approx \frac{\lambda}{D} \qquad (24.D.9)$$

ii. The Telescope

Consider a telescope whose objective lens or mirror has diameter D, and whose eyepiece is the

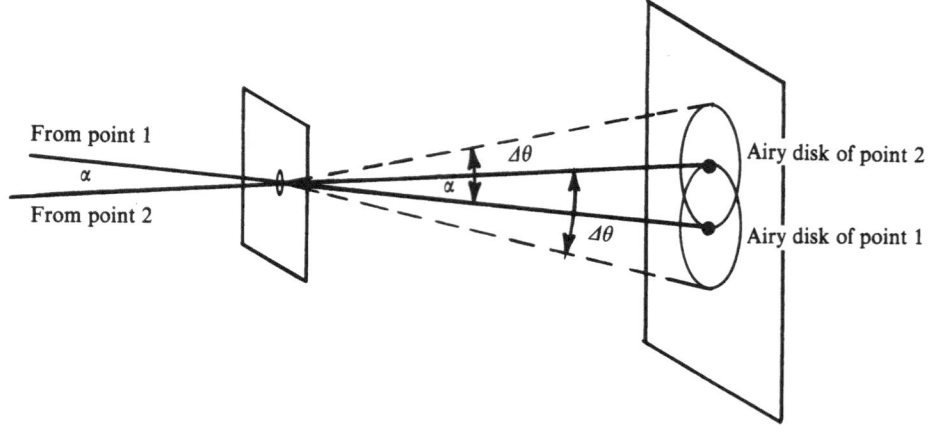

Figure 24.28. Rayleigh's criterion for barely resolved images of two points. We see that $\alpha \approx \frac{1}{2}\Delta\theta$.

best that technology can make. To within how small an angle, α, can such an instrument determine the position of a star, viewed in light of wavelength λ? The answer, equally valid for a refractor or a reflector, is

$$\boxed{\alpha \approx \frac{\lambda}{D}} \qquad (24.D.10)$$

In this formula, which reproduces (24.D.9), and which we quote without derivation, α is the angle of resolution of the telescope. When viewing a planetary surface, for example, the instrument cannot distinguish features with an angular size less than α; two stars, less than α apart, cannot be resolved and appear as a single star.

(It must be kept in mind that the large optical telescopes of today are limited not by diffraction, but by refraction through a turbulent atmosphere. Their large objectives provide light-gathering power rather than resolution. Space telescopes, on the other hand, should be able to reach their theoretical limit of resolution.)

Arrays of Several Telescopes

Formula (24.D.10) remains applicable—in order of magnitude—if D represents the largest dimension of an array of telescopes, if all these instruments could somehow be combined to give a single picture.

Example 24.9. The Very Large Array. The Y-shaped array of 27 radio telescopes shown in Fig. 24.29 is about 30 km in extent. (The radio signals received by all telescopes are combined electronically.) (a) Find the angle of resolution to be expected from the array when it is tuned to detect 6 cm waves. (b) Can detailed features be observed on a nearby radio-emitting star? Assume it has one solar diameter, $2R = 3 \times 10^{11}$ meters; take its distance from us to be $l = 10$ light years $= 10^{17}$ meters; assume 6-cm waves.

(a) Angular resolution: Equation (24.D.10) gives

$$\alpha \approx \frac{\lambda}{D} = \frac{6 \times 10^{-2}}{30 \times 10^3} \text{ rad}$$

$$\alpha \approx 2 \times 10^{-6} \text{ rad} \approx 0.4 \text{ seconds of arc}$$

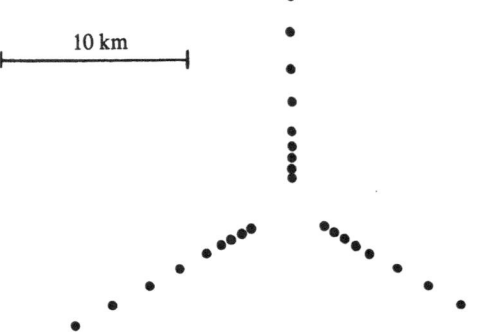

Figure 24.29. One possible configuration for the 27 track-mounted radiotelescopes of the Very Large Array. Each telescope is a "dish" 25 meters in diameter.

(b) Star features? The angular size of the star is

$$\frac{2R}{l} = \frac{3 \times 10^{11}}{10^{17}} \text{ rad} = \underline{\underline{3 \times 10^{-6} \text{ rad}}}$$

comparable to α. Thus, the Very Large Array might just be able to resolve some radio features on nearby stars.

iii. The Microscope

The resolution of a good microscope is also limited by diffraction. In contrast to the telescope, however, the microscope cannot reach an arbitrarily fine resolution through the use of larger and larger lenses, or through any other feature of its design. The ultimate resolution of a good microscope depends only on the wavelength λ used. We state without proof that

Two object points, less than a distance λ apart, cannot give separate images.

(24.D.11)

This is the most serious of all diffraction limitations; it will prevent us forever from exploring the very small (below 400 nm or so) by visible light. For example, the spatial details of atomic structure must be mapped with the help of shorter waves, like x rays.

E. Polarization

Like all electromagnetic waves, light can exhibit polarization. The origin of that phenomenon lies in the transverse nature of the wave; see Fig. 22.3 of Chap. 22. If its **E** vector vibrates in a single plane (which also contains the direction of propagation), the light wave is called **linearly polarized**; if the **E** vector randomly changes its direction of vibration, without favoring any special direction, the light is called **unpolarized**; a mixture of these kinds of waves is called **partially polarized**.

Ordinary light, say from the Sun, from a lamp, or from a flame, is predominantly unpolarized. However, after interacting with certain materials, an unpolarized light beam may emerge as one or several (less intense) polarized beams.

Some of the polarized light we encounter in daily life comes from the blue sky, or from shiny surfaces; anyone who wears polarizing sunglasses becomes aware of that fact.

i. Polarizing Filters and Malus' Law

The best-known device for polarizing light is the so-called Polaroid filter, invented by Edwin Land in 1928. It is used in sunglasses, in photography, and in many scientific applications. A typical version of that filter consists of a long-molecule plastic sheet (polyvinyl-alcohol) which has been stretched in one direction and chemically treated with iodine.

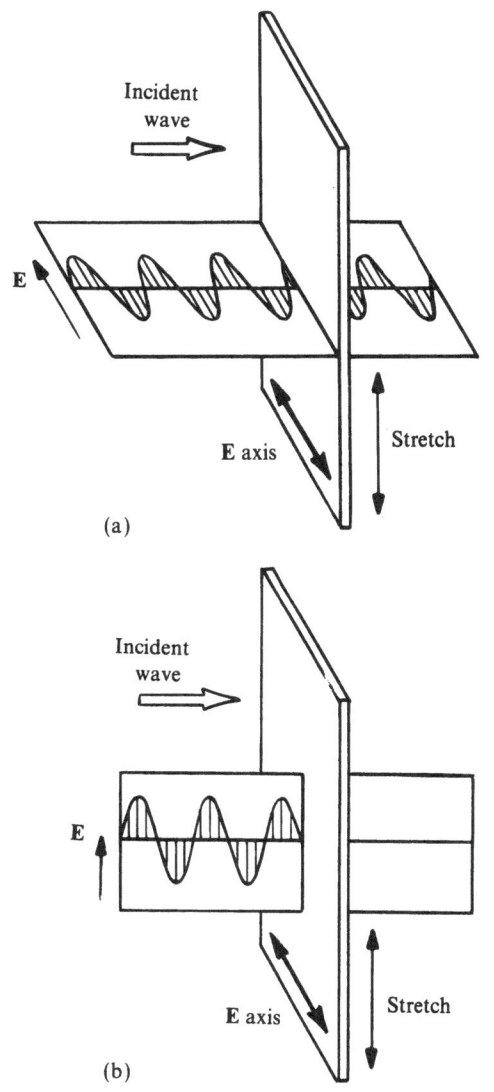

Figure 24.30. The action of a polarizing filter. Its **E** axis is labeled by a horizontal arrow ($\leftrightarrow$). (If the filter is a Polaroid sheet, its direction of stretch happens to be vertical here.)

Transmission:

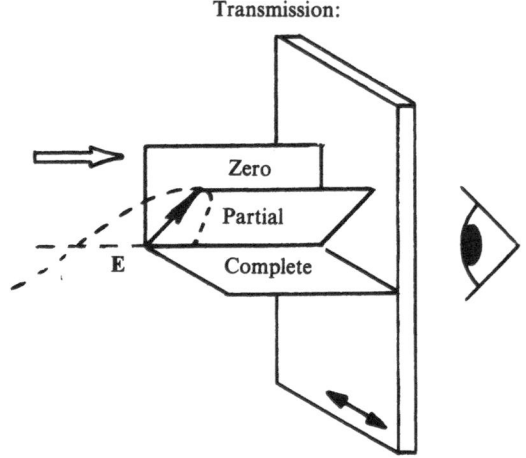

Figure 24.31. Incident **E** plane resulting in partial transmission. (The next figure will be shown from the point of view of the illustrated observer's eye.)

To a wave whose **E** vector vibrates perpendicularly across the stretched molecules, the material is transparent; but if **E** vibrates along those molecules, the wave is completely absorbed. These properties are illustrated in Fig. 24.30. The **E** plane of the transmitted wave defines what we may call the **E** axis* of the filter.

(In reality, neither the absorption nor the transmission is absolutely perfect, but we shall assume the ideal behavior. A substance that preferentially absorbs light of a certain polarization is said to exhibit **dichroism**. For this to be possible, a preferred orientation must exist in the material itself. The Polaroid plate is dichroic, and so are several kinds of crystals.)

Malus' Law

Figure 24.30 shows two extreme cases: full and zero transmission through a polarizing filter. Partial transmission is also possible; it results from an intermediate angle α for the incident polarization, as shown in Fig. 24.31. If the incident and transmitted intensities are $\mathscr{I}$ and $\mathscr{I}'$, respectively, then the transmitted fraction $\mathscr{I}'/\mathscr{I}$ depends on α through

$$\boxed{\frac{\mathscr{I}'}{\mathscr{I}} = \cos^2 \alpha} \qquad (24.\text{E}.1)$$

_____
* No standard term seems to have been agreed upon.

(Malus' law, 1808). The two cases of Fig. 24.30 correspond to $\alpha = 0$ and $\alpha = 90°$.

Explanation of Malus' Law

Figure 24.32 shows the incident **E** vector, as seen by an observer facing the incoming light in Fig. 24.31. Mathematically, he can decompose **E** into a pair of perpendicular vectors $\mathbf{E}_1$ and $\mathbf{E}_2$:

$$\mathbf{E} = \mathbf{E}_1 + \mathbf{E}_2 \qquad (24.\text{E}.2)$$

Physically, sending the obliquely polarized beam is equivalent to simultaneously sending two beams, 1 and 2, with planes of vibration along $\mathbf{E}_1$ and $\mathbf{E}_2$; beam 2 is absorbed and beam 1 survives. We see that

$$|\mathbf{E}_1| = |\mathbf{E}| \cos \alpha \qquad (24.\text{E}.3)$$

To deal with intensities we must square both sides, and then take a time average, as required by the meaning of intensity:

$$|\mathbf{E}_1|_{\text{av}}^2 = |\mathbf{E}|_{\text{av}}^2 \cos^2 \alpha \qquad (24.\text{E}.4)$$

or

$$\mathscr{I}' = \mathscr{I} \cos^2 \alpha \qquad (24.\text{E}.5)$$

as claimed in (24.E.1).

Remark on the Superposition Principle

In combining $\mathbf{E}_1$ and $\mathbf{E}_2$ we have made use of the superposition principle. In a transparent medium

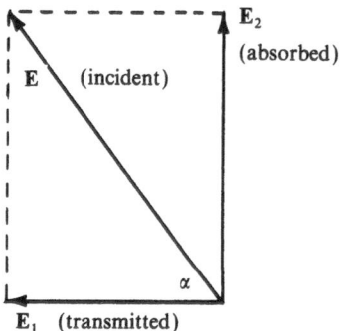

Figure 24.32. Vector decomposition of the incident **E**. After transmission, the wave has its **E** vector precisely horizontal in this illustration.

other than vacuum, that principle is still an excellent approximation. Its message is that *any light beam, polarized or not, can be viewed as a linear superposition of two linearly polarized beams whose planes of vibration are mutually perpendicular.* That is to say, at any place and time, the beam's **E** vector can always be viewed as the sum $\mathbf{E}_1 + \mathbf{E}_2$ [Eq. (24.E.2)]; the **B** vector is of course similarly decomposed.

ii. Two Polarizations Contain the Whole Energy

Figure 24.33 is a head-on view of an incoming wave with electric vector **E**, seen through a polarizing filter; the latter is successively shown in two orthogonal orientations. According to Malus' law, the transmitted intensities are

$$\mathscr{I}_{\text{horiz}} = \mathscr{I}\cos^2\alpha \quad \text{(first orientation of filter)}$$
$$\mathscr{I}_{\text{vert}} = \mathscr{I}\sin^2\alpha \quad \text{(second orientation of filter)}$$

$$(24.E.6)$$

The sum of these intensities is

$$\mathscr{I}_{\text{horiz}} + \mathscr{I}_{\text{vert}} = \mathscr{I}(\cos^2\alpha + \sin^2\alpha) = \mathscr{I}$$

$$(24.E.7)$$

The original intensity is recovered. Hence

The total energy of a wave is shared between just two (orthogonal) polarizations.

$$(24.E.8)$$

How do we select these two polarizations? The answer is, *in any way we please.* Since α is arbitrary, so are these two directions of polarization, as long as they are at right angles to one another. One of the two may contain more energy than the other; for example, if the incoming wave has a purely vertical **E**, one of the two transmitted waves in Fig. 24.33 has all the energy, the other has none.

iii. Starting from Unpolarized Light

When the incident light is unpolarized and has intensity $\mathscr{I}$, then the light emerging from the polarizing filter has the following properties:

1. It is linearly polarized.
2. Its intensity is $\frac{1}{2}\mathscr{I}$.

$$(24.E.9)$$

To understand property (1), we visualize the incoming radiation as a randomly varying mixture of waves with all angles α represented. In all cases, the same component (in Fig. 24.31, the vertical **E** vector) is extinguished.

Property (2) follows from axial symmetry. (By "axial symmetry" is meant that, without the filter, there would be no preferred value of α.) Indeed, consider once more the two filter orientations shown in Fig. 24.33. The transmitted intensities satisfy

$$\mathscr{I}_{\text{horiz}} + \mathscr{I}_{\text{vert}} = \mathscr{I} \qquad (24.E.10)$$

according to (24.E.7). On the other hand, we now have all possible α, so that the vertical and horizontal directions are equally good:

$$\mathscr{I}_{\text{horiz}} = \mathscr{I}_{\text{vert}} \qquad (24.E.11)$$

From (24.E.10) and (24.E.11) we see that $\mathscr{I}_{\text{horiz}} = \mathscr{I}_{\text{vert}} = \frac{1}{2}\mathscr{I}$, as claimed.

Figure 24.34 qualitatively illustrates some properties discussed above.

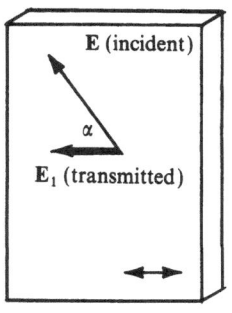

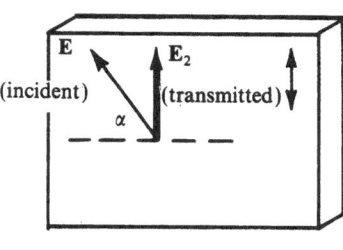

Figure 24.33. Two alternative orientations of the filter; the incident **E** is assumed fixed.

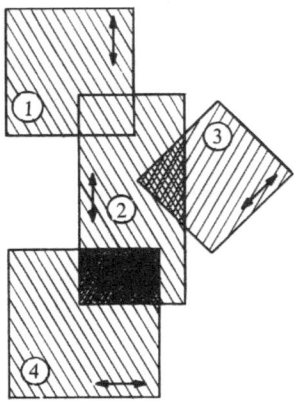

Figure 24.34. Overlapping polarizing filters in front of a uniformly lighted background. Note the E plane labels. Filters 2 and 4 are said to be **crossed**.

Example 24.10. Two Consecutive Filters. Two polarizing filters, one behind the other, have a 45° angle between their **E** axes, as shown in Fig. 24.35. An unpolarized beam, of intensity $\mathcal{I}_0 = 12$ watts/meter^2, is incident on the first filter (the "polarizer"). Find the intensity emerging from the second filter (the "analyzer").

Between the filters, the intensity is

$$\mathcal{I}_1 = \tfrac{1}{2}\mathcal{I}_0$$

see (24.E.9); the beam has become linearly polarized, with a horizontal **E** plane. Passage through the analyzer then rotates the polarization plane by 45° and reduces the intensity to

$$\mathcal{I}_2 = \mathcal{I}_1 \cos^2 45° = \left(\frac{1}{2}\mathcal{I}_0\right)\left(\frac{1}{\sqrt{2}}\right)^2$$

$$= \frac{1}{4}\mathcal{I}_0 = 3\,\frac{\text{watts}}{\text{meter}^2}$$

see (24.E.5).

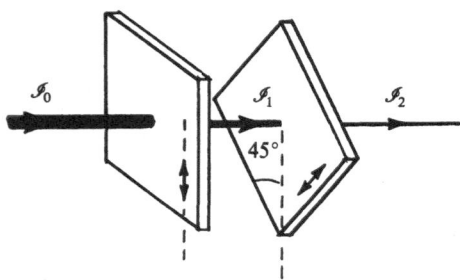

Figure 24.35. The second filter differs from the first by a 45° rotation. Ray width symbolizes intensity. (The dashed lines indicate a reference direction, here the vertical one.)

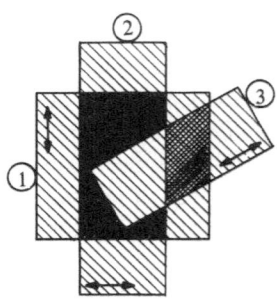

Figure 24.36. Filter 3 has been inserted between crossed filters 1 and 2. Note the E plane labels.

Examination Between Crossed Polarizing Filters

As we see, for example, in Fig. (24.34), no light at all is transmitted through a pair of crossed filters. On the other hand, the preceding example shows that a polarizing filter is capable of rotating the **E** plane of an already polarized light beam. Therefore, as in Fig. 24.36, insertion of a third filter between the crossed pair may restore transmission.

More generally, any material that affects the

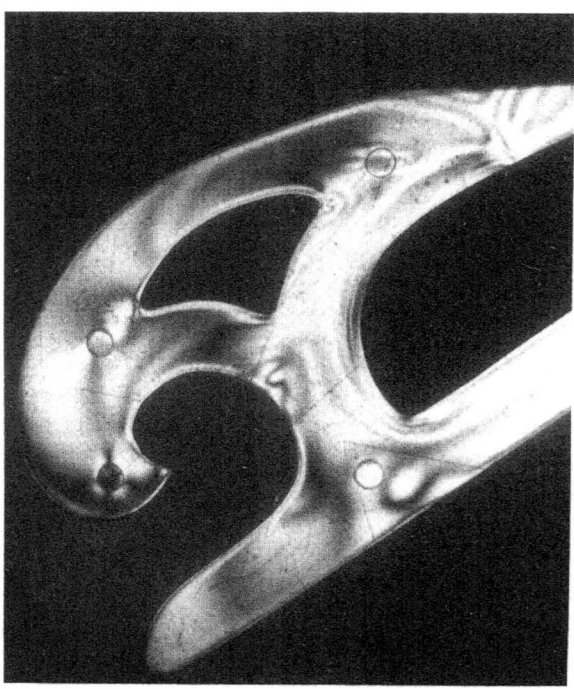

Figure 24.37. Photoelasticity: A French curve viewed between crossed polarizing filters. The manufacturing process has left permanent pressures and tensions ("stresses") within the substance.

polarization may become visible when viewed between crossed filters. Plastics under stretch or compression reveal their internal stresses in this way; see Fig. 24.37. The phenomenon is called **photoelasticity**. Its magnitude depends on the light's frequency, with a resulting color display.

Notes

i. How Much of the Incident Light is Reflected?

At a smooth dielectric surface, the incident light is partly transmitted, partly reflected (see Sec. A i); no metallic coating is assumed. What fraction of the energy is reflected? Consider the interface between two media, labeled 1 and 2; the answer is

$$\mathcal{I}_{\text{refl}} = \left(\frac{n_2 - n_1}{n_1 + n_2}\right)^2 \mathcal{I}_{\text{inc}} \qquad (24.\text{N}.1)$$

The light beam is assumed incident from either medium onto the interface, and *at right angles to it*; $\mathcal{I}_{\text{inc}}$ and $\mathcal{I}_{\text{refl}}$ are the incident and reflected intensities; n_1, n_2 are the two indices of refraction. For example, glass, with $n_2 = 1.5$, in air ($n_1 = 1$), reflects $[(1.5 - 1)/(1.5 + 1)]^2 = 4\%$ of the incident energy. No reflection occurs if $n_1 = n_2$; see Fig. 24.38. The sign of $n_2 - n_1$ is immaterial, as we can see.

Formula (24.N.1) can be derived from electromagnetism. For other than normal incidence it becomes more complicated, and involves the polarization as well as the angle of incidence.

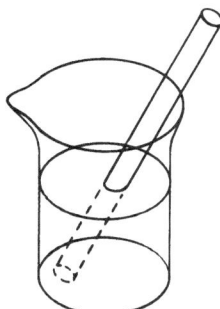

Figure 24.38. A transparent rod becomes invisible when immersed in a liquid whose index of refraction equals its own. The explanation: There is no reflection and no refraction at the rod's interface with the liquid.

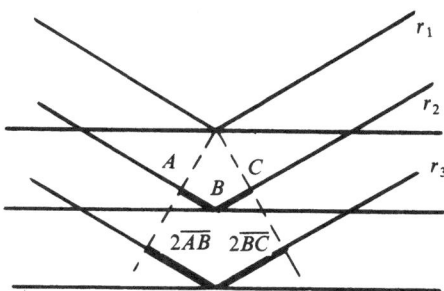

Figure 24.39. Consecutive path differences in many-layer interference.

ii. Sharp Bragg Angles from Many Atomic Layers

An example will best explain why a crystal does not reflect even at angles that are only slightly different from the Bragg angles. Suppose rays r_1, r_2 in Fig. 24.39 have a path difference

$$\overline{AB} + \overline{BC} = 1.01 \,\lambda \qquad (24.\text{N}.2)$$

(instead of λ, which would ensure constructive interference). Then rays r_1 and r_3 have a path difference $2.02 \,\lambda$, and so on; eventually we consider rays r_1 and r_{51}, which have a difference $(50\frac{1}{2})\lambda$, and therefore interfere destructively. Consequently, so do rays r_2 and r_{52}, r_3 and r_{53}, etc. With enough reflecting layers available, almost every ray has its canceling partner.

Condensed Checklist

Wavelength in a refractive medium:

$$\lambda = \frac{\lambda_0}{n} \qquad (24.\text{A}.3)$$

Thin-film interference ($n > 1$, in air):

$$s = \frac{k\lambda}{4} = \frac{k\lambda_0}{4n}$$

$$k = \begin{cases} 1, 3, 5,... & \text{(constructive)} \\ 2, 4, 6,... & \text{(destructive)} \end{cases}$$

$$(24.\text{A}.4), (24.\text{A}.5)$$

(For a coating of intermediate index, exchange "constructive" and "destructive.")

Thin wedges: Between two consecutive bright fringes or two consecutive dark fringes,

$$\Delta s = \frac{\lambda}{2} \qquad (24.\text{A}.7)$$

Mirror displacement in Michelson interferometer:

$$\Delta s = \frac{\lambda}{2} \qquad \text{(24.A.10)}$$

(between successive bright signals).

Bragg's angles of incidence:

$$\boxed{\cos\theta = \frac{k\lambda}{2s}} \qquad (k = 1, 2, 3, \dots) \qquad \text{(24.A.15)}$$

Radius of rings:

$$r = l\tan(180° - 2\theta) \qquad \text{(24.A.16)}$$

Young's fringes:

$$\boxed{\begin{array}{c} \sin\theta = \dfrac{k\lambda}{2s} \\[2mm] k = \begin{cases} 0, \pm 2, \pm 4, \dots & \text{(bright)} \\ \pm 1, \pm 3, \pm 5, \dots & \text{(dark)} \end{cases} \end{array}}$$

$$\text{(24.B.4), (24.B.5), (24.B.6)}$$

Position:

$$x \approx (l)\left(\frac{k\lambda}{2s}\right), \qquad k \text{ as above} \qquad \text{(24.B.7)}$$

(small angle)

Diffraction grating: Sec. B iv

Single slit:

$$\sin\frac{\Delta\theta}{2} = \frac{\lambda}{w} \qquad \text{(24.B.12)}$$

Coherence condition (interference from films):

Coherence length $\geq 2 \times$ film thickness   Fig. 24.23

Mutually noncoherent sources A, B:

$$\boxed{\mathcal{I} = \mathcal{I}_A + \mathcal{I}_B} \qquad \text{(24.C.1)}$$

Airy disk (small-angle approximation):

$$\Delta\theta \approx 2.44\,\frac{\lambda}{D} \qquad \text{(angle)} \qquad \text{(24.D.1)}$$

$$\Delta x \approx 2.44\,\frac{l\lambda}{D} \qquad \text{(diameter)} \qquad \text{(24.D.2)}$$

Limit of resolution:

$$\boxed{\begin{array}{lll} \alpha \approx \dfrac{\lambda}{D} & \left(\begin{array}{l}\text{Pinhole camera} \\ \text{when } l \gg D^2/\lambda\end{array}\right) & \text{(24.D.9)} \\[4mm] \alpha \approx \dfrac{\lambda}{D} & \text{(telescope)} & \text{(24.D.10)} \\[4mm] \text{Object size} \approx \lambda & \text{(microscope)} & \text{(24.D.11)} \end{array}}$$

Malus' law:

$$\mathcal{I}' = \mathcal{I}\cos^2\alpha \qquad \text{(24.E.1)}$$

Originally unpolarized beam:

$$\boxed{\mathcal{I}_{\text{polarized}} = \tfrac{1}{2}\mathcal{I}} \qquad \text{(24.E.9)}$$

True or False

1. If a surface reflects part of the incident light and transmits the rest, then two such surfaces, one behind the other, must reflect more light.

2. The wider the angle of a thin wedge, the further its interference fringes are spread apart.

3. In a Michelson interferometer, one of the two interfering waves is reflected once, the other three times. (Answer without looking at the book's diagram.)

4. In a single crystal and for x rays of given wavelength, Bragg reflection may occur at any angle of incidence.

5. The further apart two parallel slits, the further apart the fringes they produce in a given wavelength.

6. A grating spectrometer deflects blue light through a greater angle than yellow.

7. The narrower a single slit, the wider its central bright fringe in a given wavelength.

8. In a single-slit diffraction pattern, the dark fringes are about equally spaced.

9. Consider a Michelson interferometer with equal arms. By making them long enough one washes out the fringes, owing to noncoherence.

10. The interference pattern from a two-bulb chandelier is hard to observe because its fringes are too finely spaced.

11. Since a pinhole camera needs no lens, there is no limit to its resolution under visible light (although a photographic exposure may take a long time).

12. A good telescope, operating in outer space (i.e., without an intervening atmosphere) should obtain sharper images in the ultraviolet than in the visible.

13. A good polarizing filter cuts an unpolarized beam down to half its original intensity.

14. Two successive polarizing filters, identically oriented, cut an unpolarized beam down to one quarter of its original intensity.

15. Two consecutive polarizing filters, with their **E** axes at 180° to one another, transmit zero intensity.

Problems

The distance between two slits (or fringes) is understood to be measured center to center.

Section A : Interference by Reflection

*24.1. A plane wave, of wavelength 589 nm in air, is incident normally on a thin liquid film made of acetone (index of refraction 1.37) stretched in air. Find, in nanometers, the smallest three values of the film's thickness such that it reflects a maximum of light.

*24.2. (a) In Fig. 24.1a, express s in terms of the wavelength of light, λ, in the glass. (b) Do the same in Fig. 24.1b.

24.3. In Fig. 24.2, of what minimum thickness should the antireflective coating be if perpendicular reflection of wavelength 520 nm (in air) is to be suppressed? Assume an index of refraction of 1.38 for the coating.

*24.4. While mirror M_2 is moved backwards by 1.50 mm in Fig. 24.7, the observer counts a succession of 5200 bright fringes. What wavelength is being used?

24.5. In a certain Michelson interferometer, using krypton light of wavelength 605.78 nm, advancing mirror M_2 (Fig. 24.7) causes a succession of 5582 dark fringes. Over what distance was M_2 moved?

24.6. In Example 24.3, how far would the mirror move in one day? (Note: Fringe counting may be done automatically.)

24.7. Derive $\lambda = \lambda_0/n$, Eq. (24.A.3), as indicated in the text.

*24.8. Find the first-order Bragg angle for potassium chloride, KCl; assume a lattice spacing of 0.314 nm and x rays of wavelength 0.180 nm.

24.9. Two flat glass surfaces are pressed together with a hair between them, as in Fig. 24.3b. (a) In perpendicularly reflected light of wavelength 550 nm, one sees five dark fringes per millimeter, as shown. Estimate, in minutes of arc, the angle between the surfaces. (b) If 160 dark fringes are counted between the location where the plates are in contact and that of the hair, how thick is the hair?

*24.10. In the Newton's rings example, Example 24.2, what is the radius (a) of the first bright fringe? (b) Of the fifth bright fringe?

24.11. If, in Example 24.2, the wavelength were multiplied by 2, then by what factor would the radius of each fringe be multiplied?

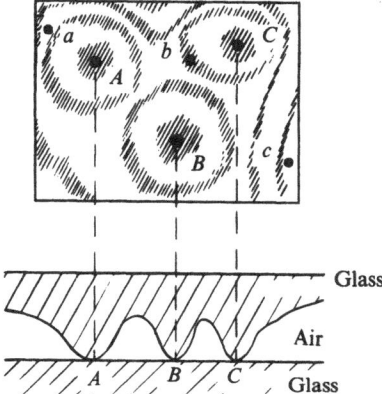

24.12 (Testing a glass surface for flatness). The adjoining figure shows interference fringes from a pair of closely spaced glass surfaces seen under perpendicularly reflected light of wavelength 590 nm. The bottom surface belongs to an "optical flat" presumed to be a perfect plane; the top surface is being tested. If A, B, C are points of contact, how wide is the gap between the two surfaces at a, b, c?

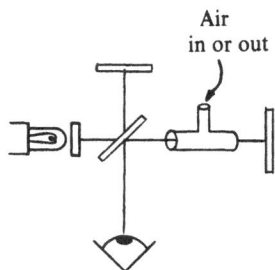

*24.13 (Measuring the index of refraction of a gas by counting fringes). A certain Michelson interferometer, using sodium yellow light ($\lambda_0 = 590$ nm), has an evacuated tube of length $l = 20$ cm, with transparent ends, in one of its arms, as shown. (a) How many wavelengths equals a roundtrip in the tube? (b) By how much does that number change after the tube is opened to the atmosphere? (Index of air: $n = 1.00030$.) (c) How many bright–dark alternations, N, are seen while this operation (very gradually) takes place? (Equivalently, this is the number of dark or bright fringes passing in front of the observer's eye.) (d) More generally, find a formula for n in terms of N, λ_0, and l.

*24.14. In Fig. 24.11, the lattice spacing for planes a is 0.281 nm (=center-to-center distance between neighboring Na and Cl ions.) Find the spacing for planes b.

24.15. A certain crystalline powder sample gives a second-order Bragg angle of 35.0°. The lattice spacing involved is 0.353 nm. (a) What are the Bragg angles of order 1, 3, etc? (b) What is the wavelength of the x rays?

*24.16. In Problem 24.8 a diffraction ring is recorded on a plane film which is at 15 cm from the sample, and normal to the incident beam. What is the ring's radius?

24.17. A thin film, of index n and of gradually varying thickness s, is observed in air under reflected light whose wavelength is λ_0 in air. The intensity received from the center of a bright fringe is $\mathscr{I}_{max}$. Find, in terms of s, λ_0, n, $\mathscr{I}_{max}$, the intensity $\mathscr{I}$ reflected from any point. (Assume approximately normal incidence.)

24.18. In the Newton's rings example, Example 24.2, what would be the radius of the third dark fringe if we were dealing with two equal lens surfaces of curvature 0.44 diopter each, arranged as shown?

*24.19. X rays of wavelength 0.210 nm are first-order diffracted by a small powdered sample, and produce a ring of radius 25.5 cm on a plane film, 30.0 cm away from the sample and normal to the incident beam. (a) What is the relevant lattice spacing? (b) What are, if any, the next observed orders of diffraction recorded on the film?

24.20. Explain why, with the geometry of Fig. 24.9, Bragg angles less then 45° do not show up on the film.

Section B: Diffraction from Slits

*24.21. What is, in terms of whole wavelengths λ, the path difference $\overline{AP} - \overline{BP}$ in Fig. 24.13?

*24.22. In Young's experimental arrangement, suppose blue light of wavelength 470 nm falls on a pair of slits separated by a distance of 0.60 mm. How far apart, in millimeters, are adjacent bright fringes on a screen 3.00 meters away? (Assume small diffraction angles; verify the consistency of that assumption.)

24.23. A collimated beam of wavelength λ is incident normally on a single slit of width 0.15 mm. If the central diffracted beam spreads over an angle of 0.35°, estimate λ.

24.24. How wide is the central fringe in the diffraction pattern of a single 1-mm-wide slit, projected 5 meters away, if the slit is illuminated normally by a plane monochromatic wave of wavelength 590 nm?

*24.25. Suppose consecutive dark fringes in Young's double slit experiment are 3.0 mm apart, the slits are 0.4 mm apart, and the screen is 2.4 meters from the slits. What is the color of the light?

24.26. Suppose Young's double slit experiment is first performed with light of wavelength 450 nm. Afterwards, it is tried once more with another wavelength, double the slit spacing, and double the distance between slits and screen. The new fringe spacing had also doubled. Find the new wavelength.

*24.27. Light of wavelength λ is incident normally on a grating having 12 000 lines per centimeter. In first order, the light is diffracted through an angle of 40.9°. What is λ, and the color of the light?

*24.28. A grating diffracts a certain monochromatic light through 8° in second order. Through what angle is the same light diffracted in fourth order?

24.29. A plane wave of wavelength 600 nm illuminates a single slit normally. The transmitted beam has a spread of 2.5 degrees of angle (i.e., most of the energy propagates within that angle). Find the width of the slit.

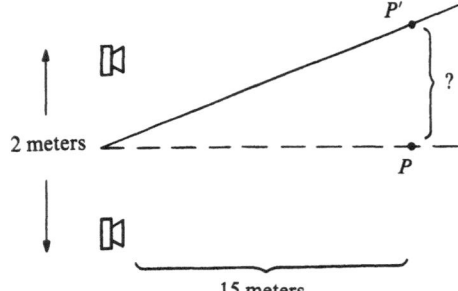

*24.30. Two small outdoor loudspeakers, arranged as shown, are connected to the same audio oscillator, and vibrate in phase at a frequency of 440 Hz (the note "A"). They are 2.0 meters apart. How close together can two listeners, P and P', sit so that P hears a loud volume while P' hears very little? Assume P and P' are 15 meters away, faced by the speakers; speed of sound in air = 340 meters/sec.

24.31. Let Young's experiment, described in Problem 24.25, be performed in water, whose index of refraction is 1.33. How is the answer affected?

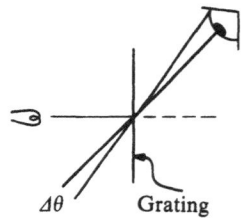

$\Delta\theta$ Grating

*24.32. A distant incandescent wire is examined through a grating that has 4500 lines per centimeter; the grating faces the wire, as shown. In first order of diffraction, the wire's visible spectrum (from $\lambda_1 = 390$ nm to $\lambda_2 = 770$ nm) is seen spread out over an angle $\Delta\theta$. Assuming a plane wave, calculate $\Delta\theta$.

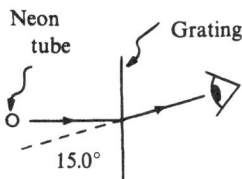

Neon tube / Grating

$15.0°$

24.33. A thin neon discharge tube, which emits a selection of well-defined wavelengths, is viewed from a large distance through a grating, as shown. One of the wavelengths, 639 nm, gives an image displaced by a 15.0° angle in first order. The grating's total width is 2.0 cm. (a) How many lines (slits) does it have? (b) In what direction is the second-order diffracted image seen in light of wavelength 614 nm?

*24.34. A grating, which transmits mixed light of wavelengths λ_1 and λ_2, causes the fourth-order λ_1 diffraction line to coincide with the fifth-order λ_2 line. If $\lambda_1 = 610$ nm (orange), find λ_2 and its color.

24.35. A certain grating diffracts light of wavelength λ by a small angle θ, in first order. Another grating, with half the width and twice the number of slits of the first one, diffracts light of wavelength λ' by an angle 2θ, in third order. What is the ratio λ'/λ?

*24.36. A certain monochromatic light is diffracted by a grating through 24.0° in third order. (a) Through what angle is it diffracted in fifth order? (b) What is the largest diffracted order?

*24.37. In the grating spectrometer of Fig. 24.16, let the grating have 5000 lines per centimeter; the transmitted light has a wavelength of 590 nm. If the lens has a focal length of 25.0 cm, how far from the undiffracted line on the film is the first-order line located?

24.38. (a) From Eqs. (24.B.10) and (24.B.11), find an expression for the position x_1 of the first-order diffraction line in a spectrometer in terms of the grating's line spacing s, the wavelength λ, and the focal length l. (Assume a small diffraction angle.) (b) What does the answer become if the order of diffraction is k instead of 1?

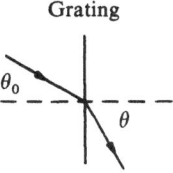

Grating

θ_0

θ

24.39. A plane grating has a spacing s between adjacent slits. A plane wave of wavelength λ is incident at an angle

θ_0, as shown. (In the text we have considered only $\theta_0 = 0$.) Derive the **grating equation** for the allowed diffraction angles θ:

$$\sin\theta - \sin\theta_0 = \frac{k\lambda}{s}, \qquad \text{where} \quad k = 0, \pm 1, \pm 2, \text{ etc.}$$

Verify that $k = 0$ gives the undeflected beam. (Note: Only two slits need be considered.)

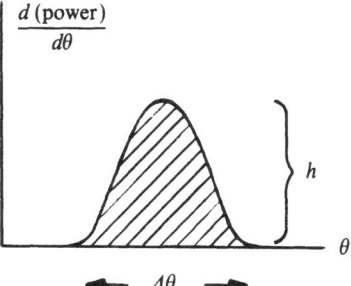

$\frac{d\,(\text{power})}{d\theta}$

h

θ

$\Delta\theta$

*24.40 (Limit of resolution of a grating). Let graphs such as those of Fig. 24.15 represent the power per unit angle transmitted by the grating in a given direction θ; see also the adjoining figure, which singles out a sample fringe. (a) Explain why the total power into the fringe equals the shaded area. (Consider that area to be proportional to the fringe width $\Delta\theta$ and to the peak's height h; do not worry about the proportionality constant, which depends on the peak's exact shape.) (b) By following the argument of Sec. B iv, show that $h\,\Delta\theta \propto N$ ($N = $ number of slits), while $h \propto N^2$. (c) From (b), how does $\Delta\theta$ depend on N? (d) What is the change in wavelength, $\Delta\lambda$, that will shift a first-order fringe by an angle $\Delta\theta$? [Use Eq. (24.B.11), assuming a small angle for simplicity.] (e) By combining (c) and (d), find how $\Delta\lambda$ depends on N; $\Delta\lambda$ is the **limit of wavelength resolution** of the grating.

Section C: Noncoherent Light

24.41. A certain nearly monochromatic light beam gives rise to interference fringes when reflected from a 0.75-mm-thick glass plate whose index of refraction is 1.55; the fringes can no longer be seen when a 0.95-mm plate of the same glass is used instead. Estimate the coherence length of the light (a) in the glass, (b) in air.

*24.42. A certain Michelson interferometer uses a light source of wavelength 486 nm whose typical coherent wave packet is 25 000 cycles long. (a) What is the maximum arm difference that still permits fringes to be seen? (b) How many fringes can be counted as one mirror is moved, from the first appearance to the vanishing of the fringes?

24.43 (Noninterference of different frequencies). Using the statement that the light intensity is proportional to the time-averaged square of the electric field amplitude [Chap. 22, Eq. (22.B.9) and statement (22.B.18)], prove that, in the superposition of two monochromatic waves, the intensities simply add. [Take the time-averaged square of the right-hand side of (24.C.2) in this chapter.]

Section D: Limits of Resolution

*24.44. A certain Earth-based laser sends to the Moon a collimated beam of diameter 5 cm and wavelength 600 nm. (a) Under the best conditions, roughly how small, in kilometers, is the illuminated spot on the Moon's surface? (Use 400 000 km for the Earth–Moon distance.) (b) What is, in seconds of arc, the angular diameter of that spot, seen from Earth? (Note: This technique is actually being used, in conjunction with Moon-based reflectors, in order to make time-delay measurements of the Earth–Moon distance.)

*24.45. What should be the diameter of a radar dish, operating with 3.0 cm waves, if it is to resolve a pair of airplanes, flying 150 meters apart at 20 km from the radar station?

24.46. It is commonly stated that a military spy satellite can "read your license plate." Whether or not this is true, and disregarding atmospheric effects, estimate the minimum diameter of the telescope objective needed to accomplish that feat. Assume a 2-cm resolution, an altitude of 200 km, and a wavelength of 500 nm, typical of white light.

*24.47. Estimate the diameter of an optical telescope whose resolution (say in 500-nm light) is the same as that of the Very Large Array in 6-cm waves; see Example 24.9. Disregard (unrealistically) atmospheric effects.

24.48. Consider a geometrically perfect orbiting telescope mirror, whose resolution is limited by diffraction only (atmospheric effects are absent). If its diameter is 2.4 meters, what is, in seconds of arc, its limiting angle of resolution in the visible (say $\lambda = 500$ nm)? (For comparison, the best ground-based telescopes have a resolution of about 1 second of arc.)

*24.49. The human ears are about 20 cm apart. Estimate roughly, in degrees, our limit of angular resolution for determining where a sound comes from, if the maximum frequency heard in that sound is 15 000 Hz. (Use 340 meters/sec = speed of sound in air. Caution: The actual limit of resolution may be much less than your result indicates.)

*24.50. Consider a pair of radiotelescopes, 8000 km apart (straight-line distance; one is in America, the other in Europe; relative phase information is broadcast from one instrument to the other). Estimate, in seconds, the angular size of a celestial object that is barely resolved in 6-cm radio waves that it emits. (This method, **very long**

baseline interferometry**, or VLBI, has been used to set an upper limit on the size of quasars.)

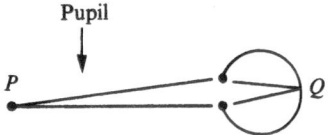

24.51. (Resolution of the eye). The adjoining sketch shows a human eye that focuses at Q, on the retina, the image of a distant object point P. According to geometrical optics, Q is an exact point, independent of pupil size. Actually, Q is blurred by diffraction into an Airy disk of diameter Δx approximated by the "pinhole camera" formula (24.D.2), even though the eye has a lens system, unlike the pinhole camera. In (24.D.2), $l \approx 2$ cm is the eyeball's diameter, $\lambda \approx 500$ nm is representative of white light, and $D \approx 5$ mm is a typical pupil diameter. (a) Find Δx. (b) Find the angular limit of resolution of the eye, considered as a telescope, from (24.D.10). [Note: For comparison with (a), the distance between neighboring light receptors in the fovea—the region of the retina where they are the most closely packed—is about 2000 nm. Would it improve image resolution if they were much closer together?]

*24.52. A sound image ("sonogram") of a live human fetus, in the womb, may be obtained as follows. The fetus is "illuminated" by ultrasonic waves, which are detected and processed electronically to reconstitute the image. In order to distinguish the size of the fetus' eyes (say, 1 cm in diameter), what minimum ultrasonic frequency must be used? (Assume 1500 meters/sec for the speed of sound in the amniotic fluid, which is mainly water.)

*24.53. Bats locate and identify flying insects by ultrasonic echoes. What minimum sound frequency must they produce in order to determine the size of a 0.2-cm beetle at close range? Use 340 meters/sec for the speed of sound in air. (Note: The human ear does not respond above 20 000 Hz.)

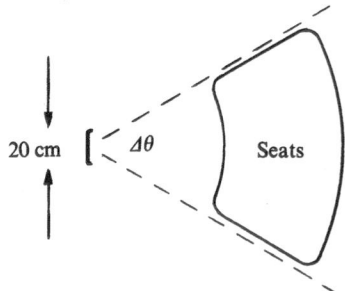

*24.54. A certain outdoor loudspeaker consists of a plane vibrating diaphragm of diameter 20 cm. Within what maximum angle $\Delta\theta$ (see the figure) should the listening public be seated so that frequencies of order 3000 Hz be about equitably distributed? (Speed of sound in air = 340 meters/sec.)

24.55. A perfectly built space telescope, with diameter 0.50 meter, has a focal length of 12 meters. If an electronic camera is mounted at the focal point, estimate what should be, in millimeters, its distance resolution in order to take full advantage of the telescope's angular resolution. Assume a wavelength of 400 nm for the incoming light.

Section E: Polarization

24.56. What fraction of the intensity of an initially unpolarized beam is transmitted by two consecutive polarizing filters whose **E** axes are at (a) 5°, (b) 30°, (c) 45°, (d) 80° to one another?

*24.57. A horizontally propagating, linearly polarized light beam, with vertical **E** plane, and with an intensity of 5.0 watts/meter^2, is incident normally on a polarizing filter whose **E** axis makes an angle α with the vertical. The emerging intensity is 3.0 watts/meter^2. Find α in degrees.

24.58. Five consecutive polarizing filters are oriented in such a way that each makes an angle of 60° with the preceding one. What fraction of the intensity of an initially unpolarized beam survives after passage through the five filters?

*24.59. An unpolarized light beam is incident on a sequence of three polarizing filters, of which the first and third are crossed. (a) If the first and second filter have an angle of 30° between their **E** axes, what fraction of the incident intensity is transmitted through the three filters? (b) What does the answer become if the 30° angle is between the second and third filter?

*24.60. A linearly polarized light beam is incident vertically on a polarizing filter that lies in a horizontal plane. The intensity emerging on the other side is 90 watts/meter^2 or 160 watts/meter^2 according to whether the filter's **E** axis points north–south or east–west. Predict the emerging intensity if the axis runs northwest to southeast. (Two answers are possible.)

Answers to True or False

1. False (not if interference is destructive).
2. False (they come closer together).
3. False (each two times).
4. False (only at the Bragg angles).
5. False (closer fringes).
6. False (the opposite).
7. True.
8. False (the central spacing has double width).
9. False (would be true for a large *difference* between arm lengths).
10. False (there is no such pattern).
11. False.
12. True.
13. True.
14. False (still 1/2).
15. False (true for 90°).

Answers or Hints to Problems

24.1. 107 nm, 322 nm, 537 nm.

24.2. (b) 2.25λ.

24.4. Hint: Use Eq. (24.A.11).

24.8. 73.3°.

24.10. (b) 2.5 mm.

24.13. (d) $n = 1 + \dfrac{N\lambda_0}{2l}$

24.14. 0.199 nm.

24.16. 9.9 cm.

24.19. (a) 0.304 nm; (b) second order.

24.21. $1 \times \lambda$.

24.22. 2.4 mm.

24.25. Green.

24.27. 546 nm, green.

24.28. 16° (a small angle means a simple proportionality).

24.30. 3.0 meters.

24.32. $20.3° - 10.1° = 10.2°$.

24.34. 488 nm.

24.36. (b) 7.

24.37. Hint: Use Eqs. (24.B.10) and (24.B.11).

24.40. (e) $\Delta\lambda \propto 1/N$.

24.42. (a) 6 mm.

24.44. (a) 12 km.

24.45. 4 meters.

24.47. About 0.25 meter.

24.49. About 7°.

24.50. About 0.0015″.

24.52. Hint: Think of a "sound microscope."

24.53. See the preceding hint.

24.54. Hint: Estimate the angular size of the Airy disk, seen from the speaker.

24.57. 39°.

24.59. 0.094.

24.60. Hint: First find the **E** plane of the incident beam.

CHAPTER 25

RELATIVITY

Our present topic is the theory of **special relativity**, formulated in 1905 by Albert Einstein. [His theory of gravitation, or **general relativity** (1916), lies outside the scope of this book.]

Every student of relativity faces two difficulties, which are easier to overcome if recognized at the outset:

1. Some intuitive notions of space and time must be abandoned.
2. Some well-learned notions of Newtonian mechanics are in need of correction—sometimes drastic correction.

Fortunately, the revisions brought about by relativity are unimportant for objects whose speeds are small compared to that of light. *Newton's physics remains approximately valid at low speeds*—exactly so in the static limit. The earlier chapters of this book must not be unlearned; they represent a well-established domain of physics—the only one many of us will ever use.

Is relativity really necessary? The answer is that countless experimental facts are known that force relativity upon us. The investigations of Albert Michelson, culminating in the historic experiment he carried out in 1887 in collaboration with Edward Morley, are the most clear-cut of all and will be our starting point.

Relativity is essential in the daily practice of physics. Material particles moving at speeds comparable to that of light are not unusual in the atomic, nuclear, and astrophysical sciences; they require relativity in order to be understood.

Selection of Topics

Even for an elementary introduction to the subject, the present discussion is by necessity very incomplete. It is oriented toward a pair of frequently used practical results, namely, the celebrated mass–energy relation ("$\mathscr{E} = Mc^2$") and the Doppler effect for light.

Some relativistic results are famous for their paradoxical aspect; they are useful in re-educating our intuition. Among these, time dilation, that is, the slowing down of events in a traveling laboratory, will be selected for discussion.

A. The Michelson–Morley Experiment

Do all light signals—in vacuum—travel at the same speed c? Maxwell's theory of electromagnetic waves asserts this to be the case. Yet, even if all stationary observers agree on their measured value of c, common sense points to speeds other than c when measured in moving laboratories; in Fig. 25.1, the flatcar, which travels at speed v, is such a laboratory. (We are assuming that the ground-based equipment has already verified Maxwell's prediction of a speed c.)

i. Is There an Ether Drift?

As an aid to common sense, we visualize the light signal as if it were a sound signal; the ground-based laboratory in Fig. 25.1 is then in motionless air; the flatcar, on the other hand, experiences a wind that brings about a modification $c \to c - v$ in the wave's speed. To remind ourselves that we are dealing with electromagnetism in vacuum, not with acoustics, we resort to the language of Michelson's own time, and speak of the vacuum as being "filled" with an

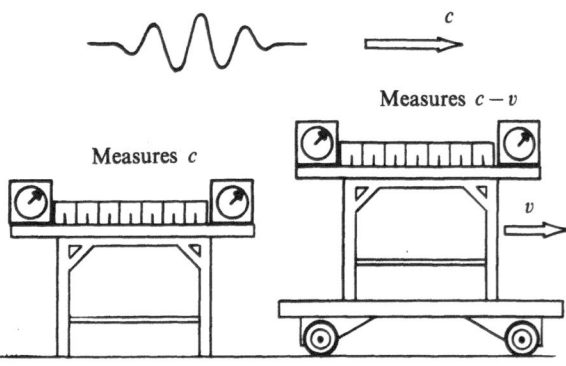

Figure 25.1. Common-sense expectation: the stationary and moving observers (clocks and meter sticks) measure different speeds for the light signal.

imaginary medium, the **ether***; the flatcar in the figure experiences, not a wind, but an "ether drift." Here we view the ether as permeating the whole Universe, including our atmosphere; its sole purpose is to play in our mind the role of a unique identifiable reference frame in which Maxwell's theory is exactly valid. Since it is only a reference frame, such an ether need not have any material properties.

What is our motion relative to the ether? This question is of great fundamental interest, and the Michelson–Morley experiment, which was designed to answer it, turned out to be one of the most basic experiments of all time.

The Earth's revolution around the Sun—at about 30 km/sec—ought to confront us with an apparent motion of the ether (an **ether drift**) of varying direction; in addition, the solar system itself is moving around the Galaxy, while the Galaxy is known to be in motion relative to a group of which it is a member.

Michelson's result was totally unexpected:

There is no ether drift at any time of year.

(25.A.1)

This would seem to imply that the Earth is standing still—at least with respect to the universal ether. Such a medieval conception is not tenable, and therefore we must assume that our common sense, and the whole notion of ether, had led us astray.

---

* Or "luminiferous ether." From the time of Maxwell to that of Michelson, the ether was taken quite seriously and became endowed with many bizarre properties, which need not concern us here. In this section, our idea of the ether is a simplified adaptation of the historical one.

But before considering how relativity resolves the paradox, we examine the details of Michelson's measurement.

ii. The Experiment

Figure 25.2 shows a Michelson interferometer with equal arm lengths l. The instrument is assumed to be stationary, while the ether moves at a constant speed v, directed along one of the arms. A monochromatic wave of frequency f issues from A. Consider now the wave's two possible round trips $M_3 M_1 M_3$ and $M_3 M_2 M_3$, each of length $2l$. We shall see that the ether drift must affect the wavelength differently in each arm; hence the length of each round trip must amount to a different number ($N_1 \neq N_2$) of wavelengths. Thus, if somehow we could "turn on" the ether's speed from zero value to its full value v, the interference effect would be similar to changing the length of one arm relative to the other; the bright fringes seen by observer ⊙ would shift by a number $|N_1 - N_2|$. We now calculate that number.

(First let us digress to point out that, ether drift or no, *the wave's frequency f is the same throughout the apparatus*. This is always the case in a steady-state situation, as Fig. 25.3 illustrates. Indeed, if we had $f_1 > f_2$ for the frequencies at two points P_1, P_2, then, per unit time, more wave fronts would be coming in at P_1 than are leaving at P_2, and a catastrophic pile-up of wave fronts would occur between these two points, in clear violation of steady-state. The presence of mirrors does not invalidate that statement.)

Figure 25.4 dissects the paths into four segments, 1–4. The respective speeds of propagation along these segments are

$$c_1 = (c^2 - v^2)^{1/2}$$
$$c_1' = (c^2 - v^2)^{1/2}$$
$$c_2 = c - v$$ (25.A.2)
$$c_2' = c + v$$

The corresponding wavelengths are $\lambda_1 = c_1/f$, $\lambda_1' = c_1'/f$, etc., so that the number of wavelengths in the round trip to M_1 is

$$N_1 = \frac{l}{\lambda_1} + \frac{l}{\lambda_1'} = \frac{lf}{c_1} + \frac{lf}{c_1'} = \frac{2lf}{(c^2 - v^2)^{1/2}}$$

(25.A.3)

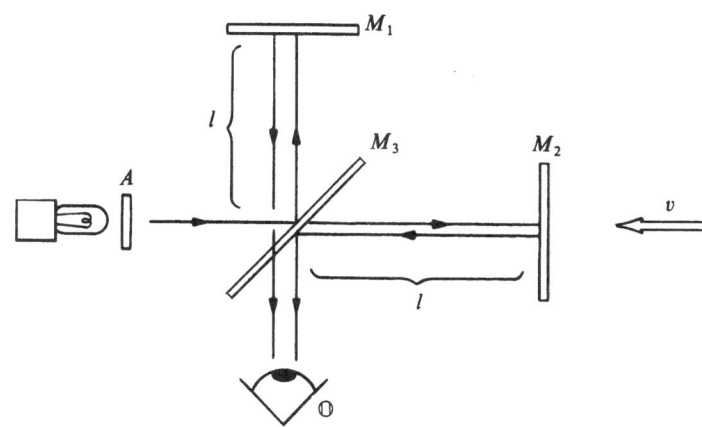

Figure 25.2. An equal-arm Michelson interferometer in a presumed ether wind of speed v.

Similarly, for the round trip to M_2,

$$N_2 = \frac{l}{\lambda_2} + \frac{l}{\lambda_2'} = \frac{lf}{c_2} + \frac{lf}{c_2'} = lf\left(\frac{1}{c-v} + \frac{1}{c+v}\right)$$

$$(25.A.4)$$

The required difference is

$$|N_1 - N_2| = lf\left(\frac{2}{(c^2 - v^2)^{1/2}} - \frac{1}{c-v} - \frac{1}{c+v}\right)$$

$$(25.A.5)$$

$$\approx \frac{lfv^2}{c^3}$$

$$(25.A.6)$$

$$\approx \frac{l}{\lambda}\frac{v^2}{c^2}$$

$$(25.A.7)$$

The small v/c approximation (25.A.6) is left as an exercise (see also the following mathematical comment); the last step, (25.A.7), uses $f = c/\lambda$, λ being the wavelength without ether drift. It is important to note that the sign of v does not affect the fringe shift.

We now come to the question of turning the ether drift on and off. This cannot be done, of course. Instead, Michelson and Morley slowly rotated their interferometer so that the two arm

directions were interchanged. The effect is the same as turning off v and turning it on again in the perpendicular direction. The overall fringe shift is then twice what is given by (25.A.7):

$$\boxed{\text{Expected fringe shift} = \frac{2l}{\lambda}\frac{v^2}{c^2}}$$

$$(25.A.8)$$

This represents the number of alternations from bright, to dark, to bright again, counted while the interferometer is being rotated by 90° in its own plane. The delicacy of the experiment is due to the smallness of the factor v^2/c^2.

Minimum Expected Shift

Even if the Sun were nearly motionless relative to the ether, we should expect at least an order of magnitude $v \approx 30$ km/sec from the Earth's orbital motion at some times of year. In the actual experiment, the interferometer arms were effectively lengthened to about $l = 11$ meters by multiple reflections from additional mirrors. With light of wavelength $\lambda = 590$ nm, Eq. (25.A.8) gives

Expected fringe shift

$$= \frac{(2)(11)(30 \times 10^3)^2}{(590 \times 10^{-9})(3 \times 10^8)^2} = 0.4$$

Thus, at least 40% of a fringe width should pass the observer's eye; in actual fact, less than 2% (consistent with zero) was ever seen. This null result has been confirmed by many subsequent researchers.

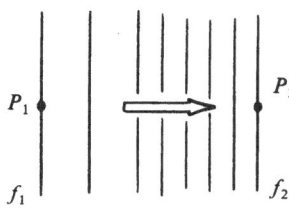

Figure 25.3. Steady state implies equal frequencies $f_1 = f_2$.

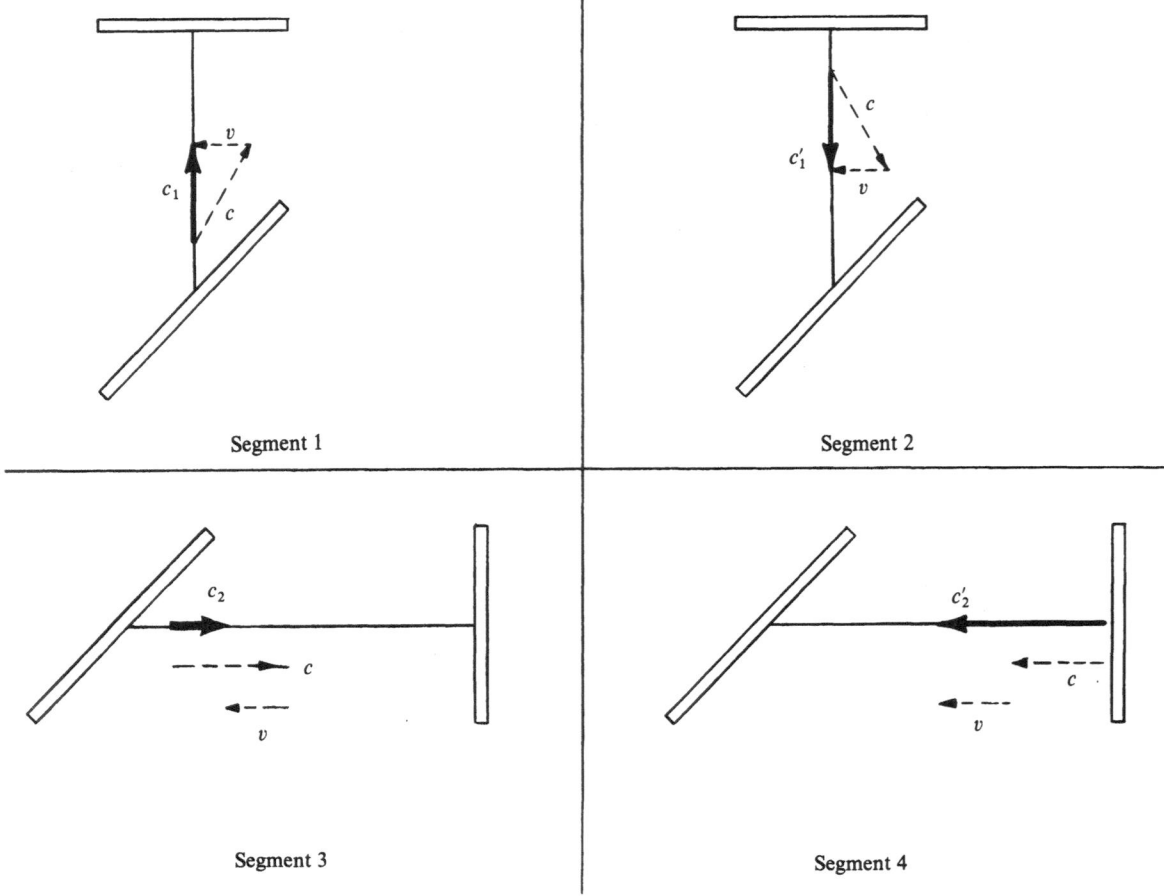

Figure 25.4. Predicted speeds c_1, c_1', c_2, c_2' in the Michelson interferometer of Fig. 25.2, as obtained from vector sums of velocities.

(Incidentally, it makes no appreciable difference whether the experiment is conducted in air or in vacuum.)

iii. A Mathematical Comment: Small-Speed Approximations

If x is a very small number,

$$|x| \ll 1 \qquad (25.A.9)$$

then the following approximation is valid:

$$\boxed{(1+x)^n \approx 1 + nx} \qquad (25.A.10)$$

This result, which is based on the so-called binomial series, is accurate to $\pm x^2$. The exponent n may be positive or negative; it need not be an integer.

Example 25.1. Approximating $1/(c^2 - v^2)^{1/2}$ for $v \ll c$.

This expression occurs in (25.A.5); we note that $v/c \ll 1$. Hence we set

$$\frac{1}{(c^2 - v^2)^{1/2}} = \frac{1}{c(1 - v^2/c^2)^{1/2}} = \frac{1}{c}\left(1 - \frac{v^2}{c^2}\right)^{-1/2}$$

$$\approx \frac{1}{c}\left[1 + \left(-\frac{1}{2}\right)\left(-\frac{v^2}{c^2}\right)\right]$$

since $n = -1/2$, $x = -v^2/c^2$. Thus, we obtain

$$\frac{1}{(c^2 - v^2)^{1/2}} \approx \frac{1}{c}\left(1 + \frac{v^2}{2c^2}\right) \qquad (25.A.11)$$

B. The Postulates of Relativity

The result of Michelson and Morley forces us to conclude that

The vacuum speed of a light signal,
as measured by a uniformly moving observer,
is independent of the observer's motion.

$$(25.B.1)$$

This is why no ether drift is ever observed, in spite of the motion of the Earth. Therefore there can be no ether at all. Light waves recognize no privileged inertial frames.

Common sense appears violated by (25.B.1), but we must keep in mind that daily experience allows us to detect only speeds that are exceedingly small compared to that of light, and hence that we lack the appropriate intuition; light signals seem to us to be transmitted instantaneously.

Accepting (25.B.1) means rejecting the nonrelativistic rule ("composition of velocities")

$$\mathbf{v}_{BA} = \mathbf{v}_{BL} - \mathbf{v}_{AL} \qquad \text{(nonrelativistic)}$$

$$(25.B.2)$$

which gives the relative velocity $\mathbf{v}_{BA}$ of an object B with respect to an object A in terms of their individual laboratory velocities $\mathbf{v}_{BL}$ and $\mathbf{v}_{AL}$. We have used that rule in connection with Fig. 25.1, and it has led to a wrong prediction—a nonexistent fringe shift.

i. Stating the Postulates

Is there an explanation for the validity of (25.B.1)? No. Special relativity adopts (25.B.1) as a basic postulate; that is to say, *it gives up any attempt at explaining* (25.B.1) in terms of simpler facts; rather, (25.B.1) itself is used in order to predict further physical results.

In addition, Einstein has assumed that

All physical laws, as tested by a uniformly
moving observer, are independent
of the observer's motion.   (25.B.3)

Galileo and Newton had already recognized this to be true in mechanics.

Inertial Frames

In both (25.B.1) and (25.B.3), a "uniformly moving observer" means an observer whose laboratory constitutes an inertial frame of reference. [We recall that (1) an inertial frame is one in which Newton's law of inertia is valid; (2) all inertial frames move uniformly and have a fixed orientation relative to each other and to the distant stars.]

Summary

The Postulates. Together, (25.B.1) and (25.B.3) are the two postulates underlying special relativity. In condensed form, they read as follows:

The speed of light is the same in all inertial frames. (25.B.4)
The laws of physics are the same in all inertial frames. (25.B.5)

[Note: (25.B.5) does not imply (25.B.4), because, while the laws do not change, the numerical data in general do. The speed of light is the exception, which requires special mention.]

The Ether. That concept has been discredited by the Michelson–Morley experiment. It amounts to a privileged frame of reference, and therefore is explicitly rejected by relativity.

Maxwell's Electromagnetism. That theory can use the same value of c in all frames, and is then in perfect accord with relativity.

Nonrelativistic Mechanics. This must be revised, except in the low-speed regime.

ii. Combining Two Velocities

At low speeds, relativistic and nonrelativistic predictions must be the same. As an illustration we examine, in one dimension, the composition of velocities (25.B.2),

$$v_{BA} = v_{BL} - v_{AL} \qquad \text{(nonrelativistic)}$$

$$(25.B.6)$$

All velocities are assumed to be in the same direction, as in Fig. 25.1.

The relativistic version of (25.B.6) turns out to be

$$\boxed{v_{BA} = \frac{v_{BL} - v_{AL}}{1 - \dfrac{v_{BL}v_{AL}}{c^2}}} \qquad (25.\text{B}.7)$$

For a proof, see Note (ii) at the end of this chapter.

First, we test (25.B.7) in the relativistic case of Fig. 25.1, where "object" B is a light signal, and A is the moving observer. Here we have

$$v_{AL} = v, \qquad v_{BL} = c \qquad (25.\text{B}.8)$$

Inserting in (25.B.7), we obtain

$$v_{BA} = \frac{c - v}{1 - cv/c^2} = c \qquad (25.\text{B}.9)$$

in happy agreement with postulate (25.B.4).

Next, we examine (25.B.7) for slow-moving objects A, B:

$$v_{AL} \ll c, \qquad v_{BL} \ll c \qquad (25.\text{B}.10)$$

We then have

$$\frac{v_{BL}v_{AL}}{c^2} \ll 1 \qquad (25.\text{B}.11)$$

and (25.B.7) reads

$$v_{BA} \approx v_{BL} - v_{AL} \qquad (25.\text{B}.12)$$

the correct nonrelativistic limit.

It follows from (25.B.7) that *no material object can ever be accelerated to the speed of light.* (For the steps involved in this deduction, see Problem 25.12.) Furthermore, it is generally accepted that

No material object, wave packet, or information-carrying signal of any kind can propagate faster than c relative to an inertial frame.

$$(25.\text{B}.13)$$

[A remark about "resultant velocities" can be use-ful at this stage. We sometimes need v_{BL} in terms of v_{AL} and v_{BA}. The nonrelativistic formula,

$$v_{BL} = v_{AL} + v_{BA} \qquad \text{(nonrelativistic)}$$
$$(25.\text{B}.14)$$

becomes, when corrected for relativity,

$$v_{BL} = \frac{v_{AL} + v_{BA}}{1 + v_{AL}v_{BA}/c^2} \qquad (25.\text{B}.15)$$

One way of obtaining this is to solve (25.B.7) for v_{BL}.]

iii. Uniqueness of c and Definition of the Meter

The speed of light in vacuum is unique in several ways. It does not depend on the velocity of the measuring apparatus; nor, as befits a wave, does it depend on the velocity of the light source. (Michelson was able to confirm this latter result experimentally as well.) Furthermore, light waves of different frequencies propagate precisely at the same speed: the vacuum shows no dispersion. (This has been established, by simultaneous radio and optical observation of flare stars, over a frequency range of as much as six orders of magnitude.)

The invariance of c is so well documented that it has given rise to a new international agreement. In 1983, the SI meter has been redefined as the distance light travels in vacuum during 1/299 792 458 of a second. In other words, the meter is chosen such that the speed of light is

$$c = 299\ 792\ 458 \text{ meters/sec} \qquad (25.\text{B}.16)$$

exactly.

(Why this particular number? The reason is that the "old" standard meter, which was defined in terms of the wavelength of a certain light emitted by a krypton-86 lamp, need not be changed in practice.)

C. The Equivalence of Mass and Energy

The most famous consequence of relativity is that *mass is a form of energy.* This will be more carefully stated below, and then proved deductively from the postulates.

i. Stating the Mass-Energy Relation

Here we focus our attention on a *system at rest*, that is, one with zero total momentum (in the non-relativistic sense). Nevertheless, the system may have component parts, like the electrons in the atoms of this page, which are in rapid motion relative to the system as a whole.

According to special relativity, a system at rest whose mass is M contains a total energy

$$\boxed{\mathscr{E} = Mc^2} \qquad (25.C.1)$$

where c is the speed of light. [We immediately note the dimensional correctness of (25.C.1); recall the formula $\frac{1}{2}mv^2$ for kinetic energy; we also note the enormity of the mass–energy conversion factor, $c^2 \approx 10^{17}$ (meters/sec)^2.] This formula needs to be interpreted carefully.

Right-Hand Side of (25.C.1)

The mass M of a system at rest (its **rest mass***) is defined in nonrelativistic terms, as in Chap. 3 of this book; it is just the system's inertia at small speeds, $M = F_{tot}/a$ (F_{tot} = total force on the system, a = acceleration); alternatively, M may be obtained from the static weight, $M = \text{weight}/g$ (g = local acceleration of gravity).

Left-Hand Side of (25.C.1)

Formula (25.C.1) makes no commitment as to the form of the energy $\mathscr{E}$; it may be, in part, the kinetic energy of the system's components, in part the mutual potential energy of these components, and in part other, unrecognized forms of energy. Two examples follow; several more will be presented in Chap. 28 (The Nucleus).

Example 25.2. Mass of a Kilocalorie. By how much does the mass of 1 kg of water increase when its temperature is raised by 1 C°?

From (25.C.1) we have, for the increases in $\mathscr{E}$ and M,

$$\Delta\mathscr{E} = (\Delta M)\, c^2 \qquad (25.C.2)$$

or

$$\Delta M = \frac{\Delta\mathscr{E}}{c^2} = \frac{1 \text{ kcal}}{c^2}$$

$$= \frac{4200 \text{ joules}}{10^{17}(\text{meters/sec})^2} \approx \underline{\underline{4 \times 10^{-14} \text{ kg}}}$$

an unobservably small amount. The relation $\mathscr{E} = Mc^2$ is not detectable in daily-life objects.

The following example, while impractical, is based on a correct and intriguing idea.

Example 25.3. Matter and Antimatter. Antimatter can be and has been produced in particle accelerators (only a few atomic particles at a time, however, and at huge expense). It can be made to undergo total annihilation together with an equal amount of ordinary matter; various kinds of radiation are then created, which, in principle, could give rise to a high-temperature source of heat. Suppose you are using a mass $m = 1$ milligram of antimatter* (about the mass of a pinhead) as fuel for your car, which needs an energy of about 10^7 joules/km. Over what distance, s, will your antimatter take you?

The energy from annihilation of a mass $2m$ (matter plus antimatter) is

$$\mathscr{E} = 2mc^2 \qquad (25.C.3)$$

If $s_0 = 1$ km requires $\mathscr{E}_0 = 10^7$ joules, then we have

$$\frac{s}{s_0} = \frac{\mathscr{E}}{\mathscr{E}_0}$$

$$s = \frac{\mathscr{E} s_0}{\mathscr{E}_0} = \frac{2mc^2 s_0}{\mathscr{E}_0}$$

$$\approx \frac{(2)(10^{-6})(10^{17})(1 \text{ km})}{10^7} = \underline{\underline{20\,000 \text{ km}}}$$

or halfway around the Earth.

---

* In modern practice, *mass* always means *rest mass*.

* Antimatter has positive mass. Negative masses do not appear to exist in Nature.

ii. Proof of $\mathscr{E} = Mc^2$

Introductory relativity often depends for its derivations on hypothetical situations, so idealized that they cannot be put into practice. These arguments are nevertheless valid because "thought experiments" are feasible *in principle*; the approach was popularized by Einstein. We shall prove the relation $\mathscr{E} = Mc^2$ on the basis of such a thought experiment.

The proof's two main ingredients are (1) the energy–momentum relation in an electromagnetic wave packet,

$$\text{Momentum} = \frac{\text{energy}}{c} \qquad (25.C.4)$$

derived in Chap. 22, and (2) the nonrelativistic definition of momentum,

$$\text{Momentum} = (\text{mass})(\text{velocity}) \qquad (25.C.5)$$

valid at small velocity.

In Fig. 25.5a, two identical electromagnetic wave packets, each of energy $\mathscr{E}/2$, are shown traveling horizontally in opposite directions inside a stationary box whose walls are perfect mirrors; the packets are simultaneously reflected from opposite walls and repeat their motion forever. There is no objection to assuming that the walls have a mass of their own, but for simplicity they are taken as massless. The box, with its two wave packets, constitutes a system at rest, with mass M and energy $\mathscr{E}$. Our task is to show that $\mathscr{E} = Mc^2$.

In part (b) of the figure we describe the system from the point of view of an observer who is moving upwards at a *small* speed u. To him, the wave packets have a downward component of velocity u; the speed of the packets is still c; the velocities make an angle θ with the horizontal such that

$$\sin \theta = \frac{u}{c} \qquad (25.C.6)$$

Parts (c) and (d) show the same situations as (a) and (b), but the labels are now *momenta* rather than speeds; from (25.C.4) we see that an energy

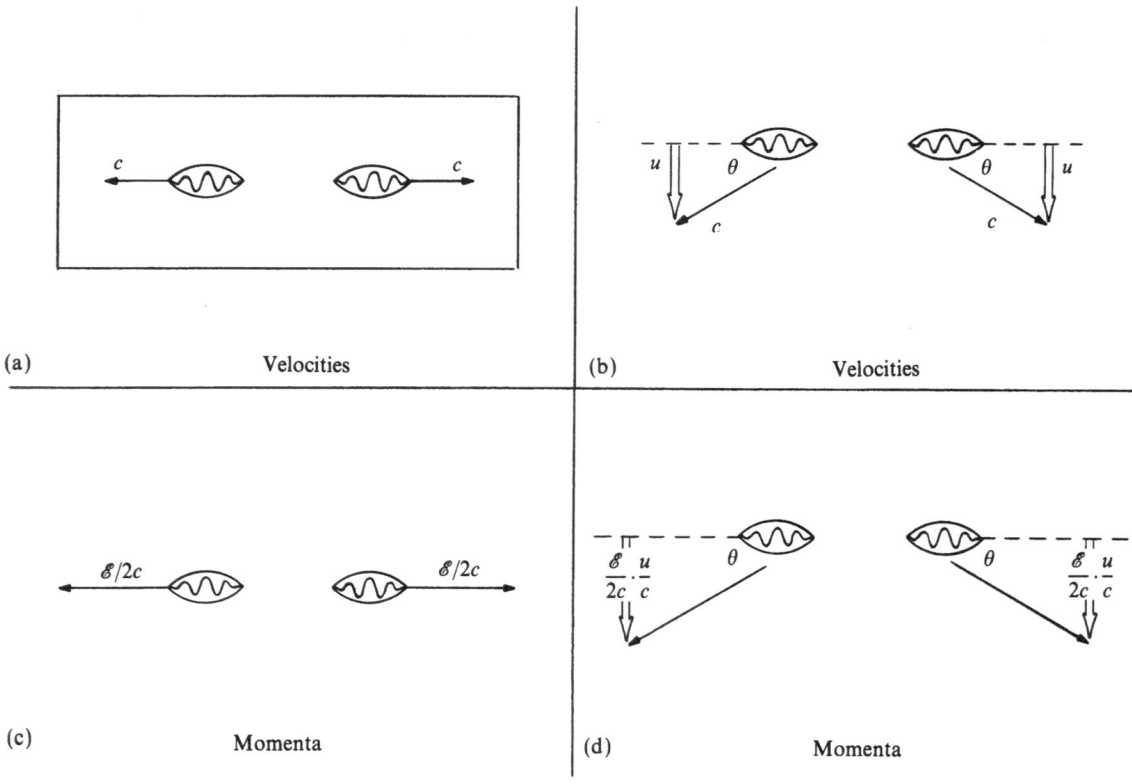

Figure 25.5. Two wave packets in a box. [The box is shown only in (a).]

$\mathscr{E}/2$ corresponds to a momentum $\mathscr{E}/2c$. For the moving observer, part (d), and for u small enough, that value is unchanged from (c). The downward component of momentum in (d) is, for each wave packet,

$$\frac{\mathscr{E}}{2c} \sin \theta = \frac{\mathscr{E}}{2c} \frac{u}{c} \qquad (25.\text{C}.7)$$

Hence the total momentum is

$$p_{\text{tot}} = \frac{\mathscr{E} u}{c^2} \qquad (25.\text{C}.8)$$

in the downward direction.

Since u is small, we have the nonrelativistic relation

$$p_{\text{tot}} = Mu \qquad (25.\text{C}.9)$$

Comparing (25.C.8) and (25.C.9) gives $\mathscr{E} = Mc^2$, as announced.

Why Is $\mathscr{E} = Mc^2$ Universal?

Two wave packets between mirrors are a very special case. However, we now recall from non-relativistic physics that, inside an isolated system at rest, neither the mass nor the energy ever changes no matter what internal events may occur. For example, our mirrors may tarnish, and the radiation be absorbed and converted to heat; neither $\mathscr{E}$ nor M will be affected. If all forms of energy are interconvertible in this way, the result is general.

D. Time Dilation

A physicist defines time as "what is measured with a clock," and then goes on to give a detailed recipe for building a clock in a standard, reproducible manner; it should function independently of environmental factors such as temperature, pressure, etc. Having achieved that aim, he might expect all clocks to agree with one another whenever they are compared. Relativity tells us that such an expectation is sometimes false.

Before we can proceed with the discussion, we must make a firm distinction (unnecessary in non-relativistic physics) between the *direct* and *indirect* comparison of two clocks. They are compared directly only when next to one another, i.e., when they occupy essentially the same location in space; we then decide to take the comparison process for granted, and do not ask any further questions about it.

If, on the other hand, two clocks are widely separated in space, then their comparison—or synchronization—must be carefully planned, and usually involves an exchange of signals.

What makes perfect clocks lose their synchronization? Consider a laboratory with many clocks, each at a fixed location in the room: if initially synchronized through some consistent procedure, they will always agree. Suppose, however, that one additional clock is carried at great speed through the laboratory in such a way that it passes by the stationary clocks one after the other. We now compare the reading on a stationary clock with that on the moving clock *as it passes by*. It turns out that *the moving clock's time lags further and further behind as it passes successive stationary clocks.* Its unit of time is "dilated." The effect is illustrated in Fig. 25.6. In what follows we describe time dilation* in more detail; we later show that it is a consequence of relativity.

i. The Time Dilation Formula

In Fig. 25.6, the laboratory, with all its clocks, is assumed to be an inertial reference frame; we choose to consider it as stationary. All the fixed clocks have been synchronized in some consistent manner, and are shown reading the same time t. If the moving clock has a constant velocity $\mathbf{v}$, it reads a time t' given by

$$\boxed{t' = \left(1 - \frac{v^2}{c^2}\right)^{1/2} t} \qquad (25.\text{D}.1)$$

(This relation assumes that the moving clock was set initially to agree with whatever stationary clock was next to it, $t = t' = 0$.) *For ordinary speeds, $v \ll c$, we see that time dilation becomes a negligible effect.* We shall see later how to derive formula (25.D.1).

Example 25.4. Deducing a Speed. Find v, as a fraction of the speed of light, in the illustration of Fig. 25.6.

_____

* Also called "time dilatation." Both usages are correct.

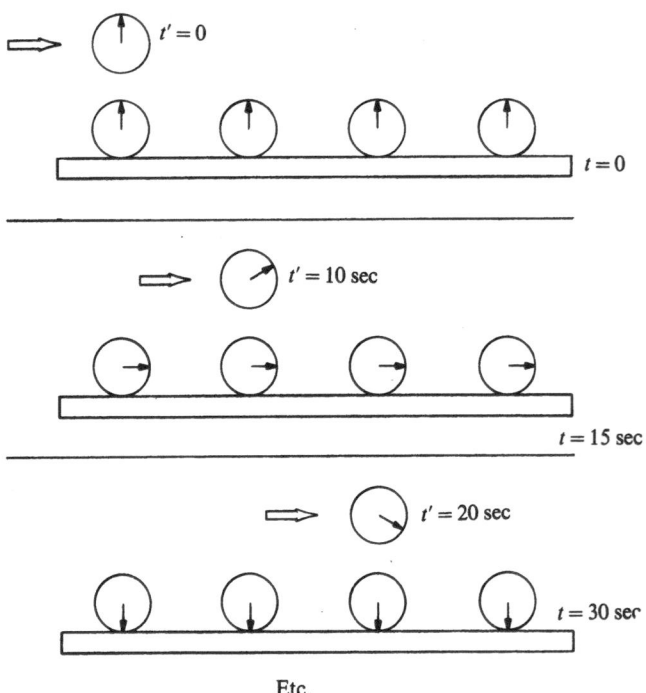

Figure 25.6. A traveling clock is compared with successive stationary clocks. The numerical values of t and t' are for illustration only.

In the figure, $t = 15$ sec implies $t' = 10$ sec. Solving (25.D.1) for v gives

$$v = \left[1 - \left(\frac{t'}{t}\right)^2\right]^{1/2} c \qquad (25.D.2)$$

$$= \left[1 - \left(\frac{10}{15}\right)^2\right]^{1/2} c \approx \underline{\underline{0.75c}}$$

about three-quarters of the speed of light.

Is the Laboratory Frame Special?

Are we stating that "clocks are slower in a moving frame than in the laboratory frame?" Such a lack of equivalence between the two inertial frames would contradict a basic postulate of relativity. Fortunately, there is no contradiction. "Clocks are slower" is a meaningless declaration unless we carefully specify the procedure for measuring a time interval. As Fig. 25.6 shows, we are using a different *procedure in both frames*. In the "stationary" frame we read the time on a succession of different clocks; in the "moving" frame we use a single clock. Thus, the asymmetry between the two results is not due to an asymmetry between the frames, but rather to the fact that we treat them differently.

Synchronizing the Laboratory Clocks

Making sure that two stationary clocks, A and B, are consistent with one another is not completely trivial if A and B are a long distance $\overline{AB}$ apart in space. Either one of the following methods is acceptable:

1. Set A to time zero, while sending a light signal from A to B. As B receives the signal, set it to a time $\overline{AB}/c$; this allows for the time lag due to distance.
2. Alternatively, bring B next to A, synchronize, and then *slowly* bring B to its final location. A time dilation effect is thereby avoided. (See also Problem 25.31.)

The Passage of Time

No specific design was given for the clocks of Fig. 25.6. They can, in principle, be mechanical, biological, or based on any phenomenon that exhibits a variation in time. This extraordinary generality implies that all processes (physical, chemical, biological including aging and even thought) are retarded according to formula (25.D.1).

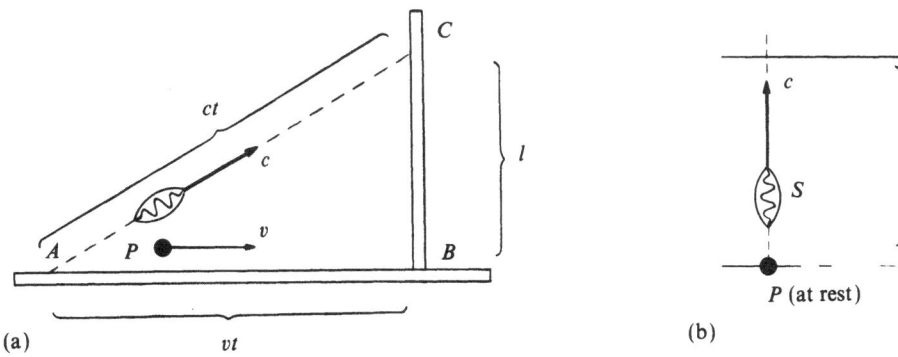

Figure 25.7. (a) Handicap race between a particle P and a light signal S. (b) The same, as seen from P.

Example 25.5. The Twin Paradox. Consider a pair of identical twins, Annette and Bernadette. At birth, Annette is sent on a round trip to a distant star at a constant 95% of the speed of light. (We assume she survives the endpoint accelerations involved.) Bernadette, who stays at home, eventually celebrates her 100th birthday ($t = 100$ years), which coincides with Annette's return. What is Annette's age, t'?

From (25.D.1) we have

$$t' = [1 - (0.95)^2]^{1/2} \, (100 \text{ years}) = \underline{31 \text{ years}}$$

Is there a fallacy involved? Here there is only one moving and one stationary clock. Why are Annette and Bernadette inequivalent?

A sketchy answer to this famous paradox is that the symmetry is not really there: Bernadette remains attached to a single inertial frame, while Annette, at the far end of her voyage, must switch from a frame with, say, velocity $+0.95c$ to one with $-0.95c$. A thorough analysis supports that explanation, as well as the fact that the age difference develops gradually—not just at the end point.*

The twin effect is now well confirmed by experiment—not on people, because actual spacecraft are too slow and people too inaccurate as clocks—but on radioactive particles. A particle does not age; it simply disintegrates at an unpredictable time. But a sample of radioactive particles does age (it *decays*) by losing a certain fraction of its members per unit time. A

sample whose speed is nearly c (inside one of today's giant accelerators) finds its decay rate spectacularly reduced, in precise agreement with (25.D.1).

ii. Theoretical Proof of Time Dilation

A thought experiment, illustrated in Fig. 25.7, will serve to establish the factor $(1 - v^2/c^2)^{1/2}$ in formula (25.D.1). In an inertial "stationary" laboratory, we race particle P (whose speed v is constant) against a light signal S (constant speed c); P and S leave point A simultaneously at time zero, and also arrive simultaneously, at time t, at the finish line BC. To compensate for the superior speed of S we aim it at an oblique angle to BC, while P travels along the perpendicular to BC. The reasoning consists of calculating, by two different methods, the final distance $l = \overline{BC}$ separating P and S. The two alternative calculations are done (a) in the laboratory frame, and (b) in a frame which moves together with P.

(a) Laboratory frame: From part (a) of Fig. 25.7, and using Pythagoras' theorem, we have

$$l = [(ct)^2 - (vt)^2]^{1/2} \qquad (25.\text{D}.3)$$

(b) Moving frame: As seen by the moving observer, P is motionless. The race lasts for a time t'. Part (b) of the figure gives

$$l = ct' \qquad (25.\text{D}.4)$$

since the speed of light is unchanged. Comparison of (25.D.3) and (25.D.4) gives

$$ct' = [(ct)^2 - (vt)^2]^{1/2}$$

*See, for example, P. A. Tipler, *Physics* (Worth, 1982), Sec. 35-7.

from which the time dilation formula (25.D.1) follows.

(In relativity, just as in nonrelativistic physics, consistency requires that a **transverse length**, i.e., one that, like *l*, is measured perpendicularly to the velocity of the observer, should not be affected by the observer's motion. We take this for granted in this derivation; the thoughtful reader may wish to derive that fact by considering two identical ribbons, each having width *l* when at rest, sliding past one another at a uniform relative speed *v*; in view of the symmetry between the two ribbons, can one of them be observed as wider than the other?)

iii. The Doppler Effect

The Acoustic Case (A Short Review)

A sound wave in air is heard with a higher pitch if source and listener are approaching one another than if both are at rest. In Chap. 14 we learned that, if the source's frequency is f while the frequency heard is f', we have

$$f' = \begin{cases} (1 + v/V)\, f & \text{(moving observer)} \quad (25.D.5) \\[2mm] \dfrac{f}{1 - v/V} & \text{(moving source)} \quad (25.D.6) \end{cases}$$

where V is the speed of sound in air; in (25.D.5), v is the observer's speed towards a stationary source, while, in (25.D.6), v is the source's speed towards a stationary observer; "stationary" means relative to the air. For small v/V, formula (25.D.6) reduces to (25.D.5).

The Electromagnetic Case

Qualitatively, light waves are affected in much the same way as sound waves. A visible source suffers an apparent **violet shift** (also called **blue shift**) if it and the observer are moving towards one another, and an apparent **red shift** in the opposite case. If light consisted of ether waves, formulas (25.D.5) and (25.D.6) would hold precisely, with V replaced by c. As it is, the ether does not exist, and the correct relativistic result is

$$\boxed{f' = \left(\frac{1 + v/c}{1 - v/c} \right)^{1/2} f} \qquad (25.D.7)$$

valid no matter whether the source or the observer is moving; they are only assumed to be approaching one another along the line of sight, in two different inertial frames with relative speed *v*.

We note that (25.D.7) happens to be a compromise (the geometric mean) between (25.D.5) and (25.D.6); for small v/c, (25.D.7) likewise becomes

$$f' \approx \left(1 + \frac{v}{c} \right) f \qquad (25.D.8)$$

In these formulas, a negative number should of course be substituted for v if approach is replaced by recession.

For a derivation of (25.D.7), see note (i) at the end of the chapter.

Shifted Spectral Lines

In order to tell whether an observed frequency is Doppler shifted, we need to know what its unshifted value is. This we fortunately do for a great variety of frequencies (known as **spectral lines***) emitted by atoms, whose unshifted lines are always available for comparison in the laboratory. Spectral lines will be discussed extensively in Chap. 27 (Atomic Structure).

Example 25.6. A Quasar's Speed of Recession. Spectral lines, characteristically emitted by hydrogen atoms, are observed in the light from **quasars** (or quasistellar objects), which are the most distant astronomical objects known. One such wavelength, $\lambda = 434$ nm (violet) is obtained from laboratory hydrogen; in a certain quasar it is seen as $\lambda' = 550$ nm (green), an extreme example of astronomical red shift. Assuming that atoms behave in the quasar as they do on Earth, and that the quasar is receding from us, along the line of sight, with a speed v, find v as a fraction of c.

Formula (25.D.7), with v replaced by $-v$, becomes

$$f' = \left(\frac{1 - v/c}{1 + v/c} \right)^{1/2} f \qquad (25.D.9)$$

---

* Because they correspond to sharp images of the slit in a spectroscope.

Solving for v, we find

$$v = \frac{1 - (f'/f)^2}{1 + (f'/f)^2} \, c \qquad (25.\text{D}.10)$$

Setting

$$\frac{f'}{f} = \frac{c/\lambda'}{c/\lambda} = \frac{\lambda}{\lambda'}$$

$$= \frac{434}{550} = 0.789$$

we get from (25.D.10)

$$v = \frac{1 - (0.789)^2}{1 + (0.789)^2} \, c = \underline{\underline{0.237c}}$$

Such enormous speeds of recession, and several larger ones, have in fact been observed, and are consistent with an expanding-universe model.

E. The Logic of This Chapter

As an aid to reviewing the preceding arguments, which are rather abstract, their logical flow is presented here. ("Frame" means "inertial frame.")

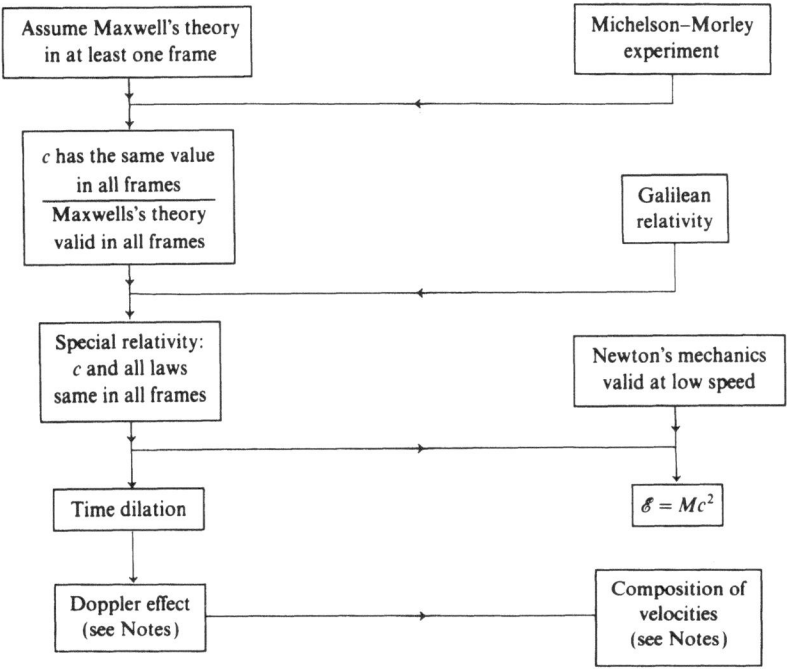

Notes

i. Deriving the Doppler Effect

Formula (25.D.7) for the Doppler-shifted frequency,

$$f' = \left(\frac{1 + v/c}{1 - v/c}\right)^{1/2} f \qquad (25.\text{N}.1)$$

is derived as in the acoustic case, but with a correction for time dilation. Consider, as in Fig. 25.8, a wave of frequency f (measured in a laboratory L), and an observer A who travels in the opposite direction at a laboratory speed v. As seen in L, the frequency $\tilde{f}$ at which A intersects the wave fronts is

$$\tilde{f} = \left(1 + \frac{v}{c}\right) f \qquad (25.\text{N}.2)$$

as would be the case for sound waves; see Sec. C iii of Chap. 14. (Relativistic effects are not involved so far, because we have not yet switched to the moving observer's point of view.) Again as seen in L, a clock traveling with A has its time slowed down by a factor $(1 - v^2/c^2)^{1/2}$; A, therefore, sees a frequency f' which is larger than $\tilde{f}$ by a factor $1/(1 - v^2/c^2)^{1/2}$;

$$f' = \frac{(1 + v/c) f}{(1 - v^2/c^2)^{1/2}} = \left(\frac{1 + v/c}{1 - v/c}\right)^{1/2} f$$

as announced in (25.N.1).

ii. Composition of Velocities

Here we derive Eq. (25.B.15),

$$v_{BL} = \frac{v_{AL} + v_{BA}}{1 + \dfrac{v_{AL} v_{BA}}{c^2}} \tag{25.N.3}$$

for the laboratory velocity of an object B which, seen from A, has a velocity v_{BA}.

The argument makes use of the Doppler formula (25.N.1). We go back to Fig. 25.8 and imagine an observer, initially at rest in L, to be set in motion, becoming attached to A. He will experience rule (25.N.1), as follows.

If an observer is given an additional speed v_{AL} relative to its former motion, the frequency it sees is multiplied by the factor

$$\left(\frac{1 + v_{AL}/c}{1 - v_{AL}/c} \right)^{1/2} \tag{25.N.4}$$

We now apply yet another change in velocity, $A \to B$. The frequency observed, first in A, and then in B, is

$$f \to \left(\frac{1 + \dfrac{v_{AL}}{c}}{1 - \dfrac{v_{AL}}{c}} \right)^{1/2} \times f$$

$$\to \left(\frac{1 + \dfrac{v_{BA}}{c}}{1 - \dfrac{v_{BA}}{c}} \right)^{1/2} \times \left(\frac{1 + \dfrac{v_{AL}}{c}}{1 - \dfrac{v_{AL}}{c}} \right)^{1/2} \times f \tag{25.N.5}$$

If the overall velocity of B relative to L is v_{BL}, we must have equivalently, for the overall change in frequency,

$$f \to \left(\frac{1 + \dfrac{v_{BL}}{c}}{1 - \dfrac{v_{BL}}{c}} \right)^{1/2} f \tag{25.N.6}$$

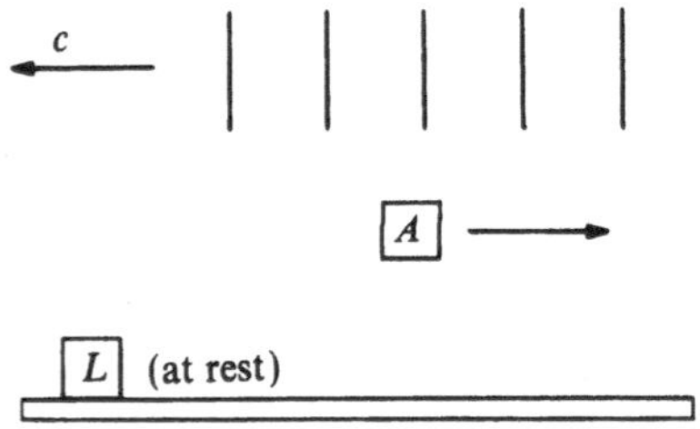

Figure 25.8. An observer successively sees a monochromatic wave from two platforms, L and A.

Squaring and comparing (25.N.5) and (25.N.6), we have

$$\frac{1 + \dfrac{v_{BL}}{c}}{1 - \dfrac{v_{BL}}{c}} = \frac{\left(1 + \dfrac{v_{BA}}{c}\right)\left(1 + \dfrac{v_{AL}}{c}\right)}{\left(1 - \dfrac{v_{BA}}{c}\right)\left(1 - \dfrac{v_{AL}}{c}\right)} \tag{25.N.7}$$

or, after some algebra, the desired result (25.N.3). [The easiest way of simplifying (25.N.7) is to use the algebraic rule that $a/b = c/d$ implies $(a - b)/(a + b) = (c - d)/(c + d)$.]

Condensed Checklist

A small-parameter approximation:

$$(1 + x)^n \approx 1 + nx \tag{25.A.10}$$

Michelson–Morley experiment:

$$\text{Expected fringe shift} = \frac{2l}{\lambda} \frac{v^2}{c^2} \tag{25.A.8}$$

Postulates of relativity:

All inertial frames exhibit the same laws and the same speed of light.

$$(25.B.4), (25.B.5)$$

Composition of velocities:

$$\boxed{v_{BA} = \frac{v_{BL} - v_{AL}}{1 - \dfrac{v_{BL} v_{AL}}{c^2}}} \quad , \quad v_{BL} = \frac{v_{AL} + v_{BA}}{1 + \dfrac{v_{AL} v_{BA}}{c^2}}$$

$$(25.B.7), (25.B.15)$$

Mass–energy relation

$$\boxed{\mathscr{E} = Mc^2} \tag{25.C.1}$$

Time dilation:

$$\boxed{t' = \left(1 - \frac{v^2}{c^2}\right)^{1/2} t} \tag{25.D.1}$$

Doppler effect:

$$f' = \left(\frac{1 + v/c}{1 - v/c}\right)^{1/2} f \approx \left(1 + \frac{v}{c}\right) f$$

$$\uparrow$$
$$\text{(25.D.7), (25.D.8)}$$
$$v \ll c$$

$$v = \frac{1 - (f'/f)^2}{1 + (f'/f)^2} c \qquad \text{(25.D.10)}$$

True or False

1. The Michelson–Morley experiment yielded only about 40% of a fringe shift, thereby disproving the existence of an ether.

2. Under the ether hypothesis, if the Earth's orbital speed around the Sun is about 30 km/sec, and if the Earth is, today, at rest relative to the ether, we should expect, six months from now, an ether wind of about 60 km/sec.

3. If a radio signal (a wave packet), traveling in space, is followed by a space ship at *nearly* the same speed, then a fast missile, sent forward from the space ship, should be able to catch up with the signal.

4. If one inertial frame, A, has a speed v relative to another inertial frame, B, then B also has speed v relative to A.

5. A top is slightly more massive when spinning than when not spinning.

6. Consider a nucleus at rest, which breaks up ("undergoes fission") into two equal parts, each of which has a high speed. The original mass of the nucleus equals the sum of the rest masses of its fragments.

7. Consider two inertial systems A and B whose relative speed is nearly c. As observed from A, the time in B is nearly "frozen"; the same statement is true for the time in A, as observed from B.

8. By speeding away from a given monochromatic light source, an observer can, in principle, receive as low a frequency as he pleases.

9. There is an absolute upper limit to the frequency that can be received from a given monochromatic light source with adjustable velocity.

Problems

Note: The speed of light is $c = 2.998 \times 10^8$ meters/sec.

Section A: The Michelson–Morley Experiment

*25.1. In the Michelson–Morley experiment, with dimensions described in Sec. A ii, suppose a fringe shift of 2 bright fringes had been observed. What speed of ether drift, as a fraction of c, would it have implied?

25.2. Under the ether-drift hypothesis, and assuming that the Earth has, at the time of measurement, an ether speed of 150 km/sec, find the expected fringe shift in a Michelson–Morley experiment performed in 450-nm light with 5.0-meter interferometer arms.

*25.3. If, believing in the ether hypothesis, we were to design a Michelson–Morley experiment such that a five-fringe shift be obtained with light of frequency 590 nm in a 120-km/sec ether drift, what effective arm length would we need for the interferometer?

*25.4. Approximate, for $v \ll c$, the following expressions:

(a) $\dfrac{1}{c - v}$; (b) $\dfrac{1}{c + v}$; (c) $(c - v)^{1/2}$;

(d) $(c + v)^{1/2}$; (e) $\left(\dfrac{c + v}{c - v}\right)^{1/2}$.

25.5. Do the small v/c algebra leading from (25.A.5) to (25.A.7).

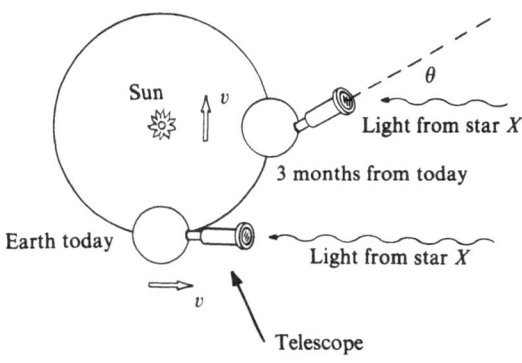

25.6 (Bradley's measurement of c). James Bradley, in 1728, deduced c from his observation of stellar *aberration*, or seasonal change in the apparent direction of a star, due to the Earth's orbital motion at right angles to the star's direction (as shown, this drawing not to scale). (Tilting one's telescope is, in this case, like tilting one's umbrella while running in vertically falling rain; a light signal plays the role of a rain drop; see Problem 2.53). Find a formula for c in terms of the aberration angle θ and the Earth's orbital speed v. [It turns out that the Sun itself may be assumed at rest without affecting the result. Assume $v \ll c$, and use (nonrelativistic) vector addition; neglect the rotational speed of the Earth's surface.]

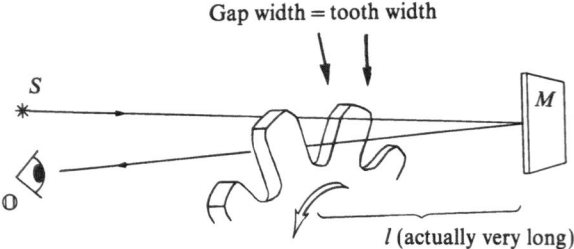

Gap width = tooth width

S

M

l (actually very long)

*25.7 (Fizeau's measurement of *c*). In 1849, Armand Fizeau used the illustrated arrangement—in a somewhat more elaborate version—to measure the speed of light. The fast rotating toothed wheel chops a light beam (coming from a source *S*) into distinct wave packets, which are reflected back from *M* a very long distance *l* away, in a neighboring town. During a signal's round trip, a tooth has replaced the gap and the returning signal is blocked; observer Ⓞ cannot see any reflected beam. (a) If the wheel has *N* teeth, and the first extinction occurs as the wheel reaches a rotational frequency *f*, express *c* in terms of *N*, *f*, and *l*. (b) By what amount would this result be affected if there were an ether drift of speed *v* along the beam's direction? (Note: The experiment's accuracy would be insufficient by far to reveal *v* even if it existed.)

Section B: The Postulates of Relativity

25.8. Show the algebra leading from v_{BA} [Eq. (25.B.7)] to v_B [Eq. (25.B.15)].

*25.9. An accelerator launches, one behind the other, particle *A* at a speed $\frac{2}{3}c$, and then particle *B* at speed $\frac{1}{3}c$. How fast is *A* relative to *B*?

25.10. A space ship from another planet is approaching Earth at a speed 0.25*c*. It then launches a missile towards the Earth at a speed 0.50*c* relative to the space ship. How fast is the missile approaching us?

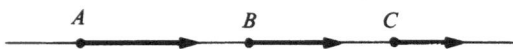

A   *B*   *C*

25.11. Three space ships, *A*, *B*, *C*, are traveling in line, as shown. Ship *B* radios to *C*: "*A* is approaching me at 0.8×10^8 meters/sec." Ship *C* itself observes that *B* is gaining on it at 2.2×10^8 meters/sec. What speed (relative to *C*) should *C* infer for *A*?

*25.12 (The speed *c* as an upper limit). Assume that some object *A* has an initial velocity $v_{AL} < c$ in the laboratory. Next, *A*'s velocity is increased by an amount v_{BA} relative to its initial motion; we also assume $v_{BA} < c$; the object is now called *B* and its resulting velocity is v_{BL} relative to the laboratory. Assuming that v_{AL}, v_{BA} are positive and in the same direction, prove that $v_{BL} < c$. [Conclusion: No object can be continuously accelerated

(i.e., by a succession of small v_{BA}) to a speed greater than *c*.]

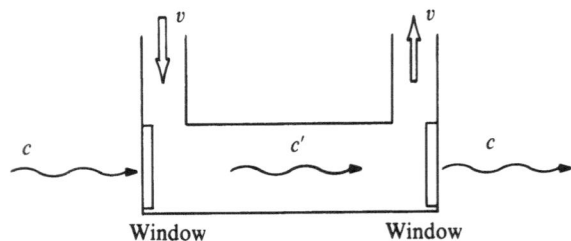

v   *v*

c   *c'*   *c*

Window   Window

*25.13 (Speed of light in moving water). A monochromatic light wave travels down a pipe in which water (index of refraction = *n*) is flowing at speed *v*, as shown. (a) Assuming that the wave fronts' speed *relative to the water* is *c/n*, find their speed *c'* relative to the laboratory. (b) Simplify your result in the case $v \ll c$. (c) Show that $c' < c$ always holds. (Note: The effect was measured interferometrically by Fizeau in 1859, long before relativity.)

Section C: The Equivalence of Mass and Energy

(For more problems on that topic, see Chap. 28, The Nucleus.)

25.14. What mass, in kilograms, is equivalent to 2×10^{20} joules (about the energy consumed in one year by the world's population)?

25.15. How much mass does a 2.5-watt flashlight radiate away in 1 hour?

*25.16. The Sun puts out, in electromagnetic radiation, a total power of about 4×10^{26} watts. Its total mass is 2×10^{30} kg. What fraction of its mass does the Sun lose in one century owing to its electromagnetic radiation alone?

25.17. In relativistic mechanics it is shown that the total energy $\mathscr{E}$ (= rest energy + kinetic energy) of a particle at speed *v* is given by $\mathscr{E} = mc^2/(1 - v^2/c^2)^{1/2}$, where *m* is the particle's rest mass. Approximate that formula for small *v/c*, and hence show that $\mathscr{E} \approx$ (rest energy) + (nonrelativistic kinetic energy).

*25.18. A cylindrical flywheel, of mass 15 kg, radius 0.20 meter, and moment of inertia 0.30 kg meter², has a rotational frequency *f*. (a) How large a value of *f* will increase the flywheel's mass by 0.04 %? (Use the nonrelativistic expression for the kinetic energy of a flywheel; assume, unrealistically, no break-up or change in shape due to stresses.) (b) What is then, as a fraction of *c*, the speed of a point on the wheel's rim?

Section D: Time Dilation

*25.19. A certain clock is traveling at a speed $\frac{1}{2}c$ relative to a set of stationary clocks. When a time of 40 sec has elapsed on the latter, how much time has elapsed on the moving clock?

25.20. If time is elapsing on a moving clock at half the rate indicated on a set of stationary clocks, find the speed of the moving clock, as a fraction of c.

25.21. Do the algebra leading from the Doppler shift (25.D.9) to the speed (25.D.10).

*25.22. How fast, as a fraction of c, is a quasar receding if its spectrum's observed frequency is half of its emitted frequency?

*25.23. By what factor is the frequency of a star's spectral line multiplied if it recedes from us (a) at a speed $\frac{1}{2}c$? (b) At a speed $0.01c$?

25.24. A certain spectral line, of frequency 434.0×10^9 Hz, is emitted by a star that is receding from us at 1.650×10^8 meters/sec. What frequency do we observe?

*25.25. How fast, in meters/sec, is a star receding if its spectrum's observed frequencies are 99.5% of its emitted frequencies?

*25.26 (A more realistic human "twin experiment"). A present-day space ship can reach, say, a speed of 10 000 meters/sec. If an astronaut keeps that speed for 30 years of Earth time until he comes back to Earth, how much less, in seconds, will he have aged than the people back home?

25.27. A sample of positive pions (a certain kind of radioactive particle), when at rest, loses half of its population in 1.8×10^{-8} sec. (This period of time is called the pion's **half-life**.) In order to extend their half-life to a full second, how close to the speed of light should the pions be made to travel?

25.28. What fraction of a second is lost in one year by a clock located at the equator, as compared with one at the pole, solely owing to the Earth's rotation, if we assume that the time-dilation formula applies?

25.29. A fictitious space ship, traveling at constant speed v, reaches the star Alpha Centauri, which is 4 light years away. Its passengers note that they have left Earth only one year before. Find v.

*25.30. Muons are radioactive particles observed in cosmic rays and in high-energy accelerator laboratories. Half the particles originally present in a sample of muons at rest have disintegrated in 2.3×10^6 sec ($=$"half-life" of the muons). (a) What is the muons' new half-life if their speed is 99.995% of the speed of light? (b) How far do they travel during that time?

*25.31. Section D-i mentions synchronization by means of clock transportation. Let two stationary clocks A and B be a distance l apart, and let us carry a clock C from A to B at a constant speed v. (a) If A and C are both set to time zero when next to one another, what do B and C each read when together? (b) Show that the discrepancy approaches zero as $v \to 0$.

*25.32. Rewrite the Doppler formula, Eq. (25.D.7), in terms of the wavelengths λ' and λ, corresponding (in vacuum) to the frequencies f' and f.

*25.33. Rewrite the approximate Doppler formula, Eq. (25.D.8), as simply as possible in terms of λ and λ'. (See also Problem 25.32.)

25.34. An optics professor, in traffic court for ignoring a red light (say $\lambda \approx 700$ nm), pleaded that, to him, it was being Doppler shifted to green (say $\lambda' \approx 550$ nm). The judge, who knew some physics, accepted the plea but convicted him for speeding instead. Estimate, in km/h, the professor's implied speed.

25.35. A certain spectral line, emitted by a distant galaxy, is found to be red-shifted in wavelength from 486.1 nm to 492.0 nm. What is the galaxy's speed of recession?

25.36. Consider the experiment of Fig. 25.7 non-relativistically (suppose c is the speed of sound, S is a sound pulse, and P is a subsonic airplane). Find the speed $c'(\neq c)$ of the sound pulse relative to the airplane, in part (b) of the figure; express your answer in terms of c and v.

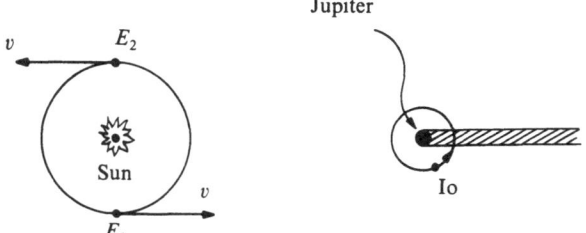

25.37 (Roemer's measurement of c). The first estimate of c was announced by Olaus Roemer in 1676; it is based on the following observation. Io, one of Jupiter's satellites, is eclipsed by Jupiter's shadow at average intervals of about 1.769 days. That interval appears shorter by about 0.01% when the Earth is at point E_1 of its orbit, as shown, and longer by the same amount when the Earth is at E_2. The effect is a Doppler shift in the number of eclipse signals per unit time. We can neglect Jupiter's orbital speed compared to ours. (a) In terms of the fractional change $(\Delta T)/T \approx 0.02\%$ in Io's apparent period, and in terms of the Earth's orbital speed $v \ll c$, find an approximate formula for c; (b) verify your result with $v \approx 30$ km/sec. (Note: the adjoining sketch is not to scale.)

Answers to True or False

1. False (there was far less than 40%, consistent with zero).
2. True.
3. False.
4. True.
5. True (it has more energy).
6. False (the original nucleus has more energy than the fragments at rest).
7. True.
8. True.
9. False.

Answers or Hints to Problems

25.1. $2.3 \times 10^{-4}c$.

25.3. 9.2 meters.

25.4. (a) $\dfrac{1}{c} + \dfrac{v}{c^2}$.

25.7. (a) $c = 4Nlf$; (b) $c(1 - v^2/c^2) = 4Nlf$, or, with $v \ll c$, $c = 4Nlf + v^2/4Nlf$.

25.9. $\frac{3}{7}c$.

25.12. Hint: Use Eq. (25.B.15); derive a contradiction from assuming $v_{BL} \geqslant c$.

25.13. Hint: Use Eq. (25.B.15); let A be the water and B the wave fronts.

25.16. 7×10^{-12}.

25.18. (a) 1×10^7 Hz.

25.19. 35 sec.

25.22. $\frac{3}{5}c$.

25.23. (a) $1/\sqrt{3}$.

25.25. Hint: Use the small v approximation.

25.26. 0.5 sec.

25.30. (b) 70 km.

25.31. (a) l/v for B, $(l/v)(1 - v^2/c^2)^{1/2}$ for C.

25.32. $\lambda' = [(1 - v/c)/(1 + v/c)]^{1/2} \lambda$.

25.33. $\lambda' \approx (1 - v/c) \lambda$.

CHAPTER 26

WAVES VERSUS PARTICLES

By now, we have accumulated massive evidence that light consists of waves. We have seen the admirable success of Maxwell's theory, which predicts, among other facts, the speed of the waves and the phenomenon of polarization. Most convincing of all are the countless optical experiments which exhibit interference and diffraction, thereby permitting accurate wavelength measurements; surely a wavelength cannot be measured unless there is a wave.

The student will therefore be legitimately confused when learning about experiments in which light shows itself to be made of particles, after all. (A particle of light is called a **photon**, or a **light quantum.***) Some of those experiments are outlined in the present chapter. They are easy to describe, and their results are simple and clear cut.

Two revolutionary statements, which seem to contradict what we have learned so far, form the gist of this chapter: *light consists of particles*; *electrons consist of waves*.

Can the wave and particle pictures be reconciled in a logical way? The answer is yes, but at a heavy price. An important traditional belief of scientists, namely, **determinism**, must be abandoned. How such a sacrifice helps resolve the wave–particle dilemma will be indicated in the last section of the chapter.

[It may comfort the puzzled reader to know that the discovery of photons induced among physicists a somewhat chaotic state of mind which persisted for many years (1905–1926).]

** Plural: **quanta**.*

Classical versus Quantum Physics

Physics is conventionally divided into a **classical** and a **quantum** domain; this nomenclature is a useful one to keep in mind. What we have surveyed up to now is **classical physics**; it consists principally of disciplines developed through 1905, and is generally understood to include relativity, its crowning achievement.

Quantum physics begins with Planck's black body theory (1900) and Einstein's photon theory (1905); it is still being vigorously developed today. We shall see, in this chapter and the next, that quantum physics is dominated by a certain numerical constant h, unimaginably small in practical units ($h = 6.6 \times 10^{-34}$ joule sec), called **Planck's constant**. As a rule of thumb, the presence of h in a formula earmarks it as a product of quantum physics, just as the presence of c (the speed of light) characterizes electromagnetic or relativistic results.

(The term **modern physics** is also frequently used; it is generally meant to include quantum physics and relativity.)

Electron Waves

Quantum physics has revised, not only our understanding of light, but our understanding of electrons (and of all other particles) as well. Just as a photon is a wave *and* a particle, so is an electron a particle *and* a wave. The wave nature of the electron has been established experimentally beyond doubt, and explains or predicts more properties of matter than classical physics has ever been able to do. Only the first direct evidence for electron waves, namely, **electron diffraction**, will be touched upon in this chapter. We leave for the next chapter a discussion of why electron waves are relevant to atomic structure.

A. Photoemission

Under certain conditions, a metal surface emits electrons when exposed to light. The effect is called **photoemission**,* and the emitted electrons are known as **photoelectrons** (they are, of course, the same as any other electrons).

In the photoemission experiment that we shall discuss, two quantities are being independently adjusted: the **frequency** of the incident light and its **intensity**. One then measures the response of two other quantities: the number of photoelectrons emitted per unit time, and the kinetic energy with which the electrons are ejected from the metal. *The experimental outcome compels us to ascribe particle-like properties to light.*

i. The Photoelectric Tube

Figure 26.1 schematically illustrates an arrangement for making a photoemission measurement. A thoroughly evacuated enclosure (**photoelectric tube**) holds two electrodes, A and B; the enclosure is preferably made of quartz in order to transmit ultraviolet radiation, which would be blocked by ordinary glass.

A beam of monochromatic light, of frequency f and intensity $\mathscr{I}$, strikes A (the **photocathode**), which responds by emitting electrons with mixed velocities $\mathbf{v}$, $\mathbf{v}'$, etc. An adjustable voltage V is maintained between A and B in such a direction as to repel the electrons away from B and make them fall back into A; V is a "retarding potential." Electrons emitted in a favorable direction, and whose kinetic energy is large enough to overcome V, reach B and complete the circuit; they give rise to a current I (the **photocurrent**). The photocurrent vanishes when V is large enough. (By convention, we consider V to be positive.)

---

* When irradiated, a material can respond electrically in several different ways. Photoemission is only one instance of these **photoelectric effects**. The photovoltaic effect, exploited in cameras' light meters and in "solar cells," is different from photoemission.

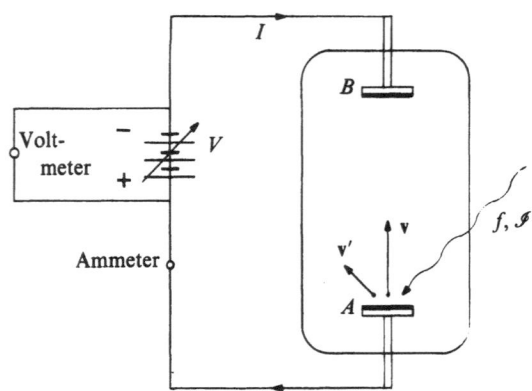

Figure 26.1. A photoelectric tube. Recall that the electrons always travel oppositely to I. Note that here I flows *against* the voltage V. In other words, the photoelectric tube acts as another source of emf, opposite to and stronger than V. Electrode A is called the **photocathode** irrespective of the sign of its potential relative to B. The arrow through the voltage supply indicates that it is adjustable.

Equality of the Two Electrodes

For the sake of simplicity, we assume that A and B face one another with identically prepared surfaces. If they did not, for example, if they were made of different metals, they would in general produce an additional (but weak) electric field in the space between them; that field would exist even under a null reading of the voltmeter in Fig. 26.1. We shall not have to worry about that effect, whose explanation* would take us too far from our subject.

Maximum Kinetic Energy

The purpose of setting up V is to measure the maximum kinetic energy, $\mathscr{K}_{\max}$, with which a photoelectron can be emitted. The lowest value of V that still prevents all electrons from reaching B is called the **stopping potential** V_{st}. In order to see how it depends on $\mathscr{K}_{\max}$ we consider, in Fig. 26.1, the fastest photoelectrons that are sent towards B. Under the effect of V_{st}, they run out of kinetic energy just short of B's surface. Their initial kinetic energy, $\mathscr{K}_{\max}$, is, at that point, completely converted to potential energy, eV_{st}:

$$\mathscr{K}_{\max} = eV_{\mathrm{st}} \qquad (26.A.1)$$

---

* See "Contact Potential Difference" in a more specialized textbook.

To summarize: Through Eq. (26.A.1), *the stopping potential reveals the maximum kinetic energy that any photoelectron can have as it leaves electrode A.*

ii. The Photoelectric Equation

To see most directly where classical physics fails, we concentrate on one innocent-looking question: "how should we adjust the incident light so as to raise $\mathcal{K}_{max}$?"

The classical prescription is: "increase the light's intensity $\mathcal{I}$." Indeed, at the electrode's surface, the E field of the light wave exerts a force $-e\mathbf{E}$ on an electron, thereby supplying the necessary energy. By increasing $\mathcal{I}$ we increase E, and $\mathcal{K}_{max}$ should therefore increase as well.

The experimental result flatly disagrees with that expectation:

> The measured $\mathcal{K}_{max}$ is independent of the light intensity $\mathcal{I}$.

(26.A.2)

Instead, $\mathcal{K}_{max}$ increases only when light of higher frequency is used:

> $\mathcal{K}_{max}$ increases with f.

(26.A.3)

Classical electromagnetism has no explanation for such a behavior.

The increase of $\mathcal{K}_{max}$ with f is accurately linear, as we can see in Fig. 26.2; $\mathcal{K}_{max}$ obeys the **photoelectric equation**

$$\mathcal{K}_{max} = h(f - f_0)$$

(26.A.4)

$$(h = \text{const}, f_0 = \text{const}, f > f_0)$$

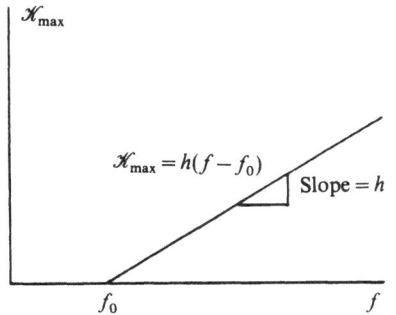

Figure 26.2. Graph of the photoelectric equation.

where h (the slope of the graph) is known as **Planck's constant**, and f_0 as the **threshold frequency**. With (26.A.1), the photoelectric equation reads

$$V_{st} = \frac{h}{e}(f - f_0)$$

(26.A.5)

Thus, assuming that $e = 1.6 \times 10^{-19}$ coulomb is already known, the values of h and f_0 are obtained directly through (26.A.5) from the experimental plot of V_{st} against f; h/e is its slope, while f_0 is its intercept on the f axis.

Planck's Constant

The number h is found to have a universal value, independent of the material used for the cathode. According to recent measurements, which utilize many other effects in addition to photoemission, we have

$$h = 6.626196 \times 10^{-34} \text{ joule sec}$$

(26.A.6)

Note the units, which follow from Eq. (26.A.4); equivalently,

$$1 \text{ joule sec} = 1 \frac{\text{kg meter}^2}{\text{sec}}$$

(26.A.7)

(The joule sec is often referred to as the SI unit of **action**, and h is the **quantum of action**.)

The Threshold Frequency

The kinetic energy $\mathcal{K}_{max}$ is, of course, necessarily positive. We therefore see from Eq. (26.A.4) that, if photoelectrons are emitted at all, the incident f must be greater than f_0.

In contrast to h, the measured value of f_0 depends strongly on the nature of the photocathode. Table 26.1 lists some threshold frequencies (and corresponding wavelengths $\lambda_0 = c/f_0$) for various materials. We see that photoemission from metals ordinarily requires ultraviolet light; the alkalis, such as sodium, potassium, and cesium, are exceptional in this regard, since they exhibit photoemission in the visible.

A considerably lower f_0, useful in infrared detectors, is featured by some well-chosen metallic compounds such as Na_2KSb (sodium-potassium-antimony), or by elaborately prepared electrode surfaces such as silver, coated with successive thin layers of silver oxide and cesium (Ag-O-Cs).

Table 26.1.   Some Photoemission Thresholds[a]

Surface material	Threshold		Response
	Frequency f_0 (10^{14} Hz)	Wavelength λ_0 (nm)	
Gold (Au)	12.5	240	
Copper (Cu)	11.0	270	
Silver (Ag)	10.3	290	Ultraviolet only
Zinc (Zn)	10.3	290	
Magnesium (Mg)	8.8	340	
Potassium (K)	5.8	520	
Sodium (Na)	5.5	540	Visible and ultraviolet
Cesium (Cs)	5.2	580	
Na_2KSb[b]	3.3	900	Infrared, visible, ultraviolet
Ag–O–Cs[b]	2.7	1100	

[a] The visible range extends between 3.90×10^{14} Hz and 7.70×10^{14} Hz, or between 770 nm and 390.nm.
[b] See text.

Example 26.1. Predicting a Stopping Potential. Find V_{st} for sodium under violet light of wavelength $\lambda = 440$ nm.

Table 26.1 gives $\lambda_0 = 540$ nm. Rewriting Eq. (26.A.5) in terms of wavelengths, we have

$$V_{st} = \frac{h}{e}\left(\frac{c}{\lambda} - \frac{c}{\lambda_0}\right)$$

or

$$V_{st} = \frac{hc}{e}\left(\frac{1}{\lambda} - \frac{1}{\lambda_0}\right) \quad (26.A.8)$$

$$= \frac{(6.6 \times 10^{-34})(3.0 \times 10^8)}{1.6 \times 10^{-19}}$$

$$\times \left(\frac{1}{440 \times 10^{-9}} - \frac{1}{540 \times 10^{-9}}\right) \text{ volt}$$

$$= \underline{0.52 \text{ volt}}$$

iii. Einstein's Theory of Photons

Photoemission had been discovered in 1887 by Heinrich Hertz—the same man who first detected radio waves. Since then, several disturbing effects, one of which we have just presented, gradually became known as experimental technique improved. Their resolution through quantum mechanics is one of the most astonishing turns of

events science has ever witnessed, and its antecedents deserve at least a brief mention, although they will not be used in the logic of this chapter.

In 1905 (the year in which he also published his special theory of relativity), Albert Einstein announced a radical solution to the photoemission puzzle. His clue came from another mystery—the familiar glow of heated objects (such as red-hot pokers). Only one theoretical model, that of Max Planck (1900) had perfectly accounted for the color of the glow at any given temperature. The amount of radiated energy, plotted wavelength by wavelength, is known as the **black-body radiation spectrum**. Planck had obtained that spectrum *by departing from the rules of classical physics*.

In effect, he had assumed that, when interacting with radiation, matter can gain or lose energy only in tiny **quanta**, or lumps. He had taken each energy quantum $\Delta\mathcal{E}_{matter}$ to be proportional to the frequency f of the radiation involved:

$$\Delta\mathcal{E}_{matter} = \pm hf \quad (26.A.9)$$

where the $\pm$ symbolizes gain or loss. Planck's argument (which we cannot reproduce here) required a numerical value $h \approx 6.6 \times 10^{-34}$ joule sec in order to fit the observed black-body spectrum.

Einstein's generalization seems quite natural to us in retrospect. On the basis of Eq. (26.A.9), he postulated that the electromagnetic radiation itself travels through space as a stream of particles, or **photons**, each with energy

$$\mathscr{E} = hf \qquad (26.A.10)$$

the radiation being assumed monochromatic in this discussion. Planck's formula, (26.A.9), simply represents the arrival or departure of a photon. *Equation (26.A.10) is among the most important in modern physics.*

We may now visualize photoemission as follows:

1. A photon of energy $\mathscr{E} = hf$ strikes the photocathode, and may penetrate it slightly.
2. The photon encounters an electron. Electrons in a piece of matter move at various speeds and have various energies; suppose the electron under consideration needs an additional energy $\mathscr{E}_{esc}$ to escape from the cathode into the vacuum.
3. The photon somehow gives all its energy $\mathscr{E}$ to the electron, and is itself annihilated in the process. If $\mathscr{E} > \mathscr{E}_{esc}$, the electron can escape, and does so with a kinetic energy $\mathscr{K} = \mathscr{E} - \mathscr{E}_{esc}$, or

$$\mathscr{K} = hf - \mathscr{E}_{esc} \qquad (26.A.11)$$

4. Consider the highest-energy electrons that reside in the cathode, and suppose that, by chance, the struck electron had been among those. Therefore, here we need to consider the minimum value of $\mathscr{E}_{esc}$, known as the cathode's **work function**, and denoted by ϕ:

$$\phi = \text{minimum value of } \mathscr{E}_{esc} \qquad (26.A.12)$$

The electron's kinetic energy $\mathscr{K}$ is then a maximum, and Eq. (26.A.11) becomes

$$\mathscr{K}_{max} = hf - \phi \qquad (26.A.13)$$

This is nothing but the photoelectric equation (26.A.4),

$$\mathscr{K}_{max} = h(f - f_0)$$

in which

$$hf_0 = \phi \qquad (26.A.14)$$

The work function is proportional to the threshold frequency.

Figure 26.3 summarizes the photoemission process, seen from the energy point of view.

(One revolutionary idea of this chapter, namely, the existence of photons, has now been outlined. The remaining discussion is one of detail, and should serve to lend some reality to photons in the reader's mind.)

Example 26.2. A Work Function in Electron Volts. Find the work function ϕ of silver, expressed in electron volts (a convenient and customary unit for this purpose).

From Table 26.1 we have, for the threshold frequency of silver,

$$f_0 = 10.3 \times 10^{14} \text{ Hz}$$

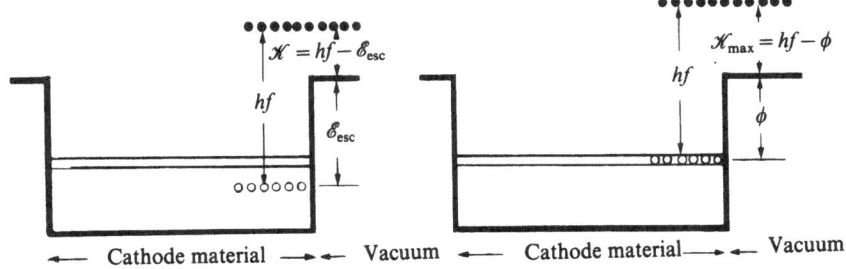

(a) TYPICAL ELECTRON   (b) HIGHEST-ENERGY ELECTRON

Figure 26.3. The photoemission process: an electron receives an energy hf, and escapes from the cathode. Energy is plotted vertically, position horizontally. Legend: ——, the potential energy well responsible for keeping the electrons inside (the zero level is arbitrary); ═══, highest energy of an electron inside the cathode; ∘∘∘, energy of an electron before the photon strikes; •••, energy of the electron after being struck.

The work function is [Eq. (26.A.14)]

$$\phi = hf_0$$

In electron volts,

$$\frac{\phi}{1 \text{ electron volt}} = \frac{hf_0}{1 \text{ electron volt}}$$

$$= \frac{(6.6 \times 10^{-34})(10.3 \times 10^{14})}{1.6 \times 10^{-19}} = \underline{\underline{4.2}}$$

The work function of silver is 4.2 electron volts.

iv. A Spectrum of Photon Energies

The basic equation (26.A.10), $\mathscr{E} = hf$, implies that any monochromatic radiation is made of photons with well-defined energy $\mathscr{E}$ calculable from the classical frequency f. The theory admits no exceptions: $\mathscr{E} = hf$ should be as true for radio "waves" (low-energy photons) as for x rays or gamma rays (high-energy photons.) It turns out, however, that radio waves behave in so many respects exactly as true waves that their particle nature is well hidden; higher-energy photons are easier to detect as particles.

Radiation that is not monochromatic is, by definition, a mixture of frequencies; hence it must contain a *mixture of photons* of different energies. A spectrograph will sort them out according to their individual energies. (In wave language, it of course "separates the frequencies.")

The energy per photon, hf, is displayed in Chap. 22, Figs. 22.13 and 22.14, for the various kinds of electromagnetic radiation.

v. Intensity of a Photon Beam

The intensity $\mathscr{I}$ of a light beam can be interpreted very simply in terms of photons. Consider a monochromatic beam of frequency f and of perpendicular cross section $\mathscr{A}_\perp$, as in Fig. 26.4. We have, by definition,

$$\mathscr{I} = \frac{\left(\begin{array}{c}\text{average energy through } \mathscr{A}_\perp \\ \text{per unit time}\end{array}\right)}{\mathscr{A}_\perp} \qquad (26.A.15)$$

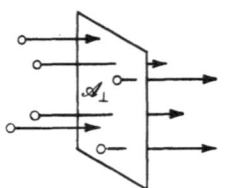

Figure 26.4. A photon beam.

or

$$\mathscr{I} = \frac{\left(\begin{array}{c}\text{average number of} \\ \text{photons per unit time}\end{array}\right)}{\mathscr{A}_V} \times \left(\begin{array}{c}\text{energy} \\ \text{per photon}\end{array}\right) \qquad (26.A.16)$$

or, more briefly,

$$\boxed{\mathscr{I} = \left(\begin{array}{c}\text{photon current} \\ \text{density}\end{array}\right)(hf)} \qquad (26.A.17)$$

Example 26.3. Estimating a Photon Flux. A certain beam of green light, with frequency $f = 5.6 \times 10^{14}$ Hz and intensity $\mathscr{I} = 0.080$ watt/meter^2, illuminates (at normal incidence) an eye pupil of area $\mathscr{A}_\perp = 0.05$ cm^2. How many photons enter the pupil per second (i.e., what is the photon flux)?

We have

$$\text{Photon flux} = \left(\begin{array}{c}\text{photon} \\ \text{current density}\end{array}\right)\mathscr{A}_\perp$$

or, using Eq. (26.A.17),

$$\text{Photon flux} = \frac{\mathscr{I}\mathscr{A}_\perp}{hf} \qquad (26.A.18)$$

$$= \frac{(0.080)(0.05 \times 10^{-4})}{(6.6 \times 10^{-34})(5.6 \times 10^{14})} \frac{\text{photons}}{\text{sec}}$$

$$\approx \underline{\underline{10^{12} \frac{\text{photons}}{\text{sec}}}}$$

Immediate Onset of the Photocurrent

When the light is turned on, the photocurrent is observed immediately, without any detectable time lag. This is true even under very weak light. Classically, one would expect the first photoelectrons to wait a long time, namely, until the energy they need to escape has been collected from the low-intensity wave. (For details, see Problem 26.13.)

On the other hand, according to photon theory, the very first photon already has enough energy to liberate an electron; a small intensity just means a longer average spacing between photons. Thus, *the absence of a time lag favors photon theory over classical electromagnetism.*

vi. Photocurrent versus Light Intensity

A third piece of evidence for the existence of photons is found in the dependence of the photocurrent I on the incident light intensity $\mathcal{I}$. The simplest assumption is that each photon acts independently of all the others in knocking an electron out of the cathode. Then the number of electrons emitted per unit time should be proportional to the photon flux; equivalently,

$$I \propto \mathcal{I} \qquad (26.A.19)$$

The photocurrent is proportional to the intensity. That fact is accurately verified in the laboratory. It cannot be predicted classically, since an increasing $\mathcal{I}$ would be expected to increase primarily the electrons' energy rather than their number.

Quantum Efficiency

The ratio $I/\mathcal{I}$, multiplied by a suitable factor, tells us the number of electrons emitted per incident photon, or **quantum efficiency**, of various materials. That number is expected to be less than 1, since some photons do not succeed in extracting an electron. (Note that the quantum efficiency does not depend on the retarding potential, which is effective only after photoemission.)

Under visible light, quantum efficiencies range from zero to as high as 30%, depending on the light's frequency and on the cathode material. High quantum efficiencies are sought for television cameras and other practical light detectors.

B. X Rays

To generate x rays in the laboratory, *we allow fast electrons to collide with the atoms of a solid target.* When an electron suddenly slows down,

stops, or changes its direction of motion, it radiates electromagnetic energy. (According to classical electromagnetism, any accelerating charge must radiate.) The radiation from colliding charges goes under the name of **bremsstrahlung** (German: Bremse = brake, Strahlung = radiation). Bremsstrahlung can, in principle, have any frequency, but its most important scientific applications are in the x-ray range. It may be considered the *inverse of photoemission,* much as buying is the inverse of selling. Here the electron's kinetic energy becomes the energy of a photon, while in photoemission the energy of a photon becomes the kinetic energy of an electron.

X rays were discovered in 1895, by Wilhelm Konrad Roentgen. The event occurred in a darkened room which contained a cathode-ray tube enclosed in black cardboard, and, some distance away, a sheet of paper coated with barium platinocyanide. [That substance was known to fluoresce when exposed (inside the tube) to cathode rays; as it happens, it fluoresces under x rays as well.] The coated paper gave off light whenever the tube was turned on. In Roentgen's experiments, the electrons' impact on the glass wall of the tube were the source of the x rays.

i. The X-Ray Tube

Figure 26.5 gives the basic design of an x-ray tube. As in Roentgen's original experiments, we have here essentially a cathode-ray tube, from which the air has been evacuated. The electron beam emerges from the heated cathode and strikes a massive metal target (the anode), often made of tungsten or molybdenum. For several reasons, including the intense hot spot due to the impinging

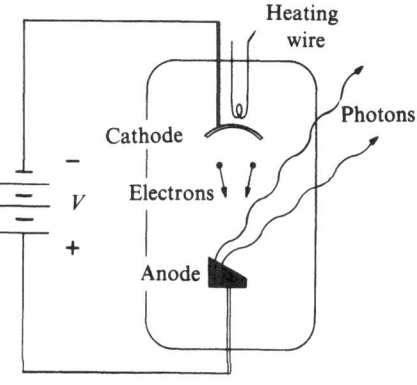

Figure 26.5. An x-ray tube.

electrons, as well as the efficiency of x-ray production, using a metal target is preferable to letting the beam strike the tube's wall, which is made of glass, quartz, or other insulator transparent to x rays.

The target is kept at a positive potential V relative to the cathode, and therefore the electrons reach the target with a kinetic energy

$$\mathscr{K} = eV \qquad (26.B.1)$$

(Their initial kinetic energy is negligible.) They typically penetrate the target somewhat, and are slowed to a stop by successive collisions with the metallic atoms. Each collision results in the creation of one or several photons, which may be absorbed by the target itself, or which may escape and be detected outside the tube.

ii. The Cutoff Frequency

How large an energy, hf, can any of these photons have? The largest possible value, hf_{max}, is one which concentrates an electron's whole kinetic energy $\mathscr{K}$ into the creation of a single photon:

$$hf_{max} = \mathscr{K} \qquad (26.B.2)$$

or, with (26.B.1),

$$\boxed{hf_{max} = eV} \qquad (26.B.3)$$

Thus, f_{max} is the so-called cutoff frequency corresponding to a tube voltage V. This relation, which was verified experimentally by W. Duane and F. L. Hunt in 1915, is another confirmation of Einstein's photon theory.

The frequency spectrum received from an x-ray tube is sketched in Fig. 26.6. What is measured in practice is the wavelength λ rather than the frequency f; we have seen in Chap. 24 how Bragg diffraction from a known crystal may be used for that purpose. The frequency is then obtained from $f = c/\lambda$. Since the cutoff frequency is a maximum, *the cutoff wavelength is a minimum:*

$$\lambda_{min} = \frac{c}{f_{max}} \qquad (26.B.4)$$

Equivalently, with Eq. (26.B.3), we have

$$\lambda_{min} = \frac{hc}{eV} \qquad (26.B.5)$$

or

$$\boxed{\lambda_{min} = \frac{1240 \text{ nm volts}}{V}} \qquad (26.B.6)$$

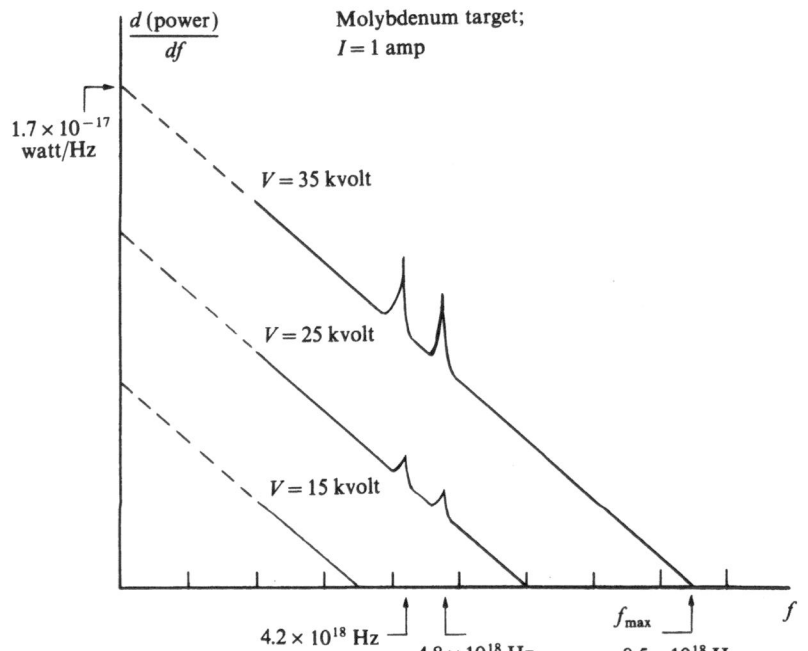

Figure 26.6. The output spectrum (in terms of power radiated per unit frequency) from an x-ray tube, for various tube voltages V. For us, the important feature of this diagram is the cutoff frequency f_{max}, which is proportional to V. The linearity of the curves results from a thick target and is only approximate. At low frequencies (dashed segments) the radiated power is difficult to measure; at higher frequencies one may find some sharp peaks which result from the detailed structure of the target atoms (here molybdenum). The total radiated power (here for a tube operated at a 1-A current) equals the area under the appropriate curve. At currents I other than 1 A, we note that the plotted quantity, $d(power)/df$, is proportional to I (V and f being fixed). For further details, see Problems 26.21–26.25.

C. The Compton Effect*

In 1923 Arthur H. Compton obtained new evidence that the photon is a true particle. He found experimentally that *a photon can collide elastically with an electron*, pretty much in billiard-ball fashion. During that collision, the two particles' combined momentum is conserved as well as their combined translational kinetic energies.

That such a scenario really takes place was inferred by Compton from wavelength measurements on x rays. Confirmation comes from photographing the track of a recoil electron after the impact of a gamma-ray photon (i.e., a very high-energy x-ray photon). A photon has a trajectory of its own; see Fig. 26.7.

In brief: Compton's clue was that *x rays increase their wavelength by being scattered.*

i. The Momentum of a Photon

Consider a collimated monochromatic light beam, of frequency f and wavelength λ; we shall view it as a stream of photons, each of energy

$$\mathscr{E} = hf \qquad (26.C.1)$$

As we shall see, it follows that *each photon also has a well-defined momentum p,* given by

$$\boxed{p = \frac{h}{\lambda}} \qquad (26.C.2)$$

(Before going on, the reader might wish to verify the dimensional correctness of that formula.)

Our method for justifying Eq. (26.C.2) is often used in quantum physics. It consists of avoiding (or postponing!) a conflict between the classical and quantum points of view.

Classically, in a traveling electromagnetic wave, we have

$$\frac{\text{Energy density}}{\text{Momentum density}} = c$$

as derived in Chap. 22. Therefore, for each particle in the beam, we must have

$$\frac{\text{Energy}}{\text{Momentum}} = c \qquad (26.C.3)$$

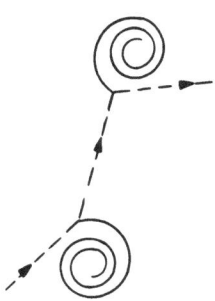

Figure 26.7. A photon's trajectory, although invisible, can *in principle* be traced from its collisions with electrons.

or, using (26.C.1) and $f = c/\lambda$,

$$p = \frac{\mathscr{E}}{c} = \frac{hf}{c} = \frac{h}{\lambda} \qquad (26.C.4)$$

as claimed.

Photons Have Zero Rest Mass

Since a photon has speed c, we must be careful in applying the nonrelativistic expression $p = mv$. In this case, $v = c$ is constant, and yet p varies from photon to photon. If we insist on writing

$$p = \text{“}m\text{”}c \qquad (26.C.5)$$

we must accept a mass "m" that also varies from photon to photon. In the limit $p \to 0$ we must have "m" $\to 0$. Hence, although there is no such thing as a photon at rest, one commonly says that "photons have zero rest mass."

Aside from motivating this remark, Eq. (26.C.5) is of doubtful utility; we shall not use it again.

Example 26.4. Photon versus Electron. A certain x-ray photon and a certain electron have the same momentum. The wavelength of the x-ray beam is $\lambda = 10^{-10}$ meter, about the size of an atom. (a) What is the speed v of the electron? (b) How does the electron's kinetic energy compare with the photon's energy?

(a) Speed of the electron: In view of Eq. (26.C.1), the equality of the momenta reads

$$mv = \frac{h}{\lambda} \qquad (26.C.6)$$

* This section can be left out without loss of continuity.

where m is the electron's mass. Hence we have

$$v = \frac{h}{\lambda m} = \frac{7 \times 10^{-34}}{(10^{-10})(9 \times 10^{-31})} \frac{\text{meters}}{\text{sec}}$$

$$\approx 10^7 \frac{\text{meters}}{\text{sec}}$$

[Important: The nonrelativistic form mv in Eq. (26.C.6) is now justified, since v is small compared to $c = 3 \times 10^8$ meters/sec.]

(b) Comparing the energies: If $f = c/\lambda$ is the x-rays' frequency, we obtain a ratio

$$\frac{\text{Electron kinetic energy}}{\text{Photon energy}} = \frac{\frac{1}{2}mv^2}{hf} = \frac{mv^2\lambda}{2hc} \quad (26.C.7)$$

The appropriate numbers may be inserted directly, but it is more satisfying to use, in (26.C.7), the value of mv taken from (26.C.6). We get

$$\frac{\text{Electron kinetic energy}}{\text{Photon energy}} = \frac{v}{2c} = \frac{10^7}{(2)(3 \times 10^8)}$$

$$\approx \underline{\underline{10^{-2}}}$$

ii. Photon–Electron Collisions

Compton's experimental arrangement is shown in Fig. (26.8). A collimated beam of x rays, with wavelength λ, strikes a target which is a sample of any substance. An incident x-ray photon can undergo four possible fates.

1. It can be unaffected by the target and continue straight through. With a thin target, this is true for most photons in the beam.
2. It can be annihilated, giving all its energy to an electron, as happens in photoemission.
3. It can be deflected by an electron without shaking it loose, even slightly, from the atom to which it is attached. The photon changes its direction, but not its energy. (Bragg diffraction is an example of such a process; we recall, incidentally, that Bragg diffraction cannot be understood except in terms of an electromagnetic wave.)
4. (Compton scattering, the case of interest). The photon can transfer *part* of its energy, and some momentum as well, to an electron.

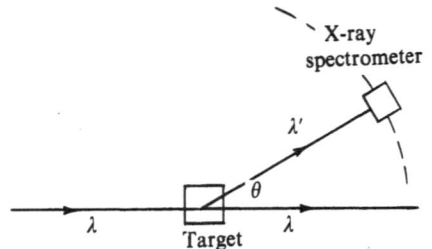

Figure 26.8. Compton's experiment. The wavelength λ' of a scattered beam is measured by an x-ray spectrometer, whose adjustable position follows the dashed curve; x rays are, in fact, scattered at all angles θ, of which one is shown being sampled by the spectrometer.

In contrast to case 2, the photon is not annihilated, but continues in a new direction. In contrast to case 3, the emerging photons ("Compton-scattered photons") have less energy than the incident ones; hence, *if they are observed as waves*, their frequency f' is lower, and their wavelength λ' longer, than the corresponding parameters f, λ of the incident beam.

iii. Compton's Formula

We now use photon theory in order to derive the *change in wavelength* suffered by Compton-scattered x rays. Given an incident beam of wavelength λ, and a scattering angle θ, as in Fig. 26.8, we must find the scattered wavelength λ' under the assumption that the incident photons make *elastic collisions* with free stationary electrons in the target.

(*Targets of Low Atomic Number.* A free electron is one that is not bound to an atom. Why are we making the "free and stationary" assumption? Mainly for the sake of getting a simple, definite answer from our calculation. Fortunately, the assumption is a good one if the target has a low atomic number; for example, it might consist of hydrogen or helium. In that case it turns out that the escape energy and the initial kinetic energy of the electrons are both negligible compared to the energy of an x-ray photon; thus, the electrons are effectively free and at rest. The next chapter will explain how to make such energy estimates.)

The energy–momentum calculation is based on Fig. 26.9; we assume initial and final frequencies f, f' for the radiation and a final speed v ($v \ll c$, non-

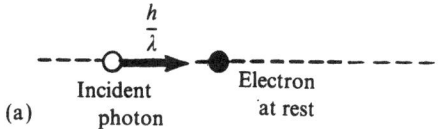

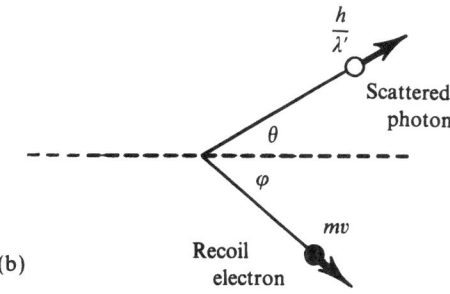

(a) Incident photon, Electron at rest

(b) Scattered photon, Recoil electron

θ, φ, mv

Figure 26.9. A photon–electron collision. (a) Initial situation. (b) Final situation: θ may have any value between zero and 180°; the electron recoils at an angle φ from the forward direction.

relativistic) for the electron. The relevant data are shown in Table 26.2.

Conservation of energy reads

$$hf = hf' + \tfrac{1}{2}mv^2 \qquad (26.C.8)$$

Conservation of x momentum:

$$\frac{hf}{c} = \frac{hf'}{c}\cos\theta + mv\cos\varphi \qquad (26.C.9)$$

Conservation of y momentum:

$$0 = \frac{hf'}{c}\sin\theta - mv\sin\varphi \qquad (26.C.10)$$

From this set of three equations (26.C.8)–(26.C.10), we eliminate the two variables v and φ; for the algebraic details, see the note at the end of

Table 26.2. Collision Data

	Photon	Electron
Initial energy	hf	0
Final energy	hf'	$\tfrac{1}{2}mv^2$
Initial momentum	$\dfrac{h}{\lambda} = \dfrac{hf}{c}$	0
Final momentum	$\dfrac{h}{\lambda'} = \dfrac{hf'}{c}$	mv

this chapter. The single resulting equation, in the nonrelativistic limit $v \ll c$, is

$$\frac{1}{f'} - \frac{1}{f} = \frac{h}{mc^2}(1 - \cos\theta) \qquad (26.C.11)$$

Introducing the wavelengths through $\lambda = c/f$, $\lambda' = c/f'$, we have

$$\boxed{\lambda' - \lambda = \frac{h}{mc}(1 - \cos\theta)} \qquad (26.C.12)$$

This is the predicted **Compton shift** in the wavelength of x rays scattered through an angle θ. That shift is in fact observed; *the energy–momentum bookkeeping of photon theory is strikingly confirmed.* Formula (26.C.12) even happens to be correct for electrons recoiling with relativistic speeds, although here we have derived it nonrelativistically.

Dimensionally, Eq. (26.C.12) is interesting. Because the left-hand side is a length, so is the factor h/mc; it is called the **Compton wavelength of the electron.** Its numerical value is

$$\frac{h}{mc} \approx \frac{6.6 \times 10^{-34}}{(9.1 \times 10^{-31})(3.0 \times 10^8)} \text{ meter}$$

$$\approx 2.4 \times 10^{-12} \text{ meter} \qquad (26.C.13)$$

quite small relative to the size of an atom ($\approx 10^{-10}$ meter). Aside from being featured in Eq. (26.C.12), that constant plays an important role in modern electron theory.

iv. How Small is a Photon?

This is an obvious question, but, because the photon is not a classical object, it has no obvious answer; the volume over which a photon interacts with matter depends on the experimental circumstances. Some experiments—which may destroy the photon—do give a very small upper limit to a photon's size. For example, a photon can be localized at one grain of a photographic emulsion, or at one photoreceptor in the eye. On a much smaller scale, it can select, as we have seen, the site of one in a dense crowd of electrons, causing photoemission or Compton scattering. A high-energy (gamma ray) photon can strike a single proton in a nucleus and eject it in a manner analogous to photoemission or Compton scattering. It is believed that there are, in principle, experiments that can concentrate the effect of a photon over as small a volume as desired; in that sense, a photon has zero size. Accurately speaking, the size of a photon is not a very meaningful concept.

D. Electron Waves

Electrons are particles. Many experiments support that conclusion; recall Millikan's measurements, in which he has counted single electrons; recall also the multitude of electron tracks recorded in high-energy laboratories, and of which Fig. 18.15b of Chap. 18 is an example. Finally, recall the electron trajectories in cathode ray tubes; Newton's laws of particle mechanics successfully predict their shape.

Nevertheless, in 1927 evidence was found in two laboratories—by Clinton Davisson and his collaborator Lester Germer, and, independently, by Sir George Paget Thomson—that *an electron beam has some properties that cannot be understood unless it is a wave.* These findings were not unexpected to some theoreticians of the period. In 1923, Prince Louis de Broglie had suggested that all particles should exhibit wave properties if given the opportunity to do so in a suitably designed experiment. We discuss, in turn, this idea—which was not immediately accepted—and its experimental confirmation.

i. De Broglie's Theory

If we put ourselves in the right frame of mind, it becomes plausible to us that particles should have wave properties. Suppose we were familiar with photons, but not with electromagnetic waves. If confronted with the interference and diffraction of light we would say, "a certain kind of particle—the photon—has wave properties." The generalization, "perhaps all kinds of particles have wave properties" is then quite natural.

De Broglie's postulates are carefully designed to agree with the already existing photon theory, and with special relativity as well; they imply the following statement:

A beam of particles, all with the same momentum

$\mathbf{p}$ *and energy* $\mathscr{E}$, *is equivalent to a plane monochromatic wave of wavelength*

$$\boxed{\lambda = \frac{h}{p}}$$ ("de Broglie wavelength") (26.D.1)

and frequency

$$f = \frac{\mathscr{E}}{h} \qquad (26.D.2)$$

($\mathscr{E}$ includes the particle's rest energy, mc^2, and its kinetic energy). Here h is once more Planck's constant, $h \approx 6.6 \times 10^{-34}$ joule sec.

These relations are clearly valid in photon theory; we recall from Eq. (26.C.2) that the wavelength of a light beam is

$$\lambda = \frac{h}{\text{photon momentum}} \qquad (26.D.3)$$

and from Eq. (26.C.1) that its frequency is

$$f = \frac{\text{photon energy}}{h} \qquad (26.D.4)$$

In this book we make no use of the energy relation (26.D.2) except in the case of photons; but the momentum relation (26.D.1), as applied to electrons, will play an important role in deducing atomic structure (next chapter). Equation (26.D.1) constitutes the second main idea of the present chapter; the first has been the existence of photons, obeying Eq. (26.A.10).

The Wave Function

The existence of a wavelength λ and frequency f implies that *something* must be vibrating. In the case of photons, the vibrating quantities are the $\mathbf{E}$ and $\mathbf{B}$ fields, which are functions of position and time. In the case of electrons, no vibrating quantity had ever been detected, and someone had to invent a new function ψ ("psi") of position and time, the so-called electron wave function. This was the achievement of Erwin Schrödinger (1926).

One simple meaning of ψ is that,

In regions of space where $\psi = 0$, we find no electrons; where ψ is large, the chance of finding an electron is high.

(26.D.5)

The interpretation of ψ will be touched upon again in the next section.

Example 26.5. Getting a Wavelength from a Voltage. A certain electron gun accelerates electrons through a potential difference $V = 80$ volts. Find the wavelength of the beam's wave function.

Each electron has a kinetic energy

$$\tfrac{1}{2}mv^2 = eV$$

and therefore has a momentum

$$p = mv$$

$$= [(2m)(\tfrac{1}{2}mv^2)]^{1/2} = [(2m)(eV)]^{1/2} \quad (26.D.6)$$

$$= [(2)(9.1 \times 10^{-31})(80)(1.6 \times 10^{-19})]^{1/2} \frac{\text{kg meter}}{\text{sec}}$$

$$= 4.9 \times 10^{-24} \frac{\text{kg meter}}{\text{sec}}$$

(Why is a nonrelativistic calculation justified?) The electrons' de Broglie wavelength is therefore, by Eq. (26.D.1),

$$\lambda = \frac{h}{p} = \frac{6.6 \times 10^{-34}}{4.9 \times 10^{-24}} \text{ meter} = \underline{1.3 \times 10^{-10} \text{ meter}}$$

comparable to an atomic diameter.

ii. The Davisson–Germer Experiment

Figure 26.10 shows the basic arrangement used by Davisson and Germer. A beam of electrons, whose momentum p is controlled by the accelerating voltage V of the electron gun, strikes a crystal (nickel in the original experiment). A movable detector measures how many electrons are reflected in various directions. The findings are as follows:

1. For most values of p, the electrons are diffusely reflected in all directions.
2. For a special value of p, a rather sharp direction is favored by the reflected electrons, as indicated by the peak in Fig. 26.10a.
3. The position of the peak, and the value of p, are consistent with a *Bragg-like diffraction of the electron's wave function from the lattice planes of the crystal*.

Example 26.6. Predicting a Gun Voltage. In one experiment, the electrons' angle of incidence with the relevant lattice planes is $\theta = 25°$, as indicated in Fig. 26.10. The lattice spacing is $s = 0.091$ nm, as independently determined by x rays. At what gun voltage, V, should we observe the diffraction peak *if de Broglie's wavelength postulate is correct?*

We treat the electron beam as we have treated x rays two chapters ago. If λ is the de Broglie wavelength, we have Bragg's condition [Eq. (24.A.15) of Chap. 24]

$$\cos \theta = \frac{k\lambda}{2s} \quad (k = 1, 2, 3, ...) \quad (26.D.7)$$

We next eliminate λ in favor of V. De Broglie's relation, $\lambda = h/p$, gives

$$\lambda = \frac{h}{(2Vme)^{1/2}} \quad (26.D.8)$$

as explained in Eq. (26.D.6). Eliminating λ from (26.D.7) and (26.D.8), and solving for V, we have

$$V = \frac{1}{2me} \left(\frac{kh}{2s \cos \theta} \right)^2 \quad (26.D.9)$$

One possible value of V corresponds to $k = 1$:

$$V = \frac{1}{(2)(9.1 \times 10^{-31})(1.6 \times 10^{-19})}$$

$$\times \left(\frac{(1)(6.6 \times 10^{-34})}{(2)(0.091 \times 10^{-9})(\cos 25°)} \right)^2 \text{ volts}$$

$$= \underline{55 \text{ volts}}$$

(close to the value measured by Davisson and Germer). For $k = 2, 3, ...$ we get $V = 212$ volts, 477 volts, etc. These voltages should also lead to the observation of a peak at $25°$.

Width of the Peak

In contrast to the Bragg diffraction of x rays, we do not get a sharp "spot" from these low-energy electrons. One reason is that they do not penetrate the target very deeply; only a few lattice planes, close to the crystal surface, are involved. We have observed before (Chap. 24, Sec. A v) that sharply

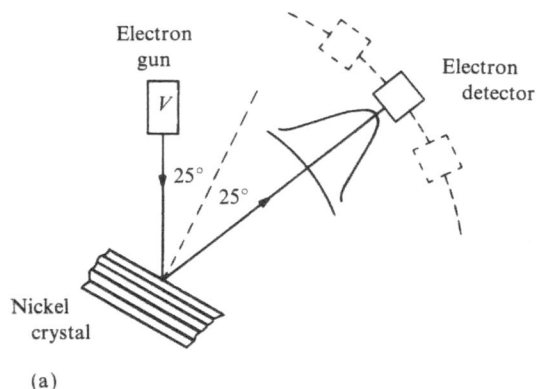

(a)

(c)

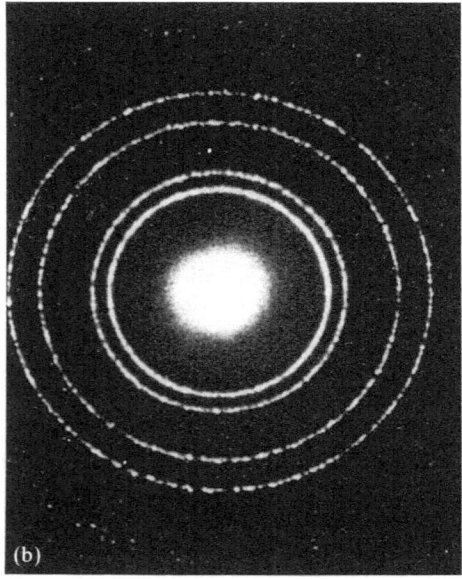

(b)

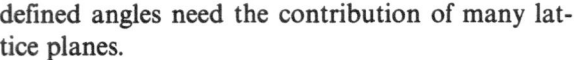

Figure 26.10. (a) The Davisson–Germer experiment. (b) The wave nature of electrons is demonstrated by the fact that they exhibit **Bragg** diffraction. This electron beam has traversed the same aluminum foil as the x rays of Fig. 24.10, Chap. 24. Note the corresponding rings in both photographs. The central bright spot is caused by the direct beam. (Courtesy Education Development Center, Inc., Newton, MA.) (c) These electrons simultaneously reveal their quantum and classical aspects. Their diffraction pattern is distorted by a magnetic field which, as we know, exerts a force on the moving electrons. [By permission, from R. B. Leighton, *Principles of Modern Physics*, p. 84 (McGraw-Hill, New York, 1959); photograph by Darwin W. Smith.]

defined angles need the contribution of many lattice planes.

Diffraction of Other Particles

Not only electrons, but much more massive particles such as neutrons and even helium atoms can exhibit wave behavior in diffraction experiments. Macroscopic particles, however, are absolutely classical in practice, as the last section of this chapter will point out in detail. No one will ever observe "golf ball waves."

E. The Role of Probability

According to classical thought, a wave cannot be a particle. A wave is a *continuous* function of posi-

tion, and can extend over an arbitrarily large region of space. A particle, on the other hand, is *discrete*. That is to say, it occupies only a pointlike region of space. Even a stream of many particles occupies only a discrete set of locations in space at any one time.

This section attempts to sketch, very roughly, how the wave–particle dilemma has been resolved sufficiently well for a consistent physical theory to emerge. That theory is **wave mechanics**. It is a special form of **quantum mechanics,** a formalism so general that it is thought applicable to all physical systems.

One essential feature of quantum mechanics is that, even assuming a faultless theory and faultless calculations, many experimental predictions are only probable, whereas classically the same type of predictions would be considered certain. Quantum mechanics assigns a well-defined, calculable probability (usually less than 100%) to each conceivable position or momentum of a particle, as well as to other numerical data.

A second feature of quantum mechanics severely limits the accuracy of some measurements, and of some predictions as well. Suppose one wishes to measure both the position and the momentum of a particle. With enough care, the position may be found with a high accuracy, but only at the cost of a low accuracy for the momentum estimate. The same limitation may be stated with the words "position" and "momentum" interchanged. In 1927 Werner Heisenberg demonstrated that such a trade-off, the **uncertainty principle**, is an unavoidable consequence of quantum mechanics; quantitatively speaking, *the product of the position and momentum uncertainties* must be larger than (approximately) Planck's constant.

That limitation applies to the initial data (position and momentum) of a trajectory, as well as to any theoretical prediction of future data. The new uncertainty is due, not to newly discovered imperfections in our apparatus, but rather to a fundamental overhaul of the cause–effect relationship. The resulting downfall of classical determinism (Note at end of Chap. 3) has shaken the scientific and philosophical outlook of our culture.

i. Probability Waves

It is not difficult to demonstrate simultaneously the wave and particle aspects of electromagnetism. Let us do Young's experiment (diffraction from two parallel slits) with very weak light, and with a sensitive photographic plate to record the fringes. A tiny crystal of silver salt in the emulsion, *when struck in a short time by just a few photons* (perhaps five or so), can develop into a silver grain, clearly seen under magnification. Each grain is a record of those photons' position.

During a short exposure, only a few photons have time to reach the emulsion; the resulting isolated dots give no hint of a wave. After a much longer exposure, the picture begins to reveal the continuous and classical aspects of the diffracted wave, as described in Chap. 24, Sec. B. Electronic image intensifiers exhibit the same phenomenon; see Fig. 26.11. We conclude that:

1. The position of a single particle does not betray the existence of a wave.
2. The largely random positions of many particles have a **probability distribution** (high probability in a bright fringe, low probability in a dark fringe) that accurately follows the electromagnetic wave pattern.
3. The wave intensity is proportional to the number of photons (recall Sec. A v); it is also proportional to $|\mathbf{E}|^2$, the square of the electric field. Therefore we have

$$\left(\begin{array}{c} \text{Probability of collecting} \\ \text{a photon per unit area} \\ \text{of the picture} \end{array}\right) \propto |\mathbf{E}|^2 \quad (26.E.1)$$

The left-hand side of (26.E.1) is known as a **probability density** (on a surface). More generally, the probability of finding a photon in any region of *three-dimensional* space is governed by a three-dimensional probability density,

$$\text{Photon probability density} \propto |\mathbf{E}|^2 \quad (26.E.2)$$

(The **B** field could be used as well.)

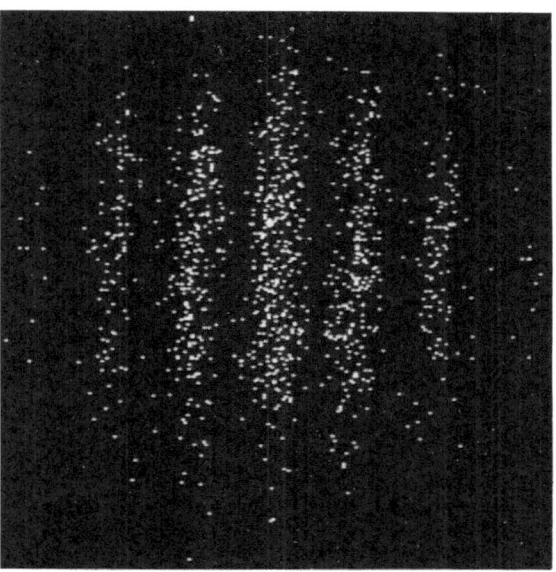

Figure 26.11. The existence of a wave is revealed by the photons' statistics. In this set of Young's fringes from a pair of pinholes, the arrival of each photon has been recorded electronically. (By permission, J. W. Goodman, Stanford University, and G. Weigelt and E. Keller, Erlangen University.)

Electron Probability Density

Turning next to electrons, we ask: How likely are we to find an electron in a given vicinity? The answer is analogous to (26.E.2) and makes use of the electron wave function ψ:

Electron probability density $\propto |\psi|^2$ (26.E.3)

The ψ function of a nonrelativistic particle obeys a famous wave equation, which differs from that of the electromagnetic field, and which was first written down (in 1926) by Erwin Schrödinger; the probabilistic nature of ψ was recognized by Max Born in the same year. These fundamental topics, whose discussion cannot be included here, are found in textbooks on modern physics.

ii. The Uncertainty Principle

Let the x coordinate of a certain particle be known to within an experimental uncertainty Δx; similarly, let the x component of that particle's momentum, p_x, be known within Δp_x. Then, in contrast to the classical situation, Δx and Δp_x cannot both be arbitrarily small. Heisenberg's **uncertainty principle** states that

$$\boxed{\Delta x\, \Delta p_x \geqslant h}$$ (26.E.4)

That is to say, *the product of these uncertainties cannot be less than Planck's constant h.* [Are the dimensions on both sides of (26.E.4) the same?] Similar relations are, of course, valid in the y and z directions. A rough, but simple argument yields (26.E.4) as follows.

The surest way to know that the particle is within a length Δx is to keep it inside a box of that size, as illustrated in Fig. 26.12. We now draw an analogy between the particle's de Broglie wave and a sound wave. The closed box can hold a standing wave of some maximum wavelength λ (the fundamental resonance mode), and harmonics of smaller wavelengths $\lambda/2$, $\lambda/3$, etc., somewhat like an organ pipe. We examine the maximum λ first. As Fig. 26.12 indicates, it obeys

$$\frac{\lambda}{2} = \Delta x$$ (26.E.5)

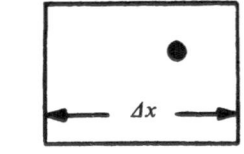

(a)

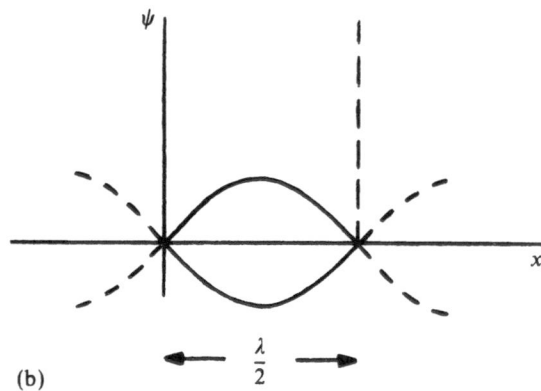

(b)

Figure 26.12. (a) A particle in a box. (b) The same, viewed as a standing de Broglie wave.

De Broglie's relation gives us the corresponding magnitude for the momentum,

$$p = \frac{h}{\lambda} = \frac{h}{2\Delta x}$$ (26.E.6)

The standing wave amounts to two superposed waves traveling in the $\pm x$ directions, and which therefore correspond to two particle momenta

$$p_x = +\frac{h}{2\Delta x}, \qquad p'_x = -\frac{h}{2\Delta x}$$ (26.E.7)

The particle's momentum, if measured, could turn out to have either of these two values; the predicted momentum therefore has an uncertainty, or spread,

$$\Delta p_x = p_x - p'_x$$

$$= \frac{h}{2\Delta x} - \left(-\frac{h}{2\Delta x}\right) = \frac{h}{\Delta x}$$ (26.E.8)

from which

$$\Delta x\, \Delta p_x = h$$ (26.E.9)

The higher harmonics have smaller λ and hence larger p and larger Δp_x; (26.E.9) becomes

$$\Delta x\, \Delta p_x \geqslant h$$

which was to be shown. (It would be natural to object that, once p_x is measured, it is exactly known. However, it turns out that the particle must be let out of its box for the purpose of an accurate momentum measurement; Δx then becomes completely washed out, preserving the uncertainty relation.)

iii. Classical Particles

Wave mechanics states that every particle has a ψ function associated with it; billiard balls, and even planets, should be no exceptions and hence should also obey the uncertainty principle. Why then is wave mechanics never observed to hold for such macroscopic objects? The answer lies in the incredible smallness of Planck's constant in terms of macroscopic units like the joule sec, as the following example shows.

Example 26.7. Wave Mechanics of a Marble. Consider a marble of mass 3 grams, rolling at a carefully measured speed $v = (2.000 + 0.004)$ cm/sec. Find (a) its de Broglie wavelength λ, and (b) the minimum uncertainty Δx in its position at any time.

_____

(a) Wavelength: We have

$$\lambda = \frac{h}{mv} \qquad (26.E.10)$$

$$= \frac{6.6 \times 10^{-34}}{(3 \times 10^{-3})(2 \times 10^{-2})} \text{ meter}$$

$$= \underline{1 \times 10^{-29} \text{ meter}}$$

This wavelength is many orders of magnitude smaller than even a single atom of the marble. The idea of observing a marble's de Broglie wavelength by diffraction or any other means is therefore absurd.

(b) Uncertainty in position: The uncertainty relation, $\Delta x\, \Delta p_x \geqslant h$, gives the minimum Δx as

$$(\Delta x)_{\min} = \frac{h}{\Delta(mv)} = \frac{h}{m\, \Delta v} \qquad (26.E.11)$$

$$= \frac{6.6 \times 10^{-34}}{(3 \times 10^{-3})(0.008 \times 10^{-2})} \text{ meter}$$

$$= \underline{3 \times 10^{-27} \text{ meter}}$$

This is such a fine precision (again to much less than an atom) that actual measuring instruments have no hope of even approaching it. Thus, the actual limitations on the measurement of a marble's position have nothing to do with wave mechanics. The conclusion from (a) and (b) is that _a marble is a classical object_. So is, a fortiori, a billiard ball or an astronomical body.

Note

Calculation of the Compton Shift

The three conservation equations (26.C.8)–(26.C.10) may be written

$$\tfrac{1}{2}mv^2 = h(f - f') \qquad (26.N.1)$$

$$mv \cos \varphi = \frac{h}{c}(f - f' \cos \theta) \qquad (26.N.2)$$

$$mv \sin \varphi = \frac{hf'}{c} \sin \theta \qquad (26.N.3)$$

To eliminate φ, square and add the corresponding sides of (26.N.2) and (26.N.3):

$$m^2v^2 = \frac{h^2}{c^2}(f^2 + f'^2 - 2ff' \cos \theta) \qquad (26.N.4)$$

To eliminate v, compare (26.N.1) and (26.N.4):

$$\frac{2mc^2}{h}(f - f') = f^2 + f'^2 - 2ff' \cos \theta \qquad (26.N.5)$$

or

$$\frac{2mc^2}{h}(f - f') = (f - f')^2 + 2ff'(1 - \cos \theta) \qquad (26.N.6)$$

(Caution: We cannot expect this result to hold if the recoil electron has a relativistic speed.)

Nonrelativistic Simplification

Equation (26.N.6) can be made much simpler on the basis of $v/c \ll 1$. Equation (26.N.1) gives

$$\frac{v^2}{c^2} = \frac{2h}{mc^2}(f - f') \ll 1 \qquad (26.N.7)$$

Multiplying both sides of this inequality by $(mc^2/2h)(f-f')$, we have

$$(f-f')^2 \ll \frac{mc^2}{2h}(f-f') \qquad (26.N.8)$$

Hence, on the right side of Eq. (26.N.6), the term $(f-f')^2$ may be neglected; we are left with

$$\frac{2mc^2}{h}(f-f') = 2ff'(1-\cos\theta) \qquad (26.N.9)$$

leading to Eq. (26.C.11) in the text.

Condensed Checklist

Planck's constant:

$$h = 6.63 \times 10^{-34}\text{ joule sec} \qquad (26.A.6)$$

Energy and momentum of a photon:

$$\boxed{\mathscr{E} = hf} \qquad (26.A.10)$$

$$p = \frac{h}{\lambda} = \frac{hf}{c} \qquad (26.C.2),\ (26.C.4)$$

$$\text{Rest mass} = 0 \qquad \text{Below (26.C.5)}$$

Stopping potential:

$$V_{st} = \frac{\mathscr{K}_{max}}{e} \qquad (26.A.1)$$

Work function:

$$\phi = \mathscr{E}_{esc,min} = hf_0$$

$$\frac{\phi}{1\text{ electron volt}} = \frac{hf_0}{(e)(1\text{ volt})} \qquad (26.A.12),\ (26.A.14)$$

$$\text{Emission thresholds:} \qquad \text{Table 26.1}$$

Photoelectric equation:

$$\boxed{\mathscr{K}_{max} = h(f-f_0) = hf - \phi} \qquad (26.A.4),\ (26.A.13)$$

$$V_{st} = \frac{h}{e}(f-f_0) = \frac{hc}{e}\left(\frac{1}{\lambda} - \frac{1}{\lambda_0}\right)$$

$$(26.A.5),\ (26.A.8)$$

$$\frac{hc}{e} = 1240\text{ nm volt}$$

Light intensity:

$$\mathscr{I} = (\text{photon current density})(hf) \qquad (26.A.17)$$

X-ray cutoff:

$$hf_{max} = eV \qquad (26.B.3)$$

$$\lambda_{min} = \frac{hc}{eV} = \frac{1240\text{ nm volt}}{V} \qquad (26.B.5),\ (26.B.6)$$

Compton effect:*

$$\lambda' - \lambda = \frac{h}{mc}(1 - \cos\theta) \qquad (26.C.12)$$

$$\frac{h}{mc} = 2.4 \times 10^{-12}\text{ meter} \qquad (26.C.13)$$

De Broglie wavelength:

$$\boxed{\lambda = \frac{h}{mv}} \qquad (v \ll c) \qquad (26.D.1)$$

Bragg's formula (reviewed):

$$\cos\theta = \frac{k\lambda}{2s} \qquad (k = 1, 2, 3,...) \qquad (26.D.7)$$

Uncertainty principle:

$$\boxed{\Delta x\,\Delta p_x \geqslant h}$$

$$\Delta x\,\Delta v_x \geqslant \frac{h}{m} \qquad (26.E.4)$$

True or False

1. Photoemission means emission of photons.

2. A photoelectron is a particle intermediate in its properties between a photon and an electron.

3. The work function is independent of the incident frequency.

4. In a given cathode material, the work function depends on the initial energy of the photoelectron.

5. In photoemission, the threshold wavelength is the smallest one at which the effect can occur.

6. The photocurrent is proportional to the incident light intensity.

7. The photoelectrons' maximum kinetic energy increases with the incident light intensity.

8. The photoelectric stopping potential decreases as the incident wavelength increases.

9. In x-ray emission, the cutoff wavelength is the smallest radiated wavelength.

* Omit if Sec. C has been omitted.

10. In x-ray emission, the cutoff wavelength is proportional to the tube voltage.

11.[†] The Compton effect arises in connection with the impact of an electron against an atom.

12.[†] Compton-scattered radiation has a smaller wavelength than the incident beam.

13. Planck's constant has the same dimensions as angular momentum.

14. The de Broglie wave of an electron is an electromagnetic wave.

15. The momentum of a neutron varies inversely with its de Broglie wavelength.

16. According to the uncertainty principle, physical predictions can always approach perfect accuracy but can never reach it.

17. The energy and momentum of a particle cannot both be measured to arbitrarily high accuracy.

Problems

Section A: Photoemission

26.1. Complete Table 26.1 by listing the work function ϕ, in electron volts, for each material.

*26.2. Calculate, in joules and in electron volts, the energy per photon in electromagnetic radiation of (a) frequency 80 MHz (radio waves), (b) wavelength 12.0 cm (radar).

26.3. A certain photoelectron is emitted with a kinetic energy of 1.5 eV. If the incident light has a frequency of 6.0×10^{14} Hz, how much energy did the electron need to escape?

*26.4. Light of frequency 5.4×10^{14} Hz, when incident on a certain photocathode, yields photoelectrons of maximum kinetic energy 0.8 eV. (a) What is the work function, in electron volts? (b) What is the threshold frequency?

26.5. When light of mixed frequency falls on a photocathode, each frequency component extracts its own photoelectrons, essentially as if the other frequencies were absent. If a magnesium cathode is exposed to the combination 5.0×10^{14} Hz, 7.0×10^{14} Hz, 9.0×10^{14} Hz, what is the maximum kinetic energy of the photoelectrons as they leave the cathode?

*26.6. A typical laser radiates 1 mW at a wavelength of 630 nm. How many photons does it emit per second?

26.7. In a series of experiments, light of wavelength 510 nm is made to fall on various photocathode

materials: (a) zinc, (b) magnesium, (c) potassium, (d) cesium. What is, in each case, the maximum kinetic energy of the photoelectrons as they leave the cathode?

*26.8. Light of wavelength 360 nm falls on one of a pair of equal electrodes. The stopping potential is 0.95 volt. (a) What is, in electron volts, the electrode's work function? (b) What is, in nanometers, its threshold wavelength?

26.9. In a certain photoelectric cell with equal electrodes, the stopping potential is 2.8 volts, and the work function is 3.6 eV. What is the wavelength of the incident light?

*26.10. A certain photoelectric cell has equal electrodes, with a work function of 4.5 eV. What is the stopping potential when ultraviolet light of wavelength 200 nm is being used?

26.11. In one photoemission experiment, the maximum kinetic energy of the photoelectrons is 2.5 eV. If the incident light has a wavelength of 290 nm, (a) what is, in electron volts, the work function of the cathode? (b) What is the threshold wavelength?

*26.12. What is, in watts/meter^2, the intensity of a blue light beam, of wavelength 470 nm, if it sends 8×10^{15} photons per second perpendicularly onto a 0.5 meter^2 screen?

26.13 (Classical time lag in photoemission). Consider a microscopic metal particle, of cross-sectional area 25 μm^2 and work function 3.8 eV. It is kept suspended in an electric field, like the oil drop in Millikan's experiment, and emits photoelectrons under a very weak ultraviolet light beam of intensity 2×10^{-10} watt/meter^2. As soon as the first electron is emitted, the particle is seen to accelerate because of its modified charge. What is the expected time lag between turning on the light and observing the acceleration? Assume a classical light wave and require an absorbed energy sufficient to liberate one electron. (This type of experiment was performed in 1914 by E. Meyer and W. Gerlach; the observed time lags were in some cases much less than a second.)

*26.14. It has been said that the dark-adapted human eye can detect a flash containing only about 100 photons striking the retina within a time less than about 0.05 sec. How far away, in kilometers, must we place a sodium lamp, emitting 10 watts of yellow light (wavelength = 590 nm) equally in all directions, in order to obtain 100 photons every 0.05 sec? Assume a diameter of 6 mm for the pupil.

26.15. At the Earth's orbit, sunlight has an intensity of about 1300 watts/meter^3. Make a very rough estimate of the number of photons striking the Earth per second, assuming a typical frequency of 4×10^{14} Hz and a terrestrial radius of 6400 km.

[†] Omit if Sec. C has been omitted.

26.16. A certain mixed light beam consists of the frequencies 4.0×10^{14} Hz and 6.0×10^{14} Hz, with respective intensities of 5.0 watts/meter^2 and 10.0 watts/meter^2. Find, *in terms of the number of photons*, the percentage composition of the beam.

Section B: X Rays

*26.17. A certain x-ray tube radiates a maximum frequency of 7.5×10^{17} Hz. At what voltage is it operating?

26.18. A certain x-ray tube radiates a minimum wavelength of 0.120 nm. At what voltage is it operating?

*26.19. (a) Verify the arithmetic leading from (26.B.5) to (26.B.6). (b) The electron gun of a certain television image tube operates at 20 kV. There is a small amount of x-ray emission, due to the electrons' impact with the screen. Find the x rays' minimum wavelength.

26.20. From the data of Fig. 26.6, estimate the power radiated per unit frequency at 30 kV and 1 A in the vicinity of 3×10^{18} Hz.

*26.21. (a) in Fig. 26.6, determine the intercepts of the 25-kV and 15-kV curves with the horizontal axis. (b) If the curves in that figure are taken to be parallel lines, calculate the corresponding intercepts with the vertical axis.

26.22. From Fig. 26.6, (a) estimate the power emitted in all frequencies for $V = 25$ kV; ignore the peaks' contributions. (b) How much is the tube's power input? (c) What is the tube's efficiency, defined as the ratio of (a) to (b)?

26.23. (a) From Fig. 26.6, estimate how much power is radiated above a frequency of 7×10^{18} Hz from a thick molybdenum target bombarded by 35-keV electrons, at a current of 0.5 A. With the same data, estimate the power radiated (b) between frequencies of 3×10^{18} Hz and 4×10^{18} Hz, and (c) between wavelengths of 0.060 nm and 0.066 nm.

*26.24. The output of a thick-target x-ray tube approximately satisfies the empirical relation (H. Kulenkampff, 1922) $d(\text{power})/df = aIZ(f_{max} - f)$, for $f < f_{max}$, where $f_{max} = eV/h$, I and V are the tube's current and voltage, Z is the atomic number of the target's material (e.g., $Z = 42$ for molybdenum), and a is an empirical constant. (a) What are the SI units of a? (b) From the intercept of the 35-kV curve with the vertical axis in Fig. 26.6, what is the value of a?

*26.25. Re-sketch the x-ray spectrum of Fig. 26.6, with the wavelength λ rather than the frequency f as the independent variable; as the dependent variable, take the power *per unit wavelength*, $|d(\text{power})/d\lambda|$.

Section C: The Compton Effect

26.26. In Eq. (26.C.12), show that h/mc has the units of length.

*26.27. In Compton scattering, what is, in nanometers, the largest possible difference between incident and scattered wavelengths?

26.28. In a certain Compton scattering experiment, the back-scattered radiation ($\theta = 180°$) was found to have a wavelength of 0.0543 nm. What was the incident wavelength λ?

*26.29. If an incident x-ray beam has a wavelength of 0.0855 nm, what is the wavelength of the Compton scattered radiation observed at 90° from the incident beam?

26.30. If a Compton shift of 3.6×10^{-12} meter in wavelength is to be observed, at what angle to the incident beam must the radiation be collected?

*26.31. In one experiment, incident photons of energy 50.00 keV are Compton scattered through a 53° angle. What is, in keV, their energy after scattering?

*26.32. Suppose that an incident x-ray beam has a wavelength of 0.0300 nm, and that the Compton scattered radiation is observed at 60° to the incident direction. Find (a) the scattered wavelength; (b) the speed of the recoil electron; (c) the angle that the recoil trajectory makes with the incident beam.

26.33. An organic molecule, of molecular weight 450, is floating at rest in empty space. It absorbs a photon of ultraviolet light corresponding to a wavelength of 180 nm. (Most of the photon's energy becomes internal energy of the molecule.) What speed, in meters/sec, does the molecule acquire? (Use 1 atomic mass unit = 1.7×10^{-27} kg.)

Sections D: Electron Waves; E: The Role of Probability

*26.34. Find the de Broglie wavelength of (a) an electron, (b) a neutron, whose momentum is 5×10^{-23} kg meter/sec.

26.35. What speed should a helium atom (mass $= 4 \times 1.7 \times 10^{-27}$ kg) have in order for its de Broglie wavelength to be as large as the atom itself (about 10^{-10} meter)?

*26.36 (Irrelevance of wave mechanics in celestial mechanics). The Earth's mass is 6×10^{24} kg and its orbital speed around the Sun is 30 km/sec. Calculate its de Broglie wavelength.

26.37. If an electron has an uncertainty of $\pm 2 \times 10^{-25}$ kg meter/sec in its momentum, estimate the minimum uncertainty in its position.

*26.38. Suppose an electron is moving in the x direction with a velocity of $(2.57 \pm 0.03) \times 10^6$ meters/sec. Estimate the minimum uncertainty of its x coordinate.

26.39. (a) If the position of a helium atom (defined as the position of its center of mass) is known within a distance of 0.5×10^{-10} meter, estimate the minimum uncertainty in its momentum. (b) Estimate the minimum uncertainty in its velocity. (Its mass is $4 \times 1.7 \times 10^{-27}$ kg.)

*26.40. Find the de Broglie wavelength of an electron whose kinetic energy is 10 keV.

26.41. Find, in electron volts, the kinetic energy of a proton whose de Broglie wavelength is 0.3×10^{-10} meter. (Proton mass $= 1.7 \times 10^{-27}$ kg.)

26.42. Find, in electron volts, the kinetic energy of an electron whose de Broglie wavelength is 3.5 nm.

*26.43. What is the minimum x-ray wavelength, λ_x, emitted by a tube in which the electron beam has a de Broglie wavelength λ_{el}? (Express λ_x in terms of λ_{el}, h, c, and the electron mass m.)

*26.44. If a second-order Bragg angle of $88.5°$ is observed in the diffraction of 35-kV electrons from a crystal, what is the relevant lattice spacing?

26.45. A first-order Bragg angle of $85.2°$ is observed when electrons of kinetic energy $\mathscr{K}$ are diffracted from a crystal of potassium chloride. If the relevant lattice planes are 0.314 nm apart, what is $\mathscr{K}$ in electron volts?

*26.46. A 15-keV electron beam is diffracted from a crystal lattice with spacing 0.12 nm. How close to 90° is the first-order Bragg angle?

26.47. If an electron has a kinetic energy of (18.0 ± 0.1) keV, what is the minimum uncertainty in its position?

26.48. In a scientific report, an author considers a proton (mass $= 1.7 \times 10^{-27}$ kg) that he claims had, at one time, a coordinate $x = 264.08$ nm and a velocity $v_x = 4.07679 \times 10^6$ meters/sec. (a) Show that these accuracies are overstated. (b) Accepting the value for v_x as given, how many significant figures make sense in x?

*26.49. If, in Compton scattering, the recoil electron has a speed of 3.0×10^7 meters/sec, find the corresponding Compton *frequency* shift of the radiation for the appropriate scattering angle.

*26.50. Consider Eqs. (26.C.8)–(26.C.10) in the case of a "head-on" collision, where the photon is sent back at $\theta = 180°$; the electron is propelled forward, at $\varphi = 0°$. In terms of the incident frequency f, find the electron's final speed v. [You may simplify your calculation (or your answer) by assuming that v is very small compared to c.]

26.51. If a photon and a (nonrelativistic) electron have the same de Broglie wavelength λ, what is the ratio of the electron's kinetic energy $\mathscr{K}$ to the photon's energy $\mathscr{E}$? (Express $\mathscr{K}/\mathscr{E}$ in terms of λ, h, c, and the electron mass m.)

26.52. If the kinetic energy of a (nonrelativistic) electron equals the energy $\mathscr{E}$ of a photon, what is the ratio $\lambda_{el}/\lambda_{phot}$ of these particles' de Broglie wavelengths? (Express your answer in terms of $\mathscr{E}$, h, c, and the electron mass m.)

*26.53. Consider a proton whose kinetic energy is $\mathscr{K} \approx 2$ MeV. If its position does not need to be known to better than 10^{-10} meter, how much more accurately, in MeV, could $\mathscr{K}$ be specified in principle?

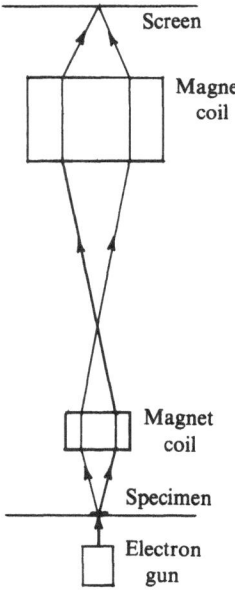

Screen

Magnet coil

Magnet coil

Specimen

Electron gun

26.54. In an electron microscope (adjoining figure), the specimen is bombarded with electrons, which are then magnetically focused onto a fluorescent screen. In this way the specimen is imaged on the screen. Since electrons are effectively point particles, it might be thought that arbitrarily small details can be picked up. In reality, the wave property of the electrons sets a **limit of resolution** λ (the electrons' de Broglie wavelength). If one wishes to resolve specimen features as small as 10 nm, what must be the minimum voltage of the electron gun in the figure?

*26.55 (Speed of an electron in a hydrogen atom). The hydrogen atom is of order 0.1 nm in diameter, and is known to contain an electron. (a) In analogy with the box argument of Sec. E ii, estimate (order of magnitude only) the uncertainty Δp in the electron's momentum. (b) Assuming, as in the box argument, that the momentum *p* itself is of the same order as Δp, estimate the electron's minimum speed in the atom. (Although very rough, this estimate is a valid one, and is confirmed by more detailed studies, as in the next chapter.)

Answers to True or False

1. False (emission of electrons).
2. False (a photoelectron is an electron).
3. True.
4. False (it is defined in terms of the highest-energy electrons).
5. False (maximum).
6. True.
7. False (independent of intensity).
8. True.
9. True.
10. False (inversely proportional).
11. False (photon against electron).
12. False (larger).
13. True.
14. False (true for a photon).
15. True.
16. False (some predictions can never approach perfect accuracy).
17. False (true for position and momentum).

Answers or Hints to Problems

26.2. Check with Fig. 22.13 of Chap. 22.

26.4. (b) 3.5×10^{14} Hz.

26.6. 3×10^{15} photons/sec.

26.8. (a) 2.5 volts.

26.10. 1.7 volts.

26.12. 6.7×10^{-3} watt/meter^2.

26.14. About 200 km.

26.17. 3.1 kV.

26.19. (b) 0.062 nm.

26.21. For 15 volts: (a) 3.6×10^{18} Hz; (b) 0.72×10^{-17} watt/Hz.

26.24. (b) $a = 4.1 \times 10^{-38}$ SI units.

26.25. Because $\lambda = c/f$, we have $|d(\text{power})/d\lambda| = (f^2/c) |d(\text{power})/df|$. A point with coordinates (x, y) in the old plot has coordinates c/x, $(x^2 y)/c$ in the new plot.

26.27. Use (26.C.12), get $(\lambda' - \lambda)_{\text{max}} = 2h/mc$.

26.29. 0.0879 nm.

26.31. Use, e.g., $(1/h) \times$ Eq. (26.C.11).

26.32. (b) use Eq. (26.C.8); (c) use Eq. (26.C.10).

26.34. Same answer for (a) and (b).

26.36. 4×10^{-63} meter.

26.38. $\pm 2.4 \times 10^{-8}$ meter.

26.40. 1.2×10^{-11} meter.

26.43. $2mc\lambda_{\text{el}}^2/h$.

26.44. 0.25 nm.

26.46. $2.4°$.

26.49. There is no need to calculate θ; use Eq. (26.C.8).

26.50. $v \approx 2hf/mc$.

26.53. To some 10^{-3} MeV accuracy.

26.55. (b) Between 10^6 meters/sec and 10^7 meters/sec.

CHAPTER 27

ATOMIC STRUCTURE

Where, inside the atom, are the electrons? How fast are they moving? What is their energy? What prevents the negative electrons from crashing into the positive nucleus? Does atomic structure explain chemistry? The answers to these questions all came within a surprisingly short period of time, roughly between the two world wars; but their research background extends to well before Isaac Newton. It is therefore important, while reading this chapter, to be aware of the wide range of ingredients it involves—ingredients to which we were introduced in the preceding chapters.

We shall approach atomic structure in three steps. First, the experimental clues: atoms emit and absorb light of selected frequencies; in a spectroscope, each frequency shows up as a **spectral line**. A characteristic set of lines (a **spectrum**) identifies a radiating atom like a signature. For example, the hydrogen spectrum looks quite different from the helium spectrum. On the basis of the spectral lines, we shall calculate the energies which an atomic electron can have; as it turns out, only certain selected energy values are possible. The electron is said to occupy **discrete energy levels**.

As a second step towards understanding atomic structure, we need a detailed scenario for the behavior of the atomic electrons. That scenario, which is technically called a **model**, must (1) agree with known physical principles, and (2) lead to the known energy levels. If conditions (1) and (2) are met, the model has a good chance of being realistic. In the case of hydrogen—the simplest of all atoms—we shall study a celebrated model due to Niels Bohr (1913). It does suffer from some inconsistencies, which, however, were eventually removed by a rigorous application of wave mechanics. (This part of the story must unfortunately be omitted.) Today's descendant of the Bohr model is, beyond any reasonable doubt, a correct description.

As our third step, we use what we have learned from the hydrogen atom in order to make sense of the other, more complex atoms. A new idea will be found essential; to state it in a somewhat inexact classical language, *the electron spins about its own axis.* Furthermore, every electron in an atom seems to know what all the other electrons in that atom are doing; no two of them ever perform the same motion at the same time. In technical language, *they cannot occupy the same state.* This bizarre-sounding rule, known as the **exclusion principle**, explains a vast number of facts; the periodic table of the elements—indeed all of chemistry—depends crucially on that principle.

A. Energy Levels

In this section we describe how an observed line spectrum reveals the energy levels which exist in an atom.

i. The Spectrum of Hydrogen

Let us pass a high-voltage discharge through a quartz tube filled with hydrogen gas at low pressure; a strong glow is observed. When examined spectroscopically, the emitted light turns out to be a mixture of many discrete frequencies. The pattern of lines is shown in Fig. 27.1; it ranges over the ultraviolet, visible, and infrared domains. The lines may be grouped into **series**; each series is

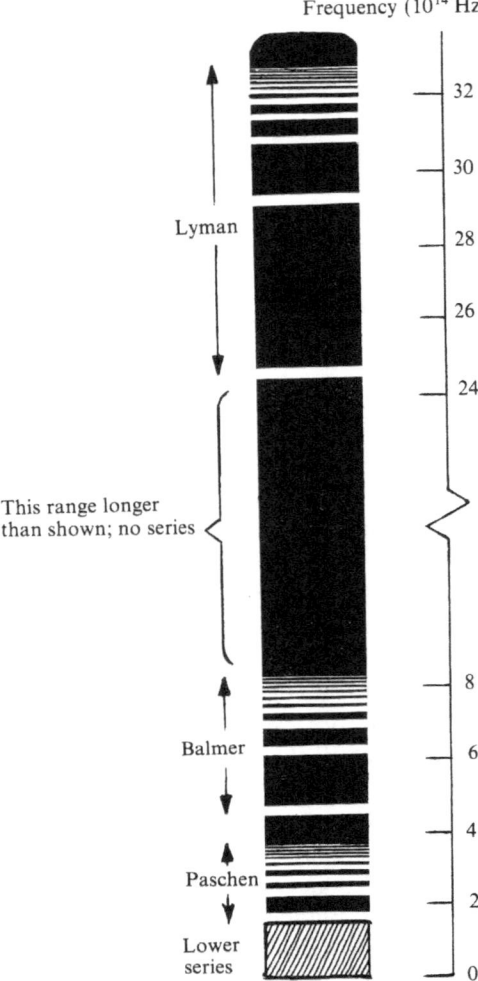

Frequency (10^{14} Hz)

Figure 27.1. The hydrogen spectrum, drawn on a uniform frequency scale. In an actual photograph, these lines would be the many images, in light of different frequencies, of a single slit. Here the thickness of a line attempts to indicate its relative brightness (which actually depends somewhat on experimental conditions). Note that the topmost three series (named Lyman, Balmer, and Paschen) are separate; the lower ones (not shown), which are very crowded, overlap among themselves and with the Paschen series.

a set of consecutive lines which crowd together at their high-frequency limit.

A simple numerical scheme, discovered empirically in 1885 by Johann Jakob Balmer, yields every frequency in the spectrum. The idea is to label each line by two positive integers n_1, n_2. The frequency f of that line is then found to obey the formula

$$\boxed{f = \left[\frac{1}{(n_1)^2} - \frac{1}{(n_2)^2}\right] f_H} \qquad (27.A.1)$$

The parameter

$$f_H = 3.2881 \times 10^{15} \text{ Hz} \qquad (27.A.2)$$

fits the whole spectrum with extraordinary accuracy.* (We must take $n_2 > n_1$, otherwise f cannot be positive. The number n_1 determines the series in which the line is found, while n_2 tells us which member of the series we are dealing with.)

Example 27.1. The Balmer Series. This name is given to the set of lines obtained from formula (27.A.1) with fixed $n_1 = 2$ and ever-increasing n_2. Determine, to three-figure accuracy, (a) the lowest four frequencies, and the corresponding wavelengths; (b) the limiting frequency and wavelength as $n_2 \to \infty$; (c) the color of these lines.

(a) Lowest frequencies: Formula (27.A.1) becomes

$$f = \left[\frac{1}{2^2} - \frac{1}{(n_2)^2}\right] f_H \qquad (27.A.3)$$

The smallest possible n_2 equals 3, and thus the lowest frequencies are for $n_2 = 3, 4, 5, 6,...$. (The corresponding spectral lines have been given the designations $H_\alpha, H_\beta, H_\gamma, H_\delta$, etc.) For $n_2 = 3$ we have

$$f = \left(\frac{1}{4} - \frac{1}{9}\right) f_H$$

$$= \left(\frac{5}{36}\right)(3.29 \times 10^{15} \text{ Hz}) \qquad (27.A.4)$$

$$= 4.57 \times 10^{14} \text{ Hz}$$

The wavelength is

$$\lambda = \frac{c}{f} = \frac{3.00 \times 10^8}{4.57 \times 10^{14}} \text{ meter} = \underline{656 \text{ nm}}$$

The other lines are calculated in a similar way, with the factor 5/36 in (27.A.4) successively replaced by 3/16, 21/100, 2/9. (Why?)

---

* In the literature, (27.A.1) is often written in terms of wavelength rather than frequency; see Problem 27.1.

(b) Series limit: As $n_2 \to \infty$, formula (27.A.3) yields

$$f \to \left(\frac{1}{4} - 0\right) f_H \tag{27.A.5}$$

$$= \left(\frac{1}{4}\right)(3.29 \times 10^{15} \text{ Hz}) = \underline{\underline{8.22 \times 10^{14} \text{ Hz}}}$$

(c) The colors are found from Fig. 22.14 of Chap. 22.

The complete results of this example are listed in Table 27.1 below.

The Spectral Series of Hydrogen

Every positive integer n_1 in formula (27.A.1) corresponds to a different whole series in the hydrogen spectrum. The most easily observed series are named as follows [see also Fig. 27.1]:

$$\begin{aligned} &\text{Lyman series } (n_1 = 1) \\ &\text{Balmer series } (n_1 = 2) \\ &\text{Paschen series } (n_1 = 3) \qquad (27.A.6) \\ &\text{Brackett series } (n_1 = 4) \\ &\text{Pfund series } (n_1 = 5) \end{aligned}$$

Only the Balmer series—the earliest to be observed and measured—has any lines in the visible. The Lyman series is entirely in the ultraviolet, all others entirely in the infrared.

ii. Line Spectra of the Chemical Elements

Any element can be made to yield a line spectrum, and hydrogen has the simplest one of all. For comparison, the iron spectrum (from a spark between iron electrodes) is partially displayed in Fig. 27.2. As another example, ordinary table salt NaCl, when sprinkled on a flame, will emit a bright orange-yellow glow, which is due spectroscopically to a pair of closely spaced lines, the "sodium D lines," at wavelengths of 589.0 nm and 589.6 nm; many other, less intense lines from Na and from Cl are present as well.

Applications

We can mention only a few of the numerous scientific uses of line spectra:

1. Determination of Atomic Structure. This is the main theme of this chapter, and in what follows we shall see how a hydrogen atom is "deciphered" through its spectrum.

2. Chemical Identification. Since each element has its distinctive pattern of lines, like a fingerprint, the element's presence in a flame or in a glow discharge is revealed spectroscopically. This method was already in use long before the origin of line spectra was understood.

The surface of the Sun and of other stars—even billions of light years away—can be analyzed chemically from their spectra. In this way, helium was discovered in the Sun more than 20 years before it was found on Earth.

3. Detection of Motion. As we recall from our discussion of the Doppler effect (Chap. 25), a recognizable, but *shifted* line spectrum denotes a velocity of the radiating atom towards or away from the observer. The velocities of stars are measured in this way. In the laboratory, a similar phenomenon is seen in line spectra from hot gases; the thermal velocities of the atoms are responsible for a range of Doppler shifts, whose superposition results in blurred ("Doppler-broadened") spectral lines.

Table 27.1. Four Balmer Lines and Series Limit

Line	H_α	H_β	H_γ	H_δ	H_∞
Frequency $(10^{14}$ Hz)	4.57	6.17	6.91	7.31	8.22
Wavelength (nm)	656	486	434	410	365
Color	Red	Blue	Violet	Violet	Ultraviolet

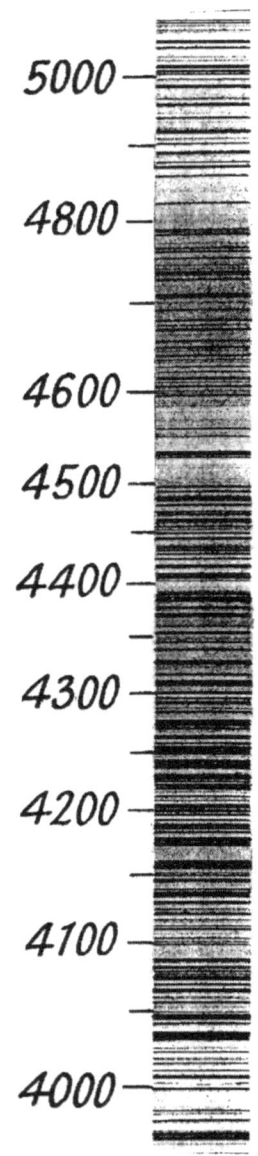

5000

4800

4600

4500

4400

4300

4200

4100

4000

Figure 27.2. A portion of the iron spectrum (actual photographic negative), giving an idea of its richness. A better resolution would reveal many more lines. The numbers refer to the vacuum wavelengths, from 400 nm to 500 nm. [By permission, from R. W. B. Pearse and A. G. Gaydon, *The Identification of Molecular Spectra* (Chapman & Hall, London, 1950).]

4. The Standard of Time. The permanence and reproducibility of a spectral line surpasses that of any tangible standard, such as a meter stick or a mechanical clock. Accordingly, the standard second (and, until recently, the standard meter) has been defined in terms of a specified line. We quote the International System definition:

> The *second* is the duration of 9 192 631 770 periods of the radiation corresponding to the transition between the two hyperfine levels of the ground state of the cesium-133 atom.

(In this definition, reference to a "transition between levels" serves to identify the spectral line —a microwave line from cesium. The terminology will be clarified in the next paragraphs, although the explanation of "hyperfine" must be left to a textbook on modern physics.)

5. The Laser. This device, in which the role of energy levels has been widely publicized, is briefly discussed in subsection (iv) further on.

iii. Levels and Transitions

Photon theory gives us the true meaning of a spectral line with frequency f: *photons of energy hf are being emitted.* When such a photon comes into existence, the emitting atom must in some way change its internal state. If its total initial and final energies are $\mathscr{E}_i$ and $\mathscr{E}_f$, conservation of energy requires

$$\boxed{hf = \mathscr{E}_i - \mathscr{E}_f} \qquad (27.A.7)$$

(We note that the internal energy $\mathscr{E}$ of an atom is partly electrostatic potential energy, and partly kinetic energy of the orbiting electrons; there is a very small amount of magnetic energy as well.)

The simplest way to visualize (27.A.7) is also the correct one. The energy of an atom is **quantized**; that is to say, the atom has a fixed, discrete set of allowed energies, two of which happen to be $\mathscr{E}_i$ and $\mathscr{E}_f$. Such discrete values are called **energy levels**, and the emission of a quantum hf implies a **transition** from level $\mathscr{E}_i$ to level $\mathscr{E}_f$. In short: *discrete spectral lines mean discrete energy levels.*

The transition is often symbolized on an **energy level diagram**, as in Fig. 27.3; we postulate that every transition releases only a *single photon*.

Energy Levels Are Often Negative

In order to specify a potential energy numerically, we must choose a reference level. We remem-

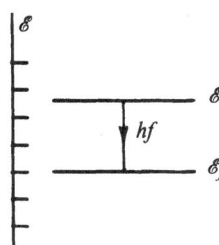

Figure 27.3. Transition between two energy levels.

ber this as a feature of gravitational problems, but it exists in atomic physics as well; here gravity is replaced by electrostatics. The total energy of an atom (potential + kinetic) depends on that zero-level convention.

Sometimes, the atom's lowest possible energy, or **ground-state energy,** is taken as its zero level; all other levels are then positive. Sometimes, however, and *almost always in the case of hydrogen*, the zero level is chosen as follows.

Imagine the atom to be completely separated into its component particles—a nucleus and one or several electrons. Let these particles be (1) far away from one another, and (2) at rest. To the extent that we can still speak of an atom, *its total energy is then defined to be zero.*

A considerable amount of work must be done on an atom in order to pull it apart in that fashion. Therefore "zero energy" really means high energy.

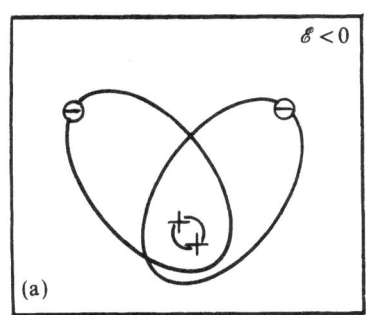

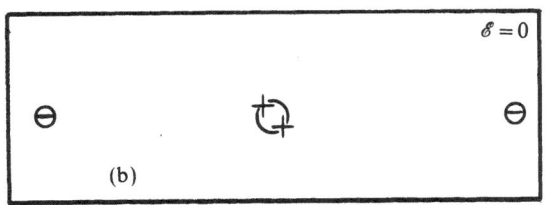

Figure 27.4. (a) A helium atom in a typical, or negative-energy state. (b) The same at zero energy. (In experimental work, however, the ground state is more often considered to have zero energy.)

Under ordinary circumstances the atom has a lower—and hence negative—energy. Figure 27.4 is a symbolic illustration of such a zero-level convention.

Example 27.2. A Set of Three Levels. Consider the three energy levels (1, 2, 3) illustrated in Fig. 27.5. (a) Which are the transitions that can result in a photon being emitted? (b) What are, in nanometers, the wavelengths of the emitted spectral lines?

(a) Possible transitions: The emission of a photon can only lower the atom's energy. Therefore the possibilities [labeled I, II, III in the figure] are $3 \to 2$, $3 \to 1$, $2 \to 1$.

(b) Wavelengths: From Eq. (27.A.1) we have

$$hf_{\mathrm{I}} = \mathscr{E}_3 - \mathscr{E}_2, \qquad hf_{\mathrm{II}} = \mathscr{E}_3 - \mathscr{E}_1, \qquad hf_{\mathrm{III}} = \mathscr{E}_2 - \mathscr{E}_1 \tag{27.A.8}$$

As a sample calculation, consider

$$f_{\mathrm{I}} = \frac{\mathscr{E}_3 - \mathscr{E}_2}{h}$$

from which we get a wavelength

$$\lambda_{\mathrm{I}} = \frac{c}{f_{\mathrm{I}}} = \frac{ch}{\mathscr{E}_3 - \mathscr{E}_2} \tag{27.A.9}$$

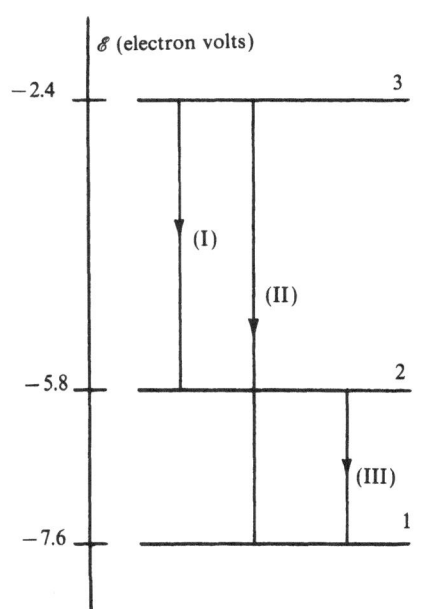

Figure 27.5. Radiative transitions among three levels.

Noting that λ_I is wanted in nanometers, and that the energies are given in electron volts [1 electron volt = (e)(1 volt)], we streamline the calculation by writing (27.A.9) in the form

$$\frac{\lambda_I}{1 \text{ nm}} = \frac{ch}{(e)(1 \text{ volt})(10^{-9} \text{ meter})}$$

$$\times \frac{1}{\dfrac{\mathscr{E}_3}{1 \text{ electron volt}} - \dfrac{\mathscr{E}_2}{1 \text{ electron volt}}}$$

(27.A.10)

$$= \frac{(3.0 \times 10^8)(6.6 \times 10^{-34})}{(1.6 \times 10^{-19})(10^{-9})}$$

$$\times \frac{1}{-2.4 - (-5.8)} = 360$$

or

$$\lambda_I = \underline{360 \text{ nm}}$$

The numerical factor in (27.A.10) is useful in many other cases:

$$\frac{ch}{(e)(1 \text{ volt})(10^{-9} \text{ meter})} = 1240 \qquad (27.A.11)$$

[Recall Eq. (26.B.6) of the preceding chapter.] In particular, for the remaining two wavelengths, we find

$$\lambda_{II} = \frac{1240}{-2.4 - (-7.6)} \text{ nm} = \underline{240 \text{ nm}}$$

$$\lambda_{III} = \frac{1240}{-5.8 - (-7.6)} \text{ nm} = \underline{690 \text{ nm}}$$

iv. Laser Radiation

Laser light is monochromatic because it originates from a transition between two sharply defined energy levels in a chosen species of atom. The popular helium–neon laser (Fig. 27.6) is a glass tube containing a low-pressure mixture of helium and neon gases. The laser transition occurs between two neon levels, shown in Fig. 27.7. The helium is needed for technical reasons that we need not go into here. The tube's input energy comes from a high-voltage discharge through the gas. Many neon atoms are excited from their ground state to level 2 in Fig. 27.7, and decay to level 1 with emission

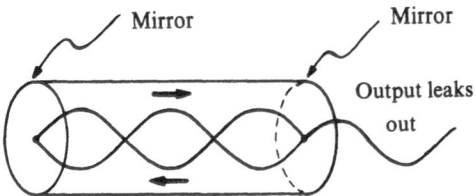

Figure 27.6. Principle of the laser. In reality, the standing wave has, not three antinodes, but perhaps a million.

of a photon of red light, whose frequency and wavelength we leave for the reader to calculate [Problem 27.12(b)].

What makes the laser different from other luminous discharges is the collective behavior of all neon atoms undergoing the transition $2 \rightarrow 1$. Their photons are emitted in such a way as to form, when taken together, a single *coherent classical wave*. That is to say, an observer outside the tube receives wave signals with the same phase from extensive portions of the gas. The resulting superposition is an almost ideal plane monochromatic wave. (Recall our earlier discussion of coherence in Sec. C of Chap. 24.)

How can distant atoms act in concert so as to radiate in phase? The answer is that they are synchronized by a common signal. The ends of the laser tube are accurately oriented mirrors which create, throughout the tube, a strong standing wave whose frequency precisely corresponds to the transition $2 \rightarrow 1$. Some radiation (the "laser beam") is

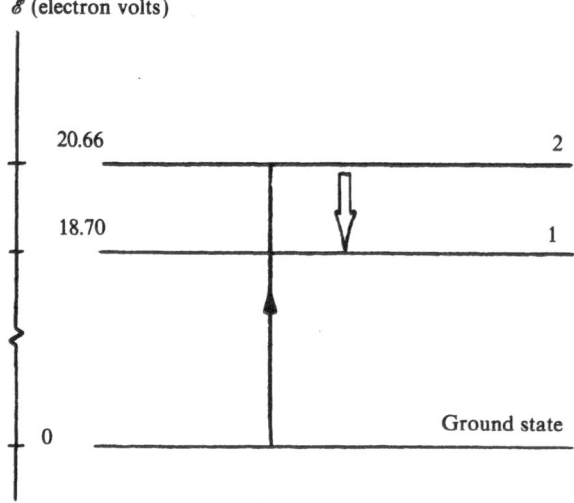

Figure 27.7. The energy levels of neon which are responsible for its laser action. Level 2 is **metastable**, which means that it hardly ever decays unless stimulated by the laser radiation itself.

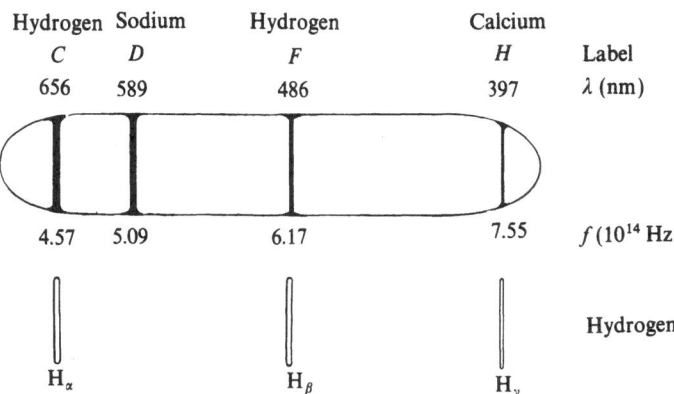

Figure 27.8. The solar spectrum (visible portion) with the most prominent Fraunhofer dark lines, revealing hydrogen, sodium, and calcium vapor in the Sun's atmosphere. The designations C, D, F, H are traditional and unrelated to chemical symbols. Three Balmer emission lines are shown for comparison; H_γ does not correspond to a conspicuous absorption line.

permitted to leak out through one of the mirrors, which is only partly reflecting.

The standing wave is what *stimulates* all neon atoms still in level 2 to radiate in unison; it also provides them with their common phase signal. (We note that the word "laser" stands for *l*ight *a*mplification by *s*timulated *e*mission of *r*adiation; the existence of stimulated emission had been predicted by Albert Einstein in 1917.)

v. Absorption Lines

In 1814 Joseph von Fraunhofer, who was using a prism of unusually high quality made by himself, noticed a large number of dark lines in the solar spectrum; the most prominent are displayed in Fig. 27.8. They are due—as Kirchhoff later deduced—to the presence of various gases in the atmosphere of the Sun; these gases are non-luminous because they are much cooler than the Sun's surface. They absorb (rather than emit) at certain frequencies; the latter are filtered out, and hence do not contribute to the observed emission spectrum. The *absorbed frequencies are identical to those that are emitted by the same gases when they are hot*. The mechanism is illustrated in Fig. 27.9. A photon can be absorbed if it has the appropriate energy, because a transition is then possible from a lower level to a higher one.

When does a given substance radiate and when does it absorb? In general, both processes occur simultaneously: some atoms radiate while others absorb. The rule of thumb is: if many atoms are in high energy levels, radiation predominates; if most are in low levels, absorption does. There are several ways of exciting atoms to higher levels—for exam-

ple, heating a gas to a high temperature; or passing an electric discharge through it with relatively little rise in temperature, as in "fluorescent" lamps.

vi. The Levels of Hydrogen

Let us go back to hydrogen, whose observed frequency spectrum is

$$f = \left[\frac{1}{(n_1)^2} - \frac{1}{(n_2)^2} \right] f_H \qquad (27.A.12)$$

see Eq. (27.A.1). Multiplication by h yields the photon energies,

$$hf = \frac{hf_H}{(n_1)^2} - \frac{hf_H}{(n_2)^2} \qquad (27.A.13)$$

We have written the right-hand side as a difference of two terms in order to suggest a transition, as in Eq. (27.A.7). What are the two relevant energy levels? (Before jumping to conclusions, we must recall that a difference, say $x - y$, is always insufficient to yield the individual values x and y;

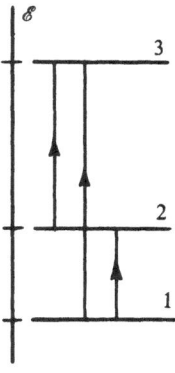

Figure 27.9. Absorptive transitions corresponding to the radiative transitions of Fig. 27.5.

furthermore, we require each of the two contributing levels to be negative.)

The simplest scheme consistent with (27.A.13) and with negative levels is to assign to a hydrogen atom the possible energies

$$\mathscr{E}_n = -\frac{hf_H}{n^2} \qquad (n = 1, 2, 3, ...)$$

$$(27.A.14)$$

This turns out to be the correct assumption; in the next section we shall justify it theoretically.

Figure 27.10 displays these levels to scale. We see that *a higher value of n goes with a higher energy.* According to (27.A.14), a transition from $n = n_2$ down to $n = n_1$ releases a photon of energy

$$hf = \mathscr{E}_{n_2} - \mathscr{E}_{n_1} = -\frac{hf_H}{n_2^2} - \left(-\frac{hf_H}{n_1^2}\right)$$

in agreement with the observed spectrum (27.A.13).

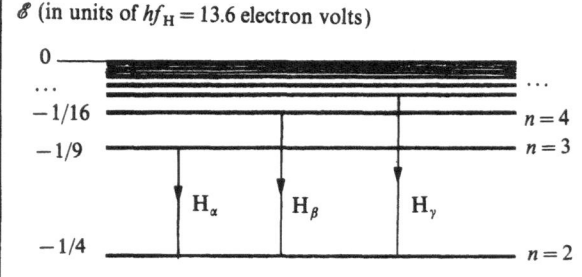

$\mathscr{E}$ (in units of $hf_H = 13.6$ electron volts)

Figure 27.10. The hydrogen levels. As an illustration, the first three Balmer transitions are shown by arrows.

Example 27.3. The Ground State Energy. What is, in electron volts, and to three figures, the lowest possible energy of a hydrogen atom?

In formula (27.A.14), we must take $n = 1$. We have

$$\mathscr{E}_1 = -hf_H \qquad (27.A.15)$$

or

$$\frac{\mathscr{E}_1}{1 \text{ electron volt}}$$

$$= -\frac{hf_H}{(e)(1 \text{ volt})}$$

$$= -\frac{(6.63 \times 10^{-34})(3.29 \times 10^{15})}{1.60 \times 10^{-19}} = -13.6$$

That is to say,

$$\mathscr{E}_1 = \underline{-13.6 \text{ electron volts}} \quad (27.A.16)$$

We recall the meaning of the minus sign: an energy of $+13.6$ electron volts must be supplied in order to split the atom (ionize it) into a proton and an electron, far apart and at rest. In short, the **ionization energy** of hydrogen is 13.6 electron volts. (It is also said that its **ionization potential** is 13.6 volts.)

B. Bohr's Model of the Hydrogen Atom

Having gained some familiarity with energy levels, we now come to our only detailed look at the physics of an atom. We focus on hydrogen, which we picture as just a pair of point particles—the massive proton and the light electron; the two are kept together by electrostatic attraction. There is excellent evidence for their pointlike, or almost pointlike nature; Note (ii) at the end of the present chapter has some comments on the electron's size; the nuclear size is discussed in the next chapter.

In order to analyze the system, we shall boldly enlist some classical mechanics as well as some wave mechanics, and as a result we shall see discrete energy levels emerge from a simple calculation, in impressive quantitative agreement with the observed spectral data. (An historical caution: The following arguments are somewhat modernized relative to Bohr's; his conclusions are reproduced exactly.)

i. Ingredients of the Model

For the sake of getting sharp predictions, Bohr intentionally made as many oversimplifications as was necessary; this is what distinguishes a "model" from a more systematic "theory." Following Bohr,

1. We consider the proton to be at rest—a fairly harmless assumption, since the proton is some 2000 times more massive than the electron.
2. We take the electron to be in uniform circular motion around the central proton. This is a serious restriction: elliptic orbits might be expected as well, since they are found in astronomy. The circular orbits will, however, make the mathematics very simple to understand.
3. We shall not hesitate to switch back and forth between classical and quantum descriptions during the discussion. (In this case such a procedure can be vindicated by a careful quantum mechanical reformulation of the problem.)

Next, we list the classical and quantum inputs of the calculation.

Classical Energy

From Newtonian mechanics we have, for the electron's total energy,

$$\mathscr{E} = \frac{1}{2} mv^2 + \left(-\frac{Ke^2}{r} \right) \qquad (27.B.1)$$

The two terms represent the electron's kinetic and potential energies. [We assume $m, v =$ mass and speed of the electron, $\pm e =$ charge on the proton or electron, $K \approx 9 \times 10^9$ (SI) = proportionality constant in Coulomb's law, $r =$ radius of the electron's orbit.] The negative sign of the potential energy is very important. As in astronomy, the $-1/r$ behavior corresponds to a $1/r^2$ attractive force.

Classical Force

Again as in astronomy, v and r are related by the centripetal force law

$$\frac{Ke^2}{r^2} = m \frac{v^2}{r} \qquad (27.B.2)$$

where the left-hand side is the electrostatic force, and the right-hand side involves the centripetal acceleration v^2/r. More simply, (27.B.2) reads

$$\frac{Ke^2}{r} = mv^2 \qquad (27.B.3)$$

or

$$|\text{potential energy}| = 2 \times (\text{kinetic energy}) \qquad (27.B.4)$$

Thus, one term of $\mathscr{E}$ [Eq. (27.B.1)] can be eliminated in favor of the other. For example, (27.B.1) may be written

$$\mathscr{E} = -\frac{Ke^2}{2r} \qquad (27.B.5)$$

displaying the fact that *the total energy is negative*; the electron is "bound" to the proton. Figure 27.11 illustrates the relation between the kinetic, potential, and total energies. So far, the discussion has been entirely classical.

The Quantization Condition

We now switch our outlook, and visualize the electron as a de Broglie wave, of wavelength

$$\lambda = \frac{h}{mv} \qquad (27.B.6)$$

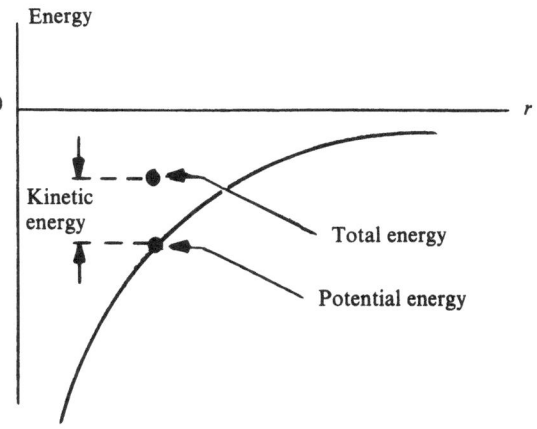

Figure 27.11. The $1/r$ potential energy curve, plotted against the orbital radius r. The heavy dots illustrate Eq. (27.B.4) in the case of a given r.

as in the preceding chapter. The wave circulates around the proton, as Fig. 27.12 indicates.

To the extent that the wave follows a path, that path should be the classical orbit; in this way the classical and wave-mechanical descriptions are in minimal conflict. Part (a) of the figure demonstrates that, in general, the wave description runs into trouble. If λ is not matched to the circumference, we cannot possibly get a smooth sine wave, or perhaps the ψ function has, at any one time, more than one value at a single point—a situation which we would not know how to interpret.

Proper matching, as in part (b) of the figure, requires that the wavelength should fit an integral number of times into the circumference,

$$\boxed{2\pi r = n\lambda} \qquad (n = 1, 2, 3,...) \quad (27.\text{B}.7)$$

or, with (27.B.6),

$$2\pi r = \frac{nh}{mv} \qquad (n = 1, 2, 3,...) \quad (27.\text{B}.8)$$

This is the **quantization condition**, and n is called a **quantum number**.

Discrete Orbits

Only those orbits exist for which n is an integer. Thus, in the Bohr model, the electron must follow one of a discrete set of concentric tracks, numbered from 1 to infinity. This feature is completely foreign to classical mechanics; the allowed astronomical orbits form a continuous set. It must be said that, in quantum mechanics, discrete orbits are only an approximation to the truth; later we shall touch upon a somewhat more realistic description.

ii. Solving for the Energy Levels

We have obtained three independent conditions for the electron's orbit. They are

$$\mathscr{E} = -\frac{Ke^2}{2r} \qquad (27.\text{B}.9)$$

$$\frac{Ke^2}{r} = mv^2 \qquad (27.\text{B}.10)$$

$$2\pi r = \frac{nh}{mv} \qquad (n = 1, 2, 3,...) \quad (27.\text{B}.11)$$

[The above are just a relisting of Eqs. (27.B.5), (27.B.3), and (27.B.8).]

For a given value of n, Eqs. (27.B.10) and (27.B.11) together yield v and r; thus, n defines the orbit. Therefore, through Eq. (27.B.9), n also defines the energy level. Finding r, v, and $\mathscr{E}$ is a simple algebraic exercise, which will be left to the reader. The results are as follows: Eqs. (27.B.10) and (27.B.11) give

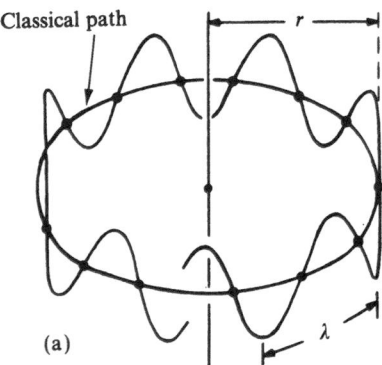

(a)

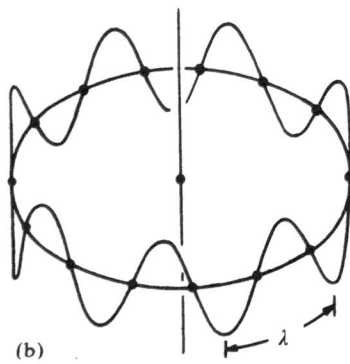

(b)

Figure 27.12. (a) Schrödinger's wave function, ψ, plotted vertically as a function of position along the classical orbit. Here the wavelength λ is improperly matched to the circumference. (b) A properly matched version of (a); in this illustration λ fits 7 times in the orbit.

$$r = \frac{h^2}{(2\pi)^2 \, Ke^2m} n^2 \qquad (27.\text{B}.12)$$

$$v = \frac{2\pi Ke^2}{h} \frac{1}{n} \qquad (27.\text{B}.13)$$

With the help of (27.B.12), (27.B.9) reads

$$\mathcal{E} = -\left(\frac{2\pi Ke^2}{h}\right)^2 \frac{m}{2}\frac{1}{n^2} \qquad (27.B.14)$$

We note immediately the hoped-for $-1/n^2$ dependence of $\mathcal{E}$, as in Eq. (27.A.12). Even more dramatically, the numerical constant in (27.B.14) agrees with the 13.6 electron volts of Eq. (27.A.16):

$$\left(\frac{2\pi Ke^2}{h}\right)^2 \frac{m}{2} = 13.6 \text{ electron volts} \qquad (27.B.15)$$

Verification is, once more, left to the reader.

We see that the electron needs a jump in its energy in order to leave its (stable) lowest orbit. If a similar feature exists in other types of atoms, we begin to understand the extraordinary stability of ordinary matter, thanks to which a desk remains a desk, or a measuring ruler stays reliable even under rough treatment.

iii. Discussion of the Bohr Model

Allowed Radii

From Eq. (27.B.12) we note the rapid expansion of the quantized orbits with increasing n. Some orbits are illustrated in Fig. 27.13. Just as $n=1$ corresponds to the lowest energy (ground state), it also corresponds to the smallest orbit, characterized by the so-called **Bohr radius**,

$$r_B = \frac{h^2}{(2\pi)^2 Ke^2 m} = 5.29 \times 10^{-11} \text{ meter}$$

$$(27.B.16)$$

This figure can be taken, roughly, as the radius of the hydrogen atom itself. It agrees fairly well with experimental estimates, although we recall that the atomic radius is not a sharp concept; its value depends somewhat on the circumstances of measurement (nearest approach in collisions, packing distance in crystals, length of a chemical bond, etc.)

In their ground state, all neutral atoms (not just

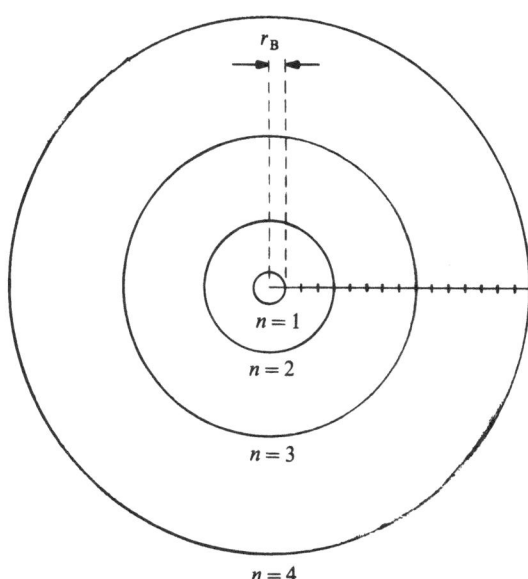

Figure 27.13. The first few Bohr orbits. The smallest circle corresponds to the stable state of the hydrogen atom, and about matches the atom's observed size.

hydrogen) have radii comparable to r_B. That size is often expressed in **Ångstrom units,**[*]

$$1 \text{ Å} = 10^{-10} \text{ meter} = 0.1 \text{ nm} \qquad (27.B.17)$$

We have

$$r_B = 0.529 \text{ Å} \qquad (27.B.18)$$

Allowed Speeds

These become smaller with increasing n; see Eq. (27.B.13). As $n \to \infty$, we find that $r \to \infty$ and $v \to 0$; the zero energy convention is fulfilled, and indeed we have $\mathcal{E} \to 0$ from (27.B.14).

In all that precedes, we have paid no attention to relativity. To see if the omission was serious, we examine the electron's speed in the fastest orbit, $n=1$. How does that speed compare with the speed of light? From Eq. (27.B.13) we have

$$\left.\frac{v}{c}\right|_{n=1} = \frac{2\pi Ke^2}{hc} \approx 7 \times 10^{-3} \qquad (27.B.19)$$

small enough to reassure us about our non-relativistic treatment.

---

[*] First used by Anders Jonas Ångstrom himself in his pioneering measurements of the solar spectrum.

Dimensionless ratios, such as (27.B.19), do not depend on any choice of units; they are often believed—rightly of wrongly—to be derivable from pure thought rather than from measurement. A precision calculation, based on the *measured constants e, h,* and *c* (we recall that *K* is numerically related to *c*) gives, for (27.B.19),

$$\boxed{\frac{2\pi Ke^2}{hc} \approx \frac{1}{137.03604}} \qquad (27.B.20)$$

This small number is known as the **fine-structure constant.** There have been mystically flavored attempts to produce it from mathematical postulates, but as of this writing none has succeeded. Because of its proportionality to e^2, the fine-structure constant is usually considered a measure of the strength with which electromagnetic fields—or better, photons—interact with the rest of matter. This strength is very modest, as we can see.

Shortcomings

The Bohr model is logically incomplete; it contradicts some of the physics we have learned in the past, and does not replace it by a systematic theory. Two discrepancies stand out in particular.

First, we have seen that, classically, an accelerating charge (the orbiting electron) necessarily radiates electromagnetic waves. Hence it ought to lose energy, somewhat like an artificial satellite grazing the upper atmosphere (in this analogy friction replaces radiation). The electron should eventually spiral into the proton, emitting a huge burst of radiation in the process. In actual fact, and also according to Bohr, nothing of the kind ever happens. The ground state ($n = 1$) is stable and can persist forever without radiating.

A second puzzle involves the discrete allowed orbits. When the atomic state decays, say from $n = 5$ to $n = 4$, it emits a photon of the appropriate energy. But the Bohr model provides no allowed orbits between $n = 5$ and $n = 4$; how can the transition occur *across this forbidden gap?*

This and many other difficulties cannot be resolved within the Bohr model itself; on the other hand, a systematically formulated quantum mechanics (or wave mechanics) can indeed be made consistent.

The Correspondence Principle

Old physical principles, which have been tested and found useful in countless different circumstances, cannot be simply discarded. We know by experiment that accelerating charges do radiate when in a *macroscopic* orbit; the rate of radiation is reliably predicted by Maxwell's theory. Therefore, as Bohr himself pointed out, large orbits ($n \to \infty$) should undergo transitions, and emit radiation, which closely resemble the classical predictions. (Problem 27.32 pursues that idea somewhat further.) The requirement that, *for large quantum numbers, quantum physics should merge with classical physics,* is called Bohr's correspondence principle.

C. Electron States and Their Four Quantum Numbers

Wave mechanics can be fairly difficult, but its results (for the hydrogen atom) are extraordinarily simple—in some respects much simpler than a classical description. This section presents the most important features of the wave-mechanical hydrogen atom; no derivation is attempted. The basic message of wave mechanics is that Bohr's discrete set of orbits must be replaced by a discrete set of wave patterns for Schrödinger's wave function ψ. These patterns are standing waves, similar to the discrete set of standing waves that can exist in an organ pipe.

Since one cannot accurately say that a standing wave pattern represents an orbit, one prefers to say that it represents a **state** (or **quantum state**) of the electron.* A state is like an orbit in that it is characterized by a well-defined energy and angular momentum. However, rather than being confined to a sharp trajectory, the electron has a certain probability for being at any given distance from the proton; certain distances, of course, are much more likely than others. Thus, one refers to a **probability cloud**, which statistically spreads the electron into a smooth charge distribution around the proton. In the quantum states to be discussed here, the probability cloud keeps a constant shape in the course of time. Hence these states are referred to as **stationary**

---

* Also considered as a state of the whole hydrogen atom, since the proton is just seen as a passive force center.

states. The ground state does not radiate because its probability cloud behaves as a motionless charge distribution. The higher-energy states, or **excited states,** do radiate in practice (one photon at a time). In our discussion, however, this fact is ignored: they are represented as stationary states as well.

In what follows we learn to classify the stationary states by a set of quantum numbers, one of which, n, determines the energy as it does in the Bohr model. The other quantum numbers are related to angular momentum.

Once we have classified the hydrogen states by their quantum numbers, it will become very easy to describe the structure of the other chemical elements, a task we leave to the next section.

i. The Probability Interpretation

At a given time, any small volume $d\mathcal{V}$ in the atom *may* contain the electron. The chance of meeting that electron in $d\mathcal{V}$ is, naturally, proportional to $d\mathcal{V}$ itself; as we remarked earlier, the chance also depends on where $d\mathcal{V}$ is located—say in the vicinity of a position $\mathbf{r}$, as in Fig. 27.14. In order to express that dependence, we introduce a function $P(\mathbf{r})$, called the **probability density**; that is to say, P is some nonnegative number that depends on $\mathbf{r}$, and whose value is such that

$$P(\mathbf{r})\, d\mathcal{V} = \text{probability that the electron}$$
$$\text{is located inside } d\mathcal{V} \text{ at } \mathbf{r}$$

$$(27.C.1)$$

This is the definition of P. That function can be obtained directly from ψ, but here we shall not be concerned with mathematical details about either P or ψ.

Figure 27.15 illustrates some electron probability densities which are known (by calculation) to occur in the hydrogen atom for various energies and angular momenta. Observe that if $n = 1$, the electron has a nonzero probability of being arbitrarily close to the proton, in contrast to the Bohr model's minimum distance r_B.

ii. Bohr's Quantization of the Angular Momentum

Let us take a fresh look at Bohr's quantization condition, Eq. (27.B.8):

$$2\pi r = \frac{nh}{mv} \qquad (27.C.2)$$

which may be written

$$mvr = \frac{nh}{2\pi} \qquad (27.C.3)$$

The left side is the **orbital angular momentum** of the electron about the proton: (27.C.3) states that

$$\boxed{\left(\begin{array}{c}\text{Orbital angular}\\ \text{momentum}\end{array}\right) = (n)\left(\frac{h}{2\pi}\right)}$$

$$(n = 1, 2, 3,...) \quad (27.C.4)$$

Thus, $h/2\pi$ is the basic quantum, or "building block," of angular momentum in the Bohr model.

[In many ways, $h/2\pi$ is a more useful number than h; it is denoted in the literature by $h/2\pi = \hbar$ ("aitch bar").]

iii. The Four Quantum Numbers

One single quantum number, n, appears so far to control two characteristics of an orbit, namely, energy and angular momentum. In a more realistic model, however, elliptic orbits are admitted, and those two quantities can be adjusted separately. Far better yet is the consistent wave-mechanical

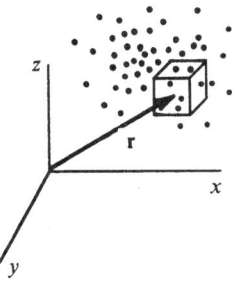

Figure 27.14. A probability cloud. The density of points is meant to represent the function $P(\mathbf{r})$. The small cube has volume $d\mathcal{V}$.

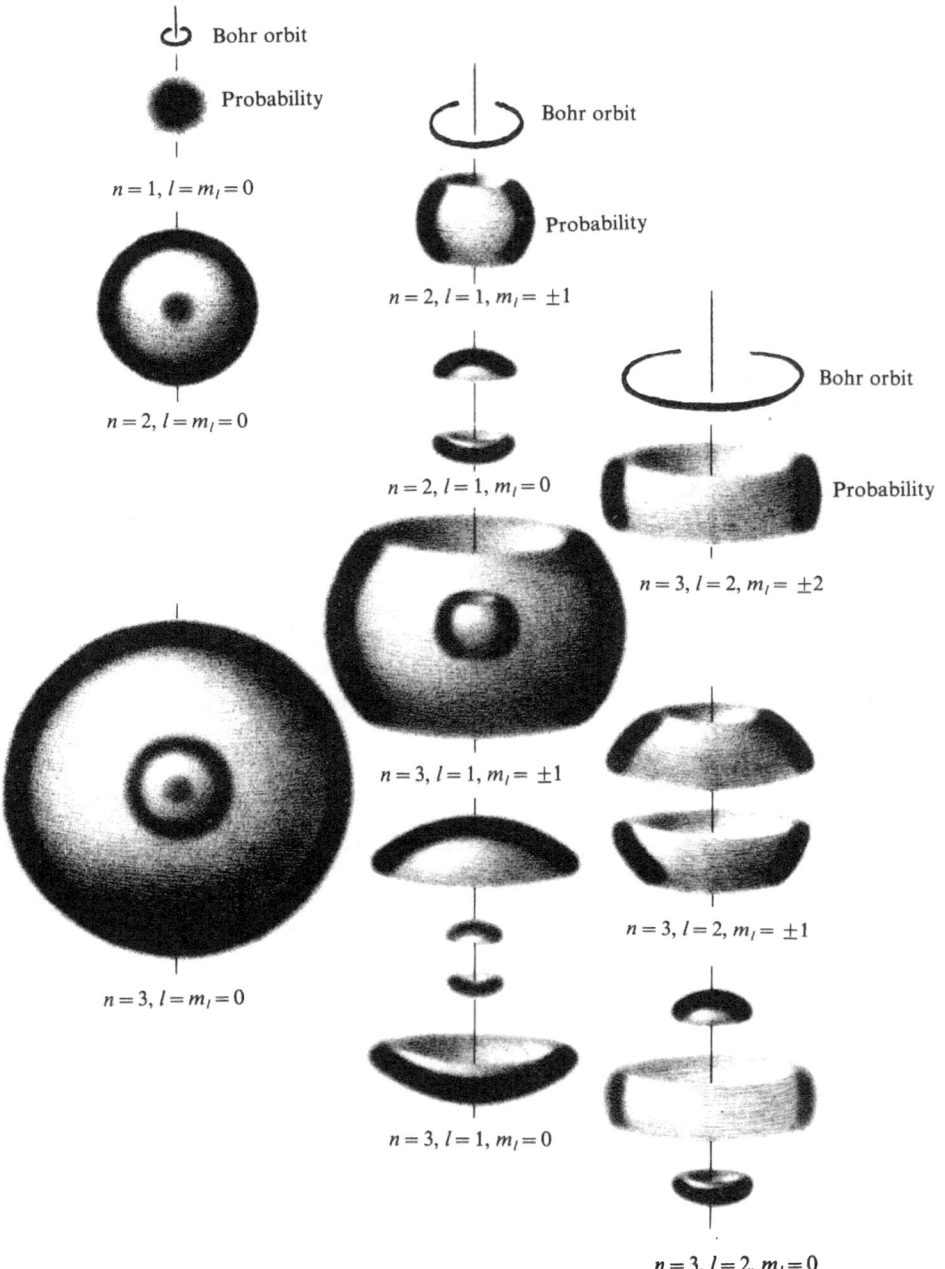

Figure 27.15. Probability cloud for the electron in several states of the hydrogen atom (an artist's conception). The Bohr orbit (a manifest oversimplification) is shown to scale for comparison in three cases. The meaning of the other configurations, and of the symbols l, m_l, is discussed further on. [By permission, from R. Eisberg and R. Resnick, *Quantum Physics*, p. 273 (Copyright © John Wiley & Sons, New York, 1974).]

theory of Erwin Schrödinger and Wolfgang Pauli. That theory, which emerged in 1925–1926, uses **four quantum numbers**—enough to specify the state of the electron in every physically meaningful aspect.

The properties we are about to list are simple —they deal mostly with small integers—but their explanation is less simple and belongs to a quantum mechanics course. With some suspension of disbelief, the reader should have no difficulty in carefully noting the results. In what follows we discuss each quantum number in turn.

1. The Principal Quantum Number

Each state has a well-defined energy, given by Bohr's formula (27.B.14),

$$\mathscr{E} = -\left(\frac{2\pi Ke^2}{h}\right)^2 \frac{m}{2}\frac{1}{n^2} = -(13.6 \text{ electron volts})\frac{1}{n^2}$$

$$(n = 1, 2, 3,...) \quad (27.C.5)$$

Here n is called the **principal quantum number**, and ranges from 1 to infinity: it governs not only the energy, but also the radial extent of the electron's probability cloud (like the radius of the orbit in the Bohr model).

2. The Orbital Quantum Number

Each state has a well-defined orbital angular momentum L about the proton; it is given by

$$\boxed{L = \sqrt{l(l+1)}\,\frac{h}{2\pi}} \qquad (l = 0, 1, 2,..., n-1)$$

$$(27.C.6)$$

Here l is the **orbital quantum number**; according to the theory, it cannot exceed $n-1$. We see that, for a given n, there are n possible values of L. Also, we are talking here about an absolute value; that is, L is defined as positive.

> **Example 27.4. Angular Momentum for $n = 3$.** List, in terms of $h/2\pi$, the possible values of L in the states with $n = 3$.
>
> From (27.C.6), with $l = 0, 1, 2$, we have
>
> $l = 0 \qquad \sqrt{(0)(1)}\,\dfrac{h}{2\pi} = \underline{\underline{0}}$
>
> $l = 1 \qquad \sqrt{(1)(2)}\,\dfrac{h}{2\pi} = \underline{\sqrt{2}\,\dfrac{h}{2\pi}}$
>
> $l = 2 \qquad \sqrt{(2)(3)}\,\dfrac{h}{2\pi} = \underline{\sqrt{6}\,\dfrac{h}{2\pi}}$

3. The Magnetic Quantum Number

Consider first a classical electron orbit, as in the Bohr model. That orbit is characterized by an angular momentum vector **L**, which one chooses to construct as in Fig. 27.16. It is drawn through the proton, and points perpendicularly to the plane of the orbit. Its magnitude $|\mathbf{L}| = L$ is, numerically, the angular momentum of the electron about the axis defined by **L**. [This construction for **L** can be shown to agree with the definition in Chap. 7, Note (ii).]

An electron state is a far cry from a classical orbit; nevertheless, it, too, has an **orbital angular momentum vector L**. The quantization rule we are about to state needs a reference direction in space, say the z axis. Experimentally, the z axis may be singled out, or "marked," by means of an external **B** or **E** field, weak enough so as not to perturb the physics we are discussing. The rule is as follows.

Each state has, for its **L** vector, a well-defined component L_z in the $+z$ direction. That component is given by

$$\boxed{L_z = m_l \frac{h}{2\pi}} \qquad (m_l = 0, \pm 1, \pm 2,..., \pm l)$$

$$(27.C.7)$$

Therefore, the **L** vector can make only certain discrete angles with the z axis. We note that, for a given l, the value of $|m_l|$ cannot exceed l, and hence there are $2l+1$ such possible angles.

Because the z direction is often taken to be that of a laboratory **B** field, m_l is referred to as the **magnetic quantum number**.

> **Example 27.5. L_z Values for $n = 3$.** (a) List, in terms of $h/2\pi$, the possible values of L_z in the states with $n = 3$. (b) Determine the angle between **L** and the $+z$ axis in the case $l = 2$, $m_l = +1$.

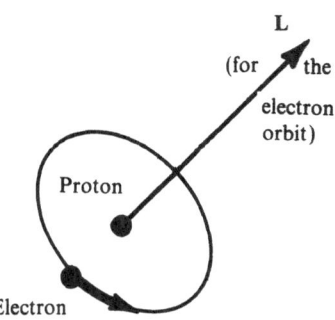

Figure 27.16. A classical illustration of **L**. (The orbit is a circular one, shown in perspective.)

(a) List of values: For $n = 3$, the possible l values are 0, 1, 2. For $l = 0$, the only possible m_l is $m_l = 0$; for $l = 1$, the values of m_l are $m_l = 0, \pm 1$. For $l = 2$, they are $m_l = 0, \pm 1, \pm 2$. We therefore have the following L_z values:

$$l = 0 \qquad L_z = \underline{\underline{0}}$$

$$l = 1 \qquad L_z = \underline{\underline{0, \pm \frac{h}{2\pi}}} \qquad (27.C.8)$$

$$l = 2 \qquad L_z = \underline{\underline{0, \pm \frac{h}{2\pi}, \pm 2\frac{h}{2\pi}}}$$

(b) Angle for $l = 2$, $m_l = +1$: Figure 27.17 shows the relevant geometry. The required angle θ follows from the $l = 2$ results above and in Example 27.4:

$$L_z = +\frac{h}{2\pi}, \qquad L = \sqrt{6}\frac{h}{2\pi},$$

$$\cos\theta = \frac{L_z}{L} = \frac{1}{\sqrt{6}}$$

$$\theta = \underline{\underline{66°}}$$

Other Components of L   Are the x and y components L_x, L_y also quantized? They are not; there exists an uncertainty principle for angular momentum, which states that

Knowledge of one component
of L (say L_z) prevents
knowledge of any other rectangular
component (L_x or L_y)   (27.C.9)

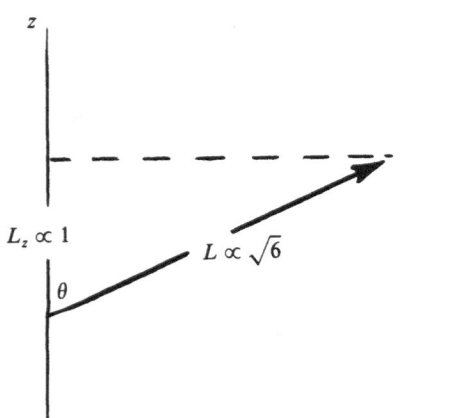

Figure 27.17. Calculating the angle between L and the z axis.

Figure 27.18 illustrates that principle for $l = 2$, Eqs. (27.C.8).

4. The Spin Quantum Number

In 1925, George Uhlenbeck and Samuel Goudsmit concluded from high-resolution spectral data that the electron behaves in some ways as a particle that spins about its own axis.

The spin angular momentum **S** of the electron exists independently of orbital motion around the nucleus. The **S** vector has all the properties which an **L** vector would have if we took $l = 1/2$ (an impossible value for l). Specifically, the magnitude of **S** is [recall formula (27.C.6)]

$$S = \sqrt{\left(\frac{1}{2}\right)\left(\frac{3}{2}\right)}\frac{h}{2\pi} = \frac{\sqrt{3}}{2}\frac{h}{2\pi} \qquad (27.C.10)$$

The z component of **S** can only assume two values,

$$\boxed{S_z = \pm\frac{1}{2}\frac{h}{2\pi}} \qquad (27.C.11)$$

corresponding to a **spin quantum number**

$$\boxed{m_s = \pm\frac{1}{2}} \qquad (27.C.12)$$

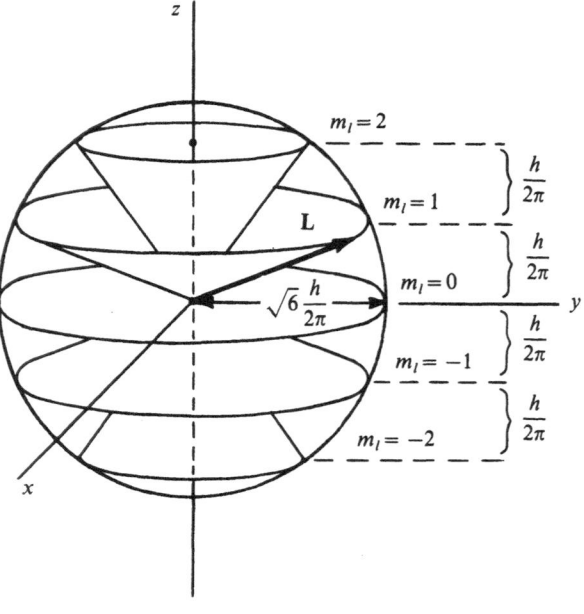

Figure 27.18. When $l = 2$, and for given m_l, the orbital angular momentum vector **L** can be anywhere on the appropriate cone.

If S_z is known, then S_x and S_y cannot be specified; see Fig. 27.19.

Simplified language is often used in connection with spin. Because of (27.C.11), one says that the electron is a "spin-half particle"; the two orientations are referred to as "spin up" or "spin down."

iv. Counting the States

In the hydrogen atom, an electron state is completely determined by the values of its four quantum numbers n, l, m_l, m_s. There are only two possible states with $n = 1$ (two versions of the ground state): they have $l = 0$, $m_l = 0$, $m_s = \pm 1/2$. The following traditional nomenclature persists in the literature. A state with $l = 0$ is called an s state, one with $l = 1$ a p state, etc.:

$$
\begin{array}{l|l}
\text{Value of } l & 0 \quad 1 \quad 2 \quad 3 \quad 4 \quad \cdots \\
\hline
\text{Name of state} & s \quad p \quad d \quad f \quad g \quad \cdots
\end{array} \quad (27.C.13)
$$

(These letters used to stand for the spectroscopic epithets "sharp," "principal," "diffuse," "fundamental"; the rest of the sequence is alphabetic.)

In naming a state, the value of n is made into a prefix: thus, a $5d$ state means a state with $n = 5$, $l = 2$; a $7g$ state is one with $n = 7$, $l = 4$, etc.

How many different states are there for a given n? Suppose first that $m_s = +1/2$. Then we tabulate the possible states in Table 27.2. The total number of states equals the arithmetic series

$$1 + 3 + 5 + \cdots + (2n - 1) = n^2 \qquad (27.C.14)$$

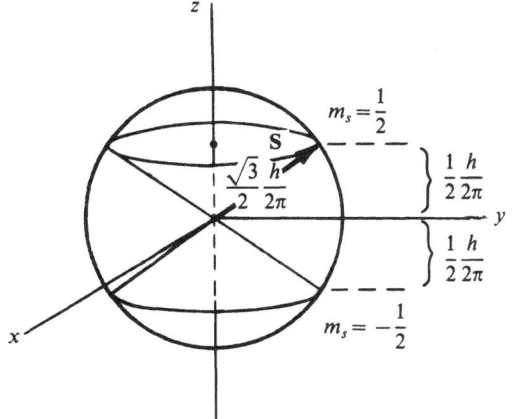

Figure 27.19. The spin vector **S** can be anywhere on the appropriate cone.

Table 27.2

l	m_l	Number of states $= 2l + 1$
0	0	1
1	$0, \pm 1$	3
2	$0, \pm 1, \pm 2$	5
$\cdots$	$\cdots$	$\cdots$
$n - 1$	$0,..., \pm(n-1)$	$2n - 1$

Thus, there are n^2 states with spin up. Similarly, there are n^2 states with spin down.

$$\text{For a given } n, \text{ the number of} \atop \text{possible states is } 2n^2. \qquad (27.C.15)$$

D. The Exclusion Principle and the Periodic Table

Having gained some insight into hydrogen, it is natural to attempt a theory of the other atoms. We shall, however, be more modest and only ask for a qualitative picture of how the electrons arrange themselves inside a given atom. The answers will turn out to be valuable—they are the key to a chemical classification of the elements.

At first glance, a many-electron atom might seem too involved for analysis. For example, in a neutral mercury atom, there are 80 electrons; each electron interacts not only with the nucleus, but with each of the 79 other electrons as well.

Fortunately, a simple description is available, after all. Help comes from two quarters. First, thinking in a global way—rather than in terms of each individual electron—turns out to be successful. Second, a remarkable quantum rule, the **exclusion principle**, ensures that the electrons keep out of each other's way to an even greater extent than might already be expected from their mutual repulsion.

The results are as follows:

1. The ground state of a neutral atom is stable, and does not radiate, because it has the lowest total energy such a system of charges can possibly have. (We already saw that statement in the context of hydrogen.) In this section, "atom" shall always mean the neutral ground state.

2. The atom consists of an almost pointlike nucleus, surrounded by the electrons' concentric probability clouds, called **shells**, which are nested inside one another somewhat like the layers of an onion. Each shell, in turn, is made of nested subshells. Probability clouds do not have sharp boundaries, and consequently there is considerable overlap between the shells (and subshells).

3. The shells and subshells are labeled by quantum numbers similar to those which exist in the hydrogen atom.

4. The chemical properties of an atom are primarily determined by the quantum numbers of its outer shell, and by how many electrons are to be found there. (Only the outer electrons are involved in chemical reactions.) Therefore a chemical classification of the elements is at the same time a classification according to the structure of the outer shell. Here our purpose is to examine, in terms of the **periodic table** of the elements, how the details work themselves out.

i. A One-Electron Problem

Consider an atom with **atomic number** Z (=number of electrons in the neutral atom; the nucleus has a charge $+Ze$). Figure 27.20 indicates how any one electron, denoted by e_1, "sees" the $Z-1$ other electrons and the nucleus. For the sake of illustration, the $Z-1$ electrons are shown forming three stationary shells (their combined probability clouds). In our simple-minded model, the electric field acting on e_1 is contributed by a *fixed*

charge distribution, amounting to those shells. In Fig. 27.20, the charge seen by e_1 is that of the nucleus ($+Ze$), shielded by the charge $(Z-1)(-e)$ of the other electrons. The overall effective charge is therefore just $+e$. If, on the other hand, e_1 were within the inner shell and close to the nucleus, it would experience the full unshielded field of the nucleus' charge $+Ze$. (We recall that a hollow spherical shell contributes zero field inside itself.) In short, the electrostatic force on e_1 does not have the simple $1/r^2$ dependence which is found in the hydrogen atom.

Any electron is as good as any other, and therefore the quantum states available to e_1 are also those available to any of the other electrons. We now turn our attention to those states. They are classified by the same quantum numbers n, l, m_l, m_s, as the hydrogen states that we have just studied. However, whereas in the Bohr model for hydrogen the energy of the electron depends only on n, in the more complicated atoms it must be assumed to depend on l as well.

The levels are shown qualitatively in Fig. 27.21.

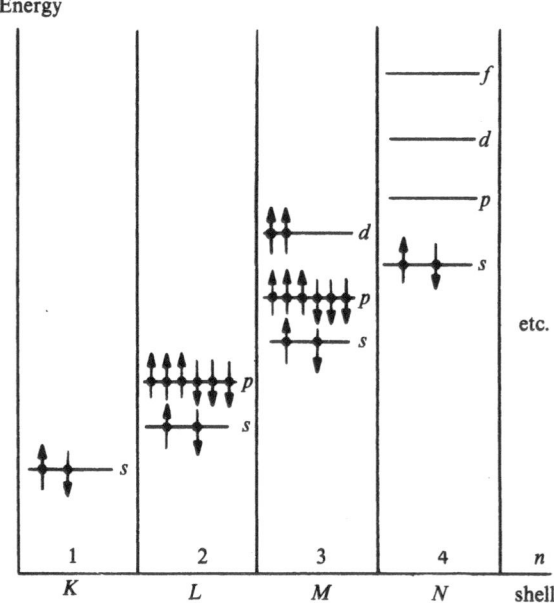

Energy

etc.

n
shell
K L M N
1   2   3   4

Figure 27.21. Energy levels (not to scale) available to an electron in a many-electron atom. As we shall see, each level can accommodate a certain number of spin-up and spin-down electrons (arrowed dots). This illustration represents the ground state of titanium ($Z=22$); each level corresponds to a subshell. [Caution: the higher energies ("excited states") of an atom cannot be inferred from the unoccupied levels: this is a very rough picture and gives about the right answer for the ground state only.]

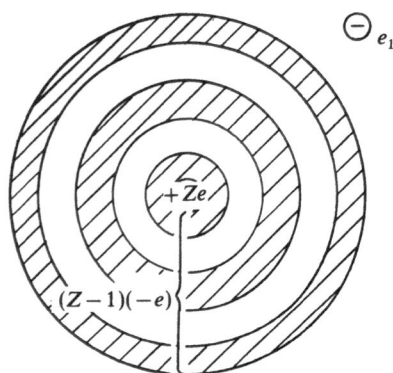

Figure 27.20. Any electron, e_1, is assumed to experience approximately a central electrostatic force due to the rest of the atom.

As we can see, the two $1s$ states ($n = 1$, $l = 0$, $m_l = 0$, $m_s = \pm 1/2$) have the lowest energy. Recalling that an electron always falls into the lowest available state, we might expect all electrons to be in $1s$ states. Yet, this does not happen; spectroscopic data (not shown here) demonstrate that there are never more than two $1s$ electrons in an atom.

ii. The Exclusion Principle

In order to make sense of the experimental evidence, Wolfgang Pauli, in 1925, postulated the following rule (**exclusion principle**):

> Two electrons cannot occupy the same quantum state in the same atom. (27.D.1)

A quantum state is fully defined by the values of n, l, m_l, and m_s. Therefore (27.D.1) stipulates that, in a single atom, *no two electrons can have the same set of quantum numbers*. For example, a helium atom cannot have, for both its electrons, the values $n = 1$, $l = 0$, $m_l = 0$, $m_s = +1/2$. It can, however, have one electron in that state, while the other electron has $n = 1$, $l = 0$, $m_l = 0$, $m_s = -1/2$; in short, the helium atom can have two $1s$ electrons with their spins oppositely lined up ("antiparallel spins"). The helium ground state does, in fact, have that configuration.

iii. Building up the Elements

Starting with hydrogen ($Z = 1$), and going up through helium ($Z = 2$), lithium ($Z = 3$), etc., we can imagine charges $+e$ being added one by one to the nucleus; every time, an electron is added to the surrounding electron cloud. That newcomer finds a suitable set of values for its quantum numbers, namely, those which give it *the lowest possible energy*; states already occupied are excluded. Usually, the electrons which were already present keep their quantum numbers unchanged.

Going back to Fig. 27.21, we can now understand why each level has the number of electrons shown. The first electron takes a $1s$ level; the next electron takes the other $1s$ level; both $1s$ states are now occupied. The third and fourth electrons must settle in the $2s$ states; the fifth electron must be in a $2p$ state. How many electrons can be added before all $2p$ states are occupied? We recall from Table 27.2 that there are three such states with spin up, and three with spin down. Hence six electrons (the fifth through the tenth) are needed to fill the $2p$

states. Thereafter one continues with the $3s$ states, etc.

Shell Nomenclature

Electrons with $n = 1$ are said to form the *K shell*, which is tightly bunched around the nucleus. Next, with increasing remoteness from the nucleus, we have the *L shell* (electrons with $n = 2$), the *M shell* (electrons with $n = 3$), and so on alphabetically. The *K* shell is not divided into subshells; the *L* shell has an s and a p subshell; the *M* shell has an s, a p, and a d subshell, etc.

The message of Fig. 27.21 is that, as we go up the atomic numbers, the shells fill up in the sequence K; $L(s, p)$; $M(s, p)$; $N(s)$; etc. That sequence is compactly displayed in Fig. 27.22. Table 27.3 lists the actual number of electrons in each subshell, as determined by years of painstaking spectroscopic studies in many laboratories. Figure 27.22 suffers a few exceptions, which amount to an occasional electron switching its subshell, and which are indicated in Table 27.3 by boxed figures; they will not affect our subsequent construction of the periodic table.

Example 27.6. A Complete Shell. How many electrons are in a complete N shell?

The N shell consists of electrons with $n = 4$. We have seen [statement (27.C.15)] that there are $2n^2$ possible such states; in this case,

$$2n^2 = (2)(4)^2 = \underline{\underline{32}}$$

Spectroscopic Notation

The shell structure of an atom is compactly represented by a list of n-values for the successive shells; each n-value is followed by the letter s, p, d, etc., indicating the subshell; and, finally, each of these letters has a superscript indicating the number of electrons occupying it.

Example 27.7. Chromium. Express, in spectroscopic notation, the shell configuration of chromium.

From Table 27.3 we have

$$Cr(Z = 24): \underline{1s^2 2s^2 2p^6 3s^2 3p^6 3d^5 4s^1}$$

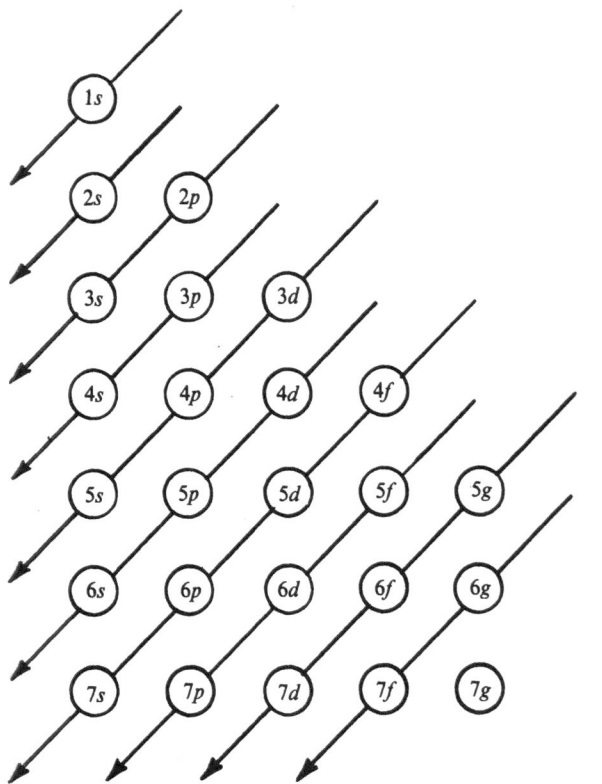

SHELL:

K

L

M

N

O

P

Q

Figure 27.22. The order in which the subshells (levels of Fig. 27.21) are filled. At this point the reader should engage in a careful, electron-by-electron comparison of Figs. 27.22 and 27.21. In the higher shells, a few exceptions to this order exist, as noted in Table 27.3 further on.

iv. The Periodic Table

Let us list all the elements in order of increasing atomic number. Many of their chemical and physical properties then also turn out to follow an orderly sequence; this is most striking in the case of their **chemical valence**.

Roughly speaking (for more precision a chemistry book should be consulted), the valence of an element X is related to how many atoms of a certain kind will bind to one atom of X. For example, hydrogen $(Z=1)$, lithium $(Z=3)$, sodium $(Z=11)$, potassium $(Z=19)$, etc., all have a valence of 1 because, with chlorine, they form the compounds HCl, LiCl, NaCl, KCl, etc. On the other hand, the elements beryllium $(Z=4)$, magnesium $(Z=12)$, calcium $(Z=20)$, etc., have valence 2 as evidenced by the compounds $BeCl_2$, $MgCl_2$, $CaCl_2$, etc. The elements can now be arranged in **periods** which display these regularities, as hinted in Fig. 27.23. The complete table is shown in Fig. 27.24.

To what do we owe the existence of this remarkable scheme? The full explanation lies in the shell structure of Table 27.3. We quote, without detail, a few highlights of that connection.

Hydrogen and the Alkali Metals (First column of Fig. 27.23) have an outer subshell consisting of a single s electron; as it happens, they easily lose that

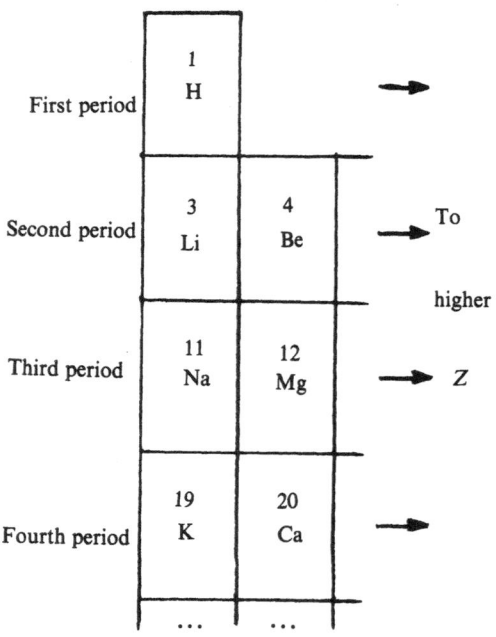

Figure 27.23. A fragment of the periodic table of elements. Each box lists Z and the chemical symbol. A vertical column has elements with the same valence.

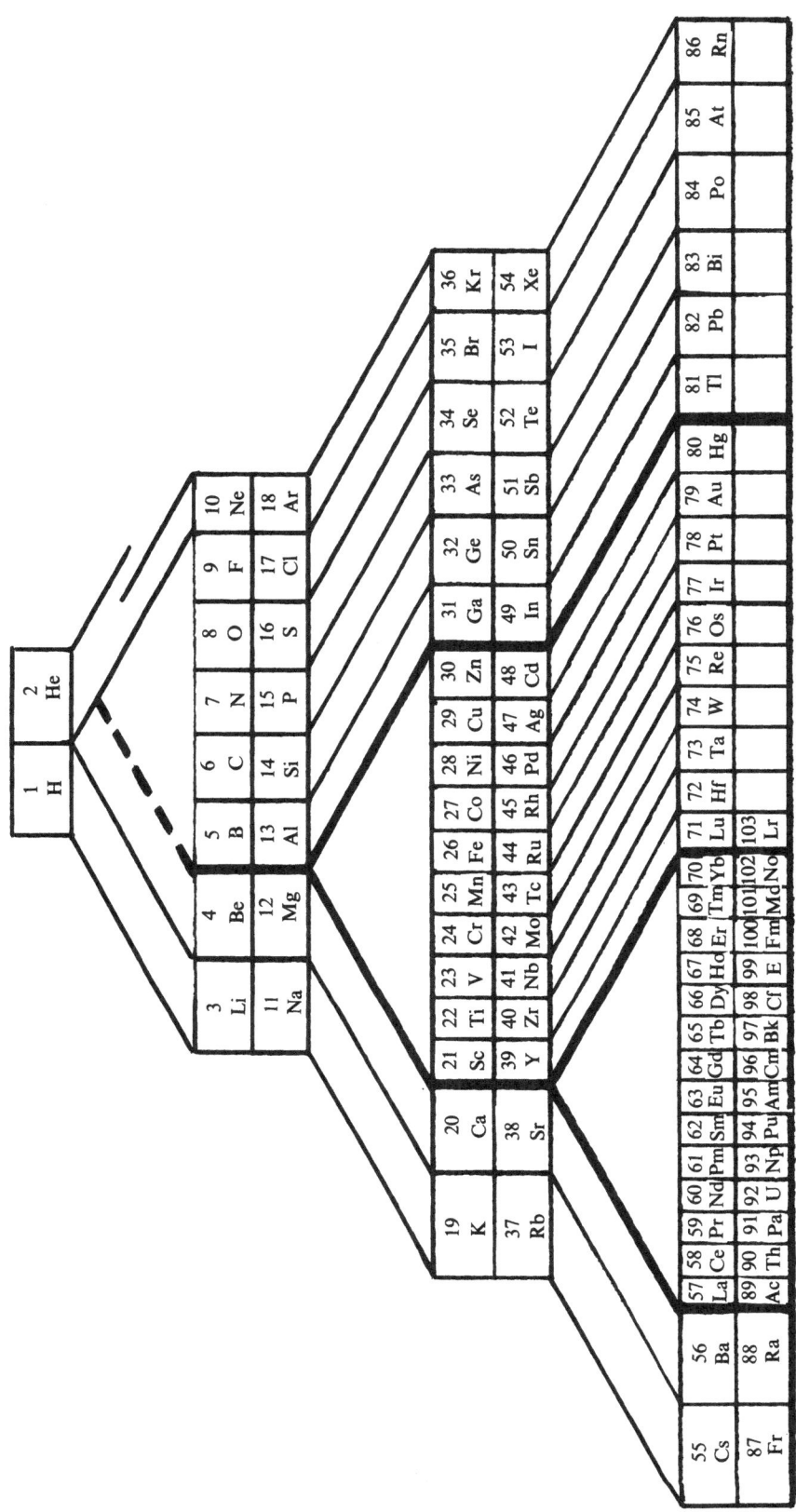

Figure 27.24. (For an alphabetic listing, with atomic weights, see Chap. 11, Table 11.1.) The complete periodic table, as arranged by Niels Bohr. The boxes list Z and the chemical symbol. The broken vertical bands represent chemical families. The reader should retrace here the two columns of Fig. 27.23. The horizontal periods expand in size because of the expanding consecutive shells; this is why the "columns" have been bent out of shape. From the shell point of view, helium (Z = 2) really belongs to the same column as Z = 4, 12, 20, etc., but as it is chemically inert, it is included with the other noble gases. The heavy lines indicate where subshells are begun and completed.

Table 27.3. Electron Configurations of the Elements*

	K	L		M			N				O				P			Q
	1s	2s	2p	3s	3p	3d	4s	4p	4d	4f	5s	5p	5d	5f	6s	6p	6d	7s
1 H	1																	
2 He	2																	
3 Li	2	1																
4 Be	2	2																
5 B	2	2	1															
6 C	2	2	2															
7 N	2	2	3															
8 O	2	2	4															
9 F	2	2	5															
10 Ne	2	2	6															
11 Na	2	2	6	1														
12 Mg	2	2	6	2														
13 Al	2	2	6	2	1													
14 Si	2	2	6	2	2													
15 P	2	2	6	2	3													
16 S	2	2	6	2	4													
17 Cl	2	2	6	2	5													
18 Ar	2	2	6	2	6													
19 K	2	2	6	2	6		1											
20 Ca	2	2	6	2	6		2											
21 Sc	2	2	6	2	6	1	2											
22 Ti	2	2	6	2	6	2	2											
23 V	2	2	6	2	6	3	2											
24 Cr	2	2	6	2	6	5	1											
25 Mn	2	2	6	2	6	5	2											
26 Fe	2	2	6	2	6	6	2											
27 Co	2	2	6	2	6	7	2											
28 Ni	2	2	6	2	6	8	2											
29 Cu	2	2	6	2	6	10	1											
30 Zn	2	2	6	2	6	10	2											
31 Ga	2	2	6	2	6	10	2	1										
32 Ge	2	2	6	2	6	10	2	2										
33 As	2	2	6	2	6	10	2	3										
34 Se	2	2	6	2	6	10	2	4										
35 Br	2	2	6	2	6	10	2	5										
36 Kr	2	2	6	2	6	10	2	6										
37 Rb	2	2	6	2	6	10	2	6			1							
38 Sr	2	2	6	2	6	10	2	6			2							
39 Y	2	2	6	2	6	10	2	6	1		2							
40 Zr	2	2	6	2	6	10	2	6	2		2							
41 Nb	2	2	6	2	6	10	2	6	4		1							
42 Mo	2	2	6	2	6	10	2	6	5		1							
43 Tc	2	2	6	2	6	10	2	6	5		2							
44 Ru	2	2	6	2	6	10	2	6	7		1							
45 Rh	2	2	6	2	6	10	2	6	8		1							
46 Pd	2	2	6	2	6	10	2	6	10									
47 Ag	2	2	6	2	6	10	2	6	10		1							
48 Cd	2	2	6	2	6	10	2	6	10		2							
49 In	2	2	6	2	6	10	2	6	10		2	1						
50 Sn	2	2	6	2	6	10	2	6	10		2	2						
51 Sb	2	2	6	2	6	10	2	6	10		2	3						
52 Te	2	2	6	2	6	10	2	6	10		2	4						

* By permission, from A. Beiser, *Concepts of Modern Physics* (McGraw–Hill, 1981).

Table 27.3 (continued)

	K	L		M			N				O				P			Q
	1s	2s	2p	3s	3p	3d	4s	4p	4d	4f	5s	5p	5d	5f	6s	6p	6d	7s
53 I	2	2	6	2	6	10	2	6	10		2	5						
54 Xe	2	2	6	2	6	10	2	6	10		2	6						
55 Cs	2	2	6	2	6	10	2	6	10		2	6			1			
56 Ba	2	2	6	2	6	10	2	6	10		2	6			2			
57 La	2	2	6	2	6	10	2	6	10		2	6	1		2			
58 Ce	2	2	6	2	6	10	2	6	10	2	2	6			2			
59 Pr	2	2	6	2	6	10	2	6	10	3	2	6			2			
60 Nd	2	2	6	2	6	10	2	6	10	4	2	6			2			
61 Pm	2	2	6	2	6	10	2	6	10	5	2	6			2			
62 Sm	2	2	6	2	6	10	2	6	10	6	2	6			2			
63 Eu	2	2	6	2	6	10	2	6	10	7	2	6			2			
64 Gd	2	2	6	2	6	10	2	6	10	7	2	6	1		2			
65 Tb	2	2	6	2	6	10	2	6	10	9	2	6			2			
66 Dy	2	2	6	2	6	10	2	6	10	10	2	6			2			
67 Ho	2	2	6	2	6	10	2	6	10	11	2	6			2			
68 Er	2	2	6	2	6	10	2	6	10	12	2	6			2			
69 Tm	2	2	6	2	6	10	2	6	10	13	2	6			2			
70 Yb	2	2	6	2	6	10	2	6	10	14	2	6			2			
71 Lu	2	2	6	2	6	10	2	6	10	14	2	6	1		2			
72 Hf	2	2	6	2	6	10	2	6	10	14	2	6	2		2			
73 Ta	2	2	6	2	6	10	2	6	10	14	2	6	3		2			
74 W	2	2	6	2	6	10	2	6	10	14	2	6	4		2			
75 Re	2	2	6	2	6	10	2	6	10	14	2	6	5		2			
76 Os	2	2	6	2	6	10	2	6	10	14	2	6	6		2			
77 Ir	2	2	6	2	6	10	2	6	10	14	2	6	7		2			
78 Pt	2	2	6	2	6	10	2	6	10	14	2	6	9		1			
79 Au	2	2	6	2	6	10	2	6	10	14	2	6	10		1			
80 Hg	2	2	6	2	6	10	2	6	10	14	2	6	10		2			
81 Tl	2	2	6	2	6	10	2	6	10	14	2	6	10		2	1		
82 Pb	2	2	6	2	6	10	2	6	10	14	2	6	10		2	2		
83 Bi	2	2	6	2	6	10	2	6	10	14	2	6	10		2	3		
84 Po	2	2	6	2	6	10	2	6	10	14	2	6	10		2	4		
85 At	2	2	6	2	6	10	2	6	10	14	2	6	10		2	5		
86 Rn	2	2	6	2	6	10	2	6	10	14	2	6	10		2	6		
87 Fr	2	2	6	2	6	10	2	6	10	14	2	6	10		2	6		1
88 Ra	2	2	6	2	6	10	2	6	10	14	2	6	10		2	6		2
89 Ac	2	2	6	2	6	10	2	6	10	14	2	6	10		2	6	1	2
90 Th	2	2	6	2	6	10	2	6	10	14	2	6	10		2	6	2	2
91 Pa	2	2	6	2	6	10	2	6	10	14	2	6	10	2	2	6	1	2
92 U	2	2	6	2	6	10	2	6	10	14	2	6	10	3	2	6	1	2
93 Np	2	2	6	2	6	10	2	6	10	14	2	6	10	4	2	6	1	2
94 Pu	2	2	6	2	6	10	2	6	10	14	2	6	10	5	2	6	1	2
95 Am	2	2	6	2	6	10	2	6	10	14	2	6	10	6	2	6	1	2
96 Cm	2	2	6	2	6	10	2	6	10	14	2	6	10	7	2	6	1	2
97 Bk	2	2	6	2	6	10	2	6	10	14	2	6	10	8	2	6	1	2
98 Cf	2	2	6	2	6	10	2	6	10	14	2	6	10	10	2	6		2
99 E	2	2	6	2	6	10	2	6	10	14	2	6	10	11	2	6		2
100 Fm	2	2	6	2	6	10	2	6	10	14	2	6	10	12	2	6		2
101 Md	2	2	6	2	6	10	2	6	10	14	2	6	10	13	2	6		2
102 No	2	2	6	2	6	10	2	6	10	14	2	6	10	14	2	6		2
103 Lr	2	2	6	2	6	10	2	6	10	14	2	6	10	14	2	6	1	2

electron, thus forming the common ions H$^+$, Li$^+$, etc.

The Halogens   F ($Z = 9$), Cl ($Z = 17$), etc., have, as their outer subshell, a set of five p electrons; one additional electron would complete the subshell. They easily gain that electron from their surroundings, forming the ions F$^-$, Cl$^-$, etc.; their valence is also equal to 1.

A Noble Gas.   In He ($Z = 2$), both electrons occupy the K shell, which is therefore complete. Helium cannot gain or lose an electron by chemical means, and has valence zero; it is chemically inert. The same can be said of the *other noble gases* Ne ($Z = 10$), Ar ($Z = 18$), etc., because their outer subshell is a complete p shell (the most stable configuration, as it turns out).

Predictions.   Science is tested, not only by its powers of explanation, but even more importantly by its powers of prediction. Several elements have had their existence and properties correctly predicted, before their discovery, from empty boxes in the periodic table. One outstanding case is man-made **technetium** (Tc, $Z = 43$), whose naturally occurring atoms have vanished in the early millennia of the Universe owing to radioactive decay. A similar situation exists for the **transuranic elements** ($Z > 92$), products of the contemporary nuclear laboratory, and whose chemical properties are indeed found to match their positions in the table. [Dmitri Mendeleyev made the first such predictions in 1871, notably in the case of germanium ($Z = 32$), and thereby established the periodic table as a research tool.]

Verification of the Atomic Number.   An important way of determining Z in the laboratory, independently of chemical properties, is outlined in Note (i) below. That determination, due to Moseley, firmly establishes physics as the basis for chemistry.

Notes

i. How Does One Tell the Atomic Number? (Moseley's Law)

The first step in constructing the periodic table is to attach to each element an atomic number, Z. How is this

done in the laboratory? Until 1913, sequencing the elements was done according to their atomic weight, which is measured chemically, and amounts to an average over the isotopes. Luckily, the charge of a nucleus is *roughly* proportional to its mass; therefore a nearly correct sequence was obtained, without any help from the modern atomic number concept. The scheme was marred by some exceptional pairs, for example cobalt ($Z = 27$, at. wt. $= 58.94$) and nickel ($Z = 28$, at. wt. $= 58.69$). But here the chemical properties took precedence over the atomic weight, and cobalt was (correctly) listed first.

The Discrete X-Ray Spectrum

In 1913 and 1914, Henry Moseley succeeded for the first time in measuring Z for many of the elements, through observation of the x-ray spectrum which they emit when used as a target in an x-ray tube (Sec. B of Chap. 26). A special case will best illustrate the reasoning.

Consider a sulfur atom, whose shell structure is crudely symbolized in Fig. 27.25. Suppose one of its K electrons has been knocked out by one of the tube's high-energy electrons; a vacancy (small white circle) is left in the K shell. The resulting atom is unstable, because an electron from one of the higher shells can "fall" (arrow) into the vacancy and release a photon, thus conserving energy. Let us restrict our attention to an L electron falling into the K vacancy. By Einstein's photoelectric equation, the emitted photon can be observed as radiation of frequency

$$f = \frac{\mathscr{E}_{LK}}{h} \qquad (27.N.1)$$

where $\mathscr{E}_{LK}$ is the L-to-K energy drop. Other well-defined frequencies result from transitions involving the M shell,

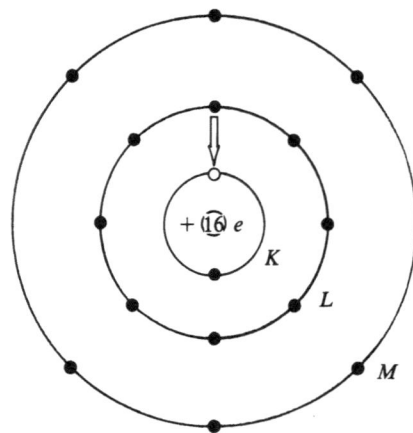

Figure 27.25. A sulfur atom, with one K electron missing.

or involving a vacancy in the L shell, etc.; to each frequency there corresponds a spectral line which belongs to the element's **discrete x ray spectrum**, found superimposed on its continuous or bremsstrahlung spectrum. [Turning back to Fig. 26.6 of Chap. 26, we see two such lines, corresponding, from left to right, to the transitions $L \rightarrow K$ ("K_α line") and $M \rightarrow K$ ("K_β line").]

Predicting f

By asking the right question, we can easily predict f if Z is known. What is the charge distribution seen by the falling electron? Figure 27.25 indicates that, underneath, it sees the nucleus (charge $= +Ze$), shielded by the probability cloud (charge $= -e$) of the one remaining K electron; we consider this combination to be a sphere of charge $(Z-1)e$, whose external **E** field is equivalent to that of a point charge at the center. Overhead, the falling electron detects nothing at all, if we assume that the hollow remaining L and M shells are spherical (a hollow spherical shell has zero internal **E** field). Therefore Fig. 27.25 reduces to Fig. 27.26, which shows a transition ($n = 2$ to $n = 1$) in a *hydrogenlike atom* whose "nucleus" has a charge $+(Z-1)e$ instead of $+e$. For hydrogen itself the energy drop would be

$$\mathscr{E}(2 \rightarrow 1) = \left(\frac{2\pi Ke^2}{h}\right)^2 \frac{m}{2}\left(\frac{1}{1^2} - \frac{1}{2^2}\right) \quad (27.N.2)$$

$$= (13.6 \text{ electron volts})\left(\frac{3}{4}\right) \quad (27.N.3)$$

as we recall from Bohr's results (27.B.14) and (27.B.15). In the present case, one of the factors e in the expression $2\pi Ke^2/h$ is still the charge of the falling electron; the other factor, which was the proton charge, must be replaced by $(Z-1)e$. We therefore find, instead of (27.N.3),

$$\mathscr{E}_{LK} = (Z-1)^2 (13.6 \text{ electron volts})\left(\frac{3}{4}\right) \quad (27.N.4)$$

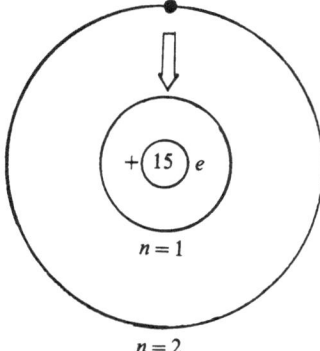

Figure 27.26. This is how the falling L electron sees the atom of Fig. 27.25.

The corresponding photon frequency is just as in hydrogen, but multiplied by $(Z-1)^2$. From Eqs. (27.A.1) and (27.A.2) we have

$$f = (Z-1)^2 \left(\tfrac{3}{4}\right)(3.29 \times 10^{15}) \text{ Hz} \quad (27.N.5)$$

It is often convenient to write (27.N.5) as a linear equation for $\sqrt{f}$ in terms of Z:

$$\boxed{\sqrt{f} = (4.97 \times 10^7)(Z-1) \text{ Hz}^{1/2}} \quad (27.N.6)$$

We conclude that, for any element, $Z-1$ is measured by *the square root of an x-ray frequency*.

Example 27.8. Molybdenum. From the spectral data of Fig. 26.6, Chap. 26, find the atomic number of molybdenum.

The figure gives $f = 4.2 \times 10^{18}$ Hz so that, from (27.N.6),

$$Z = 1 + \frac{\sqrt{f}}{4.97 \times 10^7 \text{ Hz}^{1/2}}$$

$$= 1 + \frac{(4.2 \times 10^{18})^{1/2}}{4.97 \times 10^7} = 42.2$$

pointing to the correct integer, $\underline{Z = 42}$. (We cannot expect an exact answer from the simplified model we have used.)

ii. How Small Is the Electron?*

At least three kinds of experimental evidence compel us to regard the electron, in practice, as a point.

1. Scattering of Electromagnetic Radiation

About the year 1900, J. J. Thomson (who had discovered the electron as "cathode rays") predicted, on a classical basis, how much electromagnetic energy a point electron ought to re-radiate (or "scatter") if exposed to an incoming electromagnetic wave; see Fig. 27.27.

When the experiment was performed with x rays on the electrons of hydrogen, good agreement was found with Thomson's formula, down to $\lambda \approx 2 \times 10^{-11}$ meter, about 20% of a hydrogen atom's diameter. With the advent of quantum mechanics, the deviations observed below that wavelength were shown to arise, not from a

---

* For a similar question concerning the nucleus, see the next chapter.

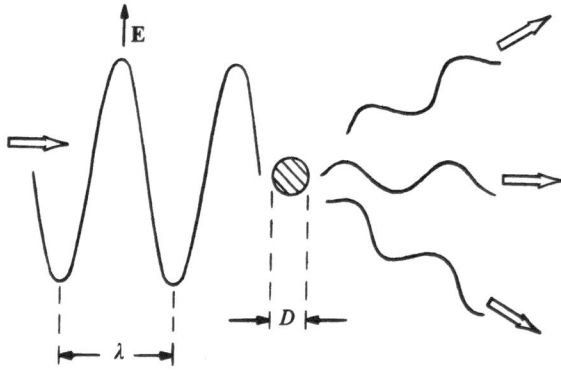

Figure 27.27. Thomson scattering. The electron (dark circle) is made to vibrate up and down by the **E** field of the incoming wave. It therefore radiates in various directions. Thomson assumed an electron of diameter D very small relative to the wavelength λ of the incoming radiation.

finite electron size, but from the fact that classical mechanics becomes a poor description. Today's scattering experiments with high energy gamma rays on electrons, when interpreted by quantum mechanics, remain consistent with a pointlike structure.

2. Atomic Spectra

The awesome success of Schrödinger's theory of atomic spectra, especially when improved by relativistic corrections, indicates that the theory's assumption of a point electron is correct.

3. Electron–Positron Collisions

In today's giant particle accelerators, electrons are made to collide with their positive counterparts, the positrons. In such a collision, the closest approach is typically 10^{-18} meter, comparable to $1/1000$ of the proton's known radius. Yet the observed recoil is still as predicted for a pair of point charges.

Condensed Checklist

Arbitrary transition:

$$\boxed{hf = \mathscr{E}_i - \mathscr{E}_f = \Delta\mathscr{E}} \qquad (27.A.7)$$

$$\frac{\lambda}{1 \text{ nm}} = \frac{1240}{\Delta\mathscr{E}/1 \text{ electron volt}} \qquad (27.A.10)$$

$$\frac{ch}{e} = 1240 \text{ nm volt} \qquad \text{(to four significant figures)}$$

$$(27.A.11)$$

Hydrogen spectrum:

$$f = \left[\frac{1}{(n_1)^2} - \frac{1}{(n_2)^2}\right] f_{\text{H}} \qquad (27.A.1)$$

$$f_{\text{H}} = 3.2881 \times 10^{15} \text{ Hz} \qquad (27.A.2)$$

Balmer lines and limit: Table 27.1 of Sec. A

Series names: (27.A.6)

Hydrogen levels:

$$\mathscr{E} = -\left(\frac{2\pi Ke^2}{h}\right)^2 \frac{m}{2} \frac{1}{n^2}$$

$$(27.B.14), (27.B.15)$$

$$\boxed{\mathscr{E} = -(13.6 \text{ electron volts})\frac{1}{n^2}}$$

Bohr orbits:

$$r = r_{\text{B}} n^2$$

$$(27.B.12), (27.B.16)$$

$$r_{\text{B}} = \text{Bohr radius} = \frac{h^2}{(2\pi)^2 Ke^2 m} = 5.29 \times 10^{-11} \text{ meter}$$

$$v = \frac{2\pi Ke^2}{h}\frac{1}{n} = \frac{c}{137}\frac{1}{n}$$

$$(27.B.13), (27.B.20)$$

Hydrogen quantum numbers:

Principal: $n = 1, 2, 3,...$ (27.C.5)

Orbital: $l = 0, 1, 2,..., n-1$ (27.C.6)

Magnetic: $m_l = 0, \pm 1,..., \pm l$ (27.C.7)

Spin: $m_s = \pm 1/2$ (27.C.12)

Names of states:

$s, p, d, f, g, h,...$ means $l = 0, 1, 2, 3, 4, 5,...$

$$(27.C.13)$$

Angular momenta:

Orbital:

$$L = \sqrt{l(l+1)}\frac{h}{2\pi} \qquad (27.C.6)$$

$$L_z = m_l\frac{h}{2\pi} \qquad (27.C.7)$$

Spin:

$$S = \sqrt{\left(\frac{1}{2}\right)\left(\frac{3}{2}\right)} \frac{h}{2\pi} \qquad (27.C.10)$$

$$S_z = m_s \frac{h}{2\pi} = \pm \frac{1}{2} \frac{h}{2\pi} \qquad (27.C.11)$$

Probability density:

$P(\mathbf{r})\, d\mathcal{V} = $ probability of being in $d\mathcal{V}$ at $\mathbf{r}$ \qquad (27.C.1)

Exclusion principle:

> The set n, l, m_l, m_s cannot be the same for two electrons in the same atom.

(27.D.1)

Electron shells:

Names: $K, L, M, N,...$ means $n = 1, 2, 3, 4,...$
\qquad Fig. 27.21 of Sec. D

Order of filling: \qquad Figs. 27.21, 27.22

Occupancy (table, for each element) \qquad Table 27.3

Notation: $1s^2 2s^2 2p^6...$ \qquad Example 27.7

Complete shell: $2n^2$ electrons \qquad Example 27.6

Periodic table: \qquad Fig. 27.24

True or False

(Try first without consulting the book's figures.)

1. The lines in a hydrogen series crowd together as frequency decreases.

2. The hydrogen spectrum does not have lines of arbitrarily high frequency.

3. On a uniform energy scale, the energy levels of hydrogen are equally spaced.

4. The hydrogen atom has a finite number of energy levels.

5. Transitions to the hydrogen ground state constitute the Balmer series.

6. With the usual zero-level convention, the ground-state energy of the hydrogen atom equals minus its ionization energy.

7. The ionization energy of hydrogen is of the order of hundreds of electron volts.

8. Successive Bohr orbits are equidistant in space.

9. In the Bohr model, the electron cannot come closer to the proton than a fixed minimum distance.

10. In a Bohr orbit, the electron's speed is proportional to the orbital radius.

11. In the Bohr model, the angular velocity of the electron is an integral multiple of a basic number.

12. The energy of a hydrogen level depends almost completely on n, hardly at all on l.

13. According to present-day theory, the orbital angular momentum of the electron in a hydrogen atom is an exact integer times $h/2\pi$.

14. The energy of an electron in neutral uranium depends on l as well as on n.

15. The exclusion principle says that, in a given atom, no two electrons can have the same value for n.

16. The P shell consists of $l = 1$ electrons only.

17. There can be as many as ten, but not more than ten d electrons in the M shell.

18. The O shell corresponds to $n = 5$.

19. An element whose outer shell is empty except for a complete s subshell is a noble gas.

Problems

Some data (all SI): $c \approx 2.9979 \times 10^8$, $K \approx 8.9875 \times 10^9$, $e \approx 1.6022 \times 10^{-19}$, $h \approx 6.6262 \times 10^{-34}$, $m_e \approx 9.1096 \times 10^{-31}$.

Section A: Energy Levels

*27.1. If, in formula (27.A.1), the left-hand side is replaced by the so-called **wave number** $1/\lambda$ ($\lambda = $ wavelength), then f_H must be replaced by another constant R_H (the **Rydberg constant** for hydrogen). Calculate R_H to five figures.

27.2. In Example 27.1, part (a), show how the quoted fractions 3/16, 21/100, 2/9 are obtained.

27.3. In formula (27.A.3), show that the frequency f increases as n_2 increases.

27.4. On the basis of (27.A.6), and of Fig. 22.14, Chap. 22, show that (a) only the Balmer series has any lines in the visible; (b) the Lyman series is entirely in the ultraviolet; (c) the other series are entirely in the infrared.

*27.5. What is, to five figures, the lowest frequency of the Paschen series?

27.6. What is, to five figures, the largest wavelength of the Brackett series?

27.7. Sketch the three spectral lines of Example 27.2 on (a) a uniform frequency scale, (b) a uniform wavelength scale.

*27.8. Consider the transition $n = 5$ to $n = 3$ in hydrogen. (a) To which series does the emitted line belong? (b) What are its frequency and its wavelength, in nanometers, to five figures?

27.9. How many spectral lines can result from transitions among five energy levels?

*27.10. When in a certain state, a helium atom can be ionized at the expense of 4.74 electron volts of energy. What minimum frequency of light is needed to achieve this?

27.11. What frequency of light is emitted by sodium atoms in a transition from a 3.192-electron volt level to a 2.105-electron volt level? (The ground state is taken as the zero level.)

*27.12. (a) Two helium levels are separated by 1.145 electron volts. What wavelength of radiation, in nanometers, is emitted or absorbed in a transition between them? (b) Find the wavelength of laser light from the transition $2 \rightarrow 1$ illustrated in Fig. 27.7.

27.13. What is the wavelength emitted by an atom in a transition from -4.92 electron volts to -7.56 electron volts?

27.14 (Photoemission from hydrogen). When hydrogen is irradiated with ultraviolet light, of wavelength 75 nm, some photons are completely absorbed, and an atomic electron is emitted. Considering the ionization energy of hydrogen, how much kinetic energy, in electron volts, is left to the escaping electron? (Neglect the fact that hydrogen occurs in H_2 molecules rather than as isolated atoms.)

*27.15. Among the many emission lines in the spectrum of sodium, three have wavelengths of 342.7 nm, 589.6 nm, 813.3 nm. (a) Find, in electron volts, the energy differences corresponding to these lines. (b) Show that transitions among only three levels can be held responsible; defining the energy of the lowest level as zero, determine, in electron volts, the other two energies. (c) Sketch the levels, approximately to scale.

27.16. A certain atom, initially in a -8.47-electron volt state, absorbs a photon corresponding to a 685-nm wavelength. What is the atom's final energy level?

*27.17. Neutral mercury has three levels (say a, b, c) such that transition $a \rightarrow b$ requires absorption of 253.6 nm ultraviolet light. Level c is 8.849 electron volts higher than level a. What wavelength of light is emitted in the transition $c \rightarrow b$?

27.18. Consider three levels, here labeled a, b, c, of the neutral helium atom. Absorption of 2058-nm radiation causes the transition $a \rightarrow b$; subsequent absorption of 667-nm radiation causes $b \rightarrow c$. (a) What is, in electron volts, the energy difference between a and c? (b) Sketch the three levels to scale.

27.19. Two closely spaced yellow emission lines, called D_1 and D_2, of wavelengths 589.59 nm and 589.00 nm, are found in the spectrum of sodium. They are due to transitions from levels with energies $\mathscr{E}_1$ and $\mathscr{E}_2$ to the ground state (energy $= 0$). Find, in electron volts, the difference $\mathscr{E}_2 - \mathscr{E}_1$.

*27.20. In a certain experiment, an ultraviolet Balmer line is measured to have a wavelength of (400 ± 4) nm. What are the two n values for the corresponding transition?

27.21. On the basis of (27.A.6), show that (a) the Lyman series does not overlap with any other; (b) ditto for the Balmer series; (c) the Paschen and Brackett series overlap.

*27.22. Find as many formulas as you can think of, different from (27.A.14), which still give negative levels *and* the observed hydrogen spectrum.

27.23. The energy liberated as heat in a chemical reaction is usually much below the ionization energy of hydrogen (13.6 electron volts per atom). On this basis, estimate, in joules, an upper limit for the energy expected from a ton (1000 kg) of explosives. (For the sake of definiteness, assume an average atomic weight of 10; use 1 atomic mass unit $\approx 10^{-27}$ kg.)

Section B: Bohr's Model of the Hydrogen Atom

Some numerical data are listed at the head of this problem section.

27.24. Show that the joule sec (the unit of action) is also the unit of angular momentum.

27.25. By inserting the appropriate numerical values, show how Eq. (27.B.15) yields 13.6 electron volts.

27.26. Verify numerically the result (27.B.16) for the Bohr radius.

27.27. Verify Eq. (27.B.20) to three significant figures.

*27.28. How many times is the Compton wavelength [Eq. (26.C.13) of Chap. 26] included in the circumference of the lowest Bohr orbit ($n = 1$) in hydrogen?

Express your answer (a) as a formula involving basic constants, (b) as a number.

27.29. Do, as simply as you can, the algebra needed to solve Eqs. (27.B.9)–(27.B.11) for r, v, and $\mathscr{E}$ [Eqs. (27.B.12)–(27.B.14)].

27.30. (a) By comparing Eqs. (27.A.14) and (27.B.14), express the constant f_H in terms of Ke^2, h, and m. (b) Verify that f_H [Eq. (27.A.2)] is rather well predicted. (Note: Taking the finite mass of the proton into account has been shown to improve the agreement even further.)

*27.31. Find, by slightly revising the Bohr-model equations, the energy levels of an arbitrary atom $(Z > 1)$ that has been multiply ionized to the extent that only one electron is left.

27.32 (Correspondence principle in the Bohr model). Assume that an electron, when in a wide circular orbit around a proton, usually falls inwards by decreasing its quantum number n one unit at a time: $\Delta n = -1$. (a) By what fraction of its value does r decrease if n changes from 1000 to 999? (b) What emitted frequency of circular motion results from Eqs. (27.B.12) and (27.B.13) when $n = 1000$? Compare with your answer to (b). (Classically, we expect the emitted wave to have the same frequency as the orbiting charge.)

27.33 (Irrelevance of the Bohr model to astronomical orbits). Consider the orbital motion of the Earth. (a) Rewrite Eqs. (27.B.9)–(27.B.11) in terms of the terrestrial and solar masses $(m_E = 6 \times 10^{24}$ kg, $m_S = 2 \times 10^{30}$ kg) and the gravitational constant, $G = 6.7 \times 10^{-11}$ newton meter^2 kg^{-2}. (b) What are the new Eqs. (27.B.12)–(27.B.14)? (c) Noting that the Earth–Sun distance is $r = 1.5 \times 10^{11}$ meters, estimate the quantum number n of the Earth's orbit. (d) By what fraction of its value would r decrease if n decreased by one unit? (e) What is the corresponding decrease of r in meters? Do we expect the Earth's orbit to be observably quantized?

Sections C: Electron States and Their Four Quantum Numbers; D: The Exclusion Principle and the Periodic Table

*27.34. In hydrogen, how many electron states are there (a) with $n = 4$ and $m_l = -2$; (b) with $n = 6$ and $m_s = +1/2$?

*27.35. What are, as in Example 27.5, the angles between $\mathbf{L}$ and the $+z$ axis for (a) $l = 2$, $m_l = 0$; (b) $l = 2$, $m_l = 2$; (c) $l = 1$, $m_l = 0$; (d) $l = 1$, $m_l = 1$?

27.36. Display, in the notation of Example 27.7, the electron configuration of technetium $(Z = 43)$, barium $(Z = 56)$, gold $(Z = 79)$, and polonium $(Z = 84)$.

27.37. Display, in the notation of Example 27.7, the electron configuration of neon $(Z = 10)$, calcium $(Z = 20)$, lanthanum $(Z = 57)$, and actinium $(Z = 89)$.

27.38. Based on Table 27.3, in what consists the similarity among the elements of the carbon $(Z = 6)$ family in Fig. 27.24? What are some differences?

27.39. Based on Table 27.3, in what consists the similarity among the elements of the oxygen $(Z = 8)$ family in Fig. 27.24? What are some differences?

*27.40. Calculate, in degrees, the angle between the spin vector $\mathbf{S}$ and the $+z$ axis when $m_s = +1/2$.

27.41. Consider an electron state with orbital angular momentum $l = 5$. (a) What is, in terms of $h/2\pi$, the magnitude of the $\mathbf{L}$ vector? (b) What is, in degrees, the smallest angle which $\mathbf{L}$ can make with the z axis? (c) How many orientations can $\mathbf{L}$ have with respect to the z axis?

*27.42. In a Bohr orbit with quantum number n, the angular momentum is $nh/2\pi$. (a) What is, in reality, the largest possible angular momentum L in a state with principal quantum number n? (Notice that $L < nh/2\pi$.) (b) How many different angles can this maximal $\mathbf{L}$ vector make with the z axis?

Answers to True or False

1. False (as frequency increases).
2. True.
3. False.
4. False.
5. False (should be Lyman).
6. True.
7. False (of the order of 10 electron volts).
8. False.
9. True (the Bohr radius).
10. False (the classical relation is used).
11. False (true for angular momentum).
12. True.
13. False [it involves $\sqrt{l(l+1)}$].
14. True (for all many-electron atoms).
15. False (all quantum numbers are involved).
16. False (true for a p subshell).

17. True.

18. True.

19. False (true for helium only).

Answers or Hints to Problems

27.1. 1.0968×10^7 meter^{-1}

27.5. 1.5984×10^{14} Hz.

27.8. (b) 1282.1 nm.

27.10. You may use (27.A.11).

27.12. (b) 633 nm.

27.15. (b) The highest is 3.618 electron volts.

27.17. 313.2 nm.

27.20. Hint: Successively try $n_2 = 6, 7, 8$; see Fig. 27.1.

27.22. Hint: Try exchanging n_1, n_2, and/or adding a constant term.

27.28. (b) 137.

27.31. Replace e^2 by Ze^2.

27.34. (a) 4; (b) 36.

27.35. (a) 90°; (d) 45°.

27.40. 55°.

27.42. (a) $\sqrt{(n-1)n}\, \dfrac{h}{2\pi}$; (b) $2n - 1$.

CHAPTER 28

THE NUCLEUS

The atom's most essential component, the **nucleus**, has almost all of the atom's mass. Through its positive charge Ze, it determines the number and behavior of the surrounding electrons.

In the preceding chapter, we have treated the nucleus as a point particle. Actually, its size, although small, is not zero. The appropriate unit to measure a nucleus is 10^{-15} meter, also called a femtometer, or a **fermi**. No nucleus has a radius larger than about 6.5 fermis; hydrogen has the smallest nucleus—the proton—with a radius of about 1 fermi, or 1/50 000 of the atomic radius itself. To quote a popular comparison: if any atom were blown up to the size of a football stadium, its nucleus would amount to no more than a marble at the center. In terms of occupied volume, all terrestrial matter, even the hardest piece of steel, or the floor on which we stand, is an almost perfect vacuum.

Aside from hydrogen, all nuclei are assemblages of protons and neutrons. These two kinds of building blocks have almost identical properties, such as mass and size, but they are distinguished by their charge: the proton's charge is positive and equal in magnitude to the electron's, while the neutron has zero charge. In addition, an isolated neutron—one that is not part of a nucleus—survives only a few minutes, after which it disintegrates into other particles. An isolated proton, on the other hand, appears to last forever.*

The existence of the nucleus, and its smallness relative to the atom as a whole, were unsuspected until Rutherford's experimental and theoretical analyses of 1908–1911. Shortly afterwards, in 1913, Mosely determined several nuclear charges by measuring the energy levels of atomic K and L electrons. [For details, see Note (i) at the end of the preceding chapter.] As to the internal structure of the nucleus, it has remained a mystery until as recently as 1932, when Chadwick discovered the neutron.

Since then, thanks to a variety of ingenious techniques, nuclear structure has gradually become almost as familiar as the surrounding atomic structure. The chief tools that have opened the new territory are of four kinds:

1. **Mass spectrometers** (see Example 18.6 of Chap. 18) yield the nuclear masses to extraordinary accuracy.
2. **Particle accelerators**, such as cyclotrons (Sec. B vii of Chap. 18), provide fast projectiles, able to penetrate the atomic electron shells and the nuclear Coulomb field.
3. **Nuclear reactors** emit large fluxes of neutrons. Being uncharged, they easily reach the nuclei of a target sample, and interact with them in many ways.
4. **Particle detectors**, such as scintillation counters and Geiger counters, are able to record the passage of a single particle. Without such detectors, the precise results of a nuclear experiment would in general not be observable.

However important all these devices, we cannot study them in this chapter, and must concentrate instead on what they have taught us. We shall first consider the **stable nuclei**, which surround us, and of which we are made; we look briefly at their com-

---

* Some recent theories predict that even an isolated proton should occasionally be seen to disintegrate, perhaps into a positron and neutral particles, but the event would be so rare as to be at the edge of observability.

position (in terms of protons and neutrons), as well as at their masses and sizes. We then discuss unstable (or **radioactive**) nuclei, which disintegrate spontaneously after existing for various lengths of time. Similar disintegrations may be induced by nuclear collisions.

The transformation of one or more nuclei into different ones is called a **nuclear reaction**. The equation describing such a process is reminiscent of a chemical equation, and we shall learn how to balance it; special attention will be paid to energy conservation and to Einstein's equation $\mathscr{E} = Mc^2$.

This chapter's concluding topics will be **fusion** and **fission**, two types of nuclear reaction that are of great technological and astrophysical interest.

A. The Stable Nuclei

The Earth has existed for billions of years, and as a result most of its radioactive minerals have decayed into stable ones; thus, the commonly encountered nuclei are stable. We start with a brief inventory of their properties.

i. Some Terminology

A typical nucleus is made of Z protons and N neutrons; Z is the **atomic number** (which defines the chemical element), and N is the **neutron number**. Both the proton and the neutron are referred to as **nucleons**; the total number of nucleons in a nucleus is called its **mass number**, and is denoted by A:

$$\boxed{A = Z + N} \qquad (28.A.1)$$

As we have seen, the exact nuclear charge is $+Ze$, and it determines the chemistry of the element; $e \approx 1.6 \times 10^{-19}$ coulomb equals the absolute charge of the electron.

A nucleus with given values of A and Z is conventionally represented by its chemical symbol (say X), preceded by A and Z as upper and lower indices:

$$_{Z}^{A}\text{X} \qquad (28.A.2)$$

Examples are

$$_{1}^{1}\text{H}, \ _{2}^{4}\text{He}, \ _{6}^{12}\text{C}, \ _{6}^{13}\text{C}, \ _{28}^{58}\text{Ni}, \ _{28}^{60}\text{Ni}, \ _{28}^{62}\text{Ni} \qquad (28.A.3)$$

etc. In the case of "heavy" hydrogen, alternative symbols are often used: $_{1}^{2}\text{D}$ ("deuterium") for $_{1}^{2}\text{H}$, and $_{1}^{3}\text{T}$ ("tritium") for $_{1}^{3}\text{H}$.

We note that the Z label is redundant, as it follows from the chemical symbol. Accordingly, one often just writes ^{1}H, ^{13}C, ^{62}Ni, etc.

The **neutron** can be symbolized by $_{0}^{1}\text{n}$.

The word **nuclide** has been coined to designate a species of nucleus with given A and Z. Thus, (28.A.3) lists seven nuclides. Nevertheless, one commonly says "nucleus" to mean "nuclide."

Two nuclides with the same Z but different A are **isotopes** of one another. In (28.A.3), for example, we see two carbon isotopes and three nickel isotopes. Naturally occurring elements are, in general, mixtures of isotopes; having the same Z, they cannot be separated by chemical means. Figure 28.1 is a stylized illustration of the two existing stable helium isotopes, $_{2}^{3}\text{He}$ and $_{2}^{4}\text{He}$.

An overview of all stable nuclides is given in Fig. 28.2. Light nuclei tend to have about as many neutrons as protons, $N \approx Z$, but in the heavier ones, N is noticeably larger than Z.

ii. Nuclear Masses

The mass of a nucleus is given *approximately* by

$$\boxed{\text{Mass} \approx (A)(1 \text{ atomic mass unit})} \qquad (28.A.4)$$

where the atomic mass unit (standard abbreviation: u) is

$$\text{u} \approx 1.660531 \times 10^{-27} \text{ kg} \qquad (28.A.5)$$

(1 u is defined as exactly 1/12 of the mass of a complete atom of carbon 12, $_{6}^{12}\text{C}$, including its six electrons).

There are two reasons why (28.A.4) is only approximate. First, the proton and neutron have

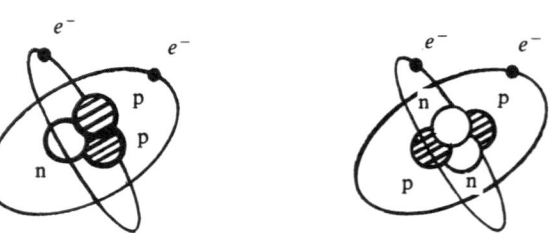

Figure 28.1. (Not to scale.) Two isotopes of helium.

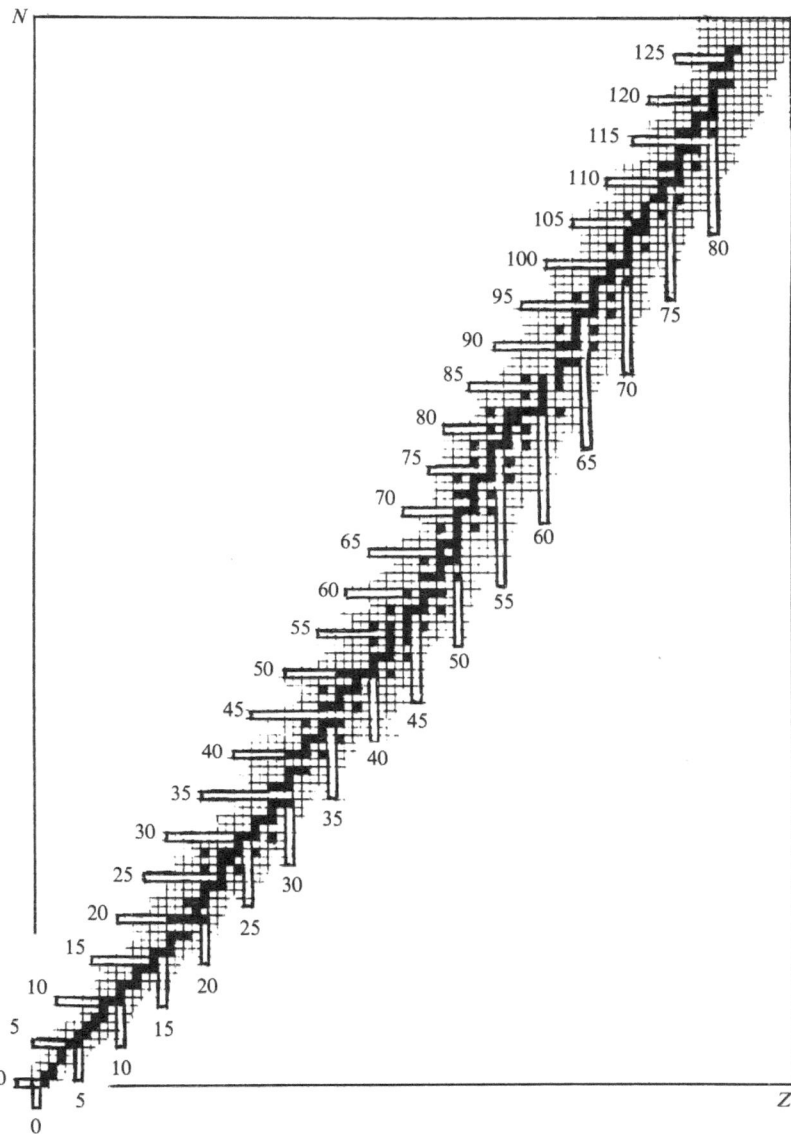

Figure 28.2. The number of protons (*Z*) and neutrons (*N*) in the stable nuclei. Each black square stands for a nuclide.

slightly different masses, neither of which is exactly 1 u. Second, as we shall see in detail further on, nuclear masses are diminished by the huge negative energy involved in binding nucleons together; that effect operates via the relativistic equivalence between mass and energy.

Accurate Masses

By means of mass spectrometry, the nuclear masses have been measured to an impressive accuracy, illustrated in Table 28.1.

iii. Nuclear Sizes

Indirect methods are necessary to measure something as small as a nucleus. Between 1908 and 1913, Ernest Rutherford pioneered the technique of bombarding a material sample with high-energy particles, and of recording, one by one, the directions in which the projectiles are scattered. These statistics are then analyzed mathematically, and give considerable information on the size of the nucleus responsible for the collisions. In Rutherford's case, the projectiles were alpha particles

Table 28.1.   Atomic Masses of the Stable Nuclides up to Carbon[a]

Atomic number, Z	Element	Mass number, A	Atomic mass[b]	Natural abundance in percent
1	H	1	1.007 825	100 $^-$
1	H, D	2	2.014 0	0.015
2	He	3	3.016 03	0.000 13
2	He	4	4.002 60	100 $^-$
3	Li	6	6.015 12	7.42
3	Li	7	7.016 00	92.58
4	Be	9	9.012 18	100
5	B	10	10.012 9	19.78
5	B	11	11.009 31	80.22
6	C	12	12	98.89
6	C	13	13.003 35	1.11

[a] Mass of proton $\approx 1.007\ 276\ 4$ u; mass of neutron $\approx 1.008\ 665$ u; mass of electron $\approx 0.000\ 548$ u; 1 atomic mass unit $= u \approx$ 1.660 531 $\times 10^{-27}$ kg $\approx 931.501\ 6$ MeV/c^2; mass of neutral $^{12}_6$C atom $= 12$ u exactly.
[b] Mass of the neutral atom, in atomic mass units.

(helium nuclei), but since then many other incident beams have been tried with great success, notably protons, neutrons, and electrons. Scattering experiments of this type are the staple of today's "big physics" experiments on many kinds of elementary particles. The Notes at the end of this chapter consider "Rutherford scattering" in one of its aspects.

Electrons of high kinetic energy (typically 200 MeV to 500 MeV), when scattered by a nucleus, not only tell us its size, but allow us to map the complete charge distribution within that nucleus. The reason is that such electrons penetrate the target and respond everywhere to its electrical forces. The inferred charge density, as a function of the distance r from the center of the nucleus, is illustrated in Fig. 28.3 in the case of gold, $^{197}_{79}$Au; the flat shape is typical of all nuclei except the proton (also shown for comparison). Here, as in all nuclear physics, a convenient length scale is the **fermi** (named after the pioneering nuclear physicist Enrico Fermi, 1901–1954):

$$\boxed{1 \text{ fermi} = 1 \text{ fm} = 10^{-15} \text{ meter}} \quad (28.A.6)$$

The overall findings are as follows:

1. Nuclei have a *fairly* well defined radius (equal to 6.2 fm for gold and 0.80 fm for the proton, as we see in Fig. 28.3).
2. Except in the proton, the charge density is almost uniform within the nucleus. Scattering

experiments using a variety of incident particles have confirmed that the nucleons themselves are rather uniformly distributed over the nuclear volume.

3. As measured by electron scattering, the nuclear radius r_{nucl} increases with mass number according to

$$r_{\text{nucl}} \approx (1.1 \text{ fm}) \sqrt[3]{A} \quad (28.A.7)$$

for $A = 3$ or more; the deuteron (i.e., the nucleus of deuterium, $^2_1$D) is an unusually loose combination and has $r_D \approx 6$ fm; it is made of one proton and one neutron.

Example 28.1. The Nuclear Mass Density. How does nuclear density depend on mass number for $A \geqslant 3$?

We have, with the help of Eqs. (28.A.4) and (28.A.7),

$$\text{Density} = \frac{\text{mass}}{\text{volume}} \approx \frac{(A)(1 \text{ u})}{(4\pi/3)\, r_{\text{nucl}}^3}$$

$$= \frac{(A)(1 \text{ u})}{(4\pi/3)\,(1.1 \text{ fm})^3\, A}$$

or

$$\text{Density} = 0.18 \frac{\text{u}}{\text{fm}^3} \quad (28.A.8)$$

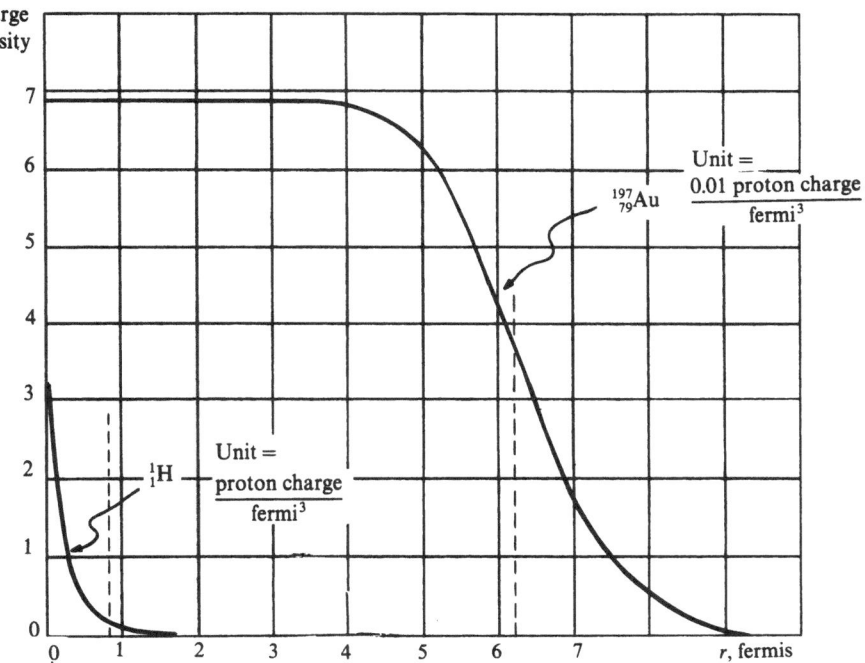

Figure 28.3. Charge density in the $^{197}_{79}$Au nucleus and in the proton; the graph for gold is scaled down by 1:100 relative to that for the proton. The approximate nuclear radii (6.2 fermis and 0.80 fermi) are indicated by the vertical dashed lines.

independent of A. [The reader should calculate in ordinary units the enormous value implied by (28.A.8).]

Nuclear Matter

Our result indicates that

> All nuclear matter
> has about the same density. $\qquad$ (28.A.9)

To put it differently, *nuclear matter is nearly incompressible,* much like a liquid under ordinary conditions; in fact, many features of nuclei have been explained by a "liquid-drop model," complete with surface tension.

(It is probable, however, that under the astonishing pressures that exist in some stellar interiors, nuclear matter is squeezed to ever higher densities, eventually resulting in black hole formation.)

iv. A Nucleon is Made of Three Quarks

Knowing about protons and neutrons has enabled physicists to predict successfully a huge number of facts. Hence those objects are often regarded in practice as basic, indivisible entities;

we, in this chapter, are taking that point of view. Yet, since 1963, it has gradually become evident that each nucleon is itself made of three pieces, which have been called **quarks** by Murray Gell-Mann, one of the theory's original proponents. Here we cannot even begin to quote the sometimes very subtle arguments leading to quarks; still, three important comments must be made:

1. It is believed that no single quark can ever be extracted from its nucleon for separate examination. Quarks are said to be **confined**.
2. Some quarks, called "up" quarks, or *u* quarks, have a charge of $+\frac{2}{3}e$, while others, called "down" quarks, or *d* quarks, have a charge of $-\frac{1}{3}e$. Nucleons comprise only *u* and *d* quarks. (Because of their bizarre fractional charges, quarks have long been considered by some to be fictitious—a convenient mathematical model without true physical counterpart.)
3. In a proton, the quarks' combined charge must amount to $+e$, while in a neutron it must be zero; no fractional charge is seen in isolation. Accordingly, the proton is a *uud* combination, with charge equal to $(2/3 + 2/3 - 1/3)e = e$ while the neutron is a *udd* combination, with charge $(2/3 - 1/3 - 1/3)e = 0$.

At present, quark research is an extensive international effort, both experimental and theoretical.

B. Radioactivity

The importance of radioactivity—natural or induced—extends over several disciplines. To mention only a few of its roles: it provides many clues to nuclear structure; it is used as a detectable label in order to trace the whereabouts of an element in complex chains of biochemical processes; it contributes to cancer therapy; it serves as an historical and geological dating mechanism; and radioactive minerals heat the Earth internally.* Radioactivity is, of course, a hazardous phenomenon, and its detrimental effects on living tissue have been well publicized.

Radioactivity was first discovered in uranium salts by Henri Becquerel (1896). The salts would fog a photographic plate even though the latter was protected by a heavy black paper wrapping. The eventual conclusion was that the uranium nuclei and their immediate decay products were emitting high-energy particles that penetrated the wrapping.

The uranium in the salts was implicated by the fact that different compounds had a fogging effect whose intensity depended on the amount of uranium, not on that of any other elements present. This chemical method of identifying radioactive elements was pioneered by Marie and Pierre Curie, and led to their discovery of radium, one of the most intensely radioactive natural substances.

Radioactivity exemplifies one aspect of quantum mechanics, namely, the role of pure chance in physics. We now know that a radioactive nucleus disintegrates spontaneously. There is no external agency or signal to trigger its decay; nor is the timing of the event based on the nucleus' past history. The disintegration just *happens*, and modern physical theory must be content to assign decay probabilities per unit time.

A radioactive decay may be described as *the emission, by a nucleus, of one or more particles*. Three cases occur predominantly: alpha (α) decay, or the emission of a helium nucleus He^{++} (an alpha particle); beta (β) decay, or the emission of an electron e^- (a beta particle); and gamma (γ) decay, or the emission of a photon (a gamma par-

ticle). One also speaks of "alpha rays," "beta rays," and "gamma rays;" the original nucleus is called the **parent**, while the nucleus that remains after the decay is the **daughter**.

Figure 28.4 is a classic illustration, used to distinguish the three modes of decay. A radioactive mixture lies in a lead-enclosed cavity; the radiations escape through a pinhole. A magnetic field is set up (into the plane of the figure) and curves the positive alpha rays to the left, the negative beta rays to the right, and the neutral gamma rays not at all. A layer of absorbing material is shown, indicating the particles' relative penetrating powers in matter. Their kinetic energies are typically of the order of 1 MeV.

In all cases, a decay must obey the **conservation of charge**:

$$\boxed{\begin{array}{l} \text{Total charge before decay} \\ \quad = \text{total charge after decay} \end{array}} \qquad (28.B.1)$$

and the **conservation of nucleons**:

$$\boxed{\begin{array}{l} \text{Number of nucleons before decay} \\ \quad = \text{number of nucleons after decay} \end{array}}$$

$$(28.B.2)$$

These two constraints limit the kinds of disintegration that can exist, and are a great simplifying factor in nuclear physics. They are true in every kind of nuclear reaction.

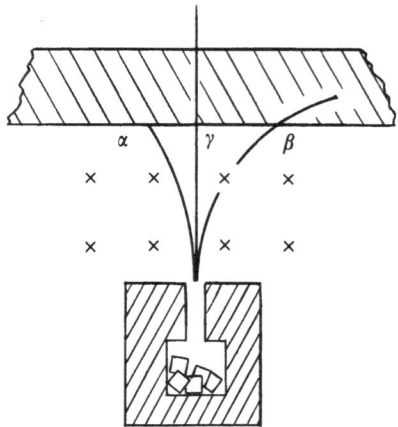

Figure 28.4. Alpha, beta, and gamma rays.

---

* Their effect on the climate is nil.

i. Alpha Decay

In this type of process, a nucleus, say $_Z^A X$, emits an alpha particle $_2^4 He$; the daughter nucleus has atomic number $Z-2$ (conservation of charge) and mass number $A-4$ (conservation of nucleons). Schematically, we have

$$_Z^A X \rightarrow _{Z-2}^{A-4} X' + _2^4 He \qquad (28.B.3)$$

For example, in the case of polonium 212 decaying to lead 208, we have

$$_{84}^{212} Po \rightarrow _{82}^{208} Pb + _2^4 He \qquad (28.B.4)$$

The alpha particle's kinetic energy is 10.54 MeV. (This is about as high as alpha energies go.) Note how superscripts and subscripts balance on both sides of the equation. *Alpha decay is seen only in heavy nuclides* (usually $A \geqslant 200$).

Penetration in Matter

Alpha particles penetrate only the thinnest manufacturable films of solid material, and are stopped by a few centimeters of air; they are said to have **low penetrating power**.

Alpha Particles as Helium Nuclei

The evidence that alphas are helium nuclei comes from many observations, notably (1) their charge-to-mass ratio in experiments where **E** and **B** fields are used, alternately or together, for their deflection; (2) their individual charge, determined from the total charge accumulated, over a length of time, on an insulated electroscope exposed to alpha decay, and divided by the total number of alpha particles (as inferred from a count of disintegrations per unit time); (3) most convincingly of all, the appearance of helium spectral lines in an originally evacuated discharge tube in which alpha particles have collected for a long time.

ii. Beta Decay

Here a nucleus, say $_Z^A X$, emits an electron, e^-; the nucleus does not change its mass number, but its atomic number becomes $Z+1$ by conservation of charge. Schematically, we have

$$_Z^A X \rightarrow _{Z+1}^A X' + e^- + \bar{v} \qquad (28.B.5)$$

The additional symbol $\bar{v}$ ("nu bar"), which appears here unexpectedly, expresses the fact that the beta particle e^- is never emitted alone; a second particle, the neutrino $\bar{v}$ (or more precisely, the **antineutrino**) is emitted simultaneously with it.

A Digression on the Neutrino

We have here a particle with zero charge and very small mass (possibly zero mass, like a photon). The neutrino cannot be detected directly except by large and elaborate equipment, available at few places in the world; its interaction with matter is almost nil, and, if a neutrino is emitted towards the ground, its chances are 10^{12} to 1 of traversing the whole Earth and emerging on the other side. All of us are, in fact, permeated by a huge but harmless neutrino flux originating from nuclear reactions inside the Sun.

In spite of its ghostlike properties, the neutrino's existence is well established. It carries measurable energy, momentum, and angular momentum away from the decaying nucleus, and indeed was originally postulated by Wolfgang Pauli in 1930 so as to account for those observably missing quantities.

Illustrations of Beta Decay

We now return to our model equation (28.B.5) for beta disintegration. One example is the decay of bismuth-210 to polonium-210:

$$_{83}^{210} Bi \rightarrow _{84}^{210} Po + e^- + \bar{v} \qquad (28.B.6)$$

The electron is emitted with, at most, a kinetic energy of 1.17 MeV—a representative figure for beta energies.

Another example is the decay of the rare isotope of hydrogen, tritium ($_1^3 H$ or $_1^3 T$):

$$_1^3 T \rightarrow _2^3 He + e^- + \bar{v} \qquad (28.B.7)$$

The end products include a rare (but stable) isotope of helium and an electron of only 0.018 MeV maximum energy.

The simplest example of all involves an isolated neutron (not found naturally but obtainable in nuclear laboratories); it decays into a proton according to

$$\, _0^1 n \rightarrow \, _1^1 H + e^- + \bar{\nu} \qquad (28.B.8)$$

with a maximum energy of 0.78 MeV for the emitted electron.

(In these illustrations, why is the electron energy quoted only as a maximum? The reason is that, although the total available energy is fixed, it can be shared in many ways among the end products of the decay. The electron achieves its maximum energy when the neutrino's share is essentially zero.)

The Creation of Particles

Does reaction (28.B.8) imply that the neutron is an unstable bound state, made of a proton ^1H, an electron e^-, and an antineutrino $\bar{\nu}$? The answer is no. Bound-state models of beta decay have been proved inconsistent or extremely implausible. A more radical description is necessary: in the process of decay, the neutron is **annihilated**, while the three decay products are **created**. (In a similar way, photons are created or annihilated when emitted or absorbed by an atom.)

Penetration in Matter

In spite of the fact that their energy is generally less than that of alpha particles, beta particles usually penetrate matter more deeply. Typical beta electrons (of the order of 1 MeV in energy) may travel through tens of meters of air before coming to a stop, or may penetrate an aluminum plate several millimeters thick.

Beta Particles as Electrons

From their deflection by **E** and **B** fields, the beta particles are found to have precisely the charge-to-mass ratio of electrons. A beam of beta particles of known energy may be simulated with cathode rays, which are known to be electrons; the beta particles and cathode rays are found to produce identical effects in matter.

Positive Beta Decay

In some cases, a positron, e^+, is emitted rather than an electron. This type of decay forms the subject of Problem 28.33.

iii. A Comment on the Strong and Weak Interactions

What binds the nucleons together is called the **strong interaction**. Is it really much stronger than the electromagnetic interaction?

The answer is that, in order to compress together, in a tiny nucleus, a large number of protons in spite of their electrostatic repulsion, the nuclear force must overwhelm the electric force. In confirmation, the approximate equality between N and Z in Fig. 28.2 shows that the proton's charge—and therefore electromagnetism—plays a minor role in nuclear structure. Indeed, if Coulomb repulsion made a nucleus unstable, stability would imply few protons (compared to neutrons).

Yet another kind of interaction, the **weak interaction**, is responsible for beta decay. One indication of its extreme weakness is the fact that the neutrino is almost undetectable. Both the strong and the weak interaction are of very short range, acting within distances comparable to or less than a nucleon radius. This feature contrasts with electromagnetism and gravity, whose $1/r^2$ forces operate across macroscopic distances.

iv. Gamma Decay

A nucleus can emit photons. Just as an atom has discrete energy levels, corresponding to the electrons' states of motion, so does the nucleus have discrete levels, related to the motion of its nucleons. A transition from a higher to a lower nuclear energy level typically results in the emission of a gamma ray photon. During this event, there is no change in either the mass number or the atomic number of the decaying nucleus. Symbolically, for an arbitrary nucleus $_Z^A X$, we have

$$\boxed{\, _Z^A X^* \rightarrow \, _Z^A X + \gamma} \qquad (28.B.9)$$

where the asterisk denotes the excited state, and γ stands for the photon. Gamma emission often follows on the heels of another decay, as in the following illustration, where magnesium-27 beta-decays to an excited state of aluminum-27:

$$\begin{aligned}{}^{27}_{12}\text{Mg} &\rightarrow {}^{27}_{13}\text{Al}^* + e^- + \bar{\nu} \\ {}^{27}_{13}\text{Al}^* &\rightarrow {}^{27}_{13}\text{Al} + \gamma\end{aligned} \qquad (28.\text{B}.10)$$

In this case the photon has an energy of 1.02 MeV. Gamma energies can reach several MeV.

Low-energy gamma rays are experimentally indistinguishable from hard x rays, and there is no question that all gamma rays consist of photons. Therefore *gamma rays are electromagnetic waves*. Their wavelength is difficult to measure by Bragg diffraction, because it is usually much smaller than any crystal lattice spacing. Hence one must resort to less direct methods, such as the Compton scattering of gamma-ray photons; what one measures here is the recoil energy of the electrons.

Penetration of Matter by Gamma Rays

Like hard x rays, gamma rays are highly penetrating; they are dangerous to life as well as to instruments. A material shield of sufficient thickness and density will attenuate the rays to any desired level. As a numerical illustration, a beam of 5.0-MeV gamma rays is attenuated to about half its original intensity after traversing a 1.5-cm thickness of lead.

v. Half-Life and Mean Life

The timing of a radioactive decay is governed by probability only. With this assertion as a starting point, we can answer the following question: given an initial sample of N_0 unstable nuclei (all identical), how many, N, remain intact after a given period of time, t? We first state the answer, and derive it afterwards. *The number of intact nuclei decreases exponentially with time*,

$$\boxed{N = N_0 e^{-\kappa t}} \qquad (28.\text{B}.11)$$

where κ is a fixed number known as the **decay constant** of the nuclide under consideration.

To see why this is so, consider an unstable nucleus that has survived up to time t. Immediately thereafter, it has a certain decay probability per unit time, κ. That number does not depend on how long the nucleus has been in existence—a nucleus

has no memory. Thus, its decay probability dP during dt is

$$dP = \kappa \, dt \qquad (28.\text{B}.12)$$

for some proportionality constant κ, which will turn out the same as in (28.B.11).

For a *large population* of nuclei, dP has a well-defined meaning: it equals the fraction, $-dN/N$, of all undecayed nuclei which subsequently decay during dt. Equation (28.B.12) reads

$$-\frac{dN}{N} = \kappa \, dt \qquad (28.\text{B}.13)$$

(Since N decreases, $-dN$ is positive.) Integrating both sides of (28.B.13) between times zero and t yields the announced decay law (28.B.11).

Mean Life

The number

$$\tau = \frac{1}{\kappa} \qquad (28.\text{B}.14)$$

has the dimensions of time, and is called the **mean life*** of the nuclei involved. Equation (28.B.11) can be written

$$\frac{N}{N_0} = e^{-t/\tau} \qquad (28.\text{B}.15)$$

We see that, after a time τ, a radioactive sample is reduced to $e^{-1} \approx 1/2.72$ of its initial population. The reason for the name "mean life" is examined in Problem 28.18.

Half-Life

The time $\tau_{1/2}$ *needed to decrease a radioactive sample by one half is called its half-life*. Since 1/2 is not as small as $1/e$, we expect $\tau_{1/2}$ to be somewhat less than τ. How much is $\tau_{1/2}$ in terms of τ?

Setting $N/N_0 = 1/2$ and $t = \tau_{1/2}$ in (28.B.15), we have

$$\frac{1}{2} = e^{-\tau_{1/2}/\tau} \qquad (28.\text{B}.16)$$

---

* Or "lifetime."

or

$$\boxed{\tau_{1/2} = (\ln 2)\tau \approx 0.70\tau}\qquad (28.B.17)$$

In nuclear tables, the half-lives are more commonly listed than the mean lives. Figure 28.5 illustrates the comparison between mean life and half-life.

The range of existing half-lives is unlimited. Stability means $\tau_{1/2} = \infty$; uranium-238 has $\tau_{1/2} = 4.5 \times 10^9$ years, comparable to the presumed age of the universe; the neutron's half-life is about 12 minutes; many nuclei have a half-life so short that their existence can only be inferred theoretically—they do not last long enough to be observed.

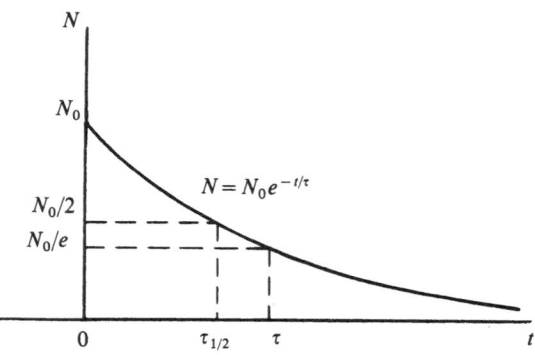

Figure 28.5. The radioactive decay curve; τ is the mean life, $\tau_{1/2} \approx 0.70\tau$ is the half-life.

Example 28.2. Waiting for 99.9 % Decay. During melting, some jewelry gold was contaminated with gold-198, a beta and gamma emitter whose half-life is 2.7 days. Suppose safety requires a reduction of the ^{198}Au content by a factor of 1/1000; how many days must one wait before using the gold?

Each half-life multiplies the ^{198}Au population by 1/2. Noting that $1000 \approx 2^{10}$, we see that after 10 half-lives the original amount is multiplied by $(1/2)^{10} \approx 1/1000$. Thus, the required time is

$$t = 10\tau_{1/2} = (10)(2.7 \text{ days}) = \underline{27 \text{ days}}$$

More formally: Setting $N/N_0 = 10^{-3}$ in Eq. (28.B.15), we get

$$10^{-3} = e^{-t/\tau}$$

$$t = 3\tau \ln 10 = 3\left(\frac{\tau_{1/2}}{\ln 2}\right)\ln 10$$

with the help of (28.B.17). Thus, we have

$$t = \frac{3 \ln 10}{\ln 2}(2.7 \text{ days}) = \underline{27 \text{ days}}$$

as before.

vi. Decay of the Activity

The **activity** R of a sample is defined as the number of its nuclei that decay per unit time,

$$\boxed{R = -\frac{dN}{dt}}\qquad (28.B.18)$$

(a positive number). Hence R is a measure of the amount of radiation emitted per unit time, and is usually of more practical importance than N. As N dwindles, so does R, according to the same exponential curve. This can be seen from Eq. (28.B.11),

$$N = N_0 e^{-\kappa t}\qquad (28.B.19)$$

Taking minus the time derivative, we have

$$-\frac{dN}{dt} = \kappa N_0 e^{-\kappa t}\qquad (28.B.20)$$

or, with $\kappa = 1/\tau$,

$$\boxed{R = \frac{N}{\tau}}\qquad (28.B.21)$$

The activity equals the sample population divided by the mean life.

Example 28.3. Sample Size from Activity. A certain amount (N atoms) of radon, $^{222}_{86}$Rn, is found to alpha-decay at a total rate (activity) $R = 450\,000$ events/sec. If the half-life of radon is $\tau_{1/2} = 3.8$ days, find N.

From (28.B.21), and noting that 1 day = 86400 sec, we have

$$N = \tau R = \frac{\tau_{1/2} R}{\ln 2}\qquad (28.B.22)$$

$$= \frac{(3.8 \text{ days})(450\,000 \text{ sec}^{-1})}{\ln 2}$$

$$= \frac{(3.8)(8.6 \times 10^4)(4.5 \times 10^5)}{\ln 2}$$

$$= \underline{2.12 \times 10^{11} \text{ (atoms)}}$$

Avogadro's Number

If, in this example, we knew the total mass of radon (perhaps by accurate weighing), then we could determine Avogadro's number. For details, see Problem 28.17.

Units of Activity

As we have seen, activity is measured in events per second, a unit also called the **becquerel** (Bq):

$$1 \text{ becquerel} = 1 \text{ Bq} = 1\, \frac{\text{decay}}{\text{sec}} = 1 \text{ sec}^{-1}$$
$$(28.B.23)$$

A more traditional (non-SI) unit, the **curie** (Ci), is given by

$$1 \text{ curie} = 1 \text{ Ci} = 3.7 \times 10^{10}\, \frac{\text{decays}}{\text{sec}} \quad (\text{exactly})$$
$$(28.B.24)$$

This is the activity of about 1 gram of radium—a very large amount. In the laboratory, typical activities range from microcuries (μCi) to millicuries (mCi).

As we have just seen, the amount of a sample may be measured by its activity; hence sample sizes are often expressed in activity units. To illustrate:

$$1 \text{ curie of radium} = 1 \text{ gram of radium}$$

or, in Example 28.3,

$$450\,000 \text{ Bq of radon} = 2.12 \times 10^{11} \text{ atoms of radon}$$

[*An important caution on half-life.* The true half-life of a nuclide shows up in its decay curve, *provided it is not constantly being replenished with undecayed nuclei.* In many substances, a chain of decays takes place, for example the successive alpha and beta disintegrations

$$^{235}_{92}\text{U} \rightarrow\ ^{231}_{90}\text{Th} \rightarrow\ ^{231}_{91}\text{Pa} \qquad (28.B.25)$$

We see that, while the ^{231}Th decays into ^{231}Pa, it is being resupplied by the ^{235}U decays.

Pure samples of ^{235}U and ^{231}Th have widely different half-lives, 7.1×10^8 years and 25.6 h, respectively; but the ^{231}Th activity in (28.B.25) exhibits a false half-life equal to that of its parent ^{235}U, namely, 7.1×10^8 years. The reason is that the ^{231}Th supply is entirely controlled by the parsimonious ^{235}U disintegrations.]

vii. Radiocarbon Dating

The concept of half-life finds one of its most elegant uses in archaeology. Consider an ancient object, such as a wooden post, that retains some carbon assimilated during its life as a tree many years ago. We shall see how to infer its present age from its carbon-14 content.

Natural carbon is about 99% carbon-12 (^{12}C) and 1% carbon-13 (^{13}C); both are stable. The atmosphere, however, contains a minute amount of beta-active carbon-14 (^{14}C); it occurs in about one out of 10^{12} molecules of atmospheric CO_2. The half-life of ^{14}C is "only" 5730 years, but those nuclei are being steadily resupplied through atmospheric nuclear reactions caused by the cosmic rays. Radiocarbon (= radioactive carbon) dating is based on the assumption that *the atmospheric ^{14}C level has been approximately constant over the past millennia.*

A living tree draws all its carbon from the atmosphere, and therefore has the same $1:10^{12}$ isotope ratio as the atmosphere. After it has been cut for lumber, say at time $t = 0$, its ^{14}C is no longer replenished; the latter then decays as

$$N = N_0 e^{-t/\tau} \qquad (28.B.26)$$

[N, N_0 = present and initial number of ^{14}C nuclei in a sample; τ = mean life = (5730 years)/ln 2]. Solving for t, we have

$$\boxed{t = \tau \ln \frac{N_0}{N} = \tau \ln \frac{R_0}{R}} \qquad (28.B.27)$$

where R is the activity corresponding to N atoms of ^{14}C; recall Eq. (28.B.21). In practice, N_0 (or R_0) can be measured from a fresh piece of wood holding the same amount of carbon as the specimen to be dated.

Example 28.4. Age from ^{14}C Activity. A 1-gram sample of charcoal (pure carbon) from an Etruscan tomb is found to emit 10.2 beta particles per minute, while a 1-gram carbon sample from fresh wood emits 14.0 beta particles per minute. What is the age of the ancient charcoal?

Equation (28.B.27) reads

$$t = \tau \ln \frac{R_0}{R}$$

$$= \frac{5730 \text{ years}}{\ln 2} \ln \frac{14 \text{ events/min}}{10.2 \text{ events/min}} = \underline{\underline{2600 \text{ years}}}$$

Limitations

Owing to the small available samples and small activities, an error of ± 100 years is usual. Furthermore, objects older than about 50 000 years no longer have enough activity to be dated quantitatively. (This limit may now be extended to perhaps 100 000 years through new methods which measure directly the number N of ^{14}C atoms rather than their activity R.) Finally, the atmospheric carbon-14 level is now believed to have fluctuated somewhat in the past.

C. Nuclear Reactions

Some nuclei decay spontaneously, as we have seen. But external bombardment can cause otherwise stable nuclei to disintegrate as well. What kind of projectile will achieve this result? Neutrons are apt to be very effective, since they are not repelled by a Coulomb barrier; but other particles —protons, deuterons, etc.—can also be used if they are energetic enough. The initial collision and subsequent breakup constitute a **nuclear reaction**. The purpose of this section is to display a few selected nuclear reactions; they will bring out the importance of energy conservation, and will serve to illustrate **fusion** and **fission**.

i. The Energy Balance

When a nucleus explodes, what is the total kinetic energy of its fragments? The amount may be predicted by a simple calculation; as an input, we use the accurately known masses of the participating nuclei. Two fundamental principles are involved: (1) **Einstein's mass–energy relation** for each nucleus:

$$\boxed{\text{Rest energy} = (\text{rest mass}) \, c^2} \quad (28.\text{C}.1)$$

and (2) **the conservation of total energy**:

$$\boxed{\begin{aligned}[\textstyle\sum(\text{rest energies}) + \sum(\text{kinetic energies})]_{\text{initial}} \\ = [\textstyle\sum(\text{rest energies}) + \sum(\text{kinetic energies})]_{\text{final}}\end{aligned}}$$
$$(28.\text{C}.2)$$

As an illustration, we consider an historic reaction—the first that involved an artificially accelerated projectile. In 1932, Sir John Cockroft and Ernest Walton exposed a sample of ordinary lithium to a beam of protons that had passed through a potential drop of 0.5×10^6 volts. The collision yields two alpha particles:

$$^7_3\text{Li} + {}^1_1\text{H} \rightarrow {}^4_2\text{He} + {}^4_2\text{He} \quad (28.\text{C}.3)$$

Note the conservation of charge and of nucleons.

The kinetic energies are as follows. Initially, we have a stationary lithium atom and a proton of energy

$$\mathscr{K} = 0.500 \text{ MeV} \quad (28.\text{C}.4)$$

The final alpha particles, taken together, have a measured kinetic energy

$$\mathscr{K}' = 17.8 \text{ MeV} \quad (28.\text{C}.5)$$

Let us now "predict" this figure from the known nuclear masses M. Table 28.1 gives

$$M(^7_3\text{Li}) = 7.01600 \text{ u}$$
$$M(^1_1\text{H}) = 1.007825 \text{ u} \quad (28.\text{C}.6)$$
$$M(^4_2\text{He}) = 4.00260 \text{ u}$$

We immediately note that *rest mass is not conserved*. Indeed, we have, for the incoming masses,

$$M(^7_3\text{Li}) + M(^1_1\text{H}) = 8.02382 \text{ u} \quad (28.\text{C}.7)$$

and for the outgoing masses

$$2M(^4_2\text{He}) = 8.00520 \text{ u} \quad (28.\text{C}.8)$$

Some mass is missing; as we can see,

$$\text{Loss of rest mass} = 0.01862 \text{ u} \quad (28.\text{C}.9)$$

Where has it gone? We shall verify that it must have been converted into some of the final kinetic energy.

Applying Eq. (28.C.2), we have

$$\mathscr{K} + [M(^7_3\text{Li}) + M(^1_1\text{H})]\, c^2 = \mathscr{K}' + 2M(^4_2\text{He})\, c^2$$

(28.C.10)

or, solving for $\mathscr{K}'$,

$$\mathscr{K}' = \mathscr{K} + (\text{loss of rest mass})\, c^2 \qquad (28.\text{C}.11)$$

Before inserting the numerical data, we note that $\mathscr{K}$ and $\mathscr{K}'$ are in MeV, while the masses are in atomic mass units. We therefore rewrite (28.C.11), divided through by 1 MeV, as

$$\frac{\mathscr{K}'}{1\ \text{MeV}} = \frac{\mathscr{K}}{1\ \text{MeV}} + \frac{(c^2)(1\ \text{u})}{1\ \text{MeV}} \frac{(\text{loss or rest mass})}{1\ \text{u}}$$

(28.C.12)

where

$$\frac{(c^2)(1\ \text{u})}{1\ \text{MeV}} = 931.5 \qquad (28.\text{C}.13)$$

that is to say, *1 atomic mass unit is equivalent to 931.5 MeV of energy.* [The numerical verification of (28.C.13) is left as an exercise.] Thus, with (28.C.9), Eq. (28.C.12) reads

$$\frac{\mathscr{K}}{1\ \text{MeV}} = 0.500 + (931.5)(0.01862) = 17.84$$

(28.C.14)

This theoretical result is in striking agreement with the measured energy (28.C.5).

The significance of the calculation (and of hundreds of others like it) lies to some extent in its predictive power, but even more fundamentally *in the confirmation of Einstein's mass–energy relation.* That relation has never been verified in chemical reactions because of the tiny mass changes involved.

Atomic versus Nuclear Mass

The masses listed in (28.C.6) are those of the *neutral atoms,* including their electrons. Does this invalidate our calculation, and should we not have used the bare nuclear masses instead? The answer is that we have made no error; to see why, consider Eq. (28.C.11), and let m be the mass of the electron. When applying that equation, we interpreted "loss of rest mass" as

Loss of rest mass = loss of *atomic* mass

$$= (\text{mass of Li nucleus} + 3m)$$
$$+ (\text{mass of proton} + m)$$
$$- 2(\text{mass of alpha particle} + 2m)$$

(28.C.15)

$$= (\text{mass of Li nucleus})$$
$$+ (\text{mass of proton})$$
$$- 2(\text{mass of alpha particle})$$
$$= \text{loss of } nuclear \text{ rest mass}$$

(28.C.16)

The electron masses cancel, and we can indifferently use *all* atomic or *all* nuclear masses.

ii. Fusion

In chemistry, some of the most spectacular reactions are **exothermal**; that is to say, they liberate heat. Such processes are very familiar; chemical fuel has been burned for hundreds of millennia, and chemical explosives have been used since antiquity.

Let us consider the reacting particles. In an exothermal reaction, *the sum of the final kinetic energies is larger than the sum of the initial kinetic energies.* (Photon energies, if any, are considered kinetic, and must be included in these two totals.)

Nuclear reactions, just like chemical ones, can be exothermal; for example, all radioactive decays fall into that category, as we have seen.

Here we shall be interested in **nuclear fusion,** which may be described as follows*:

Fusion is an exothermal reaction involving only light nuclei. (28.C.17)

("Light" means, in practice, a mass number not greater than 7.) One fusion reaction, (28.C.3), is

---

* Fusion is often defined as a reaction in which light nuclei combine to form heavier ones, but (28.C.18) is a counterexample.

already familiar to us [see also Eqs. (28.C.4) and (28.C.5)]:

$$_3^7\text{Li} + _1^1\text{H} \rightarrow _2^4\text{He} + _2^4\text{He} + (17.8 - 0.5)\text{ MeV}$$

$$(28.\text{C}.18)$$

Fusion in the Sun

When fusion takes place in bulk matter, the reaction products, with their high kinetic energy, heat their surroundings by collisions. The Sun in particular, like most other stars, derives its heat from the nuclear reactions taking place in its interior. Their overall effect is to "fuse" four protons into an alpha particle, according to

$$e^- + e^- + _1^1\text{H} + _1^1\text{H} + _1^1\text{H} + _1^1\text{H}$$
$$\rightarrow _2^4\text{He} + v + v + 26.7\text{ MeV} \qquad (28.\text{C}.19)$$

This equation compresses a great deal of physics into a single line. The left-hand side gives the false impression that the four protons and two electrons converge on one another at one time. Actually, (28.C.19) proceeds in several steps, each of which involves no more than a two-body collision.

The right-hand side exhibits neutrinos v, which have almost the same properties as antineutrons.

The 26.7 MeV of kinetic energy is unequally and variously shared between the products of the reaction; most of it serves to heat the surrounding solar matter. The neutrinos, however, escape into outer space, taking their energy along with them.

All the above may seem like nothing but theory. Actually, with the help of huge and elaborate underground installations, some of the Sun's neutrinos are clearly detected on Earth, although in smaller numbers than expected. In another star, now known as Supernova 1987A, nuclear fusion has been documented as well. On the occasion of its sudden appearance in the southern sky in 1987, a few of its neutrinos were caught in the laboratory, and they confirmed in some detail our theoretical understanding of stars and of neutrinos.

Why does fusion occur spontaneously in stars, but not on Earth, in spite of the hydrogen in sea water? The answer is that the reaction is blocked by the Coulomb repulsion between the protons. To overcome that barrier, a large initial kinetic energy is needed, which can be made available through a high temperature; hence the name **thermonuclear fusion** for a bulk fusion reaction.

Fusion on Earth

Bulk fusion requires (1) a proper choice of reacting nuclides, and (2) a sufficient temperature, lasting for a sufficient time, at a sufficient pressure. Let us look briefly at a pair of promising reactions and at the temperature that is involved.

1. The ingredients: Pure deuterium can be made to undergo thermonuclear fusion. Consider two processes, each starting with a deuteron–deuteron collision:

$$_1^2\text{D} + _1^2\text{D} \rightarrow _2^3\text{He} + _0^1\text{n} + 3.2\text{ MeV} \qquad (28.\text{C}.20)$$

$$_1^2\text{D} + _1^2\text{D} \rightarrow _1^3\text{T} + _1^1\text{H} + 4.0\text{ MeV} \qquad (28.\text{C}.21)$$

It is a matter of chance whether the nucleons rearrange themselves according to (28.C.20) or (28.C.21); both processes occur equally often. In bulk deuterium such reactions liberate an impressive amount of energy. The hydrogen bomb is one notorious application; here the necessary starting temperature is achieved through an auxiliary fission bomb (see later) incorporated in the device.

If thermonuclear fusion could take place under controlled conditions, it might some day fill the energy needs of mankind. Deuterium, although much scarcer than ordinary hydrogen, is abundant enough to last for as long as one can practically foresee. Furthermore, radioactive contaminants from fusion plants will probably be more manageable than those from present-day nuclear reactors. It should be added that many fusion reactions, besides (28.C.20) and (28.C.21), exist and hold considerable promise. The first practical reactors may well exploit a D + T reaction, which requires a lower temperature than D + D.

2. Temperature: The deuterons' electrostatic repulsion necessitates a high starting kinetic energy, and that implies a high temperature T under bulk conditions. As a numerical example we estimate very roughly, on the Kelvin scale, the value of T needed for significant fusion to occur in a deuterium plasma. (A **plasma** is a very hot, highly ionized gas.)

Example 28.5. Ignition Temperature. Approximating the deuteron as a sphere of radius $R \approx 6$ fm and charge $e \approx 1.6 \times 10^{-19}$ coulomb, and adapting formula (11.D.11) of Chap. 11 to the deuteron's mean kinetic energy,

$$\mathscr{K} = \tfrac{3}{2}kT \qquad (28.\text{C}.22)$$

where $k = 1.4 \times 10^{-23}$ joule/K, find T such that the closest approach in a typical head-on collision equals $2R$ center to center. (Such a collision then achieves nuclear contact.)

We assume equal and opposite incoming velocities for the two deuterons. At a distance $2R$, they come to rest, having converted all their kinetic energy to electrostatic potential energy:

$$2\mathcal{K} = \frac{Ke^2}{2R} \qquad (28.C.23)$$

where $K = 9 \times 10^9$ (SI) is the constant in Coulomb's law; the charged spheres interact like points. From (28.C.22) and (28.C.23), we find

$$T = \frac{2\mathcal{K}}{3k} = \frac{Ke^2}{6kR} \qquad (28.C.24)$$

$$= \frac{(9 \times 10^9)(1.6 \times 10^{-19})^2}{(6)(1.4 \times 10^{-23})(6 \times 10^{-15})} \text{ kelvins}$$

$$= \underline{4.6 \times 10^8 \text{ kelvins}}$$

about 100 000 times the surface temperature of the Sun. Actually, a more careful estimate indicates that a ten times smaller temperature, about 5×10^7 K, may be adequate.

The technical challenge of achieving such temperatures for a sufficient length of time is being mastered very slowly; the confinement of a plasma in a "magnetic bottle" was touched upon in Sec. B vii of Chap. 18.

In a thermonuclear bomb, or **hydrogen bomb**, the initial heating is done by an auxiliary nuclear explosion based on *fission*, a process that we describe further on.

Cold Fusion?

In the spring of 1989, scientific reports began to circulate claiming the achievement of deuterium fusion in an electrolytic cell at room temperature. The repulsion between the deuterons was thought to be overcome by a combination of effects: forcible crowding in a metal electrode; a quantum-mechanical phenomenon called the "tunnel effect," in which the passage of sufficient time can make a repulsive barrier penetrable; and some other, unexplained factors.

The existence of "cold" fusion would be of great practical and scientific importance. Unfortunately, after a worldwide flurry of excitement that even reached the general public, it turned out that those results had been prematurely announced and could not be independently confirmed. The episode now stands as a lesson in scientific caution.

iii. Fission

This highly exothermal and well-publicized reaction provides another way of releasing nuclear energy in bulk:

> Fission is the break-up of a heavy
> nucleus into a pair of medium-mass
> nuclei and some neutrons.   (28.C.25)

"Heavy" typically—but not always—designates an isotope of uranium (say, of mass number 233, 235, or 238), or an even heavier nuclide such as plutonium-239, while "medium" refers to mass numbers comparable to 100. Neutrons play the central role in this phenomenon:

> Fission is most easily triggered by the
> absorption of a neutron, and results
> in the emission of several neutrons   (28.C.26)

as Fig. 28.6 illustrates. For example, the fission of uranium-235 begins with the formation of a very short-lived (perhaps 10^{-14} sec) nucleus of uranium-236 according to

$$^{235}_{92}\text{U} + ^1_0\text{n} \rightarrow ^{236}_{92}\text{U}* \qquad (28.C.27)$$

(the asterisk denotes an excited state), which can then split up in many ways as governed by chance,[†] for example,

$$^{236}_{92}\text{U}* \rightarrow ^{140}_{54}\text{Xe} + ^{94}_{38}\text{Sr} + ^1_0\text{n} + ^1_0\text{n} \qquad (28.C.28)$$

$$^{236}_{92}\text{U}* \rightarrow ^{144}_{56}\text{Ba} + ^{89}_{36}\text{Kr} + ^1_0\text{n} + ^1_0\text{n} + ^1_0\text{n} \qquad (28.C.29)$$

$$^{236}_{92}\text{U}* \rightarrow ^{141}_{56}\text{Ba} + ^{92}_{36}\text{Kr} + ^1_0\text{n} + ^1_0\text{n} + ^1_0\text{n} \qquad (28.C.30)$$

and so on.

[†] Although one speaks of ^{235}U fission, it is really ^{236}U that undergoes fission.

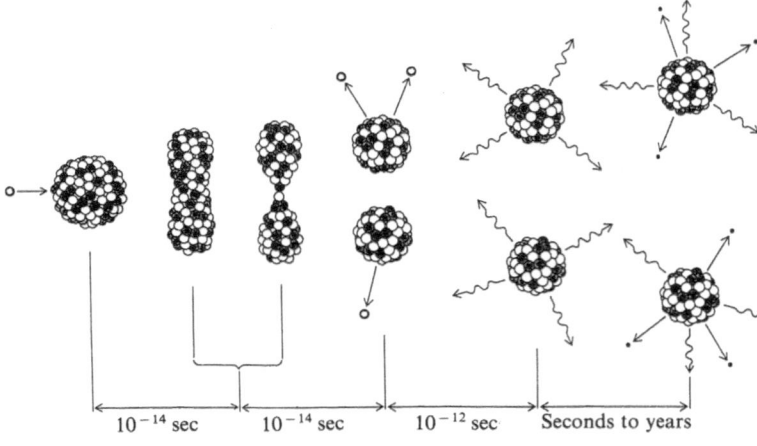

10^{-14} sec 10^{-14} sec 10^{-12} sec   Seconds to years

Figure 28.6. The approximate time sequence in the fission of $^{235}_{92}$U. Legend: ●, Protons; •, electrons; ○, neutrons; ⤳, photons (neutrinos not shown). (By permission, from R. B. Leachman, "Nuclear Fission," *Scientific American*, August 1965. Copyright © 1965 by Scientific American, Inc. All rights reserved.)

Energy Yield

The typical energy released in one such event is about 160 MeV, very large even when compared to the fusion energies mentioned earlier. In addition, most fragments are themselves radioactive, and, after some delay, liberate further energy. The total accumulated energy yield of a fission event may easily reach 200 MeV. The emitted neutrons, however, receive only a small fraction of that total—about 1 MeV or 2 MeV per neutron.

Chain Reactions

We owe large-scale fission to a special feature seen in Eqs. (28.C.28)–(28.C.30): *more than one neutron is generated*, whereas only one is absorbed in (28.C.27). In short, fission increases the number of free neutrons.

When a nucleus of ^{235}U undergoes fission, it generates, on the average, 2.5 neutrons. If each of these neutrons, in turn, could trigger a new fission reaction, the number of events would increase exponentially with time, possibly giving rise to a nuclear explosion.

(Where do the first neutrons come from? A few stray neutrons always seem to be present, possibly from the exceedingly rare spontaneous fission of uranium-238.)

Actually, fewer than those 2.5 neutrons per event are truly available. Indeed, neutrons are "wasted" in two principal ways: (1) absorption (without

fission) by other nuclei present in the reactor; and (2) direct leakage of neutrons to the outside.

1. **Absorption**. In natural uranium (the most readily available nuclear fuel) absorption is a serious problem. The metal consists of 99.3% uranium-238, and only 0.7% uranium-235. The ^{238}U rarely fissions, and it absorbs so many neutrons that not enough of them survive for ^{235}U fission to be self-sustaining in the bulk metal. In 1941–1942, Enrico Fermi circumvented the difficulty by interspersing graphite blocks ($^{12}_{6}$C) in the uranium, according to a carefully calculated geometry. Carbon nuclei do not absorb neutrons, but they do remove much of the neutrons' kinetic energy through repeated collisions: carbon is said to be a **moderator** for the neutrons. It so happens that a neutron with low kinetic energy (a "slow neutron") is selectively absorbed by ^{235}U rather than by ^{238}U, and thus has a far better chance of inducing fission. A successful, controlled reactor was the celebrated outcome of Fermi's application of these principles.

(In **nuclear bombs**, and in many controlled reactors as well, loss through absorption is compensated for by a higher ratio of ^{235}U to ^{238}U. To enrich the ^{235}U content is a major technological enterprise, because no chemical method, and few physical ones, can distinguish one isotope from the other.)

2. **Leakage**. As we recall from Sec. A, a nucleus amounts to a minuscule target, surrounded by a great expanse of vacuum. Therefore a neutron, which is essentially unaffected by the atoms' electric fields, easily escapes from a reactor without a single

collision; the neutron is then lost as far as the chain reaction is concerned.

The most direct answer to that problem consists of putting together a sufficient amount of uranium. The reactor's small surface-to-volume ratio will then reduce the importance of leakage. Thus, a sustained fission reaction requires a **critical mass** of nuclear fuel. In a bomb, the explosion is set off when several small (subcritical) pieces are suddenly brought together so as to exceed the critical amount (about 16 kg for a solid metallic sphere of ^{235}U).

In controlled reactors, a minimum mass of up to hundreds of tons is required, owing to the low ^{235}U content.

Notes

i. Rutherford Scattering (a Qualitative Description)

By the year 1908, Rutherford had determined that certain radioactive substances spontaneously emit fast positive particles (alpha particles), soon identified by him as bare helium nuclei He^{++}.

At Rutherford's suggestion, Hans Geiger and Ernest Marsden used a collimated beam of alpha particles to bombard a thin gold foil, as illustrated in Fig. 28.7. Much of the beam was transmitted undeflected through the foil, but some alpha particles were **scattered** at various angles. (Even a single deflected particle is commonly said to be "scattered.") The values of the scattering angles θ were noted event by event.

The resulting statistics, as interpreted by Rutherford in

the light of collision theory, created a sensation in the world of physics: the scattering was found to proceed as if both the alpha particle and the target nucleus were *point charges, repelling one another according to Coulomb's inverse-square law.* The general expectation, based on a model by J. J. Thomson, had been that the whole atomic volume was uniformly filled with positive charge, in which the electrons were embedded, according to a famous comparison, like raisins in a plum pudding. Figure 28.8 contrasts the Thomson and Rutherford models.

Here we focus on only one aspect of the experiment. Some alpha particles were found to be scattered almost through 180°, that is, they were bouncing back towards the source of the beam. In Thomson's model this would be impossible, because, as can be calculated, nowhere in the dilute positive charge would the electrostatic repulsion be large enough to achieve more than a minuscule deflection. In Rutherford's point-particle model, however, a close enough approach between projectile and target may give a huge $1/r^2$ force, as needed to explain the data.

(The alert reader must wonder if the electrons could spoil the experiment. Each gold atom contains 79 electrons; what is their part in deflecting the alpha particles? That part turns out to be insignificant, because an alpha particle is some 7000 times more massive than an electron.)

ii. An Upper Limit on Nuclear Radii

In one early experiment, alpha particles of kinetic energy $\mathcal{K} = 5.5$ MeV were scattered from gold foil; the

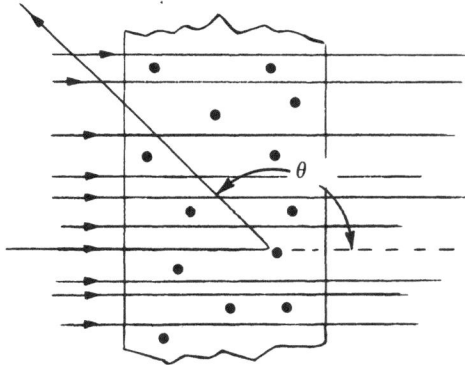

Figure 28.7. A beam of alpha particles is incident from the left on gold nuclei (black dots). One scattering event is shown, with the scattering angle θ.

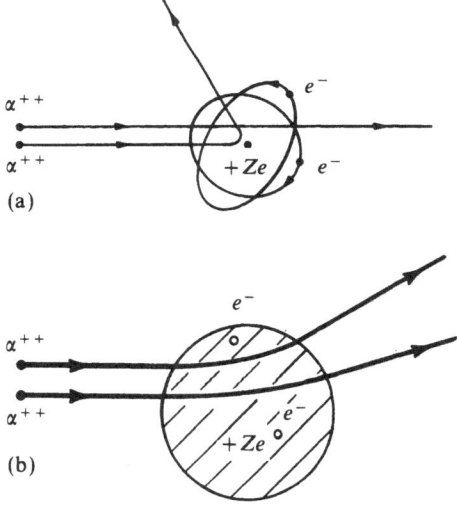

Figure 28.8. (a) The Rutherford model, with two representative electrons in orbit; (b) The Thomson model, showing a solid sphere of charge $+Ze$ and two representative "raisinlike" electrons.

observed deflections might have been due to exact point nuclei. Could a very small (but not pointlike) nucleus produce the same effect? Suppose both the alpha particle and gold nucleus are spheres, with radii a and b, respectively; we can then derive an upper limit for $a + b$ by considering the case of a **head-on collision**. Figure 28.9 shows that a $1/r^2$ interaction requires the distance of closest approach, r_{min}, not to exceed $a + b$:

$$a + b \leqslant r_{min} \qquad (28.\text{N}.1)$$

Assuming the massive gold nucleus to remain approximately motionless during the collision, we estimate r_{min} by noting that the incident kinetic energy is entirely converted to electric potential energy at closest approach:

$$\mathcal{K} = \frac{K(2e)(79e)}{r_{min}} \qquad (28.\text{N}.2)$$

where $K = 9 \times 10^9$ (SI) is the proportionality constant in Coulomb's law. With (28.N.2), inequality (28.N.1) reads

$$a + b \leqslant \frac{(2)(79)\, Ke^2}{\mathcal{K}}$$

$$a + b \leqslant \frac{(2)(79)(9 \times 10^9)(1.6 \times 10^{-19})^2}{(5.5)(1.6 \times 10^{-19})(10^6)} \text{ meter}$$

$$a + b \leqslant 4 \times 10^{-14} \text{ meter} \qquad (28.\text{N}.3)$$

Thus, *each radius a, b must surely be less than 4×10^{-14} meter.* Higher-energy experiments have further reduced that limit. We now know that the alpha particle and gold nucleus have radii of about 1.7×10^{-15} meter and 6.2×10^{-15} meter, respectively.

[Our calculation was classical; should we have used wave mechanics, assuming we knew how? The answer is that wave mechanics is indeed better justified, because

the de Broglie wavelength of the alpha particle is comparable to the range (4×10^{-14} meter) being probed. Fortunately, a correct wave mechanical calculation confirms the classical estimate.]

Summary of Rutherford's Argument

The Coulomb nature of the collision force is inferred from the observed scattering-angle statistics—an analysis not carried through in the present notes. The upper estimate on nuclear size then comes from using the Coulomb potential in a head-on collision; the actual occurrence of head-on collisions is inferred from the presence of back-scattered alpha particles.

Condensed Checklist

Nuclear composition:

$$_Z^A X : \begin{cases} A = \text{mass number} \\ Z = \text{atomic number} \end{cases} \qquad (28.\text{A}.2)$$

$$N = A - Z = \text{neutron number} \qquad (28.\text{A}.1)$$

Isotopes: same Z, different A. $\qquad (28.\text{A}.3)$

Mass:

$$\text{Nuclear mass} \approx (A)(1\text{ u}) \qquad (28.\text{A}.4)$$

1 atomic mass unit $= 1\text{ u} = 1.660531 \times 10^{-27}$ kg $\quad (28.\text{A}.5)$

Atomic mass = mass of neutral atom
Atomic mass of $_6^{12}C$ = 12 exactly $\qquad$ Table 28.1
Table, up to $_6^{13}C$:

Charge:

Nuclear charge $= +Ze$ $\quad$ (exact) $\quad$ Below (28.A.1)

$$e \approx 1.6 \times 10^{-19} \text{ coulomb}$$

Size:

$$r_{nucl} = (1.1 \text{ fm}) \sqrt[3]{A} \qquad (A > 2) \qquad (28.\text{A}.7)$$

$$1 \text{ fm} = 10^{-15} \text{ meter} \qquad (28.\text{A}.6)$$

$$\text{Density} \approx 0.18 \frac{\text{u}}{\text{fm}^3} \qquad (28.\text{A}.8)$$

Decays:

Alpha: $_Z^A X \to {}_{Z-2}^{A-4}X' + {}_2^4 He \qquad (28.\text{B}.3)$

Beta: $_Z^A X \to {}_{Z+1}^A X' + e^- + \bar{\nu} \qquad (28.\text{B}.5)$

Gamma: $_Z^A X^* \to {}_Z^A X + \gamma \qquad (28.\text{B}.9)$

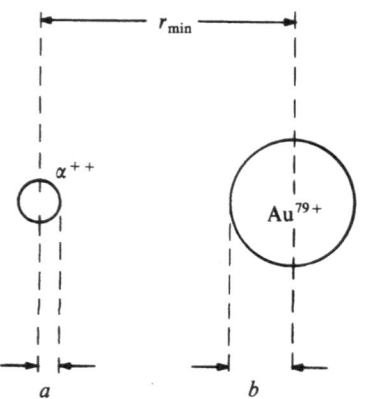

Figure 28.9. The $1/r$ potential between the two nuclei depends on the fact that they do not overlap: $r_{min} > a + b$.

Conservation laws (for all nuclear reactions):

Charge, nucleon number, energy (not rest mass).

(28.B.1), (28.B.2), (28.C.2)

Exponential decay:

$$\boxed{\frac{N}{N_0} = e^{-t/\tau}} \qquad (28.B.15)$$

Mean life $= \tau$ \qquad (28.B.14)

Half life $= \tau_{1/2} = (\ln 2)\tau \approx 0.70\tau$ \quad (28.B.17)

$$\text{Activity} = R = -\frac{dN}{dt} = \frac{N}{\tau}$$

(28.B.18), (28.B.21)

$$1 \text{ curie} = 1 \text{ Ci} = 3.7 \times 10^{10} \frac{\text{decays}}{\text{sec}} \quad \text{(exactly)}$$

(28.B.24)

Fusion:

$$X + X' \rightarrow X'' + X''' + \text{energy} \qquad (28.C.17)$$

(X, X', X'', X''' are light nuclei or neutrons)

Energy = order of 5 MeV

(28.C.20), (28.C.21)

Fission:

$$X + n \rightarrow X' + X'' + (\text{several } n) + \text{energy}$$

(X is a heavy nucleus, X', X'' are medium nuclei.)

(28.C.25), (28.C.26)

Energy = order of 100 MeV,

Energy of a neutron = order of 1 MeV

Below (28.C.30)

Energy balance:

$$\boxed{\mathscr{K}' = \mathscr{K} + (\text{loss of rest mass}) c^2} \qquad (28.C.11)$$

$$\mathscr{K}' = \mathscr{K} + (\text{loss of atomic mass}) c^2 \quad (28.C.15)$$

(The last equality not valid if a positron is created.)

Problem 28.33

1 atomic mass unit = 931.5 MeV/c^2

True or False

1. The nucleus is made of protons, neutrons, and quarks.

2. The neutron is unstable, and therefore a stable nucleus consists only of protons and electrons.

3. A large nucleus, like $^{238}_{92}$U, is almost as large as a whole hydrogen atom.

4. Nucleear sizes are of the order of 10^{-10} meter.

5. The nuclear radius increases roughly in proportion to the number of nucleons it contains.

6. After two half-lives have elapsed, a radioactive sample is completely reduced to its decay products.

7. The half-life of a nuclide equals half of its mean life.

8. After three mean lives, the activity of a radioactive sample has decayed to roughly 5 % of its initial value.

9. Of all nuclear radiations, alpha rays are the most penetrating.

10. In a nuclear reaction, the number of nucleons is usually not conserved.

11. In a nuclear reaction, the electric charge is almost, but not quite conserved.

12. A typical nuclear reaction involves kinetic energies (in the incoming or outgoing particles) of the order of 1 keV.

13. When a nucleus X splits into two nuclei Y and Z, without any other particles being emitted, conservation of mass implies that (atomic mass of X) = (atomic mass of Y) + (atomic mass of Z).

14. Fission is defined as the inverse process to fusion.

15. Fusion cannot take place except at high temperature.

16. Fission usually needs to be triggered by neutrino absorption.

Problems

Alphabetic table of elements: Table 11.1 of Chap. 11.
Partial table of masses: Table 28.1.
Periodic table: Chap. 27, Fig. 27.24.

Section A: The Stable Nuclei

*28.1. Nuclei whose numbers Z and N are both odd are called **odd–odd**; they are rather uncommon. How many stable odd–odd nuclei exist, according to Fig. 28.2?

28.2. (a) What is the specific gravity of nuclear matter? [Specific gravity of water = 1, density of water = 1000 kg/meter^3, density of nuclear matter = 0.18 u/fm^3, see Eq. (28.A.8).] (b) Some stars ("neutron stars," "pulsars") are believed to consist of close-packed neutrons. If the Sun (mass = 2×10^{30} kg) were compressed to the density of nuclear matter, what would be its radius, in kilometers? (c) What would it be for the Earth (mass $= 6 \times 10^{24}$ kg)?

*28.3. Estimate, in fermis, the radius of a uranium-238 nucleus.

28.4. How many nucleons would have to be packed together in order in order to fill a hydrogen atom (radius $\approx$ Bohr radius $\approx 0.5 \times 10^{-10}$ meter)?

*28.5. The atomic mass of gold-197 ($^{197}_{79}$Au) is 196.9666 u. Find, in atomic mass units, the mass of the $^{197}_{79}$Au *nucleus*.

Section B: Radioactivity

*28.6. Complete the following decay equations, including Z and A labels; each "..." stands for one particle.
(a) Al $\rightarrow$ ^{29}Si $+ \cdots + \bar{\nu}$; (b) ^{60}... $\rightarrow$ $_{28}$... $+ \cdots + \bar{\nu}$;
(c) ^{140}La $\rightarrow$ ... $+ e^- + ...$; (d) ^{226}Ra $\rightarrow$ ... $+ ^4$He;
(e) ^{233}... $\rightarrow$ $_{90}$... $+ ^4$He; (f) $^{235}_{92}$... $\rightarrow$ $^{231}_{90}$... $+ ...$.

28.7. How many disintegrations per second are observed in 8 µg of radium?

28.8. A certain nuclide is 80% decayed after 12 h. (a) What are its half-life and mean life? (b) What is the activity of a 3×10^{15} atom sample of that nuclide?

*28.9. Find the age of a piece of prehistoric charcoal whose ^{14}C activity is 8% that of an equal mass of fresh charcoal.

28.10. How much energy, in joules, is emitted in one year by an approximately constant 0.5-curie source radiating 8-MeV gamma rays?

28.11. During manufacture, some steel became contaminated with cobalt-60 ($^{60}_{27}$Co), a beta and gamma emitter with a half-life of 5.26 years. How many years will elapse before the activity is cut down to 10% of its initial amount?

*28.12. How many disintegrations per second are observed in a 5 kg sample of $^{238}_{92}$U, whose half-life is 4.5×10^9 years?

28.13. Given that 1 gram of radium, $^{226}_{88}$Ra, emits 3.7×10^{10} alpha particles per second (1 curie) and has a half-life of 1620 years, (a) find how many beta particles are emitted per second by 0.025 gram of $^{14}_{6}$C, whose half-life is 5730 years; and (b) what mass of polonium, $^{210}_{84}$Po, of half-life 138 days, is a 3-mCi source of alpha particles.

*28.14. What is the intensity, in watts/meter^2, of 0.2-MeV gamma rays from a 3-curie source 20 meters away?

28.15. The most energetic beta particles have kinetic energies of the order of 2 MeV. Does this imply a relativistic speed for the electron?

28.16. The most energetic alpha particles have kinetic energies of the order of 10 MeV. Do they have relativistic speeds?

*28.17 (**Avogadro's number** from radioactivity). The activity of 0.50 µg (1 µg $= 10^{-9}$ kg) of radon, $^{222}_{86}$Rn, is measured to be 77 mCi. (a) If its mean life is 5.5 days, find the number of radon atoms in the sample. (b) Find the mass, in kilograms, of a single radon atom. (c) Taking the atomic weight of radon to be 222, obtain Avogadro's number [$=$ (mass of 1 kilomole)/(mass of 1 atom) $\approx$ 6.02×10^{26}, recall Sec. B iii of Chap. 11; experimentally, the atomic weight can be found by bulk methods, for example from ideal-gas behavior.]

28.18 (**Mean life**). Consider, at time zero, a large number N_0 of identical radioactive nuclei with mean life τ. If dN of these nuclei disintegrate between times t and $t + dt$, they can be said to have had a "life" of length t. Therefore the average of all the N_0 lives is given by $(\int_0^\infty N \, dt)/N_0$. Show that this number equals τ.

*28.19. A certain 7-µC sample of ^{235}U, whose activity is essentially constant in time, keeps producing ^{231}Th. Assume a steady state, with as many ^{231}Th nuclei decaying as are produced. From the half-lives mentioned after Eq. (28.B.25), find (a) the number of ^{235}U atoms, (b) the number of ^{231}Th atoms in the sample.

Section C: Nuclear Reactions

28.20. Complete the following fission equations, including Z and A labels; each "..." stands for one particle.
(a) ^{235}U $+ n \rightarrow$ $_{35}$... $+ ^{143}$... $+ n + n + n$. (b) ^{235}U $+ n \rightarrow$ ^{95}Y $+ \cdots + n + n$.

*28.21. Complete the following equations, including all Z and A labels; each "..." stands for one particle: (a) ^{10}B $+ n \rightarrow$ ... $+ ^4$He; (b) $^{31}_{15}$... $+ \gamma \rightarrow$ ... $+ n$; (c) ^{14}N $+ 2$... $\rightarrow$ ^{17}... $+ n$; (d) ... $+ ^1$H $\rightarrow$ $^7_4$... $+ n$; (e) ^{27}Al $+ n \rightarrow$ ... $+ \gamma$; (f) $^{12}_6$... $+ \gamma \rightarrow$ ... $+ ^4_2$....

28.22. Verify that 1 atomic mass unit is equivalent to 931.5 MeV of energy, Eq. (28.C.13).

28.23. A chemical reaction delivers *of the order of* 1 electron volt of energy per atom (recall that 13.6 electron volts $=$ ionization energy of hydrogen). Examples are furnaces, combustion engines, electric batteries, etc. (a) By what approximate factor is fusion energy greater than chemical energy for a given number of moles of fuel? (b) What is this factor for fission energy?

*28.24. How much energy, in MeV, is liberated in the following fusion reactions? (a) $^3_1$T $+ ^1_1$H $\rightarrow$ $^4_2$He $+ \gamma$; (b) $^3_1$T $+ ^2_1$D $\rightarrow$ $^4_2$He $+ ^1_0$n. [Tritium is not available naturally in significant amounts, but can be generated by neutron bombardment of more plentiful materials, as in

Problem 28.25; (b) Use the atomic mass $M(_1^3T) =$ 3.01605 u, and Table 28.1; the photon has rest mass $M(\gamma) = 0.$]

28.25. How much energy, in MeV, is liberated in the following fusion reactions? (a) $_3^6Li + _1^2D \rightarrow _2^4He + _2^4He$ [Cf. Eq. (28.C.18)]; (b) $_3^6Li + _0^1n \rightarrow _2^4He + _1^3T$ [Use the atomic mass $M(_1^3T) = 3.01605$ u, and Table 28.1.]

*28.26. In the reaction $_1^1H + _0^1n \rightarrow _1^2D + \gamma$, where the initial kinetic energy is negligible, estimate the combined final kinetic energies (deuteron plus photon).

28.27. Determine, in MeV, the energy released in the fission processes (28.C.28) and (28.C.29). You may use the following atomic masses, given in atomic mass units:

^{89}Kr	88.91781
^{94}Sr	93.9154
^{140}Xe	139.921
^{144}Ba	143.923
^{235}U	235.04394

*28.28 (Fission of a medium-heavy nucleus). A high-energy particle, for example a proton, can induce fission in a medium nucleus, as in

$$_{50}^{118}Sn + _1^1H \rightarrow _{31}^{66}Ga + _{20}^{49}Ca + _0^1n + _0^1n + _0^1n + _0^1n$$

Estimate in MeV the minimum initial kinetic energy required. Use the following atomic masses, given in atomic mass units:

$_{20}^{49}Ca$	48.95567
$_{31}^{66}Ga$	65.93161
$_{50}^{118}Sn$	117.90161

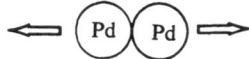

28.29. Consider the hypothetical fission process

$$_{92}^{235}U + n \rightarrow _{46}^{115}Pd + _{46}^{115}Pd + n + n + n + n$$

In the adjoining figure, the two palladium fragments are shown as spheres, just about to separate. As they fly apart, roughly how much kinetic energy do they extract from their electrostatic repulsion? (Your result will be significantly larger than the typically observed fission

energy, which is of order 100 MeV; the discrepancy is a quantum mechanical effect.[†])

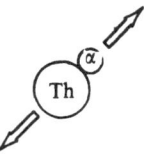

*28.30. Consider the alpha decay

$$_{92}^{238}U \rightarrow _{90}^{234}Th + _2^4He$$

The adjoining figure shows the alpha particle as it is about to separate from the thorium nucleus. Using 2 fm and 6 fm as their respective radii, find how much kinetic energy eventually results from their further separation. (The total observed energy is close to 4 MeV; the serious discrepancy with your calculation is a quantum mechanical effect.[†])

*28.31. Consider the combined alpha and gamma decay of radium into radon:

$$_{88}^{226}Ra \rightarrow _{86}^{222}Rn + _2^4He + \gamma$$

If the total kinetic energy of the products (including the photon) is 4.93 MeV, and if the atomic masses of $_{88}^{226}Ra$ and $_2^4He$ are 226.02544 u and 4.00260 u, find the atomic mass of $_{86}^{222}Rn$, in atomic mass units, to five decimal places.

28.32. Consider the solar fusion process (28.C.19). Recall that $_1^1H$ and $_2^4He$ are the (*charged*) nuclei $_1^1H^+$ and $_2^4He^{++}$. (a) Add two electrons on both sides of the reaction. Explain why the left side now has the ingredients of four *neutral* $_1^1H$ atoms, while on the right side we have the ingredients of a *neutral* $_2^4He$ atom, plus two left-over electrons. (b) Taking into account the two electron rest masses, and using *atomic* masses in your calculation, verify the 26.7 MeV energy yield. (Note: the neutrinos are considered pure energy if we assume that their rest mass is zero. Neutrino energy, which is only a small percentage of the Sun's total output, is almost unobservable.)

*28.33 (Positive beta decay). Some radioactive nuclides emit a positron e^+, which has properties identical to those of the electron except for the sign of its charge. Consider the process

$$_7^{12}N \rightarrow _6^{12}C + e^+ + \nu + 16.4 \text{ MeV}$$

(The neutrino, ν, has almost the same properties as the

[†] See "tunnel effect" in a nuclear physics textbook.

antineutrino $\bar{v}$.) From the decay energy, and from the rest masses

$$M(^{12}_{6}C) = 12.0000 \text{ u}$$

$$M(e^+) = M(e^-) = 0.0005 \text{ u}$$

$$M(v) = 0.0000 \text{ u}$$

determine the *atomic mass* of $^{12}_{7}N$, in atomic mass units, to four decimal places. [Caution: In order to deal with the masses of the neutral atoms we must add seven electrons to the system. The decay is then described by $(^{12}_{7}N)_{\text{neutral atom}} \rightarrow (^{12}_{6}C)_{\text{neutral atom}} + e^- + e^+ + v + 16.4 \text{ MeV}$. Thus, the final rest masses include two extra electron masses.]

28.34. The fusion reaction (28.C.3),

$$^{7}_{3}Li + ^{1}_{1}H \rightarrow ^{4}_{2}He + ^{4}_{2}He + 17.3 \text{ MeV}$$

has a promising energy yield, but requires a much higher temperature than a DD reaction because of the large repulsion between the nuclear charges $3e$ and e. Estimate the ignition temperature. (For simplicity consider $^{1}_{1}H$ to impinge on a stationary $^{7}_{3}Li$; assume a radius of 2.2 fm for the latter.)

Answers to True or False

1. False (just protons and neutrons, or alternatively, just quarks).
2. False (when in a nucleus, the neutron is usually stable).
3. False (still four orders of magnitude smaller).
4. False (10^{-10} meter is the atomic size).
5. False (proportional to the cube root of that number).
6. False (reduced to 1/4).
7. False (about 70% of its mean life).
8. True ($1/e^3 \approx 5\%$).
9. False.
10. False.
11. False (charge is exactly conserved as far as is known).
12. False (MeV).
13. False (there is some conversion between mass and kinetic energy; the relation would be true for atomic *number*).
14. False (they occur in heavy versus light nuclei).
15. False (the protons can be accelerated through a potential).
16. False (neutron absorption).

Answers or Hints to Problems

28.1. 8.

28.3. 6.8 fermis.

28.5. 196.9233 amu.

28.6. (a) $^{29}_{13}Al \rightarrow ^{29}_{14}Si + e^- + \bar{v}$.

28.9. About 20 000 years.

28.12. Hint: The number of atoms is about 5 kg/238 u.

28.14. Hint: The intensity decreases as (distance)$^{-2}$.

28.17. (a) 1.4×10^{15} atoms.

28.19. (b) 3×10^{10}.

28.21. (a) $^{10}_{5}B + ^{1}_{0}n \rightarrow ^{7}_{3}Li + ^{4}_{2}He$.

28.24. (a) 19.8 MeV.

28.26. 2.3 MeV.

28.28. 12 MeV.

28.30. 30 MeV.

28.31. 222.01755 u.

28.33. 12.0187 u.

ENERGY AND POWER SPECTRA

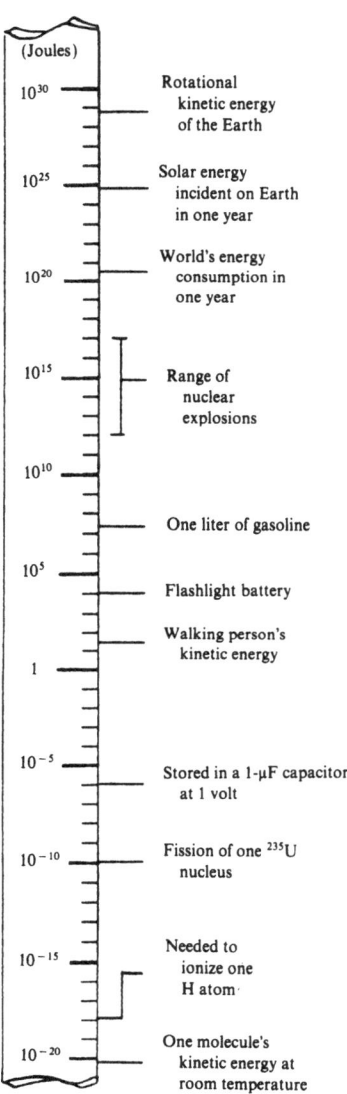

Figure I.1. Some approximate energies, on a logarithmic scale. The already huge range exhibited here could still be greatly expanded by inclusion of stellar, galactic, and cosmic energy values.

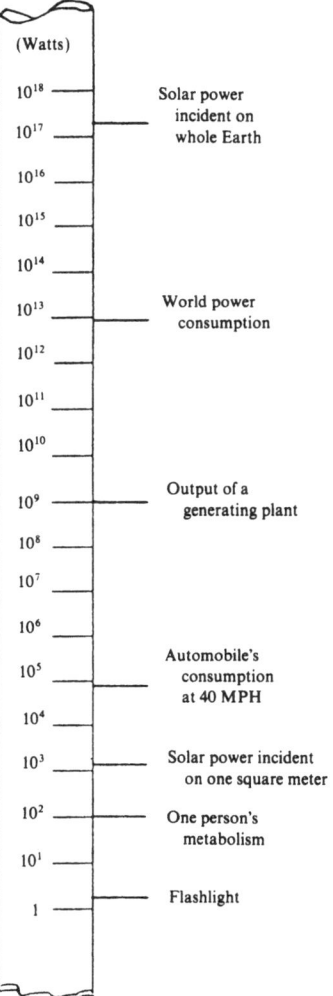

Figure I.2. Some approximate power values, on a logarithmic scale. The data are from R. H. Romer, *Energy* (W. H. Freeman & Co., San Francisco, 1976).

THE VECTOR CROSS PRODUCT

The vector cross product is often used in physics. In this book, it occurs in connection with **angular momentum** and **torque** (notes at end of Chap. 7); **magnetic forces** [Secs. B and C of Chap. 18, as well as its Note (ii)]; **magnetic fields from currents** [Sec. A of Chap. 19]; and the **energy flow in an electromagnetic wave** [Sec. B of Chap. 22]. This appendix serves as a common summary for those scattered discussions.

i. Definition

Let **A**, **B** denote two arbitrary vectors, of magnitudes A, B; see Fig. II.1. The cross product $\mathbf{A} \times \mathbf{B}$ is defined as a vector with
(1) **Magnitude**

$$|\mathbf{A} \times \mathbf{B}| = AB \sin \theta \qquad (\text{II}.1)$$

where θ is the angle between **A** and **B** when they are displayed tail to tail; and
(2) **Direction** perpendicular to the plane through **A** and **B**:

$$\mathbf{A} \times \mathbf{B} \text{ perpendicular to } \mathbf{A}$$
$$\mathbf{A} \times \mathbf{B} \text{ perpendicular to } \mathbf{B} \qquad (\text{II}.2)$$

in accordance with the *right-hand rule*. Several versions of this rule are explained in the text. They are all equivalent to that of Fig. II.1.

[We see that $\mathbf{A} \times \mathbf{B} = 0$ when **A** and **B** are mutually parallel or antiparallel (in particular, $\mathbf{A} \times \mathbf{A} = 0$); also, $|\mathbf{A} \times \mathbf{B}| = AB$ when **A** and **B** are mutually perpendicular.]

ii. Some Important Properties

Two remarkable features of $\mathbf{A} \times \mathbf{B}$ are its *noncommutativity*,

$$\mathbf{A} \times \mathbf{B} = -(\mathbf{B} \times \mathbf{A}) \qquad (\text{II}.3)$$

and its *nonassociativity*,

$$\mathbf{A} \times (\mathbf{B} \times \mathbf{C}) \neq (\mathbf{A} \times \mathbf{B}) \times \mathbf{C} \qquad (\text{II}.4)$$

(in general).

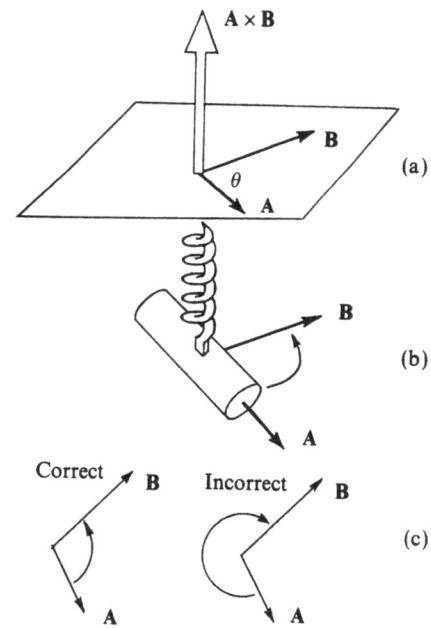

Figure II.1. (a) The cross product, $\mathbf{A} \times \mathbf{B}$. (b) The right-hand rule, here based on a right-handed corkscrew. (Are there left-handed corkscrews?) As the handle turns from **A** to **B**, the screw advances along $+(\mathbf{A} \times \mathbf{B})$. (c) The handle must turn through the smaller of the two available angles.

Illustrations

Consider the three orthogonal unit vectors **i, j, k** of Fig. II.2. We see, for example, that

$$\mathbf{i} \times \mathbf{j} = \mathbf{k}, \qquad \text{whereas} \quad \mathbf{j} \times \mathbf{i} = -\mathbf{k}$$

Also,

$$(\mathbf{i} \times \mathbf{j}) \times \mathbf{j} = \mathbf{k} \times \mathbf{j} = -\mathbf{i},$$

whereas

$$\mathbf{i} \times (\mathbf{j} \times \mathbf{j}) = \mathbf{i} \times \mathbf{0} = \mathbf{0}$$

More conventional-looking properties of the cross product are its behavior under *multiplication by an ordinary number* ("scalar") n:

$$n(\mathbf{A} \times \mathbf{B}) = (n\mathbf{A}) \times \mathbf{B} = \mathbf{A} \times (n\mathbf{B}) \qquad \text{(II.5)}$$

as well as the *distributive rule*

$$\mathbf{A} \times (\mathbf{B} + \mathbf{C}) = (\mathbf{A} \times \mathbf{B}) + (\mathbf{A} \times \mathbf{C}) \qquad \text{(II.6)}$$

and, similarly,

$$(\mathbf{B} + \mathbf{C}) \times \mathbf{A} = (\mathbf{B} \times \mathbf{A}) + (\mathbf{C} \times \mathbf{A}) \qquad \text{(II.7)}$$

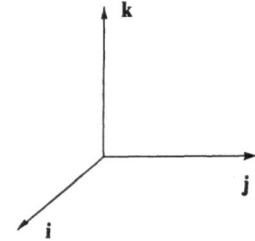

Figure II.2. Three orthogonal unit vectors.

Note how the order of factors is preserved across these equalities. We shall not go into the proofs of (II.5)–(II.7).

iii. Application of the Distributive Rule

Two uses of Eqs. (II.6), (II.7) deserve special mention.

(1) Note (iii) at the end of Chap. 7 introduces the differentiation rule (supposing that **A** and **B** both depend on time, t)

$$\frac{d}{dt}(\mathbf{A} \times \mathbf{B}) = \left(\frac{d\mathbf{A}}{dt} \times \mathbf{B}\right) + \left(\mathbf{A} \times \frac{d\mathbf{B}}{dt}\right) \qquad \text{(II.8)}$$

where the order of the factors must be preserved. The reader may derive (II.8) along the following lines. Use the definition

$$\frac{d}{dt}(\mathbf{A} \times \mathbf{B}) = \lim_{\Delta t \to 0} \frac{\Delta(\mathbf{A} \times \mathbf{B})}{\Delta t} \qquad \text{(II.9)}$$

and similarly for **A** and **B** individually. Set

$$\Delta(\mathbf{A} \times \mathbf{B}) = (\mathbf{A} + \Delta\mathbf{A}) \times (\mathbf{B} + \Delta\mathbf{B}) - (\mathbf{A} \times \mathbf{B})$$

$$\text{(II.10)}$$

and work this out with use of (II.6); neglect the $\Delta\mathbf{A} \times \Delta\mathbf{B}$ term in the limit.

(2) Note (ii) at the end of Chap. 18 points out, on the basis of the distributive rule, that the magnetic force on the drifting electrons in a wire is calculated as if all electrons had the same velocity (their average velocity).

MATHEMATICAL FORMULAS

Special Numbers

$$\sqrt{2} \approx 1.414214$$

$$\pi \approx 3.141593$$

$$e \approx 2.718282$$

Algebra

Let

$$ax^2 + bx + c = 0 \qquad (a \neq 0)$$

Then

$$x = \frac{-b \pm (b^2 - 4ac)^{1/2}}{2a}$$

$$(p + q)^n = p^n + \frac{n}{1!} p^{n-1}q + \frac{n(n-1)}{2!} p^{n-2}q^2 + \cdots$$

Small x ($x \ll 1$) Approximations

$$\frac{1}{1 \pm x} \approx 1 \mp x$$

$$(1 + x)^n \approx 1 + nx$$

$$\ln(1 + x) \approx x$$

$$\sin x \approx \tan x \approx x$$

$$\cos x \approx 1 - \tfrac{1}{2}x^2$$

Trigonometry

$$\pi = \pi \text{ rad} = 180°$$

$$\sin^2 \alpha + \cos^2 \alpha = 1$$

$$\sin 2\alpha = 2 \sin \alpha \cos \alpha$$

$$\sin^2 \alpha = \tfrac{1}{2}(1 - \cos 2\alpha)$$

$$\cos^2 \alpha = \tfrac{1}{2}(1 + \cos 2\alpha)$$

$$\sin(\alpha + \beta) = \sin \alpha \cos \beta + \cos \alpha \sin \beta$$

$$\cos(\alpha + \beta) = \cos \alpha \cos \beta - \sin \alpha \sin \beta$$

$$\cos \alpha + \cos \beta = 2 \cos \frac{\alpha + \beta}{2} \cos \frac{\alpha - \beta}{2}$$

$$\cos \alpha - \cos \beta = 2 \sin \frac{\alpha + \beta}{2} \sin \frac{\beta - \alpha}{2}$$

$$\sin \alpha + \sin \beta = 2 \sin \frac{\alpha + \beta}{2} \cos \frac{\alpha - \beta}{2}$$

$$\sin \alpha - \sin \beta = 2 \cos \frac{\alpha + \beta}{2} \sin \frac{\alpha - \beta}{2}$$

Differentiation

$$\frac{d}{dx} f(x) = f'(x)$$

$$\frac{d}{dx} f(ax) = af'(ax)$$

$$\frac{d}{dx} (fg) = f'g + fg'$$

$$\frac{d}{dx} \left(\frac{f}{g} \right) = \frac{f'g - fg'}{g^2}$$

$$\frac{d}{dx} x^n = nx^{n-1}$$

$$\frac{d}{dx} e^x = e^x$$

$$\frac{d}{dx} \ln |x| = \frac{1}{x}$$

$$\frac{d}{dx} \sin x = \cos x$$

$$\frac{d}{dx} \cos x = -\sin x$$

$$\frac{d}{dx} \tan x = \frac{1}{\cos^2 x}$$

$$\frac{d}{dx} \cot x = \frac{-1}{\sin^2 x}$$

Integration

Let

$$\frac{dF(x)}{dx} = f(x)$$

Then

$$\int f(x)\, dx = F(x) + \text{const}$$

$$\int_a^b f(x)\, dx = F(x) \Big|_a^b = F(b) - F(a)$$

$$\int x^n\, dx = \frac{1}{n+1} x^{n+1} + \text{const} \qquad (n \neq -1)$$

$$\int \frac{dx}{x} = \ln |x| + \text{const}$$

$$\int \sin x\, dx = -\cos x + \text{const}$$

$$\int \cos x\, dx = \sin x + \text{const}$$

$$\int e^x\, dx = e^x + \text{const}$$

CONVERSION FACTORS*

Conversion Factors to International-System Units

One unit of	Equals		
Acceleration			
ft/sec^2	3.048×10^{-1}	meter/sec^2	(exact)
standard g	9.80665	meters/sec^2	(exact)
Area			
acre	4.047×10^3	meters^2	
barn (nuclear cross section)	10^{-8}	meter^2	(exact)
ft^2	9.290×10^{-2}	meter^2	
Density			
lb mass/ft^3	1.602×10^1	kg/meter^3	
slug/ft^3 (British system)	5.154×10^2	kg/meter^3	
Distance, see Length			
Energy			
BTU (thermochemical)	1.054×10^3	joules	
kcal (thermochemical)	4.184×10^3	joules	(exact)
electron volt	$1.6021917 \times 10^{-19}$	joule	
erg	10^{-7}	joule	(exact)
ft lb	1.356	joules	
kW hr.	3.6×10^6	joules	(exact)
kiloton (nuclear explosion)	4.20×10^{12}	joules	
Force			
dyne	10^{-5}	newton	(exact)
lb	4.448	newtons	
Length			
Å (= Ångstrom unit)	10^{-10}	meter	(exact)
AU (= astronomical unit)	1.496×10^{11}	meters	
fermi (= femtometer)	10^{-15}	meter	(exact)
ft	3.048×10^{-1}	meter	(exact)
light year	9.46055×10^{15}	meters	
micron (= micrometer)	10^{-6}	meter	(exact)
mil	2.54×10^{-5}	meter	(exact)
mile (U.S. statute)	1.609×10^3	meters	

* For a more extensive table, see *The International System of Units*, by E. A. Mechtly (National Aeronautics and Space Publication SP-7012, 1973).

Conversion Factors to International-System Units

One unit of	Equals		
Mass			
carat	2×10^{-4}	kg	(exact)
gram	10^{-3}	kg	(exact)
lb-mass	4.536×10^{-1}	kg	
slug (British system)	1.459×10^{1}	kg	
ton (metric)	10^{3}	kg	(exact)
u (= amu = atomic mass unit = dalton)	1.66053×10^{-27}	kg	
Power			
electric horsepower	7.46×10^{2}	watts	(exact)
Pressure			
atmosphere	1.01325×10^{5}	Pa	(exact)
bar	10^{5}	Pa	(exact)
inch of mercury (60°F)	3.377×10^{3}	Pa	
lb/in.² (psi)	6.895×10^{3}	Pa	
millibar	10^{2}	Pa	(exact)
torr (=mm of Hg, 0°C)	1.333×10^{2}	Pa	
Speed			
ft/sec	3.048×10^{-1}	meter/sec	(exact)
km/h	2.778×10^{-1}	meter/sec	
mile/h (U.S. statute)	4.4704×10^{-1}	meter/sec	(exact)
Temperature			
C° (=Celsius degree)	1	K	(exact)
F° (=Fahrenheit degree)	5/9	K	(exact)
Time			
day	8.640×10^{4}	sec	(exact)
hour	3.6×10^{3}	sec	(exact)
minute	6×10^{1}	sec	(exact)
year	3.1536×10^{7}	sec	
Velocity, see Speed			
Volume			
barrel (=42 gal petroleum)	1.590×10^{-1}	meter³	
ft³	2.832×10^{-2}	meter³	
gallon (U.S., liquid)	3.785×10^{-3}	meter³	
liter	10^{-3}	meter³	(exact)
ounce (U.S., liquid)	2.957×10^{-5}	meter³	
quart (U.S., liquid)	9.464×10^{-4}	meter³	
ton (ship displacement)	2.832	meters³	

NUMERICAL CONSTANTS*

The following constants are selected for their importance in this book:

Quantity	Symbol	Value	Unit
Permittivity of the vacuum	ε_0	$8.854\,188 \times 10^{-12}$	$C^2\,N^{-1}\,m^{-2}$
Coulomb's law factor	$1/4\pi\varepsilon_0\ (=K)$	$8.987\,552 \times 10^9$	$C^{-2}\,N\,m^2$
Permeability of the vacuum (exact)	$\mu_0\ (=4\pi K')$	$4\pi \times 10^{-7}$	$A^2\,N$
Speed of light in vacuum (exact)	c	$2.997\,924\,58 \times 10^8$	$m\,sec^{-1}$
Gravitational constant	G	$6.673\,2 \times 10^{-11}$	$N\,m^2\,kg^{-2}$
Avogadro's number (for 1 kmole)	$\mathcal{N}_A$	$6.022\,169 \times 10^{26}$	
Boltzmann's constant	k	$1.380\,622 \times 10^{-23}$	$J\,K^{-1}$
Gas constant	R	$8.314\,34 \times 10^3$	$J\,kmole^{-1}\,K^{-1}$
Volume of ideal gas, standard conditions (for 1 kmole)	$\mathscr{V}_{STP}$	$2.241\,36 \times 10^1$	m^3
Atomic mass unit	$1\,u$	$1.660\,531 \times 10^{-27}$	kg
Planck's constant	h	$6.626\,196 \times 10^{-34}$	$J\,s$
	$h/2\pi\ (=\hbar)$	$1.054\,591\,9 \times 10^{-34}$	$J\,s$
Electron charge	e	$1.602\,191\,7 \times 10^{-19}$	C
Electron rest mass	m_e	$9.109\,558 \times 10^{-31}$	kg
		$5.485\,930 \times 10^{-4}$	u
Proton rest mass	m_p	$1.672\,614 \times 10^{-27}$	kg
		$1.007\,276\,61$	u
Neutron rest mass	m_n	$1.674\,920 \times 10^{-27}$	kg
		$1.008\,665\,20$	u
Electron charge to mass ratio	e/m_e	$1.758\,802\,8 \times 10^{11}$	$C\,kg^{-1}$
Standard acceleration of gravity (exact)	g_{st}	$9.806\,65$	$m\,s^{-2}$
Mass of the Earth	M_E	5.976×10^{24}	kg
Earth's radius (sea level, equatorial)	r_{Ee}	$6.378\,160 \times 10^6$	m
(Sea level, polar)	r_{Ep}	$6.356\,775 \times 10^6$	m

* After E. A. Mechtly, *The International System of Units* (National Aeronautics and Space Publication SP 7012, 1983).

THE GREEK ALPHABET

A, α	alpha	I, ι	iota	P, ρ	rho
B, β	beta	K, κ	kappa	Σ, σ	sigma
Γ, γ	gamma	Λ, λ	lambda	T, τ	tau
Δ, δ, ∂	delta	M, μ	mu	Υ, υ	upsilon
E, ε	epsilon	N, ν	nu	Φ, ϕ, φ	phi
Z, ζ	zeta	Ξ, ξ	xi	X, χ	chi
H, η	eta	O, o	omicron	Ψ, ψ	psi
Θ, θ	theta	Π, π	pi	Ω, ω	omega

APPENDIX VII

LIST OF TABLES AND SPECTRA

(Numerical)

Item	Chapter (Roman numerals refer to Appendix)	Reference label
Energies and powers	I	
Conversion factors	IV	
Numerical constants	V	
Distances	1	Fig. 1.1
Time intervals	1	Fig. 1.3
Speeds	2	Fig. 2.2
Accelerations (gravity)	2	Fig. 2.7
Masses	3	Fig. 3.4
Forces	3	Fig. 3.5
Circular frequencies	6	Fig. 6.10
Angular momenta	7	Fig. 7.33
Astronomical data	8	Table 8.1
Pressures	9	Fig. 9.7
Densities	9	Table 9.1
Temperatures	10	Figs. 10.6, 10.7
Specific heats	10	Table 10.1
	11	Fig. 11.8
Fusion and vaporization	10	Table 10.2
Thermal expansion	10	Fig. 10.11, Table 10.3
Heats of combustion	10	Table 10.4

Item	Chapter	Reference label
Atomic weights (average)	11	Table 11.1
Vibration frequencies	13	Fig. 13.26
Speed of sound	14	Table 14.1
Intensity levels (sound)	14	Table 14.2
Electric field strengths	15	Fig. 15.13
Potential differences	16	Fig. 16.11
Resistivities	17	Table 17.1
Magnetic field strengths	18	Fig. 18.8
Electromagnetic spectrum, including photon energies	22	Figs. 22.13, 22.14
Transparency of atmosphere	22	Fig. 22.16
Indices of refraction	23	Table 23.1
Dispersion in glass	23	Fig. 23.9
Photoemission thresholds	26	Table 26.1
X-ray spectrum	26	Fig. 26.6
Hydrogen spectrum	27	Fig. 27.1, Table 27.1
Fraunhofer lines	27	Fig. 27.8
Electron configuration in atoms	27	Figs. 27.21, 27.22, Table 27.3
Periodic table of elements	27	Figs. 27.23, 27.24
Stable nuclides	28	Fig. 28.2
Atomic masses up to carbon	28	Table 28.1
Charge density in nucleus	28	Fig.28.3

INDEX

A (musical note), 9, 287
Aberration, stellar, 591
Absolute value, 24; *see also* Magnitude
Absolute zero, 212
Absorption, 503
Absorption lines, 623
ac, 478
  motor, 469, 486
  power, 480
Acceleration, 17, 46
  angular, 144, 148, 163
  centripetal, 130
  gravity, 19
    spectrum, 19
  radial, 134, 136
  tangential, 134, 136
  uniform, 19
  vector, 33
Acoustic spectrum, 286
Action, line of, 152
Action, 597
  and reaction, 53, 162, 428, 439
Activity, 656
Adiabatic enclosure, 209
Adiabatic expansion, 253
Airy disc, 562
Alpha decay, 653
Alpha particles, 652, 663
Alternating current: *see* ac
Alternating voltage, 457
Altitude, maximum, 22, 23
Ammeter, 375
Ampere (unit), 372
Ampère's law, 429, 440, 490
Amplitude, 279, 281, 300
  current, 478
  voltage, 478
Angle
  deviation, 537
  incidence, 514
  reflection, 517
  refraction, 514
  total reflection, 518
Ångstrom (unit), 502, 507

Angular
  acceleration, 144, 148, 163
  coordinate, 125
  displacement, 282
  frequency, 300
  magnification, 531, 541
  momentum, 160
  momentum, conservation, 160, 166, 184
    quantized, 629
  momentum, spectrum, 163
  momentum, vector, 164
  position, 126
  velocity, 127
Annihilation, 654
Anode, 360, 362, 601
Antenna, 501
Antimatter, 583
Antineutrino, 653
Antinode, 309
Antireflective coating, 545
Archimedes' principle, 198
Astronomical data, 175, 177, 185
Astronomical unit, 177
Astronomy, 175
Atmosphere, 195, 202, 246, 269
  absorption, 503
  refraction, 538
  transmission, 503
Atom, 227
  Bohr model, 624
  Rutherford model, 663
  size, 229
  Thomson model, 663
Atomic
  mass, 229, 659
  mass unit, 229, 648
  number, 634, 640, 648
  shells, 635
  structure, 617
  table, 231
  weight, 230
Atwood's machine, 57
Avogadro's law, 232, 234
Avogadro's number, 230, 385, 657, 666

Axis, coordinate, 8
  optic, 521, 529
Azimuthal: *see* Orbital

B field: *see* Magnetic field
Back emf, 467
Balance
  Cavendish, 181, 295
  current, 433
  equal-arm, 155
  spring, 47
  torsion, 181, 295
  unequal-arm, 169
Ballistic galvanometer, 467
Ballistic pendulum, 120
Ballistics: *see* Projectiles
Balloon, 200
Balmer series, 618
Banking of road, 140
Bar (unit), 195
Bar magnet, 391, 394, 438
Barlow's wheel, 407
Battery, 371
  emf, 378
  internal resistance, 378
  See also Cell
Beam:
  collimated, 512
  electron, 360, 361, 401, 461
  light, 512
  particle, 400, 401, 663
  photon, 600, 603
  proton, 405
  splitter, 548
Beats, 287
Becquerel (unit), 657
Bel (unit), 313
Bernoulli's principle, 202
Beta decay, 653
Betatron, 461, 467
Bifilar pendulum, 283
Bifilar torsion pendulum, 171
Bimetallic thermometer, 209
Binoculars, 519

Biot–Savart law, 421
Black hole, 188
Block-and-tackle, 56
Blue shift, 588
Bohr model, 624
Bohr radius, 627
Bohr periodic table, 637
Boiling point, 219
    table, 220
Boltzmann constant, 238
Bomb, hydrogen, 661
Bomb, nuclear, 662
Boyle's law, 234
Brackett series, 619
Bradley's measurement of light speed, 591
Bragg's diffraction condition, 550
Bremsstrahlung, 601
British thermal unit (Btu), 215
Brownian motion, 228
Btu: *see* British thermal unit
Buoyancy, 198

Cable, coaxial, 431
Calorie: *see* Kilocalorie
Calorimeter, 218
Camera, 524, 562
    pinhole, 562
Capacitance, 355
Capacitor, 353, 479
    circuits: *see* CR loop, CL loop, CRL
        loop
    electrolytic, 355
    energy of, 358
    series of parallel, 357
Carbon dating, 657
Carnot cycle, 260
Cathode, 360, 599, 601
Cathode rays, 361
Cathode tube, 361
Cavendish experiment, 181, 295
Cell, 384
Celsius temperature scale, 213
Center of curvature, 521
Center of gravity, 154
Center of lens, 521
Center of mass, 111, 158, 160
    frame: *see* Collision, zero-momentum
    for various shapes, 113, 114
    motion of, 114, 117
Centigrade scale, 213
Centrifugal force, 131
Centrifuge, 133, 139
Centripetal acceleration, 130
    at the equator, 132
Centripetal force, 130
Cgs system (energy), 77
Charge, 319
    conservation of, 321, 378, 652
    electric force on, 322, 326
    electron, 324, 362, 385

Charge (*Cont.*)
    hollow spherical, 334, 363
    induced, 325, 352
    infinite sheet, 340
    long straight wire, 355
    magnetic force on, 397
    quantized, 324, 362
Charge-to-mass ratio, 400, 401
    electron, 401
    proton, 402
Charles' law, 212, 234
Chemical energy, 384
Circuit, 375
    ac, 469, 478
    dc, 371
    oscillating, 475
    resonant, 481, 483
    See also Coil; Loop (current)
Circuital law: *see* Ampère's law
Circular motion, 125
    nonuniform, 132, 134
    uniform, 127, 279
Circular aperture, 562
Circular polarization, 495
CL loop, 475, 482
Classical physics, 595, 611
Clock, 8
    cesium, 9
CLR loop, 477, 480, 481, 483
Coaxial cable, 431
Coefficient
    of expansion, 221
        table, 222
    of friction, 59, 61
    of performance, 270
Coherent wave, 560
Coil, 409, 456
    Helmholtz, 445
    search, 467
    snatch, 467
    toroidal, 448, 474
    torque on, 410
Collimated beam, 512
Collision
    elastic, 107
    head-on, 108
    inelastic, 106, 218
    off-center, 110
    zero-momentum, 107
Color, 503
Combustion, heat of, 221
    table, 222
Commutator, 412, 457
Compass, 392, 396, 411
Component of a vector, 28
Compressibility, 197, 227, 246
Compression: *see* Expansion
Compton effect, 603, 611
Compton scattering, 603
Compton wavelength, 605
Concave, 521

Conduction
    electric, 381
    thermal, 215, 265, 386
Conduction electrons, 349, 382, 408, 413
Conductivity
    electric, 382
    thermal, 386
Conductors, 321, 349
Conic section, 184
Conical pendulum, 132
Conservation
    angular momentum, 160, 166, 184
    charge, 321, 378, 652
    energy, 75, 78, 87, 91, 95, 216, 378,
        452, 605, 658
    linear momentum, 103, 605
    mass, 48
    nucleons, 652
Conservative force, 84
Constants, table, 677
Constructive interference, 544
Convection, 216
Converging lens, 521
Converging mirror, 529
Conversion table, units, 675
Convex, 521
Coordinate, 8
Core, iron, 439, 454, 458
Correspondence principle, 628, 645
Cosmic rays, 657
Coulomb (unit), 323
Coulomb's law
    electric, 322, 326
    magnetic, 392
Couple, 157
Covalent radius, 229
CR loop, 380
Creation, 654
Critical force, 165
CRL loop, 477, 480, 481, 483
Cross product, 165, 398, 408, 423, 671
Crystal, 227
    lattice, 550
Curie (unit), 657
Current, 371
    alternating: *see* ac
    amplitude, 478
    balance, 433
    density, 383
    direct: *see* dc
    direction of, 372, 379
    displacement, 489
    eddy, 453
    element, 407, 421
    induced, 451, 453
    rms, 481
    rule, 378
Curvature, 533
    center of, 521
    radius of, 521
Cutoff frequency, 602

Cycle, 279
  Carnot, 260
Cyclic process, 256
Cyclotron, 405
  frequency, 404

D'Arsonval galvanometer, 411
Ds (also dees) of cyclotron, 406
Dalton's law, 236
Damped oscillations, 477, 483
Damped resonance, 482
Damping, 291
  circuit, 477, 481
Daniell cell, 384
Data, 1
Dating, radiocarbon, 657
Daughter nucleus, 652
Davisson–Germer experiment, 607
Day, mean solar, 10
dc, 371
  circuit, 371
  generator: *see* Faraday's wheel
  motor, 412
De Broglie wavelength, 606
Decay
  alpha, 653
  beta, 653
  constant, 655
  exponential, 381, 474, 478, 656
  gamma, 654
Deceleration, 20
Decibel, 313
  table, 314
Dees of cyclotron: *see* Ds of cyclotron
Deflection, by a prism, 516, 537, 538
Density, 196
  current, 383
  energy
    electric, 359
    magnetic, 472
  lines, 329, 394
  mean, 205
  table, 198
  water, 221
Derivative, 16
  second, 18
  vector, 33, 129
Destructive interference, 544
Determinism, 64
Deuterium, 648, 660
Deuteron, 660
Deviation: *see* Deflection
Diatomic gas, 241
Dichroism, 566
Dietary calorie, 222
Diffraction, 552, 563
  Bragg, 550
  circular aperture, 562
  electron, 607
  Fraunhofer, 555
  Fresnel, 555

Diffraction (*Cont.*)
  fringes, 546, 549, 558, 562
  grating, 555
  one-slit, 558
  two-slit, 553
  Young, 553
  x-ray, 550
Dilation, time, 585
Dimensions, 22, 300
Dipole
  electric, 327, 338
  magnetic, 396, 410
Direct current: *see* dc
Direction (vector), 29
Dirod, 325
Disc dynamo: *see* Faraday's wheel
Discharge, gas, 617, 622, 623
Dispersion, 300, 516
Displacement
  angular, 282
  current, 489
  linear, 15, 273
  vector, 24
Dissipation, Joule, 375
Dissipative force, 86
Distance, 4
  spectrum, 6
Diverging
  lens, 521
  mirror, 532
Doppler effect, 306, 588, 589
Dot product, 79
Drag, magnetic, 453
Drift velocity, 382
Driving frequency, 290
Duration, 10
Dynamics, 45
Dynamo, 456, 467
  disc: *see* Faraday's wheel
Dyne, 47

E field: *see* Electric field
Earth
  weighing the, 181
  circumference of, 4
  magnetic field, 395
Eddy currents, 453
Effective
  gravity, 62, 133
  weight, 63
Efficiency of a heat engine, 258, 260, 261
Einstein's equation
  mass-energy, 583, 658
  photoelectric, 597
Elastic collision, 107
Elastic force, 273
Elastic limit, 275
Elastic potential energy, 275
Electric and magnetic fields, motion in, 399
Electric field, 326

Electric field (*Cont.*)
  energy, 359
  hollow sphere, 334
  induced, 459
  infinite sheet, 340
  line, 328
  long straight wire, 335
  motional, 450
  point charge, 326
  uniform, 331
Electric
  battery, 371
  cell, 384
  charge, 319
  conductivity, 382
  current, 371
  dipole, 327, 338
  field: *see* Electric field
  flux, 330
  force, 319, 323
  intensity: *see* Electric field
  potential, 343
  power, 374
  resistivity, 382
    table, 382
  shielding, 341
Electrical: *see* Electric
Electricity, 319
Electrode, 360, 384, 596
Electrodynamics, 491
Electrolyte, 384
Electrolytic capacitor, 355
Electromagnet, 439
Electromagnetic
  induction: *see* Magnetic induction
  theory, 491
  waves, 489
    energy, 495
    intensity, 497
    momentum, 498
    spectrum, 500, 502, 503
    speed, 413, 494
Electrometer, 356
Electromotive force
  back, 467
  battery, 378
  induced, 451, 454
  motional, 451
Electron, 163, 322
  charge, 324, 362, 385
  charge-to-mass ratio, 400, 401
  conduction, 349, 382, 408, 413
  diffraction, 607
  drift velocity, 382
  emission, 360, 596
  gun, 360
  magnetic moment, 438
  microscope, 615
  momentum, 361
  shells, 634, 638
  smallness, 641

Electron (*Cont.*)
  spin, 437
  states, 628
  volt, 360
  wave, 606
Electroscope, 321
Electrostatic
  generator, 325, 353
  induction, 325
  See also Electric
Electrostatics, 319
Elements, periodic table, 636, 637
Ellipse, 184
Elongation, 273
Emf: *see* Electromotive force
Energy, 75
  capacitor, 358
  conservation, 75, 78, 87, 91, 95, 216,
        375, 378, 452, 605, 658
  density, electric, 359
  density, magnetic, 472
  elastic, 275
  electromagnetic, 495
  fission, 661
  fusion, 659
  general definition, 95
  gravitational, 76, 185, 188
  harmonic oscillator, 283
  inductor, 472
  internal, 158, 241
  ionization, 624
  kinetic, 76, 84, 145
  level, 620
    hydrogen, 623
    mass, 583, 658
    metastable, 622
  mechanical, 76
  nuclear, 658
  potential, 76, 116, 275
  rotational, 158, 164, 241
  spectrum, 669
  translational, 158, 164, 238, 241
  vibrational, 241, 283
Engine
  heat, 256
  reversible, 259, 260
Entropy, 262
  mixing, 271
Equation of state, ideal gas, 235
Equilibrium, 50, 154
  neutral, 95
  rotational, 154
  stable, 95, 116
  thermal, 209
  translational, 154
  unstable, 95
Equipartition of energy, 239
Equipotential surface, 347
Equivalence, principle of, 19
Erect image, 525
Erg, 77

Escape, speed of, 187
Ether, 577
Excited state, 629, 654
Exclusion principle, 635
Expansion
  adiabatic, 253, 254
  free, 255, 266
  isothermal, 234, 251
  thermal, 212, 221
    table, 222
  water, 221
  work from, 249
  See also Process
Explosion, 107, 115
Exponential, 202
  decay, 381, 474, 478, 656
External force, 105
Eye, resolution of, 574
Eyepiece, 527
  Newtonian mounting, 531

Fahrenheit scale, 213
Fall
  vertical free, 18
  nonvertical free, 34
Farad, 355
Faraday (unit), 385
Faraday's ice pail, 352
Faraday's law, 454, 460
Faraday's wheel, 465
Fermi (unit), 6, 647, 650
Fermi's reactor, 662
Ferromagnetism, 438
Fiber optics, 519
Field: *see* Electric field; Magnetic field
Field ion microscope, 228, 359
Field line
  electric, 328
  magnetic, 394
Films, thin, 544, 547, 560
  nonreflective, 545
Filter, polarizing, 565
Fine-structure constant, 628
First law
  of Newton, 45
  of thermodynamics, 251
Fission, nuclear, 661
Fizeau's measurement of light speed, 592
Flow diagram (heat engine), 257
Fluid, 193
Fluorescent lamp, 623
Flux
  electric, 330
  line: *see* Field line
  magnetic, 394, 437, 455
Flywheel, 148, 161, 163
Focal length
  lens, 522, 534
  mirror, 529, 535
  thin lenses in combination, 526
Focal plane, 524

Focal point
  lens, 523
  mirror, 529, 530
Focus, 184
Foot: *see* Inch
Foot-pound, 77
Force, 46
  addition, 50
  centrifugal, 131
  centripetal, 130
  combination, 50, 156
  conservative, 84
  critical, 61
  dissipative, 86
  elastic, 273
  equilibrium, 50, 157
  external, 105
  fluid, 193
  hydrostatic, 194
  line of: *see* Field line
  nonconservative, 86
  normal, 58, 193
  restoring, 273
  shearing, 193
  spectrum, 50
  transmission of, 54
  transverse, 85
Force between
  bar magnet and coil, 425
  bar magnet and loop, 428
  capacitor plates, 356
  magnet poles, 391
  parallel wires, 427
  point charges, 322
  solenoid and loop, 438
Force constant: *see* Spring constant
Force on
  current element, 407
  moving charge, 397
Forced harmonic motion: *see* Resonance
Forcemeter, 48
Formulas, mathematical, 673
Fourier's theorem, 286
Frame
  inertial, 46
  reference, 45
Fraunhofer diffraction, 555
Fraunhofer lines, 623
Free expansion, 255, 256
Free fall, vertical, 18
Free path, mean, 247
Freezing point, 213, 214; *see also* Ice
        point; Melting point
Frequency, 127, 279, 299, 310, 500
  acoustic, spectrum, 287
  angular, 279, 300
  circular, spectrum, 133
  cutoff, 602
  cyclotron, 404
  driving, 290
  LC loop, 476

Frequency (*Cont.*)
  natural, 280, 290, 310
  resonant, 290
  threshold, 597
    table, 598
Fresnel diffraction, 555
Friction, 59, 61, 86, 89
Frictionless plane (or surface), 46, 58
Fringes, 546
  width of, 558
  Young's, 553
Fundamental frequency, 286, 310
Fusion
  heat of, 219
    table, 220
  nuclear, 659
    controlled, 660
    Sun, 660

g, 19
  spectrum, 19
Galilean relativity, 38
Galvanometer, 379, 411
  ballistic, 467
  d'Arsonval, 411
Gamma decay, 654
Gamma rays, 502
Gas
  constant, 234
  discharge, 560, 617, 622, 623
  thermometer, 211
Gases
  diatomic, 241
  heat capacity, 242, 252
  ideal, 210, 233
  kinetic theory, 236, 243
  monatomic, 241
  polyatomic, 241
Gauge, tire, 245
Gauge pressure, 195
Gauss (unit), 395
Gauss' law
  electric, 332
  magnetic, 436
Gay–Lussac's law, 235
General relativity, 188
Generator
  ac: *see* Dynamo
  dc: *see* Faraday's wheel
  electrostatic, 325, 353
  homopolar, 465
  MHD, 450
  Van de Graaff, 353
Geometrical optics, 511
Geosynchronous, 190
Governor, 140
Graph, linear, 15
Grating
  diffraction, 555
  reflection, 557
  replica, 556

Grating (*Cont.*)
  resolution of, 573
  spectrometer, 557
Gravitation, Newton's law of, 177
Gravitational
  constant, 178, 176, 181
  potential energy, 76, 185, 188
Gravity, 19, 85, 175
  acceleration of, 19
  artificial, 133
  center of, 154
  effective, 62, 133
  specific, 197
  from spheres, 179
Greek alphabet, 679
Ground state, 621
Gun
  electron, 360
  rail, 436

Half-life, 655
Harmonic motion, 278
  angular: *see* Rotational oscillation
  forced: *see* Resonance
  linear, 278
Harmonics, 285, 310
Head-to-tail method, 24
Heat, 215, 217, 251
  capacity, 218, 242, 252
  combustion, 221
    table, 222
  engine, 256
  fusion, 219
    table, 220
  mechanical equivalent, 216
  pump, 261, 270
  specific, 217
    table, 218
  vaporization, 220
Heating, Joule, 375
Heisenberg uncertainty principle, 609
Helical motion in **B** field, 406
Helmholtz coils, 445
Henry (unit), 462, 471
Hertz (unit), 128, 279
Homopolar
  generator, 465
  motor, 407
Hooke's law, 273, 280
Horizontal range, 37
Horsepower, 83
Hour, 9, 10
Hovercraft, 205
Humidity, relative, 247
Huygens'
  construction, 553
  wavelets, 553
Hydraulic jack, 206
Hydrogen
  atom, 338
  atom model, 624

Hydrogen (*Cont.*)
  bomb, 660
  levels, 623, 624, 636
  spectrum, 617
Hydrometer, 207
Hydrostatic force, 194

Ice point, 213
Ideal gas, 210, 233
Image, 519
  distance, 522
  erect, 525
  intensity, 539
  inverted, 524
  point, 520
  real, 520
  virtual, 520
Impact parameter, 110
Impedance, 480
Impulse, 119, 237
Incandescent lamp, 374, 375, 481
Inch, 5
Incidence
  angle of, 514
  plane of, 513, 517
Incident ray, 513
Inclination, magnetic, 391
Inclined plane, 58, 60
Incoherent: *see* Noncoherent
Indeterminacy: *see* Uncertainty
Index of refraction, 514
  table, 515
  gas, 571
Induced
  charge, 325, 352
  current, 451, 453
  electric field, 459
  emf, 451, 454
Inductance, 470
  mutual, 457, 462
  self-, 470
Induction
  electrostatic, 325
  Faraday's law of, 454, 460
  magnetic, 449, 459
Inductor, 470, 479
  energy in, 472
  series and parallel connection, 472
Inelastic collision, 106, 218
Inertia, 47
  law of, 45
  moment of, 144, 159, 162
    for various shapes, 146
Inertial frame, 46
Infrared, 502
Initial value, 13, 64
  projectile, 35, 36
Inner product, 79
Insulator, 321
Integral, 82

Intensity, 313
  electric: *see* Electric field
  image, 539
  level, 313
    table, 314
  light, 498, 504
  magnetic: *see* Magnetic field
  reflection, 569
  wave, 497
Interactions, weak, strong, 654
Interference, 287
  films, 544, 547, 560
  fringes, 546
  slits: *see* Diffraction
Interferometer, Michelson, 548, 578
Internal combustion, 256
Internal energy, 241
Internal reflection, total, 518
Internal resistance, 378
International System (of units), 5
Inverse-square law, 175, 177, 180
Inverted image, 524
Ion microscope, field, 359
Ionization
  energy, 624
  potential, 624
Iron
  core, 439, 458
  spectrum, 620
Irreversible, 264
Isothermal process, 234, 252, 254
Isotherms, 234, 254
Isotope, 229, 648

Jack, hydraulic, 206
Jerk, 18
Jet propulsion, 110
Joule (unit), 77
Joule's experiment, 216
Joule
  dissipation, 375
  heating, 375
Junction rule, 378
Jupiter, 183

Kelvin temperature scale, 230, 262
Kepler's rules, 182, 184
Kilocalorie, 215, 216
Kilogram, 47
Kilometer, 5
Kilomole, 232
Kilowatt-hour, 83
Kinematics, 13
Kinetic energy, 76, 158
  translational, 158, 238
Kinetic theory of gases, 236
Kirchhoff's rules, 378
Kundt tube, 318

Laminar flow: *see* Streamline flow
Laminated core, 454, 459

Lamp
  fluorescent, 623
  incandescent, 374, 375, 481
Laser, 553, 560, 622
Latent heat: *see* Fusion; Vaporization
Lateral magnification, 524
Lattice, crystal, 550
LC loop, 475, 482
LCR loop, 477, 480, 481, 483
Leiden jar, 353
Length measurement, 550
Lens, 519, 521
  center of, 521
  converging, 521
  diverging, 521
  equation, 522, 533
  focal length of, 522, 534
  focal point of, 523
  power of, 522, 526
  thin, 522, 526
Lensmaker's formula, 534
Lenz's rule, 456, 470
Levels
  energy, 620
  hydrogen, 623
  metastable, 622
Lever, 93, 168
  arm, 151
Life, mean, 655, 666
Light, speed of, 413, 492, 514, 581, 591, 592, 593
Light year, 6
Light
  bulb: *see* Lamp, incandescent
  intensity, 498, 504
  ray, 511
  wave, 500
Limit of resolution: *see* Resolution
Line, spectral, 617
Line of action, 152
Linear
  expansion, coefficient of, 221
    table, 222
  graph, 15
  motion: *see* One-dimensional motion
  polarization, 495, 565
Load, 374
Logarithmic scale, 7
Longitudinal waves, 303
Loop (current), 409
  circular, 423
  hexagonal, 446
  LC, 475, 482
  RC, 380
  rectangular, 409, 446
  RL, 473
  RLC, 477
  rule, 378
  square, 446
  torque on, 410
Loop (wave), 309

LR loop, 473
LRC loop, 477, 480, 481, 483
Lyman series, 618, 619

Macroscopic, 227
Magnet
  bar, 391, 394, 438
  permanent, 392
Magnetic and electric fields, motion in, 399
Magnetic field, 393, 398
  bar, 394
  current, 421
  curved wire, 427
  Earth, 395
  line, 394
  long straight wire, 425, 439
  loop, 423
  motion in, 402
  moving point charge, 507
  rotating, 486
  uniform, 396
Magnetic
  bottle, 406
  dipole, 396, 410
  drag, 453
  energy density, 472
  field: *see* separate entry
  flux, 394, 437, 455
  force on charge, 397
  force on current, 407
  inclination, 391
  induction, 449, 459
  intensity: *see* Magnetic field
  lens, 407
  moment, 396, 410
  monopoles, 393
  poles, 391, 392, 397, 436
  pressure, 435
  quantum number, 631
Magnetism, 391, 437
Magnetohydrodynamic generator, 540
Magnification, 524, 527, 530
  angular, 531, 541
  lateral, 524
  microscope, 527
  magnifier, 524
  telescope, 531, 541
Magnifier, 524
Magnifying glass, 524
Magnitude
  order of, 7
  vector, 24, 28
Malus' law, 565
Manometer, 194, 206
Mass, 47
  center of, 111, 158, 160
  conservation, 48
  dynamic comparison, 48
  Earth, 181
  spectrometer, 400, 404

Mass (*Cont.*)
  spectrum, 49
  standard, 47
Mass–energy relation, 583, 658
Mathematical formulas, 673
Matter, amount, 48
Maxwell
  displacement current, 489
  electromagnetic theory, 491
Mean free path, 247
Mean life, 655, 666
Measurement, 1
Mechanical
  advantage, 57
  equivalent of heat, 216
Mechanics, 45
Medium of propagation, 303
Megawatt, 83
Melting point, 219
  table, 220
Mendeleyev table, 637
Mercury
  barometer, 201
  millimeter of, 201
  thermometer, 209
Meter, 4
  defined, 582
Metric System, 4
MHD generator, 450
Michelson interferometer, 548, 578
Michelson–Morley experiment, 577
Microscope, 527
  electron, 615
  field ion, 229, 359
  resolution of, 528, 565, 615
  scanning tunneling, 228, 229
Microscopic, 227
Microwaves, 502
Millikan's experiment, 362
Minimum deflection, 538
Mirror
  converging, 529
  diverging, 532
  equation, 530
  focal length of, 529, 535
  focal point of, 529, 530
  parabolic, 531
  plane, 528
  spherical, 529
Mixing, 271
Mks: *see* Units
Moderator, 662
Modern physics, 595
Moiré pattern, 288
Mole: *see* Kilomole
Molecular
  speed, 239
  weight, 230
Molecules, 227
Moment
  arm, 152

Moment (*Cont.*)
  inertia, 144, 159, 162
    for various shapes, 146
  magnetic, 396, 410
    electron, 438
Momentum, 103
  angular, 161, 164
    conservation, 161, 166
    spectrum, 163
    vector, 165
  electromagnetic, 498
  electron, 361, 605, 607
  linear (=ordinary), 103
    conservation, 103, 605
  uncertainty, 609
Monatomic gas, 241
Monochromatic wave, 504
Monochromatic x rays, 552
Monopoles, magnetic, 393
Moon, 133, 138, 180
Moseley's law, 638
Motion
  Brownian, 228
  circular, 125
  harmonic, 278
  linear: *see* Motion, one-dimensional
  nonuniform circular, 132, 134
  one-dimensional, 15
  perpetual, 70, 171, 206
  in a plane, 30
  planetary, 182
  projectile, 34
  uniform, 13, 34
  uniform circular, 127, 279
  uniformly accelerated, 19
Motional emf, 451
Motional electric field, 450
Motor
  ac, 469, 486
  dc, 412
  homopolar, 407
Muon, 593
Mutual inductance, 457, 462
Myopia: *see* Nearsightedness

Nanometer, 6
Natural frequency, 280, 289
  string, 310
Natural period, 280
Near distance, 525
Nearsightedness, 526
Neutral equilibrium, 95
Neutrino, 653
Neutron, 647, 648
  mass, 650
  number, 648
  star, 19, 197
Newton (unit), 47
Newton
  first law, 45
  law of gravitation, 177

Newton (*Cont.*)
  pail, 136
  rings, 546
  second law, 46, 104, 117
  third law, 53, 162
Newtonian eyepiece mounting, 531
Node, 309
Nonconservative force, 86
Nonreflecting films, 545
Nonuniform circular motion, 132, 134
Normal (line), 513
Normal force, 58, 193
Nuclear
  bomb, 662
  energy, 658
  fission, 661
  fusion, 659
  reactions, 658
  reactor, 647, 662
  structure, 647
  See also Nucleus
Nucleon, 648, 651
Nucleons, conservation of, 652
Nucleus
  charge, 648
  charge density, 651
  density, 650
  mass, 648, 659
    table, 650
  size, 649, 663
  See also Nuclear
Nuclides, 648
  stable, chart, 649
Null experiment, 52
Numerical constants, table, 677

Object
  distance, 522
  point, 520
  virtual, 529
Objective, 527
Octave, 286
Ocular: *see* Eyepiece
Ohm (unit), 373
Ohm's law, 372
  thermal, 386
Oil drop experiment, 338, 362
Omegatron, 417
Optic
  axis, 521, 529
  fiber, 519
Optics
  geometrical, 511
  quantum, 595
  wave, 543
Orbit
  Bohr model, 627
  circular, 182
  elliptic, 184
  in **E** and **B** fields, 399, 402
Orbital quantum number, 631

Order of magnitude, 7
Organ pipe, 311
Oscillating circuit, 475
    damped, 477, 483
    steady-state, 478
Oscillation, 273
    rotational, 282
Overhead projector, 541
Overtones: see Harmonics

P.s.i., 195
Parabola, 16, 21, 35, 136, 184
Paraboloid, 136
Parallel
    resistors in, 376
    capacitors in, 357
    inductors in, 472
Parallel-axis theorem, 159
Parallelogram method, 25
Paraxial rays, 523
Parent nucleus, 652
Partial pressure, 236
Particle, 8, 56
    beam, 400, 401, 663
Pascal (unit), 195
Pascal's principle, 200
Paschen series, 618, 619
Pauli exclusion principle, 635
Pellat current balance, 433
Pendulum
    ballistic, 120
    bifilar, 283
    conical, 132
    physical, 295
    simple, 277, 282, 295
    torsion, 171, 295
Performance, coefficient of, 270
Period, 128, 177, 279, 280, 299
    natural, 280
Periodic table, 637
Permanent magnet, 392
Perpendicular-axis theorem, 171
Perpetual motion, 70, 171, 206
Pfund series, 619
Phase
    angle, 279
    difference, 544
    relative, 287
    shift, 478
    transition, 219
    vector, 129, 280
Phasor, 129
Photocathode, 596
Photocurrent, 596
Photoelasticity, 569
Photoelectric effects, 596
Photoelectric equation, 597, 599
Photoelectric tube, 596
Photoelectron, 596
Photoemission, 596, 644
    table of thresholds, 598

Photon, 595, 598, 605
    beam, 600, 603
    energy, 599, 600
    momentum, 605, 606
    rest mass, 603
Physical pendulum, 295
Pinch effect, 428
Pinhole, 512
    camera, 562
Pipe, vibration in, 311
Pitch, 286
Pitot tube, 204
Planck's constant, 595, 597
Plane
    frictionless, 46, 58
    incidence, 513, 517
    mirror, 528
    vibration, 494
    wave, 493
Planetary motion, 182
Plasma, 428, 660
Points, effect of, 353
Polarization of wave, 494, 565
    circular, 495
    linear, 495, 565
    partial, 565
Polarizing filter, 565
Polaroid, 565
Poles, magnetic, 391, 392, 397, 436
Polyatomic gas, 241
Position, 7
    angular, 126
    uncertainty, 609, 610
    vector, 30
Positron, 418, 654, 667
Potential
    absolute measurement, 356
    difference, 345, 354
    ionization, 624
    point charge, 343
    spherical shell, 347
    stopping, 596
Potential energy, 76, 116, 275
    elastic, 275
    gravitational, 76, 185, 188
Potentiometer, 379
Pound force, 47
Power, 82, 149
    ac, 480
    dc, 374
    from dynamo, 467
    electric, 374
    factor, 481
    lens, 522, 526
    spectrum, 669
    wave, 306, 312
Powers of ten, 3
Poynting vector, 496
Precision, 3
Prediction, 2
Pressure, 194, 305

Pressure (Cont.)
    atmospheric, 195, 197, 202, 269
    gauge, 195
    magnetic, 435
    partial, 236
    sound, 305
    spectrum, 197
Primary winding, 459
Principal quantum number, 631
Principle of equivalence, 19
Prism
    deflection by, 516, 537, 538
    dispersion by, 516
    reflecting, 519
Process (gas, summary), 256
Projectiles, 34
    initial value problem, 36
Projector, overhead, 541
Proton, 402, 405, 407, 624, 648
Pulley, 56
PV diagram, 234

Quantization
    of angular momentum, 629
    condition for hydrogen, 625
    energy, 620
Quantum, 595
    action, 597
    charge, 324, 362
    efficiency, 601
    mechanics, 608
    number, 626, 628
        magnetic, 631
        orbital, 631
        principal, 631
        spin, 632
    state, 628
    transition, 620
Quark, 324, 651

Radar, 502
Radial acceleration, 134, 136
Radian, 126
Radiation, 216; see also Decays; Rays;
        Waves
Radio waves, 501, 502
Radioactivity, 652
Radiocarbon dating, 657
Radius
    of curvature, 521
    gyration, 167
    orbit, 403, 627
Rail gun, 436
Range of a projectile, 37
Rate of change, 17
    vector, 33, 129
    See also Derivative
Ray, 511
    diagram for lens, 523
    diagram for mirror, 529
    incident, 513

Ray (*Cont.*)
  optics, 511
  reflected, 517
  refracted, 513
  See also Rays
Rayleigh's criterion, 563
Rays
  alpha, 653
  beta, 653
  cathode, 361
  cosmic, 657
  gamma, 502, 654
  Roentgen, 601
  x, 601
RC loop, 380
RCL loop, 477, 480, 481, 483
Reaction, 53, 162
Reactions, nuclear, 658
Reactor, 657, 662
Real image, 520
Red shift, 588
Reference
  frame, 45, 46
  level, 76
Reflected
  ray, 517
  intensity, 569
Reflecting
  prism, 519
  telescope, 530, 541
Reflection, 517
  angle of, 517
  grating, 557
  intensity, 569
  total internal, 518
  waves, 308
Refracted ray, 513
Refraction, 512
  angle of, 514
  atmospheric, 538
  index of, 514
    table, 515
    gas, 571
Refrigerator, 259, 261
Relative
  humidity, 247
  velocity, 32, 581, 590
Relativity, 577
  Galilean, 38
  general, 188
  postulates, 581
  special, 577
Replica grating, 556
Resistance, 373
Resistivity, 382
  table, 382
  thermal, 386
Resistor, 373
  series or parallel connection, 376
Resolution
  of camera, 563

Resolution (*Cont.*)
  electron microscope, 615
  eye, 574
  grating, 573
  light microscope, 565
  telescope, 563
Resonance, 288, 481, 483
Resultant, 50
Reversible, 259, 260, 262, 263
Revolution, 125, 128
Right-hand rule, 397, 422, 494, 671
Rigid body, 91, 96, 144
RL loop, 473
RLC loop, 477, 480, 481, 483
Rms, 239
Rocket, 52, 110
Rod, 54
Roemer's measurement of light speed, 593
Roentgen rays: *see* X rays
Rolling, 61, 158, 172
Root-mean-square, 239, 481
Rotating
  B field, 486
  drum experiment, 241
  pail, 136
Rotation, 143
Rotational
  energy, 145, 158, 164, 241
  equation of motion, 165
  equilibrium, 154
  oscillation, 282
Rough surface, 59
Rpm, 128
Rutherford
  atom model, 663
  scattering, 663
Rydberg constant, 643

Saccadic movement, 169
Sagittal theorem, 533
Satellite, 131, 139, 190
  geosynchronous, 190
Saturated vapor, 247
Scalar, 26
  product, 79
Scale, logarithmic, 7
Scanning tunneling microscope, 228, 229
Scattering, 503
  alpha particles, 663
  Compton, 603
  photons, 603
  Thomson, 642
  Rutherford, 663
Schwarzschild radius, 188
Search coil, 467
Second, 8, 10, 620
Second law
  of Newton, 46, 104, 117
  of thermodynamics, 258, 265
Secondary winding, 459
Self-inductance, 470

Series
  capacitors in, 357
  inductors in, 472
  resistors in, 376
  spectral, 617
  triboelectric, 320
Shells, atomic, 635
Shielding, electric, 341
Shift
  blue, 588
  Doppler, 588, 589
  phase, 478
  red, 588
  violet, 588
SI, 5; *see also* Units
Significant figures, 2
Simple pendulum, 277, 282, 295
Slit
  double, 553
  single, 558
Slope of a graph, 15, 16
Snatch coil, 467
Snell's law, 513
Solenoid, 432, 438, 471, 473
Sonogram, 574
Sound, 284, 304, 313
  intensity, 313
  level, 313
  frequency spectrum, 287
  power, 306
  pressure, 305
  speed, 14, 305
Special relativity, 577
  postulates, 581
Specific gravity, 197
Specific heat, 217
  table, 218
  gases, 242, 252
Specific impulse, 111
Spectra, 681
Spectral: *see* Spectrum
Spectrograph: *see* Spectrometer
Spectrometer, 504
  grating, 557
  mass, 400, 404
Spectroscope: *see* Spectrometer
Spectrum
  absorption, 623
  electromagnetic, 500, 502, 503
  hydrogen, 617
  iron, 620
  line, 617
  series, 617
  visible, 503
  x-ray, 602, 640
Speed, 14
  electromagnetic waves, 413, 494
  escape, 187
  radio signal, 14
  sound, 14, 239, 305
  spectrum, 14

Speed (*Cont.*)
  light, 14, 413, 492, 514, 581, 591, 592, 593
Spin
  of electron, 437
  quantum number, 632
Spring
  balance, 47
  constant, 273, 277
  ideal, 273
  loaded, 274, 290
  scale, 47
Stable
  equilibrium, 95, 116
  nuclides, chart, 649
Standard
  distance, 4
  length, 4
  mass, 47
  temperature and pressure, 233
Standing wave, 308, 311, 507, 610, 622
State, quantum, 628
Steam engine, 256
Stopping potential, 596
STP, 233
Streamline flow, 203
String, 54
  natural frequencies, 310
  waves in, 297
Strong interaction, 654
Sublimation, 219
Sun, 49, 50, 133, 138, 163, 176, 183, 197
Superconductivity, 382
Superposition, 178, 236, 285, 308, 311, 324, 327, 344, 347, 421
Surface
  concave, 521
  convex, 521
  equipotential, 347
  frictionless, 46, 58
  rough, 59
Symbols, 3
Synchronization, 586
Système international, 5

Tables, 681
Tangential
  acceleration, 134, 136
  velocity, 127
Telescope
  angular magnification of, 531, 541
  reflecting, 530, 541
  refracting, 540
  resolution of, 563
Temperature, 209, 212
  absolute or Kelvin, 213, 238, 262
  spectrum, 214
Tension, 54
Tesla (unit), 394
Theories, 1
Thermal conductivity, 386

Thermal expansion, 212, 221
  table, 222
Thermal resistivity, 386
Thermodynamic temperature, 262
Thermodynamics, 209
  first law, 251
  second law, 256, 265
  zeroth law, 210
Thermometer, 209.
  bimetallic, 209
  gas, 211
  mercury, 209
Thick wire, 431, 441
Thin lens, 522, 526
Thin film, 544, 547, 560
Third law of Newton, 53, 162
Thomson
  atom model, 663
  e/m determination, 401
  scattering, 642
Threshold of hearing, 314
Thrust, 52, 110
Time, 8
  flight, 22
  spectrum of intervals, 9
Time constant
  of RC loop, 380
  of RL loop, 474
Time dilation, 585
Tire gauge, 245
Toroidal coil, 448, 474
Torque, 148, 150
  bar magnet, 396
  coil, 410
  electric dipole, 338
  gravity, 153
  loop, 409
  restoring, 277
  vector, 164
Torr, 201
Torricellian vacuum, 197
Torsion
  balance, 181, 295
  pendulum, 171, 295
Torsional wave, 304
Total internal reflection, 518
Trajectory, 35; *see also* Orbit
Transient, 478
Transition, quantum, 620
Translational energy, 158, 164, 238, 241
Translational equilibrium, 154
Transmission grating: *see* Grating
Transmission
  of force, 54
  of torque, 168
Transverse force, 85
Transverse magnification: *see* Lateral magnification
Transverse waves, 303
Threshold frequency, 597
  table, 598

Triboelectric series, 320
Triple point, 212
Tritium, 648, 653
Turbine, 120, 122
Twin paradox, 587

Ultracentrifuge, 133
Ultraviolet, 502
Uncertainty, 3
  principle, 609
Uniform
  circular motion, 127, 279
  field, 331, 396
  motion, 13, 34
Uniformly accelerated motion, 19
Units, 5, 22
  British gravitational, 77
  cgs, 77
  conversion table, 675
  electric and magnetic, 412, 444, 455
  International System, 5
Universe, age, 9
Unpolarized light, 567
Unstable equilibrium, 95

Vacuum, Torricellian, 197
Van Allen belts, 407
Van de Graaff generator, 353
Van der Waals radius, 229
Vaporization, heat of, 220
Vector, 23
  acceleration, 33
  angular momentum, 164
  cross product, 165, 398, 408, 423, 671
  derivative of, 33, 129
  difference, 25
  displacement, 24
  dot product, 79
  force, 47
  momentum, 103
  phase, 129, 280
  position, 30
  sum, 25
  torque, 164
  unit, 98
  velocity, 30, 129
Velocity, 13, 30
  angular, 127
  average, 41
  drift, 382
  instantaneous, 17
  relative, 32, 581, 590
  tangential, 127
  vector, 30, 129
  See also Speed
Vertical free fall, 18
Very Large Array, 564
Very long baseline interferometry, 574
Vibration, 273
  plane of, 494
Vibrational energy, 241, 283

Violet shift, 588
Virtual image, 520
Virtual object, 529
Visible spectrum, 503
Volt, 344, 346
Voltage, 354
   amplitude, 478
   rule, 378
Voltmeter, 375

Waste, 262
Water
   changes of phase, 219
   density of, 197, 221
   waves, 303
Watt (unit), 82, 374
Wave, 492
   coherent, 560
   equation, 302
   frequency, 299
   front, 511
   function, 606
   intensity, 498, 504
   longitudinal, 303
   noncoherent, 560
   number, 643
   optics, 543

Wave (*Cont.*)
   packet, 560
   period, 299
   plane, 305, 493
   polarized, 494, 565
   power, 306, 312
   pressure, 305
   reflection, 308
   speed, 299, 300
   standing, 308, 311, 507, 610, 622
   torsional, 304
   transverse, 303
   traveling, 297
   water, 303
   See also Waves
Wavelength, 299, 500
   de Broglie, 606, 610
   threshold, table, 598
Wavelets, Huygens, 553
Waves, 492
   electromagnetic, 492
      spectrum, 502
   electron, 606
   infrared, 502
   light, 489, 543
   radar, 502
   radio, 502

Waves (*Cont.*)
   ultraviolet, 402
   See also Wave
Weak interaction, 654
Weber (unit), 395
Weighing, 51
Weight, 48
   effective, 63
Width, fringe, 558
Winch, 99
Wire, thick, 431, 441
Work, 78, 83, 149
   energy theorem, 84
   expansion, 249
   integral, 82

X rays, 502, 601
   diffraction, 550
   spectrum, 602, 640

Yard, 5
Year, 9
Young
   experiment, 553
   fringes, 553

Zero, absolute, 212